$154.95 per copy (in United States).
Price subject to change without prior notice.

0029

RSMeans

MW01676236

G

NEW FOR 2009!

Line items found in this cost data that have the icon shown above are green.

The identified items fall into a broad definition of what is considered green.

Please see page ix for more information.

Mechanical Cost Data

32nd Annual Edition

2009

R.S. Means Company, Inc.
Construction Publishers & Consultants
63 Smiths Lane
Kingston, MA 02364-3008
(781) 422-5000

Printed in the United States of America.
ISSN 0068-3531
ISBN 978-0-87629-183-2

Senior Editor
Melville J. Mossman, PE

Contributing Editors
Christopher Babbitt
Ted Baker
Barbara Balboni
Robert A. Bastoni
John H. Chiang, PE
Gary W. Christensen
David G. Drain, PE
Cheryl Elsmore
Robert J. Kuchta
Robert C. McNichols
Melville J. Mossman, PE
Jeannene D. Murphy
Stephen C. Plotner
Eugene R. Spencer
Marshall J. Stetson
Phillip R. Waier, PE

Senior Vice President & General Manager
John Ware

Vice President of Operations
Dave Walsh

Vice President of Sales & Marketing
Sev Ritchie

Marketing Director
John M. Shea

Director of Product Development
Thomas J. Dion

Engineering Manager
Bob Mewis, CCC

Production Manager
Michael Kokernak

Production Coordinator
Jill Goodman

Technical Support
Jonathan Forgit
Mary Lou Geary
Roger Hancock
Gary L. Hoitt
Genevieve Medeiros
Debbie Panarelli
Paula Reale-Camelio
Kathryn S. Rodriguez
Sheryl A. Rose

Book & Cover Design
Norman R. Forgit

PRINTED WITH SOY INK · FSC · SUSTAINABLE FORESTRY INITIATIVE

This book is printed on recycled paper (10% PCW cover, 20% PCW text), certified by the Sustainable Forestry Initiative (SFI®) and Forest Stewardship Counsil (FSC®), using soy-based printing ink. This book is recyclable.

Foreword

Our Mission

Since 1942, RSMeans has been actively engaged in construction cost publishing and consulting throughout North America.

Today, over 60 years after RSMeans began, our primary objective remains the same: to provide you, the construction and facilities professional, with the most current and comprehensive construction cost data possible.

Whether you are a contractor, an owner, an architect, an engineer, a facilities manager, or anyone else who needs a reliable construction cost estimate, you'll find this publication to be a highly useful and necessary tool.

With the constant flow of new construction methods and materials today, it's difficult to find the time to look at and evaluate all the different construction cost possibilities. In addition, because labor and material costs keep changing, last year's cost information is not a reliable basis for today's estimate or budget.

That's why so many construction professionals turn to RSMeans. We keep track of the costs for you, along with a wide range of other key information, from city cost indexes . . . to productivity rates . . . to crew composition . . . to contractor's overhead and profit rates.

RSMeans performs these functions by collecting data from all facets of the industry and organizing it in a format that is instantly accessible to you. From the preliminary budget to the detailed unit price estimate, you'll find the data in this book useful for all phases of construction cost determination.

The Staff, the Organization, and Our Services

When you purchase one of RSMeans' publications, you are, in effect, hiring the services of a full-time staff of construction and engineering professionals.

Our thoroughly experienced and highly qualified staff works daily at collecting, analyzing, and disseminating comprehensive cost information for your needs. These staff members have years of practical construction experience and engineering training prior to joining the firm. As a result, you can count on them not only for the cost figures, but also for additional background reference information that will help you create a realistic estimate.

The RSMeans organization is always prepared to help you solve construction problems through its four major divisions: Construction and Cost Data Publishing, Electronic Products and Services, Consulting and Business Solutions, and Professional Development Services.

Besides a full array of construction cost estimating books, RSMeans also publishes a number of other reference works for the construction industry. Subjects include construction estimating and project and business management; special topics such as HVAC, roofing, plumbing, and hazardous waste remediation; and a library of facility management references.

In addition, you can access all of our construction cost data electronically using *Means CostWorks®* CD or on the Web using Means CostWorks.com.

What's more, you can increase your knowledge and improve your construction estimating and management performance with an RSMeans Construction Seminar or In-House Training Program. These two-day seminar programs offer unparalleled opportunities for everyone in your organization to get updated on a wide variety of construction-related issues.

RSMeans is also a worldwide provider of construction cost management and analysis services for commercial and government owners.

In short, RSMeans can provide you with the tools and expertise for constructing accurate and dependable construction estimates and budgets in a variety of ways.

Robert Snow Means Established a Tradition of Quality That Continues Today

Robert Snow Means spent years building RSMeans, making certain he always delivered a quality product.

Today, at RSMeans, we do more than talk about the quality of our data and the usefulness of our books. We stand behind all of our data, from historical cost indexes to construction materials and techniques to current costs.

If you have any questions about our products or services, please call us toll-free at 1-800-334-3509. Our customer service representatives will be happy to assist you. You can also visit our Web site at www.rsmeans.com.

Table of Contents

Related RSMeans Products and Services

The egineers at RSMeans, suggests the following products and services as companion information resources to *RSMeans Mechanical Cost Data:*

Construction Cost Data Books

Electrical Cost Data 2009
Plumbing Cost Data 2009

Reference Books

ADA Compliance Pricing Guide, 2nd Ed.
Building Security: Strategies & Costs
Designing & Building with the IBC, 2nd Ed.
Estimating Building Costs
Estimating Handbook, 2nd Ed.
Green Building: Project Planning & Estimating, 2nd Ed.
How to Estimate with Means Data and CostWorks, 3rd Ed.
Plan Reading & Material Takeoff
Project Scheduling and Management for Construction
Mechanical Estimating Methods, 4th Ed.

Seminars and In-House Training

Mechanical & Electrical Estimating
Means CostWorks® Training
Means Data for Job Order Contracting (JOC)
Plan Reading & Material Takeoff
Scheduling & Project Management

RSMeans on the Internet

Visit RSMeans at **www.rsmeans.com.** The site contains useful interactive cost and reference material. Request or download **FREE** estimating software demos. Visit our bookstore for convenient ordering and to learn more about new publications and companion products.

RSMeans Electronic Data

Get the information found in RSMeans cost books electronically on *Means CostWorks®* CD or on the Web at Means Costworks.com

RSMeans Business Solutions

Cost Engineers and Analysts conduct facility life cycle and benchmark research studies, predictive cost modeling, as well as offer consultation services for conceptual estimating, real property management, and job order contracting. Research studies are designed with the application of proprietary cost and project data from RCD/RSMeans extensive North American databases. Analysts offer building product manufacturers qualitative and quantitative market research, as well as time/motion studies for new products, and web-based dashboards for market opportunity analysis. Clients are from both the public and private sectors, including federal, state, and municipal agencies, corporations, institutions, construction management firms, hospitals, and associations.

RSMeans for Job Order Contracting (JOC)

Best practice JOC tools for cost estimating and project management to help streamline delivery processes for renovation projects. Renovation is a $147 billion market in the U.S., and includes projects in school districts, municipalities, health care facilities, colleges and universities, and corporations.

- RSMeans Engineers consult in contracting methods and conduct JOC Facility Audits
- JOCWorks™ Software (Basic, Advanced, PRO levels)
- RSMeans Job Order Contracting Cost Data for the entire U.S.

Construction Costs for Software Applications

Over 25 unit price and assemblies cost databases are available through a number of leading estimating and facilities management software providers (listed below). For more information see the "Other RSMeans Products" pages at the back of this publication.

MeansData™ is also available to federal, state, and local government agencies as multi-year, multi-seat licenses.

- 3D International
- 4Clicks-Solutions, LLC
- Aepco, Inc.
- Applied Flow Technology
- ArenaSoft Estimating
- Ares Corporation
- Beck
- BSD – Building Systems Design, Inc.
- CMS – Construction Management Software
- Corecon Technologies, Inc.
- CorVet Systems
- Estimating Systems, Inc.
- HCSS
- Maximus Asset Solutions
- Sage Timberline Office
- Shaw Beneco Enterprises, Inc.
- US Cost, Inc.
- VFA – Vanderweil Facility Advisers
- WinEstimator, Inc.

How the Book Is Built: An Overview

The Construction Specifications Institute (CSI) and Construction Specifications Canada (CSC) have produced the 2004 edition of MasterFormat, a system of titles and numbers used extensively to organize construction information.

All unit price data in the RSMeans cost data books is now arranged in the 50-division MasterFormat 2004 system.

A Powerful Construction Tool

You have in your hands one of the most powerful construction tools available today. A successful project is built on the foundation of an accurate and dependable estimate. This book will enable you to construct just such an estimate.

For the casual user the book is designed to be:

- quickly and easily understood so you can get right to your estimate.
- filled with valuable information so you can understand the necessary factors that go into the cost estimate.

For the regular user, the book is designed to be:

- a handy desk reference that can be quickly referred to for key costs.
- a comprehensive, fully reliable source of current construction costs and productivity rates so you'll be prepared to estimate any project.
- a source book for preliminary project cost, product selections, and alternate materials and methods.

To meet all of these requirements, we have organized the book into the following clearly defined sections.

New: Quick Start

See our new "Quick Start" instructions on the following page.

Estimating with RSMeans Unit Price Cost Data

Please see these steps to complete an estimate using RSMeans unit price cost data.

How To Use the Book: The Details

This section contains an in-depth explanation of how the book is arranged and how you can use it to determine a reliable construction cost estimate. It includes information about how we develop our cost figures and how to completely prepare your estimate.

Unit Price Section

All unit price cost data has been divided into the 50 divisions according to the MasterFormat system of classification and numbering. For a listing of these divisions and an outline of their subdivisions, see the Unit Price Section Table of Contents.

Estimating tips are included at the beginning of each division.

Assemblies Section

The cost data in this section has been organized in an "Assemblies" format. These assemblies are the functional elements of a building and are arranged according to the 7 divisions of the UNIFORMAT II classification system. For a complete explanation of a typical "Assemblies" page, see "How To Use the Assemblies Cost Tables."

Reference Section

This section includes information on Equipment Rental Costs, Crew Listings, Historical Cost Indexes, City Cost Indexes, Location Factors, Reference Tables, Change Orders, Square Foot Costs, and a listing of Abbreviations.

Equipment Rental Costs: This section contains the average costs to rent and operate hundreds of pieces of construction equipment.

Crew Listings: This section lists all the crews referenced in the book. For the purposes of this book, a crew is composed of more than one trade classification and/or the addition of power equipment to any trade classification. Power equipment is included in the cost of the crew. Costs are shown both with the bare labor rates and with the installing contractor's overhead and profit added. For each, the total crew cost per eight-hour day and the composite cost per labor-hour are listed.

Historical Cost Indexes: These indexes provide you with data to adjust construction costs over time.

City Cost Indexes: All costs in this book are U.S. national averages. Costs vary because of the regional economy. You can adjust costs by CSI Division to over 316 locations throughout the U.S. and Canada by using the data in this section.

Location Factors: You can adjust total project costs to over 900 locations throughout the U.S. and Canada by using the data in this section.

Reference Tables: At the beginning of selected major classifications in the Unit Price and Assemblies sections are "reference numbers" shown in a shaded box. These numbers refer you to related information in the Reference Section. In this section, you'll find reference tables, explanations, estimating information that support how we develop the unit price data, technical data, and estimating procedures.

Change Orders: This section includes information on the factors that influence the pricing of change orders.

Square Foot Costs: This section contains costs for 59 different building types that allow you to make a rough estimate for the overall cost of a project or its major components.

Abbreviations: A listing of abbreviations used throughout this book, along with the terms they represent, is included in this section.

Index

A comprehensive listing of all terms and subjects in this book will help you quickly find what you need when you are not sure where it falls in MasterFormat.

The Scope of This Book

This book is designed to be as comprehensive and as easy to use as possible. To that end we have made certain assumptions and limited its scope in two key ways:

1. We have established material prices based on a national average.
2. We have computed labor costs based on a 30-city national average of union wage rates.

For a more detailed explanation of how the cost data is developed, see "How To Use the Book: The Details."

Project Size/Type

The material prices in Means data cost books are "contractor's prices." They are the prices that contractors can expect to pay at the lumberyards, suppliers/distributers warehouses, etc. Small orders of speciality items would be higher than the costs shown, while very large orders, such as truckload lots, would be less. The variation would depend on the size, timing, and negotiating power of the contractor. The labor costs are primarily for new construction or major renovations rather than repairs or minor alterations. **With reasonable exercise of judgment, the figures can be used for any building work.**

Absolute Essentials for a Quick Start

If you feel you are ready to use this book and don't think you will need the detailed instructions that begin on the following page, this Absolute Essentials for a Quick Start page is for you. These steps will allow you to get started estimating in a matter of minutes.

1 Scope

Think through the project that you will be estimating and identify the many individual work tasks that will need to be covered in your estimate.

2 Quantify

Determine the number of units that will be required for each work task that you identified.

3 Pricing

Locate individual Unit Price line items that match the work tasks you identified. The Unit Price Section Table of Contents that begins on page 1 and the Index in the back of the book will help you find these line items.

4 Multiply

Multiply the Total Incl O&P cost for a Unit Price line item in the book by your quantity for that item. The price you calculate will be an estimate for a completed item of work performed by a subcontractor. Keep adding line items in this manner to build your estimate.

5 Project Overhead

Include project overhead items in your estimate. These items are needed to make the job run and are typically, but not always, provided by the General Contractor. They can be found in Division 1. An alternate method of estimating project overhead costs is to apply a percentage of the total project cost.

Include rented tools not included in crews, waste, rubbish handling and cleanup.

6 Estimate Summary

Include General Contractor's markup on subcontractors, General Contractor's office overhead and profit, and sales tax on materials and equipment.

Adjust your estimate to the project's location by using the City Cost Indexes or Location Factors found in the Reference Section.

Editors' Note: We urge you to spend time reading and understanding the supporting material in the front of this book. An accurate estimate requires experience, knowledge and careful calculation. The more you know about how we at RSMeans developed the data, the more accurate your estimate will be. In addition, it is important to take into consideration the reference material in the back of the book such as Equipment Listings, Crew Listings, City Cost Indexes, Location Factors and Reference Numbers.

Estimating with RSMeans Unit Price Cost Data

Following these steps will allow you to complete an accurate estimate using RSMeans Unit Price cost data.

1 Scope out the project

- Identify the individual work tasks that will need to be covered in your estimate.
- The Unit Price data inside this book has been divided into 44 Divisions according to the CSI MasterFormat2004 – their titles are listed on the back cover of your book.
- Think through the project that you will be estimating and identify those CSI Divisions that you will need to use in your estimate.
- The Unit Price Section Table of Contents that begins on page 1 may also be helpful when scoping out your project.
- Experienced estimators find it helpful to begin an estimate with Division 2, estimating Division 1 after the full scope of the project is known.

2 Quantify

- Determine the number of units that will be required for each work task that you previously identified.
- Experienced estimators will include an allowance for waste in their quantities (waste is not included in RSMeans Unit Price line items unless so stated).

3 Price the quantities

- Use the Unit Price Table of Contents, and the Index, to locate the first individual Unit Price line item for your estimate.
- Reference Numbers indicated within a Unit Price section refer to additional information that you may find useful.
- The crew will tell you who is performing the work for that task. Crew codes are expanded in the Crew Listings in the Reference Section to include all trades and equipment that comprise the crew.
- The Daily Output is the amount of work the crew is expected to do in one day.
- The Labor-Hours value is the amount of time it will take for the average crew member to install one unit of measure.
- The abbreviated Unit designation indicates the unit of measure upon which the crew, productivity and prices are based.
- Bare Costs are shown for materials, labor, and equipment needed to complete the Unit Price line item. Bare costs do not include waste, project overhead, payroll insurance, payroll taxes, main office overhead, or profit.
- The Total Incl O&P cost is the billing rate or invoice amount of the installing contractor or subcontractor who performs the work for the Unit Price line item.

4 Multiply

- Multiply the total number of units needed for your project by the Total Incl O&P cost for the Unit Price line item.
- Be careful that the final unit of measure for your quantity of units matches the unit of measure in the Unit column in the book.
- The price you calculate will be an estimate for a completed item of work.
- Keep scoping individual tasks, determining the number of units required for those tasks, matching them up with individual Unit Price line items in the book, and multiplying quantities by Total Incl O&P costs. In this manner keep building your estimate.
- An estimate completed to this point in this manner will be priced as if a subcontractor, or set of subcontractors, will perform the work. The estimate does not yet include Project Overhead or Estimate Summary components such as General Contractor markups on subcontracted and self-performed work, General Contractor office overhead and profit, contingency, and location factor.

5 Project Overhead

- Include project overhead items from Division 1 – General Requirements.
- These are items that will be needed to make the job run. These are typically, but not always, provided by the General Contractor. They include, but are not limited to, such items as field personnel, insurance, performance bond, permits, testing, temporary utilities, field office and storage facilities, temporary scaffolding and platforms, equipment mobilization and demobilization, temporary roads and sidewalks, winter protection, temporary barricades and fencing, temporary security, temporary signs, field engineering and layout, final cleaning and commissioning.
- These items should be scoped, quantified, matched to individual Unit Price line items in Division 1, priced and added to your estimate.
- An alternate method of estimating project overhead costs is to apply a percentage of the total project cost, usually 5% to 15% with an average of 10% (see General Conditions, page ix).
- Include other project related expenses in your estimate such as:
    - Rented equipment not itemized in the Crew Listings.
    - Rubbish handling throughout the project (see 02 41 19.23).

6 Estimate Summary

- Includes sales tax on materials and equipment.
  Note: Sales tax must be added for materials in subcontracted work.
- Include the General Contractor's markup on self-performed work, usually 5% to 15% with an average of 10%.
- Include the General Contractor's markup on subcontracted work, usually 5% to 15% with an average of 10%.
- Include General Contractor's main office overhead and profit.
- RSMeans gives general guidelines on the General Contractor's main office overhead (see section 01 31 13.50 and Reference Number R013113-50).
- RSMeans gives no guidance on the General Contractor's profit.
- Markups will depend on the size of the General Contractor's operations, his projected annual revenue, the level of risk he is taking on, and on the level of competitiveness in the local area and for this project in particular.
- Include a contingency, usually 3% to 5%.
- Adjust your estimate to the project's location by using the City Cost Indexes or the Location Factors in the Reference Section.
- Look at the rules on the pages for How to Use the City Cost Indexes to see how to apply the Indexes for your location.
- When the proper Index or Factor has been identified for the project's location, convert it to a multiplier by dividing it by 100, then multiply that multiplier by your estimate total cost. Your original estimate total cost will now be adjusted up or down from the national average to a total that is appropriate for your location.

Editors' Notes:

1) *We urge you to spend time reading and understanding the supporting material in the front of this book. An accurate estimate requires experience, knowledge, and careful calculation. The more you know about how we at RSMeans developed the data, the more accurate your estimate will be. In addition, it is important to take into consideration the reference material in the back of the book such as Equipment Listings, Crew Listings, City Cost Indexes, Location Factors, and Reference Numbers.*
2) *Contractors who are bidding or are involved in JOC, DOC, SABER, or IDIQ type contracts are cautioned that workers' compensation Insurance, federal and state payroll taxes, waste, project supervision, project overhead, main office overhead, and profit are not included in bare costs. Your coefficient or multiplier must cover these costs.*

How to Use the Book: The Details

What's Behind the Numbers? The Development of Cost Data

The staff at RSMeans continuously monitors developments in the construction industry in order to ensure reliable, thorough, and up-to-date cost information.

While *overall* construction costs may vary relative to general economic conditions, price fluctuations within the industry are dependent upon many factors. Individual price variations may, in fact, be opposite to overall economic trends. Therefore, costs are continually monitored and complete updates are published yearly. Also, new items are frequently added in response to changes in materials and methods.

Costs—$ (U.S.)

All costs represent U.S. national averages and are given in U.S. dollars. The RSMeans City Cost Indexes can be used to adjust costs to a particular location. The City Cost Indexes for Canada can be used to adjust U.S. national averages to local costs in Canadian dollars. No exchange rate conversion is necessary.

The processes or products identified by **G** the green symbol in this publication have been determined to be environmentally responsible and/or resource-efficient solely by the RSMeans engineering staff. The inclusion of the green symbol does not represent compliance with any specific industry association or standard.

Material Costs

The RSMeans staff contacts manufacturers, dealers, distributors, and contractors all across the U.S. and Canada to determine national average material costs. If you have access to current material costs for your specific location, you may wish to make adjustments to reflect differences from the national average. Included within material costs are fasteners for a normal installation. RSMeans engineers use manufacturers' recommendations, written specifications and/or standard construction practice for size and spacing of fasteners. Adjustments to material costs may be required for your specific application or location. Material costs do not include sales tax.

Labor Costs

Labor costs are based on the average of wage rates from 30 major U.S. cities. Rates are determined from labor union agreements or prevailing wages for construction trades for the current year. Rates, along with overhead and profit markups, are listed on the inside back cover of this book.

- If wage rates in your area vary from those used in this book, or if rate increases are expected within a given year, labor costs should be adjusted accordingly.

Labor costs reflect productivity based on actual working conditions. In addition to actual installation, these figures include time spent during a normal workday on tasks such as, material receiving and handling, mobilization at site, site movement, breaks, and cleanup.

Productivity data is developed over an extended period so as not to be influenced by abnormal variations and reflects a typical average.

Equipment Costs

Equipment costs include not only rental, but also operating costs for equipment under normal use. The operating costs include parts and labor for routine servicing such as repair and replacement of pumps, filters, and worn lines. Normal operating expendables, such as fuel, lubricants, tires, and electricity (where applicable), are also included. Extraordinary operating expendables with highly variable wear patterns, such as diamond bits and blades, are excluded. These costs are included under materials. Equipment rental rates are obtained from industry sources throughout North America—contractors, suppliers, dealers, manufacturers, and distributors.

Equipment costs do not include operators' wages; nor do they include the cost to move equipment to a job site (mobilization) or from a job site (demobilization).

Equipment Cost/Day—The cost of power equipment required for each crew is included in the Crew Listings in the Reference Section (small tools that are considered as essential everyday tools are not listed out separately). The Crew Listings itemize specialized tools and heavy equipment along with labor trades. The daily cost of itemized equipment included in a crew is based on dividing the weekly bare rental rate by 5 (number of working days per week) and then adding the hourly operating cost times 8 (the number of hours per day). This Equipment Cost/Day is shown in the last column of the Equipment Rental Cost pages in the Reference Section.

Mobilization/Demobilization—The cost to move construction equipment from an equipment yard or rental company to the job site and back again is not included in equipment costs. Mobilization (to the site) and demobilization (from the site) costs can be found in Division 01 54 36.50. If a piece of equipment is already at the job site, it is not appropriate to utilize mobil./demob costs again in an estimate.

Overhead and Profit

Total Cost, including O&P for the *Installing Contractor* is shown in the last column on both the Unit Price and the Assemblies pages of this book. This figure is the sum of the bare material cost plus 10% for profit, the bare labor cost plus total overhead and profit, and the bare equipment cost plus 10% for profit. Details for the calculation of Overhead and Profit on labor are shown on the inside back cover and in the Reference Section of this book. (See the "How to Use the Unit Price Pages" for an example of this calculation.)

General Conditions

Cost data in this book is presented in two ways: Bare Costs and Total Cost including O&P (Overhead and Profit). General Conditions, when applicable, should also be added to the Total Cost including O&P. The costs for General Conditions are listed in Division 1 of the Unit Price Section and the Reference Section of this book. General Conditions for the *Installing Contractor* may range from 0% to 10% of the Total Cost including O&P. For the *General* or *Prime Contractor*, costs for General Conditions may range from 5% to 15% of the Total Cost, including O&P, with a figure of 10% as the most typical allowance.

Factors Affecting Costs

Costs can vary depending upon a number of variables. Here's how we have handled the main factors affecting costs.

Quality—The prices for materials and the workmanship upon which productivity is based represent sound construction work. They are also in line with U.S. government specifications.

Overtime—We have made no allowance for overtime. If you anticipate premium time or work beyond normal working hours, be sure to make an appropriate adjustment to your labor costs.

Productivity—The productivity, daily output, and labor-hour figures for each line item are based on working an eight-hour day in daylight hours in moderate temperatures. For work that extends beyond normal work hours or is performed under adverse conditions, productivity may decrease. (See the section in "How To Use the Unit Price Pages" for more on productivity.)

Size of Project—The size, scope of work, and type of construction project will have a significant impact on cost. Economies of scale can reduce costs for large projects. Unit costs can often run higher for small projects. Costs in this book are intended for the size and type of project as previously described in "How the Book Is Built: An Overview." Costs for projects of a significantly different size or type should be adjusted accordingly.

Location—Material prices in this book are for metropolitan areas. However, in dense urban areas, traffic and site storage limitations may increase costs. Beyond a 20-mile radius of large cities, extra trucking or transportation charges may also increase the material costs slightly. On the other hand, lower wage rates may be in effect. Be sure to consider both of these factors when preparing an estimate, particularly if the job site is located in a central city or remote rural location.

In addition, highly specialized subcontract items may require travel and per-diem expenses for mechanics.

Other Factors —

- season of year
- contractor management
- weather conditions
- local union restrictions
- building code requirements
- availability of:
  - adequate energy
  - skilled labor
  - building materials
- owner's special requirements/restrictions
- safety requirements
- environmental considerations

Unpredictable Factors—General business conditions influence "in-place" costs of all items. Substitute materials and construction methods may have to be employed. These may affect the installed cost and/or life cycle costs. Such factors may be difficult to evaluate and cannot necessarily be predicted on the basis of the job's location in a particular section of the country. Thus, where these factors apply, you may find significant but unavoidable cost variations for which you will have to apply a measure of judgment to your estimate.

Rounding of Costs

In general, all unit prices in excess of $5.00 have been rounded to make them easier to use and still maintain adequate precision of the results. The rounding rules we have chosen are in the following table.

Prices from . . .	Rounded to the nearest . . .
$.01 to $5.00	$.01
$5.01 to $20.00	$.05
$20.01 to $100.00	$.50
$100.01 to $300.00	$1.00
$300.01 to $1,000.00	$5.00
$1,000.01 to $10,000.00	$25.00
$10,000.01 to $50,000.00	$100.00
$50,000.01 and above	$500.00

Contingencies

Contingencies: The allowance for contingencies generally is to provide for unforeseen construction difficulties. On alterations or repair jobs, 20% is not too much. If drawings are final and only field contingencies are being considered, 2% or 3% is probably sufficient, and often nothing need be added. As far as the contract is concerned, future changes in plans will be covered by extras. The contractor should consider inflationary price trends and possible material shortages during the course of the job. If drawings are not complete or approved, or a budget cost is wanted, it is wise to add 5% to 10%. Contingencies, then, are a matter of judgment. Additional allowances are shown in Division 01 21 16.50 for contingencies and job conditions and Reference Table R012153-10 for factors to convert prices for repair and remodeling jobs.

Note: For an explanation of "Selective Demolition" versus "Removal for Replacement" see Reference Number R220105-10.

Important Estimating Considerations

One reason for listing a job size "minimum" is to ensure that the construction craftsmen are productive for 8 hours a day on a continuous basis. Otherwise, tasks that require only 6 to 7 hours could be billed as an 8-hour day, thus providing an erroneous productivity estimate. The "productivity," or daily output, of each craftsman includes mobilization and cleanup time, break time, and plan layout time, as well as an allowance to carry stock from the storage trailer or location on the job site up to 200′ into the building and onto the first or second floor. If material has to be transported over greater distances or to higher floors, an additional allowance should be considered by the estimator. An allowance has also been included in the piping and fittings installation time for a leak check and minor tightening.

Equipment installation time includes the following applicable items: positioning, leveling and securing the unit in place, connecting all associated piping, ducts, vents, etc., which shall have been estimated separately, connecting to an adjacent power source, filling/bleeding, startup, adjusting the controls up and down to assure proper response, setting the integral controls/valves/regulators/thermostats for proper operation (does not include external building type control systems, DDC systems, etc.), explaining/training owner's operator, and warranty. A reasonable breakdown of the labor costs is as follows:

1.	Movement into building, installation/ setting of equipment	35%
2.	Connecting to piping/duct/power, etc.	25%
3.	Filling/flushing/cleaning/touchup, etc.	15%
4.	Startup/running adjustments	5%
5.	Training owner's rep.	5%
6.	Warranty/call back/service	15%

Note that cranes or other lifting equipment are not included on any lines in the Mechanical divisions. For example, if a crane is required to lift a heavy piece of pipe into place high above a gym floor, or to put a rooftop unit on the roof of a four-story building, etc., it must be added. Due to the potential for extreme variation—from nothing additional required to a major crane or helicopter—we feel that including a nominal amount for "lifting contingency" would be useless and detract from the accuracy of the estimate. When using equipment rental from RSMeans, remember to include the cost of the operator(s).

Estimating Labor-Hours

The labor-hours expressed in this publication are based on Average Installation time, using an efficiency level of approximately 60–65% (see item 7), which has been found reasonable and acceptable by many contractors.

The book uses this national efficiency average to establish a consistent benchmark.

For bid situations, adjustments to this efficiency level should be the responsibility of the contractor bidding the project.

The unit labor-hour is divided in the following manner. A typical day for a journeyman might be:

1.	Study Plans	3%	14.4 min.
2.	Material Procurement	3%	14.4 min.
3.	Receiving and Storing	3%	14.4 min
4.	Mobilization	5%	24.0 min.
5.	Site Movement	5%	24.0 min.
6.	Layout and Marking	8%	38.4 min.
7.	Actual Installation	64%	307.2 min.
8.	Cleanup	3%	14.4 min.
9.	Breaks, Nonproductive	6%	28.8 min.
		100%	480.0 min.

If any of the percentages expressed in this breakdown do not apply to the particular work or project situation, then that percentage or a portion of it may be deducted from or added to labor-hours.

Final Checklist

Estimating can be a straightforward process provided you remember the basics. Here's a checklist of some of the steps you should remember to complete before finalizing your estimate.

Did you remember to . . .

- factor in the City Cost Index for your locale?
- take into consideration which items have been marked up and by how much?
- mark up the entire estimate sufficiently for your purposes?
- read the background information on techniques and technical matters that could impact your project time span and cost?
- include all components of your project in the final estimate?
- double check your figures for accuracy?
- call RSMeans if you have any questions about your estimate or the data you've found in our publications?

For more information, please see "Tips for Accurate Estimating," R011105-05 in the Reference Section.

Remember, RSMeans stands behind its publications. If you have any questions about your estimate, about the costs you've used from our books, or even about the technical aspects of the job that may affect your estimate, feel free to call the RSMeans editors at 1-800-334-3509.

Unit Price Section

Table of Contents

Table of Contents (cont.)

How to Use the Unit Price Pages

The following is a detailed explanation of a sample entry in the Unit Price Section. Next to each bold number below is the described item with the appropriate component of the sample entry following in parentheses. Some prices are listed as bare costs; others as costs that include overhead and profit of the installing contractor. In most cases, if the work is to be subcontracted, the general contractor will need to add an additional markup (RSMeans suggests using 10%) to the figures in the column "Total Incl. O&P."

1 *Division Number/Title (03 30/Cast-In-Place Concrete)*

Use the Unit Price Section Table of Contents to locate specific items. The sections are classified according to the CSI MasterFormat 2004 system.

2 *Line Numbers (03 30 53.40 3920)*

Each unit price line item has been assigned a unique 12-digit code based on the CSI MasterFormat classification.

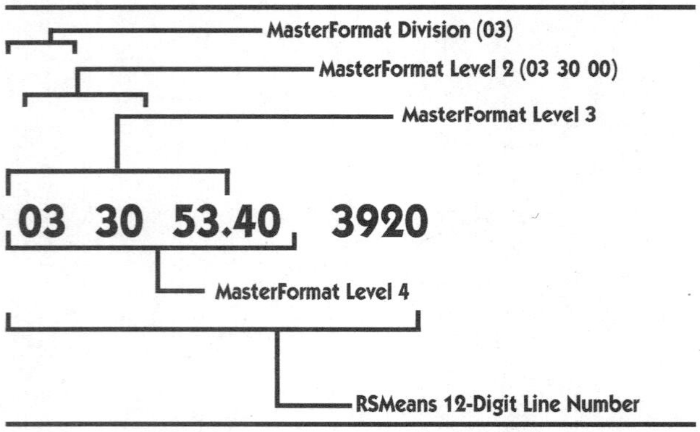

3 *Description (Concrete In Place, etc.)*

Each line item is described in detail. Sub-items and additional sizes are indented beneath the appropriate line items. The first line or two after the main item (in boldface) may contain descriptive information that pertains to all line items beneath this boldface listing.

Items which include the symbol **CN** are updated in The *RSMeans Quarterly Update Service* online. To obtain access to this service contact RSMeans customer service at 1-800-334-3509.

4 *Reference Number Information*

R033053 -50

You'll see reference numbers shown in shaded boxes at the beginning of some sections. These refer to related items in the Reference Section, visually identified by a vertical gray bar on the page edges.

The relation may be an estimating procedure that should be read before estimating, or technical information.

The "R" designates the Reference Section. The numbers refer to the MasterFormat 2004 classification system.

It is strongly recommended that you review all reference numbers that appear within the section in which you are working.

03 30 Cast-In-Place Concrete

03 30 53 – Miscellaneous Cast-In-Place Concrete

03 30 53.40 Concrete In Place			Crew	Daily Output	Labor-Hours	Unit	2009 Bare Costs Material	Labor	Equipment	Total	Total Incl O&P
0010	**CONCRETE IN PLACE**										
0020	Including forms (4 uses), reinforcing steel, concrete, placement,										
0050	and finishing unless otherwise indicated	R033053-50									
0300	Beams, 5 kip per L.F., 10' span		C-14A	15.62	12.804	C.Y.	400	515	49	964	1,300
0350	25' span		"	18.55	10.782		430	430	41	901	1,200
3850	Over 5 C.Y.		↓	75	1.493		198	57	.34	255.34	305
3900	Footings, strip, 18" x 9", unreinforced		C-14L	40	2.400		120	89.50	.65	210.15	271
3920	18" x 9", reinforced		C-14C	35	3.200		157	122	.74	279.74	365
3925	20" x 10", unreinforced		C-14L	45	2.133		117	79.50	.58	197.08	252
3930	20" x 10", reinforced		C-14C	40	2.800		147	107	.64	254.64	330
3935	24" x 12", unreinforced		C-14L	55	1.745		116	65	.47	181.47	229
3940	24" x 12", reinforced		C-14C	48	2.333	↓	147	89.50	.54	237.04	300

5 *Crew (C-14C)*

The "Crew" column designates the typical trade or crew used to install the item. If an installation can be accomplished by one trade and requires no power equipment, that trade and the number of workers are listed (for example, "2 Carpenters"). If an installation requires a composite crew, a crew code designation is listed (for example, "C-14C"). You'll find full details on all composite crews in the Crew Listings.

- For a complete list of all trades utilized in this book and their abbreviations, see the inside back cover.

Crews

Crew No.	Bare Costs		Incl. Subs O & P		Cost Per Labor-Hour	
Crew C-14C	Hr.	Daily	Hr.	Daily	Bare Costs	Incl. O&P
1 Carpenter Foreman (out)	$41.95	$335.60	$65.05	$520.40	$38.25	$59.61
6 Carpenters	39.95	1917.60	61.95	2973.60		
2 Rodmen (reinf.)	44.55	712.80	72.85	1165.60		
4 Laborers	31.60	1011.20	49.00	1568.00		
1 Cement Finisher	38.30	306.40	56.05	448.40		
1 Gas Engine Vibrator		26.00		28.60	.23	.26
112 L.H., Daily Totals		$4309.60		$6704.60	$38.48	$59.86

6 *Productivity: Daily Output (35.0)/ Labor-Hours (3.20)*

The "Daily Output" represents the typical number of units the designated crew will install in a normal 8-hour day. To find out the number of days the given crew would require to complete the installation, divide your quantity by the daily output. For example:

Quantity	÷	Daily Output	=	Duration
100 C.Y.	÷	35.0/ Crew Day	=	2.86 Crew Days

The "Labor-Hours" figure represents the number of labor-hours required to install one unit of work. To find out the number of labor-hours required for your particular task, multiply the quantity of the item times the number of labor-hours shown. For example:

Quantity	x	Productivity Rate	=	Duration
100 C.Y.	x	3.20 Labor-Hours/ C.Y.	=	320 Labor-Hours

7 *Unit (C.Y.)*

The abbreviated designation indicates the unit of measure upon which the price, production, and crew are based (C.Y. = Cubic Yard). For a complete listing of abbreviations, refer to the Abbreviations Listing in the Reference Section of this book.

8 *Bare Costs:*

Mat. (Bare Material Cost) (157)

The unit material cost is the "bare" material cost with no overhead or profit included. Costs shown reflect national average material prices for January of the current year and include delivery to the job site. No sales taxes are included.

Labor (122)

The unit labor cost is derived by multiplying bare labor-hour costs for Crew C-14C by labor-hour units. The bare labor-hour cost is found in the Crew Section under C-14C. (If a trade is listed, the hourly labor cost—the wage rate—is found on the inside back cover.)

Labor-Hour Cost Crew C-14C	x	Labor-Hour Units	=	Labor
$38.25	x	3.20		$122.00

Equip. (Equipment) (.74)

Costs for equipment included in each crew are listed in the Crew Listings in the Reference Section. Costs for small tools that are considered as essential everyday tools are covered in the "Overhead" column on the "Installing Contractor's Overhead and Profit" table that lists labor trades, base rates and markups. The unit equipment cost displayed here is derived by multiplying the bare equipment hourly cost in the Crew Listing by the unit Labor-Hours value.

Equipment Cost Crew C-14C	x	Labor-Hour Units	=	Equip.
$.23	x	3.20	=	$.74

Total (279.74)

The total of the bare costs is the arithmetic total of the three previous columns: mat., labor, and equip.

Material	+	Labor	+	Equip.	=	Total
$157	+	$122	+	$.74	=	$279.74

9 *Total Costs Including O&P*

This figure is the sum of the bare material cost plus 10% for profit; the bare labor cost plus total overhead and profit (per the inside back cover or, if a crew is listed, from the crew listings); and the bare equipment cost plus 10% for profit.

Material is Bare Material Cost + 10% = 157 + 15.70	=	$172.70
Labor for Crew C-14C = Labor-Hour Cost (59.61) x Labor-Hour Units (3.20)	=	$190.75
Equip. is Bare Equip. Cost + 10% = .74 + .07	=	$ .81
Total (Rounded)	=	$ 365

Estimating Tips

01 20 00 Price and Payment Procedures

- Allowances that should be added to estimates to cover contingencies and job conditions that are not included in the national average material and labor costs are shown in section 01 21.
- When estimating historic preservation projects (depending on the condition of the existing structure and the owner's requirements), a 15%–20% contingency or allowance is recommended, regardless of the stage of the drawings.

01 30 00 Administrative Requirements

- Before determining a final cost estimate, it is a good practice to review all the items listed in Subdivisions 01 31 and 01 32 to make final adjustments for items that may need customizing to specific job conditions.
- Requirements for initial and periodic submittals can represent a significant cost to the General Requirements of a job. Thoroughly check the submittal specifications when estimating a project to determine any costs that should be included.

01 40 00 Quality Requirements

- All projects will require some degree of Quality Control. This cost is not included in the unit cost of construction listed in each division. Depending upon the terms of the contract, the various costs of inspection and testing can be the responsibility of either the owner or the contractor. Be sure to include the required costs in your estimate.

01 50 00 Temporary Facilities and Controls

- Barricades, access roads, safety nets, scaffolding, security, and many more requirements for the execution of a safe project are elements of direct cost. These costs can easily be overlooked when preparing an estimate. When looking through the major classifications of this subdivision, determine which items apply to each division in your estimate.
- Construction Equipment Rental Costs can be found in the Reference Section in section 01 54 33. Operators' wages are not included in equipment rental costs.
- Equipment mobilization and demobilization costs are not included in equipment rental costs and must be considered separately in section 01 54 36.50.
- The cost of small tools provided by the installing contractor for his workers is covered in the "Overhead" column on the "Installing Contractor's Overhead and Profit" table that lists labor trades, base rates and markups and, therefore, is included in the "Total Incl. O&P" cost of any Unit Price line item. For those users who are constrained by contract terms to use only bare costs, there are two line items in section 01 54 39.70 for small tools as a percentage of bare labor cost. If some of those users are further constrained by contract terms to refrain from using line items from Division 1, you are advised to cover the cost of small tools within your coefficient or multiplier.

01 70 00 Execution and Closeout Requirements

- When preparing an estimate, thoroughly read the specifications to determine the requirements for Contract Closeout. Final cleaning, record documentation, operation and maintenance data, warranties and bonds, and spare parts and maintenance materials can all be elements of cost for the completion of a contract. Do not overlook these in your estimate.

Reference Numbers

Reference numbers are shown in shaded boxes at the beginning of some major classifications. These numbers refer to related items in the Reference Section. The reference information may be an estimating procedure, an alternate pricing method, or technical information.

Note: Not all subdivisions listed here necessarily appear in this publication.

01 11 Summary of Work

01 11 31 – Professional Consultants

01 11 31.10 Architectural Fees

	01 11 31.10 Architectural Fees	Crew	Daily Output	Labor-Hours	Unit	Material	Labor	Equipment	Total	Total Incl O&P
						2009 Bare Costs				
0010	**ARCHITECTURAL FEES** R011110-10									
2000	For "Greening" of building G				Project					3%

01 11 31.30 Engineering Fees

	01 11 31.30 Engineering Fees	Crew	Daily Output	Labor-Hours	Unit	Material	Labor	Equipment	Total	Total Incl O&P
0010	**ENGINEERING FEES** R011110-30									
0020	Educational planning consultant, minimum				Project					.50%
0100	Maximum				"					2.50%
0200	Electrical, minimum				Contrct					4.10%
0300	Maximum									10.10%
0400	Elevator & conveying systems, minimum									2.50%
0500	Maximum									5%
0600	Food service & kitchen equipment, minimum									8%
0700	Maximum									12%
1000	Mechanical (plumbing & HVAC), minimum									4.10%
1100	Maximum				↓					10.10%
4000	Consultant, using DOE software energy analysis, small bldg, min				SF Flr.					.25
4010	Maximum									.45
4020	Medium building, minimum									.15
4030	Maximum									.35
4040	Large building, minimum									.05
4050	Maximum				↓					.25

01 21 Allowances

01 21 16 – Contingency Allowances

01 21 16.50 Contingencies

	01 21 16.50 Contingencies	Crew	Daily Output	Labor-Hours	Unit	Material	Labor	Equipment	Total	Total Incl O&P
0010	**CONTINGENCIES**, Add to estimate									
0020	Conceptual stage				Project					20%
0050	Schematic stage									15%
0100	Preliminary working drawing stage (Design Dev.)									10%
0150	Final working drawing stage				↓					3%

01 21 53 – Factors Allowance

01 21 53.50 Factors

	01 21 53.50 Factors	Crew	Daily Output	Labor-Hours	Unit	Material	Labor	Equipment	Total	Total Incl O&P
0010	**FACTORS** Cost adjustments R012153-10									
0100	Add to construction costs for particular job requirements									
0500	Cut & patch to match existing construction, add, minimum				Costs	2%	3%			
0550	Maximum					5%	9%			
0800	Dust protection, add, minimum					1%	2%			
0850	Maximum					4%	11%			
1100	Equipment usage curtailment, add, minimum					1%	1%			
1150	Maximum					3%	10%			
1400	Material handling & storage limitation, add, minimum					1%	1%			
1450	Maximum					6%	7%			
1700	Protection of existing work, add, minimum					2%	2%			
1750	Maximum					5%	7%			
2000	Shift work requirements, add, minimum						5%			
2050	Maximum						30%			
2300	Temporary shoring and bracing, add, minimum					2%	5%			
2350	Maximum					5%	12%			
2400	Work inside prisons and high security areas, add, minimum						30%			
2450	Maximum				↓		50%			

01 21 Allowances

01 21 55 – Job Conditions Allowance

01 21 55.50 Job Conditions		Crew	Daily Output	Labor-Hours	Unit	Material	2009 Bare Costs Labor	Equipment	Total	Total Incl O&P
0010	**JOB CONDITIONS** Modifications to applicable									
0020	cost summaries									
0100	Economic conditions, favorable, deduct				Project				2%	2%
0200	Unfavorable, add								5%	5%
0300	Hoisting conditions, favorable, deduct								2%	2%
0400	Unfavorable, add								5%	5%
0700	Labor availability, surplus, deduct								1%	1%
0800	Shortage, add								10%	10%
0900	Material storage area, available, deduct								1%	1%
1000	Not available, add								2%	2%
1100	Subcontractor availability, surplus, deduct								5%	5%
1200	Shortage, add								12%	12%
1300	Work space, available, deduct								2%	2%
1400	Not available, add								5%	5%

01 21 57 – Overtime Allowance

01 21 57.50 Overtime		Crew	Daily Output	Labor-Hours	Unit	Material	Labor	Equipment	Total	Total Incl O&P
0010	**OVERTIME** for early completion of projects or where R012909-90									
0020	labor shortages exist, add to usual labor, up to				Costs		100%			

01 21 61 – Cost Indexes

01 21 61.10 Construction Cost Index		Crew	Daily Output	Labor-Hours	Unit	Material	Labor	Equipment	Total	Total Incl O&P
0010	**CONSTRUCTION COST INDEX** (Reference) over 930 zip code locations in									
0020	the U.S. and Canada, total bldg. cost, min. (Guymon, OK)				%					69.40%
0050	Average									100%
0100	Maximum (New York, NY)									130.70%
01 21 61.20 Historical Cost Indexes										
0010	**HISTORICAL COST INDEXES** (See Reference Section)									
01 21 61.30 Labor Index										
0010	**LABOR INDEX** (Reference) For over 930 zip code locations in									
0020	the U.S. and Canada, minimum (Del Rio, TX)				%		31.90%			
0050	Average						100%			
0100	Maximum (New York, NY)						164.60%			
01 21 61.50 Material Index										
0011	**MATERIAL INDEX** For over 930 zip code locations in									
0020	the U.S. and Canada, minimum (Elizabethtown, KY)				%	90.10%				
0040	Average					100%				
0060	Maximum (Ketchikan, AK)					139.60%				

01 21 63 – Taxes

01 21 63.10 Taxes		Crew	Daily Output	Labor-Hours	Unit	Material	Labor	Equipment	Total	Total Incl O&P
0010	**TAXES** R012909-80									
0020	Sales tax, State, average				%	4.91%				
0050	Maximum R012909-85					7.25%				
0200	Social Security, on first $102,000 of wages						7.65%			
0300	Unemployment, combined Federal and State, minimum						.80%			
0350	Average						6.20%			
0400	Maximum						11.76%			

01 31 Project Management and Coordination

01 31 13 – Project Coordination

01 31 13.20 Field Personnel		Crew	Daily Output	Labor-Hours	Unit	Material	2009 Bare Costs Labor	Equipment	Total	Total Incl O&P
0010	**FIELD PERSONNEL**									
0020	Clerk, average				Week		380		380	590
0100	Field engineer, minimum						895		895	1,375
0120	Average						1,165		1,165	1,800
0140	Maximum						1,350		1,350	2,100
0160	General purpose laborer, average						1,250		1,250	1,925
0180	Project manager, minimum						1,650		1,650	2,550
0200	Average						1,925		1,925	2,975
0220	Maximum						2,175		2,175	3,375
0240	Superintendent, minimum						1,600		1,600	2,475
0260	Average						1,775		1,775	2,750
0280	Maximum						2,025		2,025	3,125
0290	Timekeeper, average						1,040		1,040	1,600

01 31 13.30 Insurance		Crew	Daily Output	Labor-Hours	Unit	Material	Labor	Equipment	Total	Total Incl O&P
0010	**INSURANCE** R013113-40									
0020	Builders risk, standard, minimum				Job					.24%
0050	Maximum R013113-50									.64%
0200	All-risk type, minimum									.25%
0250	Maximum R013113-60									.62%
0400	Contractor's equipment floater, minimum				Value					.50%
0450	Maximum				"					1.50%
0600	Public liability, average				Job					2.02%
0800	Workers' compensation & employer's liability, average									
0850	by trade, carpentry, general				Payroll		17.80%			
1000	Electrical						6.46%			
1150	Insulation						14.44%			
1450	Plumbing						7.84%			
1550	Sheet metal work (HVAC)						11.57%			

01 31 13.50 General Contractor's Mark-Up		Crew	Daily Output	Labor-Hours	Unit	Material	Labor	Equipment	Total	Total Incl O&P
0010	**GENERAL CONTRACTOR'S MARK-UP** on Change Orders									
0200	Extra work, by subcontractors, add				%					10%
0250	By General Contractor, add									15%
0400	Omitted work, by subcontractors, deduct all but									5%
0450	By General Contractor, deduct all but									7.50%
0600	Overtime work, by subcontractors, add									15%
0650	By General Contractor, add									10%

01 31 13.60 Installing Contractor's Main Office Overhead		Crew	Daily Output	Labor-Hours	Unit	Material	Labor	Equipment	Total	Total Incl O&P
0010	**INSTALLING CONTRACTOR'S MAIN OFFICE OVERHEAD**									
0020	As percent of direct costs, minimum				%				5%	
0050	Average								13%	
0100	Maximum								30%	

01 31 13.80 Overhead and Profit		Crew	Daily Output	Labor-Hours	Unit	Material	Labor	Equipment	Total	Total Incl O&P
0010	**OVERHEAD & PROFIT** Allowance to add to items in this									
0020	book that do not include Subs O&P, average				%				25%	
0100	Allowance to add to items in this book that									
0110	do include Subs O&P, minimum				%					5%
0150	Average									10%
0200	Maximum									15%
0300	Typical, by size of project, under $100,000								30%	
0350	$500,000 project								25%	
0400	$2,000,000 project								20%	
0450	Over $10,000,000 project								15%	

01 31 Project Management and Coordination

01 31 13 – Project Coordination

01 31 13.90 Performance Bond

		Crew	Daily Output	Labor-Hours	Unit	Material (2009 Bare Costs)	Labor	Equipment	Total	Total Incl O&P
0010	**PERFORMANCE BOND** R013113-80									
0020	For buildings, minimum				Job					.60%
0100	Maximum				"					2.50%

01 32 Construction Progress Documentation

01 32 33 – Photographic Documentation

01 32 33.50 Photographs

		Crew	Daily Output	Labor-Hours	Unit	Material (2009 Bare Costs)	Labor	Equipment	Total	Total Incl O&P
0010	**PHOTOGRAPHS**									
0020	8" x 10", 4 shots, 2 prints ea., std. mounting				Set	475			475	520
0100	Hinged linen mounts					530			530	580
0200	8" x 10", 4 shots, 2 prints each, in color					415			415	460
0300	For I.D. slugs, add to all above					5.30			5.30	5.85
1500	Time lapse equipment, camera and projector, buy					3,775			3,775	4,175
1550	Rent per month				↓	565			565	620
1700	Cameraman and film, including processing, B.&W.				Day	1,375			1,375	1,525
1720	Color				"	1,375			1,375	1,525

01 41 Regulatory Requirements

01 41 26 – Permits

01 41 26.50 Permits

		Crew	Daily Output	Labor-Hours	Unit	Material (2009 Bare Costs)	Labor	Equipment	Total	Total Incl O&P
0010	**PERMITS**									
0020	Rule of thumb, most cities, minimum				Job					.50%
0100	Maximum				"					2%

01 45 Quality Control

01 45 23 – Testing and Inspecting Services

01 45 23.50 Testing

		Crew	Daily Output	Labor-Hours	Unit	Material (2009 Bare Costs)	Labor	Equipment	Total	Total Incl O&P
0010	**TESTING** and Inspecting Services									
9000	Thermographic testing, for bldg envelope heat loss, average 2,000 S.F.				Ea.					500

01 51 Temporary Utilities

01 51 13 – Temporary Electricity

01 51 13.80 Temporary Utilities

		Crew	Daily Output	Labor-Hours	Unit	Material (2009 Bare Costs)	Labor	Equipment	Total	Total Incl O&P
0010	**TEMPORARY UTILITIES**									
0100	Heat, incl. fuel and operation, per week, 12 hrs. per day	1 Skwk	100	.080	CSF Flr	27	3.27		30.27	34.50
0200	24 hrs. per day	"	60	.133		52	5.45		57.45	66
0350	Lighting, incl. service lamps, wiring & outlets, minimum	1 Elec	34	.235		2.63	11.05		13.68	19.35
0360	Maximum	"	17	.471		5.70	22		27.70	39.50
0400	Power for temp lighting only, per month, min/month 6.6 KWH								.75	.83
0450	Maximum/month 23.6 KWH								2.85	3.14
0600	Power for job duration incl. elevator, etc., minimum								47	51.70
0650	Maximum				↓				110	121
1000	Toilet, portable, see Equip. Rental 01 54 33 in Reference Section									

01 52 Construction Facilities

01 52 13 – Field Offices and Sheds

01 52 13.20 Office and Storage Space

	01 52 13.20 Office and Storage Space	Crew	Daily Output	Labor-Hours	Unit	2009 Bare Costs Material	Labor	Equipment	Total	Total Incl O&P
0010	**OFFICE AND STORAGE SPACE**									
0020	Trailer, furnished, no hookups, 20′ x 8′, buy	2 Skwk	1	16	Ea.	8,200	655		8,855	10,000
0250	Rent per month					163			163	179
0300	32′ x 8′, buy	2 Skwk	.70	22.857		12,200	935		13,135	15,000
0350	Rent per month					200			200	220
0400	50′ x 10′, buy	2 Skwk	.60	26.667		23,200	1,100		24,300	27,300
0450	Rent per month					281			281	310
0500	50′ x 12′, buy	2 Skwk	.50	32		27,900	1,300		29,200	32,700
0550	Rent per month					375			375	410
0700	For air conditioning, rent per month, add					41			41	45
0800	For delivery, add per mile				Mile	4.50			4.50	4.95
1000	Portable buildings, prefab, on skids, economy, 8′ x 8′	2 Carp	265	.060	S.F.	85	2.41		87.41	97
1100	Deluxe, 8′ x 12′	"	150	.107	"	95	4.26		99.26	112
1200	Storage boxes, 20′ x 8′, buy	2 Skwk	1.80	8.889	Ea.	4,675	365		5,040	5,700
1250	Rent per month					72			72	79
1300	40′ x 8′, buy	2 Skwk	1.40	11.429		6,400	465		6,865	7,775
1350	Rent per month					99			99	109

01 52 13.40 Field Office Expense

	01 52 13.40 Field Office Expense	Crew	Daily Output	Labor-Hours	Unit	Material	Labor	Equipment	Total	Total Incl O&P
0010	**FIELD OFFICE EXPENSE**									
0100	Office equipment rental average				Month	155			155	171
0120	Office supplies, average				"	85			85	93.50
0125	Office trailer rental, see Div. 01 52 13.20									
0140	Telephone bill; avg. bill/month incl. long dist.				Month	80			80	88
0160	Lights & HVAC				"	150			150	165

01 54 Construction Aids

01 54 16 – Temporary Hoists

01 54 16.50 Weekly Forklift Crew

	01 54 16.50 Weekly Forklift Crew	Crew	Daily Output	Labor-Hours	Unit	Material	Labor	Equipment	Total	Total Incl O&P
0010	**WEEKLY FORKLIFT CREW**									
0100	All-terrain forklift, 45′ lift, 35′ reach, 9000 lb. capacity	A-3P	.20	40	Month		1,550	2,225	3,775	4,800

01 54 19 – Temporary Cranes

01 54 19.50 Daily Crane Crews

	01 54 19.50 Daily Crane Crews	Crew	Daily Output	Labor-Hours	Unit	Material	Labor	Equipment	Total	Total Incl O&P
0010	**DAILY CRANE CREWS** for small jobs, portal to portal									
0100	12-ton truck-mounted hydraulic crane	A-3H	1	8	Day		340	1,025	1,365	1,625
0200	25-ton	A-3I	1	8			340	1,050	1,390	1,650
0300	40-ton	A-3J	1	8			340	1,050	1,390	1,675
0400	55-ton	A-3K	1	16			635	1,625	2,260	2,750
0500	80-ton	A-3L	1	16			635	1,850	2,485	2,975
0600	100-ton	A-3M	1	16			635	2,525	3,160	3,750

01 54 19.60 Monthly Tower Crane Crew

	01 54 19.60 Monthly Tower Crane Crew	Crew	Daily Output	Labor-Hours	Unit	Material	Labor	Equipment	Total	Total Incl O&P
0010	**MONTHLY TOWER CRANE CREW**, excludes concrete footing									
0100	Static tower crane, 130′ high, 106′ jib, 6200 lb. capaccity	A-3N	.05	176	Month		7,500	22,500	30,000	36,000

01 54 23 – Temporary Scaffolding and Platforms

01 54 23.70 Scaffolding

	01 54 23.70 Scaffolding	Crew	Daily Output	Labor-Hours	Unit	Material	Labor	Equipment	Total	Total Incl O&P
0010	**SCAFFOLDING** R015423-10									
0015	Steel tube, regular, no plank, labor only to erect & dismantle									
0090	Building exterior, wall face, 1 to 5 stories, 6′-4″ x 5′ frames	3 Carp	8	3	C.S.F.		120		120	186
0200	6 to 12 stories	4 Carp	8	4			160		160	248
0301	13 to 20 stories	5 Clab	8	5			158		158	245

01 54 23.70 Scaffolding		Crew	Daily Output	Labor-Hours	Unit	2009 Bare Costs Material	Labor	Equipment	Total	Total Incl O&P
0460	Building interior, wall face area, up to 16′ high	3 Carp	12	2	C.S.F.		80		80	124
0560	16′ to 40′ high		10	2.400			96		96	149
0800	Building interior floor area, up to 30′ high		150	.160	C.C.F.		6.40		6.40	9.90
0900	Over 30′ high	4 Carp	160	.200	"		8		8	12.40
0906	Complete system for face of walls, no plank, material only rent/mo				C.S.F.	35.50			35.50	39
0908	Interior spaces, no plank, material only rent/mo				C.C.F.	3.40			3.40	3.74
0910	Steel tubular, heavy duty shoring, buy									
0920	Frames 5′ high 2′ wide				Ea.	93			93	102
0925	5′ high 4′ wide					105			105	116
0930	6′ high 2′ wide					106			106	117
0935	6′ high 4′ wide					124			124	136
0940	Accessories									
0945	Cross braces				Ea.	18			18	19.80
0950	U-head, 8″ x 8″					21.50			21.50	24
0955	J-head, 4″ x 8″					15.80			15.80	17.40
0960	Base plate, 8″ x 8″					17.60			17.60	19.35
0965	Leveling jack					38			38	41.50
1000	Steel tubular, regular, buy									
1100	Frames 3′ high 5′ wide				Ea.	71			71	78
1150	5′ high 5′ wide					83			83	91.50
1200	6′-4″ high 5′ wide					104			104	114
1350	7′-6″ high 6′ wide					179			179	197
1500	Accessories cross braces					20.50			20.50	22.50
1550	Guardrail post					18.60			18.60	20.50
1600	Guardrail 7′ section					9.90			9.90	10.90
1650	Screw jacks & plates					27			27	29.50
1700	Sidearm brackets					43			43	47.50
1750	8″ casters					35.50			35.50	39
1800	Plank 2″ x 10″ x 16′-0″					51			51	56
1900	Stairway section					325			325	360
1910	Stairway starter bar					36.50			36.50	40
1920	Stairway inside handrail					66			66	72.50
1930	Stairway outside handrail					98			98	108
1940	Walk-thru frame guardrail					47			47	51.50
2000	Steel tubular, regular, rent/mo.									
2100	Frames 3′ high 5′ wide				Ea.	5			5	5.50
2150	5′ high 5′ wide					5			5	5.50
2200	6′-4″ high 5′ wide					5.05			5.05	5.55
2250	7′-6″ high 6′ wide					7			7	7.70
2500	Accessories, cross braces					1			1	1.10
2550	Guardrail post					1			1	1.10
2600	Guardrail 7′ section					1			1	1.10
2650	Screw jacks & plates					2			2	2.20
2700	Sidearm brackets					2			2	2.20
2750	8″ casters					8			8	8.80
2800	Outrigger for rolling tower					3			3	3.30
2850	Plank 2″ x 10″ x 16′-0″					6			6	6.60
2900	Stairway section					40			40	44
2940	Walk-thru frame guardrail					2.50			2.50	2.75
3000	Steel tubular, heavy duty shoring, rent/mo.									
3250	5′ high 2′ & 4′ wide				Ea.	5			5	5.50
3300	6′ high 2′ & 4′ wide					5			5	5.50
3500	Accessories, cross braces					1			1	1.10

01 54 Construction Aids

01 54 23 – Temporary Scaffolding and Platforms

01 54 23.70 Scaffolding		Crew	Daily Output	Labor-Hours	Unit	2009 Bare Costs Material	Labor	Equipment	Total	Total Incl O&P
3600	U-head, 8" x 8"				Ea.	1			1	1.10
3650	J-head, 4" x 8"					1			1	1.10
3700	Base plate, 8" x 8"					1			1	1.10
3750	Leveling jack					2			2	2.20
5700	Planks, 2x10x16'-0", labor only to erect & remove to 50' H	3 Carp	72	.333			13.30		13.30	20.50
5800	Over 50' high	4 Carp	80	.400			16		16	25
6000	Heavy duty shoring for elevated slab forms to 8'-2" high, floor area									
6100	Labor only to erect & dismantle	4 Carp	16	2	C.S.F.		80		80	124
6110	Materials only, rent.mo				"	29.50			29.50	32.50
6500	To 14'-8" high									
6600	Labor only to erect & dismantle	4 Carp	10	3.200	C.S.F.		128		128	198
6610	Materials only, rent/mo				"	43			43	47.50

01 54 23.75 Scaffolding Specialties		Crew	Daily Output	Labor-Hours	Unit	Material	Labor	Equipment	Total	Total Incl O&P
0010	**SCAFFOLDING SPECIALTIES**									
1200	Sidewalk bridge, heavy duty steel posts & beams, including									
1210	parapet protection & waterproofing									
1220	8' to 10' wide, 2 posts	3 Carp	15	1.600	L.F.	35	64		99	138
1230	3 posts	"	10	2.400	"	54	96		150	209
1500	Sidewalk bridge using tubular steel									
1510	scaffold frames, including planking	3 Carp	45	.533	L.F.	5.60	21.50		27.10	39
1600	For 2 uses per month, deduct from all above					50%				
1700	For 1 use every 2 months, add to all above					100%				
1900	Catwalks, 20" wide, no guardrails, 7' span, buy				Ea.	145			145	160
2000	10' span, buy					203			203	223
3720	Putlog, standard, 8' span, with hangers, buy					75			75	82.50
3730	Rent per month					10			10	11
3750	12' span, buy					113			113	124
3755	Rent per month					15			15	16.50
3760	Trussed type, 16' span, buy					260			260	286
3770	Rent per month					20			20	22
3790	22' span, buy					310			310	345
3795	Rent per month					30			30	33
3800	Rolling ladders with handrails, 30" wide, buy, 2 step					214			214	235
4000	7 step					655			655	720
4050	10 step					910			910	1,000
4100	Rolling towers, buy, 5' wide, 7' long, 10' high					1,350			1,350	1,500
4200	For 5' high added sections, to buy, add					207			207	228
4300	Complete incl. wheels, railings, outriggers,									
4350	21' high, to buy				Ea.	2,250			2,250	2,475
4400	Rent/month = 5% of purchase cost				"	7.50			7.50	8.25

01 54 36 – Equipment Mobilization

01 54 36.50 Mobilization or Demob.		Crew	Daily Output	Labor-Hours	Unit	Material	Labor	Equipment	Total	Total Incl O&P
0010	**MOBILIZATION OR DEMOB.** (One or the other, unless noted) R015433-10									
0015	Up to 25 mi haul dist (50 mi RT for mob/demob crew)									
0020	Dozer, loader, backhoe, excav., grader, paver, roller, 70 to 150 H.P.	B-34N	4	2	Ea.		64	117	181	227
0100	Above 150 HP	B-34K	3	2.667			85	261	346	420
0900	Shovel or dragline, 3/4 C.Y.	"	3.60	2.222			71	218	289	350
1100	Small equipment, placed in rear of, or towed by pickup truck	A-3A	8	1			31	16.05	47.05	65
1150	Equip up to 70 HP, on flatbed trailer behind pickup truck	A-3D	4	2			62	57.50	119.50	159
2000	Mob & demob truck-mounted crane up to 75 ton, driver only	1 Eqhv	3.60	2.222			94.50		94.50	142
2100	Crane, truck-mounted, over 75 ton	A-3E	2.50	6.400			238	51.50	289.50	415
2200	Crawler-mounted, up to 75 ton	A-3F	2	8			298	405	703	895

01 54 Construction Aids

01 54 36 – Equipment Mobilization

01 54 36.50 Mobilization or Demob.		Crew	Daily Output	Labor-Hours	Unit	Material	2009 Bare Costs Labor	Equipment	Total	Total Incl O&P
2300	Over 75 ton	A-3G	1.50	10.667	Ea.		395	610	1,005	1,275
2500	For each additional 5 miles haul distance, add						10%	10%		
3000	For large pieces of equipment, allow for assembly/knockdown									
3100	For mob/demob of micro-tunneling equip, see Div. 33 05 23.19									

01 54 39 – Construction Equipment

01 54 39.70 Small Tools

		Crew	Daily Output	Labor-Hours	Unit	Material	Labor	Equipment	Total	Total Incl O&P
0010	**SMALL TOOLS** R013113-50									
0020	As % of contractor's bare labor cost for project, minimum				Total		.50%			
0100	Maximum				"		2%			

01 55 Vehicular Access and Parking

01 55 23 – Temporary Roads

01 55 23.50 Roads and Sidewalks

		Crew	Daily Output	Labor-Hours	Unit	Material	Labor	Equipment	Total	Total Incl O&P
0010	**ROADS AND SIDEWALKS** Temporary									
0050	Roads, gravel fill, no surfacing, 4" gravel depth	B-14	715	.067	S.Y.	4.97	2.23	.41	7.61	9.35
0100	8" gravel depth	"	615	.078	"	9.95	2.59	.48	13.02	15.45
1000	Ramp, 3/4" plywood on 2" x 6" joists, 16" O.C.	2 Carp	300	.053	S.F.	1.35	2.13		3.48	4.79
1100	On 2" x 10" joists, 16" O.C.	"	275	.058	"	1.82	2.32		4.14	5.60

01 56 Temporary Barriers and Enclosures

01 56 13 – Temporary Air Barriers

01 56 13.60 Tarpaulins

		Crew	Daily Output	Labor-Hours	Unit	Material	Labor	Equipment	Total	Total Incl O&P
0010	**TARPAULINS**									
0020	Cotton duck, 10 oz. to 13.13 oz. per S.Y., minimum				S.F.	.54			.54	.59
0050	Maximum					.53			.53	.58
0100	Polyvinyl coated nylon, 14 oz. to 18 oz., minimum					.48			.48	.53
0150	Maximum					.68			.68	.75
0200	Reinforced polyethylene 3 mils thick, white					.15			.15	.17
0300	4 mils thick, white, clear or black					.20			.20	.22
0400	5.5 mils thick, clear					.21			.21	.23
0500	White, fire retardant					.35			.35	.39
0600	7.5 mils, oil resistant, fire retardant					.40			.40	.44
0700	8.5 mils, black					.53			.53	.58
0710	Woven polyethylene, 6 mils thick					.35			.35	.39
0730	Polyester reinforced w/ integral fastening system 11 mils thick					1.07			1.07	1.18
0740	Mylar polyester, non-reinforced, 7 mils thick				↓	1.17			1.17	1.29

01 56 13.90 Winter Protection

		Crew	Daily Output	Labor-Hours	Unit	Material	Labor	Equipment	Total	Total Incl O&P
0010	**WINTER PROTECTION**									
0100	Framing to close openings	2 Clab	750	.021	S.F.	.39	.67		1.06	1.48
0200	Tarpaulins hung over scaffolding, 8 uses, not incl. scaffolding		1500	.011		.25	.34		.59	.80
0250	Tarpaulin polyester reinf. w/ integral fastening system 11 mils thick		1600	.010		.80	.32		1.12	1.37
0300	Prefab fiberglass panels, steel frame, 8 uses	↓	1200	.013	↓	.85	.42		1.27	1.59

01 56 23 – Temporary Barricades

01 56 23.10 Barricades

		Crew	Daily Output	Labor-Hours	Unit	Material	Labor	Equipment	Total	Total Incl O&P
0010	**BARRICADES**									
0020	5' high, 3 rail @ 2" x 8", fixed	2 Carp	20	.800	L.F.	4.99	32		36.99	55
0150	Movable		30	.533		4.12	21.50		25.62	37.50
1000	Guardrail, wooden, 3' high, 1" x 6", on 2" x 4" posts	↓	200	.080	↓	1.02	3.20		4.22	6.10

01 56 Temporary Barriers and Enclosures

01 56 23 – Temporary Barricades

01 56 23.10 Barricades		Crew	Daily Output	Labor-Hours	Unit	2009 Bare Costs Material	Labor	Equipment	Total	Total Incl O&P
1100	2" x 6", on 4" x 4" posts	2 Carp	165	.097	L.F.	2.16	3.87		6.03	8.35
1200	Portable metal with base pads, buy					19.75			19.75	21.50
1250	Typical installation, assume 10 reuses	2 Carp	600	.027		2.50	1.07		3.57	4.40
1300	Barricade tape, polyethylene, 7 mil, 3" wide x 500' long roll				Ea.	25			25	27.50

01 56 26 – Temporary Fencing

01 56 26.50 Temporary Fencing		Crew	Daily Output	Labor-Hours	Unit	Material	Labor	Equipment	Total	Total Incl O&P
0010	**TEMPORARY FENCING**									
0020	Chain link, 11 ga, 5' high	2 Clab	400	.040	L.F.	7.25	1.26		8.51	9.95
0100	6' high		300	.053		7.75	1.69		9.44	11.15
0200	Rented chain link, 6' high, to 1000' (up to 12 mo.)		400	.040		2.69	1.26		3.95	4.92
0250	Over 1000' (up to 12 mo.)		300	.053		2.59	1.69		4.28	5.45
0350	Plywood, painted, 2" x 4" frame, 4' high	A-4	135	.178		5.05	6.80		11.85	16
0400	4" x 4" frame, 8' high	"	110	.218		9.85	8.35		18.20	23.50
0500	Wire mesh on 4" x 4" posts, 4' high	2 Carp	100	.160		9.85	6.40		16.25	20.50
0550	8' high	"	80	.200		15.05	8		23.05	29

01 56 29 – Temporary Protective Walkways

01 56 29.50 Protection		Crew	Daily Output	Labor-Hours	Unit	Material	Labor	Equipment	Total	Total Incl O&P
0010	**PROTECTION**									
0020	Stair tread, 2" x 12" planks, 1 use	1 Carp	75	.107	Tread	4.16	4.26		8.42	11.15
0100	Exterior plywood, 1/2" thick, 1 use		65	.123		1.39	4.92		6.31	9.10
0200	3/4" thick, 1 use		60	.133		2.49	5.35		7.84	11
2200	Sidewalks, 2" x 12" planks, 2 uses		350	.023	S.F.	.69	.91		1.60	2.18
2300	Exterior plywood, 2 uses, 1/2" thick		750	.011		.23	.43		.66	.91
2400	5/8" thick		650	.012		.34	.49		.83	1.13
2500	3/4" thick		600	.013		.42	.53		.95	1.29

01 58 Project Identification

01 58 13 – Temporary Project Signage

01 58 13.50 Signs		Crew	Daily Output	Labor-Hours	Unit	Material	Labor	Equipment	Total	Total Incl O&P
0010	**SIGNS**									
0020	High intensity reflectorized, no posts, buy				S.F.	21			21	23

01 74 Cleaning and Waste Management

01 74 13 – Progress Cleaning

01 74 13.20 Cleaning Up		Crew	Daily Output	Labor-Hours	Unit	Material	Labor	Equipment	Total	Total Incl O&P
0010	**CLEANING UP**									
0020	After job completion, allow, minimum				Job					.30%
0040	Maximum				"					1%
0050	Cleanup of floor area, continuous, per day, during const.	A-5	24	.750	M.S.F.	1.70	23.50	2.03	27.23	40.50
0100	Final by GC at end of job	"	11.50	1.565	"	2.71	49.50	4.23	56.44	84
1000	Mechanical demolition, see Div. 02 41 19									

01 91 Commissioning

01 91 13 – General Commissioning Requirements

01 91 13.50 Commissioning		Crew	Daily Output	Labor-Hours	Unit	Material	2009 Bare Costs Labor	Equipment	Total	Total Incl O&P
0010	**COMMISSIONING** Including documentation of design intent									
0100	performance verification, O&M, training, minimum				Project					.50%
0150	Maximum				"					.75%

01 93 Facility Maintenance

01 93 13 – Facility Maintenance Procedures

01 93 13.15 Mechanical Facilities Maintenance

	01 93 13.15 Mechanical Facilities Maintenance	Crew	Daily Output	Labor-Hours	Unit	Material	Labor	Equipment	Total	Total Incl O&P
0010	**MECHANICAL FACILITIES MAINTENANCE**									
0100	Air conditioning system maintenance									
0130	Belt, replace	1 Stpi	15	.533	Ea.		26.50		26.50	39.50
0170	Fan, clean		16	.500			24.50		24.50	37
0180	Filter, remove, clean, replace		12	.667			33		33	49.50
0190	Flexible coupling alignment, inspect		40	.200			9.85		9.85	14.80
0200	Gas leak locate and repair		4	2			98.50		98.50	148
0250	Pump packing gland, remove and replace		11	.727			36		36	54
0270	Tighten		32	.250			12.35		12.35	18.50
0290	Pump disassemble and assemble		4	2			98.50		98.50	148
0300	Air pressure regulator, disassemble, clean, assemble	1 Skwk	4	2			81.50		81.50	127
0310	Repair or replace part	"	6	1.333			54.50		54.50	84.50
0320	Purging system	1 Stpi	16	.500			24.50		24.50	37
0400	Compressor, air, remove or install fan wheel	1 Skwk	20	.400			16.35		16.35	25.50
0410	Disassemble or assemble 2 cylinder, 2 stage		4	2			81.50		81.50	127
0420	4 cylinder, 4 stage		1	8			325		325	505
0430	Repair or replace part		2	4			163		163	253
0700	Demolition, for mech. demolition see Div. 23 05 05.10 or 22 05 05.10									
0800	Ductwork, clean									
0810	Rectangular									
0820	6" G	1 Shee	187.50	.043	L.F.		2.01		2.01	3.10
0830	8" G		140.63	.057			2.69		2.69	4.13
0840	10" G		112.50	.071			3.36		3.36	5.15
0850	12" G		93.75	.085			4.03		4.03	6.20
0860	14" G		80.36	.100			4.70		4.70	7.25
0870	16" G		70.31	.114			5.35		5.35	8.25
0900	Round									
0910	4" G	1 Shee	358.10	.022	L.F.		1.05		1.05	1.62
0920	6" G		238.73	.034			1.58		1.58	2.43
0930	8" G		179.05	.045			2.11		2.11	3.25
0940	10" G		143.24	.056			2.64		2.64	4.06
0950	12" G		119.37	.067			3.16		3.16	4.87
0960	16" G		89.52	.089			4.22		4.22	6.50
1000	Expansion joint, not screwed, install or remove	1 Stpi	3	2.667	Ea.		132		132	197
1010	Repack	"	6	1.333	"		66		66	98.50
1200	Fire protection equipment									
1220	Fire hydrant, replace	Q-1	3	5.333	Ea.		234		234	350
1230	Service, lubricate, inspect, flush, clean	1 Plum	7	1.143			55.50		55.50	83.50
1240	Test	"	11	.727			35.50		35.50	53
1310	Inspect valves, pressure, nozzle,	1 Spri	4	2	System		96.50		96.50	145
1500	Instrumentation preventive maintenance, rule of thumb									
1510	Annual cost per instrument for calibration & trouble shooting									
1520	Includes flow meters, analytical transmitters, flow control									
1530	chemical feed systems, recorders and distrib. control sys.				Ea.					325

01 93 13 – Facility Maintenance Procedures

01 93 13.15 Mechanical Facilities Maintenance		Crew	Daily Output	Labor-Hours	Unit	Material	2009 Bare Costs Labor	Equipment	Total	Total Incl O&P
1800	Plumbing fixtures, for installation see Div. 22 41 00									
1801	Plumbing fixtures									
1840	Open drain with snake	1 Plum	13	.615	Ea.		30		30	45
1850	Open drain with toilet auger		16	.500			24.50		24.50	36.50
1860	Grease trap, clean		2	4			195		195	293
1870	Plaster trap, clean		6	1.333			65		65	97.50
1881	Clean commode	1 Clab	20	.400		.34	12.65		12.99	19.95
1882	Clean commode seat		48	.167		.17	5.25		5.42	8.35
1883	Clean single sink		28	.286		.22	9.05		9.27	14.25
1884	Clean double sink		20	.400		.34	12.65		12.99	19.95
1885	Clean lavatory		28	.286		.22	9.05		9.27	14.25
1886	Clean bathtub		12	.667		.68	21		21.68	33.50
1887	Clean fiberglass shower		12	.667		.68	21		21.68	33.50
1888	Clean fiberglass tub/shower		8	1		.85	31.50		32.35	50
1889	Clean faucet set		32	.250		.14	7.90		8.04	12.40
1891	Clean shower head		80	.100		.02	3.16		3.18	4.92
1893	Clean water heater		16	.500		1.19	15.80		16.99	26
1900	Relief valve, test and adjust	1 Stpi	20	.400			19.75		19.75	29.50
1910	Clean pump, heater, or motor for whirlpool	1 Clab	16	.500		.17	15.80		15.97	24.50
1920	Clean thermal cover for whirlpool	"	16	.500		1.70	15.80		17.50	26.50
2000	Repair or replace-steam trap	1 Stpi	8	1			49.50		49.50	74
2020	Y-type or bell strainer		6	1.333			66		66	98.50
2040	Water trap or vacuum breaker, screwed joints		13	.615			30.50		30.50	45.50
2100	Steam specialties, clean									
2120	Air separator with automatic trap, 1" fittings	1 Stpi	12	.667	Ea.		33		33	49.50
2130	Bucket trap, 2" pipe		7	1.143			56.50		56.50	84.50
2150	Drip leg, 2" fitting		45	.178			8.75		8.75	13.15
2200	Thermodynamic trap, 1" fittings		50	.160			7.90		7.90	11.85
2210	Thermostatic		65	.123			6.05		6.05	9.10
2240	Screen and seat in y-type strainer, plug type.		25	.320			15.80		15.80	23.50
2242	Screen and seat in y-type strainer, flange type.		12	.667			33		33	49.50
2500	Valve, replace broken handwheel		24	.333			16.45		16.45	24.50
3000	Valve, overhaul, regulator, relief, flushometer, mixing									
3040	Cold water, gas	1 Stpi	5	1.600	Ea.		79		79	118
3050	Hot water, steam		3	2.667			132		132	197
3080	Globe, gate, check up to 4" cold water, gas		10	.800			39.50		39.50	59
3090	Hot water, steam		5	1.600			79		79	118
3100	Over 4" ID hot or cold line		1.40	5.714			282		282	425
3120	Remove and replace, gate, globe or check up to 4"	Q-5	6	2.667			118		118	178
3130	Over 4"	"	2	8			355		355	535
3150	Repack up to 4"	1 Stpi	13	.615			30.50		30.50	45.50
3160	Over 4"	"	4	2			98.50		98.50	148

Estimating Tips

02 30 00 Subsurface Investigation

In preparing estimates on structures involving earthwork or foundations, all information concerning soil characteristics should be obtained. Look particularly for hazardous waste, evidence of prior dumping of debris, and previous stream beds.

02 41 00 Demolition and Structure Moving

The costs shown for selective demolition do not include rubbish handling or disposal. These items should be estimated separately using RSMeans data or other sources.

- Historic preservation often requires that the contractor remove materials from the existing structure, rehab them, and replace them. The estimator must be aware of any related measures and precautions that must be taken when doing selective demolition and cutting and patching. Requirements may include special handling and storage, as well as security.
- In addition to Subdivision 02 41 00, you can find selective demolition items in each division. Example: Roofing demolition is in Division 7.

02 80 00 Hazardous Material Disposal/ Remediation

This subdivision includes information on hazardous waste handling, asbestos remediation, lead remediation, and mold remediation. See reference R028213-20 and R028319-60 for further guidance in using these unit price lines.

Reference Numbers

Reference numbers are shown in shaded boxes at the beginning of some major classifications. These numbers refer to related items in the Reference Section. The reference information may be an estimating procedure, an alternate pricing method, or technical information.

Note: Not all subdivisions listed here necessarily appear in this publication.

02 41 Demolition

02 41 13 – Selective Site Demolition

02 41 13.17 Demolish, Remove Pavement and Curb

	02 41 13.17 Demolish, Remove Pavement and Curb	Crew	Daily Output	Labor-Hours	Unit	2009 Bare Costs Material	Labor	Equipment	Total	Total Incl O&P
0010	**DEMOLISH, REMOVE PAVEMENT AND CURB** R024119-10									
5010	Pavement removal, bituminous roads, 3" thick	B-38	690	.058	S.Y.		2.05	1.51	3.56	4.81
5050	4" to 6" thick		420	.095			3.38	2.49	5.87	7.90
5100	Bituminous driveways		640	.063			2.22	1.63	3.85	5.20
5200	Concrete to 6" thick, hydraulic hammer, mesh reinforced		255	.157			5.55	4.10	9.65	13
5300	Rod reinforced		200	.200			7.10	5.20	12.30	16.60
5400	Concrete, 7" to 24" thick, plain		33	1.212	C.Y.		43	31.50	74.50	101
5500	Reinforced		24	1.667	"		59	43.50	102.50	139
5600	With hand held air equipment, bituminous, to 6" thick	B-39	1900	.025	S.F.		.84	.10	.94	1.40
5700	Concrete to 6" thick, no reinforcing		1600	.030			1	.12	1.12	1.66
5800	Mesh reinforced		1400	.034			1.14	.14	1.28	1.90
5900	Rod reinforced		765	.063			2.08	.25	2.33	3.49

02 41 13.33 Minor Site Demolition

	02 41 13.33 Minor Site Demolition	Crew	Daily Output	Labor-Hours	Unit	Material	Labor	Equipment	Total	Total Incl O&P
0010	**MINOR SITE DEMOLITION** R024119-10									
0015	No hauling, abandon catch basin or manhole	B-6	7	3.429	Ea.		117	42	159	225
0020	Remove existing catch basin or manhole, masonry		4	6			204	73.50	277.50	395
0030	Catch basin or manhole frames and covers, stored		13	1.846			63	22.50	85.50	122
0040	Remove and reset		7	3.429			117	42	159	225
0900	Hydrants, fire, remove only	B-21A	5	8			315	120	435	605
0950	Remove and reset	"	2	20			780	300	1,080	1,500
4000	Sidewalk removal, bituminous, 2-1/2" thick	B-6	325	.074	S.Y.		2.52	.90	3.42	4.85
4050	Brick, set in mortar		185	.130			4.42	1.59	6.01	8.55
4100	Concrete, plain, 4"		160	.150			5.10	1.84	6.94	9.85
4200	Mesh reinforced		150	.160			5.45	1.96	7.41	10.50

02 41 19 – Selective Structure Demolition

02 41 19.13 Selective Building Demolition

	02 41 19.13 Selective Building Demolition	Crew	Daily Output	Labor-Hours	Unit	Material	Labor	Equipment	Total	Total Incl O&P
0010	**SELECTIVE BUILDING DEMOLITION**									
0020	Costs related to selective demolition of specific building components									
0025	are included under Common Work Results (XX 05 00)									
0030	in the component's appropriate division.									

02 41 19.19 Selective Demolition, Dump Charges

	02 41 19.19 Selective Demolition, Dump Charges	Crew	Daily Output	Labor-Hours	Unit	Material	Labor	Equipment	Total	Total Incl O&P
0010	**SELECTIVE DEMOLITION, DUMP CHARGES** R024119-10									
0020	Dump charges, typical urban city, tipping fees only									
0100	Building construction materials				Ton	100			100	110
0200	Trees, brush, lumber					75			75	82.50
0300	Rubbish only					90			90	99
0500	Reclamation station, usual charge					100			100	110

02 41 19.23 Selective Demolition, Rubbish Handling

	02 41 19.23 Selective Demolition, Rubbish Handling	Crew	Daily Output	Labor-Hours	Unit	Material	Labor	Equipment	Total	Total Incl O&P
0010	**SELECTIVE DEMOLITION, RUBBISH HANDLING** R024119-10									
0020	The following are to be added to the demolition prices									
0400	Chute, circular, prefabricated steel, 18" diameter	B-1	40	.600	L.F.	47	19.35		66.35	82
0440	30" diameter	"	30	.800	"	48.50	26		74.50	93.50
0725	Dumpster, weekly rental, 1 dump/week, 20 C.Y. capacity (8 Tons) R024119-20				Week	775			775	852.50
0800	30 C.Y. capacity (10 Tons)					1,000			1,000	1,100
0840	40 C.Y. capacity (13 Tons)					1,300			1,300	1,430
1000	Dust partition, 6 mil polyethylene, 1" x 3" frame	2 Carp	2000	.008	S.F.	.44	.32		.76	.99
1080	2" x 4" frame	"	2000	.008	"	.27	.32		.59	.80
2000	Load, haul, and dump, 50' haul	2 Clab	24	.667	C.Y.		21		21	32.50
2040	100' haul		16.50	.970			30.50		30.50	47.50
2080	Over 100' haul, add per 100 L.F.		35.50	.451			14.25		14.25	22
2120	In elevators, per 10 floors, add		140	.114			3.61		3.61	5.60

02 41 Demolition

02 41 19 – Selective Structure Demolition

02 41 19.23 Selective Demolition, Rubbish Handling		Crew	Daily Output	Labor-Hours	Unit	Material	2009 Bare Costs Labor	Equipment	Total	Total Incl O&P
3000	Loading & trucking, including 2 mile haul, chute loaded	B-16	45	.711	C.Y.		23	11.85	34.85	48.50
3040	Hand loading truck, 50′ haul	″	48	.667			21.50	11.10	32.60	45
3080	Machine loading truck	B-17	120	.267			8.95	5.15	14.10	19.40
5000	Haul, per mile, up to 8 C.Y. truck	B-34B	1165	.007			.22	.46	.68	.84
5100	Over 8 C.Y. truck	″	1550	.005			.16	.34	.50	.63
02 41 19.27 Selective Demolition, Torch Cutting										
0010	**SELECTIVE DEMOLITION, TORCH CUTTING** R024119-10									
0020	Steel, 1″ thick plate	1 Clab	360	.022	L.F.	.22	.70		.92	1.33
0040	1″ diameter bar	″	210	.038	Ea.		1.20		1.20	1.87
1000	Oxygen lance cutting, reinforced concrete walls									
1040	12″ to 16″ thick walls	1 Clab	10	.800	L.F.		25.50		25.50	39
1080	24″ thick walls	″	6	1.333	″		42		42	65.50

02 65 Underground Storage Tank Removal

02 65 10 – Underground Tank and Contaminated Soil Removal

02 65 10.30 Removal of Underground Storage Tanks		Crew	Daily Output	Labor-Hours	Unit	Material	2009 Bare Costs Labor	Equipment	Total	Total Incl O&P
0010	**REMOVAL OF UNDERGROUND STORAGE TANKS** R026510-20									
0011	Petroleum storage tanks, non-leaking									
0100	Excavate & load onto trailer									
0110	3000 gal. to 5000 gal. tank G	B-14	4	12	Ea.		400	73.50	473.50	695
0120	6000 gal to 8000 gal tank G	B-3A	3	13.333			445	289	734	1,000
0130	9000 gal to 12000 gal tank G	″	2	20			670	435	1,105	1,500
0190	Known leaking tank, add				%				100%	100%
0200	Remove sludge, water and remaining product from tank bottom									
0201	of tank with vacuum truck									
0300	3000 gal to 5000 gal tank G	A-13	5	1.600	Ea.		62.50	152	214.50	261
0310	6000 gal to 8000 gal tank G		4	2			78	190	268	325
0320	9000 gal to 12000 gal tank G		3	2.667			104	253	357	435
0390	Dispose of sludge off-site, average				Gal.				6.60	7.25
0400	Insert inert solid CO_2 "dry ice" into tank									
0401	For cleaning/transporting tanks (1.5 lbs./100 gal. cap) G	1 Clab	500	.016	Lb.	1.69	.51		2.20	2.64
1020	Haul tank to certified salvage dump, 100 miles round trip									
1023	3000 gal. to 5000 gal. tank				Ea.				725	798
1026	6000 gal. to 8000 gal. tank								860	946
1029	9,000 gal. to 12,000 gal. tank								1,160	1,276
1100	Disposal of contaminated soil to landfill									
1110	Minimum				C.Y.				130.70	144
1111	Maximum				″				392	431
1120	Disposal of contaminated soil to									
1121	bituminous concrete batch plant									
1130	Minimum				C.Y.				65.30	72
1131	Maximum				″				130.70	144
2010	Decontamination of soil on site incl poly tarp on top/bottom									
2011	Soil containment berm, and chemical treatment									
2020	Minimum G	B-11C	100	.160	C.Y.	6.90	5.85	2.94	15.69	19.75
2021	Maximum G	″	100	.160		9	5.85	2.94	17.79	22
2050	Disposal of decontaminated soil, minimum								78.20	86
2055	Maximum								160.40	176.50

02 82 Asbestos Remediation

02 82 13 – Asbestos Abatement

02 82 13.41 Asbestos Abatement Equip.

	02 82 13.41 Asbestos Abatement Equip.	Crew	Daily Output	Labor-Hours	Unit	Material	Labor	Equipment	Total	Total Incl O&P
						2009 Bare Costs				
0010	**ASBESTOS ABATEMENT EQUIP.** R028213-20									
0011	Equipment and supplies, buy									
0200	Air filtration device, 2000 C.F.M.				Ea.	795			795	875
0250	Large volume air sampling pump, minimum					305			305	335
0260	Maximum					455			455	500
0300	Airless sprayer unit, 2 gun					3,125			3,125	3,425
0350	Light stand, 500 watt					72.50			72.50	80
0400	Personal respirators									
0410	Negative pressure, 1/2 face, dual operation, min.				Ea.	25.50			25.50	28
0420	Maximum					31			31	34.50
0450	P.A.P.R., full face, minimum					124			124	137
0460	Maximum					150			150	165
0470	Supplied air, full face, incl. air line, minimum					157			157	172
0480	Maximum					239			239	263
0500	Personnel sampling pump, minimum					201			201	221
1500	Power panel, 20 unit, incl. G.F.I.					1,100			1,100	1,200
1600	Shower unit, including pump and filters					1,075			1,075	1,200
1700	Supplied air system (type C)					9,800			9,800	10,800
1750	Vacuum cleaner, HEPA, 16 gal., stainless steel, wet/dry					675			675	745
1760	55 gallon					1,700			1,700	1,875
1800	Vacuum loader, 9-18 ton/hr					90,000			90,000	99,000
1900	Water atomizer unit, including 55 gal. drum					230			230	253
2000	Worker protection, whole body, foot, head cover & gloves, plastic					6.80			6.80	7.45
2500	Respirator, single use					10.90			10.90	12
2550	Cartridge for respirator					12.30			12.30	13.55
2570	Glove bag, 7 mil, 50″ x 64″					9.20			9.20	10.10
2580	10 mil, 44″ x 60″					74.50			74.50	82
2590	6 mil, 44″ x 60″					74.50			74.50	82
3000	HEPA vacuum for work area, minimum					1,325			1,325	1,475
3050	Maximum					3,100			3,100	3,425
6000	Disposable polyethylene bags, 6 mil, 3 C.F.					1.24			1.24	1.36
6300	Disposable fiber drums, 3 C.F.					12			12	13.20
6400	Pressure sensitive caution labels, 3″ x 5″					1.19			1.19	1.31
6450	11″ x 17″					7.05			7.05	7.75
6500	Negative air machine, 1800 C.F.M.					770			770	845

02 82 13.42 Preparation of Asbestos Containment Area

	02 82 13.42 Preparation of Asbestos Containment Area	Crew	Daily Output	Labor-Hours	Unit	Material	Labor	Equipment	Total	Total Incl O&P
0010	**PREPARATION OF ASBESTOS CONTAINMENT AREA** R028213-20									
0100	Pre-cleaning, HEPA vacuum and wet wipe, flat surfaces	A-9	12000	.005	S.F.	.02	.24		.26	.39
0200	Protect carpeted area, 2 layers 6 mil poly on 3/4″ plywood	″	1000	.064		1.50	2.83		4.33	6.10
0300	Separation barrier, 2″ x 4″ @ 16″, 1/2″ plywood ea. side, 8′ high	2 Carp	400	.040		1.25	1.60		2.85	3.86
0310	12′ high		320	.050		1.40	2		3.40	4.64
0320	16′ high		200	.080		1.50	3.20		4.70	6.60
0400	Personnel decontam. chamber, 2″ x 4″ @ 16″, 3/4″ ply ea. side		280	.057		2.50	2.28		4.78	6.30
0450	Waste decontam. chamber, 2″ x 4″ studs @ 16″, 3/4″ ply ea. side		360	.044		3	1.78		4.78	6.05
0500	Cover surfaces with polyethylene sheeting									
0501	Including glue and tape									
0550	Floors, each layer, 6 mil	A-9	8000	.008	S.F.	.06	.35		.41	.61
0551	4 mil		9000	.007		.04	.31		.35	.53
0560	Walls, each layer, 6 mil		6000	.011		.06	.47		.53	.80
0561	4 mil		7000	.009		.04	.40		.44	.67
0570	For heights above 12′, add						20%			
0575	For heights above 20′, add						30%			

02 82 13.42 Preparation of Asbestos Containment Area

	02 82 13.42 Preparation of Asbestos Containment Area	Crew	Daily Output	Labor-Hours	Unit	2009 Bare Costs Material	Labor	Equipment	Total	Total Incl O&P
0580	For fire retardant poly, add					100%				
0590	For large open areas, deduct					10%	20%			
0600	Seal floor penetrations with foam firestop to 36 Sq. In.	2 Carp	200	.080	Ea.	1,275	3.20		1,278.20	1,400
0610	36 Sq. In. to 72 Sq. In.		125	.128		2,550	5.10		2,555.10	2,825
0615	72 Sq. In. to 144 Sq. In.		80	.200		5,125	8		5,133	5,625
0620	Wall penetrations, to 36 square inches		180	.089		1,275	3.55		1,278.55	1,400
0630	36 Sq. In. to 72 Sq. In.		100	.160		2,550	6.40		2,556.40	2,825
0640	72 Sq. In. to 144 Sq. In.		60	.267		5,125	10.65		5,135.65	5,650
0800	Caulk seams with latex	1 Carp	230	.035	L.F.	.15	1.39		1.54	2.32
0900	Set up neg. air machine, 1-2k C.F.M. /25 M.C.F. volume	1 Asbe	4.30	1.860	Ea.		82		82	129

02 82 13.43 Bulk Asbestos Removal

	02 82 13.43 Bulk Asbestos Removal	Crew	Daily Output	Labor-Hours	Unit	Material	Labor	Equipment	Total	Total Incl O&P
0010	**BULK ASBESTOS REMOVAL**									
0020	Includes disposable tools and 2 suits and 1 respirator filter/day/worker									
0200	Boiler insulation	A-9	480	.133	S.F.	.50	5.90		6.40	9.80
0210	With metal lath, add				%				50%	
0300	Boiler breeching or flue insulation	A-9	520	.123	S.F.	.40	5.45		5.85	8.95
0310	For active boiler, add				%				100%	
0400	Duct or AHU insulation	A-10B	440	.073	S.F.	.24	3.22		3.46	5.30
0500	Duct vibration isolation joints, up to 24 Sq. In. duct	A-9	56	1.143	Ea.	3.69	50.50		54.19	83
0520	25 Sq. In. to 48 Sq. In. duct		48	1.333		4.31	59		63.31	97
0530	49 Sq. In. to 76 Sq. In. duct		40	1.600		5.15	70.50		75.65	117
0600	Pipe insulation, air cell type, up to 4" diameter pipe		900	.071	L.F.	.23	3.14		3.37	5.15
0610	4" to 8" diameter pipe		800	.080		.26	3.53		3.79	5.85
0620	10" to 12" diameter pipe		700	.091		.30	4.04		4.34	6.70
0630	14" to 16" diameter pipe		550	.116		.38	5.15		5.53	8.45
0650	Over 16" diameter pipe		650	.098	S.F.	.32	4.35		4.67	7.15
0700	With glove bag up to 3" diameter pipe		200	.320	L.F.	2.15	14.15		16.30	24.50
1000	Pipe fitting insulation up to 4" diameter pipe		320	.200	Ea.	.65	8.85		9.50	14.55
1100	6" to 8" diameter pipe		304	.211		.68	9.30		9.98	15.30
1110	10" to 12" diameter pipe		192	.333		1.08	14.70		15.78	24
1120	14" to 16" diameter pipe		128	.500		1.62	22		23.62	36.50
1130	Over 16" diameter pipe		176	.364	S.F.	1.18	16.05		17.23	26.50
1200	With glove bag, up to 8" diameter pipe		75	.853	L.F.	6.55	37.50		44.05	66
2000	Scrape foam fireproofing from flat surface		2400	.027	S.F.	.09	1.18		1.27	1.94
2100	Irregular surfaces		1200	.053		.17	2.36		2.53	3.88
3000	Remove cementitious material from flat surface		1800	.036		.11	1.57		1.68	2.59
3100	Irregular surface		1000	.064		.15	2.83		2.98	4.59
6000	Remove contaminated soil from crawl space by hand		400	.160	C.F.	.52	7.05		7.57	11.60
6100	With large production vacuum loader	A-12	700	.091	"	.30	4.04	1.08	5.42	7.85
7000	Radiator backing, not including radiator removal	A-9	1200	.053	S.F.	.17	2.36		2.53	3.88
9000	For type C (supplied air) respirator equipment, add				%					10%

02 82 13.44 Demolition In Asbestos Contaminated Area

	02 82 13.44 Demolition In Asbestos Contaminated Area	Crew	Daily Output	Labor-Hours	Unit	Material	Labor	Equipment	Total	Total Incl O&P
0010	**DEMOLITION IN ASBESTOS CONTAMINATED AREA**									
0200	Ceiling, including suspension system, plaster and lath	A-9	2100	.030	S.F.	.10	1.35		1.45	2.22
0210	Finished plaster, leaving wire lath		585	.109		.35	4.83		5.18	7.95
0220	Suspended acoustical tile		3500	.018		.06	.81		.87	1.34
0230	Concealed tile grid system		3000	.021		.07	.94		1.01	1.56
0240	Metal pan grid system		1500	.043		.14	1.88		2.02	3.10
0250	Gypsum board		2500	.026		.08	1.13		1.21	1.86
0260	Lighting fixtures up to 2' x 4'		72	.889	Ea.	2.87	39.50		42.37	64.50
0400	Partitions, non load bearing									
0410	Plaster, lath, and studs	A-9	690	.093	S.F.	.87	4.10		4.97	7.35

02 82 13.44 Demolition In Asbestos Contaminated Area

		Crew	Daily Output	Labor-Hours	Unit	2009 Bare Costs Material	Labor	Equipment	Total	Total Incl O&P
0450	Gypsum board and studs	A-9	1390	.046	S.F.	.15	2.03		2.18	3.35
9000	For type C (supplied air) respirator equipment, add				%					10%

02 82 13.45 OSHA Testing

		Crew	Daily Output	Labor-Hours	Unit	Material	Labor	Equipment	Total	Total Incl O&P
0010	**OSHA TESTING**									
0100	Certified technician, minimum				Day				240	264
0110	Maximum				″				320	352
0200	Personal sampling, PCM analysis, NIOSH 7400, minimum	1 Asbe	8	1	Ea.	2.75	44		46.75	72
0210	Maximum	″	4	2	″	3	88		91	141
0300	Industrial hygienist, minimum				Day				320	352
0310	Maximum				″				480	528
1000	Cleaned area samples	1 Asbe	8	1	Ea.	2.63	44		46.63	72
1100	PCM air sample analysis, NIOSH 7400, minimum		8	1		30.50	44		74.50	103
1110	Maximum	↓	4	2		3.20	88		91.20	142
1200	TEM air sample analysis, NIOSH 7402, minimum								100	125
1210	Maximum				↓				400	500

02 82 13.46 Decontamination of Asbestos Containment Area

		Crew	Daily Output	Labor-Hours	Unit	Material	Labor	Equipment	Total	Total Incl O&P
0010	**DECONTAMINATION OF ASBESTOS CONTAINMENT AREA**									
0100	Spray exposed substrate with surfactant (bridging)									
0200	Flat surfaces	A-9	6000	.011	S.F.	.36	.47		.83	1.14
0250	Irregular surfaces		4000	.016	″	.31	.71		1.02	1.45
0300	Pipes, beams, and columns		2000	.032	L.F.	.56	1.41		1.97	2.83
1000	Spray encapsulate polyethylene sheeting		8000	.008	S.F.	.31	.35		.66	.89
1100	Roll down polyethylene sheeting		8000	.008	″		.35		.35	.55
1500	Bag polyethylene sheeting		400	.160	Ea.	.77	7.05		7.82	11.90
2000	Fine clean exposed substrate, with nylon brush		2400	.027	S.F.		1.18		1.18	1.85
2500	Wet wipe substrate		4800	.013			.59		.59	.92
2600	Vacuum surfaces, fine brush	↓	6400	.010	↓		.44		.44	.69
3000	Structural demolition									
3100	Wood stud walls	A-9	2800	.023	S.F.		1.01		1.01	1.58
3500	Window manifolds, not incl. window replacement		4200	.015			.67		.67	1.05
3600	Plywood carpet protection	↓	2000	.032	↓		1.41		1.41	2.21
4000	Remove custom decontamination facility	A-10A	8	3	Ea.	15	133		148	225
4100	Remove portable decontamination facility	3 Asbe	12	2	″	12.75	88		100.75	152
5000	HEPA vacuum, shampoo carpeting	A-9	4800	.013	S.F.	.05	.59		.64	.98
9000	Final cleaning of protected surfaces	A-10A	8000	.003	″		.13		.13	.21

02 82 13.47 Asbestos Waste Packaging, Handling, and Disposal

		Crew	Daily Output	Labor-Hours	Unit	Material	Labor	Equipment	Total	Total Incl O&P
0010	**ASBESTOS WASTE PACKAGING, HANDLING, AND DISPOSAL**									
0100	Collect and bag bulk material, 3 C.F. bags, by hand	A-9	400	.160	Ea.	1.24	7.05		8.29	12.40
0200	Large production vacuum loader	A-12	880	.073		.80	3.21	.86	4.87	6.90
1000	Double bag and decontaminate	A-9	960	.067		1.24	2.94		4.18	5.95
2000	Containerize bagged material in drums, per 3 C.F. drum	″	800	.080		12	3.53		15.53	18.75
3000	Cart bags 50′ to dumpster	2 Asbe	400	.040	↓		1.76		1.76	2.76
5000	Disposal charges, not including haul, minimum				C.Y.				50	55
5020	Maximum				″				170	187
5100	Remove refrigerant from system	1 Plum	40	.200	Lb.		9.75		9.75	14.65
9000	For type C (supplied air) respirator equipment, add				%					10%

02 82 13.48 Asbestos Encapsulation With Sealants

		Crew	Daily Output	Labor-Hours	Unit	Material	Labor	Equipment	Total	Total Incl O&P
0010	**ASBESTOS ENCAPSULATION WITH SEALANTS**									
0100	Ceilings and walls, minimum	A-9	21000	.003	S.F.	.27	.13		.40	.51
0110	Maximum		10600	.006	″	.42	.27		.69	.88
0300	Pipes to 12″ diameter including minor repairs, minimum		800	.080	L.F.	.37	3.53		3.90	5.95
0310	Maximum	↓	400	.160	″	1.04	7.05		8.09	12.20

02 85 Mold Remediation

02 85 16 – Mold Remediation Preparation and Containment

02 85 16.50 Preparation of Mold Containment Area		Crew	Daily Output	Labor-Hours	Unit	2009 Bare Costs Material	Labor	Equipment	Total	Total Incl O&P
0010	**PREPARATION OF MOLD CONTAINMENT AREA** R028213-20									
0100	Pre-cleaning, HEPA vacuum and wet wipe, flat surfaces	A-9	12000	.005	S.F.	.02	.24		.26	.39
0300	Separation barrier, 2" x 4" @ 16", 1/2" plywood ea. side, 8' high	2 Carp	400	.040		1.25	1.60		2.85	3.86
0310	12' high		320	.050		1.40	2		3.40	4.64
0320	16' high		200	.080		1.50	3.20		4.70	6.60
0400	Personnel decontam. chamber, 2" x 4" @ 16", 3/4" ply ea. side		280	.057		2.50	2.28		4.78	6.30
0450	Waste decontam. chamber, 2" x 4" studs @ 16", 3/4" ply each side		360	.044		3	1.78		4.78	6.05
0500	Cover surfaces with polyethylene sheeting									
0501	Including glue and tape									
0550	Floors, each layer, 6 mil	A-9	8000	.008	S.F.	.09	.35		.44	.65
0551	4 mil		9000	.007		.05	.31		.36	.55
0560	Walls, each layer, 6 mil		6000	.011		.09	.47		.56	.84
0561	4 mil		7000	.009		.07	.40		.47	.71
0570	For heights above 12', add						20%			
0575	For heights above 20', add						30%			
0580	For fire retardant poly, add					100%				
0590	For large open areas, deduct					10%	20%			
0600	Seal floor penetrations with foam firestop to 36 sq in	2 Carp	200	.080	Ea.	1,275	3.20		1,278.20	1,400
0610	36 sq in to 72 sq in		125	.128		2,550	5.10		2,555.10	2,825
0615	72 sq in to 144 sq in		80	.200		5,125	8		5,133	5,625
0620	Wall penetrations, to 36 square inches		180	.089		1,275	3.55		1,278.55	1,400
0630	36 sq. in. to 72 sq. in.		100	.160		2,550	6.40		2,556.40	2,825
0640	72 Sq. in. to 144 sq. in.		60	.267		5,125	10.65		5,135.65	5,650
0800	Caulk seams with latex caulk	1 Carp	230	.035	L.F.	.15	1.39		1.54	2.32
0900	Set up neg. air machine, 1-2k C.F.M. /25 M.C.F. volume	1 Asbe	4.30	1.860	Ea.		82		82	129

02 85 33 – Removal and Disposal of Materials with Mold

02 85 33.50 Demolition In Mold Contaminated Area		Crew	Daily Output	Labor-Hours	Unit	Material	Labor	Equipment	Total	Total Incl O&P
0010	**DEMOLITION IN MOLD CONTAMINATED AREA**									
0200	Ceiling, including suspension system, plaster and lath	A-9	2100	.030	S.F.	.10	1.35		1.45	2.22
0210	Finished plaster, leaving wire lath		585	.109		.35	4.83		5.18	7.95
0220	Suspended acoustical tile		3500	.018		.06	.81		.87	1.34
0230	Concealed tile grid system		3000	.021		.07	.94		1.01	1.56
0240	Metal pan grid system		1500	.043		.14	1.88		2.02	3.10
0250	Gypsum board		2500	.026		.08	1.13		1.21	1.86
0255	Plywood		2500	.026		.08	1.13		1.21	1.86
0260	Lighting fixtures up to 2' x 4'		72	.889	Ea.	2.87	39.50		42.37	64.50
0400	Partitions, non load bearing									
0410	Plaster, lath, and studs	A-9	690	.093	S.F.	.87	4.10		4.97	7.35
0450	Gypsum board and studs		1390	.046		.15	2.03		2.18	3.35
0465	Carpet & pad		1390	.046		.15	2.03		2.18	3.35
0600	Pipe insulation, air cell type, up to 4" diameter pipe		900	.071	L.F.	.23	3.14		3.37	5.15
0610	4" to 8" diameter pipe		800	.080		.26	3.53		3.79	5.85
0620	10" to 12" diameter pipe		700	.091		.30	4.04		4.34	6.70
0630	14" to 16" diameter pipe		550	.116		.38	5.15		5.53	8.45
0650	Over 16" diameter pipe		650	.098	S.F.	.32	4.35		4.67	7.15
9000	For type C (supplied air) respirator equipment, add				%					10%

Division Notes

		CREW	DAILY OUTPUT	LABOR-HOURS	UNIT	2009 BARE COSTS MAT.	LABOR	EQUIP.	TOTAL	TOTAL INCL O&P

Estimating Tips

General

- Carefully check all the plans and specifications. Concrete often appears on drawings other than structural drawings, including mechanical and electrical drawings for equipment pads. The cost of cutting and patching is often difficult to estimate. See Subdivision 03 81 for Concrete Cutting, Subdivision 02 41 19.16 for Cutout Demolition, Subdivision 03 05 05.10 for Concrete Demolition, and Subdivision 02 41 19.23 for Rubbish Handling (handling, loading and hauling of debris).
- Always obtain concrete prices from suppliers near the job site. A volume discount can often be negotiated, depending upon competition in the area. Remember to add for waste, particularly for slabs and footings on grade.

03 10 00 Concrete Forming and Accessories

- A primary cost for concrete construction is forming. Most jobs today are constructed with prefabricated forms. The selection of the forms best suited for the job and the total square feet of forms required for efficient concrete forming and placing are key elements in estimating concrete construction. Enough forms must be available for erection to make efficient use of the concrete placing equipment and crew.
- Concrete accessories for forming and placing depend upon the systems used. Study the plans and specifications to ensure that all special accessory requirements have been included in the cost estimate, such as anchor bolts, inserts, and hangers.
- Included within costs for forms-in-place are all necessary bracing and shoring.

03 20 00 Concrete Reinforcing

- Ascertain that the reinforcing steel supplier has included all accessories, cutting, bending, and an allowance for lapping, splicing, and waste. A good rule of thumb is 10% for lapping, splicing, and waste. Also, 10% waste should be allowed for welded wire fabric.
- The unit price items in the subdivision for Reinforcing In Place include the labor to install accessories such as beam and slab bolsters, high chairs, and bar ties and tie wire. The material cost for these accessories is not included; they may be obtained from the Accessories Division.

03 30 00 Cast-In-Place Concrete

- When estimating structural concrete, pay particular attention to requirements for concrete additives, curing methods, and surface treatments. Special consideration for climate, hot or cold, must be included in your estimate. Be sure to include requirements for concrete placing equipment, and concrete finishing.
- For accurate concrete estimating, the estimator must consider each of the following major components individually: forms, reinforcing steel, ready-mix concrete, placement of the concrete, and finishing of the top surface. For faster estimating, Subdivision 03 30 53.40 for Concrete-In-Place can be used; here, various items of concrete work are presented that include the costs of all five major components (unless specifically stated otherwise).

03 40 00 Precast Concrete
03 50 00 Cast Decks and Underlayment

- The cost of hauling precast concrete structural members is often an important factor. For this reason, it is important to get a quote from the nearest supplier. It may become economically feasible to set up precasting beds on the site if the hauling costs are prohibitive.

Reference Numbers

Reference numbers are shown in shaded boxes at the beginning of some major classifications. These numbers refer to related items in the Reference Section. The reference information may be an estimating procedure, an alternate pricing method, or technical information.

Note: Not all subdivisions listed here necessarily appear in this publication.

03 01 Maintenance of Concrete

03 01 30 – Maintenance of Cast-In-Place Concrete

03 01 30.62 Concrete Patching		Crew	Daily Output	Labor-Hours	Unit	2009 Bare Costs Material	Labor	Equipment	Total	Total Incl O&P
0010	**CONCRETE PATCHING**									
0100	Floors, 1/4" thick, small areas, regular grout	1 Cefi	170	.047	S.F.	1.07	1.80		2.87	3.81
0150	Epoxy grout	"	100	.080	"	6.85	3.06		9.91	12
2000	Walls, including chipping, cleaning and epoxy grout									
2100	1/4" deep	1 Cefi	65	.123	S.F.	7.50	4.71		12.21	15.15
2150	1/2" deep		50	.160		15.05	6.15		21.20	25.50
2200	3/4" deep	↓	40	.200	↓	22.50	7.65		30.15	36

03 11 Concrete Forming

03 11 13 – Structural Cast-In-Place Concrete Forming

03 11 13.40 Forms In Place, Equipment Foundations

		Crew	Daily Output	Labor-Hours	Unit	Material	Labor	Equipment	Total	Total Incl O&P
0010	**FORMS IN PLACE, EQUIPMENT FOUNDATIONS**									
0020	1 use	C-2	160	.300	SFCA	3.60	11.65		15.25	22
0050	2 use		190	.253		1.98	9.80		11.78	17.45
0100	3 use		200	.240		1.44	9.35		10.79	16.05
0150	4 use	↓	205	.234	↓	1.17	9.10		10.27	15.40

03 11 13.45 Forms In Place, Footings

		Crew	Daily Output	Labor-Hours	Unit	Material	Labor	Equipment	Total	Total Incl O&P
0010	**FORMS IN PLACE, FOOTINGS**									
0020	Continuous wall, plywood, 1 use	C-1	375	.085	SFCA	7.45	3.23		10.68	13.20
0050	2 use		440	.073		4.10	2.75		6.85	8.80
0100	3 use		470	.068		2.98	2.58		5.56	7.30
0150	4 use		485	.066		2.42	2.50		4.92	6.55
5000	Spread footings, job-built lumber, 1 use		305	.105		2.16	3.97		6.13	8.55
5050	2 use		371	.086		1.20	3.27		4.47	6.35
5100	3 use		401	.080		.86	3.02		3.88	5.65
5150	4 use	↓	414	.077	↓	.70	2.93		3.63	5.30

03 11 13.65 Forms In Place, Slab On Grade

		Crew	Daily Output	Labor-Hours	Unit	Material	Labor	Equipment	Total	Total Incl O&P
0010	**FORMS IN PLACE, SLAB ON GRADE**									
3000	Edge forms, wood, 4 use, on grade, to 6" high	C-1	600	.053	L.F.	.38	2.02		2.40	3.55
6000	Trench forms in floor, wood, 1 use		160	.200	SFCA	1.48	7.55		9.03	13.35
6050	2 use		175	.183		.81	6.90		7.71	11.65
6100	3 use		180	.178		.59	6.75		7.34	11.10
6150	4 use		185	.173	↓	.48	6.55		7.03	10.70
8760	Void form, corrugated fiberboard, 6" x 12", 10' long **G**	↓	240	.133	S.F.	.84	5.05		5.89	8.75

03 15 Concrete Accessories

03 15 05 – Concrete Forming Accessories

03 15 05.75 Sleeves and Chases

		Crew	Daily Output	Labor-Hours	Unit	Material	Labor	Equipment	Total	Total Incl O&P
0010	**SLEEVES AND CHASES**									
0100	Plastic, 1 use, 9" long, 2" diameter	1 Carp	100	.080	Ea.	1.41	3.20		4.61	6.50
0150	4" diameter		90	.089		4.14	3.55		7.69	10.05
0200	6" diameter		75	.107		7.30	4.26		11.56	14.65
0250	12" diameter		60	.133		21	5.35		26.35	31.50
5000	Sheet metal, 2" diameter **G**		100	.080		1.11	3.20		4.31	6.20
5100	4" diameter **G**		90	.089		1.39	3.55		4.94	7.05
5150	6" diameter **G**		75	.107		2.02	4.26		6.28	8.80
5200	12" diameter **G**		60	.133		5.20	5.35		10.55	13.95
6000	Steel pipe, 2" diameter **G**		100	.080		6.15	3.20		9.35	11.75
6100	4" diameter **G**	↓	90	.089	↓	15.50	3.55		19.05	22.50

03 15 Concrete Accessories

03 15 05 – Concrete Forming Accessories

03 15 05.75 Sleeves and Chases			Crew	Daily Output	Labor-Hours	Unit	Material	Labor	Equipment	Total	Total Incl O&P
							2009 Bare Costs				
6150	6″ diameter	G	1 Carp	75	.107	Ea.	27	4.26		31.26	36.50
6200	12″ diameter	G	↓	60	.133	↓	63.50	5.35		68.85	78

03 21 Reinforcing Steel

03 21 10 – Uncoated Reinforcing Steel

03 21 10.60 Reinforcing In Place

			Crew	Daily Output	Labor-Hours	Unit	Material	Labor	Equipment	Total	Total Incl O&P
0015	**REINFORCING IN PLACE** A615 Grade 60, incl. access. labor										
0030	Made from recycled materials										
0502	Footings, #4 to #7	G	4 Rodm	4200	.008	Lb.	.81	.34		1.15	1.45
0552	#8 to #18	G		7200	.004		.81	.20		1.01	1.21
0602	Slab on grade, #3 to #7	G	↓	4200	.008	↓	.78	.34		1.12	1.41

03 22 Welded Wire Fabric Reinforcing

03 22 05 – Uncoated Welded Wire Fabric

03 22 05.50 Welded Wire Fabric

			Crew	Daily Output	Labor-Hours	Unit	Material	Labor	Equipment	Total	Total Incl O&P
0010	**WELDED WIRE FABRIC** ASTM A185										
0030	Made from recycled materials										
0050	Sheets										
0100	6 x 6 - W1.4 x W1.4 (10 x 10) 21 lb. per C.S.F.	G	2 Rodm	35	.457	C.S.F.	18.05	20.50		38.55	53.50

03 30 Cast-In-Place Concrete

03 30 53 – Miscellaneous Cast-In-Place Concrete

03 30 53.40 Concrete In Place

		Crew	Daily Output	Labor-Hours	Unit	Material	Labor	Equipment	Total	Total Incl O&P
0010	**CONCRETE IN PLACE**									
0020	Including forms (4 uses), reinforcing steel, concrete, placement,									
0050	and finishing unless otherwise indicated									
0500	Chimney foundations, industrial, minimum	C-14C	32.22	3.476	C.Y.	166	133	.80	299.80	390
0510	Maximum	″	23.71	4.724	″	203	181	1.09	385.09	505
3540	Equipment pad, 3′ x 3′ x 6″ thick	C-14H	45	1.067	Ea.	58	42	.58	100.58	130
3550	4′ x 4′ x 6″ thick		30	1.600		84	63	.86	147.86	191
3560	5′ x 5′ x 8″ thick		18	2.667		140	105	1.44	246.44	320
3570	6′ x 6′ x 8″ thick		14	3.429		188	135	1.85	324.85	420
3580	8′ x 8′ x 10″ thick		8	6		395	236	3.24	634.24	805
3590	10′ x 10′ x 12″ thick	↓	5	9.600	↓	655	380	5.20	1,040.20	1,300
3800	Footings, spread under 1 C.Y.	C-14C	28	4	C.Y.	209	153	.92	362.92	470
3825	1 C.Y to 5 C.Y.		43	2.605		221	99.50	.60	321.10	400
3850	Over 5 C.Y.	↓	75	1.493		198	57	.34	255.34	305
3900	Footings, strip, 18″ x 9″, unreinforced	C-14L	40	2.400		120	89.50	.65	210.15	271
3920	18″ x 9″, reinforced	C-14C	35	3.200		157	122	.74	279.74	365
3925	20″ x 10″, unreinforced	C-14L	45	2.133		117	79.50	.58	197.08	252
3930	20″ x 10″, reinforced	C-14C	40	2.800		147	107	.64	254.64	330
3935	24″ x 12″, unreinforced	C-14L	55	1.745		116	65	.47	181.47	229
3940	24″ x 12″, reinforced	C-14C	48	2.333		147	89.50	.54	237.04	300
3945	36″ x 12″, unreinforced	C-14L	70	1.371		113	51	.37	164.37	203
3950	36″ x 12″, reinforced	C-14C	60	1.867		141	71.50	.43	212.93	266
4000	Foundation mat, under 10 C.Y.		38.67	2.896		255	111	.67	366.67	455
4050	Over 20 C.Y.	↓	56.40	1.986	↓	218	76	.46	294.46	360

03 30 Cast-In-Place Concrete

03 30 53 – Miscellaneous Cast-In-Place Concrete

03 30 53.40 Concrete In Place		Crew	Daily Output	Labor-Hours	Unit	2009 Bare Costs Material	Labor	Equipment	Total	Total Incl O&P
4650	Slab on grade, not including finish, 4" thick	C-14E	60.75	1.449	C.Y.	127	57	.43	184.43	230
4700	6" thick	"	92	.957	↓	121	37.50	.29	158.79	193
4751	Slab on grade, incl. troweled finish, not incl. forms									
4760	or reinforcing, over 10,000 S.F., 4" thick	C-14F	3425	.021	S.F.	1.35	.76	.01	2.12	2.63
4820	6" thick		3350	.021		1.97	.78	.01	2.76	3.34
4840	8" thick		3184	.023		2.70	.82	.01	3.53	4.19
4900	12" thick		2734	.026		4.04	.96	.01	5.01	5.90
4950	15" thick	↓	2505	.029	↓	5.10	1.04	.01	6.15	7.15

03 31 Structural Concrete

03 31 05 – Normal Weight Structural Concrete

03 31 05.35 Normal Weight Concrete, Ready Mix

		Crew	Daily Output	Labor-Hours	Unit	Material	Labor	Equipment	Total	Total Incl O&P
0010	**NORMAL WEIGHT CONCRETE, READY MIX**, delivered									
0012	Includes local aggregate, sand, Portland cement, and water									
0015	Excludes all additives and treatments									
0020	2000 psi				C.Y.	97			97	107
0100	2500 psi					98.50			98.50	108
0150	3000 psi					101			101	111
0200	3500 psi					104			104	114
0300	4000 psi					106			106	116
0350	4500 psi					109			109	120
0400	5000 psi					111			111	122
0411	6000 psi					127			127	139
0412	8000 psi					206			206	227
0413	10,000 psi					293			293	320
0414	12,000 psi					355			355	390
1000	For high early strength cement, add					10%				
1300	For winter concrete (hot water), add					5.25			5.25	5.80
1400	For hot weather concrete (ice), add					7.20			7.20	7.90
1410	For mid-range water reducer, add					3.95			3.95	4.35
1420	For high-range water reducer/superplasticizer, add					5.85			5.85	6.40
1430	For retarder, add					2			2	2.20
1440	For non-Chloride accelerator, add					4.75			4.75	5.25
1450	For Chloride accelerator, per 1%, add					2.75			2.75	3.03
1460	For fiber reinforcing, synthetic (1 Lb./C.Y.), add					6			6	6.60
1500	For Saturday delivery, add				↓	5.50			5.50	6.05
1510	For truck holding/waiting time past 1st hour per load, add				Hr.	92			92	101
1520	For short load (less than 4 C.Y.), add per load				Ea.	100			100	110
2000	For all lightweight aggregate, add				C.Y.	45%				

03 31 05.70 Placing Concrete

		Crew	Daily Output	Labor-Hours	Unit	Material	Labor	Equipment	Total	Total Incl O&P
0010	**PLACING CONCRETE**									
0020	Includes labor and equipment to place and vibrate									
1900	Footings, continuous, shallow, direct chute	C-6	120	.400	C.Y.		13.20	.43	13.63	21
1950	Pumped	C-20	150	.427			14.45	5.25	19.70	28
2000	With crane and bucket	C-7	90	.800			27.50	13.30	40.80	56.50
2100	Footings, continuous, deep, direct chute	C-6	140	.343			11.35	.37	11.72	17.80
2150	Pumped	C-20	160	.400			13.55	4.94	18.49	26.50
2200	With crane and bucket	C-7	110	.655			22.50	10.90	33.40	46
2400	Footings, spread, under 1 C.Y., direct chute	C-6	55	.873			29	.94	29.94	45
2450	Pumped	C-20	65	.985			33.50	12.15	45.65	64.50
2500	With crane and bucket	C-7	45	1.600	↓		55	26.50	81.50	113

03 31 Structural Concrete

03 31 05 – Normal Weight Structural Concrete

03 31 05.70 Placing Concrete

	03 31 05.70 Placing Concrete	Crew	Daily Output	Labor-Hours	Unit	Material	2009 Bare Costs Labor	Equipment	Total	Total Incl O&P
2600	Over 5 C.Y., direct chute	C-6	120	.400	C.Y.		13.20	.43	13.63	21
2650	Pumped	C-20	150	.427			14.45	5.25	19.70	28
2700	With crane and bucket	C-7	100	.720			24.50	11.95	36.45	50.50
2900	Foundation mats, over 20 C.Y., direct chute	C-6	350	.137			4.53	.15	4.68	7.10
2950	Pumped	C-20	400	.160			5.45	1.97	7.42	10.45
3000	With crane and bucket	C-7	300	.240			8.20	3.99	12.19	16.95

03 35 Concrete Finishing

03 35 29 – Tooled Concrete Finishing

03 35 29.30 Finishing Floors

	03 35 29.30 Finishing Floors	Crew	Daily Output	Labor-Hours	Unit	Material	Labor	Equipment	Total	Total Incl O&P
0010	**FINISHING FLOORS**									
0020	Manual screed finish	C-10	4800	.005	S.F.		.18		.18	.27
0100	Manual screed and bull float		4000	.006			.22		.22	.32
0125	Manual screed, bull float, manual float		2000	.012			.43		.43	.64
0150	Manual screed, bull float, manual float & broom finish		1850	.013			.47		.47	.70
0200	Manual screed, bull float, manual float, manual steel trowel		1265	.019			.68		.68	1.02
0250	Manual screed, bull float, machine float & trowel (walk-behind)	C-10C	1715	.014			.50	.02	.52	.78
0300	Power screed, bull float, machine float & trowel (walk-behind)	C-10D	2400	.010			.36	.05	.41	.59
0350	Power screed, bull float, machine float & trowel (ride-on)	C-10E	4000	.006			.22	.06	.28	.39

03 35 29.35 Control Joints, Saw Cut

	03 35 29.35 Control Joints, Saw Cut	Crew	Daily Output	Labor-Hours	Unit	Material	Labor	Equipment	Total	Total Incl O&P
0010	**CONTROL JOINTS, SAW CUT**									
0100	Sawcut in green concrete									
0120	1" depth	C-27	2000	.008	L.F.	.07	.31	.07	.45	.60
0140	1-1/2" depth		1800	.009		.10	.34	.08	.52	.70
0160	2" depth		1600	.010		.13	.38	.09	.60	.81
0200	Clean out control joint of debris	C-28	6000	.001			.05		.05	.07
0300	Joint sealant									
0320	Backer rod, polyethylene, 1/4" diameter	1 Cefi	460	.017	L.F.	.04	.67		.71	1.02
0340	Sealant, polyurethane									
0360	1/4" x 1/4" (308 LF/Gal)	1 Cefi	270	.030	L.F.	.18	1.13		1.31	1.85
0380	1/4" x 1/2" (154 LF/Gal)	"	255	.031	"	.35	1.20		1.55	2.15

03 54 Cast Underlayment

03 54 16 – Hydraulic Cement Underlayment

03 54 16.50 Cement Underlayment

	03 54 16.50 Cement Underlayment	Crew	Daily Output	Labor-Hours	Unit	Material	Labor	Equipment	Total	Total Incl O&P
0010	**CEMENT UNDERLAYMENT**									
2510	Underlayment, P.C based self-leveling, 4100 psi, pumped, 1/4"	C-8	20000	.003	S.F.	1.44	.10	.04	1.58	1.77
2520	1/2"		19000	.003		2.87	.10	.04	3.01	3.36
2530	3/4"		18000	.003		4.31	.11	.04	4.46	4.96
2540	1"		17000	.003		5.75	.12	.04	5.91	6.55
2550	1-1/2"		15000	.004		8.60	.13	.05	8.78	9.75
2560	Hand mix, 1/2"	C-18	4000	.002		2.87	.07	.01	2.95	3.29
2610	Topping, P.C. based self-level/dry 6100 psi, pumped, 1/4"	C-8	20000	.003		2.21	.10	.04	2.35	2.63
2620	1/2"		19000	.003		4.43	.10	.04	4.57	5.05
2630	3/4"		18000	.003		6.65	.11	.04	6.80	7.50
2660	1"		17000	.003		8.85	.12	.04	9.01	10
2670	1-1/2"		15000	.004		13.30	.13	.05	13.48	14.85
2680	Hand mix, 1/2"	C-18	4000	.002		4.43	.07	.01	4.51	5

03 63 Epoxy Grouting

03 63 05 – Grouting of Dowels and Fasteners

03 63 05.10 Epoxy Only		Crew	Daily Output	Labor-Hours	Unit	2009 Bare Costs Material	Labor	Equipment	Total	Total Incl O&P
0010	**EPOXY ONLY**									
1500	Chemical anchoring, epoxy cartridge, excludes layout, drilling, fastener									
1530	For fastener 3/4" diam. x 6" embedment	2 Skwk	72	.222	Ea.	3.54	9.10		12.64	17.95
1535	1" diam. x 8" embedment		66	.242		5.30	9.90		15.20	21
1540	1-1/4" diam. x 10" embedment		60	.267		10.60	10.90		21.50	28.50
1545	1-3/4" diam. x 12" embedment		54	.296		17.70	12.10		29.80	38
1550	14" embedment		48	.333		21	13.60		34.60	44.50
1555	2" diam. x 12" embedment		42	.381		28.50	15.55		44.05	55
1560	18" embedment		32	.500		35.50	20.50		56	70.50

03 82 Concrete Boring

03 82 13 – Concrete Core Drilling

03 82 13.10 Core Drilling

Line	Description	Crew	Daily Output	Labor-Hours	Unit	Material	Labor	Equipment	Total	Total Incl O&P
0010	**CORE DRILLING**									
0020	Reinf. conc slab, up to 6" thick, incl. bit, layout & set up									
0100	1" diameter core	B-89A	17	.941	Ea.	3.45	34	6.50	43.95	64
0150	Each added inch thick in same hole, add		1440	.011		.63	.40	.08	1.11	1.39
0300	3" diameter core		16	1		7.40	36	6.90	50.30	72
0350	Each added inch thick in same hole, add		720	.022		1.38	.81	.15	2.34	2.94
0500	4" diameter core		15	1.067		7.40	38.50	7.35	53.25	76.50
0550	Each added inch thick in same hole, add		480	.033		1.81	1.21	.23	3.25	4.11
0700	6" diameter core		14	1.143		12.20	41.50	7.85	61.55	86
0750	Each added inch thick in same hole, add		360	.044		2.17	1.61	.31	4.09	5.20
0900	8" diameter core		13	1.231		16.70	44.50	8.50	69.70	96.50
0950	Each added inch thick in same hole, add		288	.056		2.88	2.01	.38	5.27	6.70
1100	10" diameter core		12	1.333		22	48.50	9.20	79.70	109
1150	Each added inch thick in same hole, add		240	.067		3.62	2.42	.46	6.50	8.25
1300	12" diameter core		11	1.455		27	52.50	10	89.50	123
1350	Each added inch thick in same hole, add		206	.078		4.35	2.81	.54	7.70	9.75
1500	14" diameter core		10	1.600		33	58	11	102	138
1550	Each added inch thick in same hole, add		180	.089		5.85	3.22	.61	9.68	12.10
1700	18" diameter core		9	1.778		42.50	64.50	12.25	119.25	160
1750	Each added inch thick in same hole, add		144	.111		7.60	4.03	.77	12.40	15.45
1760	For horizontal holes, add to above						20%	20%		
1770	Prestressed hollow core plank, 8" thick									
1780	1" diameter core	B-89A	17.50	.914	Ea.	2.30	33	6.30	41.60	61
1790	Each added inch thick in same hole, add		3840	.004		.46	.15	.03	.64	.77
1800	3" diameter core		17	.941		4.81	34	6.50	45.31	65.50
1810	Each added inch thick in same hole, add		1920	.008		.83	.30	.06	1.19	1.44
1820	4" diameter core		16.50	.970		6.60	35	6.70	48.30	69
1830	Each added inch thick in same hole, add		1280	.013		1.21	.45	.09	1.75	2.12
1840	6" diameter core		15.50	1.032		8.10	37.50	7.10	52.70	74.50
1850	Each added inch thick in same hole, add		960	.017		1.38	.60	.11	2.09	2.59
1860	8" diameter core		15	1.067		11.50	38.50	7.35	57.35	81
1870	Each added inch thick in same hole, add		768	.021		1.93	.75	.14	2.82	3.45
1880	10" diameter core		14	1.143		15.25	41.50	7.85	64.60	89.50
1890	Each added inch thick in same hole, add		640	.025		2.01	.91	.17	3.09	3.80
1900	12" diameter core		13.50	1.185		17.85	43	8.15	69	95
1910	Each added inch thick in same hole, add		548	.029		3.22	1.06	.20	4.48	5.40
1999	Drilling, core, minimum labor/equipment charge		4	4	Job		145	27.50	172.50	256

03 82 Concrete Boring

03 82 16 – Concrete Drilling

03 82 16.10 Concrete Impact Drilling		Crew	Daily Output	Labor-Hours	Unit	2009 Bare Costs Material	Labor	Equipment	Total	Total Incl O&P
0010	**CONCRETE IMPACT DRILLING**									
0050	Up to 4" deep in conc/brick floor/wall, incl. bit & layout, no anchor									
0100	Holes, 1/4" diameter	1 Carp	75	.107	Ea.	.06	4.26		4.32	6.65
0150	For each additional inch of depth, add		430	.019		.01	.74		.75	1.17
0200	3/8" diameter		63	.127		.05	5.05		5.10	7.90
0250	For each additional inch of depth, add		340	.024		.01	.94		.95	1.47
0300	1/2" diameter		50	.160		.06	6.40		6.46	9.95
0350	For each additional inch of depth, add		250	.032		.01	1.28		1.29	2
0400	5/8" diameter		48	.167		.08	6.65		6.73	10.45
0450	For each additional inch of depth, add		240	.033		.02	1.33		1.35	2.08
0500	3/4" diameter		45	.178		.10	7.10		7.20	11.10
0550	For each additional inch of depth, add		220	.036		.03	1.45		1.48	2.28
0600	7/8" diameter		43	.186		.12	7.45		7.57	11.70
0650	For each additional inch of depth, add		210	.038		.03	1.52		1.55	2.39
0700	1" diameter		40	.200		.14	8		8.14	12.55
0750	For each additional inch of depth, add		190	.042		.04	1.68		1.72	2.65
0800	1-1/4" diameter		38	.211		.20	8.40		8.60	13.25
0850	For each additional inch of depth, add		180	.044		.05	1.78		1.83	2.81
0900	1-1/2" diameter		35	.229		.29	9.15		9.44	14.45
0950	For each additional inch of depth, add	↓	165	.048	↓	.07	1.94		2.01	3.08
1000	For ceiling installations, add						40%			

Division Notes

		CREW	DAILY OUTPUT	LABOR-HOURS	UNIT	2009 BARE COSTS MAT.	LABOR	EQUIP.	TOTAL	TOTAL INCL O&P

Estimating Tips

04 05 00 Common Work Results for Masonry

- The terms *mortar* and *grout* are often used interchangeably, and incorrectly. Mortar is used to bed masonry units, seal the entry of air and moisture, provide architectural appearance, and allow for size variations in the units. Grout is used primarily in reinforced masonry construction and is used to bond the masonry to the reinforcing steel. Common mortar types are M(2500 psi), S(1800 psi), N(750 psi), and O(350 psi), and conform to ASTM C270. Grout is either fine or coarse and conforms to ASTM C476, and in-place strengths generally exceed 2500 psi. Mortar and grout are different components of masonry construction and are placed by entirely different methods. An estimator should be aware of their unique uses and costs.
- Waste, specifically the loss/droppings of mortar and the breakage of brick and block, is included in all masonry assemblies in this division. A factor of 25% is added for mortar and 3% for brick and concrete masonry units.
- Scaffolding or staging is not included in any of the Division 4 costs. Refer to Subdivision 01 54 23 for scaffolding and staging costs.

04 20 00 Unit Masonry

- The most common types of unit masonry are brick and concrete masonry. The major classifications of brick are building brick (ASTM C62), facing brick (ASTM C216), glazed brick, fire brick, and pavers. Many varieties of texture and appearance can exist within these classifications, and the estimator would be wise to check local custom and availability within the project area. For repair and remodeling jobs, matching the existing brick may be the most important criteria.
- Brick and concrete block are priced by the piece and then converted into a price per square foot of wall. Openings less than two square feet are generally ignored by the estimator because any savings in units used is offset by the cutting and trimming required.
- It is often difficult and expensive to find and purchase small lots of historic brick. Costs can vary widely. Many design issues affect costs, selection of mortar mix, and repairs or replacement of masonry materials. Cleaning techniques must be reflected in the estimate.
- All masonry walls, whether interior or exterior, require bracing. The cost of bracing walls during construction should be included by the estimator, and this bracing must remain in place until permanent bracing is complete. Permanent bracing of masonry walls is accomplished by masonry itself, in the form of pilasters or abutting wall corners, or by anchoring the walls to the structural frame. Accessories in the form of anchors, anchor slots, and ties are used, but their supply and installation can be by different trades. For instance, anchor slots on spandrel beams and columns are supplied and welded in place by the steel fabricator, but the ties from the slots into the masonry are installed by the bricklayer. Regardless of the installation method, the estimator must be certain that these accessories are accounted for in pricing.

Reference Numbers

Reference numbers are shown in shaded boxes at the beginning of some major classifications. These numbers refer to related items in the Reference Section. The reference information may be an estimating procedure, an alternate pricing method, or technical information.

Note: Not all subdivisions listed here necessarily appear in this publication.

04 05 Common Work Results for Masonry

04 05 23 – Masonry Accessories

04 05 23.19 Vent Box		Crew	Daily Output	Labor-Hours	Unit	Material	Labor	Equipment	Total	Total Incl O&P
						2009 Bare Costs				
0010	**VENT BOX**									
0020	Extruded aluminum, 4" deep, 2-3/8" x 8-1/8"	1 Bric	30	.267	Ea.	30.50	10.80		41.30	50
0050	5" x 8-1/8"		25	.320		40	12.95		52.95	63.50
0100	2-1/4" x 25"		25	.320		69.50	12.95		82.45	96
0150	5" x 16-1/2"		22	.364		58.50	14.75		73.25	86.50
0200	6" x 16-1/2"		22	.364		81	14.75		95.75	112
0250	7-3/4" x 16-1/2"		20	.400		69	16.20		85.20	101
0400	For baked enamel finish, add					35%				
0500	For cast aluminum, painted, add					60%				
1000	Stainless steel ventilators, 6" x 6"	1 Bric	25	.320		126	12.95		138.95	158
1050	8" x 8"		24	.333		132	13.50		145.50	167
1100	12" x 12"		23	.348		153	14.10		167.10	190
1150	12" x 6"		24	.333		134	13.50		147.50	168
1200	Foundation block vent, galv., 1-1/4" thk, 8" high, 16" long, no damper		30	.267		24.50	10.80		35.30	43.50
1250	For damper, add					8.20			8.20	9

04 21 Clay Unit Masonry

04 21 13 – Brick Masonry

04 21 13.15 Chimney

		Crew	Daily Output	Labor-Hours	Unit	Material	Labor	Equipment	Total	Total Incl O&P
0010	**CHIMNEY**, excludes foundation, scaffolding, grout and reinforcing									
0100	Brick, 16" x 16", 8" flue	D-1	18.20	.879	V.L.F.	20.50	32		52.50	71.50
0150	16" x 20" with one 8" x 12" flue		16	1		32.50	36.50		69	91
0200	16" x 24" with two 8" x 8" flues		14	1.143		47.50	41.50		89	116
0250	20" x 20" with one 12" x 12" flue		13.70	1.168		38.50	42.50		81	107
0300	20" x 24" with two 8" x 12" flues		12	1.333		54	48.50		102.50	133
0350	20" x 32" with two 12" x 12" flues		10	1.600		67.50	58		125.50	163
1800	Metal, high temp. steel jacket, factory lining, 24" diam.	E-2	65	.862		210	37.50	27	274.50	325
1900	60" diameter	"	30	1.867		765	81.50	58	904.50	1,050
2100	Poured concrete, brick lining, 200' high x 10' diam.					7,275			7,275	8,000
2800	500' x 20' diameter					12,700			12,700	14,000

Estimating Tips

05 05 00 Common Work Results for Metals

- Nuts, bolts, washers, connection angles, and plates can add a significant amount to both the tonnage of a structural steel job and the estimated cost. As a rule of thumb, add 10% to the total weight to account for these accessories.
- Type 2 steel construction, commonly referred to as "simple construction," consists generally of field-bolted connections with lateral bracing supplied by other elements of the building, such as masonry walls or x-bracing. The estimator should be aware, however, that shop connections may be accomplished by welding or bolting. The method may be particular to the fabrication shop and may have an impact on the estimated cost.

05 12 23 Structural Steel

- Steel items can be obtained from two sources: a fabrication shop or a metals service center. Fabrication shops can fabricate items under more controlled conditions than can crews in the field. They are also more efficient and can produce items more economically. Metal service centers serve as a source of long mill shapes to both fabrication shops and contractors.
- Most line items in this structural steel subdivision, and most items in 05 50 00 Metal Fabrications, are indicated as being shop fabricated. The bare material cost for these shop fabricated items is the "Invoice Cost" from the shop and includes the mill base price of steel plus mill extras, transportation to the shop, shop drawings and detailing where warranted, shop fabrication and handling, sandblasting and a shop coat of primer paint, all necessary structural bolts, and delivery to the job site. The bare labor cost and bare equipment cost for these shop fabricated items is for field installation or erection.
- Line items in Subdivision 05 12 23.40 Lightweight Framing, and other items scattered in Division 5, are indicated as being field fabricated. The bare material cost for these field fabricated items is the "Invoice Cost" from the metals service center and includes the mill base price of steel plus mill extras, transportation to the metals service center, material handling, and delivery of long lengths of mill shapes to the job site. Material costs for structural bolts and welding rods should be added to the estimate. The bare labor cost and bare equipment cost for these items is for both field fabrication and field installation or erection, and include time for cutting, welding and drilling in the fabricated metal items. Drilling into concrete and fasteners to fasten field fabricated items to other work are not included and should be added to the estimate.

05 20 00 Steel Joist Framing

- In any given project the total weight of open web steel joists is determined by the loads to be supported and the design. However, economies can be realized in minimizing the amount of labor used to place the joists. This is done by maximizing the joist spacing, and therefore minimizing the number of joists required to be installed on the job. Certain spacings and locations may be required by the design, but in other cases maximizing the spacing and keeping it as uniform as possible will keep the costs down.

05 30 00 Steel Decking

- The takeoff and estimating of metal deck involves more than simply the area of the floor or roof and the type of deck specified or shown on the drawings. Many different sizes and types of openings may exist. Small openings for individual pipes or conduits may be drilled after the floor/roof is installed, but larger openings may require special deck lengths as well as reinforcing or structural support. The estimator should determine who will be supplying this reinforcing. Additionally, some deck terminations are part of the deck package, such as screed angles and pour stops, and others will be part of the steel contract, such as angles attached to structural members and cast-in-place angles and plates. The estimator must ensure that all pieces are accounted for in the complete estimate.

05 50 00 Metal Fabrications

- The most economical steel stairs are those that use common materials, standard details, and most importantly, a uniform and relatively simple method of field assembly. Commonly available A36 channels and plates are very good choices for the main stringers of the stairs, as are angles and tees for the carrier members. Risers and treads are usually made by specialty shops, and it is most economical to use a typical detail in as many places as possible. The stairs should be pre-assembled and shipped directly to the site. The field connections should be simple and straightforward to be accomplished efficiently, and with minimum equipment and labor.

Reference Numbers

Reference numbers are shown in shaded boxes at the beginning of some major classifications. These numbers refer to related items in the Reference Section. The reference information may be an estimating procedure, an alternate pricing method, or technical information. *Note:* Not all subdivisions listed here necessarily appear in this publication.

05 05 21 – Fastening Methods for Metal

05 05 21.15 Drilling Steel		Crew	Daily Output	Labor-Hours	Unit	Material	Labor	Equipment	Total	Total Incl O&P
						2009 Bare Costs				
0010	**DRILLING STEEL**									
1910	Drilling & layout for steel, up to 1/4" deep, no anchor									
1920	Holes, 1/4" diameter	1 Sswk	112	.071	Ea.	.08	3.19		3.27	5.80
1925	For each additional 1/4" depth, add		336	.024		.08	1.06		1.14	1.99
1930	3/8" diameter		104	.077		.09	3.44		3.53	6.25
1935	For each additional 1/4" depth, add		312	.026		.09	1.15		1.24	2.14
1940	1/2" diameter		96	.083		.10	3.72		3.82	6.75
1945	For each additional 1/4" depth, add		288	.028		.10	1.24		1.34	2.32
1950	5/8" diameter		88	.091		.13	4.06		4.19	7.40
1955	For each additional 1/4" depth, add		264	.030		.13	1.35		1.48	2.56
1960	3/4" diameter		80	.100		.15	4.47		4.62	8.10
1965	For each additional 1/4" depth, add		240	.033		.15	1.49		1.64	2.82
1970	7/8" diameter		72	.111		.18	4.97		5.15	9.05
1975	For each additional 1/4" depth, add		216	.037		.18	1.66		1.84	3.15
1980	1" diameter		64	.125		.20	5.60		5.80	10.15
1985	For each additional 1/4" depth, add		192	.042		.20	1.86		2.06	3.54
1990	For drilling up, add						40%			

05 05 23 – Metal Fastenings

05 05 23.05 Anchor Bolts			Crew	Daily Output	Labor-Hours	Unit	Material	Labor	Equipment	Total	Total Incl O&P
0010	**ANCHOR BOLTS**										
0015	Made from recycled materials										
0100	J-type, incl. hex nut & washer, 1/2" diameter x 6" long	G	2 Carp	70	.229	Ea.	1.01	9.15		10.16	15.25
0110	12" long	G		65	.246		1.26	9.85		11.11	16.65
0120	18" long	G		60	.267		1.64	10.65		12.29	18.30
0130	3/4" diameter x 8" long	G		50	.320		1.86	12.80		14.66	22
0140	12" long	G		45	.356		2.33	14.20		16.53	24.50
0150	18" long	G		40	.400		3.03	16		19.03	28.50
0160	1" diameter x 12" long	G		35	.457		6.90	18.25		25.15	36
0170	18" long	G		30	.533		8.30	21.50		29.80	42
0180	24" long	G		25	.640		10.15	25.50		35.65	50.50
0190	36" long	G		20	.800		13.90	32		45.90	65
0200	1-1/2" diameter x 18" long	G		22	.727		15.80	29		44.80	62.50
0210	24" long	G		16	1		18.80	40		58.80	82.50
0300	L-type, incl. hex nut & washer, 3/4" diameter x 12" long	G		45	.356		2.08	14.20		16.28	24.50
0310	18" long	G		40	.400		2.67	16		18.67	28
0320	24" long	G		35	.457		3.25	18.25		21.50	32
0330	30" long	G		30	.533		4.13	21.50		25.63	37.50
0340	36" long	G		25	.640		4.71	25.50		30.21	44.50
0350	1" diameter x 12" long	G		35	.457		4.23	18.25		22.48	33
0360	18" long	G		30	.533		5.25	21.50		26.75	39
0370	24" long	G		25	.640		6.50	25.50		32	46.50
0380	30" long	G		23	.696		7.65	28		35.65	51.50
0390	36" long	G		20	.800		8.80	32		40.80	59
0400	42" long	G		18	.889		10.65	35.50		46.15	67
0410	48" long	G		15	1.067		11.95	42.50		54.45	79
0420	1-1/4" diameter x 18" long	G		25	.640		6.80	25.50		32.30	47
0430	24" long	G		22	.727		8.10	29		37.10	54
0440	30" long	G		20	.800		9.35	32		41.35	60
0450	36" long	G		18	.889		10.65	35.50		46.15	66.50
0460	42" long	G		16	1		12.05	40		52.05	75.50
0470	48" long	G		14	1.143		13.75	45.50		59.25	86
0480	54" long	G		12	1.333		16.20	53.50		69.70	100

05 05 Common Work Results for Metals

05 05 23 – Metal Fastenings

05 05 23.05 Anchor Bolts

Line	05 05 23.05 Anchor Bolts		Crew	Daily Output	Labor-Hours	Unit	2009 Bare Costs Material	Labor	Equipment	Total	Total Incl O&P
0490	60″ long	G	2 Carp	10	1.600	Ea.	17.80	64		81.80	119
0500	1-1/2″ diameter x 18″ long	G		22	.727		10.50	29		39.50	56.50
0510	24″ long	G		19	.842		12.25	33.50		45.75	65.50
0520	30″ long	G		17	.941		13.80	37.50		51.30	73.50
0540	42″ long	G		15	1.067		18.10	42.50		60.60	86
0550	48″ long	G		13	1.231		20.50	49		69.50	99
0560	54″ long	G		11	1.455		25	58		83	118
0570	60″ long	G		9	1.778		27	71		98	140
0580	1-3/4″ diameter x 18″ long	G		20	.800		15.80	32		47.80	67
0590	24″ long	G		18	.889		18.55	35.50		54.05	75.50
0600	30″ long	G		17	.941		21.50	37.50		59	82.50
0610	36″ long	G		16	1		24.50	40		64.50	89
0620	42″ long	G		14	1.143		27.50	45.50		73	102
0630	48″ long	G		12	1.333		30.50	53.50		84	116
0640	54″ long	G		10	1.600		37.50	64		101.50	141
0650	60″ long	G		8	2		41	80		121	169
0660	2″ diameter x 24″ long	G		17	.941		23.50	37.50		61	84.50
0670	30″ long	G		15	1.067		26.50	42.50		69	95.50
0680	36″ long	G		13	1.231		29.50	49		78.50	109
0690	42″ long	G		11	1.455		32.50	58		90.50	126
0700	48″ long	G		10	1.600		37.50	64		101.50	141
0710	54″ long	G		9	1.778		44.50	71		115.50	159
0720	60″ long	G		8	2		48	80		128	177
0730	66″ long	G		7	2.286		51.50	91.50		143	199
0740	72″ long	G	↓	6	2.667		56.50	107		163.50	227
0990	For galvanized, add					↓	75%				

05 05 23.15 Chemical Anchors

Line	05 05 23.15 Chemical Anchors		Crew	Daily Output	Labor-Hours	Unit	Material	Labor	Equipment	Total	Total Incl O&P
0010	**CHEMICAL ANCHORS**										
0020	Includes layout & drilling										
1430	Chemical anchor, w/rod & epoxy cartridge, 3/4″ diam. x 9-1/2″ long		B-89A	27	.593	Ea.	7.55	21.50	4.08	33.13	46.50
1435	1″ diameter x 11-3/4″ long			24	.667		12.80	24	4.59	41.39	56.50
1440	1-1/4″ diameter x 14″ long			21	.762		23.50	27.50	5.25	56.25	75
1445	1-3/4″ diameter x 15″ long			20	.800		49.50	29	5.50	84	105
1450	18″ long			17	.941		59	34	6.50	99.50	125
1455	2″ diameter x 18″ long			16	1		78.50	36	6.90	121.40	150
1460	24″ long		↓	15	1.067	↓	102	38.50	7.35	147.85	180

05 05 23.20 Expansion Anchors

Line	05 05 23.20 Expansion Anchors		Crew	Daily Output	Labor-Hours	Unit	Material	Labor	Equipment	Total	Total Incl O&P
0010	**EXPANSION ANCHORS**										
0100	Anchors for concrete, brick or stone, no layout and drilling										
0200	Expansion shields, zinc, 1/4″ diameter, 1-5/16″ long, single	G	1 Carp	90	.089	Ea.	.28	3.55		3.83	5.80
0300	1-3/8″ long, double	G		85	.094		.33	3.76		4.09	6.20
0400	3/8″ diameter, 1-1/2″ long, single	G		85	.094		.49	3.76		4.25	6.40
0500	2″ long, double	G		80	.100		1.10	4		5.10	7.40
0600	1/2″ diameter, 2-1/16″ long, single	G		80	.100		1.11	4		5.11	7.40
0700	2-1/2″ long, double	G		75	.107		1.40	4.26		5.66	8.15
0800	5/8″ diameter, 2-5/8″ long, single	G		75	.107		1.52	4.26		5.78	8.25
0900	2-3/4″ long, double	G		70	.114		2.14	4.57		6.71	9.45
1000	3/4″ diameter, 2-3/4″ long, single	G		70	.114		1.99	4.57		6.56	9.30
1100	3-15/16″ long, double	G		65	.123		3.36	4.92		8.28	11.30
1600	Self-drilling anchor, snap-off, for 3/8″ diameter bolt	G		23	.348		1.46	13.90		15.36	23
1900	3/4″ diameter bolt	G	↓	16	.500	↓	6.30	20		26.30	38
2100	Hollow wall anchors for gypsum wall board, plaster or tile										

05 05 23.20 Expansion Anchors			Crew	Daily Output	Labor-Hours	Unit	2009 Bare Costs Material	Labor	Equipment	Total	Total Incl O&P
2300	1/8" diameter, short	G	1 Carp	160	.050	Ea.	.18	2		2.18	3.30
2400	Long	G		150	.053		.21	2.13		2.34	3.53
2500	3/16" diameter, short	G		150	.053		.36	2.13		2.49	3.70
2600	Long	G		140	.057		.47	2.28		2.75	4.06
2700	1/4" diameter, short	G		140	.057		.45	2.28		2.73	4.04
2800	Long	G		130	.062		.49	2.46		2.95	4.35
3000	Toggle bolts, bright steel, 1/8" diameter, 2" long	G		85	.094		.17	3.76		3.93	6.05
3100	4" long	G		80	.100		.20	4		4.20	6.40
3200	3/16" diameter, 3" long	G		80	.100		.22	4		4.22	6.45
3300	6" long	G		75	.107		.31	4.26		4.57	6.95
3400	1/4" diameter, 3" long	G		75	.107		.30	4.26		4.56	6.95
3500	6" long	G		70	.114		.42	4.57		4.99	7.55
3600	3/8" diameter, 3" long	G		70	.114		.68	4.57		5.25	7.85
3700	6" long	G		60	.133		1.13	5.35		6.48	9.50
3800	1/2" diameter, 4" long	G		60	.133		1.58	5.35		6.93	10
3900	6" long	G		50	.160		1.86	6.40		8.26	11.95
5000	Screw anchors for concrete, masonry,										
5100	stone & tile, no layout or drilling included										
5700	Lag screw shields, 1/4" diameter, short	G	1 Carp	90	.089	Ea.	.33	3.55		3.88	5.85
5800	Long	G		85	.094		.33	3.76		4.09	6.20
5900	3/8" diameter, short	G		85	.094		.61	3.76		4.37	6.50
6000	Long	G		80	.100		.64	4		4.64	6.90
6100	1/2" diameter, short	G		80	.100		.89	4		4.89	7.20
6200	Long	G		75	.107		1.06	4.26		5.32	7.75
6300	5/8" diameter, short	G		70	.114		1.47	4.57		6.04	8.70
6400	Long	G		65	.123		1.79	4.92		6.71	9.55
6600	Lead, #6 & #8, 3/4" long	G		260	.031		.10	1.23		1.33	2.02
6700	#10 - #14, 1-1/2" long	G		200	.040		.22	1.60		1.82	2.72
6800	#16 & #18, 1-1/2" long	G		160	.050		.28	2		2.28	3.41
6900	Plastic, #6 & #8, 3/4" long			260	.031		.03	1.23		1.26	1.94
7000	#8 & #10, 7/8" long			240	.033		.03	1.33		1.36	2.09
7100	#10 & #12, 1" long			220	.036		.04	1.45		1.49	2.29
7200	#14 & #16, 1-1/2" long			160	.050		.06	2		2.06	3.17
8000	Wedge anchors, not including layout or drilling										
8050	Carbon steel, 1/4" diameter, 1-3/4" long	G	1 Carp	150	.053	Ea.	.25	2.13		2.38	3.57
8100	3-1/4" long	G		140	.057		.32	2.28		2.60	3.90
8150	3/8" diameter, 2-1/4" long	G		145	.055		.32	2.20		2.52	3.78
8200	5" long	G		140	.057		.57	2.28		2.85	4.17
8250	1/2" diameter, 2-3/4" long	G		140	.057		.97	2.28		3.25	4.61
8300	7" long	G		125	.064		1.66	2.56		4.22	5.80
8350	5/8" diameter, 3-1/2" long	G		130	.062		1	2.46		3.46	4.91
8400	8-1/2" long	G		115	.070		2.12	2.78		4.90	6.65
8450	3/4" diameter, 4-1/4" long	G		115	.070		1.66	2.78		4.44	6.15
8500	10" long	G		95	.084		3.76	3.36		7.12	9.35
8550	1" diameter, 6" long	G		100	.080		6	3.20		9.20	11.55
8575	9" long	G		85	.094		7.80	3.76		11.56	14.45
8600	12" long	G		75	.107		8.45	4.26		12.71	15.90
8650	1-1/4" diameter, 9" long	G		70	.114		19.45	4.57		24.02	28.50
8700	12" long	G		60	.133		25	5.35		30.35	36
8750	For type 303 stainless steel, add						350%				
8800	For type 316 stainless steel, add						450%				
8950	Self-drilling concrete screw, hex washer head, 3/16" diam. x 1-3/4" long	G	1 Carp	300	.027	Ea.	.16	1.07		1.23	1.83
8960	2-1/4" long	G		250	.032		.18	1.28		1.46	2.18

05 05 23 – Metal Fastenings

05 05 23.20 Expansion Anchors

	05 05 23.20 Expansion Anchors		Crew	Daily Output	Labor-Hours	Unit	2009 Bare Costs Material	Labor	Equipment	Total	Total Incl O&P
8970	Phillips flat head, 3/16" diam. x 1-3/4" long	G	1 Carp	300	.027	Ea.	.15	1.07		1.22	1.82
8980	2-1/4" long	G		250	.032		.17	1.28		1.45	2.17

05 05 23.30 Lag Screws

	05 05 23.30 Lag Screws		Crew	Daily Output	Labor-Hours	Unit	Material	Labor	Equipment	Total	Total Incl O&P
0010	**LAG SCREWS**										
0020	Steel, 1/4" diameter, 2" long	G	1 Carp	200	.040	Ea.	.08	1.60		1.68	2.57
0100	3/8" diameter, 3" long	G		150	.053		.26	2.13		2.39	3.59
0200	1/2" diameter, 3" long	G		130	.062		.45	2.46		2.91	4.31
0300	5/8" diameter, 3" long	G		120	.067		1.12	2.66		3.78	5.35

05 05 23.35 Machine Screws

	05 05 23.35 Machine Screws		Crew	Daily Output	Labor-Hours	Unit	Material	Labor	Equipment	Total	Total Incl O&P
0010	**MACHINE SCREWS**										
0020	Steel, round head, #8 x 1" long	G	1 Carp	4.80	1.667	C	2.91	66.50		69.41	106
0110	#8 x 2" long	G		2.40	3.333		4.46	133		137.46	212
0200	#10 x 1" long	G		4	2		3.44	80		83.44	128
0300	#10 x 2" long	G		2	4		5.85	160		165.85	254

05 05 23.40 Machinery Anchors

	05 05 23.40 Machinery Anchors		Crew	Daily Output	Labor-Hours	Unit	Material	Labor	Equipment	Total	Total Incl O&P
0010	**MACHINERY ANCHORS**, heavy duty, incl. sleeve, floating base nut,										
0020	lower stud & coupling nut, fiber plug, connecting stud, washer & nut.										
0030	For flush mounted embedment in poured concrete heavy equip. pads.										
0200	Stud & bolt, 1/2" diameter	G	E-16	40	.400	Ea.	67	18.30	3.36	88.66	110
0300	5/8" diameter	G		35	.457		74.50	21	3.84	99.34	123
0500	3/4" diameter	G		30	.533		86	24.50	4.47	114.97	143
0600	7/8" diameter	G		25	.640		93.50	29.50	5.35	128.35	161
0800	1" diameter	G		20	.800		98.50	36.50	6.70	141.70	180
0900	1-1/4" diameter	G		15	1.067		131	49	8.95	188.95	241

05 05 23.50 Powder Actuated Tools and Fasteners

	05 05 23.50 Powder Actuated Tools and Fasteners		Crew	Daily Output	Labor-Hours	Unit	Material	Labor	Equipment	Total	Total Incl O&P
0010	**POWDER ACTUATED TOOLS & FASTENERS**										
0020	Stud driver, .22 caliber, buy, minimum					Ea.	224			224	246
0100	Maximum					"	360			360	395
0300	Powder charges for above, low velocity					C	8.65			8.65	9.50
0400	Standard velocity						12.30			12.30	13.55
0600	Drive pins & studs, 1/4" & 3/8" diam., to 3" long, minimum	G	1 Carp	4.80	1.667		2.88	66.50		69.38	106
0700	Maximum	G	"	4	2		11.25	80		91.25	136
0800	Pneumatic stud driver for 1/8" diameter studs					Ea.	1,550			1,550	1,725
0900	Drive pins for above, 1/2" to 3/4" long	G	1 Carp	1	8	M	355	320		675	890

05 05 23.55 Rivets

	05 05 23.55 Rivets		Crew	Daily Output	Labor-Hours	Unit	Material	Labor	Equipment	Total	Total Incl O&P
0010	**RIVETS**										
0100	Aluminum rivet & mandrel, 1/2" grip length x 1/8" diameter	G	1 Carp	4.80	1.667	C	6.15	66.50		72.65	110
0200	3/16" diameter	G		4	2		9.45	80		89.45	134
0300	Aluminum rivet, steel mandrel, 1/8" diameter	G		4.80	1.667		9.40	66.50		75.90	113
0400	3/16" diameter	G		4	2		8.60	80		88.60	133
0500	Copper rivet, steel mandrel, 1/8" diameter	G		4.80	1.667		19.90	66.50		86.40	125
1200	Steel rivet and mandrel, 1/8" diameter	G		4.80	1.667		10.45	66.50		76.95	114
1300	3/16" diameter	G		4	2		15.70	80		95.70	141
1400	Hand riveting tool, minimum					Ea.	143			143	157
1500	Maximum						273			273	300
1600	Power riveting tool, minimum						1,025			1,025	1,125
1700	Maximum						2,600			2,600	2,850

05 12 23 – Structural Steel for Buildings

05 12 23.40 Lightweight Framing

	05 12 23.40 Lightweight Framing		Crew	Daily Output	Labor-Hours	Unit	Material	Labor	Equipment	Total	Total Incl O&P
							2009 Bare Costs				
0010	**LIGHTWEIGHT FRAMING**										
0015	Made from recycled materials										
0400	Angle framing, field fabricated, 4" and larger	G	E-3	440	.055	Lb.	.87	2.47	.30	3.64	5.70
0450	Less than 4" angles	G		265	.091		.90	4.11	.51	5.52	8.85
0600	Channel framing, field fabricated, 8" and larger	G		500	.048		.90	2.18	.27	3.35	5.15
0650	Less than 8" channels	G		335	.072		.90	3.25	.40	4.55	7.25
1000	Continuous slotted channel framing system, shop fab, minimum	G	2 Sswk	2400	.007		4.65	.30		4.95	5.65
1200	Maximum	G	"	1600	.010		5.25	.45		5.70	6.60
1250	Plate & bar stock for reinforcing beams and trusses	G					1.65			1.65	1.82
1300	Cross bracing, rods, shop fabricated, 3/4" diameter	G	E-3	700	.034		1.80	1.56	.19	3.55	4.96
1310	7/8" diameter	G		850	.028		1.80	1.28	.16	3.24	4.43
1320	1" diameter	G		1000	.024		1.80	1.09	.13	3.02	4.07
1330	Angle, 5" x 5" x 3/8"	G		2800	.009		1.80	.39	.05	2.24	2.72
1350	Hanging lintels, shop fabricated, average	G		850	.028		1.80	1.28	.16	3.24	4.43
1380	Roof frames, shop fabricated, 3'-0" square, 5' span	G	E-2	4200	.013		1.80	.58	.41	2.79	3.43
1400	Tie rod, not upset, 1-1/2" to 4" diameter, with turnbuckle	G	2 Sswk	800	.020		1.95	.89		2.84	3.74
1420	No turnbuckle	G		700	.023		1.88	1.02		2.90	3.88
1500	Upset, 1-3/4" to 4" diameter, with turnbuckle	G		800	.020		1.95	.89		2.84	3.74
1520	No turnbuckle	G		700	.023		1.88	1.02		2.90	3.88

05 12 23.60 Pipe Support Framing

	05 12 23.60 Pipe Support Framing		Crew	Daily Output	Labor-Hours	Unit	Material	Labor	Equipment	Total	Total Incl O&P
0010	**PIPE SUPPORT FRAMING**										
0020	Under 10#/L.F., shop fabricated	G	E-4	3900	.008	Lb.	2.01	.37	.03	2.41	2.91
0200	10.1 to 15#/L.F.	G		4300	.007		1.98	.34	.03	2.35	2.81
0400	15.1 to 20#/L.F.	G		4800	.007		1.95	.30	.03	2.28	2.72
0600	Over 20#/L.F.	G		5400	.006		1.92	.27	.02	2.21	2.62

Division Notes

		CREW	DAILY OUTPUT	LABOR-HOURS	UNIT	2009 BARE COSTS MAT.	LABOR	EQUIP.	TOTAL	TOTAL INCL O&P

Estimating Tips

06 05 00 Common Work Results for Wood, Plastics, and Composites

- Common to any wood-framed structure are the accessory connector items such as screws, nails, adhesives, hangers, connector plates, straps, angles, and hold-downs. For typical wood-framed buildings, such as residential projects, the aggregate total for these items can be significant, especially in areas where seismic loading is a concern. For floor and wall framing, the material cost is based on 10 to 25 lbs. per MBF. Hold-downs, hangers, and other connectors should be taken off by the piece.

06 10 00 Carpentry

- Lumber is a traded commodity and therefore sensitive to supply and demand in the marketplace. Even in "budgetary" estimating of wood-framed projects, it is advisable to call local suppliers for the latest market pricing.
- Common quantity units for wood-framed projects are "thousand board feet" (MBF). A board foot is a volume of wood, 1″ x 1′ x 1′, or 144 cubic inches. Board-foot quantities are generally calculated using nominal material dimensions—dressed sizes are ignored. Board foot per lineal foot of any stick of lumber can be calculated by dividing the nominal cross-sectional area by 12. As an example, 2,000 lineal feet of 2 x 12 equates to 4 MBF by dividing the nominal area, 2 x 12, by 12, which equals 2, and multiplying by 2,000 to give 4,000 board feet. This simple rule applies to all nominal dimensioned lumber.
- Waste is an issue of concern at the quantity takeoff for any area of construction. Framing lumber is sold in even foot lengths, i.e., 10′, 12′, 14′, 16′, and depending on spans, wall heights, and the grade of lumber, waste is inevitable. A rule of thumb for lumber waste is 5%–10% depending on material quality and the complexity of the framing.
- Wood in various forms and shapes is used in many projects, even where the main structural framing is steel, concrete, or masonry. Plywood as a back-up partition material and 2x boards used as blocking and cant strips around roof edges are two common examples. The estimator should ensure that the costs of all wood materials are included in the final estimate.

06 20 00 Finish Carpentry

- It is necessary to consider the grade of workmanship when estimating labor costs for erecting millwork and interior finish. In practice, there are three grades: premium, custom, and economy. The RSMeans daily output for base and case moldings is in the range of 200 to 250 L.F. per carpenter per day. This is appropriate for most average custom-grade projects. For premium projects, an adjustment to productivity of 25%–50% should be made, depending on the complexity of the job.

Reference Numbers

Reference numbers are shown in shaded boxes at the beginning of some major classifications. These numbers refer to related items in the Reference Section. The reference information may be an estimating procedure, an alternate pricing method, or technical information.

Note: Not all subdivisions listed here necessarily appear in this publication.

06 05 23 – Wood, Plastic, and Composite Fastenings

06 05 23.10 Nails

	06 05 23.10 Nails	Crew	Daily Output	Labor-Hours	Unit	Material	Labor	Equipment	Total	Total Incl O&P
						2009 Bare Costs				
0010	**NAILS**, material only, based upon 50# box purchase									
0020	Copper nails, plain				Lb.	10.20			10.20	11.20
0400	Stainless steel, plain					6.65			6.65	7.30
0500	Box, 3d to 20d, bright					.91			.91	1
0520	Galvanized					1.67			1.67	1.84
0600	Common, 3d to 60d, plain					1.11			1.11	1.22
0700	Galvanized					1.53			1.53	1.68
0800	Aluminum					4.45			4.45	4.90
1000	Annular or spiral thread, 4d to 60d, plain					1.82			1.82	2
1200	Galvanized					1.99			1.99	2.19
1400	Drywall nails, plain					.79			.79	.87
1600	Galvanized					1.59			1.59	1.75
1800	Finish nails, 4d to 10d, plain					1.11			1.11	1.22
2000	Galvanized					1.75			1.75	1.93
2100	Aluminum					4.10			4.10	4.51
2300	Flooring nails, hardened steel, 2d to 10d, plain					2.04			2.04	2.24
2400	Galvanized					3.88			3.88	4.27
2500	Gypsum lath nails, 1-1/8", 13 ga. flathead, blued					2.64			2.64	2.90
2600	Masonry nails, hardened steel, 3/4" to 3" long, plain					1.97			1.97	2.17
2700	Galvanized					2.78			2.78	3.06
5000	Add to prices above for cement coating					.10			.10	.11
5200	Zinc or tin plating					.13			.13	.14
5500	Vinyl coated sinkers, 8d to 16d				↓	.60			.60	.66

06 05 23.20 Pneumatic Nails

	06 05 23.20 Pneumatic Nails	Crew	Daily Output	Labor-Hours	Unit	Material	Labor	Equipment	Total	Total Incl O&P
0010	**PNEUMATIC NAILS**									
0020	Framing, per carton of 5000, 2"				Ea.	40			40	44
0100	2-3/8"					45			45	49.50
0200	Per carton of 4000, 3"					40			40	44
0300	3-1/4"					40			40	44
0400	Per carton of 5000, 2-3/8", galv.					60			60	66
0500	Per carton of 4000, 3", galv.					65			65	71.50
0600	3-1/4", galv.					82			82	90
0700	Roofing, per carton of 7200, 1"					35.50			35.50	39
0800	1-1/4"					35			35	38.50
0900	1-1/2"					38.50			38.50	42.50
1000	1-3/4"				↓	47.50			47.50	52.50

06 05 23.40 Sheet Metal Screws

	06 05 23.40 Sheet Metal Screws	Crew	Daily Output	Labor-Hours	Unit	Material	Labor	Equipment	Total	Total Incl O&P
0010	**SHEET METAL SCREWS**									
0020	Steel, standard, #8 x 3/4", plain				C	3.30			3.30	3.63
0100	Galvanized					3.39			3.39	3.73
0300	#10 x 1", plain					4.25			4.25	4.68
0400	Galvanized					4.43			4.43	4.87
0600	With washers, #14 x 1", plain					11.50			11.50	12.65
0700	Galvanized					12.45			12.45	13.70
0900	#14 x 2", plain					15			15	16.50
1000	Galvanized					17.15			17.15	18.85
1500	Self-drilling, with washers, (pinch point) #8 x 3/4", plain					8			8	8.80
1600	Galvanized					8.70			8.70	9.55
1800	#10 x 3/4", plain					10			10	11
1900	Galvanized					10.95			10.95	12.05
3000	Stainless steel w/aluminum or neoprene washers, #14 x 1", plain					32			32	35
3100	#14 x 2", plain				↓	41			41	45

06 05 Common Work Results for Wood, Plastics and Composites

06 05 23 – Wood, Plastic, and Composite Fastenings

06 05 23.50 Wood Screws		Crew	Daily Output	Labor-Hours	Unit	2009 Bare Costs Material	Labor	Equipment	Total	Total Incl O&P
0010	**WOOD SCREWS**									
0020	#8 x 1" long, steel				C	5.10			5.10	5.60
0100	Brass					11.50			11.50	12.65
0200	#8, 2" long, steel					5.30			5.30	5.85
0300	Brass					19.50			19.50	21.50
0400	#10, 1" long, steel					3.71			3.71	4.08
0500	Brass					15.65			15.65	17.20
0600	#10, 2" long, steel					5.95			5.95	6.55
0700	Brass					23.50			23.50	26
0800	#10, 3" long, steel					9.95			9.95	10.95
1000	#12, 2" long, steel					7.90			7.90	8.70
1100	Brass					31.50			31.50	34.50
1500	#12, 3" long, steel					11.45			11.45	12.60
2000	#12, 4" long, steel				↓	25.50			25.50	28

06 16 Sheathing

06 16 36 – Wood Panel Product Sheathing

06 16 36.10 Sheathing		Crew	Daily Output	Labor-Hours	Unit	Material	Labor	Equipment	Total	Total Incl O&P
0010	**SHEATHING**, plywood on roofs									
0012	Plywood on roofs, CDX									
0030	5/16" thick	2 Carp	1600	.010	S.F.	.53	.40		.93	1.20
0035	Pneumatic nailed		1952	.008		.53	.33		.86	1.09
0050	3/8" thick		1525	.010		.45	.42		.87	1.14
0055	Pneumatic nailed		1860	.009		.45	.34		.79	1.02
0100	1/2" thick		1400	.011		.46	.46		.92	1.22
0105	Pneumatic nailed		1708	.009		.46	.37		.83	1.09
0200	5/8" thick		1300	.012		.67	.49		1.16	1.50
0205	Pneumatic nailed		1586	.010		.67	.40		1.07	1.37
0300	3/4" thick		1200	.013		.83	.53		1.36	1.74
0305	Pneumatic nailed		1464	.011		.83	.44		1.27	1.59
0500	Plywood on walls with exterior CDX, 3/8" thick		1200	.013		.45	.53		.98	1.32
0505	Pneumatic nailed		1488	.011		.45	.43		.88	1.16
0600	1/2" thick		1125	.014		.46	.57		1.03	1.39
0605	Pneumatic nailed		1395	.011		.46	.46		.92	1.22
0700	5/8" thick		1050	.015		.67	.61		1.28	1.68
0705	Pneumatic nailed		1302	.012		.67	.49		1.16	1.50
0800	3/4" thick		975	.016		.83	.66		1.49	1.93
0805	Pneumatic nailed	↓	1209	.013	↓	.83	.53		1.36	1.73

Division Notes

I	CREW	DAILY OUTPUT	LABOR-HOURS	UNIT	2009 BARE COSTS MAT.	LABOR	EQUIP.	TOTAL	TOTAL INCL O&P

Estimating Tips

07 10 00 Dampproofing and Waterproofing

- Be sure of the job specifications before pricing this subdivision. The difference in cost between waterproofing and dampproofing can be great. Waterproofing will hold back standing water. Dampproofing prevents the transmission of water vapor. Also included in this section are vapor retarding membranes.

07 20 00 Thermal Protection

- Insulation and fireproofing products are measured by area, thickness, volume or R-value. Specifications may give only what the specific R-value should be in a certain situation. The estimator may need to choose the type of insulation to meet that R-value.

07 30 00 Steep Slope Roofing
07 40 00 Roofing and Siding Panels

- Many roofing and siding products are bought and sold by the square. One square is equal to an area that measures 100 square feet.

This simple change in unit of measure could create a large error if the estimator is not observant. Accessories necessary for a complete installation must be figured into any calculations for both material and labor.

07 50 00 Membrane Roofing
07 60 00 Flashing and Sheet Metal
07 70 00 Roofing and Wall Specialties and Accessories

- The items in these subdivisions compose a roofing system. No one component completes the installation, and all must be estimated. Built-up or single-ply membrane roofing systems are made up of many products and installation trades. Wood blocking at roof perimeters or penetrations, parapet coverings, reglets, roof drains, gutters, downspouts, sheet metal flashing, skylights, smoke vents, and roof hatches all need to be considered along with the roofing material. Several different installation trades will need to work together on the roofing system. Inherent difficulties in the scheduling and coordination of various trades must be accounted for when estimating labor costs.

07 90 00 Joint Protection

- To complete the weather-tight shell, the sealants and caulkings must be estimated. Where different materials meet—at expansion joints, at flashing penetrations, and at hundreds of other locations throughout a construction project—they provide another line of defense against water penetration. Often, an entire system is based on the proper location and placement of caulking or sealants. The detailed drawings that are included as part of a set of architectural plans show typical locations for these materials. When caulking or sealants are shown at typical locations, this means the estimator must include them for all the locations where this detail is applicable. Be careful to keep different types of sealants separate, and remember to consider backer rods and primers if necessary.

Reference Numbers

Reference numbers are shown in shaded boxes at the beginning of some major classifications. These numbers refer to related items in the Reference Section. The reference information may be an estimating procedure, an alternate pricing method, or technical information.

Note: Not all subdivisions listed here necessarily appear in this publication.

07 65 Flexible Flashing

07 65 10 – Sheet Metal Flashing

07 65 10.10 Sheet Metal Flashing and Counter Flashing		Crew	Daily Output	Labor-Hours	Unit	2009 Bare Costs Material	Labor	Equipment	Total	Total Incl O&P
0010	**SHEET METAL FLASHING AND COUNTER FLASHING**									
0011	Including up to 4 bends									
0020	Aluminum, mill finish, .013" thick	1 Rofc	145	.055	S.F.	.67	1.89		2.56	3.92
0030	.016" thick		145	.055		.79	1.89		2.68	4.05
0060	.019" thick		145	.055		1.02	1.89		2.91	4.30
0100	.032" thick		145	.055		1.48	1.89		3.37	4.81
0200	.040" thick		145	.055		2.21	1.89		4.10	5.60
0300	.050" thick		145	.055		2.52	1.89		4.41	5.95
0400	Painted finish, add					.37			.37	.41
1600	Copper, 16 oz, sheets, under 1000 lbs.	1 Rofc	115	.070		6.80	2.38		9.18	11.45
1700	Over 4000 lbs.		155	.052		8	1.77		9.77	11.80
1900	20 oz sheets, under 1000 lbs.		110	.073		7.90	2.49		10.39	12.90
2000	Over 4000 lbs.		145	.055		8	1.89		9.89	12
2200	24 oz sheets, under 1000 lbs.		105	.076		9.60	2.61		12.21	15
2300	Over 4000 lbs.		135	.059		9.75	2.03		11.78	14.10
2500	32 oz sheets, under 1000 lbs.		100	.080		12.80	2.74		15.54	18.70
2600	Over 4000 lbs.		130	.062		20	2.11		22.11	25.50
5800	Lead, 2.5 lb. per SF, up to 12" wide		135	.059		4.14	2.03		6.17	7.95
5900	Over 12" wide		135	.059		4.14	2.03		6.17	7.95
8650	Copper, 16 oz		100	.080		5.25	2.74		7.99	10.35
8900	Stainless steel sheets, 32 ga, .010" thick		155	.052		3.79	1.77		5.56	7.15
9000	28 ga, .015" thick		155	.052		4.70	1.77		6.47	8.15
9100	26 ga, .018" thick		155	.052		5.70	1.77		7.47	9.25
9200	24 ga, .025" thick		155	.052		7.40	1.77		9.17	11.15
9400	Terne coated stainless steel, .015" thick, 28 ga		155	.052		7.35	1.77		9.12	11.10
9500	.018" thick, 26 ga		155	.052		8.30	1.77		10.07	12.10
9600	Zinc and copper alloy (brass), .020" thick		155	.052		5.15	1.77		6.92	8.65
9700	.027" thick		155	.052		6.90	1.77		8.67	10.60
9800	.032" thick		155	.052		8.05	1.77		9.82	11.85
9900	.040" thick		155	.052		9.85	1.77		11.62	13.80

07 65 13 – Laminated Sheet Flashing

07 65 13.10 Laminated Sheet Flashing		Crew	Daily Output	Labor-Hours	Unit	Material	Labor	Equipment	Total	Total Incl O&P
0010	**LAMINATED SHEET FLASHING**, Including up to 4 bends									
4300	Copper-clad stainless steel, .015" thick, under 500 lbs.	1 Rofc	115	.070	S.F.	5.25	2.38		7.63	9.75
4400	Over 2000 lbs.		155	.052		5.05	1.77		6.82	8.55
4600	.018" thick, under 500 lbs.		100	.080		6.90	2.74		9.64	12.20
4700	Over 2000 lbs.		145	.055		5.05	1.89		6.94	8.75
8550	3 ply copper and fabric, 3 oz		155	.052		2.73	1.77		4.50	6
8600	7 oz		155	.052		5.70	1.77		7.47	9.25
8700	Lead on copper and fabric, 5 oz		155	.052		2.78	1.77		4.55	6.05
8800	7 oz		155	.052		4.94	1.77		6.71	8.45

07 65 19 – Plastic Sheet Flashing

07 65 19.10 Plastic Sheet Flashing and Counter Flashing		Crew	Daily Output	Labor-Hours	Unit	Material	Labor	Equipment	Total	Total Incl O&P
0010	**PLASTIC SHEET FLASHING AND COUNTER FLASHING**									
7300	Polyvinyl chloride, black, .010" thick	1 Rofc	285	.028	S.F.	.26	.96		1.22	1.91
7400	.020" thick		285	.028		.36	.96		1.32	2.02
7600	.030" thick		285	.028		.47	.96		1.43	2.14
7700	.056" thick		285	.028		1.11	.96		2.07	2.84
7900	Black or white for exposed roofs, .060" thick		285	.028		2.45	.96		3.41	4.32
8060	PVC tape, 5" x 45 mils, for joint covers, 100 L.F./roll				Ea.	120			120	132
8850	Polyvinyl chloride, .030" thick	1 Rofc	160	.050	S.F.	.43	1.71		2.14	3.36

07 71 Roof Specialties

07 71 23 – Manufactured Gutters and Downspouts

07 71 23.10 Downspouts		Crew	Daily Output	Labor-Hours	Unit	Material	2009 Bare Costs Labor	Equipment	Total	Total Incl O&P
0010	**DOWNSPOUTS**									
0020	Aluminum 2" x 3", .020" thick, embossed	1 Shee	190	.042	L.F.	1.04	1.99		3.03	4.20
0100	Enameled		190	.042		1.59	1.99		3.58	4.81
4800	Steel, galvanized, round, corrugated, 2" or 3" diameter, 28 gauge		190	.042		1.74	1.99		3.73	4.97
4900	4" diameter, 28 gauge		145	.055		2.48	2.60		5.08	6.75
5100	5" diameter, 28 gauge		130	.062		3.68	2.90		6.58	8.50
5200	26 gauge		130	.062		2.97	2.90		5.87	7.75
5400	6" diameter, 28 gauge		105	.076		4.83	3.60		8.43	10.85
5500	26 gauge	↓	105	.076	↓	6.10	3.60		9.70	12.30

07 72 Roof Accessories

07 72 26 – Ridge Vents

07 72 26.10 Ridge Vents and Accessories

		Crew	Daily Output	Labor-Hours	Unit	Material	Labor	Equipment	Total	Total Incl O&P
0010	**RIDGE VENTS AND ACCESSORIES**									
2300	Ridge vent strip, mill finish	1 Shee	155	.052	L.F.	2.76	2.44		5.20	6.80

07 72 53 – Snow Guards

07 72 53.10 Snow Guard Options

		Crew	Daily Output	Labor-Hours	Unit	Material	Labor	Equipment	Total	Total Incl O&P
0010	**SNOW GUARD OPTIONS**									
0100	Slate & asphalt shingle roofs	1 Rofc	160	.050	Ea.	9.40	1.71		11.11	13.25
0200	Standing seam metal roofs		48	.167		14.60	5.70		20.30	25.50
0300	Surface mount for metal roofs		48	.167	↓	7.70	5.70		13.40	18.05
0400	Double rail pipe type, including pipe	↓	130	.062	L.F.	22.50	2.11		24.61	28.50

07 72 73 – Pitch Pockets

07 72 73.10 Pitch Pockets, Variable Sizes

		Crew	Daily Output	Labor-Hours	Unit	Material	Labor	Equipment	Total	Total Incl O&P
0010	**PITCH POCKETS, VARIABLE SIZES**									
0100	Adjustable, 4" to 7", welded corners, 4" deep	1 Rofc	48	.167	Ea.	12.90	5.70		18.60	24
0200	Side extenders, 6"	"	240	.033	"	1.97	1.14		3.11	4.09

07 84 Firestopping

07 84 13 – Penetration Firestopping

07 84 13.10 Firestopping

		Crew	Daily Output	Labor-Hours	Unit	Material	Labor	Equipment	Total	Total Incl O&P
0010	**FIRESTOPPING** R078413-30									
0100	Metallic piping, non insulated									
0110	Through walls, 2" diameter	1 Carp	16	.500	Ea.	12.40	20		32.40	44.50
0120	4" diameter		14	.571		18.90	23		41.90	56.50
0130	6" diameter		12	.667		25.50	26.50		52	69.50
0140	12" diameter		10	.800		45	32		77	99
0150	Through floors, 2" diameter		32	.250		7.50	10		17.50	24
0160	4" diameter		28	.286		10.80	11.40		22.20	29.50
0170	6" diameter		24	.333		14.20	13.30		27.50	36
0180	12" diameter	↓	20	.400	↓	24	16		40	51.50
0190	Metallic piping, insulated									
0200	Through walls, 2" diameter	1 Carp	16	.500	Ea.	17.55	20		37.55	50.50
0210	4" diameter		14	.571		24	23		47	62
0220	6" diameter		12	.667		30.50	26.50		57	75
0230	12" diameter		10	.800		50	32		82	105
0240	Through floors, 2" diameter		32	.250		12.70	10		22.70	29.50
0250	4" diameter	↓	28	.286	↓	15.95	11.40		27.35	35.50

07 84 Firestopping

07 84 13 – Penetration Firestopping

07 84 13.10 Firestopping		Crew	Daily Output	Labor-Hours	Unit	Material	Labor	Equipment	Total	Total Incl O&P
						2009 Bare Costs				
0260	6″ diameter	1 Carp	24	.333	Ea.	19.35	13.30		32.65	42
0270	12″ diameter	↓	20	.400	↓	24	16		40	51.50
0280	Non metallic piping, non insulated									
0290	Through walls, 2″ diameter	1 Carp	12	.667	Ea.	51	26.50		77.50	97.50
0300	4″ diameter		10	.800		64	32		96	120
0310	6″ diameter		8	1		89	40		129	160
0330	Through floors, 2″ diameter		16	.500		40	20		60	75
0340	4″ diameter		6	1.333		49.50	53.50		103	137
0350	6″ diameter	↓	6	1.333	↓	59.50	53.50		113	148
0370	Ductwork, insulated & non insulated, round									
0380	Through walls, 6″ diameter	1 Carp	12	.667	Ea.	26	26.50		52.50	70
0390	12″ diameter		10	.800		51.50	32		83.50	107
0400	18″ diameter		8	1		84	40		124	155
0410	Through floors, 6″ diameter		16	.500		14.20	20		34.20	46.50
0420	12″ diameter		14	.571		26	23		49	64
0430	18″ diameter	↓	12	.667	↓	45	26.50		71.50	91
0440	Ductwork, insulated & non insulated, rectangular									
0450	With stiffener/closure angle, through walls, 6″ x 12″	1 Carp	8	1	Ea.	21.50	40		61.50	85.50
0460	12″ x 24″		6	1.333		28.50	53.50		82	114
0470	24″ x 48″		4	2		81.50	80		161.50	214
0480	With stiffener/closure angle, through floors, 6″ x 12″		10	.800		11.65	32		43.65	62.50
0490	12″ x 24″		8	1		21	40		61	85
0500	24″ x 48″	↓	6	1.333	↓	41	53.50		94.50	128
0510	Multi trade openings									
0520	Through walls, 6″ x 12″	1 Carp	2	4	Ea.	45	160		205	298
0530	12″ x 24″	″	1	8		182	320		502	695
0540	24″ x 48″	2 Carp	1	16		730	640		1,370	1,800
0550	48″ x 96″	″	.75	21.333		2,925	850		3,775	4,550
0560	Through floors, 6″ x 12″	1 Carp	2	4		45	160		205	298
0570	12″ x 24″	″	1	8		182	320		502	695
0580	24″ x 48″	2 Carp	.75	21.333		730	850		1,580	2,125
0590	48″ x 96″	″	.50	32	↓	2,925	1,275		4,200	5,200
0600	Structural penetrations, through walls									
0610	Steel beams, W8 x 10	1 Carp	8	1	Ea.	28.50	40		68.50	93.50
0620	W12 x 14		6	1.333		45	53.50		98.50	132
0630	W21 x 44		5	1.600		90.50	64		154.50	199
0640	W36 x 135		3	2.667		220	107		327	405
0650	Bar joists, 18″ deep		6	1.333		41.50	53.50		95	128
0660	24″ deep		6	1.333		51.50	53.50		105	140
0670	36″ deep		5	1.600		77.50	64		141.50	185
0680	48″ deep	↓	4	2	↓	90.50	80		170.50	224
0690	Construction joints, floor slab at exterior wall									
0700	Precast, brick, block or drywall exterior									
0710	2″ wide joint	1 Carp	125	.064	L.F.	6.45	2.56		9.01	11.05
0720	4″ wide joint	″	75	.107	″	12.90	4.26		17.16	21
0730	Metal panel, glass or curtain wall exterior									
0740	2″ wide joint	1 Carp	40	.200	L.F.	15.30	8		23.30	29
0750	4″ wide joint	″	25	.320	″	21	12.80		33.80	43
0760	Floor slab to drywall partition									
0770	Flat joint	1 Carp	100	.080	L.F.	6.35	3.20		9.55	11.90
0780	Fluted joint		50	.160		12.90	6.40		19.30	24
0790	Etched fluted joint	↓	75	.107	↓	8.40	4.26		12.66	15.85
0800	Floor slab to concrete/masonry partition									

07 84 Firestopping

07 84 13 – Penetration Firestopping

07 84 13.10 Firestopping		Crew	Daily Output	Labor-Hours	Unit	2009 Bare Costs Material	Labor	Equipment	Total	Total Incl O&P
0810	Flat joint	1 Carp	75	.107	L.F.	14.20	4.26		18.46	22.50
0820	Fluted joint	″	50	.160	″	16.80	6.40		23.20	28.50
0830	Concrete/CMU wall joints									
0840	1″ wide	1 Carp	100	.080	L.F.	7.75	3.20		10.95	13.50
0850	2″ wide		75	.107		14.20	4.26		18.46	22.50
0860	4″ wide		50	.160		27	6.40		33.40	40
0870	Concrete/CMU floor joints									
0880	1″ wide	1 Carp	200	.040	L.F.	3.88	1.60		5.48	6.75
0890	2″ wide		150	.053		7.10	2.13		9.23	11.10
0900	4″ wide		100	.080		13.55	3.20		16.75	19.85

07 92 Joint Sealants

07 92 10 – Caulking and Sealants

07 92 10.10 Caulking and Sealant Options

Line	Description	Crew	Daily Output	Labor-Hours	Unit	Material	Labor	Equipment	Total	Total Incl O&P
0010	**CAULKING AND SEALANT OPTIONS**									
0020	Acoustical sealant, elastomeric, cartridges				Ea.	4.60			4.60	5.05
0030	Backer rod, polyethylene, 1/4″ diameter	1 Bric	4.60	1.739	C.L.F.	4.11	70.50		74.61	112
0050	1/2″ diameter		4.60	1.739		7.05	70.50		77.55	115
0070	3/4″ diameter		4.60	1.739		8.95	70.50		79.45	117
0090	1″ diameter		4.60	1.739		15.95	70.50		86.45	125
0100	Acrylic latex caulk, white									
0200	11 fl. oz cartridge				Ea.	2.14			2.14	2.35
0500	1/4″ x 1/2″	1 Bric	248	.032	L.F.	.17	1.31		1.48	2.17
0600	1/2″ x 1/2″		250	.032		.35	1.30		1.65	2.35
0800	3/4″ x 3/4″		230	.035		.79	1.41		2.20	3.01
0900	3/4″ x 1″		200	.040		1.05	1.62		2.67	3.61
1000	1″ x 1″		180	.044		1.31	1.80		3.11	4.17
1400	Butyl based, bulk				Gal.	26			26	28.50
1500	Cartridges				″	31.50			31.50	34.50
1700	Bulk, in place 1/4″ x 1/2″, 154 L.F./gal.	1 Bric	230	.035	L.F.	.17	1.41		1.58	2.33
1800	1/2″ x 1/2″, 77 L.F./gal.	″	180	.044	″	.34	1.80		2.14	3.10
2000	Latex acrylic based, bulk				Gal.	27			27	30
2100	Cartridges					33.50			33.50	36.50
2300	Polysulfide compounds, 1 component, bulk					51			51	56
2600	1 or 2 component, in place, 1/4″ x 1/4″, 308 L.F./gal.	1 Bric	145	.055	L.F.	.17	2.23		2.40	3.57
2700	1/2″ x 1/4″, 154 L.F./gal.		135	.059		.33	2.40		2.73	4.01
2900	3/4″ x 3/8″, 68 L.F./gal.		130	.062		.75	2.49		3.24	4.61
3000	1″ x 1/2″, 38 L.F./gal.		130	.062		1.35	2.49		3.84	5.25
3200	Polyurethane, 1 or 2 component				Gal.	54			54	59.50
3300	Cartridges				″	56			56	61.50
3500	Bulk, in place, 1/4″ x 1/4″	1 Bric	150	.053	L.F.	.18	2.16		2.34	3.47
3600	1/2″ x 1/4″		145	.055		.35	2.23		2.58	3.78
3800	3/4″ x 3/8″, 68 L.F./gal.		130	.062		.79	2.49		3.28	4.65
3900	1″ x 1/2″		110	.073		1.40	2.95		4.35	6
4100	Silicone rubber, bulk				Gal.	41			41	45
4200	Cartridges				″	41			41	45
4400	Neoprene gaskets, closed cell, adhesive, 1/8″ x 3/8″	1 Bric	240	.033	L.F.	.26	1.35		1.61	2.34
4500	1/4″ x 3/4″		215	.037		.58	1.51		2.09	2.93
4700	1/2″ x 1″		200	.040		1.30	1.62		2.92	3.89
4800	3/4″ x 1-1/2″		165	.048		1.44	1.96		3.40	4.56
5500	Resin epoxy coating, 2 component, heavy duty				Gal.	30.50			30.50	33.50

07 92 Joint Sealants

07 92 10 – Caulking and Sealants

07 92 10.10 Caulking and Sealant Options		Crew	Daily Output	Labor-Hours	Unit	2009 Bare Costs Material	Labor	Equipment	Total	Total Incl O&P
5800	Tapes, sealant, P.V.C. foam adhesive, 1/16" x 1/4"				C.L.F.	5.65			5.65	6.20
5900	1/16" x 1/2"					8.35			8.35	9.20
5950	1/16" x 1"					13.85			13.85	15.25
6000	1/8" x 1/2"				↓	9.35			9.35	10.30
6200	Urethane foam, 2 component, handy pack, 1 C.F.				Ea.	34			34	37
6300	50.0 C.F. pack				C.F.	17.20			17.20	18.95

Estimating Tips

08 10 00 Doors and Frames

All exterior doors should be addressed for their energy conservation.

- Most metal doors and frames look alike, but there may be significant differences among them. When estimating these items, be sure to choose the line item that most closely compares to the specification or door schedule requirements regarding:
  - — type of metal
  - — metal gauge
  - — door core material
  - — fire rating
  - — finish
- Wood and plastic doors vary considerably in price. The primary determinant is the veneer material. Lauan, birch, and oak are the most common veneers. Other variables include the following:
  - — hollow or solid core
  - — fire rating
  - — flush or raised panel
  - — finish

08 30 00 Specialty Doors and Frames

- There are many varieties of special doors, and they are usually priced per each. Add frames, hardware, or operators required for a complete installation.

08 40 00 Entrances, Storefronts, and Curtain Walls

- Glazed curtain walls consist of the metal tube framing and the glazing material. The cost data in this subdivision is presented for the metal tube framing alone or the composite wall. If your estimate requires a detailed takeoff of the framing, be sure to add the glazing cost.

08 50 00 Windows

- Most metal windows are delivered preglazed. However, some metal windows are priced without glass. Refer to 08 80 00 Glazing for glass pricing. The grade C indicates commercial grade windows, usually ASTM C-35.
- All wood windows are priced preglazed. The two glazing options priced are single pane float glass and insulating glass 1/2″ thick. Add the cost of screens and grills if required.

08 70 00 Hardware

- Hardware costs add considerably to the cost of a door. The most efficient method to determine the hardware requirements for a project is to review the door schedule.
- Door hinges are priced by the pair, with most doors requiring 1-1/2 pairs per door. The hinge prices do not include installation labor because it is included in door installation. Hinges are classified according to the frequency of use.

08 80 00 Glazing

- Different openings require different types of glass. The three most common types are:
  - — float
  - — tempered
  - — insulating
- Most exterior windows are glazed with insulating glass. Entrance doors and window walls, where the glass is less than 18″ from the floor, are generally glazed with tempered glass. Interior windows and some residential windows are glazed with float glass.
- Coastal communities are starting to address the use of impact-resistant glass.
- The insulation or 'u' value is a strong consideration.

Reference Numbers

Reference numbers are shown in shaded boxes at the beginning of some major classifications. These numbers refer to related items in the Reference Section. The reference information may be an estimating procedure, an alternate pricing method, or technical information.

Note: Not all subdivisions listed here necessarily appear in this publication.

08 34 Special Function Doors

08 34 13 – Cold Storage Doors

08 34 13.10 Doors for Cold Area Storage		Crew	Daily Output	Labor-Hours	Unit	2009 Bare Costs Material	Labor	Equipment	Total	Total Incl O&P
0010	**DOORS FOR COLD AREA STORAGE**									
0020	Single, 20 ga. galvanized steel									
0300	Horizontal sliding, 5′ x 7′, manual operation, 3.5″ thick	2 Carp	2	8	Ea.	3,100	320		3,420	3,900
0400	4″ thick		2	8		3,800	320		4,120	4,675
0500	6″ thick		2	8		3,100	320		3,420	3,925
0800	5′ x 7′, power operation, 2″ thick		1.90	8.421		5,150	335		5,485	6,200
0900	4″ thick		1.90	8.421		5,250	335		5,585	6,300
1000	6″ thick		1.90	8.421		5,975	335		6,310	7,100
1300	9′ x 10′, manual operation, 2″ insulation		1.70	9.412		4,175	375		4,550	5,150
1400	4″ insulation		1.70	9.412		4,300	375		4,675	5,300
1500	6″ insulation		1.70	9.412		5,175	375		5,550	6,275
1800	Power operation, 2″ insulation		1.60	10		7,175	400		7,575	8,525
1900	4″ insulation		1.60	10		7,350	400		7,750	8,700
2000	6″ insulation	↓	1.70	9.412	↓	8,325	375		8,700	9,725
2300	For stainless steel face, add					22%				
3000	Hinged, lightweight, 3′ x 7′-0″, galvanized 1 face, 2″ thick	2 Carp	2	8	Ea.	1,350	320		1,670	1,975
3050	4″ thick		1.90	8.421		1,675	335		2,010	2,350
3300	Aluminum doors, 3′ x 7′-0″, 4″ thick		1.90	8.421		1,225	335		1,560	1,875
3350	6″ thick		1.40	11.429		2,200	455		2,655	3,100
3600	Stainless steel, 3′ x 7′-0″, 4″ thick		1.90	8.421		1,575	335		1,910	2,250
3650	6″ thick		1.40	11.429		2,650	455		3,105	3,625
3900	Painted, 3′ x 7′-0″, 4″ thick		1.90	8.421		1,125	335		1,460	1,775
3950	6″ thick	↓	1.40	11.429	↓	2,150	455		2,605	3,075
5000	Bi-parting, electric operated									
5010	6′ x 8′ opening, galv. faces, 4″ thick for cooler	2 Carp	.80	20	Opng.	6,750	800		7,550	8,675
5050	For freezer, 4″ thick		.80	20		7,425	800		8,225	9,425
5300	For door buck framing and door protection, add		2.50	6.400		550	256		806	1,000
6000	Galvanized batten door, galvanized hinges, 4′ x 7′		2	8		1,650	320		1,970	2,325
6050	6′ x 8′		1.80	8.889		2,275	355		2,630	3,050
6500	Fire door, 3 hr., 6′ x 8′, single slide		.80	20		7,800	800		8,600	9,825
6550	Double, bi-parting	↓	.70	22.857	↓	12,000	915		12,915	14,600

08 91 Louvers

08 91 19 – Fixed Louvers

08 91 19.10 Aluminum Louvers		Crew	Daily Output	Labor-Hours	Unit	Material	Labor	Equipment	Total	Total Incl O&P
0010	**ALUMINUM LOUVERS**									
0020	Aluminum with screen, residential, 8″ x 8″	1 Carp	38	.211	Ea.	10.50	8.40		18.90	24.50
0100	12″ x 12″		38	.211		11.85	8.40		20.25	26
0200	12″ x 18″		35	.229		13.50	9.15		22.65	29
0250	14″ x 24″		30	.267		17	10.65		27.65	35
0300	18″ x 24″		27	.296		24.50	11.85		36.35	45.50
0500	24″ x 30″		24	.333		34	13.30		47.30	58
0700	Triangle, adjustable, small		20	.400		29	16		45	57
0800	Large	↓	15	.533	↓	50	21.50		71.50	88.50
1200	Extruded aluminum, see Div. 23 37 15.40									
2100	Midget, aluminum, 3/4″ deep, 1″ diameter	1 Carp	85	.094	Ea.	.77	3.76		4.53	6.70
2150	3″ diameter		60	.133		1.62	5.35		6.97	10.05
2200	4″ diameter		50	.160		3.07	6.40		9.47	13.30
2250	6″ diameter	↓	30	.267	↓	3.63	10.65		14.28	20.50

08 95 Vents

08 95 13 – Soffit Vents

08 95 13.10 Wall Louvers		Crew	Daily Output	Labor-Hours	Unit	2009 Bare Costs Material	Labor	Equipment	Total	Total Incl O&P
0010	**WALL LOUVERS**									
2330	Soffit vent, continuous, 3″ wide, aluminum, mill finish	1 Carp	200	.040	L.F.	.53	1.60		2.13	3.06
2340	Baked enamel finish		200	.040	″	.48	1.60		2.08	3.01
2400	Under eaves vent, aluminum, mill finish, 16″ x 4″		48	.167	Ea.	1.29	6.65		7.94	11.75
2500	16″ x 8″		48	.167	″	1.49	6.65		8.14	12

08 95 16 – Wall Vents

08 95 16.10 Louvers		Crew	Daily Output	Labor-Hours	Unit	Material	Labor	Equipment	Total	Total Incl O&P
0010	**LOUVERS**									
0020	Redwood, 2′-0″ diameter, full circle	1 Carp	16	.500	Ea.	159	20		179	205
0100	Half circle		16	.500		152	20		172	198
0200	Octagonal		16	.500		121	20		141	164
0300	Triangular, 5/12 pitch, 5′-0″ at base		16	.500		257	20		277	315
1100	Rectangular, 1′-4″ x 1′-8″		16	.500		25.50	20		45.50	59
1200	1′-4″ x 2′-2″		15	.533		29.50	21.50		51	65
1300	1′-9″ x 2′-2″		15	.533		35	21.50		56.50	71.50
1400	2′-3″ x 2′-2″		14	.571		47	23		70	87
1700	2′-4″ x 2′-11″		13	.615		47	24.50		71.50	89.50
2000	Aluminum, 12″ x 16″		25	.320		13.45	12.80		26.25	34.50
2010	16″ x 20″		25	.320		26	12.80		38.80	48.50
2020	24″ x 30″		25	.320		62	12.80		74.80	88
2100	6′ triangle		12	.667		147	26.50		173.50	203
7000	Vinyl gable vent, 8″ x 8″		38	.211		11.20	8.40		19.60	25.50
7020	12″ x 12″		38	.211		23.50	8.40		31.90	39
7080	12″ x 18″		35	.229		30	9.15		39.15	47
7200	18″ x 24″		30	.267		36.50	10.65		47.15	57

Division Notes

	CREW	DAILY OUTPUT	LABOR-HOURS	UNIT	2009 BARE COSTS MAT.	LABOR	EQUIP.	TOTAL	TOTAL INCL O&P

Estimating Tips

General

- Room Finish Schedule: A complete set of plans should contain a room finish schedule. If one is not available, it would be well worth the time and effort to obtain one.

09 20 00 Plaster and Gypsum Board

- Lath is estimated by the square yard plus a 5% allowance for waste. Furring, channels, and accessories are measured by the linear foot. An extra foot should be allowed for each accessory miter or stop.
- Plaster is also estimated by the square yard. Deductions for openings vary by preference, from zero deduction to 50% of all openings over 2 feet in width. The estimator should allow one extra square foot for each linear foot of horizontal interior or exterior angle located below the ceiling level. Also, double the areas of small radius work.
- Drywall accessories, studs, track, and acoustical caulking are all measured by the linear foot. Drywall taping is figured by the square foot. Gypsum wallboard is estimated by the square foot. No material deductions should be made for door or window openings under 32 S.F.

09 60 00 Flooring

- Tile and terrazzo areas are taken off on a square foot basis. Trim and base materials are measured by the linear foot. Accent tiles are listed per each. Two basic methods of installation are used. Mud set is approximately 30% more expensive than thin set. In terrazzo work, be sure to include the linear footage of embedded decorative strips, grounds, machine rubbing, and power cleanup.
- Wood flooring is available in strip, parquet, or block configuration. The latter two types are set in adhesives with quantities estimated by the square foot. The laying pattern will influence labor costs and material waste. In addition to the material and labor for laying wood floors, the estimator must make allowances for sanding and finishing these areas unless the flooring is prefinished.
- Sheet flooring is measured by the square yard. Roll widths vary, so consideration should be given to use the most economical width, as waste must be figured into the total quantity. Consider also the installation methods available, direct glue down or stretched.

09 70 00 Wall Finishes

- Wall coverings are estimated by the square foot. The area to be covered is measured, length by height of wall above baseboards, to calculate the square footage of each wall. This figure is divided by the number of square feet in the single roll which is being used. Deduct, in full, the areas of openings such as doors and windows. Where a pattern match is required allow 25%–30% waste.

09 80 00 Acoustic Treatment

- Acoustical systems fall into several categories. The takeoff of these materials should be by the square foot of area with a 5% allowance for waste. Do not forget about scaffolding, if applicable, when estimating these systems.

09 90 00 Painting and Coating

- A major portion of the work in painting involves surface preparation. Be sure to include cleaning, sanding, filling, and masking costs in the estimate.
- Protection of adjacent surfaces is not included in painting costs. When considering the method of paint application, an important factor is the amount of protection and masking required. These must be estimated separately and may be the determining factor in choosing the method of application.

Reference Numbers

Reference numbers are shown in shaded boxes at the beginning of some major classifications. These numbers refer to related items in the Reference Section. The reference information may be an estimating procedure, an alternate pricing method, or technical information.

Note: Not all subdivisions listed here necessarily appear in this publication.

09 22 Supports for Plaster and Gypsum Board

09 22 03 – Fastening Methods for Finishes

09 22 03.20 Drilling Plaster/Drywall		Crew	Daily Output	Labor-Hours	Unit	2009 Bare Costs Material	Labor	Equipment	Total	Total Incl O&P
0010	**DRILLING PLASTER/DRYWALL**									
1100	Drilling & layout for drywall/plaster walls, up to 1" deep, no anchor									
1200	Holes, 1/4" diameter	1 Carp	150	.053	Ea.	.01	2.13		2.14	3.31
1300	3/8" diameter		140	.057		.01	2.28		2.29	3.55
1400	1/2" diameter		130	.062		.01	2.46		2.47	3.82
1500	3/4" diameter		120	.067		.01	2.66		2.67	4.14
1600	1" diameter		110	.073		.02	2.91		2.93	4.53
1700	1-1/4" diameter		100	.080		.03	3.20		3.23	4.99
1800	1-1/2" diameter		90	.089		.04	3.55		3.59	5.55
1900	For ceiling installations, add						40%			

09 69 Access Flooring

09 69 13 – Rigid-Grid Access Flooring

09 69 13.10 Access Floors

Line	Description	Crew	Daily Output	Labor-Hours	Unit	Material	Labor	Equipment	Total	Total Incl O&P
0010	**ACCESS FLOORS**									
0015	Access floor package including panel, pedestal, stringers & laminate cover									
0100	Computer room, greater than 6,000 S.F.	4 Carp	750	.043	S.F.	8.65	1.70		10.35	12.15
0110	Less than 6,000 S.F.	2 Carp	375	.043		9.95	1.70		11.65	13.60
0120	Office, greater than 6,000 S.F.	4 Carp	1050	.030		4.56	1.22		5.78	6.90
0250	Panels, particle board or steel, 1250# load, no covering, under 6,000 S.F.	2 Carp	600	.027		4.24	1.07		5.31	6.30
0300	Over 6,000 S.F.		640	.025		3.72	1		4.72	5.65
0400	Aluminum, 24" panels		500	.032		31.50	1.28		32.78	36.50
0600	For carpet covering, add					8.55			8.55	9.40
0700	For vinyl floor covering, add					6.70			6.70	7.35
0900	For high pressure laminate covering, add					5.45			5.45	6
0910	For snap on stringer system, add	2 Carp	1000	.016		1.44	.64		2.08	2.57
0950	Office applications, steel or concrete panels,									
0960	no covering, over 6,000 S.F.	2 Carp	960	.017	S.F.	10.10	.67		10.77	12.15
1000	Machine cutouts after initial installation	1 Carp	50	.160	Ea.	5.05	6.40		11.45	15.45
1050	Pedestals, 6" to 12"	2 Carp	85	.188		7.95	7.50		15.45	20.50
1100	Air conditioning grilles, 4" x 12"	1 Carp	17	.471		66	18.80		84.80	102
1150	4" x 18"	"	14	.571		90.50	23		113.50	135
1200	Approach ramps, minimum	2 Carp	60	.267	S.F.	25.50	10.65		36.15	44.50
1300	Maximum	"	40	.400	"	33.50	16		49.50	61.50
1500	Handrail, 2 rail, aluminum	1 Carp	15	.533	L.F.	99.50	21.50		121	142

09 91 Painting

09 91 23 – Interior Painting

09 91 23.52 Miscellaneous, Interior

Line	Description	Crew	Daily Output	Labor-Hours	Unit	Material	Labor	Equipment	Total	Total Incl O&P
0010	**MISCELLANEOUS, INTERIOR** R099100-10									
3800	Grilles, per side, oil base, primer coat, brushwork	1 Pord	520	.015	S.F.	.10	.54		.64	.92
3850	Spray		1140	.007		.11	.25		.36	.49
3880	Paint 1 coat, brushwork		520	.015		.12	.54		.66	.94
3900	Spray		1140	.007		.13	.25		.38	.51
3920	Paint 2 coats, brushwork		325	.025		.23	.87		1.10	1.55
3940	Spray		650	.012		.26	.43		.69	.94
3950	Prime & paint 1 coat		325	.025		.22	.87		1.09	1.54
3960	Prime & paint 2 coats		270	.030		.22	1.04		1.26	1.80
4500	Louvers, one side, primer, brushwork		524	.015		.06	.54		.60	.87

09 91 23.52 Miscellaneous, Interior		Crew	Daily Output	Labor-Hours	Unit	Material	Labor	Equipment	Total	Total Incl O&P
						2009 Bare Costs				
4520	Paint one coat, brushwork	1 Pord	520	.015	S.F.	.05	.54		.59	.87
4530	Spray		1140	.007		.06	.25		.31	.44
4540	Paint two coats, brushwork		325	.025		.11	.87		.98	1.42
4550	Spray		650	.012		.12	.43		.55	.78
4560	Paint three coats, brushwork		270	.030		.16	1.04		1.20	1.73
4570	Spray		500	.016		.18	.56		.74	1.03
5000	Pipe, 1 – 4" diameter, primer or sealer coat, oil base, brushwork	2 Pord	1250	.013	L.F.	.06	.45		.51	.74
5100	Spray		2165	.007		.06	.26		.32	.45
5200	Paint 1 coat, brushwork		1250	.013		.06	.45		.51	.75
5300	Spray		2165	.007		.06	.26		.32	.45
5350	Paint 2 coats, brushwork		775	.021		.11	.73		.84	1.21
5400	Spray		1240	.013		.12	.45		.57	.82
5420	Paint 3 coats, brushwork		775	.021		.17	.73		.90	1.27
5450	5" - 8" diameter, primer or sealer coat, brushwork		620	.026		.12	.91		1.03	1.49
5500	Spray		1085	.015		.19	.52		.71	.99
5550	Paint 1 coat, brushwork		620	.026		.17	.91		1.08	1.55
5600	Spray		1085	.015		.19	.52		.71	.99
5650	Paint 2 coats, brushwork		385	.042		.22	1.46		1.68	2.44
5700	Spray		620	.026		.25	.91		1.16	1.63
5720	Paint 3 coats, brushwork		385	.042		.33	1.46		1.79	2.56
5750	9" - 12" diameter, primer or sealer coat, brushwork		415	.039		.18	1.36		1.54	2.22
5800	Spray		725	.022		.20	.78		.98	1.38
5850	Paint 1 coat, brushwork		415	.039		.17	1.36		1.53	2.22
6000	Spray		725	.022		.19	.78		.97	1.37
6200	Paint 2 coats, brushwork		260	.062		.34	2.17		2.51	3.62
6250	Spray		415	.039		.37	1.36		1.73	2.44
6270	Paint 3 coats, brushwork		260	.062		.50	2.17		2.67	3.80
6300	13" - 16" diameter, primer or sealer coat, brushwork		310	.052		.24	1.82		2.06	2.98
6350	Spray		540	.030		.26	1.04		1.30	1.85
6400	Paint 1 coat, brushwork		310	.052		.23	1.82		2.05	2.97
6450	Spray		540	.030		.26	1.04		1.30	1.84
6500	Paint 2 coats, brushwork		195	.082		.45	2.89		3.34	4.82
6550	Spray		310	.052		.50	1.82		2.32	3.27
6600	Radiators, per side, primer, brushwork	1 Pord	520	.015	S.F.	.06	.54		.60	.87
6620	Paint, one coat		520	.015		.05	.54		.59	.87
6640	Two coats		340	.024		.11	.83		.94	1.36
6660	Three coats		283	.028		.16	1		1.16	1.66

Division Notes

		CREW	DAILY OUTPUT	LABOR-HOURS	UNIT	2009 BARE COSTS MAT.	LABOR	EQUIP.	TOTAL	TOTAL INCL O&P

Estimating Tips

General

- The items in this division are usually priced per square foot or each.
- Many items in Division 10 require some type of support system or special anchors that are not usually furnished with the item. The required anchors must be added to the estimate in the appropriate division.
- Some items in Division 10, such as lockers, may require assembly before installation. Verify the amount of assembly required. Assembly can often exceed installation time.

10 20 00 Interior Specialties

- Support angles and blocking are not included in the installation of toilet compartments, shower/dressing compartments, or cubicles. Appropriate line items from Divisions 5 or 6 may need to be added to support the installations.
- Toilet partitions are priced by the stall. A stall consists of a side wall, pilaster, and door with hardware. Toilet tissue holders and grab bars are extra.
- The required acoustical rating of a folding partition can have a significant impact on costs. Verify the sound transmission coefficient rating of the panel priced to the specification requirements.
- Grab bar installation does not include supplemental blocking or backing to support the required load. When grab bars are installed at an existing facility, provisions must be made to attach the grab bars to solid structure.

Reference Numbers

Reference numbers are shown in shaded boxes at the beginning of some major classifications. These numbers refer to related items in the Reference Section. The reference information may be an estimating procedure, an alternate pricing method, or technical information.

Note: Not all subdivisions listed here necessarily appear in this publication.

10 31 Manufactured Fireplaces

10 31 13 – Manufactured Fireplace Chimneys

10 31 13.10 Fireplace Chimneys		Crew	Daily Output	Labor-Hours	Unit	2009 Bare Costs Material	Labor	Equipment	Total	Total Incl O&P
0010	**FIREPLACE CHIMNEYS**									
0500	Chimney dbl. wall, all stainless, over 8'-6", 7" diam., add to fireplace	1 Carp	33	.242	V.L.F.	70	9.70		79.70	92
0600	10" diameter, add to fireplace		32	.250		93	10		103	118
0700	12" diameter, add to fireplace		31	.258		111	10.30		121.30	138
0800	14" diameter, add to fireplace		30	.267		180	10.65		190.65	215
1000	Simulated brick chimney top, 4' high, 16" x 16"		10	.800	Ea.	244	32		276	320
1100	24" x 24"		7	1.143	"	455	45.50		500.50	570

10 31 23 – Prefabricated Fireplaces

10 31 23.10 Fireplace, Prefabricated		Crew	Daily Output	Labor-Hours	Unit	Material	Labor	Equipment	Total	Total Incl O&P
0010	**FIREPLACE, PREFABRICATED**, free standing or wall hung									
0100	With hood & screen, minimum	1 Carp	1.30	6.154	Ea.	1,175	246		1,421	1,650
0150	Average		1	8		1,625	320		1,945	2,300
0200	Maximum		.90	8.889		3,125	355		3,480	3,975
1500	Simulated logs, gas fired, 40,000 BTU, 2' long, minimum		7	1.143	Set	400	45.50		445.50	510
1600	Maximum		6	1.333		940	53.50		993.50	1,100
1700	Electric, 1,500 BTU, 1'-6" long, minimum		7	1.143		166	45.50		211.50	254
1800	11,500 BTU, maximum		6	1.333		370	53.50		423.50	495

10 35 Stoves

10 35 13 – Heating Stoves

10 35 13.10 Woodburning Stoves		Crew	Daily Output	Labor-Hours	Unit	Material	Labor	Equipment	Total	Total Incl O&P
0010	**WOODBURNING STOVES**									
0015	Cast iron, minimum	2 Carp	1.30	12.308	Ea.	1,125	490		1,615	2,000
0020	Average		1	16		1,525	640		2,165	2,675
0030	Maximum		.80	20		2,750	800		3,550	4,275
0050	For gas log lighter, add					37.50			37.50	41.50

Estimating Tips

General

- The items in this division are usually priced per square foot or each. Many of these items are purchased by the owner for installation by the contractor. Check the specifications for responsibilities and include time for receiving, storage, installation, and mechanical and electrical hookups in the appropriate divisions.
- Many items in Division 11 require some type of support system that is not usually furnished with the item. Examples of these systems include blocking for the attachment of casework and support angles for ceiling-hung projection screens. The required blocking or supports must be added to the estimate in the appropriate division.
- Some items in Division 11 may require assembly or electrical hookups. Verify the amount of assembly required or the need for a hard electrical connection and add the appropriate costs.

Reference Numbers

Reference numbers are shown in shaded boxes at the beginning of some major classifications. These numbers refer to related items in the Reference Section. The reference information may be an estimating procedure, an alternate pricing method, or technical information.

Note: Not all subdivisions listed here necessarily appear in this publication.

Division 11 - Equipment

11 11 Vehicle Service Equipment

11 11 13 – Compressed-Air Vehicle Service Equipment

11 11 13.10 Compressed Air Equipment		Crew	Daily Output	Labor-Hours	Unit	2009 Bare Costs Material	Labor	Equipment	Total	Total Incl O&P
0010	COMPRESSED AIR EQUIPMENT									
0030	Compressors, electric, 1-1/2 H.P., standard controls	L-4	1.50	16	Ea.	370	595		965	1,325
0550	Dual controls		1.50	16		695	595		1,290	1,700
0600	5 H.P., 115/230 volt, standard controls		1	24		1,900	895		2,795	3,450
0650	Dual controls		1	24		2,025	895		2,920	3,600

11 11 19 – Vehicle Lubrication Equipment

11 11 19.10 Lubrication Equipment		Crew	Daily Output	Labor-Hours	Unit	Material	Labor	Equipment	Total	Total Incl O&P
0010	LUBRICATION EQUIPMENT									
3000	Lube equipment, 3 reel type, with pumps, not including piping	L-4	.50	48	Set	7,900	1,800		9,700	11,500
3700	Pump lubrication, pneumatic, not incl. air compressor									
3710	Oil/gear lube	Q-1	9.60	1.667	Ea.	765	73		838	955
3720	Grease	"	9.60	1.667	"	920	73		993	1,100

11 11 33 – Vehicle Spray Painting Equipment

11 11 33.10 Spray Painting Equipment		Crew	Daily Output	Labor-Hours	Unit	Material	Labor	Equipment	Total	Total Incl O&P
0010	SPRAY PAINTING EQUIPMENT									
4000	Spray painting booth, 26′ long, complete	L-4	.40	60	Ea.	15,400	2,250		17,650	20,400

11 23 Commercial Laundry and Dry Cleaning Equipment

11 23 16 – Drying and Conditioning Equipment

11 23 16.13 Dryers		Crew	Daily Output	Labor-Hours	Unit	Material	Labor	Equipment	Total	Total Incl O&P
0010	**DRYERS**, Not including rough-in									
1500	Industrial, 30 lb. capacity	1 Plum	2	4	Ea.	2,975	195		3,170	3,575
1600	50 lb. capacity	"	1.70	4.706	"	3,200	229		3,429	3,875

11 23 26 – Commercial Washers and Extractors

11 23 26.13 Washers and Extractors		Crew	Daily Output	Labor-Hours	Unit	Material	Labor	Equipment	Total	Total Incl O&P
0010	**WASHERS AND EXTRACTORS**, not including rough-in									
6000	Combination washer/extractor, 20 lb. capacity	L-6	1.50	8	Ea.	5,350	385		5,735	6,475
6100	30 lb. capacity		.80	15		8,500	725		9,225	10,400
6200	50 lb. capacity		.68	17.647		9,925	850		10,775	12,200
6300	75 lb. capacity		.30	40		18,600	1,925		20,525	23,400
6350	125 lb. capacity		.16	75		24,900	3,625		28,525	32,800

11 23 33 – Coin-Operated Laundry Equipment

11 23 33.13 Coin Operated Washers and Dryers		Crew	Daily Output	Labor-Hours	Unit	Material	Labor	Equipment	Total	Total Incl O&P
0010	COIN OPERATED WASHERS AND DRYERS									
0990	Dryer, gas fired									
1000	Commercial, 30 lb. capacity, coin operated, single	1 Plum	3	2.667	Ea.	3,100	130		3,230	3,600
1100	Double stacked	"	2	4	"	6,275	195		6,470	7,200
5290	Clothes washer									
5300	Commercial, coin operated, average	1 Plum	3	2.667	Ea.	1,125	130		1,255	1,450

11 24 Maintenance Equipment

11 24 19 – Vacuum Cleaning Systems

11 24 19.10 Vacuum Cleaning		Crew	Daily Output	Labor-Hours	Unit	2009 Bare Costs Material	Labor	Equipment	Total	Total Incl O&P
0010	VACUUM CLEANING									
0020	Central, 3 inlet, residential	1 Skwk	.90	8.889	Total	925	365		1,290	1,575
0200	Commercial		.70	11.429		1,050	465		1,515	1,900
0400	5 inlet system, residential		.50	16		1,225	655		1,880	2,350
0600	7 inlet system, commercial		.40	20		1,400	815		2,215	2,800
0800	9 inlet system, residential	↓	.30	26.667		3,000	1,100		4,100	4,975
4010	Rule of thumb: First 1200 S.F., installed				↓				1,150	1,275
4020	For each additional S.F., add				S.F.					.21

11 26 Unit Kitchens

11 26 13 – Metal Unit Kitchens

11 26 13.10 Commercial Unit Kitchens

		Crew	Daily Output	Labor-Hours	Unit	Material	Labor	Equipment	Total	Total Incl O&P
0010	COMMERCIAL UNIT KITCHENS									
1500	Combination range, refrigerator and sink, 30" wide, minimum	L-1	2	8	Ea.	900	385		1,285	1,550
1550	Maximum		1	16		3,175	765		3,940	4,650
1570	60" wide, average		1.40	11.429		2,725	545		3,270	3,800
1590	72" wide, average		1.20	13.333		3,700	640		4,340	5,025
1600	Office model, 48" wide		2	8		2,750	385		3,135	3,600
1620	Refrigerator and sink only	↓	2.40	6.667	↓	2,225	320		2,545	2,900
1640	Combination range, refrigerator, sink, microwave									
1660	Oven and ice maker	L-1	.80	20	Ea.	3,950	960		4,910	5,775

11 27 Photographic Processing Equipment

11 27 13 – Darkroom Processing Equipment

11 27 13.10 Darkroom Equipment

		Crew	Daily Output	Labor-Hours	Unit	Material	Labor	Equipment	Total	Total Incl O&P
0010	DARKROOM EQUIPMENT									
0020	Developing sink, 5" deep, 24" x 48"	Q-1	2	8	Ea.	5,025	350		5,375	6,050
0050	48" x 52"		1.70	9.412		5,100	415		5,515	6,225
0200	10" deep, 24" x 48"		1.70	9.412		6,475	415		6,890	7,750
0250	24" x 108"		1.50	10.667		8,525	470		8,995	10,100
3500	Washers, round, minimum sheet 11" x 14"		2	8		3,075	350		3,425	3,900
3550	Maximum sheet 20" x 24"		1	16		3,475	700		4,175	4,875
3800	Square, minimum sheet 20" x 24"		1	16		3,125	700		3,825	4,475
3900	Maximum sheet 50" x 56"	↓	.80	20	↓	4,850	880		5,730	6,650
4500	Combination tank sink, tray sink, washers, with									
4510	Dry side tables, average	Q-1	.45	35.556	Ea.	9,600	1,550		11,150	13,000

11 31 Residential Appliances

11 31 13 – Residential Kitchen Appliances

11 31 13.13 Cooking Equipment

		Crew	Daily Output	Labor-Hours	Unit	Material	Labor	Equipment	Total	Total Incl O&P
0010	COOKING EQUIPMENT									
0020	Cooking range, 30" free standing, 1 oven, minimum	2 Clab	10	1.600	Ea.	320	50.50		370.50	430
0050	Maximum		4	4		1,775	126		1,901	2,150
0150	2 oven, minimum		10	1.600		1,450	50.50		1,500.50	1,675
0200	Maximum	↓	10	1.600	↓	1,625	50.50		1,675.50	1,875

11 31 13.23 Refrigeration Equipment

		Crew	Daily Output	Labor-Hours	Unit	Material	Labor	Equipment	Total	Total Incl O&P
0010	REFRIGERATION EQUIPMENT									

11 31 Residential Appliances

11 31 13 – Residential Kitchen Appliances

11 31 13.23 Refrigeration Equipment		Crew	Daily Output	Labor-Hours	Unit	Material	2009 Bare Costs Labor	Equipment	Total	Total Incl O&P
5200	Icemaker, automatic, 20 lb. per day	1 Plum	7	1.143	Ea.	850	55.50		905.50	1,025
5350	51 lb. per day	"	2	4	"	1,175	195		1,370	1,600

11 31 13.33 Kitchen Cleaning Equipment		Crew	Daily Output	Labor-Hours	Unit	Material	Labor	Equipment	Total	Total Incl O&P
0010	**KITCHEN CLEANING EQUIPMENT**									
2750	Dishwasher, built-in, 2 cycles, minimum	L-1	4	4	Ea.	206	192		398	515
2800	Maximum		2	8		310	385		695	910
2950	4 or more cycles, minimum		4	4		310	192		502	625
2960	Average		4	4		410	192		602	735
3000	Maximum	↓	2	8	↓	1,000	385		1,385	1,700

11 31 13.43 Waste Disposal Equipment		Crew	Daily Output	Labor-Hours	Unit	Material	Labor	Equipment	Total	Total Incl O&P
0010	**WASTE DISPOSAL EQUIPMENT**									
3300	Garbage disposal, sink type, minimum	L-1	10	1.600	Ea.	66.50	76.50		143	188
3350	Maximum	"	10	1.600	"	192	76.50		268.50	325

11 31 13.53 Kitchen Ventilation Equipment		Crew	Daily Output	Labor-Hours	Unit	Material	Labor	Equipment	Total	Total Incl O&P
0010	**KITCHEN VENTILATION EQUIPMENT**									
4150	Hood for range, 2 speed, vented, 30" wide, minimum	L-3	5	3.200	Ea.	52.50	139		191.50	271
4200	Maximum		3	5.333		810	232		1,042	1,250
4300	42" wide, minimum		5	3.200		250	139		389	490
4330	Custom		5	3.200		1,400	139		1,539	1,775
4350	Maximum	↓	3	5.333		1,725	232		1,957	2,250
4500	For ventless hood, 2 speed, add					16.55			16.55	18.20
4650	For vented 1 speed, deduct from maximum				↓	44			44	48.50

11 31 23 – Residential Laundry Appliances

11 31 23.13 Washers		Crew	Daily Output	Labor-Hours	Unit	Material	Labor	Equipment	Total	Total Incl O&P
0010	**WASHERS**									
5000	Residential, 4 cycle, average	1 Plum	3	2.667	Ea.	800	130		930	1,075
6650	Washing machine, automatic, minimum		3	2.667		315	130		445	540
6700	Maximum	↓	1	8	↓	1,100	390		1,490	1,775

11 31 23.23 Dryers		Crew	Daily Output	Labor-Hours	Unit	Material	Labor	Equipment	Total	Total Incl O&P
0010	**DRYERS**									
0500	Gas fired residential, 16 lb. capacity, average	1 Plum	3	2.667	Ea.	625	130		755	885
7450	Vent kits for dryers	1 Carp	10	.800	"	31.50	32		63.50	84.50

11 31 33 – Miscellaneous Residential Appliances

11 31 33.13 Sump Pumps		Crew	Daily Output	Labor-Hours	Unit	Material	Labor	Equipment	Total	Total Incl O&P
0010	**SUMP PUMPS**									
6400	Cellar drainer, pedestal, 1/3 H.P., molded PVC base	1 Plum	3	2.667	Ea.	119	130		249	325
6450	Solid brass	"	2	4	"	263	195		458	585
6460	Sump pump, see also Div. 22 14 29.16									

11 31 33.23 Water Heaters		Crew	Daily Output	Labor-Hours	Unit	Material	Labor	Equipment	Total	Total Incl O&P
0010	**WATER HEATERS**									
6900	Electric, glass lined, 30 gallon, minimum	L-1	5	3.200	Ea.	420	153		573	695
6950	Maximum		3	5.333		585	255		840	1,025
7100	80 gallon, minimum		2	8		760	385		1,145	1,400
7150	Maximum	↓	1	16		1,050	765		1,815	2,300
7180	Gas, glass lined, 30 gallon, minimum	2 Plum	5	3.200		565	156		721	855
7220	Maximum		3	5.333		785	260		1,045	1,250
7260	50 gallon, minimum		2.50	6.400		845	310		1,155	1,400
7300	Maximum	↓	1.50	10.667	↓	1,175	520		1,695	2,050

11 31 33.43 Air Quality		Crew	Daily Output	Labor-Hours	Unit	Material	Labor	Equipment	Total	Total Incl O&P
0010	**AIR QUALITY**									

11 31 Residential Appliances

11 31 33 – Miscellaneous Residential Appliances

11 31 33.43 Air Quality		Crew	Daily Output	Labor-Hours	Unit	Material	2009 Bare Costs Labor	Equipment	Total	Total Incl O&P
2450	Dehumidifier, portable, automatic, 15 pint				Ea.	150			150	165
2550	40 pint					179			179	197
3550	Heater, electric, built-in, 1250 watt, ceiling type, minimum	1 Elec	4	2		96	94		190	245
3600	Maximum		3	2.667		156	125		281	360
3700	Wall type, minimum		4	2		143	94		237	297
3750	Maximum		3	2.667		154	125		279	355
3900	1500 watt wall type, with blower		4	2		143	94		237	297
3950	3000 watt		3	2.667		292	125		417	505
4850	Humidifier, portable, 8 gallons per day					160			160	176
5000	15 gallons per day					193			193	212

11 41 Food Storage Equipment

11 41 13 – Refrigerated Food Storage Cases

11 41 13.10 Refrigerated Food Cases

		Crew	Daily Output	Labor-Hours	Unit	Material	Labor	Equipment	Total	Total Incl O&P
0010	**REFRIGERATED FOOD CASES**									
0030	Dairy, multi-deck, 12′ long	Q-5	3	5.333	Ea.	9,925	237		10,162	11,300
0100	For rear sliding doors, add					1,425			1,425	1,550
0200	Delicatessen case, service deli, 12′ long, single deck	Q-5	3.90	4.103		6,675	182		6,857	7,625
0300	Multi-deck, 18 S.F. shelf display		3	5.333		7,425	237		7,662	8,525
0400	Freezer, self-contained, chest-type, 30 C.F.		3.90	4.103		6,700	182		6,882	7,650
0500	Glass door, upright, 78 C.F.		3.30	4.848		9,275	215		9,490	10,500
0600	Frozen food, chest type, 12′ long		3.30	4.848		6,700	215		6,915	7,700
0700	Glass door, reach-in, 5 door		3	5.333		12,900	237		13,137	14,500
0800	Island case, 12′ long, single deck		3.30	4.848		7,600	215		7,815	8,675
0900	Multi-deck		3	5.333		16,100	237		16,337	18,100
1000	Meat case, 12′ long, single deck		3.30	4.848		5,525	215		5,740	6,400
1050	Multi-deck		3.10	5.161		9,500	229		9,729	10,700
1100	Produce, 12′ long, single deck		3.30	4.848		7,300	215		7,515	8,350
1200	Multi-deck		3.10	5.161		7,975	229		8,204	9,125

11 41 13.20 Refrigerated Food Storage Equipment

		Crew	Daily Output	Labor-Hours	Unit	Material	Labor	Equipment	Total	Total Incl O&P
0010	**REFRIGERATED FOOD STORAGE EQUIPMENT**									
2350	Cooler, reach-in, beverage, 6′ long	Q-1	6	2.667	Ea.	4,300	117		4,417	4,925
4300	Freezers, reach-in, 44 C.F.		4	4		3,150	176		3,326	3,725
4500	68 C.F.		3	5.333		4,350	234		4,584	5,125
8300	Refrigerators, reach-in type, 44 C.F.		5	3.200		6,775	140		6,915	7,650
8310	With glass doors, 68 C.F.		4	4		6,625	176		6,801	7,575

11 44 Food Cooking Equipment

11 44 13 – Commercial Ranges

11 44 13.10 Cooking Equipment

		Crew	Daily Output	Labor-Hours	Unit	Material	Labor	Equipment	Total	Total Incl O&P
0010	**COOKING EQUIPMENT**									
0020	Bake oven, gas, one section	Q-1	8	2	Ea.	4,300	88		4,388	4,875
0300	Two sections		7	2.286		10,100	100		10,200	11,400
0600	Three sections		6	2.667		13,200	117		13,317	14,700
0900	Electric convection, single deck	L-7	4	7		5,400	270		5,670	6,350
6350	Kettle, w/steam jacket, tilting, w/positive lock, SS, 20 gallons		7	4		6,675	154		6,829	7,600
6600	60 gallons		6	4.667		8,750	180		8,930	9,875

11 47 Ice Machines

11 47 10 – Commercial Ice Machines

11 47 10.10 Commercial Ice Equipment		Crew	Daily Output	Labor-Hours	Unit	2009 Bare Costs Material	Labor	Equipment	Total	Total Incl O&P
0010	**COMMERCIAL ICE EQUIPMENT**									
5800	Ice cube maker, 50 pounds per day	Q-1	6	2.667	Ea.	1,425	117		1,542	1,750
6050	500 pounds per day	"	4	4	"	3,000	176		3,176	3,575

11 48 Cleaning and Disposal Equipment

11 48 13 – Commercial Dishwashers

11 48 13.10 Dishwashers

Line	Description	Crew	Daily Output	Labor-Hours	Unit	Material	Labor	Equipment	Total	Total Incl O&P
0010	**DISHWASHERS**									
2700	Dishwasher, commercial, rack type									
2720	10 to 12 racks per hour	Q-1	3.20	5	Ea.	4,200	219		4,419	4,950
2750	Semi-automatic 38 to 50 racks per hour	"	1.30	12.308	↓	8,350	540		8,890	9,975
2800	Automatic, 190 to 230 racks per hour	L-6	.35	34.286	↓	12,400	1,650		14,050	16,200
2820	235 to 275 racks per hour	↓	.25	48	↓	27,400	2,300		29,700	33,700
2840	8,750 to 12,500 dishes per hour	↓	.10	120	↓	48,000	5,775		53,775	61,500

11 53 Laboratory Equipment

11 53 13 – Laboratory Fume Hoods

11 53 13.13 Recirculating Laboratory Fume Hoods

Line	Description	Crew	Daily Output	Labor-Hours	Unit	Material	Labor	Equipment	Total	Total Incl O&P
0010	**RECIRCULATING LABORATORY FUME HOODS**									
0600	Fume hood, with countertop & base, not including HVAC									
0610	Simple, minimum	2 Carp	5.40	2.963	L.F.	520	118		638	755
0620	Complex, including fixtures	↓	2.40	6.667	↓	1,675	266		1,941	2,275
0630	Special, maximum	↓	1.70	9.412	↓	1,825	375		2,200	2,575
0670	Service fixtures, average				Ea.	273			273	300
0680	For sink assembly with hot and cold water, add	1 Plum	1.40	5.714	"	705	279		984	1,200

11 53 13.23 Exhaust Hoods

Line	Description	Crew	Daily Output	Labor-Hours	Unit	Material	Labor	Equipment	Total	Total Incl O&P
0010	**EXHAUST HOODS**									
0650	Ductwork, minimum	2 Shee	1	16	Hood	3,575	755		4,330	5,075
0660	Maximum	"	.50	32	"	5,500	1,500		7,000	8,375

11 53 19 – Laboratory Sterilizers

11 53 19.13 Sterilizers

Line	Description	Crew	Daily Output	Labor-Hours	Unit	Material	Labor	Equipment	Total	Total Incl O&P
0010	**STERILIZERS**									
0700	Glassware washer, undercounter, minimum	L-1	1.80	8.889	Ea.	5,450	425		5,875	6,625
0710	Maximum	"	1	16	↓	12,100	765		12,865	14,600
1850	Utensil washer-sanitizer	1 Plum	2	4	↓	10,400	195		10,595	11,700

11 53 33 – Emergency Safety Appliances

11 53 33.13 Emergency Equipment

Line	Description	Crew	Daily Output	Labor-Hours	Unit	Material	Labor	Equipment	Total	Total Incl O&P
0010	**EMERGENCY EQUIPMENT**									
1400	Safety equipment, eye wash, hand held				Ea.	405			405	445
1450	Deluge shower				"	725			725	800

11 53 43 – Service Fittings and Accessories

11 53 43.13 Fittings

Line	Description	Crew	Daily Output	Labor-Hours	Unit	Material	Labor	Equipment	Total	Total Incl O&P
0010	**FITTINGS**									
1600	Sink, one piece plastic, flask wash, hose, free standing	1 Plum	1.60	5	Ea.	1,800	244		2,044	2,350
1610	Epoxy resin sink, 25" x 16" x 10"	"	2	4	"	198	195		393	510
8000	Alternate pricing method: as percent of lab furniture									
8050	Installation, not incl. plumbing & duct work				% Furn.				20%	22%

11 53 Laboratory Equipment

11 53 43 – Service Fittings and Accessories

11 53 43.13 Fittings		Crew	Daily Output	Labor-Hours	Unit	2009 Bare Costs Material	Labor	Equipment	Total	Total Incl O&P
8100	Plumbing, final connections, simple system				% Furn.				9.09%	10%
8110	Moderately complex system								13.64%	15%
8120	Complex system								18.18%	20%
8150	Electrical, simple system								9.09%	10%
8160	Moderately complex system								18.18%	20%
8170	Complex system								31.80%	35%

11 71 Medical Sterilizing Equipment

11 71 10 – Medical Sterilizers & Distillers

11 71 10.10 Sterilizers and Distillers

Line	Description	Crew	Daily Output	Labor-Hours	Unit	Material	Labor	Equipment	Total	Total Incl O&P
0010	**STERILIZERS AND DISTILLERS**									
0700	Distiller, water, steam heated, 50 gal. capacity	1 Plum	1.40	5.714	Ea.	17,600	279		17,879	19,700
5600	Sterilizers, floor loading, 26" x 62" x 42", single door, steam					151,000			151,000	166,000
5650	Double door, steam					194,000			194,000	213,500
5800	General purpose, 20" x 20" x 38", single door					14,600			14,600	16,000
6000	Portable, counter top, steam, minimum					3,625			3,625	3,975
6020	Maximum					5,650			5,650	6,225
6050	Portable, counter top, gas, 17" x 15" x 32-1/2"					37,400			37,400	41,200
6150	Manual washer/sterilizer, 16" x 16" x 26"	1 Plum	2	4		51,500	195		51,695	57,000
6200	Steam generators, electric 10 kW to 180 kW, freestanding									
6250	Minimum	1 Elec	3	2.667	Ea.	7,625	125		7,750	8,550
6300	Maximum	"	.70	11.429		27,400	535		27,935	31,000
8200	Bed pan washer-sanitizer	1 Plum	2	4		6,800	195		6,995	7,775

11 73 Patient Care Equipment

11 73 10 – Patient Treatment Equipment

11 73 10.10 Treatment Equipment

Line	Description	Crew	Daily Output	Labor-Hours	Unit	Material	Labor	Equipment	Total	Total Incl O&P
0010	**TREATMENT EQUIPMENT**									
1800	Heat therapy unit, humidified, 26" x 78" x 28"				Ea.	3,225			3,225	3,550
8400	Whirlpool bath, mobile, sst, 18" x 24" x 60"					4,275			4,275	4,725
8450	Fixed, incl. mixing valves	1 Plum	2	4		8,775	195		8,970	9,950

11 74 Dental Equipment

11 74 10 – Dental Office Equipment

11 74 10.10 Diagnostic and Treatment Equipment

Line	Description	Crew	Daily Output	Labor-Hours	Unit	Material	Labor	Equipment	Total	Total Incl O&P
0010	**DIAGNOSTIC AND TREATMENT EQUIPMENT**									
0020	Central suction system, minimum	1 Plum	1.20	6.667	Ea.	1,425	325		1,750	2,050
0100	Maximum	"	.90	8.889		3,700	435		4,135	4,725
0600	Chair, electric or hydraulic, minimum	1 Skwk	.50	16		2,675	655		3,330	3,925
0700	Maximum		.25	32		13,700	1,300		15,000	17,100
2000	Light, ceiling mounted, minimum		8	1		1,400	41		1,441	1,600
2100	Maximum		8	1		1,900	41		1,941	2,150
2200	Unit light, minimum	2 Skwk	5.33	3.002		600	123		723	850
2210	Maximum		5.33	3.002		1,625	123		1,748	2,000
2220	Track light, minimum		3.20	5		2,050	204		2,254	2,600
2230	Maximum		3.20	5		3,550	204		3,754	4,225
2300	Sterilizers, steam portable, minimum					1,225			1,225	1,350

11 74 Dental Equipment

11 74 10 – Dental Office Equipment

11 74 10.10 Diagnostic and Treatment Equipment		Crew	Daily Output	Labor-Hours	Unit	2009 Bare Costs Material	Labor	Equipment	Total	Total Incl O&P
2350	Maximum				Ea.	9,900			9,900	10,900
2600	Steam, institutional					4,250			4,250	4,675
2650	Dry heat, electric, portable, 3 trays				↓	1,150			1,150	1,275

11 78 Mortuary Equipment

11 78 13 – Mortuary Refrigerators

11 78 13.10 Mortuary and Autopsy Equipment

		Crew	Daily Output	Labor-Hours	Unit	Material	Labor	Equipment	Total	Total Incl O&P
0010	**MORTUARY AND AUTOPSY EQUIPMENT**									
0015	Autopsy table, standard	1 Plum	1	8	Ea.	8,450	390		8,840	9,850
0020	Deluxe	"	.60	13.333		16,800	650		17,450	19,400
3200	Mortuary refrigerator, end operated, 2 capacity					12,400			12,400	13,600
3300	6 capacity				↓	22,200			22,200	24,500

11 78 16 – Crematorium Equipment

11 78 16.10 Crematory

		Crew	Daily Output	Labor-Hours	Unit	Material	Labor	Equipment	Total	Total Incl O&P
0010	**CREMATORY**									
1500	Crematory, not including building, 1 place	Q-3	.20	160	Ea.	59,500	7,425		66,925	76,500
1750	2 place	"	.10	320	"	85,000	14,900		99,900	116,000

11 82 Solid Waste Handling Equipment

11 82 19 – Packaged Incinerators

11 82 19.10 Packaged Gas Fired Incinerators

		Crew	Daily Output	Labor-Hours	Unit	Material	Labor	Equipment	Total	Total Incl O&P
0010	**PACKAGED GAS FIRED INCINERATORS**									
4400	Incinerator, gas, not incl. chimney, elec. or pipe, 50#/hr., minimum	Q-3	.80	40	Ea.	11,200	1,850		13,050	15,100
4420	Maximum		.70	45.714		35,000	2,125		37,125	41,700
4440	200 lb. per hr., minimum (batch type)		.60	53.333		62,000	2,475		64,475	71,500
4460	Maximum (with feeder)		.50	64		85,000	2,975		87,975	98,000
4480	400 lb. per hr., minimum (batch type)		.30	106		68,500	4,950		73,450	83,000
4500	Maximum (with feeder)		.25	128		90,000	5,950		95,950	108,000
4520	800 lb. per hr., with feeder, minimum		.20	160		112,000	7,425		119,425	134,000
4540	Maximum		.17	188		150,000	8,750		158,750	178,000
4560	1,200 lb. per hr., with feeder, minimum		.15	213		179,500	9,900		189,400	212,000
4580	Maximum		.11	290		146,000	13,500		159,500	181,000
4600	2,000 lb. per hr., with feeder, minimum		.10	320		504,000	14,900		518,900	577,000
4620	Maximum		.05	640		357,000	29,700		386,700	437,000
4700	For heat recovery system, add, minimum		.25	128		69,000	5,950		74,950	85,000
4710	Add, maximum		.11	290		220,500	13,500		234,000	263,000
4720	For automatic ash conveyer, add		.50	64	↓	29,000	2,975		31,975	36,400
4750	Large municipal incinerators, incl. stack, minimum		.25	128	Ton/day	17,600	5,950		23,550	28,300
4850	Maximum	↓	.10	320	"	46,800	14,900		61,700	74,000

11 82 26 – Waste Compactors and Destructors

11 82 26.10 Compactors

		Crew	Daily Output	Labor-Hours	Unit	Material	Labor	Equipment	Total	Total Incl O&P
0010	**COMPACTORS**									
0020	Compactors, 115 volt, 250#/hr., chute fed	L-4	1	24	Ea.	10,900	895		11,795	13,400
0100	Hand fed		2.40	10		7,975	375		8,350	9,350
0300	Multi-bag, 230 volt, 600#/hr, chute fed		1	24		9,900	895		10,795	12,300
0400	Hand fed		1	24		8,500	895		9,395	10,700
0500	Containerized, hand fed, 2 to 6 C.Y. containers, 250#/hr.		1	24		10,500	895		11,395	13,000
0550	For chute fed, add per floor	↓	1	24	↓	1,200	895		2,095	2,700

11 82 Solid Waste Handling Equipment

11 82 26 – Waste Compactors and Destructors

11 82 26.10 Compactors		Crew	Daily Output	Labor-Hours	Unit	Material	2009 Bare Costs Labor	Equipment	Total	Total Incl O&P
1000	Heavy duty industrial compactor, 0.5 C.Y. capacity	L-4	1	24	Ea.	7,100	895		7,995	9,175
1050	1.0 C.Y. capacity		1	24		10,800	895		11,695	13,300
1100	3 C.Y. capacity		.50	48		14,800	1,800		16,600	19,000
1150	5.0 C.Y. capacity		.50	48		18,200	1,800		20,000	22,900
1200	Combination shredder/compactor (5,000 lbs./hr.)		.50	48		36,100	1,800		37,900	42,500
1400	For handling hazardous waste materials, 55 gallon drum packer, std.					16,800			16,800	18,500
1410	55 gallon drum packer w/HEPA filter					20,900			20,900	23,000
1420	55 gallon drum packer w/charcoal & HEPA filter					28,000			28,000	30,800
1430	All of the above made explosion proof, add					12,800			12,800	14,000
5800	Shredder, industrial, minimum					20,600			20,600	22,700
5850	Maximum					110,500			110,500	121,500
5900	Baler, industrial, minimum					8,250			8,250	9,075
5950	Maximum					482,500			482,500	531,000

Division Notes

		CREW	DAILY OUTPUT	LABOR-HOURS	UNIT	2009 BARE COSTS MAT.	LABOR	EQUIP.	TOTAL	TOTAL INCL O&P

Estimating Tips

General

- The items and systems in this division are usually estimated, purchased, supplied, and installed as a unit by one or more subcontractors. The estimator must ensure that all parties are operating from the same set of specifications and assumptions, and that all necessary items are estimated and will be provided. Many times the complex items and systems are covered, but the more common ones, such as excavation or a crane, are overlooked for the very reason that everyone assumes nobody could miss them. The estimator should be the central focus and be able to ensure that all systems are complete.
- Another area where problems can develop in this division is at the interface between systems. The estimator must ensure, for instance, that anchor bolts, nuts, and washers are estimated and included for the air-supported structures and pre-engineered buildings to be bolted to their foundations. Utility supply is a common area where essential items or pieces of equipment can be missed or overlooked due to the fact that each subcontractor may feel it is another's responsibility. The estimator should also be aware of certain items which may be supplied as part of a package but installed by others, and ensure that the installing contractor's estimate includes the cost of installation. Conversely, the estimator must also ensure that items are not costed by two different subcontractors, resulting in an inflated overall estimate.

13 30 00 Special Structures

- The foundations and floor slab, as well as rough mechanical and electrical, should be estimated, as this work is required for the assembly and erection of the structure. Generally, as noted in the book, the pre-engineered building comes as a shell and additional features, such as windows and doors, must be included by the estimator. Here again, the estimator must have a clear understanding of the scope of each portion of the work and all the necessary interfaces.

Reference Numbers

Reference numbers are shown in shaded boxes at the beginning of some major classifications. These numbers refer to related items in the Reference Section. The reference information may be an estimating procedure, an alternate pricing method, or technical information.

Note: Not all subdivisions listed here necessarily appear in this publication.

13 11 Swimming Pools

13 11 13 – Below-Grade Swimming Pools

13 11 13.50 Swimming Pools		Crew	Daily Output	Labor-Hours	Unit	Material	2009 Bare Costs Labor	Equipment	Total	Total Incl O&P
0010	**SWIMMING POOLS** Residential in-ground, vinyl lined, concrete									
0020	Sides including equipment, sand bottom	B-52	300	.187	SF Surf	12.60	6.90	1.65	21.15	26.50
0100	Metal or polystyrene sides	B-14	410	.117		10.55	3.88	.72	15.15	18.40
0200	Add for vermiculite bottom				↓	.81			.81	.89
0500	Gunite bottom and sides, white plaster finish									
0600	12′ x 30′ pool	B-52	145	.386	SF Surf	23.50	14.25	3.41	41.16	52
0720	16′ x 32′ pool		155	.361		21	13.30	3.19	37.49	47.50
0750	20′ x 40′ pool	↓	250	.224	↓	18.90	8.25	1.98	29.13	36
0810	Concrete bottom and sides, tile finish									
0820	12′ x 30′ pool	B-52	80	.700	SF Surf	23.50	26	6.20	55.70	73
0830	16′ x 32′ pool		95	.589		19.60	21.50	5.20	46.30	61
0840	20′ x 40′ pool	↓	130	.431	↓	15.60	15.85	3.81	35.26	46
1100	Motel, gunite with plaster finish, incl. medium									
1150	capacity filtration & chlorination	B-52	115	.487	SF Surf	29	17.95	4.30	51.25	64
1200	Municipal, gunite with plaster finish, incl. high									
1250	capacity filtration & chlorination	B-52	100	.560	SF Surf	37.50	20.50	4.95	62.95	78.50
1350	Add for formed gutters				L.F.	55			55	60.50
1360	Add for stainless steel gutters				"	163			163	179
1700	Filtration and deck equipment only, as % of total				Total				20%	20%
1800	Deck equipment, rule of thumb, 20′ x 40′ pool				SF Pool				1.18	1.30
1900	5000 S.F. pool				"				1.73	1.90
3000	Painting pools, preparation + 3 coats, 20′ x 40′ pool, epoxy	2 Pord	.33	48.485	Total	1,550	1,700		3,250	4,250
3100	Rubber base paint, 18 gallons	"	.33	48.485		1,000	1,700		2,700	3,650
3500	42′ x 82′ pool, 75 gallons, epoxy paint	3 Pord	.14	171		6,550	6,025		12,575	16,300
3600	Rubber base paint	"	.14	171	↓	4,125	6,025		10,150	13,600

13 11 46 – Swimming Pool Accessories

13 11 46.50 Swimming Pool Equipment

		Crew	Daily Output	Labor-Hours	Unit	Material	Labor	Equipment	Total	Total Incl O&P
0010	**SWIMMING POOL EQUIPMENT**									
0020	Diving stand, stainless steel, 3 meter	2 Carp	.40	40	Ea.	9,375	1,600		10,975	12,800
0300	1 meter	"	2.70	5.926	"	5,475	237		5,712	6,400
1100	Gutter system, stainless steel, with grating, stock,									
1110	contains supply and drainage system	E-1	20	1.200	L.F.	205	52	6.70	263.70	320
1120	Integral gutter and 5′ high wall system, stainless steel	"	10	2.400	"	310	104	13.40	427.40	530
2100	Lights, underwater, 12 volt, with transformer, 300 watt	1 Elec	1	8	Ea.	243	375		618	830
2200	110 volt, 500 watt, standard		1	8		201	375		576	780
2400	Low water cutoff type	↓	1	8	↓	234	375		609	815
2800	Heaters, see Div. 23 52 28.10									

13 18 Ice Rinks

13 18 13 – Ice Rink Floor Systems

13 18 13.50 Ice Skating

		Crew	Daily Output	Labor-Hours	Unit	Material	Labor	Equipment	Total	Total Incl O&P
0010	**ICE SKATING** Equipment incl. refrigeration, plumbing & cooling									
0020	coils & concrete slab, 85′ x 200′ rink									
0300	55° system, 5 mos., 100 ton				Total	596,000			596,000	655,500
0700	90° system, 12 mos., 135 ton				"	650,000			650,000	715,000
1200	Subsoil heating system (recycled from compressor), 85′ x 200′	Q-7	.27	118	Ea.	27,100	5,575		32,675	38,200
1300	Subsoil insulation, 2 lb. polystyrene with vapor barrier, 85′ x 200′	2 Carp	.14	114	"	32,600	4,575		37,175	42,900

13 18 16 – Ice Rink Dasher Boards

13 18 16.50 Ice Rink Dasher Boards

		Crew	Daily Output	Labor-Hours	Unit	Material	Labor	Equipment	Total	Total Incl O&P
0010	**ICE RINK DASHER BOARDS**									

13 18 Ice Rinks

13 18 16 – Ice Rink Dasher Boards

13 18 16.50 Ice Rink Dasher Boards		Crew	Daily Output	Labor-Hours	Unit	2009 Bare Costs Material	Labor	Equipment	Total	Total Incl O&P
1000	Dasher boards, 1/2″ H.D. polyethylene faced steel frame, 3′ acrylic									
1020	screen at sides, 5′ acrylic ends, 85′ x 200′	F-5	.06	533	Ea.	130,000	21,600		151,600	176,500
1100	Fiberglass & aluminum construction, same sides and ends	″	.06	533	″	140,000	21,600		161,600	137,500

13 21 Controlled Environment Rooms

13 21 13 – Clean Rooms

13 21 13.50 Clean Room Components

		Crew	Daily Output	Labor-Hours	Unit	Material	Labor	Equipment	Total	Total Incl O&P
0010	**CLEAN ROOM COMPONENTS**									
1100	Clean room, soft wall, 12′ x 12′, Class 100	1 Carp	.18	44.444	Ea.	15,200	1,775		16,975	19,500
1110	Class 1,000		.18	44.444		12,300	1,775		14,075	16,300
1120	Class 10,000		.21	38.095		10,800	1,525		12,325	14,200
1130	Class 100,000		.21	38.095		10,100	1,525		11,625	13,500
2800	Ceiling grid support, slotted channel struts 4′-0″ O.C., ea. way				S.F.				5.91	6.50
3000	Ceiling panel, vinyl coated foil on mineral substrate									
3020	Sealed, non-perforated				S.F.				1.27	1.40
4000	Ceiling panel seal, silicone sealant, 150 L.F./gal.	1 Carp	150	.053	L.F.	.28	2.13		2.41	3.61
4100	Two sided adhesive tape	″	240	.033	″	.13	1.33		1.46	2.20
4200	Clips, one per panel				Ea.	1.09			1.09	1.20
6000	HEPA filter, 2′x4′, 99.97% eff., 3″ dp beveled frame (silicone seal)					355			355	390
6040	6″ deep skirted frame (channel seal)					375			375	415
6100	99.99% efficient, 3″ deep beveled frame (silicone seal)					370			370	405
6140	6″ deep skirted frame (channel seal)					395			395	430
6200	99.999% efficient, 3″ deep beveled frame (silicone seal)					385			385	425
6240	6″ deep skirted frame (channel seal)					410			410	450
7000	Wall panel systems, including channel strut framing									
7020	Polyester coated aluminum, particle board				S.F.				18.18	20
7100	Porcelain coated aluminum, particle board								31.82	35
7400	Wall panel support, slotted channel struts, to 12′ high								16.36	18

13 21 26 – Cold Storage Rooms

13 21 26.50 Refrigeration

		Crew	Daily Output	Labor-Hours	Unit	Material	Labor	Equipment	Total	Total Incl O&P
0010	**REFRIGERATION**									
0020	Curbs, 12″ high, 4″ thick, concrete	2 Carp	58	.276	L.F.	4.14	11		15.14	21.50
1000	Doors, see Div. 08 34 13.10									
2400	Finishes, 2 coat portland cement plaster, 1/2″ thick	1 Plas	48	.167	S.F.	1.14	6.05		7.19	10.35
2500	For galvanized reinforcing mesh, add	1 Lath	335	.024		.76	.85		1.61	2.10
2700	3/16″ thick latex cement	1 Plas	88	.091		1.94	3.29		5.23	7.10
2900	For glass cloth reinforced ceilings, add	″	450	.018		.46	.64		1.10	1.48
3100	Fiberglass panels, 1/8″ thick	1 Carp	149.45	.054		2.57	2.14		4.71	6.15
3200	Polystyrene, plastic finish ceiling, 1″ thick		274	.029		2.37	1.17		3.54	4.42
3400	2″ thick		274	.029		2.70	1.17		3.87	4.78
3500	4″ thick		219	.037		3	1.46		4.46	5.55
3800	Floors, concrete, 4″ thick	1 Cefi	93	.086		1.13	3.29		4.42	6.05
3900	6″ thick	″	85	.094		1.68	3.60		5.28	7.15
4000	Insulation, 1″ to 6″ thick, cork				B.F.	1.10			1.10	1.21
4100	Urethane					1.08			1.08	1.19
4300	Polystyrene, regular					.73			.73	.80
4400	Bead board					.55			.55	.61
4600	Installation of above, add per layer	2 Carp	657.60	.024	S.F.	.36	.97		1.33	1.91
4700	Wall and ceiling juncture		298.90	.054	L.F.	1.77	2.14		3.91	5.25
4900	Partitions, galvanized sandwich panels, 4″ thick, stock		219.20	.073	S.F.	7.45	2.92		10.37	12.65

13 21 Controlled Environment Rooms

13 21 26 – Cold Storage Rooms

13 21 26.50 Refrigeration		Crew	Daily Output	Labor-Hours	Unit	2009 Bare Costs Material	Labor	Equipment	Total	Total Incl O&P
5000	Aluminum or fiberglass	2 Carp	219.20	.073	S.F.	8.10	2.92		11.02	13.45
5200	Prefab walk-in, 7'-6" high, aluminum, incl. door & floors,									
5210	not incl. partitions or refrigeration, 6' x 6' O.D. nominal	2 Carp	54.80	.292	SF Flr.	133	11.65		144.65	164
5500	10' x 10' O.D. nominal		82.20	.195		107	7.80		114.80	129
5700	12' x 14' O.D. nominal		109.60	.146		96	5.85		101.85	114
5800	12' x 20' O.D. nominal		109.60	.146		83.50	5.85		89.35	101
6100	For 8'-6" high, add					5%				
6300	Rule of thumb for complete units, w/o doors & refrigeration, cooler	2 Carp	146	.110		120	4.38		124.38	139
6400	Freezer		109.60	.146		142	5.85		147.85	166
6600	Shelving, plated or galvanized, steel wire type		360	.044	SF Hor.	10.55	1.78		12.33	14.35
6700	Slat shelf type		375	.043		13	1.70		14.70	16.95
6900	For stainless steel shelving, add					300%				
7000	Vapor barrier, on wood walls	2 Carp	1644	.010	S.F.	.15	.39		.54	.77
7200	On masonry walls	"	1315	.012	"	.40	.49		.89	1.19
7500	For air curtain doors, see Div. 23 34 33.10									

13 24 Special Activity Rooms

13 24 16 – Saunas

13 24 16.50 Saunas and Heaters		Crew	Daily Output	Labor-Hours	Unit	Material	Labor	Equipment	Total	Total Incl O&P
0010	**SAUNAS AND HEATERS**									
0020	Prefabricated, incl. heater & controls, 7' high, 6' x 4', C/C	L-7	2.20	12.727	Ea.	4,625	490		5,115	5,850
0050	6' x 4', C/P		2	14		4,275	540		4,815	5,525
0400	6' x 5', C/C		2	14		5,175	540		5,715	6,500
0450	6' x 5', C/P		2	14		4,800	540		5,340	6,125
0600	6' x 6', C/C		1.80	15.556		5,475	600		6,075	6,950
0650	6' x 6', C/P		1.80	15.556		5,125	600		5,725	6,550
0800	6' x 9', C/C		1.60	17.500		6,800	675		7,475	8,525
0850	6' x 9', C/P		1.60	17.500		6,300	675		6,975	8,000
1000	8' x 12', C/C		1.10	25.455		10,600	980		11,580	13,200
1050	8' x 12', C/P		1.10	25.455		9,725	980		10,705	12,200
1200	8' x 8', C/C		1.40	20		8,100	770		8,870	10,100
1250	8' x 8', C/P		1.40	20		7,600	770		8,370	9,575
1400	8' x 10', C/C		1.20	23.333		8,975	900		9,875	11,300
1450	8' x 10', C/P		1.20	23.333		8,350	900		9,250	10,600
1600	10' x 12', C/C		1	28		11,200	1,075		12,275	14,000
1650	10' x 12', C/P		1	28		10,100	1,075		11,175	12,800
2500	Heaters only (incl. above), wall mounted, to 200 C.F.					605			605	665
2750	To 300 C.F.					770			770	845
3000	Floor standing, to 720 C.F., 10,000 watts, w/controls	1 Elec	3	2.667		2,025	125		2,150	2,400
3250	To 1,000 C.F., 16,000 watts	"	3	2.667		2,525	125		2,650	2,950

13 24 26 – Steam Baths

13 24 26.50 Steam Baths and Components		Crew	Daily Output	Labor-Hours	Unit	Material	Labor	Equipment	Total	Total Incl O&P
0010	**STEAM BATHS AND COMPONENTS**									
0020	Heater, timer & head, single, to 140 C.F.	1 Plum	1.20	6.667	Ea.	1,400	325		1,725	2,050
0500	To 300 C.F.		1.10	7.273		1,625	355		1,980	2,300
1000	Commercial size, with blow-down assembly, to 800 C.F.		.90	8.889		5,200	435		5,635	6,375
1500	To 2500 C.F.		.80	10		8,025	490		8,515	9,550
2000	Multiple, motels, apts., 2 baths, w/ blow-down assm., 500 C.F.	Q-1	1.30	12.308		5,900	540		6,440	7,300
2500	4 baths	"	.70	22.857		7,450	1,000		8,450	9,700
2700	Conversion unit for residential tub, including door					4,100			4,100	4,500

13 34 Fabricated Engineered Structures

13 34 23 – Fabricated Structures

13 34 23.10 Comfort Stations		Crew	Daily Output	Labor-Hours	Unit	2009 Bare Costs Material	Labor	Equipment	Total	Total Incl O&P
0010	**COMFORT STATIONS** Prefab., stock, w/doors, windows & fixt.									
0100	Not incl. interior finish or electrical									
0300	Mobile, on steel frame, minimum				S.F.	46.50			46.50	51
0350	Maximum					75			75	82.50
0400	Permanent, including concrete slab, minimum	B-12J	50	.320		199	11.85	19.20	230.05	258
0500	Maximum	"	43	.372		288	13.80	22.50	324.30	360
0600	Alternate pricing method, mobile, minimum				Fixture	2,150			2,150	2,375
0650	Maximum					3,175			3,175	3,500
0700	Permanent, minimum	B-12J	.70	22.857		12,200	850	1,375	14,425	16,300
0750	Maximum	"	.50	32		20,400	1,175	1,925	23,500	26,300

13 34 23.16 Fabricated Control Booths		Crew	Daily Output	Labor-Hours	Unit	Material	Labor	Equipment	Total	Total Incl O&P
0010	**FABRICATED CONTROL BOOTHS**									
0100	Guard House, prefab conc w/bullet resistant doors & windows, roof & wiring									
0110	8′ x 8′, Level III	L-10	1	24	Ea.	22,000	1,075	770	23,845	26,900
0120	8′ x 8′, Level IV	"	1	24	"	27,500	1,075	770	29,345	33,000

Division Notes

I	CREW	DAILY OUTPUT	LABOR-HOURS	UNIT	2009 BARE COSTS MAT.	LABOR	EQUIP.	TOTAL	TOTAL INCL O&P

Estimating Tips

General

- Many products in Division 14 will require some type of support or blocking for installation not included with the item itself. Examples are supports for conveyors or tube systems, attachment points for lifts, and footings for hoists or cranes. Add these supports in the appropriate division.

14 10 00 Dumbwaiters
14 20 00 Elevators

- Dumbwaiters and elevators are estimated and purchased in a method similar to buying a car. The manufacturer has a base unit with standard features. Added to this base unit price will be whatever options the owner or specifications require. Increased load capacity, additional vertical travel, additional stops, higher speed, and cab finish options are items to be considered. When developing an estimate for dumbwaiters and elevators, remember that some items needed by the installers may have to be included as part of the general contract.

  Examples are:
  - — shaftway
  - — rail support brackets
  - — machine room
  - — electrical supply
  - — sill angles
  - — electrical connections
  - — pits
  - — roof penthouses
  - — pit ladders

Check the job specifications and drawings before pricing.

- Installation of elevators and handicapped lifts in historic structures can require significant additional costs. The associated structural requirements may involve cutting into and repairing finishes, mouldings, flooring, etc. The estimator must account for these special conditions.

14 30 00 Escalators and Moving Walks

- Escalators and moving walks are specialty items installed by specialty contractors. There are numerous options associated with these items. For specific options, contact a manufacturer or contractor. In a method similar to estimating dumbwaiters and elevators, you should verify the extent of general contract work and add items as necessary.

14 40 00 Lifts
14 90 00 Other Conveying Equipment

- Products such as correspondence lifts, chutes, and pneumatic tube systems, as well as other items specified in this subdivision, may require trained installers. The general contractor might not have any choice as to who will perform the installation or when it will be performed. Long lead times are often required for these products, making early decisions in scheduling necessary.

Reference Numbers

Reference numbers are shown in shaded boxes at the beginning of some major classifications. These numbers refer to related items in the Reference Section. The reference information may be an estimating procedure, an alternate pricing method, or technical information.

Note: Not all subdivisions listed here necessarily appear in this publication.

14 91 Facility Chutes

14 91 33 – Laundry and Linen Chutes

14 91 33.10 Chutes		Crew	Daily Output	Labor-Hours	Unit	Material	Labor	Equipment	Total	Total Incl O&P
						2009 Bare Costs				
0011	**CHUTES**, linen, trash or refuse									
0050	Aluminized steel, 16 ga., 18″ diameter	2 Shee	3.50	4.571	Floor	1,100	216		1,316	1,550
0100	24″ diameter		3.20	5		1,175	236		1,411	1,650
0200	30″ diameter		3	5.333		1,325	252		1,577	1,825
0300	36″ diameter		2.80	5.714		1,500	270		1,770	2,075
0400	Galvanized steel, 16 ga., 18″ diameter		3.50	4.571		1,000	216		1,216	1,425
0500	24″ diameter		3.20	5		1,125	236		1,361	1,625
0600	30″ diameter		3	5.333		1,275	252		1,527	1,775
0700	36″ diameter		2.80	5.714		1,500	270		1,770	2,075
0800	Stainless steel, 18″ diameter		3.50	4.571		2,350	216		2,566	2,900
0900	24″ diameter		3.20	5		2,550	236		2,786	3,175
1000	30″ diameter		3	5.333		3,100	252		3,352	3,775
1005	36″ diameter		2.80	5.714		3,450	270		3,720	4,225
1200	Linen chute bottom collector, aluminized steel		4	4	Ea.	1,400	189		1,589	1,850
1300	Stainless steel		4	4		1,800	189		1,989	2,275
1500	Refuse, bottom hopper, aluminized steel, 18″ diameter		3	5.333		890	252		1,142	1,375
1600	24″ diameter		3	5.333		1,075	252		1,327	1,550
1800	36″ diameter		3	5.333		1,800	252		2,052	2,350

14 91 82 – Trash Chutes

14 91 82.10 Trash Chutes and Accessories		Crew	Daily Output	Labor-Hours	Unit	Material	Labor	Equipment	Total	Total Incl O&P
0010	**TRASH CHUTES AND ACCESSORIES**									
2900	Package chutes, spiral type, minimum	2 Shee	4.50	3.556	Floor	2,425	168		2,593	2,925
3000	Maximum	″	1.50	10.667	″	6,325	505		6,830	7,750

14 92 Pneumatic Tube Systems

14 92 10 – Conventional, Automatic and Computer Controlled Pneumatic Tube Systems

14 92 10.10 Pneumatic Tube Systems		Crew	Daily Output	Labor-Hours	Unit	Material	Labor	Equipment	Total	Total Incl O&P
0010	**PNEUMATIC TUBE SYSTEMS**									
0020	100′ long, single tube, 2 stations, stock									
0100	3″ diameter	2 Stpi	.12	133	Total	6,600	6,575		13,175	17,100
0300	4″ diameter	″	.09	177	″	7,450	8,775		16,225	21,400
0400	Twin tube, two stations or more, conventional system									
0600	2-1/2″ round	2 Stpi	62.50	.256	L.F.	13.30	12.65		25.95	33.50
0700	3″ round		46	.348		15.25	17.15		32.40	43
0900	4″ round		49.60	.323		16.90	15.90		32.80	42.50
1000	4″ x 7″ oval		37.60	.426		24.50	21		45.50	58.50
1050	Add for blower		2	8	System	5,200	395		5,595	6,325
1110	Plus for each round station, add		7.50	2.133	Ea.	585	105		690	805
1150	Plus for each oval station, add		7.50	2.133	″	585	105		690	805
1200	Alternate pricing method: base cost, minimum		.75	21.333	Total	6,375	1,050		7,425	8,600
1300	Maximum		.25	64	″	12,700	3,150		15,850	18,700
1500	Plus total system length, add, minimum		93.40	.171	L.F.	8.20	8.45		16.65	22
1600	Maximum		37.60	.426	″	24.50	21		45.50	58
1800	Completely automatic system, 4″ round, 15 to 50 stations		.29	55.172	Station	20,100	2,725		22,825	26,200
2200	51 to 144 stations		.32	50		15,600	2,475		18,075	20,900
2400	6″ round or 4″ x 7″ oval, 15 to 50 stations		.24	66.667		25,200	3,300		28,500	32,600
2800	51 to 144 stations		.23	69.565		21,100	3,425		24,525	28,400

Estimating Tips

22 10 00 Plumbing Piping and Pumps

This subdivision is primarily basic pipe and related materials. The pipe may be used by any of the mechanical disciplines, i.e., plumbing, fire protection, heating, and air conditioning.

- The labor adjustment factors listed in Subdivision 22 01 02.20 apply throughout Divisions 21, 22, and 23. CAUTION: the correct percentage may vary for the same items. For example, the percentage add for the basic pipe installation should be based on the maximum height that the craftsman must install for that particular section. If the pipe is to be located 14′ above the floor but it is suspended on threaded rod from beams, the bottom flange of which is 18′ high (4′ rods), then the height is actually 18′ and the add is 20%. The pipe coverer, however, does not have to go above the 14′, and so his or her add should be 10%.
- Most pipe is priced first as straight pipe with a joint (coupling, weld, etc.) every 10′ and a hanger usually every 10′. There are exceptions with hanger spacing such as for cast iron pipe (5′) and plastic pipe (3 per 10′). Following each type of pipe there are several lines listing sizes and the amount to be subtracted to delete couplings and hangers. This is for pipe that is to be buried or supported together on trapeze hangers. The reason that the couplings are deleted is that these runs are usually long, and frequently longer lengths of pipe are used. By deleting the couplings, the estimator is expected to look up and add back the correct reduced number of couplings.
- When preparing an estimate, it may be necessary to approximate the fittings. Fittings usually run between 25% and 50% of the cost of the pipe. The lower percentage is for simpler runs, and the higher number is for complex areas, such as mechanical rooms.
- For historic restoration projects, the systems must be as invisible as possible, and pathways must be sought for pipes, conduit, and ductwork. While installations in accessible spaces (such as basements and attics) are relatively straightforward to estimate, labor costs may be more difficult to determine when delivery systems must be concealed.

22 40 00 Plumbing Fixtures

- Plumbing fixture costs usually require two lines: the fixture itself and its "rough-in, supply, and waste."
- In the Assemblies Section (Plumbing D2010) for the desired fixture, the System Components Group at the center of the page shows the fixture on the first line. The rest of the list (fittings, pipe, tubing, etc.) will total up to what we refer to in the Unit Price section as "Rough-in, supply, waste, and vent." Note that for most fixtures we allow a nominal 5′ of tubing to reach from the fixture to a main or riser.
- Remember that gas- and oil-fired units need venting.

Reference Numbers

Reference numbers are shown in shaded boxes at the beginning of some major classifications. These numbers refer to related items in the Reference Section. The reference information may be an estimating procedure, an alternate pricing method, or technical information.

Note: Not all subdivisions listed here necessarily appear in this publication.

Note: **Trade Service,** *in part, has been used as a reference source for some of the material prices used in Division 22.*

22 01 Operation and Maintenance of Plumbing

22 01 02 – Labor Adjustments

22 01 02.10 Boilers, General		Crew	Daily Output	Labor-Hours	Unit	Material	2009 Bare Costs Labor	Equipment	Total	Total Incl O&P
0010	**BOILERS, GENERAL**, Prices do not include flue piping, elec. wiring,									
0020	gas or oil piping, boiler base, pad, or tankless unless noted									
0100	Boiler H.P.: 10 KW = 34 lbs/steam/hr = 33,475 BTU/hr. R235000-50									
0150	To convert SFR to BTU rating: Hot water, 150 x SFR;									
0160	Forced hot water, 180 x SFR; steam, 240 x SFR									

22 01 02.20 Labor Adjustment Factors		Crew	Daily Output	Labor-Hours	Unit	Material	Labor	Equipment	Total	Total Incl O&P
0010	**LABOR ADJUSTMENT FACTORS**, (For Div. 22 and 23)									
1000	Add to labor for elevated installation (Above floor level)									
1080	10′ to 14.5′ high R221113-70						10%			
1100	15′ to 19.5′ high						20%			
1120	20′ to 24.5′ high						25%			
1140	25′ to 29.5′ high						35%			
1160	30′ to 34.5′ high						40%			
1180	35′ to 39.5′ high						50%			
1200	40′ and higher						55%			
2000	Add to labor for crawl space									
2100	3′ high						40%			
2140	4′ high						30%			
3000	Add to labor for multi-story building									
3100	Add per floor for floors 3 thru 19						2%			
3140	Add per floor for floors 20 and up						4%			
4000	Add to labor for working in existing occupied buildings									
4100	Hospital						35%			
4140	Office building						25%			
4180	School						20%			
4220	Factory or warehouse						15%			
4260	Multi dwelling						15%			
5000	Add to labor, miscellaneous									
5100	Cramped shaft						35%			
5140	Congested area						15%			
5180	Excessive heat or cold						30%			
9000	Labor factors, The above are reasonable suggestions, however									
9010	each project should be evaluated for its own peculiarities.									
9100	Other factors to be considered are:									
9140	Movement of material and equipment through finished areas									
9180	Equipment room									
9220	Attic space									
9260	No service road									
9300	Poor unloading/storage area									
9340	Congested site area/heavy traffic									

22 05 05 – Selective Plumbing Demolition

22 05 05.10 Plumbing Demolition		Crew	Daily Output	Labor-Hours	Unit	Material	2009 Bare Costs Labor	Equipment	Total	Total Incl O&P
0010	**PLUMBING DEMOLITION** R220105-10									
0400	Air compressor, up thru 2 H.P.	Q-1	10	1.600	Ea.		70		70	105
0410	3 H.P. thru 7-1/2 H.P.		5.60	2.857			125		125	188
0420	10 H.P. thru 15 H.P.		1.40	11.429			500		500	755
0430	20 H.P. thru 30 H.P.	Q-2	1.30	18.462			840		840	1,250
0500	Backflow preventer, up thru 2" diameter	1 Plum	17	.471			23		23	34.50
0510	2-1/2" thru 3" diameter	Q-1	10	1.600			70		70	105
0520	4" thru 6" diameter	"	5	3.200			140		140	211
0530	8" thru 10" diameter	Q-2	3	8			365		365	545
0700	Carriers and supports									
0710	Fountains, sinks, lavatories and urinals	1 Plum	14	.571	Ea.		28		28	42
0720	Water closets	"	12	.667	"		32.50		32.50	49
0730	Grinder pump or sewage ejector system									
0732	Simplex	Q-1	7	2.286	Ea.		100		100	151
0734	Duplex	"	2.80	5.714			251		251	375
0738	Hot water dispenser	1 Plum	36	.222			10.85		10.85	16.25
0740	Hydrant, wall		26	.308			15		15	22.50
0744	Ground		12	.667			32.50		32.50	49
0760	Cleanouts and drains, up thru 4" pipe diameter		10	.800			39		39	58.50
0764	5" thru 8" pipe diameter	Q-1	10	1.600			70		70	105
0780	Industrial safety fixtures	1 Plum	8	1			49		49	73
1020	Fixtures, including 10' piping									
1100	Bath tubs, cast iron	1 Plum	4	2	Ea.		97.50		97.50	146
1120	Fiberglass		6	1.333			65		65	97.50
1140	Steel		5	1.600			78		78	117
1150	Bidet	Q-1	7	2.286			100		100	151
1200	Lavatory, wall hung	1 Plum	10	.800			39		39	58.50
1220	Counter top		8	1			49		49	73
1300	Sink, single compartment		8	1			49		49	73
1320	Double compartment		7	1.143			55.50		55.50	83.50
1340	Shower, stall and receptor	Q-1	6	2.667			117		117	176
1350	Group	"	7	2.286			100		100	151
1400	Water closet, floor mounted	1 Plum	8	1			49		49	73
1420	Wall mounted	"	7	1.143			55.50		55.50	83.50
1440	Wash fountain, 36" diameter	Q-2	8	3			137		137	205
1442	54" diameter	"	7	3.429			156		156	234
1500	Urinal, floor mounted	1 Plum	4	2			97.50		97.50	146
1520	Wall mounted	"	7	1.143			55.50		55.50	83.50
1590	Whirl pool or hot tub	Q-1	2.60	6.154			270		270	405
1600	Water fountains, free standing	1 Plum	8	1			49		49	73
1620	Wall or deck mounted		6	1.333			65		65	97.50
1800	Medical gas specialties		8	1			49		49	73
1810	Plumbing demo, Floor drain, Remove		12	.667			32.50		32.50	49
1900	Piping fittings, single connection, up thru 1-1/2" diameter		30	.267			13		13	19.50
1910	2" thru 4" diameter		14	.571			28		28	42
1980	Pipe hanger/support removal		80	.100			4.88		4.88	7.30
1990	Glass pipe with fittings, 1" thru 3" diameter		200	.040	L.F.		1.95		1.95	2.93
1992	4" thru 6" diameter		150	.053			2.60		2.60	3.90
2000	Piping, metal, up thru 1-1/2" diameter		200	.040			1.95		1.95	2.93
2050	2" thru 3-1/2" diameter		150	.053			2.60		2.60	3.90
2100	4" thru 6" diameter	2 Plum	100	.160			7.80		7.80	11.70
2150	8" thru 14" diameter	"	60	.267			13		13	19.50
2153	16" thru 20" diameter	Q-18	70	.343			15.80	.80	16.60	24.50

22 05 05 – Selective Plumbing Demolition

22 05 05.10 Plumbing Demolition		Crew	Daily Output	Labor-Hours	Unit	Material	2009 Bare Costs Labor	Equipment	Total	Total Incl O&P
2155	24″ thru 26″ diameter	Q-18	55	.436	L.F.		20	1.01	21.01	31
2156	30″ thru 36″ diameter		40	.600			27.50	1.39	28.89	43
2160	Plastic pipe with fittings, up thru 1-1/2″ diameter	1 Plum	250	.032			1.56		1.56	2.34
2162	2″ thru 3″ diameter	″	200	.040			1.95		1.95	2.93
2164	4″ thru 6″ diameter	Q-1	200	.080			3.51		3.51	5.25
2166	8″ thru 14″ diameter		150	.107			4.68		4.68	7
2168	16″ diameter		100	.160			7		7	10.55
2170	Prison fixtures, lavatory or sink		18	.889	Ea.		39		39	58.50
2172	Shower		5.60	2.857			125		125	188
2174	Urinal or water closet		13	1.231			54		54	81
2180	Pumps, all fractional horse-power		12	1.333			58.50		58.50	88
2184	1 H.P. thru 5 H.P.		6	2.667			117		117	176
2186	7-1/2 H.P. thru 15 H.P.		2.50	6.400			281		281	420
2188	20 H.P. thru 25 H.P.	Q-2	4	6			273		273	410
2190	30 H.P. thru 60 H.P.		.80	30			1,375		1,375	2,050
2192	75 H.P. thru 100 H.P.		.60	40			1,825		1,825	2,725
2194	150 H.P.		.50	48			2,175		2,175	3,275
2198	Pump, sump or submersible	1 Plum	12	.667			32.50		32.50	49
2200	Receptors and interceptors, up thru 20 GPM	″	8	1			49		49	73
2204	25 thru 100 GPM	Q-1	6	2.667			117		117	176
2208	125 thru 300 GPM	″	2.40	6.667			293		293	440
2211	325 thru 500 GPM	Q-2	2.60	9.231			420		420	630
2212	Deduct for salvage, aluminum scrap				Ton				1,474	1,620
2214	Brass scrap								3,440	3,780
2216	Copper scrap								5,114	5,620
2218	Lead scrap								764	840
2220	Steel scrap								205	225
2230	Temperature maintenance cable	1 Plum	1200	.007	L.F.		.33		.33	.49
2250	Water heater, 40 gal.	″	6	1.333	Ea.		65		65	97.50
3100	Tanks, water heaters and liquid containers									
3110	Up thru 45 gallons	Q-1	22	.727	Ea.		32		32	48
3120	50 thru 120 gallons		14	1.143			50		50	75.50
3130	130 thru 240 gallons		7.60	2.105			92.50		92.50	139
3140	250 thru 500 gallons		5.40	2.963			130		130	195
3150	600 thru 1000 gallons	Q-2	1.60	15			685		685	1,025
3160	1100 thru 2000 gallons		.70	34.286			1,550		1,550	2,350
3170	2100 thru 4000 gallons		.50	48			2,175		2,175	3,275
6000	Remove and reset fixtures, minimum	1 Plum	6	1.333			65		65	97.50
6100	Maximum	″	4	2			97.50		97.50	146
9100	Valve, metal valves, strainers and similar, up thru 1-1/2″ diameter	1 Stpi	28	.286			14.10		14.10	21
9110	2″ thru 3″ diameter	Q-1	11	1.455			64		64	96
9120	4″ thru 6″ diameter	″	8	2			88		88	132
9130	8″ thru 14″ diameter	Q-2	8	3			137		137	205
9140	16″ thru 20″ diameter		2	12			545		545	820
9150	24″ diameter		1.20	20			910		910	1,375
9200	Valve, plastic, up thru 1-1/2″ diameter	1 Plum	42	.190			9.30		9.30	13.95
9210	2″ thru 3″ diameter		15	.533			26		26	39
9220	4″ thru 6″ diameter		12	.667			32.50		32.50	49
9300	Vent flashing and caps		55	.145			7.10		7.10	10.65
9350	Water filter, commercial, 1″ thru 1-1/2″	Q-1	2	8			350		350	525
9360	2″ thru 2-1/2″	″	1.60	10			440		440	660
9402	Water heaters									
9410	Up thru 245 GPH	Q-1	2.40	6.667	Ea.		293		293	440

22 05 05 – Selective Plumbing Demolition

22 05 05.10 Plumbing Demolition		Crew	Daily Output	Labor-Hours	Unit	Material	2009 Bare Costs Labor	Equipment	Total	Total Incl O&P
9420	250 thru 756 GPH	Q-1	1.60	10	Ea.		440		440	660
9430	775 thru 1640 GPH	↓	.80	20			880		880	1,325
9440	1650 thru 4000 GPH	Q-2	.50	48			2,175		2,175	3,275
9470	Water softener	Q-1	2	8	↓		350		350	525

22 05 23 – General-Duty Valves for Plumbing Piping

22 05 23.10 Valves, Brass

		Crew	Daily Output	Labor-Hours	Unit	Material	Labor	Equipment	Total	Total Incl O&P
0010	**VALVES, BRASS**									
0032	For motorized valves, see Div. 23 09 53.10									
0500	Gas cocks, threaded									
0510	1/4"	1 Plum	26	.308	Ea.	8.65	15		23.65	32
0520	3/8"		24	.333		8.65	16.25		24.90	34
0530	1/2"		24	.333		8.65	16.25		24.90	34
0540	3/4"		22	.364		13.65	17.75		31.40	41.50
0550	1"		19	.421		17.95	20.50		38.45	50.50
0560	1-1/4"		15	.533		30	26		56	72
0570	1-1/2"		13	.615		38.50	30		68.50	87.50
0580	2"	↓	11	.727	↓	64.50	35.50		100	124
0672	For larger sizes use lubricated plug valve, Div. 23 05 23.70									

22 05 23.20 Valves, Bronze

		Crew	Daily Output	Labor-Hours	Unit	Material	Labor	Equipment	Total	Total Incl O&P
0010	**VALVES, BRONZE** R220523-90									
1020	Angle, 150 lb., rising stem, threaded									
1030	1/8"	1 Plum	24	.333	Ea.	87	16.25		103.25	120
1040	1/4"		24	.333		90	16.25		106.25	124
1050	3/8"		24	.333		90	16.25		106.25	124
1060	1/2"		22	.364		90	17.75		107.75	126
1070	3/4"		20	.400		123	19.50		142.50	165
1080	1"		19	.421		177	20.50		197.50	226
1090	1-1/4"		15	.533		227	26		253	289
1100	1-1/2"		13	.615		296	30		326	370
1110	2"	↓	11	.727	↓	480	35.50		515.50	580
1380	Ball, 150 psi, threaded									
1400	1/4"	1 Plum	24	.333	Ea.	10.30	16.25		26.55	36
1430	3/8"		24	.333		10.30	16.25		26.55	36
1450	1/2"		22	.364		10.30	17.75		28.05	38
1460	3/4"		20	.400		16.95	19.50		36.45	48
1470	1"		19	.421		21.50	20.50		42	54.50
1480	1-1/4"		15	.533		39.50	26		65.50	82.50
1490	1-1/2"		13	.615		51	30		81	101
1500	2"		11	.727		61.50	35.50		97	121
1510	2-1/2"		9	.889		207	43.50		250.50	293
1520	3"	↓	8	1	↓	315	49		364	420
1600	Butterfly, 175 psi, full port, solder or threaded ends									
1610	Stainless steel disc and stem									
1620	1/4"	1 Plum	24	.333	Ea.	11.60	16.25		27.85	37.50
1630	3/8"		24	.333		11.60	16.25		27.85	37.50
1640	1/2"		22	.364		12.20	17.75		29.95	40
1650	3/4"		20	.400		19.75	19.50		39.25	51
1660	1"		19	.421		24	20.50		44.50	57.50
1670	1-1/4"		15	.533		38	26		64	81
1680	1-1/2"		13	.615		50	30		80	100
1690	2"	↓	11	.727	↓	62.50	35.50		98	122
1750	Check, swing, class 150, regrinding disc, threaded									

22 05 23.20 Valves, Bronze		Crew	Daily Output	Labor-Hours	Unit	2009 Bare Costs Material	Labor	Equipment	Total	Total Incl O&P
1800	1/8"	1 Plum	24	.333	Ea.	42	16.25		58.25	70.50
1830	1/4"		24	.333		42	16.25		58.25	70.50
1840	3/8"		24	.333		44.50	16.25		60.75	73
1850	1/2"		24	.333		44.50	16.25		60.75	73
1860	3/4"		20	.400		64	19.50		83.50	99.50
1870	1"		19	.421		91.50	20.50		112	132
1880	1-1/4"		15	.533		132	26		158	184
1890	1-1/2"		13	.615		154	30		184	214
1900	2"		11	.727		226	35.50		261.50	300
1910	2-1/2"	Q-1	15	1.067		505	47		552	630
1920	3"	"	13	1.231		675	54		729	825
2000	For 200 lb, add					5%	10%			
2040	For 300 lb, add					15%	15%			
2060	Check swing, 300#, sweat, 3/8" size	1 Plum	24	.333	Ea.	29	16.25		45.25	56.50
2070	1/2"		24	.333		29	16.25		45.25	56.50
2080	3/4"		20	.400		36.50	19.50		56	70
2090	1"		19	.421		51	20.50		71.50	87
2100	1-1/4"		15	.533		66	26		92	112
2110	1-1/2"		13	.615		87.50	30		117.50	142
2120	2"		11	.727		132	35.50		167.50	199
2130	2-1/2"	Q-1	15	1.067		315	47		362	415
2140	3"	"	13	1.231		420	54		474	540
2350	Check, lift, class 150, horizontal composition disc, threaded									
2430	1/4"	1 Plum	24	.333	Ea.	126	16.25		142.25	163
2440	3/8"		24	.333		160	16.25		176.25	201
2450	1/2"		24	.333		139	16.25		155.25	178
2460	3/4"		20	.400		170	19.50		189.50	217
2470	1"		19	.421		241	20.50		261.50	296
2480	1-1/4"		15	.533		325	26		351	395
2490	1-1/2"		13	.615		385	30		415	470
2500	2"		11	.727		645	35.50		680.50	765
2850	Gate, N.R.S., soldered, 125 psi									
2900	3/8"	1 Plum	24	.333	Ea.	36.50	16.25		52.75	65
2920	1/2"		24	.333		31	16.25		47.25	58.50
2940	3/4"		20	.400		35.50	19.50		55	68.50
2950	1"		19	.421		50.50	20.50		71	86.50
2960	1-1/4"		15	.533		80.50	26		106.50	128
2970	1-1/2"		13	.615		85.50	30		115.50	140
2980	2"		11	.727		122	35.50		157.50	187
2990	2-1/2"	Q-1	15	1.067		305	47		352	405
3000	3"	"	13	1.231		400	54		454	520
3350	Threaded, class 150									
3410	1/4"	1 Plum	24	.333	Ea.	52.50	16.25		68.75	82.50
3420	3/8"		24	.333		52.50	16.25		68.75	82.50
3430	1/2"		24	.333		49	16.25		65.25	78.50
3440	3/4"		20	.400		56	19.50		75.50	91.50
3450	1"		19	.421		71.50	20.50		92	110
3460	1-1/4" size		15	.533		96	26		122	144
3470	1-1/2"		13	.615		138	30		168	196
3480	2"		11	.727		164	35.50		199.50	234
3490	2-1/2"	Q-1	15	1.067		385	47		432	490
3500	3" size	"	13	1.231		540	54		594	675
3600	Gate, flanged, 150 lb.									

22 05 23.20 Valves, Bronze		Crew	Daily Output	Labor-Hours	Unit	2009 Bare Costs Material	Labor	Equipment	Total	Total Incl O&P
3610	1"	1 Plum	7	1.143	Ea.	1,025	55.50		1,080.50	1,200
3620	1-1/2"		6	1.333		1,275	65		1,340	1,500
3630	2"		5	1.600		1,900	78		1,978	2,200
3634	2-1/2"	Q-1	5	3.200		2,925	140		3,065	3,425
3640	3"	"	4.50	3.556		3,425	156		3,581	3,975
3850	Rising stem, soldered, 300 psi									
3900	3/8"	1 Plum	24	.333	Ea.	83.50	16.25		99.75	117
3920	1/2"		24	.333		83.50	16.25		99.75	117
3940	3/4"		20	.400		91	19.50		110.50	130
3950	1"		19	.421		132	20.50		152.50	176
3960	1-1/4"		15	.533		183	26		209	240
3970	1-1/2"		13	.615		215	30		245	282
3980	2"		11	.727		350	35.50		385.50	440
3990	2-1/2"	Q-1	15	1.067		780	47		827	930
4000	3"	"	13	1.231		1,100	54		1,154	1,300
4250	Threaded, class 150									
4310	1/4"	1 Plum	24	.333	Ea.	47	16.25		63.25	76
4320	3/8"		24	.333		47	16.25		63.25	76
4330	1/2"		24	.333		44	16.25		60.25	73
4340	3/4"		20	.400		51	19.50		70.50	85.50
4350	1"		19	.421		68.50	20.50		89	106
4360	1-1/4"		15	.533		92	26		118	140
4370	1-1/2"		13	.615		117	30		147	173
4380	2"		11	.727		158	35.50		193.50	226
4390	2-1/2"	Q-1	15	1.067		365	47		412	475
4400	3"	"	13	1.231		510	54		564	640
4500	For 300 psi, threaded, add					100%	15%			
4540	For chain operated type, add					15%				
4850	Globe, class 150, rising stem, threaded									
4920	1/4"	1 Plum	24	.333	Ea.	67	16.25		83.25	98
4940	3/8"		24	.333		67	16.25		83.25	98
4950	1/2"		24	.333		67	16.25		83.25	98
4960	3/4"		20	.400		90	19.50		109.50	129
4970	1"		19	.421		141	20.50		161.50	186
4980	1-1/4"		15	.533		224	26		250	286
4990	1-1/2"		13	.615		273	30		303	345
5000	2"		11	.727		410	35.50		445.50	505
5010	2-1/2"	Q-1	15	1.067		825	47		872	975
5020	3"	"	13	1.231		1,175	54		1,229	1,375
5120	For 300 lb. threaded, add					50%	15%			
5130	Globe, 300 lb., sweat, 3/8" size	1 Plum	24	.333	Ea.	83.50	16.25		99.75	117
5140	1/2"		24	.333		85.50	16.25		101.75	119
5150	3/4"		20	.400		117	19.50		136.50	158
5160	1"		19	.421		161	20.50		181.50	208
5170	1-1/4"		15	.533		256	26		282	320
5180	1-1/2"		13	.615		310	30		340	385
5190	2"		11	.727		455	35.50		490.50	555
5200	2-1/2"	Q-1	15	1.067		890	47		937	1,050
5210	3"	"	13	1.231		1,150	54		1,204	1,325
5600	Relief, pressure & temperature, self-closing, ASME, threaded									
5640	3/4"	1 Plum	28	.286	Ea.	118	13.95		131.95	151
5650	1"		24	.333		180	16.25		196.25	224
5660	1-1/4"		20	.400		345	19.50		364.50	410

22 05 23 – General-Duty Valves for Plumbing Piping

22 05 23.20	Valves, Bronze	Crew	Daily Output	Labor-Hours	Unit	2009 Bare Costs Material	Labor	Equipment	Total	Total Incl O&P
5670	1-1/2"	1 Plum	18	.444	Ea.	660	21.50		681.50	760
5680	2"		16	.500		720	24.50		744.50	825
5950	Pressure, poppet type, threaded									
6000	1/2"	1 Plum	30	.267	Ea.	33	13		46	55.50
6040	3/4"	"	28	.286	"	35	13.95		48.95	59.50
6400	Pressure, water, ASME, threaded									
6440	3/4"	1 Plum	28	.286	Ea.	70	13.95		83.95	98
6450	1"		24	.333		149	16.25		165.25	189
6460	1-1/4"		20	.400		236	19.50		255.50	289
6470	1-1/2"		18	.444		325	21.50		346.50	395
6480	2"		16	.500		470	24.50		494.50	555
6490	2-1/2"		15	.533		2,250	26		2,276	2,525
6900	Reducing, water pressure									
6920	300 psi to 25-75 psi, threaded or sweat									
6940	1/2"	1 Plum	24	.333	Ea.	214	16.25		230.25	261
6950	3/4"		20	.400		214	19.50		233.50	266
6960	1"		19	.421		330	20.50		350.50	395
6970	1-1/4"		15	.533		595	26		621	695
6980	1-1/2"		13	.615		895	30		925	1,025
6990	2"		11	.727		1,300	35.50		1,335.50	1,475
7100	For built-in by-pass or 10-35 psi, add					10.75			10.75	11.80
7700	High capacity, 250 psi to 25-75 psi, threaded									
7740	1/2"	1 Plum	24	.333	Ea.	214	16.25		230.25	261
7780	3/4"		20	.400		214	19.50		233.50	266
7790	1"		19	.421		330	20.50		350.50	395
7800	1-1/4"		15	.533		595	26		621	695
7810	1-1/2"		13	.615		895	30		925	1,025
7820	2"		11	.727		1,300	35.50		1,335.50	1,475
7830	2-1/2"		9	.889		1,850	43.50		1,893.50	2,125
7840	3"		8	1		2,025	49		2,074	2,300
7850	3" flanged (iron body)	Q-1	10	1.600		2,100	70		2,170	2,400
7860	4" flanged (iron body)	"	8	2		3,450	88		3,538	3,925
7920	For higher pressure, add					25%				
8000	Silent check, bronze trim									
8010	Compact wafer type, for 125 or 150 lb. flanges									
8020	1-1/2"	1 Plum	11	.727	Ea.	247	35.50		282.50	325
8021	2"	"	9	.889		261	43.50		304.50	355
8022	2-1/2"	Q-1	9	1.778		275	78		353	415
8023	3"		8	2		305	88		393	465
8024	4"		5	3.200		525	140		665	790
8025	5"	Q-2	6	4		725	182		907	1,075
8026	6"		5	4.800		820	218		1,038	1,225
8027	8"		4.50	5.333		1,750	243		1,993	2,300
8028	10"		4	6		3,200	273		3,473	3,925
8029	12"		3	8		7,250	365		7,615	8,525
8050	For 250 or 300 lb flanges, thru 6" no change									
8051	For 8" and 10", add				Ea.	40%	10%			
8060	Full flange wafer type, 150 lb.									
8061	1"	1 Plum	14	.571	Ea.	260	28		288	330
8062	1-1/4"		12	.667		271	32.50		303.50	345
8063	1-1/2"		11	.727		325	35.50		360.50	415
8064	2"		9	.889		440	43.50		483.50	550
8065	2-1/2"	Q-1	9	1.778		495	78		573	660

22 05 23.20 Valves, Bronze		Crew	Daily Output	Labor-Hours	Unit	2009 Bare Costs Material	Labor	Equipment	Total	Total Incl O&P
8066	3″	Q-1	8	2	Ea.	550	88		638	735
8067	4″		5	3.200		810	140		950	1,100
8068	5″	Q-2	6	4		1,050	182		1,232	1,425
8069	6″	″	5	4.800		1,375	218		1,593	1,850
8080	For 300 lb., add					40%	10%			
8100	Globe type, 150 lb.									
8110	2″	1 Plum	9	.889	Ea.	510	43.50		553.50	630
8111	2-1/2″	Q-1	9	1.778		630	78		708	810
8112	3″		8	2		755	88		843	960
8113	4″		5	3.200		1,050	140		1,190	1,350
8114	5″	Q-2	6	4		1,275	182		1,457	1,675
8115	6″	″	5	4.800		1,775	218		1,993	2,300
8130	For 300 lb., add					20%	10%			
8140	Screwed end type, 250 lb.									
8141	1/2″	1 Plum	24	.333	Ea.	46	16.25		62.25	75
8142	3/4″		20	.400		46	19.50		65.50	80
8143	1″		19	.421		52.50	20.50		73	89
8144	1-1/4″		15	.533		72.50	26		98.50	119
8145	1-1/2″		13	.615		81.50	30		111.50	135
8146	2″		11	.727		110	35.50		145.50	174
8350	Tempering, water, sweat connections									
8400	1/2″	1 Plum	24	.333	Ea.	71	16.25		87.25	103
8440	3/4″	″	20	.400	″	87	19.50		106.50	125
8650	Threaded connections									
8700	1/2″	1 Plum	24	.333	Ea.	87	16.25		103.25	120
8740	3/4″		20	.400		370	19.50		389.50	440
8750	1″		19	.421		410	20.50		430.50	480
8760	1-1/4″		15	.533		650	26		676	755
8770	1-1/2″		13	.615		695	30		725	810
8780	2″		11	.727		1,200	35.50		1,235.50	1,375
22 05 23.60 Valves, Plastic										
0010	**VALVES, PLASTIC** R220523-90									
1100	Angle, PVC, threaded									
1110	1/4″	1 Plum	26	.308	Ea.	61	15		76	89.50
1120	1/2″		26	.308		61	15		76	89.50
1130	3/4″		25	.320		72.50	15.60		88.10	104
1140	1″		23	.348		88	16.95		104.95	123
1150	Ball, PVC, socket or threaded, single union									
1230	1/2″	1 Plum	26	.308	Ea.	20.50	15		35.50	45
1240	3/4″		25	.320		23.50	15.60		39.10	49.50
1250	1″		23	.348		29.50	16.95		46.45	58
1260	1-1/4″		21	.381		39.50	18.55		58.05	71.50
1270	1-1/2″		20	.400		47	19.50		66.50	81.50
1280	2″		17	.471		68.50	23		91.50	110
1290	2-1/2″	Q-1	26	.615		288	27		315	355
1300	3″		24	.667		350	29.50		379.50	430
1310	4″		20	.800		590	35		625	705
1360	For PVC, flanged, add					100%	15%			
1450	Double union 1/2″	1 Plum	26	.308		30	15		45	55.50
1460	3/4″		25	.320		35.50	15.60		51.10	62.50
1470	1″		23	.348		42	16.95		58.95	72
1480	1-1/4″		21	.381		58.50	18.55		77.05	92.50

22 05 23 – General-Duty Valves for Plumbing Piping

22 05 23.60 Valves, Plastic		Crew	Daily Output	Labor-Hours	Unit	2009 Bare Costs Material	Labor	Equipment	Total	Total Incl O&P
1490	1-1/2"	1 Plum	20	.400	Ea.	70.50	19.50		90	107
1500	2"	↓	17	.471	↓	93	23		116	137
1650	CPVC, socket or threaded, single union									
1700	1/2"	1 Plum	26	.308	Ea.	39	15		54	65
1720	3/4"		25	.320		49	15.60		64.60	77
1730	1"		23	.348		58	16.95		74.95	89.50
1750	1-1/4"		21	.381		93	18.55		111.55	130
1760	1-1/2"		20	.400		93	19.50		112.50	132
1770	2"	↓	17	.471		128	23		151	176
1780	3"	Q-1	24	.667		380	29.50		409.50	465
1840	For CPVC, flanged, add					65%	15%			
1880	For true union, socket or threaded, add				↓	50%	5%			
2050	Polypropylene, threaded									
2100	1/4"	1 Plum	26	.308	Ea.	39	15		54	65.50
2120	3/8"		26	.308		39	15		54	65.50
2130	1/2"		26	.308		39	15		54	65.50
2140	3/4"		25	.320		49	15.60		64.60	77
2150	1"		23	.348		57.50	16.95		74.45	89
2160	1-1/4"		21	.381		84	18.55		102.55	121
2170	1-1/2"		20	.400		97	19.50		116.50	136
2180	2"	↓	17	.471		131	23		154	179
2190	3"	Q-1	24	.667		345	29.50		374.50	425
2200	4"	"	20	.800	↓	575	35		610	690
2550	PVC, three way, socket or threaded									
2600	1/2"	1 Plum	26	.308	Ea.	48	15		63	75
2640	3/4"		25	.320		54.50	15.60		70.10	83.50
2650	1"		23	.348		58.50	16.95		75.45	90
2660	1-1/2"		20	.400		119	19.50		138.50	161
2670	2"	↓	17	.471		159	23		182	209
2680	3"	Q-1	24	.667		345	29.50		374.50	425
2740	For flanged, add				↓	60%	15%			
3150	Ball check, PVC, socket or threaded									
3200	1/4"	1 Plum	26	.308	Ea.	32	15		47	58
3220	3/8"		26	.308		32	15		47	58
3240	1/2"		26	.308		33.50	15		48.50	59.50
3250	3/4"		25	.320		36	15.60		51.60	63
3260	1"		23	.348		45	16.95		61.95	75
3270	1-1/4"		21	.381		75.50	18.55		94.05	111
3280	1-1/2"		20	.400		75.50	19.50		95	113
3290	2"	↓	17	.471		103	23		126	148
3310	3"	Q-1	24	.667		285	29.50		314.50	360
3320	4"	"	20	.800		405	35		440	500
3360	For PVC, flanged, add				↓	50%	15%			
3750	CPVC, socket or threaded									
3800	1/2"	1 Plum	26	.308	Ea.	47	15		62	74
3840	3/4"		25	.320		56	15.60		71.60	85
3850	1"		23	.348		66	16.95		82.95	98
3860	1-1/2"		20	.400		114	19.50		133.50	155
3870	2"	↓	17	.471		153	23		176	204
3880	3"	Q-1	24	.667		415	29.50		444.50	500
3920	4"	"	20	.800		555	35		590	665
3930	For CPVC, flanged, add				↓	40%	15%			
4340	Polypropylene, threaded									

22 05 23 – General-Duty Valves for Plumbing Piping

22 05 23.60 Valves, Plastic		Crew	Daily Output	Labor-Hours	Unit	Material	2009 Bare Costs Labor	Equipment	Total	Total Incl O&P
4360	1/2"	1 Plum	26	.308	Ea.	33.50	15		48.50	59.50
4400	3/4"		25	.320		39	15.60		54.60	66.50
4440	1"		23	.348		50.50	16.95		67.45	81
4450	1-1/2"		20	.400		97	19.50		116.50	136
4460	2"		17	.471		122	23		145	169
4500	For polypropylene flanged, add					200%	15%			
4850	Foot valve, PVC, socket or threaded									
4900	1/2"	1 Plum	34	.235	Ea.	57	11.45		68.45	79.50
4930	3/4"		32	.250		64	12.20		76.20	89
4940	1"		28	.286		83	13.95		96.95	113
4950	1-1/4"		27	.296		160	14.45		174.45	198
4960	1-1/2"		26	.308		160	15		175	199
4970	2"		24	.333		186	16.25		202.25	229
4980	3"		20	.400		445	19.50		464.50	520
4990	4"		18	.444		780	21.50		801.50	890
5000	For flanged, add					25%	10%			
5050	CPVC, socket or threaded									
5060	1/2"	1 Plum	34	.235	Ea.	59.50	11.45		70.95	82.50
5070	3/4"		32	.250		69.50	12.20		81.70	95
5080	1"		28	.286		85	13.95		98.95	115
5090	1-1/4"		27	.296		136	14.45		150.45	171
5100	1-1/2"		26	.308		136	15		151	172
5110	2"		24	.333		174	16.25		190.25	217
5120	3"		20	.400		355	19.50		374.50	420
5130	4"		18	.444		645	21.50		666.50	745
5140	For flanged, add					25%	10%			
5280	Needle valve, PVC, threaded									
5300	1/4"	1 Plum	26	.308	Ea.	41	15		56	67.50
5340	3/8"		26	.308		47.50	15		62.50	74.50
5360	1/2"		26	.308		47.50	15		62.50	74.50
5380	For polypropylene, add					10%				
5800	Y check, PVC, socket or threaded									
5820	1/2"	1 Plum	26	.308	Ea.	68	15		83	97
5840	3/4"		25	.320		78	15.60		93.60	110
5850	1"		23	.348		79.50	16.95		96.45	113
5860	1-1/4"		21	.381		125	18.55		143.55	166
5870	1-1/2"		20	.400		137	19.50		156.50	180
5880	2"		17	.471		170	23		193	222
5890	2-1/2"		15	.533		360	26		386	435
5900	3"	Q-1	24	.667		425	29.50		454.50	515
5910	4"	"	20	.800		585	35		620	700
5960	For PVC flanged, add					45%	15%			
6350	Y sediment strainer, PVC, socket or threaded									
6400	1/2"	1 Plum	26	.308	Ea.	38.50	15		53.50	64.50
6440	3/4"		24	.333		41.50	16.25		57.75	70.50
6450	1"		23	.348		50.50	16.95		67.45	81
6460	1-1/4"		21	.381		83	18.55		101.55	120
6470	1-1/2"		20	.400		83	19.50		102.50	121
6480	2"		17	.471		102	23		125	147
6490	2-1/2"		15	.533		249	26		275	315
6500	3"	Q-1	24	.667		249	29.50		278.50	320
6510	4"	"	20	.800		415	35		450	510
6560	For PVC, flanged, add					55%	15%			

22 05 29 – Hangers and Supports for Plumbing Piping and Equipment

22 05 29.10 Hangers & Supp. for Plumb'g/HVAC Pipe/Equip.		Crew	Daily Output	Labor-Hours	Unit	2009 Bare Costs Material	Labor	Equipment	Total	Total Incl O&P
0010	HANGERS AND SUPPORTS FOR PLUMB'G/HVAC PIPE/EQUIP.									
0011	TYPE numbers per MSS-SP58									
0050	Brackets									
0060	Beam side or wall, malleable iron, TYPE 34									
0070	3/8" threaded rod size	1 Plum	48	.167	Ea.	2.19	8.15		10.34	14.60
0080	1/2" threaded rod size		48	.167		4.22	8.15		12.37	16.85
0090	5/8" threaded rod size		48	.167		5.60	8.15		13.75	18.40
0100	3/4" threaded rod size		48	.167		6.75	8.15		14.90	19.60
0110	7/8" threaded rod size		48	.167		7.85	8.15		16	21
0120	For concrete installation, add						30%			
0150	Wall, welded steel, medium, TYPE 32									
0160	0 size, 12" wide, 18" deep	1 Plum	34	.235	Ea.	177	11.45		188.45	211
0170	1 size, 18" wide, 24" deep		34	.235		210	11.45		221.45	248
0180	2 size, 24" wide, 30" deep		34	.235		278	11.45		289.45	320
0200	Beam attachment, welded, TYPE 22									
0202	3/8"	Q-15	80	.200	Ea.	8.15	8.80	.70	17.65	23
0203	1/2"		76	.211		8.15	9.25	.73	18.13	23.50
0204	5/8"		72	.222		8.55	9.75	.77	19.07	25
0205	3/4"		68	.235		9.35	10.30	.82	20.47	26.50
0206	7/8"		64	.250		13.55	10.95	.87	25.37	32.50
0207	1"		56	.286		19.35	12.55	.99	32.89	41.50
0300	Clamps									
0310	C-clamp, for mounting on steel beam flange, w/locknut, TYPE 23									
0320	3/8" threaded rod size	1 Plum	160	.050	Ea.	1.81	2.44		4.25	5.65
0330	1/2" threaded rod size		160	.050		2.30	2.44		4.74	6.20
0340	5/8" threaded rod size		160	.050		3.81	2.44		6.25	7.85
0350	3/4" threaded rod size		160	.050		5.80	2.44		8.24	10.05
0352	7/8" threaded rod size		140	.057		18.55	2.79		21.34	24.50
0400	High temperature to 1050° F, alloy steel									
0410	4" pipe size	Q-1	106	.151	Ea.	26.50	6.60		33.10	39
0420	6" pipe size		106	.151		44.50	6.60		51.10	59
0430	8" pipe size		97	.165		50.50	7.25		57.75	66.50
0440	10" pipe size		84	.190		68	8.35		76.35	87.50
0450	12" pipe size		72	.222		79	9.75		88.75	101
0460	14" pipe size		64	.250		165	10.95		175.95	198
0470	16" pipe size		56	.286		188	12.55		200.55	226
0480	Beam clamp, flange type, TYPE 25									
0482	For 3/8" bolt	1 Plum	48	.167	Ea.	3.68	8.15		11.83	16.25
0483	For 1/2" bolt		44	.182		6.05	8.85		14.90	19.95
0484	For 5/8" bolt		40	.200		6.05	9.75		15.80	21.50
0485	For 3/4" bolt		36	.222		6.05	10.85		16.90	23
0486	For 1" bolt		32	.250		6.05	12.20		18.25	25
0500	I-beam, for mounting on bottom flange, strap iron, TYPE 21									
0510	2" flange size	1 Plum	96	.083	Ea.	8.85	4.06		12.91	15.80
0520	3" flange size		95	.084		10.50	4.11		14.61	17.70
0530	4" flange size		93	.086		6.20	4.19		10.39	13.15
0540	5" flange size		92	.087		6.50	4.24		10.74	13.50
0550	6" flange size		90	.089		7.85	4.33		12.18	15.10
0560	7" flange size		88	.091		8.25	4.43		12.68	15.75
0570	8" flange size		86	.093		8.85	4.53		13.38	16.50
0600	One hole, vertical mounting, malleable iron									
0610	1/2" pipe size	1 Plum	160	.050	Ea.	.73	2.44		3.17	4.46

22 05 29.10 Hangers & Supp. for Plumb'g/HVAC Pipe/Equip.		Crew	Daily Output	Labor-Hours	Unit	2009 Bare Costs Material	Labor	Equipment	Total	Total Incl O&P
0620	3/4" pipe size	1 Plum	145	.055	Ea.	.88	2.69		3.57	5
0630	1" pipe size		136	.059		1.30	2.87		4.17	5.75
0640	1-1/4" pipe size		128	.063		1.46	3.05		4.51	6.20
0650	1-1/2" pipe size		120	.067		1.64	3.25		4.89	6.70
0660	2" pipe size		112	.071		2.37	3.48		5.85	7.85
0670	2-1/2" pipe size		104	.077		4.62	3.75		8.37	10.75
0680	3" pipe size		96	.083		6.55	4.06		10.61	13.30
0690	3-1/2" pipe size		90	.089		10.65	4.33		14.98	18.25
0700	4" pipe size	↓	84	.095	↓	13.75	4.64		18.39	22
0750	Riser or extension pipe, carbon steel, TYPE 8									
0756	1/2" pipe size	1 Plum	52	.154	Ea.	3.05	7.50		10.55	14.60
0760	3/4" pipe size		48	.167		3.05	8.15		11.20	15.55
0770	1" pipe size		47	.170		3.08	8.30		11.38	15.85
0780	1-1/4" pipe size		46	.174		3.84	8.50		12.34	16.90
0790	1-1/2" pipe size		45	.178		4.09	8.65		12.74	17.50
0800	2" pipe size		43	.186		4.25	9.05		13.30	18.30
0810	2-1/2" pipe size		41	.195		4.48	9.50		13.98	19.20
0820	3" pipe size		40	.200		4.74	9.75		14.49	19.85
0830	3-1/2" pipe size		39	.205		9.05	10		19.05	25
0840	4" pipe size		38	.211		6	10.25		16.25	22
0850	5" pipe size		37	.216		8.95	10.55		19.50	25.50
0860	6" pipe size		36	.222		10.25	10.85		21.10	27.50
0870	8" pipe size		34	.235		16.55	11.45		28	35.50
0880	10" pipe size		32	.250		23	12.20		35.20	43.50
0890	12" pipe size	↓	28	.286	↓	28	13.95		41.95	52
0900	For plastic coating 3/4" to 4", add					190%				
0910	For copper plating 3/4" to 4", add					58%				
0950	Two piece, complete, carbon steel, medium weight, TYPE 4									
0960	1/2" pipe size	Q-1	137	.117	Ea.	2.23	5.10		7.33	10.15
0970	3/4" pipe size		134	.119		2.23	5.25		7.48	10.30
0980	1" pipe size		132	.121		2.27	5.30		7.57	10.50
0990	1-1/4" pipe size		130	.123		2.94	5.40		8.34	11.35
1000	1-1/2" pipe size		126	.127		2.94	5.55		8.49	11.60
1010	2" pipe size		124	.129		3.42	5.65		9.07	12.25
1020	2-1/2" pipe size		120	.133		3.62	5.85		9.47	12.80
1030	3" pipe size		117	.137		4.02	6		10.02	13.40
1040	3-1/2" pipe size		114	.140		5.60	6.15		11.75	15.40
1050	4" pipe size		110	.145		5.60	6.40		12	15.75
1060	5" pipe size		106	.151		9.20	6.60		15.80	20
1070	6" pipe size		104	.154		14.05	6.75		20.80	25.50
1080	8" pipe size		100	.160		16.80	7		23.80	29
1090	10" pipe size		96	.167		30.50	7.30		37.80	44.50
1100	12" pipe size		89	.180		35	7.90		42.90	50.50
1110	14" pipe size		82	.195		67.50	8.55		76.05	87.50
1120	16" pipe size	↓	68	.235		73	10.30		83.30	96
1130	For galvanized, add				↓	45%				
1150	Insert, concrete									
1160	Wedge type, carbon steel body, malleable iron nut									
1170	1/4" threaded rod size	1 Plum	96	.083	Ea.	1.10	4.06		5.16	7.30
1180	3/8" threaded rod size		96	.083		1.15	4.06		5.21	7.35
1190	1/2" threaded rod size		96	.083		1.18	4.06		5.24	7.40
1200	5/8" threaded rod size		96	.083		1.20	4.06		5.26	7.40
1210	3/4" threaded rod size	↓	96	.083	↓	1.42	4.06		5.48	7.65

22 05 29 – Hangers and Supports for Plumbing Piping and Equipment

22 05 29.10 Hangers & Supp. for Plumb'g/HVAC Pipe/Equip.		Crew	Daily Output	Labor-Hours	Unit	2009 Bare Costs Material	Labor	Equipment	Total	Total Incl O&P
1220	7/8" threaded rod size	1 Plum	96	.083	Ea.	1.91	4.06		5.97	8.20
1230	For galvanized, add					.60			.60	.66
1250	Pipe guide sized for insulation									
1260	No. 1, 1" pipe size, 1" thick insulation	1 Stpi	26	.308	Ea.	118	15.20		133.20	153
1270	No. 2, 1-1/4"-2" pipe size, 1" thick insulation		23	.348		138	17.15		155.15	177
1280	No. 3, 1-1/4"-2" pipe size, 1-1/2" thick insulation		21	.381		138	18.80		156.80	179
1290	No. 4, 2-1/2"-3-1/2" pipe size, 1-1/2" thick insulation		18	.444		138	22		160	184
1300	No. 5, 4"-5" pipe size, 1-1/2" thick insulation		16	.500		155	24.50		179.50	208
1310	No. 6, 5"-6" pipe size, 2" thick insulation	Q-5	21	.762		171	34		205	239
1320	No. 7, 8" pipe size, 2" thick insulation		16	1		300	44.50		344.50	395
1330	No. 8, 10" pipe size, 2" thick insulation		12	1.333		570	59		629	720
1340	No. 9, 12" pipe size, 2" thick insulation	Q-6	17	1.412		570	65		635	730
1350	No. 10, 12"-14" pipe size, 2-1/2" thick insulation		16	1.500		745	69		814	925
1360	No. 11, 16" pipe size, 2-1/2" thick insulation		10.50	2.286		745	105		850	980
1370	No. 12, 16"-18" pipe size, 3" thick insulation		9	2.667		785	123		908	1,050
1380	No. 13, 20" pipe size, 3" thick insulation		7.50	3.200		785	147		932	1,075
1390	No. 14, 24" pipe size, 3" thick insulation		7	3.429		1,000	158		1,158	1,325
1400	Bands									
1410	Adjustable band, carbon steel, for non-insulated pipe, TYPE 7									
1420	1/2" pipe size	Q-1	142	.113	Ea.	.49	4.94		5.43	7.95
1430	3/4" pipe size		140	.114		.49	5		5.49	8.10
1440	1" pipe size		137	.117		.49	5.10		5.59	8.25
1450	1-1/4" pipe size		134	.119		.55	5.25		5.80	8.45
1460	1-1/2" pipe size		131	.122		.57	5.35		5.92	8.70
1470	2" pipe size		129	.124		.57	5.45		6.02	8.80
1480	2-1/2" pipe size		125	.128		1.02	5.60		6.62	9.55
1490	3" pipe size		122	.131		1.12	5.75		6.87	9.90
1500	3-1/2" pipe size		119	.134		1.83	5.90		7.73	10.85
1510	4" pipe size		114	.140		1.83	6.15		7.98	11.25
1520	5" pipe size		110	.145		3.04	6.40		9.44	12.95
1530	6" pipe size		108	.148		3.17	6.50		9.67	13.25
1540	8" pipe size		104	.154		3.79	6.75		10.54	14.30
1550	For copper plated, add					50%				
1560	For galvanized, add					30%				
1570	For plastic coating, add					30%				
1600	Adjusting nut malleable iron, steel band, TYPE 9									
1610	1/2" pipe size, galvanized band	Q-1	137	.117	Ea.	.86	5.10		5.96	8.65
1620	3/4" pipe size, galvanized band		135	.119		.86	5.20		6.06	8.75
1630	1" pipe size, galvanized band		132	.121		.90	5.30		6.20	9
1640	1-1/4" pipe size, galvanized band		129	.124		.90	5.45		6.35	9.15
1650	1-1/2" pipe size, galvanized band		126	.127		.90	5.55		6.45	9.35
1660	2" pipe size, galvanized band		124	.129		.95	5.65		6.60	9.55
1670	2-1/2" pipe size, galvanized band		120	.133		1.67	5.85		7.52	10.65
1680	3" pipe size, galvanized band		117	.137		1.67	6		7.67	10.85
1690	3-1/2" pipe size, galvanized band		114	.140		2.79	6.15		8.94	12.30
1700	4" pipe size, cadmium plated band		110	.145		2.88	6.40		9.28	12.75
1740	For plastic coated band, add					35%				
1750	For completely copper coated, add					45%				
1800	Clevis, adjustable, carbon steel, for non-insulated pipe, TYPE 1									
1810	1/2" pipe size	Q-1	137	.117	Ea.	1.05	5.10		6.15	8.85
1820	3/4" pipe size		135	.119		1.05	5.20		6.25	8.95
1830	1" pipe size		132	.121		1.09	5.30		6.39	9.20
1840	1-1/4" pipe size		129	.124		1.17	5.45		6.62	9.45

22 05 29.10 Hangers & Supp. for Plumb'g/HVAC Pipe/Equip.		Crew	Daily Output	Labor-Hours	Unit	2009 Bare Costs Material	Labor	Equipment	Total	Total Incl O&P
1850	1-1/2" pipe size	Q-1	126	.127	Ea.	1.24	5.55		6.79	9.70
1860	2" pipe size		124	.129		1.48	5.65		7.13	10.15
1870	2-1/2" pipe size		120	.133		2.33	5.85		8.18	11.35
1880	3" pipe size		117	.137		2.90	6		8.90	12.20
1890	3-1/2" pipe size		114	.140		3.08	6.15		9.23	12.65
1900	4" pipe size		110	.145		3.54	6.40		9.94	13.50
1910	5" pipe size		106	.151		4.85	6.60		11.45	15.30
1920	6" pipe size		104	.154		5.65	6.75		12.40	16.35
1930	8" pipe size		100	.160		9.30	7		16.30	21
1940	10" pipe size		96	.167		16.75	7.30		24.05	29.50
1950	12" pipe size		89	.180		19.95	7.90		27.85	34
1960	14" pipe size		82	.195		32.50	8.55		41.05	48.50
1970	16" pipe size		68	.235		46	10.30		56.30	66.50
1971	18" pipe size		54	.296		57.50	13		70.50	82.50
1972	20" pipe size	↓	38	.421	↓	120	18.50		138.50	161
1980	For galvanized, add					66%				
1990	For copper plated 1/2" to 4", add					77%				
2000	For light weight 1/2" to 4", deduct					13%				
2010	Insulated pipe type, 3/4" to 12" pipe, add					180%				
2020	Insulated pipe type, chrome-moly U-strap, add					530%				
2250	Split ring, malleable iron, for non-insulated pipe, TYPE 11									
2260	1/2" pipe size	Q-1	137	.117	Ea.	5.60	5.10		10.70	13.85
2270	3/4" pipe size		135	.119		3.79	5.20		8.99	11.95
2280	1" pipe size		132	.121		4.01	5.30		9.31	12.40
2290	1-1/4" pipe size		129	.124		4.98	5.45		10.43	13.65
2300	1-1/2" pipe size		126	.127		6.05	5.55		11.60	15
2310	2" pipe size		124	.129		6.85	5.65		12.50	16.05
2320	2-1/2" pipe size		120	.133		9.95	5.85		15.80	19.70
2330	3" pipe size		117	.137		12.35	6		18.35	22.50
2340	3-1/2" pipe size		114	.140		13.35	6.15		19.50	24
2350	4" pipe size		110	.145		12.90	6.40		19.30	24
2360	5" pipe size		106	.151		17.15	6.60		23.75	29
2370	6" pipe size		104	.154		38.50	6.75		45.25	52
2380	8" pipe size	↓	100	.160	↓	62.50	7		69.50	79
2390	For copper plated, add					8%				
2500	Washer, flat steel									
2502	3/8"	1 Plum	240	.033	Ea.	.10	1.62		1.72	2.55
2503	1/2"		220	.036		.14	1.77		1.91	2.81
2504	5/8"		200	.040		.35	1.95		2.30	3.32
2505	3/4"		180	.044		.51	2.17		2.68	3.81
2506	7/8"		160	.050		1	2.44		3.44	4.76
2507	1"		140	.057		1.10	2.79		3.89	5.40
2508	1-1/4"	↓	120	.067	↓	1.36	3.25		4.61	6.40
2520	Nut, steel, hex									
2522	3/8"	1 Plum	200	.040	Ea.	.33	1.95		2.28	3.29
2523	1/2"		180	.044		.47	2.17		2.64	3.77
2524	5/8"		160	.050		.99	2.44		3.43	4.75
2525	3/4"		140	.057		1.45	2.79		4.24	5.80
2526	7/8"		120	.067		1.83	3.25		5.08	6.90
2527	1"		100	.080		2.37	3.90		6.27	8.45
2528	1-1/4"	↓	80	.100	↓	2.98	4.88		7.86	10.60
2532	Turnbuckle, TYPE 13									
2534	3/8"	1 Plum	80	.100	Ea.	14.45	4.88		19.33	23

22 05 29 – Hangers and Supports for Plumbing Piping and Equipment

22 05 29.10	Hangers & Supp. for Plumb'g/HVAC Pipe/Equip.	Crew	Daily Output	Labor-Hours	Unit	2009 Bare Costs Material	Labor	Equipment	Total	Total Incl O&P
2535	1/2"	1 Plum	72	.111	Ea.	22	5.40		27.40	32
2536	5/8"		64	.125		25	6.10		31.10	36.50
2537	3/4"		56	.143		38	6.95		44.95	52
2538	7/8"		48	.167		47	8.15		55.15	63.50
2539	1"		40	.200		65	9.75		74.75	86
2540	1-1/4"		32	.250		75.50	12.20		87.70	101
2544	Eye rod, welded									
2546	3/8"	1 Plum	56	.143	Ea.	10.70	6.95		17.65	22
2547	1/2"		48	.167		15.85	8.15		24	29.50
2548	5/8"		40	.200		24	9.75		33.75	41
2549	3/4"		32	.250		28	12.20		40.20	49.50
2550	7/8"		24	.333		37	16.25		53.25	65.50
2551	1"		16	.500		47	24.50		71.50	88
2556	Eye rod, welded, linked									
2558	3/8"	1 Plum	56	.143	Ea.	61.50	6.95		68.45	78
2559	1/2"		48	.167		82.50	8.15		90.65	103
2560	5/8"		40	.200		118	9.75		127.75	145
2561	3/4"		32	.250		138	12.20		150.20	170
2562	7/8"		24	.333		190	16.25		206.25	234
2563	1"		16	.500		241	24.50		265.50	300
2650	Rods, carbon steel									
2660	Continuous thread									
2670	1/4" thread size	1 Plum	144	.056	L.F.	.14	2.71		2.85	4.21
2680	3/8" thread size		144	.056		.23	2.71		2.94	4.31
2690	1/2" thread size		144	.056		.46	2.71		3.17	4.57
2700	5/8" thread size		144	.056		.73	2.71		3.44	4.86
2710	3/4" thread size		144	.056		1.10	2.71		3.81	5.25
2720	7/8" thread size		144	.056		1.78	2.71		4.49	6
2721	1" thread size	Q-1	160	.100		2.37	4.39		6.76	9.20
2722	1-1/8" thread size	"	120	.133		11.50	5.85		17.35	21.50
2725	1/4" thread size, bright finish	1 Plum	144	.056		1.21	2.71		3.92	5.40
2726	1/2" thread size, bright finish	"	144	.056		2.07	2.71		4.78	6.35
2730	For galvanized, add					40%				
2750	Both ends machine threaded 18" length									
2760	3/8" thread size	1 Plum	240	.033	Ea.	3.94	1.62		5.56	6.75
2770	1/2" thread size		240	.033		7.55	1.62		9.17	10.75
2780	5/8" thread size		240	.033		10.85	1.62		12.47	14.40
2790	3/4" thread size		240	.033		16.70	1.62		18.32	21
2800	7/8" thread size		240	.033		20	1.62		21.62	25
2810	1" thread size		240	.033		28	1.62		29.62	33
2820	Rod couplings									
2821	3/8"	1 Plum	60	.133	Ea.	1.25	6.50		7.75	11.15
2822	1/2"		54	.148		1.67	7.20		8.87	12.70
2823	5/8"		48	.167		1.91	8.15		10.06	14.30
2824	3/4"		44	.182		2.61	8.85		11.46	16.15
2825	7/8"		40	.200		4.37	9.75		14.12	19.45
2826	1"		34	.235		5.90	11.45		17.35	23.50
2827	1-1/8"		30	.267		18.95	13		31.95	40.50
2860	Pipe hanger assy, adj. clevis, saddle, rod, clamp, insul. allowance									
2864	1/2" pipe size	Q-5	35	.457	Ea.	29	20.50		49.50	62.50
2866	3/4" pipe size		34.80	.460		30	20.50		50.50	63.50
2868	1" pipe size		34.60	.462		27	20.50		47.50	61
2869	1-1/4" pipe size		34.30	.466		27.50	20.50		48	61

22 05 29.10 Hangers & Supp. for Plumb'g/HVAC Pipe/Equip.		Crew	Daily Output	Labor-Hours	Unit	2009 Bare Costs Material	Labor	Equipment	Total	Total Incl O&P
2870	1-1/2" pipe size	Q-5	33.90	.472	Ea.	28.50	21		49.50	62.50
2872	2" pipe size		33.30	.480		29.50	21.50		51	64.50
2874	2-1/2" pipe size		32.30	.495		36.50	22		58.50	73.50
2876	3" pipe size		31.20	.513		42.50	23		65.50	80.50
2880	4" pipe size		30.70	.521		45.50	23		68.50	85
2884	6" pipe size		29.80	.537		55.50	24		79.50	97
2888	8" pipe size		28	.571		78	25.50		103.50	124
2892	10" pipe size		25.20	.635		86	28		114	137
2896	12" pipe size		23.20	.690		138	30.50		168.50	198
2900	Rolls									
2910	Adjustable yoke, carbon steel with CI roll, TYPE 43									
2918	2" pipe size	Q-1	140	.114	Ea.	8.95	5		13.95	17.40
2920	2-1/2" pipe size		137	.117		10.60	5.10		15.70	19.35
2930	3" pipe size		131	.122		11.85	5.35		17.20	21
2940	3-1/2" pipe size		124	.129		15.75	5.65		21.40	26
2950	4" pipe size		117	.137		15.75	6		21.75	26.50
2960	5" pipe size		110	.145		18.40	6.40		24.80	29.50
2970	6" pipe size		104	.154		23	6.75		29.75	35
2980	8" pipe size		96	.167		34	7.30		41.30	48.50
2990	10" pipe size		80	.200		40.50	8.80		49.30	58
3000	12" pipe size		68	.235		60	10.30		70.30	81.50
3010	14" pipe size		56	.286		166	12.55		178.55	201
3020	16" pipe size		48	.333		213	14.65		227.65	256
3050	Chair, carbon steel with CI roll									
3060	2" pipe size	1 Plum	68	.118	Ea.	13.95	5.75		19.70	24
3070	2-1/2" pipe size		65	.123		15.10	6		21.10	25.50
3080	3" pipe size		62	.129		15.45	6.30		21.75	26.50
3090	3-1/2" pipe size		60	.133		19.75	6.50		26.25	31.50
3100	4" pipe size		58	.138		22	6.70		28.70	34.50
3110	5" pipe size		56	.143		24	6.95		30.95	37
3120	6" pipe size		53	.151		33.50	7.35		40.85	48
3130	8" pipe size		50	.160		45	7.80		52.80	61
3140	10" pipe size		48	.167		56.50	8.15		64.65	74.50
3150	12" pipe size		46	.174		82.50	8.50		91	103
3170	Single pipe roll, (see line 2650 for rods), TYPE 41, 1" pipe size	Q-1	137	.117		11.20	5.10		16.30	20
3180	1-1/4" pipe size		131	.122		11.20	5.35		16.55	20.50
3190	1-1/2" pipe size		129	.124		11.20	5.45		16.65	20.50
3200	2" pipe size		124	.129		11.90	5.65		17.55	21.50
3210	2-1/2" pipe size		118	.136		12.80	5.95		18.75	23
3220	3" pipe size		115	.139		13.50	6.10		19.60	24
3230	3-1/2" pipe size		113	.142		15.70	6.20		21.90	26.50
3240	4" pipe size		112	.143		17.15	6.25		23.40	28.50
3250	5" pipe size		110	.145		19.05	6.40		25.45	30.50
3260	6" pipe size		101	.158		24.50	6.95		31.45	37.50
3270	8" pipe size		90	.178		38.50	7.80		46.30	53.50
3280	10" pipe size		80	.200		45.50	8.80		54.30	63
3290	12" pipe size		68	.235		65.50	10.30		75.80	87.50
3291	14" pipe size		56	.286		109	12.55		121.55	138
3292	16" pipe size		48	.333		147	14.65		161.65	184
3293	18" pipe size		40	.400		159	17.55		176.55	201
3294	20" pipe size		35	.457		217	20		237	269
3296	24" pipe size		30	.533		345	23.50		368.50	415
3297	30" pipe size		25	.640		680	28		708	790

22 05 29.10 Hangers & Supp. for Plumb'g/HVAC Pipe/Equip.		Crew	Daily Output	Labor-Hours	Unit	2009 Bare Costs Material	Labor	Equipment	Total	Total Incl O&P
3298	36" pipe size	Q-1	20	.800	Ea.	835	35		870	975
3300	Saddles (add vertical pipe riser, usually 3" diameter)									
3310	Pipe support, complete, adjust., CI saddle, TYPE 36									
3320	2-1/2" pipe size	1 Plum	96	.083	Ea.	118	4.06		122.06	135
3330	3" pipe size		88	.091		119	4.43		123.43	138
3340	3-1/2" pipe size		79	.101		120	4.94		124.94	139
3350	4" pipe size		68	.118		156	5.75		161.75	180
3360	5" pipe size		64	.125		158	6.10		164.10	183
3370	6" pipe size		59	.136		160	6.60		166.60	187
3380	8" pipe size		53	.151		170	7.35		177.35	198
3390	10" pipe size		50	.160		188	7.80		195.80	219
3400	12" pipe size		48	.167		195	8.15		203.15	227
3450	For standard pipe support, one piece, CI deduct					34%				
3460	For stanchion support, CI with steel yoke, deduct					60%				
3550	Insulation shield 1" thick, 1/2" pipe size, TYPE 40	1 Asbe	100	.080	Ea.	4.89	3.53		8.42	10.95
3560	3/4" pipe size		100	.080		6.10	3.53		9.63	12.25
3570	1" pipe size		98	.082		6.10	3.60		9.70	12.35
3580	1-1/4" pipe size		98	.082		6.50	3.60		10.10	12.80
3590	1-1/2" pipe size		96	.083		6.50	3.67		10.17	12.90
3600	2" pipe size		96	.083		6.75	3.67		10.42	13.15
3610	2-1/2" pipe size		94	.085		6.95	3.75		10.70	13.55
3620	3" pipe size		94	.085		7.45	3.75		11.20	14.05
3630	2" thick, 3-1/2" pipe size		92	.087		8.30	3.83		12.13	15.15
3640	4" pipe size		92	.087		9.10	3.83		12.93	16
3650	5" pipe size		90	.089		14.85	3.92		18.77	22.50
3660	6" pipe size		90	.089		16.95	3.92		20.87	25
3670	8" pipe size		88	.091		23.50	4.01		27.51	32
3680	10" pipe size		88	.091		34.50	4.01		38.51	44.50
3690	12" pipe size		86	.093		39.50	4.10		43.60	50
3700	14" pipe size		86	.093		79.50	4.10		83.60	94
3710	16" pipe size		84	.095		84.50	4.20		88.70	99.50
3720	18" pipe size		84	.095		93.50	4.20		97.70	110
3730	20" pipe size		82	.098		103	4.30		107.30	120
3732	24" pipe size		80	.100		125	4.41		129.41	144
3733	30" pipe size		78	.103		390	4.52		394.52	435
3735	36" pipe size		75	.107		590	4.70		594.70	655
3750	Covering protection saddle, TYPE 39									
3760	1" covering size									
3770	3/4" pipe size	1 Plum	68	.118	Ea.	8.55	5.75		14.30	18
3780	1" pipe size		68	.118		8.55	5.75		14.30	18
3790	1-1/4" pipe size		68	.118		8.55	5.75		14.30	18
3800	1-1/2" pipe size		66	.121		9.30	5.90		15.20	19.05
3810	2" pipe size		66	.121		9.30	5.90		15.20	19.05
3820	2-1/2" pipe size		64	.125		9.30	6.10		15.40	19.35
3830	3" pipe size		64	.125		13.35	6.10		19.45	24
3840	3-1/2" pipe size		62	.129		14.25	6.30		20.55	25
3850	4" pipe size		62	.129		14.25	6.30		20.55	25
3860	5" pipe size		60	.133		14.25	6.50		20.75	25.50
3870	6" pipe size		60	.133		17	6.50		23.50	28.50
3900	1-1/2" covering size									
3910	3/4" pipe size	1 Plum	68	.118	Ea.	10.70	5.75		16.45	20.50
3920	1" pipe size		68	.118		10.70	5.75		16.45	20.50
3930	1-1/4" pipe size		68	.118		10.70	5.75		16.45	20.50

22 05 29.10 Hangers & Supp. for Plumb'g/HVAC Pipe/Equip.		Crew	Daily Output	Labor-Hours	Unit	Material	2009 Bare Costs Labor	Equipment	Total	Total Incl O&P
3940	1-1/2" pipe size	1 Plum	66	.121	Ea.	10.70	5.90		16.60	20.50
3950	2" pipe size		66	.121		11.60	5.90		17.50	21.50
3960	2-1/2" pipe size		64	.125		13.70	6.10		19.80	24
3970	3" pipe size		64	.125		13.70	6.10		19.80	24
3980	3-1/2" pipe size		62	.129		21.50	6.30		27.80	33
3990	4" pipe size		62	.129		13.90	6.30		20.20	24.50
4000	5" pipe size		60	.133		13.90	6.50		20.40	25
4010	6" pipe size		60	.133		19.90	6.50		26.40	32
4020	8" pipe size		58	.138		25	6.70		31.70	37.50
4022	10" pipe size		56	.143		25	6.95		31.95	38
4024	12" pipe size	↓	54	.148	↓	36.50	7.20		43.70	51.50
4028	2" covering size									
4029	2-1/2" pipe size	1 Plum	62	.129	Ea.	13.15	6.30		19.45	24
4032	3" pipe size		60	.133		15.45	6.50		21.95	26.50
4033	4" pipe size		58	.138		15.45	6.70		22.15	27
4034	6" pipe size		56	.143		22	6.95		28.95	35
4035	8" pipe size		54	.148		25	7.20		32.20	38.50
4080	10" pipe size		58	.138		25	6.70		31.70	37.50
4090	12" pipe size		56	.143		41	6.95		47.95	55.50
4100	14" pipe size		56	.143		41	6.95		47.95	55.50
4110	16" pipe size		54	.148		58	7.20		65.20	74.50
4120	18" pipe size		54	.148		58	7.20		65.20	74.50
4130	20" pipe size		52	.154		64	7.50		71.50	82
4150	24" pipe size		50	.160		75	7.80		82.80	94
4160	30" pipe size		48	.167		82	8.15		90.15	103
4180	36" pipe size	↓	45	.178	↓	90.50	8.65		99.15	113
4186	2-1/2" covering size									
4187	3" pipe size	1 Plum	58	.138	Ea.	15.05	6.70		21.75	26.50
4188	4" pipe size		56	.143		16.35	6.95		23.30	28.50
4189	6" pipe size		52	.154		25	7.50		32.50	39
4190	8" pipe size		48	.167		30.50	8.15		38.65	45.50
4191	10" pipe size		44	.182		33.50	8.85		42.35	50
4192	12" pipe size		40	.200		44.50	9.75		54.25	63.50
4193	14" pipe size		36	.222		44.50	10.85		55.35	65.50
4194	16" pipe size		32	.250		65.50	12.20		77.70	90.50
4195	18" pipe size	↓	28	.286	↓	70	13.95		83.95	98.50
4200	Sockets									
4210	Rod end, malleable iron, TYPE 16									
4220	1/4" thread size	1 Plum	240	.033	Ea.	1.18	1.62		2.80	3.74
4230	3/8" thread size		240	.033		1.27	1.62		2.89	3.84
4240	1/2" thread size		230	.035		1.52	1.70		3.22	4.21
4250	5/8" thread size		225	.036		3	1.73		4.73	5.90
4260	3/4" thread size		220	.036		4.89	1.77		6.66	8.05
4270	7/8" thread size	↓	210	.038		9.60	1.86		11.46	13.35
4290	Strap, 1/2" pipe size, TYPE 26	Q-1	142	.113		2.86	4.94		7.80	10.55
4300	3/4" pipe size		140	.114		2.92	5		7.92	10.75
4310	1" pipe size		137	.117		2.98	5.10		8.08	11
4320	1-1/4" pipe size		134	.119		3.35	5.25		8.60	11.55
4330	1-1/2" pipe size		131	.122		4.16	5.35		9.51	12.65
4340	2" pipe size		129	.124		4.31	5.45		9.76	12.90
4350	2-1/2" pipe size		125	.128		7.50	5.60		13.10	16.70
4360	3" pipe size		122	.131		8.25	5.75		14	17.70
4370	3-1/2" pipe size	↓	119	.134	↓	9.25	5.90		15.15	19

22 05 Common Work Results for Plumbing

22 05 29 – Hangers and Supports for Plumbing Piping and Equipment

	22 05 29.10 Hangers & Supp. for Plumb'g/HVAC Pipe/Equip.	Crew	Daily Output	Labor-Hours	Unit	2009 Bare Costs Material	Labor	Equipment	Total	Total Incl O&P
4380	4" pipe size	Q-1	114	.140	Ea.	10.30	6.15		16.45	20.50
4400	U-bolt, carbon steel									
4410	Standard, with nuts, TYPE 42									
4420	1/2" pipe size	1 Plum	160	.050	Ea.	1.28	2.44		3.72	5.05
4430	3/4" pipe size		158	.051		1.28	2.47		3.75	5.10
4450	1" pipe size		152	.053		1.43	2.57		4	5.40
4460	1-1/4" pipe size		148	.054		1.65	2.63		4.28	5.75
4470	1-1/2" pipe size		143	.056		1.81	2.73		4.54	6.10
4480	2" pipe size		139	.058		1.85	2.81		4.66	6.25
4490	2-1/2" pipe size		134	.060		3.06	2.91		5.97	7.75
4500	3" pipe size		128	.063		3.21	3.05		6.26	8.10
4510	3-1/2" pipe size		122	.066		3.36	3.20		6.56	8.50
4520	4" pipe size		117	.068		3.39	3.33		6.72	8.75
4530	5" pipe size		114	.070		3.70	3.42		7.12	9.20
4540	6" pipe size		111	.072		6.45	3.51		9.96	12.35
4550	8" pipe size		109	.073		7.25	3.58		10.83	13.30
4560	10" pipe size		107	.075		12.90	3.65		16.55	19.65
4570	12" pipe size		104	.077		17.25	3.75		21	24.50
4580	For plastic coating on 1/2" thru 6" size, add					150%				
4700	U-hook, carbon steel, requires mounting screws or bolts									
4710	3/4" thru 2" pipe size									
4720	6" long	1 Plum	96	.083	Ea.	.39	4.06		4.45	6.55
4730	8" long		96	.083		.51	4.06		4.57	6.65
4740	10" long		96	.083		.58	4.06		4.64	6.75
4750	12" long		96	.083		.66	4.06		4.72	6.85
4760	For copper plated, add					50%				
8000	Pipe clamp, plastic, 1/2" CTS	1 Plum	80	.100		.22	4.88		5.10	7.55
8010	3/4" CTS		73	.110		.23	5.35		5.58	8.25
8020	1" CTS		68	.118		.52	5.75		6.27	9.15
8080	Economy clamp, 1/4" CTS		175	.046		.05	2.23		2.28	3.40
8090	3/8" CTS		168	.048		.07	2.32		2.39	3.56
8100	1/2" CTS		160	.050		.07	2.44		2.51	3.74
8110	3/4" CTS		145	.055		.14	2.69		2.83	4.19
8200	Half clamp, 1/2" CTS		80	.100		.08	4.88		4.96	7.40
8210	3/4" CTS		73	.110		.12	5.35		5.47	8.15
8300	Suspension clamp, 1/2" CTS		80	.100		.22	4.88		5.10	7.55
8310	3/4" CTS		73	.110		.23	5.35		5.58	8.25
8320	1" CTS		68	.118		.52	5.75		6.27	9.15
8400	Insulator, 1/2" CTS		80	.100		.36	4.88		5.24	7.70
8410	3/4" CTS		73	.110		.37	5.35		5.72	8.40
8420	1" CTS		68	.118		.38	5.75		6.13	9
8500	J hook clamp with nail 1/2" CTS		240	.033		.13	1.62		1.75	2.58
8501	3/4" CTS		240	.033		.13	1.62		1.75	2.58
8800	Wire cable support system									
8810	Cable with hook terminal and locking device									
8830	2 mm, (.079 ") dia cable, (100 lb. cap.)									
8840	1 m, (3.3') length, with hook	1 Shee	96	.083	Ea.	2.36	3.93		6.29	8.65
8850	2 m, (6.6') length, with hook		84	.095		2.72	4.50		7.22	9.90
8860	3 m, (9.9') length, with hook		72	.111		3.09	5.25		8.34	11.45
8870	5 m, (16.4') length, with hook	Q-9	60	.267		3.87	11.35		15.22	21.50
8880	10 m, (32.8') length, with hook	"	30	.533		5.60	22.50		28.10	41
8900	3mm, (.118 ") dia cable, (200 lb. cap.)									
8910	1 m, (3.3') length, with hook	1 Shee	96	.083	Ea.	3.02	3.93		6.95	9.35

22 05 29 – Hangers and Supports for Plumbing Piping and Equipment

22 05 29.10 Hangers & Supp. for Plumb'g/HVAC Pipe/Equip.		Crew	Daily Output	Labor-Hours	Unit	2009 Bare Costs Material	Labor	Equipment	Total	Total Incl O&P
8920	2 m, (6.6') length, with hook	1 Shee	84	.095	Ea.	3.40	4.50		7.90	10.65
8930	3 m, (9.9') length, with hook		72	.111		3.62	5.25		8.87	12.05
8940	5 m, (16.4') length, with hook	Q-9	60	.267		4.62	11.35		15.97	22.50
8950	10 m, (32.8') length, with hook	"	30	.533		6.65	22.50		29.15	42.50
9000	Cable system accessories									
9010	Anchor bolt, 3/8", with nut	1 Shee	140	.057	Ea.	.90	2.70		3.60	5.15
9020	Air duct corner protector		160	.050		.41	2.36		2.77	4.08
9030	Air duct support attachment		140	.057		.15	2.70		2.85	4.32
9040	Flange clip, hammer-on style									
9044	For flange thickness 3/32" - 9/64", 160 lb. cap.	1 Shee	180	.044	Ea.	.21	2.10		2.31	3.46
9048	For flange thickness 1/8" - 1/4", 200 lb. cap.		160	.050		.21	2.36		2.57	3.86
9052	For flange thickness 5/16" - 1/2", 200 lb. cap.		150	.053		.44	2.52		2.96	4.35
9056	For flange thickness 9/16" - 3/4", 200 lb. cap.		140	.057		.58	2.70		3.28	4.79
9060	Wire insulation protection tube		180	.044	L.F.	.22	2.10		2.32	3.47
9070	Wire cutter				Ea.	25			25	27.50

22 05 48 – Vibration and Seismic Controls for Plumbing Piping and Equipment

22 05 48.40 Vibration Absorbers		Crew	Daily Output	Labor-Hours	Unit	Material	Labor	Equipment	Total	Total Incl O&P
0010	**VIBRATION ABSORBERS**									
0100	Hangers, neoprene flex									
0200	10 – 120 lb. capacity				Ea.	16.55			16.55	18.20
0220	75 – 550 lb. capacity					21.50			21.50	23.50
0240	250 – 1100 lb. capacity					44			44	48
0260	1000 – 4000 lb. capacity					69.50			69.50	76
0500	Spring flex, 60 lb. capacity					23			23	25.50
0520	450 lb. capacity					29			29	31.50
0540	900 lb. capacity					38.50			38.50	42
0560	1100 – 1300 lb. capacity					38.50			38.50	42
0600	Rubber in shear									
0610	45 – 340 lb., up to 1/2" rod size	1 Stpi	22	.364	Ea.	14.30	17.95		32.25	43
0620	130 – 700 lb., up to 3/4" rod size		20	.400		31	19.75		50.75	63.50
0630	50 – 1000 lb., up to 3/4" rod size		18	.444		36.50	22		58.50	73
1000	Mounts, neoprene, 45 – 380 lb. capacity					11			11	12.10
1020	250 – 1100 lb. capacity					29			29	31.50
1040	1000 – 4000 lb. capacity					64.50			64.50	71
1100	Spring flex, 60 lb. capacity					45			45	49.50
1120	165 lb. capacity					45			45	49.50
1140	260 lb. capacity					45			45	49.50
1160	450 lb. capacity					59.50			59.50	65.50
1180	600 lb. capacity					59.50			59.50	65.50
1200	750 lb. capacity					59.50			59.50	65.50
1220	900 lb. capacity					62			62	68
1240	1100 lb. capacity					65.50			65.50	72
1260	1300 lb. capacity					65.50			65.50	72
1280	1500 lb. capacity					96			96	105
1300	1800 lb. capacity					103			103	113
1320	2200 lb. capacity					101			101	111
1340	2600 lb. capacity					103			103	113
1399	Spring type									
1400	2 Piece									
1410	50 – 1000 lb.	1 Stpi	12	.667	Ea.	130	33		163	193
1420	1100 – 1600 lb.	"	12	.667	"	141	33		174	205
1500	Double spring open									

22 05 48 – Vibration and Seismic Controls for Plumbing Piping and Equipment

22 05 48.40 Vibration Absorbers		Crew	Daily Output	Labor-Hours	Unit	2009 Bare Costs Material	Labor	Equipment	Total	Total Incl O&P
1510	150 – 450 lb.	1 Stpi	24	.333	Ea.	89	16.45		105.45	123
1520	500 – 1000 lb.		24	.333		89	16.45		105.45	123
1530	1100 – 1600 lb.		24	.333		89	16.45		105.45	123
1540	1700 – 2400 lb.		24	.333		97	16.45		113.45	131
1550	2500 – 3400 lb.		24	.333		165	16.45		181.45	207
2000	Pads, cork rib, 18" x 18" x 1", 10-50 psi					104			104	114
2020	18" x 36" x 1", 10-50 psi					208			208	229
2100	Shear flexible pads, 18" x 18" x 3/8", 20-70 psi					49.50			49.50	54
2120	18" x 36" x 3/8", 20-70 psi					104			104	114
2150	Laminated neoprene and cork									
2160	1" thick	1 Stpi	16	.500	S.F.	54	24.50		78.50	96.50
3000	Note overlap in capacities due to deflections									

22 05 53 – Identification for Plumbing Piping and Equipment

22 05 53.10 Piping System Identification Labels		Crew	Daily Output	Labor-Hours	Unit	Material	Labor	Equipment	Total	Total Incl O&P
0010	**PIPING SYSTEM IDENTIFICATION LABELS**									
0100	Indicate contents and flow direction									
0106	Pipe markers									
0110	Plastic snap around									
0114	1/2" pipe	1 Plum	80	.100	Ea.	6	4.88		10.88	13.90
0116	3/4" pipe		80	.100		6	4.88		10.88	13.90
0118	1" pipe		80	.100		6	4.88		10.88	13.90
0120	2" pipe		75	.107		7.10	5.20		12.30	15.60
0122	3" pipe		70	.114		11	5.55		16.55	20.50
0124	4" pipe		60	.133		11	6.50		17.50	22
0126	6" pipe		60	.133		11.60	6.50		18.10	22.50
0128	8" pipe		56	.143		15.40	6.95		22.35	27.50
0130	10" pipe		56	.143		15.40	6.95		22.35	27.50
0200	Over 10" pipe size		50	.160		18.70	7.80		26.50	32
1110	Self adhesive									
1114	1" pipe	1 Plum	80	.100	Ea.	3.86	4.88		8.74	11.55
1116	2" pipe		75	.107		5.10	5.20		10.30	13.40
1118	3" pipe		70	.114		9.60	5.55		15.15	18.90
1120	4" pipe		60	.133		9.60	6.50		16.10	20.50
1122	6" pipe		60	.133		10.70	6.50		17.20	21.50
1124	8" pipe		56	.143		14.60	6.95		21.55	26.50
1126	10" pipe		56	.143		14.60	6.95		21.55	26.50
1200	Over 10" pipe size		50	.160		17.80	7.80		25.60	31.50
2000	Valve tags									
2010	Numbered plus identifying legend									
2100	Brass, 2" diameter	1 Plum	40	.200	Ea.	2.70	9.75		12.45	17.60
2200	Plastic, 1-1/2" Dia.	"	40	.200	"	3.10	9.75		12.85	18.05

22 07 Plumbing Insulation

22 07 16 – Plumbing Equipment Insulation

22 07 16.10 Insulation for Plumbing Equipment		Crew	Daily Output	Labor-Hours	Unit	2009 Bare Costs Material	Labor	Equipment	Total	Total Incl O&P
0010	**INSULATION FOR PLUMBING EQUIPMENT**									
2900	Domestic water heater wrap kit									
2920	1-1/2" with vinyl jacket, 20-60 gal. G	1 Plum	8	1	Ea.	16.05	49		65.05	90.50

22 07 19 – Plumbing Piping Insulation

22 07 19.10 Piping Insulation		Crew	Daily Output	Labor-Hours	Unit	Material	Labor	Equipment	Total	Total Incl O&P
0010	**PIPING INSULATION**									
0100	Rule of thumb, as a percentage of total mechanical costs				Job				10%	
0110	Insulation req'd is based on the surface size/area to be covered									
2930	Insulated protectors, (ADA)									
2935	For exposed piping under sinks or lavatories									
2940	Vinyl coated foam, velcro tabs									
2945	P Trap, 1-1/4" or 1-1/2"	1 Plum	32	.250	Ea.	18.50	12.20		30.70	39
2960	Valve and supply cover									
2965	1/2", 3/8", and 7/16" pipe size	1 Plum	32	.250	Ea.	17.60	12.20		29.80	37.50
2970	Extension drain cover									
2975	1-1/4", or 1-1/2" pipe size	1 Plum	32	.250	Ea.	19.30	12.20		31.50	40
2980	Tailpiece offset (wheelchair)									
2985	1-1/4" pipe size	1 Plum	32	.250	Ea.	21.50	12.20		33.70	42
4000	Pipe covering (price copper tube one size less than IPS)									
4280	Cellular glass, closed cell foam, all service jacket, sealant,									
4281	working temp. (-450° F to +900° F), 0 water vapor transmission									
4284	1" wall,									
4286	1/2" iron pipe size G	Q-14	120	.133	L.F.	1.82	5.30		7.12	10.30
4300	1-1/2" wall,									
4301	1" iron pipe size G	Q-14	105	.152	L.F.	2.93	6.05		8.98	12.70
4304	2-1/2" iron pipe size G		90	.178		4.91	7.05		11.96	16.45
4306	3" iron pipe size G		85	.188		4.99	7.45		12.44	17.20
4308	4" iron pipe size G		70	.229		6.60	9.05		15.65	21.50
4310	5" iron pipe size G		65	.246		7.60	9.75		17.35	23.50
4320	2" wall,									
4322	1" iron pipe size G	Q-14	100	.160	L.F.	4.35	6.35		10.70	14.75
4324	2-1/2" iron pipe size G		85	.188		6.45	7.45		13.90	18.80
4326	3" iron pipe size G		80	.200		6.60	7.95		14.55	19.75
4328	4" iron pipe size G		65	.246		7.40	9.75		17.15	23.50
4330	5" iron pipe size G		60	.267		8.90	10.60		19.50	26.50
4332	6" iron pipe size G		50	.320		10.55	12.70		23.25	31.50
4336	8" iron pipe size G		40	.400		13.60	15.90		29.50	40
4338	10" iron pipe size G		35	.457		14.05	18.15		32.20	44
4350	2-1/2" wall,									
4360	12" iron pipe size G	Q-14	32	.500	L.F.	22.50	19.85		42.35	55.50
4362	14" iron pipe size G	"	28	.571	"	24	22.50		46.50	62
4370	3" wall,									
4378	6" iron pipe size G	Q-14	48	.333	L.F.	14.05	13.25		27.30	36
4380	8" iron pipe size G		38	.421		16.40	16.70		33.10	44
4382	10" iron pipe size G		33	.485		21.50	19.25		40.75	54
4384	16" iron pipe size G		25	.640		28	25.50		53.50	71
4386	18" iron pipe size G		22	.727		30.50	29		59.50	79
4388	20" iron pipe size G		20	.800		36	32		68	89.50
4400	3-1/2" wall,									
4412	12" iron pipe size G	Q-14	27	.593	L.F.	31.50	23.50		55	72
4414	14" iron pipe size G	"	25	.640	"	33.50	25.50		59	77
4430	4" wall,									

22 07 19 – Plumbing Piping Insulation

22 07 19.10 Piping Insulation			Crew	Daily Output	Labor-Hours	Unit	2009 Bare Costs Material	Labor	Equipment	Total	Total Incl O&P
4446	16" iron pipe size	G	Q-14	22	.727	L.F.	34	29		63	82.50
4448	18" iron pipe size	G		20	.800		44	32		76	98.50
4450	20" iron pipe size	G		18	.889		45.50	35.50		81	106
4480	Fittings, average with fabric and mastic										
4484	1" wall,										
4486	1/2" iron pipe size	G	1 Asbe	40	.200	Ea.	3.23	8.80		12.03	17.35
4500	1-1/2" wall,										
4502	1" iron pipe size	G	1 Asbe	38	.211	Ea.	5.55	9.30		14.85	20.50
4504	2-1/2" iron pipe size	G		32	.250		6.90	11.05		17.95	25
4506	3" iron pipe size	G		30	.267		7.60	11.75		19.35	27
4508	4" iron pipe size	G		28	.286		11.70	12.60		24.30	32.50
4510	5" iron pipe size	G		24	.333		12.90	14.70		27.60	37
4520	2" wall,										
4522	1" iron pipe size	G	1 Asbe	36	.222	Ea.	6.45	9.80		16.25	22.50
4524	2-1/2" iron pipe size	G		30	.267		7.60	11.75		19.35	27
4526	3" iron pipe size	G		28	.286		9.15	12.60		21.75	30
4528	4" iron pipe size	G		24	.333		12.30	14.70		27	36.50
4530	5" iron pipe size	G		22	.364		15.50	16.05		31.55	42
4532	6" iron pipe size	G		20	.400		18.15	17.65		35.80	47.50
4536	8" iron pipe size	G		12	.667		27.50	29.50		57	76.50
4538	10" iron pipe size	G		8	1		34	44		78	107
4550	2-1/2" wall,										
4560	12" iron pipe size	G	1 Asbe	6	1.333	Ea.	58.50	59		117.50	157
4562	14" iron pipe size	G	"	4	2	"	68	88		156	213
4570	3" wall,										
4578	6" iron pipe size	G	1 Asbe	16	.500	Ea.	21	22		43	57.50
4580	8" iron pipe size	G		10	.800		31	35.50		66.50	89.50
4582	10" iron pipe size	G		6	1.333		44	59		103	141
4584	16" iron pipe size	G		2	4		105	176		281	390
4586	18" iron pipe size	G		2	4		136	176		312	425
4588	20" iron pipe size	G		2	4		160	176		336	455
4600	3-1/2" wall,										
4612	12" iron pipe size	G	1 Asbe	4	2	Ea.	76	88		164	222
4614	14" iron pipe size	G	"	2	4	"	84.50	176		260.50	370
4630	4" wall,										
4646	16" iron pipe size	G	1 Asbe	2	4	Ea.	126	176		302	415
4648	18" iron pipe size	G		2	4		155	176		331	445
4650	20" iron pipe size	G		2	4		180	176		356	475
4900	Calcium silicate, with cover										
5100	1" wall, 1/2" iron pipe size	G	Q-14	170	.094	L.F.	2.56	3.74		6.30	8.65
5130	3/4" iron pipe size	G		170	.094		2.58	3.74		6.32	8.70
5140	1" iron pipe size	G		170	.094		2.50	3.74		6.24	8.60
5150	1-1/4" iron pipe size	G		165	.097		2.56	3.85		6.41	8.85
5160	1-1/2" iron pipe size	G		165	.097		2.60	3.85		6.45	8.90
5170	2" iron pipe size	G		160	.100		2.98	3.97		6.95	9.50
5180	2-1/2" iron pipe size	G		160	.100		3.17	3.97		7.14	9.70
5190	3" iron pipe size	G		150	.107		3.43	4.23		7.66	10.40
5200	4" iron pipe size	G		140	.114		4.25	4.54		8.79	11.80
5210	5" iron pipe size	G		135	.119		4.48	4.71		9.19	12.30
5220	6" iron pipe size	G		130	.123		4.76	4.89		9.65	12.90
5280	1-1/2" wall, 1/2" iron pipe size	G		150	.107		2.77	4.23		7	9.70
5310	3/4" iron pipe size	G		150	.107		2.82	4.23		7.05	9.75
5320	1" iron pipe size	G		150	.107		3.07	4.23		7.30	10.05

22 07 19.10 Piping Insulation			Crew	Daily Output	Labor-Hours	Unit	Material	Labor	Equipment	Total	Total Incl O&P
							2009 Bare Costs				
5330	1-1/4" iron pipe size	G	Q-14	145	.110	L.F.	3.30	4.38		7.68	10.50
5340	1-1/2" iron pipe size	G		145	.110		3.54	4.38		7.92	10.75
5350	2" iron pipe size	G		140	.114		3.90	4.54		8.44	11.40
5360	2-1/2" iron pipe size	G		140	.114		4.25	4.54		8.79	11.80
5370	3" iron pipe size	G		135	.119		4.47	4.71		9.18	12.25
5380	4" iron pipe size	G		125	.128		5.15	5.10		10.25	13.65
5390	5" iron pipe size	G		120	.133		5.85	5.30		11.15	14.70
5400	6" iron pipe size	G		110	.145		6	5.75		11.75	15.65
5402	8" iron pipe size	G		95	.168		7.90	6.70		14.60	19.20
5404	10" iron pipe size	G		85	.188		10.30	7.45		17.75	23
5406	12" iron pipe size	G		80	.200		12.20	7.95		20.15	26
5408	14" iron pipe size	G		75	.213		13.70	8.45		22.15	28.50
5410	16" iron pipe size	G		70	.229		15.35	9.05		24.40	31
5412	18" iron pipe size	G		65	.246		16.95	9.75		26.70	34
5460	2" wall, 1/2" iron pipe size	G		135	.119		4.27	4.71		8.98	12.05
5490	3/4" iron pipe size	G		135	.119		4.49	4.71		9.20	12.30
5500	1" iron pipe size	G		135	.119		4.73	4.71		9.44	12.55
5510	1-1/4" iron pipe size	G		130	.123		5.05	4.89		9.94	13.25
5520	1-1/2" iron pipe size	G		130	.123		5.30	4.89		10.19	13.45
5530	2" iron pipe size	G		125	.128		5.55	5.10		10.65	14.05
5540	2-1/2" iron pipe size	G		125	.128		6.65	5.10		11.75	15.25
5550	3" iron pipe size	G		120	.133		6.70	5.30		12	15.65
5560	4" iron pipe size	G		115	.139		7.75	5.50		13.25	17.15
5570	5" iron pipe size	G		110	.145		8.80	5.75		14.55	18.75
5580	6" iron pipe size	G		105	.152		9.65	6.05		15.70	20
5581	8" iron pipe size	G		95	.168		11.45	6.70		18.15	23
5582	10" iron pipe size	G		85	.188		14.30	7.45		21.75	27.50
5583	12" iron pipe size	G		80	.200		15.95	7.95		23.90	30
5584	14" iron pipe size	G		75	.213		17.50	8.45		25.95	32.50
5585	16" iron pipe size	G		70	.229		19.30	9.05		28.35	35
5586	18" iron pipe size	G		65	.246		21	9.75		30.75	38.50
5587	3" wall, 1-1/4" iron pipe size	G		100	.160		7.85	6.35		14.20	18.55
5588	1-1/2" iron pipe size	G		100	.160		7.90	6.35		14.25	18.65
5589	2" iron pipe size	G		95	.168		8.20	6.70		14.90	19.50
5590	2-1/2" iron pipe size	G		95	.168		9.55	6.70		16.25	21
5591	3" iron pipe size	G		90	.178		9.65	7.05		16.70	21.50
5592	4" iron pipe size	G		85	.188		12.20	7.45		19.65	25
5593	6" iron pipe size	G		80	.200		14.90	7.95		22.85	29
5594	8" iron pipe size	G		75	.213		17.70	8.45		26.15	32.50
5595	10" iron pipe size	G		65	.246		21.50	9.75		31.25	39
5596	12" iron pipe size	G		60	.267		23.50	10.60		34.10	42.50
5597	14" iron pipe size	G		55	.291		26.50	11.55		38.05	47
5598	16" iron pipe size	G		50	.320		29	12.70		41.70	52
5599	18" iron pipe size	G		45	.356		32	14.10		46.10	57
5600	Calcium silicate, no cover										
5720	1" wall, 1/2" iron pipe size	G	Q-14	180	.089	L.F.	2.33	3.53		5.86	8.10
5740	3/4" iron pipe size	G		180	.089		2.33	3.53		5.86	8.10
5750	1" iron pipe size	G		180	.089		2.23	3.53		5.76	8
5760	1-1/4" iron pipe size	G		175	.091		2.27	3.63		5.90	8.20
5770	1-1/2" iron pipe size	G		175	.091		2.29	3.63		5.92	8.20
5780	2" iron pipe size	G		170	.094		2.64	3.74		6.38	8.75
5790	2-1/2" iron pipe size	G		170	.094		2.79	3.74		6.53	8.90
5800	3" iron pipe size	G		160	.100		3	3.97		6.97	9.50

22 07 19.10 Piping Insulation		Crew	Daily Output	Labor-Hours	Unit	2009 Bare Costs Material	Labor	Equipment	Total	Total Incl O&P
5810	4" iron pipe size	G	Q-14	150	.107	L.F.	3.75	4.23		7.98	10.80
5820	5" iron pipe size	G		145	.110		3.90	4.38		8.28	11.15
5830	6" iron pipe size	G		140	.114		4.10	4.54		8.64	11.60
5900	1-1/2" wall, 1/2" iron pipe size	G		160	.100		2.46	3.97		6.43	8.90
5920	3/4" iron pipe size	G		160	.100		2.50	3.97		6.47	8.95
5930	1" iron pipe size	G		160	.100		2.73	3.97		6.70	9.20
5940	1-1/4" iron pipe size	G		155	.103		2.94	4.10		7.04	9.65
5950	1-1/2" iron pipe size	G		155	.103		3.16	4.10		7.26	9.90
5960	2" iron pipe size	G		150	.107		3.48	4.23		7.71	10.50
5970	2-1/2" iron pipe size	G		150	.107		3.80	4.23		8.03	10.85
5980	3" iron pipe size	G		145	.110		3.97	4.38		8.35	11.20
5990	4" iron pipe size	G		135	.119		4.60	4.71		9.31	12.40
6000	5" iron pipe size	G		130	.123		5.20	4.89		10.09	13.35
6010	6" iron pipe size	G		120	.133		5.30	5.30		10.60	14.10
6020	7" iron pipe size	G		115	.139		6.20	5.50		11.70	15.50
6030	8" iron pipe size	G		105	.152		7	6.05		13.05	17.20
6040	9" iron pipe size	G		100	.160		8.40	6.35		14.75	19.20
6050	10" iron pipe size	G		95	.168		9.30	6.70		16	20.50
6060	12" iron pipe size	G		90	.178		11	7.05		18.05	23
6070	14" iron pipe size	G		85	.188		12.45	7.45		19.90	25.50
6080	16" iron pipe size	G		80	.200		13.95	7.95		21.90	28
6090	18" iron pipe size	G		75	.213		15.40	8.45		23.85	30
6120	2" wall, 1/2" iron pipe size	G		145	.110		3.89	4.38		8.27	11.15
6140	3/4" iron pipe size	G		145	.110		4.10	4.38		8.48	11.35
6150	1" iron pipe size	G		145	.110		4.32	4.38		8.70	11.60
6160	1-1/4" iron pipe size	G		140	.114		4.63	4.54		9.17	12.20
6170	1-1/2" iron pipe size	G		140	.114		4.83	4.54		9.37	12.40
6180	2" iron pipe size	G		135	.119		5.05	4.71		9.76	12.95
6190	2-1/2" iron pipe size	G		135	.119		6.10	4.71		10.81	14.05
6200	3" iron pipe size	G		130	.123		6.10	4.89		10.99	14.40
6210	4" iron pipe size	G		125	.128		7.10	5.10		12.20	15.75
6220	5" iron pipe size	G		120	.133		8.10	5.30		13.40	17.20
6230	6" iron pipe size	G		115	.139		8.85	5.50		14.35	18.35
6240	7" iron pipe size	G		110	.145		9.55	5.75		15.30	19.60
6250	8" iron pipe size	G		105	.152		10.50	6.05		16.55	21
6260	9" iron pipe size	G		100	.160		11.90	6.35		18.25	23
6270	10" iron pipe size	G		95	.168		13.20	6.70		19.90	25
6280	12" iron pipe size	G		90	.178		14.70	7.05		21.75	27
6290	14" iron pipe size	G		85	.188		16.15	7.45		23.60	29.50
6300	16" iron pipe size	G		80	.200		17.80	7.95		25.75	32
6310	18" iron pipe size	G		75	.213		19.40	8.45		27.85	35
6320	20" iron pipe size	G		65	.246		24	9.75		33.75	42
6330	22" iron pipe size	G		60	.267		26.50	10.60		37.10	45.50
6340	24" iron pipe size	G		55	.291		27	11.55		38.55	48
6360	3" wall, 1/2" iron pipe size	G		115	.139		7.05	5.50		12.55	16.40
6380	3/4" iron pipe size	G		115	.139		7.10	5.50		12.60	16.45
6390	1" iron pipe size	G		115	.139		7.15	5.50		12.65	16.50
6400	1-1/4" iron pipe size	G		110	.145		7.25	5.75		13	17.05
6410	1-1/2" iron pipe size	G		110	.145		7.30	5.75		13.05	17.10
6420	2" iron pipe size	G		105	.152		7.55	6.05		13.60	17.80
6430	2-1/2" iron pipe size	G		105	.152		8.85	6.05		14.90	19.25
6440	3" iron pipe size	G		100	.160		8.95	6.35		15.30	19.80
6450	4" iron pipe size	G	↓	95	.168	↓	11.40	6.70		18.10	23

22 07 19.10 Piping Insulation			Crew	Daily Output	Labor-Hours	Unit	2009 Bare Costs Material	Labor	Equipment	Total	Total Incl O&P
6460	5" iron pipe size	G	Q-14	90	.178	L.F.	12.40	7.05		19.45	24.50
6470	6" iron pipe size	G		90	.178		13.95	7.05		21	26.50
6480	7" iron pipe size	G		85	.188		15.50	7.45		22.95	29
6490	8" iron pipe size	G		85	.188		16.60	7.45		24.05	30
6500	9" iron pipe size	G		80	.200		18.80	7.95		26.75	33
6510	10" iron pipe size	G		75	.213		20	8.45		28.45	35.50
6520	12" iron pipe size	G		70	.229		22	9.05		31.05	38.50
6530	14" iron pipe size	G		65	.246		25	9.75		34.75	43
6540	16" iron pipe size	G		60	.267		27.50	10.60		38.10	46.50
6550	18" iron pipe size	G		55	.291		30	11.55		41.55	51
6560	20" iron pipe size	G		50	.320		36	12.70		48.70	59.50
6570	22" iron pipe size	G		45	.356		39	14.10		53.10	64.50
6580	24" iron pipe size	G		40	.400		42	15.90		57.90	71
6600	Fiberglass, with all service jacket										
6640	1/2" wall, 1/2" iron pipe size	G	Q-14	250	.064	L.F.	.73	2.54		3.27	4.78
6660	3/4" iron pipe size	G		240	.067		.83	2.65		3.48	5.05
6670	1" iron pipe size	G		230	.070		.86	2.76		3.62	5.30
6680	1-1/4" iron pipe size	G		220	.073		.91	2.89		3.80	5.50
6690	1-1/2" iron pipe size	G		220	.073		1.05	2.89		3.94	5.70
6700	2" iron pipe size	G		210	.076		1.12	3.02		4.14	5.95
6710	2-1/2" iron pipe size	G		200	.080		1.17	3.18		4.35	6.25
6840	1" wall, 1/2" iron pipe size	G		240	.067		.89	2.65		3.54	5.15
6860	3/4" iron pipe size	G		230	.070		.97	2.76		3.73	5.40
6870	1" iron pipe size	G		220	.073		1.04	2.89		3.93	5.65
6880	1-1/4" iron pipe size	G		210	.076		1.12	3.02		4.14	5.95
6890	1-1/2" iron pipe size	G		210	.076		1.21	3.02		4.23	6.05
6900	2" iron pipe size	G		200	.080		1.31	3.18		4.49	6.40
6910	2-1/2" iron pipe size	G		190	.084		1.49	3.34		4.83	6.90
6920	3" iron pipe size	G		180	.089		1.59	3.53		5.12	7.30
6930	3-1/2" iron pipe size	G		170	.094		1.73	3.74		5.47	7.75
6940	4" iron pipe size	G		150	.107		2.11	4.23		6.34	8.95
6950	5" iron pipe size	G		140	.114		2.38	4.54		6.92	9.70
6960	6" iron pipe size	G		120	.133		2.52	5.30		7.82	11.05
6970	7" iron pipe size	G		110	.145		3	5.75		8.75	12.35
6980	8" iron pipe size	G		100	.160		4.12	6.35		10.47	14.50
6990	9" iron pipe size	G		90	.178		4.34	7.05		11.39	15.80
7000	10" iron pipe size	G		90	.178		4.37	7.05		11.42	15.85
7010	12" iron pipe size	G		80	.200		4.76	7.95		12.71	17.70
7020	14" iron pipe size	G		80	.200		5.50	7.95		13.45	18.50
7030	16" iron pipe size	G		70	.229		7	9.05		16.05	22
7040	18" iron pipe size	G		70	.229		7.75	9.05		16.80	23
7050	20" iron pipe size	G		60	.267		8.65	10.60		19.25	26
7060	24" iron pipe size	G		60	.267		10.55	10.60		21.15	28
7080	1-1/2" wall, 1/2" iron pipe size	G		230	.070		1.67	2.76		4.43	6.15
7100	3/4" iron pipe size	G		220	.073		1.68	2.89		4.57	6.35
7110	1" iron pipe size	G		210	.076		1.80	3.02		4.82	6.70
7120	1-1/4" iron pipe size	G		200	.080		1.96	3.18		5.14	7.15
7130	1-1/2" iron pipe size	G		200	.080		2.05	3.18		5.23	7.25
7140	2" iron pipe size	G		190	.084		2.26	3.34		5.60	7.75
7150	2-1/2" iron pipe size	G		180	.089		2.42	3.53		5.95	8.20
7160	3" iron pipe size	G		170	.094		2.54	3.74		6.28	8.65
7170	3-1/2" iron pipe size	G		160	.100		2.78	3.97		6.75	9.25
7180	4" iron pipe size	G		140	.114		2.88	4.54		7.42	10.25

22 07 Plumbing Insulation

22 07 19 – Plumbing Piping Insulation

22 07 19.10 Piping Insulation			Crew	Daily Output	Labor-Hours	Unit	2009 Bare Costs Material	Labor	Equipment	Total	Total Incl O&P
7190	5" iron pipe size	G	Q-14	130	.123	L.F.	3.23	4.89		8.12	11.20
7200	6" iron pipe size	G		110	.145		3.41	5.75		9.16	12.80
7210	7" iron pipe size	G		100	.160		3.79	6.35		10.14	14.10
7220	8" iron pipe size	G		90	.178		4.78	7.05		11.83	16.30
7230	9" iron pipe size	G		85	.188		4.98	7.45		12.43	17.20
7240	10" iron pipe size	G		80	.200		5.15	7.95		13.10	18.10
7250	12" iron pipe size	G		75	.213		5.85	8.45		14.30	19.70
7260	14" iron pipe size	G		70	.229		6.70	9.05		15.75	21.50
7270	16" iron pipe size	G		65	.246		8.75	9.75		18.50	25
7280	18" iron pipe size	G		60	.267		9.85	10.60		20.45	27.50
7290	20" iron pipe size	G		55	.291		10.05	11.55		21.60	29
7300	24" iron pipe size	G		50	.320		12.30	12.70		25	33.50
7320	2" wall, 1/2" iron pipe size	G		220	.073		2.60	2.89		5.49	7.40
7340	3/4" iron pipe size	G		210	.076		2.68	3.02		5.70	7.70
7350	1" iron pipe size	G		200	.080		2.86	3.18		6.04	8.15
7360	1-1/4" iron pipe size	G		190	.084		3.02	3.34		6.36	8.55
7370	1-1/2" iron pipe size	G		190	.084		3.14	3.34		6.48	8.70
7380	2" iron pipe size	G		180	.089		3.32	3.53		6.85	9.20
7390	2-1/2" iron pipe size	G		170	.094		3.57	3.74		7.31	9.80
7400	3" iron pipe size	G		160	.100		3.79	3.97		7.76	10.35
7410	3-1/2" iron pipe size	G		150	.107		4.10	4.23		8.33	11.15
7420	4" iron pipe size	G		130	.123		4.42	4.89		9.31	12.50
7430	5" iron pipe size	G		120	.133		5.05	5.30		10.35	13.85
7440	6" iron pipe size	G		100	.160		5.20	6.35		11.55	15.70
7450	7" iron pipe size	G		90	.178		5.95	7.05		13	17.55
7460	8" iron pipe size	G		80	.200		6.35	7.95		14.30	19.40
7470	9" iron pipe size	G		75	.213		6.95	8.45		15.40	21
7480	10" iron pipe size	G		70	.229		7.60	9.05		16.65	22.50
7490	12" iron pipe size	G		65	.246		8.50	9.75		18.25	24.50
7500	14" iron pipe size	G		60	.267		10.55	10.60		21.15	28
7510	16" iron pipe size	G		55	.291		11.60	11.55		23.15	31
7520	18" iron pipe size	G		50	.320		12.95	12.70		25.65	34
7530	20" iron pipe size	G		45	.356		14.65	14.10		28.75	38
7540	24" iron pipe size	G		40	.400		15.75	15.90		31.65	42.50
7560	2-1/2" wall, 1/2" iron pipe size	G		210	.076		3.08	3.02		6.10	8.15
7562	3/4" iron pipe size	G		200	.080		3.22	3.18		6.40	8.50
7564	1" iron pipe size	G		190	.084		3.35	3.34		6.69	8.95
7566	1-1/4" iron pipe size	G		185	.086		3.47	3.43		6.90	9.20
7568	1-1/2" iron pipe size	G		180	.089		3.65	3.53		7.18	9.55
7570	2" iron pipe size	G		170	.094		3.83	3.74		7.57	10.05
7572	2-1/2" iron pipe size	G		160	.100		4.41	3.97		8.38	11.05
7574	3" iron pipe size	G		150	.107		4.64	4.23		8.87	11.75
7576	3-1/2" iron pipe size	G		140	.114		5.05	4.54		9.59	12.65
7578	4" iron pipe size	G		120	.133		5.35	5.30		10.65	14.15
7580	5" iron pipe size	G		110	.145		7.10	5.75		12.85	16.85
7582	6" iron pipe size	G		90	.178		7.65	7.05		14.70	19.45
7584	7" iron pipe size	G		80	.200		7.70	7.95		15.65	21
7586	8" iron pipe size	G		70	.229		8.15	9.05		17.20	23
7588	9" iron pipe size	G		65	.246		8.95	9.75		18.70	25
7590	10" iron pipe size	G		60	.267		9.75	10.60		20.35	27.50
7592	12" iron pipe size	G		55	.291		11.80	11.55		23.35	31
7594	14" iron pipe size	G		50	.320		13.30	12.70		26	34.50
7596	16" iron pipe size	G		45	.356		15.25	14.10		29.35	39

22 07 19.10 Piping Insulation		Crew	Daily Output	Labor-Hours	Unit	Material	Labor	Equipment	Total	Total Incl O&P
7598	18" iron pipe size	G	Q-14	40	.400	L.F.	16.55	15.90		32.45	43
7602	24" iron pipe size	G		30	.533		22	21		43	57
7620	3" wall, 1/2" iron pipe size	G		200	.080		3.94	3.18		7.12	9.30
7622	3/4" iron pipe size	G		190	.084		4.17	3.34		7.51	9.85
7624	1" iron pipe size	G		180	.089		4.41	3.53		7.94	10.40
7626	1-1/4" iron pipe size	G		175	.091		4.50	3.63		8.13	10.65
7628	1-1/2" iron pipe size	G		170	.094		4.70	3.74		8.44	11
7630	2" iron pipe size	G		160	.100		5.10	3.97		9.07	11.80
7632	2-1/2" iron pipe size	G		150	.107		5.30	4.23		9.53	12.50
7634	3" iron pipe size	G		140	.114		5.65	4.54		10.19	13.30
7636	3-1/2" iron pipe size	G		130	.123		6.20	4.89		11.09	14.50
7638	4" iron pipe size	G		110	.145		6.70	5.75		12.45	16.40
7640	5" iron pipe size	G		100	.160		7.55	6.35		13.90	18.25
7642	6" iron pipe size	G		80	.200		8.10	7.95		16.05	21.50
7644	7" iron pipe size	G		70	.229		9.25	9.05		18.30	24.50
7646	8" iron pipe size	G		60	.267		10.05	10.60		20.65	27.50
7648	9" iron pipe size	G		55	.291		10.85	11.55		22.40	30
7650	10" iron pipe size	G		50	.320		11.65	12.70		24.35	32.50
7652	12" iron pipe size	G		45	.356		14.50	14.10		28.60	38
7654	14" iron pipe size	G		40	.400		16.10	15.90		32	42.50
7656	16" iron pipe size	G		35	.457		18.25	18.15		36.40	48.50
7658	18" iron pipe size	G		32	.500		19.55	19.85		39.40	52.50
7660	20" iron pipe size	G		30	.533		21.50	21		42.50	56.50
7662	24" iron pipe size	G		28	.571		27	22.50		49.50	65
7664	26" iron pipe size	G		26	.615		29.50	24.50		54	71
7666	30" iron pipe size	G	↓	24	.667	↓	34	26.50		60.50	79
7800	For fiberglass with standard canvas jacket, deduct						5%				
7802	For fittings, add 3 L.F. for each fitting										
7804	plus 4 L.F. for each flange of the fitting										
7810	Finishes										
7814	For single layer of felt, add						10%	10%			
7816	For roofing paper, 45 lb to 55 lb, add						25%	10%			
7820	Polyethylene tubing flexible closed cell foam, UV resistant										
7828	Standard temperature (-90° F to +212° F)										
7830	3/8" wall, 1/8" iron pipe size	G	1 Asbe	130	.062	L.F.	.17	2.71		2.88	4.44
7831	1/4" iron pipe size	G		130	.062		.18	2.71		2.89	4.45
7832	3/8" iron pipe size	G		130	.062		.20	2.71		2.91	4.47
7833	1/2" iron pipe size	G		126	.063		.22	2.80		3.02	4.63
7834	3/4" iron pipe size	G		122	.066		.26	2.89		3.15	4.82
7835	1" iron pipe size	G		120	.067		.29	2.94		3.23	4.93
7836	1-1/4" iron pipe size	G		118	.068		.36	2.99		3.35	5.10
7837	1-1/2" iron pipe size	G		118	.068		.53	2.99		3.52	5.25
7838	2" iron pipe size	G		116	.069		.73	3.04		3.77	5.55
7839	2-1/2" iron pipe size	G		114	.070		1.03	3.09		4.12	6
7840	3" iron pipe size	G		112	.071		1.20	3.15		4.35	6.25
7842	1/2" wall, 1/8" iron pipe size	G		120	.067		.25	2.94		3.19	4.89
7843	1/4" iron pipe size	G		120	.067		.26	2.94		3.20	4.90
7844	3/8" iron pipe size	G		120	.067		.29	2.94		3.23	4.93
7845	1/2" iron pipe size	G		118	.068		.33	2.99		3.32	5.05
7846	3/4" iron pipe size	G		116	.069		.37	3.04		3.41	5.20
7847	1" iron pipe size	G		114	.070		.42	3.09		3.51	5.30
7848	1-1/4" iron pipe size	G		112	.071		.49	3.15		3.64	5.50
7849	1-1/2" iron pipe size	G	↓	110	.073	↓	.70	3.21		3.91	5.80

22 07 19.10 Piping Insulation		Crew	Daily Output	Labor-Hours	Unit	Material	Labor	Equipment	Total	Total Incl O&P
						2009 Bare Costs				
7850	2" iron pipe size	G	1 Asbe	108	.074	L.F.	.84	3.27		4.11	6
7851	2-1/2" iron pipe size	G		106	.075		.95	3.33		4.28	6.25
7852	3" iron pipe size	G		104	.077		1.44	3.39		4.83	6.90
7853	3-1/2" iron pipe size	G		102	.078		1.65	3.46		5.11	7.20
7854	4" iron pipe size	G		100	.080		1.88	3.53		5.41	7.60
7855	3/4" wall, 1/8" iron pipe size	G		110	.073		.38	3.21		3.59	5.45
7856	1/4" iron pipe size	G		110	.073		.40	3.21		3.61	5.50
7857	3/8" iron pipe size	G		108	.074		.46	3.27		3.73	5.60
7858	1/2" iron pipe size	G		106	.075		.55	3.33		3.88	5.80
7859	3/4" iron pipe size	G		104	.077		.67	3.39		4.06	6.05
7860	1" iron pipe size	G		102	.078		.77	3.46		4.23	6.25
7861	1-1/4" iron pipe size	G		100	.080		1.05	3.53		4.58	6.70
7862	1-1/2" iron pipe size	G		100	.080		1.37	3.53		4.90	7.05
7863	2" iron pipe size	G		98	.082		1.52	3.60		5.12	7.30
7864	2-1/2" iron pipe size	G		96	.083		1.62	3.67		5.29	7.55
7865	3" iron pipe size	G		94	.085		2.46	3.75		6.21	8.60
7866	3-1/2" iron pipe size	G		92	.087		2.77	3.83		6.60	9.05
7867	4" iron pipe size	G		90	.089		3.12	3.92		7.04	9.60
7868	1" wall, 1/4" iron pipe size	G		100	.080		.81	3.53		4.34	6.45
7869	3/8" iron pipe size	G		98	.082		.82	3.60		4.42	6.55
7870	1/2" iron pipe size	G		96	.083		1.02	3.67		4.69	6.85
7871	3/4" iron pipe size	G		94	.085		1.21	3.75		4.96	7.25
7872	1" iron pipe size	G		92	.087		1.47	3.83		5.30	7.60
7873	1-1/4" iron pipe size	G		90	.089		1.68	3.92		5.60	8
7874	1-1/2" iron pipe size	G		90	.089		2.10	3.92		6.02	8.45
7875	2" iron pipe size	G		88	.091		2.52	4.01		6.53	9.05
7876	2-1/2" iron pipe size	G		86	.093		2.85	4.10		6.95	9.60
7877	3" iron pipe size	G		84	.095		3.59	4.20		7.79	10.55
7878	Contact cement, quart can	G				Ea.	9.20			9.20	10.10
7879	Rubber tubing, flexible closed cell foam										
7880	3/8" wall, 1/4" iron pipe size	G	1 Asbe	120	.067	L.F.	.55	2.94		3.49	5.20
7900	3/8" iron pipe size	G		120	.067		.68	2.94		3.62	5.35
7910	1/2" iron pipe size	G		115	.070		.70	3.07		3.77	5.60
7920	3/4" iron pipe size	G		115	.070		.85	3.07		3.92	5.75
7930	1" iron pipe size	G		110	.073		.98	3.21		4.19	6.15
7940	1-1/4" iron pipe size	G		110	.073		1.20	3.21		4.41	6.35
7950	1-1/2" iron pipe size	G		110	.073		1.30	3.21		4.51	6.50
8100	1/2" wall, 1/4" iron pipe size	G		90	.089		.44	3.92		4.36	6.65
8120	3/8" iron pipe size	G		90	.089		.49	3.92		4.41	6.70
8130	1/2" iron pipe size	G		89	.090		.54	3.96		4.50	6.80
8140	3/4" iron pipe size	G		89	.090		.60	3.96		4.56	6.85
8150	1" iron pipe size	G		88	.091		.67	4.01		4.68	7.05
8160	1-1/4" iron pipe size	G		87	.092		.77	4.06		4.83	7.20
8170	1-1/2" iron pipe size	G		87	.092		.94	4.06		5	7.40
8180	2" iron pipe size	G		86	.093		1.22	4.10		5.32	7.80
8190	2-1/2" iron pipe size	G		86	.093		1.54	4.10		5.64	8.15
8200	3" iron pipe size	G		85	.094		1.72	4.15		5.87	8.40
8210	3-1/2" iron pipe size	G		85	.094		2.39	4.15		6.54	9.15
8220	4" iron pipe size	G		80	.100		2.55	4.41		6.96	9.70
8230	5" iron pipe size	G		80	.100		3.51	4.41		7.92	10.75
8240	6" iron pipe size	G		75	.107		3.51	4.70		8.21	11.20
8300	3/4" wall, 1/4" iron pipe size	G		90	.089		.68	3.92		4.60	6.90
8320	3/8" iron pipe size	G		90	.089		.75	3.92		4.67	7

22 07 19.10 Piping Insulation			Crew	Daily Output	Labor-Hours	Unit	2009 Bare Costs Material	Labor	Equipment	Total	Total Incl O&P
8330	1/2" iron pipe size	G	1 Asbe	89	.090	L.F.	.90	3.96		4.86	7.20
8340	3/4" iron pipe size	G		89	.090		1.09	3.96		5.05	7.40
8350	1" iron pipe size	G		88	.091		1.25	4.01		5.26	7.70
8360	1-1/4" iron pipe size	G		87	.092		1.68	4.06		5.74	8.20
8370	1-1/2" iron pipe size	G		87	.092		1.90	4.06		5.96	8.45
8380	2" iron pipe size	G		86	.093		2.23	4.10		6.33	8.90
8390	2-1/2" iron pipe size	G		86	.093		2.98	4.10		7.08	9.75
8400	3" iron pipe size	G		85	.094		3.40	4.15		7.55	10.25
8410	3-1/2" iron pipe size	G		85	.094		4.05	4.15		8.20	10.95
8420	4" iron pipe size	G		80	.100		4.31	4.41		8.72	11.65
8430	5" iron pipe size	G		80	.100		5.20	4.41		9.61	12.60
8440	6" iron pipe size	G		80	.100		6.30	4.41		10.71	13.80
8444	1" wall, 1/2" iron pipe size	G		86	.093		1.71	4.10		5.81	8.35
8445	3/4" iron pipe size	G		84	.095		2.07	4.20		6.27	8.90
8446	1" iron pipe size	G		84	.095		2.41	4.20		6.61	9.25
8447	1-1/4" iron pipe size	G		82	.098		2.72	4.30		7.02	9.75
8448	1-1/2" iron pipe size	G		82	.098		3.16	4.30		7.46	10.25
8449	2" iron pipe size	G		80	.100		4.23	4.41		8.64	11.55
8450	2-1/2" iron pipe size	G	↓	80	.100	↓	5.50	4.41		9.91	12.95
8456	Rubber insulation tape, 1/8" x 2" x 30'	G				Ea.	11.40			11.40	12.50
8460	Polyolefin tubing, flexible closed cell foam, UV stabilized, work										
8462	temp. -165° F to +210° F, 0 water vapor transmission										
8464	3/8" wall, 1/8" iron pipe size	G	1 Asbe	140	.057	L.F.	.35	2.52		2.87	4.34
8466	1/4" iron pipe size	G		140	.057		.37	2.52		2.89	4.36
8468	3/8" iron pipe size	G		140	.057		.40	2.52		2.92	4.39
8470	1/2" iron pipe size	G		136	.059		.45	2.59		3.04	4.56
8472	3/4" iron pipe size	G		132	.061		.51	2.67		3.18	4.75
8474	1" iron pipe size	G		130	.062		.60	2.71		3.31	4.91
8476	1-1/4" iron pipe size	G		128	.063		.73	2.76		3.49	5.10
8478	1-1/2" iron pipe size	G		128	.063		.91	2.76		3.67	5.30
8480	2" iron pipe size	G		126	.063		1.01	2.80		3.81	5.50
8482	2-1/2" iron pipe size	G		123	.065		1.47	2.87		4.34	6.10
8484	3" iron pipe size	G		121	.066		1.91	2.92		4.83	6.65
8486	4" iron pipe size	G		118	.068		3.14	2.99		6.13	8.15
8500	1/2" wall, 1/8" iron pipe size	G		130	.062		.51	2.71		3.22	4.81
8502	1/4" iron pipe size	G		130	.062		.55	2.71		3.26	4.86
8504	3/8" iron pipe size	G		130	.062		.59	2.71		3.30	4.90
8506	1/2" iron pipe size	G		128	.063		.64	2.76		3.40	5
8508	3/4" iron pipe size	G		126	.063		.74	2.80		3.54	5.20
8510	1" iron pipe size	G		123	.065		.86	2.87		3.73	5.45
8512	1-1/4" iron pipe size	G		121	.066		1.01	2.92		3.93	5.70
8514	1-1/2" iron pipe size	G		119	.067		1.22	2.96		4.18	6
8516	2" iron pipe size	G		117	.068		1.43	3.02		4.45	6.30
8518	2-1/2" iron pipe size	G		114	.070		1.92	3.09		5.01	6.95
8520	3" iron pipe size	G		112	.071		2.40	3.15		5.55	7.60
8522	4" iron pipe size	G		110	.073		3.33	3.21		6.54	8.70
8534	3/4" wall, 1/8" iron pipe size	G		120	.067		.78	2.94		3.72	5.45
8536	1/4" iron pipe size	G		120	.067		.83	2.94		3.77	5.50
8538	3/8" iron pipe size	G		117	.068		.95	3.02		3.97	5.80
8540	1/2" iron pipe size	G		114	.070		1.05	3.09		4.14	6
8542	3/4" iron pipe size	G		112	.071		1.37	3.15		4.52	6.45
8544	1" iron pipe size	G		110	.073		1.60	3.21		4.81	6.80
8546	1-1/4" iron pipe size	G	↓	108	.074	↓	2.13	3.27		5.40	7.45

22 07 19 – Plumbing Piping Insulation

22 07 19.10 Piping Insulation		Crew	Daily Output	Labor-Hours	Unit	Material	2009 Bare Costs Labor	Equipment	Total	Total Incl O&P
8548	1-1/2" iron pipe size	G	1 Asbe	108	.074	L.F.	2.50	3.27		5.77	7.85
8550	2" iron pipe size	G		106	.075		2.92	3.33		6.25	8.40
8552	2-1/2" iron pipe size	G		104	.077		3.68	3.39		7.07	9.35
8554	3" iron pipe size	G		102	.078		4.43	3.46		7.89	10.25
8556	4" iron pipe size	G		100	.080		5.85	3.53		9.38	11.95
8570	1" wall, 1/8" iron pipe size	G		110	.073		1.42	3.21		4.63	6.60
8572	1/4" iron pipe size	G		108	.074		1.48	3.27		4.75	6.75
8574	3/8" iron pipe size	G		106	.075		1.57	3.33		4.90	6.95
8576	1/2" iron pipe size	G		104	.077		2.06	3.39		5.45	7.55
8578	3/4" iron pipe size	G		102	.078		2.38	3.46		5.84	8
8580	1" iron pipe size	G		100	.080		2.70	3.53		6.23	8.50
8582	1-1/4" iron pipe size	G		97	.082		3.09	3.64		6.73	9.10
8584	1-1/2" iron pipe size	G		97	.082		4.05	3.64		7.69	10.15
8586	2" iron pipe size	G		95	.084		5.30	3.71		9.01	11.60
8588	2-1/2" iron pipe size	G		93	.086		5.30	3.79		9.09	11.75
8590	3" iron pipe size	G	↓	91	.088	↓	6.20	3.88		10.08	12.85
8606	Contact adhesive (R-320)	G				Qt.	13			13	14.30
8608	Contact adhesive (R-320)	G				Gal.	45.50			45.50	50
8610	NOTE: Preslit/preglued vs unslit, same price										

22 07 19.30 Piping Insulation Protective Jacketing, PVC		Crew	Daily Output	Labor-Hours	Unit	Material	Labor	Equipment	Total	Total Incl O&P
0010	**PIPING INSULATION PROTECTIVE JACKETING, PVC**									
0100	PVC, white, 48" lengths cut from roll goods									
0120	20 mil thick									
0140	Size based on OD of insulation									
0150	1-1/2" ID	Q-14	270	.059	L.F.	.26	2.35		2.61	3.98
0152	2" ID		260	.062		.32	2.44		2.76	4.18
0154	2-1/2" ID		250	.064		.40	2.54		2.94	4.42
0156	3" ID		240	.067		.47	2.65		3.12	4.67
0158	3-1/2" ID		230	.070		.54	2.76		3.30	4.92
0160	4" ID		220	.073		.60	2.89		3.49	5.20
0162	4-1/2" ID		210	.076		.67	3.02		3.69	5.50
0164	5" ID		200	.080		.75	3.18		3.93	5.80
0166	5-1/2" ID		190	.084		.82	3.34		4.16	6.15
0168	6" ID		180	.089		.88	3.53		4.41	6.50
0170	6-1/2" ID		175	.091		.95	3.63		4.58	6.75
0172	7" ID		170	.094		1.03	3.74		4.77	7
0174	7-1/2" ID		164	.098		1.10	3.87		4.97	7.25
0176	8" ID		161	.099		1.18	3.95		5.13	7.50
0178	8-1/2" ID		158	.101		1.24	4.02		5.26	7.65
0180	9" ID		155	.103		1.31	4.10		5.41	7.85
0182	9-1/2" ID		152	.105		1.38	4.18		5.56	8.05
0184	10" ID		149	.107		1.46	4.26		5.72	8.30
0186	10-1/2" ID		146	.110		1.52	4.35		5.87	8.45
0188	11" ID		143	.112		1.59	4.44		6.03	8.70
0190	11-1/2" ID		140	.114		1.66	4.54		6.20	8.95
0192	12" ID		137	.117		1.74	4.64		6.38	9.15
0194	12-1/2" ID		134	.119		1.80	4.74		6.54	9.45
0195	13" ID		132	.121		1.88	4.81		6.69	9.60
0196	13-1/2" ID		132	.121		1.95	4.81		6.76	9.70
0198	14" ID		130	.123		2.02	4.89		6.91	9.85
0200	15" ID		128	.125		2.15	4.96		7.11	10.15
0202	16" ID	↓	126	.127	↓	2.30	5.05		7.35	10.45

22 07 19.30 Piping Insulation Protective Jacketing, PVC		Crew	Daily Output	Labor-Hours	Unit	Material (2009 Bare Costs)	Labor (2009 Bare Costs)	Equipment (2009 Bare Costs)	Total (2009 Bare Costs)	Total Incl O&P
0204	17" ID	Q-14	124	.129	L.F.	2.43	5.10		7.53	10.70
0206	18" ID		122	.131		2.58	5.20		7.78	11
0208	19" ID		120	.133		2.72	5.30		8.02	11.30
0210	20" ID		118	.136		2.86	5.40		8.26	11.60
0212	21" ID		116	.138		3	5.50		8.50	11.90
0214	22" ID		114	.140		3.14	5.55		8.69	12.20
0216	23" ID		112	.143		3.28	5.65		8.93	12.50
0218	24" ID		110	.145		3.43	5.75		9.18	12.80
0220	25" ID		108	.148		3.56	5.90		9.46	13.10
0222	26" ID		106	.151		3.71	6		9.71	13.50
0224	27" ID		104	.154		3.84	6.10		9.94	13.75
0226	28" ID		102	.157		3.99	6.25		10.24	14.15
0228	29" ID		100	.160		4.13	6.35		10.48	14.50
0230	30" ID		98	.163		4.27	6.50		10.77	14.85
0300	For colors, add				Ea.	10%				
1000	30 mil thick									
1010	Size based on OD of insulation									
1020	2" ID	Q-14	260	.062	L.F.	.50	2.44		2.94	4.38
1022	2-1/2" ID		250	.064		.60	2.54		3.14	4.64
1024	3" ID		240	.067		.72	2.65		3.37	4.94
1026	3-1/2" ID		230	.070		.82	2.76		3.58	5.25
1028	4" ID		220	.073		.92	2.89		3.81	5.55
1030	4-1/2" ID		210	.076		1.03	3.02		4.05	5.85
1032	5" ID		200	.080		1.13	3.18		4.31	6.20
1034	5-1/2" ID		190	.084		1.24	3.34		4.58	6.60
1036	6" ID		180	.089		1.34	3.53		4.87	7
1038	6-1/2" ID		175	.091		1.46	3.63		5.09	7.30
1040	7" ID		170	.094		1.56	3.74		5.30	7.55
1042	7-1/2" ID		164	.098		1.68	3.87		5.55	7.90
1044	8" ID		161	.099		1.78	3.95		5.73	8.15
1046	8-1/2" ID		158	.101		1.89	4.02		5.91	8.40
1048	9" ID		155	.103		1.99	4.10		6.09	8.60
1050	9-1/2" ID		152	.105		2.11	4.18		6.29	8.85
1052	10" ID		149	.107		2.21	4.26		6.47	9.15
1054	10-1/2" ID		146	.110		2.32	4.35		6.67	9.35
1056	11" ID		143	.112		2.42	4.44		6.86	9.60
1058	11-1/2" ID		140	.114		2.53	4.54		7.07	9.90
1060	12" ID		137	.117		2.63	4.64		7.27	10.15
1062	12-1/2" ID		134	.119		2.74	4.74		7.48	10.45
1063	13" ID		132	.121		2.86	4.81		7.67	10.70
1064	13-1/2" ID		132	.121		2.96	4.81		7.77	10.80
1066	14" ID		130	.123		3.07	4.89		7.96	11.05
1068	15" ID		128	.125		3.28	4.96		8.24	11.40
1070	16" ID		126	.127		3.49	5.05		8.54	11.75
1072	17" ID		124	.129		3.71	5.10		8.81	12.15
1074	18" ID		122	.131		3.92	5.20		9.12	12.45
1076	19" ID		120	.133		4.13	5.30		9.43	12.85
1078	20" ID		118	.136		4.35	5.40		9.75	13.25
1080	21" ID		116	.138		4.56	5.50		10.06	13.60
1082	22" ID		114	.140		4.77	5.55		10.32	14
1084	23" ID		112	.143		4.98	5.65		10.63	14.40
1086	24" ID		110	.145		5.20	5.75		10.95	14.80
1088	25" ID		108	.148		5.40	5.90		11.30	15.15

22 07 19.30 Piping Insulation Protective Jacketing, PVC		Crew	Daily Output	Labor-Hours	Unit	2009 Bare Costs Material	Labor	Equipment	Total	Total Incl O&P
1090	26" ID	Q-14	106	.151	L.F.	5.65	6		11.65	15.60
1092	27" ID		104	.154		5.85	6.10		11.95	16
1094	28" ID		102	.157		6.05	6.25		12.30	16.40
1096	29" ID		100	.160		6.30	6.35		12.65	16.85
1098	30" ID		98	.163		6.50	6.50		13	17.30
1300	For colors, add				Ea.	10%				
2000	PVC, white, fitting covers									
2020	Fiberglass insulation inserts included with sizes 1-3/4" thru 9-3/4"									
2030	Size is based on OD of insulation									
2040	90° Elbow fitting									
2060	1-3/4"	Q-14	135	.119	Ea.	.49	4.71		5.20	7.90
2062	2"		130	.123		.60	4.89		5.49	8.30
2064	2-1/4"		128	.125		.66	4.96		5.62	8.55
2068	2-1/2"		126	.127		.74	5.05		5.79	8.70
2070	2-3/4"		123	.130		.82	5.15		5.97	9
2072	3"		120	.133		.83	5.30		6.13	9.20
2074	3-3/8"		116	.138		.95	5.50		6.45	9.65
2076	3-3/4"		113	.142		1.08	5.60		6.68	10
2078	4-1/8"		110	.145		1.34	5.75		7.09	10.50
2080	4-3/4"		105	.152		1.61	6.05		7.66	11.25
2082	5-1/4"		100	.160		1.88	6.35		8.23	12
2084	5-3/4"		95	.168		2.52	6.70		9.22	13.25
2086	6-1/4"		90	.178		4.24	7.05		11.29	15.70
2088	6-3/4"		87	.184		4.49	7.30		11.79	16.40
2090	7-1/4"		85	.188		5.60	7.45		13.05	17.90
2092	7-3/4"		83	.193		5.80	7.65		13.45	18.40
2094	8-3/4"		80	.200		7.60	7.95		15.55	21
2096	9-3/4"		77	.208		10.15	8.25		18.40	24
2098	10-7/8"		74	.216		11.30	8.60		19.90	26
2100	11-7/8"		71	.225		12.95	8.95		21.90	28.50
2102	12-7/8"		68	.235		16.70	9.35		26.05	33
2104	14-1/8"		66	.242		18.55	9.60		28.15	35.50
2106	15-1/8"		64	.250		20	9.95		29.95	37.50
2108	16-1/8"		63	.254		22	10.10		32.10	40
2110	17-1/8"		62	.258		24.50	10.25		34.75	43
2112	18-1/8"		61	.262		30.50	10.40		40.90	50
2114	19-1/8"		60	.267		42.50	10.60		53.10	63.50
2116	20-1/8"		59	.271		55	10.75		65.75	77.50
2200	45° Elbow fitting									
2220	1-3/4" thru 9-3/4" same price as 90° Elbow fitting									
2320	10-7/8"	Q-14	74	.216	Ea.	11.30	8.60		19.90	26
2322	11-7/8"		71	.225		12.50	8.95		21.45	28
2324	12-7/8"		68	.235		13.90	9.35		23.25	30
2326	14-1/8"		66	.242		15.85	9.60		25.45	32.50
2328	15-1/8"		64	.250		17.05	9.95		27	34.50
2330	16-1/8"		63	.254		19.50	10.10		29.60	37.50
2332	17-1/8"		62	.258		22	10.25		32.25	40
2334	18-1/8"		61	.262		27	10.40		37.40	46
2336	19-1/8"		60	.267		36.50	10.60		47.10	56.50
2338	20-1/8"		59	.271		41.50	10.75		52.25	62.50
2400	Tee fitting									
2410	1-3/4"	Q-14	96	.167	Ea.	.93	6.60		7.53	11.35
2412	2"		94	.170		1.03	6.75		7.78	11.75

22 07 19 – Plumbing Piping Insulation

22 07 19.30 Piping Insulation Protective Jacketing, PVC		Crew	Daily Output	Labor-Hours	Unit	Material	Labor	Equipment	Total	Total Incl O&P
						2009 Bare Costs				
2414	2-1/4"	Q-14	91	.176	Ea.	1.15	7		8.15	12.20
2416	2-1/2"		88	.182		1.25	7.20		8.45	12.70
2418	2-3/4"		85	.188		1.37	7.45		8.82	13.20
2420	3"		82	.195		1.49	7.75		9.24	13.80
2422	3-3/8"		79	.203		1.71	8.05		9.76	14.50
2424	3-3/4"		76	.211		1.96	8.35		10.31	15.25
2426	4-1/8"		73	.219		2.30	8.70		11	16.20
2428	4-3/4"		70	.229		2.88	9.05		11.93	17.35
2430	5-1/4"		67	.239		3.45	9.50		12.95	18.65
2432	5-3/4"		63	.254		4.57	10.10		14.67	21
2434	6-1/4"		60	.267		6	10.60		16.60	23
2436	6-3/4"		59	.271		7.45	10.75		18.20	25
2438	7-1/4"		57	.281		12.05	11.15		23.20	30.50
2440	7-3/4"		54	.296		13.20	11.75		24.95	33
2442	8-3/4"		52	.308		16.05	12.20		28.25	37
2444	9-3/4"		50	.320		18.95	12.70		31.65	41
2446	10-7/8"		48	.333		19.15	13.25		32.40	41.50
2448	11-7/8"		47	.340		21.50	13.50		35	44.50
2450	12-7/8"		46	.348		23.50	13.80		37.30	47.50
2452	14-1/8"		45	.356		25.50	14.10		39.60	50
2454	15-1/8"		44	.364		27.50	14.45		41.95	53
2456	16-1/8"		43	.372		30	14.75		44.75	56
2458	17-1/8"		42	.381		32	15.10		47.10	58.50
2460	18-1/8"		41	.390		35	15.50		50.50	63
2462	19-1/8"		40	.400		38.50	15.90		54.40	67
2464	20-1/8"		39	.410		41.50	16.30		57.80	71
4000	Mechanical grooved fitting cover, including insert									
4020	90° Elbow fitting									
4030	3/4" & 1"	Q-14	140	.114	Ea.	6.75	4.54		11.29	14.55
4040	1-1/4" & 1-1/2"		135	.119		7.05	4.71		11.76	15.10
4042	2"		130	.123		7.75	4.89		12.64	16.15
4044	2-1/2"		125	.128		8.60	5.10		13.70	17.40
4046	3"		120	.133		9.65	5.30		14.95	18.95
4048	3-1/2"		115	.139		11.20	5.50		16.70	21
4050	4"		110	.145		12.50	5.75		18.25	23
4052	5"		100	.160		15.45	6.35		21.80	27
4054	6"		90	.178		22	7.05		29.05	35
4056	8"		80	.200		25	7.95		32.95	40
4058	10"		75	.213		32	8.45		40.45	48.50
4060	12"		68	.235		46	9.35		55.35	65
4062	14"		65	.246		57	9.75		66.75	78
4064	16"		63	.254		77.50	10.10		87.60	101
4066	18"		61	.262		106	10.40		116.40	133
4100	45° Elbow fitting									
4120	3/4" & 1"	Q-14	140	.114	Ea.	6.45	4.54		10.99	14.20
4130	1-1/4" & 1-1/2"		135	.119		6.75	4.71		11.46	14.80
4140	2"		130	.123		7.10	4.89		11.99	15.45
4142	2-1/2"		125	.128		7.85	5.10		12.95	16.60
4144	3"		120	.133		8.65	5.30		13.95	17.85
4146	3-1/2"		115	.139		10.10	5.50		15.60	19.80
4148	4"		110	.145		10.45	5.75		16.20	20.50
4150	5"		100	.160		13.95	6.35		20.30	25.50
4152	6"		90	.178		19.40	7.05		26.45	32.50

22 07 19.30 Piping Insulation Protective Jacketing, PVC

	22 07 19.30 Piping Insulation Protective Jacketing, PVC	Crew	Daily Output	Labor-Hours	Unit	2009 Bare Costs Material	Labor	Equipment	Total	Total Incl O&P
4154	8"	Q-14	80	.200	Ea.	22	7.95		29.95	36.50
4156	10"		75	.213		28	8.45		36.45	44
4158	12"		68	.235		39.50	9.35		48.85	58
4160	14"		65	.246		44.50	9.75		54.25	64.50
4162	16"		63	.254		65.50	10.10		75.60	88.50
4164	18"	↓	61	.262	↓	94	10.40		104.40	119
4200	Tee fitting									
4220	3/4" & 1"	Q-14	93	.172	Ea.	9.30	6.85		16.15	21
4230	1-1/4" & 1-1/2"		90	.178		9.90	7.05		16.95	22
4240	2"		87	.184		10.95	7.30		18.25	23.50
4242	2-1/2"		84	.190		12.20	7.55		19.75	25.50
4244	3"		80	.200		13.10	7.95		21.05	27
4246	3-1/2"		77	.208		15.25	8.25		23.50	29.50
4248	4"		73	.219		17.95	8.70		26.65	33.50
4250	5"		67	.239		21	9.50		30.50	38
4252	6"		60	.267		25	10.60		35.60	44
4254	8"		54	.296		31	11.75		42.75	52.50
4256	10"		50	.320		39.50	12.70		52.20	63.50
4258	12"		46	.348		52.50	13.80		66.30	79.50
4260	14"		43	.372		67	14.75		81.75	96.50
4262	16"		42	.381		89	15.10		104.10	122
4264	18"	↓	41	.390	↓	118	15.50		133.50	154

22 07 19.40 Pipe Insulation Protective Jacketing, Aluminum

	22 07 19.40 Pipe Insulation Protective Jacketing, Aluminum	Crew	Daily Output	Labor-Hours	Unit	Material	Labor	Equipment	Total	Total Incl O&P
0010	**PIPE INSULATION PROTECTIVE JACKETING, ALUMINUM**									
0100	Metal roll jacketing									
0120	Aluminum with polykraft moisture barrier									
0140	Smooth, based on OD of insulation, .010" thick									
0180	1/2" ID	Q-14	220	.073	L.F.	.26	2.89		3.15	4.81
0190	3/4" ID		215	.074		.35	2.95		3.30	5
0200	1" ID		210	.076		.43	3.02		3.45	5.20
0210	1-1/4" ID		205	.078		.51	3.10		3.61	5.40
0220	1-1/2" ID		202	.079		.59	3.14		3.73	5.60
0230	1-3/4" ID		199	.080		.67	3.19		3.86	5.75
0240	2" ID		195	.082		.75	3.26		4.01	5.95
0250	2-1/4" ID		191	.084		.83	3.33		4.16	6.10
0260	2-1/2" ID		187	.086		.92	3.40		4.32	6.30
0270	2-3/4" ID		184	.087		1	3.45		4.45	6.50
0280	3" ID		180	.089		1.07	3.53		4.60	6.75
0290	3-1/4" ID		176	.091		1.16	3.61		4.77	6.95
0300	3-1/2" ID		172	.093		1.24	3.69		4.93	7.15
0310	3-3/4" ID		169	.095		1.32	3.76		5.08	7.35
0320	4" ID		165	.097		1.40	3.85		5.25	7.60
0330	4-1/4" ID		161	.099		1.48	3.95		5.43	7.85
0340	4-1/2" ID		157	.102		1.56	4.05		5.61	8.05
0350	4-3/4" ID		154	.104		1.64	4.12		5.76	8.25
0360	5" ID		150	.107		1.73	4.23		5.96	8.55
0370	5-1/4" ID		146	.110		1.80	4.35		6.15	8.80
0380	5-1/2" ID		143	.112		1.88	4.44		6.32	9
0390	5-3/4" ID		139	.115		1.97	4.57		6.54	9.30
0400	6" ID		135	.119		2.05	4.71		6.76	9.60
0410	6-1/4" ID		133	.120		2.13	4.78		6.91	9.85
0420	6-1/2" ID	↓	131	.122	↓	2.21	4.85		7.06	10.05

22 07 19.40 Pipe Insulation Protective Jacketing, Aluminum		Crew	Daily Output	Labor-Hours	Unit	2009 Bare Costs Material	Labor	Equipment	Total	Total Incl O&P
0430	7" ID	Q-14	128	.125	L.F.	2.37	4.96		7.33	10.40
0440	7-1/4" ID		125	.128		2.45	5.10		7.55	10.65
0450	7-1/2" ID		123	.130		2.54	5.15		7.69	10.90
0460	8" ID		121	.132		2.69	5.25		7.94	11.15
0470	8-1/2" ID		119	.134		2.86	5.35		8.21	11.50
0480	9" ID		116	.138		3.02	5.50		8.52	11.90
0490	9-1/2" ID		114	.140		3.18	5.55		8.73	12.25
0500	10" ID		112	.143		3.35	5.65		9	12.60
0510	10-1/2" ID		110	.145		3.50	5.75		9.25	12.90
0520	11" ID		107	.150		3.67	5.95		9.62	13.35
0530	11-1/2" ID		105	.152		3.83	6.05		9.88	13.70
0540	12" ID		103	.155		3.99	6.15		10.14	14.05
0550	12-1/2" ID		100	.160		4.16	6.35		10.51	14.55
0560	13" ID		99	.162		4.31	6.40		10.71	14.80
0570	14" ID		98	.163		4.64	6.50		11.14	15.25
0580	15" ID		96	.167		4.97	6.60		11.57	15.80
0590	16" ID		95	.168		5.30	6.70		12	16.30
0600	17" ID		93	.172		5.60	6.85		12.45	16.85
0610	18" ID		92	.174		5.85	6.90		12.75	17.20
0620	19" ID		90	.178		6.25	7.05		13.30	17.95
0630	20" ID		89	.180		6.60	7.15		13.75	18.45
0640	21" ID		87	.184		6.90	7.30		14.20	19.05
0650	22" ID		86	.186		7.25	7.40		14.65	19.50
0660	23" ID		84	.190		7.55	7.55		15.10	20
0670	24" ID		83	.193		7.90	7.65		15.55	20.50
0710	For smooth .020" thick, add					27%	10%			
0720	For smooth .024" thick, add					52%	20%			
0730	For smooth .032" thick, add					104%	33%			
0800	For stucco embossed, add					1%				
0820	For corrugated, add					2.50%				
0900	White aluminum with polysurlyn moisture barrier									
0910	Smooth, % is an add to polykraft lines of same thickness									
0940	For smooth .016" thick, add				L.F.	35%				
0960	For smooth .024" thick, add				"	22%				
1000	Aluminum fitting covers									
1010	Size is based on OD of insulation									
1020	90 LR elbow, 2 piece									
1100	1-1/2"	Q-14	140	.114	Ea.	3.05	4.54		7.59	10.45
1110	1-3/4"		135	.119		3.05	4.71		7.76	10.70
1120	2"		130	.123		3.83	4.89		8.72	11.85
1130	2-1/4"		128	.125		3.83	4.96		8.79	12
1140	2-1/2"		126	.127		3.83	5.05		8.88	12.10
1150	2-3/4"		123	.130		3.83	5.15		8.98	12.30
1160	3"		120	.133		4.16	5.30		9.46	12.90
1170	3-1/4"		117	.137		4.16	5.45		9.61	13.10
1180	3-1/2"		115	.139		4.91	5.50		10.41	14.05
1190	3-3/4"		113	.142		5.05	5.60		10.65	14.35
1200	4"		110	.145		5.25	5.75		11	14.85
1210	4-1/4"		108	.148		5.40	5.90		11.30	15.15
1220	4-1/2"		106	.151		5.40	6		11.40	15.35
1230	4-3/4"		104	.154		5.40	6.10		11.50	15.50
1240	5"		102	.157		6.10	6.25		12.35	16.45
1250	5-1/4"		100	.160		7.05	6.35		13.40	17.70

22 07 19.40 Pipe Insulation Protective Jacketing, Aluminum		Crew	Daily Output	Labor-Hours	Unit	2009 Bare Costs Material	Labor	Equipment	Total	Total Incl O&P
1260	5-1/2"	Q-14	97	.165	Ea.	8.80	6.55		15.35	19.90
1270	5-3/4"		95	.168		8.80	6.70		15.50	20
1280	6"		92	.174		7.50	6.90		14.40	19.05
1290	6-1/4"		90	.178		7.50	7.05		14.55	19.30
1300	6-1/2"		87	.184		9	7.30		16.30	21.50
1310	7"		85	.188		10.95	7.45		18.40	24
1320	7-1/4"		84	.190		10.95	7.55		18.50	24
1330	7-1/2"		83	.193		16.10	7.65		23.75	29.50
1340	8"		82	.195		12	7.75		19.75	25.50
1350	8-1/2"		80	.200		22	7.95		29.95	37
1360	9"		78	.205		22	8.15		30.15	37.50
1370	9-1/2"		77	.208		16.30	8.25		24.55	31
1380	10"		76	.211		16.30	8.35		24.65	31
1390	10-1/2"		75	.213		16.65	8.45		25.10	31.50
1400	11"		74	.216		16.65	8.60		25.25	32
1410	11-1/2"		72	.222		18.95	8.80		27.75	35
1420	12"		71	.225		18.95	8.95		27.90	35
1430	12-1/2"		69	.232		33.50	9.20		42.70	51.50
1440	13"		68	.235		33.50	9.35		42.85	51.50
1450	14"		66	.242		45	9.60		54.60	65
1460	15"		64	.250		47.50	9.95		57.45	67.50
1470	16"	↓	63	.254	↓	51.50	10.10		61.60	73
2000	45 Elbow, 2 piece									
2010	2-1/2"	Q-14	126	.127	Ea.	3.17	5.05		8.22	11.40
2020	2-3/4"		123	.130		3.17	5.15		8.32	11.60
2030	3"		120	.133		3.58	5.30		8.88	12.25
2040	3-1/4"		117	.137		3.58	5.45		9.03	12.45
2050	3-1/2"		115	.139		4.08	5.50		9.58	13.15
2060	3-3/4"		113	.142		4.08	5.60		9.68	13.30
2070	4"		110	.145		4.15	5.75		9.90	13.60
2080	4-1/4"		108	.148		4.73	5.90		10.63	14.40
2090	4-1/2"		106	.151		4.73	6		10.73	14.60
2100	4-3/4"		104	.154		4.73	6.10		10.83	14.75
2110	5"		102	.157		5.55	6.25		11.80	15.85
2120	5-1/4"		100	.160		5.55	6.35		11.90	16.05
2130	5-1/2"		97	.165		5.85	6.55		12.40	16.65
2140	6"		92	.174		5.85	6.90		12.75	17.20
2150	6-1/2"		87	.184		8.05	7.30		15.35	20.50
2160	7"		85	.188		8.05	7.45		15.50	20.50
2170	7-1/2"		83	.193		8.10	7.65		15.75	21
2180	8"		82	.195		8.10	7.75		15.85	21
2190	8-1/2"		80	.200		10.10	7.95		18.05	23.50
2200	9"		78	.205		10.10	8.15		18.25	24
2210	9-1/2"		77	.208		14.05	8.25		22.30	28.50
2220	10"		76	.211		14.05	8.35		22.40	28.50
2230	10-1/2"		75	.213		13.10	8.45		21.55	27.50
2240	11"		74	.216		13.10	8.60		21.70	28
2250	11-1/2"		72	.222		15.75	8.80		24.55	31
2260	12"		71	.225		15.75	8.95		24.70	31.50
2270	13"		68	.235		18.50	9.35		27.85	35
2280	14"		66	.242		22.50	9.60		32.10	40
2290	15"		64	.250		35	9.95		44.95	54
2300	16"	↓	63	.254	↓	38	10.10		48.10	57.50

22 07 19.40 Pipe Insulation Protective Jacketing, Aluminum		Crew	Daily Output	Labor-Hours	Unit	2009 Bare Costs Material	Labor	Equipment	Total	Total Incl O&P
2310	17"	Q-14	62	.258	Ea.	37	10.25		47.25	56.50
2320	18"		61	.262		42	10.40		52.40	63
2330	19"		60	.267		51.50	10.60		62.10	73
2340	20"		59	.271		50	10.75		60.75	72
2350	21"		58	.276		54	10.95		64.95	76
3000	Tee, 4 piece									
3010	2-1/2"	Q-14	88	.182	Ea.	24.50	7.20		31.70	38.50
3020	2-3/4"		86	.186		24.50	7.40		31.90	38.50
3030	3"		84	.190		28	7.55		35.55	42.50
3040	3-1/4"		82	.195		28	7.75		35.75	42.50
3050	3-1/2"		80	.200		29	7.95		36.95	44.50
3060	4"		78	.205		29.50	8.15		37.65	45.50
3070	4-1/4"		76	.211		31	8.35		39.35	47
3080	4-1/2"		74	.216		31	8.60		39.60	47.50
3090	4-3/4"		72	.222		31	8.80		39.80	48
3100	5"		70	.229		32	9.05		41.05	49.50
3110	5-1/4"		68	.235		32	9.35		41.35	50
3120	5-1/2"		66	.242		33.50	9.60		43.10	52
3130	6"		64	.250		33.50	9.95		43.45	52.50
3140	6-1/2"		60	.267		37	10.60		47.60	57
3150	7"		58	.276		37	10.95		47.95	57.50
3160	7-1/2"		56	.286		41	11.35		52.35	63
3170	8"		54	.296		41	11.75		52.75	63.50
3180	8-1/2"		52	.308		42	12.20		54.20	65
3190	9"		50	.320		42	12.70		54.70	66
3200	9-1/2"		49	.327		31	12.95		43.95	55
3210	10"		48	.333		31	13.25		44.25	55
3220	10-1/2"		47	.340		33	13.50		46.50	57.50
3230	11"		46	.348		33	13.80		46.80	58
3240	11-1/2"		45	.356		35	14.10		49.10	60.50
3250	12"		44	.364		35	14.45		49.45	61
3260	13"		43	.372		37	14.75		51.75	63.50
3270	14"		42	.381		38.50	15.10		53.60	66
3280	15"		41	.390		41.50	15.50		57	70.50
3290	16"		40	.400		43	15.90		58.90	72
3300	17"		39	.410		49	16.30		65.30	79.50
3310	18"		38	.421		51	16.70		67.70	82
3320	19"		37	.432		56.50	17.15		73.65	89
3330	20"		36	.444		57.50	17.65		75.15	91
3340	22"		35	.457		73.50	18.15		91.65	109
3350	23"		34	.471		75.50	18.70		94.20	113
3360	24"		31	.516		77.50	20.50		98	117

22 07 19.50 Pipe Insulation Protective Jacketing, St. Stl.

		Crew	Daily Output	Labor-Hours	Unit	Material	Labor	Equipment	Total	Total Incl O&P
0010	**PIPE INSULATION PROTECTIVE JACKETING, STAINLESS STEEL**									
0100	Metal roll jacketing									
0120	Type 304 with moisture barrier									
0140	Smooth, based on OD of insulation, .010" thick									
0260	2-1/2" ID	Q-14	250	.064	L.F.	2.70	2.54		5.24	6.95
0270	2-3/4" ID		245	.065		2.94	2.59		5.53	7.30
0280	3" ID		240	.067		3.18	2.65		5.83	7.65
0290	3-1/4" ID		235	.068		3.42	2.70		6.12	8
0300	3-1/2" ID		230	.070		3.66	2.76		6.42	8.35

22 07 19.50 Pipe Insulation Protective Jacketing, St. Stl.		Crew	Daily Output	Labor-Hours	Unit	Material	2009 Bare Costs Labor	Equipment	Total	Total Incl O&P
0310	3-3/4" ID	Q-14	225	.071	L.F.	3.90	2.82		6.72	8.70
0320	4" ID		220	.073		4.14	2.89		7.03	9.05
0330	4-1/4" ID		215	.074		4.38	2.95		7.33	9.45
0340	4-1/2" ID		210	.076		4.62	3.02		7.64	9.85
0350	5" ID		200	.080		5.10	3.18		8.28	10.60
0360	5-1/2" ID		190	.084		5.55	3.34		8.89	11.40
0370	6" ID		180	.089		6.05	3.53		9.58	12.20
0380	6-1/2" ID		175	.091		6.55	3.63		10.18	12.90
0390	7" ID		170	.094		7	3.74		10.74	13.55
0400	7-1/2" ID		164	.098		7.50	3.87		11.37	14.30
0410	8" ID		161	.099		7.95	3.95		11.90	14.95
0420	8-1/2" ID		158	.101		8.45	4.02		12.47	15.60
0430	9" ID		155	.103		8.95	4.10		13.05	16.20
0440	9-1/2" ID		152	.105		9.40	4.18		13.58	16.90
0450	10" ID		149	.107		9.90	4.26		14.16	17.60
0460	10-1/2" ID		146	.110		10.35	4.35		14.70	18.20
0470	11" ID		143	.112		10.85	4.44		15.29	18.90
0480	12" ID		137	.117		11.80	4.64		16.44	20.50
0490	13" ID		132	.121		12.75	4.81		17.56	21.50
0500	14" ID	↓	130	.123		13.70	4.89		18.59	23
0700	For smooth .016" thick, add				↓	45%	33%			
1000	Stainless steel, Type 316, fitting covers									
1010	Size is based on OD of insulation									
1020	90 LR Elbow, 2 piece									
1100	1-1/2"	Q-14	126	.127	Ea.	19.55	5.05		24.60	29.50
1110	2-3/4"		123	.130		12.50	5.15		17.65	22
1120	3"		120	.133		13.05	5.30		18.35	22.50
1130	3-1/4"		117	.137		13.05	5.45		18.50	23
1140	3-1/2"		115	.139		13.75	5.50		19.25	24
1150	3-3/4"		113	.142		14.50	5.60		20.10	25
1160	4"		110	.145		15.95	5.75		21.70	26.50
1170	4-1/4"		108	.148		21	5.90		26.90	32.50
1180	4-1/2"		106	.151		21	6		27	33
1190	5"		102	.157		21.50	6.25		27.75	34
1200	5-1/2"		97	.165		32.50	6.55		39.05	46
1210	6"		92	.174		35.50	6.90		42.40	50
1220	6-1/2"		87	.184		49	7.30		56.30	65
1230	7"		85	.188		49	7.45		56.45	65
1240	7-1/2"		83	.193		57	7.65		64.65	74.50
1250	8"		80	.200		57	7.95		64.95	75
1260	8-1/2"		80	.200		59	7.95		66.95	77.50
1270	9"		78	.205		89.50	8.15		97.65	111
1280	9-1/2"		77	.208		88	8.25		96.25	110
1290	10"		76	.211		88	8.35		96.35	110
1300	10-1/2"		75	.213		105	8.45		113.45	129
1310	11"		74	.216		101	8.60		109.60	124
1320	12"		71	.225		114	8.95		122.95	140
1330	13"		68	.235		158	9.35		167.35	189
1340	14"	↓	66	.242	↓	159	9.60		168.60	190
2000	45 Elbow, 2 piece									
2010	2-1/2"	Q-14	126	.127	Ea.	10.40	5.05		15.45	19.35
2020	2-3/4"		123	.130		10.40	5.15		15.55	19.55
2030	3"	↓	120	.133	↓	11.20	5.30		16.50	20.50

22 07 Plumbing Insulation

22 07 19 – Plumbing Piping Insulation

22 07 19.50 Pipe Insulation Protective Jacketing, St. Stl.		Crew	Daily Output	Labor-Hours	Unit	2009 Bare Costs Material	Labor	Equipment	Total	Total Incl O&P
2040	3-1/4"	Q-14	117	.137	Ea.	11.20	5.45		16.65	21
2050	3-1/2"		115	.139		11.35	5.50		16.85	21
2060	3-3/4"		113	.142		11.35	5.60		16.95	21.50
2070	4"		110	.145		14.60	5.75		20.35	25
2080	4-1/4"		108	.148		20.50	5.90		26.40	31.50
2090	4-1/2"		106	.151		20.50	6		26.50	32
2100	4-3/4"		104	.154		20.50	6.10		26.60	32
2110	5"		102	.157		21	6.25		27.25	33
2120	5-1/2"		97	.165		21	6.55		27.55	33.50
2130	6"		92	.174		24.50	6.90		31.40	38
2140	6-1/2"		87	.184		24.50	7.30		31.80	38.50
2150	7"		85	.188		42	7.45		49.45	57.50
2160	7-1/2"		83	.193		42.50	7.65		50.15	59
2170	8"		82	.195		42.50	7.75		50.25	59
2180	8-1/2"		80	.200		50	7.95		57.95	67.50
2190	9"		78	.205		50	8.15		58.15	68
2200	9-1/2"		77	.208		58.50	8.25		66.75	77.50
2210	10"		76	.211		58.50	8.35		66.85	77.50
2220	10-1/2"		75	.213		70.50	8.45		78.95	91
2230	11"		74	.216		70.50	8.60		79.10	91
2240	12"		71	.225		76.50	8.95		85.45	98
2250	13"	↓	68	.235	↓	90	9.35		99.35	114

22 11 Facility Water Distribution

22 11 13 – Facility Water Distribution Piping

22 11 13.14 Pipe, Brass

		Crew	Daily Output	Labor-Hours	Unit	Material	Labor	Equipment	Total	Total Incl O&P
0010	**PIPE, BRASS**, Plain end									
0900	Field threaded, coupling & clevis hanger assembly 10' O.C.									
0920	Regular weight									
1120	1/2" diameter	1 Plum	48	.167	L.F.	4.62	8.15		12.77	17.30
1140	3/4" diameter		46	.174		6.30	8.50		14.80	19.65
1160	1" diameter	↓	43	.186		9.30	9.05		18.35	24
1180	1-1/4" diameter	Q-1	72	.222		14.30	9.75		24.05	30.50
1200	1-1/2" diameter		65	.246		17.10	10.80		27.90	35
1220	2" diameter		53	.302		24	13.25		37.25	46.50
1240	2-1/2" diameter		41	.390		37.50	17.10		54.60	67
1260	3" diameter	↓	31	.516		53	22.50		75.50	92
1300	4" diameter	Q-2	37	.649	↓	92	29.50		121.50	146
1930	To delete coupling & hanger, subtract									
1940	1/8" diam. to 1/2" diam.					14%	46%			
1950	3/4" diam. to 1-1/2" diam.					8%	47%			
1960	2" diam. to 4" diam.					10%	37%			

22 11 13.16 Pipe Fittings, Brass

		Crew	Daily Output	Labor-Hours	Unit	Material	Labor	Equipment	Total	Total Incl O&P
0010	**PIPE FITTINGS, BRASS,** Rough bronze, threaded.									
1000	Standard wt., 90° Elbow									
1040	1/8"	1 Plum	13	.615	Ea.	12.90	30		42.90	59
1060	1/4"		13	.615		12.90	30		42.90	59
1080	3/8"		13	.615		12.90	30		42.90	59
1100	1/2"		12	.667		12.90	32.50		45.40	63
1120	3/4"		11	.727		17.25	35.50		52.75	72
1140	1"	↓	10	.800	↓	28	39		67	89.50

22 11 13.16 Pipe Fittings, Brass		Crew	Daily Output	Labor-Hours	Unit	2009 Bare Costs Material	Labor	Equipment	Total	Total Incl O&P
1160	1-1/4"	Q-1	17	.941	Ea.	45.50	41.50		87	112
1180	1-1/2"		16	1		56	44		100	128
1200	2"		14	1.143		90.50	50		140.50	175
1220	2-1/2"		11	1.455		219	64		283	335
1240	3"		8	2		335	88		423	500
1260	4"	Q-2	11	2.182		680	99.50		779.50	895
1280	5"		8	3		1,875	137		2,012	2,250
1300	6"		7	3.429		2,750	156		2,906	3,275
1500	45° Elbow, 1/8"	1 Plum	13	.615		15.85	30		45.85	62.50
1540	1/4"		13	.615		15.85	30		45.85	62.50
1560	3/8"		13	.615		15.85	30		45.85	62.50
1580	1/2"		12	.667		15.85	32.50		48.35	66.50
1600	3/4"		11	.727		22.50	35.50		58	78
1620	1"		10	.800		38.50	39		77.50	101
1640	1-1/4"	Q-1	17	.941		61	41.50		102.50	129
1660	1-1/2"		16	1		77	44		121	151
1680	2"		14	1.143		124	50		174	213
1700	2-1/2"		11	1.455		237	64		301	355
1720	3"		8	2		365	88		453	530
1740	4"	Q-2	11	2.182		825	99.50		924.50	1,050
1760	5"		8	3		1,525	137		1,662	1,875
1780	6"		7	3.429		2,100	156		2,256	2,550
2000	Tee, 1/8"	1 Plum	9	.889		15.15	43.50		58.65	81.50
2040	1/4"		9	.889		15.15	43.50		58.65	81.50
2060	3/8"		9	.889		15.15	43.50		58.65	81.50
2080	1/2"		8	1		15.15	49		64.15	89.50
2100	3/4"		7	1.143		21.50	55.50		77	107
2120	1"		6	1.333		39	65		104	141
2140	1-1/4"	Q-1	10	1.600		67	70		137	179
2160	1-1/2"		9	1.778		75.50	78		153.50	200
2180	2"		8	2		125	88		213	270
2200	2-1/2"		7	2.286		298	100		398	480
2220	3"		5	3.200		455	140		595	710
2240	4"	Q-2	7	3.429		1,125	156		1,281	1,475
2260	5"		5	4.800		2,350	218		2,568	2,925
2280	6"		4	6		3,675	273		3,948	4,450
2500	Coupling, 1/8"	1 Plum	26	.308		10.80	15		25.80	34.50
2540	1/4"		22	.364		10.80	17.75		28.55	38.50
2560	3/8"		18	.444		10.80	21.50		32.30	44.50
2580	1/2"		15	.533		10.80	26		36.80	51
2600	3/4"		14	.571		15.15	28		43.15	58.50
2620	1"		13	.615		26	30		56	73.50
2640	1-1/4"	Q-1	22	.727		43	32		75	95.50
2660	1-1/2"		20	.800		56	35		91	114
2680	2"		18	.889		93	39		132	161
2700	2-1/2"		14	1.143		158	50		208	250
2720	3"		10	1.600		219	70		289	345
2740	4"	Q-2	12	2		450	91		541	630
2760	5"		10	2.400		845	109		954	1,100
2780	6"		9	2.667		1,200	121		1,321	1,500
3000	Union, 125 lb									
3020	1/8"	1 Plum	12	.667	Ea.	36.50	32.50		69	89.50
3040	1/4"		12	.667		36.50	32.50		69	89.50

22 11 13.16 Pipe Fittings, Brass		Crew	Daily Output	Labor-Hours	Unit	Material	Labor	Equipment	Total	Total Incl O&P
						2009 Bare Costs				
3060	3/8"	1 Plum	12	.667	Ea.	36.50	32.50		69	89.50
3080	1/2"		11	.727		36.50	35.50		72	93.50
3100	3/4"		10	.800		50.50	39		89.50	114
3120	1"		9	.889		66.50	43.50		110	139
3140	1-1/4"	Q-1	16	1		96.50	44		140.50	172
3160	1-1/2"		15	1.067		115	47		162	196
3180	2"		13	1.231		176	54		230	275
3200	2-1/2"		10	1.600		485	70		555	635
3220	3"		7	2.286		750	100		850	975
3240	4"	Q-2	10	2.400		2,350	109		2,459	2,750
3320	For 250 lb. (navy pattern), add					100%				

22 11 13.23 Pipe, Copper		Crew	Daily Output	Labor-Hours	Unit	Material	Labor	Equipment	Total	Total Incl O&P
0010	**PIPE, COPPER**, Solder joints R221113-50									
0100	Solder									
0120	Solder, lead free, roll				Lb.	15.35			15.35	16.85
1000	Type K tubing, couplings & clevis hanger assemblies 10' O.C.									
1100	1/4" diameter	1 Plum	84	.095	L.F.	3.43	4.64		8.07	10.70
1120	3/8" diameter		82	.098		4.29	4.76		9.05	11.85
1140	1/2" diameter		78	.103		5	5		10	13
1160	5/8" diameter		77	.104		6.40	5.05		11.45	14.65
1180	3/4" diameter		74	.108		9.15	5.25		14.40	18
1200	1" diameter		66	.121		12.10	5.90		18	22
1220	1-1/4" diameter		56	.143		15.10	6.95		22.05	27
1240	1-1/2" diameter		50	.160		19.70	7.80		27.50	33
1260	2" diameter		40	.200		30.50	9.75		40.25	48
1280	2-1/2" diameter	Q-1	60	.267		45.50	11.70		57.20	67.50
1300	3" diameter		54	.296		63.50	13		76.50	89.50
1320	3-1/2" diameter		42	.381		85	16.70		101.70	119
1330	4" diameter		38	.421		107	18.50		125.50	146
1340	5" diameter		32	.500		238	22		260	295
1360	6" diameter	Q-2	38	.632		345	28.50		373.50	425
1380	8" diameter	"	34	.706		615	32		647	725
1390	For other than full hard temper, add					13%				
1440	For silver solder, add						15%			
1800	For medical clean, (oxygen class), add					12%				
1950	To delete cplgs. & hngrs., 1/4"-1" pipe, subtract					27%	60%			
1960	1-1/4"-3" pipe, subtract					14%	52%			
1970	3-1/2"-5" pipe, subtract					10%	60%			
1980	6"-8" pipe, subtract					19%	53%			
2000	Type L tubing, couplings & clevis hanger assemblies 10' O.C.									
2100	1/4" diameter	1 Plum	88	.091	L.F.	2.23	4.43		6.66	9.10
2120	3/8" diameter		84	.095		3.26	4.64		7.90	10.55
2140	1/2" diameter		81	.099		3.83	4.82		8.65	11.45
2160	5/8" diameter		79	.101		5.45	4.94		10.39	13.40
2180	3/4" diameter		76	.105		6	5.15		11.15	14.30
2200	1" diameter		68	.118		9.05	5.75		14.80	18.55
2220	1-1/4" diameter		58	.138		12.50	6.70		19.20	24
2240	1-1/2" diameter		52	.154		16.10	7.50		23.60	29
2260	2" diameter		42	.190		25	9.30		34.30	41.50
2280	2-1/2" diameter	Q-1	62	.258		38	11.30		49.30	59
2300	3" diameter		56	.286		51.50	12.55		64.05	76
2320	3-1/2" diameter		43	.372		70	16.35		86.35	102

22 11 13 – Facility Water Distribution Piping

22 11 13.23 Pipe, Copper		Crew	Daily Output	Labor-Hours	Unit	Material	Labor	Equipment	Total	Total Incl O&P
						2009 Bare Costs				
2340	4" diameter	Q-1	39	.410	L.F.	87	18		105	123
2360	5" diameter	↓	34	.471		186	20.50		206.50	235
2380	6" diameter	Q-2	40	.600		256	27.50		283.50	325
2400	8" diameter	"	36	.667		450	30.50		480.50	540
2410	For other than full hard temper, add				↓	21%				
2590	For silver solder, add						15%			
2900	For medical clean, (oxygen class), add					12%				
2940	To delete cplgs. & hngrs., 1/4"-1" pipe, subtract					37%	63%			
2960	1-1/4"-3" pipe, subtract					12%	53%			
2970	3-1/2"-5" pipe, subtract					12%	63%			
2980	6"-8" pipe, subtract					24%	55%			
3000	Type M tubing, couplings & clevis hanger assemblies 10' O.C.									
3100	1/4" diameter	1 Plum	90	.089	L.F.	2.62	4.33		6.95	9.40
3120	3/8" diameter		87	.092		2.65	4.48		7.13	9.65
3140	1/2" diameter		84	.095		2.79	4.64		7.43	10
3160	5/8" diameter		81	.099		4.45	4.82		9.27	12.15
3180	3/4" diameter		78	.103		4.41	5		9.41	12.35
3200	1" diameter		70	.114		6.95	5.55		12.50	15.95
3220	1-1/4" diameter		60	.133		10.25	6.50		16.75	21
3240	1-1/2" diameter		54	.148		14	7.20		21.20	26.50
3260	2" diameter	↓	44	.182		22	8.85		30.85	37.50
3280	2-1/2" diameter	Q-1	64	.250		33	10.95		43.95	53
3300	3" diameter		58	.276		44.50	12.10		56.60	66.50
3320	3-1/2" diameter		45	.356		62.50	15.60		78.10	92
3340	4" diameter		40	.400		79.50	17.55		97.05	114
3360	5" diameter	↓	36	.444		186	19.50		205.50	234
3370	6" diameter	Q-2	42	.571		256	26		282	320
3380	8" diameter	"	38	.632	↓	450	28.50		478.50	540
3440	For silver solder, add						15%			
3960	To delete cplgs. & hngrs., 1/4"-1" pipe, subtract					35%	65%			
3970	1-1/4"-3" pipe, subtract					19%	56%			
3980	3-1/2"-5" pipe, subtract					13%	65%			
3990	6"-8" pipe, subtract					28%	58%			
4000	Type DWV tubing, couplings & clevis hanger assemblies 10' O.C.									
4100	1-1/4" diameter	1 Plum	60	.133	L.F.	10.90	6.50		17.40	22
4120	1-1/2" diameter		54	.148		13.70	7.20		20.90	26
4140	2" diameter	↓	44	.182		18.45	8.85		27.30	34
4160	3" diameter	Q-1	58	.276		32.50	12.10		44.60	54
4180	4" diameter		40	.400		57.50	17.55		75.05	90
4200	5" diameter	↓	36	.444		165	19.50		184.50	211
4220	6" diameter	Q-2	42	.571		235	26		261	298
4240	8" diameter	"	38	.632	↓	575	28.50		603.50	675
4730	To delete cplgs. & hngrs., 1-1/4"-2" pipe, subtract					16%	53%			
4740	3"-4" pipe, subtract					13%	60%			
4750	5"-8" pipe, subtract					23%	58%			
5200	ACR tubing, type L, hard temper, cleaned and									
5220	capped, no couplings or hangers									
5240	3/8" OD				L.F.	2.06			2.06	2.27
5250	1/2" OD					3.26			3.26	3.59
5260	5/8" OD					3.97			3.97	4.37
5270	3/4" OD					5.60			5.60	6.15
5280	7/8" OD					6.20			6.20	6.85
5290	1-1/8" OD				↓	9			9	9.90

22 11 13.23 Pipe, Copper		Crew	Daily Output	Labor-Hours	Unit	2009 Bare Costs Material	Labor	Equipment	Total	Total Incl O&P
5300	1-3/8" OD				L.F.	12.05			12.05	13.25
5310	1-5/8" OD					15.45			15.45	17
5320	2-1/8" OD					24.50			24.50	27
5330	2-5/8" OD					36.50			36.50	40
5340	3-1/8" OD					49			49	53.50
5350	3-5/8" OD					64			64	70
5360	4-1/8" OD				↓	81			81	89
5380	ACR tubing, type L, hard, cleaned and capped									
5381	No couplings or hangers									
5384	3/8"	1 Stpi	160	.050	L.F.	2.06	2.47		4.53	5.95
5385	1/2"		160	.050		3.26	2.47		5.73	7.30
5386	5/8"		160	.050		3.97	2.47		6.44	8.05
5387	3/4"		130	.062		5.60	3.04		8.64	10.70
5388	7/8"		130	.062		6.20	3.04		9.24	11.40
5389	1-1/8"		115	.070		9	3.43		12.43	15.05
5390	1-3/8"		100	.080		12.05	3.95		16	19.15
5391	1-5/8"		90	.089		15.45	4.39		19.84	23.50
5392	2-1/8"	↓	80	.100		24.50	4.94		29.44	34.50
5393	2-5/8"	Q-5	125	.128		36.50	5.70		42.20	48.50
5394	3-1/8"		105	.152		49	6.75		55.75	63.50
5395	4-1/8"	↓	95	.168	↓	81	7.50		88.50	100
5800	Refrigeration tubing, dryseal, 60′ coils									
5840	1/8" OD				Coil	41.50			41.50	45.50
5850	3/16" OD					48.50			48.50	53
5860	1/4" OD					57.50			57.50	63
5870	5/16" OD					74.50			74.50	82
5880	3/8" OD					85			85	93.50
5890	1/2" OD					112			112	123
5900	5/8" OD					151			151	166
5910	3/4" OD					182			182	200
5920	7/8" OD					272			272	299
5930	1-1/8" OD					390			390	430
5940	1-3/8" OD					600			600	660
5950	1-5/8" OD				↓	760			760	835
9400	Sub assemblies used in assembly systems									
9410	Chilled water unit, coil connections per unit under 10 ton	Q-5	.80	20	System	895	890		1,785	2,300
9420	Chilled water unit, coil connections per unit 10 ton and up		1	16		1,375	710		2,085	2,500
9430	Chilled water dist. piping per ton, less than 61 ton systems	↓	26	.615		21.50	27.50		49	64.50
9440	Chilled water dist. piping per ton, 61 through 120 ton systems	Q-6	31	.774		35.50	35.50		71	92.50
9450	Chilled water dist. piping/ton, 135 ton systems and up	Q-8	25.40	1.260		47.50	59	2.19	108.69	143
9510	Refrigerant piping/ton of cooling for remote condensers	Q-5	2	8		500	355		855	1,075
9520	Refrigerant piping per ton up to 10 ton w/remote condensing unit		2.40	6.667		202	296		498	665
9530	Refrigerant piping per ton, 20 ton w/remote condensing unit		2	8		310	355		665	875
9540	Refrigerant piping per ton, 40 ton w/remote condensing unit	↓	1.90	8.421		415	375		790	1,025
9550	Refrigerant piping per ton, 75-80 ton w/remote condensing unit	Q-6	2.40	10		615	460		1,075	1,375
9560	Refrigerant piping per ton, 100 ton w/remote condensing unit	"	2.20	10.909	↓	820	505		1,325	1,650

22 11 13.25 Pipe Fittings, Copper		Crew	Daily Output	Labor-Hours	Unit	Material	Labor	Equipment	Total	Total Incl O&P
0010	**PIPE FITTINGS, COPPER,** Wrought unless otherwise noted									
0040	Solder joints, copper x copper									
0070	90° elbow, 1/4"	1 Plum	22	.364	Ea.	3.89	17.75		21.64	31
0090	3/8"		22	.364		3.69	17.75		21.44	30.50
0100	1/2"	↓	20	.400	↓	1.23	19.50		20.73	31

22 11 Facility Water Distribution

22 11 13 – Facility Water Distribution Piping

22 11 13.25 Pipe Fittings, Copper		Crew	Daily Output	Labor-Hours	Unit	2009 Bare Costs Material	Labor	Equipment	Total	Total Incl O&P
0110	5/8"	1 Plum	19	.421	Ea.	4.34	20.50		24.84	36
0120	3/4"		19	.421		2.76	20.50		23.26	34
0130	1"		16	.500		6.80	24.50		31.30	44
0140	1-1/4"		15	.533		10.25	26		36.25	50.50
0150	1-1/2"		13	.615		16.10	30		46.10	63
0160	2"		11	.727		29	35.50		64.50	85
0170	2-1/2"	Q-1	13	1.231		58.50	54		112.50	145
0180	3"		11	1.455		78	64		142	182
0190	3-1/2"		10	1.600		272	70		342	405
0200	4"		9	1.778		200	78		278	335
0210	5"		6	2.667		835	117		952	1,100
0220	6"	Q-2	9	2.667		1,125	121		1,246	1,400
0230	8"	"	8	3		4,125	137		4,262	4,725
0250	45° elbow, 1/4"	1 Plum	22	.364		7.30	17.75		25.05	34.50
0270	3/8"		22	.364		5.90	17.75		23.65	33
0280	1/2"		20	.400		2.26	19.50		21.76	32
0290	5/8"		19	.421		11.30	20.50		31.80	43.50
0300	3/4"		19	.421		3.96	20.50		24.46	35.50
0310	1"		16	.500		9.95	24.50		34.45	47.50
0320	1-1/4"		15	.533		13.45	26		39.45	54
0330	1-1/2"		13	.615		16.20	30		46.20	63
0340	2"		11	.727		27	35.50		62.50	83
0350	2-1/2"	Q-1	13	1.231		57.50	54		111.50	145
0360	3"		13	1.231		85.50	54		139.50	175
0370	3-1/2"		10	1.600		144	70		214	263
0380	4"		9	1.778		182	78		260	315
0390	5"		6	2.667		705	117		822	950
0400	6"	Q-2	9	2.667		1,125	121		1,246	1,400
0410	8"	"	8	3		3,775	137		3,912	4,350
0450	Tee, 1/4"	1 Plum	14	.571		8.15	28		36.15	51
0470	3/8"		14	.571		6.25	28		34.25	49
0480	1/2"		13	.615		2.11	30		32.11	47.50
0490	5/8"		12	.667		13.55	32.50		46.05	64
0500	3/4"		12	.667		5.10	32.50		37.60	54.50
0510	1"		10	.800		15.70	39		54.70	76
0520	1-1/4"		9	.889		21.50	43.50		65	88.50
0530	1-1/2"		8	1		33	49		82	110
0540	2"		7	1.143		51.50	55.50		107	140
0550	2-1/2"	Q-1	8	2		104	88		192	246
0560	3"		7	2.286		159	100		259	325
0570	3-1/2"		6	2.667		460	117		577	680
0580	4"		5	3.200		385	140		525	630
0590	5"		4	4		1,250	176		1,426	1,650
0600	6"	Q-2	6	4		1,725	182		1,907	2,175
0610	8"	"	5	4.800		6,625	218		6,843	7,625
0612	Tee, reducing on the outlet, 1/4"	1 Plum	15	.533		13.15	26		39.15	53.50
0613	3/8"		15	.533		12.35	26		38.35	52.50
0614	1/2"		14	.571		10.85	28		38.85	54
0615	5/8"		13	.615		19.40	30		49.40	66.50
0616	3/4"		12	.667		7	32.50		39.50	56.50
0617	1"		11	.727		15.75	35.50		51.25	70.50
0618	1-1/4"		10	.800		29.50	39		68.50	91
0619	1-1/2"		9	.889		31.50	43.50		75	100

22 11 13.25 Pipe Fittings, Copper		Crew	Daily Output	Labor-Hours	Unit	2009 Bare Costs Material	Labor	Equipment	Total	Total Incl O&P
0620	2"	1 Plum	8	1	Ea.	44.50	49		93.50	122
0621	2-1/2"	Q-1	9	1.778		119	78		197	247
0622	3"		8	2		133	88		221	279
0623	4"		6	2.667		245	117		362	445
0624	5"		5	3.200		1,250	140		1,390	1,575
0625	6"	Q-2	7	3.429		1,625	156		1,781	2,025
0626	8"	"	6	4		6,625	182		6,807	7,575
0630	Tee, reducing on the run, 1/4"	1 Plum	15	.533		16.45	26		42.45	57
0631	3/8"		15	.533		22	26		48	63
0632	1/2"		14	.571		14.90	28		42.90	58.50
0633	5/8"		13	.615		22	30		52	69
0634	3/4"		12	.667		6	32.50		38.50	55.50
0635	1"		11	.727		18.70	35.50		54.20	73.50
0636	1-1/4"		10	.800		29.50	39		68.50	91
0637	1-1/2"		9	.889		49.50	43.50		93	120
0638	2"		8	1		72	49		121	152
0639	2-1/2"	Q-1	9	1.778		157	78		235	289
0640	3"		8	2		231	88		319	385
0641	4"		6	2.667		510	117		627	740
0642	5"		5	3.200		1,200	140		1,340	1,500
0643	6"	Q-2	7	3.429		1,825	156		1,981	2,225
0644	8"	"	6	4		6,525	182		6,707	7,450
0650	Coupling, 1/4"	1 Plum	24	.333		.90	16.25		17.15	25.50
0670	3/8"		24	.333		1.19	16.25		17.44	26
0680	1/2"		22	.364		.94	17.75		18.69	27.50
0690	5/8"		21	.381		2.81	18.55		21.36	31
0700	3/4"		21	.381		1.87	18.55		20.42	30
0710	1"		18	.444		3.73	21.50		25.23	36.50
0715	1-1/4"		17	.471		6.65	23		29.65	42
0716	1-1/2"		15	.533		8.75	26		34.75	48.50
0718	2"		13	.615		14.65	30		44.65	61
0721	2-1/2"	Q-1	15	1.067		31	47		78	104
0722	3"		13	1.231		47	54		101	133
0724	3-1/2"		8	2		89.50	88		177.50	231
0726	4"		7	2.286		98.50	100		198.50	259
0728	5"		6	2.667		242	117		359	440
0731	6"	Q-2	8	3		400	137		537	645
0732	8"	"	7	3.429		1,300	156		1,456	1,650
0741	Coupling, reducing, concentric									
0743	1/2"	1 Plum	23	.348	Ea.	2.61	16.95		19.56	28.50
0745	3/4"		21.50	.372		4.19	18.15		22.34	31.50
0747	1"		19.50	.410		7.50	20		27.50	38.50
0748	1-1/4"		18	.444		9	21.50		30.50	42.50
0749	1-1/2"		16	.500		15.20	24.50		39.70	53
0751	2"		14	.571		23.50	28		51.50	67.50
0752	2-1/2"		13	.615		57	30		87	108
0753	3"	Q-1	14	1.143		70.50	50		120.50	153
0755	4"	"	8	2		150	88		238	297
0757	5"	Q-2	7.50	3.200		550	146		696	825
0759	6"		7	3.429		855	156		1,011	1,175
0761	8"		6.50	3.692		1,400	168		1,568	1,800
0771	Cap, sweat									
0773	1/2"	1 Plum	40	.200	Ea.	.90	9.75		10.65	15.65

22 11 13 – Facility Water Distribution Piping

22 11 13.25 Pipe Fittings, Copper		Crew	Daily Output	Labor-Hours	Unit	Material	Labor	Equipment	Total	Total Incl O&P
0775	3/4"	1 Plum	38	.211	Ea.	1.67	10.25		11.92	17.25
0777	1"		32	.250		3.96	12.20		16.16	22.50
0778	1-1/4"		29	.276		5.30	13.45		18.75	26
0779	1-1/2"		26	.308		7.75	15		22.75	31
0781	2"		22	.364		14.30	17.75		32.05	42.50
0791	Flange, sweat									
0793	3"	Q-1	22	.727	Ea.	138	32		170	200
0795	4"		18	.889		148	39		187	222
0797	5"		12	1.333		360	58.50		418.50	485
0799	6"	Q-2	18	1.333		375	60.50		435.50	505
0801	8"	"	16	1.500		1,300	68.50		1,368.50	1,525
0850	Unions, 1/4"	1 Plum	21	.381		24	18.55		42.55	54.50
0870	3/8"		21	.381		24	18.55		42.55	54.50
0880	1/2"		19	.421		12.90	20.50		33.40	45
0890	5/8"		18	.444		55	21.50		76.50	93
0900	3/4"		18	.444		16.15	21.50		37.65	50.50
0910	1"		15	.533		28	26		54	69.50
0920	1-1/4"		14	.571		48	28		76	95
0930	1-1/2"		12	.667		63.50	32.50		96	119
0940	2"		10	.800		108	39		147	178
0950	2-1/2"	Q-1	12	1.333		230	58.50		288.50	340
0960	3"	"	10	1.600		615	70		685	780
0980	Adapter, copper x male IPS, 1/4"	1 Plum	20	.400		11.95	19.50		31.45	42.50
0990	3/8"		20	.400		6.35	19.50		25.85	36.50
1000	1/2"		18	.444		2.58	21.50		24.08	35.50
1010	3/4"		17	.471		4.33	23		27.33	39.50
1020	1"		15	.533		11.20	26		37.20	51.50
1030	1-1/4"		13	.615		16.55	30		46.55	63
1040	1-1/2"		12	.667		19	32.50		51.50	70
1050	2"		11	.727		32	35.50		67.50	88.50
1060	2-1/2"	Q-1	10.50	1.524		123	67		190	236
1070	3"		10	1.600		163	70		233	284
1080	3-1/2"		9	1.778		251	78		329	395
1090	4"		8	2		243	88		331	400
1200	5"		6	2.667		1,150	117		1,267	1,425
1210	6"	Q-2	8.50	2.824		1,400	128		1,528	1,750
1214	Adapter, copper x female IPS									
1216	1/2"	1 Plum	18	.444	Ea.	4.09	21.50		25.59	37
1218	3/4"		17	.471		5.60	23		28.60	40.50
1220	1"		15	.533		13	26		39	53.50
1221	1-1/4"		13	.615		19	30		49	66
1222	1-1/2"		12	.667		29.50	32.50		62	81.50
1224	2"		11	.727		40.50	35.50		76	97.50
1250	Cross, 1/2"		10	.800		17.45	39		56.45	77.50
1260	3/4"		9.50	.842		34	41		75	99
1270	1"		8	1		57.50	49		106.50	137
1280	1-1/4"		7.50	1.067		82.50	52		134.50	169
1290	1-1/2"		6.50	1.231		118	60		178	220
1300	2"		5.50	1.455		223	71		294	350
1310	2-1/2"	Q-1	6.50	2.462		515	108		623	725
1320	3"	"	5.50	2.909		645	128		773	900
1500	Tee fitting, mechanically formed, (Type 1)									
1520	1/2" run size, 3/8" to 1/2" branch size	1 Plum	80	.100	Ea.		4.88		4.88	7.30

Material, Labor, Equipment and Total columns: 2009 Bare Costs

22 11 13 – Facility Water Distribution Piping

22 11 13.25 Pipe Fittings, Copper		Crew	Daily Output	Labor-Hours	Unit	Material	2009 Bare Costs Labor	Equipment	Total	Total Incl O&P
1530	3/4" run size, 3/8" to 3/4" branch size	1 Plum	60	.133	Ea.		6.50		6.50	9.75
1540	1" run size, 3/8" to 1" branch size		54	.148			7.20		7.20	10.85
1550	1-1/4" run size, 3/8" to 1-1/4" branch size		48	.167			8.15		8.15	12.20
1560	1-1/2" run size, 3/8" to 1-1/2" branch size		40	.200			9.75		9.75	14.65
1570	2" run size, 3/8" to 2" branch size		35	.229			11.15		11.15	16.70
1580	2-1/2" run size, 1/2" to 2" branch size		32	.250			12.20		12.20	18.30
1590	3" run size, 1" to 2" branch size		26	.308			15		15	22.50
1600	4" run size, 1" to 2" branch size		24	.333			16.25		16.25	24.50
1640	Tee fitting, mechanically formed, (Type 2)									
1650	2-1/2" run size, 2-1/2" branch size	1 Plum	12.50	.640	Ea.		31		31	47
1660	3" run size, 2-1/2" to 3" branch size		12	.667			32.50		32.50	49
1670	3-1/2" run size, 2-1/2" to 3-1/2" branch size		11	.727			35.50		35.50	53
1680	4" run size, 2-1/2" to 4" branch size		10.50	.762			37		37	55.50
1698	5" run size, 2" to 4" branch size		9.50	.842			41		41	61.50
1700	6" run size, 2" to 4" branch size		8.50	.941			46		46	69
1710	8" run size, 2" to 4" branch size		7	1.143			55.50		55.50	83.50
1800	ACR fittings, OD size									
1802	Tee, straight									
1808	5/8"	1 Stpi	12	.667	Ea.	2.11	33		35.11	52
1810	3/4"		12	.667		13.55	33		46.55	64.50
1812	7/8"		10	.800		5.10	39.50		44.60	54.50
1813	1"		10	.800		37	39.50		76.50	99.50
1814	1-1/8"		10	.800		15.70	39.50		55.20	76.50
1816	1-3/8"		9	.889		21.50	44		65.50	89.50
1818	1-5/8"		8	1		33	49.50		82.50	111
1820	2-1/8"		7	1.143		51.50	56.50		108	141
1822	2-5/8"	Q-5	8	2		104	89		193	247
1824	3-1/8"		7	2.286		159	102		261	325
1826	4-1/8"		5	3.200		385	142		527	635
1830	90° elbow									
1836	5/8"	1 Stpi	19	.421	Ea.	4.13	21		25.13	35.50
1838	3/4"		19	.421		7.50	21		28.50	39.50
1840	7/8"		16	.500		7.45	24.50		31.95	45
1842	1-1/8"		16	.500		9.95	24.50		34.45	48
1844	1-3/8"		15	.533		10.25	26.50		36.75	51
1846	1-5/8"		13	.615		16	30.50		46.50	63
1848	2-1/8"		11	.727		29	36		65	86
1850	2-5/8"	Q-5	13	1.231		58.50	54.50		113	146
1852	3-1/8"		11	1.455		78	64.50		142.50	183
1854	4-1/8"		9	1.778		200	79		279	340
1860	Coupling									
1866	5/8"	1 Stpi	21	.381	Ea.	.94	18.80		19.74	29
1868	3/4"		21	.381		2.83	18.80		21.63	31
1870	7/8"		18	.444		1.87	22		23.87	35
1871	1"		18	.444		6.35	22		28.35	40
1872	1-1/8"		18	.444		3.78	22		25.78	37
1874	1-3/8"		17	.471		6.65	23		29.65	42.50
1876	1-5/8"		15	.533		8.75	26.50		35.25	49
1878	2-1/8"		13	.615		14.65	30.50		45.15	61.50
1880	2-5/8"	Q-5	15	1.067		33	47.50		80.50	108
1882	3-1/8"		13	1.231		48	54.50		102.50	135
1884	4-1/8"		7	2.286		114	102		216	277
2000	DWV, solder joints, copper x copper									

22 11 Facility Water Distribution

22 11 13 – Facility Water Distribution Piping

22 11 13.25 Pipe Fittings, Copper		Crew	Daily Output	Labor-Hours	Unit	2009 Bare Costs Material	Labor	Equipment	Total	Total Incl O&P
2030	90° Elbow, 1-1/4"	1 Plum	13	.615	Ea.	11.90	30		41.90	58
2050	1-1/2"		12	.667		17.85	32.50		50.35	68.50
2070	2"		10	.800		23.50	39		62.50	84
2090	3"	Q-1	10	1.600		57.50	70		127.50	169
2100	4"	"	9	1.778		281	78		359	425
2150	45° Elbow, 1-1/4"	1 Plum	13	.615		10.95	30		40.95	57
2170	1-1/2"		12	.667		9.05	32.50		41.55	59
2180	2"		10	.800		21	39		60	81.50
2190	3"	Q-1	10	1.600		44.50	70		114.50	154
2200	4"	"	9	1.778		193	78		271	330
2250	Tee, Sanitary, 1-1/4"	1 Plum	9	.889		23.50	43.50		67	91
2270	1-1/2"		8	1		29.50	49		78.50	105
2290	2"		7	1.143		34	55.50		89.50	121
2310	3"	Q-1	7	2.286		126	100		226	290
2330	4"	"	6	2.667		320	117		437	530
2400	Coupling, 1-1/4"	1 Plum	14	.571		5.55	28		33.55	48
2420	1-1/2"		13	.615		6.90	30		36.90	52.50
2440	2"		11	.727		9.60	35.50		45.10	63.50
2460	3"	Q-1	11	1.455		18.65	64		82.65	117
2480	4"	"	10	1.600		59.50	70		129.50	171
2602	Traps, see Div. 22 13 16.60									
3500	Compression joint fittings									
3510	As used for plumbing and oil burner work									
3520	Fitting price includes nuts and sleeves									
3540	Sleeve, 1/8"				Ea.	.14			.14	.15
3550	3/16"					.14			.14	.15
3560	1/4"					.05			.05	.06
3570	5/16"					.14			.14	.15
3580	3/8"					.28			.28	.31
3600	1/2"					.38			.38	.42
3620	Nut, 1/8"					.20			.20	.22
3630	3/16"					.21			.21	.23
3640	1/4"					.21			.21	.23
3650	5/16"					.31			.31	.34
3660	3/8"					.33			.33	.36
3670	1/2"					.61			.61	.67
3710	Union, 1/8"	1 Plum	26	.308		1.25	15		16.25	24
3720	3/16"		24	.333		1.20	16.25		17.45	26
3730	1/4"		24	.333		1.32	16.25		17.57	26
3740	5/16"		23	.348		1.69	16.95		18.64	27.50
3750	3/8"		22	.364		1.70	17.75		19.45	28.50
3760	1/2"		22	.364		2.62	17.75		20.37	29.50
3780	5/8"		21	.381		2.90	18.55		21.45	31
3820	Union tee, 1/8"		17	.471		5	23		28	40
3830	3/16"		16	.500		3.09	24.50		27.59	40
3840	1/4"		15	.533		2.79	26		28.79	42
3850	5/16"		15	.533		3.93	26		29.93	43.50
3860	3/8"		15	.533		3.75	26		29.75	43
3870	1/2"		15	.533		6.35	26		32.35	46
3910	Union elbow, 1/4"		24	.333		1.61	16.25		17.86	26.50
3920	5/16"		23	.348		2.94	16.95		19.89	28.50
3930	3/8"		22	.364		2.84	17.75		20.59	29.50
3940	1/2"		22	.364		4.48	17.75		22.23	31.50

22 11 13.25 Pipe Fittings, Copper		Crew	Daily Output	Labor-Hours	Unit	Material	Labor	Equipment	Total	Total Incl O&P
						2009 Bare Costs				
3980	Female connector, 1/8"	1 Plum	26	.308	Ea.	1.07	15		16.07	23.50
4000	3/16" x 1/8"		24	.333		1.38	16.25		17.63	26
4010	1/4" x 1/8"		24	.333		1.22	16.25		17.47	26
4020	1/4"		24	.333		1.65	16.25		17.90	26.50
4030	3/8" x 1/4"		22	.364		1.65	17.75		19.40	28.50
4040	1/2" x 3/8"		22	.364		2.63	17.75		20.38	29.50
4050	5/8" x 1/2"		21	.381		3.46	18.55		22.01	32
4090	Male connector, 1/8"		26	.308		.82	15		15.82	23.50
4100	3/16" x 1/8"		24	.333		.85	16.25		17.10	25.50
4110	1/4" x 1/8"		24	.333		.88	16.25		17.13	25.50
4120	1/4"		24	.333		.88	16.25		17.13	25.50
4130	5/16" x 1/8"		23	.348		1.04	16.95		17.99	26.50
4140	5/16" x 1/4"		23	.348		1.17	16.95		18.12	27
4150	3/8" x 1/8"		22	.364		1.22	17.75		18.97	28
4160	3/8" x 1/4"		22	.364		1.41	17.75		19.16	28
4170	3/8"		22	.364		1.40	17.75		19.15	28
4180	3/8" x 1/2"		22	.364		1.93	17.75		19.68	28.50
4190	1/2" x 3/8"		22	.364		1.88	17.75		19.63	28.50
4200	1/2"		22	.364		2.11	17.75		19.86	29
4210	5/8" x 1/2"		21	.381		2.42	18.55		20.97	30.50
4240	Male elbow, 1/8"		26	.308		1.69	15		16.69	24.50
4250	3/16" x 1/8"		24	.333		1.86	16.25		18.11	26.50
4260	1/4" x 1/8"		24	.333		1.41	16.25		17.66	26
4270	1/4"		24	.333		1.61	16.25		17.86	26.50
4280	3/8" x 1/4"		22	.364		2.03	17.75		19.78	28.50
4290	3/8"		22	.364		3.50	17.75		21.25	30.50
4300	1/2" x 1/4"		22	.364		3.37	17.75		21.12	30
4310	1/2" x 3/8"		22	.364		5.30	17.75		23.05	32.50
4340	Female elbow, 1/8"		26	.308		3.62	15		18.62	26.50
4350	1/4" x 1/8"		24	.333		2.06	16.25		18.31	27
4360	1/4"		24	.333		2.71	16.25		18.96	27.50
4370	3/8" x 1/4"		22	.364		2.67	17.75		20.42	29.50
4380	1/2" x 3/8"		22	.364		4.55	17.75		22.30	31.50
4390	1/2"		22	.364		6.15	17.75		23.90	33.50
4420	Male run tee, 1/4" x 1/8"		15	.533		2.53	26		28.53	42
4430	5/16" x 1/8"		15	.533		5.35	26		31.35	45
4440	3/8" x 1/4"		15	.533		3.90	26		29.90	43.50
4480	Male branch tee, 1/4" x 1/8"		15	.533		2.93	26		28.93	42
4490	1/4"		15	.533		3.15	26		29.15	42.50
4500	3/8" x 1/4"		15	.533		3.67	26		29.67	43
4510	1/2" x 3/8"		15	.533		5.70	26		31.70	45.50
4520	1/2"		15	.533		9.15	26		35.15	49
4800	Flare joint fittings									
4810	Refrigeration fittings									
4820	Flare joint nuts and labor not incl. in price. Add 1 nut per jnt.									
4830	90° Elbow, 1/4"				Ea.	1.80			1.80	1.98
4840	3/8"					2.81			2.81	3.09
4850	1/2"					4.07			4.07	4.48
4860	5/8"					5.95			5.95	6.55
4870	3/4"					8.05			8.05	8.85
5030	Tee, 1/4"					2.28			2.28	2.51
5040	5/16"					3.15			3.15	3.47
5050	3/8"					3.37			3.37	3.71

22 11 13 – Facility Water Distribution Piping

22 11 13.25 Pipe Fittings, Copper		Crew	Daily Output	Labor-Hours	Unit	Material	Labor	Equipment	Total	Total Incl O&P
						2009 Bare Costs				
5060	1/2"				Ea.	5.15			5.15	5.65
5070	5/8"					8			8	8.80
5080	3/4"					13.10			13.10	14.40
5140	Union, 3/16"					.80			.80	.88
5150	1/4"					.54			.54	.59
5160	5/16"					1			1	1.10
5170	3/8"					1.04			1.04	1.14
5180	1/2"					1.55			1.55	1.71
5190	5/8"					2.48			2.48	2.73
5200	3/4"					6.20			6.20	6.80
5260	Long flare nut, 3/16"	1 Stpi	42	.190		1.22	9.40		10.62	15.45
5270	1/4"		41	.195		.67	9.65		10.32	15.20
5280	5/16"		40	.200		1.22	9.85		11.07	16.15
5290	3/8"		39	.205		1.34	10.10		11.44	16.65
5300	1/2"		38	.211		2.34	10.40		12.74	18.15
5310	5/8"		37	.216		4.07	10.65		14.72	20.50
5320	3/4"		34	.235		9.80	11.60		21.40	28
5380	Short flare nut, 3/16"		42	.190		.47	9.40		9.87	14.60
5390	1/4"		41	.195		.47	9.65		10.12	14.95
5400	5/16"		40	.200		.56	9.85		10.41	15.40
5410	3/8"		39	.205		.74	10.10		10.84	16
5420	1/2"		38	.211		1.10	10.40		11.50	16.80
5430	5/8"		36	.222		1.54	10.95		12.49	18.15
5440	3/4"		34	.235		2.97	11.60		14.57	20.50
5500	90° Elbow flare by MIPS, 1/4"					1.66			1.66	1.83
5510	3/8"					1.83			1.83	2.01
5520	1/2"					2.93			2.93	3.22
5530	5/8"					7.75			7.75	8.50
5540	3/4"					8.75			8.75	9.65
5600	Flare by FIPS, 1/4"					4.85			4.85	5.35
5610	3/8"					17.70			17.70	19.45
5620	1/2"					4.75			4.75	5.25
5670	Flare by sweat, 1/4"					1.86			1.86	2.05
5680	3/8"					2.15			2.15	2.37
5690	1/2"					3.83			3.83	4.21
5700	5/8"					14.80			14.80	16.30
5760	Tee flare by IPS, 1/4"					2.59			2.59	2.85
5770	3/8"					3.53			3.53	3.88
5780	1/2"					5.40			5.40	5.95
5790	5/8"					6.95			6.95	7.60
5850	Connector, 1/4"					.86			.86	.95
5860	3/8"					1.11			1.11	1.22
5870	1/2"					1.40			1.40	1.54
5880	5/8"					1.55			1.55	1.71
5890	3/4"					5.10			5.10	5.65
5950	Seal cap, 1/4"					.34			.34	.37
5960	3/8"					.66			.66	.73
5970	1/2"					.74			.74	.81
5980	5/8"					1.11			1.11	1.22
5990	3/4"					3.26			3.26	3.59
6000	Water service fittings									
6010	Flare joints nut and labor are included in the fitting price.									
6020	90° Elbow, C x C, 3/8"	1 Plum	19	.421	Ea.	43	20.50		63.50	78.50

22 11 13.25 Pipe Fittings, Copper		Crew	Daily Output	Labor-Hours	Unit	Material	Labor	Equipment	Total	Total Incl O&P
						2009 Bare Costs				
6030	1/2"	1 Plum	18	.444	Ea.	43	21.50		64.50	80
6040	3/4"		16	.500		55	24.50		79.50	97
6050	1"		15	.533		104	26		130	154
6080	2"		10	.800		360	39		399	455
6090	90° Elbow, C x MPT, 3/8"		19	.421		31	20.50		51.50	65
6100	1/2"		18	.444		31	21.50		52.50	66.50
6110	3/4"		16	.500		36	24.50		60.50	76
6120	1"		15	.533		98	26		124	147
6130	1-1/4"		13	.615		202	30		232	267
6140	1-1/2"		12	.667		202	32.50		234.50	271
6150	2"		10	.800		264	39		303	350
6160	90° Elbow, C x FPT, 3/8"		19	.421		33.50	20.50		54	67.50
6170	1/2"		18	.444		33.50	21.50		55	69
6180	3/4"		16	.500		41.50	24.50		66	82
6190	1"		15	.533		95.50	26		121.50	144
6200	1-1/4"		13	.615		106	30		136	162
6210	1-1/2"		12	.667		410	32.50		442.50	500
6220	2"		10	.800		590	39		629	710
6230	Tee, C x C x C, 3/8"		13	.615		60	30		90	111
6240	1/2"		12	.667		63	32.50		95.50	119
6250	3/4"		11	.727		74	35.50		109.50	135
6260	1"		10	.800		131	39		170	203
6330	Tube nut, C x nut seat, 3/8"		40	.200		12.75	9.75		22.50	28.50
6340	1/2"		38	.211		12.75	10.25		23	29.50
6350	3/4"		34	.235		12.75	11.45		24.20	31.50
6360	1"		32	.250		23	12.20		35.20	44
6380	Coupling, C x C, 3/8"		19	.421		38.50	20.50		59	73.50
6390	1/2"		18	.444		38.50	21.50		60	75
6400	3/4"		16	.500		49	24.50		73.50	90.50
6410	1"		15	.533		90.50	26		116.50	139
6420	1-1/4"		12	.667		147	32.50		179.50	210
6430	1-1/2"		12	.667		222	32.50		254.50	293
6440	2"		10	.800		335	39		374	425
6450	Adapter, C x FPT, 3/8"		19	.421		29.50	20.50		50	63
6460	1/2"		18	.444		29.50	21.50		51	64.50
6470	3/4"		16	.500		35.50	24.50		60	75.50
6480	1"		15	.533		75.50	26		101.50	122
6490	1-1/4"		13	.615		160	30		190	221
6500	1-1/2"		12	.667		160	32.50		192.50	225
6510	2"		10	.800		209	39		248	288
6520	Adapter, C x MPT, 3/8"		19	.421		26	20.50		46.50	60
6530	1/2"		18	.444		26	21.50		47.50	61.50
6540	3/4"		16	.500		36	24.50		60.50	76
6550	1"		15	.533		63.50	26		89.50	109
6560	1-1/4"		13	.615		151	30		181	211
6570	1-1/2"		12	.667		151	32.50		183.50	215
6580	2"	↓	10	.800	↓	215	39		254	296
6992	Polybutylene/polyethylene pipe, See Div. 22 11 13.76 for plastic ftng.									
7000	Insert type Brass/copper, 100 psi @ 180° F, CTS									
7010	Adapter MPT 3/8" x 3/8" CTS	1 Plum	29	.276	Ea.	1.60	13.45		15.05	22
7020	1/2" x 1/2"		26	.308		1.60	15		16.60	24.50
7030	3/4" x 1/2"		26	.308		2.73	15		17.73	25.50
7040	3/4" x 3/4"	↓	25	.320	↓	2.36	15.60		17.96	26

22 11 13.25 Pipe Fittings, Copper		Crew	Daily Output	Labor-Hours	Unit	Material	Labor	Equipment	Total	Total Incl O&P
						2009 Bare Costs				
7050	Adapter CTS 1/2" x 1/2" sweat	1 Plum	24	.333	Ea.	.60	16.25		16.85	25
7060	3/4" x 3/4" sweat		22	.364		.75	17.75		18.50	27.50
7070	Coupler center set 3/8" CTS		25	.320		.78	15.60		16.38	24.50
7080	1/2" CTS		23	.348		.60	16.95		17.55	26
7090	3/4" CTS		22	.364		.75	17.75		18.50	27.50
7100	Elbow 90°, copper 3/8"		25	.320		1.71	15.60		17.31	25.50
7110	1/2" CTS		23	.348		1.50	16.95		18.45	27
7120	3/4" CTS		22	.364		1.33	17.75		19.08	28
7130	Tee copper 3/8" CTS		17	.471		2.03	23		25.03	36.50
7140	1/2" CTS		15	.533		1.39	26		27.39	40.50
7150	3/4" CTS		14	.571		2.14	28		30.14	44.50
7160	3/8" x 3/8" x 1/2"		16	.500		1.39	24.50		25.89	38
7170	1/2" x 3/8" x 1/2"		15	.533		1.39	26		27.39	40.50
7180	3/4" x 1/2" x 3/4"		14	.571		1.39	28		29.39	43.50

22 11 13.27 Pipe/Tube, Grooved Joint for Copper		Crew	Daily Output	Labor-Hours	Unit	Material	Labor	Equipment	Total	Total Incl O&P
0010	**PIPE/TUBE, GROOVED JOINT FOR COPPER**									
4000	Fittings: coupling material required at joints not incl. in fitting price.									
4001	Add 1 selected coupling, material only, per joint for installed price.									
4010	Coupling, rigid style									
4018	2" diameter	1 Plum	50	.160	Ea.	12.15	7.80		19.95	25
4020	2-1/2" diameter	Q-1	80	.200		13.90	8.80		22.70	28.50
4022	3" diameter		67	.239		15.35	10.50		25.85	32.50
4024	4" diameter		50	.320		23.50	14.05		37.55	47
4026	5" diameter		40	.400		40	17.55		57.55	70.50
4028	6" diameter	Q-2	50	.480		53	22		75	91.50
4100	Elbow, 90° or 45°									
4108	2" diameter	1 Plum	25	.320	Ea.	22	15.60		37.60	47.50
4110	2-1/2" diameter	Q-1	40	.400		24	17.55		41.55	53
4112	3" diameter		33	.485		33.50	21.50		55	69
4114	4" diameter		25	.640		78	28		106	128
4116	5" diameter		20	.800		222	35		257	297
4118	6" diameter	Q-2	25	.960		355	43.50		398.50	455
4200	Tee									
4208	2" diameter	1 Plum	17	.471	Ea.	36.50	23		59.50	74.50
4210	2-1/2" diameter	Q-1	27	.593		39	26		65	81.50
4212	3" diameter		22	.727		58	32		90	112
4214	4" diameter		17	.941		127	41.50		168.50	202
4216	5" diameter		13	1.231		360	54		414	475
4218	6" diameter	Q-2	17	1.412		440	64		504	580
4300	Reducer, concentric									
4310	3" x 2-1/2" diameter	Q-1	35	.457	Ea.	31.50	20		51.50	65
4312	4" x 2-1/2" diameter		32	.500		65.50	22		87.50	105
4314	4" x 3" diameter		29	.552		65.50	24		89.50	109
4316	5" x 3" diameter		25	.640		185	28		213	246
4318	5" x 4" diameter		22	.727		185	32		217	252
4320	6" x 3" diameter	Q-2	28	.857		201	39		240	280
4322	6" x 4" diameter		26	.923		201	42		243	284
4324	6" x 5" diameter		24	1		201	45.50		246.50	290
4350	Flange, w/groove gasket									
4351	ANSI class 125 and 150									
4355	2" diameter	1 Plum	23	.348	Ea.	97	16.95		113.95	132
4356	2-1/2" diameter	Q-1	37	.432		101	19		120	140

22 11 13.27 Pipe/Tube, Grooved Joint for Copper		Crew	Daily Output	Labor-Hours	Unit	Material	Labor	Equipment	Total	Total Incl O&P
						2009 Bare Costs				
4358	3″ diameter	Q-1	31	.516	Ea.	106	22.50		128.50	150
4360	4″ diameter		23	.696		116	30.50		146.50	173
4362	5″ diameter		19	.842		156	37		193	227
4364	6″ diameter	Q-2	23	1.043		170	47.50		217.50	259

22 11 13.44 Pipe, Steel		Crew	Daily Output	Labor-Hours	Unit	Material	Labor	Equipment	Total	Total Incl O&P
0010	**PIPE, STEEL** R221113-50									
0020	All pipe sizes are to Spec. A-53 unless noted otherwise R221113-70									
0032	Schedule 10, see Div. 22 11 13.48 0500									
0050	Schedule 40, threaded, with couplings, and clevis hanger									
0060	assemblies sized for covering, 10′ O.C.									
0540	Black, 1/4″ diameter	1 Plum	66	.121	L.F.	2.17	5.90		8.07	11.25
0550	3/8″ diameter		65	.123		2.46	6		8.46	11.70
0560	1/2″ diameter		63	.127		2.23	6.20		8.43	11.75
0570	3/4″ diameter		61	.131		2.64	6.40		9.04	12.50
0580	1″ diameter		53	.151		3.87	7.35		11.22	15.30
0590	1-1/4″ diameter	Q-1	89	.180		4.89	7.90		12.79	17.25
0600	1-1/2″ diameter		80	.200		5.75	8.80		14.55	19.50
0610	2″ diameter		64	.250		7.70	10.95		18.65	25
0620	2-1/2″ diameter		50	.320		12.05	14.05		26.10	34.50
0630	3″ diameter		43	.372		15.50	16.35		31.85	41.50
0640	3-1/2″ diameter		40	.400		25.50	17.55		43.05	54.50
0650	4″ diameter		36	.444		23	19.50		42.50	54.50
0809	A-106, gr. A/B, seamless w/cplgs. & clevis hanger assemblies									
0811	1/4″ diameter	1 Plum	66	.121	L.F.	7.10	5.90		13	16.65
0812	3/8″ diameter		65	.123		8.70	6		14.70	18.55
0813	1/2″ diameter		63	.127		6.70	6.20		12.90	16.65
0814	3/4″ diameter		61	.131		8.25	6.40		14.65	18.65
0815	1″ diameter		53	.151		9.15	7.35		16.50	21
0816	1-1/4″ diameter	Q-1	89	.180		11.20	7.90		19.10	24
0817	1-1/2″ diameter		80	.200		14.35	8.80		23.15	29
0819	2″ diameter		64	.250		14.75	10.95		25.70	32.50
0821	2-1/2″ diameter		50	.320		19.05	14.05		33.10	42
0822	3″ diameter		43	.372		25	16.35		41.35	52
0823	4″ diameter		36	.444		34	19.50		53.50	66.50
1220	To delete coupling & hanger, subtract									
1230	1/4″ diam. to 3/4″ diam.					31%	56%			
1240	1″ diam. to 1-1/2″ diam.					23%	51%			
1250	2″ diam. to 4″ diam.					23%	41%			
1280	All pipe sizes are to Spec. A-53 unless noted otherwise									
1281	Schedule 40, threaded, with couplings and clevis hanger									
1282	assemblies sized for covering, 10′ O. C.									
1290	Galvanized, 1/4″ diameter	1 Plum	66	.121	L.F.	2.36	5.90		8.26	11.45
1300	3/8″ diameter		65	.123		3.31	6		9.31	12.65
1310	1/2″ diameter		63	.127		3.11	6.20		9.31	12.70
1320	3/4″ diameter		61	.131		3.76	6.40		10.16	13.75
1330	1″ diameter		53	.151		5.10	7.35		12.45	16.70
1340	1-1/4″ diameter	Q-1	89	.180		6.55	7.90		14.45	19.05
1350	1-1/2″ diameter		80	.200		7.80	8.80		16.60	21.50
1360	2″ diameter		64	.250		10.35	10.95		21.30	28
1370	2-1/2″ diameter		50	.320		17.60	14.05		31.65	40.50
1380	3″ diameter		43	.372		22	16.35		38.35	48.50
1390	3-1/2″ diameter		40	.400		28	17.55		45.55	57.50

22 11 13.44 Pipe, Steel		Crew	Daily Output	Labor-Hours	Unit	Material	2009 Bare Costs Labor	Equipment	Total	Total Incl O&P
1400	4" diameter	Q-1	36	.444	L.F.	32.50	19.50		52	65
1750	To delete coupling & hanger, subtract									
1760	1/4" diam. to 3/4" diam.					31%	56%			
1770	1" diam. to 1-1/2" diam.					23%	51%			
1780	2" diam. to 4" diam.					23%	41%			
2000	Welded, sch. 40, on yoke & roll hanger assys, sized for covering, 10' O.C.									
2040	Black, 1" diameter	Q-15	93	.172	L.F.	4.30	7.55	.60	12.45	16.75
2050	1-1/4" diameter		84	.190		5.55	8.35	.66	14.56	19.40
2060	1-1/2" diameter		76	.211		6.25	9.25	.73	16.23	21.50
2070	2" diameter		61	.262		7.95	11.50	.91	20.36	27
2080	2-1/2" diameter		47	.340		9.85	14.95	1.18	25.98	34.50
2090	3" diameter		43	.372		15.55	16.35	1.29	33.19	43
2100	3-1/2" diameter		39	.410		17.45	18	1.43	36.88	48
2110	4" diameter		37	.432		18.10	19	1.50	38.60	50
2120	5" diameter		32	.500		24.50	22	1.74	48.24	62
2130	6" diameter	Q-16	36	.667		31	30.50	1.55	63.05	81
2140	8" diameter		29	.828		46	37.50	1.92	85.42	110
2150	10" diameter		24	1		67	45.50	2.32	114.82	145
2160	12" diameter		19	1.263		80.50	57.50	2.93	140.93	178
2170	14" diameter, (two rod roll type hanger for 14" dia. and up)		15	1.600		91	73	3.71	167.71	213
2180	16" diameter, (two rod roll type hanger)		13	1.846		104	84	4.28	192.28	245
2190	18" diameter, (two rod roll type hanger)		11	2.182		121	99.50	5.05	225.55	288
2200	20" diameter, (two rod roll type hanger)		9	2.667		145	121	6.20	272.20	350
2220	24" diameter, (two rod roll type hanger)		8	3		201	137	6.95	344.95	435
2345	Sch. 40, A-53, gr. A/B, ERW, welded w/hngrs.									
2346	2" diameter	Q-15	61	.262	L.F.	6.50	11.50	.91	18.91	25.50
2347	2-1/2" diameter		47	.340		8.20	14.95	1.18	24.33	33
2348	3" diameter		43	.372		11.90	16.35	1.29	29.54	39
2349	4" diameter		38	.421		13.40	18.50	1.47	33.37	44
2350	6" diameter	Q-16	37	.649		45	29.50	1.50	76	95.50
2351	8" diameter		29	.828		58.50	37.50	1.92	97.92	123
2352	10" diameter		24	1		83	45.50	2.32	130.82	162
2363	.375" wall, A-53, gr. A/B, ERW, welded w/hngrs.									
2364	12" diameter	Q-16	20	1.200	L.F.	99	54.50	2.78	156.28	194
2365	14" diameter		17	1.412		109	64	3.28	176.28	220
2366	16" diameter		14	1.714		146	78	3.98	227.98	282
2560	To delete hanger, subtract									
2570	1" diam. to 1-1/2" diam.					15%	34%			
2580	2" diam. to 3-1/2" diam.					9%	21%			
2590	4" diam. to 12" diam.					5%	12%			
2596	14" diam. to 36" diam.					3%	10%			
2640	Galvanized, 1" diameter	Q-15	93	.172	L.F.	6.15	7.55	.60	14.30	18.80
2650	1-1/4" diameter		84	.190		8	8.35	.66	17.01	22
2660	1-1/2" diameter		76	.211		9.15	9.25	.73	19.13	25
2670	2" diameter		61	.262		11.75	11.50	.91	24.16	31
2680	2-1/2" diameter		47	.340		17.45	14.95	1.18	33.58	43
2690	3" diameter		43	.372		22.50	16.35	1.29	40.14	50.50
2700	3-1/2" diameter		39	.410		24.50	18	1.43	43.93	55.50
2710	4" diameter		37	.432		29	19	1.50	49.50	62
2720	5" diameter		32	.500		33	22	1.74	56.74	71.50
2730	6" diameter	Q-16	36	.667		38.50	30.50	1.55	70.55	89.50
2740	8" diameter		29	.828		44.50	37.50	1.92	83.92	108
2750	10" diameter		24	1		72	45.50	2.32	119.82	150

22 11 13.44 Pipe, Steel		Crew	Daily Output	Labor-Hours	Unit	2009 Bare Costs Material	Labor	Equipment	Total	Total Incl O&P
2760	12" diameter	Q-16	19	1.263	L.F.	91	57.50	2.93	151.43	190
3160	To delete hanger, subtract									
3170	1" diam. to 1-1/2" diam.					30%	34%			
3180	2" diam. to 3-1/2" diam.					19%	21%			
3190	4" diam. to 12" diam.					10%	12%			
3250	Flanged, 150 lb. weld neck, on yoke & roll hangers									
3260	sized for covering, 10' O.C.									
3290	Black, 1" diameter	Q-15	70	.229	L.F.	8.45	10.05	.80	19.30	25
3300	1-1/4" diameter		64	.250		9.70	10.95	.87	21.52	28
3310	1-1/2" diameter		58	.276		10.45	12.10	.96	23.51	30.50
3320	2" diameter		45	.356		12.60	15.60	1.24	29.44	39
3330	2-1/2" diameter		36	.444		14.95	19.50	1.55	36	47.50
3340	3" diameter		32	.500		21	22	1.74	44.74	58
3350	3-1/2" diameter		29	.552		25	24	1.92	50.92	66
3360	4" diameter		26	.615		26	27	2.14	55.14	71.50
3370	5" diameter		21	.762		36	33.50	2.65	72.15	92.50
3380	6" diameter	Q-16	25	.960		44	43.50	2.23	89.73	116
3390	8" diameter		19	1.263		66.50	57.50	2.93	126.93	163
3400	10" diameter		16	1.500		109	68.50	3.48	180.98	226
3410	12" diameter		14	1.714		130	78	3.98	211.98	263
3470	For 300 lb. flanges, add					63%				
3480	For 600 lb. flanges, add					310%				
3960	To delete flanges & hanger, subtract									
3970	1" diam. to 2" diam.					76%	65%			
3980	2-1/2" diam. to 4" diam.					62%	59%			
3990	5" diam. to 12" diam.					60%	46%			
4040	Galvanized, 1" diameter	Q-15	70	.229	L.F.	10.35	10.05	.80	21.20	27.50
4050	1-1/4" diameter		64	.250		12.15	10.95	.87	23.97	31
4060	1-1/2" diameter		58	.276		13.35	12.10	.96	26.41	34
4070	2" diameter		45	.356		16.10	15.60	1.24	32.94	42.50
4080	2-1/2" diameter		36	.444		22.50	19.50	1.55	43.55	55.50
4090	3" diameter		32	.500		27.50	22	1.74	51.24	65.50
4100	3-1/2" diameter		29	.552		30.50	24	1.92	56.42	72
4110	4" diameter		26	.615		37	27	2.14	66.14	83.50
4120	5" diameter		21	.762		44	33.50	2.65	80.15	101
4130	6" diameter	Q-16	25	.960		49.50	43.50	2.23	95.23	122
4140	8" diameter		19	1.263		57	57.50	2.93	117.43	153
4150	10" diameter		16	1.500		92	68.50	3.48	163.98	207
4160	12" diameter		14	1.714		116	78	3.98	197.98	248
4220	For 300 lb. flanges, add					65%				
4240	For 600 lb. flanges, add					350%				
4660	To delete flanges & hanger, subtract									
4670	1" diam. to 2" diam.					73%	65%			
4680	2-1/2" diam. to 4" diam.					58%	59%			
4690	5" diam. to 12" diam.					43%	46%			
4750	Schedule 80, threaded, with couplings, and clevis hanger assemblies									
4760	sized for covering, 10' O.C.									
4790	Black, 1/4" diameter	1 Plum	54	.148	L.F.	2.81	7.20		10.01	13.95
4800	3/8" diameter		53	.151		3.60	7.35		10.95	15
4810	1/2" diameter		52	.154		3.32	7.50		10.82	14.90
4820	3/4" diameter		50	.160		4.05	7.80		11.85	16.15
4830	1" diameter		45	.178		5.55	8.65		14.20	19.10
4840	1-1/4" diameter	Q-1	75	.213		7.35	9.35		16.70	22

22 11 13.44 Pipe, Steel		Crew	Daily Output	Labor-Hours	Unit	2009 Bare Costs Material	Labor	Equipment	Total	Total Incl O&P
4850	1-1/2" diameter	Q-1	69	.232	L.F.	8.60	10.15		18.75	25
4860	2" diameter		56	.286		11.70	12.55		24.25	31.50
4870	2-1/2" diameter		44	.364		18.10	15.95		34.05	44
4880	3" diameter		38	.421		24	18.50		42.50	54
4890	3-1/2" diameter		35	.457		28.50	20		48.50	61.50
4900	4" diameter		32	.500		26	22		48	61.50
5061	A-106, gr. A/B seamless with cplgs. & clevis hanger assemblies, 1/4" dia.	1 Plum	63	.127		6.60	6.20		12.80	16.55
5062	3/8" diameter		62	.129		7.75	6.30		14.05	17.95
5063	1/2" diameter		61	.131		6.75	6.40		13.15	17
5064	3/4" diameter		57	.140		7.85	6.85		14.70	18.90
5065	1" diameter		51	.157		9.40	7.65		17.05	22
5066	1-1/4" diameter	Q-1	85	.188		12.50	8.25		20.75	26
5067	1-1/2" diameter		77	.208		15.50	9.10		24.60	31
5071	2" diameter		61	.262		18.10	11.50		29.60	37
5072	2-1/2" diameter		48	.333		19.30	14.65		33.95	43.50
5073	3" diameter		41	.390		25.50	17.10		42.60	53.50
5074	4" diameter		35	.457		37.50	20		57.50	71
5430	To delete coupling & hanger, subtract									
5440	1/4" diam. to 1/2" diam.					31%	54%			
5450	3/4" diam. to 1-1/2" diam.					28%	49%			
5460	2" diam. to 4" diam.					21%	40%			
5510	Galvanized, 1/4" diameter	1 Plum	54	.148	L.F.	4.20	7.20		11.40	15.45
5520	3/8" diameter		53	.151		4.23	7.35		11.58	15.70
5530	1/2" diameter		52	.154		4.34	7.50		11.84	16
5540	3/4" diameter		50	.160		5.25	7.80		13.05	17.50
5550	1" diameter		45	.178		7.35	8.65		16	21
5560	1-1/4" diameter	Q-1	75	.213		9.85	9.35		19.20	25
5570	1-1/2" diameter		69	.232		11.60	10.15		21.75	28
5580	2" diameter		56	.286		16.05	12.55		28.60	36.50
5590	2-1/2" diameter		44	.364		25	15.95		40.95	51.50
5600	3" diameter		38	.421		33	18.50		51.50	63.50
5610	3-1/2" diameter		35	.457		35.50	20		55.50	69
5620	4" diameter		32	.500		46.50	22		68.50	84
5930	To delete coupling & hanger, subtract									
5940	1/4" diam. to 1/2" diam.					31%	54%			
5950	3/4" diam. to 1-1/2" diam.					28%	49%			
5960	2" diam. to 4" diam.					21%	40%			
6000	Welded, on yoke & roller hangers									
6010	sized for covering, 10' O.C.									
6040	Black, 1" diameter	Q-15	85	.188	L.F.	6	8.25	.66	14.91	19.70
6050	1-1/4" diameter		79	.203		7.90	8.90	.70	17.50	23
6060	1-1/2" diameter		72	.222		9.15	9.75	.77	19.67	25.50
6070	2" diameter		57	.281		11.80	12.30	.98	25.08	32.50
6080	2-1/2" diameter		44	.364		17.75	15.95	1.27	34.97	45
6090	3" diameter		40	.400		23.50	17.55	1.39	42.44	53.50
6100	3-1/2" diameter		34	.471		27.50	20.50	1.64	49.64	63.50
6110	4" diameter		33	.485		24.50	21.50	1.69	47.69	61
6120	5" diameter		26	.615		50.50	27	2.14	79.64	98.50
6130	6" diameter	Q-16	30	.800		48.50	36.50	1.86	86.86	110
6140	8" diameter		25	.960		84	43.50	2.23	129.73	160
6150	10" diameter		20	1.200		150	54.50	2.78	207.28	250
6160	12" diameter		15	1.600		231	73	3.71	307.71	365
6540	To delete hanger, subtract									

22 11 13.44 Pipe, Steel		Crew	Daily Output	Labor-Hours	Unit	Material	Labor	Equipment	Total	Total Incl O&P
						2009 Bare Costs				
6550	1" diam. to 1-1/2" diam.					30%	14%			
6560	2" diam. to 3" diam.					23%	9%			
6570	3-1/2" diam. to 5" diam.					12%	6%			
6580	6" diam. to 12" diam.					10%	4%			
6610	Galvanized, 1" diameter	Q-15	85	.188	L.F.	7.75	8.25	.66	16.66	21.50
6650	1-1/4" diameter		79	.203		10.35	8.90	.70	19.95	25.50
6660	1-1/2" diameter		72	.222		12.10	9.75	.77	22.62	29
6670	2" diameter		57	.281		16.10	12.30	.98	29.38	37.50
6680	2-1/2" diameter		44	.364		24.50	15.95	1.27	41.72	52.50
6690	3" diameter		40	.400		32	17.55	1.39	50.94	63.50
6700	3-1/2" diameter		34	.471		34.50	20.50	1.64	56.64	71
6710	4" diameter		33	.485		45	21.50	1.69	68.19	83.50
6720	5" diameter		26	.615		77	27	2.14	106.14	127
6730	6" diameter	Q-16	30	.800		84.50	36.50	1.86	122.86	150
6740	8" diameter		25	.960		127	43.50	2.23	172.73	208
6750	10" diameter		20	1.200		237	54.50	2.78	294.28	345
6760	12" diameter		15	1.600		340	73	3.71	416.71	490
7150	To delete hanger, subtract									
7160	1" diam. to 1-1/2" diam.					26%	14%			
7170	2" diam. to 3" diam.					16%	9%			
7180	3-1/2" diam. to 5" diam.					10%	6%			
7190	6" diam. to 12" diam.					7%	4%			
7250	Flanged, 300 lb. weld neck, on yoke & roll hangers									
7260	sized for covering, 10' O.C.									
7290	Black, 1" diameter	Q-15	66	.242	L.F.	11.10	10.65	.84	22.59	29
7300	1-1/4" diameter		61	.262		13	11.50	.91	25.41	32.50
7310	1-1/2" diameter		54	.296		14.30	13	1.03	28.33	36.50
7320	2" diameter		42	.381		17.65	16.70	1.33	35.68	46
7330	2-1/2" diameter		33	.485		24.50	21.50	1.69	47.69	61
7340	3" diameter		29	.552		31	24	1.92	56.92	72.50
7350	3-1/2" diameter		24	.667		40	29.50	2.32	71.82	90.50
7360	4" diameter		23	.696		37	30.50	2.42	69.92	89
7370	5" diameter		19	.842		69	37	2.93	108.93	135
7380	6" diameter	Q-16	23	1.043		66	47.50	2.42	115.92	147
7390	8" diameter		17	1.412		115	64	3.28	182.28	226
7400	10" diameter		14	1.714		205	78	3.98	286.98	345
7410	12" diameter		12	2		300	91	4.64	395.64	470
7470	For 600 lb. flanges, add					100%				
7940	To delete flanges & hanger, subtract									
7950	1" diam. to 1-1/2" diam.					75%	66%			
7960	2" diam. to 3" diam.					62%	60%			
7970	3-1/2" diam. to 5" diam.					54%	66%			
7980	6" diam. to 12" diam.					55%	62%			
8040	Galvanized, 1" diameter	Q-15	66	.242	L.F.	12.90	10.65	.84	24.39	31
8050	1-1/4" diameter		61	.262		15.50	11.50	.91	27.91	35.50
8060	1-1/2" diameter		54	.296		17.25	13	1.03	31.28	39.50
8070	2" diameter		42	.381		22	16.70	1.33	40.03	50.50
8080	2-1/2" diameter		33	.485		31.50	21.50	1.69	54.69	68.50
8090	3" diameter		29	.552		40	24	1.92	65.92	82.50
8100	3-1/2" diameter		24	.667		47	29.50	2.32	78.82	98.50
8110	4" diameter		23	.696		57.50	30.50	2.42	90.42	112
8120	5" diameter		19	.842		95.50	37	2.93	135.43	164
8130	6" diameter	Q-16	23	1.043		102	47.50	2.42	151.92	187

22 11 13.44 Pipe, Steel		Crew	Daily Output	Labor-Hours	Unit	Material	2009 Bare Costs Labor	Equipment	Total	Total Incl O&P
8140	8" diameter	Q-16	17	1.412	L.F.	158	64	3.28	225.28	274
8150	10" diameter		14	1.714		291	78	3.98	372.98	440
8160	12" diameter		12	2		410	91	4.64	505.64	590
8240	For 600 lb. flanges, add					100%				
8900	To delete flanges & hangers, subtract									
8910	1" diam. to 1-1/2" diam.					72%	66%			
8920	2" diam. to 3" diam.					59%	60%			
8930	3-1/2" diam. to 5" diam.					51%	66%			
8940	6" diam. to 12" diam.					49%	62%			
9000	Threading pipe labor, one end, all schedules through 80									
9010	1/4" through 3/4" pipe size	1 Plum	80	.100	Ea.		4.88		4.88	7.30
9020	1" through 2" pipe size		73	.110			5.35		5.35	8
9030	2-1/2" pipe size		53	.151			7.35		7.35	11.05
9040	3" pipe size		50	.160			7.80		7.80	11.70
9050	3-1/2" pipe size	Q-1	89	.180			7.90		7.90	11.85
9060	4" pipe size		73	.219			9.60		9.60	14.45
9070	5" pipe size		53	.302			13.25		13.25	19.90
9080	6" pipe size		46	.348			15.25		15.25	23
9090	8" pipe size		29	.552			24		24	36.50
9100	10" pipe size		21	.762			33.50		33.50	50
9110	12" pipe size		13	1.231			54		54	81
9200	Welding labor per joint									
9210	Schedule 40,									
9230	1/2" pipe size	Q-15	32	.500	Ea.		22	1.74	23.74	35
9240	3/4" pipe size		27	.593			26	2.06	28.06	41.50
9250	1" pipe size		23	.696			30.50	2.42	32.92	48.50
9260	1-1/4" pipe size		20	.800			35	2.78	37.78	55.50
9270	1-1/2" pipe size		19	.842			37	2.93	39.93	58.50
9280	2" pipe size		16	1			44	3.48	47.48	70
9290	2-1/2" pipe size		13	1.231			54	4.28	58.28	85.50
9300	3" pipe size		12	1.333			58.50	4.64	63.14	93
9310	4" pipe size		10	1.600			70	5.55	75.55	111
9320	5" pipe size		9	1.778			78	6.20	84.20	124
9330	6" pipe size		8	2			88	6.95	94.95	140
9340	8" pipe size		5	3.200			140	11.15	151.15	223
9350	10" pipe size		4	4			176	13.90	189.90	278
9360	12" pipe size		3	5.333			234	18.55	252.55	370
9370	14" pipe size		2.60	6.154			270	21.50	291.50	430
9380	16" pipe size		2.20	7.273			320	25.50	345.50	510
9390	18" pipe size		2	8			350	28	378	555
9400	20" pipe size		1.80	8.889			390	31	421	620
9410	22" pipe size		1.70	9.412			415	33	448	655
9420	24" pipe size		1.50	10.667			470	37	507	740
9450	Schedule 80,									
9460	1/2" pipe size	Q-15	27	.593	Ea.		26	2.06	28.06	41.50
9470	3/4" pipe size		23	.696			30.50	2.42	32.92	48.50
9480	1" pipe size		20	.800			35	2.78	37.78	55.50
9490	1-1/4" pipe size		19	.842			37	2.93	39.93	58.50
9500	1-1/2" pipe size		18	.889			39	3.09	42.09	62
9510	2" pipe size		15	1.067			47	3.71	50.71	74
9520	2-1/2" pipe size		12	1.333			58.50	4.64	63.14	93
9530	3" pipe size		11	1.455			64	5.05	69.05	102
9540	4" pipe size		8	2			88	6.95	94.95	140

22 11 13 – Facility Water Distribution Piping

22 11 13.44 Pipe, Steel		Crew	Daily Output	Labor-Hours	Unit	Material	2009 Bare Costs Labor	Equipment	Total	Total Incl O&P
9550	5" pipe size	Q-15	6	2.667	Ea.		117	9.30	126.30	186
9560	6" pipe size		5	3.200			140	11.15	151.15	223
9570	8" pipe size		4	4			176	13.90	189.90	278
9580	10" pipe size		3	5.333			234	18.55	252.55	370
9590	12" pipe size		2	8			350	28	378	555
9600	14" pipe size	Q-16	2.60	9.231			420	21.50	441.50	655
9610	16" pipe size		2.30	10.435			475	24	499	735
9620	18" pipe size		2	12			545	28	573	850
9630	20" pipe size		1.80	13.333			605	31	636	945
9640	22" pipe size		1.60	15			685	35	720	1,075
9650	24" pipe size		1.50	16			730	37	767	1,150

22 11 13.45 Pipe Fittings, Steel, Threaded		Crew	Daily Output	Labor-Hours	Unit	Material	Labor	Equipment	Total	Total Incl O&P
0010	**PIPE FITTINGS, STEEL, THREADED**									
0020	Cast Iron									
0040	Standard weight, black									
0060	90° Elbow, straight									
0070	1/4"	1 Plum	16	.500	Ea.	5.80	24.50		30.30	43
0080	3/8"		16	.500		8.40	24.50		32.90	46
0090	1/2"		15	.533		3.69	26		29.69	43
0100	3/4"		14	.571		3.84	28		31.84	46
0110	1"		13	.615		4.54	30		34.54	50
0120	1-1/4"	Q-1	22	.727		6.45	32		38.45	55
0130	1-1/2"		20	.800		8.90	35		43.90	62.50
0140	2"		18	.889		13.95	39		52.95	74
0150	2-1/2"		14	1.143		33.50	50		83.50	113
0160	3"		10	1.600		54.50	70		124.50	165
0170	3-1/2"		8	2		149	88		237	295
0180	4"		6	2.667		101	117		218	288
0250	45° Elbow, straight									
0260	1/4"	1 Plum	16	.500	Ea.	6.95	24.50		31.45	44
0270	3/8"		16	.500		7.50	24.50		32	45
0280	1/2"		15	.533		5.65	26		31.65	45
0300	3/4"		14	.571		5.65	28		33.65	48.50
0320	1"		13	.615		6.65	30		36.65	52.50
0330	1-1/4"	Q-1	22	.727		8.95	32		40.95	58
0340	1-1/2"		20	.800		14.80	35		49.80	69
0350	2"		18	.889		17.10	39		56.10	77.50
0360	2-1/2"		14	1.143		44.50	50		94.50	125
0370	3"		10	1.600		70.50	70		140.50	183
0380	3-1/2"		8	2		156	88		244	305
0400	4"		6	2.667		146	117		263	335
0500	Tee, straight									
0510	1/4"	1 Plum	10	.800	Ea.	9.10	39		48.10	68.50
0520	3/8"		10	.800		8.85	39		47.85	68
0530	1/2"		9	.889		5.75	43.50		49.25	71.50
0540	3/4"		9	.889		6.70	43.50		50.20	72.50
0550	1"		8	1		5.95	49		54.95	79.50
0560	1-1/4"	Q-1	14	1.143		10.85	50		60.85	87.50
0570	1-1/2"		13	1.231		14.15	54		68.15	96.50
0580	2"		11	1.455		19.70	64		83.70	118
0590	2-1/2"		9	1.778		51	78		129	173
0600	3"		6	2.667		79	117		196	263

22 11 Facility Water Distribution

22 11 13 – Facility Water Distribution Piping

22 11 13.45 Pipe Fittings, Steel, Threaded		Crew	Daily Output	Labor-Hours	Unit	2009 Bare Costs Material	Labor	Equipment	Total	Total Incl O&P
0610	3-1/2"	Q-1	5	3.200	Ea.	158	140		298	385
0620	4"		4	4		153	176		329	430
0660	Tee, reducing, run or outlet									
0661	1/2"	1 Plum	9	.889	Ea.	12.90	43.50		56.40	79
0662	3/4"		9	.889		11.20	43.50		54.70	77.50
0663	1"		8	1		12.25	49		61.25	86.50
0664	1-1/4"	Q-1	14	1.143		10.90	50		60.90	87.50
0665	1-1/2"		13	1.231		19.50	54		73.50	103
0666	2"		11	1.455		22.50	64		86.50	121
0667	2-1/2"		9	1.778		67	78		145	191
0668	3"		6	2.667		133	117		250	320
0669	3-1/2"		5	3.200		171	140		311	400
0670	4"		4	4		243	176		419	530
0674	Reducer, concentric									
0675	3/4"	1 Plum	18	.444	Ea.	10.40	21.50		31.90	44
0676	1"	"	15	.533		7	26		33	46.50
0677	1-1/4"	Q-1	26	.615		20	27		47	62.50
0678	1-1/2"		24	.667		32	29.50		61.50	79
0679	2"		21	.762		37	33.50		70.50	91
0680	2-1/2"		18	.889		53	39		92	117
0681	3"		14	1.143		73.50	50		123.50	156
0682	3-1/2"		12	1.333		139	58.50		197.50	240
0683	4"		10	1.600		150	70		220	270
0687	Reducer, eccentric									
0688	3/4"	1 Plum	16	.500	Ea.	21.50	24.50		46	60.50
0689	1"	"	14	.571		23.50	28		51.50	67.50
0690	1-1/4"	Q-1	25	.640		37.50	28		65.50	83
0691	1-1/2"		22	.727		49.50	32		81.50	103
0692	2"		20	.800		69	35		104	129
0693	2-1/2"		16	1		95	44		139	171
0694	3"		12	1.333		150	58.50		208.50	253
0695	3-1/2"		10	1.600		209	70		279	335
0696	4"		9	1.778		271	78		349	415
0700	Standard weight, galvanized cast iron									
0720	90° Elbow, straight									
0730	1/4"	1 Plum	16	.500	Ea.	9.10	24.50		33.60	46.50
0740	3/8"		16	.500		9.10	24.50		33.60	46.50
0750	1/2"		15	.533		6.90	26		32.90	46.50
0760	3/4"		14	.571		10.20	28		38.20	53
0770	1"		13	.615		11.75	30		41.75	58
0780	1-1/4"	Q-1	22	.727		18.30	32		50.30	68
0790	1-1/2"		20	.800		25	35		60	80
0800	2"		18	.889		37	39		76	99.50
0810	2-1/2"		14	1.143		76	50		126	160
0820	3"		10	1.600		116	70		186	232
0830	3-1/2"		8	2		185	88		273	335
0840	4"		6	2.667		212	117		329	410
0900	45° Elbow, straight									
0910	1/4"	1 Plum	16	.500	Ea.	13	24.50		37.50	51
0920	3/8"		16	.500		14	24.50		38.50	52
0930	1/2"		15	.533		10.35	26		36.35	50.50
0940	3/4"		14	.571		11.90	28		39.90	55
0950	1"		13	.615		12.45	30		42.45	58.50

	22 11 13.45 Pipe Fittings, Steel, Threaded	Crew	Daily Output	Labor-Hours	Unit	2009 Bare Costs Material	Labor	Equipment	Total	Total Incl O&P
0960	1-1/4"	Q-1	22	.727	Ea.	16.70	32		48.70	66.50
0970	1-1/2"		20	.800		31	35		66	86.50
0980	2"		18	.889		43.50	39		82.50	106
0990	2-1/2"		14	1.143		85	50		135	169
1000	3"		10	1.600		136	70		206	255
1010	3-1/2"		8	2		217	88		305	370
1020	4"		6	2.667		294	117		411	500
1100	Tee, straight									
1110	1/4"	1 Plum	10	.800	Ea.	16.85	39		55.85	77
1120	3/8"		10	.800		11.60	39		50.60	71.50
1130	1/2"		9	.889		10.70	43.50		54.20	77
1140	3/4"		9	.889		14.65	43.50		58.15	81
1150	1"		8	1		15.90	49		64.90	90.50
1160	1-1/4"	Q-1	14	1.143		28	50		78	106
1170	1-1/2"		13	1.231		37	54		91	122
1180	2"		11	1.455		46	64		110	147
1190	2-1/2"		9	1.778		99	78		177	226
1200	3"		6	2.667		256	117		373	455
1210	3-1/2"		5	3.200		263	140		403	500
1220	4"		4	4		294	176		470	590
1300	Extra heavy weight, black									
1310	Couplings, steel straight									
1320	1/4"	1 Plum	19	.421	Ea.	2.34	20.50		22.84	33.50
1330	3/8"		19	.421		2.56	20.50		23.06	34
1340	1/2"		19	.421		3.46	20.50		23.96	35
1350	3/4"		18	.444		3.70	21.50		25.20	36.50
1360	1"		15	.533		4.70	26		30.70	44
1370	1-1/4"	Q-1	26	.615		7.55	27		34.55	49
1380	1-1/2"		24	.667		7.55	29.50		37.05	52.50
1390	2"		21	.762		11.60	33.50		45.10	63
1400	2-1/2"		18	.889		17.05	39		56.05	77.50
1410	3"		14	1.143		20.50	50		70.50	98
1420	3-1/2"		12	1.333		27	58.50		85.50	118
1430	4"		10	1.600		32	70		102	140
1510	90° Elbow, straight									
1520	1/2"	1 Plum	15	.533	Ea.	18.25	26		44.25	59
1530	3/4"		14	.571		19.30	28		47.30	63
1540	1"		13	.615		23.50	30		53.50	70.50
1550	1-1/4"	Q-1	22	.727		35	32		67	86
1560	1-1/2"		20	.800		43	35		78	100
1580	2"		18	.889		53	39		92	117
1590	2-1/2"		14	1.143		124	50		174	213
1600	3"		10	1.600		164	70		234	286
1610	4"		6	2.667		345	117		462	555
1650	45° Elbow, straight									
1660	1/2"	1 Plum	15	.533	Ea.	25.50	26		51.50	67
1670	3/4"		14	.571		25	28		53	69.50
1680	1"		13	.615		30	30		60	78
1690	1-1/4"	Q-1	22	.727		49.50	32		81.50	103
1700	1-1/2"		20	.800		54.50	35		89.50	113
1710	2"		18	.889		78	39		117	144
1720	2-1/2"		14	1.143		134	50		184	224
1800	Tee, straight									

22 11 Facility Water Distribution

22 11 13 – Facility Water Distribution Piping

22 11 13.45 Pipe Fittings, Steel, Threaded		Crew	Daily Output	Labor-Hours	Unit	Material	Labor	Equipment	Total	Total Incl O&P
						2009 Bare Costs				
1810	1/2"	1 Plum	9	.889	Ea.	29	43.50		72.50	96.50
1820	3/4"		9	.889		29	43.50		72.50	96.50
1830	1"		8	1		35	49		84	111
1840	1-1/4"	Q-1	14	1.143		52	50		102	133
1850	1-1/2"		13	1.231		66.50	54		120.50	155
1860	2"		11	1.455		83	64		147	187
1870	2-1/2"		9	1.778		177	78		255	310
1880	3"		6	2.667		241	117		358	440
1890	4"		4	4		465	176		641	775
4000	Standard weight, black									
4010	Couplings, steel straight, merchants									
4030	1/4"	1 Plum	19	.421	Ea.	.63	20.50		21.13	31.50
4040	3/8"		19	.421		.75	20.50		21.25	32
4050	1/2"		19	.421		.80	20.50		21.30	32
4060	3/4"		18	.444		1.03	21.50		22.53	33.50
4070	1"		15	.533		1.44	26		27.44	40.50
4080	1-1/4"	Q-1	26	.615		1.84	27		28.84	42.50
4090	1-1/2"		24	.667		2.33	29.50		31.83	46.50
4100	2"		21	.762		3.33	33.50		36.83	53.50
4110	2-1/2"		18	.889		9.50	39		48.50	69
4120	3"		14	1.143		13.30	50		63.30	90
4130	3-1/2"		12	1.333		23.50	58.50		82	114
4140	4"		10	1.600		23.50	70		93.50	131
4166	Plug, 1/4"	1 Plum	38	.211		1.84	10.25		12.09	17.40
4167	3/8"		38	.211		1.84	10.25		12.09	17.40
4168	1/2"		38	.211		1.63	10.25		11.88	17.20
4169	3/4"		32	.250		4.94	12.20		17.14	24
4170	1"		30	.267		5.25	13		18.25	25.50
4171	1-1/4"	Q-1	52	.308		6.05	13.50		19.55	27
4172	1-1/2"		48	.333		8.60	14.65		23.25	31.50
4173	2"		42	.381		11.10	16.70		27.80	37
4176	2-1/2"		36	.444		16.40	19.50		35.90	47.50
4180	4"		20	.800		28.50	35		63.50	83.50
4200	Standard weight, galvanized									
4210	Couplings, steel straight, merchants									
4230	1/4"	1 Plum	19	.421	Ea.	.73	20.50		21.23	32
4240	3/8"		19	.421		.93	20.50		21.43	32
4250	1/2"		19	.421		.99	20.50		21.49	32
4260	3/4"		18	.444		1.24	21.50		22.74	34
4270	1"		15	.533		1.73	26		27.73	41
4280	1-1/4"	Q-1	26	.615		2.21	27		29.21	43
4290	1-1/2"		24	.667		2.75	29.50		32.25	47
4300	2"		21	.762		4.11	33.50		37.61	54.50
4310	2-1/2"		18	.889		11.65	39		50.65	71.50
4320	3"		14	1.143		15.45	50		65.45	92.50
4330	3-1/2"		12	1.333		27.50	58.50		86	118
4340	4"		10	1.600		27.50	70		97.50	135
4370	Plug, galvanized, square head									
4374	1/2"	1 Plum	38	.211	Ea.	3.95	10.25		14.20	19.75
4375	3/4"		32	.250		3.95	12.20		16.15	22.50
4376	1"		30	.267		3.95	13		16.95	24
4377	1-1/4"	Q-1	52	.308		6.60	13.50		20.10	28
4378	1-1/2"		48	.333		8.90	14.65		23.55	32

22 11 13.45 Pipe Fittings, Steel, Threaded		Crew	Daily Output	Labor-Hours	Unit	2009 Bare Costs Material	Labor	Equipment	Total	Total Incl O&P
4379	2″	Q-1	42	.381	Ea.	11.05	16.70		27.75	37
4380	2-1/2″		36	.444		23.50	19.50		43	55
4381	3″		28	.571		30	25		55	70.50
4382	4″		20	.800		80	35		115	141
4700	Nipple, black									
4710	1/2″ x 4″ long	1 Plum	19	.421	Ea.	1.33	20.50		21.83	32.50
4712	3/4″ x 4″ long		18	.444		1.62	21.50		23.12	34.50
4714	1″ x 4″ long		15	.533		2.24	26		28.24	41.50
4716	1-1/4″ x 4″ long	Q-1	26	.615		2.81	27		29.81	43.50
4718	1-1/2″ x 4″ long		24	.667		3.33	29.50		32.83	47.50
4720	2″ x 4″ long		21	.762		4.64	33.50		38.14	55
4722	2-1/2″ x 4″ long		18	.889		12.65	39		51.65	72.50
4724	3″ x 4″ long		14	1.143		16.05	50		66.05	93
4726	4″ x 4″ long		10	1.600		21.50	70		91.50	129
4800	Nipple, galvanized									
4810	1/2″ x 4″ long	1 Plum	19	.421	Ea.	1.64	20.50		22.14	33
4812	3/4″ x 4″ long		18	.444		2.03	21.50		23.53	34.50
4814	1″ x 4″ long		15	.533		2.74	26		28.74	42
4816	1-1/4″ x 4″ long	Q-1	26	.615		3.37	27		30.37	44
4818	1-1/2″ x 4″ long		24	.667		4.26	29.50		33.76	48.50
4820	2″ x 4″ long		21	.762		5.45	33.50		38.95	56
4822	2-1/2″ x 4″ long		18	.889		14.25	39		53.25	74
4824	3″ x 4″ long		14	1.143		18.90	50		68.90	96.50
4826	4″ x 4″ long		10	1.600		25.50	70		95.50	133
5000	Malleable iron, 150 lb.									
5020	Black									
5040	90° elbow, straight									
5060	1/4″	1 Plum	16	.500	Ea.	2.94	24.50		27.44	39.50
5070	3/8″		16	.500		2.94	24.50		27.44	39.50
5080	1/2″		15	.533		2.94	26		28.94	42
5090	3/4″		14	.571		3.37	28		31.37	45.50
5100	1″		13	.615		4.27	30		34.27	49.50
5110	1-1/4″	Q-1	22	.727		7.05	32		39.05	56
5120	1-1/2″		20	.800		9.25	35		44.25	62.50
5130	2″		18	.889		16.70	39		55.70	77
5140	2-1/2″		14	1.143		35.50	50		85.50	115
5150	3″		10	1.600		52	70		122	163
5160	3-1/2″		8	2		139	88		227	285
5170	4″		6	2.667		112	117		229	299
5250	45° elbow, straight									
5270	1/4″	1 Plum	16	.500	Ea.	4.43	24.50		28.93	41.50
5280	3/8″		16	.500		4.43	24.50		28.93	41.50
5290	1/2″		15	.533		3.36	26		29.36	42.50
5300	3/4″		14	.571		4.16	28		32.16	46.50
5310	1″		13	.615		5.25	30		35.25	51
5320	1-1/4″	Q-1	22	.727		9.25	32		41.25	58
5330	1-1/2″		20	.800		11.45	35		46.45	65
5340	2″		18	.889		17.35	39		56.35	77.50
5350	2-1/2″		14	1.143		50	50		100	131
5360	3″		10	1.600		65	70		135	177
5370	3-1/2″		8	2		124	88		212	268
5380	4″		6	2.667		128	117		245	315
5450	Tee, straight									

22 11 13 – Facility Water Distribution Piping

22 11 13.45 Pipe Fittings, Steel, Threaded		Crew	Daily Output	Labor-Hours	Unit	Material	Labor	Equipment	Total	Total Incl O&P
						2009 Bare Costs				
5470	1/4"	1 Plum	10	.800	Ea.	4.27	39		43.27	63
5480	3/8"		10	.800		4.27	39		43.27	63
5490	1/2"		9	.889		2.72	43.50		46.22	68
5500	3/4"		9	.889		3.91	43.50		47.41	69.50
5510	1"		8	1		6.70	49		55.70	80.50
5520	1-1/4"	Q-1	14	1.143		10.85	50		60.85	87.50
5530	1-1/2"		13	1.231		13.50	54		67.50	96
5540	2"		11	1.455		23	64		87	122
5550	2-1/2"		9	1.778		49.50	78		127.50	172
5560	3"		6	2.667		73	117		190	257
5570	3-1/2"		5	3.200		164	140		304	390
5580	4"		4	4		176	176		352	455
5601	Tee, reducing, on outlet									
5602	1/2"	1 Plum	9	.889	Ea.	6.15	43.50		49.65	72
5603	3/4"		9	.889		6.75	43.50		50.25	72.50
5604	1"		8	1		10.75	49		59.75	85
5605	1-1/4"	Q-1	14	1.143		18.40	50		68.40	95.50
5606	1-1/2"		13	1.231		19	54		73	102
5607	2"		11	1.455		26	64		90	125
5608	2-1/2"		9	1.778		75	78		153	200
5609	3"		6	2.667		102	117		219	288
5610	3-1/2"		5	3.200		228	140		368	460
5611	4"		4	4		228	176		404	515
5650	Coupling									
5670	1/4"	1 Plum	19	.421	Ea.	3.65	20.50		24.15	35
5680	3/8"		19	.421		3.65	20.50		24.15	35
5690	1/2"		19	.421		2.81	20.50		23.31	34
5700	3/4"		18	.444		3.30	21.50		24.80	36
5710	1"		15	.533		4.93	26		30.93	44.50
5720	1-1/4"	Q-1	26	.615		6.40	27		33.40	47.50
5730	1-1/2"		24	.667		8.65	29.50		38.15	53.50
5740	2"		21	.762		12.80	33.50		46.30	64
5750	2-1/2"		18	.889		35.50	39		74.50	97.50
5760	3"		14	1.143		48	50		98	128
5770	3-1/2"		12	1.333		96	58.50		154.50	194
5780	4"		10	1.600		96	70		166	211
5840	Reducer, concentric, 1/4"	1 Plum	19	.421		3.84	20.50		24.34	35
5850	3/8"		19	.421		4.93	20.50		25.43	36.50
5860	1/2"		19	.421		3.89	20.50		24.39	35.50
5870	3/4"		16	.500		4.60	24.50		29.10	41.50
5880	1"		15	.533		7.25	26		33.25	47
5890	1-1/4"	Q-1	26	.615		8.70	27		35.70	50
5900	1-1/2"		24	.667		12.45	29.50		41.95	57.50
5910	2"		21	.762		18	33.50		51.50	70
5911	2-1/2"		18	.889		40	39		79	103
5912	3"		14	1.143		49.50	50		99.50	130
5913	3-1/2"		12	1.333		150	58.50		208.50	253
5914	4"		10	1.600		114	70		184	230
5981	Bushing, 1/4"	1 Plum	19	.421		3.82	20.50		24.32	35
5982	3/8"		19	.421		4.20	20.50		24.70	35.50
5983	1/2"		19	.421		4.20	20.50		24.70	35.50
5984	3/4"		16	.500		4.04	24.50		28.54	41
5985	1"		15	.533		5.90	26		31.90	45.50

22 11 13.45 Pipe Fittings, Steel, Threaded		Crew	Daily Output	Labor-Hours	Unit	2009 Bare Costs Material	Labor	Equipment	Total	Total Incl O&P
5986	1-1/4"	Q-1	26	.615	Ea.	7.35	27		34.35	48.50
5987	1-1/2"		24	.667		8.30	29.50		37.80	53
5988	2"		21	.762		8.95	33.50		42.45	60
5989	Cap, 1/4"	1 Plum	38	.211		2.71	10.25		12.96	18.40
5991	3/8"		38	.211		2.71	10.25		12.96	18.40
5992	1/2"		38	.211		2.06	10.25		12.31	17.65
5993	3/4"		32	.250		2.79	12.20		14.99	21.50
5994	1"		30	.267		3.37	13		16.37	23
5995	1-1/4"	Q-1	52	.308		4.44	13.50		17.94	25.50
5996	1-1/2"		48	.333		6.10	14.65		20.75	28.50
5997	2"		42	.381		8.90	16.70		25.60	35
6000	For galvanized elbows, tees, and couplings add					20%				
6058	For galvanized reducers, caps and bushings add					20%				
6100	90° Elbow, galvanized, 150 lb., reducing									
6110	3/4" x 1/2"	1 Plum	15.40	.519	Ea.	5.85	25.50		31.35	44.50
6112	1" x 3/4"		14	.571		7.50	28		35.50	50.50
6114	1" x 1/2"		14.50	.552		7.95	27		34.95	49.50
6116	1-1/4" x 1"	Q-1	24.20	.661		12.60	29		41.60	57.50
6118	1-1/4" x 3/4"		25.40	.630		15.15	27.50		42.65	58
6120	1-1/4" x 1/2"		26.20	.611		16.05	27		43.05	57.50
6122	1-1/2" x 1-1/4"		21.60	.741		20	32.50		52.50	71
6124	1-1/2" x 1"		23.50	.681		20	30		50	67
6126	1-1/2" x 3/4"		24.60	.650		20	28.50		48.50	65
6128	2" x 1-1/2"		20.50	.780		23.50	34.50		58	77.50
6130	2" x 1-1/4"		21	.762		26.50	33.50		60	79.50
6132	2" x 1"		22.80	.702		28	31		59	77
6134	2" x 3/4"		23.90	.669		28.50	29.50		58	75.50
6136	2-1/2" x 2"		12.30	1.301		79.50	57		136.50	173
6138	2-1/2" x 1-1/2"		12.50	1.280		89	56		145	182
6140	3" x 2-1/2"		8.60	1.860		141	81.50		222.50	278
6142	3" x 2"		11.80	1.356		126	59.50		185.50	228
6144	4" x 3"		8.20	1.951		320	85.50		405.50	480
6160	90° Elbow, black, 150 lb., reducing									
6170	1" x 3/4"	1 Plum	14	.571	Ea.	5.15	28		33.15	47.50
6174	1-1/2" x 1"	Q-1	23.50	.681		12.15	30		42.15	58.50
6178	1-1/2" x 3/4"		24.60	.650		13.90	28.50		42.40	58.50
6182	2" x 1-1/2"		20.50	.780		17.60	34.50		52.10	71
6186	2" x 1"		22.80	.702		20	31		51	68
6190	2" x 3/4"		23.90	.669		21	29.50		50.50	67.50
6194	2-1/2" x 2"		12.30	1.301		48.50	57		105.50	139
7000	Union, with brass seat									
7010	1/4"	1 Plum	15	.533	Ea.	14.45	26		40.45	55
7020	3/8"		15	.533		9.95	26		35.95	50
7030	1/2"		14	.571		9.05	28		37.05	52
7040	3/4"		13	.615		10.35	30		40.35	56.50
7050	1"		12	.667		13.50	32.50		46	64
7060	1-1/4"	Q-1	21	.762		19.35	33.50		52.85	71.50
7070	1-1/2"		19	.842		24	37		61	82
7080	2"		17	.941		28	41.50		69.50	93
7090	2-1/2"		13	1.231		83.50	54		137.50	173
7100	3"		9	1.778		101	78		179	228
7120	Union, galvanized									
7124	1/2"	1 Plum	14	.571	Ea.	11.95	28		39.95	55

22 11 13.45 Pipe Fittings, Steel, Threaded		Crew	Daily Output	Labor-Hours	Unit	2009 Bare Costs Material	Labor	Equipment	Total	Total Incl O&P
7125	3/4"	1 Plum	13	.615	Ea.	13.75	30		43.75	60
7126	1"		12	.667		18	32.50		50.50	69
7127	1-1/4"	Q-1	21	.762		26	33.50		59.50	78.50
7128	1-1/2"		19	.842		31.50	37		68.50	90
7129	2"		17	.941		36	41.50		77.50	102
7130	2-1/2"		13	1.231		126	54		180	219
7131	3"		9	1.778		176	78		254	310
7500	Malleable iron, 300 lb									
7520	Black									
7540	90° Elbow, straight, 1/4"	1 Plum	16	.500	Ea.	10.30	24.50		34.80	48
7560	3/8"		16	.500		9.15	24.50		33.65	46.50
7570	1/2"		15	.533		11.90	26		37.90	52
7580	3/4"		14	.571		13.45	28		41.45	57
7590	1"		13	.615		17.15	30		47.15	64
7600	1-1/4"	Q-1	22	.727		25	32		57	75.50
7610	1-1/2"		20	.800		29.50	35		64.50	85
7620	2"		18	.889		42.50	39		81.50	105
7630	2-1/2"		14	1.143		110	50		160	197
7640	3"		10	1.600		126	70		196	244
7650	4"		6	2.667		289	117		406	495
7700	45° Elbow, straight, 1/4"	1 Plum	16	.500		15.35	24.50		39.85	53.50
7720	3/8"		16	.500		15.35	24.50		39.85	53.50
7730	1/2"		15	.533		17.15	26		43.15	58
7740	3/4"		14	.571		19	28		47	63
7750	1"		13	.615		21	30		51	68
7760	1-1/4"	Q-1	22	.727		33.50	32		65.50	85
7770	1-1/2"		20	.800		44	35		79	101
7780	2"		18	.889		66.50	39		105.50	132
7790	2-1/2"		14	1.143		144	50		194	235
7800	3"		10	1.600		190	70		260	315
7810	4"		6	2.667		430	117		547	650
7850	Tee, straight, 1/4"	1 Plum	10	.800		13	39		52	73
7870	3/8"		10	.800		13.75	39		52.75	73.50
7880	1/2"		9	.889		17.55	43.50		61.05	84.50
7890	3/4"		9	.889		18.90	43.50		62.40	86
7900	1"		8	1		23	49		72	98
7910	1-1/4"	Q-1	14	1.143		32.50	50		82.50	112
7920	1-1/2"		13	1.231		38	54		92	123
7930	2"		11	1.455		56	64		120	158
7940	2-1/2"		9	1.778		133	78		211	263
7950	3"		6	2.667		194	117		311	390
7960	4"		4	4		540	176		716	860
8050	Couplings, straight, 1/4"	1 Plum	19	.421		9.30	20.50		29.80	41.50
8070	3/8"		19	.421		9.30	20.50		29.80	41.50
8080	1/2"		19	.421		10.40	20.50		30.90	42.50
8090	3/4"		18	.444		11.95	21.50		33.45	45.50
8100	1"		15	.533		13.65	26		39.65	54
8110	1-1/4"	Q-1	26	.615		16.25	27		43.25	58.50
8120	1-1/2"		24	.667		24.50	29.50		54	71
8130	2"		21	.762		35	33.50		68.50	88.50
8140	2-1/2"		18	.889		63	39		102	128
8150	3"		14	1.143		90.50	50		140.50	175
8160	4"		10	1.600		113	70		183	229

22 11 13.45 Pipe Fittings, Steel, Threaded		Crew	Daily Output	Labor-Hours	Unit	2009 Bare Costs Material	Labor	Equipment	Total	Total Incl O&P
8200	Galvanized									
8220	90° Elbow, straight, 1/4"	1 Plum	16	.500	Ea.	18.45	24.50		42.95	57
8222	3/8"		16	.500		18.85	24.50		43.35	57
8224	1/2"		15	.533		22.50	26		48.50	63.50
8226	3/4"		14	.571		25	28		53	69.50
8228	1"		13	.615		32.50	30		62.50	80.50
8230	1-1/4"	Q-1	22	.727		50.50	32		82.50	104
8232	1-1/2"		20	.800		55	35		90	113
8234	2"		18	.889		92.50	39		131.50	161
8236	2-1/2"		14	1.143		198	50		248	294
8238	3"		10	1.600		243	70		313	370
8240	4"		6	2.667		705	117		822	950
8280	45° Elbow, straight									
8282	1/2"	1 Plum	15	.533	Ea.	33.50	26		59.50	75.50
8284	3/4"		14	.571		38.50	28		66.50	84.50
8286	1"		13	.615		42	30		72	91.50
8288	1-1/4"	Q-1	22	.727		65.50	32		97.50	120
8290	1-1/2"		20	.800		84.50	35		119.50	146
8292	2"		18	.889		117	39		156	188
8310	Tee, straight, 1/4"	1 Plum	10	.800		26.50	39		65.50	87.50
8312	3/8"		10	.800		27	39		66	88
8314	1/2"		9	.889		33.50	43.50		77	102
8316	3/4"		9	.889		37	43.50		80.50	106
8318	1"		8	1		45.50	49		94.50	123
8320	1-1/4"	Q-1	14	1.143		65	50		115	147
8322	1-1/2"		13	1.231		67.50	54		121.50	156
8324	2"		11	1.455		108	64		172	215
8326	2-1/2"		9	1.778		330	78		408	480
8328	3"		6	2.667		385	117		502	600
8330	4"		4	4		965	176		1,141	1,350
8380	Couplings, straight, 1/4"	1 Plum	19	.421		12.10	20.50		32.60	44.50
8382	3/8"		19	.421		17.45	20.50		37.95	50
8384	1/2"		19	.421		17.90	20.50		38.40	50.50
8386	3/4"		18	.444		20	21.50		41.50	54.50
8388	1"		15	.533		27	26		53	68.50
8390	1-1/4"	Q-1	26	.615		35	27		62	79
8392	1-1/2"		24	.667		49	29.50		78.50	98
8394	2"		21	.762		59.50	33.50		93	116
8396	2-1/2"		18	.889		152	39		191	227
8398	3"		14	1.143		181	50		231	275
8399	4"		10	1.600		189	70		259	315
8529	Black									
8530	Reducer, concentric, 1/4"	1 Plum	19	.421	Ea.	12.30	20.50		32.80	44.50
8531	3/8"		19	.421		12.30	20.50		32.80	44.50
8532	1/2"		17	.471		14.80	23		37.80	51
8533	3/4"		16	.500		20.50	24.50		45	59
8534	1"		15	.533		24.50	26		50.50	66
8535	1-1/4"	Q-1	26	.615		36.50	27		63.50	81
8536	1-1/2"		24	.667		38	29.50		67.50	86
8537	2"		21	.762		55	33.50		88.50	111
8550	Cap, 1/4"	1 Plum	38	.211		9.55	10.25		19.80	26
8551	3/8"		38	.211		9.55	10.25		19.80	26
8552	1/2"		34	.235		9.45	11.45		20.90	27.50

22 11 13.45 Pipe Fittings, Steel, Threaded		Crew	Daily Output	Labor-Hours	Unit	2009 Bare Costs Material	Labor	Equipment	Total	Total Incl O&P
8553	3/4"	1 Plum	32	.250	Ea.	11.95	12.20		24.15	31.50
8554	1"		30	.267		15.65	13		28.65	37
8555	1-1/4"	Q-1	52	.308		17.35	13.50		30.85	39.50
8556	1-1/2"		48	.333		25.50	14.65		40.15	50
8557	2"		42	.381		35	16.70		51.70	64
8570	Plug, 1/4"	1 Plum	38	.211		2.09	10.25		12.34	17.70
8571	3/8"		38	.211		2.09	10.25		12.34	17.70
8572	1/2"		34	.235		2.80	11.45		14.25	20.50
8573	3/4"		32	.250		3.48	12.20		15.68	22
8574	1"		30	.267		5.35	13		18.35	25.50
8575	1-1/4"	Q-1	52	.308		10.80	13.50		24.30	32.50
8576	1-1/2"		48	.333		12.55	14.65		27.20	36
8577	2"		42	.381		19.85	16.70		36.55	47
9500	Union with brass seat, 1/4"	1 Plum	15	.533		21	26		47	62
9530	3/8"		15	.533		21.50	26		47.50	62.50
9540	1/2"		14	.571		17.25	28		45.25	61
9550	3/4"		13	.615		19.10	30		49.10	66
9560	1"		12	.667		25	32.50		57.50	76.50
9570	1-1/4"	Q-1	21	.762		40.50	33.50		74	95
9580	1-1/2"		19	.842		42.50	37		79.50	103
9590	2"		17	.941		53	41.50		94.50	120
9600	2-1/2"		13	1.231		170	54		224	268
9610	3"		9	1.778		220	78		298	360
9620	4"		5	3.200		700	140		840	980
9630	Union, all iron, 1/4"	1 Plum	15	.533		32.50	26		58.50	75
9650	3/8"		15	.533		32.50	26		58.50	75
9660	1/2"		14	.571		34.50	28		62.50	80
9670	3/4"		13	.615		37	30		67	85.50
9680	1"		12	.667		45.50	32.50		78	99
9690	1-1/4"	Q-1	21	.762		69.50	33.50		103	127
9700	1-1/2"		19	.842		85	37		122	149
9710	2"		17	.941		107	41.50		148.50	180
9720	2-1/2"		13	1.231		190	54		244	290
9730	3"		9	1.778		360	78		438	515
9750	For galvanized unions, add					15%				
9757	Forged steel, 3000 lb.									
9758	Black									
9760	90° Elbow, 1/4"	1 Plum	16	.500	Ea.	15.35	24.50		39.85	53.50
9761	3/8"		16	.500		15.35	24.50		39.85	53.50
9762	1/2"		15	.533		12.45	26		38.45	52.50
9763	3/4"		14	.571		14.50	28		42.50	58
9764	1"		13	.615		21	30		51	68
9765	1-1/4"	Q-1	22	.727		42.50	32		74.50	94.50
9766	1-1/2"		20	.800		54	35		89	112
9767	2"		18	.889		65	39		104	130
9780	45° Elbow 1/4"	1 Plum	16	.500		19.50	24.50		44	58
9781	3/8"		16	.500		19.50	24.50		44	58
9782	1/2"		15	.533		19.20	26		45.20	60
9783	3/4"		14	.571		22	28		50	66.50
9784	1"		13	.615		30	30		60	78.50
9785	1-1/4"	Q-1	22	.727		42	32		74	94
9786	1-1/2"		20	.800		60	35		95	119
9787	2"		18	.889		83	39		122	150

22 11 13.45 Pipe Fittings, Steel, Threaded		Crew	Daily Output	Labor-Hours	Unit	Material	Labor	Equipment	Total	Total Incl O&P
						2009 Bare Costs				
9800	Tee, 1/4"	1 Plum	10	.800	Ea.	18.65	39		57.65	79
9801	3/8"		10	.800		18.65	39		57.65	79
9802	1/2"		9	.889		16.80	43.50		60.30	83.50
9803	3/4"		9	.889		22.50	43.50		66	89.50
9804	1"	↓	8	1		29	49		78	105
9805	1-1/4"	Q-1	14	1.143		58	50		108	140
9806	1-1/2"		13	1.231		67.50	54		121.50	156
9807	2"	↓	11	1.455		132	64		196	242
9820	Reducer, concentric, 1/4"	1 Plum	19	.421		10.35	20.50		30.85	42.50
9821	3/8"		19	.421		10.80	20.50		31.30	43
9822	1/2"		17	.471		10.80	23		33.80	46.50
9823	3/4"		16	.500		12.80	24.50		37.30	50.50
9824	1"	↓	15	.533		16.65	26		42.65	57.50
9825	1-1/4"	Q-1	26	.615		27.50	27		54.50	71
9826	1-1/2"		24	.667		30	29.50		59.50	77
9827	2"	↓	21	.762		43.50	33.50		77	98
9840	Cap, 1/4"	1 Plum	38	.211		5.75	10.25		16	21.50
9841	3/8"		38	.211		5.75	10.25		16	21.50
9842	1/2"		34	.235		5.60	11.45		17.05	23.50
9843	3/4"		32	.250		7.80	12.20		20	27
9844	1"	↓	30	.267		11.95	13		24.95	32.50
9845	1-1/4"	Q-1	52	.308		18.90	13.50		32.40	41.50
9846	1-1/2"		48	.333		22	14.65		36.65	46.50
9847	2"	↓	42	.381		32.50	16.70		49.20	60.50
9860	Plug, 1/4"	1 Plum	38	.211		2.54	10.25		12.79	18.20
9861	3/8"		38	.211		2.67	10.25		12.92	18.35
9862	1/2"		34	.235		2.80	11.45		14.25	20.50
9863	3/4"		32	.250		3.48	12.20		15.68	22
9864	1"	↓	30	.267		5.35	13		18.35	25.50
9865	1-1/4"	Q-1	52	.308		10.80	13.50		24.30	32.50
9866	1-1/2"		48	.333		12.55	14.65		27.20	36
9867	2"	↓	42	.381		19.85	16.70		36.55	47
9880	Union, bronze seat, 1/4"	1 Plum	15	.533		41.50	26		67.50	85
9881	3/8"		15	.533		41.50	26		67.50	85
9882	1/2"		14	.571		40	28		68	86
9883	3/4"		13	.615		53.50	30		83.50	104
9884	1"	↓	12	.667		60.50	32.50		93	116
9885	1-1/4"	Q-1	21	.762		117	33.50		150.50	179
9886	1-1/2"		19	.842		121	37		158	189
9887	2"	↓	17	.941		143	41.50		184.50	219
9900	Coupling, 1/4"	1 Plum	19	.421		5.55	20.50		26.05	37
9901	3/8"		19	.421		5.55	20.50		26.05	37
9902	1/2"		17	.471		4.81	23		27.81	40
9903	3/4"		16	.500		6.15	24.50		30.65	43.50
9904	1"	↓	15	.533		10.90	26		36.90	51
9905	1-1/4"	Q-1	26	.615		18.15	27		45.15	60.50
9906	1-1/2"		24	.667		23.50	29.50		53	70
9907	2"	↓	21	.762	↓	28	33.50		61.50	81

22 11 13.47 Pipe Fittings, Steel		Crew	Daily Output	Labor-Hours	Unit	Material	Labor	Equipment	Total	Total Incl O&P
0010	**PIPE FITTINGS, STEEL,** Flanged, Welded & Special									
0020	Flanged joints, C.I., standard weight, black. One gasket & bolt									
0040	set, mat'l only, required at each joint, not included (see line 0620)									

22 11 13 – Facility Water Distribution Piping

22 11 13.47 Pipe Fittings, Steel		Crew	Daily Output	Labor-Hours	Unit	Material	Labor	Equipment	Total	Total Incl O&P
						2009 Bare Costs				
0060	90° Elbow, straight, 1-1/2" pipe size	Q-1	14	1.143	Ea.	320	50		370	425
0080	2" pipe size		13	1.231		194	54		248	294
0090	2-1/2" pipe size		12	1.333		210	58.50		268.50	320
0100	3" pipe size		11	1.455		175	64		239	288
0110	4" pipe size		8	2		216	88		304	370
0120	5" pipe size	↓	7	2.286		510	100		610	715
0130	6" pipe size	Q-2	9	2.667		340	121		461	550
0140	8" pipe size		8	3		585	137		722	850
0150	10" pipe size		7	3.429		1,275	156		1,431	1,625
0160	12" pipe size	↓	6	4	↓	2,575	182		2,757	3,125
0171	90° Elbow, reducing									
0172	2-1/2" by 2" pipe size	Q-1	12	1.333	Ea.	615	58.50		673.50	770
0173	3" by 2-1/2" pipe size		11	1.455		650	64		714	810
0174	4" by 3" pipe size		8	2		480	88		568	660
0175	5" by 3" pipe size	↓	7	2.286		990	100		1,090	1,250
0176	6" by 4" pipe size	Q-2	9	2.667		595	121		716	835
0177	8" by 6" pipe size		8	3		870	137		1,007	1,150
0178	10" by 8" pipe size		7	3.429		1,650	156		1,806	2,050
0179	12" by 10" pipe size	↓	6	4		3,175	182		3,357	3,750
0200	45° Elbow, straight, 1-1/2" pipe size	Q-1	14	1.143		385	50		435	495
0220	2" pipe size		13	1.231		281	54		335	390
0230	2-1/2" pipe size		12	1.333		299	58.50		357.50	420
0240	3" pipe size		11	1.455		291	64		355	415
0250	4" pipe size		8	2		330	88		418	490
0260	5" pipe size	↓	7	2.286		785	100		885	1,025
0270	6" pipe size	Q-2	9	2.667		535	121		656	770
0280	8" pipe size		8	3		785	137		922	1,075
0290	10" pipe size		7	3.429		1,650	156		1,806	2,050
0300	12" pipe size	↓	6	4	↓	2,525	182		2,707	3,050
0310	Cross, straight									
0311	2-1/2" pipe size	Q-1	6	2.667	Ea.	615	117		732	850
0312	3" pipe size		5	3.200		645	140		785	920
0313	4" pipe size		4	4		845	176		1,021	1,200
0314	5" pipe size	↓	3	5.333		1,775	234		2,009	2,300
0315	6" pipe size	Q-2	5	4.800		1,775	218		1,993	2,275
0316	8" pipe size		4	6		2,850	273		3,123	3,525
0317	10" pipe size		3	8		3,375	365		3,740	4,275
0318	12" pipe size	↓	2	12		5,150	545		5,695	6,500
0350	Tee, straight, 1-1/2" pipe size	Q-1	10	1.600		365	70		435	505
0370	2" pipe size		9	1.778		212	78		290	350
0380	2-1/2" pipe size		8	2		310	88		398	470
0390	3" pipe size		7	2.286		217	100		317	390
0400	4" pipe size		5	3.200		330	140		470	570
0410	5" pipe size	↓	4	4		880	176		1,056	1,225
0420	6" pipe size	Q-2	6	4		480	182		662	800
0430	8" pipe size		5	4.800		815	218		1,033	1,225
0440	10" pipe size		4	6		2,200	273		2,473	2,825
0450	12" pipe size	↓	3	8	↓	3,450	365		3,815	4,350
0459	Tee, reducing on outlet									
0460	2-1/2" by 2" pipe size	Q-1	8	2	Ea.	590	88		678	775
0461	3" by 2-1/2" pipe size		7	2.286		645	100		745	855
0462	4" by 3" pipe size		5	3.200		680	140		820	955
0463	5" by 4" pipe size	↓	4	4	↓	1,425	176		1,601	1,850

22 11 13.47 Pipe Fittings, Steel		Crew	Daily Output	Labor-Hours	Unit	Material	2009 Bare Costs Labor	Equipment	Total	Total Incl O&P
0464	6" by 4" pipe size	Q-2	6	4	Ea.	655	182		837	995
0465	8" by 6" pipe size		5	4.800		1,025	218		1,243	1,450
0466	10" by 8" pipe size		4	6		2,325	273		2,598	2,975
0467	12" by 10" pipe size		3	8		4,025	365		4,390	4,975
0476	Reducer, concentric									
0477	3" by 2-1/2"	Q-1	12	1.333	Ea.	390	58.50		448.50	515
0478	4" by 3"		9	1.778		435	78		513	595
0479	5" by 4"		8	2		675	88		763	870
0480	6" by 4"	Q-2	10	2.400		585	109		694	805
0481	8" by 6"		9	2.667		735	121		856	990
0482	10" by 8"		8	3		1,525	137		1,662	1,875
0483	12" by 10"		7	3.429		2,625	156		2,781	3,100
0492	Reducer, eccentric									
0493	4" by 3"	Q-1	8	2	Ea.	710	88		798	910
0494	5" by 4"	"	7	2.286		1,050	100		1,150	1,300
0495	6" by 4"	Q-2	9	2.667		670	121		791	915
0496	8" by 6"		8	3		845	137		982	1,125
0497	10" by 8"		7	3.429		2,150	156		2,306	2,600
0498	12" by 10"		6	4		2,925	182		3,107	3,500
0500	For galvanized elbows and tees, add					100%				
0520	For extra heavy weight elbows and tees, add					140%				
0620	Gasket and bolt set, 150#, 1/2" pipe size	1 Plum	20	.400		2.16	19.50		21.66	32
0622	3/4" pipe size		19	.421		2.82	20.50		23.32	34
0624	1" pipe size		18	.444		2.31	21.50		23.81	35
0626	1-1/4" pipe size		17	.471		2.43	23		25.43	37
0628	1-1/2" pipe size		15	.533		2.58	26		28.58	42
0630	2" pipe size		13	.615		4.69	30		34.69	50
0640	2-1/2" pipe size		12	.667		4.84	32.50		37.34	54.50
0650	3" pipe size		11	.727		5	35.50		40.50	58.50
0660	3-1/2" pipe size		9	.889		8.15	43.50		51.65	74
0670	4" pipe size		8	1		9	49		58	83
0680	5" pipe size		7	1.143		14.80	55.50		70.30	100
0690	6" pipe size		6	1.333		14.80	65		79.80	114
0700	8" pipe size		5	1.600		16.95	78		94.95	136
0710	10" pipe size		4.50	1.778		30.50	86.50		117	164
0720	12" pipe size		4.20	1.905		32.50	93		125.50	175
0730	14" pipe size		4	2		34.50	97.50		132	184
0740	16" pipe size		3	2.667		36	130		166	235
0750	18" pipe size		2.70	2.963		70	144		214	294
0760	20" pipe size		2.30	3.478		111	170		281	375
0780	24" pipe size		1.90	4.211		142	205		347	465
0790	26" pipe size		1.60	5		192	244		436	575
0810	30" pipe size		1.40	5.714		375	279		654	830
0830	36" pipe size		1.10	7.273		700	355		1,055	1,300
0850	For 300 lb gasket set, add					40%				
2000	Flanged unions, 125 lb., black, 1/2" pipe size	1 Plum	17	.471	Ea.	44	23		67	82.50
2040	3/4" pipe size		17	.471		57.50	23		80.50	98
2050	1" pipe size		16	.500		56	24.50		80.50	98
2060	1-1/4" pipe size	Q-1	28	.571		66	25		91	111
2070	1-1/2" pipe size		27	.593		61	26		87	106
2080	2" pipe size		26	.615		72	27		99	120
2090	2-1/2" pipe size		24	.667		97.50	29.50		127	151
2100	3" pipe size		22	.727		110	32		142	169

22 11 13.47 Pipe Fittings, Steel		Crew	Daily Output	Labor-Hours	Unit	2009 Bare Costs Material	Labor	Equipment	Total	Total Incl O&P
2110	3-1/2" pipe size	Q-1	18	.889	Ea.	187	39		226	264
2120	4" pipe size		16	1		150	44		194	231
2130	5" pipe size		14	1.143		350	50		400	460
2140	6" pipe size	Q-2	19	1.263		330	57.50		387.50	445
2150	8" pipe size	"	16	1.500		730	68.50		798.50	900
2200	For galvanized unions, add					150%				
2290	Threaded flange									
2300	Cast iron									
2310	Black, 125 lb., per flange									
2320	1" pipe size	1 Plum	27	.296	Ea.	27	14.45		41.45	51
2330	1-1/4" pipe size	Q-1	44	.364		32	15.95		47.95	59.50
2340	1-1/2" pipe size		40	.400		29.50	17.55		47.05	58.50
2350	2" pipe size		36	.444		30	19.50		49.50	62.50
2360	2-1/2" pipe size		28	.571		34.50	25		59.50	75.50
2370	3" pipe size		20	.800		44.50	35		79.50	102
2380	3-1/2" pipe size		16	1		62.50	44		106.50	135
2390	4" pipe size		12	1.333		60	58.50		118.50	154
2400	5" pipe size		10	1.600		85	70		155	199
2410	6" pipe size	Q-2	14	1.714		96	78		174	223
2420	8" pipe size		12	2		151	91		242	305
2430	10" pipe size		10	2.400		268	109		377	460
2440	12" pipe size		8	3		575	137		712	835
2460	For galvanized flanges, add					95%				
2490	Blind flange									
2492	Cast iron									
2494	Black, 125 lb., per flange									
2496	1" pipe size	1 Plum	27	.296	Ea.	42.50	14.45		56.95	68.50
2500	1-1/2" pipe size	Q-1	40	.400		48	17.55		65.55	79
2502	2" pipe size		36	.444		54	19.50		73.50	89
2504	2-1/2" pipe size		28	.571		59.50	25		84.50	103
2506	3" pipe size		20	.800		71.50	35		106.50	131
2508	4" pipe size		12	1.333		91	58.50		149.50	188
2510	5" pipe size		10	1.600		148	70		218	268
2512	6" pipe size	Q-2	14	1.714		161	78		239	294
2514	8" pipe size		12	2		253	91		344	415
2516	10" pipe size		10	2.400		375	109		484	575
2518	12" pipe size		8	3		680	137		817	955
2520	For galvanized flanges, add					80%				
2570	Threaded flange									
2580	Forged steel,									
2590	Black 150 lb., per flange									
2600	1/2" pipe size	1 Plum	30	.267	Ea.	21	13		34	42.50
2610	3/4" pipe size		28	.286		21	13.95		34.95	44
2620	1" pipe size		27	.296		21	14.45		35.45	44.50
2630	1-1/4" pipe size	Q-1	44	.364		21	15.95		36.95	47
2640	1-1/2" pipe size		40	.400		21	17.55		38.55	49.50
2650	2" pipe size		36	.444		21	19.50		40.50	52.50
2660	2-1/2" pipe size		28	.571		26	25		51	66
2670	3" pipe size		20	.800		26	35		61	81
2690	4" pipe size		12	1.333		32	58.50		90.50	124
2700	5" pipe size		10	1.600		46.50	70		116.50	157
2710	6" pipe size	Q-2	14	1.714		54	78		132	177
2720	8" pipe size		12	2		88.50	91		179.50	235

22 11 13.47 Pipe Fittings, Steel		Crew	Daily Output	Labor-Hours	Unit	2009 Bare Costs Material	Labor	Equipment	Total	Total Incl O&P
2730	10″ pipe size	Q-2	10	2.400	Ea.	166	109		275	345
2860	Black 300 lb., per flange									
2870	1/2″ pipe size	1 Plum	30	.267	Ea.	23	13		36	45
2880	3/4″ pipe size		28	.286		23	13.95		36.95	46.50
2890	1″ pipe size		27	.296		23	14.45		37.45	47
2900	1-1/4″ pipe size	Q-1	44	.364		23	15.95		38.95	49.50
2910	1-1/2″ pipe size		40	.400		23	17.55		40.55	52
2920	2″ pipe size		36	.444		24.50	19.50		44	56.50
2930	2-1/2″ pipe size		28	.571		33.50	25		58.50	74.50
2940	3″ pipe size		20	.800		37.50	35		72.50	93.50
2960	4″ pipe size		12	1.333		53	58.50		111.50	146
2970	6″ pipe size	Q-2	14	1.714		102	78		180	229
3000	Weld joint, butt, carbon steel, standard weight									
3040	90° elbow, long radius									
3050	1/2″ pipe size	Q-15	16	1	Ea.	27.50	44	3.48	74.98	100
3060	3/4″ pipe size		16	1		27.50	44	3.48	74.98	100
3070	1″ pipe size		16	1		12.80	44	3.48	60.28	84
3080	1-1/4″ pipe size		14	1.143		12.80	50	3.98	66.78	94
3090	1-1/2″ pipe size		13	1.231		12.80	54	4.28	71.08	100
3100	2″ pipe size		10	1.600		14.40	70	5.55	89.95	127
3110	2-1/2″ pipe size		8	2		16.95	88	6.95	111.90	158
3120	3″ pipe size		7	2.286		20.50	100	7.95	128.45	182
3130	4″ pipe size		5	3.200		34	140	11.15	185.15	261
3136	5″ pipe size		4	4		75	176	13.90	264.90	360
3140	6″ pipe size	Q-16	5	4.800		74.50	218	11.15	303.65	425
3150	8″ pipe size		3.75	6.400		141	291	14.85	446.85	605
3160	10″ pipe size		3	8		283	365	18.55	666.55	875
3170	12″ pipe size		2.50	9.600		420	435	22.50	877.50	1,150
3180	14″ pipe size		2	12		625	545	28	1,198	1,525
3190	16″ pipe size		1.50	16		840	730	37	1,607	2,075
3191	18″ pipe size		1.25	19.200		1,125	875	44.50	2,044.50	2,575
3192	20″ pipe size		1.15	20.870		1,525	950	48.50	2,523.50	3,150
3194	24″ pipe size		1.02	23.529		2,225	1,075	54.50	3,354.50	4,100
3200	45° Elbow, long									
3210	1/2″ pipe size	Q-15	16	1	Ea.	38.50	44	3.48	85.98	112
3220	3/4″ pipe size		16	1		38.50	44	3.48	85.98	112
3230	1″ pipe size		16	1		13.75	44	3.48	61.23	85
3240	1-1/4″ pipe size		14	1.143		13.75	50	3.98	67.73	95
3250	1-1/2″ pipe size		13	1.231		13.75	54	4.28	72.03	101
3260	2″ pipe size		10	1.600		13.75	70	5.55	89.30	126
3270	2-1/2″ pipe size		8	2		16	88	6.95	110.95	157
3280	3″ pipe size		7	2.286		16.95	100	7.95	124.90	178
3290	4″ pipe size		5	3.200		30.50	140	11.15	181.65	257
3296	5″ pipe size		4	4		46.50	176	13.90	236.40	330
3300	6″ pipe size	Q-16	5	4.800		60	218	11.15	289.15	410
3310	8″ pipe size		3.75	6.400		100	291	14.85	405.85	560
3320	10″ pipe size		3	8		200	365	18.55	583.55	785
3330	12″ pipe size		2.50	9.600		283	435	22.50	740.50	990
3340	14″ pipe size		2	12		380	545	28	953	1,275
3341	16″ pipe size		1.50	16		685	730	37	1,452	1,900
3342	18″ pipe size		1.25	19.200		975	875	44.50	1,894.50	2,425
3343	20″ pipe size		1.15	20.870		1,000	950	48.50	1,998.50	2,575
3345	24″ pipe size		1.05	22.857		1,525	1,050	53	2,628	3,275

22 11 Facility Water Distribution

22 11 13 – Facility Water Distribution Piping

	22 11 13.47 Pipe Fittings, Steel	Crew	Daily Output	Labor-Hours	Unit	2009 Bare Costs Material	Labor	Equipment	Total	Total Incl O&P
3346	26" pipe size	Q-16	.85	28.235	Ea.	1,700	1,275	65.50	3,040.50	3,850
3347	30" pipe size		.45	53.333		1,850	2,425	124	4,399	5,825
3349	36" pipe size		.38	63.158		2,050	2,875	147	5,072	6,700
3350	Tee, straight									
3352	For reducing tees and concentrics see starting line 4600									
3360	1/2" pipe size	Q-15	10	1.600	Ea.	68.50	70	5.55	144.05	187
3370	3/4" pipe size		10	1.600		68.50	70	5.55	144.05	187
3380	1" pipe size		10	1.600		34	70	5.55	109.55	149
3390	1-1/4" pipe size		9	1.778		42.50	78	6.20	126.70	171
3400	1-1/2" pipe size		8	2		42.50	88	6.95	137.45	187
3410	2" pipe size		6	2.667		34	117	9.30	160.30	224
3420	2-1/2" pipe size		5	3.200		46.50	140	11.15	197.65	274
3430	3" pipe size		4	4		51.50	176	13.90	241.40	335
3440	4" pipe size		3	5.333		72.50	234	18.55	325.05	450
3446	5" pipe size		2.50	6.400		120	281	22.50	423.50	575
3450	6" pipe size	Q-16	3	8		130	365	18.55	513.55	710
3460	8" pipe size		2.50	9.600		228	435	22.50	685.50	930
3470	10" pipe size		2	12		450	545	28	1,023	1,350
3480	12" pipe size		1.60	15		630	685	35	1,350	1,750
3481	14" pipe size		1.30	18.462		1,125	840	43	2,008	2,550
3482	16" pipe size		1	24		1,250	1,100	55.50	2,405.50	3,075
3483	18" pipe size		.80	30		1,875	1,375	69.50	3,319.50	4,200
3484	20" pipe size		.75	32		3,025	1,450	74	4,549	5,600
3486	24" pipe size		.70	34.286		3,925	1,550	79.50	5,554.50	6,750
3487	26" pipe size		.55	43.636		4,325	1,975	101	6,401	7,825
3488	30" pipe size		.30	80		4,725	3,650	186	8,561	10,900
3490	36" pipe size		.25	96		5,225	4,375	223	9,823	12,500
3491	Eccentric reducer, 1-1/2" pipe size	Q-15	14	1.143		87.50	50	3.98	141.48	176
3492	2" pipe size		11	1.455		124	64	5.05	193.05	239
3493	2-1/2" pipe size		9	1.778		157	78	6.20	241.20	296
3494	3" pipe size		8	2		167	88	6.95	261.95	325
3495	4" pipe size		6	2.667		183	117	9.30	309.30	385
3496	6" pipe size	Q-16	5	4.800		320	218	11.15	549.15	690
3497	8" pipe size		4	6		365	273	13.90	651.90	825
3498	10" pipe size		3	8		425	365	18.55	808.55	1,025
3499	12" pipe size		2.50	9.600		605	435	22.50	1,062.50	1,350
3501	Cap, 1-1/2" pipe size	Q-15	28	.571		15.35	25	1.99	42.34	56.50
3502	2" pipe size		22	.727		15.05	32	2.53	49.58	67.50
3503	2-1/2" pipe size		18	.889		16.30	39	3.09	58.39	80
3504	3" pipe size		16	1		16.30	44	3.48	63.78	88
3505	4" pipe size		12	1.333		23.50	58.50	4.64	86.64	119
3506	6" pipe size	Q-16	10	2.400		40	109	5.55	154.55	214
3507	8" pipe size		8	3		60.50	137	6.95	204.45	279
3508	10" pipe size		6	4		109	182	9.30	300.30	405
3509	12" pipe size		5	4.800		165	218	11.15	394.15	525
3511	14" pipe size		4	6		220	273	13.90	506.90	670
3512	16" pipe size		4	6		257	273	13.90	543.90	710
3513	18" pipe size		3	8		350	365	18.55	733.55	950
3517	Weld joint, butt, carbon steel, extra strong									
3519	90° elbow, long									
3520	1/2" pipe size	Q-15	13	1.231	Ea.	35	54	4.28	93.28	124
3530	3/4" pipe size		12	1.333		35	58.50	4.64	98.14	132
3540	1" pipe size		11	1.455		16.95	64	5.05	86	120

22 11 13.47 Pipe Fittings, Steel		Crew	Daily Output	Labor-Hours	Unit	Material	Labor	Equipment	Total	Total Incl O&P
						2009 Bare Costs				
3550	1-1/4" pipe size	Q-15	10	1.600	Ea.	16.95	70	5.55	92.50	130
3560	1-1/2" pipe size		9	1.778		16.95	78	6.20	101.15	142
3570	2" pipe size		8	2		16.95	88	6.95	111.90	158
3580	2-1/2" pipe size		7	2.286		24.50	100	7.95	132.45	187
3590	3" pipe size		6	2.667		30.50	117	9.30	156.80	220
3600	4" pipe size		4	4		50.50	176	13.90	240.40	335
3606	5" pipe size		3.50	4.571		107	201	15.90	323.90	435
3610	6" pipe size	Q-16	4.50	5.333		127	243	12.35	382.35	520
3620	8" pipe size		3.50	6.857		244	310	15.90	569.90	755
3630	10" pipe size		2.50	9.600		545	435	22.50	1,002.50	1,275
3640	12" pipe size		2.25	10.667		630	485	25	1,140	1,450
3650	45° Elbow, long									
3660	1/2" pipe size	Q-15	13	1.231	Ea.	38.50	54	4.28	96.78	128
3670	3/4" pipe size		12	1.333		38.50	58.50	4.64	101.64	135
3680	1" pipe size		11	1.455		17.60	64	5.05	86.65	121
3690	1-1/4" pipe size		10	1.600		17.60	70	5.55	93.15	131
3700	1-1/2" pipe size		9	1.778		17.60	78	6.20	101.80	143
3710	2" pipe size		8	2		17.60	88	6.95	112.55	159
3720	2-1/2" pipe size		7	2.286		39	100	7.95	146.95	203
3730	3" pipe size		6	2.667		22.50	117	9.30	148.80	211
3740	4" pipe size		4	4		36	176	13.90	225.90	320
3746	5" pipe size		3.50	4.571		86	201	15.90	302.90	410
3750	6" pipe size	Q-16	4.50	5.333		99	243	12.35	354.35	490
3760	8" pipe size		3.50	6.857		172	310	15.90	497.90	680
3770	10" pipe size		2.50	9.600		330	435	22.50	787.50	1,050
3780	12" pipe size		2.25	10.667		495	485	25	1,005	1,300
3800	Tee, straight									
3810	1/2" pipe size	Q-15	9	1.778	Ea.	85.50	78	6.20	169.70	218
3820	3/4" pipe size		8.50	1.882		81.50	82.50	6.55	170.55	221
3830	1" pipe size		8	2		45	88	6.95	139.95	189
3840	1-1/4" pipe size		7	2.286		42	100	7.95	149.95	206
3850	1-1/2" pipe size		6	2.667		51	117	9.30	177.30	242
3860	2" pipe size		5	3.200		42	140	11.15	193.15	270
3870	2-1/2" pipe size		4	4		70	176	13.90	259.90	355
3880	3" pipe size		3.50	4.571		65.50	201	15.90	282.40	390
3890	4" pipe size		2.50	6.400		106	281	22.50	409.50	560
3896	5" pipe size		2.25	7.111		228	310	25	563	750
3900	6" pipe size	Q-16	2.25	10.667		186	485	25	696	960
3910	8" pipe size		2	12		350	545	28	923	1,225
3920	10" pipe size		1.75	13.714		600	625	32	1,257	1,625
3930	12" pipe size		1.50	16		910	730	37	1,677	2,150
4000	Eccentric reducer, 1-1/2" pipe size	Q-15	10	1.600		85	70	5.55	160.55	205
4010	2" pipe size		9	1.778		231	78	6.20	315.20	380
4020	2-1/2" pipe size		8	2		158	88	6.95	252.95	315
4030	3" pipe size		7	2.286		203	100	7.95	310.95	385
4040	4" pipe size		5	3.200		297	140	11.15	448.15	550
4046	5" pipe size		4.70	3.404		625	149	11.85	785.85	920
4050	6" pipe size	Q-16	4.50	5.333		415	243	12.35	670.35	840
4060	8" pipe size		3.50	6.857		460	310	15.90	785.90	995
4070	10" pipe size		2.50	9.600		585	435	22.50	1,042.50	1,325
4080	12" pipe size		2.25	10.667		640	485	25	1,150	1,450
4090	14" pipe size		2.10	11.429		1,250	520	26.50	1,796.50	2,175
4100	16" pipe size		1.90	12.632		1,250	575	29.50	1,854.50	2,275

22 11 13.47 Pipe Fittings, Steel		Crew	Daily Output	Labor-Hours	Unit	Material	Labor	Equipment	Total	Total Incl O&P
						2009 Bare Costs				
4151	Cap, 1-1/2" pipe size	Q-15	24	.667	Ea.	19.50	29.50	2.32	51.32	68
4152	2" pipe size		18	.889		15.05	39	3.09	57.14	78.50
4153	2-1/2" pipe size		16	1		21	44	3.48	68.48	93
4154	3" pipe size		14	1.143		23	50	3.98	76.98	105
4155	4" pipe size		10	1.600		32	70	5.55	107.55	146
4156	6" pipe size	Q-16	9	2.667		64.50	121	6.20	191.70	260
4157	8" pipe size		7	3.429		97	156	7.95	260.95	350
4158	10" pipe size		5	4.800		149	218	11.15	378.15	505
4159	12" pipe size		4	6		199	273	13.90	485.90	645
4190	Weld fittings, reducing, standard weight									
4200	Welding ring w/spacer pins, 2" pipe size				Ea.	2.96			2.96	3.26
4210	2-1/2" pipe size					3.58			3.58	3.94
4220	3" pipe size					3.70			3.70	4.07
4230	4" pipe size					3.85			3.85	4.24
4236	5" pipe size					4.53			4.53	4.98
4240	6" pipe size					4.53			4.53	4.98
4250	8" pipe size					5.35			5.35	5.90
4260	10" pipe size					6.55			6.55	7.20
4270	12" pipe size					7.50			7.50	8.25
4280	14" pipe size					8.75			8.75	9.60
4290	16" pipe size					9.85			9.85	10.85
4300	18" pipe size					11.40			11.40	12.50
4310	20" pipe size					12.05			12.05	13.30
4330	24" pipe size					14.05			14.05	15.45
4340	26" pipe size					18.75			18.75	20.50
4350	30" pipe size					18.75			18.75	20.50
4370	36" pipe size					21			21	23
4600	Tee, reducing on outlet									
4601	2-1/2" x 2" pipe size	Q-15	5	3.200	Ea.	62.50	140	11.15	213.65	292
4602	3" x 2-1/2" pipe size		4	4		73	176	13.90	262.90	360
4604	4" x 3" pipe size		3	5.333		76	234	18.55	328.55	455
4605	5" x 4" pipe size		2.50	6.400		165	281	22.50	468.50	625
4606	6" x 5" pipe size	Q-16	3	8		229	365	18.55	612.55	820
4607	8" x 6" pipe size		2.50	9.600		284	435	22.50	741.50	995
4608	10" x 8" pipe size		2	12		530	545	28	1,103	1,425
4609	12" x 10" pipe size		1.60	15		780	685	35	1,500	1,925
4610	16" x 12" pipe size		1.50	16		1,300	730	37	2,067	2,575
4611	14" x 12" pipe size		1.52	15.789		1,200	720	36.50	1,956.50	2,425
4618	Reducer, concentric									
4619	2-1/2" by 2" pipe size	Q-15	10	1.600	Ea.	26.50	70	5.55	102.05	140
4620	3" by 2-1/2" pipe size		9	1.778		22	78	6.20	106.20	148
4621	3-1/2" by 3" pipe size		8	2		56.50	88	6.95	151.45	202
4622	4" by 2-1/2" pipe size		7	2.286		34	100	7.95	141.95	197
4623	5" by 3" pipe size		7	2.286		70.50	100	7.95	178.45	237
4624	6" by 4" pipe size	Q-16	6	4		58.50	182	9.30	249.80	350
4625	8" by 6" pipe size		5	4.800		77	218	11.15	306.15	425
4626	10" by 8" pipe size		4	6		149	273	13.90	435.90	590
4627	12" by 10" pipe size		3	8		185	365	18.55	568.55	770
4660	Reducer, eccentric									
4662	3" x 2" pipe size	Q-15	8	2	Ea.	36.50	88	6.95	131.45	180
4664	4" x 3" pipe size		6	2.667		44.50	117	9.30	170.80	235
4666	4" x 2" pipe size		6	2.667		54.50	117	9.30	180.80	246
4670	6" x 4" pipe size	Q-16	5	4.800		92	218	11.15	321.15	445

22 11 13.47 Pipe Fittings, Steel		Crew	Daily Output	Labor-Hours	Unit	2009 Bare Costs Material	Labor	Equipment	Total	Total Incl O&P
4672	6" x 3" pipe size	Q-16	5	4.800	Ea.	144	218	11.15	373.15	500
4676	8" x 6" pipe size		4	6		117	273	13.90	403.90	555
4678	8" x 4" pipe size		4	6		214	273	13.90	500.90	660
4682	10" x 8" pipe size		3	8		153	365	18.55	536.55	735
4684	10" x 6" pipe size		3	8		271	365	18.55	654.55	865
4688	12" x 10" pipe size		2.50	9.600		240	435	22.50	697.50	945
4690	12" x 8" pipe size		2.50	9.600		375	435	22.50	832.50	1,100
4691	14" x 12" pipe size		2.20	10.909		420	495	25.50	940.50	1,250
4693	16" x 14" pipe size		1.80	13.333		645	605	31	1,281	1,650
4694	16" x 12" pipe size		2	12		855	545	28	1,428	1,800
4696	18" x 16" pipe size		1.60	15		1,175	685	35	1,895	2,375
5000	Weld joint, socket, forged steel, 3000 lb., schedule 40 pipe									
5010	90° elbow, straight									
5020	1/4" pipe size	Q-15	22	.727	Ea.	18.75	32	2.53	53.28	71.50
5030	3/8" pipe size		22	.727		18.75	32	2.53	53.28	71.50
5040	1/2" pipe size		20	.800		10.45	35	2.78	48.23	67
5050	3/4" pipe size		20	.800		11.10	35	2.78	48.88	68
5060	1" pipe size		20	.800		14.10	35	2.78	51.88	71
5070	1-1/4" pipe size		18	.889		27	39	3.09	69.09	91.50
5080	1-1/2" pipe size		16	1		30	44	3.48	77.48	103
5090	2" pipe size		12	1.333		46	58.50	4.64	109.14	144
5100	2-1/2" pipe size		10	1.600		131	70	5.55	206.55	255
5110	3" pipe size		8	2		230	88	6.95	324.95	395
5120	4" pipe size		6	2.667		595	117	9.30	721.30	840
5130	45° Elbow, straight									
5134	1/4" pipe size	Q-15	22	.727	Ea.	18.85	32	2.53	53.38	72
5135	3/8" pipe size		22	.727		18.85	32	2.53	53.38	72
5136	1/2" pipe size		20	.800		13.95	35	2.78	51.73	71
5137	3/4" pipe size		20	.800		16.20	35	2.78	53.98	73.50
5140	1" pipe size		20	.800		21	35	2.78	58.78	79
5150	1-1/4" pipe size		18	.889		29	39	3.09	71.09	94
5160	1-1/2" pipe size		16	1		35.50	44	3.48	82.98	109
5170	2" pipe size		12	1.333		57.50	58.50	4.64	120.64	156
5180	2-1/2" pipe size		10	1.600		150	70	5.55	225.55	277
5190	3" pipe size		8	2		250	88	6.95	344.95	415
5200	4" pipe size		6	2.667		495	117	9.30	621.30	725
5250	Tee, straight									
5254	1/4" pipe size	Q-15	15	1.067	Ea.	20.50	47	3.71	71.21	97
5255	3/8" pipe size		15	1.067		20.50	47	3.71	71.21	97
5256	1/2" pipe size		13	1.231		12.95	54	4.28	71.23	100
5257	3/4" pipe size		13	1.231		15.70	54	4.28	73.98	103
5260	1" pipe size		13	1.231		21.50	54	4.28	79.78	109
5270	1-1/4" pipe size		12	1.333		33	58.50	4.64	96.14	130
5280	1-1/2" pipe size		11	1.455		44	64	5.05	113.05	150
5290	2" pipe size		8	2		63.50	88	6.95	158.45	209
5300	2-1/2" pipe size		6	2.667		182	117	9.30	308.30	385
5310	3" pipe size		5	3.200		435	140	11.15	586.15	705
5320	4" pipe size		4	4		700	176	13.90	889.90	1,050
5350	For reducing sizes, add					60%				
5450	Couplings									
5451	1/4" pipe size	Q-15	23	.696	Ea.	11.65	30.50	2.42	44.57	61.50
5452	3/8" pipe size		23	.696		11.65	30.50	2.42	44.57	61.50
5453	1/2" pipe size		21	.762		5.40	33.50	2.65	41.55	59

22 11 13 – Facility Water Distribution Piping

22 11 13.47 Pipe Fittings, Steel		Crew	Daily Output	Labor-Hours	Unit	2009 Bare Costs Material	Labor	Equipment	Total	Total Incl O&P
5454	3/4" pipe size	Q-15	21	.762	Ea.	6.95	33.50	2.65	43.10	60.50
5460	1" pipe size		20	.800		7.70	35	2.78	45.48	64
5470	1-1/4" pipe size		20	.800		13.60	35	2.78	51.38	70.50
5480	1-1/2" pipe size		18	.889		15.05	39	3.09	57.14	78.50
5490	2" pipe size		14	1.143		24	50	3.98	77.98	106
5500	2-1/2" pipe size		12	1.333		54	58.50	4.64	117.14	152
5510	3" pipe size		9	1.778		121	78	6.20	205.20	257
5520	4" pipe size		7	2.286		182	100	7.95	289.95	360
5570	Union, 1/4" pipe size		21	.762		25	33.50	2.65	61.15	80.50
5571	3/8" pipe size		21	.762		25	33.50	2.65	61.15	80.50
5572	1/2" pipe size		19	.842		21.50	37	2.93	61.43	82
5573	3/4" pipe size		19	.842		24.50	37	2.93	64.43	85.50
5574	1" pipe size		19	.842		31.50	37	2.93	71.43	93
5575	1-1/4" pipe size		17	.941		51.50	41.50	3.28	96.28	123
5576	1-1/2" pipe size		15	1.067		55	47	3.71	105.71	135
5577	2" pipe size		11	1.455		76	64	5.05	145.05	185
5600	Reducer, 1/4" pipe size		23	.696		27.50	30.50	2.42	60.42	78.50
5601	3/8" pipe size		23	.696		28	30.50	2.42	60.92	79.50
5602	1/2" pipe size		21	.762		19.10	33.50	2.65	55.25	74
5603	3/4" pipe size		21	.762		19.15	33.50	2.65	55.30	74
5604	1" pipe size		21	.762		21	33.50	2.65	57.15	76.50
5605	1-1/4" pipe size		19	.842		35	37	2.93	74.93	97
5607	1-1/2" pipe size		17	.941		41	41.50	3.28	85.78	111
5608	2" pipe size		13	1.231		45.50	54	4.28	103.78	136
5612	Cap, 1/4" pipe size		46	.348		11.15	15.25	1.21	27.61	36.50
5613	3/8" pipe size		46	.348		11.15	15.25	1.21	27.61	36.50
5614	1/2" pipe size		42	.381		6.90	16.70	1.33	24.93	34
5615	3/4" pipe size		42	.381		8.05	16.70	1.33	26.08	35.50
5616	1" pipe size		42	.381		12.20	16.70	1.33	30.23	40
5617	1-1/4" pipe size		38	.421		14.20	18.50	1.47	34.17	44.50
5618	1-1/2" pipe size		34	.471		19.65	20.50	1.64	41.79	54.50
5619	2" pipe size		26	.615		30.50	27	2.14	59.64	76.50
5630	T-O-L, 1/4" pipe size, nozzle		23	.696		6.10	30.50	2.42	39.02	55.50
5631	3/8" pipe size, nozzle		23	.696		6.10	30.50	2.42	39.02	55.50
5632	1/2" pipe size, nozzle		22	.727		6.10	32	2.53	40.63	57.50
5633	3/4" pipe size, nozzle		21	.762		7	33.50	2.65	43.15	60.50
5634	1" pipe size, nozzle		20	.800		8.15	35	2.78	45.93	64.50
5635	1-1/4" pipe size, nozzle		18	.889		12.95	39	3.09	55.04	76
5636	1-1/2" pipe size, nozzle		16	1		12.95	44	3.48	60.43	84
5637	2" pipe size, nozzle		12	1.333		14.75	58.50	4.64	77.89	109
5638	2-1/2" pipe size, nozzle		10	1.600		49.50	70	5.55	125.05	166
5639	4" pipe size, nozzle		6	2.667		112	117	9.30	238.30	310
5640	W-O-L, 1/4" pipe size, nozzle		23	.696		12.60	30.50	2.42	45.52	62.50
5641	3/8" pipe size, nozzle		23	.696		12.60	30.50	2.42	45.52	62.50
5642	1/2" pipe size, nozzle		22	.727		12.60	32	2.53	47.13	64.50
5643	3/4" pipe size, nozzle		21	.762		13.30	33.50	2.65	49.45	67.50
5644	1" pipe size, nozzle		20	.800		13.90	35	2.78	51.68	71
5645	1-1/4" pipe size, nozzle		18	.889		16.50	39	3.09	58.59	80
5646	1-1/2" pipe size, nozzle		16	1		16.50	44	3.48	63.98	88
5647	2" pipe size, nozzle		12	1.333		16.65	58.50	4.64	79.79	111
5648	2-1/2" pipe size, nozzle		10	1.600		38	70	5.55	113.55	153
5649	3" pipe size, nozzle		8	2		42	88	6.95	136.95	186
5650	4" pipe size, nozzle	↓	6	2.667	↓	53	117	9.30	179.30	244

22 11 13.47 Pipe Fittings, Steel		Crew	Daily Output	Labor-Hours	Unit	2009 Bare Costs Material	Labor	Equipment	Total	Total Incl O&P
5651	5" pipe size, nozzle	Q-15	5	3.200	Ea.	130	140	11.15	281.15	365
5652	6" pipe size, nozzle		4	4		149	176	13.90	338.90	440
5653	8" pipe size, nozzle		3	5.333		285	234	18.55	537.55	685
5654	10" pipe size, nozzle		2.60	6.154		520	270	21.50	811.50	1,000
5655	12" pipe size, nozzle		2.20	7.273		985	320	25.50	1,330.50	1,575
5656	14" pipe size, nozzle		2	8		2,350	350	28	2,728	3,125
5674	S-O-L, 1/4" pipe size, outlet		23	.696		7.55	30.50	2.42	40.47	57
5675	3/8" pipe size, outlet		23	.696		7.55	30.50	2.42	40.47	57
5676	1/2" pipe size, outlet		22	.727		7.55	32	2.53	42.08	59
5677	3/4" pipe size, outlet		21	.762		7.65	33.50	2.65	43.80	61.50
5678	1" pipe size, outlet		20	.800		8.50	35	2.78	46.28	65
5679	1-1/4" pipe size, outlet		18	.889		14.20	39	3.09	56.29	77.50
5680	1-1/2" pipe size, outlet		16	1		14.20	44	3.48	61.68	85.50
5681	2" pipe size, outlet		12	1.333		15.60	58.50	4.64	78.74	110
6000	Weld-on flange, forged steel									
6020	Slip-on, 150 lb. flange (welded front and back)									
6050	1/2" pipe size	Q-15	18	.889	Ea.	14	39	3.09	56.09	77.50
6060	3/4" pipe size		18	.889		14	39	3.09	56.09	77.50
6070	1" pipe size		17	.941		14	41.50	3.28	58.78	81
6080	1-1/4" pipe size		16	1		14	44	3.48	61.48	85
6090	1-1/2" pipe size		15	1.067		14	47	3.71	64.71	89.50
6100	2" pipe size		12	1.333		14	58.50	4.64	77.14	109
6110	2-1/2" pipe size		10	1.600		24.50	70	5.55	100.05	138
6120	3" pipe size		9	1.778		21.50	78	6.20	105.70	148
6130	3-1/2" pipe size		7	2.286		25.50	100	7.95	133.45	188
6140	4" pipe size		6	2.667		24.50	117	9.30	150.80	213
6150	5" pipe size		5	3.200		42	140	11.15	193.15	270
6160	6" pipe size	Q-16	6	4		39.50	182	9.30	230.80	325
6170	8" pipe size		5	4.800		63.50	218	11.15	292.65	410
6180	10" pipe size		4	6		104	273	13.90	390.90	540
6190	12" pipe size		3	8		153	365	18.55	536.55	735
6191	14" pipe size		2.50	9.600		209	435	22.50	666.50	910
6192	16" pipe size		1.80	13.333		345	605	31	981	1,325
6200	300 lb. flange									
6210	1/2" pipe size	Q-15	17	.941	Ea.	18.50	41.50	3.28	63.28	86
6220	3/4" pipe size		17	.941		18.50	41.50	3.28	63.28	86
6230	1" pipe size		16	1		18.50	44	3.48	65.98	90.50
6240	1-1/4" pipe size		13	1.231		18.50	54	4.28	76.78	106
6250	1-1/2" pipe size		12	1.333		18.50	58.50	4.64	81.64	114
6260	2" pipe size		11	1.455		22	64	5.05	91.05	126
6270	2-1/2" pipe size		9	1.778		27.50	78	6.20	111.70	154
6280	3" pipe size		7	2.286		32	100	7.95	139.95	195
6290	4" pipe size		6	2.667		46.50	117	9.30	172.80	238
6300	5" pipe size		4	4		70	176	13.90	259.90	355
6310	6" pipe size	Q-16	5	4.800		74.50	218	11.15	303.65	425
6320	8" pipe size		4	6		128	273	13.90	414.90	565
6330	10" pipe size		3.40	7.059		202	320	16.40	538.40	720
6340	12" pipe size		2.80	8.571		266	390	19.90	675.90	900
6400	Welding neck, 150 lb. flange									
6410	1/2" pipe size	Q-15	40	.400	Ea.	19.90	17.55	1.39	38.84	50
6420	3/4" pipe size		36	.444		19.90	19.50	1.55	40.95	53
6430	1" pipe size		32	.500		19.90	22	1.74	43.64	57
6440	1-1/4" pipe size		29	.552		19.90	24	1.92	45.82	60.50

22 11 13.47 Pipe Fittings, Steel		Crew	Daily Output	Labor-Hours	Unit	Material	Labor	Equipment	Total	Total Incl O&P
						2009 Bare Costs				
6450	1-1/2" pipe size	Q-15	26	.615	Ea.	19.90	27	2.14	49.04	65
6460	2" pipe size		20	.800		21.50	35	2.78	59.28	79
6470	2-1/2" pipe size		16	1		23.50	44	3.48	70.98	95.50
6480	3" pipe size		14	1.143		25	50	3.98	78.98	107
6500	4" pipe size		10	1.600		34.50	70	5.55	110.05	149
6510	5" pipe size		8	2		48.50	88	6.95	143.45	193
6520	6" pipe size	Q-16	10	2.400		54.50	109	5.55	169.05	230
6530	8" pipe size		7	3.429		85	156	7.95	248.95	335
6540	10" pipe size		6	4		143	182	9.30	334.30	440
6550	12" pipe size		5	4.800		206	218	11.15	435.15	570
6551	14" pipe size		4.50	5.333		278	243	12.35	533.35	685
6552	16" pipe size		3	8		425	365	18.55	808.55	1,025
6553	18" pipe size		2.50	9.600		620	435	22.50	1,077.50	1,350
6554	20" pipe size		2.30	10.435		720	475	24	1,219	1,525
6556	24" pipe size		2	12		1,025	545	28	1,598	1,975
6557	26" pipe size		1.70	14.118		1,100	640	33	1,773	2,225
6558	30" pipe size		.90	26.667		1,275	1,225	62	2,562	3,300
6559	36" pipe size		.75	32		1,450	1,450	74	2,974	3,850
6560	300 lb. flange									
6570	1/2" pipe size	Q-15	36	.444	Ea.	24.50	19.50	1.55	45.55	58
6580	3/4" pipe size		34	.471		24.50	20.50	1.64	46.64	60
6590	1" pipe size		30	.533		24.50	23.50	1.86	49.86	64
6600	1-1/4" pipe size		28	.571		24.50	25	1.99	51.49	66.50
6610	1-1/2" pipe size		24	.667		24.50	29.50	2.32	56.32	73.50
6620	2" pipe size		18	.889		27	39	3.09	69.09	91.50
6630	2-1/2" pipe size		14	1.143		31.50	50	3.98	85.48	114
6640	3" pipe size		12	1.333		36.50	58.50	4.64	99.64	133
6650	4" pipe size		8	2		58	88	6.95	152.95	204
6660	5" pipe size		7	2.286		85.50	100	7.95	193.45	254
6670	6" pipe size	Q-16	9	2.667		93	121	6.20	220.20	291
6680	8" pipe size		6	4		159	182	9.30	350.30	455
6690	10" pipe size		5	4.800		258	218	11.15	487.15	625
6700	12" pipe size		4	6		330	273	13.90	616.90	790
6710	14" pipe size		3.50	6.857		760	310	15.90	1,085.90	1,325
6720	16" pipe size		2	12		860	545	28	1,433	1,800
7740	Plain ends for plain end pipe, mechanically coupled									
7750	Cplg & labor required at joints not included, add 1 per									
7760	joint for installed price, see line 9180									
7770	Malleable iron, painted, unless noted otherwise									
7800	90° Elbow 1"				Ea.	68.50			68.50	75.50
7810	1-1/2"					81.50			81.50	89.50
7820	2"					122			122	134
7830	2-1/2"					142			142	157
7840	3"					148			148	163
7860	4"					168			168	184
7870	5" welded steel					193			193	212
7880	6"					244			244	268
7890	8" welded steel					460			460	505
7900	10" welded steel					605			605	665
7910	12" welded steel					675			675	740
7970	45° Elbow 1"					41			41	45
7980	1-1/2"					61.50			61.50	67.50
7990	2"					129			129	142

22 11 13 – Facility Water Distribution Piping

22 11 13.47 Pipe Fittings, Steel		Crew	Daily Output	Labor-Hours	Unit	Material	Labor	Equipment	Total	Total Incl O&P
						2009 Bare Costs				
8000	2-1/2"				Ea.	129			129	142
8010	3"					148			148	163
8030	4"					154			154	169
8040	5" welded steel					193			193	212
8050	6"					221			221	243
8060	8"					259			259	285
8070	10" welded steel					278			278	305
8080	12" welded steel					505			505	555
8140	Tee, straight 1"					77			77	84.50
8150	1-1/2"					99			99	109
8160	2"					99			99	109
8170	2-1/2"					129			129	142
8180	3"					199			199	219
8200	4"					286			286	315
8210	5" welded steel					390			390	430
8220	6"					340			340	375
8230	8" welded steel					490			490	540
8240	10" welded steel					760			760	835
8250	12" welded steel				▼	935			935	1 025
8340	Segmentally welded steel, painted									
8390	Wye 2"				Ea.	124			124	136
8400	2-1/2"					124			124	136
8410	3"					139			139	152
8430	4"					207			207	228
8440	5"					251			251	277
8450	6"					345			345	380
8460	8"					450			450	495
8470	10"					660			660	730
8480	12"					1,025			1,025	1,125
8540	Wye, lateral 2"					134			134	148
8550	2-1/2"					155			155	170
8560	3"					183			183	202
8580	4"					253			253	278
8590	5"					430			430	475
8600	6"					445			445	485
8610	8"					745			745	820
8620	10"					770			770	850
8630	12"					1,425			1,425	1,550
8690	Cross, 2"					141			141	155
8700	2-1/2"					141			141	155
8710	3"					168			168	185
8730	4"					230			230	253
8740	5"					325			325	360
8750	6"					430			430	475
8760	8"					555			555	610
8770	10"					805			805	890
8780	12"					1,175			1,175	1,275
8800	Tees, reducing 2" x 1"					84			84	92.50
8810	2" x 1-1/2"					84			84	92.50
8820	3" x 1"					84			84	92.50
8830	3" x 1-1/2"					84			84	92.50
8840	3" x 2"					84			84	92.50
8850	4" x 1"				▼	121			121	133

22 11 13.47 Pipe Fittings, Steel		Crew	Daily Output	Labor-Hours	Unit	2009 Bare Costs Material	Labor	Equipment	Total	Total Incl O&P
8860	4" x 1-1/2"				Ea.	121			121	133
8870	4" x 2"					121			121	133
8880	4" x 2-1/2"					125			125	138
8890	4" x 3"					125			125	138
8900	6" x 2"					199			199	218
8910	6" x 3"					214			214	235
8920	6" x 4"					260			260	286
8930	8" x 2"					271			271	298
8940	8" x 3"					277			277	305
8950	8" x 4"					277			277	305
8960	8" x 5"					286			286	315
8970	8" x 6"					286			286	315
8980	10" x 4"					410			410	450
8990	10" x 6"					410			410	450
9000	10" x 8"					435			435	480
9010	12" x 6"					630			630	695
9020	12" x 8"					635			635	695
9030	12" x 10"				↓	655			655	725
9080	Adapter nipples 3" long									
9090	1"				Ea.	11.40			11.40	12.50
9100	1-1/2"					11.40			11.40	12.50
9110	2"					11.40			11.40	12.50
9120	2-1/2"					13.25			13.25	14.60
9130	3"					16.25			16.25	17.85
9140	4"					27			27	29.50
9150	6"				↓	72			72	79.50
9180	Coupling, mechanical, plain end pipe to plain end pipe or fitting									
9190	1"	Q-1	29	.552	Ea.	41	24		65	82
9200	1-1/2"		28	.571		41	25		66	83
9210	2"		27	.593		41	26		67	84.50
9220	2-1/2"		26	.615		41	27		68	86
9230	3"		25	.640		60.50	28		88.50	109
9240	3-1/2"		24	.667		70.50	29.50		100	122
9250	4"	↓	22	.727		70.50	32		102.50	126
9260	5"	Q-2	28	.857		100	39		139	169
9270	6"		24	1		122	45.50		167.50	203
9280	8"		19	1.263		216	57.50		273.50	325
9290	10"		16	1.500		281	68.50		349.50	410
9300	12"	↓	12	2	↓	350	91		441	525
9310	Outlets for precut holes through pipe wall									
9331	Strapless type, with gasket									
9332	4" to 8" pipe x 1/2"	1 Plum	13	.615	Ea.	45	30		75	94.50
9333	4" to 8" pipe x 3/4"		13	.615		48.50	30		78.50	98
9334	10" pipe and larger x 1/2"		11	.727		45	35.50		80.50	103
9335	10" pipe and larger x 3/4"	↓	11	.727	↓	48.50	35.50		84	106
9341	Thermometer wells with gasket									
9342	4" to 8" pipe, 6" stem	1 Plum	14	.571	Ea.	68.50	28		96.50	117
9343	8" pipe and larger, 6" stem	"	13	.615	"	68.50	30		98.50	120
9940	For galvanized fittings for plain end pipe, add					20%				

22 11 13.48 Pipe, Fittings and Valves, Steel, Grooved-Joint	
0010	**PIPE, FITTINGS AND VALVES, STEEL, GROOVED-JOINT**
0012	Fittings are ductile iron. Steel fittings noted.

22 11 13.48 Pipe, Fittings and Valves, Steel, Grooved-Joint		Crew	Daily Output	Labor-Hours	Unit	2009 Bare Costs Material	Labor	Equipment	Total	Total Incl O&P
0020	Pipe includes coupling & clevis type hanger assemblies, 10′ O.C.									
0500	Schedule 10, black									
0550	2″ diameter	1 Plum	43	.186	L.F.	5.30	9.05		14.35	19.45
0560	2-1/2″ diameter	Q-1	61	.262		6.90	11.50		18.40	25
0570	3″ diameter		55	.291		8.20	12.75		20.95	28
0580	3-1/2″ diameter		53	.302		10.60	13.25		23.85	31.50
0590	4″ diameter		49	.327		10.90	14.35		25.25	33.50
0600	5″ diameter		40	.400		14.20	17.55		31.75	42
0610	6″ diameter	Q-2	46	.522		18.40	23.50		41.90	55.50
0620	8″ diameter	″	41	.585		24.50	26.50		51	67
0700	To delete couplings & hangers, subtract									
0710	2″ diam. to 5″ diam.					25%	20%			
0720	6″ diam. to 8″ diam.					27%	15%			
1000	Schedule 40, black									
1040	3/4″ diameter	1 Plum	71	.113	L.F.	3.53	5.50		9.03	12.15
1050	1″ diameter		63	.127		4.23	6.20		10.43	13.95
1060	1-1/4″ diameter		58	.138		5.45	6.70		12.15	16.10
1070	1-1/2″ diameter		51	.157		6.30	7.65		13.95	18.40
1080	2″ diameter		40	.200		8.05	9.75		17.80	23.50
1090	2-1/2″ diameter	Q-1	57	.281		10.05	12.30		22.35	29.50
1100	3″ diameter		50	.320		16.10	14.05		30.15	38.50
1110	4″ diameter		45	.356		19.10	15.60		34.70	44.50
1120	5″ diameter		37	.432		26.50	19		45.50	57.50
1130	6″ diameter	Q-2	42	.571		33.50	26		59.50	76
1140	8″ diameter		37	.649		51.50	29.50		81	102
1150	10″ diameter		31	.774		77.50	35		112.50	139
1160	12″ diameter		27	.889		82.50	40.50		123	152
1170	14″ diameter		20	1.200		94.50	54.50		149	186
1180	16″ diameter		17	1.412		112	64		176	221
1190	18″ diameter		14	1.714		129	78		207	259
1200	20″ diameter		12	2		156	91		247	310
1210	24″ diameter		10	2.400		189	109		298	370
1740	To delete coupling & hanger, subtract									
1750	3/4″ diam. to 2″ diam.					65%	27%			
1760	2-1/2″ diam. to 5″ diam.					41%	18%			
1770	6″ diam. to 12″ diam.					31%	13%			
1780	14″ diam. to 24″ diam.					35%	10%			
1800	Galvanized									
1840	3/4″ diameter	1 Plum	71	.113	L.F.	4.63	5.50		10.13	13.35
1850	1″ diameter		63	.127		6.10	6.20		12.30	16
1860	1-1/4″ diameter		58	.138		7.85	6.70		14.55	18.75
1870	1-1/2″ diameter		51	.157		9.20	7.65		16.85	21.50
1880	2″ diameter		40	.200		11.85	9.75		21.60	27.50
1890	2-1/2″ diameter	Q-1	57	.281		17.95	12.30		30.25	38
1900	3″ diameter		50	.320		23	14.05		37.05	46.50
1910	4″ diameter		45	.356		30	15.60		45.60	56.50
1920	5″ diameter		37	.432		34.50	19		53.50	66.50
1930	6″ diameter	Q-2	42	.571		40.50	26		66.50	83.50
1940	8″ diameter		37	.649		48.50	29.50		78	97.50
1950	10″ diameter		31	.774		71.50	35		106.50	132
1960	12″ diameter		27	.889		93	40.50		133.50	163
2540	To delete coupling & hanger, subtract									
2550	3/4″ diam. to 2″ diam.					36%	27%			

22 11 13.48 Pipe, Fittings and Valves, Steel, Grooved-Joint		Crew	Daily Output	Labor-Hours	Unit	Material	Labor	Equipment	Total	Total Incl O&P
						2009 Bare Costs				
2560	2-1/2" diam. to 5" diam.					19%	18%			
2570	6" diam. to 12" diam.					14%	13%			
2600	Schedule 80, black									
2610	3/4" diameter	1 Plum	65	.123	L.F.	4.67	6		10.67	14.15
2650	1" diameter		61	.131		5.95	6.40		12.35	16.15
2660	1-1/4" diameter		55	.145		7.80	7.10		14.90	19.25
2670	1-1/2" diameter		49	.163		9.20	7.95		17.15	22
2680	2" diameter		38	.211		11.95	10.25		22.20	28.50
2690	2-1/2" diameter	Q-1	54	.296		18.05	13		31.05	39.50
2700	3" diameter		48	.333		24	14.65		38.65	48
2710	4" diameter		44	.364		25.50	15.95		41.45	52
2720	5" diameter		35	.457		52.50	20		72.50	87.50
2730	6" diameter	Q-2	40	.600		48	27.50		75.50	94
2740	8" diameter		35	.686		85.50	31		116.50	141
2750	10" diameter		29	.828		150	37.50		187.50	222
2760	12" diameter		24	1		229	45.50		274.50	320
3240	To delete coupling & hanger, subtract									
3250	3/4" diam. to 2" diam.					30%	25%			
3260	2-1/2" diam. to 5" diam.					14%	17%			
3270	6" diam. to 12" diam.					12%	12%			
3300	Galvanized									
3310	3/4" diameter	1 Plum	65	.123	L.F.	5.90	6		11.90	15.45
3350	1" diameter		61	.131		7.75	6.40		14.15	18.10
3360	1-1/4" diameter		55	.145		10.25	7.10		17.35	22
3370	1-1/2" diameter		46	.174		12.20	8.50		20.70	26
3380	2" diameter		38	.211		16.15	10.25		26.40	33
3390	2-1/2" diameter	Q-1	54	.296		24.50	13		37.50	46.50
3400	3" diameter		48	.333		32.50	14.65		47.15	58
3410	4" diameter		44	.364		46	15.95		61.95	74.50
3420	5" diameter		35	.457		77.50	20		97.50	116
3430	6" diameter	Q-2	40	.600		83.50	27.50		111	133
3440	8" diameter		35	.686		127	31		158	187
3450	10" diameter		29	.828		235	37.50		272.50	315
3460	12" diameter		24	1		335	45.50		380.50	440
3920	To delete coupling & hanger, subtract									
3930	3/4" diam. to 2" diam.					30%	25%			
3940	2-1/2" diam. to 5" diam.					15%	17%			
3950	6" diam. to 12" diam.					11%	12%			
3990	Fittings: coupling material required at joints not incl. in fitting price.									
3994	Add 1 selected coupling, material only, per joint for installed price.									
4000	Elbow, 90° or 45°, painted									
4030	3/4" diameter	1 Plum	50	.160	Ea.	36.50	7.80		44.30	51.50
4040	1" diameter		50	.160		19.35	7.80		27.15	33
4050	1-1/4" diameter		40	.200		19.35	9.75		29.10	36
4060	1-1/2" diameter		33	.242		19.35	11.80		31.15	39.50
4070	2" diameter		25	.320		19.35	15.60		34.95	45
4080	2-1/2" diameter	Q-1	40	.400		19.35	17.55		36.90	48
4090	3" diameter		33	.485		34.50	21.50		56	69.50
4100	4" diameter		25	.640		37.50	28		65.50	83
4110	5" diameter		20	.800		90	35		125	152
4120	6" diameter	Q-2	25	.960		106	43.50		149.50	182
4130	8" diameter		21	1.143		220	52		272	320
4140	10" diameter		18	1.333		400	60.50		460.50	530

22 11 13 – Facility Water Distribution Piping

22 11 13.48 Pipe, Fittings and Valves, Steel, Grooved-Joint		Crew	Daily Output	Labor-Hours	Unit	Material	2009 Bare Costs Labor	Equipment	Total	Total Incl O&P
4150	12″ diameter	Q-2	15	1.600	Ea.	640	73		713	815
4170	14″ diameter		12	2		835	91		926	1,050
4180	16″ diameter		11	2.182		1,075	99.50		1,174.50	1,350
4190	18″ diameter	Q-3	15	2.133		1,375	99		1,474	1,675
4200	20″ diameter		13	2.462		1,825	114		1,939	2,175
4210	24″ diameter		11	2.909		2,625	135		2,760	3,100
4250	For galvanized elbows, add					26%				
4690	Tee, painted									
4700	3/4″ diameter	1 Plum	38	.211	Ea.	39	10.25		49.25	58.50
4740	1″ diameter		33	.242		30	11.80		41.80	51
4750	1-1/4″ diameter		27	.296		30	14.45		44.45	54.50
4760	1-1/2″ diameter		22	.364		30	17.75		47.75	59.50
4770	2″ diameter		17	.471		30	23		53	67.50
4780	2-1/2″ diameter	Q-1	27	.593		30	26		56	72
4790	3″ diameter		22	.727		41.50	32		73.50	93.50
4800	4″ diameter		17	.941		63.50	41.50		105	132
4810	5″ diameter		13	1.231		148	54		202	244
4820	6″ diameter	Q-2	17	1.412		171	64		235	285
4830	8″ diameter		14	1.714		375	78		453	530
4840	10″ diameter		12	2		780	91		871	995
4850	12″ diameter		10	2.400		1,100	109		1,209	1,375
4851	14″ diameter		9	2.667		1,125	121		1,246	1,400
4852	16″ diameter		8	3		1,275	137		1,412	1,500
4853	18″ diameter	Q-3	11	2.909		1,575	135		1,710	1,950
4854	20″ diameter		10	3.200		2,275	149		2,424	2,725
4855	24″ diameter		8	4		3,450	186		3,636	4,075
4900	For galvanized tees, add					24%				
4906	Couplings, rigid style, painted									
4908	1″ diameter	1 Plum	100	.080	Ea.	14.50	3.90		18.40	22
4909	1-1/4″ diameter		100	.080		14.50	3.90		18.40	22
4910	1-1/2″ diameter		67	.119		14.50	5.80		20.30	24.50
4912	2″ diameter		50	.160		14.80	7.80		22.60	28
4914	2-1/2″ diameter	Q-1	80	.200		17	8.80		25.80	32
4916	3″ diameter		67	.239		19.90	10.50		30.40	38
4918	4″ diameter		50	.320		28	14.05		42.05	52
4920	5″ diameter		40	.400		36	17.55		53.55	66.50
4922	6″ diameter	Q-2	50	.480		48.50	22		70.50	86
4924	8″ diameter		42	.571		76	26		102	123
4926	10″ diameter		35	.686		135	31		166	196
4928	12″ diameter		32	.750		151	34		185	217
4930	14″ diameter		24	1		192	45.50		237.50	281
4931	16″ diameter		20	1.200		250	54.50		304.50	355
4932	18″ diameter		18	1.333		289	60.50		349.50	410
4933	20″ diameter		16	1.500		395	68.50		463.50	535
4934	24″ diameter		13	1.846		505	84		589	685
4940	Flexible, standard, painted									
4950	3/4″ diameter	1 Plum	100	.080	Ea.	10.40	3.90		14.30	17.30
4960	1″ diameter		100	.080		10.40	3.90		14.30	17.30
4970	1-1/4″ diameter		80	.100		13.80	4.88		18.68	22.50
4980	1-1/2″ diameter		67	.119		15.15	5.80		20.95	25.50
4990	2″ diameter		50	.160		16	7.80		23.80	29.50
5000	2-1/2″ diameter	Q-1	80	.200		18.90	8.80		27.70	34
5010	3″ diameter		67	.239		21	10.50		31.50	39

22 11 Facility Water Distribution

22 11 13 – Facility Water Distribution Piping

22 11 13.48 Pipe, Fittings and Valves, Steel, Grooved-Joint		Crew	Daily Output	Labor-Hours	Unit	2009 Bare Costs Material	Labor	Equipment	Total	Total Incl O&P
5020	3-1/2" diameter	Q-1	57	.281	Ea.	30.50	12.30		42.80	52
5030	4" diameter		50	.320		30.50	14.05		44.55	54.50
5040	5" diameter		40	.400		46	17.55		63.55	77.50
5050	6" diameter	Q-2	50	.480		54.50	22		76.50	93
5070	8" diameter		42	.571		89	26		115	137
5090	10" diameter		35	.686		147	31		178	209
5110	12" diameter		32	.750		168	34		202	236
5120	14" diameter		24	1		198	45.50		243.50	287
5130	16" diameter		20	1.200		260	54.50		314.50	370
5140	18" diameter		18	1.333		305	60.50		365.50	425
5150	20" diameter		16	1.500		405	68.50		473.50	545
5160	24" diameter		13	1.846		525	84		609	700
5176	Lightweight style, painted									
5178	1-1/2" diameter	1 Plum	67	.119	Ea.	13.25	5.80		19.05	23.50
5180	2" diameter	"	50	.160		13.60	7.80		21.40	26.50
5182	2-1/2" diameter	Q-1	80	.200		15.70	8.80		24.50	30.50
5184	3" diameter		67	.239		18.25	10.50		28.75	36
5186	3-1/2" diameter		57	.281		25.50	12.30		37.80	46.50
5188	4" diameter		50	.320		25.50	14.05		39.55	49
5190	5" diameter		40	.400		36.50	17.55		54.05	67
5192	6" diameter	Q-2	50	.480		44	22		66	81.50
5194	8" diameter		42	.571		69	26		95	115
5196	10" diameter		35	.686		184	31		215	250
5198	12" diameter		32	.750		206	34		240	277
5200	For galvanized couplings, add					33%				
5220	Tee, reducing, painted									
5225	2" x 1-1/2" diameter	Q-1	38	.421	Ea.	64	18.50		82.50	98
5226	2-1/2" x 2" diameter		28	.571		64	25		89	108
5227	3" x 2-1/2" diameter		23	.696		56.50	30.50		87	108
5228	4" x 3" diameter		18	.889		76.50	39		115.50	143
5229	5" x 4" diameter		15	1.067		164	47		211	250
5230	6" x 4" diameter	Q-2	18	1.333		180	60.50		240.50	289
5231	8" x 6" diameter		15	1.600		375	73		448	525
5232	10" x 8" diameter		13	1.846		495	84		579	665
5233	12" x 10" diameter		11	2.182		750	99.50		849.50	975
5234	14" x 12" diameter		10	2.400		815	109		924	1,075
5235	16" x 12" diameter		9	2.667		1,000	121		1,121	1,275
5236	18" x 12" diameter	Q-3	12	2.667		1,225	124		1,349	1,500
5237	18" x 16" diameter		11	2.909		1,525	135		1,660	1,875
5238	20" x 16" diameter		10	3.200		2,000	149		2,149	2,425
5239	24" x 20" diameter		9	3.556		3,150	165		3,315	3,725
5240	Reducer, concentric, painted									
5241	2-1/2" x 2" diameter	Q-1	43	.372	Ea.	23	16.35		39.35	49.50
5242	3" x 2-1/2" diameter		35	.457		27.50	20		47.50	60
5243	4" x 3" diameter		29	.552		33	24		57	73
5244	5" x 4" diameter		22	.727		46	32		78	98.50
5245	6" x 4" diameter	Q-2	26	.923		53	42		95	122
5246	8" x 6" diameter		23	1.043		138	47.50		185.50	224
5247	10" x 8" diameter		20	1.200		282	54.50		336.50	390
5248	12" x 10" diameter		16	1.500		505	68.50		573.50	655
5255	Eccentric, painted									
5256	2-1/2" x 2" diameter	Q-1	42	.381	Ea.	47.50	16.70		64.20	77.50
5257	3" x 2-1/2" diameter		34	.471		54.50	20.50		75	91

22 11 13.48 Pipe, Fittings and Valves, Steel, Grooved-Joint		Crew	Daily Output	Labor-Hours	Unit	2009 Bare Costs Material	Labor	Equipment	Total	Total Incl O&P
5258	4" x 3" diameter	Q-1	28	.571	Ea.	66.50	25		91.50	111
5259	5" x 4" diameter	↓	21	.762		90.50	33.50		124	150
5260	6" x 4" diameter	Q-2	25	.960		105	43.50		148.50	182
5261	8" x 6" diameter		22	1.091		213	49.50		262.50	310
5262	10" x 8" diameter		19	1.263		585	57.50		642.50	730
5263	12" x 10" diameter	↓	15	1.600	↓	805	73		878	995
5270	Coupling, reducing, painted									
5272	2" x 1-1/2" diameter	1 Plum	52	.154	Ea.	22.50	7.50		30	36.50
5274	2-1/2" x 2" diameter	Q-1	82	.195		29.50	8.55		38.05	45.50
5276	3" x 2" diameter		69	.232		33.50	10.15		43.65	52.50
5278	4" x 2" diameter		52	.308		53.50	13.50		67	79
5280	5" x 4" diameter	↓	42	.381		60.50	16.70		77.20	91.50
5282	6" x 4" diameter	Q-2	52	.462		91	21		112	132
5284	8" x 6" diameter	"	44	.545	↓	137	25		162	187
5290	Outlet coupling, painted									
5294	1-1/2" x 1" pipe size	1 Plum	65	.123	Ea.	29	6		35	40.50
5296	2" x 1" pipe size	"	48	.167		29.50	8.15		37.65	44.50
5298	2-1/2" x 1" pipe size	Q-1	78	.205		45.50	9		54.50	63.50
5300	2-1/2" x 1-1/4" pipe size	1 Plum	70	.114		51.50	5.55		57.05	65.50
5302	3" x 1" pipe size	Q-1	65	.246		58	10.80		68.80	80
5304	4" x 3/4" pipe size		48	.333		65	14.65		79.65	93
5306	4" x 1-1/2" pipe size	↓	46	.348		92	15.25		107.25	124
5308	6" x 1-1/2" pipe size	Q-2	44	.545	↓	130	25		155	180
5750	Flange, w/groove gasket, black steel									
5754	See Div. 22 11 13.44 0620 for gasket & bolt set									
5760	ANSI class 125 and 150, painted									
5780	2" pipe size	1 Plum	23	.348	Ea.	65	16.95		81.95	97
5790	2-1/2" pipe size	Q-1	37	.432		80.50	19		99.50	117
5800	3" pipe size		31	.516		87	22.50		109.50	130
5820	4" pipe size		23	.696		116	30.50		146.50	174
5830	5" pipe size	↓	19	.842		135	37		172	205
5840	6" pipe size	Q-2	23	1.043		147	47.50		194.50	234
5850	8" pipe size		17	1.412		167	64		231	280
5860	10" pipe size		14	1.714		264	78		342	405
5870	12" pipe size		12	2		345	91		436	515
5880	14" pipe size		10	2.400		770	109		879	1,000
5890	16" pipe size		9	2.667		895	121		1,016	1,150
5900	18" pipe size		6	4		1,100	182		1,282	1,475
5910	20" pipe size		5	4.800		1,325	218		1,543	1,775
5920	24" pipe size	↓	4.50	5.333	↓	1,700	243		1,943	2,225
5940	ANSI class 350, painted									
5946	2" pipe size	1 Plum	23	.348	Ea.	81	16.95		97.95	115
5948	2-1/2" pipe size	Q-1	37	.432		94	19		113	132
5950	3" pipe size		31	.516		129	22.50		151.50	176
5952	4" pipe size		23	.696		171	30.50		201.50	235
5954	5" pipe size	↓	19	.842		195	37		232	270
5956	6" pipe size	Q-2	23	1.043		227	47.50		274.50	320
5958	8" pipe size		17	1.412		260	64		324	385
5960	10" pipe size	↓	14	1.714		415	78		493	570
5962	12" pipe size	1 Plum	12	.667	↓	445	32.50		477.50	535
6100	Cross, painted									
6110	2" diameter	1 Plum	12.50	.640	Ea.	54	31		85	107
6112	2-1/2" diameter	Q-1	20	.800	↓	54	35		89	112

22 11 13.48	Pipe, Fittings and Valves, Steel, Grooved-Joint	Crew	Daily Output	Labor-Hours	Unit	2009 Bare Costs Material	Labor	Equipment	Total	Total Incl O&P
6114	3″ diameter	Q-1	15.50	1.032	Ea.	96.50	45.50		142	174
6116	4″ diameter		12.50	1.280		160	56		216	262
6118	6″ diameter	Q-2	12.50	1.920		420	87.50		507.50	595
6120	8″ diameter		10.50	2.286		555	104		659	765
6122	10″ diameter		9	2.667		945	121		1,066	1,225
6124	12″ diameter		7.50	3.200		1,375	146		1,521	1,725
7400	Suction diffuser									
7402	Grooved end inlet x flanged outlet									
7410	3″ x 3″	Q-1	50	.320	Ea.	610	14.05		624.05	690
7412	4″ x 4″		38	.421		825	18.50		843.50	940
7414	5″ x 5″		30	.533		970	23.50		993.50	1,100
7416	6″ x 6″	Q-2	38	.632		1,225	28.50		1,253.50	1,400
7418	8″ x 8″		27	.889		2,275	40.50		2,315.50	2,550
7420	10″ x 10″		20	1.200		3,075	54.50		3,129.50	3,475
7422	12″ x 12″		16	1.500		5,075	68.50		5,143.50	5,700
7424	14″ x 14″		15	1.600		6,425	73		6,498	7,150
7426	16″ x 14″		14	1.714		6,600	78		6,678	7,375
7500	Strainer, tee type, painted									
7506	2″ pipe size	1 Plum	38	.211	Ea.	410	10.25		420.25	465
7508	2-1/2″ pipe size	Q-1	62	.258		430	11.30		441.30	490
7510	3″ pipe size		50	.320		485	14.05		499.05	555
7512	4″ pipe size		38	.421		550	18.50		568.50	635
7514	5″ pipe size		30	.533		795	23.50		818.50	910
7516	6″ pipe size	Q-2	38	.632		860	28.50		888.50	995
7518	8″ pipe size		27	.889		1,325	40.50		1,365.50	1,500
7520	10″ pipe size		20	1.200		1,950	54.50		2,004.50	2,200
7522	12″ pipe size		16	1.500		2,500	68.50		2,568.50	2,850
7524	14″ pipe size		15	1.600		8,650	73		8,723	9,625
7526	16″ pipe size		14	1.714		10,800	78		10,878	11,900
7570	Expansion joint, max. 3″ travel									
7572	2″ diameter	1 Plum	38	.211	Ea.	445	10.25		455.25	505
7574	3″ diameter	Q-1	50	.320		510	14.05		524.05	585
7576	4″ diameter	″	38	.421		675	18.50		693.50	770
7578	6″ diameter	Q-2	38	.632		1,050	28.50		1,078.50	1,200
7800	Ball valve w/handle, carbon steel trim									
7810	1-1/2″ pipe size	1 Plum	50	.160	Ea.	117	7.80		124.80	141
7812	2″ pipe size	″	38	.211		130	10.25		140.25	158
7814	2-1/2″ pipe size	Q-1	62	.258		283	11.30		294.30	325
7816	3″ pipe size		50	.320		460	14.05		474.05	525
7818	4″ pipe size		38	.421		710	18.50		728.50	810
7820	6″ pipe size	Q-2	30	.800		2,175	36.50		2,211.50	2,450
7830	With gear operator									
7834	2-1/2″ pipe size	Q-1	62	.258	Ea.	575	11.30		586.30	645
7836	3″ pipe size		50	.320		815	14.05		829.05	915
7838	4″ pipe size		38	.421		1,025	18.50		1,043.50	1,175
7840	6″ pipe size	Q-2	30	.800		2,450	36.50		2,486.50	2,750
7870	Check valve									
7874	2-1/2″ pipe size	Q-1	62	.258	Ea.	196	11.30		207.30	233
7876	3″ pipe size		50	.320		232	14.05		246.05	276
7878	4″ pipe size		38	.421		245	18.50		263.50	297
7880	5″ pipe size		30	.533		405	23.50		428.50	485
7882	6″ pipe size	Q-2	38	.632		485	28.50		513.50	575
7884	8″ pipe size		27	.889		660	40.50		700.50	785

22 11 13.48 Pipe, Fittings and Valves, Steel, Grooved-Joint		Crew	Daily Output	Labor-Hours	Unit	Material	Labor	Equipment	Total	Total Incl O&P
						2009 Bare Costs				
7886	10″ pipe size	Q-2	20	1.200	Ea.	1,900	54.50		1,954.50	2,175
7888	12″ pipe size	↓	16	1.500	↓	2,250	68.50		2,318.50	2,575
7900	Plug valve, balancing, w/lever operator									
7906	3″ pipe size	Q-1	50	.320	Ea.	455	14.05		469.05	525
7908	4″ pipe size	″	38	.421		560	18.50		578.50	650
7909	6″ pipe size	Q-2	30	.800	↓	860	36.50		896.50	1,000
7916	With gear operator									
7920	3″ pipe size	Q-1	50	.320	Ea.	865	14.05		879.05	975
7922	4″ pipe size	″	38	.421		900	18.50		918.50	1,025
7924	6″ pipe size	Q-2	38	.632		1,200	28.50		1,228.50	1,375
7926	8″ pipe size		27	.889		1,875	40.50		1,915.50	2,100
7928	10″ pipe size		20	1.200		2,550	54.50		2,604.50	2,875
7930	12″ pipe size	↓	16	1.500	↓	3,875	68.50		3,943.50	4,350
8000	Butterfly valve, 2 position handle, with standard trim									
8010	1-1/2″ pipe size	1 Plum	50	.160	Ea.	173	7.80		180.80	202
8020	2″ pipe size	″	38	.211		173	10.25		183.25	205
8030	3″ pipe size	Q-1	50	.320		249	14.05		263.05	294
8050	4″ pipe size	″	38	.421		273	18.50		291.50	330
8070	6″ pipe size	Q-2	38	.632		550	28.50		578.50	650
8080	8″ pipe size		27	.889		790	40.50		830.50	930
8090	10″ pipe size	↓	20	1.200	↓	1,400	54.50		1,454.50	1,625
8200	With stainless steel trim									
8240	1-1/2″ pipe size	1 Plum	50	.160	Ea.	220	7.80		227.80	253
8250	2″ pipe size	″	38	.211		220	10.25		230.25	256
8270	3″ pipe size	Q-1	50	.320		295	14.05		309.05	345
8280	4″ pipe size	″	38	.421		320	18.50		338.50	380
8300	6″ pipe size	Q-2	38	.632		600	28.50		628.50	705
8310	8″ pipe size		27	.889		1,575	40.50		1,615.50	1,775
8320	10″ pipe size		20	1.200		2,850	54.50		2,904.50	3,200
8322	12″ pipe size		16	1.500		3,275	68.50		3,343.50	3,700
8324	14″ pipe size		15	1.600		3,975	73		4,048	4,475
8326	16″ pipe size	↓	14	1.714		5,725	78		5,803	6,425
8328	18″ pipe size	Q-3	12	2.667		7,050	124		7,174	7,925
8330	20″ pipe size		11	2.909		8,700	135		8,835	9,750
8332	24″ pipe size	↓	10	3.200	↓	11,600	149		11,749	13,000
8336	Note: sizes 8″ up w/manual gear operator									
9000	Cut one groove, labor									
9010	3/4″ pipe size	Q-1	152	.105	Ea.		4.62		4.62	6.95
9020	1″ pipe size		140	.114			5		5	7.55
9030	1-1/4″ pipe size		124	.129			5.65		5.65	8.50
9040	1-1/2″ pipe size		114	.140			6.15		6.15	9.25
9050	2″ pipe size		104	.154			6.75		6.75	10.15
9060	2-1/2″ pipe size		96	.167			7.30		7.30	11
9070	3″ pipe size		88	.182			8		8	11.95
9080	3-1/2″ pipe size		83	.193			8.45		8.45	12.70
9090	4″ pipe size		78	.205			9		9	13.50
9100	5″ pipe size		72	.222			9.75		9.75	14.65
9110	6″ pipe size		70	.229			10.05		10.05	15.05
9120	8″ pipe size		54	.296			13		13	19.50
9130	10″ pipe size		38	.421			18.50		18.50	27.50
9140	12″ pipe size		30	.533			23.50		23.50	35
9150	14″ pipe size		20	.800			35		35	52.50
9160	16″ pipe size	↓	19	.842	↓		37		37	55.50

22 11 13.48 Pipe, Fittings and Valves, Steel, Grooved-Joint		Crew	Daily Output	Labor-Hours	Unit	Material	2009 Bare Costs Labor	Equipment	Total	Total Incl O&P
9170	18" pipe size	Q-1	18	.889	Ea.		39		39	58.50
9180	20" pipe size		17	.941			41.50		41.50	62
9190	24" pipe size		15	1.067			47		47	70
9210	Roll one groove									
9220	3/4" pipe size	Q-1	266	.060	Ea.		2.64		2.64	3.96
9230	1" pipe size		228	.070			3.08		3.08	4.62
9240	1-1/4" pipe size		200	.080			3.51		3.51	5.25
9250	1-1/2" pipe size		178	.090			3.94		3.94	5.90
9260	2" pipe size		116	.138			6.05		6.05	9.10
9270	2-1/2" pipe size		110	.145			6.40		6.40	9.60
9280	3" pipe size		100	.160			7		7	10.55
9290	3-1/2" pipe size		94	.170			7.45		7.45	11.20
9300	4" pipe size		86	.186			8.15		8.15	12.25
9310	5" pipe size		84	.190			8.35		8.35	12.55
9320	6" pipe size		80	.200			8.80		8.80	13.15
9330	8" pipe size		66	.242			10.65		10.65	15.95
9340	10" pipe size		58	.276			12.10		12.10	18.15
9350	12" pipe size		46	.348			15.25		15.25	23
9360	14" pipe size		30	.533			23.50		23.50	35
9370	16" pipe size		28	.571			25		25	37.50
9380	18" pipe size		27	.593			26		26	39
9390	20" pipe size		25	.640			28		28	42
9400	24" pipe size		23	.696			30.50		30.50	46

22 11 13.60 Tubing Fittings, Stainless Steel		Crew	Daily Output	Labor-Hours	Unit	Material	Labor	Equipment	Total	Total Incl O&P
0010	**TUBING FITTINGS, STAINLESS STEEL**									
5010	Type 304, no joints, no hangers									
5020	.035 wall									
5021	1/4"	1 Plum	160	.050	L.F.	3.20	2.44		5.64	7.20
5022	3/8"		160	.050		4.05	2.44		6.49	8.10
5023	1/2"		160	.050		5.20	2.44		7.64	9.40
5024	5/8"		160	.050		6.40	2.44		8.84	10.70
5025	3/4"		133	.060		7.70	2.93		10.63	12.85
5026	7/8"		133	.060		9.15	2.93		12.08	14.45
5027	1"		114	.070		10.95	3.42		14.37	17.20
5040	.049 wall									
5041	1/4"	1 Plum	160	.050	L.F.	3.77	2.44		6.21	7.80
5042	3/8"		160	.050		5.20	2.44		7.64	9.40
5043	1/2"		160	.050		5.80	2.44		8.24	10
5044	5/8"		160	.050		7.80	2.44		10.24	12.25
5045	3/4"		133	.060		8.30	2.93		11.23	13.55
5046	7/8"		133	.060		9.75	2.93		12.68	15.15
5047	1"		114	.070		10.35	3.42		13.77	16.55
5060	.065 wall									
5061	1/4"	1 Plum	160	.050	L.F.	4.23	2.44		6.67	8.30
5062	3/8"		160	.050		5.85	2.44		8.29	10.05
5063	1/2"		160	.050		7.05	2.44		9.49	11.40
5064	5/8"		160	.050		8.60	2.44		11.04	13.10
5065	3/4"		133	.060		9.10	2.93		12.03	14.40
5066	7/8"		133	.060		11.75	2.93		14.68	17.30
5067	1"		114	.070		11.50	3.42		14.92	17.80
5210	Type 316									
5220	.035 wall									

22 11 Facility Water Distribution

22 11 13 – Facility Water Distribution Piping

22 11 13.60 Tubing Fittings, Stainless Steel		Crew	Daily Output	Labor-Hours	Unit	Material	Labor	Equipment	Total	Total Incl O&P
						2009 Bare Costs				
5221	1/4"	1 Plum	160	.050	L.F.	3.99	2.44		6.43	8.05
5222	3/8"		160	.050		4.98	2.44		7.42	9.15
5223	1/2"		160	.050		8.20	2.44		10.64	12.70
5224	5/8"		160	.050		10.20	2.44		12.64	14.85
5225	3/4"		133	.060		13.05	2.93		15.98	18.75
5226	7/8"		133	.060		17.70	2.93		20.63	24
5227	1"		114	.070		19.50	3.42		22.92	26.50
5240	.049 wall									
5241	1/4"	1 Plum	160	.050	L.F.	5.05	2.44		7.49	9.20
5242	3/8"		160	.050		5.60	2.44		8.04	9.80
5243	1/2"		160	.050		7.50	2.44		9.94	11.90
5244	5/8"		160	.050		10.15	2.44		12.59	14.80
5245	3/4"		133	.060		10.30	2.93		13.23	15.75
5246	7/8"		133	.060		14.80	2.93		17.73	20.50
5247	1"		114	.070		14.90	3.42		18.32	21.50
5260	.065 wall									
5261	1/4"	1 Plum	160	.050	L.F.	5.05	2.44		7.49	9.20
5262	3/8"		160	.050		8.50	2.44		10.94	13
5263	1/2"		160	.050		9.30	2.44		11.74	13.90
5264	5/8"		160	.050		11.35	2.44		13.79	16.10
5265	3/4"		133	.060		12	2.93		14.93	17.60
5266	7/8"		133	.060		17.35	2.93		20.28	23.50
5267	1"		114	.070		18.45	3.42		21.87	25.50
22 11 13.61 Tubing Fittings, Stainless Steel										
0010	**TUBING FITTINGS, STAINLESS STEEL**									
8200	Tube fittings, compression type									
8202	Type 316									
8204	90° elbow									
8206	1/4"	1 Plum	24	.333	Ea.	14.20	16.25		30.45	40
8207	3/8"		22	.364		17.35	17.75		35.10	45.50
8208	1/2"		22	.364		28.50	17.75		46.25	58
8209	5/8"		21	.381		32	18.55		50.55	63
8210	3/4"		21	.381		50.50	18.55		69.05	83.50
8211	7/8"		20	.400		77.50	19.50		97	115
8212	1"		20	.400		96.50	19.50		116	136
8220	Union tee									
8222	1/4"	1 Plum	15	.533	Ea.	19.95	26		45.95	61
8224	3/8"		15	.533		25.50	26		51.50	67.50
8225	1/2"		15	.533		39.50	26		65.50	82.50
8226	5/8"		14	.571		44.50	28		72.50	91
8227	3/4"		14	.571		59.50	28		87.50	108
8228	7/8"		13	.615		96	30		126	151
8229	1"		13	.615		127	30		157	185
8234	Union									
8236	1/4"	1 Plum	24	.333	Ea.	9.75	16.25		26	35.50
8237	3/8"		22	.364		12.90	17.75		30.65	40.50
8238	1/2"		22	.364		21	17.75		38.75	49.50
8239	5/8"		21	.381		27	18.55		45.55	58
8240	3/4"		21	.381		33.50	18.55		52.05	65
8241	7/8"		20	.400		55	19.50		74.50	90
8242	1"		20	.400		58.50	19.50		78	93.50
8250	Male connector									

22 11 13.61 Tubing Fittings, Stainless Steel		Crew	Daily Output	Labor-Hours	Unit	2009 Bare Costs Material	Labor	Equipment	Total	Total Incl O&P
8252	1/4" x 1/4"	1 Plum	24	.333	Ea.	6.30	16.25		22.55	31.50
8253	3/8" x 3/8"		22	.364		9.75	17.75		27.50	37.50
8254	1/2" x 1/2"		22	.364		14.50	17.75		32.25	42.50
8256	3/4" x 3/4"		21	.381		22.50	18.55		41.05	52.50
8258	1" x 1"		20	.400		39	19.50		58.50	72

22 11 13.64 Pipe, Stainless Steel		Crew	Daily Output	Labor-Hours	Unit	Material	Labor	Equipment	Total	Total Incl O&P
0010	**PIPE, STAINLESS STEEL**									
0020	Welded, with clevis type hanger assemblies, 10' O.C.									
0500	Schedule 5, type 304									
0540	1/2" diameter	Q-15	128	.125	L.F.	7.30	5.50	.44	13.24	16.80
0550	3/4" diameter		116	.138		9.35	6.05	.48	15.88	19.95
0560	1" diameter		103	.155		11.75	6.80	.54	19.09	24
0570	1-1/4" diameter		93	.172		16.50	7.55	.60	24.65	30
0580	1-1/2" diameter		85	.188		22.50	8.25	.66	31.41	37.50
0590	2" diameter		69	.232		37	10.15	.81	47.96	57
0600	2-1/2" diameter		53	.302		41	13.25	1.05	55.30	66
0610	3" diameter		48	.333		50	14.65	1.16	65.81	78
0620	4" diameter		44	.364		63.50	15.95	1.27	80.72	95
0630	5" diameter		36	.444		130	19.50	1.55	151.05	174
0640	6" diameter	Q-16	42	.571		120	26	1.33	147.33	171
0650	8" diameter		34	.706		185	32	1.64	218.64	253
0660	10" diameter		26	.923		256	42	2.14	300.14	345
0670	12" diameter		21	1.143		335	52	2.65	389.65	450
0700	To delete hangers, subtract									
0710	1/2" diam. to 1-1/2" diam.					8%	19%			
0720	2" diam. to 5" diam.					4%	9%			
0730	6" diam. to 12" diam.					3%	4%			
0750	For small quantities, add				L.F.	10%				
1250	Schedule 5, type 316									
1290	1/2" diameter	Q-15	128	.125	L.F.	9.55	5.50	.44	15.49	19.25
1300	3/4" diameter		116	.138		12.25	6.05	.48	18.78	23
1310	1" diameter		103	.155		16.75	6.80	.54	24.09	29.50
1320	1-1/4" diameter		93	.172		20.50	7.55	.60	28.65	34.50
1330	1-1/2" diameter		85	.188		29	8.25	.66	37.91	45
1340	2" diameter		69	.232		48	10.15	.81	58.96	68.50
1350	2-1/2" diameter		53	.302		56.50	13.25	1.05	70.80	83
1360	3" diameter		48	.333		71.50	14.65	1.16	87.31	102
1370	4" diameter		44	.364		89	15.95	1.27	106.22	123
1380	5" diameter		36	.444		171	19.50	1.55	192.05	219
1390	6" diameter	Q-16	42	.571		166	26	1.33	193.33	222
1400	8" diameter		34	.706		250	32	1.64	283.64	325
1410	10" diameter		26	.923		360	42	2.14	404.14	460
1420	12" diameter		21	1.143		460	52	2.65	514.65	590
1490	For small quantities, add					10%				
1940	To delete hanger, subtract									
1950	1/2" diam. to 1-1/2" diam.					5%	19%			
1960	2" diam. to 5" diam.					3%	9%			
1970	6" diam. to 12" diam.					2%	4%			
2000	Schedule 10, type 304									
2040	1/4" diameter	Q-15	131	.122	L.F.	4.46	5.35	.43	10.24	13.40
2050	3/8" diameter		128	.125		4.69	5.50	.44	10.63	13.90
2060	1/2" diameter		125	.128		4.93	5.60	.45	10.98	14.40

22 11 13.64 Pipe, Stainless Steel		Crew	Daily Output	Labor-Hours	Unit	Material	Labor	Equipment	Total	Total Incl O&P
						2009 Bare Costs				
2070	3/4" diameter	Q-15	113	.142	L.F.	7.35	6.20	.49	14.04	17.90
2080	1" diameter		100	.160		9.85	7	.56	17.41	22
2090	1-1/4" diameter		91	.176		9.90	7.70	.61	18.21	23
2100	1-1/2" diameter		83	.193		10.85	8.45	.67	19.97	25.50
2110	2" diameter		67	.239		13.25	10.50	.83	24.58	31
2120	2-1/2" diameter		51	.314		17.40	13.75	1.09	32.24	41
2130	3" diameter		46	.348		21.50	15.25	1.21	37.96	48
2140	4" diameter		42	.381		31.50	16.70	1.33	49.53	61.50
2150	5" diameter		35	.457		42.50	20	1.59	64.09	79
2160	6" diameter	Q-16	40	.600		46	27.50	1.39	74.89	93
2170	8" diameter		33	.727		79.50	33	1.69	114.19	139
2180	10" diameter		25	.960		110	43.50	2.23	155.73	189
2190	12" diameter		21	1.143		138	52	2.65	192.65	233
2250	For small quantities, add					10%				
2650	To delete hanger, subtract									
2660	1/4" diam. to 3/4" diam.					9%	22%			
2670	1" diam. to 2" diam.					4%	15%			
2680	2-1/2" diam. to 5" diam.					3%	8%			
2690	6" diam. to 12" diam.					3%	4%			
2750	Schedule 10, type 316									
2790	1/4" diameter	Q-15	131	.122	L.F.	5.30	5.35	.43	11.08	14.30
2800	3/8" diameter		128	.125		6.25	5.50	.44	12.19	15.65
2810	1/2" diameter		125	.128		6.55	5.60	.45	12.60	16.20
2820	3/4" diameter		113	.142		8.60	6.20	.49	15.29	19.30
2830	1" diameter		100	.160		10.45	7	.56	18.01	22.50
2840	1-1/4" diameter		91	.176		15	7.70	.61	23.31	29
2850	1-1/2" diameter		83	.193		20.50	8.45	.67	29.62	36
2860	2" diameter		67	.239		18.60	10.50	.83	29.93	37
2870	2-1/2" diameter		51	.314		27.50	13.75	1.09	42.34	51.50
2880	3" diameter		46	.348		32	15.25	1.21	48.46	59.50
2890	4" diameter		42	.381		43	16.70	1.33	61.03	74
2900	5" diameter		35	.457		60	20	1.59	81.59	98
2910	6" diameter	Q-16	40	.600		68.50	27.50	1.39	97.39	118
2920	8" diameter		33	.727		117	33	1.69	151.69	179
2930	10" diameter		25	.960		154	43.50	2.23	199.73	237
2940	12" diameter		21	1.143		193	52	2.65	247.65	293
2990	For small quantities, add					10%				
3430	To delete hanger, subtract									
3440	1/4" diam. to 3/4" diam.					6%	22%			
3450	1" diam. to 2" diam.					3%	15%			
3460	2-1/2" diam. to 5" diam.					2%	8%			
3470	6" diam. to 12" diam.					2%	4%			
3500	Threaded, couplings and clevis hanger assemblies, 10' O.C.									
3520	Schedule 40, type 304									
3540	1/4" diameter	1 Plum	54	.148	L.F.	6.80	7.20		14	18.35
3550	3/8" diameter		53	.151		7.10	7.35		14.45	18.85
3560	1/2" diameter		52	.154		7.65	7.50		15.15	19.65
3570	3/4" diameter		51	.157		9.60	7.65		17.25	22
3580	1" diameter		45	.178		13.55	8.65		22.20	28
3590	1-1/4" diameter	Q-1	76	.211		18.70	9.25		27.95	34.50
3600	1-1/2" diameter		69	.232		22	10.15		32.15	39.50
3610	2" diameter		57	.281		29.50	12.30		41.80	51
3620	2-1/2" diameter		44	.364		54.50	15.95		70.45	84

22 11 Facility Water Distribution

22 11 13 – Facility Water Distribution Piping

	22 11 13.64 Pipe, Stainless Steel	Crew	Daily Output	Labor-Hours	Unit	2009 Bare Costs Material	Labor	Equipment	Total	Total Incl O&P
3630	3" diameter	Q-1	38	.421	L.F.	71.50	18.50		90	106
3640	4" diameter	Q-2	51	.471		92.50	21.50		114	134
3740	For small quantities, add					10%				
4200	To delete couplings & hangers, subtract									
4210	1/4" diam. to 3/4" diam.					15%	56%			
4220	1" diam. to 2" diam.					18%	49%			
4230	2-1/2" diam. to 4" diam.					34%	40%			
4250	Schedule 40, type 316									
4290	1/4" diameter	1 Plum	54	.148	L.F.	8.45	7.20		15.65	20
4300	3/8" diameter		53	.151		8.90	7.35		16.25	21
4310	1/2" diameter		52	.154		12.25	7.50		19.75	24.50
4320	3/4" diameter		51	.157		15.95	7.65		23.60	29
4330	1" diameter		45	.178		23	8.65		31.65	38.50
4340	1-1/4" diameter	Q-1	76	.211		33	9.25		42.25	50
4350	1-1/2" diameter		69	.232		36	10.15		46.15	55
4360	2" diameter		57	.281		51.50	12.30		63.80	75.50
4370	2-1/2" diameter		44	.364		86	15.95		101.95	119
4380	3" diameter		38	.421		113	18.50		131.50	152
4390	4" diameter	Q-2	51	.471		154	21.50		175.50	201
4490	For small quantities, add					10%				
4900	To delete couplings & hangers, subtract									
4910	1/4" diam. to 3/4" diam.					12%	56%			
4920	1" diam. to 2" diam.					14%	49%			
4930	2-1/2" diam. to 4" diam.					27%	40%			
5000	Schedule 80, type 304									
5040	1/4" diameter	1 Plum	53	.151	L.F.	15.60	7.35		22.95	28
5050	3/8" diameter		52	.154		16.85	7.50		24.35	30
5060	1/2" diameter		51	.157		21.50	7.65		29.15	35.50
5070	3/4" diameter		48	.167		25.50	8.15		33.65	40
5080	1" diameter		43	.186		30.50	9.05		39.55	47
5090	1-1/4" diameter	Q-1	73	.219		40	9.60		49.60	58.50
5100	1-1/2" diameter		67	.239		50	10.50		60.50	71.50
5110	2" diameter		54	.296		67.50	13		80.50	93.50
5190	For small quantities, add					10%				
5700	To delete couplings & hangers, subtract									
5710	1/4" diam. to 3/4" diam.					10%	53%			
5720	1" diam. to 2" diam.					14%	47%			
5750	Schedule 80, type 316									
5790	1/4" diameter	1 Plum	53	.151	L.F.	18.15	7.35		25.50	31
5800	3/8" diameter		52	.154		19.35	7.50		26.85	33
5810	1/2" diameter		51	.157		25.50	7.65		33.15	39.50
5820	3/4" diameter		48	.167		34.50	8.15		42.65	50
5830	1" diameter		43	.186		41	9.05		50.05	58.50
5840	1-1/4" diameter	Q-1	73	.219		58.50	9.60		68.10	79
5850	1-1/2" diameter		67	.239		66	10.50		76.50	88.50
5860	2" diameter		54	.296		94	13		107	123
5950	For small quantities, add					10%				
7000	To delete couplings & hangers, subtract									
7010	1/4" diam. to 3/4" diam.					9%	53%			
7020	1" diam. to 2" diam.					14%	47%			
8000	Weld joints with clevis type hanger assemblies, 10' O.C.									
8010	Schedule 40, type 304									
8050	1/8" pipe size	Q-15	126	.127	L.F.	5.50	5.55	.44	11.49	14.90

22 11 13.64 Pipe, Stainless Steel		Crew	Daily Output	Labor-Hours	Unit	Material	Labor	Equipment	Total	Total Incl O&P
						2009 Bare Costs				
8060	1/4" pipe size	Q-15	125	.128	L.F.	5.65	5.60	.45	11.70	15.20
8070	3/8" pipe size		122	.131		5.70	5.75	.46	11.91	15.45
8080	1/2" pipe size		118	.136		5.80	5.95	.47	12.22	15.85
8090	3/4" pipe size		109	.147		7.15	6.45	.51	14.11	18.10
8100	1" pipe size		95	.168		9.60	7.40	.59	17.59	22.50
8110	1-1/4" pipe size		86	.186		12.35	8.15	.65	21.15	26.50
8120	1-1/2" pipe size		78	.205		14.85	9	.71	24.56	30.50
8130	2" pipe size		62	.258		18.45	11.30	.90	30.65	38.50
8140	2-1/2" pipe size		49	.327		28.50	14.35	1.14	43.99	54.50
8150	3" pipe size		44	.364		36	15.95	1.27	53.22	65
8160	3-1/2" pipe size		44	.364		38.50	15.95	1.27	55.72	67.50
8170	4" pipe size		39	.410		40.50	18	1.43	59.93	73
8180	5" pipe size		32	.500		69	22	1.74	92.74	111
8190	6" pipe size	Q-16	37	.649		84	29.50	1.50	115	139
8200	8" pipe size		29	.828		118	37.50	1.92	157.42	139
8210	10" pipe size		24	1		199	45.50	2.32	246.82	290
8220	12" pipe size		20	1.200		226	54.50	2.78	283.28	335
8300	Schedule 40, type 316									
8310	1/8" pipe size	Q-15	126	.127	L.F.	7.45	5.55	.44	13.44	17.05
8320	1/4" pipe size		125	.128		7.05	5.60	.45	13.10	16.70
8330	3/8" pipe size		122	.131		7.25	5.75	.46	13.46	17.10
8340	1/2" pipe size		118	.136		10	5.95	.47	16.42	20.50
8350	3/4" pipe size		109	.147		13	6.45	.51	19.96	24.50
8360	1" pipe size		95	.168		18.40	7.40	.59	26.39	32.50
8370	1-1/4" pipe size		86	.186		25	8.15	.65	33.80	41
8380	1-1/2" pipe size		78	.205		27.50	9	.71	37.21	45
8390	2" pipe size		62	.258		38	11.30	.90	50.20	60
8400	2-1/2" pipe size		49	.327		55	14.35	1.14	70.49	83.50
8410	3" pipe size		44	.364		70.50	15.95	1.27	87.72	103
8420	3-1/2" pipe size		44	.364		83	15.95	1.27	100.22	117
8430	4" pipe size		39	.410		91	18	1.43	110.43	129
8440	5" pipe size		32	.500		123	22	1.74	146.74	171
8450	6" pipe size	Q-16	37	.649		144	29.50	1.50	175	205
8460	8" pipe size		29	.828		270	37.50	1.92	309.42	355
8470	10" pipe size		24	1		360	45.50	2.32	407.82	465
8480	12" pipe size		20	1.200		490	54.50	2.78	547.28	625
8500	Schedule 80, type 304									
8510	1/4" pipe size	Q-15	110	.145	L.F.	14.25	6.40	.51	21.16	26
8520	3/8" pipe size		109	.147		15.35	6.45	.51	22.31	27
8530	1/2" pipe size		106	.151		19.80	6.60	.53	26.93	32.50
8540	3/4" pipe size		96	.167		23	7.30	.58	30.88	37
8550	1" pipe size		87	.184		26.50	8.05	.64	35.19	42
8560	1-1/4" pipe size		81	.198		30	8.65	.69	39.34	47
8570	1-1/2" pipe size		74	.216		38.50	9.50	.75	48.75	57
8580	2" pipe size		58	.276		51.50	12.10	.96	64.56	75
8590	2-1/2" pipe size		46	.348		106	15.25	1.21	122.46	140
8600	3" pipe size		41	.390		142	17.10	1.36	160.46	183
8610	4" pipe size		33	.485		223	21.50	1.69	246.19	279
8630	6" pipe size	Q-16	30	.800		490	36.50	1.86	528.36	595
8640	Schedule 80, type 316									
8650	1/4" pipe size	Q-15	110	.145	L.F.	16.60	6.40	.51	23.51	28.50
8660	3/8" pipe size		109	.147		17.70	6.45	.51	24.66	29.50
8670	1/2" pipe size		106	.151		23	6.60	.53	30.13	36

22 11 13.64 Pipe, Stainless Steel		Crew	Daily Output	Labor-Hours	Unit	Material	Labor	Equipment	Total	Total Incl O&P
						2009 Bare Costs				
8680	3/4" pipe size	Q-15	96	.167	L.F.	31	7.30	.58	38.88	45.50
8690	1" pipe size		87	.184		35	8.05	.64	43.69	51.50
8700	1-1/4" pipe size		81	.198		45.50	8.65	.69	54.84	64
8710	1-1/2" pipe size		74	.216		51	9.50	.75	61.25	71
8720	2" pipe size		58	.276		73	12.10	.96	86.06	99.50
8730	2-1/2" pipe size		46	.348		197	15.25	1.21	213.46	241
8740	3" pipe size		41	.390		230	17.10	1.36	248.46	280
8760	4" pipe size		33	.485		320	21.50	1.69	343.19	385
8770	6" pipe size	Q-16	30	.800		865	36.50	1.86	903.36	1,000
9100	Threading pipe labor, sst, one end, schedules 40 & 80									
9110	1/4" through 3/4" pipe size	1 Plum	61.50	.130	Ea.		6.35		6.35	9.50
9120	1" through 2" pipe size		55.90	.143			7		7	10.45
9130	2-1/2" pipe size		41.50	.193			9.40		9.40	14.10
9140	3" pipe size		38.50	.208			10.15		10.15	15.20
9150	3-1/2" pipe size	Q-1	68.40	.234			10.25		10.25	15.40
9160	4" pipe size		73	.219			9.60		9.60	14.45
9170	5" pipe size		40.70	.393			17.25		17.25	26
9180	6" pipe size		35.40	.452			19.85		19.85	30
9190	8" pipe size		22.30	.717			31.50		31.50	47.50
9200	10" pipe size		16.10	.994			43.50		43.50	65.50
9210	12" pipe size		12.30	1.301			57		57	85.50
9250	Welding labor per joint for stainless steel									
9260	Schedule 5 and 10									
9270	1/4" pipe size	Q-15	36	.444	Ea.		19.50	1.55	21.05	31
9280	3/8" pipe size		35	.457			20	1.59	21.59	32
9290	1/2" pipe size		35	.457			20	1.59	21.59	32
9300	3/4" pipe size		28	.571			25	1.99	26.99	39.50
9310	1" pipe size		25	.640			28	2.23	30.23	44.50
9320	1-1/4" pipe size		22	.727			32	2.53	34.53	51
9330	1-1/2" pipe size		21	.762			33.50	2.65	36.15	53
9340	2" pipe size		18	.889			39	3.09	42.09	62
9350	2-1/2" pipe size		12	1.333			58.50	4.64	63.14	93
9360	3" pipe size		9.73	1.644			72	5.70	77.70	114
9370	4" pipe size		7.37	2.171			95.50	7.55	103.05	151
9380	5" pipe size		6.15	2.602			114	9.05	123.05	181
9390	6" pipe size		5.71	2.802			123	9.75	132.75	196
9400	8" pipe size		3.69	4.336			190	15.10	205.10	305
9410	10" pipe size		2.91	5.498			241	19.15	260.15	380
9420	12" pipe size		2.31	6.926			305	24	329	480
9500	Schedule 40									
9510	1/4" pipe size	Q-15	28	.571	Ea.		25	1.99	26.99	39.50
9520	3/8" pipe size		27	.593			26	2.06	28.06	41.50
9530	1/2" pipe size		25.40	.630			27.50	2.19	29.69	44
9540	3/4" pipe size		22.22	.720			31.50	2.51	34.01	50.50
9550	1" pipe size		20.25	.790			34.50	2.75	37.25	55
9560	1-1/4" pipe size		18.82	.850			37.50	2.96	40.46	59.50
9570	1-1/2" pipe size		17.78	.900			39.50	3.13	42.63	63
9580	2" pipe size		15.09	1.060			46.50	3.69	50.19	74
9590	2-1/2" pipe size		7.96	2.010			88	7	95	140
9600	3" pipe size		6.43	2.488			109	8.65	117.65	174
9610	4" pipe size		4.88	3.279			144	11.40	155.40	229
9620	5" pipe size		4.26	3.756			165	13.05	178.05	261
9630	6" pipe size		3.77	4.244			186	14.75	200.75	295

22 11 13.64 Pipe, Stainless Steel

	22 11 13.64 Pipe, Stainless Steel	Crew	Daily Output	Labor-Hours	Unit	Material	2009 Bare Costs Labor	Equipment	Total	Total Incl O&P
9640	8" pipe size	Q-15	2.44	6.557	Ea.		288	23	311	455
9650	10" pipe size		1.92	8.333			365	29	394	580
9660	12" pipe size		1.52	10.526			460	36.50	496.50	735
9750	Schedule 80									
9760	1/4" pipe size	Q-15	21.55	.742	Ea.		32.50	2.58	35.08	52
9770	3/8" pipe size		20.75	.771			34	2.68	36.68	54
9780	1/2" pipe size		19.54	.819			36	2.85	38.85	57
9790	3/4" pipe size		17.09	.936			41	3.26	44.26	65
9800	1" pipe size		15.58	1.027			45	3.57	48.57	71.50
9810	1-1/4" pipe size		14.48	1.105			48.50	3.85	52.35	77
9820	1-1/2" pipe size		13.68	1.170			51.50	4.07	55.57	81.50
9830	2" pipe size		11.61	1.378			60.50	4.80	65.30	96.50
9840	2-1/2" pipe size		6.12	2.614			115	9.10	124.10	182
9850	3" pipe size		4.94	3.239			142	11.25	153.25	225
9860	4" pipe size		3.75	4.267			187	14.85	201.85	297
9870	5" pipe size		3.27	4.893			215	17.05	232.05	340
9880	6" pipe size		2.90	5.517			242	19.20	261.20	385
9890	8" pipe size		1.87	8.556			375	30	405	600
9900	10" pipe size		1.48	10.811			475	37.50	512.50	750
9910	12" pipe size		1.17	13.675			600	47.50	647.50	955
9920	Schedule 160, 1/2" pipe size		17	.941			41.50	3.28	44.78	65.50
9930	3/4" pipe size		14.81	1.080			47.50	3.76	51.26	75
9940	1" pipe size		13.50	1.185			52	4.12	56.12	82.50
9950	1-1/4" pipe size		12.55	1.275			56	4.44	60.44	89
9960	1-1/2" pipe size		11.85	1.350			59.50	4.70	64.20	94
9970	2" pipe size		10	1.600			70	5.55	75.55	111
9980	3" pipe size		4.28	3.738			164	13	177	260
9990	4" pipe size		3.25	4.923			216	17.15	233.15	345

22 11 13.66 Pipe Fittings, Stainless Steel

	22 11 13.66 Pipe Fittings, Stainless Steel	Crew	Daily Output	Labor-Hours	Unit	Material	Labor	Equipment	Total	Total Incl O&P
0010	**PIPE FITTINGS, STAINLESS STEEL**									
0100	Butt weld joint, schedule 5, type 304									
0120	90° Elbow, long									
0140	1/2"	Q-15	17.50	.914	Ea.	18	40	3.18	61.18	83.50
0150	3/4"		14	1.143		18	50	3.98	71.98	99.50
0160	1"		12.50	1.280		19	56	4.45	79.45	110
0170	1-1/4"		11	1.455		26	64	5.05	95.05	130
0180	1-1/2"		10.50	1.524		22	67	5.30	94.30	130
0190	2"		9	1.778		26	78	6.20	110.20	152
0200	2-1/2"		6	2.667		60	117	9.30	186.30	252
0210	3"		4.86	3.292		66	144	11.45	221.45	300
0220	3-1/2"		4.27	3.747		165	164	13.05	342.05	445
0230	4"		3.69	4.336		90	190	15.10	295.10	400
0240	5"		3.08	5.195		350	228	18.10	596.10	745
0250	6"	Q-16	4.29	5.594		270	255	13	538	690
0260	8"		2.76	8.696		570	395	20	985	1,250
0270	10"		2.18	11.009		880	500	25.50	1,405.50	1,750
0280	12"		1.73	13.873		1,250	630	32	1,912	2,350
0320	For schedule 5, type 316, add					30%				
0600	45° Elbow, long									
0620	1/2"	Q-15	17.50	.914	Ea.	18	40	3.18	61.18	83.50
0630	3/4"		14	1.143		18	50	3.98	71.98	99.50
0640	1"		12.50	1.280		19	56	4.45	79.45	110

22 11 13.66 Pipe Fittings, Stainless Steel		Crew	Daily Output	Labor-Hours	Unit	Material	Labor	Equipment	Total	Total Incl O&P
						2009 Bare Costs				
0650	1-1/4"	Q-15	11	1.455	Ea.	26	64	5.05	95.05	130
0660	1-1/2"		10.50	1.524		22	67	5.30	94.30	130
0670	2"		9	1.778		26	78	6.20	110.20	152
0680	2-1/2"		6	2.667		60	117	9.30	186.30	252
0690	3"		4.86	3.292		66	144	11.45	221.45	300
0700	3-1/2"		4.27	3.747		165	164	13.05	342.05	445
0710	4"		3.69	4.336		76	190	15.10	281.10	385
0720	5"		3.08	5.195		280	228	18.10	526.10	670
0730	6"	Q-16	4.29	5.594		190	255	13	458	605
0740	8"		2.76	8.696		400	395	20	815	1,050
0750	10"		2.18	11.009		700	500	25.50	1,225.50	1,550
0760	12"		1.73	13.873		875	630	32	1,537	1,950
0800	For schedule 5, type 316, add					25%				
1100	Tee, straight									
1130	1/2"	Q-15	11.66	1.372	Ea.	54	60	4.78	118.78	155
1140	3/4"		9.33	1.715		54	75.50	5.95	135.45	179
1150	1"		8.33	1.921		57	84.50	6.70	148.20	196
1160	1-1/4"		7.33	2.183		46	96	7.60	149.60	203
1170	1-1/2"		7	2.286		45	100	7.95	152.95	209
1180	2"		6	2.667		47	117	9.30	173.30	238
1190	2-1/2"		4	4		113	176	13.90	302.90	400
1200	3"		3.24	4.938		89	217	17.20	323.20	440
1210	3-1/2"		2.85	5.614		236	246	19.55	501.55	650
1220	4"		2.46	6.504		128	285	22.50	435.50	595
1230	5"		2	8		405	350	28	783	1,000
1240	6"	Q-16	2.85	8.421		320	385	19.55	724.55	945
1250	8"		1.84	13.043		680	595	30.50	1,305.50	1,675
1260	10"		1.45	16.552		1,100	755	38.50	1,893.50	2,375
1270	12"		1.15	20.870		1,525	950	48.50	2,523.50	3,150
1320	For schedule 5, type 316, add					25%				
2000	Butt weld joint, schedule 10, type 304									
2020	90° elbow, long									
2040	1/2"	Q-15	17	.941	Ea.	13.50	41.50	3.28	58.28	80.50
2050	3/4"		14	1.143		13.50	50	3.98	67.48	94.50
2060	1"		12.50	1.280		14.25	56	4.45	74.70	105
2070	1-1/4"		11	1.455		19.50	64	5.05	88.55	123
2080	1-1/2"		10.50	1.524		16.50	67	5.30	88.80	124
2090	2"		9	1.778		19.50	78	6.20	103.70	145
2100	2-1/2"		6	2.667		45	117	9.30	171.30	236
2110	3"		4.86	3.292		42	144	11.45	197.45	276
2120	3-1/2"		4.27	3.747		124	164	13.05	301.05	395
2130	4"		3.69	4.336		70.50	190	15.10	275.60	380
2140	5"		3.08	5.195		263	228	18.10	509.10	650
2150	6"	Q-16	4.29	5.594		203	255	13	471	615
2160	8"		2.76	8.696		430	395	20	845	1,075
2170	10"		2.18	11.009		660	500	25.50	1,185.50	1,500
2180	12"		1.73	13.873		940	630	32	1,602	2,000
2500	45° elbow, long									
2520	1/2"	Q-15	17.50	.914	Ea.	13.50	40	3.18	56.68	78.50
2530	3/4"		14	1.143		13.50	50	3.98	67.48	94.50
2540	1"		12.50	1.280		14.25	56	4.45	74.70	105
2550	1-1/4"		11	1.455		19.50	64	5.05	88.55	123
2560	1-1/2"		10.50	1.524		16.50	67	5.30	88.80	124

22 11 Facility Water Distribution

22 11 13 – Facility Water Distribution Piping

22 11 13.66 Pipe Fittings, Stainless Steel		Crew	Daily Output	Labor-Hours	Unit	2009 Bare Costs Material	Labor	Equipment	Total	Total Incl O&P
2570	2″	Q-15	9	1.778	Ea.	19.50	78	6.20	103.70	145
2580	2-1/2″		6	2.667		45	117	9.30	171.30	236
2590	3″		4.86	3.292		34	144	11.45	189.45	267
2600	3-1/2″		4.27	3.747		124	164	13.05	301.05	395
2610	4″		3.69	4.336		57	190	15.10	262.10	365
2620	5″		3.08	5.195		210	228	18.10	456.10	590
2630	6″	Q-16	4.29	5.594		143	255	13	411	550
2640	8″		2.76	8.696		300	395	20	715	945
2650	10″		2.18	11.009		525	500	25.50	1,050.50	1,350
2660	12″		1.73	13.873		655	630	32	1,317	1,700
2670	Reducer, concentric									
2674	1″ x 3/4″	Q-15	13.25	1.208	Ea.	29.50	53	4.20	86.70	116
2676	2″ x 1-1/2″	″	9.75	1.641		24	72	5.70	101.70	141
2678	6″ x 4″	Q-16	4.91	4.888		72	222	11.35	305.35	425
2680	Caps									
2682	1″	Q-15	25	.640	Ea.	30	28	2.23	60.23	77.50
2684	1-1/2″		21	.762		36	33.50	2.65	72.15	92.50
2685	2″		18	.889		34	39	3.09	76.09	99
2686	4″		7.38	2.168		54	95	7.55	156.55	211
2687	6″	Q-16	8.58	2.797		86.50	127	6.50	220	293
3000	Tee, straight									
3030	1/2″	Q-15	11.66	1.372	Ea.	40.50	60	4.78	105.28	140
3040	3/4″		9.33	1.715		40.50	75.50	5.95	121.95	164
3050	1″		8.33	1.921		43	84.50	6.70	134.20	180
3060	1-1/4″		7.33	2.183		52.50	96	7.60	156.10	210
3070	1-1/2″		7	2.286		34	100	7.95	141.95	197
3080	2″		6	2.667		35.50	117	9.30	161.80	225
3090	2-1/2″		4	4		85	176	13.90	274.90	370
3100	3″		3.24	4.938		67	217	17.20	301.20	415
3110	3-1/2″		2.85	5.614		98.50	246	19.55	364.05	500
3120	4″		2.46	6.504		96	285	22.50	403.50	560
3130	5″		2	8		305	350	28	683	890
3140	6″	Q-16	2.85	8.421		240	385	19.55	644.55	860
3150	8″		1.84	13.043		510	595	30.50	1,135.50	1,475
3151	10″		1.45	16.552		825	755	38.50	1,618.50	2,075
3152	12″		1.15	20.870		1,150	950	48.50	2,148.50	2,725
3154	For schedule 10, type 316, add					25%				
3281	Butt weld joint, schedule 40, type 304									
3284	90° Elbow, long, 1/2″	Q-15	12.70	1.260	Ea.	15.75	55.50	4.38	75.63	105
3288	3/4″		11.10	1.441		15.75	63.50	5	84.25	118
3289	1″		10.13	1.579		16.50	69.50	5.50	91.50	128
3290	1-1/4″		9.40	1.702		22	74.50	5.90	102.40	143
3300	1-1/2″		8.89	1.800		17.25	79	6.25	102.50	145
3310	2″		7.55	2.119		25	93	7.35	125.35	175
3320	2-1/2″		3.98	4.020		47.50	176	14	237.50	330
3330	3″		3.21	4.984		61	219	17.35	297.35	415
3340	3-1/2″		2.83	5.654		225	248	19.65	492.65	640
3350	4″		2.44	6.557		105	288	23	416	570
3360	5″		2.13	7.512		355	330	26	711	915
3370	6″	Q-16	2.83	8.481		310	385	19.65	714.65	940
3380	8″		1.83	13.115		615	595	30.50	1,240.50	1,600
3390	10″		1.44	16.667		1,300	760	38.50	2,098.50	2,625
3400	12″		1.14	21.053		1,650	960	49	2,659	3,300

22 11 13 – Facility Water Distribution Piping

22 11 13.66 Pipe Fittings, Stainless Steel		Crew	Daily Output	Labor-Hours	Unit	2009 Bare Costs Material	Labor	Equipment	Total	Total Incl O&P
3410	For schedule 40, type 316, add					25%				
3460	45° Elbow, long, 1/2"	Q-15	12.70	1.260	Ea.	15.75	55.50	4.38	75.63	105
3470	3/4"		11.10	1.441		15.75	63.50	5	84.25	118
3480	1"		10.13	1.579		16.50	69.50	5.50	91.50	128
3490	1-1/4"		9.40	1.702		22	74.50	5.90	102.40	143
3500	1-1/2"		8.89	1.800		17.25	79	6.25	102.50	145
3510	2"		7.55	2.119		25	93	7.35	125.35	175
3520	2-1/2"		3.98	4.020		47.50	176	14	237.50	330
3530	3"		3.21	4.984		48	219	17.35	284.35	400
3540	3-1/2"		2.83	5.654		225	248	19.65	492.65	640
3550	4"		2.44	6.557		75	288	23	386	540
3560	5"		2.13	7.512		248	330	26	604	795
3570	6"	Q-16	2.83	8.481		218	385	19.65	622.65	840
3580	8"		1.83	13.115		430	595	30.50	1,055.50	1,400
3590	10"		1.44	16.667		910	760	38.50	1,708.50	2,200
3600	12"		1.14	21.053		1,150	960	49	2,159	2,750
3610	For schedule 40, type 316, add					25%				
3660	Tee, straight 1/2"	Q-15	8.46	1.891	Ea.	40.50	83	6.60	130.10	177
3670	3/4"		7.40	2.162		40.50	95	7.50	143	195
3680	1"		6.74	2.374		43	104	8.25	155.25	212
3690	1-1/4"		6.27	2.552		100	112	8.90	220.90	288
3700	1-1/2"		5.92	2.703		46	119	9.40	174.40	239
3710	2"		5.03	3.181		68.50	140	11.05	219.55	296
3720	2-1/2"		2.65	6.038		86.50	265	21	372.50	520
3730	3"		2.14	7.477		87	330	26	443	615
3740	3-1/2"		1.88	8.511		164	375	29.50	568.50	775
3750	4"		1.62	9.877		164	435	34.50	633.50	870
3760	5"		1.42	11.268		380	495	39	914	1,200
3770	6"	Q-16	1.88	12.766		335	580	29.50	944.50	1,275
3780	8"		1.22	19.672		735	895	45.50	1,675.50	2,200
3790	10"		.96	25		1,425	1,150	58	2,633	3,350
3800	12"		.76	31.579		1,900	1,425	73.50	3,398.50	4,325
3810	For schedule 40, type 316, add					25%				
3820	Tee, reducing on outlet, 3/4" x 1/2"	Q-15	7.73	2.070	Ea.	54	91	7.20	152.20	203
3822	1" x 1/2"		7.24	2.210		58.50	97	7.70	163.20	219
3824	1" x 3/4"		6.96	2.299		54	101	8	163	219
3826	1-1/4" x 1"		6.43	2.488		132	109	8.65	249.65	320
3828	1-1/2" x 1/2"		6.58	2.432		79.50	107	8.45	194.95	257
3830	1-1/2" x 3/4"		6.35	2.520		74.50	111	8.75	194.25	257
3832	1-1/2" x 1"		6.18	2.589		57	114	9	180	242
3834	2" x 1"		5.50	2.909		102	128	10.10	240.10	315
3836	2" x 1-1/2"		5.30	3.019		85.50	132	10.50	228	305
3838	2-1/2" x 2"		3.15	5.079		129	223	17.70	369.70	495
3840	3" x 1-1/2"		2.72	5.882		131	258	20.50	409.50	550
3842	3" x 2"		2.65	6.038		109	265	21	395	545
3844	4" x 2"		2.10	7.619		272	335	26.50	633.50	830
3846	4" x 3"		1.77	9.040		197	395	31.50	623.50	845
3848	5" x 4"		1.48	10.811		455	475	37.50	967.50	1,250
3850	6" x 3"	Q-16	2.19	10.959		505	500	25.50	1,030.50	1,325
3852	6" x 4"		2.04	11.765		435	535	27.50	997.50	1,325
3854	8" x 4"		1.46	16.438		1,025	750	38	1,813	2,300
3856	10" x 8"		.69	34.783		1,725	1,575	80.50	3,380.50	4,350
3858	12" x 10"		.55	43.636		2,275	1,975	101	4,351	5,600

22 11 13.66 Pipe Fittings, Stainless Steel		Crew	Daily Output	Labor-Hours	Unit	Material	Labor	Equipment	Total	Total Incl O&P
						2009 Bare Costs				
3950	Reducer, concentric, 3/4" x 1/2"	Q-15	11.85	1.350	Ea.	30	59.50	4.70	94.20	127
3952	1" x 3/4"		10.60	1.509		31.50	66	5.25	102.75	140
3954	1-1/4" x 3/4"		10.19	1.570		77.50	69	5.45	151.95	194
3956	1-1/4" x 1"		9.76	1.639		39	72	5.70	116.70	157
3958	1-1/2" x 3/4"		9.88	1.619		65.50	71	5.65	142.15	185
3960	1-1/2" x 1"		9.47	1.690		52.50	74	5.90	132.40	175
3962	2" x 1"		8.65	1.850		22.50	81	6.45	109.95	154
3964	2" x 1-1/2"		8.16	1.961		26.50	86	6.80	119.30	166
3966	2-1/2" x 1"		5.71	2.802		113	123	9.75	245.75	320
3968	2-1/2" x 2"		5.21	3.071		58	135	10.70	203.70	277
3970	3" x 1"		4.88	3.279		78	144	11.40	233.40	315
3972	3" x 1-1/2"		4.72	3.390		40	149	11.80	200.80	280
3974	3" x 2"		4.51	3.548		37.50	156	12.35	205.85	289
3976	4" x 2"		3.69	4.336		53.50	190	15.10	258.60	360
3978	4" x 3"		2.77	5.776		40	253	20	313	445
3980	5" x 3"		2.56	6.250		237	274	22	533	695
3982	5" x 4"	↓	2.27	7.048		191	310	24.50	525.50	700
3984	6" x 3"	Q-16	3.57	6.723		122	305	15.60	442.60	610
3986	6" x 4"		3.19	7.524		101	340	17.45	458.45	645
3988	8" x 4"		2.44	9.836		315	450	23	788	1,050
3990	8" x 6"		2.22	10.811		236	490	25	751	1,025
3992	10" x 6"		1.91	12.565		440	570	29	1,039	1,375
3994	10" x 8"		1.61	14.907		365	680	34.50	1,079.50	1,475
3995	12" x 6"		1.63	14.724		675	670	34	1,379	1,775
3996	12" x 8"		1.41	17.021		620	775	39.50	1,434.50	1,875
3997	12" x 10"	↓	1.27	18.898	↓	435	860	44	1,339	1,825
4000	Socket weld joint, 3000 lb., type 304									
4100	90° Elbow									
4140	1/4"	Q-15	13.47	1.188	Ea.	42	52	4.13	98.13	129
4150	3/8"		12.97	1.234		54.50	54	4.29	112.79	146
4160	1/2"		12.21	1.310		56.50	57.50	4.56	118.56	154
4170	3/4"		10.68	1.498		68.50	65.50	5.20	139.20	179
4180	1"		9.74	1.643		103	72	5.70	180.70	227
4190	1-1/4"		9.05	1.768		179	77.50	6.15	262.65	320
4200	1-1/2"		8.55	1.871		218	82	6.50	306.50	370
4210	2"	↓	7.26	2.204	↓	350	96.50	7.65	454.15	540
4300	45° Elbow									
4340	1/4"	Q-15	13.47	1.188	Ea.	78.50	52	4.13	134.63	169
4350	3/8"		12.97	1.234		78.50	54	4.29	136.79	172
4360	1/2"		12.21	1.310		78.50	57.50	4.56	140.56	178
4370	3/4"		10.68	1.498		89.50	65.50	5.20	160.20	202
4380	1"		9.74	1.643		129	72	5.70	206.70	256
4390	1-1/4"		9.05	1.768		213	77.50	6.15	296.65	355
4400	1-1/2"		8.55	1.871		213	82	6.50	301.50	365
4410	2"	↓	7.26	2.204	↓	390	96.50	7.65	494.15	585
4500	Tee									
4540	1/4"	Q-15	8.97	1.784	Ea.	66	78.50	6.20	150.70	196
4550	3/8"		8.64	1.852		66	81.50	6.45	153.95	202
4560	1/2"		8.13	1.968		81	86.50	6.85	174.35	227
4570	3/4"		7.12	2.247		93	98.50	7.80	199.30	260
4580	1"		6.48	2.469		125	108	8.60	241.60	310
4590	1-1/4"		6.03	2.653		223	116	9.25	348.25	430
4600	1-1/2"	↓	5.69	2.812	↓	320	123	9.80	452.80	545

22 11 13.66 Pipe Fittings, Stainless Steel		Crew	Daily Output	Labor-Hours	Unit	2009 Bare Costs Material	Labor	Equipment	Total	Total Incl O&P
4610	2″	Q-15	4.83	3.313	Ea.	485	145	11.55	641.55	760
5000	Socket weld joint, 3000 lb., type 316									
5100	90° Elbow									
5140	1/4″	Q-15	13.47	1.188	Ea.	51	52	4.13	107.13	139
5150	3/8″		12.97	1.234		60	54	4.29	118.29	152
5160	1/2″		12.21	1.310		72	57.50	4.56	134.06	171
5170	3/4″		10.68	1.498		95	65.50	5.20	165.70	208
5180	1″		9.74	1.643		135	72	5.70	212.70	262
5190	1-1/4″		9.05	1.768		240	77.50	6.15	323.65	385
5200	1-1/2″		8.55	1.871		271	82	6.50	359.50	430
5210	2″		7.26	2.204		460	96.50	7.65	564.15	665
5300	45° Elbow									
5340	1/4″	Q-15	13.47	1.188	Ea.	103	52	4.13	159.13	197
5350	3/8″		12.97	1.234		103	54	4.29	161.29	200
5360	1/2″		12.21	1.310		103	57.50	4.56	165.06	206
5370	3/4″		10.68	1.498		115	65.50	5.20	185.70	230
5380	1″		9.74	1.643		173	72	5.70	250.70	305
5390	1-1/4″		9.05	1.768		247	77.50	6.15	330.65	395
5400	1-1/2″		8.55	1.871		278	82	6.50	366.50	435
5410	2″		7.26	2.204		415	96.50	7.65	519.15	610
5500	Tee									
5540	1/4″	Q-15	8.97	1.784	Ea.	69.50	78.50	6.20	154.20	200
5550	3/8″		8.64	1.852		85	81.50	6.45	172.95	223
5560	1/2″		8.13	1.968		94.50	86.50	6.85	187.85	242
5570	3/4″		7.12	2.247		117	98.50	7.80	223.30	286
5580	1″		6.48	2.469		179	108	8.60	295.60	370
5590	1-1/4″		6.03	2.653		284	116	9.25	409.25	495
5600	1-1/2″		5.69	2.812		400	123	9.80	532.80	635
5610	2″		4.83	3.313		640	145	11.55	796.55	935
5700	For socket weld joint, 6000 lb., type 304 and 316, add					100%				
6000	Threaded companion flange									
6010	Stainless steel, 150 lb., type 304									
6020	1/2″ diam.	1 Plum	30	.267	Ea.	35	13		48	58
6030	3/4″ diam.		28	.286		39	13.95		52.95	64
6040	1″ diam.		27	.296		43	14.45		57.45	69
6050	1-1/4″ diam.	Q-1	44	.364		55	15.95		70.95	84.50
6060	1-1/2″ diam.		40	.400		55	17.55		72.55	87
6070	2″ diam.		36	.444		72	19.50		91.50	109
6080	2-1/2″ diam.		28	.571		100	25		125	148
6090	3″ diam.		20	.800		105	35		140	169
6110	4″ diam.		12	1.333		144	58.50		202.50	246
6130	6″ diam.	Q-2	14	1.714		259	78		337	400
6140	8″ diam.	″	12	2		500	91		591	685
6150	For type 316 add					40%				
6260	Weld flanges, stainless steel, type 304									
6270	Slip on, 150 lb. (welded, front and back)									
6280	1/2″ diam.	Q-15	12.70	1.260	Ea.	31	55.50	4.38	90.88	122
6290	3/4″ diam.		11.11	1.440		32	63	5	100	136
6300	1″ diam.		10.13	1.579		35	69.50	5.50	110	149
6310	1-1/4″ diam.		9.41	1.700		48	74.50	5.90	128.40	172
6320	1-1/2″ diam.		8.89	1.800		48	79	6.25	133.25	179
6330	2″ diam.		7.55	2.119		61	93	7.35	161.35	215
6340	2-1/2″ diam.		3.98	4.020		86	176	14	276	375

22 11 13.66 Pipe Fittings, Stainless Steel		Crew	Daily Output	Labor-Hours	Unit	2009 Bare Costs Material	Labor	Equipment	Total	Total Incl O&P
6350	3″ diam.	Q-15	3.21	4.984	Ea.	92	219	17.35	328.35	450
6370	4″ diam.		2.44	6.557		125	288	23	436	595
6390	6″ diam.	Q-16	1.89	12.698		190	580	29.50	799.50	1,100
6400	8″ diam.	″	1.22	19.672		360	895	45.50	1,300.50	1,800
6410	For type 316, add					40%				
6530	Weld neck 150 lb.									
6540	1/2″ diam.	Q-15	25.40	.630	Ea.	16.85	27.50	2.19	46.54	62.50
6550	3/4″ diam.		22.22	.720		20.50	31.50	2.51	54.51	73
6560	1″ diam.		20.25	.790		25	34.50	2.75	62.25	82.50
6570	1-1/4″ diam.		18.82	.850		43	37.50	2.96	83.46	107
6580	1-1/2″ diam.		17.78	.900		38.50	39.50	3.13	81.13	105
6590	2″ diam.		15.09	1.060		40	46.50	3.69	90.19	118
6600	2-1/2″ diam.		7.96	2.010		64	88	7	159	210
6610	3″ diam.		6.43	2.488		69.50	109	8.65	187.15	250
6630	4″ diam.		4.88	3.279		107	144	11.40	262.40	345
6640	5″ diam.		4.26	3.756		148	165	13.05	326.05	425
6650	6″ diam.	Q-16	5.66	4.240		189	193	9.85	391.85	510
6652	8″ diam.		3.65	6.575		330	299	15.25	644.25	825
6654	10″ diam.		2.88	8.333		465	380	19.35	864.35	1,100
6656	12″ diam.		2.28	10.526		720	480	24.50	1,224.50	1,550
6670	For type 316 add					23%				
7000	Threaded joint, 150 lb., type 304									
7030	90° elbow									
7040	1/8″	1 Plum	13	.615	Ea.	24	30		54	71.50
7050	1/4″		13	.615		24	30		54	71.50
7070	3/8″		13	.615		26.50	30		56.50	74
7080	1/2″		12	.667		23.50	32.50		56	75
7090	3/4″		11	.727		27	35.50		62.50	82.50
7100	1″		10	.800		35.50	39		74.50	98
7110	1-1/4″	Q-1	17	.941		55.50	41.50		97	123
7120	1-1/2″		16	1		61	44		105	133
7130	2″		14	1.143		88	50		138	172
7140	2-1/2″		11	1.455		243	64		307	365
7150	3″		8	2		335	88		423	500
7160	4″	Q-2	11	2.182		450	99.50		549.50	645
7180	45° elbow									
7190	1/8″	1 Plum	13	.615	Ea.	35.50	30		65.50	84
7200	1/4″		13	.615		35.50	30		65.50	84
7210	3/8″		13	.615		33.50	30		63.50	82
7220	1/2″		12	.667		34	32.50		66.50	86.50
7230	3/4″		11	.727		26.50	35.50		62	82.50
7240	1″		10	.800		40.50	39		79.50	103
7250	1-1/4″	Q-1	17	.941		52	41.50		93.50	120
7260	1-1/2″		16	1		76.50	44		120.50	150
7270	2″		14	1.143		95.50	50		145.50	181
7280	2-1/2″		11	1.455		345	64		409	470
7290	3″		8	2		500	88		588	680
7300	4″	Q-2	11	2.182		905	99.50		1,004.50	1,150
7320	Tee, straight									
7330	1/8″	1 Plum	9	.889	Ea.	37	43.50		80.50	106
7340	1/4″		9	.889		37	43.50		80.50	106
7350	3/8″		9	.889		37.50	43.50		81	106
7360	1/2″		8	1		35.50	49		84.50	112

22 11 13 – Facility Water Distribution Piping

22 11 13.66 Pipe Fittings, Stainless Steel		Crew	Daily Output	Labor-Hours	Unit	2009 Bare Costs Material	Labor	Equipment	Total	Total Incl O&P
7370	3/4"	1 Plum	7	1.143	Ea.	38	55.50		93.50	125
7380	1"		6.50	1.231		44.50	60		104.50	139
7390	1-1/4"	Q-1	11	1.455		75	64		139	179
7400	1-1/2"		10	1.600		92	70		162	206
7410	2"		9	1.778		114	78		192	243
7420	2-1/2"		7	2.286		350	100		450	535
7430	3"		5	3.200		530	140		670	795
7440	4"	Q-2	7	3.429		1,300	156		1,456	1,675
7460	Coupling, straight									
7470	1/8"	1 Plum	19	.421	Ea.	9.70	20.50		30.20	41.50
7480	1/4"		19	.421		11.50	20.50		32	43.50
7490	3/8"		19	.421		13.80	20.50		34.30	46
7500	1/2"		19	.421		18.35	20.50		38.85	51
7510	3/4"		18	.444		24.50	21.50		46	59.50
7520	1"		15	.533		39.50	26		65.50	82.50
7530	1-1/4"	Q-1	26	.615		63.50	27		90.50	111
7540	1-1/2"		24	.667		71	29.50		100.50	122
7550	2"		21	.762		112	33.50		145.50	173
7560	2-1/2"		18	.889		260	39		299	345
7570	3"		14	1.143		355	50		405	465
7580	4"	Q-2	16	1.500		520	68.50		588.50	675
7600	Reducer, concentric, 1/2"	1 Plum	12	.667		18.35	32.50		50.85	69
7610	3/4"		11	.727		24.50	35.50		60	80
7612	1"		10	.800		42	39		81	105
7614	1-1/4"	Q-1	17	.941		89.50	41.50		131	161
7616	1-1/2"		16	1		98	44		142	174
7618	2"		14	1.143		115	50		165	202
7620	2-1/2"		11	1.455		380	64		444	510
7622	3"		8	2		420	88		508	595
7624	4"	Q-2	11	2.182		700	99.50		799.50	920
7710	Union									
7720	1/8"	1 Plum	12	.667	Ea.	45	32.50		77.50	98.50
7730	1/4"		12	.667		45	32.50		77.50	98.50
7740	3/8"		12	.667		53	32.50		85.50	107
7750	1/2"		11	.727		66	35.50		101.50	126
7760	3/4"		10	.800		90.50	39		129.50	158
7770	1"		9	.889		131	43.50		174.50	209
7780	1-1/4"	Q-1	16	1		305	44		349	400
7790	1-1/2"		15	1.067		330	47		377	435
7800	2"		13	1.231		420	54		474	540
7810	2-1/2"		10	1.600		790	70		860	975
7820	3"		7	2.286		1,050	100		1,150	1,300
7830	4"	Q-2	10	2.400		1,450	109		1,559	1,775
7838	Caps									
7840	1/2"	1 Plum	24	.333	Ea.	12.20	16.25		28.45	38
7841	3/4"		22	.364		17.85	17.75		35.60	46
7842	1"		20	.400		28.50	19.50		48	61
7843	1-1/2"	Q-1	32	.500		68	22		90	108
7844	2"	"	28	.571		86	25		111	132
7845	4"	Q-2	22	1.091		340	49.50		389.50	445
7850	For 150 lb., type 316, add					25%				
8750	Threaded joint, 2000 lb., type 304									
8770	90° Elbow									

22 11 13.66 Pipe Fittings, Stainless Steel		Crew	Daily Output	Labor-Hours	Unit	2009 Bare Costs Material	Labor	Equipment	Total	Total Incl O&P
8780	1/8"	1 Plum	13	.615	Ea.	31	30		61	79
8790	1/4"		13	.615		31	30		61	79
8800	3/8"		13	.615		38	30		68	87
8810	1/2"		12	.667		51.50	32.50		84	106
8820	3/4"		11	.727		60	35.50		95.50	119
8830	1"		10	.800		79.50	39		118.50	146
8840	1-1/4"	Q-1	17	.941		129	41.50		170.50	204
8850	1-1/2"		16	1		208	44		252	294
8860	2"		14	1.143		292	50		342	395
8880	45° Elbow									
8890	1/8"	1 Plum	13	.615	Ea.	64	30		94	116
8900	1/4"		13	.615		64	30		94	116
8910	3/8"		13	.615		80.50	30		110.50	134
8920	1/2"		12	.667		80.50	32.50		113	138
8930	3/4"		11	.727		88.50	35.50		124	151
8940	1"		10	.800		106	39		145	175
8950	1-1/4"	Q-1	17	.941		184	41.50		225.50	265
8960	1-1/2"		16	1		247	44		291	340
8970	2"		14	1.143		285	50		335	390
8990	Tee, straight									
9000	1/8"	1 Plum	9	.889	Ea.	40	43.50		83.50	109
9010	1/4"		9	.889		40	43.50		83.50	109
9020	3/8"		9	.889		51	43.50		94.50	122
9030	1/2"		8	1		65.50	49		114.50	145
9040	3/4"		7	1.143		74.50	55.50		130	166
9050	1"		6.50	1.231		103	60		163	203
9060	1-1/4"	Q-1	11	1.455		173	64		237	287
9070	1-1/2"		10	1.600		289	70		359	420
9080	2"		9	1.778		400	78		478	555
9100	For couplings and unions use 3000 lb., type 304									
9120	2000 lb., type 316									
9130	90° Elbow									
9140	1/8"	1 Plum	13	.615	Ea.	38.50	30		68.50	87
9150	1/4"		13	.615		38.50	30		68.50	87
9160	3/8"		13	.615		45	30		75	94.50
9170	1/2"		12	.667		58	32.50		90.50	113
9180	3/4"		11	.727		71	35.50		106.50	132
9190	1"		10	.800		85.50	39		124.50	153
9200	1-1/4"	Q-1	17	.941		199	41.50		240.50	281
9210	1-1/2"		16	1		221	44		265	310
9220	2"		14	1.143		375	50		425	490
9240	45° Elbow									
9250	1/8"	1 Plum	13	.615	Ea.	81.50	30		111.50	135
9260	1/4"		13	.615		81.50	30		111.50	135
9270	3/8"		13	.615		81.50	30		111.50	135
9280	1/2"		12	.667		81.50	32.50		114	139
9300	3/4"		11	.727		91.50	35.50		127	154
9310	1"		10	.800		138	39		177	211
9320	1-1/4"	Q-1	17	.941		198	41.50		239.50	279
9330	1-1/2"		16	1		270	44		314	365
9340	2"		14	1.143		330	50		380	440
9360	Tee, straight									
9370	1/8"	1 Plum	9	.889	Ea.	49	43.50		92.50	119

22 11 Facility Water Distribution

22 11 13 – Facility Water Distribution Piping

22 11 13.66 Pipe Fittings, Stainless Steel		Crew	Daily Output	Labor-Hours	Unit	2009 Bare Costs Material	Labor	Equipment	Total	Total Incl O&P
9380	1/4"	1 Plum	9	.889	Ea.	49	43.50		92.50	119
9390	3/8"		9	.889		59.50	43.50		103	131
9400	1/2"		8	1		74.50	49		123.50	155
9410	3/4"		7	1.143		92	55.50		147.50	185
9420	1"		6.50	1.231		140	60		200	244
9430	1-1/4"	Q-1	11	1.455		227	64		291	345
9440	1-1/2"		10	1.600		325	70		395	465
9450	2"		9	1.778		520	78		598	685
9470	For couplings and unions use 3000 lb., type 316									
9490	3000 lb., type 304									
9510	Coupling									
9520	1/8"	1 Plum	19	.421	Ea.	12.15	20.50		32.65	44.50
9530	1/4"		19	.421		13.20	20.50		33.70	45.50
9540	3/8"		19	.421		15.40	20.50		35.90	48
9550	1/2"		19	.421		18.55	20.50		39.05	51.50
9560	3/4"		18	.444		24.50	21.50		46	59.50
9570	1"		15	.533		42	26		68	85
9580	1-1/4"	Q-1	26	.615		102	27		129	153
9590	1-1/2"		24	.667		119	29.50		148.50	175
9600	2"		21	.762		157	33.50		190.50	222
9620	Union									
9630	1/8"	1 Plum	12	.667	Ea.	93	32.50		125.50	151
9640	1/4"		12	.667		93	32.50		125.50	151
9650	3/8"		12	.667		99.50	32.50		132	158
9660	1/2"		11	.727		99.50	35.50		135	162
9670	3/4"		10	.800		122	39		161	193
9680	1"		9	.889		188	43.50		231.50	272
9690	1-1/4"	Q-1	16	1		345	44		389	445
9700	1-1/2"		15	1.067		390	47		437	500
9710	2"		13	1.231		525	54		579	655
9730	3000 lb., type 316									
9750	Coupling									
9770	1/8"	1 Plum	19	.421	Ea.	13.70	20.50		34.20	46
9780	1/4"		19	.421		15.45	20.50		35.95	48
9790	3/8"		19	.421		16.55	20.50		37.05	49
9800	1/2"		19	.421		23.50	20.50		44	57
9810	3/4"		18	.444		34.50	21.50		56	70.50
9820	1"		15	.533		57.50	26		83.50	103
9830	1-1/4"	Q-1	26	.615		129	27		156	183
9840	1-1/2"		24	.667		151	29.50		180.50	210
9850	2"		21	.762		209	33.50		242.50	279
9870	Union									
9880	1/8"	1 Plum	12	.667	Ea.	102	32.50		134.50	161
9890	1/4"		12	.667		102	32.50		134.50	161
9900	3/8"		12	.667		118	32.50		150.50	179
9910	1/2"		11	.727		118	35.50		153.50	183
9920	3/4"		10	.800		155	39		194	229
9930	1"		9	.889		244	43.50		287.50	335
9940	1-1/4"	Q-1	16	1		425	44		469	535
9950	1-1/2"		15	1.067		525	47		572	650
9960	2"		13	1.231		670	54		724	815

22 11 13.74 Pipe, Plastic		Crew	Daily Output	Labor-Hours	Unit	2009 Bare Costs Material	Labor	Equipment	Total	Total Incl O&P
0010	**PIPE, PLASTIC**									
0020	Fiberglass reinforced, couplings 10′ O.C., clevis hanger assy's, 3 per 10′									
0080	General service									
0120	2″ diameter	Q-1	59	.271	L.F.	5.90	11.90		17.80	24.50
0140	3″ diameter		52	.308		8.05	13.50		21.55	29.50
0150	4″ diameter		48	.333		10.85	14.65		25.50	34
0160	6″ diameter	↓	39	.410		19.60	18		37.60	48.50
0170	8″ diameter	Q-2	49	.490		32.50	22.50		55	69
0180	10″ diameter		41	.585		49	26.50		75.50	94
0190	12″ diameter	↓	36	.667	↓	63.50	30.50		94	116
0600	PVC, high impact/pressure, cplgs. 10′ O.C., clevis hanger assy's, 3 per 10′									
1020	Schedule 80									
1070	1/2″ diameter	1 Plum	50	.160	L.F.	3.03	7.80		10.83	15.05
1080	3/4″ diameter		47	.170		3.84	8.30		12.14	16.65
1090	1″ diameter		43	.186		5.15	9.05		14.20	19.25
1100	1-1/4″ diameter		39	.205		6.90	10		16.90	22.50
1110	1-1/2″ diameter	↓	34	.235		8.15	11.45		19.60	26
1120	2″ diameter	Q-1	55	.291		10.75	12.75		23.50	31
1140	3″ diameter		50	.320		22	14.05		36.05	45
1150	4″ diameter		46	.348		31.50	15.25		46.75	58
1170	6″ diameter	↓	38	.421	↓	71.50	18.50		90	106
1730	To delete coupling & hangers, subtract									
1740	1/4″ diam. to 1/2″ diam.					62%	80%			
1750	3/4″ diam. to 1-1/4″ diam.					58%	73%			
1760	1-1/2″ diam. to 6″ diam.					40%	57%			
1770	8″ diam. to 12″ diam.					34%	50%			
1800	PVC, couplings 10′ O.C., clevis hanger assy's, 3 per 10′									
1820	Schedule 40									
1860	1/2″ diameter	1 Plum	54	.148	L.F.	1.43	7.20		8.63	12.40
1870	3/4″ diameter		51	.157		1.67	7.65		9.32	13.30
1880	1″ diameter		46	.174		1.96	8.50		10.46	14.85
1890	1-1/4″ diameter		42	.190		2.49	9.30		11.79	16.70
1900	1-1/2″ diameter	↓	36	.222		2.73	10.85		13.58	19.25
1910	2″ diameter	Q-1	59	.271		3.33	11.90		15.23	21.50
1920	2-1/2″ diameter		56	.286		4.93	12.55		17.48	24.50
1930	3″ diameter		53	.302		6.60	13.25		19.85	27
1940	4″ diameter		48	.333		9.15	14.65		23.80	32
1950	5″ diameter		43	.372		12.85	16.35		29.20	38.50
1960	6″ diameter	↓	39	.410		16.95	18		34.95	45.50
1970	8″ diameter	Q-2	48	.500		24.50	23		47.50	61
1980	10″ diameter		43	.558		55	25.50		80.50	98.50
1990	12″ diameter		42	.571		66	26		92	112
2000	14″ diameter		31	.774		116	35		151	180
2010	16″ diameter	↓	23	1.043	↓	166	47.50		213.50	255
2340	To delete coupling & hangers, subtract									
2360	1/2″ diam. to 1-1/4″ diam.					65%	74%			
2370	1-1/2″ diam. to 6″ diam.					44%	57%			
2380	8″ diam. to 12″ diam.					41%	53%			
2390	14″ diam. to 16″ diam.					48%	45%			
2420	Schedule 80									
2440	1/4″ diameter	1 Plum	58	.138	L.F.	1.09	6.70		7.79	11.30
2450	3/8″ diameter	↓	55	.145	↓	1.09	7.10		8.19	11.85

22 11 13 – Facility Water Distribution Piping

22 11 13.74 Pipe, Plastic		Crew	Daily Output	Labor-Hours	Unit	Material	Labor	Equipment	Total	Total Incl O&P
						2009 Bare Costs				
2460	1/2" diameter	1 Plum	50	.160	L.F.	1.52	7.80		9.32	13.35
2470	3/4" diameter		47	.170		1.79	8.30		10.09	14.40
2480	1" diameter		43	.186		2.14	9.05		11.19	15.95
2490	1-1/4" diameter		39	.205		2.76	10		12.76	18.05
2500	1-1/2" diameter		34	.235		3.07	11.45		14.52	20.50
2510	2" diameter	Q-1	55	.291		3.75	12.75		16.50	23.50
2520	2-1/2" diameter		52	.308		4.60	13.50		18.10	25.50
2530	3" diameter		50	.320		7.65	14.05		21.70	29.50
2540	4" diameter		46	.348		10.75	15.25		26	35
2550	5" diameter		42	.381		15.25	16.70		31.95	42
2560	6" diameter		38	.421		24	18.50		42.50	54
2570	8" diameter	Q-2	47	.511		30.50	23		53.50	68.50
2580	10" diameter		42	.571		64.50	26		90.50	110
2590	12" diameter		38	.632		80	28.50		108.50	131
2830	To delete coupling & hangers, subtract									
2840	1/4" diam. to 1/2" diam.					66%	80%			
2850	3/4" diam. to 1-1/4" diam.					61%	73%			
2860	1-1/2" diam. to 6" diam.					41%	57%			
2870	8" diam. to 12" diam.					31%	50%			
2900	Schedule 120									
2910	1/2" diameter	1 Plum	50	.160	L.F.	1.65	7.80		9.45	13.50
2950	3/4" diameter		47	.170		1.95	8.30		10.25	14.60
2960	1" diameter		43	.186		2.38	9.05		11.43	16.20
2970	1-1/4" diameter		39	.205		3.10	10		13.10	18.40
2980	1-1/2" diameter		33	.242		3.49	11.80		15.29	21.50
2990	2" diameter	Q-1	54	.296		4.38	13		17.38	24.50
3000	2-1/2" diameter		52	.308		7.15	13.50		20.65	28.50
3010	3" diameter		49	.327		8.90	14.35		23.25	31.50
3020	4" diameter		45	.356		13.40	15.60		29	38
3030	6" diameter		37	.432		25.50	19		44.50	56.50
3240	To delete coupling & hangers, subtract									
3250	1/2" diam. to 1-1/4" diam.					52%	74%			
3260	1-1/2" diam. to 4" diam.					30%	57%			
3270	6" diam.					17%	50%			
3300	PVC, pressure, couplings 10' O.C., clevis hanger assy's, 3 per 10'									
3310	SDR 26, 160 psi									
3350	1-1/4" diameter	1 Plum	42	.190	L.F.	2.07	9.30		11.37	16.20
3360	1-1/2" diameter	"	36	.222		2.27	10.85		13.12	18.75
3370	2" diameter	Q-1	59	.271		2.71	11.90		14.61	21
3380	2-1/2" diameter		56	.286		3.97	12.55		16.52	23
3390	3" diameter		53	.302		5.60	13.25		18.85	26
3400	4" diameter		48	.333		8.10	14.65		22.75	31
3420	6" diameter		39	.410		16.30	18		34.30	45
3430	8" diameter	Q-2	48	.500		25	23		48	61.50
3660	To delete coupling & clevis hanger assy's, subtract									
3670	1-1/4" diam.					63%	68%			
3680	1-1/2" diam. to 4" diam.					48%	57%			
3690	6" diam. to 8" diam.					60%	54%			
3720	SDR 21, 200 psi, 1/2" diameter	1 Plum	54	.148	L.F.	1.21	7.20		8.41	12.20
3740	3/4" diameter		51	.157		1.48	7.65		9.13	13.10
3750	1" diameter		46	.174		1.64	8.50		10.14	14.50
3760	1-1/4" diameter		42	.190		2.15	9.30		11.45	16.30
3770	1-1/2" diameter		36	.222		2.38	10.85		13.23	18.85

22 11 13.74 Pipe, Plastic		Crew	Daily Output	Labor-Hours	Unit	2009 Bare Costs Material	Labor	Equipment	Total	Total Incl O&P
3780	2″ diameter	Q-1	59	.271	L.F.	2.90	11.90		14.80	21
3790	2-1/2″ diameter		56	.286		5.15	12.55		17.70	24.50
3800	3″ diameter		53	.302		6	13.25		19.25	26.50
3810	4″ diameter		48	.333		9.90	14.65		24.55	33
3830	6″ diameter		39	.410		17.85	18		35.85	46.50
3840	8″ diameter	Q-2	48	.500		27.50	23		50.50	64.50
4000	To delete coupling & hangers, subtract									
4010	1/2″ diam. to 3/4″ diam.					71%	77%			
4020	1″ diam. to 1-1/4″ diam.					63%	70%			
4030	1-1/2″ diam. to 6″ diam.					44%	57%			
4040	8″ diam.					46%	54%			
4100	DWV type, schedule 40, couplings 10′ O.C., clevis hanger assy's, 3 per 10′									
4120	ABS									
4140	1-1/4″ diameter	1 Plum	42	.190	L.F.	1.58	9.30		10.88	15.70
4150	1-1/2″ diameter	″	36	.222		1.72	10.85		12.57	18.15
4160	2″ diameter	Q-1	59	.271		2.21	11.90		14.11	20.50
4170	3″ diameter		53	.302		4.94	13.25		18.19	25.50
4180	4″ diameter		48	.333		8	14.65		22.65	31
4190	6″ diameter		39	.410		16.30	18		34.30	45
4194	To delete coupling & hangers, subtract									
4196	1-1/4″ diam.					56%	62%			
4198	1-1/2″ diam. to 6″ diam.					50%	52%			
4210	ABS, schedule 40, foam core type									
4212	Plain end black									
4214	1-1/2″ diameter	1 Plum	39	.205	L.F.	1.76	10		11.76	16.95
4216	2″ diameter	Q-1	62	.258		2.17	11.30		13.47	19.40
4218	3″ diameter		56	.286		4.15	12.55		16.70	23.50
4220	4″ diameter		51	.314		5.75	13.75		19.50	27
4222	6″ diameter		42	.381		11.95	16.70		28.65	38
4240	To delete coupling & hangers, subtract									
4244	1-1/2″ diam. to 6″ diam.					43%	48%			
4400	PVC									
4410	1-1/4″ diameter	1 Plum	42	.190	L.F.	1.85	9.30		11.15	16
4420	1-1/2″ diameter	″	36	.222		1.82	10.85		12.67	18.25
4460	2″ diameter	Q-1	59	.271		2.20	11.90		14.10	20.50
4470	3″ diameter		53	.302		4.21	13.25		17.46	24.50
4480	4″ diameter		48	.333		5.75	14.65		20.40	28.50
4490	6″ diameter		39	.410		10.55	18		28.55	38.50
4500	8″ diameter	Q-2	48	.500		17.50	23		40.50	53.50
4510	To delete coupling & hangers, subtract									
4520	1-1/4″ diam. to 1-1/2″ diam.					48%	60%			
4530	2″ diam. to 8″ diam.					42%	54%			
4550	PVC, schedule 40, foam core type									
4552	Plain end, white									
4554	1-1/2″ diameter	1 Plum	39	.205	L.F.	1.56	10		11.56	16.70
4556	2″ diameter	Q-1	62	.258		1.88	11.30		13.18	19.05
4558	3″ diameter		56	.286		3.60	12.55		16.15	23
4560	4″ diameter		51	.314		4.83	13.75		18.58	26
4562	6″ diameter		42	.381		8.70	16.70		25.40	34.50
4564	8″ diameter	Q-2	51	.471		14.60	21.50		36.10	48
4568	10″ diameter		48	.500		19.25	23		42.25	55
4570	12″ diameter		46	.522		24	23.50		47.50	62
4580	To delete coupling & hangers, subtract									

22 11 Facility Water Distribution

22 11 13 – Facility Water Distribution Piping

22 11 13.74 Pipe, Plastic		Crew	Daily Output	Labor-Hours	Unit	2009 Bare Costs Material	Labor	Equipment	Total	Total Incl O&P
4582	1-1/2" dia to 2" dia					58%	54%			
4584	3" dia to 12" dia					46%	42%			
4800	PVC, clear pipe, cplgs. 10' O.C., clevis hanger assy's 3 per 10', Sched. 40									
4840	1/4" diameter	1 Plum	59	.136	L.F.	1.52	6.60		8.12	11.55
4850	3/8" diameter		56	.143		1.79	6.95		8.74	12.40
4860	1/2" diameter		54	.148		2.31	7.20		9.51	13.40
4870	3/4" diameter		51	.157		2.85	7.65		10.50	14.60
4880	1" diameter		46	.174		3.94	8.50		12.44	17.05
4890	1-1/4" diameter		42	.190		5.10	9.30		14.40	19.55
4900	1-1/2" diameter		36	.222		5.95	10.85		16.80	23
4910	2" diameter	Q-1	59	.271		7.80	11.90		19.70	26.50
4920	2-1/2" diameter		56	.286		12.30	12.55		24.85	32.50
4930	3" diameter		53	.302		15.90	13.25		29.15	37.50
4940	3-1/2" diameter		50	.320		22.50	14.05		36.55	46
4950	4" diameter		48	.333		23.50	14.65		38.15	48
5250	To delete coupling & hangers, subtract									
5260	1/4" diam. to 3/8" diam.					60%	81%			
5270	1/2" diam. to 3/4" diam.					41%	77%			
5280	1" diam. to 1-1/2" diam.					26%	67%			
5290	2" diam. to 4" diam.					16%	58%			
5360	CPVC, couplings 10' O.C., clevis hanger assy's, 3 per 10'									
5380	Schedule 40									
5460	1/2" diameter	1 Plum	54	.148	L.F.	2.84	7.20		10.04	13.95
5470	3/4" diameter		51	.157		3.37	7.65		11.02	15.15
5480	1" diameter		46	.174		4.13	8.50		12.63	17.25
5490	1-1/4" diameter		42	.190		5.05	9.30		14.35	19.50
5500	1-1/2" diameter		36	.222		5.70	10.85		16.55	22.50
5510	2" diameter	Q-1	59	.271		7.05	11.90		18.95	25.50
5520	2-1/2" diameter		56	.286		11.50	12.55		24.05	31.50
5530	3" diameter		53	.302		14.20	13.25		27.45	35.50
5540	4" diameter		48	.333		23.50	14.65		38.15	47.50
5550	6" diameter		43	.372		37	16.35		53.35	65
5730	To delete coupling & hangers, subtract									
5740	1/2" diam. to 3/4" diam.					37%	77%			
5750	1" diam. to 1-1/4" diam.					27%	70%			
5760	1-1/2" diam. to 3" diam.					21%	57%			
5770	4" diam. to 6" diam.					16%	57%			
5800	Schedule 80									
5860	1/2" diameter	1 Plum	50	.160	L.F.	2.99	7.80		10.79	15
5870	3/4" diameter		47	.170		3.60	8.30		11.90	16.40
5880	1" diameter		43	.186		4.45	9.05		13.50	18.50
5890	1-1/4" diameter		39	.205		5.55	10		15.55	21
5900	1-1/2" diameter		34	.235		6.30	11.45		17.75	24
5910	2" diameter	Q-1	55	.291		8.05	12.75		20.80	28
5920	2-1/2" diameter		52	.308		12.85	13.50		26.35	34.50
5930	3" diameter		50	.320		16.15	14.05		30.20	39
5940	4" diameter		46	.348		26.50	15.25		41.75	52
5950	6" diameter		38	.421		44	18.50		62.50	76
5960	8" diameter	Q-2	47	.511		80.50	23		103.50	124
6060	To delete couplings & hangers, subtract									
6070	1/2" diam. to 3/4" diam.					44%	77%			
6080	1" diam. to 1-1/4" diam.					32%	71%			
6090	1-1/2" diam. to 4" diam.					25%	58%			

22 11 13.74 Pipe, Plastic

	22 11 13.74 Pipe, Plastic	Crew	Daily Output	Labor-Hours	Unit	Material	Labor	Equipment	Total	Total Incl O&P
6100	6" diam. to 8" diam.					20%	53%			
6240	CTS, 1/2" diameter	1 Plum	54	.148	L.F.	1.35	7.20		8.55	12.35
6250	3/4" diameter		51	.157		1.93	7.65		9.58	13.60
6260	1" diameter		46	.174		3.21	8.50		11.71	16.25
6270	1 1/4"		42	.190		4.42	9.30		13.72	18.80
6280	1 1/2" diameter		36	.222		5.65	10.85		16.50	22.50
6290	2" diameter	Q-1	59	.271		9.10	11.90		21	28
6370	To delete coupling & hangers, subtract									
6380	1/2" diam.					51%	79%			
6390	3/4" diam.					40%	76%			
6392	1" thru 2" diam.					72%	68%			
7280	Polyethylene, flexible, no couplings or hangers									
7282	Note: For labor costs add 25% to the couplings and fittings labor total.									
7300	SDR 15, 100 psi									
7310	3/4" diameter				L.F.	.32			.32	.35
7350	1" diameter					.43			.43	.47
7360	1-1/4" diameter					.78			.78	.86
7370	1-1/2" diameter					.95			.95	1.05
7380	2" diameter					1.55			1.55	1.71
7700	SDR 9, 160 psi									
7710	1/2" diameter				L.F.	.43			.43	.47
7750	3/4" diameter					.43			.43	.47
7760	1" diameter					.65			.65	.72
7770	1-1/4" diameter					1.16			1.16	1.28
7780	1-1/2" diameter					1.62			1.62	1.78
7790	2" diameter					2.68			2.68	2.95
8120	SDR 9, 200 psi									
8150	3/4" diameter				L.F.	.38			.38	.42
8160	1" diameter					.63			.63	.69
8170	1-1/4" diameter					.95			.95	1.05
8420	SDR 7, 250 psi									
8440	3/4" diameter				L.F.	.55			.55	.61
8450	1" diameter					.88			.88	.97
8460	1-1/4" diameter					1.55			1.55	1.71

(2009 Bare Costs: Material, Labor, Equipment, Total)

22 11 13.76 Pipe Fittings, Plastic

	22 11 13.76 Pipe Fittings, Plastic	Crew	Daily Output	Labor-Hours	Unit	Material	Labor	Equipment	Total	Total Incl O&P
0010	**PIPE FITTINGS, PLASTIC**									
0030	Epoxy resin, fiberglass reinforced, general service									
0090	Elbow, 90°, 2"	Q-1	33.10	.483	Ea.	79.50	21		100.50	120
0100	3"		20.80	.769		91.50	34		125.50	152
0110	4"		16.50	.970		126	42.50		168.50	202
0120	6"		10.10	1.584		182	69.50		251.50	305
0130	8"	Q-2	9.30	2.581		335	117		452	545
0140	10"		8.50	2.824		425	128		553	660
0150	12"		7.60	3.158		605	144		749	885
0160	45° Elbow, same as 90°									
0170	Elbow, 90°, flanged									
0172	2"	Q-1	23	.696	Ea.	121	30.50		151.50	179
0173	3"		16	1		138	44		182	218
0174	4"		13	1.231		180	54		234	279
0176	6"		8	2		325	88		413	490
0177	8"	Q-2	9	2.667		590	121		711	830
0178	10"		7	3.429		805	156		961	1,125

22 11 13.76 Pipe Fittings, Plastic		Crew	Daily Output	Labor-Hours	Unit	Material	Labor	Equipment	Total	Total Incl O&P
						2009 Bare Costs				
0179	12"	Q-2	5	4.800	Ea.	1,075	218		1,293	1,525
0186	Elbow, 45°, flanged									
0188	2"	Q-1	23	.696	Ea.	121	30.50		151.50	179
0189	3"		16	1		138	44		182	218
0190	4"		13	1.231		180	54		234	279
0192	6"	↓	8	2		325	88		413	490
0193	8"	Q-2	9	2.667		560	121		681	795
0194	10"		7	3.429		710	156		866	1,025
0195	12"	↓	5	4.800		910	218		1,128	1,325
0290	Tee, 2"	Q-1	20	.800		43.50	35		78.50	101
0300	3"		13.90	1.151		52.50	50.50		103	134
0310	4"		11	1.455		72	64		136	176
0320	6"	↓	6.70	2.388		207	105		312	385
0330	8"	Q-2	6.20	3.871		1,300	176		1,476	1,700
0340	10"		5.70	4.211		1,475	192		1,667	1,925
0350	12"	↓	5.10	4.706	↓	1,800	214		2,014	2,300
0352	Tee, flanged									
0354	2"	Q-1	17	.941	Ea.	163	41.50		204.50	242
0355	3"		10	1.600		216	70		286	345
0356	4"		8	2		242	88		330	400
0358	6"	↓	5	3.200		415	140		555	670
0359	8"	Q-2	6	4		790	182		972	1,150
0360	10"		5	4.800		1,150	218		1,368	1,575
0361	12"	↓	4	6	↓	1,575	273		1,848	2,150
0365	Wye, flanged									
0367	2"	Q-1	17	.941	Ea.	325	41.50		366.50	415
0368	3"		10	1.600		445	70		515	595
0369	4"		8	2		595	88		683	780
0371	6"	↓	5	3.200		830	140		970	1,125
0372	8"	Q-2	6	4		1,375	182		1,557	1,775
0373	10"		5	4.800		2,125	218		2,343	2,675
0374	12"	↓	4	6	↓	3,100	273		3,373	3,800
0380	Couplings									
0410	2"	Q-1	33.10	.483	Ea.	14	21		35	47.50
0420	3"		20.80	.769		15	34		49	67
0430	4"		16.50	.970		20	42.50		62.50	86
0440	6"	↓	10.10	1.584		48	69.50		117.50	157
0450	8"	Q-2	9.30	2.581		81.50	117		198.50	266
0460	10"		8.50	2.824		120	128		248	325
0470	12"	↓	7.60	3.158	↓	160	144		304	390
0473	High corrosion resistant couplings, add					30%				
0474	Reducer, concentric, flanged									
0475	2" x 1-1/2"	Q-1	30	.533	Ea.	315	23.50		338.50	380
0476	3" x 2"		24	.667		340	29.50		369.50	420
0477	4" x 3"		19	.842		385	37		422	475
0479	6" x 4"	↓	15	1.067		385	47		432	490
0480	8" x 6"	Q-2	16	1.500		630	68.50		698.50	795
0481	10" x 8"		13	1.846		585	84		669	770
0482	12" x 10"	↓	11	2.182	↓	905	99.50		1,004.50	1,150
0486	Adapter, bell x male or female									
0488	2"	Q-1	28	.571	Ea.	15.60	25		40.60	54.50
0489	3"		20	.800		23	35		58	78
0491	4"	↓	17	.941	↓	31	41.50		72.50	96

22 11 13.76 Pipe Fittings, Plastic		Crew	Daily Output	Labor-Hours	Unit	Material	Labor	Equipment	Total	Total Incl O&P
						2009 Bare Costs				
0492	6"	Q-1	12	1.333	Ea.	65	58.50		123.50	160
0493	8"	Q-2	15	1.600		94	73		167	213
0494	10"	"	11	2.182		136	99.50		235.50	299
0528	Flange									
0532	2"	Q-1	46	.348	Ea.	23	15.25		38.25	48.50
0533	3"		32	.500		28	22		50	64
0534	4"		26	.615		37	27		64	81
0536	6"		16	1		70	44		114	143
0537	8"	Q-2	18	1.333		107	60.50		167.50	209
0538	10"		14	1.714		152	78		230	284
0539	12"		10	2.400		191	109		300	375
2100	PVC schedule 80, socket joint									
2110	90° elbow, 1/2"	1 Plum	30.30	.264	Ea.	2.08	12.85		14.93	21.50
2130	3/4"		26	.308		2.66	15		17.66	25.50
2140	1"		22.70	.352		3.98	17.20		21.18	30.50
2150	1-1/4"		20.20	.396		5.75	19.30		25.05	35.50
2160	1-1/2"		18.20	.440		6.15	21.50		27.65	39
2170	2"	Q-1	33.10	.483		7.45	21		28.45	40
2180	3"		20.80	.769		19.50	34		53.50	72
2190	4"		16.50	.970		29.50	42.50		72	96.50
2200	6"		10.10	1.584		84.50	69.50		154	197
2210	8"	Q-2	9.30	2.581		233	117		350	430
2250	45° elbow, 1/2"	1 Plum	30.30	.264		3.92	12.85		16.77	23.50
2270	3/4"		26	.308		5.95	15		20.95	29
2280	1"		22.70	.352		9	17.20		26.20	36
2290	1-1/4"		20.20	.396		11.45	19.30		30.75	41.50
2300	1-1/2"		18.20	.440		13.50	21.50		35	47
2310	2"	Q-1	33.10	.483		17.50	21		38.50	51.50
2320	3"		20.80	.769		22.50	34		56.50	75
2330	4"		16.50	.970		80.50	42.50		123	153
2340	6"		10.10	1.584		102	69.50		171.50	216
2350	8"	Q-2	9.30	2.581		220	117		337	420
2400	Tee, 1/2"	1 Plum	20.20	.396		5.85	19.30		25.15	35.50
2420	3/4"		17.30	.462		6.15	22.50		28.65	41
2430	1"		15.20	.526		7.70	25.50		33.20	47
2440	1-1/4"		13.50	.593		21	29		50	67
2450	1-1/2"		12.10	.661		21	32		53	72
2460	2"	Q-1	20	.800		26.50	35		61.50	81.50
2470	3"		13.90	1.151		36	50.50		86.50	116
2480	4"		11	1.455		41.50	64		105.50	142
2490	6"		6.70	2.388		142	105		247	315
2500	8"	Q-2	6.20	3.871		296	176		472	590
2510	Flange, socket, 150 lb., 1/2"	1 Plum	55.60	.144		11.40	7		18.40	23
2514	3/4"		47.60	.168		12.20	8.20		20.40	25.50
2518	1"		41.70	.192		13.55	9.35		22.90	29
2522	1-1/2"		33.30	.240		14.30	11.70		26	33.50
2526	2"	Q-1	60.60	.264		19	11.60		30.60	38.50
2530	4"		30.30	.528		41	23		64	80
2534	6"		18.50	.865		64.50	38		102.50	128
2538	8"	Q-2	17.10	1.404		115	64		179	223
2550	Coupling, 1/2"	1 Plum	30.30	.264		3.77	12.85		16.62	23.50
2570	3/4"		26	.308		5.10	15		20.10	28
2580	1"		22.70	.352		5.25	17.20		22.45	32

22 11 13.76 Pipe Fittings, Plastic		Crew	Daily Output	Labor-Hours	Unit	2009 Bare Costs Material	Labor	Equipment	Total	Total Incl O&P
2590	1-1/4"	1 Plum	20.20	.396	Ea.	8	19.30		27.30	38
2600	1-1/2"		18.20	.440		8.60	21.50		30.10	41.50
2610	2"	Q-1	33.10	.483		9.25	21		30.25	42
2620	3"		20.80	.769		22.50	34		56.50	75.50
2630	4"		16.50	.970		32.50	42.50		75	100
2640	6"		10.10	1.584		70.50	69.50		140	182
2650	8"	Q-2	9.30	2.581		96	117		213	281
2660	10"		8.50	2.824		330	128		458	555
2670	12"		7.60	3.158		380	144		524	630
2700	PVC (white), schedule 40, socket joints									
2760	90° elbow, 1/2"	1 Plum	33.30	.240	Ea.	.41	11.70		12.11	18
2770	3/4"		28.60	.280		.45	13.65		14.10	21
2780	1"		25	.320		.82	15.60		16.42	24.50
2790	1-1/4"		22.20	.360		1.44	17.55		18.99	28
2800	1-1/2"		20	.400		1.55	19.50		21.05	31
2810	2"	Q-1	36.40	.440		2.42	19.30		21.72	31.50
2820	2-1/2"		26.70	.599		7.40	26.50		33.90	47.50
2830	3"		22.90	.699		8.85	30.50		39.35	55.50
2840	4"		18.20	.879		15.80	38.50		54.30	75.50
2850	5"		12.10	1.322		41	58		99	132
2860	6"		11.10	1.441		50	63.50		113.50	151
2870	8"	Q-2	10.30	2.330		129	106		235	300
2980	45° elbow, 1/2"	1 Plum	33.30	.240		.67	11.70		12.37	18.30
2990	3/4"		28.60	.280		1.04	13.65		14.69	21.50
3000	1"		25	.320		1.25	15.60		16.85	25
3010	1-1/4"		22.20	.360		1.73	17.55		19.28	28.50
3020	1-1/2"		20	.400		2.18	19.50		21.68	32
3030	2"	Q-1	36.40	.440		2.84	19.30		22.14	32
3040	2-1/2"		26.70	.599		7.40	26.50		33.90	47.50
3050	3"		22.90	.699		11.50	30.50		42	58.50
3060	4"		18.20	.879		20.50	38.50		59	80.50
3070	5"		12.10	1.322		41	58		99	132
3080	6"		11.10	1.441		51	63.50		114.50	151
3090	8"	Q-2	10.30	2.330		122	106		228	294
3180	Tee, 1/2"	1 Plum	22.20	.360		.50	17.55		18.05	27
3190	3/4"		19	.421		.58	20.50		21.08	31.50
3200	1"		16.70	.479		1.08	23.50		24.58	36
3210	1-1/4"		14.80	.541		1.68	26.50		28.18	41.50
3220	1-1/2"		13.30	.602		2.06	29.50		31.56	46.50
3230	2"	Q-1	24.20	.661		2.99	29		31.99	47
3240	2-1/2"		17.80	.899		9.85	39.50		49.35	70
3250	3"		15.20	1.053		12.95	46		58.95	84
3260	4"		12.10	1.322		23.50	58		81.50	113
3270	5"		8.10	1.975		56.50	86.50		143	193
3280	6"		7.40	2.162		79	95		174	229
3290	8"	Q-2	6.80	3.529		183	161		344	440
3380	Coupling, 1/2"	1 Plum	33.30	.240		.27	11.70		11.97	17.85
3390	3/4"		28.60	.280		.37	13.65		14.02	21
3400	1"		25	.320		.66	15.60		16.26	24
3410	1-1/4"		22.20	.360		.89	17.55		18.44	27.50
3420	1-1/2"		20	.400		.95	19.50		20.45	30.50
3430	2"	Q-1	36.40	.440		1.45	19.30		20.75	30.50
3440	2-1/2"		26.70	.599		3.20	26.50		29.70	43

22 11 13.76 Pipe Fittings, Plastic		Crew	Daily Output	Labor-Hours	Unit	Material	2009 Bare Costs Labor	Equipment	Total	Total Incl O&P
3450	3"	Q-1	22.90	.699	Ea.	5	30.50		35.50	51.50
3460	4"		18.20	.879		7.25	38.50		45.75	66
3470	5"		12.10	1.322		13.30	58		71.30	102
3480	6"		11.10	1.441		23	63.50		86.50	120
3490	8"	Q-2	10.30	2.330		43	106		149	206
3710	Reducing insert, schedule 40, socket weld									
3712	3/4"	1 Plum	31.50	.254	Ea.	.42	12.40		12.82	19.05
3713	1"		27.50	.291		.77	14.20		14.97	22.50
3715	1-1/2"		22	.364		1.10	17.75		18.85	27.50
3716	2"	Q-1	40	.400		1.82	17.55		19.37	28.50
3717	4"		20	.800		9.60	35		44.60	63
3718	6"		12.20	1.311		24	57.50		81.50	113
3719	8"	Q-2	11.30	2.124		93.50	96.50		190	248
3730	Reducing insert, socket weld x female/male thread									
3732	1/2"	1 Plum	38.30	.209	Ea.	1.92	10.20		12.12	17.40
3733	3/4"		32.90	.243		1.19	11.85		13.04	19.10
3734	1"		28.80	.278		1.67	13.55		15.22	22.50
3736	1-1/2"		23	.348		2.99	16.95		19.94	29
3737	2"	Q-1	41.90	.382		3.20	16.75		19.95	28.50
3738	4"	"	20.90	.766		12.15	33.50		45.65	64
3742	Male adapter, socket weld x male thread									
3744	1/2"	1 Plum	38.30	.209	Ea.	.37	10.20		10.57	15.70
3745	3/4"		32.90	.243		.41	11.85		12.26	18.25
3746	1"		28.80	.278		.73	13.55		14.28	21.50
3748	1-1/2"		23	.348		1.19	16.95		18.14	27
3749	2"	Q-1	41.90	.382		1.50	16.75		18.25	26.50
3750	4"	"	20.90	.766		8.95	33.50		42.45	60.50
3754	Female adapter, socket weld x female thread									
3756	1/2"	1 Plum	38.30	.209	Ea.	.45	10.20		10.65	15.80
3757	3/4"		32.90	.243		.58	11.85		12.43	18.45
3758	1"		28.80	.278		.67	13.55		14.22	21
3760	1-1/2"		23	.348		1.19	16.95		18.14	27
3761	2"	Q-1	41.90	.382		1.59	16.75		18.34	27
3762	4"	"	20.90	.766		9	33.50		42.50	60.50
3800	PVC, schedule 80, socket joints									
3810	Reducing insert									
3812	3/4"	1 Plum	28.60	.280	Ea.	1.22	13.65		14.87	22
3813	1"		25	.320		3.47	15.60		19.07	27.50
3815	1-1/2"		20	.400		7.45	19.50		26.95	37.50
3816	2"	Q-1	36.40	.440		10.60	19.30		29.90	40.50
3817	4"		18.20	.879		40.50	38.50		79	103
3818	6"		11.10	1.441		56	63.50		119.50	157
3819	8"	Q-2	10.20	2.353		325	107		432	515
3830	Reducing insert, socket weld x female/male thread									
3832	1/2"	1 Plum	34.80	.230	Ea.	7.60	11.20		18.80	25
3833	3/4"		29.90	.268		4.66	13.05		17.71	24.50
3834	1"		26.10	.307		7.30	14.95		22.25	30.50
3836	1-1/2"		20.90	.383		9.15	18.65		27.80	38
3837	2"	Q-1	38	.421		13.45	18.50		31.95	42.50
3838	4"	"	19	.842		65	37		102	127
3844	Adapter, male socket x male thread									
3846	1/2"	1 Plum	34.80	.230	Ea.	2.94	11.20		14.14	20
3847	3/4"		29.90	.268		3.24	13.05		16.29	23

22 11 13.76 Pipe Fittings, Plastic		Crew	Daily Output	Labor-Hours	Unit	Material	Labor (2009 Bare Costs)	Equipment	Total	Total Incl O&P
3848	1″	1 Plum	26.10	.307	Ea.	5.60	14.95		20.55	28.50
3850	1-1/2″		20.90	.383		9.40	18.65		28.05	38.50
3851	2″	Q-1	38	.421		13.60	18.50		32.10	42.50
3852	4″	″	19	.842		30.50	37		67.50	89
3860	Adapter, female socket x female thread									
3862	1/2″	1 Plum	34.80	.230	Ea.	3.53	11.20		14.73	20.50
3863	3/4″		29.90	.268		5.25	13.05		18.30	25.50
3864	1″		26.10	.307		7.70	14.95		22.65	31
3866	1-1/2″		20.90	.383		15.35	18.65		34	45
3867	2″	Q-1	38	.421		27	18.50		45.50	57
3868	4″	″	19	.842		81.50	37		118.50	146
3872	Union, socket joints									
3874	1/2″	1 Plum	25.80	.310	Ea.	7.70	15.10		22.80	31
3875	3/4″		22.10	.362		9.80	17.65		27.45	37.50
3876	1″		19.30	.415		11.20	20		31.20	43
3878	1-1/2″		15.50	.516		25	25		50	65.50
3879	2″	Q-1	28.10	.569		34	25		59	75
3888	Cap									
3890	1/2″	1 Plum	54.50	.147	Ea.	3.69	7.15		10.84	14.80
3891	3/4″		46.70	.171		3.89	8.35		12.24	16.85
3892	1″		41	.195		6.90	9.50		16.40	22
3894	1-1/2″		32.80	.244		8.35	11.90		20.25	27
3895	2″	Q-1	59.50	.269		22	11.80		33.80	41.50
3896	4″		30	.533		66	23.50		89.50	108
3897	6″		18.20	.879		164	38.50		202.50	239
3898	8″	Q-2	16.70	1.437		212	65.50		277.50	330
4500	DWV, ABS, non pressure, socket joints									
4540	1/4 Bend, 1-1/4″	1 Plum	20.20	.396	Ea.	2.78	19.30		22.08	32
4560	1-1/2″	″	18.20	.440		1.96	21.50		23.46	34
4570	2″	Q-1	33.10	.483		3.04	21		24.04	35.50
4580	3″		20.80	.769		7.30	34		41.30	58.50
4590	4″		16.50	.970		14.05	42.50		56.55	79.50
4600	6″		10.10	1.584		66.50	69.50		136	178
4650	1/8 Bend, same as 1/4 Bend									
4800	Tee, sanitary									
4820	1-1/4″	1 Plum	13.50	.593	Ea.	3.31	29		32.31	47
4830	1-1/2″	″	12.10	.661		2.67	32		34.67	51.50
4840	2″	Q-1	20	.800		4.12	35		39.12	57
4850	3″		13.90	1.151		11.35	50.50		61.85	88.50
4860	4″		11	1.455		20.50	64		84.50	119
4862	Tee, sanitary, reducing, 2″ x 1-1/2″		22	.727		3.94	32		35.94	52.50
4864	3″ x 2″		15.30	1.046		8.25	46		54.25	78
4868	4″ x 3″		12.10	1.322		19.20	58		77.20	108
4870	Combination Y and 1/8 bend									
4872	1-1/2″	1 Plum	12.10	.661	Ea.	6.35	32		38.35	55.50
4874	2″	Q-1	20	.800		7.55	35		42.55	61
4876	3″		13.90	1.151		16.35	50.50		66.85	94
4878	4″		11	1.455		33	64		97	132
4880	3″ x 1-1/2″		15.50	1.032		15.55	45.50		61.05	85
4882	4″ x 3″		12.10	1.322		26.50	58		84.50	116
4900	Wye, 1-1/4″	1 Plum	13.50	.593		3.82	29		32.82	47.50
4902	1-1/2″	″	12.10	.661		3.91	32		35.91	53
4904	2″	Q-1	20	.800		5.50	35		40.50	58.50

	22 11 13.76 Pipe Fittings, Plastic	Crew	Daily Output	Labor-Hours	Unit	2009 Bare Costs Material	Labor	Equipment	Total	Total Incl O&P
4906	3″	Q-1	13.90	1.151	Ea.	12.60	50.50		63.10	90
4908	4″		11	1.455		27	64		91	126
4910	6″		6.70	2.388		83	105		188	249
4918	3″ x 1-1/2″		15.50	1.032		9.15	45.50		54.65	78
4920	4″ x 3″		12.10	1.322		21	58		79	110
4922	6″ x 4″		6.90	2.319		83	102		185	245
4930	Double Wye, 1-1/2″	1 Plum	9.10	.879		8.95	43		51.95	74.50
4932	2″	Q-1	16.60	.964		10.70	42.50		53.20	75.50
4934	3″		10.40	1.538		27.50	67.50		95	132
4936	4″		8.25	1.939		56	85		141	190
4940	2″ x 1-1/2″		16.80	.952		10.70	42		52.70	74.50
4942	3″ x 2″		10.60	1.509		20.50	66		86.50	122
4944	4″ x 3″		8.45	1.893		44.50	83		127.50	174
4946	6″ x 4″		7.25	2.207		87	97		184	241
4950	Reducer bushing, 2″ x 1-1/2″		36.40	.440		1.48	19.30		20.78	30.50
4952	3″ x 1-1/2″		27.30	.586		5.55	25.50		31.05	44.50
4954	4″ x 2″		18.20	.879		11.35	38.50		49.85	70.50
4956	6″ x 4″		11.10	1.441		30.50	63.50		94	129
4960	Couplings, 1-1/2″	1 Plum	18.20	.440		.90	21.50		22.40	33
4962	2″	Q-1	33.10	.483		1.33	21		22.33	33.50
4963	3″		20.80	.769		3.54	34		37.54	54.50
4964	4″		16.50	.970		6.30	42.50		48.80	71
4966	6″		10.10	1.584		25	69.50		94.50	132
4970	2″ x 1-1/2″		33.30	.480		2.46	21		23.46	34
4972	3″ x 1-1/2″		21	.762		6.80	33.50		40.30	57.50
4974	4″ x 3″		16.70	.958		12.05	42		54.05	76.50
4978	Closet flange, 4″	1 Plum	32	.250		5.75	12.20		17.95	24.50
4980	4″ x 3″	″	34	.235		6.45	11.45		17.90	24.50
5000	DWV, PVC, schedule 40, socket joints									
5040	1/4 bend, 1-1/4″	1 Plum	20.20	.396	Ea.	5.70	19.30		25	35.50
5060	1-1/2″	″	18.20	.440		1.90	21.50		23.40	34
5070	2″	Q-1	33.10	.483		2.89	21		23.89	35
5080	3″		20.80	.769		7.30	34		41.30	58.50
5090	4″		16.50	.970		13.75	42.50		56.25	79
5100	6″		10.10	1.584		63.50	69.50		133	174
5105	8″	Q-2	9.30	2.581		92.50	117		209.50	278
5106	10″	″	8.50	2.824		206	128		334	420
5110	1/4 bend, long sweep, 1-1/2″	1 Plum	18.20	.440		3.41	21.50		24.91	36
5112	2″	Q-1	33.10	.483		3.68	21		24.68	36
5114	3″		20.80	.769		8.45	34		42.45	60
5116	4″		16.50	.970		16.15	42.50		58.65	82
5150	1/8 bend, 1-1/4″	1 Plum	20.20	.396		2.68	19.30		21.98	32
5170	1-1/2″	″	18.20	.440		1.36	21.50		22.86	33.50
5180	2″	Q-1	33.10	.483		2.10	21		23.10	34.50
5190	3″		20.80	.769		5.75	34		39.75	57
5200	4″		16.50	.970		9.90	42.50		52.40	75
5210	6″		10.10	1.584		41.50	69.50		111	150
5215	8″	Q-2	9.30	2.581		69.50	117		186.50	253
5216	10″		8.50	2.824		173	128		301	385
5217	12″		7.60	3.158		238	144		382	480
5250	Tee, sanitary 1-1/4″	1 Plum	13.50	.593		4.10	29		33.10	48
5254	1-1/2″	″	12.10	.661		2.57	32		34.57	51.50
5255	2″	Q-1	20	.800		3.78	35		38.78	56.50

22 11 13.76 Pipe Fittings, Plastic		Crew	Daily Output	Labor-Hours	Unit	2009 Bare Costs Material	Labor	Equipment	Total	Total Incl O&P
5256	3"	Q-1	13.90	1.151	Ea.	9.70	50.50		60.20	86.50
5257	4"		11	1.455		16.85	64		80.85	115
5259	6"		6.70	2.388		73.50	105		178.50	238
5261	8"	Q-2	6.20	3.871		184	176		360	465
5276	Tee, sanitary, reducing									
5281	2" x 1-1/2" x 1-1/2"	Q-1	23	.696	Ea.	3.32	30.50		33.82	49.50
5282	2" x 1-1/2" x 2"		22	.727		4.02	32		36.02	52.50
5283	2" x 2" x 1-1/2"		22	.727		3.34	32		35.34	51.50
5284	3" x 3" x 1-1/2"		15.50	1.032		6.15	45.50		51.65	75
5285	3" x 3" x 2"		15.30	1.046		7.25	46		53.25	77
5286	4" x 4" x 1-1/2"		12.30	1.301		17.45	57		74.45	105
5287	4" x 4" x 2"		12.20	1.311		15.75	57.50		73.25	104
5288	4" x 4" x 3"		12.10	1.322		21.50	58		79.50	111
5291	6" x 6" x 4"		6.90	2.319		71	102		173	232
5294	Tee, double sanitary									
5295	1-1/2"	1 Plum	9.10	.879	Ea.	5.25	43		48.25	70.50
5296	2"	Q-1	16.60	.964		7.70	42.50		50.20	72
5297	3"		10.40	1.538		21.50	67.50		89	125
5298	4"		8.25	1.939		34.50	85		119.50	166
5303	Wye, reducing									
5304	2" x 1-1/2" x 1-1/2"	Q-1	23	.696	Ea.	6.50	30.50		37	53
5305	2" x 2" x 1-1/2"		22	.727		5.65	32		37.65	54.50
5306	3" x 3" x 2"		15.30	1.046		17.95	46		63.95	89
5307	4" x 4" x 2"		12.20	1.311		12.30	57.50		69.80	100
5309	4" x 4" x 3"		12.10	1.322		17.50	58		75.50	106
5314	Combination Y & 1/8 bend, 1-1/2"	1 Plum	12.10	.661		5.60	32		37.60	54.50
5315	2"	Q-1	20	.800		7.40	35		42.40	60.50
5317	3"		13.90	1.151		17.90	50.50		68.40	95.50
5318	4"		11	1.455		32	64		96	131
5319	6"		6.70	2.388		141	105		246	310
5320	8"	Q-2	6.20	3.871		225	176		401	510
5324	Combination Y & 1/8 bend, reducing									
5325	2" x 2" x 1-1/2"	Q-1	22	.727	Ea.	8	32		40	57
5327	3" x 3" x 1-1/2"		15.50	1.032		13.85	45.50		59.35	83.50
5328	3" x 3" x 2"		15.30	1.046		10.65	46		56.65	80.50
5329	4" x 4" x 2"		12.20	1.311		17.75	57.50		75.25	106
5331	Wye, 1-1/4"	1 Plum	13.50	.593		5.25	29		34.25	49.50
5332	1-1/2"	"	12.10	.661		3.93	32		35.93	53
5333	2"	Q-1	20	.800		4.60	35		39.60	57.50
5334	3"		13.90	1.151		11.85	50.50		62.35	89
5335	4"		11	1.455		22	64		86	120
5336	6"		6.70	2.388		68	105		173	232
5337	8"	Q-2	6.20	3.871		185	176		361	465
5338	10"		5.70	4.211		355	192		547	680
5339	12"		5.10	4.706		575	214		789	950
5341	2" x 1-1/2"	Q-1	22	.727		5.65	32		37.65	54.50
5342	3" x 1-1/2"		15.50	1.032		7.95	45.50		53.45	76.50
5343	4" x 3"		12.10	1.322		17.50	58		75.50	106
5344	6" x 4"		6.90	2.319		50	102		152	208
5345	8" x 6"	Q-2	6.40	3.750		151	171		322	420
5347	Double wye, 1-1/2"	1 Plum	9.10	.879		8.90	43		51.90	74.50
5348	2"	Q-1	16.60	.964		9.55	42.50		52.05	74
5349	3"		10.40	1.538		24.50	67.50		92	128

22 11 13.76 Pipe Fittings, Plastic		Crew	Daily Output	Labor-Hours	Unit	2009 Bare Costs Material	Labor	Equipment	Total	Total Incl O&P
5350	4"	Q-1	8.25	1.939	Ea.	50	85		135	183
5354	2" x 1-1/2"		16.80	.952		8.75	42		50.75	72
5355	3" x 2"		10.60	1.509		18.35	66		84.35	120
5356	4" x 3"		8.45	1.893		39.50	83		122.50	169
5357	6" x 4"		7.25	2.207		100	97		197	255
5374	Coupling, 1-1/4"	1 Plum	20.20	.396		2.48	19.30		21.78	31.50
5376	1-1/2"	"	18.20	.440		.66	21.50		22.16	32.50
5378	2"	Q-1	33.10	.483		.89	21		21.89	33
5380	3"		20.80	.769		3.25	34		37.25	54
5390	4"		16.50	.970		5.60	42.50		48.10	70
5400	6"		10.10	1.584		18.25	69.50		87.75	124
5402	8"	Q-2	9.30	2.581		45	117		162	225
5404	2" x 1-1/2"	Q-1	33.30	.480		1.97	21		22.97	33.50
5406	3" x 1-1/2"		21	.762		6.25	33.50		39.75	57
5408	4" x 3"		16.70	.958		10.10	42		52.10	74
5410	Reducer bushing, 2" x 1-1/4"		36.50	.438		2.51	19.25		21.76	32
5411	2" x 1-1/2"		36.40	.440		1.21	19.30		20.51	30.50
5412	3" x 1-1/2"		27.30	.586		6.05	25.50		31.55	45
5413	3" x 2"		27.10	.590		2.96	26		28.96	42.50
5414	4" x 2"		18.20	.879		11.25	38.50		49.75	70.50
5415	4" x 3"		16.70	.958		5.75	42		47.75	69.50
5416	6" x 4"		11.10	1.441		30	63.50		93.50	128
5418	8" x 6"	Q-2	10.20	2.353		73	107		180	242
5425	Closet flange 4"	Q-1	32	.500		6.55	22		28.55	40.50
5426	4" x 3"	"	34	.471		5.80	20.50		26.30	37.50
5450	Solvent cement for PVC, industrial grade, per quart				Qt.	17.25			17.25	19
5500	CPVC, Schedule 80, threaded joints									
5540	90° Elbow, 1/4"	1 Plum	32	.250	Ea.	10.20	12.20		22.40	29.50
5560	1/2"		30.30	.264		5.90	12.85		18.75	26
5570	3/4"		26	.308		8.85	15		23.85	32.50
5580	1"		22.70	.352		12.40	17.20		29.60	39.50
5590	1-1/4"		20.20	.396		24	19.30		43.30	55.50
5600	1-1/2"		18.20	.440		26	21.50		47.50	60.50
5610	2"	Q-1	33.10	.483		34.50	21		55.50	70
5620	2-1/2"		24.20	.661		107	29		136	162
5630	3"		20.80	.769		115	34		149	177
5640	4"		16.50	.970		180	42.50		222.50	262
5650	6"		10.10	1.584		212	69.50		281.50	335
5700	45° Elbow same as 90° Elbow									
5850	Tee, 1/4"	1 Plum	22	.364	Ea.	19.80	17.75		37.55	48.50
5870	1/2"		20.20	.396		19.80	19.30		39.10	51
5880	3/4"		17.30	.462		28.50	22.50		51	65.50
5890	1"		15.20	.526		30.50	25.50		56	72.50
5900	1-1/4"		13.50	.593		31	29		60	77.50
5910	1-1/2"		12.10	.661		32.50	32		64.50	84
5920	2"	Q-1	20	.800		36	35		71	92
5930	2-1/2"		16.20	.988		177	43.50		220.50	260
5940	3"		13.90	1.151		206	50.50		256.50	300
5950	4"		11	1.455		485	64		549	630
5960	6"		6.70	2.388		580	105		685	795
6000	Coupling, 1/4"	1 Plum	32	.250		13	12.20		25.20	32.50
6020	1/2"		30.30	.264		13	12.85		25.85	33.50
6030	3/4"		26	.308		15.60	15		30.60	39.50

22 11 13.76 Pipe Fittings, Plastic		Crew	Daily Output	Labor-Hours	Unit	Material	2009 Bare Costs Labor	Equipment	Total	Total Incl O&P
6040	1"	1 Plum	22.70	.352	Ea.	17.25	17.20		34.45	45
6050	1-1/4"		20.20	.396		21	19.30		40.30	52
6060	1-1/2"		18.20	.440		22.50	21.50		44	56.50
6070	2"	Q-1	33.10	.483		26.50	21		47.50	61
6080	2-1/2"		24.20	.661		47	29		76	95.50
6090	3"		20.80	.769		55	34		89	111
6100	4"		16.50	.970		112	42.50		154.50	187
6110	6"		10.10	1.584		153	69.50		222.50	272
6120	8"	Q-2	9.30	2.581		325	117		442	535
6200	CTS, 100 psi at 180° F, hot and cold water									
6230	90° Elbow, 1/2"	1 Plum	20	.400	Ea.	.08	19.50		19.58	29.50
6250	3/4"		19	.421		.17	20.50		20.67	31
6251	1"		16	.500		.58	24.50		25.08	37
6252	1-1/4"		15	.533		.99	26		26.99	40
6253	1-1/2"		14	.571		1.59	28		29.59	44
6254	2"	Q-1	23	.696		3.43	30.50		33.93	50
6260	45° Elbow, 1/2"	1 Plum	20	.400		.27	19.50		19.77	30
6280	3/4"		19	.421		.27	20.50		20.77	31.50
6281	1"		16	.500		.53	24.50		25.03	37
6282	1-1/4"		15	.533		1.04	26		27.04	40
6283	1-1/2"		14	.571		1.65	28		29.65	44
6284	2"	Q-1	23	.696		3.48	30.50		33.98	50
6290	Tee, 1/2"	1 Plum	13	.615		.14	30		30.14	45
6310	3/4"		12	.667		.26	32.50		32.76	49.50
6311	1"		11	.727		1.20	35.50		36.70	54.50
6312	1-1/4"		10	.800		1.60	39		40.60	60.50
6313	1-1/2"		10	.800		2.60	39		41.60	61.50
6314	2"	Q-1	17	.941		4.22	41.50		45.72	66.50
6320	Coupling, 1/2"	1 Plum	22	.364		.10	17.75		17.85	26.50
6340	3/4"		21	.381		.14	18.55		18.69	28
6341	1"		18	.444		.50	21.50		22	33
6342	1-1/4"		17	.471		.62	23		23.62	35
6343	1-1/2"		16	.500		.88	24.50		25.38	37.50
6344	2"	Q-1	28	.571		1.75	25		26.75	39.50
6360	Solvent cement for CPVC, commercial grade, per quart				Qt.	17			17	18.70
7550	Union, schedule 40, socket joints, 1/2"	1 Plum	19	.421	Ea.	3.97	20.50		24.47	35.50
7560	3/4"		18	.444		4.52	21.50		26.02	37.50
7570	1"		15	.533		4.65	26		30.65	44
7580	1-1/4"		14	.571		14	28		42	57.50
7590	1-1/2"		13	.615		15	30		45	61.50
7600	2"	Q-1	20	.800		21.50	35		56.50	76.50
7992	Polybutyl/polyethyl pipe, for copper fittings see Div. 22 11 13.25 7000									
8000	Compression type, PVC, 160 psi cold water									
8010	Coupling, 3/4" CTS	1 Plum	21	.381	Ea.	2.76	18.55		21.31	31
8020	1" CTS		18	.444		3.43	21.50		24.93	36.50
8030	1-1/4" CTS		17	.471		4.83	23		27.83	40
8040	1-1/2" CTS		16	.500		6.60	24.50		31.10	44
8050	2" CTS		15	.533		9.15	26		35.15	49
8060	Female adapter, 3/4" FPT x 3/4" CTS		23	.348		3.12	16.95		20.07	29
8070	3/4" FPT x 1" CTS		21	.381		3.80	18.55		22.35	32
8080	1" FPT x 1" CTS		20	.400		4.98	19.50		24.48	35
8090	1-1/4" FPT x 1-1/4" CTS		18	.444		6.50	21.50		28	39.50
8100	1-1/2" FPT x 1-1/2" CTS		16	.500		7.45	24.50		31.95	44.50

22 11 13.76 Pipe Fittings, Plastic		Crew	Daily Output	Labor-Hours	Unit	Material	Labor	Equipment	Total	Total Incl O&P
						2009 Bare Costs				
8110	2" FPT x 2" CTS	1 Plum	13	.615	Ea.	9.05	30		39.05	55
8130	Male adapter, 3/4" MPT x 3/4" CTS		23	.348		3.02	16.95		19.97	29
8140	3/4" MPT x 1" CTS		21	.381		3.25	18.55		21.80	31.50
8150	1" MPT x 1" CTS		20	.400		4.22	19.50		23.72	34
8160	1-1/4" MPT x 1-1/4" CTS		18	.444		5.70	21.50		27.20	39
8170	1-1/2" MPT x 1-1/2" CTS		16	.500		6.85	24.50		31.35	44
8180	2" MPT x 2" CTS		13	.615		8.90	30		38.90	55
8200	Spigot adapter, 3/4" IPS x 3/4" CTS		23	.348		2.15	16.95		19.10	28
8210	3/4" IPS x 1" CTS		21	.381		2.76	18.55		21.31	31
8220	1" IPS x 1" CTS		20	.400		2.76	19.50		22.26	32.50
8230	1-1/4" IPS x 1-1/4" CTS		18	.444		4.30	21.50		25.80	37
8240	1-1/2" IPS x 1-1/2" CTS		16	.500		4.33	24.50		28.83	41.50
8250	2" IPS x 2" CTS	↓	13	.615	↓	5.40	30		35.40	51
8270	Price includes insert stiffeners									
8280	250 psi is same price as 160 psi									
8300	Insert type, nylon, 160 & 250 psi, cold water									
8310	Clamp ring stainless steel, 3/4" IPS	1 Plum	115	.070	Ea.	1.95	3.39		5.34	7.25
8320	1" IPS		107	.075		1.61	3.65		5.26	7.20
8330	1-1/4" IPS		101	.079		2	3.86		5.86	8
8340	1-1/2" IPS		95	.084		2.18	4.11		6.29	8.55
8350	2" IPS		85	.094		2.50	4.59		7.09	9.65
8370	Coupling, 3/4" IPS		22	.364		.74	17.75		18.49	27.50
8390	1-1/4" IPS		18	.444		1.15	21.50		22.65	34
8400	1-1/2" IPS		17	.471		1.36	23		24.36	36
8410	2" IPS		16	.500		2.57	24.50		27.07	39.50
8430	Elbow, 90°, 3/4" IPS		22	.364		1.49	17.75		19.24	28
8440	1" IPS		19	.421		1.64	20.50		22.14	33
8450	1-1/4" IPS		18	.444		1.83	21.50		23.33	34.50
8460	1-1/2" IPS		17	.471		2.17	23		25.17	37
8470	2" IPS		16	.500		3.02	24.50		27.52	40
8490	Male adapter, 3/4" IPS x 3/4" MPT		25	.320		.74	15.60		16.34	24.50
8500	1" IPS x 1" MPT		21	.381		.77	18.55		19.32	29
8510	1-1/4" IPS x 1-1/4" MPT		20	.400		1.22	19.50		20.72	31
8520	1-1/2" IPS x 1-1/2" MPT		18	.444		1.36	21.50		22.86	34
8530	2" IPS x 2" MPT		15	.533		2.60	26		28.60	42
8550	Tee, 3/4" IPS		14	.571		1.44	28		29.44	43.50
8560	1" IPS		13	.615		1.87	30		31.87	47
8570	1-1/4" IPS		12	.667		2.93	32.50		35.43	52
8580	1-1/2" IPS		11	.727		3.32	35.50		38.82	56.50
8590	2" IPS	↓	10	.800	↓	6.55	39		45.55	65.50
8610	Insert type, PVC, 100 psi @ 180° F, hot & cold water									
8620	Coupler, male, 3/8" CTS x 3/8" MPT	1 Plum	29	.276	Ea.	.61	13.45		14.06	20.50
8630	3/8" CTS x 1/2" MPT		28	.286		.61	13.95		14.56	21.50
8640	1/2" CTS x 1/2" MPT		27	.296		.61	14.45		15.06	22
8650	1/2" CTS x 3/4" MPT		26	.308		1.93	15		16.93	24.50
8660	3/4" CTS x 1/2" MPT		25	.320		1.73	15.60		17.33	25.50
8670	3/4" CTS x 3/4" MPT		25	.320		.74	15.60		16.34	24.50
8700	Coupling, 3/8" CTS x 1/2" CTS		25	.320		2.86	15.60		18.46	26.50
8710	1/2" CTS		23	.348		3.41	16.95		20.36	29.50
8730	3/4" CTS		22	.364		5.40	17.75		23.15	32.50
8750	Elbow 90°, 3/8" CTS		25	.320		3.90	15.60		19.50	28
8760	1/2" CTS		23	.348		5.60	16.95		22.55	31.50
8770	3/4" CTS	↓	22	.364	↓	6.25	17.75		24	33.50

22 11 13 – Facility Water Distribution Piping

22 11 13.76 Pipe Fittings, Plastic		Crew	Daily Output	Labor-Hours	Unit	2009 Bare Costs Material	Labor	Equipment	Total	Total Incl O&P
8800	Rings, crimp, copper, 3/8" CTS	1 Plum	120	.067	Ea.	.16	3.25		3.41	5.05
8810	1/2" CTS		117	.068		.14	3.33		3.47	5.15
8820	3/4" CTS		115	.070		.19	3.39		3.58	5.30
8850	Reducer tee, 3/8" x 3/8" x 1/2" CTS		17	.471		1.18	23		24.18	36
8860	1/2" x 3/8" x 1/2" CTS		15	.533		1.18	26		27.18	40.50
8870	3/4" x 1/2" x 1/2" CTS		14	.571		1.82	28		29.82	44
8890	3/4" x 3/4" x 1/2" CTS		14	.571		1.82	28		29.82	44
8900	3/4" x 1/2" x 3/8" CTS		14	.571		1.82	28		29.82	44
8930	Tee, 3/8" CTS		17	.471		1.40	23		24.40	36
8940	1/2" CTS		15	.533		1.58	26		27.58	40.50
8950	3/4" CTS	↓	14	.571	↓	1.71	28		29.71	44
8960	Copper rings included in fitting price									
9000	Flare type, assembled, acetal, hot & cold water									
9010	Coupling, 1/4" & 3/8" CTS	1 Plum	24	.333	Ea.	2.90	16.25		19.15	27.50
9020	1/2" CTS		22	.364		3.32	17.75		21.07	30
9030	3/4" CTS		21	.381		4.92	18.55		23.47	33.50
9040	1" CTS		18	.444		6.25	21.50		27.75	39.50
9050	Elbow 90°, 1/4" CTS		26	.308		3.23	15		18.23	26
9060	3/8" CTS		24	.333		3.46	16.25		19.71	28.50
9070	1/2" CTS		22	.364		4.10	17.75		21.85	31
9080	3/4" CTS		21	.381		6.25	18.55		24.80	35
9090	1" CTS		18	.444		7.90	21.50		29.40	41
9110	Tee ,1/4" & 3/8" CTS		15	.533		3.57	26		29.57	43
9120	1/2" CTS		14	.571		4.69	28		32.69	47
9130	3/4" CTS		13	.615		7.15	30		37.15	53
9140	1" CTS	↓	12	.667	↓	9.55	32.50		42.05	59.50
9552	For plastic hangers see Div. 22 05 29.10 8000									
9562	For copper/brass fittings see Div. 22 11 13.25 7000									

22 11 13.78 Pipe, High Density Polyethylene Plastic (HDPE)		Crew	Daily Output	Labor-Hours	Unit	Material	Labor	Equipment	Total	Total Incl O&P
0010	**PIPE, HIGH DENSITY POLYETHYLENE PLASTIC (HDPE)**									
0020	Not incl. hangers, trenching, backfill, hoisting or digging equipment.									
0030	Standard length is 40', add a weld for each joint									
0040	Single wall									
0050	Straight									
0054	1" diameter DR 11				L.F.	.53			.53	.58
0058	1-1/2" diameter DR 11					.66			.66	.73
0062	2" diameter DR 11					1.10			1.10	1.21
0066	3" diameter DR 11					1.32			1.32	1.45
0070	3" diameter DR 17					1.05			1.05	1.16
0074	4" diameter DR 11					2.20			2.20	2.42
0078	4" diameter DR 17					2.20			2.20	2.42
0082	6" diameter DR 11					5.50			5.50	6.05
0086	6" diameter DR 17					3.63			3.63	3.99
0090	8" diameter DR 11					9.15			9.15	10.05
0094	8" diameter DR 26					4.29			4.29	4.72
0098	10" diameter DR 11					14.30			14.30	15.75
0102	10" diameter DR 26					6.60			6.60	7.25
0106	12" diameter DR 11					21			21	23
0110	12" diameter DR 26					9.90			9.90	10.90
0114	16" diameter DR 11					32			32	35
0118	16" diameter DR 26					14.30			14.30	15.75
0122	18" diameter DR 11				↓	40.50			40.50	45

22 11 13.78 Pipe, High Density Polyethylene Plastic (HDPE)		Crew	Daily Output	Labor-Hours	Unit	Material	Labor	Equipment	Total	Total Incl O&P
						2009 Bare Costs				
0126	18" diameter DR 26				L.F.	18.70			18.70	20.50
0130	20" diameter DR 11					49.50			49.50	54.50
0134	20" diameter DR 26					22			22	24
0138	22" diameter DR 11					60.50			60.50	66.50
0142	22" diameter DR 26					27.50			27.50	30.50
0146	24" diameter DR 11					71.50			71.50	78.50
0150	24" diameter DR 26					32			32	35
0154	28" diameter DR 17					66			66	72.50
0158	28" diameter DR 26					44			44	48.50
0162	30" diameter DR 21					61.50			61.50	68
0166	30" diameter DR 26					50.50			50.50	55.50
0170	36" diameter DR 26					50.50			50.50	55.50
0174	42" diameter DR 26					98			98	108
0178	48" diameter DR 26					129			129	142
0182	54" diameter DR 26				↓	162			162	178
0300	90° Elbow									
0304	1" diameter DR 11				Ea.	4.40			4.40	4.84
0308	1-1/2" diameter DR 11					5.50			5.50	6.05
0312	2" diameter DR 11					5.50			5.50	6.05
0316	3" diameter DR 11					11			11	12.10
0320	3" diameter DR 17					11			11	12.10
0324	4" diameter DR 11					15.40			15.40	16.95
0328	4" diameter DR 17					15.40			15.40	16.95
0332	6" diameter DR 11					35			35	38.50
0336	6" diameter DR 17					35			35	38.50
0340	8" diameter DR 11					87			87	95.50
0344	8" diameter DR 26					77			77	84.50
0348	10" diameter DR 11					325			325	360
0352	10" diameter DR 26					300			300	330
0356	12" diameter DR 11					345			345	375
0360	12" diameter DR 26					310			310	340
0364	16" diameter DR 11					405			405	445
0368	16" diameter DR 26					400			400	440
0372	18" diameter DR 11					510			510	560
0376	18" diameter DR 26					480			480	530
0380	20" diameter DR 11					600			600	660
0384	20" diameter DR 26					580			580	640
0388	22" diameter DR 11					620			620	680
0392	22" diameter DR 26					600			600	660
0396	24" diameter DR 11					700			700	770
0400	24" diameter DR 26					680			680	750
0404	28" diameter DR 17					850			850	935
0408	28" diameter DR 26					800			800	880
0412	30" diameter DR 17					1,200			1,200	1,325
0416	30" diameter DR 26					1,100			1,100	1,200
0420	36" diameter DR 26					1,400			1,400	1,550
0424	42" diameter DR 26					1,800			1,800	1,975
0428	48" diameter DR 26					2,100			2,100	2,300
0432	54" diameter DR 26				↓	5,000			5,000	5,500
0500	45° Elbow									
0512	2" diameter DR 11				Ea.	4.40			4.40	4.84
0516	3" diameter DR 11					11			11	12.10
0520	3" diameter DR 17				↓	11			11	12.10

22 11 13.78 Pipe, High Density Polyethylene Plastic (HDPE)		Crew	Daily Output	Labor-Hours	Unit	Material	2009 Bare Costs Labor	Equipment	Total	Total Incl O&P
0524	4" diameter DR 11				Ea.	15.40			15.40	16.95
0528	4" diameter DR 17					15.40			15.40	16.95
0532	6" diameter DR 11					35			35	38.50
0536	6" diameter DR 17					35			35	38.50
0540	8" diameter DR 11					87			87	95.50
0544	8" diameter DR 26					50.50			50.50	55.50
0548	10" diameter DR 11					325			325	360
0552	10" diameter DR 26					300			300	330
0556	12" diameter DR 11					345			345	375
0560	12" diameter DR 26					310			310	340
0564	16" diameter DR 11					180			180	198
0568	16" diameter DR 26					170			170	187
0572	18" diameter DR 11					191			191	210
0576	18" diameter DR 26					175			175	193
0580	20" diameter DR 11					290			290	320
0584	20" diameter DR 26					275			275	305
0588	22" diameter DR 11					400			400	440
0592	22" diameter DR 26					380			380	420
0596	24" diameter DR 11					490			490	540
0600	24" diameter DR 26					470			470	515
0604	28" diameter DR 17					555			555	610
0608	28" diameter DR 26					540			540	595
0612	30" diameter DR 17					690			690	760
0616	30" diameter DR 26					670			670	735
0620	36" diameter DR 26					850			850	935
0624	42" diameter DR 26					1,100			1,100	1,200
0628	48" diameter DR 26					1,200			1,200	1,300
0632	54" diameter DR 26				↓	1,600			1,600	1,750
0700	Tee									
0704	1" diameter DR 11				Ea.	5.75			5.75	6.35
0708	1-1/2" diameter DR 11					8.05			8.05	8.85
0712	2" diameter DR 11					6.90			6.90	7.60
0716	3" diameter DR 11					12.65			12.65	13.90
0720	3" diameter DR 17					12.65			12.65	13.90
0724	4" diameter DR 11					18.40			18.40	20
0728	4" diameter DR 17					18.40			18.40	20
0732	6" diameter DR 11					46			46	50.50
0736	6" diameter DR 17					46			46	50.50
0740	8" diameter DR 11					114			114	125
0744	8" diameter DR 17					114			114	125
0748	10" diameter DR 11					340			340	375
0752	10" diameter DR 17					340			340	375
0756	12" diameter DR 11					455			455	500
0760	12" diameter DR 17					455			455	500
0764	16" diameter DR 11					239			239	263
0768	16" diameter DR 17					197			197	217
0772	18" diameter DR 11					335			335	370
0776	18" diameter DR 17					277			277	305
0780	20" diameter DR 11					410			410	455
0784	20" diameter DR 17					335			335	370
0788	22" diameter DR 11					530			530	580
0792	22" diameter DR 17					415			415	460
0796	24" diameter DR 11				↓	655			655	720

22 11 13.78 Pipe, High Density Polyethylene Plastic (HDPE)		Crew	Daily Output	Labor-Hours	Unit	Material	Labor	Equipment	Total	Total Incl O&P
						2009 Bare Costs				
0800	24" diameter DR 17				Ea.	550			550	605
0804	28" diameter DR 17					1,075			1,075	1,175
0812	30" diameter DR 17					1,225			1,225	1,350
0820	36" diameter DR 17					2,025			2,025	2,225
0824	42" diameter DR 26					2,250			2,250	2,475
0828	48" diameter DR 26				↓	2,425			2,425	2,650
1000	Flange adptr, w/back-up ring and 1/2 cost of plated bolt set									
1004	1" diameter DR 11				Ea.	22			22	24
1008	1-1/2" diameter DR 11					22			22	24
1012	2" diameter DR 11					13.80			13.80	15.20
1016	3" diameter DR 11					16.10			16.10	17.70
1020	3" diameter DR 17					16.10			16.10	17.70
1024	4" diameter DR 11					22			22	24
1028	4" diameter DR 17					22			22	24
1032	6" diameter DR 11					31			31	34
1036	6" diameter DR 17					31			31	34
1040	8" diameter DR 11					45			45	49.50
1044	8" diameter DR 26					45			45	49.50
1048	10" diameter DR 11					71.50			71.50	78.50
1052	10" diameter DR 26					71.50			71.50	78.50
1056	12" diameter DR 11					105			105	115
1060	12" diameter DR 26					105			105	115
1064	16" diameter DR 11					224			224	247
1068	16" diameter DR 26					224			224	247
1072	18" diameter DR 11					291			291	320
1076	18" diameter DR 26					291			291	320
1080	20" diameter DR 11					405			405	445
1084	20" diameter DR 26					405			405	445
1088	22" diameter DR 11					440			440	480
1092	22" diameter DR 26					435			435	480
1096	24" diameter DR 17					475			475	525
1100	24" diameter DR 32.5					475			475	525
1104	28" diameter DR 15.5					645			645	710
1108	28" diameter DR 32.5					645			645	710
1112	30" diameter DR 11					750			750	820
1116	30" diameter DR 21					750			750	820
1120	36" diameter DR 26					815			815	900
1124	42" diameter DR 26					930			930	1,025
1128	48" diameter DR 26					1,150			1,150	1,250
1132	54" diameter DR 26				↓	1,375			1,375	1,525
1200	Reducer									
1208	2" x 1-1/2" diameter DR 11				Ea.	6.60			6.60	7.25
1212	3" x 2" diameter DR 11					6.60			6.60	7.25
1216	4" x 2" diameter DR 11					7.70			7.70	8.45
1220	4" x 3" diameter DR 11					9.90			9.90	10.90
1224	6" x 4" diameter DR 11					23			23	25.50
1228	8" x 6" diameter DR 11					35			35	38.50
1232	10" x 8" diameter DR 11					60.50			60.50	66.50
1236	12" x 8" diameter DR 11					99			99	109
1240	12" x 10" diameter DR 11					79			79	87
1244	14" x 12" diameter DR 11					88			88	97
1248	16" x 14" diameter DR 11					112			112	123
1252	18" x 16" diameter DR 11				↓	136			136	150

22 11 Facility Water Distribution

22 11 13 – Facility Water Distribution Piping

22 11 13.78	Pipe, High Density Polyethylene Plastic (HDPE)	Crew	Daily Output	Labor-Hours	Unit	Material	2009 Bare Costs Labor	Equipment	Total	Total Incl O&P
1256	20″ x 18″ diameter DR 11				Ea.	271			271	298
1260	22″ x 20″ diameter DR 11					330			330	365
1264	24″ x 22″ diameter DR 11					375			375	410
1268	26″ x 24″ diameter DR 11					440			440	485
1272	28″ x 24″ diameter DR 11					560			560	615
1276	32″ x 28″ diameter DR 17					725			725	800
1280	36″ x 32″ diameter DR 17					990			990	1,100
4000	Welding labor per joint, not including welding machine									
4010	Pipe joint size (cost based on thickest wall for each dia.)									
4030	1″ pipe size	4 Skwk	273	.117	Ea.		4.79		4.79	7.40
4040	1-1/2″ pipe size		175	.183			7.45		7.45	11.55
4050	2″ pipe size		128	.250			10.20		10.20	15.80
4060	3″ pipe size		100	.320			13.05		13.05	20
4070	4″ pipe size		77	.416			17		17	26.50
4080	6″ pipe size	5 Skwk	63	.635			26		26	40
4090	8″ pipe size		48	.833			34		34	52.50
4100	10″ pipe size		40	1			41		41	63.50
4110	12″ pipe size	6 Skwk	41	1.171			48		48	74
4120	16″ pipe size		34	1.412			57.50		57.50	89.50
4130	18″ pipe size		32	1.500			61.50		61.50	95
4140	20″ pipe size	8 Skwk	37	1.730			70.50		70.50	109
4150	22″ pipe size		35	1.829			74.50		74.50	116
4160	24″ pipe size		34	1.882			77		77	119
4170	28″ pipe size		33	1.939			79		79	123
4180	30″ pipe size		32	2			81.50		81.50	127
4190	36″ pipe size		31	2.065			84.50		84.50	131
4200	42″ pipe size		30	2.133			87		87	135
4210	48″ pipe size	9 Skwk	33	2.182			89		89	138
4220	54″ pipe size	″	31	2.323			95		95	147
4300	Note: Cost for set up each time welder is moved.									
4301	Add 50% of a weld cost									
4310	Welder usually remains stationary with pipe moved through it.									
4340	Weld machine, rental per day based on dia. capacity									
4350	1″ thru 2″ diameter				Ea.			40.25	40.25	44.25
4360	3″ thru 4″ diameter							46	46	50.50
4370	6″ thru 8″ diameter							103	103	113
4380	10″ thru 12″ diameter							178	178	196
4390	16″ thru 18″ diameter							259	259	285
4400	20″ thru 24″ diameter							500	500	550
4410	28″ thru 32″ diameter							545	545	600
4420	36″ diameter							570	570	627
4430	42″ thru 54″ diameter							890	890	980
5000	Dual wall contained pipe									
5040	Straight									
5054	1″ DR 11 x 3″ DR 11				L.F.	8.10			8.10	8.90
5058	1″ DR 11 x 4″ DR 11					8.65			8.65	9.50
5062	1-1/2″ DR 11 x 4″ DR 17					8.90			8.90	9.80
5066	2″ DR 11 x 4″ DR 17					9.60			9.60	10.55
5070	2″ DR 11 x 6″ DR 17					14.35			14.35	15.75
5074	3″ DR 11 x 6″ DR 17					16			16	17.60
5078	3″ DR 11 x 6″ DR 26					13.75			13.75	15.10
5086	3″ DR 17 x 8″ DR 17					20.50			20.50	22.50
5090	4″ DR 11 x 8″ DR 17					24			24	26

22 11 13.78 Pipe, High Density Polyethylene Plastic (HDPE)		Crew	Daily Output	Labor-Hours	Unit	Material	Labor	Equipment	Total	Total Incl O&P
						2009 Bare Costs				
5094	4" DR 17 x 8" DR 26				L.F.	18.80			18.80	20.50
5098	6" DR 11 x 10" DR 17					37			37	40.50
5102	6" DR 17 x 10" DR 26					27.50			27.50	30.50
5106	6" DR 26 x 10" DR 26					26			26	29
5110	8" DR 17 x 12" DR 26					39			39	43
5114	8" DR 26 x 12" DR 32.5					33			33	36.50
5118	10" DR 17 x 14" DR 26					54.50			54.50	60
5122	10" DR 17 x 16" DR 26					60			60	66
5126	10" DR 26 x 16" DR 26					54			54	59.50
5130	12" DR 26 x 16" DR 26					61.50			61.50	68
5134	12" DR 17 x 18" DR 26					78.50			78.50	86.50
5138	12" DR 26 x 18" DR 26					70.50			70.50	77.50
5142	14" DR 26 x 20" DR 32.5					77.50			77.50	85
5146	16" DR 26 x 22" DR 32.5					117			117	128
5150	18" DR 26 x 24" DR 32.5					107			107	118
5154	20" DR 32.5 x 28" DR 32.5					139			139	153
5158	22" DR 32.5 x 30" DR 32.5					139			139	153
5162	24" DR 32.5 x 32" DR 32.5					177			177	195
5166	36" DR 32.5 x 42" DR 32.5					299			299	330
5300	Force transfer coupling									
5354	1" DR 11 x 3" DR 11				Ea.	248			248	273
5358	1" DR 11 x 4" DR 17					261			261	287
5362	1-1/2" DR 11 x 4" DR 17					273			273	300
5366	2" DR 11 x 4" DR 17					285			285	315
5370	2" DR 11 x 6" DR 17					415			415	455
5374	3" DR 11 x 6" DR 17					415			415	455
5378	3" DR 11 x 6" DR 26					415			415	455
5382	3" DR 11 x 8" DR 11					440			440	480
5386	3" DR 11 x 8" DR 17					440			440	480
5390	4" DR 11 x 8" DR 17					440			440	480
5394	4" DR 17 x 8" DR 26					385			385	425
5398	6" DR 11 x 10" DR 17					555			555	610
5402	6" DR 17 x 10" DR 26					460			460	505
5406	6" DR 26 x 10" DR 26					460			460	505
5410	8" DR 17 x 12" DR 26					610			610	670
5414	8" DR 26 x 12" DR 32.5					490			490	540
5418	10" DR 17 x 14" DR 26					795			795	875
5422	10" DR 17 x 16" DR 26					795			795	875
5426	10" DR 26 x 16" DR 26					795			795	875
5430	12" DR 26 x 16" DR 26					1,000			1,000	1,100
5434	12" DR 17 x 18" DR 26					1,025			1,025	1,125
5438	12" DR 26 x 18" DR 26					1,025			1,025	1,125
5442	14" DR 26 x 20" DR 32.5					1,075			1,075	1,200
5446	16" DR 26 x 22" DR 32.5					1,125			1,125	1,225
5450	18" DR 26 x 24" DR 32.5					1,200			1,200	1,325
5454	20" DR 32.5 x 28" DR 32.5					1,475			1,475	1,625
5458	22" DR 32.5 x 30" DR 32.5					1,625			1,625	1,775
5462	24" DR 32.5 x 32" DR 32.5					1,750			1,750	1,925
5466	36" DR 32.5 x 42" DR 32.5					2,925			2,925	3,200
5600	90° Elbow									
5654	1" DR 11 x 3" DR 11				Ea.	205			205	226
5658	1" DR 11 x 4" DR 17					196			196	216
5662	1-1/2" DR 11 x 4" DR 17					215			215	236

22 11 13 – Facility Water Distribution Piping

22 11 13.78 Pipe, High Density Polyethylene Plastic (HDPE)		Crew	Daily Output	Labor-Hours	Unit	Material	Labor	Equipment	Total	Total Incl O&P
5666	2″ DR 11 x 4″ DR 17				Ea.	227			227	250
5670	2″ DR 11 x 6″ DR 17					289			289	320
5674	3″ DR 11 x 6″ DR 17					330			330	365
5678	3″ DR 17 x 6″ DR 26					264			264	290
5682	3″ DR 17 x 8″ DR 11					535			535	590
5686	3″ DR 17 x 8″ DR 17					415			415	455
5690	4″ DR 11 x 8″ DR 17					490			490	535
5694	4″ DR 17 x 8″ DR 26					390			390	425
5698	6″ DR 11 x 10″ DR 17					650			650	715
5702	6″ DR 17 x 10″ DR 26					500			500	550
5706	6″ DR 26 x 10″ DR 26					465			465	515
5710	8″ DR 17 x 12″ DR 26					740			740	815
5714	8″ DR 26 x 12″ DR 32.5					685			685	755
5718	10″ DR 17 x 14″ DR 26					1,175			1,175	1,275
5722	10″ DR 17 x 16″ DR 26					1,125			1,125	1,225
5726	10″ DR 26 x 16″ DR 26					1,025			1,025	1,125
5730	12″ DR 26 x 16″ DR 26					1,150			1,150	1,250
5734	12″ DR 17 x 18″ DR 26					1,375			1,375	1,500
5738	12″ DR 26 x 18″ DR 26					1,250			1,250	1,375
5742	14″ DR 26 x 20″ DR 32.5					1,650			1,650	1,825
5746	16″ DR 26 x 22″ DR 32.5					1,675			1,675	1,825
5750	18″ DR 26 x 24″ DR 32.5					2,050			2,050	2,250
5754	20″ DR 32.5 x 28″ DR 32.5					2,350			2,350	2,600
5758	22″ DR 32.5 x 30″ DR 32.5					3,200			3,200	3,525
5762	24″ DR 32.5 x 32″ DR 32.5					3,125			3,125	3,425
5766	36″ DR 32.5 x 42″ DR 32.5				↓	5,075			5,075	5,575
5800	45° Elbow									
5804	1″ DR 11 x 3″ DR 11				Ea.	123			123	135
5808	1″ DR 11 x 4″ DR 17					122			122	134
5812	1-1/2″ DR 11 x 4″ DR 17					131			131	144
5816	2″ DR 11 x 4″ DR 17					143			143	157
5820	2″ DR 11 x 6″ DR 17					179			179	197
5824	3″ DR 11 x 6″ DR 17					199			199	219
5828	3″ DR 17 x 6″ DR 26					165			165	181
5832	3″ DR 17 x 8″ DR 11					305			305	335
5836	3″ DR 17 x 8″ DR 17					244			244	268
5840	4″ DR 11 x 8″ DR 17					285			285	315
5844	4″ DR 17 x 8″ DR 26					232			232	255
5848	6″ DR 11 x 10″ DR 17					375			375	415
5852	6″ DR 17 x 10″ DR 26					295			295	325
5856	6″ DR 26 x 10″ DR 26					278			278	305
5860	8″ DR 17 x 12″ DR 26					450			450	495
5864	8″ DR 26 x 12″ DR 32.5					420			420	460
5868	10″ DR 17 x 14″ DR 26					700			700	770
5872	10″ DR 17 x 16″ DR 26					685			685	755
5876	10″ DR 26 x 16″ DR 26					645			645	710
5880	12″ DR 26 x 16″ DR 26					725			725	800
5884	12″ DR 17 x 18″ DR 26					855			855	940
5888	12″ DR 26 x 18″ DR 26					795			795	875
5892	14″ DR 26 x 20″ DR 32.5					1,025			1,025	1,125
5896	16″ DR 26 x 22″ DR 32.5					1,075			1,075	1,175
5900	18″ DR 26 x 24″ DR 32.5					1,275			1,275	1,400
5904	20″ DR 32.5 x 28″ DR 32.5				↓	1,550			1,550	1,725

Material, Labor, Equipment and Total are 2009 Bare Costs.

22 11 13.78 Pipe, High Density Polyethylene Plastic (HDPE)		Crew	Daily Output	Labor-Hours	Unit	2009 Bare Costs Material	Labor	Equipment	Total	Total Incl O&P
5908	22" DR 32.5 x 30" DR 32.5				Ea.	1,975			1,975	2,175
5912	24" DR 32.5 x 32" DR 32.5					1,975			1,975	2,150
5916	36" DR 32.5 x 42" DR 32.5					3,050			3,050	3,375
6000	Access port with 4" riser									
6050	1" DR 11 x 4" DR 17				Ea.	268			268	295
6054	1-1/2" DR 11 x 4" DR 17					274			274	300
6058	2" DR 11 x 6" DR 17					335			335	365
6062	3" DR 11 x 6" DR 17					335			335	370
6066	3" DR 17 x 6" DR 26					330			330	360
6070	3" DR 17 x 8" DR 11					375			375	415
6074	3" DR 17 x 8" DR 17					360			360	395
6078	4" DR 11 x 8" DR 17					370			370	410
6082	4" DR 17 x 8" DR 26					355			355	390
6086	6" DR 11 x 10" DR 17					465			465	510
6090	6" DR 17 x 10" DR 26					430			430	470
6094	6" DR 26 x 10" DR 26					420			420	460
6098	8" DR 17 x 12" DR 26					450			450	495
6102	8" DR 26 x 12" DR 32.5					430			430	470
6200	End termination with vent plug									
6204	1" DR 11 x 3" DR 11				Ea.	222			222	244
6208	1" DR 11 x 4" DR 17					237			237	261
6212	1-1/2" DR 11 x 4" DR 17					237			237	261
6216	2" DR 11 x 4" DR 17					258			258	284
6220	2" DR 11 x 6" DR 17					294			294	325
6224	3" DR 11 x 6" DR 17					294			294	325
6228	3" DR 17 x 6" DR 26					294			294	325
6232	3" DR 17 x 8" DR 11					405			405	445
6236	3" DR 17 x 8" DR 17					405			405	445
6240	4" DR 11 x 8" DR 17					405			405	445
6244	4" DR 17 x 8" DR 26					355			355	390
6248	6" DR 11 x 10" DR 17					485			485	530
6252	6" DR 17 x 10" DR 26					395			395	435
6256	6" DR 26 x 10" DR 26					395			395	435
6260	8" DR 17 x 12" DR 26					590			590	645
6264	8" DR 26 x 12" DR 32.5					470			470	515
6268	10" DR 17 x 14" DR 26					610			610	670
6272	10" DR 17 x 16" DR 26					610			610	670
6276	10" DR 26 x 16" DR 26					610			610	670
6280	12" DR 26 x 16" DR 26					610			610	670
6284	12" DR 17 x 18" DR 26					805			805	885
6288	12" DR 26 x 18" DR 26					805			805	885
6292	14" DR 26 x 20" DR 32.5					820			820	900
6296	16" DR 26 x 22" DR 32.5					950			950	1,050
6300	18" DR 26 x 24" DR 32.5					980			980	1,075
6304	20" DR 32.5 x 28" DR 32.5					1,250			1,250	1,375
6308	22" DR 32.5 x 30" DR 32.5					1,375			1,375	1,525
6312	24" DR 32.5 x 32" DR 32.5					1,525			1,525	1,675
6316	36" DR 32.5 x 42" DR 32.5					2,225			2,225	2,450
6600	Tee									
6604	1" DR 11 x 3" DR 11				Ea.	226			226	249
6608	1" DR 11 x 4" DR 17					280			280	310
6612	1-1/2" DR 11 x 4" DR 17					310			310	345
6616	2" DR 11 x 4" DR 17					340			340	375

22 11 13 – Facility Water Distribution Piping

22 11 13.78 Pipe, High Density Polyethylene Plastic (HDPE)		Crew	Daily Output	Labor-Hours	Unit	2009 Bare Costs Material	Labor	Equipment	Total	Total Incl O&P
6620	2″ DR 11 x 6″ DR 17				Ea.	415			415	455
6624	3″ DR 11 x 6″ DR 17					475			475	525
6628	3″ DR 17 x 6″ DR 26					465			465	510
6632	3″ DR 17 x 8″ DR 11					560			560	620
6636	3″ DR 17 x 8″ DR 17					535			535	585
6640	4″ DR 11 x 8″ DR 17					580			580	640
6644	4″ DR 17 x 8″ DR 26					560			560	615
6648	6″ DR 11 x 10″ DR 17					730			730	800
6652	6″ DR 17 x 10″ DR 26					690			690	760
6656	6″ DR 26 x 10″ DR 26					680			680	750
6660	8″ DR 17 x 12″ DR 26					680			680	750
6664	8″ DR 26 x 12″ DR 32.5					890			890	980
6668	10″ DR 17 x 14″ DR 26					1,025			1,025	1,125
6672	10″ DR 17 x 16″ DR 26					1,075			1,075	1,200
6676	10″ DR 26 x 16″ DR 26					1,075			1,075	1,200
6680	12″ DR 26 x 16″ DR 26					1,175			1,175	1,275
6684	12″ DR 17 x 18″ DR 26					1,175			1,175	1,275
6688	12″ DR 26 x 18″ DR 26					1,300			1,300	1,425
6692	14″ DR 26 x 20″ DR 32.5					1,700			1,700	1,875
6696	16″ DR 26 x 22″ DR 32.5					2,100			2,100	2,300
6700	18″ DR 26 x 24″ DR 32.5					2,275			2,275	2,500
6704	20″ DR 32.5 x 28″ DR 32.5					2,875			2,875	3,150
6708	22″ DR 32.5 x 30″ DR 32.5					3,400			3,400	3,750
6712	24″ DR 32.5 x 32″ DR 32.5					4,075			4,075	4,500
6716	36″ DR 32.5 x 42″ DR 32.5				▼	8,325			8,325	9,150
6800	Wye									
6816	2″ DR 11 x 4″ DR 17				Ea.	430			430	475
6820	2″ DR 11 x 6″ DR 17					490			490	535
6824	3″ DR 11 x 6″ DR 17					495			495	540
6828	3″ DR 17 x 6″ DR 26					475			475	525
6832	3″ DR 17 x 8″ DR 11					600			600	660
6836	3″ DR 17 x 8″ DR 17					560			560	615
6840	4″ DR 11 x 8″ DR 17					610			610	670
6844	4″ DR 17 x 8″ DR 26					580			580	640
6848	6″ DR 11 x 10″ DR 17					755			755	830
6852	6″ DR 17 x 10″ DR 26					705			705	775
6856	6″ DR 26 x 10″ DR 26					700			700	770
6860	8″ DR 17 x 12″ DR 26					1,000			1,000	1,100
6864	8″ DR 26 x 12″ DR 32.5					965			965	1,075
6868	10″ DR 17 x 14″ DR 26					1,200			1,200	1,325
6872	10″ DR 17 x 16″ DR 26					1,275			1,275	1,400
6876	10″ DR 26 x 16″ DR 26					1,275			1,275	1,400
6880	12″ DR 26 x 16″ DR 26					1,425			1,425	1,550
6884	12″ DR 17 x 18″ DR 26					1,600			1,600	1,750
6888	12″ DR 26 x 18″ DR 26					1,625			1,625	1,775
6892	14″ DR 26 x 20″ DR 32.5					1,975			1,975	2,150
6896	16″ DR 26 x 22″ DR 32.5					2,825			2,825	3,125
6900	18″ DR 26 x 24″ DR 32.5					3,325			3,325	3,675
6904	20″ DR 32.5 x 28″ DR 32.5					4,175			4,175	4,600
6908	22″ DR 32.5 x 30″ DR 32.5					4,750			4,750	5,225
6912	24″ DR 32.5 x 32″ DR 32.5				▼	5,425			5,425	5,950
9000	Welding labor per joint, not including welding machine									
9010	Pipe joint size, outer pipe (cost based on the thickest walls)									

22 11 13 – Facility Water Distribution Piping

22 11 13.78 Pipe, High Density Polyethylene Plastic (HDPE)		Crew	Daily Output	Labor-Hours	Unit	2009 Bare Costs Material	Labor	Equipment	Total	Total Incl O&P
9020	Straight pipe									
9050	3″ pipe size	4 Skwk	96	.333	Ea.		13.60		13.60	21
9060	4″ pipe size	″	77	.416			17		17	26.50
9070	6″ pipe size	5 Skwk	60	.667			27		27	42
9080	8″ pipe size	″	40	1			41		41	63.50
9090	10″ pipe size	6 Skwk	41	1.171			48		48	74
9100	12″ pipe size		39	1.231			50.50		50.50	78
9110	14″ pipe size		38	1.263			51.50		51.50	80
9120	16″ pipe size		35	1.371			56		56	86.50
9130	18″ pipe size	8 Skwk	45	1.422			58		58	90
9140	20″ pipe size		42	1.524			62.50		62.50	96.50
9150	22″ pipe size		40	1.600			65.50		65.50	101
9160	24″ pipe size		38	1.684			69		69	107
9170	28″ pipe size		37	1.730			70.50		70.50	109
9180	30″ pipe size		36	1.778			72.50		72.50	112
9190	32″ pipe size		35	1.829			74.50		74.50	116
9200	42″ pipe size		32	2			81.50		81.50	127
9300	Note: Cost for set up each time welder is moved.									
9301	Add 100% of weld labor cost									
9310	For handling between fitting welds add 50% of weld labor cost									
9320	Welder usually remains stationary with pipe moved through it									
9360	Weld machine, rental per day based on dia. capacity									
9380	3″ thru 4″ diameter				Ea.			63.50	63.50	70
9390	6″ thru 8″ diameter							207	207	228
9400	10″ thru 12″ diameter							310	310	341
9410	14″ thru 18″ diameter							420	420	462
9420	20″ thru 24″ diameter							748	748	823
9430	28″ thru 32″ diameter							880	880	968
9440	42″ thru 58″ diameter							910	910	1,000

22 11 19 – Domestic Water Piping Specialties

22 11 19.10 Flexible Connectors		Crew	Daily Output	Labor-Hours	Unit	Material	Labor	Equipment	Total	Total Incl O&P
0010	**FLEXIBLE CONNECTORS**, Corrugated, 7/8″ O.D., 1/2″ I.D.									
0050	Gas, seamless brass, steel fittings									
0200	12″ long	1 Plum	36	.222	Ea.	15.10	10.85		25.95	33
0220	18″ long		36	.222		18.35	10.85		29.20	36.50
0240	24″ long		34	.235		22	11.45		33.45	41.50
0260	30″ long		34	.235		24	11.45		35.45	43.50
0280	36″ long		32	.250		26.50	12.20		38.70	47.50
0320	48″ long		30	.267		33.50	13		46.50	56.50
0340	60″ long		30	.267		40	13		53	63.50
0360	72″ long		30	.267		46	13		59	70
2000	Water, copper tubing, dielectric separators									
2100	12″ long	1 Plum	36	.222	Ea.	14	10.85		24.85	31.50
2220	15″ long		36	.222		15.45	10.85		26.30	33.50
2240	18″ long		36	.222		16.85	10.85		27.70	35
2260	24″ long		34	.235		20.50	11.45		31.95	39.50

22 11 19.14 Flexible Metal Hose		Crew	Daily Output	Labor-Hours	Unit	Material	Labor	Equipment	Total	Total Incl O&P
0010	**FLEXIBLE METAL HOSE**, Connectors, standard lengths									
0100	Bronze braided, bronze ends									
0120	3/8″ diameter x 12″	1 Stpi	26	.308	Ea.	18.30	15.20		33.50	43
0140	1/2″ diameter x 12″		24	.333		17.40	16.45		33.85	43.50
0160	3/4″ diameter x 12″		20	.400		25.50	19.75		45.25	57.50

22 11 19 – Domestic Water Piping Specialties

22 11 19.14 Flexible Metal Hose		Crew	Daily Output	Labor-Hours	Unit	Material	Labor	Equipment	Total	Total Incl O&P
						2009 Bare Costs				
0180	1" diameter x 18"	1 Stpi	19	.421	Ea.	33	21		54	67
0200	1-1/2" diameter x 18"		13	.615		52	30.50		82.50	103
0220	2" diameter x 18"		11	.727		63	36		99	123
1000	Carbon steel ends									
1020	1/4" diameter x 12"	1 Stpi	28	.286	Ea.	13.60	14.10		27.70	36
1040	3/8" diameter x 12"		26	.308		13.95	15.20		29.15	38.50
1060	1/2" diameter x 12"		24	.333		15.40	16.45		31.85	41.50
1080	1/2" diameter x 24"		24	.333		34.50	16.45		50.95	62
1100	1/2" diameter x 36"		24	.333		40.50	16.45		56.95	69
1120	3/4" diameter x 12"		20	.400		25	19.75		44.75	57
1140	3/4" diameter x 24"		20	.400		24.50	19.75		44.25	56.50
1160	3/4" diameter x 36"		20	.400		51	19.75		70.75	85.50
1180	1" diameter x 18"		19	.421		31.50	21		52.50	65.50
1200	1" diameter x 30"		19	.421		51	21		72	87
1220	1" diameter x 36"		19	.421		60	21		81	97
1240	1-1/4" diameter x 18"		15	.533		43	26.50		69.50	87
1260	1-1/4" diameter x 36"		15	.533		79.50	26.50		106	127
1280	1-1/2" diameter x 18"		13	.615		61	30.50		91.50	113
1300	1-1/2" diameter x 36"		13	.615		87.50	30.50		118	142
1320	2" diameter x 24"		11	.727		88.50	36		124.50	152
1340	2" diameter x 36"		11	.727		109	36		145	174
1360	2-1/2" diameter x 24"		9	.889		202	44		246	289
1380	2-1/2" diameter x 36"		9	.889		263	44		307	355
1400	3" diameter x 24"		7	1.143		279	56.50		335.50	390
1420	3" diameter x 36"		7	1.143		375	56.50		431.50	495
2000	Carbon steel braid, carbon steel solid ends									
2100	1/2" diameter x 12"	1 Stpi	24	.333	Ea.	34	16.45		50.45	62
2120	3/4" diameter x 12"		20	.400		43	19.75		62.75	77
2140	1" diameter x 12"		19	.421		48.50	21		69.50	84
2160	1-1/4" diameter x 12"		15	.533		54	26.50		80.50	98.50
2180	1-1/2" diameter x 12"		13	.615		55.50	30.50		86	107
3000	Stainless steel braid, welded on carbon steel ends									
3100	1/2" diameter x 12"	1 Stpi	24	.333	Ea.	43.50	16.45		59.95	72
3120	3/4" diameter x 12"		20	.400		55	19.75		74.75	90
3140	3/4" diameter x 24"		20	.400		63.50	19.75		83.25	99
3160	3/4" diameter x 36"		20	.400		72	19.75		91.75	109
3180	1" diameter x 12"		19	.421		65.50	21		86.50	103
3200	1" diameter x 24"		19	.421		77.50	21		98.50	116
3220	1" diameter x 36"		19	.421		90.50	21		111.50	131
3240	1-1/4" diameter x 12"		15	.533		98.50	26.50		125	149
3260	1-1/4" diameter x 24"		15	.533		104	26.50		130.50	154
3280	1-1/4" diameter x 36"		15	.533		118	26.50		144.50	169
3300	1-1/2" diameter x 12"		13	.615		103	30.50		133.50	159
3320	1-1/2" diameter x 24"		13	.615		115	30.50		145.50	173
3340	1-1/2" diameter x 36"		13	.615		135	30.50		165.50	194
3400	Metal stainless steel braid, over corrugated stainless steel, flanged ends									
3410	150 PSI									
3420	1/2" diameter x 12"	1 Stpi	24	.333	Ea.	144	16.45		160.45	183
3430	1" diameter x 12"		20	.400		180	19.75		199.75	228
3440	1-1/2" diameter x 12"		15	.533		204	26.50		230.50	264
3450	2-1/2" diameter x 9"		12	.667		75	33		108	132
3460	3" diameter x 9"		9	.889		90	44		134	165
3470	4" diameter x 9"		7	1.143		110	56.50		166.50	206

22 11 Facility Water Distribution

22 11 19 – Domestic Water Piping Specialties

22 11 19.14 Flexible Metal Hose		Crew	Daily Output	Labor-Hours	Unit	2009 Bare Costs Material	Labor	Equipment	Total	Total Incl O&P
3480	4" diameter x 30"	1 Stpi	5	1.600	Ea.	325	79		404	475
3490	4" diameter x 36"		4.80	1.667		350	82.50		432.50	510
3500	6" diameter x 11"		5	1.600		183	79		262	320
3510	6" diameter x 36"		3.80	2.105		505	104		609	715
3520	8" diameter x 12"		4	2		360	98.50		458.50	545
3530	10" diameter x 13"		3	2.667		490	132		622	735
3540	12" diameter x 14"	Q-5	4	4		670	178		848	1,000
6000	Molded rubber with helical wire reinforcement									
6010	150 PSI									
6020	1-1/2" diameter x 12"	1 Stpi	15	.533	Ea.	82	26.50		108.50	130
6030	2" diameter x 12"		12	.667		234	33		267	305
6040	3" diameter x 12"		8	1		244	49.50		293.50	340
6050	4" diameter x 12"		6	1.333		310	66		376	440
6060	6" diameter x 18"		4	2		465	98.50		563.50	660
6070	8" diameter x 24"		3	2.667		650	132		782	910
6080	10" diameter x 24"		2	4		790	197		987	1,175
6090	12" diameter x 24"	Q-5	3	5.333		900	237		1,137	1,350
7000	Molded teflon with stainless steel flanges									
7010	150 PSI									
7020	2-1/2" diameter x 3-3/16"	Q-1	7.80	2.051	Ea.	1,350	90		1,440	1,600
7030	3" diameter x 3-5/8"		6.50	2.462		1,575	108		1,683	1,900
7040	4" diameter x 3-5/8"		5	3.200		2,050	140		2,190	2,450
7050	6" diameter x 4"		4.30	3.721		2,875	163		3,038	3,425
7060	8" diameter x 6"		3.80	4.211		4,575	185		4,760	5,325

22 11 19.18 Mixing Valve		Crew	Daily Output	Labor-Hours	Unit	Material	Labor	Equipment	Total	Total Incl O&P
0010	**MIXING VALVE**, Automatic, water tempering.									
0040	1/2" size	1 Stpi	19	.421	Ea.	435	21		456	510
0050	3/4" size		18	.444		435	22		457	515
0100	1" size		16	.500		625	24.50		649.50	720
0120	1-1/4" size		13	.615		905	30.50		935.50	1,050
0140	1-1/2" size		10	.800		1,075	39.50		1,114.50	1,225
0160	2" size		8	1		1,350	49.50		1,399.50	1,575
0170	2-1/2" size		6	1.333		1,350	66		1,416	1,600
0180	3" size		4	2		3,125	98.50		3,223.50	3,600
0190	4" size		3	2.667		3,125	132		3,257	3,650

22 11 19.22 Pressure Reducing Valve		Crew	Daily Output	Labor-Hours	Unit	Material	Labor	Equipment	Total	Total Incl O&P
0010	**PRESSURE REDUCING VALVE**, Steam, pilot operated.									
0100	Threaded, iron body									
0200	1-1/2" size	1 Stpi	8	1	Ea.	1,000	49.50		1,049.50	1,175
0220	2" size	"	5	1.600	"	1,325	79		1,404	1,575
1000	Flanged, iron body, 125 lb. flanges									
1020	2" size	1 Stpi	8	1	Ea.	1,450	49.50		1,499.50	1,675
1040	2-1/2" size	"	4	2		1,750	98.50		1,848.50	2,075
1060	3" size	Q-5	4.50	3.556		1,775	158		1,933	2,175
1080	4" size	"	3	5.333		2,200	237		2,437	2,775
1500	For 250 lb. flanges, add					5%				

22 11 19.26 Pressure Regulators		Crew	Daily Output	Labor-Hours	Unit	Material	Labor	Equipment	Total	Total Incl O&P
0010	**PRESSURE REGULATORS**									
0100	Gas appliance regulators									
0106	Main burner and pilot applications									
0108	Rubber seat poppet type									
0109	1/8" pipe size	1 Stpi	24	.333	Ea.	13.70	16.45		30.15	39.50

22 11 19 – Domestic Water Piping Specialties

22 11 19.26 Pressure Regulators		Crew	Daily Output	Labor-Hours	Unit	Material	2009 Bare Costs Labor	Equipment	Total	Total Incl O&P
0110	1/4" pipe size	1 Stpi	24	.333	Ea.	15.75	16.45		32.20	42
0112	3/8" pipe size		24	.333		16.75	16.45		33.20	43
0113	1/2" pipe size		24	.333		18.50	16.45		34.95	45
0114	3/4" pipe size		20	.400		22	19.75		41.75	53.50
0122	Lever action type									
0123	3/8" pipe size	1 Stpi	24	.333	Ea.	21	16.45		37.45	47.50
0124	1/2" pipe size		24	.333		21	16.45		37.45	47.50
0125	3/4" pipe size		20	.400		41.50	19.75		61.25	75.50
0126	1" pipe size		19	.421		41.50	21		62.50	77
0132	Double diaphragm type									
0133	3/8" pipe size	1 Stpi	24	.333	Ea.	30	16.45		46.45	57.50
0134	1/2" pipe size		24	.333		42.50	16.45		58.95	71
0135	3/4" pipe size		20	.400		71	19.75		90.75	108
0136	1" pipe size		19	.421		86.50	21		107.50	126
0137	1-1/4" pipe size		15	.533		300	26.50		326.50	370
0138	1-1/2" pipe size		13	.615		555	30.50		585.50	655
0139	2" pipe size		11	.727		555	36		591	665
0140	2-1/2" pipe size	Q-5	15	1.067		1,075	47.50		1,122.50	1,250
0141	3" pipe size		13	1.231		1,075	54.50		1,129.50	1,250
0142	4" pipe size (flanged)		8	2		1,900	89		1,989	2,225
0160	Main burner only									
0162	Straight-thru-flow design									
0163	1/2" pipe size	1 Stpi	24	.333	Ea.	34.50	16.45		50.95	62.50
0164	3/4" pipe size		20	.400		47	19.75		66.75	81.50
0165	1" pipe size		19	.421		67	21		88	105
0166	1-1/4" pipe size		15	.533		67	26.50		93.50	113
0200	Oil, light, hot water, ordinary steam, threaded									
0220	Bronze body, 1/4" size	1 Stpi	24	.333	Ea.	173	16.45		189.45	216
0230	3/8" size		24	.333		187	16.45		203.45	231
0240	1/2" size		24	.333		233	16.45		249.45	281
0250	3/4" size		20	.400		278	19.75		297.75	335
0260	1" size		19	.421		425	21		446	495
0270	1-1/4" size		15	.533		570	26.50		596.50	665
0280	1-1/2" size		13	.615		610	30.50		640.50	720
0290	2" size		11	.727		1,100	36		1,136	1,275
0320	Iron body, 1/4" size		24	.333		136	16.45		152.45	174
0330	3/8" size		24	.333		160	16.45		176.45	200
0340	1/2" size		24	.333		169	16.45		185.45	211
0350	3/4" size		20	.400		205	19.75		224.75	256
0360	1" size		19	.421		268	21		289	325
0370	1-1/4" size		15	.533		370	26.50		396.50	450
0380	1-1/2" size		13	.615		415	30.50		445.50	500
0390	2" size		11	.727		600	36		636	715
0500	Oil, heavy, viscous fluids, threaded									
0520	Bronze body, 3/8" size	1 Stpi	24	.333	Ea.	298	16.45		314.45	355
0530	1/2" size		24	.333		385	16.45		401.45	445
0540	3/4" size		20	.400		425	19.75		444.75	500
0550	1" size		19	.421		535	21		556	615
0560	1-1/4" size		15	.533		750	26.50		776.50	865
0570	1-1/2" size		13	.615		860	30.50		890.50	990
0600	Iron body, 3/8" size		24	.333		237	16.45		253.45	286
0620	1/2" size		24	.333		288	16.45		304.45	340
0630	3/4" size		20	.400		325	19.75		344.75	390

22 11 19.26 Pressure Regulators		Crew	Daily Output	Labor-Hours	Unit	Material	2009 Bare Costs Labor	Equipment	Total	Total Incl O&P
0640	1" size	1 Stpi	19	.421	Ea.	385	21		406	455
0650	1-1/4" size		15	.533		555	26.50		581.50	655
0660	1-1/2" size		13	.615		595	30.50		625.50	700
0800	Process steam, wet or super heated, monel trim, threaded									
0820	Bronze body, 1/4" size	1 Stpi	24	.333	Ea.	625	16.45		641.45	715
0830	3/8" size		24	.333		625	16.45		641.45	715
0840	1/2" size		24	.333		720	16.45		736.45	815
0850	3/4" size		20	.400		845	19.75		864.75	960
0860	1" size		19	.421		1,075	21		1,096	1,200
0870	1-1/4" size		15	.533		1,325	26.50		1,351.50	1,500
0880	1-1/2" size		13	.615		1,600	30.50		1,630.50	1,800
0890	2" size		11	.727		2,325	36		2,361	2,600
0920	Iron body, max 125 PSIG press out, 1/4" size		24	.333		136	16.45		152.45	174
0930	3/8" size		24	.333		120	16.45		136.45	157
0940	1/2" size		24	.333		127	16.45		143.45	164
0950	3/4" size		20	.400		154	19.75		173.75	199
0960	1" size		19	.421		201	21		222	252
0970	1-1/4" size		15	.533		279	26.50		305.50	345
0980	1-1/2" size		13	.615		310	30.50		340.50	385
0990	2" size		11	.727		450	36		486	550
1000	Flanged, Class 125									
1006	2-1/2" size	Q-5	10	1.600	Ea.	1,425	71		1,496	1,650
1008	3" size		9	1.778		1,475	79		1,554	1,750
1010	4" size		8	2		1,875	89		1,964	2,200
1020	6" size		7	2.286		4,775	102		4,877	5,400
3000	Steam, high capacity, bronze body, stainless steel trim									
3020	Threaded, 1/2" diameter	1 Stpi	24	.333	Ea.	1,300	16.45		1,316.45	1,450
3030	3/4" diameter		24	.333		1,300	16.45		1,316.45	1,450
3040	1" diameter		19	.421		1,450	21		1,471	1,625
3060	1-1/4" diameter		15	.533		1,600	26.50		1,626.50	1,825
3080	1-1/2" diameter		13	.615		1,825	30.50		1,855.50	2,075
3100	2" diameter		11	.727		2,250	36		2,286	2,525
3120	2-1/2" diameter	Q-5	12	1.333		2,825	59		2,884	3,200
3140	3" diameter	"	11	1.455		3,200	64.50		3,264.50	3,625
3500	Flanged connection, iron body, 125 lb. W.S.P.									
3520	3" diameter	Q-5	11	1.455	Ea.	3,525	64.50		3,589.50	3,975
3540	4" diameter	"	5	3.200	"	4,425	142		4,567	5,100
9002	For water pressure regulators, see Div. 22 05 23.20									

22 11 19.30 Pressure and Temperature Safety Plug		Crew	Daily Output	Labor-Hours	Unit	Material	Labor	Equipment	Total	Total Incl O&P
0010	**PRESSURE & TEMPERATURE SAFETY PLUG**									
1000	3/4" external thread, 3/8" diam. element									
1020	Carbon steel									
1050	7-1/2" insertion	1 Stpi	32	.250	Ea.	70	12.35		82.35	95
1120	304 stainless steel									
1150	7-1/2" insertion	1 Stpi	32	.250	Ea.	65.50	12.35		77.85	90.50
1220	316 Stainless steel									
1250	7-1/2" insertion	1 Stpi	32	.250	Ea.	68.50	12.35		80.85	93.50

22 11 19.34 Sleeves and Escutcheons		Crew	Daily Output	Labor-Hours	Unit	Material	Labor	Equipment	Total	Total Incl O&P
0010	**SLEEVES & ESCUTCHEONS**									
0100	Pipe sleeve									
0110	Steel, w/water stop, 12" long, with link seal									
0120	2" diam. for 1/2" carrier pipe	1 Plum	8.40	.952	Ea.	45	46.50		91.50	119

22 11 19.34 Sleeves and Escutcheons		Crew	Daily Output	Labor-Hours	Unit	2009 Bare Costs Material	Labor	Equipment	Total	Total Incl O&P
0130	2-1/2" diam. for 3/4" carrier pipe	1 Plum	8	1	Ea.	51	49		100	129
0140	2-1/2" diam. for 1" carrier pipe		8	1		48	49		97	126
0150	3" diam. for 1-1/4" carrier pipe		7.20	1.111		62	54		116	150
0160	3-1/2" diam. for 1-1/2" carrier pipe		6.80	1.176		63	57.50		120.50	155
0170	4" diam. for 2" carrier pipe		6	1.333		68	65		133	173
0180	4" diam. for 2-1/2" carrier pipe		6	1.333		68	65		133	172
0190	5" diam. for 3" carrier pipe		5.40	1.481		81.50	72		153.50	198
0200	6" diam. for 4" carrier pipe		4.80	1.667		92	81.50		173.50	223
0210	10" diam. for 6" carrier pipe	Q-1	8	2		159	88		247	305
0220	12" diam. for 8" carrier pipe		7.20	2.222		222	97.50		319.50	390
0230	14" diam. for 10" carrier pipe		6.40	2.500		256	110		366	445
0240	16" diam. for 12" carrier pipe		5.80	2.759		292	121		413	500
0250	18" diam. for 14" carrier pipe		5.20	3.077		395	135		530	640
0260	24" diam. for 18" carrier pipe		4	4		670	176		846	1,000
0270	24" diam. for 20" carrier pipe		4	4		585	176		761	905
0280	30" diam. for 24" carrier pipe		3.20	5		905	219		1,124	1,325
0500	Wall sleeve									
0510	Ductile iron with rubber gasket seal									
0520	3"	1 Plum	8.40	.952	Ea.	700	46.50		746.50	840
0530	4"		7.20	1.111		750	54		804	905
0540	6"		6	1.333		920	65		985	1,125
0550	8"		4	2		1,125	97.50		1,222.50	1,400
0560	10"		3	2.667		1,375	130		1,505	1,725
0570	12"		2.40	3.333		1,625	163		1,788	2,050
5000	Escutcheon									
5100	Split ring, pipe									
5110	Chrome plated									
5120	1/2"	1 Plum	160	.050	Ea.	1.05	2.44		3.49	4.82
5130	3/4"		160	.050		1.09	2.44		3.53	4.86
5140	1"		135	.059		1.17	2.89		4.06	5.60
5150	1-1/2"		115	.070		1.76	3.39		5.15	7.05
5160	2"		100	.080		1.85	3.90		5.75	7.90
5170	4"		80	.100		4.11	4.88		8.99	11.80
5180	6"		68	.118		6.85	5.75		12.60	16.15
5400	Shallow flange type									
5410	Chrome plated steel									
5420	1/2" CTS	1 Plum	180	.044	Ea.	.28	2.17		2.45	3.56
5430	3/4" CTS		180	.044		.28	2.17		2.45	3.56
5440	1/2" IPS		180	.044		.26	2.17		2.43	3.54
5450	3/4" IPS		180	.044		.28	2.17		2.45	3.56
5460	1" IPS		175	.046		.46	2.23		2.69	3.85
5470	1-1/2" IPS		170	.047		.72	2.29		3.01	4.23
5480	2" IPS		160	.050		.86	2.44		3.30	4.61

22 11 19.38 Water Supply Meters		Crew	Daily Output	Labor-Hours	Unit	Material	Labor	Equipment	Total	Total Incl O&P
0010	**WATER SUPPLY METERS**									
1000	Detector, serves dual systems such as fire and domestic or									
1020	process water, wide range cap., UL and FM approved									
1100	3" mainline x 2" by-pass, 400 GPM	Q-1	3.60	4.444	Ea.	6,250	195		6,445	7,175
1140	4" mainline x 2" by-pass, 700 GPM	"	2.50	6.400		6,250	281		6,531	7,300
1180	6" mainline x 3" by-pass, 1600 GPM	Q-2	2.60	9.231		9,550	420		9,970	11,100
1220	8" mainline x 4" by-pass, 2800 GPM		2.10	11.429		14,200	520		14,720	16,400
1260	10" mainline x 6" by-pass, 4400 GPM		2	12		20,300	545		20,845	23,100

22 11 19 – Domestic Water Piping Specialties

22 11 19.38 Water Supply Meters		Crew	Daily Output	Labor-Hours	Unit	Material	Labor	Equipment	Total	Total Incl O&P
						2009 Bare Costs				
1300	10"x12" mainlines x 6" by-pass,5400 GPM	Q-2	1.70	14.118	Ea.	27,500	640		28,140	31,200
2000	Domestic/commercial, bronze									
2020	Threaded									
2060	5/8" diameter, to 20 GPM	1 Plum	16	.500	Ea.	42	24.50		66.50	82.50
2080	3/4" diameter, to 30 GPM		14	.571		76.50	28		104.50	126
2100	1" diameter, to 50 GPM	↓	12	.667	↓	116	32.50		148.50	177
2300	Threaded/flanged									
2340	1-1/2" diameter, to 100 GPM	1 Plum	8	1	Ea.	284	49		333	385
2360	2" diameter, to 160 GPM	"	6	1.333	"	385	65		450	525
2600	Flanged, compound									
2640	3" diameter, 320 GPM	Q-1	3	5.333	Ea.	2,625	234		2,859	3,225
2660	4" diameter, to 500 GPM		1.50	10.667		4,200	470		4,670	5,300
2680	6" diameter, to 1,000 GPM		1	16		6,675	700		7,375	8,400
2700	8" diameter, to 1,800 GPM	↓	.80	20	↓	10,500	880		11,380	12,800
7000	Turbine									
7260	Flanged									
7300	2" diameter, to 160 GPM	1 Plum	7	1.143	Ea.	510	55.50		565.50	650
7320	3" diameter, to 450 GPM	Q-1	3.60	4.444		1,075	195		1,270	1,475
7340	4" diameter, to 650 GPM	"	2.50	6.400		1,775	281		2,056	2,375
7360	6" diameter, to 1800 GPM	Q-2	2.60	9.231		3,150	420		3,570	4,100
7380	8" diameter, to 2500 GPM		2.10	11.429		5,000	520		5,520	6,300
7400	10" diameter, to 5500 GPM	↓	1.70	14.118	↓	6,775	640		7,415	8,425

22 11 19.42 Backflow Preventers		Crew	Daily Output	Labor-Hours	Unit	Material	Labor	Equipment	Total	Total Incl O&P
0010	**BACKFLOW PREVENTERS**, Includes valves									
0020	and four test cocks, corrosion resistant, automatic operation									
1000	Double check principle									
1010	Threaded, with ball valves									
1020	3/4" pipe size	1 Plum	16	.500	Ea.	223	24.50		247.50	282
1030	1" pipe size		14	.571		257	28		285	325
1040	1-1/2" pipe size		10	.800		670	39		709	800
1050	2" pipe size	↓	7	1.143	↓	800	55.50		855.50	965
1080	Threaded, with gate valves									
1100	3/4" pipe size	1 Plum	16	.500	Ea.	860	24.50		884.50	980
1120	1" pipe size		14	.571		865	28		893	995
1140	1-1/2" pipe size		10	.800		1,125	39		1,164	1,275
1160	2" pipe size	↓	7	1.143	↓	1,375	55.50		1,430.50	1,575
1200	Flanged, valves are gate									
1210	3" pipe size	Q-1	4.50	3.556	Ea.	2,000	156		2,156	2,425
1220	4" pipe size	"	3	5.333		2,450	234		2,684	3,050
1230	6" pipe size	Q-2	3	8		3,625	365		3,990	4,525
1240	8" pipe size		2	12		6,575	545		7,120	8,075
1250	10" pipe size	↓	1	24	↓	9,775	1,100		10,875	12,400
1300	Flanged, valves are OS&Y									
1370	1" pipe size	1 Plum	5	1.600	Ea.	925	78		1,003	1,150
1374	1-1/2" pipe size		5	1.600		1,175	78		1,253	1,425
1378	2" pipe size	↓	4.80	1.667		1,425	81.50		1,506.50	1,700
1380	3" pipe size	Q-1	4.50	3.556		3,525	156		3,681	4.100
1400	4" pipe size	"	3	5.333		3,925	234		4,159	4.675
1420	6" pipe size	Q-2	3	8		6,125	365		6,490	7.300
1430	8" pipe size	"	2	12	↓	12,100	545		12,645	14 100
4000	Reduced pressure principle									
4100	Threaded, bronze, valves are ball									

22 11 19.42 Backflow Preventers		Crew	Daily Output	Labor-Hours	Unit	Material	Labor	Equipment	Total	Total Incl O&P
4120	3/4" pipe size	1 Plum	16	.500	Ea.	330	24.50		354.50	400
4140	1" pipe size		14	.571		355	28		383	430
4150	1-1/4" pipe size		12	.667		610	32.50		642.50	720
4160	1-1/2" pipe size		10	.800		670	39		709	795
4180	2" pipe size		7	1.143		750	55.50		805.50	910
5000	Flanged, bronze, valves are OS&Y									
5060	2-1/2" pipe size	Q-1	5	3.200	Ea.	3,200	140		3,340	3,725
5080	3" pipe size		4.50	3.556		3,350	156		3,506	3,925
5100	4" pipe size		3	5.333		4,225	234		4,459	5,000
5120	6" pipe size	Q-2	3	8		6,100	365		6,465	7,275
5200	Flanged, iron, valves are gate									
5210	2-1/2" pipe size	Q-1	5	3.200	Ea.	2,250	140		2,390	2,675
5220	3" pipe size		4.50	3.556		2,350	156		2,506	2,800
5230	4" pipe size		3	5.333		3,175	234		3,409	3,850
5240	6" pipe size	Q-2	3	8		4,475	365		4,840	5,475
5250	8" pipe size		2	12		8,025	545		8,570	9,650
5260	10" pipe size		1	24		11,300	1,100		12,400	14,200
5600	Flanged, iron, valves are OS&Y									
5660	2-1/2" pipe size	Q-1	5	3.200	Ea.	1,750	140		1,890	2,125
5680	3" pipe size		4.50	3.556		2,700	156		2,856	3,175
5700	4" pipe size		3	5.333		3,450	234		3,684	4,150
5720	6" pipe size	Q-2	3	8		4,875	365		5,240	5,925
5740	8" pipe size		2	12		8,600	545		9,145	10,300
5760	10" pipe size		1	24		11,500	1,100		12,600	14,400

22 11 19.50 Vacuum Breakers		Crew	Daily Output	Labor-Hours	Unit	Material	Labor	Equipment	Total	Total Incl O&P
0010	**VACUUM BREAKERS**									
0013	See also backflow preventers Div. 22 11 19.42									
1000	Anti-siphon continuous pressure type									
1010	Max. 150 PSI - 210° F									
1020	Bronze body									
1030	1/2" size	1 Stpi	24	.333	Ea.	143	16.45		159.45	182
1040	3/4" size		20	.400		143	19.75		162.75	187
1050	1" size		19	.421		148	21		169	194
1060	1-1/4" size		15	.533		291	26.50		317.50	360
1070	1-1/2" size		13	.615		360	30.50		390.50	440
1080	2" size		11	.727		370	36		406	460
1200	Max. 125 PSI with atmospheric vent									
1210	Brass, in-line construction									
1220	1/4" size	1 Stpi	24	.333	Ea.	57.50	16.45		73.95	88
1230	3/8" size	"	24	.333		57.50	16.45		73.95	88
1260	For polished chrome finish, add					13%				
2000	Anti-siphon, non-continuous pressure type									
2010	Hot or cold water 125 PSI - 210° F									
2020	Bronze body									
2030	1/4" size	1 Stpi	24	.333	Ea.	36	16.45		52.45	64
2040	3/8" size		24	.333		36	16.45		52.45	64
2050	1/2" size		24	.333		41	16.45		57.45	69.50
2060	3/4" size		20	.400		48.50	19.75		68.25	83
2070	1" size		19	.421		75.50	21		96.50	114
2080	1-1/4" size		15	.533		133	26.50		159.50	186
2090	1-1/2" size		13	.615		137	30.50		167.50	197
2100	2" size		11	.727		243	36		279	320

22 11 Facility Water Distribution

22 11 19 – Domestic Water Piping Specialties

22 11 19.50 Vacuum Breakers		Crew	Daily Output	Labor-Hours	Unit	2009 Bare Costs Material	Labor	Equipment	Total	Total Incl O&P
2110	2-1/2" size	1 Stpi	8	1	Ea.	695	49.50		744.50	840
2120	3" size		6	1.333		925	66		991	1,125
2150	For polished chrome finish, add					50%				

22 11 19.54 Water Hammer Arresters/Shock Absorbers

Line	Description	Crew	Daily Output	Labor-Hours	Unit	Material	Labor	Equipment	Total	Total Incl O&P
0010	**WATER HAMMER ARRESTERS/SHOCK ABSORBERS**									
0490	Copper									
0500	3/4" male I.P.S. For 1 to 11 fixtures	1 Plum	12	.667	Ea.	17	32.50		49.50	67.50
0600	1" male I.P.S., For 12 to 32 fixtures		8	1		41.50	49		90.50	119
0700	1-1/4" male I.P.S. For 33 to 60 fixtures		8	1		44	49		93	122
0800	1-1/2" male I.P.S. For 61 to 113 fixtures		8	1		63	49		112	143
0900	2" male I.P.S.For 114 to 154 fixtures		8	1		92	49		141	174
1000	2-1/2" male I.P.S. For 155 to 330 fixtures		4	2		285	97.50		382.50	460
4000	Bellows type									
4010	3/4" FNPT, to 11 fixture units	1 Plum	10	.800	Ea.	183	39		222	261
4020	1" FNPT, to 32 fixture units		8.80	.909		370	44.50		414.50	470
4030	1" FNPT, to 60 fixture units		8.80	.909		555	44.50		599.50	675
4040	1" FNPT, to 113 fixture units		8.80	.909		1,400	44.50		1,444.50	1,600
4050	1" FNPT, to 154 fixture units		6.60	1.212		1,575	59		1,634	1,825
4060	1-1/2" FNPT, to 300 fixture units		5	1.600		1,925	78		2,003	2,250

22 11 19.64 Hydrants

Line	Description	Crew	Daily Output	Labor-Hours	Unit	Material	Labor	Equipment	Total	Total Incl O&P
0010	**HYDRANTS**									
0050	Wall type, moderate climate, bronze, encased									
0200	3/4" IPS connection	1 Plum	16	.500	Ea.	535	24.50		559.50	625
0300	1" IPS connection	"	14	.571		615	28		643	715
0500	Anti-siphon type, 3/4" connection					465			465	510
1000	Non-freeze, bronze, exposed									
1100	3/4" IPS connection, 4" to 9" thick wall	1 Plum	14	.571	Ea.	365	28		393	440
1120	10" to 14" thick wall		12	.667		475	32.50		507.50	570
1140	15" to 19" thick wall		12	.667		475	32.50		507.50	570
1160	20" to 24" thick wall		10	.800		480	39		519	590
1200	For 1" IPS connection, add					15%	10%			
1240	For 3/4" adapter type vacuum breaker, add				Ea.	45.50			45.50	50
1280	For anti-siphon type, add				"	95			95	105
2000	Non-freeze bronze, encased, anti-siphon type									
2100	3/4" IPS connection, 5" to 9" thick wall	1 Plum	14	.571	Ea.	920	28		948	1,050
2120	10" to 14" thick wall		12	.667		950	32.50		982.50	1,100
2140	15" to 19" thick wall		12	.667		995	32.50		1,027.50	1,150
2160	20" to 24" thick wall		10	.800		1,025	39		1,064	1,175
2200	For 1" IPS connection, add					10%	10%			
3000	Ground box type, bronze frame, 3/4" IPS connection									
3080	Non-freeze, all bronze, polished face, set flush									
3100	2 feet depth of bury	1 Plum	8	1	Ea.	685	49		734	830
3120	3 feet depth of bury		8	1		735	49		784	885
3140	4 feet depth of bury		8	1		790	49		839	940
3160	5 feet depth of bury		7	1.143		845	55.50		900.50	1,025
3180	6 feet depth of bury		7	1.143		895	55.50		950.50	1,075
3200	7 feet depth of bury		6	1.333		950	65		1,015	1,150
3220	8 feet depth of bury		5	1.600		1,000	78		1,078	1,225
3240	9 feet depth of bury		4	2		1,050	97.50		1,147.50	1,300
3260	10 feet depth of bury		4	2		1,100	97.50		1,197.50	1,375
3400	For 1" IPS connection, add					15%	10%			
3450	For 1-1/4" IPS connection, add					325%	14%			

22 11 19 – Domestic Water Piping Specialties

22 11 19.64 Hydrants		Crew	Daily Output	Labor-Hours	Unit	2009 Bare Costs Material	Labor	Equipment	Total	Total Incl O&P
3500	For 1-1/2″ connection, add				Ea.	370%	18%			
3550	For 2″ connection, add					445%	24%			
3600	For tapped drain port in box, add					61.50			61.50	68
4000	Non-freeze, CI body, bronze frame & scoriated cover									
4010	with hose storage									
4100	2 feet depth of bury	1 Plum	7	1.143	Ea.	1,250	55.50		1,305.50	1,475
4120	3 feet depth of bury		7	1.143		1,325	55.50		1,380.50	1,525
4140	4 feet depth of bury		7	1.143		1,350	55.50		1,405.50	1,575
4160	5 feet depth of bury		6.50	1.231		1,375	60		1,435	1,625
4180	6 feet depth of bury		6	1.333		1,400	65		1,465	1,650
4200	7 feet depth of bury		5.50	1.455		1,450	71		1,521	1,700
4220	8 feet depth of bury		5	1.600		1,500	78		1,578	1,775
4240	9 feet depth of bury		4.50	1.778		1,550	86.50		1,636.50	1,850
4260	10 feet depth of bury		4	2		1,600	97.50		1,697.50	1,925
4280	For 1″ IPS connection, add					277			277	305
4300	For tapped drain port in box, add					61.50			61.50	68
5000	Moderate climate, all bronze, polished face									
5020	and scoriated cover, set flush									
5100	3/4″ IPS connection	1 Plum	16	.500	Ea.	475	24.50		499.50	555
5120	1″ IPS connection	″	14	.571		585	28		613	680
5200	For tapped drain port in box, add					61.50			61.50	68
6000	Ground post type, all non-freeze, all bronze, aluminum casing									
6010	guard, exposed head, 3/4″ IPS connection									
6100	2 feet depth of bury	1 Plum	8	1	Ea.	670	49		719	810
6120	3 feet depth of bury		8	1		725	49		774	870
6140	4 feet depth of bury		8	1		795	49		844	950
6160	5 feet depth of bury		7	1.143		835	55.50		890.50	1,000
6180	6 feet depth of bury		7	1.143		895	55.50		950.50	1,075
6200	7 feet depth of bury		6	1.333		955	65		1,020	1,150
6220	8 feet depth of bury		5	1.600		1,025	78		1,103	1,250
6240	9 feet depth of bury		4	2		1,075	97.50		1,172.50	1,325
6260	10 feet depth of bury		4	2		1,150	97.50		1,247.50	1,400
6300	For 1″ IPS connection, add					40%	10%			
6350	For 1-1/4″ IPS connection, add					140%	14%			
6400	For 1-1/2″ IPS connection, add					225%	18%			
6450	For 2″ IPS connection, add					315%	24%			

22 11 23 – Domestic Water Pumps

22 11 23.10 General Utility Pumps		Crew	Daily Output	Labor-Hours	Unit	Material	Labor	Equipment	Total	Total Incl O&P
0010	**GENERAL UTILITY PUMPS**									
2000	Single stage									
3000	Double suction,									
3190	75 HP, to 2500 GPM	Q-3	.28	114	Ea.	15,600	5,300		20,900	25,100
3220	100 HP, to 3000 GPM		.26	123		19,600	5,725		25,325	30,100
3240	150 HP, to 4000 GPM		.24	133		30,200	6,200		36,400	42,500
4000	Centrifugal, end suction, mounted on base									
4010	Horizontal mounted, with drip proof motor, rated @ 100′ head									
4020	Vertical split case, single stage									
4040	100 GPM, 5 HP, 1-1/2″ discharge	Q-1	1.70	9.412	Ea.	2,875	415		3,290	3,775
4050	200 GPM, 10 HP, 2″ discharge		1.30	12.308		3,475	540		4,015	4,625
4060	250 GPM, 10 HP, 3″ discharge		1.28	12.500		3,875	550		4,425	5,100
4070	300 GPM, 15 HP, 2″ discharge	Q-2	1.56	15.385		4,075	700		4,775	5,525
4080	500 GPM, 20 HP, 4″ discharge		1.44	16.667		4,925	760		5,685	6,575

22 11 23.10 General Utility Pumps

Line	22 11 23.10 General Utility Pumps	Crew	Daily Output	Labor-Hours	Unit	2009 Bare Costs Material	Labor	Equipment	Total	Total Incl O&P
4090	750 GPM, 30 HP, 4" discharge	Q-2	1.20	20	Ea.	5,100	910		6,010	6,975
4100	1050 GPM, 40 HP, 5" discharge		1	24		6,075	1,100		7,175	8,350
4110	1500 GPM, 60 HP, 6" discharge		.60	40		10,600	1,825		12,425	14,300
4120	2000 GPM, 75 HP, 6" discharge		.50	48		10,600	2,175		12,775	14,900
4130	3000 GPM, 100 HP, 8" discharge		.40	60		14,800	2,725		17,525	20,400
4200	Horizontal split case, single stage									
4210	100 GPM, 7.5 HP, 1-1/2" discharge	Q-1	1.70	9.412	Ea.	3,250	415		3,665	4,200
4220	250 GPM, 15 HP, 2-1/2" discharge	"	1.30	12.308		6,025	540		6,565	7,425
4230	500 GPM, 20 HP, 4" discharge	Q-2	1.60	15		7,625	685		8,310	9,400
4240	750 GPM, 25 HP, 5" discharge		1.54	15.584		7,875	710		8,585	9,725
4250	1000 GPM, 40 HP, 5" discharge		1.20	20		10,300	910		11,210	12,800
4260	1500 GPM, 50 HP, 6" discharge	Q-3	1.42	22.535		12,800	1,050		13,850	15,600
4270	2000 GPM, 75 HP, 8" discharge		1.14	28.070		19,600	1,300		20,900	23,500
4280	3000 GPM, 100 HP, 10" discharge		.96	33.333		20,500	1,550		22,050	24,800
4290	3500 GPM, 150 HP, 10" discharge		.86	37.209		23,100	1,725		24,825	28,000
4300	4000 GPM, 200 HP, 10" discharge		.66	48.485		34,500	2,250		36,750	41,400
4330	Horizontal split case, two stage, 500' head									
4340	100 GPM, 40 HP, 1-1/2" discharge	Q-2	1.70	14.118	Ea.	13,500	640		14,140	15,900
4350	200 GPM, 50 HP, 1-1/2" discharge	"	1.44	16.667		13,500	760		14,260	16,100
4360	300 GPM, 75 HP, 2" discharge	Q-3	1.57	20.382		19,500	945		20,445	22,800
4370	400 GPM, 100 HP, 3" discharge		1.14	28.070		19,500	1,300		20,800	23,400
4380	800 GPM, 200 HP, 4" discharge		.86	37.209		30,800	1,725		32,525	36,500
5000	Centrifugal, in-line									
5006	Vertical mount, iron body, 125 lb. flgd, 3550 RPM TEFC mtr									
5010	Single stage									
5020	50 GPM, 3 HP, 1-1/2" discharge	Q-1	2.30	6.957	Ea.	990	305		1,295	1,525
5030	75 GPM, 5 HP, 1-1/2" discharge		1.60	10		1,075	440		1,515	1,825
5040	100 GPM, 7.5 HP, 1-1/2" discharge		1.30	12.308		1,325	540		1,865	2,250
5050	125 GPM, 10 HP, 1-1/2" discharge	Q-2	1.70	14.118		1,525	640		2,165	2,650
5060	150 GPM, 15 HP, 1-1/2" discharge		1.60	15		2,575	685		3,260	3,850
5080	250 GPM, 40 HP, 2" discharge		1.40	17.143		3,825	780		4,605	5,375
5090	300 GPM, 50 HP, 3" discharge		1.30	18.462		4,100	840		4,940	5,750
5100	400 GPM, 75 HP, 3" discharge	Q-3	.60	53.333		5,200	2,475		7,675	9,425
5110	600 GPM, 100 HP, 3" discharge	"	.50	64		5,475	2,975		8,450	10,500

22 11 23.11 Miscellaneous Pumps

Line	22 11 23.11 Miscellaneous Pumps	Crew	Daily Output	Labor-Hours	Unit	Material	Labor	Equipment	Total	Total Incl O&P
0010	**MISCELLANEOUS PUMPS**									
0020	Water pump, portable, gasoline powered									
0100	6000 GPH, 2" discharge	Q-1	11	1.455	Ea.	660	64		724	825
0110	8000 GPH, 2" discharge		10.50	1.524		600	67		667	760
0120	10,000 GPH, 2" discharge		10	1.600		1,225	70		1,295	1,450
0500	Pump, propylene body, housing and impeller									
0510	22 GPM, 1/3 HP, 40' HD	1 Plum	5	1.600	Ea.	440	78		518	595
0520	33 GPM, 1/2 HP, 40' HD	Q-1	5	3.200		490	140		630	750
0530	53 GPM, 3/4 HP, 40' HD	"	4	4		890	176		1,066	1,250
0600	Rotary pump, CI									
0614	202 GPH, 3/4 HP, 1" discharge	Q-1	4.50	3.556	Ea.	930	156		1,086	1,250
0618	277 GPH, 1 HP, 1" discharge		4	4		930	176		1,106	1,300
0624	1100 GPH, 1.5 HP, 1-1/4" discharge		3.60	4.444		2,325	195		2,520	2,850
0628	1900 GPH, 2 HP, 1-1/4" discharge		3.20	5		3,200	219		3,419	3,825
1000	Turbine pump, CI									
1010	50 GPM, 2 HP, 3" discharge	Q-1	.80	20	Ea.	3,125	880		4,005	4,775
1020	100 GPM, 3 HP, 4" discharge	Q-2	.96	25		4,925	1,150		6,075	7,125

22 11 Facility Water Distribution

22 11 23 – Domestic Water Pumps

22 11 23.11 Miscellaneous Pumps		Crew	Daily Output	Labor-Hours	Unit	2009 Bare Costs Material	Labor	Equipment	Total	Total Incl O&P
1030	250 GPM, 15 HP, 6" discharge	Q-2	.94	25.532	Ea.	6,100	1,150		7,250	8,475
1040	500 GPM, 25 HP, 6" discharge	Q-3	1.22	26.230		6,150	1,225		7,375	8,600
1050	1000 GPM, 50 HP, 8" discharge		1.14	28.070		8,500	1,300		9,800	11,300
1060	2000 GPM, 100 HP, 10" discharge		1	32		10,100	1,475		11,575	13,300
1070	3000 GPM, 150 HP, 10" discharge		.80	40		16,200	1,850		18,050	20,600
1080	4000 GPM, 200 HP, 12" discharge		.70	45.714		17,700	2,125		19,825	22,700
1090	6000 GPM, 300 HP, 14" discharge		.60	53.333		27,000	2,475		29,475	33,400
1100	10,000 GPM, 300 HP, 18" discharge		.58	55.172		31,200	2,550		33,750	38,300
2000	Centrifugal stainless steel pumps									
2100	100 GPM, 100' TDH, 3 HP	Q-1	1.80	8.889	Ea.	8,900	390		9,290	10,400
2130	250 GPM, 100' TDH, 10 HP	"	1.28	12.500		12,500	550		13,050	14,500
2160	500 GPM, 100' TDH, 20 HP	Q-2	1.44	16.667		14,300	760		15,060	16,900
2200	Vertical turbine stainless steel pumps									
2220	100 GPM, 100' TDH, 7.5 HP	Q-2	.95	25.263	Ea.	49,900	1,150		51,050	56,500
2240	250 GPM, 100' TDH, 15 HP		.94	25.532		55,000	1,150		56,150	63,000
2260	500 GPM, 100' TDH, 20 HP		.93	25.806		60,500	1,175		61,675	68,500
2280	750 GPM, 100' TDH, 30 HP	Q-3	1.19	26.891		66,000	1,250		67,250	74,500
2300	1000 GPM, 100' TDH, 40 HP	"	1.16	27.586		71,500	1,275		72,775	80,500
2320	100 GPM, 200' TDH, 15 HP	Q-2	.94	25.532		76,500	1,150		77,650	86,500
2340	250 GPM, 200' TDH, 25 HP	Q-3	1.22	26.230		82,000	1,225		83,225	92,000
2360	500 GPM, 200' TDH, 40 HP		1.16	27.586		87,500	1,275		88,775	98,000
2380	750 GPM, 200' TDH, 60 HP		1.13	28.319		89,000	1,325		90,325	100,000
2400	1000 GPM, 200' TDH, 75 HP		1.12	28.571		91,000	1,325		92,325	102,000

22 11 23.13 Domestic-Water Packaged Booster Pumps		Crew	Daily Output	Labor-Hours	Unit	Material	Labor	Equipment	Total	Total Incl O&P
0010	**DOMESTIC-WATER PACKAGED BOOSTER PUMPS**									
0200	Pump system, with diaphragm tank, control, press. switch									
0300	1 HP pump	Q-1	1.30	12.308	Ea.	5,050	540		5,590	6,375
0400	1-1/2 HP pump		1.25	12.800		5,125	560		5,685	6,475
0420	2 HP pump		1.20	13.333		5,225	585		5,810	6,625
0440	3 HP pump		1.10	14.545		5,325	640		5,965	6,800
0460	5 HP pump	Q-2	1.50	16		5,875	730		6,605	7,575
0480	7-1/2 HP pump		1.42	16.901		6,550	770		7,320	8,350
0500	10 HP pump		1.34	17.910		6,850	815		7,665	8,750
2000	Pump system, variable speed, base, controls, starter									
2010	Duplex, 100' head									
2020	400 GPM, 7-1/2HP, 4" discharge	Q-2	.70	34.286	Ea.	48,400	1,550		49,950	55,500
2025	Triplex, 100' head									
2030	1000 GPM, 15HP, 6" discharge	Q-2	.50	48	Ea.	69,000	2,175		71,175	79,500
2040	1700 GPM, 30HP, 6" discharge	"	.30	80	"	78,000	3,650		81,650	91,000

22 12 Facility Potable-Water Storage Tanks

22 12 23 – Facility Indoor Potable-Water Storage Tanks

22 12 23.13 Facility Steel, Indoor Pot.-Water Storage Tanks		Crew	Daily Output	Labor-Hours	Unit	Material	Labor	Equipment	Total	Total Incl O&P
0010	**FACILITY STEEL, INDOOR POT.-WATER STORAGE TANKS**									
2000	Galvanized steel, 15 gal., 14" diam., 26" LOA	1 Plum	12	.667	Ea.	990	32.50		1,022.50	1,125
2060	30 gal., 14" diam. x 49" LOA		11	.727		1,175	35.50		1,210.50	1,325
2080	80 gal., 20" diam. x 64" LOA		9	.889		1,825	43.50		1,868.50	2,075
2100	135 gal., 24" diam. x 75" LOA		6	1.333		2,650	65		2,715	3,025
2120	240 gal., 30" diam. x 86" LOA		4	2		5,200	97.50		5,297.50	5,875
2140	300 gal., 36" diam. x 76" LOA		3	2.667		7,375	130		7,505	8,300
2160	400 gal., 36" diam. x 100" LOA	Q-1	4	4		9,075	176		9,251	10,200

22 12 Facility Potable-Water Storage Tanks

22 12 23 – Facility Indoor Potable-Water Storage Tanks

22 12 23.13 Facility Steel, Indoor Pot.-Water Storage Tanks		Crew	Daily Output	Labor-Hours	Unit	2009 Bare Costs Material	Labor	Equipment	Total	Total Incl O&P
2180	500 gal., 36" diam., x 126" LOA	Q-1	3	5.333	Ea.	11,200	234		11,434	12,800
3000	Glass lined, P.E., 80 gal., 20" diam. x 60" LOA	1 Plum	9	.889		2,450	43.50		2,493.50	2,775
3060	140 gal., 24" diam. x 80" LOA		6	1.333		3,400	65		3,465	3,850
3080	225 gal., 30" diam. x 78" LOA		4	2		4,250	97.50		4,347.50	4,825
3100	325 gal., 36" diam. x 81" LOA		3	2.667		5,425	130		5,555	6,150
3120	460 gal., 42" diam. x 84" LOA	Q-1	4	4		5,700	176		5,876	6,550
3140	605 gal., 48" diam. x 87" LOA		3	5.333		10,900	234		11,134	12,400
3160	740 gal., 54" diam. x 91" LOA		3	5.333		12,300	234		12,534	13,900
3180	940 gal., 60" diam. x 93" LOA		2.50	6.400		14,200	281		14,481	16,000
3200	1330 gal., 66" diam. x 107" LOA		2	8		19,400	350		19,750	21,900
3220	1615 gal., 72" diam. x 110" LOA		1.50	10.667		22,100	470		22,570	25,000
3240	2285 gal., 84" diam. x 128" LOA		1	16		27,300	700		28,000	31,100
3260	3440 gal., 96" diam. x 157" LOA	Q-2	1.50	16		40,400	730		41,130	45,600

22 13 Facility Sanitary Sewerage

22 13 16 – Sanitary Waste and Vent Piping

22 13 16.60 Traps		Crew	Daily Output	Labor-Hours	Unit	Material	Labor	Equipment	Total	Total Incl O&P
0010	**TRAPS**									
4700	Copper, drainage, drum trap									
4800	3" x 5" solid, 1-1/2" pipe size	1 Plum	16	.500	Ea.	97.50	24.50		122	144
4840	3" x 6" swivel, 1-1/2" pipe size	"	16	.500	"	155	24.50		179.50	207
5100	P trap, standard pattern									
5200	1-1/4" pipe size	1 Plum	18	.444	Ea.	72	21.50		93.50	112
5240	1-1/2" pipe size		17	.471		69.50	23		92.50	111
5260	2" pipe size		15	.533		107	26		133	157
5280	3" pipe size		11	.727		258	35.50		293.50	335
5340	With cleanout and slip joint									
5360	1-1/4" pipe size	1 Plum	18	.444	Ea.	66.50	21.50		88	106
5400	1-1/2" pipe size		17	.471		106	23		129	152
5420	2" pipe size		15	.533		171	26		197	227
5460	For swivel, 1-1/2"					155			155	170
6710	ABS DWV P trap, solvent weld joint									
6720	1-1/2" pipe size	1 Plum	18	.444	Ea.	6.75	21.50		28.25	40
6722	2" pipe size		17	.471		9.15	23		32.15	44.50
6724	3" pipe size		15	.533		35	26		61	77.50
6726	4" pipe size		14	.571		72	28		100	121
6732	PVC DWV P trap, solvent weld joint									
6733	1-1/2" pipe size	1 Plum	18	.444	Ea.	5.45	21.50		26.95	38.50
6734	2" pipe size		17	.471		7.40	23		30.40	42.50
6735	3" pipe size		15	.533		25	26		51	66.50
6736	4" pipe size		14	.571		57	28		85	105
6760	PP DWV, dilution trap, 1-1/2" pipe size		16	.500		158	24.50		182.50	211
6770	P trap, 1-1/2" pipe size		17	.471		39	23		62	77.50
6780	2" pipe size		16	.500		58	24.50		82.50	101
6790	3" pipe size		14	.571		106	28		134	159
6800	4" pipe size		13	.615		179	30		209	242
6830	S trap, 1-1/2" pipe size		16	.500		32	24.50		56.50	71.50
6840	2" pipe size		15	.533		48	26		74	92
6850	Universal trap, 1-1/2" pipe size		14	.571		64	28		92	113
6860	PVC DWV hub x hub, basin trap, 1-1/4" pipe size		18	.444		8.95	21.50		30.45	42.50
6870	Sink P trap, 1-1/2" pipe size		18	.444		8.95	21.50		30.45	42.50

22 13 Facility Sanitary Sewerage

22 13 16 – Sanitary Waste and Vent Piping

22 13 16.60 Traps		Crew	Daily Output	Labor-Hours	Unit	Material	Labor	Equipment	Total	Total Incl O&P
						2009 Bare Costs				
6880	Tubular S trap, 1-1/2" pipe size	1 Plum	17	.471	Ea.	16.55	23		39.55	52.50
6890	PVC sch. 40 DWV, drum trap									
6900	1-1/2" pipe size	1 Plum	16	.500	Ea.	20.50	24.50		45	59
6910	P trap, 1-1/2" pipe size		18	.444		5.05	21.50		26.55	38
6920	2" pipe size		17	.471		6.65	23		29.65	42
6930	3" pipe size		15	.533		23	26		49	64.50
6940	4" pipe size		14	.571		53	28		81	100
6950	P trap w/clean out, 1-1/2" pipe size		18	.444		8.35	21.50		29.85	41.50
6960	2" pipe size		17	.471		14.20	23		37.20	50
6970	P trap adjustable, 1-1/2" pipe size		17	.471		6.45	23		29.45	41.50
6980	P trap adj. w/union & cleanout, 1-1/2" pipe size	↓	16	.500	↓	14.90	24.50		39.40	53

22 13 16.80 Vent Flashing and Caps

		Crew	Daily Output	Labor-Hours	Unit	Material	Labor	Equipment	Total	Total Incl O&P
0010	**VENT FLASHING AND CAPS**									
0120	Vent caps									
0140	Cast iron									
0180	2-1/2" - 3-5/8" pipe	1 Plum	21	.381	Ea.	40	18.55		58.55	72
0190	4" - 4-1/8" pipe	"	19	.421	"	48	20.50		68.50	84
0900	Vent flashing									
1000	Aluminum with lead ring									
1020	1-1/4" pipe	1 Plum	20	.400	Ea.	9.90	19.50		29.40	40.50
1030	1-1/2" pipe		20	.400		10.60	19.50		30.10	41
1040	2" pipe		18	.444		10.60	21.50		32.10	44
1050	3" pipe		17	.471		11.70	23		34.70	47.50
1060	4" pipe	↓	16	.500	↓	14.15	24.50		38.65	52
1350	Copper with neoprene ring									
1400	1-1/4" pipe	1 Plum	20	.400	Ea.	19.95	19.50		39.45	51.50
1430	1-1/2" pipe		20	.400		19.95	19.50		39.45	51.50
1440	2" pipe		18	.444		21	21.50		42.50	55.50
1450	3" pipe		17	.471		24.50	23		47.50	61.50
1460	4" pipe	↓	16	.500	↓	27	24.50		51.50	66.50
2000	Galvanized with neoprene ring									
2020	1-1/4" pipe	1 Plum	20	.400	Ea.	3.07	19.50		22.57	33
2030	1-1/2" pipe		20	.400		12.10	19.50		31.60	43
2040	2" pipe		18	.444		9.10	21.50		30.60	42.50
2050	3" pipe		17	.471		9.90	23		32.90	45.50
2060	4" pipe	↓	16	.500	↓	12.85	24.50		37.35	50.50
2980	Neoprene, one piece									
3000	1-1/4" pipe	1 Plum	24	.333	Ea.	6.30	16.25		22.55	31.50
3030	1-1/2" pipe		24	.333		6.90	16.25		23.15	32
3040	2" pipe		23	.348		7.25	16.95		24.20	33.50
3050	3" pipe		21	.381		8	18.55		26.55	37
3060	4" pipe	↓	20	.400	↓	10.80	19.50		30.30	41.50
4000	Lead, 4#, 8" skirt, VTR									
4100	2" pipe	1 Plum	18	.444	Ea.	38	21.50		59.50	74
4110	3" pipe		17	.471		45	23		68	84
4120	4" pipe		16	.500		51.50	24.50		76	93
4130	6" pipe	↓	14	.571	↓	73.50	28		101.50	123

22 13 26 – Sanitary Waste Separators

22 13 26.10 Separators

		Crew	Daily Output	Labor-Hours	Unit	Material	Labor	Equipment	Total	Total Incl O&P
0010	**SEPARATORS**, Entrainment eliminator, steel body, 150 PSIG.									
0100	1/4" size	1 Stpi	24	.333	Ea.	197	16.45		213.45	242
0120	1/2" size	↓	24	.333	↓	207	16.45		223.45	253

RSMeans Business Solutions

Leaders in Cost Engineering for the Design, Construction, and Facilities Management Industry...

Market Analysis and Third Party Validation – RSMeans conducts research studies from numerous perspectives, including escalation costs in local markets, cost estimates for legal damages, and trend research for industry associations, as well as product adoption and productivity studies for building product manufacturers. Means is often called upon to produce "third party validation" construction cost estimates for owners, architects, and boards of directors.

Predictive Cost Models for Facility Owners – From early planning, cost modeling is driven by close attention to standards, codes, methods, and business processes that impact cost and operational effectiveness. Predictive cost models offer cost control and standardization for facility owners, particularly in repetitive building environments. RSMeans custom models are based upon actual architectural plans and historical budgets, with cost data and location factors adjusted to regions anywhere in North America.

Custom Cost Databases and Web-Based Dashboards – Custom applications of the RSMeans cost databases use the building blocks of construction cost engineering, including a standard code of accounts, activity descriptions, material, labor, and equipment costs. RSMeans has in-house technical expertise for rapid deployment of cost databases, online cost calculators, and web-based analytic dashboards.

Job Order Contracting and FacilitiesWorks™ – A delivery method for construction renovation to reduce cycle times and overhead costs. RSMeans offers consulting in JOC management practices, FacilitiesWorks™ software for renovation cost estimating and project management, and on-site training programs. JOC is available to government agencies, colleges and universities, school districts, municipalities, and health care systems.

Life Cycle Cost Analysis and Sustainability – Building upon its extensive facility cost databases, RSMeans offers life cycle cost studies and sustainability models to analyze total cost of ownership (TCO) for high performance buildings. Analyses include capital costs, life cycle and operations costs, as well as studies specific to integrated building automation, energy efficiency and security systems. Cost savings realized in operations, maintenance and replacement are analyzed enterprise-wide.

CORPORATE • HEALTH CARE • EDUCATION • GOVERNMENT • MUNICIPAL • RETAIL • ASSOCIATIONS • BUILDING PRODUCT MANUFACTURERS • LEGAL

For more information contact RSMeans Business Solutions at 781-422-5101 or visit our website at www.rsmeans.com or email consulting@rsmeans.com

RSMeans

22 13 26 – Sanitary Waste Separators

22 13 26.10 Separators		Crew	Daily Output	Labor-Hours	Unit	Material	2009 Bare Costs Labor	Equipment	Total	Total Incl O&P
0140	3/4" size	1 Stpi	20	.400	Ea.	221	19.75		240.75	273
0160	1" size		19	.421		291	21		312	350
0180	1-1/4" size		15	.533		330	26.50		356.50	405
0200	1-1/2" size		13	.615		365	30.50		395.50	445
0220	2" size		11	.727		405	36		441	500
0240	2-1/2" size	Q-5	15	1.067		1,650	47.50		1,697.50	1,875
0260	3" size		13	1.231		1,950	54.50		2,004.50	2,225
0280	4" size		10	1.600		2,200	71		2,271	2,525
0300	5" size		6	2.667		2,575	118		2,693	3,000
0320	6" size		3	5.333		2,850	237		3,087	3,475
0340	8" size	Q-6	4.40	5.455		3,525	251		3,776	4,250
0360	10" size	"	4	6		5,125	276		5,401	6,050
1000	For 300 PSIG, add					15%				

22 13 29 – Sanitary Sewerage Pumps

22 13 29.13 Wet-Pit-Mounted, Vertical Sewerage Pumps

		Crew	Daily Output	Labor-Hours	Unit	Material	Labor	Equipment	Total	Total Incl O&P
0010	**WET-PIT-MOUNTED, VERTICAL SEWERAGE PUMPS**									
0020	Controls incl. alarm/disconnect panel w/wire. Excavation not included									
0260	Simplex, 9 GPM at 60 PSIG, 91 gal. tank				Ea.	2,825			2,825	3,100
0300	Unit with manway, 26" I.D., 18" high					3,250			3,250	3,575
0340	26" I.D., 36" high					3,400			3,400	3,725
0380	43" I.D., 4' high					3,450			3,450	3,800
0600	Simplex, 9 GPM at 60 PSIG, 150 gal. tank, indoor					3,100			3,100	3,400
0700	Unit with manway, 26" I.D., 36" high					3,750			3,750	4,125
0740	26" I.D., 4' high					3,875			3,875	4,250
2000	Duplex, 18 GPM at 60 PSIG, 150 gal. tank, indoor					5,400			5,400	5,950
2060	Unit with manway, 43" I.D., 4' high					7,125			7,125	7,825
2400	For core only					1,525			1,525	1,675

22 13 29.14 Sewage Ejector Pumps

		Crew	Daily Output	Labor-Hours	Unit	Material	Labor	Equipment	Total	Total Incl O&P
0010	**SEWAGE EJECTOR PUMPS**, With operating and level controls									
0100	Simplex system incl. tank, cover, pump 15' head									
0500	37 gal PE tank, 12 GPM, 1/2 HP, 2" discharge	Q-1	3.20	5	Ea.	440	219		659	815
0510	3" discharge		3.10	5.161		480	226		706	865
0530	87 GPM, .7 HP, 2" discharge		3.20	5		670	219		889	1,075
0540	3" discharge		3.10	5.161		725	226		951	1,150
0600	45 gal. coated stl tank, 12 GPM, 1/2 HP, 2" discharge		3	5.333		780	234		1,014	1,200
0610	3" discharge		2.90	5.517		810	242		1,052	1,250
0630	87 GPM, .7 HP, 2" discharge		3	5.333		1,000	234		1,234	1,450
0640	3" discharge		2.90	5.517		1,050	242		1,292	1,525
0660	134 GPM, 1 HP, 2" discharge		2.80	5.714		1,075	251		1,326	1,575
0680	3" discharge		2.70	5.926		1,150	260		1,410	1,650
0700	70 gal. PE tank, 12 GPM, 1/2 HP, 2" discharge		2.60	6.154		850	270		1,120	1,350
0710	3" discharge		2.40	6.667		910	293		1,203	1,450
0730	87 GPM, 0.7 HP, 2" discharge		2.50	6.400		1,100	281		1,381	1,625
0740	3" discharge		2.30	6.957		1,150	305		1,455	1,725
0760	134 GPM, 1 HP, 2" discharge		2.20	7.273		1,200	320		1,520	1,775
0770	3" discharge		2	8		1,275	350		1,625	1,925
0800	75 gal. coated stl. tank, 12 GPM, 1/2 HP, 2 " discharge		2.40	6.667		935	293		1,228	1,475
0810	3" discharge		2.20	7.273		970	320		1,290	1,550
0830	87 GPM, .7 HP, 2" discharge		2.30	6.957		1,175	305		1,480	1,750
0840	3" discharge		2.10	7.619		1,225	335		1,560	1,850
0860	134 GPM, 1 HP, 2" discharge		2	8		1,250	350		1,600	1,900
0880	3" discharge		1.80	8.889		1,325	390		1,715	2,025

22 13 Facility Sanitary Sewerage

22 13 29 – Sanitary Sewerage Pumps

22 13 29.14 Sewage Ejector Pumps		Crew	Daily Output	Labor-Hours	Unit	Material	Labor	Equipment	Total	Total Incl O&P
						2009 Bare Costs				
1040	Duplex system incl. tank, covers, pumps									
1060	110 gal. fiberglass tank, 24 GPM, 1/2 HP, 2" discharge	Q-1	1.60	10	Ea.	1,700	440		2,140	2,525
1080	3" discharge		1.40	11.429		1,800	500		2,300	2,725
1100	174 GPM, .7 HP, 2" discharge		1.50	10.667		2,200	470		2,670	3,100
1120	3" discharge		1.30	12.308		2,275	540		2,815	3,300
1140	268 GPM, 1 HP, 2" discharge		1.20	13.333		2,375	585		2,960	3,500
1160	3" discharge		1	16		2,475	700		3,175	3,750
1260	135 gal. coated stl. tank, 24 GPM, 1/2 HP, 2" discharge	Q-2	1.70	14.118		1,750	640		2,390	2,875
2000	3" discharge		1.60	15		1,850	685		2,535	3,050
2640	174 GPM, .7 HP, 2" discharge		1.60	15		2,275	685		2,960	3,525
2660	3" discharge		1.50	16		2,400	730		3,130	3,725
2700	268 GPM, 1 HP, 2" discharge		1.30	18.462		2,475	840		3,315	3,975
3040	3" discharge		1.10	21.818		2,625	995		3,620	4,400
3060	275 gal. coated stl. tank, 24 GPM, 1/2 HP, 2" discharge		1.50	16		2,175	730		2,905	3,500
3080	3" discharge		1.40	17.143		2,200	780		2,980	3,600
3100	174 GPM, .7 HP, 2" discharge		1.40	17.143		2,800	780		3,580	4,250
3120	3" discharge		1.30	18.462		2,975	840		3,815	4,525
3140	268 GPM, 1 HP, 2" discharge		1.10	21.818		3,075	995		4,070	4,875
3160	3" discharge		.90	26.667		3,250	1,225		4,475	5,400
3260	Pump system accessories, add									
3300	Alarm horn and lights, 115V mercury switch	Q-1	8	2	Ea.	93	88		181	234
3340	Switch, mag. contactor, alarm bell, light, 3 level control		5	3.200		455	140		595	710
3380	Alternator, mercury switch activated		4	4		825	176		1,001	1,175

22 14 Facility Storm Drainage

22 14 29 – Sump Pumps

22 14 29.13 Wet-Pit-Mounted, Vertical Sump Pumps		Crew	Daily Output	Labor-Hours	Unit	Material	Labor	Equipment	Total	Total Incl O&P
0010	**WET-PIT-MOUNTED, VERTICAL SUMP PUMPS**									
0400	Molded PVC base, 21 GPM at 15' head, 1/3 HP	1 Plum	5	1.600	Ea.	119	78		197	248
0800	Iron base, 21 GPM at 15' head, 1/3 HP		5	1.600		150	78		228	282
1200	Solid brass, 21 GPM at 15' head, 1/3 HP		5	1.600		263	78		341	405

22 14 29.16 Submersible Sump Pumps		Crew	Daily Output	Labor-Hours	Unit	Material	Labor	Equipment	Total	Total Incl O&P
0010	**SUBMERSIBLE SUMP PUMPS**									
7000	Sump pump, automatic									
7100	Plastic, 1-1/4" discharge, 1/4 HP	1 Plum	6	1.333	Ea.	136	65		201	247
7140	1/3 HP		5	1.600		163	78		241	297
7160	1/2 HP		5	1.600		201	78		279	340
7180	1-1/2" discharge, 1/2 HP		4	2		223	97.50		320.50	390
7500	Cast iron, 1-1/4" discharge, 1/4 HP		6	1.333		156	65		221	269
7540	1/3 HP		6	1.333		184	65		249	300
7560	1/2 HP		5	1.600		222	78		300	360

22 15 13 – General Service Compressed-Air Piping

22 15 13.10 Compressor Accessories		Crew	Daily Output	Labor-Hours	Unit	Material	Labor	Equipment	Total	Total Inc O&P
						2009 Bare Costs				
0010	**COMPRESSOR ACCESSORIES**									
1700	Refrigerated air dryers with ambient air filters									
1710	10 CFM	Q-5	8	2	Ea.	700	89		789	905
1720	25 CFM		6.60	2.424		1,875	108		1,983	2,200
1730	50 CFM		6.20	2.581		2,200	115		2,315	2,600
1740	75 CFM		5.80	2.759		2,625	123		2,748	3,050
1750	100 CFM	↓	5.60	2.857	↓	2,875	127		3,002	3,350
4000	Couplers, air line, sleeve type									
4010	Female, connection size NPT									
4020	1/4"	1 Stpi	38	.211	Ea.	6.30	10.40		16.70	22.50
4030	3/8"		36	.222		7.10	10.95		18.05	24.50
4040	1/2"		35	.229		14.40	11.30		25.70	33
4050	3/4"	↓	34	.235	↓	14.40	11.60		26	33.50
4100	Male									
4110	1/4"	1 Stpi	38	.211	Ea.	6.35	10.40		16.75	22.50
4120	3/8"		36	.222		7.10	10.95		18.05	24.50
4130	1/2"		35	.229		14.40	11.30		25.70	33
4140	3/4"	↓	34	.235	↓	14.40	11.60		26	33.50
4150	Coupler, combined male and female halves									
4160	1/2"	1 Stpi	17	.471	Ea.	29	23		52	66.50
4170	3/4"	"	15	.533	"	29	26.50		55.50	71

22 15 19 – General Service Packaged Air Compressors and Receivers

22 15 19.10 Air Compressors		Crew	Daily Output	Labor-Hours	Unit	Material	Labor	Equipment	Total	Total Inc O&P
0010	**AIR COMPRESSORS**									
5250	Air, reciprocating air cooled, splash lubricated, tank mounted									
5300	Single stage, 1 phase, 140 psi									
5303	1/2 HP, 30 gal tank	1 Stpi	3	2.667	Ea.	1,250	132		1,382	1,575
5305	3/4 HP, 30 gal tank		2.60	3.077		1,275	152		1,427	1,625
5307	1 HP, 30 gal tank	↓	2.20	3.636		1,625	179		1,804	2,075
5309	2 HP, 30 gal tank	Q-5	4	4		2,075	178		2,253	2,575
5310	3 HP, 30 gal tank		3.60	4.444		2,500	197		2,697	3,050
5314	3 HP, 60 gal tank		3.50	4.571		3,200	203		3,403	3,800
5320	5 HP, 60 gal tank		3.20	5		3,275	222		3,497	3,925
5330	5 HP, 80 gal tank		3	5.333		3,550	237		3,787	4,275
5340	7.5 HP, 80 gal tank	↓	2.60	6.154	↓	5,725	273		5,998	6,675
5600	2 stage pkg., 3 phase									
5650	6 CFM at 125 psi 1-1/2 HP, 60 gal tank	Q-5	3	5.333	Ea.	2,750	237		2,987	3,375
5670	10.9 CFM at 125 psi, 3 HP, 80 gal tank		1.50	10.667		3,375	475		3,850	4,425
5680	38.7 CFM at 125 psi, 10 HP, 120 gal tank	↓	.60	26.667		5,375	1,175		6,550	7,675
5690	105 CFM at 125 psi, 25 HP, 250 gal tank	Q-6	.60	40	↓	11,000	1,850		12,850	15,000
5800	With single stage pump									
5850	8.3 CFM at 125 psi, 2 HP, 80 gal tank	Q-6	3.50	6.857	Ea.	3,225	315		3,540	4,000
5860	38.7 CFM at 125 psi, 10 HP, 120 gal tank	"	.90	26.667	"	5,800	1,225		7,025	8,225
6000	Reciprocating, 2 stage, tank mtd, 3 Ph., Cap. rated @175 PSIG									
6050	Pressure lubricated, hvy duty, 9.7 CFM, 3 HP, 120 Gal. tank	Q-5	1.30	12.308	Ea.	5,025	545		5,570	6,350
6054	5 CFM, 1-1/2 HP, 80 gal tank		2.80	5.714		2,925	254		3,179	3,575
6056	6.4 CFM, 2 HP, 80 gal tank		2	8		3,225	355		3,580	4,050
6058	8.1 CFM, 3 HP, 80 gal tank		1.70	9.412		3,375	420		3,795	4,350
6059	14.8 CFM, 5 HP, 80 gal tank		1	16		3,550	710		4,260	5,030
6060	16.5 CFM, 5 HP, 120 gal tank		1	16		5,075	710		5,785	6,675
6063	13 CFM, 6 HP, 80 gal tank		.90	17.778		5,475	790		6,265	7,230
6066	19.8 CFM, 7.5 HP, 80 gal tank	↓	.80	20	↓	5,775	890		6,665	7,675

22 15 General Service Compressed-Air Systems

22 15 19 – General Service Packaged Air Compressors and Receivers

22 15 19.10 Air Compressors		Crew	Daily Output	Labor-Hours	Unit	2009 Bare Costs Material	Labor	Equipment	Total	Total Incl O&P
6070	25.8 CFM, 7-1/2 HP, 120 gal. tank	Q-5	.80	20	Ea.	7,575	890		8,465	9,650
6078	34.8 CFM, 10 HP, 80 gal. tank		.70	22.857		7,100	1,025		8,125	9,325
6080	34.8 CFM, 10 HP, 120 gal. tank		.60	26.667		7,450	1,175		8,625	9,975
6090	53.7 CFM, 15 HP, 120 gal. tank	Q-6	.80	30		9,350	1,375		10,725	12,400
6100	76.7 CFM, 20 HP, 120 gal. tank		.70	34.286		11,600	1,575		13,175	15,200
6104	76.7 CFM, 20 HP, 240 gal tank		.68	35.294		12,600	1,625		14,225	16,400
6110	90.1 CFM, 25 HP, 120 gal. tank		.63	38.095		12,100	1,750		13,850	15,900
6120	101 CFM, 30 HP, 120 gal. tank		.57	42.105		13,000	1,950		14,950	17,200
6130	101 CFM, 30 HP, 250 gal. tank		.52	46.154		14,100	2,125		16,225	18,700
6200	Oil-less, 13.6 CFM, 5 HP, 120 gal. tank	Q-5	.88	18.182		14,500	810		15,310	17,100
6210	13.6 CFM, 5 HP, 250 gal. tank		.80	20		15,900	890		16,790	18,800
6220	18.2 CFM, 7.5 HP, 120 gal. tank		.73	21.918		15,000	975		15,975	18,000
6230	18.2 CFM, 7.5 HP, 250 gal. tank		.67	23.881		16,100	1,050		17,150	19,300
6250	30.5 CFM, 10 HP, 120 gal. tank		.57	28.070		17,400	1,250		18,650	21,100
6260	30.5 CFM, 10 HP, 250 gal. tank		.53	30.189		18,500	1,350		19,850	22,400
6270	41.3 CFM, 15 HP, 120 gal. tank	Q-6	.70	34.286		19,000	1,575		20,575	23,300
6280	41.3 CFM, 15 HP, 250 gal. tank	"	.67	35.821		20,100	1,650		21,750	24,600

22 31 Domestic Water Softeners

22 31 13 – Residential Domestic Water Softeners

22 31 13.10 Residential Water Softeners		Crew	Daily Output	Labor-Hours	Unit	Material	Labor	Equipment	Total	Total Incl O&P
0010	**RESIDENTIAL WATER SOFTENERS**									
7350	Water softener, automatic, to 30 grains per gallon	2 Plum	5	3.200	Ea.	410	156		566	685
7400	To 100 grains per gallon	"	4	4	"	680	195		875	1,050

22 35 Domestic Water Heat Exchangers

22 35 30 – Water Heating by Steam

22 35 30.10 Water Heating Transfer Package		Crew	Daily Output	Labor-Hours	Unit	Material	Labor	Equipment	Total	Total Incl O&P
0010	**WATER HEATING TRANSFER PACKAGE**, Complete controls,									
0020	expansion tank, converter, air separator									
1000	Hot water, 180° F enter, 200° F leaving, 15# steam									
1010	One pump system, 28 GPM	Q-6	.75	32	Ea.	15,600	1,475		17,075	19,400
1020	35 GPM		.70	34.286		17,200	1,575		18,775	21,300
1040	55 GPM		.65	36.923		19,500	1,700		21,200	24,100
1060	130 GPM		.55	43.636		24,200	2,000		26,200	29,600
1080	255 GPM		.40	60		32,900	2,775		35,675	40,400
1100	550 GPM		.30	80		41,600	3,675		45,275	51,500
1120	800 GPM		.25	96		49,400	4,425		53,825	61,000
1220	Two pump system, 28 GPM		.70	34.286		21,300	1,575		22,875	25,800
1240	35 GPM		.65	36.923		25,300	1,700		27,000	30,400
1260	55 GPM		.60	40		25,800	1,850		27,650	31,200
1280	130 GPM		.50	48		33,100	2,200		35,300	39,700
1300	255 GPM		.35	68.571		44,600	3,150		47,750	54,000
1320	550 GPM		.25	96		56,000	4,425		60,425	68,000
1340	800 GPM		.20	120		65,000	5,525		70,525	80,000

22 41 Residential Plumbing Fixtures

22 41 39 – Residential Faucets, Supplies and Trim

22 41 39.10 Faucets and Fittings		Crew	Daily Output	Labor-Hours	Unit	Material	Labor	Equipment	Total	Total Incl O&P
						2009 Bare Costs				
0010	FAUCETS AND FITTINGS									
5000	Sillcock, compact, brass, IPS or copper to hose	1 Plum	24	.333	Ea.	8.05	16.25		24.30	33.50
6000	Stop and waste valves, bronze									
6100	Angle, solder end 1/2"	1 Plum	24	.333	Ea.	7.05	16.25		23.30	32.50
6110	3/4"		20	.400		7.75	19.50		27.25	38
6300	Straightway, solder end 3/8"		24	.333		6.15	16.25		22.40	31.50
6310	1/2"		24	.333		2.81	16.25		19.06	27.50
6320	3/4"		20	.400		3.48	19.50		22.98	33.50
6330	1"		19	.421		19.95	20.50		40.45	53
6400	Straightway, threaded 3/8"		24	.333		7.15	16.25		23.40	32.50
6410	1/2"		24	.333		3.82	16.25		20.07	28.50
6420	3/4"		20	.400		6.35	19.50		25.85	36.50
6430	1"	↓	19	.421	↓	12.60	20.50		33.10	45

22 51 Swimming Pool Plumbing Systems

22 51 19 – Swimming Pool Water Treatment Equipment

22 51 19.50 Swimming Pool Filtration Equipment		Crew	Daily Output	Labor-Hours	Unit	Material	Labor	Equipment	Total	Total Incl O&P
0010	SWIMMING POOL FILTRATION EQUIPMENT									
0900	Filter system, sand or diatomite type, incl. pump, 6,000 gal./hr.	2 Plum	1.80	8.889	Total	1,475	435		1,910	2,250
1020	Add for chlorination system, 800 S.F. pool		3	5.333	Ea.	78.50	260		338.50	475
1040	5,000 S.F. pool	↓	3	5.333	"	1,950	260		2,210	2,525

22 52 Fountain Plumbing Systems

22 52 16 – Fountain Pumps

22 52 16.10 Fountain Water Pumps		Crew	Daily Output	Labor-Hours	Unit	Material	Labor	Equipment	Total	Total Incl O&P
0010	FOUNTAIN WATER PUMPS									
0100	Pump w/controls									
0200	Single phase, 100' cord, 1/2 H.P. pump	2 Skwk	4.40	3.636	Ea.	1,800	149		1,949	2,200
0300	3/4 H.P. pump		4.30	3.721		2,050	152		2,202	2,475
0400	1 H.P. pump		4.20	3.810		3,800	156		3,956	4,425
0500	1-1/2 H.P. pump		4.10	3.902		3,500	159		3,659	4,100
0600	2 H.P. pump		4	4		4,725	163		4,888	5,450
0700	Three phase, 200' cord, 5 H.P. pump		3.90	4.103		5,000	168		5,168	5,750
0800	7-1/2 H.P. pump		3.80	4.211		8,500	172		8,672	9,625
0900	10 H.P. pump		3.70	4.324		9,275	177		9,452	10,500
1000	15 H.P. pump	↓	3.60	4.444	↓	11,300	182		11,482	12,800

22 52 33 – Fountain Ancillary

22 52 33.10 Fountain Miscellaneous		Crew	Daily Output	Labor-Hours	Unit	Material	Labor	Equipment	Total	Total Incl O&P
0010	FOUNTAIN MISCELLANEOUS									
1100	Nozzles, minimum	2 Skwk	8	2	Ea.	150	81.50		231.50	292
1200	Maximum		8	2		300	81.50		381.50	455
1300	Lights w/mounting kits, 200 watt		18	.889		1,000	36.50		1,036.50	1,150
1400	300 watt		18	.889		1,200	36.50		1,236.50	1,375
1500	500 watt		18	.889		1,300	36.50		1,336.50	1,475
1600	Color blender	↓	12	1.333	↓	530	54.50		584.50	670

22 62 Vacuum Systems for Laboratory and Healthcare Facilities

22 62 19 – Vacuum Equipment for Laboratory and Healthcare Facilities

22 62 19.70 Healthcare Vacuum Equipment		Crew	Daily Output	Labor-Hours	Unit	2009 Bare Costs Material	Labor	Equipment	Total	Total Incl O&P
0010	**HEALTHCARE VACUUM EQUIPMENT**									
0100	Medical, with receiver									
0110	Duplex									
0120	20 SCFM	Q-1	1.14	14.035	Ea.	15,700	615		16,315	18,100
0130	60 SCFM, 10 HP	Q-2	1.20	20	"	27,100	910		28,010	31,300
0200	Triplex									
0220	180 SCFM	Q-2	.86	27.907	Ea.	51,000	1,275		52,275	58,500
0300	Dental oral									
0310	Duplex									
0330	165 SCFM with 77 Gal separator	Q-2	1.30	18.462	Ea.	46,300	840		47,140	52,500
1100	Vacuum system									
1110	Vacuum outlet alarm panel	1 Plum	3.20	2.500	Ea.	935	122		1,057	1,200

22 63 Gas Systems for Laboratory and Healthcare Facilities

22 63 13 – Gas Piping for Laboratory and Healthcare Facilities

22 63 13.70 Healthcare Gas Piping		Crew	Daily Output	Labor-Hours	Unit	Material	Labor	Equipment	Total	Total Incl O&P
0010	**HEALTHCARE GAS PIPING**									
1000	Nitrogen or oxygen system									
1010	Cylinder manifold									
1020	5 cylinder	1 Plum	.80	10	Ea.	5,025	490		5,515	6,250
1026	10 cylinder	"	.40	20	"	6,100	975		7,075	8,175
3000	Outlets and valves									
3010	Recessed, wall mounted									
3012	Single outlet	1 Plum	3.20	2.500	Ea.	60	122		182	249
3100	Ceiling outlet									
3200	Zone valve with box									
3210	2"	1 Plum	3.20	2.500	Ea.	420	122		542	645
4000	Alarm panel, medical gases and vacuum									
4010	Alarm panel	1 Plum	3.20	2.500	Ea.	935	122		1,057	1,200

Estimating Tips

The labor adjustment factors listed in Subdivision 22 01 02.20 also apply to Division 23.

23 10 00 Facility Fuel Systems

- The prices in this subdivision for above- and below-ground storage tanks do not include foundations or hold-down slabs. The estimator should refer to Divisions 3 and 31 for foundation system pricing. In addition to the foundations, required tank accessories, such as tank gauges, leak detection devices, and additional manholes and piping, must be added to the tank prices.

23 50 00 Central Heating Equipment

- When estimating the cost of an HVAC system, check to see who is responsible for providing and installing the temperature control system. It is possible to overlook controls, assuming that they would be included in the electrical estimate.
- When looking up a boiler, be careful on specified capacity. Some manufacturers rate their products on output while others use input.
- Include HVAC insulation for pipe, boiler, and duct (wrap and liner).
- Be careful when looking up mechanical items to get the correct pressure rating and connection type (thread, weld, flange).

23 70 00 Central HVAC Equipment

- Combination heating and cooling units are sized by the air conditioning requirements. (See Reference No. R236000-20 for preliminary sizing guide.)
- A ton of air conditioning is nominally 400 CFM.
- Rectangular duct is taken off by the linear foot for each size, but its cost is usually estimated by the pound. Remember that SMACNA standards now base duct on internal pressure.
- Prefabricated duct is estimated and purchased like pipe: straight sections and fittings.
- Note that cranes or other lifting equipment are not included on any lines in Division 23. For example, if a crane is required to lift a heavy piece of pipe into place high above a gym floor, or to put a rooftop unit on the roof of a four-story building, etc., it must be added. Due to the potential for extreme variation—from nothing additional required to a major crane or helicopter—we feel that including a nominal amount for "lifting contingency" would be useless and detract from the accuracy of the estimate. When using equipment rental cost data from RSMeans, do not forget to include the cost of the operator(s).

Reference Numbers

Reference numbers are shown in shaded boxes at the beginning of some major classifications. These numbers refer to related items in the Reference Section. The reference information may be an estimating procedure, an alternate pricing method, or technical information.

Note: Not all subdivisions listed here necessarily appear in this publication.

Note: **Trade Service,** *in part, has been used as a reference source for some of the material prices used in Division 23.*

23 05 Common Work Results for HVAC

23 05 05 – Selective HVAC Demolition

23 05 05.10 HVAC Demolition		Crew	Daily Output	Labor-Hours	Unit	2009 Bare Costs Material	Labor	Equipment	Total	Total Incl O&P
0010	**HVAC DEMOLITION** R220105-10									
0100	Air conditioner, split unit, 3 ton	Q-5	2	8	Ea.		355		355	535
0150	Package unit, 3 ton	Q-6	3	8			370		370	555
0190	Rooftop, self contained, up to 5 ton	1 Plum	1.20	6.667			325		325	490
0250	Air curtain	Q-9	20	.800	L.F.		34		34	52.50
0254	Air filters, up thru 16,000 CFM		20	.800	Ea.		34		34	52.50
0256	20,000 thru 60,000 CFM		16	1			42.50		42.50	65.50
0297	Boiler blowdown	Q-5	8	2			89		89	133
0298	Boilers									
0300	Electric, up thru 148 kW	Q-19	2	12	Ea.		545		545	815
0310	150 thru 518 kW	"	1	24			1,075		1,075	1,625
0320	550 thru 2000 kW	Q-21	.40	80			3,700		3,700	5,550
0330	2070 kW and up	"	.30	106			4,950		4,950	7,400
0340	Gas and/or oil, up thru 150 MBH	Q-7	2.20	14.545			685		685	1,025
0350	160 thru 2000 MBH		.80	40			1,875		1,875	2,825
0360	2100 thru 4500 MBH		.50	64			3,000		3,000	4,525
0370	4600 thru 7000 MBH		.30	106			5,025		5,025	7,525
0380	7100 thru 12,000 MBH		.16	200			9,400		9,400	14,100
0390	12,200 thru 25,000 MBH		.12	266			12,500		12,500	18,800
0400	Central station air handler unit, up thru 15 ton	Q-5	1.60	10			445		445	665
0410	17.5 thru 30 ton	"	.80	20			890		890	1,325
0430	Computer room unit									
0434	Air cooled split, up thru 10 ton	Q-5	.67	23.881	Ea.		1,050		1,050	1,600
0436	12 thru 23 ton		.53	30.189			1,350		1,350	2,025
0440	Chilled water, up thru 10 ton		1.30	12.308			545		545	820
0444	12 thru 23 ton		1	16			710		710	1,075
0450	Glycol system, up thru 10 ton		.53	30.189			1,350		1,350	2,025
0454	12 thru 23 ton		.40	40			1,775		1,775	2,675
0460	Water cooled, not including condenser, up thru 10 ton		.80	20			890		890	1,325
0464	12 thru 23 ton		.60	26.667			1,175		1,175	1,775
0600	Condenser, up thru 50 ton		1	16			710		710	1,075
0610	51 thru 100 ton	Q-6	.80	30			1,375		1,375	2,075
0620	101 thru 1000 ton	"	.16	150			6,900		6,900	10,400
0660	Condensing unit, up thru 10 ton	Q-5	1.25	12.800			570		570	855
0670	11 thru 50 ton	"	.40	40			1,775		1,775	2,675
0680	60 thru 100 ton	Q-6	.30	80			3,675		3,675	5,525
0700	Cooling tower, up thru 400 ton		.80	30			1,375		1,375	2,075
0710	450 thru 600 ton		.53	45.283			2,075		2,075	3,125
0720	700 thru 1300 ton		.40	60			2,775		2,775	4,150
0780	Dehumidifier, up thru 155 lb./hr.	Q-1	8	2			88		88	132
0790	240 lb./hr. and up	"	2	8			350		350	525
1560	Ductwork									
1570	Metal, steel, sst, fabricated	Q-9	1000	.016	Lb.		.68		.68	1.05
1580	Aluminum, fabricated		485	.033	"		1.40		1.40	2.16
1590	Spiral, prefabricated		400	.040	L.F.		1.70		1.70	2.62
1600	Fiberglass, prefabricated		400	.040			1.70		1.70	2.62
1610	Flex, prefabricated		500	.032			1.36		1.36	2.09
1620	Glass fiber reinforced plastic, prefabricated		280	.057			2.43		2.43	3.74
1630	Diffusers, registers or grills, up thru 20" max dimension	1 Shee	50	.160	Ea.		7.55		7.55	11.60
1640	21 thru 36" max dimension		36	.222			10.50		10.50	16.15
1650	Above 36" max dimension		30	.267			12.60		12.60	19.35
1700	Evaporator, up thru 12,000 BTUH	Q-5	5.30	3.019			134		134	201
1710	12,500 thru 30,000 BTUH	"	2.70	5.926			263		263	395

23 05 05 – Selective HVAC Demolition

23 05 05.10 HVAC Demolition		Crew	Daily Output	Labor-Hours	Unit	Material	2009 Bare Costs Labor	Equipment	Total	Total Incl O&P
1720	31,000 BTUH and up	Q-6	1.50	16	Ea.		735		735	1,100
1730	Evaporative cooler, up thru 5 H.P.	Q-9	2.70	5.926			252		252	385
1740	10 thru 30 H.P.	"	.67	23.881			1,025		1,025	1,550
1750	Exhaust systems									
1760	Exhaust components	1 Shee	8	1	System		47		47	72.50
1770	Weld fume hoods	"	20	.400	Ea.		18.90		18.90	29
2120	Fans, up thru 1 H.P. or 2000 CFM	Q-9	8	2			85		85	131
2124	1-1/2 thru 10 H.P. or 20,000 CFM		5.30	3.019			128		128	197
2128	15 thru 30 H.P. or above 20,000 CFM		4	4			170		170	262
2150	Fan coil air conditioner, chilled water, up thru 7.5 ton	Q-5	14	1.143			51		51	76
2154	Direct expansion, up thru 10 ton		8	2			89		89	133
2158	11 thru 30 ton		2	8			355		355	535
2170	Flue shutter damper	Q-9	8	2			85		85	131
2200	Furnace, electric	Q-20	2	10			435		435	665
2300	Gas or oil, under 120 MBH	Q-9	4	4			170		170	262
2340	Over 120 MBH	"	3	5.333			227		227	350
2730	Heating and ventilating unit	Q-5	2.70	5.926			263		263	395
2740	Heater, electric, wall, baseboard and quartz	1 Elec	10	.800			37.50		37.50	56
2750	Heater, electric, unit, cabinet, fan and convector	"	8	1			47		47	70
2760	Heat exchanger, shell and tube type	Q-5	1.60	10			445		445	665
2770	Plate type	Q-6	.60	40			1,850		1,850	2,775
2810	Heat pump									
2820	Air source, split, up thru 10 ton	Q-5	.90	17.778	Ea.		790		790	1,175
2830	15 thru 25 ton	Q-6	.80	30			1,375		1,375	2,075
2850	Single package, up thru 12 ton	Q-5	1	16			710		710	1,075
2860	Water source, up thru 15 ton	"	.90	17.778			790		790	1,175
2870	20 thru 50 ton	Q-6	.80	30			1,375		1,375	2,075
2910	Heat recovery package, up thru 20,000 CFM	Q-5	2	8			355		355	535
2920	25,000 CFM and up		1.20	13.333			590		590	890
2930	Heat transfer package, up thru 130 GPM		.80	20			890		890	1,325
2934	255 thru 800 GPM		.42	38.095			1,700		1,700	2,550
2940	Humidifier		10.60	1.509			67		67	101
2961	Hydronic unit heaters, up thru 200 MBH		14	1.143			51		51	76
2962	Above 200 MBH		8	2			89		89	133
2964	Valance units		32	.500			22		22	33.50
2966	Radiant floor heating									
2967	System valves, controls, manifolds	Q-5	16	1	Ea.		44.50		44.50	66.50
2968	Per room distribution		8	2			89		89	133
2970	Hydronic heating, baseboard radiation		16	1			44.50		44.50	66.50
2976	Convectors and free standing radiators		18	.889			39.50		39.50	59.50
2980	Induced draft fan, up thru 1 H.P.	Q-9	4.60	3.478			148		148	227
2984	1-1/2 H.P. thru 7-1/2 H.P.	"	2.20	7.273			310		310	475
2988	Infra-red unit	Q-5	16	1			44.50		44.50	66.50
2992	Louvers	1 Shee	46	.174	S.F.		8.20		8.20	12.65
3000	Mechanical equipment, light items. Unit is weight, not cooling.	Q-5	.90	17.778	Ton		790		790	1,175
3600	Heavy items	"	1.10	14.545	"		645		645	970
3700	Deduct for salvage (when applicable), minimum				Job				73	80
3710	Maximum				"				455	500
3720	Make up air unit, up thru 6000 CFM	Q-5	3	5.333	Ea.		237		237	355
3730	6500 thru 30,000 CFM	"	1.60	10			445		445	665
3740	35,000 thru 75,000 CFM	Q-6	1	24			1,100		1,100	1 650
3800	Mixing boxes, constant and VAV	Q-9	18	.889			38		38	58
4000	Packaged terminal air conditioner, up thru 18,000 BTUH	Q-5	8	2			89		89	133

23 05 Common Work Results for HVAC

23 05 05 – Selective HVAC Demolition

23 05 05.10 HVAC Demolition		Crew	Daily Output	Labor-Hours	Unit	Material	2009 Bare Costs Labor	Equipment	Total	Total Incl O&P
4010	24,000 thru 48,000 BTUH	Q-5	2.80	5.714	Ea.		254		254	380
5000	Refrigerant compressor, reciprocating or scroll									
5010	Up thru 5 ton	1 Stpi	6	1.333	Ea.		66		66	98.50
5020	5.08 thru 10 ton	Q-5	6	2.667			118		118	178
5030	15 thru 50 ton	″	.40	40			1,775		1,775	2,675
5040	60 thru 130 ton	Q-6	.48	50			2,300		2,300	3,450
5100	Roof top air conditioner, up thru 10 ton	Q-5	1.40	11.429			510		510	760
5110	12 thru 40 ton	Q-6	1	24			1,100		1,100	1,650
5120	50 thru 140 ton		.50	48			2,200		2,200	3,325
5130	150 thru 300 ton		.30	80			3,675		3,675	5,525
6000	Self contained single package air conditioner, up thru 10 ton	Q-5	1.60	10			445		445	665
6010	15 thru 60 ton	Q-6	1.20	20			920		920	1,375
6100	Space heaters, up thru 200 MBH	Q-5	10	1.600			71		71	107
6110	Over 200 MBH		5	3.200			142		142	213
6200	Split ductless, both sections		8	2			89		89	133
6300	Steam condensate meter	1 Stpi	11	.727			36		36	54
6600	Thru-the-wall air conditioner	L-2	8	2			70.50		70.50	109
7000	Vent chimney, prefabricated, up thru 12″ diameter	Q-9	94	.170	V.L.F.		7.25		7.25	11.15
7010	14″ thru 36″ diameter		40	.400			17		17	26
7020	38″ thru 48″ diameter		32	.500			21		21	32.50
7030	54″ thru 60″ diameter	Q-10	14	1.714			75.50		75.50	116
7400	Ventilators, up thru 14″ neck diameter	Q-9	58	.276	Ea.		11.70		11.70	18.05
7410	16″ thru 50″ neck diameter		40	.400			17		17	26
7450	Relief vent, up thru 24″ x 96″		22	.727			31		31	47.50
7460	48″ x 60″ thru 96″ x 144″		10	1.600			68		68	105
8000	Water chiller up thru 10 ton	Q-5	2.50	6.400			284		284	425
8010	15 thru 100 ton	Q-6	.48	50			2,300		2,300	3,450
8020	110 thru 500 ton	Q-7	.29	110			5,175		5,175	7,775
8030	600 thru 1000 ton		.23	139			6,550		6,550	9,825
8040	1100 ton and up		.20	160			7,525		7,525	11,300
8400	Window air conditioner	1 Carp	16	.500			20		20	31

23 05 23 – General-Duty Valves for HVAC Piping

23 05 23.30 Valves, Iron Body

		Crew	Daily Output	Labor-Hours	Unit	Material	Labor	Equipment	Total	Total Incl O&P
0010	**VALVES, IRON BODY** R220523-90									
0022	For grooved joint, see Div. 22 11 13.48									
0100	Angle, 125 lb.									
0110	Flanged									
0116	2″	1 Plum	5	1.600	Ea.	900	78		978	1,100
0118	4″	Q-1	3	5.333		1,500	234		1,734	2,000
0120	6″	Q-2	3	8		2,925	365		3,290	3,750
0122	8″	″	2.50	9.600		5,225	435		5,660	6,400
0560	Butterfly, lug type, pneumatic operator, 2″ size	1 Stpi	14	.571		192	28		220	254
0570	3″	Q-1	8	2		205	88		293	360
0580	4″	″	5	3.200		230	140		370	465
0590	6″	Q-2	5	4.800		320	218		538	680
0600	8″		4.50	5.333		410	243		653	815
0610	10″		4	6		520	273		793	980
0620	12″		3	8		820	365		1,185	1,450
0630	14″		2.30	10.435		1,325	475		1,800	2,175
0640	18″		1.50	16		2,175	730		2,905	3,500
0650	20″		1	24		2,650	1,100		3,750	4,550
0790	Butterfly, lug type, gear operated, 2″ size, 200 lb. except noted	1 Plum	14	.571		254	28		282	320

23 05 23.30 Valves, Iron Body		Crew	Daily Output	Labor-Hours	Unit	2009 Bare Costs Material	Labor	Equipment	Total	Total Incl O&P
0800	2-1/2"	Q-1	9	1.778	Ea.	255	78		333	400
0810	3"		8	2		268	88		356	425
0820	4"		5	3.200		305	140		445	545
0830	5"	Q-2	5	4.800		375	218		593	745
0840	6"		5	4.800		420	218		638	790
0850	8"		4.50	5.333		540	243		783	960
0860	10"		4	6		715	273		988	1,200
0870	12"		3	8		1,050	365		1,415	1,700
0880	14", 150 lb.		2.30	10.435		1,975	475		2,450	2,875
0890	16", 150 lb.		1.75	13.714		2,525	625		3,150	3,725
0900	18", 150 lb.		1.50	16		3,625	730		4,355	5,075
0910	20", 150 lb.		1	24		4,575	1,100		5,675	6,675
0930	24", 150 lb.		.75	32		8,200	1,450		9,650	11,200
1020	Butterfly, wafer type, gear actuator, 200 lb.									
1030	2"	1 Plum	14	.571	Ea.	155	28		183	212
1040	2-1/2"	Q-1	9	1.778		159	78		237	292
1050	3"		8	2		165	88		253	315
1060	4"		5	3.200		188	140		328	415
1070	5"	Q-2	5	4.800		249	218		467	605
1080	6"		5	4.800		281	218		499	640
1090	8"		4.50	5.333		320	243		563	715
1100	10"		4	6		470	273		743	925
1110	12"		3	8		650	365		1,015	1,250
1200	Wafer type, lever actuator, 200 lb.									
1220	2"	1 Plum	14	.571	Ea.	113	28		141	166
1230	2-1/2"	Q-1	9	1.778		116	78		194	244
1240	3"		8	2		122	88		210	266
1250	4"		5	3.200		149	140		289	375
1260	5"	Q-2	5	4.800		207	218		425	560
1270	6"		5	4.800		250	218		468	605
1280	8"		4.50	5.333		370	243		613	770
1290	10"		4	6		515	273		788	975
1300	12"		3	8		930	365		1,295	1,575
1600	Gate, threaded, 125 lb.									
1603	1-1/2"	1 Plum	13	.615	Ea.	310	30		340	390
1604	2"		11	.727		745	35.50		780.50	875
1605	2-1/2"		10	.800		810	39		849	950
1606	3"		8	1		965	49		1,014	1,150
1607	4"		5	1.600		1,350	78		1,428	1,625
1650	Gate, 125 lb., N.R.S.									
2150	Flanged									
2200	2"	1 Plum	5	1.600	Ea.	495	78		573	660
2240	2-1/2"	Q-1	5	3.200		510	140		650	770
2260	3"		4.50	3.556		570	156		726	860
2280	4"		3	5.333		815	234		1,049	1,250
2290	5"	Q-2	3.40	7.059		1,400	320		1,720	2,000
2300	6"		3	8		1,400	365		1,765	2,075
2320	8"		2.50	9.600		2,375	435		2,810	3,275
2340	10"		2.20	10.909		4,200	495		4,695	5,350
2360	12"		1.70	14.118		5,750	640		6,390	7,300
2370	14"		1.30	18.462		7,475	840		8,315	9,475
2380	16"		1	24		10,400	1,100		11,500	13,200
2420	For 250 lb flanged, add					200%	10%			

23 05 23.30 Valves, Iron Body		Crew	Daily Output	Labor-Hours	Unit	Material	2009 Bare Costs Labor	Equipment	Total	Total Incl O&P
3500	OS&Y, 125 lb., (225# noted), threaded									
3504	3/4" (225 lb.)	1 Plum	18	.444	Ea.	440	21.50		461.50	520
3505	1" (225 lb.)		16	.500		500	24.50		524.50	585
3506	1-1/2" (225 lb.)		13	.615		730	30		760	845
3507	2"		11	.727		875	35.50		910.50	1,025
3550	OS&Y, 125 lb., flanged									
3600	2"	1 Plum	5	1.600	Ea.	310	78		388	455
3640	2-1/2"	Q-1	5	3.200		320	140		460	560
3660	3"		4.50	3.556		355	156		511	630
3670	3-1/2"		3	5.333		405	234		639	795
3680	4"		3	5.333		510	234		744	910
3690	5"	Q-2	3.40	7.059		840	320		1,160	1,400
3700	6"		3	8		840	365		1,205	1,475
3720	8"		2.50	9.600		1,500	435		1,935	2,300
3740	10"		2.20	10.909		2,725	495		3,220	3,750
3760	12"		1.70	14.118		3,650	640		4,290	5,000
3770	14"		1.30	18.462		6,975	840		7,815	8,900
3780	16"		1	24		10,400	1,100		11,500	13,200
3790	18"		.80	30		13,500	1,375		14,875	16,900
3800	20"		.60	40		19,500	1,825		21,325	24,200
3830	24"		.50	48		27,800	2,175		29,975	33,900
3900	For 175 lb, flanged, add					200%	10%			
4350	Globe, OS&Y									
4540	Class 125, flanged									
4550	2"	1 Plum	5	1.600	Ea.	630	78		708	810
4560	2-1/2"	Q-1	5	3.200		635	140		775	910
4570	3"		4.50	3.556		775	156		931	1,075
4580	4"		3	5.333		1,100	234		1,334	1,575
4590	5"	Q-2	3.40	7.059		2,025	320		2,345	2,700
4600	6"		3	8		2,025	365		2,390	2,775
4610	8"		2.50	9.600		3,950	435		4,385	5,000
4612	10"		2.20	10.909		5,925	495		6,420	7,275
4614	12"		1.70	14.118		6,450	640		7,090	8,050
5040	Class 250, flanged									
5050	2"	1 Plum	4.50	1.778	Ea.	1,000	86.50		1,086.50	1,250
5060	2-1/2"	Q-1	4.50	3.556		1,325	156		1,481	1,675
5070	3"		4	4		1,350	176		1,526	1,775
5080	4"		2.70	5.926		2,000	260		2,260	2,600
5090	5"	Q-2	3	8		3,575	365		3,940	4,475
5100	6"		2.70	8.889		3,575	405		3,980	4,525
5110	8"		2.20	10.909		6,075	495		6,570	7,425
5120	10"		2	12		9,275	545		9,820	11,000
5130	12"		1.60	15		13,900	685		14,585	16,300
5240	Valve sprocket rim w/chain, for 2" valve	1 Stpi	30	.267		108	13.15		121.15	139
5250	2-1/2" valve		27	.296		108	14.60		122.60	141
5260	3-1/2" valve		25	.320		108	15.80		123.80	143
5270	6" valve		20	.400		108	19.75		127.75	149
5280	8" valve		18	.444		144	22		166	191
5290	12" valve		16	.500		144	24.50		168.50	195
5300	16" valve		12	.667		199	33		232	268
5310	20" valve		10	.800		199	39.50		238.50	277
5320	36" valve		8	1		375	49.50		424.50	490
5450	Swing check, 125 lb., threaded									

23 05 23.30 Valves, Iron Body		Crew	Daily Output	Labor-Hours	Unit	Material	Labor	Equipment	Total	Total Incl O&P
						2009 Bare Costs				
5470	1"	1 Plum	13	.615	Ea.	560	30		590	660
5500	2"	"	11	.727		635	35.50		670.50	755
5540	2-1/2"	Q-1	15	1.067		645	47		692	775
5550	3"		13	1.231		715	54		769	865
5560	4"		10	1.600		1,225	70		1,295	1,450
5950	Flanged									
5994	1"	1 Plum	7	1.143	Ea.	139	55.50		194.50	237
5998	1-1/2"		6	1.333		197	65		262	315
6000	2"		5	1.600		230	78		308	370
6040	2-1/2"	Q-1	5	3.200		258	140		398	495
6050	3"		4.50	3.556		276	156		432	540
6060	4"		3	5.333		435	234		669	825
6070	6"	Q-2	3	8		740	365		1,105	1,350
6080	8"		2.50	9.600		1,400	435		1,835	2,175
6090	10"		2.20	10.909		2,375	495		2,870	3,375
6100	12"		1.70	14.118		3,850	640		4,490	5,225
6110	18"		1.30	18.462		11,800	840		12,640	14,300
6114	24"		.75	32		22,200	1,450		23,650	26,600
6160	For 250 lb flanged, add					200%	20%			
6600	Silent check, bronze trim									
6610	Compact wafer type, for 125 or 150 lb. flanges									
6630	1-1/2"	1 Plum	11	.727	Ea.	107	35.50		142.50	170
6640	2"	"	9	.889		127	43.50		170.50	205
6650	2-1/2"	Q-1	9	1.778		139	78		217	270
6660	3"		8	2		150	88		238	297
6670	4"		5	3.200		191	140		331	420
6680	5"	Q-2	6	4		249	182		431	545
6690	6"		6	4		335	182		517	640
6700	8"		4.50	5.333		595	243		838	1,025
6710	10"		4	6		1,025	273		1,298	1,525
6720	12"		3	8		1,975	365		2,340	2,700
6740	For 250 or 300 lb. flanges, thru 6" no change									
6741	For 8" and 10", add				Ea.	11%	10%			
6750	Twin disc									
6752	2"	1 Plum	9	.889	Ea.	220	43.50		263.50	305
6754	4"	Q-1	5	3.200		360	140		500	605
6756	6"	Q-2	5	4.800		535	218		753	915
6758	8"		4.50	5.333		815	243		1,058	1,275
6760	10"		4	6		1,300	273		1,573	1,825
6762	12"		3	8		1,700	365		2,065	2,400
6764	18"		1.50	16		7,300	730		8,030	9,150
6766	24"		.75	32		10,300	1,450		11,750	13,500
6800	Full flange type, 150 lb.									
6810	1"	1 Plum	14	.571	Ea.	125	28		153	180
6811	1-1/4"		12	.667		132	32.50		164.50	194
6812	1-1/2"		11	.727		152	35.50		187.50	220
6813	2"		9	.889		175	43.50		218.50	257
6814	2-1/2"	Q-1	9	1.778		212	78		290	350
6815	3"		8	2		219	88		307	375
6816	4"		5	3.200		272	140		412	510
6817	5"	Q-2	6	4		430	182		612	745
6818	6"		5	4.800		565	218		783	950
6819	8"		4.50	5.333		950	243		1,193	1,425

23 05 23 – General-Duty Valves for HVAC Piping

23 05 23.30 Valves, Iron Body		Crew	Daily Output	Labor-Hours	Unit	Material	2009 Bare Costs Labor	Equipment	Total	Total Incl O&P
6820	10"	Q-2	4	6	Ea.	1,200	273		1,473	1,725
6840	For 250 lb., add					30%	10%			
6900	Globe type, 125 lb.									
6910	2"	1 Plum	9	.889	Ea.	242	43.50		285.50	330
6911	2-1/2"	Q-1	9	1.778		267	78		345	410
6912	3"		8	2		288	88		376	445
6913	4"		5	3.200		385	140		525	635
6914	5"	Q-2	6	4		495	182		677	820
6915	6"		5	4.800		625	218		843	1,025
6916	8"		4.50	5.333		1,150	243		1,393	1,625
6917	10"		4	6		1,425	273		1,698	1,975
6918	12"		3	8		2,400	365		2,765	3,175
6919	14"		2.30	10.435		3,250	475		3,725	4,275
6920	16"		1.75	13.714		4,700	625		5,325	6,075
6921	18"		1.50	16		6,775	730		7,505	8,550
6922	20"		1	24		8,225	1,100		9,325	10,700
6923	24"		.75	32		10,100	1,450		11,550	13,400
6940	For 250 lb., add					40%	10%			
6980	Screwed end type, 125 lb.									
6981	1"	1 Plum	19	.421	Ea.	46	20.50		66.50	81.50
6982	1-1/4"		15	.533		60	26		86	105
6983	1-1/2"		13	.615		74	30		104	126
6984	2"		11	.727		101	35.50		136.50	164

23 05 23.50 Valves, Multipurpose		Crew	Daily Output	Labor-Hours	Unit	Material	Labor	Equipment	Total	Total Incl O&P
0010	**VALVES, MULTIPURPOSE**									
0100	Functions as a shut off, balancing, check & metering valve									
1000	Cast iron body									
1010	Threaded									
1020	1-1/2" size	1 Stpi	11	.727	Ea.	310	36		346	400
1030	2" size	"	8	1		380	49.50		429.50	490
1040	2-1/2" size	Q-5	5	3.200		410	142		552	665
1050	3" size	"	4.50	3.556		475	158		633	755
1200	Flanged									
1210	3" size	Q-5	4.20	3.810	Ea.	475	169		644	775
1220	4" size	"	3	5.333		955	237		1,192	1,400
1230	5" size	Q-6	3.80	6.316		1,125	291		1,416	1,675
1240	6" size		3	8		1,400	370		1,770	2,100
1250	8" size		2.50	9.600		2,300	440		2,740	3,200
1260	10" size		2.20	10.909		3,275	505		3,780	4,350
1270	12" size		2.10	11.429		8,975	525		9,500	10,700
1280	14" size		2	12		12,500	555		13,055	14,500

23 05 23.70 Valves, Semi-Steel		Crew	Daily Output	Labor-Hours	Unit	Material	Labor	Equipment	Total	Total Incl O&P
0010	**VALVES, SEMI-STEEL** R220523-90									
1020	Lubricated plug valve, threaded, 200 psi									
1030	1/2"	1 Plum	18	.444	Ea.	73	21.50		94.50	113
1040	3/4"		16	.500		73	24.50		97.50	117
1050	1"		14	.571		93	28		121	144
1060	1-1/4"		12	.667		110	32.50		142.50	170
1070	1-1/2"		11	.727		119	35.50		154.50	184
1080	2"		8	1		142	49		191	229
1090	2-1/2"	Q-1	5	3.200		218	140		358	450
1100	3"	"	4.50	3.556		268	156		424	530

23 05 23.70 Valves, Semi-Steel

	23 05 23.70 Valves, Semi-Steel	Crew	Daily Output	Labor-Hours	Unit	Material	2009 Bare Costs Labor	Equipment	Total	Total Incl O&P
6990	Flanged, 200 psi									
7000	2″	1 Plum	8	1	Ea.	176	49		225	267
7010	2-1/2″	Q-1	5	3.200		253	140		393	490
7020	3″		4.50	3.556		305	156		461	570
7030	4″		3	5.333		385	234		619	770
7036	5″		2.50	6.400		840	281		1,121	1,350
7040	6″	Q-2	3	8		630	365		995	1,250
7050	8″		2.50	9.600		1,225	435		1,660	2,000
7060	10″		2.20	10.909		1,950	495		2,445	2,875
7070	12″		1.70	14.118		4,425	640		5,065	5,825

23 05 23.80 Valves, Steel

	23 05 23.80 Valves, Steel	Crew	Daily Output	Labor-Hours	Unit	Material	2009 Bare Costs Labor	Equipment	Total	Total Incl O&P
0010	**VALVES, STEEL** R220523-90									
0800	Cast									
1350	Check valve, swing type, 150 lb., flanged									
1370	1″	1 Plum	10	.800	Ea.	510	39		549	620
1400	2″	″	8	1		640	49		689	780
1440	2-1/2″	Q-1	5	3.200		740	140		880	1,025
1450	3″		4.50	3.556		755	156		911	1,075
1460	4″		3	5.333		1,075	234		1,309	1,550
1470	6″	Q-2	3	8		1,700	365		2,065	2,425
1480	8″		2.50	9.600		2,775	435		3,210	3,700
1490	10″		2.20	10.909		4,100	495		4,595	5,275
1500	12″		1.70	14.118		5,850	640		6,490	7,425
1510	14″		1.30	18.462		8,450	840		9,290	10,600
1520	16″		1	24		9,600	1,100		10,700	12,300
1548	For 600 lb., flanged, add					110%	20%			
1571	300 lb., 2″	1 Plum	7.40	1.081		680	52.50		732.50	830
1572	2-1/2″	Q-1	4.20	3.810		1,025	167		1,192	1,375
1573	3″		4	4		1,025	176		1,201	1,400
1574	4″		2.80	5.714		1,325	251		1,576	1,850
1575	6″	Q-2	2.90	8.276		2,575	375		2,950	3,425
1576	8″		2.40	10		3,725	455		4,180	4,775
1577	10″		2.10	11.429		5,525	520		6,045	6,850
1578	12″		1.60	15		7,700	685		8,385	9,500
1579	14″		1.20	20		11,700	910		12,610	14,200
1581	16″		.90	26.667		15,000	1,225		16,225	18,400
1950	Gate valve, 150 lb., flanged									
2000	2″	1 Plum	8	1	Ea.	685	49		734	830
2040	2-1/2″	Q-1	5	3.200		970	140		1,110	1,275
2050	3″		4.50	3.556		970	156		1,126	1,300
2060	4″		3	5.333		1,200	234		1,434	1,675
2070	6″	Q-2	3	8		1,800	365		2,165	2,525
2080	8″		2.50	9.600		1,875	435		2,310	2,725
2090	10″		2.20	10.909		4,275	495		4,770	5,450
2100	12″		1.70	14.118		5,800	640		6,440	7,350
2110	14″		1.30	18.462		9,225	840		10,065	11,400
2120	16″		1	24		11,900	1,100		13,000	14,800
2130	18″		.80	30		14,500	1,375		15,875	18,000
2140	20″		.60	40		17,400	1,825		19,225	21,800
2650	300 lb., flanged									
2700	2″	1 Plum	7.40	1.081	Ea.	940	52.50		992.50	1,100
2740	2-1/2″	Q-1	4.20	3.810		1,275	167		1,442	1,650

23 05 23.80 Valves, Steel		Crew	Daily Output	Labor-Hours	Unit	Material	Labor	Equipment	Total	Total Incl O&P
						2009 Bare Costs				
2750	3"	Q-1	4	4	Ea.	1,275	176		1,451	1,675
2760	4"	↓	2.80	5.714		1,775	251		2,026	2,325
2770	6"	Q-2	2.90	8.276		2,975	375		3,350	3,850
2780	8"		2.40	10		4,675	455		5,130	5,800
2790	10"		2.10	11.429		6,250	520		6,770	7,650
2800	12"		1.60	15		9,025	685		9,710	11,000
2810	14"		1.20	20		17,100	910		18,010	20,200
2820	16"		.90	26.667		21,500	1,225		22,725	25,400
2830	18"		.70	34.286		30,600	1,550		32,150	36,100
2840	20"	↓	.50	48	↓	34,100	2,175		36,275	40,800
3650	Globe valve, 150 lb., flanged									
3700	2"	1 Plum	8	1	Ea.	865	49		914	1,025
3740	2-1/2"	Q-1	5	3.200		1,100	140		1,240	1,400
3750	3"		4.50	3.556		1,100	156		1,256	1,425
3760	4"	↓	3	5.333		1,600	234		1,834	2,100
3770	6"	Q-2	3	8		2,525	365		2,890	3,325
3780	8"		2.50	9.600		4,625	435		5,060	5,725
3790	10"		2.20	10.909		8,200	495		8,695	9,750
3800	12"	↓	1.70	14.118	↓	11,000	640		11,640	13,100
4080	300 lb., flanged									
4100	2"	1 Plum	7.40	1.081	Ea.	1,150	52.50		1,202.50	1,350
4140	2-1/2"	Q-1	4.20	3.810		1,575	167		1,742	1,975
4150	3"		4	4		1,575	176		1,751	2,000
4160	4"	↓	2.80	5.714		2,175	251		2,426	2,775
4170	6"	Q-2	2.90	8.276		3,925	375		4,300	4,900
4180	8"		2.40	10		6,425	455		6,880	7,750
4190	10"		2.10	11.429		12,100	520		12,620	14,100
4200	12"	↓	1.60	15	↓	14,000	685		14,685	16,400
4680	600 lb., flanged									
4700	2"	1 Plum	7	1.143	Ea.	1,600	55.50		1,655.50	1,850
4740	2-1/2"	Q-1	4	4		2,400	176		2,576	2,925
4750	3"		3.60	4.444		2,400	195		2,595	2,950
4760	4"	↓	2.50	6.400		3,875	281		4,156	4,700
4770	6"	Q-2	2.60	9.231		8,075	420		8,495	9,500
4780	8"	"	2.10	11.429	↓	13,100	520		13,620	15,300
4800	Silent check, 316 S.S. trim									
4810	Full flange type, 150 lb.									
4811	1"	1 Plum	14	.571	Ea.	340	28		368	415
4812	1-1/4"		12	.667		375	32.50		407.50	460
4813	1-1/2"		11	.727		410	35.50		445.50	505
4814	2"	↓	9	.889		495	43.50		538.50	610
4815	2-1/2"	Q-1	9	1.778		595	78		673	770
4816	3"		8	2		695	88		783	895
4817	4"	↓	5	3.200		945	140		1,085	1,250
4818	5"	Q-2	6	4		1,175	182		1,357	1,575
4819	6"		5	4.800		1,525	218		1,743	2,000
4820	8"		4.50	5.333		2,225	243		2,468	2,825
4821	10"	↓	4	6		3,125	273		3,398	3,850
4840	For 300 lb., add					20%	10%			
4860	For 600 lb., add				↓	50%	15%			
4900	Globe type, 150 lb.									
4910	2"	1 Plum	9	.889	Ea.	590	43.50		633.50	710
4911	2-1/2"	Q-1	9	1.778	↓	705	78		783	890

23 05 23 – General-Duty Valves for HVAC Piping

23 05 23.80 Valves, Steel		Crew	Daily Output	Labor-Hours	Unit	2009 Bare Costs Material	Labor	Equipment	Total	Total Incl O&P
4912	3"	Q-1	8	2	Ea.	790	88		878	995
4913	4"		5	3.200		1,050	140		1,190	1,350
4914	5"	Q-2	6	4		1,325	182		1,507	1,725
4915	6"		5	4.800		1,625	218		1,843	2,100
4916	8"		4.50	5.333		2,275	243		2,518	2,875
4917	10"		4	6		4,000	273		4,273	4,800
4918	12"		3	8		4,825	365		5,190	5,850
4919	14"		2.30	10.435		7,300	475		7,775	8,725
4920	16"		1.75	13.714		10,200	625		10,825	12,100
4921	18"		1.50	16		10,700	730		11,430	12,900
4940	For 300 lb., add					30%	10%			
4960	For 600 lb., add					60%	15%			
5150	Forged									
5340	Ball valve, 800 psi, threaded, 1/4" size	1 Plum	24	.333	Ea.	50.50	16.25		66.75	80.50
5350	3/8"		24	.333		50.50	16.25		66.75	80
5360	1/2"		24	.333		63	16.25		79.25	93.50
5370	3/4"		20	.400		83.50	19.50		103	122
5380	1"		19	.421		104	20.50		124.50	145
5390	1-1/4"		15	.533		138	26		164	190
5400	1-1/2"		13	.615		181	30		211	244
5410	2"		11	.727		239	35.50		274.50	315
5460	Ball valve, 800 lb., socket weld, 1/4" size	Q-15	19	.842		62.50	37	2.93	102.43	127
5470	3/8"		19	.842		62.50	37	2.93	102.43	127
5480	1/2"		19	.842		62.50	37	2.93	102.43	127
5490	3/4"		19	.842		83	37	2.93	122.93	150
5500	1"		15	1.067		103	47	3.71	153.71	188
5510	1-1/4"		13	1.231		138	54	4.28	196.28	238
5520	1-1/2"		11	1.455		180	64	5.05	249.05	300
5530	2"		8.50	1.882		255	82.50	6.55	344.05	410
5550	Ball valve, 150 lb., flanged									
5560	4"	Q-1	3	5.333	Ea.	955	234		1,189	1,400
5570	6"	Q-2	3	8		2,000	365		2,365	2,750
5580	8"	"	2.50	9.600		2,900	435		3,335	3,850
5650	Check valve, class 800, horizontal, socket									
5652	1/4"	Q-15	19	.842	Ea.	76.50	37	2.93	116.43	143
5654	3/8"		19	.842		76.50	37	2.93	116.43	143
5656	1/2"		19	.842		76.50	37	2.93	116.43	143
5658	3/4"		19	.842		82	37	2.93	121.93	149
5660	1"		15	1.067		96.50	47	3.71	147.21	180
5662	1-1/4"		13	1.231		189	54	4.28	247.28	294
5664	1-1/2"		11	1.455		189	64	5.05	258.05	310
5666	2"		8.50	1.882		265	82.50	6.55	354.05	420
5698	Threaded									
5700	1/4"	1 Plum	24	.333	Ea.	76.50	16.25		92.75	109
5720	3/8"		24	.333		76.50	16.25		92.75	109
5730	1/2"		24	.333		76.50	16.25		92.75	109
5740	3/4"		20	.400		82	19.50		101.50	120
5750	1"		19	.421		96.50	20.50		117	137
5760	1-1/4"		15	.533		189	26		215	247
5770	1-1/2"		13	.615		189	30		219	253
5780	2"		11	.727		265	35.50		300.50	345
5840	For class 150, flanged, add					100%	15%			
5860	For class 300, flanged, add					120%	20%			

23 05 23.80 Valves, Steel		Crew	Daily Output	Labor-Hours	Unit	Material	2009 Bare Costs Labor	Equipment	Total	Total Incl O&P
6100	Gate, class 800, OS&Y, socket									
6102	3/8"	Q-15	19	.842	Ea.	53.50	37	2.93	93.43	117
6103	1/2"		19	.842		53.50	37	2.93	93.43	117
6104	3/4"		19	.842		58.50	37	2.93	98.43	123
6105	1"		15	1.067		70.50	47	3.71	121.21	152
6106	1-1/4"		13	1.231		134	54	4.28	192.28	233
6107	1-1/2"		11	1.455		134	64	5.05	203.05	249
6108	2"		8.50	1.882		174	82.50	6.55	263.05	325
6118	Threaded									
6120	3/8"	1 Plum	24	.333	Ea.	53.50	16.25		69.75	83
6130	1/2"		24	.333		53.50	16.25		69.75	83
6140	3/4"		20	.400		58.50	19.50		78	93.50
6150	1"		19	.421		70.50	20.50		91	109
6160	1-1/4"		15	.533		134	26		160	186
6170	1-1/2"		13	.615		134	30		164	192
6180	2"		11	.727		174	35.50		209.50	245
6260	For OS&Y, flanged, add					100%	20%			
6700	Globe, OS&Y, class 800, socket									
6710	1/4"	Q-15	19	.842	Ea.	82	37	2.93	121.93	149
6720	3/8"		19	.842		82	37	2.93	121.93	149
6730	1/2"		19	.842		82	37	2.93	121.93	149
6740	3/4"		19	.842		93.50	37	2.93	133.43	162
6750	1"		15	1.067		122	47	3.71	172.71	209
6760	1-1/4"		13	1.231		240	54	4.28	298.28	350
6770	1-1/2"		11	1.455		240	64	5.05	309.05	365
6780	2"		8.50	1.882		305	82.50	6.55	394.05	470
6860	For OS&Y, flanged, add					300%	20%			
6880	Threaded									
6882	1/4"	1 Plum	24	.333	Ea.	82	16.25		98.25	115
6884	3/8"		24	.333		82	16.25		98.25	115
6886	1/2"		24	.333		82	16.25		98.25	115
6888	3/4"		20	.400		93.50	19.50		113	133
6890	1"		19	.421		122	20.50		142.50	166
6892	1-1/4"		15	.533		240	26		266	305
6894	1-1/2"		13	.615		240	30		270	310
6896	2"		11	.727		305	35.50		340.50	395

23 05 23.90 Valves, Stainless Steel		Crew	Daily Output	Labor-Hours	Unit	Material	Labor	Equipment	Total	Total Incl O&P
0010	**VALVES, STAINLESS STEEL** R220523-90									
1610	Ball, threaded 1/4"	1 Stpi	24	.333	Ea.	40.50	16.45		56.95	69
1620	3/8"		24	.333		40.50	16.45		56.95	69
1630	1/2"		22	.364		40.50	17.95		58.45	71.50
1640	3/4"		20	.400		67.50	19.75		87.25	104
1650	1"		19	.421		82	21		103	122
1660	1-1/4		15	.533		156	26.50		182.50	212
1670	1-1/2"		13	.615		163	30.50		193.50	226
1680	2"		11	.727		212	36		248	287
1700	Check, 200 lb., threaded									
1710	1/4"	1 Plum	24	.333	Ea.	126	16.25		142.25	164
1720	1/2"		22	.364		126	17.75		143.75	166
1730	3/4"		20	.400		138	19.50		157.50	182
1750	1"		19	.421		176	20.50		196.50	225
1760	1-1/2"		13	.615		340	30		370	415

23 05 23.90 Valves, Stainless Steel		Crew	Daily Output	Labor-Hours	Unit	Material	Labor	Equipment	Total	Total Incl O&P
						2009 Bare Costs				
1770	2"	1 Plum	11	.727	Ea.	570	35.50		605.50	680
1800	150 lb., flanged									
1810	2-1/2"	Q-1	5	3.200	Ea.	1,375	140		1,515	1,700
1820	3"		4.50	3.556		1,375	156		1,531	1,725
1830	4"		3	5.333		2,025	234		2,259	2,575
1840	6"	Q-2	3	8		3,625	365		3,990	4,550
1850	8"	"	2.50	9.600		6,050	435		6,485	7,325
2100	Gate, OS&Y, 150 lb., flanged									
2120	1/2"	1 Plum	18	.444	Ea.	400	21.50		421.50	475
2140	3/4"		16	.500		425	24.50		449.50	505
2150	1"		14	.571		525	28		553	620
2160	1-1/2"		11	.727		730	35.50		765.50	860
2170	2"		8	1		820	49		869	975
2180	2-1/2"	Q-1	5	3.200		1,350	140		1,490	1,700
2190	3"		4.50	3.556		2,000	156		2,156	2,425
2200	4"		3	5.333		2,000	234		2,234	2,550
2205	5"		2.80	5.714		3,450	251		3,701	4,175
2210	6"	Q-2	3	8		3,450	365		3,815	4,350
2220	8"		2.50	9.600		5,550	435		5,985	6,750
2230	10"		2.30	10.435		9,100	475		9,575	10,700
2240	12"		1.90	12.632		11,700	575		12,275	13,700
2260	For 300 lb., flanged, add					120%	15%			
2600	600 lb., flanged									
2620	1/2"	1 Plum	16	.500	Ea.	173	24.50		197.50	227
2640	3/4"		14	.571		187	28		215	248
2650	1"		12	.667		225	32.50		257.50	297
2660	1-1/2"		10	.800		360	39		399	455
2670	2"		7	1.143		495	55.50		550.50	630
2680	2-1/2"	Q-1	4	4		4,800	176		4,976	5,550
2690	3"	"	3.60	4.444		4,800	195		4,995	5,575
3100	Globe, OS&Y, 150 lb., flanged									
3120	1/2"	1 Plum	18	.444	Ea.	400	21.50		421.50	475
3140	3/4"		16	.500		445	24.50		469.50	525
3150	1"		14	.571		555	28		583	650
3160	1-1/2"		11	.727		825	35.50		860.50	960
3170	2"		8	1		1,000	49		1,049	1,175
3180	2-1/2"	Q-1	5	3.200		2,150	140		2,290	2,575
3190	3"		4.50	3.556		2,150	156		2,306	2,600
3200	4"		3	5.333		2,525	234		2,759	3,150
3210	6"	Q-2	3	8		4,450	365		4,815	5,450
5000	Silent check, 316 S.S. body and trim									
5010	Compact wafer type, 300 lb.									
5020	1"	1 Plum	14	.571	Ea.	620	28		648	720
5021	1-1/4"		12	.667		670	32.50		702.50	785
5022	1-1/2"		11	.727		690	35.50		725.50	815
5023	2"		9	.889		790	43.50		833.50	935
5024	2-1/2"	Q-1	9	1.778		1,075	78		1,153	1,325
5025	3"	"	8	2		1,250	88		1,338	1,500
5100	Full flange wafer type, 150 lb.									
5110	1"	1 Plum	14	.571	Ea.	435	28		463	520
5111	1-1/4"		12	.667		465	32.50		497.50	565
5112	1-1/2"		11	.727		535	35.50		570.50	640
5113	2"		9	.889		700	43.50		743.50	835

23 05 23 – General-Duty Valves for HVAC Piping

23 05 23.90 Valves, Stainless Steel		Crew	Daily Output	Labor-Hours	Unit	Material	2009 Bare Costs Labor	Equipment	Total	Total Incl O&P
5114	2-1/2"	Q-1	9	1.778	Ea.	820	78		898	1,025
5115	3"		8	2		935	88		1,023	1,150
5116	4"		5	3.200		1,400	140		1,540	1,725
5117	5"	Q-2	6	4		1,625	182		1,807	2,050
5118	6"		5	4.800		2,025	218		2,243	2,550
5119	8"		4.50	5.333		3,400	243		3,643	4,100
5120	10"		4	6		4,700	273		4,973	5,575
5200	Globe type, 300 lb.									
5210	2"	1 Plum	9	.889	Ea.	795	43.50		838.50	940
5211	2-1/2"	Q-1	9	1.778		980	78		1,058	1,200
5212	3"		8	2		1,200	88		1,288	1,450
5213	4"		5	3.200		1,700	140		1,840	2,075
5214	5"	Q-2	6	4		2,050	182		2,232	2,525
5215	6"		5	4.800		2,650	218		2,868	3,250
5216	8"		4.50	5.333		3,750	243		3,993	4,500
5217	10"		4	6		5,325	273		5,598	6,250
5300	Screwed end type, 300 lb.									
5310	1/2"	1 Plum	24	.333	Ea.	67.50	16.25		83.75	98.50
5311	3/4"		20	.400		80	19.50		99.50	118
5312	1"		19	.421		93	20.50		113.50	134
5313	1-1/4"		15	.533		124	26		150	175
5314	1-1/2"		13	.615		134	30		164	192
5315	2"		11	.727		166	35.50		201.50	235

23 05 93 – Testing, Adjusting, and Balancing for HVAC

23 05 93.10 Balancing, Air		Crew	Daily Output	Labor-Hours	Unit	Material	Labor	Equipment	Total	Total Incl O&P
0010	**BALANCING, AIR**, (Subcontractor's quote incl. material and labor.) R230500-10									
0900	Heating and ventilating equipment									
1000	Centrifugal fans, utility sets				Ea.				317.34	348.72
1100	Heating and ventilating unit								476	523.08
1200	In-line fan								476	523.08
1300	Propeller and wall fan								89.91	98.80
1400	Roof exhaust fan								211.56	232.48
2000	Air conditioning equipment, central station								687.56	755.56
2100	Built-up low pressure unit								634.67	697.44
2200	Built-up high pressure unit								740.45	813.68
2300	Built-up high pressure dual duct								1,163.56	1,278.64
2400	Built-up variable volume								1,375.12	1,511.12
2500	Multi-zone A.C. and heating unit								476	523.08
2600	For each zone over one, add								105.78	116.24
2700	Package A.C. unit								264.45	290.60
2800	Rooftop heating and cooling unit								370.22	406.84
3000	Supply, return, exhaust, registers & diffusers, avg. height ceiling								63.47	69.74
3100	High ceiling								95.20	104.62
3200	Floor height								52.89	58.12
3300	Off mixing box								42.31	46.50
3500	Induction unit								68.76	75.56
3600	Lab fume hood								317.34	348.72
3700	Linear supply								158.67	174.36
3800	Linear supply high								185.11	203.42
4000	Linear return								52.89	58.12
4100	Light troffers								63.47	69.74
4200	Moduline - master								63.47	69.74

23 05 93.10 Balancing, Air		Crew	Daily Output	Labor-Hours	Unit	Material	2009 Bare Costs Labor	Equipment	Total	Total Incl O&P
4300	Moduline - slaves				Ea.				31.73	34.87
4400	Regenerators								423.11	464.96
4500	Taps into ceiling plenums								79.33	87.18
4600	Variable volume boxes				↓				63.47	69.74

23 05 93.20 Balancing, Water		Crew	Daily Output	Labor-Hours	Unit	Material	Labor	Equipment	Total	Total Incl O&P
0010	**BALANCING, WATER**, (Subcontractor's quote incl. material and labor.)									
0050	Air cooled condenser				Ea.				188.68	207.34
0080	Boiler								379.52	417.05
0100	Cabinet unit heater R230500-10								64.69	71.09
0200	Chiller								458.22	503.54
0300	Convector								53.91	59.24
0400	Converter								269.54	296.20
0500	Cooling tower								350.40	385.06
0600	Fan coil unit, unit ventilator								97.04	106.63
0700	Fin tube and radiant panels								107.82	118.48
0800	Main and duct re-heat coils								99.73	109.59
0810	Heat exchanger								99.73	109.59
0900	Main balancing cocks								80.86	88.86
1000	Pumps								237.20	260.66
1100	Unit heater				↓				75.47	82.94

23 05 93.50 Piping, Testing		Crew	Daily Output	Labor-Hours	Unit	Material	Labor	Equipment	Total	Total Incl O&P
0010	**PIPING, TESTING**									
0100	Nondestructive testing									
0110	Nondestructive hydraulic pressure test, isolate & 1 hr. hold									
0120	1" - 4" pipe									
0140	0 – 250 L.F.	1 Stpi	1.33	6.015	Ea.		297		297	445
0160	250 – 500 L.F.	"	.80	10			495		495	740
0180	500 – 1000 L.F.	Q-5	1.14	14.035			625		625	935
0200	1000 – 2000 L.F.	"	.80	20	↓		890		890	1,325
0300	6" - 10" pipe									
0320	0 – 250 L.F.	Q-5	1	16	Ea.		710		710	1,075
0340	250 – 500 L.F.		.73	21.918			975		975	1,450
0360	500 – 1000 L.F.		.53	30.189			1,350		1,350	2,025
0380	1000 – 2000 L.F.	↓	.38	42.105	↓		1,875		1,875	2,800
1000	Pneumatic pressure test, includes soaping joints									
1120	1" - 4" pipe									
1140	0 – 250 L.F.	Q-5	2.67	5.993	Ea.	7	266		273	410
1160	250 – 500 L.F.		1.33	12.030		14.05	535		549.05	815
1180	500 – 1000 L.F.		.80	20		21	890		911	1,350
1200	1000 – 2000 L.F.	↓	.50	32	↓	28	1,425		1,453	2,150
1300	6" - 10" pipe									
1320	0 – 250 L.F.	Q-5	1.33	12.030	Ea.	7	535		542	810
1340	250 – 500 L.F.		.67	23.881		14.05	1,050		1,064.05	1,625
1360	500 – 1000 L.F.		.40	40		28	1,775		1,803	2,700
1380	1000 – 2000 L.F.	↓	.25	64	↓	35	2,850		2,885	4,325
2000	X-Ray of welds									
2110	2" diam.	1 Stpi	8	1	Ea.	10.20	49.50		59.70	85
2120	3" diam.		8	1		10.20	49.50		59.70	85
2130	4" diam.		8	1		15.30	49.50		64.80	91
2140	6" diam.		8	1		15.30	49.50		64.80	91
2150	8" diam.		6.60	1.212		15.30	60		75.30	107
2160	10" diam.	↓	6	1.333	↓	20.50	66		86.50	121

23 05 Common Work Results for HVAC

23 05 93 – Testing, Adjusting, and Balancing for HVAC

23 05 93.50 Piping, Testing		Crew	Daily Output	Labor-Hours	Unit	2009 Bare Costs Material	Labor	Equipment	Total	Total Incl O&P
3000	Liquid penetration of welds									
3110	2" diam.	1 Stpi	14	.571	Ea.	2.87	28		30.87	45.50
3120	3" diam.		13.60	.588		2.87	29		31.87	46.50
3130	4" diam.		13.40	.597		2.87	29.50		32.37	47
3140	6" diam.		13.20	.606		2.87	30		32.87	48
3150	8" diam.		13	.615		4.31	30.50		34.81	50
3160	10" diam.	↓	12.80	.625	↓	4.31	31		35.31	51

23 07 HVAC Insulation

23 07 13 – Duct Insulation

23 07 13.10 Duct Thermal Insulation

			Crew	Daily Output	Labor-Hours	Unit	Material	Labor	Equipment	Total	Total Incl O&P
0010	**DUCT THERMAL INSULATION**										
0100	Rule of thumb, as a percentage of total mechanical costs					Job				10%	
0110	Insulation req'd is based on the surface size/area to be covered										
3000	Ductwork										
3020	Blanket type, fiberglass, flexible										
3140	FSK vapor barrier wrap, .75 lb. density										
3160	1" thick	G	Q-14	350	.046	S.F.	.17	1.81		1.98	3.03
3170	1-1/2" thick	G		320	.050		.20	1.99		2.19	3.33
3180	2" thick	G		300	.053		.26	2.12		2.38	3.61
3190	3" thick	G		260	.062		.39	2.44		2.83	4.26
3200	4" thick	G		242	.066		.56	2.63		3.19	4.73
3212	Vinyl Jacket, .75 lb. density, 1-1/2" thick	G	↓	320	.050	↓	.20	1.99		2.19	3.33
3280	Unfaced, 1 lb. density										
3310	1" thick	G	Q-14	360	.044	S.F.	.16	1.76		1.92	2.94
3320	1-1/2" thick	G		330	.048		.19	1.92		2.11	3.23
3330	2" thick	G	↓	310	.052	↓	.26	2.05		2.31	3.50
3400	FSK facing, 1 lb. density										
3420	1-1/2" thick	G	Q-14	310	.052	S.F.	.17	2.05		2.22	3.40
3430	2" thick	G		300	.053		.19	2.12		2.31	3.53
3470	1-1/2" thick	G		300	.053		.25	2.12		2.37	3.60
3480	2" thick	G	↓	290	.055	↓	.32	2.19		2.51	3.78
3730	Sheet insulation										
3760	Polyethylene foam, closed cell, UV resistant										
3770	Standard temperature (-90° F to +212° F)										
3771	1/4" thick	G	Q-14	450	.036	S.F.	1.30	1.41		2.71	3.64
3772	3/8" thick	G		440	.036		1.86	1.44		3.30	4.31
3773	1/2" thick	G		420	.038		2.28	1.51		3.79	4.88
3774	3/4" thick	G		400	.040		3.26	1.59		4.85	6.10
3775	1" thick	G		380	.042		4.41	1.67		6.08	7.45
3776	1-1/2" thick	G		360	.044		6.95	1.76		8.71	10.40
3777	2" thick	G		340	.047		9.35	1.87		11.22	13.20
3778	2-1/2" thick	G	↓	320	.050	↓	11.80	1.99		13.79	16.10
3779	Adhesive (see line 7878)										
3780	Foam, rubber										
3782	1" thick	G	1 Stpi	50	.160	S.F.	3.43	7.90		11.33	15.60
3795	Finishes										
3800	Stainless steel woven mesh		Q-14	100	.160	S.F.	.84	6.35		7.19	10.85
3820	18 oz. fiberglass cloth, pasted on			170	.094		.77	3.74		4.51	6.70
3900	8 oz. canvas, pasted on		↓	180	.089	↓	.49	3.53		4.02	6.10
7878	Contact cement, quart can					Ea.	9.20			9.20	10.10

23 07 HVAC Insulation

23 07 13 – Duct Insulation

23 07 13.10 Duct Thermal Insulation		Crew	Daily Output	Labor-Hours	Unit	Material	2009 Bare Costs Labor	Equipment	Total	Total Incl O&P
9600	Minimum labor/equipment charge	1 Stpi	4	2	Job		98.50		98.50	148

23 07 16 – HVAC Equipment Insulation

23 07 16.10 HVAC Equipment Thermal Insulation

			Crew	Daily Output	Labor-Hours	Unit	Material	Labor	Equipment	Total	Total Incl O&P
0010	**HVAC EQUIPMENT THERMAL INSULATION**										
0100	Rule of thumb, as a percentage of total mechanical costs					Job				10%	
0110	Insulation req'd is based on the surface size/area to be covered										
1000	Boiler, 1-1/2" calcium silicate only	G	Q-14	110	.145	S.F.	2.65	5.75		8.40	11.95
1020	Plus 2" fiberglass	G	"	80	.200	"	3.51	7.95		11.46	16.30
2000	Breeching, 2" calcium silicate										
2020	Rectangular	G	Q-14	42	.381	S.F.	5.20	15.10		20.30	29
2040	Round	G	"	38.70	.413	"	5.45	16.40		21.85	31.50
2300	Calcium silicate block, + 200° F to + 1200° F										
2310	On irregular surfaces, valves and fittings										
2340	1" thick	G	Q-14	30	.533	S.F.	2.35	21		23.35	35.50
2360	1-1/2" thick	G		25	.640		2.65	25.50		28.15	43
2380	2" thick	G		22	.727		3.48	29		32.48	49
2400	3" thick	G	↓	18	.889	↓	5.35	35.50		40.85	61.50
2410	On plane surfaces										
2420	1" thick	G	Q-14	126	.127	S.F.	2.35	5.05		7.40	10.50
2430	1-1/2" thick	G		120	.133		2.65	5.30		7.95	11.20
2440	2" thick	G		100	.160		3.48	6.35		9.83	13.80
2450	3" thick	G	↓	70	.229	↓	5.35	9.05		14.40	20

23 09 Instrumentation and Control for HVAC

23 09 13 – Instrumentation and Control Devices for HVAC

23 09 13.60 Water Level Controls

		Crew	Daily Output	Labor-Hours	Unit	Material	Labor	Equipment	Total	Total Incl O&P
0010	**WATER LEVEL CONTROLS**									
1000	Electric water feeder	1 Stpi	12	.667	Ea.	229	33		262	300
2000	Feeder cut-off combination									
2100	Steam system up to 5000 sq. ft.	1 Stpi	12	.667	Ea.	490	33		523	590
2200	Steam system above 5000 sq. ft.		12	.667		605	33		638	715
2300	Steam and hot water, high pressure	↓	10	.800	↓	710	39.50		749.50	840
3000	Low water cut-off for hot water boiler, 50 psi maximum									
3100	1" top & bottom equalizing pipes, manual reset	1 Stpi	14	.571	Ea.	275	28		303	345
3200	1" top & bottom equalizing pipes		14	.571		275	28		303	345
3300	2-1/2" side connection for nipple-to-boiler	↓	14	.571	↓	248	28		276	315
4000	Low water cut-off for low pressure steam with quick hook-up ftgs.									
4100	For installation in gauge glass tappings	1 Stpi	16	.500	Ea.	198	24.50		222.50	255
4200	Built-in type, 2-1/2" tap - 3-1/8" insertion		16	.500		166	24.50		190.50	219
4300	Built-in type, 2-1/2" tap - 1-3/4" insertion		16	.500		176	24.50		200.50	231
4400	Side connection to 2-1/2" tapping		16	.500		189	24.50		213.50	245
5000	Pump control, low water cut-off and alarm switch	↓	14	.571	↓	425	28		453	515
9000	Water gauges, complete									
9010	Rough brass, wheel type									
9020	125 PSI at 350° F									
9030	3/8" pipe size	1 Stpi	11	.727	Ea.	49	36		85	108
9040	1/2" pipe size	"	10	.800	"	53.50	39.50		93	118
9060	200 PSI at 400° F									
9070	3/8" pipe size	1 Stpi	11	.727	Ea.	61	36		97	121
9080	1/2" pipe size	↓	10	.800	↓	61	39.50		100.50	126

23 09 13 – Instrumentation and Control Devices for HVAC

23 09 13.60 Water Level Controls		Crew	Daily Output	Labor-Hours	Unit	Material	2009 Bare Costs Labor	Equipment	Total	Total Incl O&P
9090	3/4" pipe size	1 Stpi	9	.889	Ea.	77.50	44		121.50	152
9130	Rough brass, chain lever type									
9140	250 PSI at 400° F									
9200	Polished brass, wheel type									
9210	200 PSI at 400° F									
9220	3/8" pipe size	1 Stpi	11	.727	Ea.	117	36		153	183
9230	1/2" pipe size	↓	10	.800	↓	118	39.50		157.50	189
9240	3/4" pipe size	↓	9	.889	↓	125	44		169	204
9260	Polished brass, chain lever type									
9270	250 PSI at 400° F									
9280	1/2" pipe size	1 Stpi	10	.800	Ea.	168	39.50		207.50	244
9290	3/4" pipe size	"	9	.889	"	183	44		227	267
9400	Bronze, high pressure, ASME									
9410	1/2" pipe size	1 Stpi	10	.800	Ea.	212	39.50		251.50	292
9420	3/4" pipe size	"	9	.889	"	220	44		264	310
9460	Chain lever type									
9470	1/2" pipe size	1 Stpi	10	.800	Ea.	273	39.50		312.50	360
9480	3/4" pipe size	"	9	.889	"	288	44		332	380
9500	316 stainless steel, high pressure, ASME									
9510	500 PSI at 450° F									
9520	1/2" pipe size	1 Stpi	10	.800	Ea.	655	39.50		694.50	780
9530	3/4" pipe size	"	9	.889	"	635	44		679	765

23 09 23 – Direct-Digital Control System for HVAC

23 09 23.10 Control Components/DDC Systems		Crew	Daily Output	Labor-Hours	Unit	Material	Labor	Equipment	Total	Total Incl O&P
0010	**CONTROL COMPONENTS/DDC SYSTEMS** (Sub's quote incl. M & L)									
0100	Analog inputs									
0110	Sensors (avg. 50' run in 1/2" EMT)									
0120	Duct temperature				Ea.				360	396
0130	Space temperature				↓				577	634.64
0140	Duct humidity, +/- 3%				↓				605.32	665.85
0150	Space humidity +/- 2%				↓				926.90	1,019.59
0160	Duct static pressure				↓				491.82	541
0170	C.F.M./Transducer				↓				662.07	728.28
0172	Water temp. (see Div. 23 21 20 for well tap add)				↓				567.64	624.24
0174	Water flow (see Div. 23 21 20 for circuit sensor add)				↓				2,080.80	2,288.88
0176	Water pressure differential (see Div. 23 21 20 for tap add)				↓				851.24	936.36
0177	Steam flow (see Div. 23 21 20 for circuit sensor add)				↓				2,080.80	2,288.88
0178	Steam pressure (see Div. 23 21 20 for tap add)				↓				889.06	977.97
0180	K.W./Transducer				↓				1,182.27	1,300.50
0182	K.W.H. totalization (not incl. elec. meter pulse xmtr.)				↓				543.38	597.72
0190	Space static pressure				↓				927.27	1,020
1000	Analog outputs (avg. 50' run in 1/2" EMT)									
1010	P/I Transducer				Ea.				548.57	603.43
1020	Analog output, matl. in MUX				↓				264.83	291.31
1030	Pneumatic (not incl. control device)				↓				558	613.83
1040	Electric (not incl control device)				↓				331	364.14
2000	Status (Alarms)									
2100	Digital inputs (avg. 50' run in 1/2" EMT)									
2110	Freeze				Ea.				378.33	416.16
2120	Fire				↓				340.49	374.54
2130	Differential pressure, (air)				↓				520.20	572.22
2140	Differential pressure, (water)				↓				850	935

23 09 23 – Direct-Digital Control System for HVAC

23 09 23.10 Control Components/DDC Systems		Crew	Daily Output	Labor-Hours	Unit	2009 Bare Costs Material	Labor	Equipment	Total	Total Incl O&P
2150	Current sensor				Ea.				378.33	416.16
2160	Duct high temperature thermostat								496.55	546.21
2170	Duct smoke detector								614.78	676.26
2200	Digital output (avg. 50′ run in 1/2″ EMT)									
2210	Start/stop				Ea.				296.73	326.40
2220	On/off (maintained contact)				"				510	561
3000	Controller M.U.X. panel, incl. function boards									
3100	48 point				Ea.				4,587.22	5,045.94
3110	128 point				"				6,290	6,918.66
3200	D.D.C. controller (avg. 50′ run in conduit)									
3210	Mechanical room									
3214	16 point controller (incl. 120v/1ph power supply)				Ea.				1,937.45	3,121.20
3229	32 point controller (incl. 120v/1ph power supply)				"				4,729	5,202
3230	Includes software programming and checkout									
3260	Space									
3266	V.A.V. terminal box (incl. space temp. sensor)				Ea.				733	806.31
3280	Host computer (avg. 50′ run in conduit)									
3281	Package complete with PC, keyboard,									
3282	printer, color CRT, modem, basic software				Ea.				8,512.36	9,363.60
4000	Front end costs									
4100	Computer (P.C.)/software program				Ea.				5,675	6,242.40
4200	Color graphics software								3,405	3,745.44
4300	Color graphics slides								426	468.18
4350	Additional dot matrix printer								851	936.36
4400	Communications trunk cable				L.F.				3.31	3.64
4500	Engineering labor, (not incl. dftg.)				Point				71.88	79.07
4600	Calibration labor								71.88	79.07
4700	Start-up, checkout labor								108.76	119.64
4800	Drafting labor, as req'd									
5000	Communications bus (data transmission cable)									
5010	#18 twisted shielded pair in 1/2″ EMT conduit				C.L.F.				331.04	364.14
8000	Applications software									
8050	Basic maintenance manager software (not incl. data base entry)				Ea.				1,702	1,872.72
8100	Time program				Point				5.95	6.55
8120	Duty cycle								11.86	13.05
8140	Optimum start/stop								35.94	39.53
8160	Demand limiting								17.78	19.56
8180	Enthalpy program								35.94	39.53
8200	Boiler optimization				Ea.				1,064	1,170.45
8220	Chiller optimization				"				1,419	1,560.60
8240	Custom applications									
8260	Cost varies with complexity									

23 09 33 – Electric and Electronic Control System for HVAC

23 09 33.10 Electronic Control Systems

		Crew	Daily Output	Labor-Hours	Unit	Material	Labor	Equipment	Total	Total Incl O&P
0010	**ELECTRONIC CONTROL SYSTEMS** R230500-10									
0020	For electronic costs, add to Div. 23 09 43.10				Ea.					15%

23 09 43 – Pneumatic Control System for HVAC

23 09 43.10 Pneumatic Control Systems

		Crew	Daily Output	Labor-Hours	Unit	Material	Labor	Equipment	Total	Total Incl O&P
0010	**PNEUMATIC CONTROL SYSTEMS**									
0011	Including a nominal 50 Ft. of tubing. Add control panelboard if req'd.									
0100	Heating and ventilating, split system R230500-10									
0200	Mixed air control, economizer cycle, panel readout, tubing									

23 09 43 – Pneumatic Control System for HVAC

23 09 43.10 Pneumatic Control Systems			Crew	Daily Output	Labor-Hours	Unit	2009 Bare Costs Material	Labor	Equipment	Total	Total Incl O&P
0220	Up to 10 tons	G	Q-19	.68	35.294	Ea.	3,575	1,600		5,175	6,325
0240	For 10 to 20 tons	G		.63	37.915		3,825	1,725		5,550	6,775
0260	For over 20 tons	G		.58	41.096		4,150	1,850		6,000	7,325
0270	Enthalpy cycle, up to 10 tons			.50	48.387		3,950	2,200		6,150	7,625
0280	For 10 to 20 tons			.46	52.174		4,250	2,350		6,600	8,200
0290	For over 20 tons			.42	56.604		4,625	2,575		7,200	8,900
0300	Heating coil, hot water, 3 way valve,										
0320	Freezestat, limit control on discharge, readout		Q-5	.69	23.088	Ea.	2,650	1,025		3,675	4,475
0500	Cooling coil, chilled water, room										
0520	Thermostat, 3 way valve		Q-5	2	8	Ea.	1,175	355		1,530	1,825
0600	Cooling tower, fan cycle, damper control,										
0620	Control system including water readout in/out at panel		Q-19	.67	35.821	Ea.	4,700	1,625		6,325	7,600
1000	Unit ventilator, day/night operation,										
1100	freezestat, ASHRAE, cycle 2		Q-19	.91	26.374	Ea.	2,600	1,200		3,800	4,625
2000	Compensated hot water from boiler, valve control,										
2100	readout and reset at panel, up to 60 GPM		Q-19	.55	43.956	Ea.	4,875	2,000		6,875	8,325
2120	For 120 GPM			.51	47.059		5,200	2,125		7,325	8,925
2140	For 240 GPM			.49	49.180		5,450	2,225		7,675	9,325
3000	Boiler room combustion air, damper to 5 SF, controls			1.37	17.582		2,350	795		3,145	3,775
3500	Fan coil, heating and cooling valves, 4 pipe control system			3	8		1,050	360		1,410	1,725
3600	Heat exchanger system controls			.86	27.907		2,275	1,275		3,550	4,400
3900	Multizone control (one per zone), includes thermostat, damper										
3910	motor and reset of discharge temperature		Q-5	.51	31.373	Ea.	2,350	1,400		3,750	4,675
4000	Pneumatic thermostat, including controlling room radiator valve		"	2.43	6.593		705	293		998	1,225
4040	Program energy saving optimizer	G	Q-19	1.21	19.786		5,875	895		6,770	7,800
4060	Pump control system		"	3	8		1,075	360		1,435	1,750
4080	Reheat coil control system, not incl coil		Q-5	2.43	6.593		920	293		1,213	1,450
4500	Air supply for pneumatic control system										
4600	Tank mounted duplex compressor, starter, alternator,										
4620	piping, dryer, PRV station and filter										
4630	1/2 HP		Q-19	.68	35.139	Ea.	8,725	1,600		10,325	12,000
4640	3/4 HP			.64	37.383		9,150	1,700		10,850	12,600
4650	1 HP			.61	39.539		10,000	1,800		11,800	13,700
4660	1-1/2 HP			.58	41.739		10,700	1,900		12,600	14,500
4680	3 HP			.55	43.956		14,500	2,000		16,500	19,000
4690	5 HP			.42	57.143		25,300	2,575		27,875	31,800
4800	Main air supply, includes 3/8" copper main and labor		Q-5	1.82	8.791	C.L.F.	293	390		683	910
4810	If poly tubing used, deduct										30%
7000	Static pressure control for air handling unit, includes pressure										
7010	sensor, receiver controller, readout and damper motors		Q-19	.64	37.383	Ea.	6,950	1,700		8,650	10,200
7020	If return air fan requires control, add										70%
8600	VAV boxes, incl. thermostat, damper motor, reheat coil & tubing		Q-5	1.46	10.989		1,100	490		1,590	1,925
8610	If no reheat coil, deduct										204
9400	Sub assemblies for assembly systems										

23 09 53 – Pneumatic and Electric Control System for HVAC

23 09 53.10 Control Components		Crew	Daily Output	Labor-Hours	Unit	Material	Labor	Equipment	Total	Total Incl O&P
0010	**CONTROL COMPONENTS**									
0500	Aquastats									
0508	Immersion type									
0510	High/low limit, breaks contact w/temp rise	1 Stpi	4	2	Ea.	292	98.50		390.50	470
0514	Sequencing, break 2 switches w/temp rise		4	2		219	98.50		317.50	390
0518	Circulating, makes contact w/temp rise		4	2		199	98.50		297.50	365

23 09 53.10 Control Components		Crew	Daily Output	Labor-Hours	Unit	2009 Bare Costs Material	Labor	Equipment	Total	Total Incl O&P
0600	Carbon monoxide detector system									
0606	Panel	1 Stpi	4	2	Ea.	1,350	98.50		1,448.50	1,650
0610	Sensor	"	7.30	1.096	"	785	54		839	945
0700	Controller, receiver									
0730	Pneumatic, panel mount, single input	1 Plum	8	1	Ea.	410	49		459	525
0740	With conversion mounting bracket		8	1		410	49		459	525
0750	Dual input, with control point adjustment	↓	7	1.143		565	55.50		620.50	705
0850	Electric, single snap switch	1 Elec	4	2		390	94		484	570
0860	Dual snap switches		3	2.667		525	125		650	765
0870	Humidity controller		8	1		210	47		257	300
0880	Load limiting controller		8	1		775	47		822	925
0890	Temperature controller	↓	8	1	↓	435	47		482	550
0900	Control panel readout									
0910	Panel up to 12 indicators	1 Stpi	1.20	6.667	Ea.	264	330		594	785
0914	Panel up to 24 indicators		.86	9.302		355	460		815	1,075
0918	Panel up to 48 indicators	↓	.50	16	↓	860	790		1,650	2,125
1000	Enthalpy control, boiler water temperature control									
1010	governed by outdoor temperature, with timer	1 Elec	3	2.667	Ea.	330	125		455	550
1300	Energy control/monitor									
1320	BTU computer meter and controller	1 Stpi	1	8	Ea.	1,425	395		1,820	2,150
1600	Flow meters									
1610	Gas	1 Stpi	4	2	Ea.	103	98.50		201.50	261
1620	Liquid	"	4	2	"	103	98.50		201.50	261
1640	Freezestat									
1644	20' sensing element, adjustable	1 Stpi	3.50	2.286	Ea.	101	113		214	280
2000	Gauges, pressure or vacuum									
2100	2" diameter dial	1 Stpi	32	.250	Ea.	22	12.35		34.35	42.50
2200	2-1/2" diameter dial		32	.250		24	12.35		36.35	45
2300	3-1/2" diameter dial		32	.250		35.50	12.35		47.85	58
2400	4-1/2" diameter dial	↓	32	.250	↓	52.50	12.35		64.85	76.50
2700	Flanged iron case, black ring									
2800	3-1/2" diameter dial	1 Stpi	32	.250	Ea.	102	12.35		114.35	131
2900	4-1/2" diameter dial		32	.250		105	12.35		117.35	135
3000	6" diameter dial	↓	32	.250	↓	167	12.35		179.35	202
3010	Steel case, 0 – 300 psi									
3012	2" diam. dial	1 Stpi	16	.500	Ea.	24	24.50		48.50	63.50
3014	4" diam. dial	"	16	.500	"	28.50	24.50		53	68.50
3020	Aluminum case, 0 – 300 psi									
3022	3-1/2" diam. dial	1 Stpi	16	.500	Ea.	37.50	24.50		62	78.50
3024	4-1/2" diam. dial		16	.500		134	24.50		158.50	185
3026	6" diam. dial		16	.500		211	24.50		235.50	269
3028	8-1/2" diam. dial	↓	16	.500	↓	305	24.50		329.50	370
3030	Brass case, 0 – 300 psi									
3032	2" diam. dial	1 Stpi	16	.500	Ea.	34.50	24.50		59	75
3034	4-1/2" diam. dial	"	16	.500	"	88	24.50		112.50	134
3040	Steel case, high pressure, 0 -10,000 psi									
3042	4-1/2" diam. dial	1 Stpi	16	.500	Ea.	205	24.50		229.50	263
3044	6-1/2" diam. dial		16	.500		211	24.50		235.50	269
3046	8-1/2" diam. dial	↓	16	.500	↓	305	24.50		329.50	370
3080	Pressure gauge, differential, magnehelic									
3084	0 – 2" W.C., with air filter kit	1 Stpi	6	1.333	Ea.	91	66		157	199
3300	For compound pressure-vacuum, add					18%				
3350	Humidistat									

23 09 53.10 Control Components		Crew	Daily Output	Labor-Hours	Unit	Material	2009 Bare Costs Labor	Equipment	Total	Total Incl O&P
3360	Pneumatic operation									
3361	Room humidistat, direct acting	1 Stpi	12	.667	Ea.	289	33		322	370
3362	Room humidistat, reverse acting		12	.667		289	33		322	370
3363	Room humidity transmitter		17	.471		310	23		333	375
3364	Duct mounted controller		12	.667		340	33		373	420
3365	Duct mounted transmitter		12	.667		310	33		343	390
3366	Humidity indicator, 3-1/2"		28	.286		123	14.10		137.10	156
3390	Electric operated	1 Shee	8	1		69.50	47		116.50	149
3400	Relays									
3430	Pneumatic/electric	1 Plum	16	.500	Ea.	310	24.50		334.50	375
3440	Pneumatic proportioning		8	1		209	49		258	305
3450	Pneumatic switching		12	.667		142	32.50		174.50	205
3460	Selector, 3 point		6	1.333		101	65		166	209
3470	Pneumatic time delay		8	1		264	49		313	365
3500	Sensor, air operated									
3520	Humidity	1 Plum	16	.500	Ea.	340	24.50		364.50	405
3540	Pressure		16	.500		58	24.50		82.50	101
3560	Temperature		12	.667		152	32.50		184.50	216
3600	Electric operated									
3620	Humidity	1 Elec	8	1	Ea.	77	47		124	155
3650	Pressure		8	1		1,500	47		1,547	1,725
3680	Temperature		10	.800		93.50	37.50		131	159
3700	Switches									
3710	Minimum position									
3720	Electrical	1 Stpi	4	2	Ea.	19.45	98.50		117.95	170
3730	Pneumatic	"	4	2	"	69	98.50		167.50	224
4000	Thermometers									
4100	Dial type, 3-1/2" diameter, vapor type, union connection	1 Stpi	32	.250	Ea.	152	12.35		164.35	186
4120	Liquid type, union connection		32	.250		295	12.35		307.35	345
4130	Remote reading, 15' capillary		32	.250		172	12.35		184.35	208
4500	Stem type, 6-1/2" case, 2" stem, 1/2" NPT		32	.250		40.50	12.35		52.85	63
4520	4" stem, 1/2" NPT		32	.250		50	12.35		62.35	73
4600	9" case, 3-1/2" stem, 3/4" NPT		28	.286		81.50	14.10		95.60	111
4620	6" stem, 3/4" NPT		28	.286		96	14.10		110.10	127
4640	8" stem, 3/4" NPT		28	.286		109	14.10		123.10	141
4660	12" stem, 1" NPT		26	.308		152	15.20		167.20	190
4670	Bi-metal, dial type, steel case brass stem									
4672	2" dial, 4" - 9" stem	1 Stpi	16	.500	Ea.	39.50	24.50		64	80
4673	2-1/2" dial, 4" - 9" stem		16	.500		39.50	24.50		64	80
4674	3-1/2" dial, 4" - 9" stem		16	.500		44	24.50		68.50	85
4680	Mercury filled, industrial, union connection type									
4682	Angle stem, 7" scale	1 Stpi	16	.500	Ea.	59.50	24.50		84	102
4683	9" scale		16	.500		109	24.50		133.50	157
4684	12" scale		16	.500		141	24.50		165.50	192
4686	Straight stem, 7" scale		16	.500		96	24.50		120.50	142
4687	9" scale		16	.500		90	24.50		114.50	136
4688	12" scale		16	.500		110	24.50		134.50	158
4690	Mercury filled, industrial, separable socket type									
4692	Angle stem, with socket, 7" scale	1 Stpi	16	.500	Ea.	52	24.50		76.50	94
4693	9" scale		16	.500		59	24.50		83.50	102
4694	12" scale		16	.500		84.50	24.50		109	130
4696	Straight stem, with socket, 7" scale		16	.500		25	24.50		49.50	64.50
4697	9" scale		16	.500		32	24.50		56.50	72

23 09 53.10 Control Components		Crew	Daily Output	Labor-Hours	Unit	Material	2009 Bare Costs Labor	Equipment	Total	Total Incl O&P
4698	12″ scale	1 Stpi	16	.500	Ea.	98	24.50		122.50	145
5000	Thermostats									
5030	Manual	1 Shee	8	1	Ea.	27	47		74	102
5040	1 set back, electric, timed G		8	1		95	47		142	177
5050	2 set back, electric, timed G		8	1		209	47		256	300
5100	Locking cover					21			21	23
5200	24 hour, automatic, clock G	1 Shee	8	1		115	47		162	200
5220	Electric, low voltage, 2 wire	1 Elec	13	.615		18	29		47	63
5230	3 wire		10	.800		22	37.50		59.50	80.50
5236	Heating/cooling, low voltage, with clock		8	1		140	47		187	225
5240	Pneumatic									
5250	Single temp., single pressure	1 Stpi	8	1	Ea.	200	49.50		249.50	293
5251	Dual pressure		8	1		280	49.50		329.50	385
5252	Dual temp., dual pressure		8	1		257	49.50		306.50	355
5253	Reverse acting w/averaging element		8	1		198	49.50		247.50	292
5254	Heating-cooling w/deadband		8	1		445	49.50		494.50	565
5255	Integral w/piston top valve actuator		8	1		177	49.50		226.50	269
5256	Dual temp., dual pressure		8	1		181	49.50		230.50	273
5257	Low limit, 8′ averaging element		8	1		171	49.50		220.50	262
5258	Room single temp. proportional		8	1		71.50	49.50		121	153
5260	Dual temp, direct acting for VAV fan		8	1		267	49.50		316.50	365
5262	Capillary tube type, 20′, -30° F to +100° F		8	1		161	49.50		210.50	251
5300	Transmitter, pneumatic									
5320	Temperature averaging element	Q-1	8	2	Ea.	116	88		204	259
5350	Pressure differential	1 Plum	7	1.143		975	55.50		1,030.50	1,150
5370	Humidity, duct		8	1		310	49		359	415
5380	Room		12	.667		310	32.50		342.50	390
5390	Temperature, with averaging element		6	1.333		164	65		229	278
5500	Timer/time clocks									
5510	7 day, 12 hr. battery	1 Stpi	2.60	3.077	Ea.	157	152		309	400
5600	Recorders									
5610	Hydrograph humidity, wall mtd., aluminum case, grad. chart									
5612	8″	1 Stpi	1.60	5	Ea.	615	247		862	1,050
5614	10″	″	1.60	5	″	1,025	247		1,272	1,500
5620	Thermo-hydrograph, wall mtd., grad. chart, 5′ SS probe									
5622	8″	1 Stpi	1.60	5	Ea.	530	247		777	950
5624	10″	″	1.60	5	″	890	247		1,137	1,350
5630	Time of operation, wall mtd., 1 pen cap, 24 hr. clock									
5632	6″	1 Stpi	1.60	5	Ea.	420	247		667	830
5634	8″	″	1.60	5	″	630	247		877	1,050
5640	Pressure & vacuum, bourdon tube type									
5642	Flush mount, 1 pen									
5644	4″	1 Stpi	1.60	5	Ea.	425	247		672	840
5645	6″		1.60	5		420	247		667	830
5646	8″		1.60	5		700	247		947	1,150
5648	10″		1.60	5		615	247		862	1,050
6000	Valves, motorized zone									
6100	Sweat connections, 1/2″ C x C	1 Stpi	20	.400	Ea.	140	19.75		159.75	184
6110	3/4″ C x C		20	.400		141	19.75		160.75	185
6120	1″ C x C		19	.421		183	21		204	233
6140	1/2″ C x C, with end switch, 2 wire		20	.400		139	19.75		158.75	183
6150	3/4″ C x C, with end switch, 2 wire		20	.400		148	19.75		167.75	193
6160	1″ C x C, with end switch, 2 wire		19	.421		169	21		190	217

23 09 53 – Pneumatic and Electric Control System for HVAC

23 09 53.10 Control Components		Crew	Daily Output	Labor-Hours	Unit	Material	Labor	Equipment	Total	Total Incl O&P
						2009 Bare Costs				
7090	Valves, motor controlled, including actuator									
7100	Electric motor actuated									
7200	Brass, two way, screwed									
7210	1/2" pipe size	L-6	36	.333	Ea.	202	16.05		218.05	247
7220	3/4" pipe size		30	.400		258	19.25		277.25	310
7230	1" pipe size		28	.429		300	20.50		320.50	360
7240	1-1/2" pipe size		19	.632		395	30.50		425.50	480
7250	2" pipe size		16	.750		650	36		686	770
7350	Brass, three way, screwed									
7360	1/2" pipe size	L-6	33	.364	Ea.	254	17.50		271.50	305
7370	3/4" pipe size		27	.444		296	21.50		317.50	355
7380	1" pipe size		25.50	.471		385	22.50		407.50	460
7384	1-1/4" pipe size		21	.571		490	27.50		517.50	580
7390	1-1/2" pipe size		17	.706		560	34		594	665
7400	2" pipe size		14	.857		745	41.50		786.50	880
7550	Iron body, two way, flanged									
7560	2-1/2" pipe size	L-6	4	3	Ea.	1,075	145		1,220	1,400
7570	3" pipe size		3	4		1,150	193		1,343	1,575
7580	4" pipe size		2	6		2,025	289		2,314	2,650
7850	Iron body, three way, flanged									
7860	2-1/2" pipe size	L-6	3	4	Ea.	1,125	193		1,318	1,525
7870	3" pipe size		2.50	4.800		1,250	231		1,481	1,750
7880	4" pipe size		2	6		1,625	289		1,914	2,200
8000	Pneumatic, air operated									
8050	Brass, two way, screwed									
8060	1/2" pipe size, class 250	1 Plum	24	.333	Ea.	164	16.25		180.25	205
8070	3/4" pipe size, class 250		20	.400		196	19.50		215.50	246
8080	1" pipe size, class 250		19	.421		229	20.50		249.50	283
8090	1-1/4" pipe size, class 125		15	.533		285	26		311	355
8100	1-1/2" pipe size, class 125		13	.615		365	30		395	445
8110	2" pipe size, class 125		11	.727		420	35.50		455.50	520
8180	Brass, three way, screwed									
8190	1/2" pipe size, class 250	1 Plum	22	.364	Ea.	180	17.75		197.75	225
8200	3/4" pipe size, class 250		18	.444		223	21.50		244.50	278
8210	1" pipe size, class 250		17	.471		262	23		285	325
8214	1-1/4" pipe size, class 250		14	.571		365	28		393	445
8220	1-1/2" pipe size, class 125		11	.727		435	35.50		470.50	535
8230	2" pipe size, class 125		9	.889		525	43.50		568.50	640
8450	Iron body, two way, flanged									
8460	2-1/2" pipe size, 250 lb. flanges	Q-1	5	3.200	Ea.	1,700	140		1,840	2,075
8470	3" pipe size, 250 lb. flanges		4.50	3.556		1,800	156		1,956	2,200
8480	4" pipe size, 250 lb. flanges		3	5.333		2,100	234		2,334	2,650
8510	2-1/2" pipe size, class 125		5	3.200		890	140		1,030	1,200
8520	3" pipe size, class 125		4.50	3.556		975	156		1,131	1,300
8530	4" pipe size, class 125		3	5.333		1,450	234		1,684	1,950
8540	5" pipe size, class 125	Q-2	3.40	7.059		2,950	320		3,270	3,725
8550	6" pipe size, class 125	"	3	8		3,400	365		3,765	4,275
8560	Iron body, three way, flanged									
8570	2-1/2" pipe size, class 125	Q-1	4.50	3.556	Ea.	960	156		1,116	1,275
8580	3" pipe size, class 125		4	4		1,100	176		1,276	1,500
8590	4" pipe size, class 125		2.50	6.400		2,225	281		2,506	2,875
8600	6" pipe size, class 125	Q-2	3	8		3,400	365		3,765	4,275
9005	Pneumatic system misc. components									

23 09 53.10 Control Components		Crew	Daily Output	Labor-Hours	Unit	Material	Labor	Equipment	Total	Total Incl O&P
						2009 Bare Costs				
9010	Adding/subtracting repeater	1 Stpi	20	.400	Ea.	156	19.75		175.75	202
9020	Adjustable ratio network		16	.500		147	24.50		171.50	199
9030	Comparator		20	.400		300	19.75		319.75	360
9040	Cumulator									
9041	Air switching	1 Stpi	20	.400	Ea.	142	19.75		161.75	186
9042	Averaging		20	.400		128	19.75		147.75	170
9043	2:1 ratio		20	.400		288	19.75		307.75	345
9044	Sequencing		20	.400		90.50	19.75		110.25	129
9045	Two-position		16	.500		177	24.50		201.50	231
9046	Two-position pilot		16	.500		340	24.50		364.50	405
9050	Damper actuator									
9051	For smoke control	1 Stpi	8	1	Ea.	134	49.50		183.50	221
9052	For ventilation	"	8	1	"	193	49.50		242.50	286
9053	Series duplex pedestal mounted									
9054	For inlet vanes on fans, compressors	1 Stpi	6	1.333	Ea.	1,475	66		1,541	1,725
9060	Enthalpy logic center		16	.500		287	24.50		311.50	350
9070	High/low pressure selector		22	.364		57	17.95		74.95	90
9080	Signal limiter		20	.400		90.50	19.75		110.25	129
9081	Transmitter		28	.286		34.50	14.10		48.60	59
9090	Optimal start									
9092	Mass temperature transmitter	1 Stpi	18	.444	Ea.	135	22		157	182
9100	Step controller with positioner, time delay restrictor									
9110	recycler air valve, switches and cam settings									
9112	With cabinet									
9113	6 points	1 Stpi	4	2	Ea.	885	98.50		983.50	1,125
9114	6 points, 6 manual sequences		2	4		775	197		972	1,150
9115	8 points		3	2.667		965	132		1,097	1,275
9200	Pressure controller and switches									
9210	High static pressure limit	1 Stpi	8	1	Ea.	202	49.50		251.50	296
9220	Pressure transmitter		8	1		213	49.50		262.50	310
9230	Differential pressure transmitter		8	1		835	49.50		884.50	995
9240	Static pressure transmitter		8	1		835	49.50		884.50	995
9250	Proportional-only control, single input		6	1.333		410	66		476	550
9260	Proportional plus integral, single input		6	1.333		610	66		676	770
9270	Proportional-only, dual input		6	1.333		565	66		631	720
9280	Proportional plus integral, dual input		6	1.333		745	66		811	920
9281	Time delay for above proportional units		16	.500		143	24.50		167.50	195
9290	Proportional/2 position controller units		6	1.333		430	66		496	575
9300	Differential pressure control direct/reverse		6	1.333		365	66		431	500
9310	Air pressure reducing valve									
9311	1/8" size	1 Stpi	18	.444	Ea.	31.50	22		53.50	67.50
9315	Precision valve, 1/4" size		17	.471		176	23		199	228
9320	Air flow controller		12	.667		164	33		197	230
9330	Booster relay volume amplifier		18	.444		97	22		119	140
9340	Series restrictor, straight		60	.133		4.99	6.60		11.59	15.35
9341	T-fitting		40	.200		7.25	9.85		17.10	23
9342	In-line adjustable		48	.167		32	8.25		40.25	48
9350	Diode tee		40	.200		16.70	9.85		26.55	33
9351	Restrictor tee		40	.200		16.70	9.85		26.55	33
9360	Pneumatic gradual switch		8	1		279	49.50		328.50	380
9361	Selector switch		8	1		101	49.50		150.50	185
9370	Electro-pneumatic motor driven servo		10	.800		880	39.50		919.50	1,025
9390	Fan control switch and mounting base		16	.500		97.50	24.50		122	144

23 09 53.10 Control Components		Crew	Daily Output	Labor-Hours	Unit	2009 Bare Costs Material	Labor	Equipment	Total	Total Incl O&P
9400	Circulating pump sequencer	1 Stpi	14	.571	Ea.	1,150	28		1,178	1,325
9410	Pneumatic tubing, fittings and accessories									
9414	Tubing, urethane									
9415	1/8″ OD x 1/16″ ID	1 Stpi	120	.067	L.F.	.66	3.29		3.95	5.65
9416	1/4″ OD x 1/8″ ID		115	.070		1.30	3.43		4.73	6.60
9417	5/32″ OD x 3/32″ ID		110	.073		.74	3.59		4.33	6.20
9420	Coupling, straight									
9422	Barb x barb									
9423	1/4″ x 5/32″	1 Stpi	160	.050	Ea.	.74	2.47		3.21	4.51
9424	1/4″ x 1/4″		158	.051		.38	2.50		2.88	4.17
9425	3/8″ x 3/8″		154	.052		.51	2.56		3.07	4.41
9426	3/8″ x 1/4″		150	.053		.51	2.63		3.14	4.51
9427	1/2″ x 1/4″		148	.054		4.73	2.67		7.40	9.20
9428	1/2″ x 3/8″		144	.056		.99	2.74		3.73	5.20
9429	1/2″ x 1/2″		140	.057		.65	2.82		3.47	4.95
9440	Tube x tube									
9441	1/4″ x 1/4″	1 Stpi	100	.080	Ea.	2.03	3.95		5.98	8.15
9442	3/8″ x 1/4″		96	.083		3.77	4.11		7.88	10.30
9443	3/8″ x 3/8″		92	.087		2.52	4.29		6.81	9.20
9444	1/2″ x 3/8″		88	.091		4.05	4.49		8.54	11.20
9445	1/2″ x 1/2″		84	.095		4.62	4.70		9.32	12.15
9450	Elbow coupling									
9452	Barb x barb									
9454	1/4″ x 1/4″	1 Stpi	158	.051	Ea.	.87	2.50		3.37	4.71
9455	3/8″ x 3/8″		154	.052		2.37	2.56		4.93	6.45
9456	1/2″ x 1/2″		140	.057		1.32	2.82		4.14	5.70
9460	Tube x tube									
9462	1/2″ x 1/2″	1 Stpi	84	.095	Ea.	6.40	4.70		11.10	14.10
9470	Tee coupling									
9472	Barb x barb x barb									
9474	1/4″ x 1/4″ x 5/32″	1 Stpi	108	.074	Ea.	1.31	3.66		4.97	6.95
9475	1/4″ x 1/4″ x 1/4″		104	.077		.89	3.80		4.69	6.70
9476	3/8″ x 3/8″ x 5/32″		102	.078		2.84	3.87		6.71	8.90
9477	3/8″ x 3/8″ x 1/4″		99	.081		1.31	3.99		5.30	7.45
9478	3/8″ x 3/8″ x 3/8″		98	.082		1.40	4.03		5.43	7.60
9479	1/2″ x 1/2″ x 1/4″		96	.083		3.57	4.11		7.68	10.10
9480	1/2″ x 1/2″ x 3/8″		95	.084		2.97	4.16		7.13	9.50
9481	1/2″ x 1/2″ x 1/2″		92	.087		1.59	4.29		5.88	8.20
9484	Tube x tube x tube									
9485	1/4″ x 1/4″ x 1/4″	1 Stpi	66	.121	Ea.	4.03	6		10.03	13.45
9486	3/8″ x 1/4″ x 1/4″		64	.125		12.30	6.15		18.45	23
9487	3/8″ x 3/8″ x 1/4″		61	.131		5.85	6.45		12.30	16.10
9488	3/8″ x 3/8″ x 3/8″		60	.133		6.10	6.60		12.70	16.60
9489	1/2″ x 1/2″ x 3/8″		58	.138		12.75	6.80		19.55	24.50
9490	1/2″ x 1/2″ x 1/2″		56	.143		8.60	7.05		15.65	20
9492	Needle valve									
9494	Tube x tube									
9495	1/4″ x 1/4″	1 Stpi	86	.093	Ea.	7.15	4.59		11.74	14.80
9496	3/8″ x 3/8″	″	82	.098	″	10.60	4.81		15.41	18.90
9600	Electronic system misc. components									
9610	Electric motor damper actuator	1 Elec	8	1	Ea.	605	47		652	735
9620	Damper position indicator		10	.800		114	37.50		151.50	182
9630	Pneumatic-electronic transducer		12	.667		59.50	31.50		91	112

23 09 Instrumentation and Control for HVAC

23 09 53 – Pneumatic and Electric Control System for HVAC

23 09 53.10 Control Components		Crew	Daily Output	Labor-Hours	Unit	2009 Bare Costs Material	Labor	Equipment	Total	Total Incl O&P
9700	Modulating step controller									
9701	2-5 steps	1 Elec	5.30	1.509	Ea.	1,975	71		2,046	2,275
9702	6-10 steps		3.20	2.500		2,550	118		2,668	3,000
9703	11-15 steps		2.70	2.963		3,100	139		3,239	3,600
9704	16-20 steps	↓	1.60	5	↓	3,600	235		3,835	4,300
9710	Staging network									
9740	Accessory unit regulator									
9750	Accessory power supply	1 Elec	18	.444	Ea.	175	21		196	223
9760	Step down transformer	"	16	.500	"	181	23.50		204.50	234

23 11 Facility Fuel Piping

23 11 13 – Facility Fuel-Oil Piping

23 11 13.10 Fuel Oil Specialties

		Crew	Daily Output	Labor-Hours	Unit	Material	Labor	Equipment	Total	Total Incl O&P
0010	**FUEL OIL SPECIALTIES**									
0020	Foot valve, single poppet, metal to metal construction									
0040	Bevel seat, 1/2" diameter	1 Stpi	20	.400	Ea.	59	19.75		78.75	94.50
0060	3/4" diameter		18	.444		57	22		79	95.50
0080	1" diameter		16	.500		24	24.50		48.50	63.50
0100	1-1/4" diameter		15	.533		88	26.50		114.50	137
0120	1-1/2" diameter		13	.615		37	30.50		67.50	86
0140	2" diameter	↓	11	.727	↓	38.50	36		74.50	96
0160	Foot valve, double poppet, metal to metal construction									
0164	1" diameter	1 Stpi	15	.533	Ea.	134	26.50		160.50	187
0166	1-1/2" diameter	"	12	.667	"	161	33		194	227
0400	Fuel fill box, flush type									
0408	Nonlocking, watertight									
0410	1-1/2" diameter	1 Stpi	12	.667	Ea.	19.40	33		52.40	71
0440	2" diameter		10	.800		16.75	39.50		56.25	77.50
0450	2-1/2" diameter		9	.889		56	44		100	128
0460	3" diameter		7	1.143		54	56.50		110.50	144
0470	4" diameter	↓	5	1.600	↓	75	79		154	201
0500	Locking inner cover									
0510	2" diameter	1 Stpi	8	1	Ea.	82.50	49.50		132	165
0520	2-1/2" diameter		7	1.143		110	56.50		166.50	206
0530	3" diameter		5	1.600		118	79		197	248
0540	4" diameter	↓	4	2	↓	152	98.50		250.50	315
0600	Fuel system components									
0620	Spill container	1 Stpi	4	2	Ea.	585	98.50		683.50	790
0640	Fill adapter, 4", straight drop		8	1		47	49.50		96.50	126
0680	Fill cap, 4"		30	.267		31.50	13.15		44.65	55
0700	Extractor fitting, 4" x 1-1/2"		8	1		310	49.50		359.50	415
0740	Vapor hose adapter, 4"	↓	8	1		90	49.50		139.50	173
0760	Wood gage stick, 10'					11			11	12.10
1000	Oil filters, 3/8" IPT., 20 gal. per hour	1 Stpi	20	.400		19.25	19.75		39	50.50
1020	32 gal. per hour		18	.444		34.50	22		56.50	71
1040	40 gal. per hour		16	.500		37	24.50		61.50	78
1060	50 gal. per hour	↓	14	.571	↓	38.50	28		66.50	84.50
2000	Remote tank gauging system, self contained									
2100	Single tank kit/8 sensors inputs w/printer	1 Stpi	2.50	3.200	Ea.	3,275	158		3,433	3,825
2120	Two tank kit/8 sensors inputs w/printer		2	4		4,425	197		4,622	5,150
3000	Valve, ball check, globe type, 3/8" diameter	↓	24	.333	↓	11.10	16.45		27.55	36.50

23 11 13.10 Fuel Oil Specialties		Crew	Daily Output	Labor-Hours	Unit	2009 Bare Costs Material	Labor	Equipment	Total	Total Incl O&P
3500	Fusible, 3/8" diameter	1 Stpi	24	.333	Ea.	11.40	16.45		27.85	37
3600	1/2" diameter		24	.333		27.50	16.45		43.95	55
3610	3/4" diameter		20	.400		61	19.75		80.75	96.50
3620	1" diameter		19	.421		175	21		196	224
4000	Nonfusible, 3/8" diameter		24	.333		19.25	16.45		35.70	45.50
4500	Shutoff, gate type, lever handle, spring-fusible kit									
4520	1/4" diameter	1 Stpi	14	.571	Ea.	30	28		58	75.50
4540	3/8" diameter		12	.667		29	33		62	81.50
4560	1/2" diameter		10	.800		43	39.50		82.50	107
4570	3/4" diameter		8	1		74	49.50		123.50	156
4580	Lever handle, requires weight and fusible kit									
4600	1" diameter	1 Stpi	9	.889	Ea.	164	44		208	246
4620	1-1/4" diameter		8	1		189	49.50		238.50	282
4640	1-1/2" diameter		7	1.143		244	56.50		300.50	355
4660	2" diameter		6	1.333		305	66		371	435
4680	For fusible link, weight and braided wire, add					5%				
5000	Vent alarm, whistling signal					26			26	28.50
5500	Vent protector/breather, 1-1/4" diameter	1 Stpi	32	.250		10.40	12.35		22.75	30
5520	1-1/2" diameter		32	.250		12	12.35		24.35	31.50
5540	2" diameter		32	.250		21	12.35		33.35	41.50
5560	3" diameter		28	.286		33.50	14.10		47.60	58
5580	4" diameter		24	.333		37.50	16.45		53.95	66
5600	Dust cap, breather, 2"		40	.200		11.70	9.85		21.55	27.50
8000	Fuel oil and tank heaters									
8020	Electric, capacity rated at 230 volts									
8040	Immersion element in steel manifold									
8060	96 GPH at 50° F rise	Q-5	6.40	2.500	Ea.	870	111		981	1,125
8070	128 GPH at 50° F rise		6.20	2.581		900	115		1,015	1,150
8080	160 GPH at 50° F rise		5.90	2.712		1,000	120		1,120	1,275
8090	192 GPH at 50° F rise		5.50	2.909		1,100	129		1,229	1,400
8100	240 GPH at 50° F rise		5.10	3.137		1,250	139		1,389	1,575
8110	288 GPH at 50° F rise		4.60	3.478		1,425	155		1,580	1,775
8120	384 GPH at 50° F rise		3.10	5.161		1,650	229		1,879	2,175
8130	480 GPH at 50° F rise		2.30	6.957		2,075	310		2,385	2,750
8140	576 GPH at 50° F rise		2.10	7.619		2,075	340		2,415	2,775
8300	Suction stub, immersion type									
8320	75" long, 750 watts	1 Stpi	14	.571	Ea.	590	28		618	695
8330	99" long, 2000 watts		12	.667		830	33		863	965
8340	123" long, 3000 watts		10	.800		895	39.50		934.50	1,050
8660	Steam, cross flow, rated at 5 PSIG									
8680	42 GPH	Q-5	7	2.286	Ea.	1,375	102		1,477	1,650
8690	73 GPH		6.70	2.388		1,600	106		1,706	1,900
8700	112 GPH		6.20	2.581		1,650	115		1,765	2,000
8710	158 GPH		5.80	2.759		1,800	123		1,923	2,150
8720	187 GPH		4	4		2,475	178		2,653	3,000
8730	270 GPH		3.60	4.444		2,675	197		2,872	3,225
8740	365 GPH	Q-6	4.90	4.898		3,125	226		3,351	3,800
8750	635 GPH		3.70	6.486		3,950	299		4,249	4,800
8760	845 GPH		2.50	9.600		5,800	440		6,240	7,050
8770	1420 GPH		1.60	15		9,425	690		10,115	11,400
8780	2100 GPH		1.10	21.818		12,900	1,000		13,900	15,700

23 11 Facility Fuel Piping

23 11 23 – Facility Natural-Gas Piping

23 11 23.10 Gas Meters		Crew	Daily Output	Labor-Hours	Unit	2009 Bare Costs Material	Labor	Equipment	Total	Total Incl O&P
0010	**GAS METERS**									
4000	Residential									
4010	Gas meter, residential, 3/4" pipe size	1 Plum	14	.571	Ea.	131	28		159	186
4020	Gas meter, residential, 1" pipe size		12	.667		181	32.50		213.50	248
4030	Gas meter, residential, 1-1/4" pipe size	↓	10	.800	↓	198	39		237	277

23 12 Facility Fuel Pumps

23 12 13 – Facility Fuel-Oil Pumps

23 12 13.10 Pump and Motor Sets

		Crew	Daily Output	Labor-Hours	Unit	Material	Labor	Equipment	Total	Total Incl O&P
0010	**PUMP AND MOTOR SETS**									
1810	Light fuel and diesel oils									
1820	20 GPH 1/3 HP	Q-5	6	2.667	Ea.	790	118		908	1,050
1830	27 GPH, 1/3 HP		6	2.667		790	118		908	1,050
1840	80 GPH, 1/3 HP		5	3.200		790	142		932	1,075
1850	145 GPH, 1/2 HP		4	4		835	178		1,013	1,175
1860	277 GPH, 1 HP		4	4		930	178		1,108	1,300
1870	700 GPH, 1-1/2 HP		3	5.333		1,925	237		2,162	2,475
1880	1000 GPH, 2 HP		3	5.333		2,325	237		2,562	2,900
1890	1800 GPH, 5 HP	↓	1.80	8.889	↓	3,200	395		3,595	4,100

23 12 16 – Facility Gasoline Dispensing Pumps

23 12 16.20 Fuel Dispensing Equipment

		Crew	Daily Output	Labor-Hours	Unit	Material	Labor	Equipment	Total	Total Incl O&P
0010	**FUEL DISPENSING EQUIPMENT**									
1100	Product dispenser with vapor recovery for 6 nozzles, installed, not									
1110	including piping to storage tanks				Ea.	20,800			20,800	22,900

23 13 Facility Fuel-Storage Tanks

23 13 13 – Facility Underground Fuel-Oil, Storage Tanks

23 13 13.09 Single-Wall Steel Fuel-Oil Tanks

		Crew	Daily Output	Labor-Hours	Unit	Material	Labor	Equipment	Total	Total Incl O&P
0010	**SINGLE-WALL STEEL FUEL-OIL TANKS**									
5000	Steel underground, sti-P3, set in place, not incl. hold-down bars.									
5500	Excavation, pad, pumps and piping not included									
5510	Single wall, 500 gallon capacity, 7 gauge shell	Q-5	2.70	5.926	Ea.	1,975	263		2,238	2,575
5520	1,000 gallon capacity, 7 gauge shell	"	2.50	6.400		3,150	284		3,434	3,900
5530	2,000 gallon capacity, 1/4" thick shell	Q-7	4.60	6.957		5,125	325		5,450	6,125
5535	2,500 gallon capacity, 7 gauge shell	Q-5	3	5.333		5,650	237		5,887	6,575
5540	5,000 gallon capacity, 1/4" thick shell	Q-7	3.20	10		8,250	470		8,720	9,775
5560	10,000 gallon capacity, 1/4" thick shell		2	16		10,200	750		10,950	12,300
5580	15,000 gallon capacity, 5/16" thick shell		1.70	18.824		12,700	885		13,585	15,200
5600	20,000 gallon capacity, 5/16" thick shell		1.50	21.333		21,100	1,000		22,100	24,700
5610	25,000 gallon capacity, 3/8" thick shell		1.30	24.615		25,700	1,150		26,850	30,000
5620	30,000 gallon capacity, 3/8" thick shell		1.10	29.091		31,700	1,375		33,075	37,000
5630	40,000 gallon capacity, 3/8" thick shell		.90	35.556		39,600	1,675		41,275	46,100
5640	50,000 gallon capacity, 3/8" thick shell	↓	.80	40	↓	49,500	1,875		51,375	57,500

23 13 13.13 Dbl-Wall Stl, Undgrnd Fuel-Oil, Stor. Tanks

		Crew	Daily Output	Labor-Hours	Unit	Material	Labor	Equipment	Total	Total Incl O&P
0010	**DOUBLE-WALL STEEL, UNDERGROUND FUEL-OIL, STORAGE TANKS**									
6200	Steel, underground, 360°, double wall, U.L. listed,									
6210	with sti-P3 corrosion protection,									
6220	(dielectric coating, cathodic protection, electrical									

23 13 13 – Facility Underground Fuel-Oil, Storage Tanks

23 13 13.13 Dbl-Wall Stl, Undgrnd Fuel-Oil, Stor. Tanks	Crew	Daily Output	Labor-Hours	Unit	Material	Labor	Equipment	Total	Total Incl O&P
6230 isolation) 30 year warranty,									
6240 not incl. manholes or hold-downs.									
6250 500 gallon capacity	Q-5	2.40	6.667	Ea.	3,900	296		4,196	4,750
6260 1,000 gallon capactiy	"	2.25	7.111		5,700	315		6,015	6,750
6270 2,000 gallon capacity	Q-7	4.16	7.692		7,475	360		7,835	8,775
6280 3,000 gallon capacity		3.90	8.205		9,350	385		9,735	10,900
6290 4,000 gallon capacity		3.64	8.791		11,500	415		11,915	13,300
6300 5,000 gallon capacity		2.91	10.997		14,100	515		14,615	16,300
6310 6,000 gallon capacity		2.42	13.223		16,000	620		16,620	18,500
6320 8,000 gallon capacity		2.08	15.385		17,800	725		18,525	20,700
6330 10,000 gallon capacity		1.82	17.582		21,900	825		22,725	25,400
6340 12,000 gallon capacity		1.70	18.824		23,700	885		24,585	27,400
6350 15,000 gallon capacity		1.33	24.060		29,000	1,125		30,125	33,600
6360 20,000 gallon capacity		1.33	24.060		37,600	1,125		38,725	43,100
6370 25,000 gallon capacity		1.16	27.586		60,000	1,300		61,300	68,000
6380 30,000 gallon capacity		1.03	31.068		72,500	1,450		73,950	81,500
6390 40,000 gallon capacity		.80	40		93,000	1,875		94,875	105,000
6395 50,000 gallon capacity		.73	43.836	↓	113,000	2,050		115,050	127,000
6400 For hold-downs 500-2000 gal, add		16	2	Set	151	94		245	305
6410 For hold-downs 3000-6000 gal, add		12	2.667		305	125		430	525
6420 For hold-downs 8000-12,000 gal, add		11	2.909		365	137		502	610
6430 For hold-downs 15,000 gal, add		9	3.556		530	167		697	830
6440 For hold-downs 20,000 gal, add		8	4		610	188		798	950
6450 For hold-downs 20,000 gal plus, add	↓	6	5.333	↓	790	251		1,041	1,250
6500 For manways, add				Ea.	1,500			1,500	1,650
6600 In place with hold-downs									
6652 550 gallon capacity	Q-5	1.84	8.696	Ea.	4,050	385		4,435	5,050

23 13 13.23 Glass-Fiber-Reinfcd-Plastic, Fuel-Oil, Storage	Crew	Daily Output	Labor-Hours	Unit	Material	Labor	Equipment	Total	Total Incl O&P
0010 **GLASS-FIBER-REINFCD-PLASTIC, UNDERGRND FUEL-OIL, STORAGE**									
0210 Fiberglass, underground, single wall, U.L. listed, not including									
0220 manway or hold-down strap									
0225 550 gallon capacity	Q-5	2.67	5.993	Ea.	3,000	266		3,266	3,700
0230 1,000 gallon capacity	"	2.46	6.504		3,650	289		3,939	4,450
0240 2,000 gallon capacity	Q-7	4.57	7.002		5,750	330		6,080	6,825
0250 4,000 gallon capacity		3.55	9.014		8,150	425		8,575	9,600
0260 6,000 gallon capacity		2.67	11.985		9,050	565		9,615	10,800
0270 8,000 gallon capacity		2.29	13.974		10,800	655		11,455	12,800
0280 10,000 gallon capacity		2	16		12,200	750		12,950	14,500
0282 12,000 gallon capacity		1.88	17.021		20,500	800		21,300	23,800
0284 15,000 gallon capacity		1.68	19.048		22,600	895		23,495	26,200
0290 20,000 gallon capacity		1.45	22.069		24,400	1,025		25,425	28,400
0300 25,000 gallon capacity		1.28	25		36,100	1,175		37,275	41,500
0320 30,000 gallon capacity		1.14	28.070		43,500	1,325		44,825	49,800
0340 40,000 gallon capacity		.89	35.955		62,000	1,700		63,700	70,500
0360 48,000 gallon capacity	↓	.81	39.506		73,000	1,850		74,850	83,000
0500 For manway, fittings and hold-downs, add					20%	15%			
0600 For manways, add					1,800			1,800	1,975
1000 For helical heating coil, add	Q-5	2.50	6.400	↓	3,650	284		3,934	4,425
1020 Fiberglass, underground, double wall, U.L. listed									
1030 includes manways, not incl. hold-down straps									
1040 600 gallon capacity	Q-5	2.42	6.612	Ea.	6,650	294		6,944	7,750
1050 1,000 gallon capacity	"	2.25	7.111	↓	8,350	315		8,665	9,675

Column heading note: Material, Labor, Equipment and Total are under "2009 Bare Costs".

23 13 13 – Facility Underground Fuel-Oil, Storage Tanks

23 13 13.23 Glass-Fiber-Reinfcd-Plastic, Fuel-Oil, Storage		Crew	Daily Output	Labor-Hours	Unit	2009 Bare Costs Material	Labor	Equipment	Total	Total Incl O&P
1060	2,500 gallon capacity	Q-7	4.16	7.692	Ea.	12,700	360		13,060	14,500
1070	3,000 gallon capacity		3.90	8.205		13,700	385		14,085	15,700
1080	4,000 gallon capacity		3.64	8.791		15,500	415		15,915	17,600
1090	6,000 gallon capacity		2.42	13.223		15,700	620		16,320	18,200
1100	8,000 gallon capacity		2.08	15.385		20,100	725		20,825	23,200
1110	10,000 gallon capacity		1.82	17.582		23,000	825		23,825	26,600
1120	12,000 gallon capacity		1.70	18.824		28,300	885		29,185	32,500
1122	15,000 gallon capacity		1.52	21.053		36,700	990		37,690	41,800
1124	20,000 gallon capacity		1.33	24.060		43,900	1,125		45,025	50,000
1126	25,000 gallon capacity		1.16	27.586		58,500	1,300		59,800	66,500
1128	30,000 gallon capacity	↓	1.03	31.068		70,500	1,450		71,950	79,500
1140	For hold-down straps, add				↓	2%	10%			
1150	For hold-downs 500-4000 gal., add	Q-7	16	2	Set	420	94		514	600
1160	For hold-downs 5000-15000 gal., add		8	4		835	188		1,023	1,200
1170	For hold-downs 20,000 gal., add		5.33	6.004		1,250	282		1,532	1,800
1180	For hold-downs 25,000 gal., add		4	8		1,675	375		2,050	2,425
1190	For hold-downs 30,000 gal., add	↓	2.60	12.308	↓	2,500	580		3,080	3,625
2210	Fiberglass, underground, single wall, U.L. listed, including									
2220	hold-down straps, no manways									
2225	550 gallon capacity	Q-5	2	8	Ea.	3,400	355		3,755	4,275
2230	1,000 gallon capacity	"	1.88	8.511		4,075	380		4,455	5,050
2240	2,000 gallon capacity	Q-7	3.55	9.014		6,150	425		6,575	7,400
2250	4,000 gallon capacity		2.90	11.034		8,575	520		9,095	10,200
2260	6,000 gallon capacity		2	16		9,900	750		10,650	12,000
2270	8,000 gallon capacity		1.78	17.978		11,600	845		12,445	14,100
2280	10,000 gallon capacity		1.60	20		13,000	940		13,940	15,700
2282	12,000 gallon capacity		1.52	21.053		21,400	990		22,390	25,000
2284	15,000 gallon capacity		1.39	23.022		23,400	1,075		24,475	27,400
2290	20,000 gallon capacity		1.14	28.070		25,600	1,325		26,925	30,200
2300	25,000 gallon capacity		.96	33.333		37,700	1,575		39,275	43,900
2320	30,000 gallon capacity	↓	.80	40	↓	46,000	1,875		47,875	53,500
3020	Fiberglass, underground, double wall, U.L. listed									
3030	includes manways and hold-down straps									
3040	600 gallon capacity	Q-5	1.86	8.602	Ea.	7,050	380		7,430	8,350
3050	1,000 gallon capacity	"	1.70	9.412		8,775	420		9,195	10,300
3060	2,500 gallon capacity	Q-7	3.29	9.726		13,100	455		13,555	15,100
3070	3,000 gallon capacity		3.13	10.224		14,100	480		14,580	16,300
3080	4,000 gallon capacity		2.93	10.922		15,900	515		16,415	18,300
3090	6,000 gallon capacity		1.86	17.204		16,500	810		17,310	19,400
3100	8,000 gallon capacity		1.65	19.394		21,000	910		21,910	24,500
3110	10,000 gallon capacity		1.48	21.622		23,900	1,025		24,925	27,700
3120	12,000 gallon capacity		1.40	22.857		29,200	1,075		30,275	33,700
3122	15,000 gallon capacity		1.28	25		37,500	1,175		38,675	43,100
3124	20,000 gallon capacity		1.06	30.189		45,200	1,425		46,625	52,000
3126	25,000 gallon capacity		.90	35.556		60,500	1,675		62,175	69,000
3128	30,000 gallon capacity	↓	.74	43.243	↓	73,000	2,025		75,025	83,000

23 13 23 – Facility Aboveground Fuel-Oil, Storage Tanks

23 13 23.13 Vert., Steel, Abvgrnd Fuel-Oil, Stor. Tanks		Crew	Daily Output	Labor-Hours	Unit	Material	Labor	Equipment	Total	Total Incl O&P
0010	**VERTICAL, STEEL, ABOVEGROUND FUEL-OIL, STORAGE TANKS**									
4000	Fixed roof oil storage tanks, steel, (1 BBL=42 GAL w/ foundation 3'D x 1'W)									
4200	5,000 barrels				Ea.				157,273	173,000
4300	24,000 barrels				↓				269,091	296,000

23 13 Facility Fuel-Storage Tanks

23 13 23 – Facility Aboveground Fuel-Oil, Storage Tanks

23 13 23.13 Vert., Steel, Abvgrnd Fuel-Oil, Stor. Tanks		Crew	Daily Output	Labor-Hours	Unit	2009 Bare Costs Material	Labor	Equipment	Total	Total Incl O&P
4500	56,000 barrels				Ea.				535,000	588,500
4600	110,000 barrels								777,273	855,000
4800	143,000 barrels								996,818	1,096,500
4900	224,000 barrels								1,495,909	1,645,500
5100	Floating roof gasoline tanks, steel, 5,000 barrels								172,727	190,000
5200	25,000 barrels								295,455	325,000
5400	55,000 barrels								590,909	650,000
5500	100,000 barrels								854,545	940,000
5700	150,000 barrels								1,095,455	1,205,000
5800	225,000 barrels								1,645,455	1,810,000

23 13 23.16 Horizontal, Stl, Abvgrd Fuel-Oil, Storage Tanks		Crew	Daily Output	Labor-Hours	Unit	Material	Labor	Equipment	Total	Total Incl O&P
0010	**HORIZONTAL, STEEL, ABOVEGROUND FUEL-OIL, STORAGE TANKS**									
3000	Steel, storage, above ground, including cradles, coating,									
3020	fittings, not including foundation, pumps or piping									
3040	Single wall, interior, 275 gallon	Q-5	5	3.200	Ea.	385	142		527	635
3060	550 gallon	"	2.70	5.926		2,275	263		2,538	2,925
3080	1,000 gallon	Q-7	5	6.400		2,875	300		3,175	3,600
3100	1,500 gallon		4.75	6.737		4,275	315		4,590	5,200
3120	2,000 gallon		4.60	6.957		4,925	325		5,250	5,900
3140	5,000 gallon		3.20	10		6,400	470		6,870	7,750
3150	10,000 gallon		2	16		18,400	750		19,150	21,400
3160	15,000 gallon		1.70	18.824		21,000	885		21,885	24,400
3170	20,000 gallon		1.45	22.069		26,300	1,025		27,325	30,500
3180	25,000 gallon		1.30	24.615		31,900	1,150		33,050	36,800
3190	30,000 gallon		1.10	29.091		36,100	1,375		37,475	41,800
3320	Double wall, 500 gallon capacity	Q-5	2.40	6.667		2,300	296		2,596	2,975
3330	2000 gallon capacity	Q-7	4.15	7.711		8,250	360		8,610	9,625
3340	4000 gallon capacity		3.60	8.889		14,700	420		15,120	16,800
3350	6000 gallon capacity		2.40	13.333		17,400	625		18,025	20,000
3360	8000 gallon capacity		2	16		22,300	750		23,050	25,600
3370	10000 gallon capacity		1.80	17.778		25,000	835		25,835	28,800
3380	15000 gallon capacity		1.50	21.333		37,900	1,000		38,900	43,200
3390	20000 gallon capacity		1.30	24.615		43,200	1,150		44,350	49,200
3400	25000 gallon capacity		1.15	27.826		52,500	1,300		53,800	59,500
3410	30000 gallon capacity		1	32		57,500	1,500		59,000	66,000

23 21 Hydronic Piping and Pumps

23 21 20 – Hydronic HVAC Piping Specialties

23 21 20.10 Air Control		Crew	Daily Output	Labor-Hours	Unit	Material	Labor	Equipment	Total	Total Incl O&P
0010	**AIR CONTROL**									
0030	Air separator, with strainer									
0040	2" diameter	Q-5	6	2.667	Ea.	750	118		868	1,000
0080	2-1/2" diameter		5	3.200		850	142		992	1,150
0100	3" diameter		4	4		1,300	178		1,478	1,700
0120	4" diameter		3	5.333		1,875	237		2,112	2,425
0130	5" diameter	Q-6	3.60	6.667		2,375	305		2,680	3,075
0140	6" diameter		3.40	7.059		2,850	325		3,175	3,625
0160	8" diameter		3	8		4,275	370		4,645	5,250
0180	10" diameter		2.20	10.909		6,625	505		7,130	8,025
0200	12" diameter		1.70	14.118		11,100	650		11,750	13,200
0210	14" diameter		1.30	18.462		13,200	850		14,050	15,800

23 21 20.10 Air Control		Crew	Daily Output	Labor-Hours	Unit	2009 Bare Costs Material	Labor	Equipment	Total	Total Incl O&P
0220	16″ diameter	Q-6	1	24	Ea.	19,900	1,100		21,000	23,600
0230	18″ diameter		.80	30		26,200	1,375		27,575	30,900
0240	20″ diameter		.60	40		29,300	1,850		31,150	35,000
0300	Without strainer									
0310	2″ diameter	Q-5	6	2.667	Ea.	580	118		698	820
0320	2-1/2″ diameter		5	3.200		695	142		837	980
0330	3″ diameter		4	4		975	178		1,153	1,350
0340	4″ diameter		3	5.333		1,525	237		1,762	2,025
0350	5″ diameter		2.40	6.667		2,075	296		2,371	2,725
0360	6″ diameter	Q-6	3.40	7.059		2,375	325		2,700	3,125
0370	8″ diameter		3	8		3,275	370		3,645	4,150
0380	10″ diameter		2.20	10.909		4,825	505		5,330	6,050
0390	12″ diameter		1.70	14.118		7,425	650		8,075	9,125
0400	14″ diameter		1.30	18.462		11,100	850		11,950	13,500
0410	16″ diameter		1	24		15,500	1,100		16,600	18,800
0420	18″ diameter		.80	30		20,200	1,375		21,575	24,400
0430	20″ diameter		.60	40		24,300	1,850		26,150	29,500
1000	Micro-bubble separator for total air removal, closed loop system									
1010	Requires bladder type tank in system.									
1020	Water (hot or chilled) or glycol system									
1030	Threaded									
1040	3/4″ diameter	1 Stpi	20	.400	Ea.	82.50	19.75		102.25	121
1050	1″ diameter		19	.421		92	21		113	132
1060	1-1/4″ diameter		16	.500		128	24.50		152.50	178
1070	1-1/2″ diameter		13	.615		166	30.50		196.50	229
1080	2″ diameter		11	.727		810	36		846	945
1090	2-1/2″ diameter	Q-5	15	1.067		900	47.50		947.50	1,050
1100	3″ diameter		13	1.231		1,300	54.50		1,354.50	1,500
1110	4″ diameter		10	1.600		1,400	71		1,471	1,625
1230	Flanged									
1250	2″ diameter	1 Stpi	8	1	Ea.	1,025	49.50		1,074.50	1,225
1260	2-1/2″ diameter	Q-5	5	3.200		1,125	142		1,267	1,450
1270	3″ diameter		4.50	3.556		1,500	158		1,658	1,875
1280	4″ diameter		3	5.333		1,675	237		1,912	2,175
1290	5″ diameter	Q-6	3.40	7.059		2,675	325		3,000	3,450
1300	6″ diameter	″	3	8		3,250	370		3,620	4,125
1400	Larger sizes available									
1590	With extended tank dirt catcher									
1600	Flanged									
1620	2″ diameter	1 Stpi	7.40	1.081	Ea.	1,350	53.50		1,403.50	1,575
1630	2-1/2″ diameter	Q-5	4.20	3.810		1,425	169		1,594	1,825
1640	3″ diameter		4	4		1,875	178		2,053	2,325
1650	4″ diameter		2.80	5.714		2,025	254		2,279	2,600
1660	5″ diameter	Q-6	3.10	7.742		3,000	355		3,355	3,800
1670	6″ diameter	″	2.90	8.276		3,675	380		4,055	4,625
1800	With drain/dismantle/cleaning access flange									
1810	Flanged									
1820	2″ diameter	1 Stpi	7	1.143	Ea.	2,125	56.50		2,181.50	2,425
1830	2-1/2″ diameter	Q-5	4	4		2,300	178		2,478	2,800
1840	3″ diameter		3.60	4.444		3,100	197		3,297	3,700
1850	4″ diameter		2.50	6.400		3,425	284		3,709	4,200
1860	5″ diameter	Q-6	2.80	8.571		5,100	395		5,495	6,225
1870	6″ diameter	″	2.60	9.231		6,650	425		7,075	7,975

23 21 20 – Hydronic HVAC Piping Specialties

23 21 20.10 Air Control		Crew	Daily Output	Labor-Hours	Unit	2009 Bare Costs Material	Labor	Equipment	Total	Total Incl O&P
2000	Boiler air fitting, separator									
2010	1-1/4"	Q-5	22	.727	Ea.	175	32.50		207.50	242
2020	1-1/2"		20	.800		335	35.50		370.50	425
2021	2"		18	.889		580	39.50		619.50	700
2400	Compression tank air fitting									
2410	For tanks 9" to 24" diameter	1 Stpi	15	.533	Ea.	56.50	26.50		83	102
2420	For tanks 100 gallon or larger	"	12	.667	"	215	33		248	286

23 21 20.14 Air Purging Scoop		Crew	Daily Output	Labor-Hours	Unit	Material	Labor	Equipment	Total	Total Incl O&P
0010	**AIR PURGING SCOOP**, With tappings.									
0020	For air vent and expansion tank connection									
0100	1" pipe size, threaded	1 Stpi	19	.421	Ea.	18.85	21		39.85	52
0110	1-1/4" pipe size, threaded		15	.533		18.85	26.50		45.35	60.50
0120	1-1/2" pipe size, threaded		13	.615		41.50	30.50		72	91
0130	2" pipe size, threaded		11	.727		47.50	36		83.50	106
0140	2-1/2" pipe size, threaded	Q-5	15	1.067		101	47.50		148.50	182
0150	3" pipe size, threaded		13	1.231		132	54.50		186.50	228
0160	4" 150 lb. flanges		3	5.333		296	237		533	680

23 21 20.18 Automatic Air Vent		Crew	Daily Output	Labor-Hours	Unit	Material	Labor	Equipment	Total	Total Incl O&P
0010	**AUTOMATIC AIR VENT**									
0020	Cast iron body, stainless steel internals, float type									
0060	1/2" NPT inlet, 300 psi	1 Stpi	12	.667	Ea.	88	33		121	146
0140	3/4" NPT inlet, 300 psi		12	.667		88	33		121	146
0180	1/2" NPT inlet, 250 psi		10	.800		275	39.50		314.50	360
0220	3/4" NPT inlet, 250 psi		10	.800		275	39.50		314.50	360
0260	1" NPT inlet, 250 psi		10	.800		410	39.50		449.50	515
0340	1-1/2" NPT inlet, 250 psi	Q-5	12	1.333		870	59		929	1,050
0380	2" NPT inlet, 250 psi	"	12	1.333		870	59		929	1,050
0600	Forged steel body, stainless steel internals, float type									
0640	1/2" NPT inlet, 750 psi	1 Stpi	12	.667	Ea.	920	33		953	1,050
0680	3/4" NPT inlet, 750 psi		12	.667		920	33		953	1,050
0760	3/4" NPT inlet, 1000 psi		10	.800		1,375	39.50		1,414.50	1,575
0800	1" NPT inlet, 1000 psi	Q-5	12	1.333		1,375	59		1,434	1,625
0880	1-1/2" NPT inlet, 1000 psi		10	1.600		3,875	71		3,946	4,375
0920	2" NPT inlet, 1000 psi		10	1.600		3,875	71		3,946	4,375
1100	Formed steel body, noncorrosive									
1110	1/8" NPT inlet 150 psi	1 Stpi	32	.250	Ea.	9.65	12.35		22	29
1120	1/4" NPT inlet 150 psi		32	.250		32.50	12.35		44.85	54.50
1130	3/4" NPT inlet 150 psi		32	.250		32.50	12.35		44.85	54.50
1300	Chrome plated brass, automatic/manual, for radiators									
1310	1/8" NPT inlet, nickel plated brass	1 Stpi	32	.250	Ea.	5.65	12.35		18	24.50

23 21 20.22 Circuit Sensor		Crew	Daily Output	Labor-Hours	Unit	Material	Labor	Equipment	Total	Total Incl O&P
0010	**CIRCUIT SENSOR**, Flow meter									
0020	Metering stations									
0040	Wafer orifice insert type									
0060	2-1/2" pipe size	Q-5	12	1.333	Ea.	164	59		223	269
0100	3" pipe size		11	1.455		180	64.50		244.50	295
0140	4" pipe size		8	2		205	89		294	360
0180	5" pipe size		7.30	2.192		262	97.50		359.50	435
0220	6" pipe size		6.40	2.500		310	111		421	505
0260	8" pipe size	Q-6	5.30	4.528		435	209		644	790
0280	10" pipe size		4.60	5.217		505	240		745	915
0360	12" pipe size		4.20	5.714		835	263		1,098	1,325

23 21 Hydronic Piping and Pumps

23 21 20 – Hydronic HVAC Piping Specialties

23 21 20.22 Circuit Sensor		Crew	Daily Output	Labor-Hours	Unit	2009 Bare Costs Material	Labor	Equipment	Total	Total Incl O&P
2000	In-line probe type									
2200	Copper, soldered									
2210	1/2″ size	1 Stpi	24	.333	Ea.	147	16.45		163.45	187
2220	3/4″ size		20	.400		153	19.75		172.75	198
2230	1″ size		19	.421		173	21		194	221
2240	1-1/4″ size		15	.533		182	26.50		208.50	240
2250	1-1/2″ size		13	.615		216	30.50		246.50	284
2260	2″ size		11	.727		241	36		277	320
2270	2-1/2″ size	Q-5	15	1.067		254	47.50		301.50	350
2280	3″ size	″	13	1.231		275	54.50		329.50	385
2500	Brass, threaded									
2510	1″ size	1 Stpi	19	.421	Ea.	182	21		203	231
2520	1-1/4″ size		15	.533		201	26.50		227.50	261
2530	1-1/2″ size		13	.615		241	30.50		271.50	310
2540	2″ size		11	.727		254	36		290	335
2550	2-1/2″ size	Q-5	15	1.067		275	47.50		322.50	375
2560	3″ size	″	13	1.231		282	54.50		336.50	390
2700	Steel, threaded									
2710	1″ size	1 Stpi	19	.421	Ea.	174	21		195	223
2720	1-1/4″ size		15	.533		189	26.50		215.50	248
2730	1-1/2″ size		13	.615		215	30.50		245.50	282
2740	2″ size		11	.727		239	36		275	315
2750	2-1/2″ size	Q-5	15	1.067		254	47.50		301.50	350
2760	3″ size	″	13	1.231		281	54.50		335.50	390
3000	Pitot tube probe type									
3100	Weld-on mounting with probe									
3110	2″ size	Q-17	18	.889	Ea.	197	39.50	3.09	239.59	279
3120	2-1/2″ size		18	.889		201	39.50	3.09	243.59	284
3130	3″ size		18	.889		213	39.50	3.09	255.59	297
3140	4″ size		18	.889		282	39.50	3.09	324.59	375
3150	5″ size		17	.941		315	42	3.28	360.28	415
3160	6″ size		17	.941		410	42	3.28	455.28	515
3170	8″ size		16	1		585	44.50	3.48	632.98	710
3180	10″ size		15	1.067		630	47.50	3.71	681.21	770
3190	12″ size		15	1.067		675	47.50	3.71	726.21	820
3200	14″ size		14	1.143		775	51	3.98	829.98	930
3210	16″ size		14	1.143		915	51	3.98	969.98	1,075
3220	18″ size		13	1.231		975	54.50	4.28	1,033.78	1,150
3230	20″ size		13	1.231		1,050	54.50	4.28	1,108.78	1,250
3240	24″ size		12	1.333		1,125	59	4.64	1,188.64	1,350
3400	Weld-on wet tap with probe									
3410	2″ size, 5/16″ probe dia.	Q-17	14	1.143	Ea.	475	51	3.98	529.98	605
3420	2-1/2″ size, 5/16″ probe dia.		14	1.143		480	51	3.98	534.98	610
3430	3″ size, 5/16″ probe dia.		14	1.143		515	51	3.98	569.98	645
3440	4″ size, 5/16″ probe dia.		14	1.143		530	51	3.98	584.98	660
3450	5″ size, 3/8″ probe dia.		13	1.231		550	54.50	4.28	608.78	690
3460	6″ size, 3/8″ probe dia.		13	1.231		585	54.50	4.28	643.78	730
3470	8″ size, 1/2″ probe dia.		12	1.333		610	59	4.64	673.64	765
3480	10″ size, 3/4″ probe dia.		11	1.455		670	64.50	5.05	739.55	840
3490	12″ size, 3/4″ probe dia.		11	1.455		710	64.50	5.05	779.55	885
3500	14″ size, 1″ probe dia.		10	1.600		805	71	5.55	881.55	1,000
3510	16″ size, 1″ probe dia.		10	1.600		955	71	5.55	1,031.55	1,175
3520	18″ size, 1″ probe dia.		9	1.778		1,025	79	6.20	1,110.20	1,250

23 21 20 – Hydronic HVAC Piping Specialties

23 21 20.22 Circuit Sensor		Crew	Daily Output	Labor-Hours	Unit	Material	Labor (2009 Bare Costs)	Equipment (2009 Bare Costs)	Total (2009 Bare Costs)	Total Incl O&P
3530	20" size, 1" probe dia.	Q-17	8	2	Ea.	1,100	89	6.95	1,195.95	1,350
3540	24" size, 1" probe dia.	↓	8	2	↓	1,225	89	6.95	1,320.95	1,500
3700	Clamp-on wet tap with probe									
3710	2" size, 5/16" probe dia.	Q-5	16	1	Ea.	560	44.50		604.50	680
3720	2-1/2" size, 5/16" probe dia.		16	1		595	44.50		639.50	720
3730	3" size, 5/16" probe dia.		15	1.067		595	47.50		642.50	725
3740	4" size, 5/16" probe dia.		15	1.067		690	47.50		737.50	830
3750	5" size, 3/8" probe dia.		14	1.143		730	51		781	880
3760	6" size, 3/8" probe dia.		14	1.143		775	51		826	925
3770	8" size, 1/2" probe dia.		13	1.231		850	54.50		904.50	1,025
3780	10" size, 3/4" probe dia.		12	1.333		890	59		949	1,075
3790	12" size, 3/4" probe dia.		12	1.333		1,025	59		1,084	1,225
3800	14" size, 1" probe dia.		11	1.455		1,350	64.50		1,414.50	1,575
3810	16" size, 1" probe dia.		11	1.455		1,550	64.50		1,614.50	1,800
3820	18" size, 1" probe dia.		10	1.600		1,675	71		1,746	1,950
3830	20" size, 1" probe dia		9	1.778		1,900	79		1,979	2,225
3840	24" size, 1" probe dia.	↓	9	1.778	↓	1,975	79		2,054	2,300
4000	Wet tap drills									
4010	3/8" dia. for 5/16" probe				Ea.	115			115	126
4020	1/2" dia. for 3/8" probe					119			119	131
4030	5/8" dia for 1/2" probe					141			141	156
4040	7/8" dia for 3/4" probe					184			184	203
4050	1-3/16" dia for 1" probe				↓	197			197	216
4100	Wet tap punch									
4110	5/16" dia for 5/16" probe				Ea.	118			118	130
4120	3/8" dia for 3/8" probe					127			127	140
4130	1/2" dia for 1/2" probe				↓	153			153	168
4150	Note: labor for wet tap drill or punch									
4160	is included with the mounting and probe assembly									
9000	Readout instruments									
9200	Gauges									
9220	GPM gauge				Ea.	1,925			1,925	2,125
9230	Dual gauge kit				"	2,000			2,000	2,200

23 21 20.26 Circuit Setter		Crew	Daily Output	Labor-Hours	Unit	Material	Labor	Equipment	Total	Total Incl O&P
0010	**CIRCUIT SETTER**, Balance valve									
0012	Bronze body, soldered									
0013	1/2" size	1 Stpi	24	.333	Ea.	56.50	16.45		72.95	86.50
0014	3/4" size	"	20	.400	"	59.50	19.75		79.25	94.50
0018	Threaded									
0019	1/2" pipe size	1 Stpi	22	.364	Ea.	58	17.95		75.95	91
0020	3/4" pipe size		20	.400		62	19.75		81.75	98
0040	1" pipe size		18	.444		80.50	22		102.50	122
0050	1-1/4" pipe size		15	.533		119	26.50		145.50	171
0060	1-1/2" pipe size		12	.667		140	33		173	204
0080	2" pipe size	↓	10	.800		200	39.50		239.50	279
0100	2-1/2" pipe size	Q-5	15	1.067		450	47.50		497.50	565
0120	3" pipe size	"	10	1.600	↓	635	71		706	805
0130	Cast iron body, flanged									
0136	3" pipe size, flanged	Q-5	4	4	Ea.	635	178		813	965
0140	4" pipe size	"	3	5.333		955	237		1,192	1,400
0200	For differential meter, accurate to 1%, add					735			735	810
0300	For differential meter 400 psi/200° F continuous, add				↓	1,825			1,825	2,000

23 21 20.30 Cocks, Drains and Specialties		Crew	Daily Output	Labor-Hours	Unit	2009 Bare Costs Material	Labor	Equipment	Total	Total Incl O&P
0010	**COCKS, DRAINS AND SPECIALTIES**									
1000	Boiler drain									
1010	Pipe thread to hose									
1020	Bronze									
1030	1/2" size	1 Stpi	36	.222	Ea.	6.95	10.95		17.90	24
1040	3/4" size	"	34	.235	"	7.55	11.60		19.15	26
1100	Solder to hose									
1110	Bronze									
1120	1/2" size	1 Stpi	46	.174	Ea.	6.25	8.60		14.85	19.75
1130	3/4" size	"	44	.182	"	6.65	8.95		15.60	21
1600	With built-in vacuum breaker									
1610	1/2" IP or solder	1 Stpi	36	.222	Ea.	25.50	10.95		36.45	45
1630	With tamper proof vacuum breaker									
1640	1/2" IP or solder	1 Stpi	36	.222	Ea.	29	10.95		39.95	48.50
1650	3/4" IP or solder	"	34	.235	"	29.50	11.60		41.10	50
3000	Cocks									
3010	Air, lever or tee handle									
3020	Bronze, single thread									
3030	1/8" size	1 Stpi	52	.154	Ea.	7.65	7.60		15.25	19.80
3040	1/4" size		46	.174		8.10	8.60		16.70	22
3050	3/8" size		40	.200		8.40	9.85		18.25	24
3060	1/2" size	↓	36	.222	↓	9.65	10.95		20.60	27
3100	Bronze, double thread									
3110	1/8" size	1 Stpi	26	.308	Ea.	9.60	15.20		24.80	33.50
3120	1/4" size		22	.364		10.05	17.95		28	38
3130	3/8" size		18	.444		10.45	22		32.45	44.50
3140	1/2" size	↓	15	.533	↓	12.75	26.50		39.25	53.50
4000	Steam									
4010	Bronze									
4020	1/8" size	1 Stpi	26	.308	Ea.	28	15.20		43.20	54
4030	1/4" size		22	.364		28	17.95		45.95	58
4040	3/8" size		18	.444		35.50	22		57.50	72
4050	1/2" size		15	.533		37.50	26.50		64	80.50
4060	3/4" size		14	.571		44	28		72	91
4070	1" size		13	.615		57.50	30.50		88	109
4080	1-1/4" size		11	.727		88.50	36		124.50	151
4090	1-1/2" size		10	.800		114	39.50		153.50	184
4100	2" size	↓	9	.889	↓	188	44		232	273
4300	Bronze, 3 way									
4320	1/4" size	1 Stpi	15	.533	Ea.	42	26.50		68.50	85.50
4330	3/8" size		12	.667		42	33		75	95.50
4340	1/2" size		10	.800		42	39.50		81.50	105
4350	3/4" size		9.50	.842		72	41.50		113.50	142
4360	1" size		8.50	.941		86.50	46.50		133	165
4370	1-1/4" size		7.50	1.067		130	52.50		182.50	222
4380	1-1/2" size		6.50	1.231		151	60.50		211.50	257
4390	2" size	↓	6	1.333	↓	223	66		289	345
4500	Gauge cock, brass									
4510	1/4" FPT	1 Stpi	24	.333	Ea.	8.10	16.45		24.55	33.50
4512	1/4" MPT	"	24	.333	"	10.15	16.45		26.60	35.50
4600	Pigtail, steam syphon									
4604	1/4"	1 Stpi	24	.333	Ea.	15.15	16.45		31.60	41

23 21 Hydronic Piping and Pumps

23 21 20 – Hydronic HVAC Piping Specialties

23 21 20.30 Cocks, Drains and Specialties		Crew	Daily Output	Labor-Hours	Unit	2009 Bare Costs Material	Labor	Equipment	Total	Total Incl O&P
4650	Snubber valve									
4654	1/4″	1 Stpi	22	.364	Ea.	11	17.95		28.95	39
4660	Nipple, black steel									
4664	1/4″ x 3″	1 Stpi	37	.216	Ea.	.84	10.65		11.49	16.90

23 21 20.34 Dielectric Unions

No.	Description	Crew	Daily Output	Labor-Hours	Unit	Material	Labor	Equipment	Total	Total Incl O&P
0010	**DIELECTRIC UNIONS**, Standard gaskets for water and air.									
0020	250 psi maximum pressure									
0280	Female IPT to sweat, straight									
0300	1/2″ pipe size	1 Plum	24	.333	Ea.	4.25	16.25		20.50	29
0340	3/4″ pipe size		20	.400		4.79	19.50		24.29	35
0360	1″ pipe size		19	.421		6.60	20.50		27.10	38.50
0380	1-1/4″ pipe size		15	.533		10.25	26		36.25	50.50
0400	1-1/2″ pipe size		13	.615		15.40	30		45.40	62
0420	2″ pipe size		11	.727		21	35.50		56.50	76
0580	Female IPT to brass pipe thread, straight									
0600	1/2″ pipe size	1 Plum	24	.333	Ea.	9.40	16.25		25.65	35
0640	3/4″ pipe size		20	.400		10.35	19.50		29.85	41
0660	1″ pipe size		19	.421		19.25	20.50		39.75	52
0680	1-1/4″ pipe size		15	.533		23.50	26		49.50	65
0700	1-1/2″ pipe size		13	.615		34.50	30		64.50	82.50
0720	2″ pipe size		11	.727		67.50	35.50		103	127
0780	Female IPT to female IPT, straight									
0800	1/2″ pipe size	1 Plum	24	.333	Ea.	8.80	16.25		25.05	34
0840	3/4″ pipe size		20	.400		9.90	19.50		29.40	40.50
0860	1″ pipe size		19	.421		13.25	20.50		33.75	45.50
0880	1-1/4″ pipe size		15	.533		17.90	26		43.90	58.50
0900	1-1/2″ pipe size		13	.615		27.50	30		57.50	75
0920	2″ pipe size		11	.727		40	35.50		75.50	97
2000	175 psi maximum pressure									
2180	Female IPT to sweat									
2240	2″ pipe size	1 Plum	9	.889	Ea.	116	43.50		159.50	193
2260	2-1/2″ pipe size	Q-1	15	1.067		126	47		173	209
2280	3″ pipe size		14	1.143		173	50		223	266
2300	4″ pipe size		11	1.455		460	64		524	600
2480	Female IPT to brass pipe									
2500	1-1/2″ pipe size	1 Plum	11	.727	Ea.	145	35.50		180.50	212
2540	2″ pipe size	″	9	.889		173	43.50		216.50	255
2560	2-1/2″ pipe size	Q-1	15	1.067		250	47		297	345
2580	3″ pipe size		14	1.143		290	50		340	395
2600	4″ pipe size		11	1.455		520	64		584	670

23 21 20.38 Expansion Couplings

No.	Description	Crew	Daily Output	Labor-Hours	Unit	Material	Labor	Equipment	Total	Total Incl O&P
0010	**EXPANSION COUPLINGS**, Hydronic									
0100	Copper to copper, sweat									
0120	3/4″ diameter	1 Stpi	20	.400	Ea.	32	19.75		51.75	65
0140	1″ diameter		19	.421		41	21		62	76
0160	1-1/4″ diameter		15	.533		55.50	26.50		82	101
1000	Baseboard riser fitting, 5″ stub by coupling 12″ long									
1020	1/2″ diameter	1 Stpi	24	.333	Ea.	15.60	16.45		32.05	41.50
1040	3/4″ diameter		20	.400		22	19.75		41.75	54
1060	1″ diameter		19	.421		31	21		52	65
1080	1-1/4″ diameter		15	.533		40.50	26.50		67	84
1180	9″ Stub by tubing 8″ long									

23 21 20.38 Expansion Couplings		Crew	Daily Output	Labor-Hours	Unit	2009 Bare Costs Material	Labor	Equipment	Total	Total Incl O&P
1200	1/2" diameter	1 Stpi	24	.333	Ea.	14	16.45		30.45	40
1220	3/4" diameter		20	.400		18.55	19.75		38.30	50
1240	1" diameter		19	.421		25.50	21		46.50	59.50
1260	1-1/4" diameter	↓	15	.533	↓	36	26.50		62.50	79

23 21 20.42 Expansion Joints		Crew	Daily Output	Labor-Hours	Unit	Material	Labor	Equipment	Total	Total Incl O&P
0010	**EXPANSION JOINTS**									
0100	Bellows type, neoprene cover, flanged spool									
0140	6" face to face, 1-1/4" diameter	1 Stpi	11	.727	Ea.	248	36		284	325
0160	1-1/2" diameter	"	10.60	.755		248	37.50		285.50	330
0180	2" diameter	Q-5	13.30	1.203		251	53.50		304.50	355
0190	2-1/2" diameter		12.40	1.290		260	57.50		317.50	370
0200	3" diameter		11.40	1.404		291	62.50		353.50	415
0210	4" diameter		8.40	1.905		315	84.50		399.50	470
0220	5" diameter		7.60	2.105		385	93.50		478.50	565
0230	6" diameter		6.80	2.353		395	105		500	590
0240	8" diameter		5.40	2.963		460	132		592	705
0250	10" diameter		5	3.200		635	142		777	915
0260	12" diameter		4.60	3.478		730	155		885	1,025
0480	10" face to face, 2" diameter		13	1.231		360	54.50		414.50	475
0500	2-1/2" diameter		12	1.333		380	59		439	510
0520	3" diameter		11	1.455		390	64.50		454.50	520
0540	4" diameter		8	2		440	89		529	620
0560	5" diameter		7	2.286		525	102		627	730
0580	6" diameter		6	2.667		545	118		663	730
0600	8" diameter		5	3.200		650	142		792	930
0620	10" diameter		4.60	3.478		720	155		875	1,025
0640	12" diameter		4	4		890	178		1,068	1,250
0660	14" diameter		3.80	4.211		1,100	187		1,287	1,475
0680	16" diameter		2.90	5.517		1,275	245		1,520	1,775
0700	18" diameter		2.50	6.400		1,425	284		1,709	2,000
0720	20" diameter		2.10	7.619		1,500	340		1,840	2,150
0740	24" diameter		1.80	8.889		1,750	395		2,145	2,525
0760	26" diameter		1.40	11.429		1,975	510		2,485	2,900
0780	30" diameter		1.20	13.333		2,200	590		2,790	3,325
0800	36" diameter	↓	1	16	↓	2,725	710		3,435	4,075
1000	Bellows with internal sleeves and external covers									
1010	Stainless steel, 150 lb.									
1020	With male threads									
1030	1/2" diameter	1 Stpi	22	.364	Ea.	132	17.95		149.95	173
1040	3/4" diameter		20	.400		132	19.75		151.75	176
1050	1" diameter		19	.421		140	21		161	185
1060	1-1/4" diameter		16	.500		146	24.50		170.50	198
1070	1-1/2" diameter		13	.615		178	30.50		208.50	242
1080	2" diameter	↓	11	.727	↓	228	36		264	305
1110	With flanged ends									
1120	1-1/4" diameter	1 Stpi	12	.667	Ea.	430	33		463	520
1130	1-1/2" diameter		11	.727		430	36		466	525
1140	2" diameter	↓	9	.889		685	44		729	820
1150	3" diameter	Q-5	8	2		690	89		779	895
1160	4" diameter	"	5	3.200		740	142		882	1,025
1170	5" diameter	Q-6	6	4		790	184		974	1,150
1180	6" diameter	↓	5	4.800	↓	865	221		1,086	1,275

23 21 Hydronic Piping and Pumps

23 21 20 – Hydronic HVAC Piping Specialties

23 21 20.42 Expansion Joints		Crew	Daily Output	Labor-Hours	Unit	2009 Bare Costs Material	Labor	Equipment	Total	Total Incl O&P
1190	8″ diameter	Q-6	4.50	5.333	Ea.	1,025	246		1,271	1,525
1200	10″ diameter		4	6		1,275	276		1,551	1,825
1210	12″ diameter		3.60	6.667		1,625	305		1,930	2,225

23 21 20.46 Expansion Tanks		Crew	Daily Output	Labor-Hours	Unit	Material	Labor	Equipment	Total	Total Incl O&P
0010	**EXPANSION TANKS**									
1400	Plastic, corrosion resistant, see Plumbing Cost Data									
1507	Fiberglass and steel single / double wall storage, see Div. 23 13 13									
1512	Tank leak detection systems, see Div. 28 33 33.50									
2000	Steel, liquid expansion, ASME, painted, 15 gallon capacity	Q-5	17	.941	Ea.	440	42		482	550
2020	24 gallon capacity		14	1.143		470	51		521	595
2040	30 gallon capacity		12	1.333		495	59		554	635
2060	40 gallon capacity		10	1.600		580	71		651	740
2080	60 gallon capacity		8	2		695	89		784	900
2100	80 gallon capacity		7	2.286		745	102		847	970
2120	100 gallon capacity		6	2.667		1,000	118		1,118	1,275
2130	120 gallon capacity		5	3.200		1,075	142		1,217	1,400
2140	135 gallon capacity		4.50	3.556		1,125	158		1,283	1,450
2150	175 gallon capacity		4	4		1,750	178		1,928	2,200
2160	220 gallon capacity		3.60	4.444		2,000	197		2,197	2,500
2170	240 gallon capacity		3.30	4.848		2,075	215		2,290	2,600
2180	305 gallon capacity		3	5.333		2,925	237		3,162	3,575
2190	400 gallon capacity		2.80	5.714		3,600	254		3,854	4,350
2360	Galvanized									
2370	15 gallon capacity	Q-5	17	.941	Ea.	780	42		822	920
2380	24 gallon capacity		14	1.143		895	51		946	1,050
2390	30 gallon capacity		12	1.333		930	59		989	1,125
2400	40 gallon capacity		10	1.600		1,100	71		1,171	1,300
2410	60 gallon capacity		8	2		1,250	89		1,339	1,500
2420	80 gallon capacity		7	2.286		1,450	102		1,552	1,725
2430	100 gallon capacity		6	2.667		1,875	118		1,993	2,250
2440	120 gallon capacity		5	3.200		2,000	142		2,142	2,425
2450	135 gallon capacity		4.50	3.556		2,125	158		2,283	2,550
2460	175 gallon capacity		4	4		3,450	178		3,628	4,075
2470	220 gallon capacity		3.60	4.444		3,925	197		4,122	4,625
2480	240 gallon capacity		3.30	4.848		4,125	215		4,340	4,850
2490	305 gallon capacity		3	5.333		5,975	237		6,212	6,925
2500	400 gallon capacity		2.80	5.714		7,350	254		7,604	8,450
3000	Steel ASME expansion, rubber diaphragm, 19 gal. cap. accept.		12	1.333		1,925	59		1,984	2,225
3020	31 gallon capacity		8	2		2,150	89		2,239	2,500
3040	61 gallon capacity		6	2.667		3,025	118		3,143	3,500
3060	79 gallon capacity		5	3.200		3,075	142		3,217	3,600
3080	119 gallon capacity		4	4		3,250	178		3,428	3,850
3100	158 gallon capacity		3.80	4.211		4,525	187		4,712	5,250
3120	211 gallon capacity		3.30	4.848		5,225	215		5,440	6,075
3140	317 gallon capacity		2.80	5.714		6,825	254		7,079	7,875
3160	422 gallon capacity		2.60	6.154		10,100	273		10,373	11,500
3180	528 gallon capacity		2.40	6.667		11,100	296		11,396	12,600

23 21 20.50 Float Valves		Crew	Daily Output	Labor-Hours	Unit	Material	Labor	Equipment	Total	Total Incl O&P
0010	**FLOAT VALVES**									
0020	With ball and bracket									
0030	Single seat, threaded									
0040	Brass body									

23 21 20 – Hydronic HVAC Piping Specialties

23 21 20.50 Float Valves		Crew	Daily Output	Labor-Hours	Unit	2009 Bare Costs Material	Labor	Equipment	Total	Total Incl O&P
0050	1/2"	1 Stpi	11	.727	Ea.	69	36		105	130
0060	3/4"		9	.889		76.50	44		120.50	151
0070	1"		7	1.143		101	56.50		157.50	196
0080	1-1/2"		4.50	1.778		149	87.50		236.50	296
0090	2"	↓	3.60	2.222	↓	152	110		262	330
0300	For condensate receivers, CI, in-line mount									
0320	1" inlet	1 Stpi	7	1.143	Ea.	124	56.50		180.50	221
0360	For condensate receiver, CI, external float, flanged tank mount									
0370	3/4" inlet	1 Stpi	5	1.600	Ea.	103	79		182	231
23 21 20.54 Flow Check Control										
0010	**FLOW CHECK CONTROL**									
0100	Bronze body, soldered									
0110	3/4" size	1 Stpi	20	.400	Ea.	49.50	19.75		69.25	84
0120	1" size	"	19	.421	"	59.50	21		80.50	96.50
0200	Cast iron body, threaded									
0210	3/4" size	1 Stpi	20	.400	Ea.	38	19.75		57.75	71.50
0220	1" size		19	.421		43	21		64	78.50
0230	1-1/4" size		15	.533		52.50	26.50		79	97
0240	1-1/2" size		13	.615		80	30.50		110.50	134
0250	2" size	↓	11	.727	↓	115	36		151	180
0300	Flanged inlet, threaded outlet									
0310	2-1/2" size	1 Stpi	8	1	Ea.	257	49.50		306.50	355
0320	3" size	Q-5	9	1.778		305	79		384	455
0330	4" size	"	7	2.286	↓	595	102		697	805
23 21 20.58 Hydronic Heating Control Valves										
0010	**HYDRONIC HEATING CONTROL VALVES**									
0050	Hot water, nonelectric, thermostatic									
0100	Radiator supply, 1/2" diameter	1 Stpi	24	.333	Ea.	52.50	16.45		68.95	82
0120	3/4" diameter		20	.400		54	19.75		73.75	89
0140	1" diameter		19	.421		68.50	21		89.50	107
0160	1-1/4" diameter	↓	15	.533	↓	98	26.50		124.50	148
0500	For low pressure steam, add					25%				
1000	Manual, radiator supply									
1010	1/2" pipe size, angle union	1 Stpi	24	.333	Ea.	34.50	16.45		50.95	62.50
1020	3/4" pipe size, angle union		20	.400		44	19.75		63.75	78
1030	1" pipe size, angle union	↓	19	.421	↓	57	21		78	94
1100	Radiator, balancing, straight, sweat connections									
1110	1/2" pipe size	1 Stpi	24	.333	Ea.	13.35	16.45		29.80	39
1120	3/4" pipe size		20	.400		18.70	19.75		38.45	50
1130	1" pipe size	↓	19	.421	↓	30.50	21		51.50	64.50
1200	Steam, radiator, supply									
1210	1/2" pipe size, angle union	1 Stpi	24	.333	Ea.	34.50	16.45		50.95	62
1220	3/4" pipe size, angle union		20	.400		37	19.75		56.75	70.50
1230	1" pipe size, angle union		19	.421		43	21		64	78.50
1240	1-1/4" pipe size, angle union	↓	15	.533	↓	55	26.50		81.50	100
8000	System balancing and shut-off									
8020	Butterfly, quarter turn, calibrated, threaded or solder									
8040	Bronze, -30° F to +350° F, pressure to 175 psi									
8060	1/2" size	1 Stpi	22	.364	Ea.	12.20	17.95		30.15	40.50
8070	3/4" size		20	.400		19.75	19.75		39.50	51
8080	1" size		19	.421		24	21		45	57.50
8090	1-1/4" size	↓	15	.533	↓	38	26.50		64.50	81.50

23 21 20.58 Hydronic Heating Control Valves		Crew	Daily Output	Labor-Hours	Unit	Material	Labor	Equipment	Total	Total Incl O&P
						2009 Bare Costs				
8100	1-1/2" size	1 Stpi	13	.615	Ea.	50	30.50		80.50	101
8110	2" size	↓	11	.727	↓	62.50	36		98.50	123
23 21 20.62 Liquid Drainers										
0010	**LIQUID DRAINERS**									
0100	Guided lever type									
0110	Cast iron body, threaded									
0120	1/2" pipe size	1 Stpi	22	.364	Ea.	88	17.95		105.95	124
0130	3/4" pipe size		20	.400		234	19.75		253.75	287
0140	1" pipe size		19	.421		385	21		406	455
0150	1-1/2" pipe size		11	.727		815	36		851	950
0160	2" pipe size	↓	8	1	↓	815	49.50		864.50	970
0200	Forged steel body, threaded									
0210	1/2" pipe size	1 Stpi	18	.444	Ea.	870	22		892	990
0220	3/4" pipe size		16	.500		870	24.50		894.50	990
0230	1" pipe size	↓	14	.571		1,300	28		1,328	1,475
0240	1-1/2" pipe size	Q-5	10	1.600		3,800	71		3,871	4,275
0250	2" pipe size	"	8	2	↓	3,800	89		3,889	4,300
0300	Socket weld									
0310	1/2" pipe size	Q-17	16	1	Ea.	1,000	44.50	3.48	1,047.98	1,175
0320	3/4" pipe size		14	1.143		1,000	51	3.98	1,054.98	1,175
0330	1" pipe size		12	1.333		1,450	59	4.64	1,513.64	1,700
0340	1-1/2" pipe size		9	1.778		1,450	79	6.20	1,535.20	1,725
0350	2" pipe size	↓	7	2.286	↓	3,950	102	7.95	4,059.95	4,475
0400	Flanged									
0410	1/2" pipe size	1 Stpi	12	.667	Ea.	1,725	33		1,758	1,950
0420	3/4" pipe size		10	.800		1,725	39.50		1,764.50	1,950
0430	1" pipe size	↓	7	1.143		2,225	56.50		2,281.50	2,500
0440	1-1/2" pipe size	Q-5	9	1.778		4,900	79		4,979	5,525
0450	2" pipe size	"	7	2.286	↓	4,900	102		5,002	5,550
1000	Fixed lever and snap action type									
1100	Cast iron body, threaded									
1110	1/2" pipe size	1 Stpi	22	.364	Ea.	137	17.95		154.95	178
1120	3/4" pipe size		20	.400		365	19.75		384.75	430
1130	1" pipe size	↓	19	.421	↓	365	21		386	430
1200	Forged steel body, threaded									
1210	1/2" pipe size	1 Stpi	18	.444	Ea.	940	22		962	1,050
1220	3/4" pipe size		16	.500		940	24.50		964.50	1,050
1230	1" pipe size		14	.571		2,325	28		2,353	2,600
1240	1-1/4" pipe size	↓	12	.667	↓	2,325	33		2,358	2,600
2000	High pressure high leverage type									
2100	Forged steel body, threaded									
2110	1/2" pipe size	1 Stpi	18	.444	Ea.	1,850	22		1,872	2,050
2120	3/4" pipe size		16	.500		1,850	24.50		1,874.50	2,050
2130	1" pipe size		14	.571		3,100	28		3,128	3,450
2140	1-1/4" pipe size	↓	12	.667		3,100	33		3,133	3,450
2150	1-1/2" pipe size	Q-5	10	1.600		4,650	71		4,721	5,225
2160	2" pipe size	"	8	2	↓	4,650	89		4,739	5,250
2200	Socket weld									
2210	1/2" pipe size	Q-17	16	1	Ea.	2,000	44.50	3.48	2,047.98	2,275
2220	3/4" pipe size		14	1.143		2,000	51	3.98	2,054.98	2,275
2230	1" pipe size		12	1.333		3,250	59	4.64	3,313.64	3,675
2240	1-1/4" pipe size	↓	11	1.455	↓	3,250	64.50	5.05	3,319.55	3,675

23 21 20.62 Liquid Drainers

	23 21 20.62 Liquid Drainers	Crew	Daily Output	Labor-Hours	Unit	Material	2009 Bare Costs Labor	Equipment	Total	Total Incl O&P
2250	1-1/2" pipe size	Q-17	9	1.778	Ea.	4,800	79	6.20	4,885.20	5,400
2260	2" pipe size		7	2.286		4,800	102	7.95	4,909.95	5,425
2300	Flanged									
2310	1/2" pipe size	1 Stpi	12	.667	Ea.	2,675	33		2,708	3,000
2320	3/4" pipe size		10	.800		2,675	39.50		2,714.50	3,000
2330	1" pipe size		7	1.143		4,225	56.50		4,281.50	4,725
2340	1-1/4" pipe size		6	1.333		4,225	66		4,291	4,750
2350	1-1/2" pipe size	Q-5	9	1.778		5,775	79		5,854	6,475
2360	2" pipe size	"	7	2.286		5,775	102		5,877	6,500
3000	Guided lever dual gravity type									
3100	Cast iron body, threaded									
3110	1/2" pipe size	1 Stpi	22	.364	Ea.	340	17.95		357.95	395
3120	3/4" pipe size		20	.400		340	19.75		359.75	400
3130	1" pipe size		19	.421		490	21		511	570
3140	1-1/2" pipe size		11	.727		950	36		986	1,[illegible]00
3150	2" pipe size		8	1		950	49.50		999.50	1,[illegible]25
3200	Forged steel body, threaded									
3210	1/2" pipe size	1 Stpi	18	.444	Ea.	1,000	22		1,022	1,125
3220	3/4" pipe size		16	.500		1,000	24.50		1,024.50	1,125
3230	1" pipe size		14	.571		1,500	28		1,528	1,700
3240	1-1/2" pipe size	Q-5	10	1.600		3,975	71		4,046	4,475
3250	2" pipe size	"	8	2		3,975	89		4,064	4,500
3300	Socket weld									
3310	1/2" pipe size	Q-17	16	1	Ea.	1,150	44.50	3.48	1,197.98	1,325
3320	3/4" pipe size		14	1.143		1,150	51	3.98	1,204.98	1,325
3330	1" pipe size		12	1.333		1,650	59	4.64	1,713.64	1,900
3340	1-1/2" pipe size		9	1.778		4,125	79	6.20	4,210.20	4,575
3350	2" pipe size		7	2.286		4,125	102	7.95	4,234.95	4,700
3400	Flanged									
3410	1/2" pipe size	1 Stpi	12	.667	Ea.	1,825	33		1,858	2,075
3420	3/4" pipe size		10	.800		1,825	39.50		1,864.50	2,075
3430	1" pipe size		7	1.143		2,400	56.50		2,456.50	2,725
3440	1-1/2" pipe size	Q-5	9	1.778		5,100	79		5,179	5,725
3450	2" pipe size	"	7	2.286		5,100	102		5,202	5,750

23 21 20.66 Monoflow Tee Fitting

	23 21 20.66 Monoflow Tee Fitting	Crew	Daily Output	Labor-Hours	Unit	Material	Labor	Equipment	Total	Total Incl O&P
0010	**MONOFLOW TEE FITTING**									
1100	For one pipe hydronic, supply and return									
1110	Copper, soldered									
1120	3/4" x 1/2" size	1 Stpi	13	.615	Ea.	16.15	30.50		46.65	63.50
1130	1" x 1/2" size		12	.667		16.15	33		49.15	67.50
1140	1" x 3/4" size		11	.727		16.15	36		52.15	72
1150	1-1/4" x 1/2" size		11	.727		16.90	36		52.90	72.50
1160	1-1/4" x 3/4" size		10	.800		16.90	39.50		56.40	77.50
1170	1-1/2" x 3/4" size		10	.800		23.50	39.50		63	85
1180	1-1/2" x 1" size		9	.889		23.50	44		67.50	92
1190	2" x 3/4" size		9	.889		37	44		81	107
1200	2" x 1" size		8	1		37	49.50		86.50	115

23 21 20.70 Steam Traps

	23 21 20.70 Steam Traps	Crew	Daily Output	Labor-Hours	Unit	Material	Labor	Equipment	Total	Total Incl O&P
0010	**STEAM TRAPS**									
0030	Cast iron body, threaded									
0040	Inverted bucket									
0050	1/2" pipe size	1 Stpi	12	.667	Ea.	128	33		161	190

23 21 Hydronic Piping and Pumps

23 21 20 – Hydronic HVAC Piping Specialties

23 21 20.70 Steam Traps		Crew	Daily Output	Labor-Hours	Unit	Material	2009 Bare Costs Labor	Equipment	Total	Total Incl O&P
0070	3/4" pipe size	1 Stpi	10	.800	Ea.	222	39.50		261.50	305
0100	1" pipe size		9	.889		340	44		384	440
0120	1-1/4" pipe size		8	1		515	49.50		564.50	645
0130	1-1/2" pipe size		7	1.143		650	56.50		706.50	800
0140	2" pipe size	↓	6	1.333	↓	1,025	66		1,091	1,225
0200	With thermic vent & check valve									
0202	1/2" pipe size	1 Stpi	12	.667	Ea.	150	33		183	215
0204	3/4" pipe size		10	.800		254	39.50		293.50	340
0206	1" pipe size		9	.889		370	44		414	470
0208	1-1/4" pipe size		8	1		560	49.50		609.50	690
0210	1-1/2" pipe size		7	1.143		620	56.50		676.50	770
0212	2" pipe size	↓	6	1.333	↓	950	66		1,016	1,150
1000	Float & thermostatic, 15 psi									
1010	3/4" pipe size	1 Stpi	16	.500	Ea.	103	24.50		127.50	150
1020	1" pipe size		15	.533		124	26.50		150.50	176
1030	1-1/4" pipe size		13	.615		150	30.50		180.50	211
1040	1-1/2" pipe size		9	.889		218	44		262	305
1060	2" pipe size	↓	6	1.333		400	66		466	540
1100	For vacuum breaker add				↓	54.50			54.50	60
1110	Differential condensate controller									
1120	With union and manual metering valve									
1130	1/2" pipe size	1 Stpi	24	.333	Ea.	480	16.45		496.45	555
1140	3/4" pipe size		20	.400		560	19.75		579.75	645
1150	1" pipe size		19	.421		750	21		771	855
1160	1-1/4" pipe size		15	.533		1,000	26.50		1,026.50	1,175
1170	1-1/2" pipe size		13	.615		1,225	30.50		1,255.50	1,400
1180	2" pipe size	↓	11	.727	↓	1,600	36		1,636	1,800
1290	Brass body, threaded									
1300	Thermostatic, angle union, 25 psi									
1310	1/2" pipe size	1 Stpi	24	.333	Ea.	49.50	16.45		65.95	79
1320	3/4" pipe size		20	.400		75	19.75		94.75	112
1330	1" pipe size	↓	19	.421	↓	111	21		132	153
1990	Carbon steel body, threaded									
2000	Controlled disc, integral strainer & blow down									
2010	Threaded									
2020	3/8" pipe size	1 Stpi	24	.333	Ea.	215	16.45		231.45	262
2030	1/2" pipe size		24	.333		215	16.45		231.45	262
2040	3/4" pipe size		20	.400		289	19.75		308.75	350
2050	1" pipe size	↓	19	.421	↓	370	21		391	440
2100	Socket weld									
2110	3/8" pipe size	Q-17	17	.941	Ea.	257	42	3.28	302.28	350
2120	1/2" pipe size		16	1		257	44.50	3.48	304.98	350
2130	3/4" pipe size		14	1.143		335	51	3.98	389.98	450
2140	1" pipe size	↓	12	1.333	↓	420	59	4.64	483.64	560
2170	Flanged, 600 lb. ASA									
2180	3/8" pipe size	1 Stpi	12	.667	Ea.	580	33		613	690
2190	1/2" pipe size		12	.667		580	33		613	690
2200	3/4" pipe size		10	.800		690	39.50		729.50	820
2210	1" pipe size	↓	7	1.143	↓	820	56.50		876.50	990
5000	Forged steel body									
5010	Inverted bucket									
5020	Threaded									
5030	1/2" pipe size	1 Stpi	12	.667	Ea.	385	33		418	475

23 21 Hydronic Piping and Pumps

23 21 20 – Hydronic HVAC Piping Specialties

23 21 20.70 Steam Traps		Crew	Daily Output	Labor-Hours	Unit	Material	Labor	Equipment	Total	Total Incl O&P
						2009 Bare Costs				
5040	3/4″ pipe size	1 Stpi	10	.800	Ea.	665	39.50		704.50	790
5050	1″ pipe size		9	.889		995	44		1,039	1,175
5060	1-1/4″ pipe size		8	1		1,450	49.50		1,499.50	1,675
5070	1-1/2″ pipe size		7	1.143		2,050	56.50		2,106.50	2,325
5080	2″ pipe size		6	1.333		3,600	66		3,666	4,075
5100	Socket weld									
5110	1/2″ pipe size	Q-17	16	1	Ea.	465	44.50	3.48	512.98	580
5120	3/4″ pipe size		14	1.143		745	51	3.98	799.98	900
5130	1″ pipe size		12	1.333		1,075	59	4.64	1,138.64	1,275
5140	1-1/4″ pipe size		10	1.600		1,525	71	5.55	1,601.55	1,800
5150	1-1/2″ pipe size		9	1.778		2,125	79	6.20	2,210.20	2,475
5160	2″ pipe size		7	2.286		3,700	102	7.95	3,809.95	4,225
5200	Flanged									
5210	1/2″ pipe size	1 Stpi	12	.667	Ea.	710	33		743	835
5220	3/4″ pipe size		10	.800		1,175	39.50		1,214.50	1,325
5230	1″ pipe size		7	1.143		1,575	56.50		1,631.50	1,825
5240	1-1/4″ pipe size		6	1.333		2,175	66		2,241	2,475
5250	1-1/2″ pipe size	Q-5	9	1.778		2,750	79		2,829	3,150
5260	2″ pipe size	″	7	2.286		4,325	102		4,427	4,900
6000	Stainless steel body									
6010	Inverted bucket									
6020	Threaded									
6030	1/2″ pipe size	1 Stpi	12	.667	Ea.	127	33		160	189
6040	3/4″ pipe size		10	.800		166	39.50		205.50	242
6050	1″ pipe size		9	.889		590	44		634	715
6100	Threaded with thermic vent									
6110	1/2″ pipe size	1 Stpi	12	.667	Ea.	144	33		177	209
6120	3/4″ pipe size		10	.800		186	39.50		225.50	263
6130	1″ pipe size		9	.889		615	44		659	740
6200	Threaded with check valve									
6210	1/2″ pipe size	1 Stpi	12	.667	Ea.	181	33		214	249
6220	3/4″ pipe size	″	10	.800	″	186	39.50		225.50	263
6300	Socket weld									
6310	1/2″ pipe size	Q-17	16	1	Ea.	146	44.50	3.48	193.98	231
6320	3/4″ pipe size	″	14	1.143	″	190	51	3.98	244.98	289

23 21 20.74 Strainers, Basket Type

		Crew	Daily Output	Labor-Hours	Unit	Material	Labor	Equipment	Total	Total Incl O&P
0010	**STRAINERS, BASKET TYPE**, Perforated stainless steel basket									
0100	Brass or monel available									
2000	Simplex style									
2300	Bronze body									
2320	Screwed, 3/8″ pipe size	1 Stpi	22	.364	Ea.	127	17.95		144.95	167
2340	1/2″ pipe size		20	.400		129	19.75		148.75	172
2360	3/4″ pipe size		17	.471		133	23		156	181
2380	1″ pipe size		15	.533		173	26.50		199.50	230
2400	1-1/4″ pipe size		13	.615		260	30.50		290.50	330
2420	1-1/2″ pipe size		12	.667		268	33		301	345
2440	2″ pipe size		10	.800		400	39.50		439.50	500
2460	2-1/2″ pipe size	Q-5	15	1.067		570	47.50		617.50	695
2480	3″ pipe size	″	14	1.143		825	51		876	980
2600	Flanged, 2″ pipe size	1 Stpi	6	1.333		605	66		671	765
2620	2-1/2″ pipe size	Q-5	4.50	3.556		950	158		1,108	1,275
2640	3″ pipe size		3.50	4.571		1,075	203		1,278	1,475

23 21 20.74 Strainers, Basket Type		Crew	Daily Output	Labor-Hours	Unit	2009 Bare Costs Material	Labor	Equipment	Total	Total Incl O&P
2660	4" pipe size	Q-5	3	5.333	Ea.	1,800	237		2,037	2,325
2680	5" pipe size	Q-6	3.40	7.059		2,775	325		3,100	3,550
2700	6" pipe size		3	8		3,525	370		3,895	4,425
2710	8" pipe size		2.50	9.600		5,725	440		6,165	6,975
3600	Iron body									
3700	Screwed, 3/8" pipe size	1 Stpi	22	.364	Ea.	88.50	17.95		106.45	125
3720	1/2" pipe size		20	.400		88.50	19.75		108.25	127
3740	3/4" pipe size		17	.471		113	23		136	159
3760	1" pipe size		15	.533		115	26.50		141.50	166
3780	1-1/4" pipe size		13	.615		150	30.50		180.50	211
3800	1-1/2" pipe size		12	.667		166	33		199	232
3820	2" pipe size		10	.800		198	39.50		237.50	277
3840	2-1/2" pipe size	Q-5	15	1.067		265	47.50		312.50	360
3860	3" pipe size	"	14	1.143		320	51		371	430
4000	Flanged, 2" pipe size	1 Stpi	6	1.333		266	66		332	390
4020	2-1/2" pipe size	Q-5	4.50	3.556		360	158		518	630
4040	3" pipe size		3.50	4.571		375	203		578	720
4060	4" pipe size		3	5.333		575	237		812	990
4080	5" pipe size	Q-6	3.40	7.059		875	325		1,200	1,450
4100	6" pipe size		3	8		1,125	370		1,495	1,775
4120	8" pipe size		2.50	9.600		2,025	440		2,465	2,900
4140	10" pipe size		2.20	10.909		4,450	505		4,955	5,650
6000	Cast steel body									
6400	Screwed, 1" pipe size	1 Stpi	15	.533	Ea.	185	26.50		211.50	244
6410	1-1/4" pipe size		13	.615		278	30.50		308.50	350
6420	1-1/2" pipe size		12	.667		278	33		311	355
6440	2" pipe size		10	.800		380	39.50		419.50	480
6460	2-1/2" pipe size	Q-5	15	1.067		540	47.50		587.50	665
6480	3" pipe size	"	14	1.143		705	51		756	850
6560	Flanged, 2" pipe size	1 Stpi	6	1.333		560	66		626	715
6580	2-1/2" pipe size	Q-5	4.50	3.556		840	158		998	1,150
6600	3" pipe size		3.50	4.571		900	203		1,103	1,300
6620	4" pipe size		3	5.333		1,275	237		1,512	1,750
6640	6" pipe size	Q-6	3	8		2,350	370		2,720	3,125
6660	8" pipe size	"	2.50	9.600		3,775	440		4,215	4,825
7000	Stainless steel body									
7200	Screwed, 1" pipe size	1 Stpi	15	.533	Ea.	280	26.50		306.50	345
7210	1-1/4" pipe size		13	.615		430	30.50		460.50	520
7220	1-1/2" pipe size		12	.667		430	33		463	525
7240	2" pipe size		10	.800		645	39.50		684.50	765
7260	2-1/2" pipe size	Q-5	15	1.067		910	47.50		957.50	1,075
7280	3" pipe size	"	14	1.143		1,250	51		1,301	1,450
7400	Flanged, 2" pipe size	1 Stpi	6	1.333		920	66		986	1,125
7420	2-1/2" pipe size	Q-5	4.50	3.556		1,675	158		1,833	2,075
7440	3" pipe size		3.50	4.571		1,725	203		1,928	2,200
7460	4" pipe size		3	5.333		2,700	237		2,937	3,325
7480	6" pipe size	Q-6	3	8		4,700	370		5,070	5,725
7500	8" pipe size	"	2.50	9.600		6,850	440		7,290	8,200
8100	Duplex style									
8200	Bronze body									
8240	Screwed, 3/4" pipe size	1 Stpi	16	.500	Ea.	895	24.50		919.50	1,025
8260	1" pipe size		14	.571		895	28		923	1,025
8280	1-1/4" pipe size		12	.667		1,750	33		1,783	1,975

23 21 20 – Hydronic HVAC Piping Specialties

23 21 20.74 Strainers, Basket Type		Crew	Daily Output	Labor-Hours	Unit	Material	2009 Bare Costs Labor	Equipment	Total	Total Incl O&P
8300	1-1/2" pipe size	1 Stpi	11	.727	Ea.	1,750	36		1,786	1,975
8320	2" pipe size	↓	9	.889		2,825	44		2,869	3,200
8340	2-1/2" pipe size	Q-5	14	1.143		3,650	51		3,701	4,100
8420	Flanged, 2" pipe size	1 Stpi	6	1.333		3,050	66		3,116	3,450
8440	2-1/2" pipe size	Q-5	4.50	3.556		4,200	158		4,358	4,850
8460	3" pipe size		3.50	4.571		4,575	203		4,778	5,350
8480	4" pipe size	↓	3	5.333		6,575	237		6,812	7,575
8500	5" pipe size	Q-6	3.40	7.059		14,600	325		14,925	16,600
8520	6" pipe size	"	3	8	↓	14,600	370		14,970	16,700
8700	Iron body									
8740	Screwed, 3/4" pipe size	1 Stpi	16	.500	Ea.	525	24.50		549.50	610
8760	1" pipe size		14	.571		525	28		553	620
8780	1-1/4" pipe size		12	.667		925	33		958	1,075
8800	1-1/2" pipe size		11	.727		925	36		961	1,075
8820	2" pipe size	↓	9	.889		1,550	44		1,594	1,300
8840	2-1/2" pipe size	Q-5	14	1.143		1,725	51		1,776	1,975
9000	Flanged, 2" pipe size	1 Stpi	6	1.333		1,675	66		1,741	1,925
9020	2-1/2" pipe size	Q-5	4.50	3.556		1,775	158		1,933	2,175
9040	3" pipe size		3.50	4.571		1,950	203		2,153	2,425
9060	4" pipe size	↓	3	5.333		3,275	237		3,512	3,950
9080	5" pipe size	Q-6	3.40	7.059		6,500	325		6,825	7,650
9100	6" pipe size		3	8		6,500	370		6,870	7,700
9120	8" pipe size		2.50	9.600		11,300	440		11,740	13,200
9140	10" pipe size		2.20	10.909		15,800	505		16,305	18,100
9160	12" pipe size		1.70	14.118		17,500	650		18,150	20,200
9170	14" pipe size		1.40	17.143		21,200	790		21,990	24,500
9180	16" pipe size	↓	1	24	↓	26,600	1,100		27,700	30,900
9300	Cast steel body									
9340	Screwed, 1" pipe size	1 Stpi	14	.571	Ea.	1,100	28		1,128	1,275
9360	1-1/2" pipe size		11	.727		1,800	36		1,836	2,025
9380	2" pipe size		9	.889		2,375	44		2,419	2,675
9460	Flanged, 2" pipe size	↓	6	1.333		2,550	66		2,616	2,900
9480	2-1/2" pipe size	Q-5	4.50	3.556		4,100	158		4,258	4,725
9500	3" pipe size		3.50	4.571		4,475	203		4,678	5,225
9520	4" pipe size	↓	3	5.333		5,525	237		5,762	6,425
9540	6" pipe size	Q-6	3	8		10,300	370		10,670	12,000
9560	8" pipe size	"	2.50	9.600	↓	22,800	440		23,240	25,800
9700	Stainless steel body									
9740	Screwed, 1" pipe size	1 Stpi	14	.571	Ea.	1,425	28		1,453	1,625
9760	1-1/2" pipe size		11	.727		2,175	36		2,211	2,425
9780	2" pipe size		9	.889		3,075	44		3,119	3,450
9860	Flanged, 2" pipe size	↓	6	1.333		3,300	66		3,366	3,725
9880	2-1/2" pipe size	Q-5	4.50	3.556		5,525	158		5,683	6,300
9900	3" pipe size		3.50	4.571		6,025	203		6,228	6,925
9920	4" pipe size	↓	3	5.333		7,850	237		8,087	8,975
9940	6" pipe size	Q-6	3	8		12,000	370		12,370	13,800
9960	8" pipe size	"	2.50	9.600	↓	33,500	440		33,940	37,500

23 21 20.76 Strainers, Y Type, Bronze Body		Crew	Daily Output	Labor-Hours	Unit	Material	Labor	Equipment	Total	Total Incl O&P
0010	**STRAINERS, Y TYPE, BRONZE BODY**									
0050	Screwed, 150 lb., 1/4" pipe size	1 Stpi	24	.333	Ea.	14.55	16.45		31	40.50
0070	3/8" pipe size		24	.333		19.20	16.45		35.65	45.50
0100	1/2" pipe size	↓	20	.400	↓	19.20	19.75		38.95	50.50

23 21 20 – Hydronic HVAC Piping Specialties

23 21 20.76 Strainers, Y Type, Bronze Body		Crew	Daily Output	Labor-Hours	Unit	2009 Bare Costs Material	Labor	Equipment	Total	Total Incl O&P
0120	3/4" pipe size	1 Stpi	19	.421	Ea.	21.50	21		42.50	55
0140	1" pipe size		17	.471		23	23		46	60
0150	1-1/4" pipe size		15	.533		51.50	26.50		78	96.50
0160	1-1/2" pipe size		14	.571		49.50	28		77.50	97
0180	2" pipe size		13	.615		66	30.50		96.50	118
0182	3" pipe size		12	.667		590	33		623	700
0200	300 lb., 2-1/2" pipe size	Q-5	17	.941		350	42		392	450
0220	3" pipe size		16	1		695	44.50		739.50	830
0240	4" pipe size		15	1.067		1,575	47.50		1,622.50	1,825
0500	For 300 lb rating 1/4" thru 2", add					15%				
1000	Flanged, 150 lb., 1-1/2" pipe size	1 Stpi	11	.727	Ea.	365	36		401	455
1020	2" pipe size	"	8	1		440	49.50		489.50	560
1030	2-1/2" pipe size	Q-5	5	3.200		650	142		792	930
1040	3" pipe size		4.50	3.556		805	158		963	1,125
1060	4" pipe size		3	5.333		1,225	237		1,462	1,700
1080	5" pipe size	Q-6	3.40	7.059		1,225	325		1,550	1,850
1100	6" pipe size		3	8		2,325	370		2,695	3,100
1106	8" pipe size		2.60	9.231		2,550	425		2,975	3,475
1500	For 300 lb rating, add					40%				

23 21 20.78 Strainers, Y Type, Iron Body		Crew	Daily Output	Labor-Hours	Unit	Material	Labor	Equipment	Total	Total Incl O&P
0010	**STRAINERS, Y TYPE, IRON BODY**									
0050	Screwed, 250 lb., 1/4" pipe size	1 Stpi	20	.400	Ea.	7.75	19.75		27.50	38
0070	3/8" pipe size		20	.400		7.75	19.75		27.50	38
0100	1/2" pipe size		20	.400		7.75	19.75		27.50	38
0120	3/4" pipe size		18	.444		9.45	22		31.45	43.50
0140	1" pipe size		16	.500		12.90	24.50		37.40	51
0150	1-1/4" pipe size		15	.533		17.20	26.50		43.70	58.50
0160	1-1/2" pipe size		12	.667		21	33		54	72.50
0180	2" pipe size		8	1		32	49.50		81.50	109
0200	2-1/2" pipe size	Q-5	12	1.333		174	59		233	280
0220	3" pipe size		11	1.455		187	64.50		251.50	305
0240	4" pipe size		5	3.200		315	142		457	565
0500	For galvanized body, add					50%				
1000	Flanged, 125 lb., 1-1/2" pipe size	1 Stpi	11	.727	Ea.	101	36		137	165
1020	2" pipe size	"	8	1		106	49.50		155.50	190
1030	2-1/2" pipe size	Q-5	5	3.200		128	142		270	355
1040	3" pipe size		4.50	3.556		142	158		300	395
1060	4" pipe size		3	5.333		181	237		418	555
1080	5" pipe size	Q-6	3.40	7.059		283	325		608	800
1100	6" pipe size		3	8		283	370		653	865
1120	8" pipe size		2.50	9.600		580	440		1,020	1,300
1140	10" pipe size		2	12		1,100	555		1,655	2,050
1160	12" pipe size		1.70	14.118		1,675	650		2,325	2,800
1170	14" pipe size		1.30	18.462		3,075	850		3,925	4,650
1180	16" pipe size		1	24		4,350	1,100		5,450	6,425
1500	For 250 lb rating, add					20%				
2000	For galvanized body, add					50%				
2500	For steel body, add					40%				

23 21 20.80 Suction Diffusers		Crew	Daily Output	Labor-Hours	Unit	Material	Labor	Equipment	Total	Total Incl O&P
0010	**SUCTION DIFFUSERS**									
0100	Cast iron body with integral straightening vanes, strainer									
1000	Flanged									

23 21 20.80 Suction Diffusers

	23 21 20.80 Suction Diffusers	Crew	Daily Output	Labor-Hours	Unit	Material	2009 Bare Costs Labor	Equipment	Total	Total Inc O&P
1010	2″ inlet, 1-1/2″ pump side	1 Stpi	6	1.333	Ea.	250	66		316	375
1020	2″ pump side	″	5	1.600		259	79		338	405
1030	3″ inlet, 2″ pump side	Q-5	6.50	2.462		279	109		388	470
1040	2-1/2″ pump side		5.30	3.019		465	134		599	710
1050	3″ pump side		4.50	3.556		470	158		628	750
1060	4″ inlet, 3″ pump side		3.50	4.571		550	203		753	910
1070	4″ pump side		3	5.333		635	237		872	1,050
1080	5″ inlet, 4″ pump side	Q-6	3.80	6.316		725	291		1,016	1,225
1090	5″ pump side		3.40	7.059		865	325		1,190	1,450
1100	6″ inlet, 4″ pump side		3.30	7.273		755	335		1,090	1,325
1110	5″ pump side		3.10	7.742		910	355		1,265	1,525
1120	6″ pump side		3	8		955	370		1,325	1,600
1130	8″ inlet, 6″ pump side		2.70	8.889		1,025	410		1,435	1,750
1140	8″ pump side		2.50	9.600		1,775	440		2,215	2,625
1150	10″ inlet, 6″ pump side		2.40	10		2,475	460		2,935	3,400
1160	8″ pump side		2.30	10.435		2,775	480		3,255	3,775
1170	10″ pump side		2.20	10.909		2,975	505		3,480	4,025
1180	12″ inlet, 8″ pump side		2	12		3,150	555		3,705	4,300
1190	10″ pump side		1.80	13.333		3,475	615		4,090	4,750
1200	12″ pump side		1.70	14.118		3,925	650		4,575	5,275
1210	14″ inlet, 12″ pump side		1.50	16		5,075	735		5,810	6,700
1220	14″ pump side		1.30	18.462		5,175	850		6,025	6,975

23 21 20.84 Thermoflo Indicator

	23 21 20.84 Thermoflo Indicator	Crew	Daily Output	Labor-Hours	Unit	Material	Labor	Equipment	Total	Total Inc O&P
0010	**THERMOFLO INDICATOR**, For balancing									
1000	Sweat connections, 1-1/4″ pipe size	1 Stpi	12	.667	Ea.	505	33		538	605
1020	1-1/2″ pipe size		10	.800		515	39.50		554.50	625
1040	2″ pipe size		8	1		540	49.50		589.50	670
1060	2-1/2″ pipe size		7	1.143		820	56.50		876.50	990
2000	Flange connections, 3″ pipe size	Q-5	5	3.200		980	142		1,122	1,300
2020	4″ pipe size		4	4		1,175	178		1,353	1,575
2030	5″ pipe size		3.50	4.571		1,475	203		1,678	1,925
2040	6″ pipe size		3	5.333		1,575	237		1,812	2,100
2060	8″ pipe size		2	8		1,925	355		2,280	2,650

23 21 20.88 Venturi Flow

	23 21 20.88 Venturi Flow	Crew	Daily Output	Labor-Hours	Unit	Material	Labor	Equipment	Total	Total Inc O&P
0010	**VENTURI FLOW**, Measuring device									
0050	1/2″ diameter	1 Stpi	24	.333	Ea.	228	16.45		244.45	276
0100	3/4″ diameter		20	.400		211	19.75		230.75	262
0120	1″ diameter		19	.421		225	21		246	279
0140	1-1/4″ diameter		15	.533		280	26.50		306.50	350
0160	1-1/2″ diameter		13	.615		293	30.50		323.50	365
0180	2″ diameter		11	.727		299	36		335	385
0200	2-1/2″ diameter	Q-5	16	1		410	44.50		454.50	515
0220	3″ diameter		14	1.143		425	51		476	540
0240	4″ diameter		11	1.455		635	64.50		699.50	790
0260	5″ diameter	Q-6	4	6		835	276		1,111	1,325
0280	6″ diameter		3.50	6.857		925	315		1,240	1,500
0300	8″ diameter		3	8		1,175	370		1,545	1,850
0320	10″ diameter		2	12		2,800	555		3,355	3,900
0330	12″ diameter		1.80	13.333		3,825	615		4,440	5,150
0340	14″ diameter		1.60	15		4,350	690		5,040	5,825
0350	16″ diameter		1.40	17.143		4,975	790		5,765	6,650
0500	For meter, add					1,675			1,675	1,850

23 21 20 – Hydronic HVAC Piping Specialties

23 21 20.94 Weld End Ball Joints		Crew	Daily Output	Labor-Hours	Unit	2009 Bare Costs Material	Labor	Equipment	Total	Total Incl O&P
0010	**WELD END BALL JOINTS**, Steel									
0050	2-1/2" diameter	Q-17	13	1.231	Ea.	575	54.50	4.28	633.78	715
0100	3" diameter		12	1.333		675	59	4.64	738.64	840
0120	4" diameter		11	1.455		1,075	64.50	5.05	1,144.55	1,300
0140	5" diameter	Q-18	14	1.714		1,200	79	3.98	1,282.98	1,450
0160	6" diameter		12	2		1,950	92	4.64	2,046.64	2,300
0180	8" diameter		9	2.667		2,900	123	6.20	3,029.20	3,375
0200	10" diameter		8	3		3,775	138	6.95	3,919.95	4,375
0220	12" diameter		6	4		5,025	184	9.30	5,218.30	5,800

23 21 20.98 Zone Valves		Crew	Daily Output	Labor-Hours	Unit	Material	Labor	Equipment	Total	Total Incl O&P
0010	**ZONE VALVES**									
1000	Bronze body									
1010	2 way, 125 psi									
1020	Standard, 65' pump head									
1030	1/2" soldered	1 Stpi	24	.333	Ea.	102	16.45		118.45	137
1040	3/4" soldered		20	.400		102	19.75		121.75	142
1050	1" soldered		19	.421		106	21		127	148
1060	1-1/4" soldered		15	.533		124	26.50		150.50	176
1200	High head, 125' pump head									
1210	1/2" soldered	1 Stpi	24	.333	Ea.	135	16.45		151.45	173
1220	3/4" soldered		20	.400		139	19.75		158.75	183
1230	1" soldered		19	.421		152	21		173	198
1300	Geothermal, 150' pump head									
1310	3/4" soldered	1 Stpi	20	.400	Ea.	160	19.75		179.75	206
1320	1" soldered		19	.421		170	21		191	218
1330	3/4" threaded		20	.400		160	19.75		179.75	206
3000	3 way, 125 psi									
3200	Bypass, 65' pump head									
3210	1/2" soldered	1 Stpi	24	.333	Ea.	141	16.45		157.45	180
3220	3/4" soldered		20	.400		145	19.75		164.75	189
3230	1" soldered		19	.421		166	21		187	213
9000	Transformer, for up to 3 zone valves		20	.400		20.50	19.75		40.25	52

23 21 23 – Hydronic Pumps

23 21 23.13 In-Line Centrifugal Hydronic Pumps		Crew	Daily Output	Labor-Hours	Unit	Material	Labor	Equipment	Total	Total Incl O&P
0010	**IN-LINE CENTRIFUGAL HYDRONIC PUMPS**									
0600	Bronze, sweat connections, 1/40 HP, in line									
0640	3/4" size	Q-1	16	1	Ea.	172	44		216	255
1000	Flange connection, 3/4" to 1-1/2" size									
1040	1/12 HP	Q-1	6	2.667	Ea.	445	117		562	665
1060	1/8 HP		6	2.667		750	117		867	1,000
1100	1/3 HP		6	2.667		855	117		972	1,125
1140	2" size, 1/6 HP		5	3.200		1,100	140		1,240	1,400
1180	2-1/2" size, 1/4 HP		5	3.200		1,375	140		1,515	1,700
1220	3" size, 1/4 HP		4	4		1,425	176		1,601	1,850
1260	1/3 HP		4	4		1,750	176		1,926	2,200
1300	1/2 HP		4	4		1,800	176		1,976	2,250
1340	3/4 HP		4	4		1,950	176		2,126	2,425
1380	1 HP		4	4		3,175	176		3,351	3,750
2000	Cast iron, flange connection									
2040	3/4" to 1-1/2" size, in line, 1/12 HP	Q-1	6	2.667	Ea.	287	117		404	490
2060	1/8 HP		6	2.667		480	117		597	700

23 21 23 – Hydronic Pumps

23 21 23.13 In-Line Centrifugal Hydronic Pumps		Crew	Daily Output	Labor-Hours	Unit	Material	2009 Bare Costs Labor	Equipment	Total	Total Incl O&P
2100	1/3 HP	Q-1	6	2.667	Ea.	535	117		652	765
2140	2" size, 1/6 HP		5	3.200		585	140		725	855
2180	2-1/2" size, 1/4 HP		5	3.200		740	140		880	1,025
2220	3" size, 1/4 HP		4	4		755	176		931	1,100
2260	1/3 HP		4	4		1,025	176		1,201	1,400
2300	1/2 HP		4	4		1,050	176		1,226	1,425
2340	3/4 HP		4	4		1,225	176		1,401	1,625
2380	1 HP		4	4		1,750	176		1,926	2,200
2600	For non-ferrous impeller, add					3%				
3000	High head, bronze impeller									
3030	1-1/2" size 1/2 HP	Q-1	5	3.200	Ea.	880	140		1,020	1,175
3040	1-1/2" size 3/4 HP		5	3.200		940	140		1,080	1,225
3050	2" size 1 HP		4	4		1,125	176		1,301	1,525
3090	2" size 1-1/2 HP		4	4		1,300	176		1,476	1,725
4000	Close coupled, end suction, bronze impeller									
4040	1-1/2" size, 1-1/2 HP, to 40 GPM	Q-1	3	5.333	Ea.	1,625	234		1,859	2,125
4090	2" size, 2 HP, to 50 GPM		3	5.333		1,925	234		2,159	2,450
4100	2" size, 3 HP, to 90 GPM		2.30	6.957		2,000	305		2,305	2,650
4190	2-1/2" size, 3 HP, to 150 GPM		2	8		2,175	350		2,525	2,900
4300	3" size, 5 HP, to 225 GPM		1.80	8.889		2,475	390		2,865	3,300
4410	3" size, 10 HP, to 350 GPM		1.60	10		3,325	440		3,765	4,300
4420	4" size, 7-1/2 HP, to 350 GPM		1.60	10		3,225	440		3,665	4,200
4520	4" size, 10 HP, to 600 GPM	Q-2	1.70	14.118		3,700	640		4,340	5,025
4530	5" size, 15 HP, to 1000 GPM		1.70	14.118		3,700	640		4,340	5,050
4610	5" size, 20 HP, to 1350 GPM		1.50	16		3,925	730		4,655	5,425
4620	5" size, 25 HP, to 1550 GPM		1.50	16		5,375	730		6,105	7,000
5000	Base mounted, bronze impeller, coupling guard									
5040	1-1/2" size, 1-1/2 HP, to 40 GPM	Q-1	2.30	6.957	Ea.	3,100	305		3,405	3,875
5090	2" size, 2 HP, to 50 GPM		2.30	6.957		3,400	305		3,705	4,175
5100	2" size, 3 HP, to 90 GPM		2	8		3,425	350		3,775	4,275
5190	2-1/2" size, 3 HP, to 150 GPM		1.80	8.889		3,700	390		4,090	4,650
5300	3" size, 5 HP, to 225 GPM		1.60	10		3,775	440		4,215	4,800
5410	4" size, 5 HP, to 350 GPM		1.50	10.667		4,025	470		4,495	5,125
5420	4" size, 7-1/2 HP, to 350 GPM		1.50	10.667		4,150	470		4,620	5,275
5520	5" size, 10 HP, to 600 GPM	Q-2	1.60	15		4,275	685		4,960	5,725
5530	5" size, 15 HP, to 1000 GPM		1.60	15		4,425	685		5,110	5,900
5610	6" size, 20 HP, to 1350 GPM		1.40	17.143		5,750	780		6,530	7,500
5620	6" size, 25 HP, to 1550 GPM		1.40	17.143		5,900	780		6,680	7,675
5800	The above pump capacities are based on 1800 RPM,									
5810	at a 60 foot head. Increasing the RPM									
5820	or decreasing the head will increase the GPM.									

23 22 Steam and Condensate Piping and Pumps

23 22 13 – Steam and Condensate Heating Piping

23 22 13.23 Aboveground Steam and Condensate Piping		Crew	Daily Output	Labor-Hours	Unit	2009 Bare Costs Material	Labor	Equipment	Total	Total Incl O&P
0010	ABOVEGROUND STEAM AND CONDENSATE HEATING PIPING									
0020	Condensate meter									
0100	500 lb. per hour	1 Stpi	14	.571	Ea.	2,300	28		2,328	2,575
0140	1500 lb. per hour		7	1.143		2,475	56.50		2,531.50	2,800
0160	3000 lb. per hour		5	1.600		3,325	79		3,404	3,775
0200	12,000 lb. per hour	Q-5	3.50	4.571		4,450	203		4,653	5,175

23 22 23 – Steam Condensate Pumps

23 22 23.10 Condensate Return System

		Crew	Daily Output	Labor-Hours	Unit	Material	Labor	Equipment	Total	Total Incl O&P
0010	CONDENSATE RETURN SYSTEM									
2000	Simplex									
2010	With pump, motor, CI receiver, float switch									
2020	3/4 HP, 15 GPM	Q-1	1.80	8.889	Ea.	5,225	390		5,615	6,325
2100	Duplex									
2110	With 2 pumps and motors, CI receiver, float switch, alternator									
2120	3/4 HP, 15 GPM, 15 Gal CI rcvr	Q-1	1.40	11.429	Ea.	5,700	500		6,200	7,025
2130	1 HP, 25 GPM		1.20	13.333		6,850	585		7,435	8,400
2140	1-1/2 HP, 45 GPM		1	16		7,850	700		8,550	9,675
2150	1-1/2 HP, 60 GPM		1	16		8,925	700		9,625	10,900

23 23 Refrigerant Piping

23 23 13 – Refrigerant Piping Valves

23 23 13.10 Valves

		Crew	Daily Output	Labor-Hours	Unit	Material	Labor	Equipment	Total	Total Incl O&P
0010	VALVES									
8100	Check valve, soldered									
8110	5/8"	1 Stpi	36	.222	Ea.	79.50	10.95		90.45	103
8114	7/8"		26	.308		108	15.20		123.20	142
8118	1-1/8"		18	.444		129	22		151	175
8122	1-3/8"		14	.571		179	28		207	239
8126	1-5/8"		13	.615		172	30.50		202.50	235
8130	2-1/8"		12	.667		256	33		289	330
8134	2-5/8"	Q-5	22	.727		1,625	32.50		1,657.50	1,850
8138	3-1/8"	"	20	.800		1,950	35.50		1,985.50	2,200
8500	Refrigeration valve, packless, soldered									
8510	1/2"	1 Stpi	38	.211	Ea.	29.50	10.40		39.90	48
8514	5/8"		36	.222		31	10.95		41.95	50.50
8518	7/8"		26	.308		273	15.20		288.20	325
8520	Packed, soldered									
8522	1-1/8"	1 Stpi	18	.444	Ea.	161	22		183	210
8526	1-3/8"		14	.571		293	28		321	365
8530	1-5/8"		13	.615		315	30.50		345.50	390
8534	2-1/8"		12	.667		500	33		533	600
8538	2-5/8"	Q-5	22	.727		680	32.50		712.50	795
8542	3-1/8"		20	.800		785	35.50		820.50	920
8546	4-1/8"		18	.889		1,375	39.50		1,414.50	1,575
8600	Solenoid valve, flange/solder									
8610	1/2"	1 Stpi	38	.211	Ea.	144	10.40		154.40	174
8614	5/8"		36	.222		178	10.95		188.95	211
8618	3/4"		30	.267		225	13.15		238.15	267
8622	7/8"		26	.308		283	15.20		298.20	335
8626	1-1/8"		18	.444		370	22		392	440

23 23 Refrigerant Piping

23 23 13 – Refrigerant Piping Valves

23 23 13.10 Valves		Crew	Daily Output	Labor-Hours	Unit	2009 Bare Costs Material	Labor	Equipment	Total	Total Incl O&P
8630	1-3/8"	1 Stpi	14	.571	Ea.	445	28		473	535
8634	1-5/8"		13	.615		630	30.50		660.50	740
8638	2-1/8"		12	.667		690	33		723	810
8800	Thermostatic valve, flange/solder									
8810	1/2 – 3 ton, 3/8" x 5/8"	1 Stpi	9	.889	Ea.	131	44		175	210
8814	4 – 5 ton, 1/2" x 7/8"		7	1.143		131	56.50		187.50	229
8818	6 – 8 ton, 5/8" x 7/8"		5	1.600		131	79		210	262
8822	7 – 12 ton, 7/8" x 1-1/8"		4	2		153	98.50		251.50	315
8826	15 – 20 ton, 7/8" x 1-3/8"		3.20	2.500		153	123		276	355

23 23 16 – Refrigerant Piping Specialties

23 23 16.10 Refrigerant Piping Component Specialties

Line	Description	Crew	Daily Output	Labor-Hours	Unit	Material	Labor	Equipment	Total	Total Incl O&P
0010	**REFRIGERANT PIPING COMPONENT SPECIALTIES**									
0600	Accumulator									
0610	3/4"	1 Stpi	8.80	.909	Ea.	70.50	45		115.50	145
0614	7/8"		6.40	1.250		80.50	61.50		142	181
0618	1-1/8"		4.80	1.667		111	82.50		193.50	245
0622	1-3/8"		4	2		145	98.50		243.50	310
0626	1-5/8"		3.20	2.500		157	123		280	355
0630	2-1/8"		2.40	3.333		395	165		560	680
0700	Condensate drip/drain pan		24	.333		116	16.45		132.45	152
1000	Filter dryer									
1010	Replaceable core type, solder									
1020	1/2"	1 Stpi	20	.400	Ea.	108	19.75		127.75	149
1030	5/8"		19	.421		142	21		163	187
1040	7/8"		18	.444		142	22		164	189
1050	1-1/8"		15	.533		148	26.50		174.50	203
1060	1-3/8"		14	.571		148	28		176	206
1070	1-5/8"		12	.667		152	33		185	218
1080	2-1/8"		10	.800		155	39.50		194.50	229
1090	2-5/8"		9	.889		605	44		649	730
1100	3-1/8"		8	1		630	49.50		679.50	765
1200	Sealed in-line, solder									
1210	1/4", 3 cubic inches	1 Stpi	22	.364	Ea.	8.10	17.95		26.05	36
1220	3/8", 5 cubic inches		21	.381		7.40	18.80		26.20	36
1230	1/2", 9 cubic inches		20	.400		12	19.75		31.75	42.50
1240	1/2", 16 cubic inches		20	.400		15.60	19.75		35.35	46.50
1250	5/8", 16 cubic inches		19	.421		16.35	21		37.35	49
1260	5/8", 30 cubic inches		19	.421		26.50	21		47.50	60
1270	7/8", 30 cubic inches		18	.444		28	22		50	64
1280	7/8", 41 cubic inches		17	.471		37.50	23		60.50	76.50
1290	1-1/8", 60 cubic inches		15	.533		52	26.50		78.50	97
4000	P-Trap, suction line, solder									
4010	5/8"	1 Stpi	19	.421	Ea.	52	21		73	88.50
4020	3/4"		19	.421		52	21		73	88.50
4030	7/8"		18	.444		56	22		78	94.50
4040	1-1/8"		15	.533		59	26.50		85.50	105
4050	1-3/8"		14	.571		107	28		135	161
4060	1-5/8"		12	.667		163	33		196	229
4070	2-1/8"		10	.800		335	39.50		374.50	425
5000	Sightglass									
5010	Moisture and liquid indicator, solder									
5020	1/4"	1 Stpi	22	.364	Ea.	19.30	17.95		37.25	48

23 23 Refrigerant Piping

23 23 16 – Refrigerant Piping Specialties

23 23 16.10 Refrigerant Piping Component Specialties		Crew	Daily Output	Labor-Hours	Unit	2009 Bare Costs Material	Labor	Equipment	Total	Total Incl O&P
5030	3/8"	1 Stpi	21	.381	Ea.	18.40	18.80		37.20	48.50
5040	1/2"		20	.400		22.50	19.75		42.25	54
5050	5/8"		19	.421		24	21		45	57
5060	7/8"		18	.444		42	22		64	79
5070	1-1/8"		15	.533		43	26.50		69.50	87
5080	1-3/8"		14	.571		32.50	28		60.50	78.50
5090	1-5/8"		12	.667		38	33		71	91.50
5100	2-1/8"	↓	10	.800	↓	47.50	39.50		87	111
7400	Vacuum pump set									
7410	Two stage, high vacuum continuous duty	1 Stpi	8	1	Ea.	3,000	49.50		3,049.50	3,375

23 23 23 – Refrigerants

23 23 23.10 Anti-Freeze

		Crew	Daily Output	Labor-Hours	Unit	Material	Labor	Equipment	Total	Total Incl O&P
0010	**ANTI-FREEZE**, Inhibited									
0900	Ethylene glycol concentrated									
1000	55 gallon drums, small quantities				Gal.	12.10			12.10	13.30
1200	Large quantities					12			12	13.20
2000	Propylene glycol, for solar heat, small quantities					13.85			13.85	15.25
2100	Large quantities				↓	13.75			13.75	15.15

23 23 23.20 Refrigerant

		Crew	Daily Output	Labor-Hours	Unit	Material	Labor	Equipment	Total	Total Incl O&P
0010	**REFRIGERANT**									
4420	Refrigerant, R-22, 30 lb. disposable cylinder				Lb.	17.45			17.45	19.15
4440	Refrigerant, R-507, 25 lb. disposable cylinder				"	10.50			10.50	11.55

23 31 HVAC Ducts and Casings

23 31 13 – Metal Ducts

23 31 13.13 Rectangular Metal Ducts

		Crew	Daily Output	Labor-Hours	Unit	Material	Labor	Equipment	Total	Total Incl O&P
0010	**RECTANGULAR METAL DUCTS** R233100-20									
0020	Fabricated rectangular, includes fittings, joints, supports,									
0030	allowance for flexible connections, no insulation R233100-30									
0031	NOTE: Fabrication and installation are combined									
0040	as LABOR cost. Approx. 25% fittings assumed. R233100-40									
0042	Fabrication/Inst. is to commercial quality standards									
0043	(SMACNA or equiv.) for structure, sealing, leak testing, etc. R233100-50									
0050	Add to labor for elevated installation									
0051	of fabricated ductwork									
0052	10' to 15' high						6%			
0053	15' to 20' high						12%			
0054	20' to 25' high						15%			
0055	25' to 30' high						21%			
0056	30' to 35' high						24%			
0057	35' to 40' high						30%			
0058	Over 40' high						33%			
0072	For duct insulation see Div. 23 07 13.10 3000									
0100	Aluminum, alloy 3003-H14, under 100 lb.	Q-10	75	.320	Lb.	3.72	14.10		17.82	25.50
0110	100 to 500 lb.		80	.300		2.48	13.20		15.68	23
0120	500 to 1,000 lb.		95	.253		2.31	11.15		13.46	19.70
0140	1,000 to 2,000 lb.		120	.200		2.15	8.80		10.95	15.90
0150	2,000 to 5,000 lb.		130	.185		2.15	8.15		10.30	14.85
0160	Over 5,000 lb.		145	.166		2.15	7.30		9.45	13.55
0500	Galvanized steel, under 200 lb.	↓	235	.102	↓	1.12	4.50		5.62	8.15

23 31 13.13 Rectangular Metal Ducts

	23 31 13.13 Rectangular Metal Ducts	Crew	Daily Output	Labor-Hours	Unit	Material	2009 Bare Costs Labor	Equipment	Total	Total Ind O&P
0520	200 to 500 lb.	Q-10	245	.098	Lb.	.99	4.32		5.31	7.75
0540	500 to 1,000 lb.		255	.094		.97	4.15		5.12	7.45
0560	1,000 to 2,000 lb.		265	.091		.97	3.99		4.96	7.20
0570	2,000 to 5,000 lb.		275	.087		.95	3.84		4.79	6.95
0580	Over 5,000 lb.		285	.084		.93	3.71		4.64	6.70
0600	For large quantities special prices available from supplier									
1000	Stainless steel, type 304, under 100 lb.	Q-10	165	.145	Lb.	4.35	6.40		10.75	14.65
1020	100 to 500 lb.		175	.137		4.02	6.05		10.07	13.70
1030	500 to 1,000 lb.		190	.126		3.68	5.55		9.23	12.60
1040	1,000 to 2,000 lb.		200	.120		3.65	5.30		8.95	12.15
1050	2,000 to 5,000 lb.		225	.107		3.31	4.70		8.01	10.90
1060	Over 5,000 lb.		235	.102		3.24	4.50		7.74	10.45
1080	Note: Minimum order cost exceeds per lb. cost for min. wt.									
1100	For medium pressure ductwork, add				Lb.		15%			
1200	For high pressure ductwork, add						40%			
1210	For welded ductwork, add						85%			
1220	For 30% fittings, add						11%			
1224	For 40% fittings, add						34%			
1228	For 50% fittings, add						56%			
1232	For 60% fittings, add						79%			
1236	For 70% fittings, add						101%			
1240	For 80% fittings, add						124%			
1244	For 90% fittings, add						147%			
1248	For 100% fittings, add						169%			
1252	Note: Fittings add includes time for detailing and installation.									
1276	Duct tape, 2" x 60 yd roll				Ea.	5.20			5.20	5.70

23 31 13.16 Round and Flat-Oval Spiral Ducts

	23 31 13.16 Round and Flat-Oval Spiral Ducts	Crew	Daily Output	Labor-Hours	Unit	Material	Labor	Equipment	Total	Total Ind O&P
0010	**ROUND AND FLAT-OVAL SPIRAL DUCTS**									
1280	Add to labor for elevated installation									
1282	of prefabricated (purchased) ductwork									
1283	10′ to 15′ high						10%			
1284	15′ to 20′ high						20%			
1285	20′ to 25′ high						25%			
1286	25′ to 30′ high						35%			
1287	30′ to 35′ high						40%			
1288	35′ to 40′ high						50%			
1289	Over 40′ high						55%			
5400	Spiral preformed, steel, galv., straight lengths, Max 10" spwg.									
5410	4" diameter, 26 ga.	Q-9	360	.044	L.F.	1.39	1.89		3.28	4.44
5416	5" diameter, 26 ga.		320	.050		1.77	2.12		3.89	5.20
5420	6" diameter, 26 ga.		280	.057		2.02	2.43		4.45	5.95
5425	7" diameter, 26 ga.		240	.067		2.40	2.83		5.23	7
5430	8" diameter, 26 ga.		200	.080		2.66	3.40		6.06	8.20
5440	10" diameter, 26 ga.		160	.100		4.30	4.25		8.55	11.30
5450	12" diameter, 26 ga.		120	.133		5.20	5.65		10.85	14.40
5460	14" diameter, 26 ga.		80	.200		6.05	8.50		14.55	19.80
5480	16" diameter, 24 ga.		60	.267		6.95	11.35		18.30	25
5490	18" diameter, 24 ga.		50	.320		9.50	13.60		23.10	31.50
5500	20" diameter, 24 ga.	Q-10	65	.369		10.50	16.25		26.75	36.50
5510	22" diameter, 24 ga.		60	.400		11.50	17.60		29.10	39.50
5520	24" diameter, 24 ga.		55	.436		12.50	19.20		31.70	43.50
5540	30" diameter, 22 ga.		45	.533		18.45	23.50		41.95	56.50

23 31 13.16 Round and Flat-Oval Spiral Ducts		Crew	Daily Output	Labor-Hours	Unit	2009 Bare Costs Material	Labor	Equipment	Total	Total Incl O&P
5600	36″ diameter, 22 ga.	Q-10	40	.600	L.F.	29.50	26.50		56	73
5800	Connector, 4″ diameter	Q-9	100	.160	Ea.	2.44	6.80		9.24	13.15
5810	5″ diameter		94	.170		2.61	7.25		9.86	14
5820	6″ diameter		88	.182		2.81	7.70		10.51	15
5840	8″ diameter		78	.205		3.63	8.70		12.33	17.40
5860	10″ diameter		70	.229		5.70	9.70		15.40	21.50
5880	12″ diameter		50	.320		6.90	13.60		20.50	28.50
5900	14″ diameter		44	.364		8	15.45		23.45	33
5920	16″ diameter		40	.400		9	17		26	36
5930	18″ diameter		37	.432		10.05	18.35		28.40	39.50
5940	20″ diameter		34	.471		10.90	20		30.90	43
5950	22″ diameter		31	.516		11.90	22		33.90	46.50
5960	24″ diameter		28	.571		12.75	24.50		37.25	51.50
5980	30″ diameter		22	.727		16.60	31		47.60	66
6000	36″ diameter		18	.889		25	38		63	85.50
6300	Elbow, 45°, 4″ diameter		60	.267		2.75	11.35		14.10	20.50
6310	5″ diameter		52	.308		3.54	13.05		16.59	24
6320	6″ diameter		44	.364		3.87	15.45		19.32	28.50
6340	8″ diameter		28	.571		5.10	24.50		29.60	43
6360	10″ diameter		18	.889		7.75	38		45.75	66.50
6380	12″ diameter		13	1.231		10.05	52.50		62.55	91.50
6400	14″ diameter		11	1.455		18.85	62		80.85	116
6420	16″ diameter		10	1.600		31	68		99	139
6430	18″ diameter		9.60	1.667		38	71		109	151
6440	20″ diameter	Q-10	14	1.714		48	75.50		123.50	169
6450	22″ diameter		13	1.846		57.50	81.50		139	189
6460	24″ diameter		12	2		64.50	88		152.50	207
6480	30″ diameter		9	2.667		99	117		216	290
6500	36″ diameter		7	3.429		120	151		271	365
6600	Elbow, 90°, 4″ diameter	Q-9	60	.267		4.71	11.35		16.06	22.50
6610	5″ diameter		52	.308		5.40	13.05		18.45	26
6620	6″ diameter		44	.364		6.20	15.45		21.65	31
6625	7″ diameter		36	.444		6.55	18.90		25.45	36
6630	8″ diameter		28	.571		7.60	24.50		32.10	46
6640	10″ diameter		18	.889		10.85	38		48.85	70
6650	12″ diameter		13	1.231		14.35	52.50		66.85	96.50
6660	14″ diameter		11	1.455		27	62		89	125
6670	16″ diameter		10	1.600		42.50	68		110.50	152
6676	18″ diameter		9.60	1.667		55.50	71		126.50	170
6680	20″ diameter	Q-10	14	1.714		74	75.50		149.50	197
6684	22″ diameter		13	1.846		115	81.50		196.50	252
6690	24″ diameter		12	2		129	88		217	277
6700	30″ diameter		9	2.667		198	117		315	400
6710	36″ diameter		7	3.429		240	151		391	495
6720	End cap									
6722	4″	Q-9	110	.145	Ea.	2.43	6.20		8.63	12.15
6724	6″		94	.170		3.24	7.25		10.49	14.70
6725	8″		88	.182		4.64	7.70		12.34	17
6726	10″		78	.205		6.75	8.70		15.45	21
6727	12″		70	.229		9	9.70		18.70	25
6728	14″		50	.320		12.40	13.60		26	34.50
6729	16″		44	.364		14.90	15.45		30.35	40.50
6730	18″		40	.400		17.45	17		34.45	45

23 31 13.16 Round and Flat-Oval Spiral Ducts		Crew	Daily Output	Labor-Hours	Unit	2009 Bare Costs Material	Labor	Equipment	Total	Total Incl O&P
6731	20″	Q-9	37	.432	Ea.	19	18.35		37.35	49.50
6732	22″		34	.471		21	20		41	54.50
6733	24″		30	.533		25	22.50		47.50	62.50
6800	Reducing coupling, 6″ x 4″		46	.348		7	14.80		21.80	30
6820	8″ x 6″		40	.400		8.60	17		25.60	35.50
6840	10″ x 8″		32	.500		10.90	21		31.90	44.50
6860	12″ x 10″		24	.667		15.60	28.50		44.10	60.50
6880	14″ x 12″		20	.800		25.50	34		59.50	81
6900	16″ x 14″		18	.889		37.50	38		75.50	99
6920	18″ x 16″		16	1		46.50	42.50		89	117
6940	20″ x 18″	Q-10	24	1		56.50	44		100.50	130
6950	22″ x 20″		23	1.043		66	46		112	144
6960	24″ x 22″		22	1.091		76	48		124	158
6980	30″ x 28″		18	1.333		113	58.50		171.50	215
7000	36″ x 34″		16	1.500		152	66		218	259
7100	Tee, 90°, 4″ diameter	Q-9	40	.400		6.15	17		23.15	33
7110	5″ diameter		34.60	.462		8.20	19.65		27.85	39
7120	6″ diameter		30	.533		8.85	22.50		31.35	45
7130	8″ diameter		19	.842		11.55	36		47.55	67.50
7140	10″ diameter		12	1.333		16.45	56.50		72.95	105
7150	12″ diameter		9	1.778		21	75.50		96.50	139
7160	14″ diameter		7	2.286		40	97		137	193
7170	16″ diameter		6.70	2.388		56.50	101		157.50	218
7176	18″ diameter		6	2.667		78	113		191	260
7180	20″ diameter	Q-10	8.70	2.759		86	122		208	282
7190	24″ diameter		7	3.429		103	151		254	345
7200	30″ diameter		6	4		126	176		302	410
7210	36″ diameter		5	4.800		160	211		371	500
7350	Lateral 45°, reducing									
7354	6″ x 6″ x 4″	Q-9	30	.533	Ea.	11.35	22.50		33.85	47.50
7358	8″ x 8″ x 4″		19	.842		14	36		50	70.50
7362	10″ x 10″ x 4″		12	1.333		19.60	56.50		76.10	109
7366	12″ x 12″ x 4″		9	1.778		22	75.50		97.50	141
7370	14″ x 14″ x 4″		7	2.286		31.50	97		128.50	184
7374	16″ x 16″ x 4″		6.70	2.388		50	101		151	211
7378	18″ x 18″ x 4″		6	2.667		63	113		176	244
7382	20″ x 20″ x 4″	Q-10	8.70	2.759		68.50	122		190.50	263
7386	22″ x 22″ x 4″		8	3		77	132		209	288
7390	24″ x 24″ x 4″		7	3.429		82.50	151		233.50	325
7400	Steel, PVC coated both sides, straight lengths									
7410	4″ diameter, 26 ga.	Q-9	360	.044	L.F.	2.52	1.89		4.41	5.70
7416	5″ diameter, 26 ga.		240	.067		3.14	2.83		5.97	7.80
7420	6″ diameter, 26 ga.		280	.057		3.49	2.43		5.92	7.60
7440	8″ diameter, 26 ga.		200	.080		4.82	3.40		8.22	10.55
7460	10″ diameter, 26 ga.		160	.100		6.50	4.25		10.75	13.70
7480	12″ diameter, 26 ga.		120	.133		8.05	5.65		13.70	17.55
7500	14″ diameter, 26 ga.		80	.200		11	8.50		19.50	25
7520	16″ diameter, 24 ga.		60	.267		15.55	11.35		26.90	34.50
7540	18″ diameter, 24 ga.		45	.356		16.75	15.10		31.85	42
7560	20″ diameter, 24 ga.	Q-10	65	.369		19.80	16.25		36.05	47
7580	24″ diameter, 24 ga.		55	.436		25	19.20		44.20	57
7600	30″ diameter, 22 ga.		45	.533		36.50	23.50		60	76
7620	36″ diameter, 22 ga.		40	.600		44.50	26.50		71	89.50

23 31 13.16	Round and Flat-Oval Spiral Ducts	Crew	Daily Output	Labor-Hours	Unit	2009 Bare Costs Material	Labor	Equipment	Total	Total Incl O&P
7890	Connector, 4″ diameter	Q-9	100	.160	Ea.	4.81	6.80		11.61	15.75
7896	5″ diameter		83	.193		5.15	8.20		13.35	18.25
7900	6″ diameter		88	.182		5.40	7.70		13.10	17.85
7920	8″ diameter		78	.205		6.85	8.70		15.55	21
7940	10″ diameter		70	.229		9.40	9.70		19.10	25.50
7960	12″ diameter		50	.320		12	13.60		25.60	34
7980	14″ diameter		44	.364		15.45	15.45		30.90	41
8000	16″ diameter		40	.400		17.10	17		34.10	45
8020	18″ diameter		37	.432		19.30	18.35		37.65	49.50
8040	20″ diameter		34	.471		21	20		41	54
8060	24″ diameter		28	.571		23	24.50		47.50	63
8080	30″ diameter		22	.727		30.50	31		61.50	81
8100	36″ diameter		18	.889		41	38		79	104
8390	Elbow, 45°, 4″ diameter		60	.267		6.05	11.35		17.40	24
8396	5″ diameter		36	.444		7.30	18.90		26.20	37
8400	6″ diameter		44	.364		7.95	15.45		23.40	33
8420	8″ diameter		28	.571		10.50	24.50		35	49
8440	10″ diameter		18	.889		15.25	38		53.25	75
8460	12″ diameter		13	1.231		22.50	52.50		75	105
8480	14″ diameter		11	1.455		40.50	62		102.50	140
8500	16″ diameter		10	1.600		58	68		126	169
8520	18″ diameter		9	1.778		67	75.50		142.50	190
8540	20″ diameter	Q-10	13	1.846		89.50	81.50		171	224
8560	24″ diameter		11	2.182		113	96		209	273
8580	30″ diameter		9	2.667		174	117		291	375
8600	36″ diameter		7	3.429		212	151		363	465
8800	Elbow, 90°, 4″ diameter	Q-9	60	.267		10.05	11.35		21.40	28.50
8804	5″ diameter		52	.308		11.45	13.05		24.50	32.50
8806	6″ diameter		44	.364		12.55	15.45		28	38
8808	8″ diameter		28	.571		15.65	24.50		40.15	54.50
8810	10″ diameter		18	.889		21	38		59	81
8812	12″ diameter		13	1.231		29	52.50		81.50	113
8814	14″ diameter		11	1.455		52	62		114	152
8816	16″ diameter		10	1.600		81	68		149	194
8818	18″ diameter		9.60	1.667		99	71		170	218
8820	20″ diameter	Q-10	14	1.714		126	75.50		201.50	255
8822	24″ diameter		12	2		227	88		315	385
8824	30″ diameter		9	2.667		350	117		467	565
8826	36″ diameter		7	3.429		425	151		576	695
8850	Tee, 90°, 4″ diameter	Q-9	40	.400		11.30	17		28.30	38.50
8854	5″ diameter		34.60	.462		14.90	19.65		34.55	46.50
8856	6″ diameter		30	.533		16.45	22.50		38.95	53
8858	8″ diameter		19	.842		21	36		57	78.50
8860	10″ diameter		12	1.333		27.50	56.50		84	118
8862	12″ diameter		9	1.778		39	75.50		114.50	159
8864	14″ diameter		7	2.286		68.50	97		165.50	224
8866	16″ diameter		6.70	2.388		99.50	101		200.50	265
8868	18″ diameter		6	2.667		144	113		257	330
8870	20″ diameter	Q-10	8.70	2.759		155	122		277	360
8872	24″ diameter		7	3.429		214	151		365	465
8874	30″ diameter		6	4		267	176		443	565
8876	36″ diameter		5	4.800		340	211		551	695
9000	Reducing coupling, tapered offset									

23 31 13 – Metal Ducts

23 31 13.16 Round and Flat-Oval Spiral Ducts		Crew	Daily Output	Labor-Hours	Unit	Material	2009 Bare Costs Labor	Equipment	Total	Total Incl O&P
9020	Diameter x any smaller diameter									
9040	6″ x	Q-9	46	.348	Ea.	12.50	14.80		27.30	36.50
9060	8″ x		40	.400		15.45	17		32.45	43
9080	10″ x		32	.500		18.65	21		39.65	53
9100	12″ x		24	.667		26	28.50		54.50	72
9120	14″ x		20	.800		41.50	34		75.50	98.50
9140	16″ x		18	.889		63.50	38		101.50	128
9160	18″ x		16	1		77	42.50		119.50	150
9180	20″ x	Q-10	24	1		94.50	44		138.50	172
9200	24″ x		22	1.091		128	48		176	215
9220	30″ x		18	1.333		181	58.50		239.50	290
9240	36″ x		16	1.500		236	66		302	360
9400	Lateral 45°, straight barrel									
9410	4″ size	Q-9	40	.400	Ea.	12.95	17		29.95	40.50
9416	5″ size		34.60	.462		17.10	19.65		36.75	49
9420	6″ size		30	.533		18.95	22.50		41.45	56
9440	8″ size		19	.842		24.50	36		60.50	81.50
9460	10″ size		12	1.333		35	56.50		91.50	126
9480	12″ size		9	1.778		45	75.50		120.50	166
9500	14″ size		7	2.286		78.50	97		175.50	236
9520	16″ size		6.70	2.388		114	101		215	282
9540	18″ size		6	2.667		166	113		279	355
9560	20″ size	Q-10	8.70	2.759		179	122		301	385
9580	24″ size		7	3.429		246	151		397	505
9600	30″ size		6	4		305	176		481	610
9620	36″ size		5	4.800		390	211		601	755
9640	For reducing lateral, add					40%				
9650	Lateral 45°, reducing									
9654	6″ x 6″ x 4″	Q-9	30	.533	Ea.	18.95	22.50		41.45	56
9658	8″ x 8″ x 4″		19	.842		24.50	36		60.50	81.50
9662	10″ x 10″ x 4″		12	1.333		32	56.50		88.50	122
9666	12″ x 12″ x 4″		9	1.778		39	75.50		114.50	159
9670	14″ x 14″ x 4″		7	2.286		52.50	97		149.50	207
9674	16″ x 16″ x 4″		6.70	2.388		89	101		190	254
9678	18″ x 18″ x 4″		6	2.667		124	113		237	310
9682	20″ x 20″ x 4″	Q-10	8.70	2.759		132	122		254	335
9686	22″ x 22″ x 4″		8	3		180	132		312	400
9690	24″ x 24″ x 4″		7	3.429		193	151		344	445
9700	Register boot, elbow or perimeter									
9710	6″ duct diameter	Q-9	60	.267	Ea.	26.50	11.35		37.85	47
9720	8″ duct diameter		56	.286		32	12.15		44.15	53.50
9730	10″ duct diameter		45	.356		43.50	15.10		58.60	71.50
9740	12″ duct diameter		36	.444		43.50	18.90		62.40	77

23 31 13.17 Round Grease Duct		Crew	Daily Output	Labor-Hours	Unit	Material	Labor	Equipment	Total	Total Incl O&P
0010	**ROUND GREASE DUCT**									
1280	Add to labor for elevated installation									
1282	of prefabricated (purchased) ductwork									
1283	10′ to 15′ high						10%			
1284	15′ to 20′ high						20%			
1285	20′ to 25′ high						25%			
1286	25′ to 30′ high						35%			
1287	30′ to 35′ high						40%			

23 31 13.17 Round Grease Duct		Crew	Daily Output	Labor-Hours	Unit	Material	2009 Bare Costs Labor	Equipment	Total	Total Incl O&P
1288	35′ to 40′ high						50%			
1289	Over 40′ high						55%			
4020	Zero clearance, 2 hr. fire rated, UL listed, 3″ double wall									
4030	304 interior, aluminized exterior, all sizes are ID									
4032	For 316 interior and aluminized exterior, add					6%				
4034	For 304 interior and 304 exterior, add					21%				
4036	For 316 interior and 316 exterior, add					40%				
4040	Straight									
4050	5″	Q-9	36.48	.439	L.F.	62.50	18.65		81.15	97
4052	6″		34.20	.468		79	19.85		98.85	117
4054	8″		29.64	.540		90.50	23		113.50	135
4056	10″		27.36	.585		102	25		127	150
4058	12″		25.10	.637		116	27		143	170
4060	14″		23.94	.668		134	28.50		162.50	191
4062	16″		22.80	.702		140	30		170	200
4064	18″	↓	21.66	.739		172	31.50		203.50	239
4066	20″	Q-10	20.52	1.170		198	51.50		249.50	298
4068	22″		19.38	1.238		225	54.50		279.50	330
4070	24″		18.24	1.316		256	58		314	370
4072	26″		17.67	1.358		271	60		331	390
4074	28″		17.10	1.404		285	62		347	410
4076	30″		15.96	1.504		310	66		376	440
4078	32″		15.39	1.559		335	68.50		403.50	475
4080	36″	↓	14.25	1.684	↓	375	74		449	525
4100	Adjustable section, 30″ long									
4102	5″	Q-9	18.24	.877	Ea.	196	37.50		233.50	273
4104	6″		17.10	.936		245	40		285	330
4106	8″		14.82	1.080		250	46		296	345
4108	10″		13.68	1.170		320	49.50		369.50	425
4110	12″		12.54	1.276		365	54		419	485
4112	14″		11.97	1.337		420	57		477	550
4114	16″		11.40	1.404		435	59.50		494.50	570
4116	18″	↓	10.83	1.477		535	63		598	685
4118	20″	Q-10	10.26	2.339		620	103		723	840
4120	22″		9.69	2.477		705	109		814	945
4122	24″		9.12	2.632		800	116		916	1,050
4124	26″		8.83	2.718		845	120		965	1,125
4126	28″		8.55	2.807		890	124		1,014	1,175
4128	30″		8.26	2.906		965	128		1,093	1,275
4130	32″		7.98	3.008		1,050	132		1,182	1,350
4132	36″	↓	6.84	3.509	↓	1,150	155		1,305	1,500
4140	Tee, 90°, grease									
4142	5″	Q-9	14.25	1.123	Ea.	520	47.50		567.50	645
4144	6″		13.68	1.170		610	49.50		659.50	745
4146	8″		12.54	1.276		700	54		754	855
4148	10″		11.97	1.337		820	57		877	990
4150	12″		11.40	1.404		955	59.50		1,014.50	1,150
4152	14″		10.26	1.559		1,125	66.50		1,191.50	1,350
4154	16″		9.12	1.754		1,275	74.50		1,349.50	1,525
4156	18″		7.98	2.005		1,500	85		1,585	1,775
4158	20″		9.69	1.651		1,675	70		1,745	1,950
4160	22″		8.26	1.937		1,875	82.50		1,957.50	2,200
4162	24″	↓	6.84	2.339	↓	2,075	99.50		2,174.50	2,425

23 31 13.17 Round Grease Duct		Crew	Daily Output	Labor-Hours	Unit	2009 Bare Costs Material	Labor	Equipment	Total	Total Incl O&P
4164	26"	Q-9	6.55	2.443	Ea.	2,350	104		2,454	2,725
4166	28"	Q-10	6.27	3.828		2,600	169		2,769	3,100
4168	30"		5.98	4.013		3,075	177		3,252	3,675
4170	32"		5.70	4.211		3,550	185		3,735	4,200
4172	36"		5.13	4.678		4,150	206		4,356	4,875
4180	Cleanout Tee Cap									
4182	5"	Q-9	22.80	.702	Ea.	152	30		182	214
4184	6"		21.09	.759		180	32		212	248
4186	8"		19.38	.826		183	35		218	255
4188	10"		18.24	.877		201	37.50		238.50	280
4190	12"		17.10	.936		210	40		250	292
4192	14"		15.96	1.003		242	42.50		284.50	330
4194	16"		14.25	1.123		265	47.50		312.50	365
4196	18"		13.68	1.170		299	49.50		348.50	405
4198	20"	Q-10	15.39	1.559		325	68.50		393.50	460
4200	22"		12.54	1.914		365	84.50		449.50	530
4202	24"		11.97	2.005		410	88.50		498.50	585
4204	26"		11.40	2.105		455	92.50		547.50	650
4206	28"		10.83	2.216		505	97.50		602.50	705
4208	30"		10.26	2.339		560	103		663	780
4210	32"		9.69	2.477		620	109		729	850
4212	36"		8.55	2.807		770	124		894	1,025
4220	Elbow, 45°									
4222	5"	Q-9	18.24	.877	Ea.	284	37.50		321.50	370
4224	6"		17.10	.936		335	40		375	425
4226	8"		14.82	1.080		380	46		426	485
4228	10"		13.68	1.170		425	49.50		474.50	540
4230	12"		12.54	1.276		485	54		539	615
4232	14"		11.97	1.337		555	57		612	700
4234	16"		11.40	1.404		625	59.50		684.50	780
4236	18"		10.83	1.477		710	63		773	875
4238	20"	Q-10	10.26	2.339		810	103		913	1,050
4240	22"		9.69	2.477		915	109		1,024	1,175
4242	24"		9.12	2.632		1,050	116		1,166	1,325
4244	26"		8.83	2.718		1,100	120		1,220	1,400
4246	28"		8.55	2.807		1,175	124		1,299	1,500
4248	30"		8.26	2.906		1,375	128		1,503	1,725
4250	32"		7.98	3.008		1,575	132		1,707	1,950
4252	36"		6.84	3.509		2,225	155		2,380	2,700
4260	Elbow, 90°									
4262	5"	Q-9	18.24	.877	Ea.	450	37.50		487.50	555
4264	6"		17.10	.936		530	40		570	640
4266	8"		14.82	1.080		580	46		626	710
4268	10"		13.68	1.170		670	49.50		719.50	810
4270	12"		12.54	1.276		760	54		814	920
4272	14"		11.97	1.337		890	57		947	1,075
4274	16"		11.40	1.404		990	59.50		1,049.50	1,200
4276	18"		10.83	1.477		1,150	63		1,213	1,375
4278	20"	Q-10	10.26	2.339		1,275	103		1,378	1,575
4280	22"		9.69	2.477		1,450	109		1,559	1,750
4282	24"		9.12	2.632		1,600	116		1,716	1,925
4284	26"		8.83	2.718		1,800	120		1,920	2,150
4286	28"		8.55	2.807		2,000	124		2,124	2,400

23 31 13.17 Round Grease Duct		Crew	Daily Output	Labor-Hours	Unit	Material	2009 Bare Costs Labor	Equipment	Total	Total Incl O&P
4288	30″	Q-10	8.26	2.906	Ea.	2,325	128		2,453	2,750
4290	32″		7.98	3.008		2,650	132		2,782	3,125
4292	36″		6.84	3.509		3,250	155		3,405	3,825
4300	Support strap									
4302	5″	Q-9	27	.593	Ea.	22.50	25		47.50	63
4304	6″		25.50	.627		22.50	26.50		49	65.50
4306	8″		22	.727		23.50	31		54.50	73
4308	10″		20.50	.780		23.50	33		56.50	76.50
4310	12″		18.80	.851		23.50	36		59.50	81
4312	14″		17.90	.894		23.50	38		61.50	84
4314	16″		17	.941		23.50	40		63.50	87
4316	18″		16.20	.988		24.50	42		66.50	91.50
4318	20″	Q-10	15.40	1.558		25.50	68.50		94	134
4320	22″	″	14.50	1.655		25.50	73		98.50	140
4340	Plate Support Assemaly									
4342	5″	Q-9	17.10	.936	Ea.	106	40		146	177
4344	6″		14.82	1.080		124	46		170	207
4346	8″		12.54	1.276		139	54		193	237
4348	10″		11.40	1.404		152	59.50		211.50	260
4350	12″		10.26	1.559		158	66.50		224.50	276
4352	14″		9.69	1.651		193	70		263	320
4354	16″		9.12	1.754		201	74.50		275.50	335
4356	18″		8.55	1.871		214	79.50		293.50	355
4358	20″	Q-10	9.12	2.632		225	116		341	425
4360	22″		8.55	2.807		232	124		356	445
4362	24″		7.98	3.008		238	132		370	465
4364	26″		7.69	3.121		292	137		429	530
4366	28″		7.41	3.239		330	143		473	585
4368	30″		7.12	3.371		335	148		483	600
4370	32″		6.84	3.509		380	155		535	655
4372	36″		5.70	4.211		475	185		660	805
4380	Wall Guide Assembly									
4382	5″	Q-9	20.20	.792	Ea.	124	33.50		157.50	188
4384	6″		19.10	.838		135	35.50		170.50	203
4386	8″		16.50	.970		141	41		182	219
4388	10″		15.30	1.046		148	44.50		192.50	232
4390	12″		14.10	1.135		155	48		203	244
4392	14″		13.40	1.194		167	50.50		217.50	262
4394	16″		12.75	1.255		169	53.50		222.50	268
4396	18″		12.10	1.322		177	56		233	282
4398	20″	Q-10	11.50	2.087		186	92		278	345
4400	22″		10.90	2.202		191	97		288	360
4402	24″		10.20	2.353		193	104		297	370
4404	26″		9.90	2.424		201	107		308	385
4406	28″		9.60	2.500		216	110		326	410
4408	30″		9.30	2.581		225	114		339	420
4410	32″		9	2.667		235	117		352	440
4412	36″		7.60	3.158		256	139		395	495
4420	Hood Transition, Flanged or Unflanged									
4422	5″	Q-9	16.60	.964	Ea.	42.50	41		83.50	110
4424	6″		15.60	1.026		50	43.50		93.50	122
4426	8″		13.50	1.185		57	50.50		107.50	140
4428	10″		12.50	1.280		63.50	54.50		118	154

23 31 13.17 Round Grease Duct		Crew	Daily Output	Labor-Hours	Unit	Material	2009 Bare Costs Labor	Equipment	Total	Total Incl O&P
4430	12"	Q-9	11.40	1.404	Ea.	73.50	59.50		133	173
4432	14"		10.90	1.468		84.50	62.50		147	189
4434	16"		10.40	1.538		94.50	65.50		160	205
4436	18"		9.80	1.633		109	69.50		178.50	227
4438	20"	Q-10	9.40	2.553		127	112		239	315
4440	22"		8.80	2.727		149	120		269	350
4442	24"		8.30	2.892		168	127		295	380
4444	26"		8.10	2.963		179	131		310	400
4446	28"		7.80	3.077		193	136		329	420
4448	30"		7.50	3.200		208	141		349	445
4450	32"		7.30	3.288		218	145		363	465
4452	36"		6.20	3.871		264	171		435	550
4460	Fan Adapter									
4462	5"	Q-9	16	1	Ea.	240	42.50		282.50	330
4464	6"		15	1.067		290	45.50		335.50	390
4466	8"		13	1.231		325	52.50		377.50	435
4468	10"		12	1.333		360	56.50		416.50	485
4470	12"		11	1.455		415	62		477	555
4472	14"		10.50	1.524		480	64.50		544.50	630
4474	16"		10	1.600		550	68		618	705
4476	18"		9.50	1.684		625	71.50		696.50	795
4478	20"	Q-10	9	2.667		730	117		847	980
4480	22"		8.50	2.824		850	124		974	1,125
4482	24"		8	3		875	132		1,007	1,175
4484	26"		7.70	3.117		905	137		1,042	1,200
4486	28"		7.50	3.200		925	141		1,066	1,250
4488	30"		7.20	3.333		955	147		1,102	1,275
4490	32"		7	3.429		985	151		1,136	1,300
4492	36"		6	4		1,025	176		1,201	1,400
4500	Tapered Increaser/Reducer									
4510	5" x 10" dia	Q-9	13.68	1.170	Ea.	465	49.50		514.50	585
4520	5" x 20" dia		10.30	1.553		725	66		791	900
4530	6" x 10" dia		13.40	1.194		465	50.50		515.50	590
4540	6" x 20" dia		10.10	1.584		765	67.50		832.50	945
4550	8" x 10" dia		13.20	1.212		440	51.50		491.50	560
4560	8" x 20" dia		9.90	1.616		765	68.50		833.50	950
4570	10" x 20" dia		10.60	1.509		825	64		889	1,000
4580	10" x 28" dia	Q-10	8.60	2.791		1,125	123		1,248	1,450
4590	12" x 20" dia		8.80	2.727		870	120		990	1,150
4600	14" x 28" dia		8.40	2.857		1,300	126		1,426	1,625
4610	16" x 20" dia		8.60	2.791		870	123		993	1,150
4620	16" x 30" dia		8.30	2.892		1,425	127		1,552	1,775
4630	18" x 24" dia		8.80	2.727		1,050	120		1,170	1,325
4640	18" x 30" dia		8.10	2.963		1,425	131		1,556	1,775
4650	20" x 24" dia		8.30	2.892		995	127		1,122	1,300
4660	20" x 32" dia		8	3		1,525	132		1,657	1,875
4670	24" mm x 26" dia		8	3		995	132		1,127	1,300
4680	24" mm x 36" dia		7.30	3.288		1,750	145		1,895	2,150
4690	28" x 30" dia		7.90	3.038		1,200	134		1,334	1,525
4700	28" x 36" dia		7.10	3.380		1,750	149		1,899	2,150
4710	30" x 36" dia		6.90	3.478		1,800	153		1,953	2,200
4730	Many intermediate standard sizes of tapered fittings are available									

23 31 13 – Metal Ducts

23 31 13.19 Metal Duct Fittings		Crew	Daily Output	Labor-Hours	Unit	Material	Labor	Equipment	Total	Total Incl O&P
						2009 Bare Costs				
0010	**METAL DUCT FITTINGS**									
0050	Air extractors, 12" x 4"	1 Shee	24	.333	Ea.	15.65	15.75		31.40	41
0100	8" x 6"		22	.364		15.65	17.15		32.80	43.50
0120	12" x 6"		21	.381		21.50	18		39.50	51.50
0140	16" x 6"		20	.400		27	18.90		45.90	59
0160	24" x 6"		18	.444		45.50	21		66.50	82.50
0180	12" x 8"		20	.400		24.50	18.90		43.40	55.50
0200	20" x 8"		16	.500		35.50	23.50		59	75.50
0240	18" x 10"		14	.571		34.50	27		61.50	79.50
0260	24" x 10"		12	.667		44.50	31.50		76	97.50
0280	24" x 12"		10	.800		48	38		86	111
0300	30" x 12"		8	1		60	47		107	139
0350	Duct collar, spin-in type									
0352	Without damper									
0354	Round									
0356	4"	1 Shee	32	.250	Ea.	2.01	11.80		13.81	20.50
0357	5"		30	.267		2.01	12.60		14.61	21.50
0358	6"		28	.286		2.01	13.50		15.51	23
0359	7"		26	.308		2.31	14.50		16.81	25
0360	8"		24	.333		2.31	15.75		18.06	26.50
0361	9"		22	.364		3.49	17.15		20.64	30.50
0362	10"		20	.400		3.49	18.90		22.39	33
0363	12"		17	.471		4.48	22		26.48	39
2000	Fabrics for flexible connections, with metal edge		100	.080	L.F.	2.13	3.78		5.91	8.15
2100	Without metal edge		160	.050	"	1.53	2.36		3.89	5.30

23 31 16 – Nonmetal Ducts

23 31 16.13 Fibrous-Glass Ducts

		Crew	Daily Output	Labor-Hours	Unit	Material	Labor	Equipment	Total	Total Incl O&P
0010	**FIBROUS-GLASS DUCTS** R233100-20									
1280	Add to labor for elevated installation									
1282	of prefabricated (purchased) ductwork									
1283	10' to 15' high						10%			
1284	15' to 20' high						20%			
1285	20' to 25' high						25%			
1286	25' to 30' high						35%			
1287	30' to 35' high						40%			
1288	35' to 40' high						50%			
1289	Over 40' high						55%			
3490	Rigid fiberglass duct board, foil reinf. kraft facing									
3500	Rectangular, 1" thick, alum. faced, (FRK), std. weight	Q-10	350	.069	SF Surf	.73	3.02		3.75	5.45

23 31 16.16 Thermoset Fiberglass-Reinforced Plastic Ducts

		Crew	Daily Output	Labor-Hours	Unit	Material	Labor	Equipment	Total	Total Incl O&P
0010	**THERMOSET FIBERGLASS-REINFORCED PLASTIC DUCTS** R233100-20									
1280	Add to labor for elevated installation									
1282	of prefabricated (purchased) ductwork									
1283	10' to 15' high						10%			
1284	15' to 20' high						20%			
1285	20' to 25' high						25%			
1286	25' to 30' high						35%			
1287	30' to 35' high						40%			
1288	35' to 40' high						50%			
1289	Over 40' high						55%			
3550	Rigid fiberglass reinforced plastic, FM approved									

23 31 HVAC Ducts and Casings

23 31 16 – Nonmetal Ducts

23 31 16.16 Thermoset Fiberglass-Reinforced Plastic Ducts		Crew	Daily Output	Labor-Hours	Unit	Material	2009 Bare Costs Labor	Equipment	Total	Total Incl O&P
3552	for acid fume and smoke exhaust system, nonflammable									
3554	Straight, 4″ diameter	Q-9	200	.080	L.F.	10.60	3.40		14	16.90
3555	6″ diameter		146	.110		13.90	4.66		18.56	22.50
3556	8″ diameter		106	.151		16.85	6.40		23.25	28.50
3557	10″ diameter		85	.188		20	8		28	34.50
3558	12″ diameter		68	.235		23.50	10		33.50	41.50
3559	14″ diameter		50	.320		35	13.60		48.60	59.50
3560	16″ diameter		40	.400		39	17		56	69
3561	18″ diameter		31	.516		43.50	22		65.50	81
3562	20″ diameter	Q-10	44	.545		48	24		72	89.50
3563	22″ diameter		40	.600		51.50	26.50		78	97.50
3564	24″ diameter		37	.649		55.50	28.50		84	105
3565	26″ diameter		34	.706		60	31		91	114
3566	28″ diameter		28	.857		64.50	38		102.50	129
3567	30″ diameter		24	1		68.50	44		112.50	144
3568	32″ diameter		22.60	1.062		73	47		120	152
3569	34″ diameter		21.60	1.111		77	49		126	160
3570	36″ diameter		20.60	1.165		81.50	51.50		133	169
3571	38″ diameter		19.80	1.212		125	53.50		178.50	220
3572	42″ diameter		18.30	1.311		140	58		198	243
3573	46″ diameter		17	1.412		155	62		217	266
3574	50″ diameter		15.90	1.509		168	66.50		234.50	287
3575	54″ diameter		14.90	1.611		185	71		256	315
3576	60″ diameter		13.60	1.765		212	77.50		289.50	355
3577	64″ diameter		12.80	1.875		310	82.50		392.50	465
3578	68″ diameter		12.20	1.967		330	86.50		416.50	500
3579	72″ diameter		11.90	2.017		355	89		444	525
3584	Note: joints are cemented with									
3586	fiberglass resin, included in material cost.									
3590	Elbow, 90°, 4″ diameter	Q-9	20.30	.788	Ea.	51	33.50		84.50	108
3591	6″ diameter		13.80	1.159		71.50	49.50		121	155
3592	8″ diameter		10	1.600		79	68		147	192
3593	10″ diameter		7.80	2.051		103	87		190	248
3594	12″ diameter		6.50	2.462		131	105		236	305
3595	14″ diameter		5.50	2.909		161	124		285	365
3596	16″ diameter		4.80	3.333		191	142		333	430
3597	18″ diameter		4.40	3.636		274	154		428	540
3598	20″ diameter	Q-10	5.90	4.068		289	179		468	590
3599	22″ diameter		5.40	4.444		355	196		551	690
3600	24″ diameter		4.90	4.898		410	216		626	780
3601	26″ diameter		4.60	5.217		415	230		645	810
3602	28″ diameter		4.20	5.714		515	252		767	955
3603	30″ diameter		3.90	6.154		575	271		846	1,050
3604	32″ diameter		3.70	6.486		640	286		926	1,[illegible]50
3605	34″ diameter		3.45	6.957		755	305		1,060	1,300
3606	36″ diameter		3.20	7.500		865	330		1,195	1,475
3607	38″ diameter		3.07	7.818		1,225	345		1,570	1,875
3608	42″ diameter		2.70	8.889		1,475	390		1,865	2,225
3609	46″ diameter		2.54	9.449		1,725	415		2,140	2,550
3610	50″ diameter		2.33	10.300		2,000	455		2,455	2,900
3611	54″ diameter		2.14	11.215		2,300	495		2,795	3,275
3612	60″ diameter		1.94	12.371		3,050	545		3,595	4,200
3613	64″ diameter		1.82	13.187		4,425	580		5,005	5,775

23 31 16.16 Thermoset Fiberglass-Reinforced Plastic Ducts		Crew	Daily Output	Labor-Hours	Unit	2009 Bare Costs Material	Labor	Equipment	Total	Total Incl O&P
3614	68" diameter	Q-10	1.71	14.035	Ea.	4,925	620		5,545	6,375
3615	72" diameter	↓	1.62	14.815		5,450	655		6,105	6,975
3626	Elbow, 45°, 4" diameter	Q-9	22.30	.717		49	30.50		79.50	101
3627	6" diameter		15.20	1.053		68	44.50		112.50	144
3628	8" diameter		11	1.455		72.50	62		134.50	175
3629	10" diameter		8.60	1.860		76.50	79		155.50	206
3630	12" diameter		7.20	2.222		83.50	94.50		178	237
3631	14" diameter		6.10	2.623		102	111		213	284
3632	16" diameter		5.28	3.030		122	129		251	330
3633	18" diameter	↓	4.84	3.306		143	140		283	375
3634	20" diameter	Q-10	6.50	3.692		165	163		328	430
3635	22" diameter		5.94	4.040		226	178		404	520
3636	24" diameter		5.40	4.444		257	196		453	585
3637	26" diameter		5.06	4.743		291	209		500	640
3638	28" diameter		4.62	5.195		315	229		544	695
3639	30" diameter		4.30	5.581		365	246		611	780
3640	32" diameter		4.07	5.897		405	260		665	845
3641	34" diameter		3.80	6.316		480	278		758	955
3642	36" diameter		3.52	6.818		550	300		850	1,075
3643	38" diameter		3.38	7.101		775	315		1,090	1,325
3644	42" diameter		2.97	8.081		925	355		1,280	1,575
3645	46" diameter		2.80	8.571		1,075	380		1,455	1,775
3646	50" diameter		2.56	9.375		1,250	415		1,665	2,000
3647	54" diameter		2.35	10.213		1,425	450		1,875	2,275
3648	60" diameter		2.13	11.268		1,925	495		2,420	2,875
3649	64" diameter		2	12		2,775	530		3,305	3,875
3650	68" diameter		1.88	12.766		3,075	560		3,635	4,250
3651	72" diameter	↓	1.78	13.483		3,400	595		3,995	4,650
3660	Tee, 90°, 4" diameter	Q-9	14.06	1.138		28	48.50		76.50	106
3661	6" diameter		9.52	1.681		43	71.50		114.50	157
3662	8" diameter		6.98	2.292		58.50	97.50		156	215
3663	10" diameter		5.45	2.936		76.50	125		201.50	276
3664	12" diameter		4.60	3.478		95.50	148		243.50	330
3665	14" diameter		3.86	4.145		139	176		315	425
3666	16" diameter		3.40	4.706		165	200		365	490
3667	18" diameter	↓	3.05	5.246		196	223		419	560
3668	20" diameter	Q-10	4.12	5.825		229	257		486	645
3669	22" diameter		3.78	6.349		262	280		542	720
3670	24" diameter		3.46	6.936		295	305		600	795
3671	26" diameter		3.20	7.500		335	330		665	880
3672	28" diameter		2.96	8.108		375	355		730	960
3673	30" diameter		2.75	8.727		420	385		805	1,050
3674	32" diameter		2.54	9.449		465	415		880	1,150
3675	34" diameter		2.40	10		505	440		945	1,250
3676	36" diameter		2.27	10.573		555	465		1,020	1,325
3677	38" diameter		2.15	11.163		835	490		1,325	1,675
3678	42" diameter		1.91	12.565		1,000	555		1,555	1,950
3679	46" diameter		1.78	13.483		1,175	595		1,770	2,225
3680	50" diameter		1.63	14.724		1,375	650		2,025	2,525
3681	54" diameter		1.50	16		1,600	705		2,305	2,825
3682	60" diameter		1.35	17.778		2,000	785		2,785	3,400
3683	64" diameter		1.27	18.898		3,000	830		3,830	4,575
3684	68" diameter	↓	1.20	20	↓	3,350	880		4,230	5,025

23 31 16 – Nonmetal Ducts

23 31 16.16 Thermoset Fiberglass-Reinforced Plastic Ducts		Crew	Daily Output	Labor-Hours	Unit	Material	2009 Bare Costs Labor	Equipment	Total	Total Incl O&P
3685	72" diameter	Q-10	1.13	21.239	Ea.	3,750	935		4,685	5,575
3690	For Y @ 45°, add					22%				
3700	Blast gate, 4" diameter	Q-9	22.30	.717		136	30.50		166.50	196
3701	6" diameter		15.20	1.053		154	44.50		198.50	238
3702	8" diameter		11	1.455		207	62		269	320
3703	10" diameter		8.60	1.860		236	79		315	380
3704	12" diameter		7.20	2.222		289	94.50		383.50	460
3705	14" diameter		6.10	2.623		320	111		431	520
3706	16" diameter		5.28	3.030		355	129		484	590
3707	18" diameter		4.84	3.306		390	140		530	645
3708	20" diameter	Q-10	6.50	3.692		420	163		583	710
3709	22" diameter		5.94	4.040		455	178		633	775
3710	24" diameter		5.40	4.444		485	196		681	835
3711	26" diameter		5.06	4.743		520	209		729	895
3712	28" diameter		4.62	5.195		560	229		789	965
3713	30" diameter		4.30	5.581		595	246		841	1,025
3714	32" diameter		4.07	5.897		650	260		910	1,125
3715	34" diameter		3.80	6.316		710	278		988	1,200
3716	36" diameter		3.52	6.818		765	300		1,065	1,300
3717	38" diameter		3.38	7.101		1,025	315		1,340	1,600
3718	42" diameter		2.97	8.081		1,175	355		1,530	1,850
3719	46" diameter		2.80	8.571		1,350	380		1,730	2,075
3720	50" diameter		2.56	9.375		1,550	415		1,965	2,325
3721	54" diameter		2.35	10.213		1,750	450		2,200	2,625
3722	60" diameter		2.13	11.268		2,050	495		2,545	3,025
3723	64" diameter		2	12		2,275	530		2,805	3,325
3724	68" diameter		1.88	12.766		2,500	560		3,060	3,625
3725	72" diameter		1.78	13.483		2,750	595		3,345	3,950
3740	Reducers, 4" diameter	Q-9	24.50	.653		34	27.50		61.50	80
3741	6" diameter		16.70	.958		50.50	40.50		91	119
3742	8" diameter		12	1.333		68.50	56.50		125	163
3743	10" diameter		9.50	1.684		84.50	71.50		156	203
3744	12" diameter		7.90	2.025		101	86		187	244
3745	14" diameter		6.70	2.388		118	101		219	286
3746	16" diameter		5.80	2.759		135	117		252	330
3747	18" diameter		5.30	3.019		153	128		281	365
3748	20" diameter	Q-10	7.20	3.333		169	147		316	410
3749	22" diameter		6.50	3.692		273	163		436	550
3750	24" diameter		5.90	4.068		295	179		474	600
3751	26" diameter		5.60	4.286		320	189		509	640
3752	28" diameter		5.10	4.706		350	207		557	705
3753	30" diameter		4.70	5.106		375	225		600	755
3754	32" diameter		4.50	5.333		400	235		635	800
3755	34" diameter		4.20	5.714		425	252		677	850
3756	36" diameter		3.90	6.154		450	271		721	910
3757	38" diameter		3.70	6.486		680	286		966	1,175
3758	42" diameter		3.30	7.273		750	320		1,070	1,325
3759	46" diameter		3.10	7.742		820	340		1,160	1,425
3760	50" diameter		2.80	8.571		890	380		1,270	1,550
3761	54" diameter		2.60	9.231		965	405		1,370	1,675
3762	60" diameter		2.34	10.256		1,075	450		1,525	1,900
3763	64" diameter		2.20	10.909		1,675	480		2,155	2,600
3764	68" diameter		2.06	11.650		1,775	515		2,290	2,750

23 31 16.16 Thermoset Fiberglass-Reinforced Plastic Ducts		Crew	Daily Output	Labor-Hours	Unit	2009 Bare Costs Material	Labor	Equipment	Total	Total Incl O&P
3765	72″ diameter	Q-10	1.96	12.245	Ea.	1,900	540		2,440	2,900
3780	Flange, per each, 4″ diameter	Q-9	40.60	.394		82	16.75		98.75	117
3781	6″ diameter		27.70	.578		96	24.50		120.50	144
3782	8″ diameter		20.50	.780		120	33		153	183
3783	10″ diameter		16.30	.982		144	41.50		185.50	222
3784	12″ diameter		13.90	1.151		165	49		214	258
3785	14″ diameter		11.30	1.416		191	60		251	305
3786	16″ diameter		9.70	1.649		215	70		285	345
3787	18″ diameter		8.30	1.928		238	82		320	390
3788	20″ diameter	Q-10	11.30	2.124		262	93.50		355.50	430
3789	22″ diameter		10.40	2.308		286	102		388	470
3790	24″ diameter		9.50	2.526		310	111		421	510
3791	26″ diameter		8.80	2.727		335	120		455	555
3792	28″ diameter		8.20	2.927		360	129		489	595
3793	30″ diameter		7.60	3.158		380	139		519	630
3794	32″ diameter		7.20	3.333		405	147		552	670
3795	34″ diameter		6.80	3.529		430	155		585	710
3796	36″ diameter		6.40	3.750		455	165		620	755
3797	38″ diameter		6.10	3.934		480	173		653	790
3798	42″ diameter		5.50	4.364		520	192		712	870
3799	46″ diameter		5	4.800		570	211		781	955
3800	50″ diameter		4.60	5.217		620	230		850	1,025
3801	54″ diameter		4.20	5.714		665	252		917	1,125
3802	60″ diameter		3.80	6.316		740	278		1,018	1,250
3803	64″ diameter		3.60	6.667		785	294		1,079	1,300
3804	68″ diameter		3.40	7.059		830	310		1,140	1,400
3805	72″ diameter		3.20	7.500		880	330		1,210	1,475

23 31 16.19 PVC Ducts

		Crew	Daily Output	Labor-Hours	Unit	Material	Labor	Equipment	Total	Total Incl O&P
0010	**PVC DUCTS**									
4000	Rigid plastic, corrosive fume resistant PVC									
4020	Straight, 6″ diameter	Q-9	220	.073	L.F.	7.50	3.09		10.59	13
4030	7″ diameter		160	.100		11.10	4.25		15.35	18.80
4040	8″ diameter		160	.100		11.10	4.25		15.35	18.80
4050	9″ diameter		140	.114		12.55	4.86		17.41	21.50
4060	10″ diameter		120	.133		14	5.65		19.65	24
4070	12″ diameter		100	.160		16.35	6.80		23.15	28.50
4080	14″ diameter		70	.229		21	9.70		30.70	38.50
4090	16″ diameter		62	.258		24.50	10.95		35.45	44
4100	18″ diameter		58	.276		38	11.70		49.70	60
4110	20″ diameter	Q-10	75	.320		49	14.10		63.10	75.50
4120	22″ diameter		60	.400		56	17.60		73.60	88.50
4130	24″ diameter		55	.436		63	19.20		82.20	99
4140	26″ diameter		42	.571		71.50	25		96.50	117
4150	28″ diameter		35	.686		77	30		107	131
4160	30″ diameter		28	.857		83	38		121	150
4170	36″ diameter		23	1.043		99.50	46		145.50	180
4180	42″ diameter		20	1.200		116	53		169	209
4200	48″ diameter		18	1.333		132	58.50		190.50	237
4250	Coupling, 6″ diameter	Q-9	88	.182	Ea.	9.45	7.70		17.15	22.50
4260	7″ diameter		82	.195		10.10	8.30		18.40	24
4270	8″ diameter		78	.205		10.55	8.70		19.25	25
4280	9″ diameter		74	.216		11.45	9.20		20.65	27

23 31 HVAC Ducts and Casings

23 31 16 – Nonmetal Ducts

23 31 16.19 PVC Ducts		Crew	Daily Output	Labor-Hours	Unit	2009 Bare Costs Material	Labor	Equipment	Total	Total Incl O&P
4290	10″ diameter	Q-9	70	.229	Ea.	11.45	9.70		21.15	27.50
4300	12″ diameter		55	.291		12.30	12.35		24.65	32.50
4310	14″ diameter		44	.364		14.95	15.45		30.40	40.50
4320	16″ diameter		40	.400		17.70	17		34.70	45.50
4330	18″ diameter		38	.421		20.50	17.90		38.40	50
4340	20″ diameter	Q-10	55	.436		35.50	19.20		54.70	68.50
4350	22″ diameter		50	.480		41	21		62	77.50
4360	24″ diameter		45	.533		46.50	23.50		70	87
4370	26″ diameter		39	.615		49	27		76	95
4380	28″ diameter		33	.727		51	32		83	106
4390	30″ diameter		27	.889		56	39		95	122
4400	36″ diameter		25	.960		88.50	42.50		131	162
4410	42″ diameter		23	1.043		99	46		145	180
4420	48″ diameter		21	1.143		115	50.50		165.50	205
4470	Elbow, 90°, 6″ diameter	Q-9	44	.364		74.50	15.45		89.95	106
4480	7″ diameter		36	.444		80	18.90		98.90	117
4490	8″ diameter		28	.571		80	24.50		104.50	126
4500	9″ diameter		22	.727		110	31		141	169
4510	10″ diameter		18	.889		130	38		168	201
4520	12″ diameter		15	1.067		177	45.50		222.50	264
4530	14″ diameter		11	1.455		242	62		304	360
4540	16″ diameter		10	1.600		292	68		360	425
4550	18″ diameter	Q-10	15	1.600		315	70.50		385.50	460
4560	20″ diameter		14	1.714		335	75.50		410.50	485
4570	22″ diameter		13	1.846		425	81.50		506.50	595
4580	24″ diameter		12	2		480	88		568	665
4590	26″ diameter		11	2.182		550	96		646	755
4600	28″ diameter		10	2.400		650	106		756	880
4610	30″ diameter		9	2.667		770	117		887	1,025
4620	36″ diameter		7	3.429		1,050	151		1,201	1,375
4630	42″ diameter		5	4.800		1,450	211		1,661	1,925
4640	48″ diameter		4	6		1,900	264		2,164	2,500
4750	Elbow 45°, use 90° and deduct					45%				

23 33 Air Duct Accessories

23 33 13 – Dampers

23 33 13.13 Volume-Control Dampers		Crew	Daily Output	Labor-Hours	Unit	Material	Labor	Equipment	Total	Total Incl O&P
0010	**VOLUME-CONTROL DAMPERS**									
5990	Multi-blade dampers, opposed blade, 8″ x 6″	1 Shee	24	.333	Ea.	20	15.75		35.75	46
5994	8″ x 8″		22	.364		21	17.15		38.15	49.50
5996	10″ x 10″		21	.381		24.50	18		42.50	54.50
6000	12″ x 12″		21	.381		27.50	18		45.50	57.50
6020	12″ x 18″		18	.444		36.50	21		57.50	73
6030	14″ x 10″		20	.400		26.50	18.90		45.40	58.50
6031	14″ x 14″		17	.471		32.50	22		54.50	69.50
6032	16″ x 10″		18	.444		29	21		50	64
6033	16″ x 12″		17	.471		32.50	22		54.50	69.50
6035	16″ x 16″		16	.500		40.50	23.50		64	81
6036	18″ x 14″		16	.500		40.50	23.50		64	81
6037	18″ x 16″		15	.533		44.50	25		69.50	88
6038	18″ x 18″		15	.533		48	25		73	92

23 33 13.13 Volume-Control Dampers		Crew	Daily Output	Labor-Hours	Unit	Material	Labor	Equipment	Total	Total Incl O&P
						2009 Bare Costs				
6040	18" x 24"	1 Shee	12	.667	Ea.	62	31.50		93.50	117
6060	18" x 28"		10	.800		72	38		110	137
6068	20" x 6"		18	.444		26	21		47	61
6069	20" x 8"		16	.500		30	23.50		53.50	70
6070	20" x 16"		14	.571		48	27		75	94.50
6071	20" x 18"		13	.615		52	29		81	102
6072	20" x 20"		13	.615		57.50	29		86.50	108
6073	22" x 6"		20	.400		26.50	18.90		45.40	58.50
6074	22" x 18"		14	.571		57.50	27		84.50	105
6075	22" x 20"		12	.667		62.50	31.50		94	118
6076	24" x 16"		11	.727		57	34.50		91.50	116
6077	22" x 22"		10	.800		67	38		105	132
6078	24" x 20"		8	1		67	47		114	146
6080	24" x 24"		8	1		78.50	47		125.50	159
6100	24" x 28"		6	1.333		92	63		155	198
6110	26" x 26"		6	1.333		86.50	63		149.50	192
6120	28" x 28"	Q-9	11	1.455		98	62		160	203
6130	30" x 18"		10	1.600		83.50	68		151.50	197
6132	30" x 24"		7	2.286		104	97		201	264
6133	30" x 30"		6.60	2.424		124	103		227	295
6135	32" x 32"		6.40	2.500		144	106		250	320
6150	34" x 34"		6	2.667		161	113		274	350
6151	36" x 12"		10	1.600		65.50	68		133.50	177
6152	36" x 16"		8	2		89.50	85		174.50	229
6156	36" x 32"		6.30	2.540		161	108		269	345
6157	36" x 34"		6.20	2.581		168	110		278	355
6158	36" x 36"		6	2.667		176	113		289	370
6160	44" x 28"		5.80	2.759		181	117		298	380
6180	48" x 36"		5.60	2.857		237	121		358	450
6200	56" x 36"		5.40	2.963		287	126		413	510
6220	60" x 36"		5.20	3.077		305	131		436	535
6240	60" x 44"		5	3.200		360	136		496	605
7500	Variable volume modulating motorized damper, incl. elect. mtr.									
7504	8" x 6"	1 Shee	15	.533	Ea.	109	25		134	159
7506	10" x 6"		14	.571		109	27		136	162
7510	10" x 10"		13	.615		112	29		141	169
7520	12" x 12"		12	.667		117	31.50		148.50	178
7522	12" x 16"		11	.727		118	34.50		152.50	183
7524	16" x 10"		12	.667		115	31.50		146.50	176
7526	16" x 14"		10	.800		120	38		158	189
7528	16" x 18"		9	.889		125	42		167	202
7540	18" x 12"		10	.800		120	38		158	190
7542	18" x 18"		8	1		127	47		174	213
7544	20" x 14"		8	1		130	47		177	216
7546	20" x 18"		7	1.143		136	54		190	233
7560	24" x 12"		8	1		132	47		179	218
7562	24" x 18"		7	1.143		143	54		197	240
7568	28" x 10"		7	1.143		132	54		186	228
7580	28" x 16"		6	1.333		149	63		212	261
7590	30" x 14"		5	1.600		140	75.50		215.50	270
7600	30" x 18"		4	2		180	94.50		274.50	345
7610	30" x 24"		3.80	2.105		239	99.50		338.50	415
7690	48" x 48"	Q-9	6	2.667		740	113		853	985

23 33 13.13 Volume-Control Dampers		Crew	Daily Output	Labor-Hours	Unit	Material	Labor	Equipment	Total	Total Incl O&P
						2009 Bare Costs				
7694	6′ x 14′ w/2 motors	Q-9	3	5.333	Ea.	3,175	227		3,402	3,825
7700	For thermostat, add	1 Shee	8	1		40.50	47		87.50	117
7800	For transformer 40 VA capacity, add	"	16	.500		18	23.50		41.50	56.50
8000	Multi-blade dampers, parallel blade									
8100	8″ x 8″	1 Shee	24	.333	Ea.	71	15.75		86.75	102
8120	12″ x 8″		22	.364		71	17.15		88.15	105
8140	16″ x 10″		20	.400		91.50	18.90		110.40	129
8160	18″ x 12″		18	.444		104	21		125	147
8180	22″ x 12″		15	.533		106	25		131	155
8200	24″ x 16″		11	.727		116	34.50		150.50	181
8220	28″ x 16″		10	.800		130	38		168	201
8240	30″ x 16″		8	1		134	47		181	221
8260	30″ x 18″		7	1.143		159	54		213	258
8400	Round damper, butterfly, vol. control w/lever lock reg.									
8410	6″ diam.	1 Shee	22	.364	Ea.	23	17.15		40.15	52
8412	7″ diam.		21	.381		24	18		42	53.50
8414	8″ diam.		20	.400		24.50	18.90		43.40	56
8416	9″ diam.		19	.421		25.50	19.85		45.35	58.50
8418	10″ diam.		18	.444		28	21		49	63.50
8420	12″ diam.		16	.500		29	23.50		52.50	68
8422	14″ diam.		14	.571		37.50	27		64.50	82.50
8424	16″ diam.		13	.615		40.50	29		69.50	89
8426	18″ diam.		12	.667		49.50	31.50		81	103
8428	20″ diam.		11	.727		60.50	34.50		95	120
8430	24″ diam.		10	.800		75.50	38		113.50	141
8432	30″ diam.		9	.889		106	42		148	181
8434	36″ diam.		8	1		141	47		188	228
8500	Round motor operated damper									
8510	6″ dia.	1 Shee	22	.364	Ea.	57	17.15		74.15	89
8512	7″ dia.		21	.381		60	18		78	93
8514	8″ dia.		20	.400		62	18.90		80.90	97
8516	10″ dia.		18	.444		68.50	21		89.50	108
8518	12″ dia.		16	.500		73.50	23.50		97	118
8520	14″ dia.		14	.571		83.50	27		110.50	134

23 33 13.16 Fire Dampers		Crew	Daily Output	Labor-Hours	Unit	Material	Labor	Equipment	Total	Total Incl O&P
0010	**FIRE DAMPERS**									
3000	Fire damper, curtain type, 1-1/2 hr rated, vertical, 6″ x 6″	1 Shee	24	.333	Ea.	21	15.75		36.75	47
3020	8″ x 6″		22	.364		21	17.15		38.15	49.50
3021	8″ x 8″		22	.364		21	17.15		38.15	49.50
3030	10″ x 10″		22	.364		21	17.15		38.15	49.50
3040	12″ x 6″		22	.364		21	17.15		38.15	49.50
3044	12″ x 12″		20	.400		21	18.90		39.90	52
3060	20″ x 6″		18	.444		24	21		45	59
3080	12″ x 8″		22	.364		24	17.15		41.15	53
3100	24″ x 8″		16	.500		29	23.50		52.50	68.50
3120	12″ x 10″		21	.381		21	18		39	50.50
3140	24″ x 10″		15	.533		31	25		56	73
3160	36″ x 10″		12	.667		39.50	31.50		71	92
3180	16″ x 12″		20	.400		23	18.90		41.90	54.50
3200	24″ x 12″		13	.615		32.50	29		61.50	80
3220	48″ x 12″		10	.800		49.50	38		87.50	113
3238	14″ x 14″		19	.421		28.50	19.85		48.35	62

23 33 13 – Dampers

23 33 13.16 Fire Dampers		Crew	Daily Output	Labor-Hours	Unit	2009 Bare Costs Material	Labor	Equipment	Total	Total Incl O&P
3240	16" x 14"	1 Shee	18	.444	Ea.	28.50	21		49.50	64
3260	24" x 14"		12	.667		35	31.50		66.50	87
3280	30" x 14"		11	.727		42	34.50		76.50	99
3298	16" x 16"		18	.444		32.50	21		53.50	68
3300	18" x 16"		17	.471		34.50	22		56.50	72
3320	24" x 16"		11	.727		37.50	34.50		72	94
3340	36" x 16"		10	.800		50	38		88	113
3356	18" x 18"		16	.500		36	23.50		59.50	76
3360	24" x 18"		10	.800		43.50	38		81.50	106
3380	48" x 18"		8	1		62.50	47		109.50	142
3398	20" x 20"		10	.800		37.50	38		75.50	99
3400	24" x 20"		8	1		40.50	47		87.50	118
3420	36" x 20"		7	1.143		53.50	54		107.50	142
3440	24" x 22"		7	1.143		42	54		96	129
3460	30" x 22"		6	1.333		44.50	63		107.50	146
3478	24" x 24"		8	1		43	47		90	120
3480	26" x 24"		7	1.143		45.50	54		99.50	133
3484	32" x 24"	Q-9	13	1.231		53.50	52.50		106	139
3500	48" x 24"		12	1.333		70	56.50		126.50	164
3520	28" x 26"		13	1.231		51	52.50		103.50	137
3540	30" x 28"		12	1.333		56.50	56.50		113	150
3560	48" x 30"		11	1.455		82.50	62		144.50	186
3580	48" x 36"		10	1.600		90	68		158	204
3590	32" x 40"		10	1.600		71.50	68		139.50	184
3600	44" x 44"		10	1.600		94	68		162	208
3620	48" x 48"		8	2		124	85		209	267
3700	U.L. label included in above									
3800	For horizontal operation, add				Ea.	20%				
3900	For cap for blades out of air stream, add					20%				
4000	For 10" 22 ga., U.L. approved sleeve, add					35%				
4100	For 10" 22 ga., U.L. sleeve, 100% free area, add					65%				
4200	For oversize openings group dampers									
4210	Fire damper, round, Type A									
4214	4" diam.	1 Shee	20	.400	Ea.	55	18.90		73.90	89.50
4216	5" diam.		16	.500		49.50	23.50		73	91
4218	6" diam.		14.30	.559		49.50	26.50		76	95
4220	7" diam.		13.30	.602		49.50	28.50		78	98
4222	8" diam.		11.40	.702		49.50	33		82.50	106
4224	9" diam.		10.80	.741		54	35		89	114
4226	10" diam.		10	.800		56	38		94	120
4228	12" diam.		9.50	.842		61.50	40		101.50	129
4230	14" diam.		8.20	.976		77.50	46		123.50	157
4232	16" diam.		7.60	1.053		77.50	49.50		127	162
4234	18" diam.		7.20	1.111		93	52.50		145.50	183
4236	20" diam.		6.40	1.250		97.50	59		156.50	198
4238	22" diam.		6.10	1.311		106	62		168	212
4240	24" diam.		5.70	1.404		110	66.50		176.50	223
4242	26" diam.		5.50	1.455		124	68.50		192.50	243
4244	28" diam.		5.40	1.481		131	70		201	252
4246	30" diam.		5.30	1.509		136	71.50		207.50	260
4500	Fire/smoke combination damper, louver type, UL									
4506	6" x 6"	1 Shee	24	.333	Ea.	36	15.75		51.75	63.50
4510	8" x 8"		22	.364		126	17.15		143.15	166

23 33 13.16 Fire Dampers		Crew	Daily Output	Labor-Hours	Unit	Material	2009 Bare Costs Labor	Equipment	Total	Total Incl O&P
4520	16" x 8"	1 Shee	20	.400	Ea.	138	18.90		156.90	131
4540	18" x 8"		18	.444		138	21		159	135
4560	20" x 8"		16	.500		148	23.50		171.50	199
4580	10" x 10"		21	.381		143	18		161	135
4600	24" x 10"		15	.533		168	25		193	224
4620	30" x 10"		12	.667		173	31.50		204.50	240
4640	12" x 12"		20	.400		149	18.90		167.90	193
4660	18" x 12"		18	.444		152	21		173	200
4680	24" x 12"		13	.615		160	29		189	221
4700	30" x 12"		11	.727		166	34.50		200.50	236
4720	14" x 14"		17	.471		160	22		182	211
4740	16" x 14"		18	.444		165	21		186	214
4760	20" x 14"		14	.571		168	27		195	227
4780	24" x 14"		12	.667		172	31.50		203.50	238
4800	30" x 14"		10	.800		182	38		220	258
4820	16" x 16"		16	.500		178	23.50		201.50	232
4840	20" x 16"		14	.571		182	27		209	243
4860	24" x 16"		11	.727		188	34.50		222.50	260
4880	30" x 16"		8	1		196	47		243	289
4900	18" x 18"		15	.533		176	25		201	232
5000	24" x 18"		10	.800		188	38		226	265
5020	36" x 18"		7	1.143		206	54		260	310
5040	20" x 20"		13	.615		196	29		225	261
5060	24" x 20"		8	1		201	47		248	294
5080	30" x 20"		8	1		211	47		258	305
5100	36" x 20"		7	1.143		222	54		276	325
5120	24" x 24"		8	1		217	47		264	310
5130	30" x 24"		7.60	1.053		206	49.50		255.50	305
5140	36" x 24"	Q-9	12	1.333		237	56.50		293.50	350
5141	36" x 30"		10	1.600		250	68		318	380
5142	40" x 36"		8	2		445	85		530	620
5143	48" x 48"		4	4		870	170		1,040	1,225
5150	Damper operator motor, 24 or 120 volt	1 Shee	16	.500		231	23.50		254.50	291

23 33 13.25 Exhaust Vent Damper		Crew	Daily Output	Labor-Hours	Unit	Material	Labor	Equipment	Total	Total Incl O&P
0010	**EXHAUST VENT DAMPER**, Auto, OB with elect. actuator.									
1110	6" x 6"	1 Shee	24	.333	Ea.	59	15.75		74.75	89
1114	8" x 6"		23	.348		60.50	16.40		76.90	92
1116	8" x 8"		22	.364		62.50	17.15		79.65	95.50
1118	10" x 6"		23	.348		62.50	16.40		78.90	94.50
1120	10" x 10"		22	.364		67	17.15		84.15	100
1122	12" x 6"		22	.364		63.50	17.15		80.65	96
1124	12" x 10"		21	.381		70.50	18		88.50	105
1126	12" x 12"		20	.400		74	18.90		92.90	111
1128	14" x 6"		21	.381		65.50	18		83.50	99.50
1130	14" x 10"		19	.421		74	19.85		93.85	112
1132	16" x 6"		21	.381		66	18		84	101
1134	16" x 8"		20	.400		73.50	18.90		92.40	110
1136	18" x 6"		20	.400		70	18.90		88.90	106
1138	18" x 8"		19	.421		74	19.85		93.85	112

23 33 13.28 Splitter Damper Assembly		Crew	Daily Output	Labor-Hours	Unit	Material	Labor	Equipment	Total	Total Incl O&P
0009	**SPLITTER DAMPER ASSEMBLY**									
0010	Self locking, 1' rod	1 Shee	24	.333	Ea.	21.50	15.75		37.25	48

23 33 Air Duct Accessories

23 33 13 – Dampers

23 33 13.28 Splitter Damper Assembly		Crew	Daily Output	Labor-Hours	Unit	Material	Labor	Equipment	Total	Total Incl O&P
						2009 Bare Costs				
7020	3′ rod	1 Shee	22	.364	Ea.	28	17.15		45.15	57.50
7040	4′ rod		20	.400		31.50	18.90		50.40	63.50
7060	6′ rod	↓	18	.444	↓	38	21		59	74.50

23 33 13.32 Relief Damper

Code	Description	Crew	Daily Output	Labor-Hours	Unit	Material	Labor	Equipment	Total	Total Incl O&P
0010	**RELIEF DAMPER**, Electronic bypass with tight seal									
8310	8″ x 6″	1 Shee	22	.364	Ea.	163	17.15		180.15	207
8314	10″ x 6″		22	.364		163	17.15		180.15	207
8318	10″ x 10″		21	.381		170	18		188	215
8322	12″ x 12″		20	.400		173	18.90		191.90	219
8326	12″ x 16″		19	.421		180	19.85		199.85	229
8330	16″ x 10″		20	.400		189	18.90		207.90	237
8334	16″ x 14″		18	.444		194	21		215	246
8338	16″ x 18″		17	.471		211	22		233	266
8342	18″ x 12″		18	.444		211	21		232	265
8346	18″ x 18″		15	.533		199	25		224	258
8350	20″ x 14″		14	.571		196	27		223	258
8354	20″ x 18″		13	.615		205	29		234	270
8358	24″ x 12″		13	.615		200	29		229	265
8362	24″ x 18″		10	.800		156	38		194	229
8363	24″ x 24″		10	.800		218	38		256	298
8364	24″ x 36″		9	.889		259	42		301	350
8365	24″ x 48″		6	1.333		290	63		353	415
8366	28″ x 10″		13	.615		144	29		173	203
8370	28″ x 16″		10	.800		221	38		259	300
8374	30″ x 14″		11	.727		151	34.50		185.50	219
8378	30″ x 18″		8	1		228	47		275	325
8382	30″ x 24″		6	1.333		242	63		305	365
8390	46″ x 36″		4	2		320	94.50		414.50	500
8394	48″ x 48″		3	2.667		350	126		476	580
8396	54″ x 36″	↓	2	4	↓	370	189		559	695

23 33 19 – Duct Silencers

23 33 19.10 Duct Silencers

Code	Description	Crew	Daily Output	Labor-Hours	Unit	Material	Labor	Equipment	Total	Total Incl O&P
0009	**DUCT SILENCERS**									
0010	Silencers, noise control for air flow, duct				MCFM	54			54	59.50
9004	Duct sound trap, packaged									
9010	12″ x 18″ x 36″, 2000 CFM	1 Shee	12	.667	Ea.	420	31.50		451.50	510
9011	24″ x 18″ x 36″, 9000 CFM	Q-9	80	.200		480	8.50		488.50	545
9012	24″ x 24″ x 36″, 9000 CFM		60	.267		720	11.35		731.35	810
9013	24″ x 30″ x 36″, 9000 CFM		50	.320		770	13.60		783.60	865
9014	24″ x 36″ x 36″, 9000 CFM		40	.400		960	17		977	1,075
9015	24″ x 48″ x 36″, 9000 CFM		28	.571		1,250	24.50		1,274.50	1,425
9018	36″ x 18″ x 36″, 6300 CFM		42	.381		960	16.20		976.20	1,075
9019	36″ x 36″ x 36″, 6300 CFM		21	.762		1,925	32.50		1,957.50	2,175
9020	36″ x 48″ x 36″, 6300 CFM		14	1.143		2,500	48.50		2,548.50	2,825
9021	36″ x 60″ x 36″, 6300 CFM		13	1.231		3,175	52.50		3,227.50	3,575
9022	48″ x 48″ x 36″, 6300 CFM		12	1.333		3,375	56.50		3,431.50	3,775
9023	24″ x 24″ x 60″, 6000 CFM		3.60	4.444		1,100	189		1,289	1,525
9026	24″ x 30″ x 60″, 7000 CFM		3.50	4.571		1,250	194		1,444	1,675
9030	24″ x 36″ x 60″, 8000 CFM		3.40	4.706		1,375	200		1,575	1,800
9034	24″ x 48″ x 60″, 11,000 CFM		3.30	4.848		1,175	206		1,381	1,625
9038	36″ x 18″ x 60″, 6300 CFM		3.50	4.571		920	194		1,114	1,325
9042	36″ x 36″ x 60″, 12,000 CFM	↓	3.30	4.848	↓	1,475	206		1,681	1,950

23 33 Air Duct Accessories

23 33 19 – Duct Silencers

23 33 19.10 Duct Silencers		Crew	Daily Output	Labor-Hours	Unit	Material	Labor	Equipment	Total	Total Incl O&P
						2009 Bare Costs				
9046	36″ x 48″ x 60″, 17,000 CFM	Q-9	3.10	5.161	Ea.	1,950	219		2,169	2,450
9050	36″ x 60″ x 60″, 20,000 CFM		2.80	5.714		2,375	243		2,618	2,975
9054	48″ x 48″ x 60″, 23,000 CFM		2.60	6.154		2,725	261		2,986	3,400
9056	48″ x 60″ x 60″, 23,000 CFM	↓	2.10	7.619	↓	3,425	325		3,750	4,250

23 33 23 – Turning Vanes

23 33 23.13 Air Turning Vanes

		Crew	Daily Output	Labor-Hours	Unit	Material	Labor	Equipment	Total	Total Incl O&P
0010	**AIR TURNING VANES**									
9400	Turning vane components									
9410	Turning vane rail	1 Shee	160	.050	L.F.	.77	2.36		3.13	4.48
9420	Double thick, factory fab. vane		300	.027		1.12	1.26		2.38	3.17
9428	12″ high set		170	.047		1.57	2.22		3.79	5.15
9432	14″ high set		160	.050		1.82	2.36		4.18	5.65
9434	16″ high set		150	.053		2.08	2.52		4.60	6.15
9436	18″ high set		144	.056		2.34	2.62		4.96	6.60
9438	20″ high set		138	.058		2.60	2.74		5.34	7.05
9440	22″ high set		130	.062		2.86	2.90		5.76	7.60
9442	24″ high set		124	.065		3.12	3.05		6.17	8.10
9444	26″ high set		116	.069		3.38	3.26		6.64	8.70
9446	30″ high set	↓	112	.071	↓	3.90	3.37		7.27	9.50

23 33 33 – Duct-Mounting Access Doors

23 33 33.13 Duct Access Doors

		Crew	Daily Output	Labor-Hours	Unit	Material	Labor	Equipment	Total	Total Incl O&P
0010	**DUCT ACCESS DOORS**									
1000	Duct access door, insulated, 6″ x 6″	1 Shee	14	.571	Ea.	14.55	27		41.55	57.50
1020	10″ x 10″		11	.727		17.60	34.50		52.10	72.50
1040	12″ x 12″		10	.800		18.80	38		56.80	78.50
1050	12″ x 18″		9	.889		33.50	42		75.50	102
1060	16″ x 12″		9	.889		26.50	42		68.50	94
1070	18″ x 18″		8	1		33.50	47		80.50	109
1074	24″ x 18″		8	1		40	47		87	117
1080	24″ x 24″	↓	8	1	↓	42.50	47		89.50	119

23 33 46 – Flexible Ducts

23 33 46.10 Flexible Air Ducts

		Crew	Daily Output	Labor-Hours	Unit	Material	Labor	Equipment	Total	Total Incl O&P
0010	**FLEXIBLE AIR DUCTS** R233100-20									
1280	Add to labor for elevated installation									
1282	of prefabricated (purchased) ductwork									
1283	10′ to 15′ high						10%			
1284	15′ to 20′ high						20%			
1285	20′ to 25′ high						25%			
1286	25′ to 30′ high						35%			
1287	30′ to 35′ high						40%			
1288	35′ to 40′ high						50%			
1289	Over 40′ high						55%			
1300	Flexible, coated fiberglass fabric on corr. resist. metal helix									
1400	pressure to 12″ (WG) UL-181									
1500	Non-insulated, 3″ diameter	Q-9	400	.040	L.F.	.99	1.70		2.69	3.71
1520	4″ diameter		360	.044		.99	1.89		2.88	4
1540	5″ diameter		320	.050		1.21	2.12		3.33	4.60
1560	6″ diameter		280	.057		1.49	2.43		3.92	5.40
1580	7″ diameter		240	.067		1.76	2.83		4.59	6.30
1600	8″ diameter		200	.080		2.04	3.40		5.44	7.50
1620	9″ diameter	↓	180	.089	↓	2.42	3.78		6.20	8.45

23 33 46 – Flexible Ducts

23 33 46.10 Flexible Air Ducts			Crew	Daily Output	Labor-Hours	Unit	2009 Bare Costs Material	Labor	Equipment	Total	Total Incl O&P
1640	10" diameter		Q-9	160	.100	L.F.	2.53	4.25		6.78	9.35
1660	12" diameter			120	.133		3.08	5.65		8.73	12.10
1680	14" diameter			80	.200		3.74	8.50		12.24	17.20
1700	16" diameter			60	.267		5.35	11.35		16.70	23.50
1800	For adjustable clamps, add					Ea.	.69			.69	.76
1900	Insulated, 1" thick, PE jacket, 3" diameter	G	Q-9	380	.042	L.F.	2.09	1.79		3.88	5.05
1910	4" diameter	G		340	.047		2.09	2		4.09	5.40
1920	5" diameter	G		300	.053		2.09	2.27		4.36	5.80
1940	6" diameter	G		260	.062		2.31	2.61		4.92	6.55
1960	7" diameter	G		220	.073		2.70	3.09		5.79	7.75
1980	8" diameter	G		180	.089		2.92	3.78		6.70	9
2000	9" diameter	G		160	.100		3.30	4.25		7.55	10.20
2020	10" diameter	G		140	.114		3.41	4.86		8.27	11.20
2040	12" diameter	G		100	.160		4.24	6.80		11.04	15.10
2060	14" diameter	G		80	.200		5.10	8.50		13.60	18.75
2080	16" diameter	G		60	.267		6.40	11.35		17.75	24.50
2100	18" diameter	G		45	.356		8.10	15.10		23.20	32.50
2120	20" diameter	G	Q-10	65	.369		8.85	16.25		25.10	35
2500	Insulated, heavy duty, coated fiberglass fabric										
2520	4" diameter	G	Q-9	340	.047	L.F.	3.36	2		5.36	6.80
2540	5" diameter	G		300	.053		3.58	2.27		5.85	7.45
2560	6" diameter	G		260	.062		4.02	2.61		6.63	8.45
2580	7" diameter	G		220	.073		4.46	3.09		7.55	9.65
2600	8" diameter	G		180	.089		5.05	3.78		8.83	11.35
2620	9" diameter	G		160	.100		5.40	4.25		9.65	12.50
2640	10" diameter	G		140	.114		6.55	4.86		11.41	14.65
2660	12" diameter	G		100	.160		7	6.80		13.80	18.15
2680	14" diameter	G		80	.200		9.15	8.50		17.65	23
2700	16" diameter	G		60	.267		13.70	11.35		25.05	32.50
2720	18" diameter	G		45	.356		16.80	15.10		31.90	42
2800	Flexible, aluminum, pressure to 12" (WG) UL-181										
2820	Non-insulated										
2830	3" diameter		Q-9	400	.040	L.F.	1.11	1.70		2.81	3.84
2831	4" diameter			360	.044		1.32	1.89		3.21	4.36
2832	5" diameter			320	.050		1.71	2.12		3.83	5.15
2833	6" diameter			280	.057		2.16	2.43		4.59	6.10
2834	7" diameter			240	.067		2.61	2.83		5.44	7.25
2835	8" diameter			200	.080		2.99	3.40		6.39	8.55
2836	9" diameter			180	.089		3.55	3.78		7.33	9.70
2837	10" diameter			160	.100		4.17	4.25		8.42	11.15
2838	12" diameter			120	.133		5.15	5.65		10.80	14.35
2839	14" diameter			80	.200		6.35	8.50		14.85	20
2840	15" diameter			70	.229		8.05	9.70		17.75	24
2841	16" diameter			60	.267		9.85	11.35		21.20	28.50
2842	18" diameter			45	.356		11	15.10		26.10	35.50
2843	20" diameter		Q-10	65	.369		12.15	16.25		28.40	38.50
2880	Insulated, 1" thick with 3/4 lb., PE jacket										
2890	3" diameter	G	Q-9	380	.042	L.F.	1.91	1.79		3.70	4.85
2891	4" diameter	G		340	.047		2.20	2		4.20	5.50
2892	5" diameter	G		300	.053		3.03	2.27		5.30	6.80
2893	6" diameter	G		260	.062		3.35	2.61		5.96	7.70
2894	7" diameter	G		220	.073		4.13	3.09		7.22	9.30
2895	8" diameter	G		180	.089		4.42	3.78		8.20	10.65

23 33 46 – Flexible Ducts

23 33 46.10 Flexible Air Ducts			Crew	Daily Output	Labor-Hours	Unit	2009 Bare Costs Material	Labor	Equipment	Total	Total Incl O&P
2896	9" diameter	G	Q-9	160	.100	L.F.	5	4.25		9.25	12.05
2897	10" diameter	G		140	.114		6.15	4.86		11.01	14.20
2898	12" diameter	G		100	.160		6.75	6.80		13.55	17.85
2899	14" diameter	G		80	.200		8.20	8.50		16.70	22
2900	15" diameter	G		70	.229		10.60	9.70		20.30	26.50
2901	16" diameter	G		60	.267		12.20	11.35		23.55	31
2902	18" diameter	G	↓	45	.356	↓	14.30	15.10		29.40	39
3000	Transitions, 3" deep, square to round										
3010	6" x 6" to round		Q-9	44	.364	Ea.	8.25	15.45		23.70	33
3014	6" x 12" to round			42	.381		15.40	16.20		31.60	42
3018	8" x 8" to round			40	.400		9.90	17		26.90	37
3022	9" x 9" to round			38	.421		9.90	17.90		27.80	38.50
3026	9" x 12" to round			37	.432		19.80	18.35		38.15	50.50
3030	9" x 15" to round			36	.444		23	18.90		41.90	54.50
3034	10" x 10" to round			34	.471		12.10	20		32.10	44.50
3038	10" x 22" to round			32	.500		28	21		49	63.50
3042	12" x 12" to round			30	.533		12.10	22.50		34.60	48.50
3046	12" x 18" to round			28	.571		28	24.50		52.50	68.50
3050	12" x 24" to round			26	.615		28	26		54	71
3054	15" x 15" to round			25	.640		15.40	27		42.40	59
3058	15" x 18" to round			24	.667		31	28.50		59.50	77.50
3062	18" x 18" to round			22	.727		19.25	31		50.25	68.50
3066	21" x 21" to round			21	.762		33	32.50		65.50	86.50
3070	22" x 22" to round			20	.800		43	34		77	99.50
3074	24" x 24" to round		↓	18	.889	↓	43	38		81	105

23 33 53 – Duct Liners

23 33 53.10 Duct Liner Board			Crew	Daily Output	Labor-Hours	Unit	Material	Labor	Equipment	Total	Total Incl O&P
0010	**DUCT LINER BOARD**										
3340	Board type fiberglass liner, FSK, 1-1/2 lb. density										
3344	1" thick	G	Q-14	150	.107	S.F.	.54	4.23		4.77	7.25
3345	1-1/2" thick	G		130	.123		.59	4.89		5.48	8.30
3346	2" thick	G		120	.133		.69	5.30		5.99	9.05
3348	3" thick	G		110	.145		.89	5.75		6.64	10.05
3350	4" thick	G		100	.160		1.09	6.35		7.44	11.15
3356	3 lb. density, 1" thick	G		150	.107		.73	4.23		4.96	7.45
3358	1-1/2" thick	G		130	.123		.92	4.89		5.81	8.65
3360	2" thick	G		120	.133		1.12	5.30		6.42	9.55
3362	2-1/2" thick	G		110	.145		1.32	5.75		7.07	10.50
3364	3" thick	G		100	.160		1.52	6.35		7.87	11.60
3366	4" thick	G		90	.178		1.90	7.05		8.95	13.15
3370	6 lb. density, 1" thick	G		140	.114		1.03	4.54		5.57	8.25
3374	1-1/2" thick	G		120	.133		1.39	5.30		6.69	9.85
3378	2" thick	G	↓	100	.160	↓	1.73	6.35		8.08	11.85
3490	Board type, fiberglass liner, 3 lb. density										
3500	Fire resistant, black pigmented, 1 side										
3520	1" thick	G	Q-14	150	.107	S.F.	.73	4.23		4.96	7.45
3540	1-1/2" thick	G		130	.123		.92	4.89		5.81	8.65
3560	2" thick	G	↓	120	.133	↓	1.12	5.30		6.42	9.55
3600	FSK vapor barrier										
3620	1" thick	G	Q-14	150	.107	S.F.	.73	4.23		4.96	7.45
3630	1-1/2" thick	G		130	.123		.92	4.89		5.81	8.65
3640	2" thick	G	↓	120	.133	↓	1.12	5.30		6.42	9.55

23 33 Air Duct Accessories

23 33 53 – Duct Liners

23 33 53.10 Duct Liner Board		Crew	Daily Output	Labor-Hours	Unit	Material	Labor	Equipment	Total	Total Incl O&P
						2009 Bare Costs				
3680	No finish									
3700	1" thick G	Q-14	170	.094	S.F.	.38	3.74		4.12	6.25
3710	1-1/2" thick G		140	.114		.57	4.54		5.11	7.75
3720	2" thick G		130	.123		.77	4.89		5.66	8.50
3940	Board type, non-fibrous foam									
3950	Temperature, bacteria and fungi resistant									
3960	1" thick G	Q-14	150	.107	S.F.	2.25	4.23		6.48	9.15
3970	1-1/2" thick G		130	.123		3	4.89		7.89	10.95
3980	2" thick G		120	.133		3.60	5.30		8.90	12.25

23 34 HVAC Fans

23 34 13 – Axial HVAC Fans

23 34 13.10 Axial Flow HVAC Fans		Crew	Daily Output	Labor-Hours	Unit	Material	Labor	Equipment	Total	Total Incl O&P
0010	**AXIAL FLOW HVAC FANS** R233100-10									
0020	Air conditioning and process air handling									
0030	Axial flow, compact, low sound, 2.5" S.P. R233400-10									
0050	3,800 CFM, 5 HP	Q-20	3.40	5.882	Ea.	4,375	255		4,630	5,200
0080	6,400 CFM, 5 HP		2.80	7.143		4,875	310		5,185	5,850
0100	10,500 CFM, 7-1/2 HP		2.40	8.333		6,075	360		6,435	7,225
0120	15,600 CFM, 10 HP		1.60	12.500		7,650	540		8,190	9,250
0140	23,000 CFM, 15 HP		.70	28.571		11,700	1,250		12,950	14,800
0160	28,000 CFM, 20 HP		.40	50		13,100	2,175		15,275	17,700
0500	Axial flow, constant speed									
0505	Direct drive, 1/8" S.P.									
0510	12", 1060 CFM, 1/6 HP	Q-20	3	6.667	Ea.	545	289		834	1,050
0514	12", 2095 CFM, 1/2 HP		3	6.667		640	289		929	1,150
0518	16", 2490 CFM, 1/3 HP		2.80	7.143		640	310		950	1,175
0522	20", 4130 CFM, 3/4 HP		2.60	7.692		765	335		1,100	1,350
0526	22", 4700 CFM, 3/4 HP		2.60	7.692		1,125	335		1,460	1,750
0530	24", 5850 CFM, 1 HP		2.50	8		1,175	345		1,520	1,825
0534	24", 7925 CFM, 1-1/2 HP		2.40	8.333		1,425	360		1,785	2,125
0538	30", 10640 CFM, 2 HP		2.20	9.091		1,850	395		2,245	2,650
0542	30", 14765 CFM, 2-1/2 HP		2.10	9.524		2,150	415		2,565	3,000
0546	36", 16780 CFM, 2 HP		2.10	9.524		2,225	415		2,640	3,075
0550	36", 22920 CFM, 5 HP		1.80	11.111		2,450	480		2,930	3,400
0560	Belt drive, 1/8" S.P.									
0562	15", 2800 CFM, 1/3 HP	Q-20	3.20	6.250	Ea.	735	271		1,006	1,225
0564	15", 3400 CFM, 1/2 HP		3	6.667		750	289		1,039	1,275
0568	18", 3280 CFM, 1/3 HP		2.80	7.143		790	310		1,100	1,350
0572	18", 3900 CFM, 1/2 HP		2.80	7.143		790	310		1,100	1,350
0576	18", 5250 CFM, 1 HP		2.70	7.407		875	320		1,195	1,450
0584	24", 6430 CFM, 1 HP		2.50	8		1,075	345		1,420	1,725
0588	24", 8860 CFM, 2 HP		2.40	8.333		1,125	360		1,485	1,800
0592	30", 9250 CFM, 1 HP		2.20	9.091		1,350	395		1,745	2,075
0596	30", 16900 CFM, 5 HP		2	10		1,475	435		1,910	2,300
0604	36", 14475 CFM, 2 HP		2.30	8.696		1,650	375		2,025	2,400
0608	36", 20080 CFM, 5 HP		1.80	11.111		1,750	480		2,230	2,650
0612	36", 14475 CFM, 7-1/2 HP		1.60	12.500		2,000	540		2,540	3,025
0616	42", 29000 CFM, 7-1/2 HP		1.40	14.286		2,725	620		3,345	3,950
1000	Return air fan									
1010	9200 CFM	Q-20	1.40	14.286	Ea.	3,575	620		4,195	4,875

23 34 HVAC Fans

23 34 13 – Axial HVAC Fans

23 34 13.10 Axial Flow HVAC Fans		Crew	Daily Output	Labor-Hours	Unit	2009 Bare Costs Material	Labor	Equipment	Total	Total Incl O&P
1020	13,200 CFM	Q-20	1.30	15.385	Ea.	5,125	665		5,790	6,650
1030	16,500 CFM		1.20	16.667		6,400	725		7,125	8,125
1040	19,500 CFM		1	20		7,550	870		8,420	9,625
1500	Vaneaxial, low pressure, 2000 CFM, 1/2 HP		3.60	5.556		1,825	241		2,066	2,375
1520	4,000 CFM, 1 HP		3.20	6.250		2,100	271		2,371	2,725
1540	8,000 CFM, 2 HP		2.80	7.143		2,675	310		2,985	3,400
1560	16,000 CFM, 5 HP	↓	2.40	8.333	↓	3,850	360		4,210	4,775

23 34 14 – Blower HVAC Fans

23 34 14.10 Blower Type HVAC Fans		Crew	Daily Output	Labor-Hours	Unit	Material	Labor	Equipment	Total	Total Incl O&P
0010	**BLOWER TYPE HVAC FANS**									
2000	Blowers, direct drive with motor, complete									
2020	1045 CFM @ .5" S.P., 1/5 HP	Q-20	18	1.111	Ea.	208	48		256	305
2040	1385 CFM @ .5" S.P., 1/4 HP		18	1.111		202	48		250	296
2060	1640 CFM @ .5" S.P., 1/3 HP		18	1.111		232	48		280	330
2080	1760 CFM @ .5" S.P., 1/2 HP	↓	18	1.111	↓	233	48		281	330
2090	4 speed									
2100	1164 to 1739 CFM @ .5" S.P., 1/3 HP	Q-20	16	1.250	Ea.	264	54		318	375
2120	1467 to 2218 CFM @ 1.0" S.P., 3/4 HP	"	14	1.429	"	340	62		402	470
2500	Ceiling fan, right angle, extra quiet, 0.10" S.P.									
2520	95 CFM	Q-20	20	1	Ea.	228	43.50		271.50	320
2540	210 CFM		19	1.053		270	45.50		315.50	365
2560	385 CFM		18	1.111		345	48		393	450
2580	885 CFM		16	1.250		675	54		729	825
2600	1,650 CFM		13	1.538		935	66.50		1,001.50	1,125
2620	2,960 CFM	↓	11	1.818		1,250	79		1,329	1,500
2640	For wall or roof cap, add	1 Shee	16	.500		228	23.50		251.50	288
2660	For straight thru fan, add					10%				
2680	For speed control switch, add	1 Elec	16	.500	↓	125	23.50		148.50	172
7500	Utility set, steel construction, pedestal, 1/4" S.P.									
7520	Direct drive, 150 CFM, 1/8 HP	Q-20	6.40	3.125	Ea.	735	136		871	1,000
7540	485 CFM, 1/6 HP		5.80	3.448		925	150		1,075	1,250
7560	1950 CFM, 1/2 HP		4.80	4.167		1,075	181		1,256	1,475
7580	2410 CFM, 3/4 HP		4.40	4.545		2,000	197		2,197	2,500
7600	3328 CFM, 1-1/2 HP	↓	3	6.667	↓	2,225	289		2,514	2,900
7680	V-belt drive, drive cover, 3 phase									
7700	800 CFM, 1/4 HP	Q-20	6	3.333	Ea.	740	145		885	1,025
7720	1,300 CFM, 1/3 HP		5	4		775	174		949	1,125
7740	2,000 CFM, 1 HP		4.60	4.348		915	189		1,104	1,300
7760	2,900 CFM, 3/4 HP		4.20	4.762		1,250	207		1,457	1,675
7780	3,600 CFM, 3/4 HP		4	5		1,525	217		1,742	2,000
7800	4,800 CFM, 1 HP		3.50	5.714		1,800	248		2,048	2,350
7820	6,700 CFM, 1-1/2 HP		3	6.667		2,225	289		2,514	2,900
7830	7,500 CFM, 2 HP		2.50	8		3,000	345		3,345	3,850
7840	11,000 CFM, 3 HP		2	10		4,025	435		4,460	5,100
7860	13,000 CFM, 3 HP		1.60	12.500		4,075	540		4,615	5,325
7880	15,000 CFM, 5 HP		1	20		4,225	870		5,095	5,975
7900	17,000 CFM, 7-1/2 HP		.80	25		4,525	1,075		5,600	6,625
7920	20,000 CFM, 7-1/2 HP	↓	.80	25	↓	5,400	1,075		6,475	7,575

23 34 16 – Centrifugal HVAC Fans

23 34 16.10 Centrifugal Type HVAC Fans		Crew	Daily Output	Labor-Hours	Unit	Material	Labor	Equipment	Total	Total Incl O&P
0010	**CENTRIFUGAL TYPE HVAC FANS**									
0200	In-line centrifugal, supply/exhaust booster									

23 34 16 – Centrifugal HVAC Fans

23 34 16.10 Centrifugal Type HVAC Fans		Crew	Daily Output	Labor-Hours	Unit	2009 Bare Costs Material	Labor	Equipment	Total	Total Incl O&P
0220	aluminum wheel/hub, disconnect switch, 1/4" S.P.									
0240	500 CFM, 10" diameter connection	Q-20	3	6.667	Ea.	1,125	289		1,414	1,675
0260	1,380 CFM, 12" diameter connection		2	10		1,175	435		1,610	1,975
0280	1,520 CFM, 16" diameter connection		2	10		1,300	435		1,735	2,100
0300	2,560 CFM, 18" diameter connection		1	20		1,400	870		2,270	2,875
0320	3,480 CFM, 20" diameter connection		.80	25		1,675	1,075		2,750	3,475
0326	5,080 CFM, 20" diameter connection		.75	26.667		1,800	1,150		2,950	3,775
0340	7,500 CFM, 22" diameter connection		.70	28.571		2,150	1,250		3,400	4,275
0350	10,000 CFM, 27" diameter connection		.65	30.769		2,625	1,325		3,950	4,925
3500	Centrifugal, airfoil, motor and drive, complete									
3520	1000 CFM, 1/2 HP	Q-20	2.50	8	Ea.	1,525	345		1,870	2,225
3540	2,000 CFM, 1 HP		2	10		1,675	435		2,110	2,525
3560	4,000 CFM, 3 HP		1.80	11.111		2,350	480		2,830	3,300
3580	8,000 CFM, 7-1/2 HP		1.40	14.286		3,625	620		4,245	4,950
3600	12,000 CFM, 10 HP		1	20		4,375	870		5,245	6,150
4000	Single width, belt drive, not incl. motor, capacities									
4020	at 2000 FPM, 2.5" S.P. for indicated motor									
4040	6900 CFM, 5 HP	Q-9	2.40	6.667	Ea.	3,200	283		3,483	3,925
4060	10,340 CFM, 7-1/2 HP		2.20	7.273		4,450	310		4,760	5,350
4080	15,320 CFM, 10 HP		2	8		5,225	340		5,565	6,275
4100	22,780 CFM, 15 HP		1.80	8.889		7,775	380		8,155	9,125
4120	33,840 CFM, 20 HP		1.60	10		10,300	425		10,725	12,100
4140	41,400 CFM, 25 HP		1.40	11.429		13,700	485		14,185	15,800
4160	50,100 CFM, 30 HP		.80	20		17,400	850		18,250	20,400
4200	Double width wheel, 12,420 CFM, 7.5 HP		2.20	7.273		4,550	310		4,860	5,500
4220	18,620 CFM, 15 HP		2	8		6,525	340		6,865	7,700
4240	27,580 CFM, 20 HP		1.80	8.889		7,975	380		8,355	9,350
4260	40,980 CFM, 25 HP		1.50	10.667		12,500	455		12,955	14,400
4280	60,920 CFM, 40 HP		1	16		16,800	680		17,480	19,600
4300	74,520 CFM, 50 HP		.80	20		19,900	850		20,750	23,200
4320	90,160 CFM, 50 HP		.70	22.857		27,300	970		28,270	31,500
4340	110,300 CFM, 60 HP		.50	32		36,900	1,350		38,250	42,700
4360	134,960 CFM, 75 HP		.40	40		49,100	1,700		50,800	56,500
4500	Corrosive fume resistant, plastic									
4600	roof ventilators, centrifugal, V belt drive, motor									
4620	1/4" S.P., 250 CFM, 1/4 HP	Q-20	6	3.333	Ea.	3,350	145		3,495	3,925
4640	895 CFM, 1/3 HP		5	4		3,650	174		3,824	4,300
4660	1630 CFM, 1/2 HP		4	5		4,325	217		4,542	5,075
4680	2240 CFM, 1 HP		3	6.667		4,500	289		4,789	5,400
4700	3810 CFM, 2 HP		2	10		5,000	435		5,435	6,175
4710	5000 CFM, 2 HP		1.80	11.111		10,500	480		10,980	12,300
4715	8000 CFM, 5 HP		1.40	14.286		11,700	620		12,320	13,800
4720	11760 CFM, 5 HP		1	20		11,900	870		12,770	14,400
4740	18810 CFM, 10 HP		.70	28.571		12,300	1,250		13,550	15,400
4800	For intermediate capacity, motors may be varied									
4810	For explosion proof motor, add				Ea.	15%				
5000	Utility set, centrifugal, V belt drive, motor									
5020	1/4" S.P., 1200 CFM, 1/4 HP	Q-20	6	3.333	Ea.	4,075	145		4,220	4,700
5040	1520 CFM, 1/3 HP		5	4		4,075	174		4,249	4,750
5060	1850 CFM, 1/2 HP		4	5		4,100	217		4,317	4,825
5080	2180 CFM, 3/4 HP		3	6.667		4,150	289		4,439	5,000
5100	1/2" S.P., 3600 CFM, 1 HP		2	10		6,000	435		6,435	7,275
5120	4250 CFM, 1-1/2 HP		1.60	12.500		6,075	540		6,615	7,525

23 34 16 – Centrifugal HVAC Fans

23 34 16.10 Centrifugal Type HVAC Fans		Crew	Daily Output	Labor-Hours	Unit	Material	2009 Bare Costs Labor	Equipment	Total	Total Incl O&P
5140	4800 CFM, 2 HP	Q-20	1.40	14.286	Ea.	6,150	620		6,770	7,725
5160	6920 CFM, 5 HP		1.30	15.385		6,350	665		7,015	8,000
5180	7700 CFM, 7-1/2 HP	↓	1.20	16.667	↓	6,550	725		7,275	8,300
5200	For explosion proof motor, add					15%				
5300	Fume exhauster without hose and intake nozzle									
5310	630 CFM, 1-1/2 HP	Q-20	2	10	Ea.	775	435		1,210	1,525
5320	Hose extension kit 5′	"	20	1	"	57.50	43.50		101	130
5500	Fans, industrial exhauster, for air which may contain granular matl.									
5520	1000 CFM, 1-1/2 HP	Q-20	2.50	8	Ea.	2,475	345		2,820	3,250
5540	2000 CFM, 3 HP		2	10		3,050	435		3,485	4,050
5560	4000 CFM, 7-1/2 HP		1.80	11.111		4,350	480		4,830	5,500
5580	8000 CFM, 15 HP		1.40	14.286		5,625	620		6,245	7,150
5600	12,000 CFM, 30 HP	↓	1	20	↓	9,150	870		10,020	11,400
7000	Roof exhauster, centrifugal, aluminum housing, 12″ galvanized									
7020	curb, bird screen, back draft damper, 1/4″ S.P.									
7100	Direct drive, 320 CFM, 11″ sq. damper	Q-20	7	2.857	Ea.	510	124		634	755
7120	600 CFM, 11″ sq. damper		6	3.333		520	145		665	790
7140	815 CFM, 13″ sq. damper		5	4		520	174		694	835
7160	1450 CFM, 13″ sq. damper		4.20	4.762		665	207		872	1,050
7180	2050 CFM, 16″ sq. damper		4	5		665	217		882	1,050
7200	V-belt drive, 1650 CFM, 12″ sq. damper		6	3.333		915	145		1,060	1,225
7220	2750 CFM, 21″ sq. damper		5	4		1,075	174		1,249	1,450
7230	3500 CFM, 21″ sq. damper		4.50	4.444		1,175	193		1,368	1,600
7240	4910 CFM, 23″ sq. damper		4	5		1,450	217		1,667	1,925
7260	8525 CFM, 28″ sq. damper		3	6.667		1,825	289		2,114	2,450
7280	13,760 CFM, 35″ sq. damper		2	10		2,500	435		2,935	3,425
7300	20,558 CFM, 43″ sq. damper	↓	1	20		5,300	870		6,170	7,175
7320	For 2 speed winding, add					15%				
7340	For explosionproof motor, add					375			375	415
7360	For belt driven, top discharge, add				↓	15%				
7400	Roof mounted kitchen exhaust, aluminum, centrifugal									
7410	Direct drive, 2 speed, temp to 200° F									
7412	1/12 HP, 9-3/4″	Q-20	8	2.500	Ea.	515	108		623	730
7414	1/3 HP, 12-5/8″		7	2.857		635	124		759	890
7416	1/2 HP, 13-1/2″		6	3.333		700	145		845	990
7418	3/4 HP, 15″	↓	5	4	↓	1,000	174		1,174	1,375
7424	Belt drive, temp to 250° F									
7426	3/4 HP, 20″	Q-20	5	4	Ea.	2,200	174		2,374	2,700
7428	1-1/2 HP, 24-1/2″		4	5		2,425	217		2,642	3,000
7430	1-1/2 HP, 30″	↓	4	5	↓	3,225	217		3,442	3,875
7450	Upblast, propeller, w/BDD, BS, + C									
7454	30,300 CFM @ 3/8″ S.P., 5 HP	Q-20	1.60	12.500	Ea.	3,725	540		4,265	4,925
7458	36,000 CFM @ 1/2″ S.P., 15 HP	"	1.60	12.500	"	5,300	540		5,840	6,650
8500	Wall exhausters, centrifugal, auto damper, 1/8″ S.P.									
8520	Direct drive, 610 CFM, 1/20 HP	Q-20	14	1.429	Ea.	330	62		392	460
8540	796 CFM, 1/12 HP		13	1.538		675	66.50		741.50	845
8560	822 CFM, 1/6 HP		12	1.667		840	72.50		912.50	1,025
8580	1,320 CFM, 1/4 HP		12	1.667		985	72.50		1,057.50	1,175
8600	1756 CFM, 1/4 HP		11	1.818		990	79		1,069	1,200
8620	1983 CFM, 1/4 HP		10	2		1,000	87		1,087	1,250
8640	2900 CFM, 1/2 HP		9	2.222		1,075	96.50		1,171.50	1,325
8660	3307 CFM, 3/4 HP		8	2.500		1,175	108		1,283	1,450
8670	5940 CFM, 1 HP	↓	7	2.857	↓	1,300	124		1,424	1,625

23 34 HVAC Fans

23 34 16 – Centrifugal HVAC Fans

23 34 16.10 Centrifugal Type HVAC Fans		Crew	Daily Output	Labor-Hours	Unit	2009 Bare Costs Material	Labor	Equipment	Total	Total Incl O&P
9500	V-belt drive, 3 phase									
9520	2,800 CFM, 1/4 HP	Q-20	9	2.222	Ea.	1,500	96.50		1,596.50	1,800
9540	3,740 CFM, 1/2 HP		8	2.500		1,550	108		1,658	1,900
9560	4400 CFM, 3/4 HP		7	2.857		1,575	124		1,699	1,950
9580	5700 CFM, 1-1/2 HP	↓	6	3.333	↓	1,650	145		1,795	2,025

23 34 23 – HVAC Power Ventilators

23 34 23.10 HVAC Power Circulators and Ventilators		Crew	Daily Output	Labor-Hours	Unit	Material	Labor	Equipment	Total	Total Incl O&P
0010	**HVAC POWER CIRCULATORS AND VENTILATORS**									
3000	Paddle blade air circulator, 3 speed switch									
3020	42", 5,000 CFM high, 3000 CFM low G	1 Elec	2.40	3.333	Ea.	89	157		246	330
3040	52", 6,500 CFM high, 4000 CFM low G	"	2.20	3.636	"	96.50	171		267.50	360
3100	For antique white motor, same cost									
3200	For brass plated motor, same cost									
3300	For light adaptor kit, add G				Ea.	29			29	31.50
3310	Industrial grade, reversible, 4 blade									
3312	5500 CFM	Q-20	5	4	Ea.	128	174		302	405
3314	7600 CFM		5	4		139	174		313	420
3316	21,015 CFM	↓	4	5	↓	179	217		396	525
6000	Propeller exhaust, wall shutter, 1/4" S.P.									
6020	Direct drive, two speed									
6100	375 CFM, 1/10 HP	Q-20	10	2	Ea.	410	87		497	585
6120	730 CFM, 1/7 HP		9	2.222		455	96.50		551.50	645
6140	1000 CFM, 1/8 HP		8	2.500		635	108		743	865
6160	1890 CFM, 1/4 HP		7	2.857		645	124		769	900
6180	3275 CFM, 1/2 HP		6	3.333		660	145		805	945
6200	4720 CFM, 1 HP	↓	5	4	↓	1,025	174		1,199	1,400
6300	V-belt drive, 3 phase									
6320	6175 CFM, 3/4 HP	Q-20	5	4	Ea.	855	174		1,029	1,200
6340	7500 CFM, 3/4 HP		5	4		900	174		1,074	1,250
6360	10,100 CFM, 1 HP		4.50	4.444		1,100	193		1,293	1,500
6380	14,300 CFM, 1-1/2 HP		4	5		1,300	217		1,517	1,750
6400	19,800 CFM, 2 HP		3	6.667		1,475	289		1,764	2,075
6420	26,250 CFM, 3 HP		2.60	7.692		1,775	335		2,110	2,450
6440	38,500 CFM, 5 HP		2.20	9.091		2,100	395		2,495	2,900
6460	46,000 CFM, 7-1/2 HP		2	10		2,250	435		2,685	3,150
6480	51,500 CFM, 10 HP	↓	1.80	11.111	↓	2,325	480		2,805	3,300
6490	V-belt drive, 115V., residential, whole house									
6500	Ceiling-wall, 5200 CFM, 1/4 HP, 30" x 30"	1 Shee	6	1.333	Ea.	445	63		508	585
6510	7500 CFM, 1/3 HP, 36"x36"		5	1.600		465	75.50		540.50	630
6520	10,500 CFM, 1/3 HP, 42"x42"		5	1.600		490	75.50		565.50	650
6530	13,200 CFM, 1/3 HP, 48" x 48"		4	2		540	94.50		634.50	740
6540	15,445 CFM, 1/2 HP, 48" x 48"		4	2		560	94.50		654.50	760
6550	17,025 CFM, 1/2 HP, 54" x 54"	↓	4	2	↓	1,075	94.50		1,169.50	1,350
6560	For two speed motor, add					20%				
6570	Shutter, automatic, ceiling/wall									
6580	30" x 30"	1 Shee	8	1	Ea.	145	47		192	233
6590	36" x 36"		8	1		166	47		213	256
6600	42" x 42"		8	1		208	47		255	300
6610	48" x 48"		7	1.143		225	54		279	330
6620	54" x 54"		6	1.333		297	63		360	420
6630	Timer, shut off, to 12 Hr.	↓	20	.400	↓	48	18.90		66.90	81.50
6650	Residential, bath exhaust, grille, back draft damper									

23 34 23.10 HVAC Power Circulators and Ventilators		Crew	Daily Output	Labor-Hours	Unit	2009 Bare Costs Material	Labor	Equipment	Total	Total Incl O&P
6660	50 CFM	Q-20	24	.833	Ea.	40.50	36		76.50	97.50
6670	110 CFM		22	.909		66.50	39.50		106	134
6672	180 CFM		22	.909		138	39.50		177.50	212
6673	210 CFM		22	.909		185	39.50		224.50	264
6674	260 CFM		22	.909		207	39.50		246.50	289
6675	300 CFM		22	.909		282	39.50		321.50	370
6680	Light combination, squirrel cage, 100 watt, 70 CFM	↓	24	.833	↓	73.50	36		109.50	136
6700	Light/heater combination, ceiling mounted									
6710	70 CFM, 1450 watt	Q-20	24	.833	Ea.	89	36		125	153
6800	Heater combination, recessed, 70 CFM		24	.833		42	36		78	102
6820	With 2 infrared bulbs		23	.870		64	37.50		101.50	128
6840	Wall mount, 170 CFM		22	.909		212	39.50		251.50	295
6846	Ceiling mount, 180 CFM		22	.909		215	39.50		254.50	297
6900	Kitchen exhaust, grille, complete, 160 CFM		22	.909		76	39.50		115.50	145
6910	180 CFM		20	1		66.50	43.50		110	140
6920	270 CFM		18	1.111		117	48		165	202
6930	350 CFM	↓	16	1.250	↓	91	54		145	183
6940	Residential roof jacks and wall caps									
6944	Wall cap with back draft damper									
6946	3" & 4" dia. round duct	1 Shee	11	.727	Ea.	16.15	34.50		50.65	71
6948	6" dia. round duct	"	11	.727	"	39	34.50		73.50	96
6958	Roof jack with bird screen and back draft damper									
6960	3" & 4" dia. round duct	1 Shee	11	.727	Ea.	15	34.50		49.50	69.50
6962	3-1/4" x 10" rectangular duct	"	10	.800	"	28.50	38		66.50	89.50
6980	Transition									
6982	3-1/4" x 10" to 6" dia. round	1 Shee	20	.400	Ea.	18.55	18.90		37.45	49.50
8020	Attic, roof type									
8030	Aluminum dome, damper & curb									
8040	6" diameter, 300 CFM	1 Elec	16	.500	Ea.	380	23.50		403.50	455
8050	7" diameter, 450 CFM		15	.533		415	25		440	495
8060	9" diameter, 900 CFM		14	.571		455	27		482	540
8080	12" diameter, 1000 CFM (gravity)		10	.800		470	37.50		507.50	575
8090	16" diameter, 1500 CFM (gravity)		9	.889		570	42		612	685
8100	20" diameter, 2500 CFM (gravity)		8	1		695	47		742	835
8110	26" diameter, 4000 CFM (gravity)		7	1.143		845	53.50		898.50	1,000
8120	32" diameter, 6500 CFM (gravity)		6	1.333		1,150	62.50		1,212.50	1,375
8130	38" diameter, 8000 CFM (gravity)		5	1.600		1,725	75		1,800	2,000
8140	50" diameter, 13,000 CFM (gravity)	↓	4	2	↓	2,500	94		2,594	2,900
8160	Plastic, ABS dome									
8180	1050 CFM	1 Elec	14	.571	Ea.	139	27		166	192
8200	1600 CFM	"	12	.667	"	208	31.50		239.50	275
8240	Attic, wall type, with shutter, one speed									
8250	12" diameter, 1000 CFM	1 Elec	14	.571	Ea.	300	27		327	370
8260	14" diameter, 1500 CFM		12	.667		325	31.50		356.50	400
8270	16" diameter, 2000 CFM	↓	9	.889	↓	365	42		407	465
8290	Whole house, wall type, with shutter, one speed									
8300	30" diameter, 4800 CFM	1 Elec	7	1.143	Ea.	785	53.50		838.50	945
8310	36" diameter, 7000 CFM		6	1.333		855	62.50		917.50	1,025
8320	42" diameter, 10,000 CFM		5	1.600		960	75		1,035	1,150
8330	48" diameter, 16,000 CFM	↓	4	2		1,200	94		1,294	1,450
8340	For two speed, add				↓	71.50			71.50	79
8350	Whole house, lay-down type, with shutter, one speed									
8360	30" diameter, 4500 CFM	1 Elec	8	1	Ea.	835	47		882	990

23 34 HVAC Fans

23 34 23 – HVAC Power Ventilators

23 34 23.10 HVAC Power Circulators and Ventilators		Crew	Daily Output	Labor-Hours	Unit	2009 Bare Costs Material	Labor	Equipment	Total	Total Incl O&P
8370	36" diameter, 6500 CFM	1 Elec	7	1.143	Ea.	900	53.50		953.50	1,075
8380	42" diameter, 9000 CFM		6	1.333		990	62.50		1,052.50	1,175
8390	48" diameter, 12,000 CFM		5	1.600		1,125	75		1,200	1,325
8440	For two speed, add					54			54	59.50
8450	For 12 hour timer switch, add	1 Elec	32	.250		54	11.75		65.75	77

23 34 33 – Air Curtains

23 34 33.10 Air Barrier Curtains

		Crew	Daily Output	Labor-Hours	Unit	Material	Labor	Equipment	Total	Total Incl O&P
0010	**AIR BARRIER CURTAINS**, Incl. motor starters, transformers,									
0050	door switches & temperature controls									
0100	Shipping and receiving doors, unheated, minimal wind stoppage									
0150	8' high, multiples of 3' wide	2 Shee	6	2.667	L.F.	410	126		536	650
0160	5' wide		10	1.600		350	75.50		425.50	500
0210	10' high, multiples of 4' wide		8	2		440	94.50		534.50	625
0250	12' high, 3'-6" wide		7	2.286		194	108		302	380
0260	12' wide		6	2.667		250	126		376	470
0350	16' high, 3'-6" wide		7	2.286		194	108		302	380
0360	12' wide		6	2.667		250	126		376	470
1500	Customer entrance doors, unheated, minimal wind stoppage									
1550	10' high, multiples of 3' wide	2 Shee	6	2.667	L.F.	305	126		431	530
1560	5' wide		10	1.600		350	75.50		425.50	500
1650	Maximum wind stoppage, 12' high, multiples of 4' wide		8	2		340	94.50		434.50	520
1700	Heated, minimal wind stoppage, electric heat									
1750	8' high, multiples of 3' wide	2 Shee	6	2.667	L.F.	380	126		506	615
1850	10' high, multiples of 3' wide		6	2.667		815	126		941	1,100
1860	Multiples of 5' wide		10	1.600		535	75.50		610.50	700
1950	Maximum wind stoppage, steam heat									
1960	12' high, multiples of 4' wide	2 Shee	8	2	L.F.	655	94.50		749.50	865
2000	Walk-in coolers and freezers, ambient air, minimal wind stoppage									
2050	8' high, multiples of 3' wide	2 Shee	6	2.667	L.F.	360	126		486	590
2060	Multiples of 5' wide		10	1.600		238	75.50		313.50	380
2250	Maximum wind stoppage, 12' high, multiples of 3' wide		6	2.667		222	126		348	440
2450	Conveyor openings or service windows, unheated, 5' high		5	3.200		228	151		379	485
2460	Heated, electric, 5' high, 2'-6" wide		5	3.200		146	151		297	395

23 35 Special Exhaust Systems

23 35 16 – Engine Exhaust Systems

23 35 16.10 Engine Exhaust Removal Systems

		Crew	Daily Output	Labor-Hours	Unit	Material	Labor	Equipment	Total	Total Incl O&P
0010	**ENGINE EXHAUST REMOVAL SYSTEMS** D3090–320									
0500	Engine exhaust, garage, in-floor system									
0510	Single tube outlet assemblies									
0520	For transite pipe ducting, self-storing tube									
0530	3" tubing adapter plate	1 Shee	16	.500	Ea.	217	23.50		240.50	276
0540	4" tubing adapter plate		16	.500		217	23.50		240.50	276
0550	5" tubing adapter plate		16	.500		217	23.50		240.50	276
0600	For vitrified tile ducting									
0610	3" tubing adapter plate, self-storing tube	1 Shee	16	.500	Ea.	217	23.50		240.50	276
0620	4" tubing adapter plate, self-storing tube		16	.500		217	23.50		240.50	276
0660	5" tubing adapter plate, self-storing tube		16	.500		217	23.50		240.50	276
0800	Two tube outlet assemblies									
0810	For transite pipe ducting, self-storing tube									

23 35 Special Exhaust Systems

23 35 16 – Engine Exhaust Systems

23 35 16.10 Engine Exhaust Removal Systems		Crew	Daily Output	Labor-Hours	Unit	2009 Bare Costs Material	Labor	Equipment	Total	Total Incl O&P
0820	3″ tubing, dual exhaust adapter plate	1 Shee	16	.500	Ea.	217	23.50		240.50	275
0850	For vitrified tile ducting									
0860	3″ tubing, dual exhaust, self-storing tube	1 Shee	16	.500	Ea.	217	23.50		240.50	275
0870	3″ tubing, double outlet, non-storing tubes	″	16	.500	″	217	23.50		240.50	275
0900	Accessories for metal tubing, (overhead systems also)									
0910	Adapters, for metal tubing end									
0920	3″ tail pipe type				Ea.	48.50			48.50	53
0930	4″ tail pipe type				↓	48.50			48.50	53
0940	5″ tail pipe type				↓	48.50			48.50	53
0990	5″ diesel stack type				↓	340			340	370
1000	6″ diesel stack type				↓	340			340	370
1100	Bullnose (guide) required for in-floor assemblies									
1110	3″ tubing size				Ea.	24			24	26.50
1120	4″ tubing size				↓	24			24	26.50
1130	5″ tubing size				↓	24			24	26.50
1150	Plain rings, for tubing end									
1160	3″ tubing size				Ea.	11.50			11.50	12.65
1170	4″ tubing size				″	14.50			14.50	15.95
1200	Tubing, galvanized, flexible, (for overhead systems also)									
1210	3″ ID				L.F.	6.45			6.45	7.10
1220	4″ ID				↓	7.80			7.80	8.55
1230	5″ ID				↓	9.05			9.05	9.95
1240	6″ ID				↓	10.40			10.40	11.40
1250	Stainless steel, flexible, (for overhead system, also)									
1260	3″ ID				L.F.	13.40			13.40	14.75
1270	4″ ID				↓	17.45			17.45	19.20
1280	5″ ID				↓	21			21	23
1290	6″ ID				↓	25			25	27.50
1500	Engine exhaust, garage, overhead components, for neoprene tubing									
1510	Alternate metal tubing & accessories see above									
1550	Adapters, for neoprene tubing end									
1560	3″ tail pipe, adjustable, neoprene				Ea.	42.50			42.50	46.50
1570	3″ tail pipe, heavy wall neoprene				↓	54.50			54.50	60
1580	4″ tail pipe, heavy wall neoprene				↓	58			58	64
1590	5″ tail pipe, heavy wall neoprene				↓	78.50			78.50	86.50
1650	Connectors, tubing									
1660	3″ interior, aluminum				Ea.	11.50			11.50	12.65
1670	4″ interior, aluminum				↓	14.50			14.50	15.95
1710	5″ interior, neoprene				↓	36			36	40
1750	3″ spiralock, neoprene				↓	11.50			11.50	12.65
1760	4″ spiralock, neoprene				↓	14.50			14.50	15.95
1780	Y for 3″ ID tubing, neoprene, dual exhaust				↓	30			30	33
1790	Y for 4″ ID tubing, aluminum, dual exhaust				↓	54.50			54.50	60
1850	Elbows, aluminum, splice into tubing for strap									
1860	3″ neoprene tubing size				Ea.	36			36	40
1870	4″ neoprene tubing size				↓	42.50			42.50	46.50
1900	Flange assemblies, connect tubing to overhead duct				↓	40			40	44
2000	Hardware and accessories									
2020	Cable, galvanized, 1/8″ diameter				L.F.	.42			.42	.46
2040	Cleat, tie down cable or rope				Ea.	4.83			4.83	5.30
2060	Pulley				↓	6.65			6.65	7.30
2080	Pulley hook, universal				↓	4.83			4.83	5.30
2100	Rope, nylon, 1/4″ diameter				L.F.	.36			.36	.40

23 35 16 – Engine Exhaust Systems

23 35 16.10 Engine Exhaust Removal Systems		Crew	Daily Output	Labor-Hours	Unit	2009 Bare Costs Material	Labor	Equipment	Total	Total Incl O&P
2120	Winch, 1" diameter				Ea.	103			103	113
2150	Lifting strap, mounts on neoprene									
2160	3" tubing size				Ea.	24			24	26.50
2170	4" tubing size					24			24	26.50
2180	5" tubing size					24			24	26.50
2190	6" tubing size					24			24	26.50
2200	Tubing, neoprene, 11' lengths									
2210	3" ID				L.F.	7.25			7.25	8
2220	4" ID					9.05			9.05	9.95
2230	5" ID					16.30			16.30	17.95
2500	Engine exhaust, thru-door outlet									
2510	3" tube size	1 Carp	16	.500	Ea.	42.50	20		62.50	77.50
2530	4" tube size	"	16	.500	"	42.50	20		62.50	77.50
3000	Tubing, exhaust, flex hose, with									
3010	coupler, damper and tail pipe adapter									
3020	Neoprene									
3040	3" x 20'	1 Shee	6	1.333	Ea.	230	63		293	350
3050	4" x 15'		5.40	1.481		230	70		300	360
3060	4" x 20'		5	1.600		275	75.50		350.50	420
3070	5" x 15'		4.40	1.818		390	86		476	555
3100	Galvanized									
3110	3" x 20'	1 Shee	6	1.333	Ea.	209	63		272	325
3120	4" x 17'		5.60	1.429		217	67.50		284.50	345
3130	4" x 20'		5	1.600		240	75.50		315.50	380
3140	5" x 17'		4.60	1.739		268	82		350	420
8000	Blower, for tailpipe exhaust system									
8010	Direct drive									
8012	495 CFM, 1/3 HP	Q-9	5	3.200	Ea.	1,700	136		1,836	2,075
8014	1445 CFM, 3/4 HP		4	4		2,100	170		2,270	2,550
8016	1840 CFM, 1-1/2 HP		3	5.333		2,500	227		2,727	3,100
8030	Beltdrive									
8032	1400 CFM, 1 HP	Q-9	4	4	Ea.	2,100	170		2,270	2,550
8034	2023 CFM, 1-1/2 HP		3	5.333		2,850	227		3,077	3,475
8036	2750 CFM, 2 HP		2	8		3,400	340		3,740	4,275
8038	4400 CFM, 3 HP		1.80	8.889		5,000	380		5,380	6,075
8040	7060 CFM, 5 HP		1.40	11.429		7,500	485		7,985	9,000

23 35 43 – Welding Fume Elimination Systems

23 35 43.10 Welding Fume Elimination System Components		Crew	Daily Output	Labor-Hours	Unit	Material	Labor	Equipment	Total	Total Incl O&P
0010	**WELDING FUME ELIMINATION SYSTEM COMPONENTS**									
7500	Welding fume elimination accessories for garage exhaust systems									
7600	Cut off (blast gate)									
7610	3" tubing size, 3" x 6" opening	1 Shee	24	.333	Ea.	19.30	15.75		35.05	45.50
7620	4" tubing size, 4" x 8" opening		24	.333		21.50	15.75		37.25	48
7630	5" tubing size, 5" x 10" opening		24	.333		29	15.75		44.75	56
7640	6" tubing size		24	.333		34	15.75		49.75	61
7650	8" tubing size		24	.333		48.50	15.75		64.25	77
7700	Hoods, magnetic, with handle & screen									
7710	3" tubing size, 3" x 6" opening	1 Shee	24	.333	Ea.	90.50	15.75		106.25	124
7720	4" tubing size, 4" x 8" opening		24	.333		90.50	15.75		106.25	124
7730	5" tubing size, 5" x 10" opening		24	.333		90.50	15.75		106.25	124

23 36 13 – Constant-Air-Volume Units

23 36 13.10 Constant Volume Mixing Boxes		Crew	Daily Output	Labor-Hours	Unit	2009 Bare Costs Material	Labor	Equipment	Total	Total Incl O&P
0010	**CONSTANT VOLUME MIXING BOXES**									
5180	Mixing box, includes electric or pneumatic motor									
5192	Recommend use with silencer, see Div. 23 33 19.10 0010									
5200	Constant volume, 150 to 270 CFM	Q-9	12	1.333	Ea.	620	56.50		676.50	765
5210	270 to 600 CFM		11	1.455		640	62		702	795
5230	550 to 1000 CFM		9	1.778		640	75.50		715.50	815
5240	1000 to 1600 CFM		8	2		655	85		740	850
5250	1300 to 1900 CFM		6	2.667		670	113		783	900
5260	550 to 2640 CFM		5.60	2.857		735	121		856	995
5270	650 to 3120 CFM		5.20	3.077		785	131		916	1,050

23 36 16 – Variable-Air-Volume Units

23 36 16.10 Variable Volume Mixing Boxes		Crew	Daily Output	Labor-Hours	Unit	Material	Labor	Equipment	Total	Total Incl O&P
0010	**VARIABLE VOLUME MIXING BOXES**									
5180	Mixing box, includes electric or pneumatic motor									
5192	Recommend use with attenuator, see Div. 23 33 19.10 0010									
5500	VAV Cool only, pneumatic, pressure independent 300 to 600 CFM	Q-9	11	1.455	Ea.	400	62		462	535
5510	500 to 1000 CFM		9	1.778		410	75.50		485.50	565
5520	800 to 1600 CFM		9	1.778		425	75.50		500.50	580
5530	1100 to 2000 CFM		8	2		435	85		520	605
5540	1500 to 3000 CFM		7	2.286		460	97		557	660
5550	2000 to 4000 CFM		6	2.667		465	113		578	650
5560	For electric, w/thermostat, pressure dependent, add					23			23	25.50
5600	VAV Cool & HW coils, damper, actuator and t'stat									
5610	200 CFM	Q-9	11	1.455	Ea.	700	62		762	865
5620	400 CFM		10	1.600		705	68		773	880
5630	600 CFM		10	1.600		705	68		773	880
5640	800 CFM		8	2		725	85		810	930
5650	1000 CFM		8	2		725	85		810	930
5660	1250 CFM		6	2.667		795	113		908	1,050
5670	1500 CFM		6	2.667		795	113		908	1,050
5680	2000 CFM		4	4		865	170		1,035	1,200
5684	3000 CFM		3.80	4.211		940	179		1,119	1,300
5700	VAV Cool only, fan powered, damper, actuator, thermostat									
5710	200 CFM	Q-9	10	1.600	Ea.	955	68		1,023	1,150
5720	400 CFM		9	1.778		990	75.50		1,065.50	1,225
5730	600 CFM		9	1.778		990	75.50		1,065.50	1,225
5740	800 CFM		7	2.286		1,050	97		1,147	1,325
5750	1000 CFM		7	2.286		1,050	97		1,147	1,325
5760	1250 CFM		5	3.200		1,150	136		1,286	1,450
5770	1500 CFM		5	3.200		1,150	136		1,286	1,450
5780	2000 CFM		4	4		1,275	170		1,445	1,650
5800	VAV Fan powr'd, cooling, with HW coils, dampers, actuators, t'stat									
5810	200 CFM	Q-9	10	1.600	Ea.	1,150	68		1,218	1,375
5820	400 CFM		9	1.778		1,200	75.50		1,275.50	1,425
5830	600 CFM		9	1.778		1,200	75.50		1,275.50	1,425
5840	800 CFM		7	2.286		1,250	97		1,347	1,525
5850	1000 CFM		7	2.286		1,250	97		1,347	1,525
5860	1250 CFM		5	3.200		1,425	136		1,561	1,775
5870	1500 CFM		5	3.200		1,425	136		1,561	1,775
5880	2000 CFM		3.50	4.571		1,525	194		1,719	2,000

23 37 13 – Diffusers, Registers, and Grilles

23 37 13.10 Diffusers		Crew	Daily Output	Labor-Hours	Unit	Material	Labor	Equipment	Total	Total Incl O&P
						2009 Bare Costs				
0010	**DIFFUSERS**, Aluminum, opposed blade damper unless noted R233100-10									
0100	Ceiling, linear, also for sidewall									
0120	2" wide R233100-30	1 Shee	32	.250	L.F.	39.50	11.80		51.30	61.50
0140	3" wide		30	.267		45	12.60		57.60	69
0160	4" wide R233700-60		26	.308		51.50	14.50		66	79.50
0180	6" wide		24	.333		63.50	15.75		79.25	94
0200	8" wide		22	.364		75	17.15		92.15	109
0220	10" wide		20	.400		85	18.90		103.90	123
0240	12" wide		18	.444		91	21		112	133
0260	For floor or sill application, add					15%				
0500	Perforated, 24" x 24" lay-in panel size, 6" x 6"	1 Shee	16	.500	Ea.	56	23.50		79.50	98
0520	8" x 8"		15	.533		57	25		82	102
0530	9" x 9"		14	.571		61	27		88	109
0540	10" x 10"		14	.571		63	27		90	111
0560	12" x 12"		12	.667		64	31.50		95.50	119
0580	15" x 15"		11	.727		127	34.50		161.50	193
0590	16" x 16"		11	.727		106	34.50		140.50	170
0600	18" x 18"		10	.800		112	38		150	181
0610	20" x 20"		10	.800		129	38		167	200
0620	24" x 24"		9	.889		161	42		203	242
1000	Rectangular, 1 to 4 way blow, 6" x 6"		16	.500		45	23.50		68.50	86
1010	8" x 8"		15	.533		54	25		79	98.50
1014	9" x 9"		15	.533		59.50	25		84.50	105
1016	10" x 10"		15	.533		64	25		89	110
1020	12" x 6"		15	.533		65.50	25		90.50	111
1040	12" x 9"		14	.571		71.50	27		98.50	120
1060	12" x 12"		12	.667		72	31.50		103.50	128
1070	14" x 6"		13	.615		71	29		100	123
1074	14" x 14"		12	.667		101	31.50		132.50	160
1080	18" x 12"		11	.727		126	34.50		160.50	192
1120	15" x 15"		10	.800		103	38		141	171
1140	18" x 15"		9	.889		148	42		190	228
1150	18" x 18"		9	.889		126	42		168	204
1160	21" x 21"		8	1		180	47		227	271
1170	24" x 12"		10	.800		134	38		172	205
1180	24" x 24"		7	1.143		256	54		310	365
1500	Round, butterfly damper, steel, diffuser size, 6" diameter		18	.444		22	21		43	56.50
1520	8" diameter		16	.500		23.50	23.50		47	62.50
1540	10" diameter		14	.571		29	27		56	73.50
1560	12" diameter		12	.667		38.50	31.50		70	91
1580	14" diameter		10	.800		49	38		87	112
1600	18" diameter		9	.889		117	42		159	194
1610	20" diameter		9	.889		125	42		167	203
1620	22" diameter		8	1		137	47		184	224
1630	24" diameter		8	1		150	47		197	238
1640	28" diameter		7	1.143		183	54		237	285
1650	32" diameter		6	1.333		205	63		268	325
1700	Round, steel, adjustable core, annular segmented damper, diffuser size									
1704	6" diameter	1 Shee	18	.444	Ea.	77	21		98	117
1708	8" diameter		16	.500		82.50	23.50		106	128
1712	10" diameter		14	.571		93	27		120	144
1716	12" diameter		12	.667		103	31.50		134.50	163
1720	14" diameter		10	.800		115	38		153	184

23 37 Air Outlets and Inlets

23 37 13 – Diffusers, Registers, and Grilles

23 37 13.10 Diffusers		Crew	Daily Output	Labor-Hours	Unit	Material	Labor	Equipment	Total	Total Incl O&P
						2009 Bare Costs				
1724	18″ diameter	1 Shee	9	.889	Ea.	175	42		217	258
1728	20″ diameter		9	.889		203	42		245	288
1732	24″ diameter		8	1		315	47		362	425
1736	30″ diameter		7	1.143		550	54		604	690
1740	36″ diameter		6	1.333		670	63		733	835
2000	T bar mounting, 24″ x 24″ lay-in frame, 6″ x 6″		16	.500		42.50	23.50		66	83.50
2020	8″ x 8″		14	.571		44.50	27		71.50	90
2040	12″ x 12″		12	.667		52	31.50		83.50	106
2060	16″ x 16″		11	.727		70	34.50		104.50	130
2080	18″ x 18″		10	.800		78	38		116	144
2500	Combination supply and return									
2520	21″ x 21″ supply, 15″ x 15″ return	Q-9	10	1.600	Ea.	305	68		373	440
2540	24″ x 24″ supply, 18″ x 18″ return		9.50	1.684		385	71.50		456.50	535
2560	27″ x 27″ supply, 18″ x 18″ return		9	1.778		450	75.50		525.50	610
2580	30″ x 30″ supply, 21″ x 21″ return		8.50	1.882		570	80		650	750
2600	33″ x 33″ supply, 24″ x 24″ return		8	2		665	85		750	860
2620	36″ x 36″ supply, 24″ x 24″ return		7.50	2.133		780	90.50		870.50	1,000
3000	Baseboard, white enameled steel									
3100	18″ long	1 Shee	20	.400	Ea.	11.75	18.90		30.65	42
3120	24″ long		18	.444		23	21		44	57.50
3140	48″ long		16	.500		42	23.50		65.50	82.50
3400	For matching return, deduct					10%				
4000	Floor, steel, adjustable pattern									
4100	2″ x 10″	1 Shee	34	.235	Ea.	15.75	11.10		26.85	34.50
4120	2″ x 12″		32	.250		17.30	11.80		29.10	37
4140	2″ x 14″		30	.267		19.60	12.60		32.20	41
4200	4″ x 10″		28	.286		18.20	13.50		31.70	41
4220	4″ x 12″		26	.308		20	14.50		34.50	44.50
4240	4″ x 14″		25	.320		22.50	15.10		37.60	48.50
4260	6″ x 10″		26	.308		27.50	14.50		42	52.50
4280	6″ x 12″		24	.333		28.50	15.75		44.25	55.50
4300	6″ x 14″		22	.364		30	17.15		47.15	59.50
4500	Linear bar diffuser in wall, sill or ceiling									
4510	2″ wide	1 Shee	32	.250	L.F.	35	11.80		46.80	56.50
4512	3″ wide		30	.267		37.50	12.60		50.10	61
4514	4″ wide		26	.308		46.50	14.50		61	73.50
4516	6″ wide		24	.333		57.50	15.75		73.25	87
4520	10″ wide		20	.400		76.50	18.90		95.40	114
4522	12″ wide		18	.444		86	21		107	127
5000	Sidewall, aluminum, 3 way dispersion									
5100	8″ x 4″	1 Shee	24	.333	Ea.	8.65	15.75		24.40	33.50
5110	8″ x 6″		24	.333		9.25	15.75		25	34
5120	10″ x 4″		22	.364		9.10	17.15		26.25	36.50
5130	10″ x 6″		22	.364		9.50	17.15		26.65	37
5140	10″ x 8″		20	.400		13.25	18.90		32.15	43.50
5160	12″ x 4″		18	.444		9.75	21		30.75	43.50
5170	12″ x 6″		17	.471		10.80	22		32.80	46
5180	12″ x 8″		16	.500		14	23.50		37.50	52
5200	14″ x 4″		15	.533		10.85	25		35.85	51
5220	14″ x 6″		14	.571		12	27		39	54.50
5240	14″ x 8″		13	.615		15.60	29		44.60	61.50
5260	16″ x 6″		12.50	.640		17.15	30		47.15	65.50
6000	For steel diffusers instead of aluminum, deduct					10%				

23 37 13 – Diffusers, Registers, and Grilles

23 37 13.30 Grilles		Crew	Daily Output	Labor-Hours	Unit	2009 Bare Costs Material	Labor	Equipment	Total	Total Incl O&P
0010	**GRILLES**									
0020	Aluminum									
0100	Air supply, single deflection, adjustable									
0120	8" x 4"	1 Shee	30	.267	Ea.	17.70	12.60		30.30	39
0140	8" x 8"		28	.286		21.50	13.50		35	44.50
0160	10" x 4"		24	.333		19.50	15.75		35.25	45.50
0180	10" x 10"		23	.348		25	16.40		41.40	53
0200	12" x 6"		23	.348		23	16.40		39.40	51
0220	12" x 12"		22	.364		30.50	17.15		47.65	60
0230	14" x 6"		23	.348		25.50	16.40		41.90	54
0240	14" x 8"		23	.348		27.50	16.40		43.90	56
0260	14" x 14"		22	.364		37	17.15		54.15	67.50
0270	18" x 8"		23	.348		31.50	16.40		47.90	60.50
0280	18" x 10"		23	.348		36.50	16.40		52.90	65.50
0300	18" x 18"		21	.381		56.50	18		74.50	90
0320	20" x 12"		22	.364		41.50	17.15		58.65	72
0340	20" x 20"		21	.381		68.50	18		86.50	103
0350	24" x 8"		20	.400		43	18.90		61.90	76.50
0360	24" x 14"		18	.444		56	21		77	94
0380	24" x 24"		15	.533		97	25		122	146
0400	30" x 8"		20	.400		51	18.90		69.90	85
0420	30" x 10"		19	.421		67.50	19.85		87.35	105
0440	30" x 12"		18	.444		62	21		83	101
0460	30" x 16"		17	.471		55.50	22		77.50	95
0480	30" x 18"		17	.471		90	22		112	133
0500	30" x 30"		14	.571		160	27		187	218
0520	36" x 12"		17	.471		77	22		99	119
0540	36" x 16"		16	.500		98	23.50		121.50	145
0560	36" x 18"		15	.533		109	25		134	158
0570	36" x 20"		15	.533		119	25		144	170
0580	36" x 24"		14	.571		154	27		181	212
0600	36" x 28"		13	.615		189	29		218	253
0620	36" x 30"		12	.667		208	31.50		239.50	278
0640	36" x 32"		12	.667		228	31.50		259.50	300
0660	36" x 34"		11	.727		248	34.50		282.50	325
0680	36" x 36"		11	.727		270	34.50		304.50	350
0700	For double deflecting, add					70%				
1000	Air return, 6" x 6"	1 Shee	26	.308		17.05	14.50		31.55	41.50
1020	10" x 6"		24	.333		17.05	15.75		32.80	43
1040	14" x 6"		23	.348		19.80	16.40		36.20	47.50
1060	10" x 8"		23	.348		19.80	16.40		36.20	47.50
1080	16" x 8"		22	.364		21.50	17.15		38.65	50
1100	12" x 12"		22	.364		23	17.15		40.15	52
1120	24" x 12"		18	.444		32	21		53	67.50
1140	30" x 12"		16	.500		41.50	23.50		65	82
1160	14" x 14"		22	.364		27.50	17.15		44.65	57
1180	16" x 16"		22	.364		31	17.15		48.15	60.50
1200	18" x 18"		21	.381		35	18		53	66
1220	24" x 18"		16	.500		40	23.50		63.50	80.50
1240	36" x 18"		15	.533		61.50	25		86.50	107
1260	24" x 24"		15	.533		47.50	25		72.50	91
1280	36" x 24"		14	.571		69	27		96	117

23 37 Air Outlets and Inlets

23 37 13 – Diffusers, Registers, and Grilles

23 37 13.30 Grilles		Crew	Daily Output	Labor-Hours	Unit	Material	Labor	Equipment	Total	Total Incl O&P
						2009 Bare Costs				
1300	48″ x 24″	1 Shee	12	.667	Ea.	118	31.50		149.50	179
1320	48″ x 30″		11	.727		153	34.50		187.50	221
1340	36″ x 36″		13	.615		92	29		121	146
1360	48″ x 36″		11	.727		190	34.50		224.50	262
1380	48″ x 48″		8	1		234	47		281	330
2000	Door grilles, 12″ x 12″		22	.364		52.50	17.15		69.65	84
2020	18″ x 12″		22	.364		59.50	17.15		76.65	92
2040	24″ x 12″		18	.444		64.50	21		85.50	104
2060	18″ x 18″		18	.444		80	21		101	120
2080	24″ x 18″		16	.500		91	23.50		114.50	137
2100	24″ x 24″		15	.533		125	25		150	177
3000	Filter grille with filter, 12″ x 12″		24	.333		45.50	15.75		61.25	74
3020	18″ x 12″		20	.400		64	18.90		82.90	99
3040	24″ x 18″		18	.444		75	21		96	115
3060	24″ x 24″		16	.500		86.50	23.50		110	132
3080	30″ x 24″		14	.571		108	27		135	161
3100	30″ x 30″		13	.615		127	29		156	185
3950	Eggcrate, framed, 6″ x 6″ opening		26	.308		18.70	14.50		33.20	43
3954	8″ x 8″ opening		24	.333		19.80	15.75		35.55	46
3960	10″ x 10″ opening		23	.348		21.50	16.40		37.90	49
3970	12″ x 12″ opening		22	.364		26	17.15		43.15	55
3980	14″ x 14″ opening		22	.364		27	17.15		44.15	56
3984	16″ x 16″ opening		21	.381		39.50	18		57.50	71
3990	18″ x 18″ opening		21	.381		45.50	18		63.50	77.50
4020	22″ x 22″ opening		17	.471		59.50	22		81.50	99.50
4040	24″ x 24″ opening		15	.533		75	25		100	122
4044	28″ x 28″ opening		14	.571		116	27		143	169
4048	36″ x 36″ opening		12	.667		132	31.50		163.50	194
4050	48″ x 24″ opening		12	.667		168	31.50		199.50	234
4060	Eggcrate, lay-in, T-bar system									
4070	48″ x 24″ sheet	1 Shee	40	.200	Ea.	90	9.45		99.45	114
5000	Transfer grille, vision proof, 8″ x 4″		30	.267		32.50	12.60		45.10	55.50
5020	8″ x 6″		28	.286		32.50	13.50		46	57
5040	8″ x 8″		26	.308		36.50	14.50		51	62.50
5060	10″ x 6″		24	.333		34	15.75		49.75	61.50
5080	10″ x 10″		23	.348		43	16.40		59.40	72.50
5090	12″ x 6″		23	.348		36.50	16.40		52.90	65.50
5100	12″ x 10″		22	.364		47	17.15		64.15	78
5110	12″ x 12″		22	.364		51	17.15		68.15	82.50
5120	14″ x 10″		22	.364		51	17.15		68.15	82.50
5140	16″ x 8″		21	.381		51	18		69	83.50
5160	18″ x 12″		20	.400		66.50	18.90		85.40	102
5170	18″ x 18″		21	.381		93.50	18		111.50	131
5180	20″ x 12″		20	.400		71.50	18.90		90.40	108
5200	20″ x 20″		19	.421		113	19.85		132.85	155
5220	24″ x 12″		17	.471		87	22		109	130
5240	24″ x 24″		16	.500		155	23.50		178.50	207
5260	30″ x 6″		15	.533		71.50	25		96.50	118
5280	30″ x 8″		15	.533		83.50	25		108.50	131
5300	30″ x 12″		14	.571		101	27		128	153
5320	30″ x 16″		13	.615		124	29		153	181
5340	30″ x 20″		12	.667		156	31.50		187.50	221
5360	30″ x 24″		11	.727		192	34.50		226.50	265

23 37 Air Outlets and Inlets

23 37 13 – Diffusers, Registers, and Grilles

23 37 13.30 Grilles		Crew	Daily Output	Labor-Hours	Unit	Material	Labor	Equipment	Total	Total Incl O&P
						2009 Bare Costs				
5380	30″ x 30″	1 Shee	10	.800	Ea.	254	38		292	335
6000	For steel grilles instead of aluminum in above, deduct					10%				
6200	Plastic, eggcrate, lay-in, T-bar system									
6210	48″ x 24″ sheet	1 Shee	50	.160	Ea.	22.50	7.55		30.05	36.50
6250	Steel door louver									
6270	With fire link, steel only									
6350	12″ x 18″	1 Shee	22	.364	Ea.	187	17.15		204.15	233
6410	12″ x 24″		21	.381		214	18		232	263
6470	18″ x 18″		20	.400		209	18.90		227.90	259
6530	18″ x 24″		19	.421		240	19.85		259.85	295
6600	18″ x 36″		17	.471		279	22		301	340
6660	24″ x 24″		15	.533		242	25		267	305
6720	24″ x 30″		15	.533		279	25		304	345
6780	30″ x 30″		13	.615		325	29		354	405
6840	30″ x 36″		12	.667		380	31.50		411.50	465
6900	36″ x 36″		10	.800		435	38		473	540
23 37 13.60 Registers										
0010	**REGISTERS**									
0980	Air supply									
1000	Ceiling/wall, O.B. damper, anodized aluminum									
1010	One or two way deflection, adj. curved face bars									
1014	6″ x 6″	1 Shee	24	.333	Ea.	30.50	15.75		46.25	57.50
1020	8″ x 4″		26	.308		30.50	14.50		45	56
1040	8″ x 8″		24	.333		38	15.75		53.75	66
1060	10″ x 6″		20	.400		35	18.90		53.90	67.50
1080	10″ x 10″		19	.421		47	19.85		66.85	82.50
1100	12″ x 6″		19	.421		35.50	19.85		55.35	69.50
1120	12″ x 12″		18	.444		44.50	21		65.50	81.50
1140	14″ x 8″		17	.471		48.50	22		70.50	87.50
1160	14″ x 14″		18	.444		65	21		86	104
1170	16″ x 16″		17	.471		72	22		94	114
1180	18″ x 8″		18	.444		54.50	21		75.50	92.50
1200	18″ x 18″		17	.471		99	22		121	143
1220	20″ x 4″		19	.421		50	19.85		69.85	85.50
1240	20″ x 6″		18	.444		50	21		71	87.50
1260	20″ x 8″		18	.444		54	21		75	92
1280	20″ x 20″		17	.471		120	22		142	167
1290	22″ x 22″		15	.533		138	25		163	191
1300	24″ x 4″		17	.471		61.50	22		83.50	102
1320	24″ x 6″		16	.500		61.50	23.50		85	104
1340	24″ x 8″		13	.615		67.50	29		96.50	119
1350	24″ x 18″		12	.667		125	31.50		156.50	187
1360	24″ x 24″		11	.727		170	34.50		204.50	240
1380	30″ x 4″		16	.500		77	23.50		100.50	121
1400	30″ x 6″		15	.533		77	25		102	124
1420	30″ x 8″		14	.571		88	27		115	139
1440	30″ x 24″		12	.667		219	31.50		250.50	290
1460	30″ x 30″		10	.800		281	38		319	370
1504	4 way deflection, adjustable curved face bars									
1510	6″ x 6″	1 Shee	26	.308	Ea.	36.50	14.50		51	63
1514	8″ x 8″		24	.333		45.50	15.75		61.25	74.50
1518	10″ x 10″		19	.421		56.50	19.85		76.35	93

23 37 13.60 Registers		Crew	Daily Output	Labor-Hours	Unit	2009 Bare Costs Material	Labor	Equipment	Total	Total Incl O&P
1522	12" x 6"	1 Shee	19	.421	Ea.	42.50	19.85		62.35	77
1526	12" x 12"		18	.444		53.50	21		74.50	91.50
1530	14" x 14"		18	.444		78	21		99	119
1534	16" x 16"		17	.471		86.50	22		108.50	130
1538	18" x 18"		17	.471		119	22		141	165
1542	22" x 22"	↓	16	.500	↓	166	23.50		189.50	220
1980	One way deflection, adj. vert. or horiz. face bars									
1990	6" x 6"	1 Shee	26	.308	Ea.	32	14.50		46.50	57.50
2000	8" x 4"		26	.308		31.50	14.50		46	57
2020	8" x 8"		24	.333		37	15.75		52.75	65
2040	10" x 6"		20	.400		37.50	18.90		56.40	70
2060	10" x 10"		19	.421		43.50	19.85		63.35	78.50
2080	12" x 6"		19	.421		40.50	19.85		60.35	75
2100	12" x 12"		18	.444		52.50	21		73.50	90.50
2120	14" x 8"		17	.471		48.50	22		70.50	87
2140	14" x 12"		18	.444		58	21		79	96
2160	14" x 14"		18	.444		65.50	21		86.50	105
2180	16" x 6"		18	.444		44	21		65	80.50
2200	16" x 12"		18	.444		64	21		85	103
2220	16" x 16"		17	.471		81	22		103	123
2240	18" x 8"		18	.444		56	21		77	94.50
2260	18" x 12"		17	.471		68.50	22		90.50	109
2280	18" x 18"		16	.500		99.50	23.50		123	146
2300	20" x 10"		18	.444		67	21		88	106
2320	20" x 16"		16	.500		96.50	23.50		120	143
2340	20" x 20"		16	.500		122	23.50		145.50	171
2360	24" x 12"		15	.533		84	25		109	132
2380	24" x 16"		14	.571		112	27		139	165
2400	24" x 20"		12	.667		140	31.50		171.50	203
2420	24" x 24"		10	.800		171	38		209	246
2440	30" x 12"		13	.615		109	29		138	165
2460	30" x 16"		13	.615		132	29		161	191
2480	30" x 24"		11	.727		221	34.50		255.50	296
2500	30" x 30"		9	.889		283	42		325	375
2520	36" x 12"		11	.727		135	34.50		169.50	202
2540	36" x 24"		10	.800		272	38		310	360
2560	36" x 36"	↓	8	1		455	47		502	575
2600	For 2 way deflect., adj. vert. or horiz. face bars, add					40%				
2700	Above registers in steel instead of aluminum, deduct				↓	10%				
3000	Baseboard, hand adj. damper, enameled steel									
3012	8" x 6"	1 Shee	26	.308	Ea.	12.85	14.50		27.35	36.50
3020	10" x 6"		24	.333		14	15.75		29.75	39.50
3040	12" x 5"		23	.348		15.20	16.40		31.60	42
3060	12" x 6"		23	.348		15.20	16.40		31.60	42
3080	12" x 8"		22	.364		22	17.15		39.15	50.50
3100	14" x 6"	↓	20	.400	↓	16.45	18.90		35.35	47
4000	Floor, toe operated damper, enameled steel									
4020	4" x 8"	1 Shee	32	.250	Ea.	24	11.80		35.80	44
4040	4" x 12"		26	.308		28	14.50		42.50	53
4060	6" x 8"		28	.286		25.50	13.50		39	49
4080	6" x 14"		22	.364		32.50	17.15		49.65	62
4100	8" x 10"		22	.364		29	17.15		46.15	58.50
4120	8" x 16"	↓	20	.400	↓	48.50	18.90		67.40	82.50

23 37 13.60 Registers		Crew	Daily Output	Labor-Hours	Unit	Material	Labor	Equipment	Total	Total Incl O&P
						2009 Bare Costs				
4140	10″ x 10″	1 Shee	20	.400	Ea.	35	18.90		53.90	67.50
4160	10″ x 16″		18	.444		53	21		74	90.50
4180	12″ x 12″		18	.444		43	21		64	79.50
4200	12″ x 24″		16	.500		148	23.50		171.50	200
4220	14″ x 14″		16	.500		122	23.50		145.50	171
4240	14″ x 20″	↓	15	.533	↓	163	25		188	218
4300	Spiral pipe supply register									
4310	Aluminum, double deflection, w/damper extractor									
4320	4″ x 12″, for 6″ thru 10″ diameter duct	1 Shee	25	.320	Ea.	57	15.10		72.10	86.50
4330	4″ x 18″, for 6″ thru 10″ diameter duct		18	.444		71	21		92	111
4340	6″ x 12″, for 8″ thru 12″ diameter duct		19	.421		62.50	19.85		82.35	99
4350	6″ x 18″, for 8″ thru 12″ diameter duct		18	.444		81	21		102	122
4360	6″ x 24″, for 8″ thru 12″ diameter duct		16	.500		100	23.50		123.50	147
4370	6″ x 30″, for 8″ thru 12″ diameter duct		15	.533		127	25		152	179
4380	8″ x 18″, for 10″ thru 14″ diameter duct		18	.444		86.50	21		107.50	128
4390	8″ x 24″, for 10″ thru 14″ diameter duct		15	.533		108	25		133	158
4400	8″ x 30″, for 10″ thru 14″ diameter duct		14	.571		144	27		171	200
4410	10″ x 24″, for 12″ thru 18″ diameter duct		13	.615		119	29		148	176
4420	10″ x 30″, for 12″ thru 18″ diameter duct		12	.667		157	31.50		188.50	221
4430	10″ x 36″, for 12″ thru 18″ diameter duct	↓	11	.727	↓	195	34.50		229.50	267
4980	Air return									
5000	Ceiling or wall, fixed 45° face blades									
5010	Adjustable O.B. damper, anodized aluminum									
5020	4″ x 8″	1 Shee	26	.308	Ea.	23	14.50		37.50	48
5040	6″ x 8″		24	.333		26.50	15.75		42.25	53
5060	6″ x 10″		19	.421		26.50	19.85		46.35	59.50
5070	6″ x 12″		19	.421		31	19.85		50.85	64.50
5080	6″ x 16″		18	.444		34	21		55	70
5100	8″ x 10″		19	.421		31	19.85		50.85	64.50
5120	8″ x 12″		16	.500		32.50	23.50		56	72
5140	10″ x 10″		18	.444		32.50	21		53.50	68
5160	10″ x 16″		17	.471		40	22		62	78
5170	12″ x 12″		18	.444		35	21		56	71
5180	12″ x 18″		18	.444		46	21		67	83.50
5184	12″ x 24″		14	.571		53	27		80	99.50
5200	12″ x 30″		12	.667		70.50	31.50		102	126
5210	12″ x 36″		10	.800		96.50	38		134.50	164
5220	16″ x 16″		17	.471		49	22		71	88
5240	18″ x 18″		16	.500		58.50	23.50		82	101
5250	18″ x 24″		13	.615		69.50	29		98.50	121
5254	18″ x 30″		11	.727		107	34.50		141.50	170
5260	18″ x 36″		10	.800		127	38		165	197
5270	20″ x 20″		10	.800		67	38		105	132
5274	20″ x 24″		12	.667		80	31.50		111.50	136
5280	24″ x 24″		11	.727		89	34.50		123.50	151
5290	24″ x 30″		10	.800		99	38		137	167
5300	24″ x 36″		8	1		140	47		187	227
5320	24″ x 48″		6	1.333		209	63		272	325
5340	30″ x 30″	↓	6.50	1.231		164	58		222	270
5344	30″ x 120″	Q-9	6	2.667		705	113		818	950
5360	36″ x 36″	1 Shee	6	1.333		310	63		373	440
5370	40″ x 32″	″	6	1.333	↓	420	63		483	560
5400	Ceiling or wall, removable-reversible core, single deflection									

23 37 13 – Diffusers, Registers, and Grilles

23 37 13.60 Registers

23 37 13.60 Registers		Crew	Daily Output	Labor-Hours	Unit	2009 Bare Costs Material	Labor	Equipment	Total	Total Incl O&P
5402	Adjustable O.B. damper, radiused frame with border, aluminum									
5410	8" x 4"	1 Shee	26	.308	Ea.	29.50	14.50		44	55
5414	8" x 6"		24	.333		29.50	15.75		45.25	56.50
5418	10" x 6"		19	.421		37.50	19.85		57.35	71.50
5422	10" x 10"		18	.444		42.50	21		63.50	79.50
5426	12" x 6"		19	.421		37.50	19.85		57.35	71.50
5430	12" x 12"		18	.444		48	21		69	85.50
5434	16" x 16"		16	.500		72	23.50		95.50	116
5438	18" x 18"		17	.471		91	22		113	134
5442	20" x 20"		16	.500		109	23.50		132.50	157
5446	24" x 12"		15	.533		72	25		97	119
5450	24" x 18"		13	.615		109	29		138	165
5454	24" x 24"		11	.727		152	34.50		186.50	220
5458	30" x 12"		13	.615		96	29		125	151
5462	30" x 18"		12.50	.640		139	30		169	200
5466	30" x 24"		11	.727		195	34.50		229.50	267
5470	30" x 30"		9	.889		243	42		285	330
5474	36" x 12"		11	.727		120	34.50		154.50	185
5478	36" x 18"		10.50	.762		168	36		204	241
5482	36" x 24"		10	.800		235	38		273	315
5486	36" x 30"		9	.889		315	42		357	410
5490	36" x 36"		8	1		405	47		452	520
6000	For steel construction instead of aluminum, deduct					10%				

23 37 15 – Louvers

23 37 15.40 HVAC Louvers

23 37 15.40 HVAC Louvers		Crew	Daily Output	Labor-Hours	Unit	Material	Labor	Equipment	Total	Total Incl O&P
0010	**HVAC LOUVERS**									
0100	Aluminum, extruded, with screen, mill finish									
1002	Brick vent, see also Div. 04 05 23.19									
1100	Standard, 4" deep, 8" wide, 5" high	1 Shee	24	.333	Ea.	34.50	15.75		50.25	61.50
1200	Modular, 4" deep, 7-3/4" wide, 5" high		24	.333		36	15.75		51.75	63.50
1300	Speed brick, 4" deep, 11-5/8" wide, 3-7/8" high		24	.333		36	15.75		51.75	63.50
1400	Fuel oil brick, 4" deep, 8" wide, 5" high		24	.333		61.50	15.75		77.25	92
2000	Cooling tower and mechanical equip., screens, light weight		40	.200	S.F.	15.40	9.45		24.85	31.50
2020	Standard weight		35	.229		41	10.80		51.80	61.50
2500	Dual combination, automatic, intake or exhaust		20	.400		56	18.90		74.90	90.50
2520	Manual operation		20	.400		41.50	18.90		60.40	75
2540	Electric or pneumatic operation		20	.400		41.50	18.90		60.40	75
2560	Motor, for electric or pneumatic		14	.571	Ea.	480	27		507	570
3000	Fixed blade, continuous line									
3100	Mullion type, stormproof	1 Shee	28	.286	S.F.	41.50	13.50		55	67
3200	Stormproof		28	.286		41.50	13.50		55	67
3300	Vertical line		28	.286		49.50	13.50		63	75
3500	For damper to use with above, add					50%	30%			
3520	Motor, for damper, electric or pneumatic	1 Shee	14	.571	Ea.	480	27		507	570
4000	Operating, 45°, manual, electric or pneumatic		24	.333	S.F.	41.50	15.75		57.25	70
4100	Motor, for electric or pneumatic		14	.571	Ea.	480	27		507	570
4200	Penthouse, roof		56	.143	S.F.	24.50	6.75		31.25	37.50
4300	Walls		40	.200		58	9.45		67.45	78.50
5000	Thinline, under 4" thick, fixed blade		40	.200		24	9.45		33.45	41
5010	Finishes, applied by mfr. at additional cost, available in colors									
5020	Prime coat only, add				S.F.	3.30			3.30	3.63
5040	Baked enamel finish coating, add					6.10			6.10	6.70

23 37 15 – Louvers

23 37 15.40 HVAC Louvers		Crew	Daily Output	Labor-Hours	Unit	Material	Labor	Equipment	Total	Total Incl O&P
						2009 Bare Costs				
5060	Anodized finish, add				S.F.	6.60			6.60	7.25
5080	Duranodic finish, add					12			12	13.20
5100	Fluoropolymer finish coating, add					18.90			18.90	21
9980	For small orders (under 10 pieces), add					25%				

23 37 23 – HVAC Gravity Ventilators

23 37 23.10 HVAC Gravity Air Ventilators		Crew	Daily Output	Labor-Hours	Unit	Material	Labor	Equipment	Total	Total Incl O&P
0010	**HVAC GRAVITY AIR VENTILATORS**, Incl. base and damper									
1280	Rotary ventilators, wind driven, galvanized									
1300	4″ neck diameter	Q-9	20	.800	Ea.	35	34		69	91
1320	5″ neck diameter		18	.889		39	38		77	101
1340	6″ neck diameter		16	1		39	42.50		81.50	109
1360	8″ neck diameter		14	1.143		40.50	48.50		89	119
1380	10″ neck diameter		12	1.333		52.50	56.50		109	145
1400	12″ neck diameter		10	1.600		56.50	68		124.50	168
1420	14″ neck diameter		10	1.600		78.50	68		146.50	192
1440	16″ neck diameter		9	1.778		95	75.50		170.50	221
1460	18″ neck diameter, 1700 CFM		9	1.778		111	75.50		186.50	238
1480	20″ neck diameter, 2100 CFM		8	2		123	85		208	266
1500	24″ neck diameter, 3,100 CFM		8	2		165	85		250	315
1520	30″ neck diameter, 4500 CFM		7	2.286		345	97		442	530
1540	36″ neck diameter, 5,500 CFM		6	2.667		485	113		598	710
1600	For aluminum, add					300%				
1620	For stainless steel, add					600%				
1630	For copper, add					600%				
2000	Stationary, gravity, syphon, galvanized									
2100	3″ neck diameter, 40 CFM	Q-9	24	.667	Ea.	31.50	28.50		60	78.50
2120	4″ neck diameter, 50 CFM		20	.800		34.50	34		68.50	90.50
2140	5″ neck diameter, 58 CFM		18	.889		35	38		73	96.50
2160	6″ neck diameter, 66 CFM		16	1		30	42.50		72.50	98.50
2180	7″ neck diameter, 86 CFM		15	1.067		38	45.50		83.50	112
2200	8″ neck diameter, 110 CFM		14	1.143		38	48.50		86.50	117
2220	10″ neck diameter, 140 CFM		12	1.333		49.50	56.50		106	142
2240	12″ neck diameter, 160 CFM		10	1.600		63.50	68		131.50	175
2260	14″ neck diameter, 250 CFM		10	1.600		80	68		148	193
2280	16″ neck diameter, 380 CFM		9	1.778		107	75.50		182.50	233
2300	18″ neck diameter, 500 CFM		9	1.778		142	75.50		217.50	272
2320	20″ neck diameter, 625 CFM		8	2		150	85		235	296
2340	24″ neck diameter, 900 CFM		8	2		173	85		258	320
2360	30″ neck diameter, 1375 CFM		7	2.286		395	97		492	585
2380	36″ neck diameter, 2,000 CFM		6	2.667		540	113		653	770
2400	42″ neck diameter, 3000 CFM		4	4		765	170		935	1,100
2410	48″ neck diameter, 4000 CFM		3	5.333		1,025	227		1,252	1,475
2500	For aluminum, add					300%				
2520	For stainless steel, add					600%				
2530	For copper, add					600%				
3000	Rotating chimney cap, galvanized, 4″ neck diameter	Q-9	20	.800		27	34		61	82.50
3020	5″ neck diameter		18	.889		28	38		66	88.50
3040	6″ neck diameter		16	1		28.50	42.50		71	96.50
3060	7″ neck diameter		15	1.067		30.50	45.50		76	103
3080	8″ neck diameter		14	1.143		32	48.50		80.50	110
3100	10″ neck diameter		12	1.333		41	56.50		97.50	132
3600	Stationary chimney rain cap, galvanized, 3″ neck diameter		24	.667		24	28.50		52.50	69.50

23 37 Air Outlets and Inlets

23 37 23 – HVAC Gravity Ventilators

23 37 23.10 HVAC Gravity Air Ventilators		Crew	Daily Output	Labor-Hours	Unit	2009 Bare Costs Material	Labor	Equipment	Total	Total Incl O&P
3620	4″ neck diameter	Q-9	20	.800	Ea.	24.50	34		58.50	79
3640	6″ neck diameter		16	1		25.50	42.50		68	93.50
3680	8″ neck diameter		14	1.143		27.50	48.50		76	105
3700	10″ neck diameter		12	1.333		34.50	56.50		91	125
3720	12″ neck diameter		10	1.600		57.50	68		125.50	168
3740	14″ neck diameter		10	1.600		61.50	68		129.50	173
3760	16″ neck diameter		9	1.778		75.50	75.50		151	199
3770	18″ neck diameter		9	1.778		109	75.50		184.50	236
3780	20″ neck diameter		8	2		146	85		231	291
3790	24″ neck diameter		8	2		170	85		255	320
4200	Stationary mushroom, aluminum, 16″ orifice diameter		10	1.600		510	68		578	665
4220	26″ orifice diameter		6.15	2.602		755	111		866	1,000
4230	30″ orifice diameter		5.71	2.802		1,100	119		1,219	1,400
4240	38″ orifice diameter		5	3.200		1,575	136		1,711	1,950
4250	42″ orifice diameter		4.70	3.404		2,100	145		2,245	2,525
4260	50″ orifice diameter		4.44	3.604		2,500	153		2,653	2,975
5000	Relief vent									
5500	Rectangular, aluminum, galvanized curb									
5510	intake/exhaust, 0.033″ SP									
5580	500 CFM, 12″ x 12″	Q-9	8.60	1.860	Ea.	590	79		669	770
5600	600 CFM, 12″x16″		8	2		665	85		750	860
5620	750 CFM, 12″ x 20″		7.20	2.222		715	94.50		809.50	930
5640	1000 CFM, 12″ x 24″		6.60	2.424		750	103		853	985
5660	1500 CFM, 12″x 36″		5.80	2.759		995	117		1,112	1 275
5680	3000 CFM, 20″ x 42″		4	4		1,325	170		1,495	1.700
5700	5000 CFM, 20″ x 72″		3	5.333		1,900	227		2,127	2.425
5720	6000 CFM, 24″ x 72″		2.30	6.957		2,050	296		2,346	2.700
5740	10,000 CFM, 48″ x 60″		1.80	8.889		2,600	380		2,980	3,425
5760	12,000 CFM, 48″ x 72″		1.60	10		3,075	425		3,500	4,025
5780	13,750 CFM, 60″ x 66″		1.40	11.429		3,400	485		3,885	4,475
5800	15,000 CFM, 60″ x 72″		1.30	12.308		3,625	525		4,150	4,800
5820	18,000 CFM, 72″ x 72″		1.20	13.333		4,125	565		4,690	5,425
5880	Size is throat area, volume is at 500 FPM									
7000	Note: sizes based on exhaust. Intake, with 0.125″ SP									
7100	loss, approximately twice listed capacity.									

23 38 Ventilation Hoods

23 38 13 – Commercial-Kitchen Hoods

23 38 13.10 Hood and Ventilation Equipment		Crew	Daily Output	Labor-Hours	Unit	Material	Labor	Equipment	Total	Total Incl O&P
0010	**HOOD AND VENTILATION EQUIPMENT**									
7950	Hood fire protection system, minimum	Q-1	3	5.333	Ea.	4,100	234		4,334	4,850
8050	Maximum	″	1	16	″	31,600	700		32,300	35,900

23 38 13.20 Kitchen Ventilation		Crew	Daily Output	Labor-Hours	Unit	Material	Labor	Equipment	Total	Total Incl O&P
0010	**KITCHEN VENTILATION**, Commercial									
1010	Heat reclaim unit, air to air									
1100	Heat pipe exchanger									
1110	Combined supply/exhaust air volume									
1120	2.5 to 6.0 MCFM	Q-10	2.80	8.571	MCFM	6,775	380		7,155	8,025
1130	6 to 16 MCFM		5	4.800		4,150	211		4,361	4,900
1140	16 to 22 MCFM		6	4		3,250	176		3,426	3,850
2010	Packaged ventilation									

23 38 Ventilation Hoods

23 38 13 – Commercial-Kitchen Hoods

23 38 13.20 Kitchen Ventilation		Crew	Daily Output	Labor-Hours	Unit	2009 Bare Costs Material	Labor	Equipment	Total	Total Incl O&P
2100	Combined supply/exhaust air volume									
2110	With ambient supply									
2120	3 to 7 MCFM	Q-10	5	4.800	MCFM	1,225	211		1,436	1,675
2130	7 to 12 MCFM		7.10	3.380		815	149		964	1,125
2140	12 to 22 MCFM	↓	10.70	2.243	↓	575	99		674	785
2400	With tempered supply									
2410	3 to 7 MCFM	Q-10	5	4.800	MCFM	2,125	211		2,336	2,650
2420	7 to 12 MCFM		7.10	3.380		1,375	149		1,524	1,725
2430	12 to 22 MCFM	↓	10.70	2.243	↓	1,000	99		1,099	1,250
6000	Exhaust hoods									
6100	Centrifugal grease extraction									
6110	Water wash type									
6120	Per foot of length, minimum	Q-10	7.10	3.380	L.F.	1,575	149		1,724	1,950
6130	Maximum	"	7.10	3.380	"	2,700	149		2,849	3,175
6200	Non water wash type									
6210	Per foot of length	Q-10	13.40	1.791	L.F.	1,150	79		1,229	1,400
6400	Island style ventilator									
6410	Water wash type									
6420	Per foot of length, minimum	Q-10	7.10	3.380	L.F.	2,100	149		2,249	2,525
6430	Maximum	"	7.10	3.380	"	3,500	149		3,649	4,075
6460	Non water wash type									
6470	Per foot of length, minimum	Q-10	13.40	1.791	L.F.	910	79		989	1,125
6480	Maximum	"	13.40	1.791	"	1,850	79		1,929	2,175

23 41 Particulate Air Filtration

23 41 13 – Panel Air Filters

23 41 13.10 Panel Type Air Filters		Crew	Daily Output	Labor-Hours	Unit	Material	Labor	Equipment	Total	Total Incl O&P
0010	**PANEL TYPE AIR FILTERS**									
2950	Mechanical media filtration units									
3000	High efficiency type, with frame, non-supported G				MCFM	45			45	49.50
3100	Supported type G				"	60			60	66
5500	Throwaway glass or paper media type				Ea.	3.35			3.35	3.69

23 41 16 – Renewable-Media Air Filters

23 41 16.10 Disposable Media Air Filters		Crew	Daily Output	Labor-Hours	Unit	Material	Labor	Equipment	Total	Total Incl O&P
0010	**DISPOSABLE MEDIA AIR FILTERS**									
5000	Renewable disposable roll				MCFM	250			250	275
5800	Filter, bag type									
5810	90-95% efficiency									
5820	24" x 12" x 29", .75-1.25 MCFM	1 Shee	1.90	4.211	Ea.	43.50	199		242.50	355
5830	24" x 24" x 29", 1.5-2.5 MCFM	"	1.90	4.211	"	79	199		278	390
5850	80-85% efficiency									
5860	24" x 12" x 29", .75-1.25 MCFM	1 Shee	1.90	4.211	Ea.	39	199		238	350
5870	24" x 24" x 29", 1.5-2.5 MCFM	"	1.90	4.211	"	71	199		270	385

23 41 19 – Washable Air Filters

23 41 19.10 Permanent Air Filters		Crew	Daily Output	Labor-Hours	Unit	Material	Labor	Equipment	Total	Total Incl O&P
0010	**PERMANENT AIR FILTERS**									
4500	Permanent washable G				MCFM	20			20	22

23 41 Particulate Air Filtration

23 41 23 – Extended Surface Filters

23 41 23.10 Expanded Surface Filters		Crew	Daily Output	Labor-Hours	Unit	2009 Bare Costs Material	Labor	Equipment	Total	Total Incl O&P
0010	**EXPANDED SURFACE FILTERS**									
4000	Medium efficiency, extended surface G				MCFM	5.50			5.50	6.05

23 41 33 – High-Efficiency Particulate Filtration

23 41 33.10 HEPA Filters

Line	Description	Crew	Daily Output	Labor-Hours	Unit	Material	Labor	Equipment	Total	Total Incl O&P
0010	**HEPA FILTERS**									
6000	HEPA filter complete w/particle board,									
6010	kraft paper frame, separator material									
6020	95% DOP efficiency									
6030	12" x 12" x 6", 150 CFM	1 Shee	3.70	2.162	Ea.	44	102		146	206
6034	24" x 12" x 6", 375 CFM		1.90	4.211		73.50	199		272.50	335
6038	24" x 18" x 6", 450 CFM		1.90	4.211		90	199		289	405
6042	24" x 24" x 6", 700 CFM		1.90	4.211		101	199		300	415
6046	12" x 12" x 12", 250 CFM		3.70	2.162		74.50	102		176.50	239
6050	24" x 12" x 12", 500 CFM		1.90	4.211		105	199		304	420
6054	24" x 18" x 12", 875 CFM		1.90	4.211		146	199		345	455
6058	24" x 24" x 12", 1000 CFM		1.90	4.211		158	199		357	430
6100	99% DOP efficiency									
6110	12" x 12" x 6", 150 CFM	1 Shee	3.70	2.162	Ea.	46	102		148	208
6114	24" x 12" x 6", 325 CFM		1.90	4.211		76	199		275	390
6118	24" x 18" x 6", 550 CFM		1.90	4.211		93	199		292	405
6122	24" x 24" x 6", 775 CFM		1.90	4.211		105	199		304	420
6124	24" x 48" x 6", 775 CFM		1.30	6.154		700	290		990	1,225
6126	12" x 12" x 12", 250 CFM		3.60	2.222		77	105		182	246
6130	24" x 12" x 12", 500 CFM		1.90	4.211		109	199		308	425
6134	24" x 18" x 12", 775 CFM		1.90	4.211		152	199		351	470
6138	24" x 24" x 12", 1100 CFM		1.90	4.211		164	199		363	485
6500	HEPA filter housing, 14 ga. galv. sheet metal									
6510	12" x 12" x 6"	1 Shee	2.50	3.200	Ea.	660	151		811	955
6514	24" x 12" x 6"		2	4		740	189		929	1,100
6518	12" x 12" x 12"		2.40	3.333		660	157		817	965
6522	14" x 12" x 12"		2.30	3.478		660	164		824	980
6526	24" x 18" x 6"		1.90	4.211		940	199		1,139	1,325
6530	24" x 24" x 6"		1.80	4.444		855	210		1,065	1,275
6534	24" x 48" x 6"		1.70	4.706		1,150	222		1,372	1,600
6538	24" x 72" x 6"		1.60	5		1,475	236		1,711	2,000
6542	24" x 18" x 12"		1.80	4.444		940	210		1,150	1,350
6546	24" x 24" x 12"		1.70	4.706		1,050	222		1,272	1,500
6550	24" x 48" x 12"		1.60	5		1,225	236		1,461	1,725
6554	24" x 72" x 12"		1.50	5.333		1,525	252		1,777	2,075
6558	48" x 48" x 6"	Q-9	2.80	5.714		1,700	243		1,943	2,225
6562	48" x 72" x 6"		2.60	6.154		2,025	261		2,286	2,625
6566	48" x 96" x 6"		2.40	6.667		2,575	283		2,858	3,250
6570	48" x 48" x 12"		2.70	5.926		1,700	252		1,952	2,225
6574	48" x 72" x 12"		2.50	6.400		2,025	272		2,297	2,650
6578	48" x 96" x 12"		2.30	6.957		2,575	296		2,871	3,275
6582	114" x 72" x 12"		2	8		3,550	340		3,890	4,425

23 42 Gas-Phase Air Filtration

23 42 13 – Activated-Carbon Air Filtration

23 42 13.10 Charcoal Type Air Filtration		Crew	Daily Output	Labor-Hours	Unit	Material	2009 Bare Costs Labor	Equipment	Total	Total Incl O&P
0010	**CHARCOAL TYPE AIR FILTRATION**									
0050	Activated charcoal type, full flow				MCFM	600			600	660
0060	Full flow, impregnated media 12" deep					225			225	248
0070	HEPA filter & frame for field erection					300			300	330
0080	HEPA filter-diffuser, ceiling install.					275			275	305

23 42 16 – Chemically-Impregnated Adsorption Air Filtration

23 42 16.10 Chemical Adsorption Air Filtration		Crew	Daily Output	Labor-Hours	Unit	Material	Labor	Equipment	Total	Total Incl O&P
0010	**CHEMICAL ADSORPTION AIR FILTRATION**									
0500	Chemical media filtration type									
1100	Industrial air fume & odor scrubber unit w/pump & motor									
1110	corrosion resistant PVC construction									
1120	Single pack filter, horizontal type									
1130	500 CFM	Q-9	14	1.143	Ea.	5,200	48.50		5,248.50	5,800
1140	1000 CFM		11	1.455		6,100	62		6,162	6,825
1150	2000 CFM		8	2		7,400	85		7,485	8,275
1160	3000 CFM		7	2.286		8,575	97		8,672	9,575
1170	5000 CFM		5	3.200		11,200	136		11,336	12,500
1180	8000 CFM		4	4		15,600	170		15,770	17,500
1190	12,000 CFM		3	5.333		20,700	227		20,927	23,200
1200	16,000 CFM		2.50	6.400		25,600	272		25,872	28,500
1210	20,000 CFM		2	8		32,100	340		32,440	35,800
1220	26,000 CFM		1.50	10.667		36,900	455		37,355	41,300
1230	30,000 CFM	Q-10	2	12		42,000	530		42,530	46,900
1240	40,000 CFM	"	1.50	16		55,000	705		55,705	61,500
1250	50,000 CFM	Q-11	2	16		67,000	720		67,720	74,500
1260	55,000 CFM		1.80	17.778		73,000	800		73,800	81,000
1270	60,000 CFM		1.50	21.333		78,500	960		79,460	88,000
1300	Double pack filter, horizontal type									
1310	500 CFM	Q-9	10	1.600	Ea.	6,775	68		6,843	7,550
1320	1000 CFM		8	2		8,125	85		8,210	9,050
1330	2000 CFM		6	2.667		9,775	113		9,888	11,000
1340	3000 CFM		5	3.200		11,700	136		11,836	13,100
1350	5000 CFM		4	4		14,900	170		15,070	16,600
1360	8000 CFM		3	5.333		19,600	227		19,827	22,000
1370	12,000 CFM		2.50	6.400		26,600	272		26,872	29,700
1380	16,000 CFM		2	8		34,800	340		35,140	38,700
1390	20,000 CFM		1.50	10.667		42,100	455		42,555	47,000
1400	26,000 CFM		1	16		48,800	680		49,480	54,500
1410	30,000 CFM	Q-10	1.50	16		57,500	705		58,205	64,000
1420	40,000 CFM	"	1.30	18.462		73,500	815		74,315	82,000
1430	50,000 CFM	Q-11	1.50	21.333		91,500	960		92,460	102,500
1440	55,000 CFM		1.30	24.615		100,000	1,100		101,100	111,500
1450	60,000 CFM		1	32		108,500	1,450		109,950	121,500
1500	Single pack filter, vertical type									
1510	500 CFM	Q-9	24	.667	Ea.	4,975	28.50		5,003.50	5,525
1520	1000 CFM		18	.889		5,875	38		5,913	6,500
1530	2000 CFM		12	1.333		7,050	56.50		7,106.50	7,825
1540	3000 CFM		9	1.778		8,225	75.50		8,300.50	9,175
1550	5000 CFM		6	2.667		11,000	113		11,113	12,300
1560	8000 CFM		4	4		15,600	170		15,770	17,500
1570	12,000 CFM		3	5.333		21,200	227		21,427	23,800
1580	16,000 CFM		2	8		24,800	340		25,140	27,800

23 42 Gas-Phase Air Filtration

23 42 16 – Chemically-Impregnated Adsorption Air Filtration

23 42 16.10 Chemical Adsorption Air Filtration		Crew	Daily Output	Labor-Hours	Unit	2009 Bare Costs Material	Labor	Equipment	Total	Total Inc O&P
1590	20,000 CFM	Q-9	1.80	8.889	Ea.	31,300	380		31,680	35,100
1600	24,000 CFM	↓	1.60	10	↓	35,000	425		35,425	39,200
1650	Double pack filter, vertical type									
1660	500 CFM	Q-9	22	.727	Ea.	6,450	31		6,481	7,150
1670	1000 CFM		16	1		7,700	42.50		7,742.50	8,525
1680	2000 CFM		10	1.600		9,225	68		9,293	10,300
1690	3000 CFM		7	2.286		10,900	97		10,997	12,000
1700	5000 CFM		5	3.200		14,200	136		14,336	15,800
1710	8000 CFM		3	5.333		19,400	227		19,627	21,800
1720	12,000 CFM		2.50	6.400		26,300	272		26,572	29,300
1730	16,000 CFM		2	8		33,000	340		33,340	36,800
1740	20,000 CFM		1.50	10.667		41,200	455		41,655	46,100
1750	24,000 CFM	↓	1	16	↓	46,000	680		46,680	51,500
1800	Inlet or outlet transition, horizontal									
1810	Single pack to 12,000 CFM, add					4%				
1820	Single pack to 30,000 CFM, add					6%				
1830	Single pack to 60,000 CFM, add					8%				
1840	Double pack to 12,000 CFM, add					3%				
1850	Double pack to 30,000 CFM, add					5%				
1860	Double pack to 60,000 CFM, add					6%				
1870	Inlet or outlet transition, vertical									
1880	Single pack to 5000 CFM, add					2%				
1890	Single pack to 24,000 CFM, add					3%				
1900	Double pack to 24,000 CFM, add					2%				

23 43 Electronic Air Cleaners

23 43 13 – Washable Electronic Air Cleaners

23 43 13.10 Electronic Air Cleaners

		Crew	Daily Output	Labor-Hours	Unit	Material	Labor	Equipment	Total	Total Inc O&P
0010	**ELECTRONIC AIR CLEANERS**									
2000	Electronic air cleaner, duct mounted									
2150	400 – 1000 CFM	1 Shee	2.30	3.478	Ea.	1,000	164		1,164	1,350
2200	1000 – 1400 CFM		2.20	3.636		1,225	172		1,397	1,625
2250	1400 – 2000 CFM		2.10	3.810		1,400	180		1,580	1,825
2260	2000 – 2500 CFM	↓	2	4	↓	1,500	189		1,689	1,950

23 51 Breechings, Chimneys, and Stacks

23 51 13 – Draft Control Devices

23 51 13.13 Draft-Induction Fans

		Crew	Daily Output	Labor-Hours	Unit	Material	Labor	Equipment	Total	Total Inc O&P
0010	**DRAFT-INDUCTION FANS**									
1000	Breeching installation									
1800	Hot gas, 600° F, variable pitch pulley and motor									
1840	6" diam. inlet, 1/4 H.P., 1 phase, 400 CFM	Q-9	6	2.667	Ea.	1,500	113		1,613	1,825
1850	7" diam. inlet, 1/4 H.P., 1 phase, 800 CFM		5	3.200		1,600	136		1,736	1,950
1860	8" diam. inlet, 1/4 H.P., 1phase, 1120 CFM		4	4		2,325	170		2,495	2,800
1870	9" diam. inlet, 3/4 H.P., 1 phase, 1440 CFM		3.60	4.444		2,750	189		2,939	3,325
1880	10" diam. inlet, 3/4 H.P., 1 phase, 2000 CFM		3.30	4.848		3,100	206		3,306	3,725
1900	12" diam. inlet, 3/4 H.P., 3 phase, 2960 CFM		3	5.333		3,175	227		3,402	3,850
1910	14" diam. inlet, 1 H.P., 3 phase, 4160 CFM		2.60	6.154		3,750	261		4,011	4,525
1920	16" diam. inlet, 2 H.P., 3 phase, 6720 CFM	↓	2.30	6.957	↓	4,100	296		4,396	4,950

23 51 Breechings, Chimneys, and Stacks

23 51 13 – Draft Control Devices

23 51 13.13 Draft-Induction Fans		Crew	Daily Output	Labor-Hours	Unit	2009 Bare Costs Material	Labor	Equipment	Total	Total Incl O&P
1940	18" diam. inlet, 3 H.P., 3 phase, 9120 CFM	Q-9	2	8	Ea.	4,925	340		5,265	5,950
1950	20" diam. inlet, 3 H.P., 3 phase, 9760 CFM		1.50	10.667		5,375	455		5,830	6,600
1960	22" diam. inlet, 5 H.P., 3 phase, 13,360 CFM		1	16		6,300	680		6,980	7,975
1980	24" diam. inlet, 7-1/2 H.P., 3 phase, 17,760 CFM		.80	20		7,750	850		8,600	9,825
2300	For multi-blade damper at fan inlet, add					20%				
3600	Chimney-top installation									
3700	6" size	1 Shee	8	1	Ea.	690	47		737	830
3740	8" size		7	1.143		690	54		744	845
3750	10" size		6.50	1.231		905	58		963	1,075
3780	13" size		6	1.333		930	63		993	1,125
3880	For speed control switch, add					54.50			54.50	60
3920	For thermal fan control, add					54.50			54.50	60
5500	Flue shutter damper for draft control,									
5510	parallel blades									
5550	8" size	Q-9	8	2	Ea.	520	85		605	705
5560	9" size		7.50	2.133		565	90.50		655.50	760
5570	10" size		7	2.286		585	97		682	790
5580	12" size		6.50	2.462		640	105		745	865
5590	14" size		6	2.667		725	113		838	975
5600	16" size		5.50	2.909		790	124		914	1,050
5610	18" size		5	3.200		865	136		1,001	1,150
5620	20" size		4.50	3.556		960	151		1,111	1,275
5630	22" size		4	4		1,050	170		1,220	1,400
5640	24" size		3.50	4.571		1,150	194		1,344	1,575
5650	27" size		3	5.333		1,175	227		1,402	1,650
5660	30" size		2.50	6.400		1,225	272		1,497	1,775
5670	32" size		2	8		1,275	340		1,615	1,950
5680	36" size		1.50	10.667		1,375	455		1,830	2,225

23 51 13.16 Vent Dampers		Crew	Daily Output	Labor-Hours	Unit	Material	Labor	Equipment	Total	Total Incl O&P
0010	**VENT DAMPERS**									
5000	Vent damper, bi-metal, gas, 3" diameter	Q-9	24	.667	Ea.	38	28.50		66.50	85.50
5010	4" diameter		24	.667		38	28.50		66.50	85.50
5020	5" diameter		23	.696		38	29.50		67.50	87.50
5030	6" diameter		22	.727		38	31		69	89.50
5040	7" diameter		21	.762		40.50	32.50		73	94.50
5050	8" diameter		20	.800		44	34		78	101

23 51 13.19 Barometric Dampers		Crew	Daily Output	Labor-Hours	Unit	Material	Labor	Equipment	Total	Total Incl O&P
0010	**BAROMETRIC DAMPERS**									
1000	Barometric, gas fired system only, 6" size for 5" and 6" pipes	1 Shee	20	.400	Ea.	54	18.90		72.90	88.50
1020	7" size, for 6" and 7" pipes		19	.421		59	19.85		78.85	95.50
1040	8" size, for 7" and 8" pipes		18	.444		74.50	21		95.50	115
1060	9" size, for 8" and 9" pipes		16	.500		87	23.50		110.50	133
2000	All fuel, oil, oil/gas, coal									
2020	10" for 9" and 10" pipes	1 Shee	15	.533	Ea.	125	25		150	177
2040	12" for 11" and 12" pipes		15	.533		163	25		188	219
2060	14" for 13" and 14" pipes		14	.571		212	27		239	275
2080	16" for 15" and 16" pipes		13	.615		297	29		326	370
2100	18" for 17" and 18" pipes		12	.667		395	31.50		426.50	485
2120	20" for 19" and 21" pipes		10	.800		470	38		508	580
2140	24" for 22" and 25" pipes	Q-9	12	1.333		580	56.50		636.50	720
2160	28" for 26" and 30" pipes		10	1.600		715	68		783	895
2180	32" for 31" and 34" pipes		8	2		915	85		1,000	1,125

23 51 13 – Draft Control Devices

23 51 13.19 Barometric Dampers		Crew	Daily Output	Labor-Hours	Unit	2009 Bare Costs Material	Labor	Equipment	Total	Total Incl O&P
3260	For thermal switch for above, add	1 Shee	24	.333	Ea.	52.50	15.75		68.25	82

23 51 23 – Gas Vents

23 51 23.10 Gas Chimney Vents

23 51 23.10 Gas Chimney Vents		Crew	Daily Output	Labor-Hours	Unit	Material	Labor	Equipment	Total	Total Incl O&P
0010	**GAS CHIMNEY VENTS,** Prefab metal, U.L. listed									
0020	Gas, double wall, galvanized steel									
0080	3″ diameter	Q-9	72	.222	V.L.F.	5.35	9.45		14.80	20.50
0100	4″ diameter		68	.235		6.70	10		16.70	23
0120	5″ diameter		64	.250		7.80	10.60		18.40	25
0140	6″ diameter		60	.267		9.10	11.35		20.45	27.50
0160	7″ diameter		56	.286		13.35	12.15		25.50	33.50
0180	8″ diameter		52	.308		14.85	13.05		27.90	36.50
0200	10″ diameter		48	.333		34	14.15		48.15	59
0220	12″ diameter		44	.364		45	15.45		60.45	73.50
0240	14″ diameter		42	.381		76	16.20		92.20	109
0260	16″ diameter		40	.400		103	17		120	139
0280	18″ diameter		38	.421		133	17.90		150.90	174
0300	20″ diameter	Q-10	36	.667		139	29.50		168.50	198
0320	22″ diameter		34	.706		199	31		230	267
0340	24″ diameter		32	.750		245	33		278	320
0360	26″ diameter		31	.774		294	34		328	380
0380	28″ diameter		30	.800		310	35		345	395
0400	30″ diameter		28	.857		330	38		368	420
0420	32″ diameter		27	.889		380	39		419	480
0440	34″ diameter		26	.923		435	40.50		475.50	540
0460	36″ diameter		25	.960		460	42.50		502.50	570
0480	38″ diameter		24	1		500	44		544	620
0500	40″ diameter		23	1.043		560	46		606	685
0520	42″ diameter		22	1.091		585	48		633	720
0540	44″ diameter		21	1.143		645	50.50		695.50	790
0560	46″ diameter		20	1.200		710	53		763	865
0580	48″ diameter		19	1.263		790	55.50		845.50	950
0600	For 4″, 5″ and 6″ oval, add					50%				
0650	Gas, double wall, galvanized steel, fittings									
0660	Elbow 45°, 3″ diameter	Q-9	36	.444	Ea.	13.25	18.90		32.15	43.50
0670	4″ diameter		34	.471		15.75	20		35.75	48.50
0680	5″ diameter		32	.500		18.50	21		39.50	53
0690	6″ diameter		30	.533		22	22.50		44.50	59.50
0700	7″ diameter		28	.571		32	24.50		56.50	72.50
0710	8″ diameter		26	.615		37	26		63	80.50
0720	10″ diameter		24	.667		98	28.50		126.50	152
0730	12″ diameter		22	.727		124	31		155	185
0740	14″ diameter		21	.762		174	32.50		206.50	241
0750	16″ diameter		20	.800		218	34		252	293
0760	18″ diameter		19	.842		281	36		317	365
0770	20″ diameter	Q-10	18	1.333		330	58.50		388.50	455
0780	22″ diameter		17	1.412		510	62		572	655
0790	24″ diameter		16	1.500		635	66		701	800
0800	26″ diameter		16	1.500		1,250	66		1,316	1,475
0810	28″ diameter		15	1.600		1,275	70.50		1,345.50	1,500
0820	30″ diameter		14	1.714		1,350	75.50		1,425.50	1,625
0830	32″ diameter		14	1.714		1,525	75.50		1,600.50	1,800
0840	34″ diameter		13	1.846		1,675	81.50		1,756.50	1,975

23 51 23.10 Gas Chimney Vents		Crew	Daily Output	Labor-Hours	Unit	Material	Labor	Equipment	Total	Total Incl O&P
						2009 Bare Costs				
0850	36" diameter	Q-10	12	2	Ea.	1,725	88		1,813	2,000
0860	38" diameter		12	2		1,850	88		1,938	2,150
0870	40" diameter		11	2.182		1,950	96		2,046	2,300
0880	42" diameter		11	2.182		2,050	96		2,146	2,400
0890	44" diameter		10	2.400		2,200	106		2,306	2,600
0900	46" diameter		10	2.400		2,350	106		2,456	2,750
0910	48" diameter		10	2.400		2,450	106		2,556	2,850
0916	Adjustable length									
0918	3" diameter, to 12"	Q-9	36	.444	Ea.	10.90	18.90		29.80	41
0920	4" diameter, to 12"		34	.471		12.90	20		32.90	45
0924	6" diameter, to 12"		30	.533		16.60	22.50		39.10	53.50
0928	8" diameter, to 12"		26	.615		24.50	26		50.50	67
0930	10" diameter, to 18"		24	.667		99	28.50		127.50	153
0932	12" diameter, to 18"		22	.727		121	31		152	181
0936	16" diameter, to 18"		20	.800		251	34		285	330
0938	18" diameter, to 18"		19	.842		277	36		313	360
0944	24" diameter, to 18"	Q-10	16	1.500		495	66		561	645
0950	Elbow 90°, adjustable, 3" diameter	Q-9	36	.444		22	18.90		40.90	53.50
0960	4" diameter		34	.471		27	20		47	60.50
0970	5" diameter		32	.500		31.50	21		52.50	67
0980	6" diameter		30	.533		37.50	22.50		60	76.50
0990	7" diameter		28	.571		59.50	24.50		84	103
1010	8" diameter		26	.615		63.50	26		89.50	110
1020	Wall thimble, 4 to 7" adjustable, 3" diameter		36	.444		14.05	18.90		32.95	44.50
1022	4" diameter		34	.471		15.65	20		35.65	48.50
1024	5" diameter		32	.500		14.35	21		35.35	48.50
1026	6" diameter		30	.533		21.50	22.50		44	58.50
1028	7" diameter		28	.571		25	24.50		49.50	65
1030	8" diameter		26	.615		26	26		52	68.50
1040	Roof flashing, 3" diameter		36	.444		24.50	18.90		43.40	56
1050	4" diameter		34	.471		28	20		48	62
1060	5" diameter		32	.500		29.50	21		50.50	65
1070	6" diameter		30	.533		31.50	22.50		54	69.50
1080	7" diameter		28	.571		53	24.50		77.50	95.50
1090	8" diameter		26	.615		54	26		80	99.50
1100	10" diameter		24	.667		83.50	28.50		112	135
1110	12" diameter		22	.727		106	31		137	165
1120	14" diameter		20	.800		124	34		158	190
1130	16" diameter		18	.889		154	38		192	227
1140	18" diameter		16	1		208	42.50		250.50	295
1150	20" diameter	Q-10	18	1.333		270	58.50		328.50	390
1160	22" diameter		14	1.714		340	75.50		415.50	490
1170	24" diameter		12	2		410	88		498	585
1200	Tee, 3" diameter	Q-9	27	.593		33	25		58	74.50
1210	4" diameter		26	.615		34	26		60	77.50
1220	5" diameter		25	.640		37	27		64	83
1230	6" diameter		24	.667		40	28.50		68.50	87.50
1240	7" diameter		23	.696		44	29.50		73.50	94
1250	8" diameter		22	.727		60.50	31		91.50	114
1260	10" diameter		21	.762		176	32.50		208.50	244
1270	12" diameter		20	.800		212	34		246	286
1280	14" diameter		18	.889		345	38		383	440
1290	16" diameter		16	1		490	42.50		532.50	605

23 51 23.10 Gas Chimney Vents		Crew	Daily Output	Labor-Hours	Unit	2009 Bare Costs Material	Labor	Equipment	Total	Total Incl O&P
1300	18″ diameter	Q-9	14	1.143	Ea.	610	48.50		658.50	750
1310	20″ diameter	Q-10	17	1.412		830	62		892	1,000
1320	22″ diameter		13	1.846		1,025	81.50		1,106.50	1,250
1330	24″ diameter		12	2		1,200	88		1,288	1,450
1340	26″ diameter		12	2		1,725	88		1,813	2,025
1350	28″ diameter		11	2.182		1,925	96		2,021	2,275
1360	30″ diameter		10	2.400		2,100	106		2,206	2,475
1370	32″ diameter		10	2.400		2,425	106		2,531	2,850
1380	34″ diameter		9	2.667		2,450	117		2,567	2,875
1390	36″ diameter		9	2.667		2,925	117		3,042	3,400
1400	38″ diameter		8	3		2,950	132		3,082	3,425
1410	40″ diameter		7	3.429		3,225	151		3,376	3,775
1420	42″ diameter		6	4		3,500	176		3,676	4,125
1430	44″ diameter		5	4.800		3,800	211		4,011	4,525
1440	46″ diameter		5	4.800		4,175	211		4,386	4,925
1450	48″ diameter		5	4.800		4,450	211		4,661	5,225
1460	Tee cap, 3″ diameter	Q-9	45	.356		2.34	15.10		17.44	26
1470	4″ diameter		42	.381		2.63	16.20		18.83	28
1480	5″ diameter		40	.400		3.48	17		20.48	30
1490	6″ diameter		37	.432		4.06	18.35		22.41	33
1500	7″ diameter		35	.457		6.55	19.40		25.95	37.50
1510	8″ diameter		34	.471		7.75	20		27.75	39.50
1520	10″ diameter		32	.500		15.95	21		36.95	50
1530	12″ diameter		30	.533		21.50	22.50		44	58.50
1540	14″ diameter		28	.571		44	24.50		68.50	85.50
1550	16″ diameter		25	.640		51	27		78	98
1560	18″ diameter		24	.667		57.50	28.50		86	107
1570	20″ diameter	Q-10	27	.889		61.50	39		100.50	128
1580	22″ diameter		22	1.091		70.50	48		118.50	152
1590	24″ diameter		21	1.143		79.50	50.50		130	165
1600	26″ diameter		20	1.200		138	53		191	233
1610	28″ diameter		19	1.263		325	55.50		380.50	440
1620	30″ diameter		18	1.333		375	58.50		433.50	500
1630	32″ diameter		17	1.412		435	62		497	575
1640	34″ diameter		16	1.500		440	66		506	580
1650	36″ diameter		15	1.600		450	70.50		520.50	605
1660	38″ diameter		14	1.714		460	75.50		535.50	620
1670	40″ diameter		14	1.714		470	75.50		545.50	635
1680	42″ diameter		13	1.846		485	81.50		566.50	660
1690	44″ diameter		12	2		490	88		578	675
1700	46″ diameter		12	2		500	88		588	685
1710	48″ diameter		11	2.182		505	96		601	705
1750	Top, 3″ diameter	Q-9	46	.348		12.30	14.80		27.10	36
1760	4″ diameter		44	.364		16.20	15.45		31.65	42
1770	5″ diameter		42	.381		18.10	16.20		34.30	45
1780	6″ diameter		40	.400		22.50	17		39.50	50.50
1790	7″ diameter		38	.421		33	17.90		50.90	64
1800	8″ diameter		36	.444		43.50	18.90		62.40	76.50
1810	10″ diameter		34	.471		109	20		129	151
1820	12″ diameter		32	.500		165	21		186	214
1830	14″ diameter		30	.533		188	22.50		210.50	242
1840	16″ diameter		28	.571		340	24.50		364.50	415
1850	18″ diameter		26	.615		600	26		626	695

23 51 23.10 Gas Chimney Vents		Crew	Daily Output	Labor-Hours	Unit	2009 Bare Costs Material	Labor	Equipment	Total	Total Incl O&P
1860	20" diameter	Q-10	28	.857	Ea.	665	38		703	790
1870	22" diameter		22	1.091		905	48		953	1,075
1880	24" diameter		20	1.200		1,125	53		1,178	1,325
1882	26" diameter		19	1.263		1,700	55.50		1,755.50	1,950
1884	28" diameter		18	1.333		1,950	58.50		2,008.50	2,225
1886	30" diameter		17	1.412		2,175	62		2,237	2,500
1888	32" diameter		16	1.500		2,300	66		2,366	2,625
1890	34" diameter		15	1.600		2,450	70.50		2,520.50	2,800
1892	36" diameter		14	1.714		2,625	75.50		2,700.50	3,025
1900	Gas, double wall, galvanized steel, oval									
1904	4" x 1'	Q-9	68	.235	V.L.F.	19.65	10		29.65	37
1906	5"		64	.250		54.50	10.60		65.10	76.50
1908	5"/6" x 1'		60	.267		54.50	11.35		65.85	77.50
1910	Oval fittings									
1912	Adjustable length									
1914	4" diameter to 12" lg	Q-9	34	.471	Ea.	20.50	20		40.50	54
1916	5" diameter to 12" lg		32	.500		38	21		59	74
1918	5"/6" diameter to 12"		30	.533		40	22.50		62.50	79
1920	Elbow 45°									
1922	4"	Q-9	34	.471	Ea.	29.50	20		49.50	63.50
1924	5"		32	.500		68	21		89	108
1926	5"/6"		30	.533		64.50	22.50		87	106
1930	Elbow 45°, flat									
1932	4"	Q-9	34	.471	Ea.	36	20		56	70.50
1934	5"		32	.500		68	21		89	108
1936	5"/6"		30	.533		68	22.50		90.50	110
1940	Top									
1942	4"	Q-9	44	.364	Ea.	27	15.45		42.45	53.50
1944	5"		42	.381		18.65	16.20		34.85	45.50
1946	5"/6"		40	.400		45.50	17		62.50	76
1950	Adjustable flashing									
1952	4"	Q-9	34	.471	Ea.	16.10	20		36.10	49
1954	5"		32	.500		47.50	21		68.50	84.50
1956	5"/6"		30	.533		34	22.50		56.50	72.50
1960	Tee									
1962	4"	Q-9	26	.615	Ea.	53	26		79	98.50
1964	5"		25	.640		102	27		129	154
1966	5"/6"		24	.667		103	28.50		131.50	158
1970	Tee with short snout									
1972	4"	Q-9	26	.615	Ea.	53	26		79	98.50

23 51 26 – All-Fuel Vent Chimneys

23 51 26.10 All-Fuel Vent Chimneys, Press. Tight, Dbl. Wall		Crew	Daily Output	Labor-Hours	Unit	Material	Labor	Equipment	Total	Total Incl O&P
0010	**ALL-FUEL VENT CHIMNEYS, PRESSURE TIGHT, DOUBLE WALL**									
3200	All fuel, pressure tight, double wall, U.L. listed, 1400° F.									
3210	304 stainless steel liner, aluminized steel outer jacket									
3220	6" diameter	Q-9	60	.267	L.F.	47	11.35		58.35	69.50
3221	8" diameter		52	.308		58	13.05		71.05	83.50
3222	10" diameter		48	.333		64.50	14.15		78.65	93
3223	12" diameter		44	.364		73.50	15.45		88.95	105
3224	14" diameter		42	.381		81	16.20		97.20	114
3225	16" diameter		40	.400		89.50	17		106.50	125
3226	18" diameter		38	.421		101	17.90		118.90	139

23 51 26 – All-Fuel Vent Chimneys

23 51 26.10 All-Fuel Vent Chimneys, Press. Tight, Dbl. Wall		Crew	Daily Output	Labor-Hours	Unit	2009 Bare Costs Material	Labor	Equipment	Total	Total Ind O&P
3227	20" diameter	Q-10	36	.667	L.F.	115	29.50		144.50	172
3228	24" diameter		32	.750		148	33		181	214
3229	28" diameter		30	.800		170	35		205	241
3230	32" diameter		27	.889		191	39		230	271
3231	36" diameter		25	.960		224	42.50		266.50	310
3232	42" diameter		22	1.091		258	48		306	355
3233	48" diameter		19	1.263		300	55.50		355.50	415
3260	For 316 stainless steel liner add					6%				
3280	All fuel, pressure tight, double wall fittings									
3284	304 stainless steel inner, aluminized steel jacket									
3288	Adjustable 18"/30" section									
3291	5" diameter	Q-9	32	.500	Ea.	121	21		142	166
3292	6" diameter		30	.533		153	22.50		175.50	203
3293	8" diameter		26	.615		176	26		202	234
3294	10" diameter		24	.667		198	28.50		226.50	262
3295	12" diameter		22	.727		225	31		256	296
3296	14" diameter		21	.762		260	32.50		292.50	335
3297	16" diameter		20	.800		270	34		304	350
3298	18" diameter		19	.842		335	36		371	420
3299	20" diameter	Q-10	18	1.333		385	58.50		443.50	515
3300	24" diameter		16	1.500		495	66		561	645
3301	28" diameter		15	1.600		555	70.50		625.50	720
3302	32" diameter		14	1.714		650	75.50		725.50	830
3303	36" diameter		12	2		680	88		768	885
3304	42" diameter		11	2.182		745	96		841	970
3305	48" diameter		10	2.400		875	106		981	1,125
3326	For 316 stainless steel liner, add					15%				
3350	Elbow 90° fixed									
3353	5" diameter	Q-9	32	.500	Ea.	227	21		248	282
3354	6" diameter		30	.533		266	22.50		288.50	330
3355	8" diameter		26	.615		294	26		320	365
3356	10" diameter		24	.667		335	28.50		363.50	415
3357	12" diameter		22	.727		380	31		411	470
3358	14" diameter		21	.762		445	32.50		477.50	540
3359	16" diameter		20	.800		500	34		534	605
3360	18" diameter		19	.842		580	36		616	695
3361	20" diameter	Q-10	18	1.333		650	58.50		708.50	805
3362	24" diameter		16	1.500		805	66		871	985
3363	28" diameter		15	1.600		1,000	70.50		1,070.50	1,200
3364	32" diameter		14	1.714		1,350	75.50		1,425.50	1,600
3365	36" diameter		12	2		1,575	88		1,663	1,875
3366	42" diameter		11	2.182		2,400	96		2,496	2,800
3367	48" diameter		10	2.400		3,025	106		3,131	3,500
3380	For 316 stainless steel liner, add					20%				
3400	Elbow 45°									
3404	6" diameter	Q-9	30	.533	Ea.	168	22.50		190.50	219
3405	8" diameter		26	.615		190	26		216	249
3406	10" diameter		24	.667		213	28.50		241.50	279
3407	12" diameter		22	.727		243	31		274	315
3408	14" diameter		21	.762		280	32.50		312.50	360
3409	16" diameter		20	.800		315	34		349	400
3410	18" diameter		19	.842		355	36		391	445
3411	20" diameter	Q-10	18	1.333		410	58.50		468.50	540

23 51 26 – All-Fuel Vent Chimneys

23 51 26.10 All-Fuel Vent Chimneys, Press. Tight, Dbl. Wall		Crew	Daily Output	Labor-Hours	Unit	2009 Bare Costs Material	Labor	Equipment	Total	Total Incl O&P
3412	24" diameter	Q-10	16	1.500	Ea.	525	66		591	675
3413	28" diameter		15	1.600		595	70.50		665.50	765
3414	32" diameter		14	1.714		795	75.50		870.50	990
3415	36" diameter		12	2		1,125	88		1,213	1,350
3416	42" diameter		11	2.182		1,525	96		1,621	1,825
3417	48" diameter		10	2.400		2,050	106		2,156	2,425
3430	For 316 stainless steel liner, add					20%				
3450	Tee 90°									
3454	6" diameter	Q-9	24	.667	Ea.	175	28.50		203.50	237
3455	8" diameter		22	.727		202	31		233	270
3456	10" diameter		21	.762		236	32.50		268.50	310
3457	12" diameter		20	.800		275	34		309	355
3458	14" diameter		18	.889		325	38		363	420
3459	16" diameter		16	1		365	42.50		407.50	470
3460	18" diameter		14	1.143		430	48.50		478.50	550
3461	20" diameter	Q-10	17	1.412		485	62		547	630
3462	24" diameter		12	2		600	88		688	790
3463	28" diameter		11	2.182		745	96		841	970
3464	32" diameter		10	2.400		1,025	106		1,131	1,300
3465	36" diameter		9	2.667		1,200	117		1,317	1,500
3466	42" diameter		6	4		1,900	176		2,076	2,375
3467	48" diameter		5	4.800		2,425	211		2,636	3,000
3480	For Tee Cap, add					35%	20%			
3500	For 316 stainless steel liner, add					22%				
3520	Plate support, galvanized									
3524	6" diameter	Q-9	26	.615	Ea.	91	26		117	140
3525	8" diameter		22	.727		103	31		134	162
3526	10" diameter		20	.800		112	34		146	177
3527	12" diameter		18	.889		117	38		155	187
3528	14" diameter		17	.941		143	40		183	219
3529	16" diameter		16	1		150	42.50		192.50	231
3530	18" diameter		15	1.067		158	45.50		203.50	244
3531	20" diameter	Q-10	16	1.500		168	66		234	286
3532	24" diameter		14	1.714		176	75.50		251.50	310
3533	28" diameter		13	1.846		216	81.50		297.50	360
3534	32" diameter		12	2		281	88		369	445
3535	36" diameter		10	2.400		350	106		456	550
3536	42" diameter		9	2.667		620	117		737	860
3537	48" diameter		8	3		745	132		877	1,025
3570	Bellows, lined, 316 stainless steel only									
3574	6" diameter	Q-9	30	.533	Ea.	440	22.50		462.50	515
3575	8" diameter		26	.615		475	26		501	565
3576	10" diameter		24	.667		525	28.50		553.50	620
3577	12" diameter		22	.727		580	31		611	690
3578	14" diameter		21	.762		655	32.50		687.50	770
3579	16" diameter		20	.800		725	34		759	850
3580	18" diameter		19	.842		805	36		841	945
3581	20" diameter	Q-10	18	1.333		910	58.50		968.50	1,100
3582	24" diameter		16	1.500		1,150	66		1,216	1,350
3583	28" diameter		15	1.600		1,275	70.50		1,345.50	1,500
3584	32" diameter		14	1.714		1,475	75.50		1,550.50	1,750
3585	36" diameter		12	2		1,700	88		1,788	2,000
3586	42" diameter		11	2.182		2,525	96		2,621	2,950

23 51 26.10 All-Fuel Vent Chimneys, Press. Tight, Dbl. Wall		Crew	Daily Output	Labor-Hours	Unit	2009 Bare Costs Material	Labor	Equipment	Total	Total Incl O&P
3587	48″ diameter	Q-10	10	2.400	Ea.	2,625	106		2,731	3,075
3600	Ventilated roof thimble, 304 stainless steel									
3620	6″ diameter	Q-9	26	.615	Ea.	225	26		251	287
3624	8″ diameter		22	.727		241	31		272	315
3625	10″ diameter		20	.800		260	34		294	340
3626	12″ diameter		18	.889		281	38		319	370
3627	14″ diameter		17	.941		300	40		340	390
3628	16″ diameter		16	1		320	42.50		362.50	420
3629	18″ diameter		15	1.067		345	45.50		390.50	450
3630	20″ diameter	Q-10	16	1.500		370	66		436	510
3631	24″ diameter		14	1.714		420	75.50		495.50	575
3632	28″ diameter		13	1.846		480	81.50		561.50	655
3633	32″ diameter		12	2		540	88		628	730
3634	36″ diameter		10	2.400		620	106		726	845
3635	42″ diameter		9	2.667		820	117		937	1,075
3636	48″ diameter		8	3		1,025	132		1,157	1,350
3650	For 316 stainless steel, add					10%				
3670	Exit cone, 316 stainless steel only									
3674	6″ diameter	Q-9	46	.348	Ea.	178	14.80		192.80	219
3675	8″ diameter		42	.381		186	16.20		202.20	230
3676	10″ diameter		40	.400		202	17		219	248
3677	12″ diameter		38	.421		211	17.90		228.90	260
3678	14″ diameter		37	.432		234	18.35		252.35	287
3679	16″ diameter		36	.444		258	18.90		276.90	315
3680	18″ diameter		35	.457		315	19.40		334.40	380
3681	20″ diameter	Q-10	28	.857		380	38		418	475
3682	24″ diameter		26	.923		485	40.50		525.50	600
3683	28″ diameter		25	.960		590	42.50		632.50	715
3684	32″ diameter		24	1		735	44		779	875
3685	36″ diameter		22	1.091		880	48		928	1,050
3686	42″ diameter		21	1.143		1,125	50.50		1,175.50	1,325
3687	48″ diameter		20	1.200		1,450	53		1,503	1,675
3720	Roof support assembly, incl. 30″ pipe sect, 304 stainless steel									
3724	6″ diameter	Q-9	25	.640	Ea.	259	27		286	325
3725	8″ diameter		21	.762		282	32.50		314.50	360
3726	10″ diameter		19	.842		310	36		346	395
3727	12″ diameter		17	.941		330	40		370	420
3728	14″ diameter		16	1		370	42.50		412.50	470
3729	16″ diameter		15	1.067		390	45.50		435.50	500
3730	18″ diameter		14	1.143		420	48.50		468.50	540
3731	20″ diameter	Q-10	15	1.600		450	70.50		520.50	605
3732	24″ diameter		13	1.846		500	81.50		581.50	675
3733	28″ diameter		12	2		590	88		678	780
3734	32″ diameter		11	2.182		705	96		801	925
3735	36″ diameter		9	2.667		855	117		972	1,125
3736	42″ diameter		8	3		1,350	132		1,482	1,675
3737	48″ diameter		7	3.429		1,650	151		1,801	2,050
3750	For 316 stainless steel, add					20%				
3770	Stack cap, 316 stainless steel only									
3773	5″ diameter	Q-9	48	.333	Ea.	169	14.15		183.15	208
3774	6″ diameter		46	.348		198	14.80		212.80	241
3775	8″ diameter		42	.381		229	16.20		245.20	276
3776	10″ diameter		40	.400		270	17		287	325

23 51 26 – All-Fuel Vent Chimneys

23 51 26.10 All-Fuel Vent Chimneys, Press. Tight, Dbl. Wall		Crew	Daily Output	Labor-Hours	Unit	Material	Labor	Equipment	Total	Total Incl O&P
						2009 Bare Costs				
3777	12″ diameter	Q-9	38	.421	Ea.	315	17.90		332.90	375
3778	14″ diameter		37	.432		370	18.35		388.35	435
3779	16″ diameter		36	.444		420	18.90		438.90	495
3780	18″ diameter		35	.457		475	19.40		494.40	550
3781	20″ diameter	Q-10	28	.857		540	38		578	655
3782	24″ diameter		26	.923		655	40.50		695.50	785
3783	28″ diameter		25	.960		710	42.50		752.50	850
3784	32″ diameter		24	1		820	44		864	970
3785	36″ diameter		22	1.091		1,000	48		1,048	1,175
3786	42″ diameter		21	1.143		1,225	50.50		1,275.50	1,425
3787	48″ diameter		20	1.200		1,450	53		1,503	1,675
3800	Round storm collar									
3806	5″ diameter	Q-9	32	.500	Ea.	26.50	21		47.50	61.50

23 51 26.30 All-Fuel Vent Chimneys, Double Wall, St. Stl.		Crew	Daily Output	Labor-Hours	Unit	Material	Labor	Equipment	Total	Total Incl O&P
0010	**ALL-FUEL VENT CHIMNEYS, DOUBLE WALL, STAINLESS STEEL**									
7800	All fuel, double wall, stainless steel, 6″ diameter	Q-9	60	.267	V.L.F.	56.50	11.35		67.85	79.50
7802	7″ diameter		56	.286		73	12.15		85.15	99
7804	8″ diameter		52	.308		86.50	13.05		99.55	115
7806	10″ diameter		48	.333		125	14.15		139.15	159
7808	12″ diameter		44	.364		167	15.45		182.45	208
7810	14″ diameter		42	.381		219	16.20		235.20	266
8000	All fuel, double wall, stainless steel fittings									
8010	Roof support 6″ diameter	Q-9	30	.533	Ea.	103	22.50		125.50	149
8020	7″ diameter		28	.571		117	24.50		141.50	166
8030	8″ diameter		26	.615		130	26		156	183
8040	10″ diameter		24	.667		150	28.50		178.50	209
8050	12″ diameter		22	.727		181	31		212	247
8060	14″ diameter		21	.762		229	32.50		261.50	300
8100	Elbow 15°, 6″ diameter		30	.533		165	22.50		187.50	217
8120	7″ diameter		28	.571		186	24.50		210.50	243
8140	8″ diameter		26	.615		212	26		238	273
8160	10″ diameter		24	.667		278	28.50		306.50	350
8180	12″ diameter		22	.727		335	31		366	420
8200	14″ diameter		21	.762		405	32.50		437.50	495
8300	Insulated tee with insulated tee cap, 6″ diameter		30	.533		156	22.50		178.50	207
8340	7″ diameter		28	.571		205	24.50		229.50	263
8360	8″ diameter		26	.615		231	26		257	294
8380	10″ diameter		24	.667		325	28.50		353.50	405
8400	12″ diameter		22	.727		455	31		486	550
8420	14″ diameter		21	.762		600	32.50		632.50	710
8500	Joist shield, 6″ diameter		30	.533		47.50	22.50		70	87.50
8510	7″ diameter		28	.571		51.50	24.50		76	94.50
8520	8″ diameter		26	.615		63.50	26		89.50	110
8530	10″ diameter		24	.667		86	28.50		114.50	138
8540	12″ diameter		22	.727		107	31		138	166
8550	14″ diameter		21	.762		133	32.50		165.50	196
8600	Round top, 6″ diameter		30	.533		53	22.50		75.50	93
8620	7″ diameter		28	.571		72	24.50		96.50	117
8640	8″ diameter		26	.615		97	26		123	147
8660	10″ diameter		24	.667		180	28.50		208.50	242
8680	12″ diameter		22	.727		249	31		280	320
8700	14″ diameter		21	.762		330	32.50		362.50	415

23 51 26 – All-Fuel Vent Chimneys

23 51 26.30 All-Fuel Vent Chimneys, Double Wall, St. Stl.		Crew	Daily Output	Labor-Hours	Unit	2009 Bare Costs Material	Labor	Equipment	Total	Total Incl O&P
8800	Adjustable roof flashing, 6" diameter	Q-9	30	.533	Ea.	63	22.50		85.50	104
8820	7" diameter		28	.571		72	24.50		96.50	117
8840	8" diameter		26	.615		78	26		104	126
8860	10" diameter		24	.667		100	28.50		128.50	154
8880	12" diameter		22	.727		130	31		161	190
8900	14" diameter		21	.762		162	32.50		194.50	228

23 51 33 – Insulated Sectional Chimneys

23 51 33.10 Prefabricated Insulated Sectional Chimneys		Crew	Daily Output	Labor-Hours	Unit	Material	Labor	Equipment	Total	Total Incl O&P
0010	**PREFABRICATED INSULATED SECTIONAL CHIMNEYS**									
9000	High temp. (2000° F), steel jacket, acid resistant refractory lining									
9010	11 ga. galvanized jacket, U.L. listed									
9020	Straight section, 48" long, 10" diameter	Q-10	13.30	1.805	Ea.	400	79.50		479.50	560
9030	12" diameter		11.20	2.143		440	94.50		534.50	625
9040	18" diameter		7.40	3.243		615	143		758	895
9050	24" diameter		4.60	5.217		840	230		1,070	1,275
9060	30" diameter		3.70	6.486		1,075	286		1,361	1,625
9070	36" diameter		2.70	8.889		1,325	390		1,715	2,050
9080	42" diameter	Q-11	3.10	10.323		1,625	465		2,090	2,525
9090	48" diameter		2.70	11.852		1,925	535		2,460	2,925
9100	54" diameter		2.20	14.545		2,325	655		2,980	3,550
9110	60" diameter		2	16		3,050	720		3,770	4,450
9120	Tee section, 10" diameter	Q-10	4.40	5.455		855	240		1,095	1,300
9130	12" diameter		3.70	6.486		900	286		1,186	1,425
9140	18" diameter		2.40	10		1,225	440		1,665	2,025
9150	24" diameter		1.50	16		1,675	705		2,380	2,925
9160	30" diameter		1.20	20		2,350	880		3,230	3,950
9170	36" diameter		.80	30		3,975	1,325		5,300	6,400
9180	42" diameter	Q-11	1	32		5,225	1,450		6,675	7,975
9190	48" diameter		.90	35.556		6,775	1,600		8,375	9,900
9200	54" diameter		.70	45.714		9,900	2,050		11,950	14,100
9210	60" diameter		.60	53.333		12,500	2,400		14,900	17,500
9220	Cleanout pier section, 10" diameter	Q-10	3.50	6.857		655	300		955	1,200
9230	12" diameter		2.50	9.600		690	425		1,115	1,400
9240	18" diameter		1.90	12.632		885	555		1,440	1,825
9250	24" diameter		1.30	18.462		1,125	815		1,940	2,475
9260	30" diameter		1	24		1,375	1,050		2,425	3,125
9270	36" diameter		.75	32		1,625	1,400		3,025	3,950
9280	42" diameter	Q-11	.90	35.556		2,050	1,600		3,650	4,725
9290	48" diameter		.75	42.667		2,475	1,925		4,400	5,675
9300	54" diameter		.60	53.333		2,950	2,400		5,350	6,950
9310	60" diameter		.50	64		3,775	2,875		6,650	8,575
9320	For drain, add					53%				
9330	Elbow, 30° and 45°, 10" diameter	Q-10	6.60	3.636		555	160		715	855
9340	12" diameter		5.60	4.286		600	189		789	945
9350	18" diameter		3.70	6.486		900	286		1,186	1,425
9360	24" diameter		2.30	10.435		1,350	460		1,810	2,175
9370	30" diameter		1.85	12.973		1,850	570		2,420	2,900
9380	36" diameter		1.30	18.462		2,225	815		3,040	3,700
9390	42" diameter	Q-11	1.60	20		3,050	900		3,950	4,750
9400	48" diameter		1.35	23.704		3,900	1,075		4,975	5,925
9410	54" diameter		1.10	29.091		5,050	1,300		6,350	7,575
9420	60" diameter		1	32		6,150	1,450		7,600	9,000

23 51 33 – Insulated Sectional Chimneys

23 51 33.10 Prefabricated Insulated Sectional Chimneys		Crew	Daily Output	Labor-Hours	Unit	2009 Bare Costs Material	Labor	Equipment	Total	Total Incl O&P
9430	For 60° and 90° elbow, add				Ea.	112%				
9440	End cap, 10" diameter	Q-10	26	.923		955	40.50		995.50	1,125
9450	12" diameter		22	1.091		1,100	48		1,148	1,275
9460	18" diameter		15	1.600		1,350	70.50		1,420.50	1,575
9470	24" diameter		9	2.667		1,650	117		1,767	2,000
9480	30" diameter		7.30	3.288		2,050	145		2,195	2,475
9490	36" diameter		5	4.800		2,350	211		2,561	2,925
9500	42" diameter	Q-11	6.30	5.079		2,950	228		3,178	3,575
9510	48" diameter		5.30	6.038		3,475	271		3,746	4,225
9520	54" diameter		4.40	7.273		3,975	325		4,300	4,850
9530	60" diameter		3.90	8.205		4,675	370		5,045	5,725
9540	Increaser (1 diameter), 10" diameter	Q-10	6.60	3.636		475	160		635	765
9550	12" diameter		5.60	4.286		505	189		694	845
9560	18" diameter		3.70	6.486		740	286		1,026	1,250
9570	24" diameter		2.30	10.435		1,050	460		1,510	1,875
9580	30" diameter		1.85	12.973		1,325	570		1,895	2,325
9590	36" diameter		1.30	18.462		1,450	815		2,265	2,850
9592	42" diameter	Q-11	1.60	20		1,600	900		2,500	3,125
9594	48" diameter		1.35	23.704		1,925	1,075		3,000	3,750
9596	54" diameter		1.10	29.091		2,325	1,300		3,625	4,575
9600	For expansion joints, add to straight section					7%				
9610	For 1/4" hot rolled steel jacket, add					157%				
9620	For 2950° F very high temperature, add					89%				
9630	26 ga. aluminized jacket, straight section, 48" long									
9640	10" diameter	Q-10	15.30	1.569	V.L.F.	237	69		306	365
9650	12" diameter		12.90	1.860		252	82		334	405
9660	18" diameter		8.50	2.824		365	124		489	595
9670	24" diameter		5.30	4.528		520	199		719	875
9680	30" diameter		4.30	5.581		665	246		911	1,125
9690	36" diameter		3.10	7.742		845	340		1,185	1,450
9700	Accessories (all models)									
9710	Guy band, 10" diameter	Q-10	32	.750	Ea.	74	33		107	132
9720	12" diameter		30	.800		78	35		113	140
9730	18" diameter		26	.923		102	40.50		142.50	175
9740	24" diameter		24	1		149	44		193	231
9750	30" diameter		20	1.200		173	53		226	272
9760	36" diameter		18	1.333		191	58.50		249.50	300
9770	42" diameter	Q-11	22	1.455		345	65.50		410.50	480
9780	48" diameter		20	1.600		380	72		452	525
9790	54" diameter		16	2		420	90		510	600
9800	60" diameter		12	2.667		455	120		575	685
9810	Draw band, galv. stl., 11 gauge, 10" diameter	Q-10	32	.750		65	33		98	123
9820	12" diameter		30	.800		69	35		104	130
9830	18" diameter		26	.923		86	40.50		126.50	157
9840	24" diameter		24	1		110	44		154	189
9850	30" diameter		20	1.200		127	53		180	221
9860	36" diameter		18	1.333		146	58.50		204.50	251
9870	42" diameter	Q-11	22	1.455		196	65.50		261.50	315
9880	48" diameter		20	1.600		251	72		323	390
9890	54" diameter		16	2		291	90		381	460
9900	60" diameter		12	2.667		370	120		490	590
9910	Draw band, aluminized stl., 26 gauge, 10" diameter	Q-10	32	.750		18.95	33		51.95	72
9920	12" diameter		30	.800		18.95	35		53.95	75

23 51 Breechings, Chimneys, and Stacks

23 51 33 – Insulated Sectional Chimneys

23 51 33.10 Prefabricated Insulated Sectional Chimneys		Crew	Daily Output	Labor-Hours	Unit	2009 Bare Costs Material	Labor	Equipment	Total	Total Incl O&P
9930	18" diameter	Q-10	26	.923	Ea.	24	40.50		64.50	89
9940	24" diameter		24	1		28	44		72	98.50
9950	30" diameter		20	1.200		32	53		85	117
9960	36" diameter		18	1.333		43	58.50		101.50	138
9962	42" diameter	Q-11	22	1.455		43	65.50		108.50	148
9964	48" diameter		20	1.600		51	72		123	167
9966	54" diameter		16	2		51	90		141	194
9968	60" diameter		12	2.667		56	120		176	247

23 52 Heating Boilers

23 52 13 – Electric Boilers

23 52 13.10 Electric Boilers, ASME

		Crew	Daily Output	Labor-Hours	Unit	Material	Labor	Equipment	Total	Total Incl O&P
0010	**ELECTRIC BOILERS, ASME**, Standard controls and trim. R235000-30									
1000	Steam, 6 KW, 20.5 MBH	Q-19	1.20	20	Ea.	3,375	905		4,280	5,075
1040	9 KW, 30.7 MBH D3020-102		1.20	20		3,525	905		4,430	5,225
1060	18 KW, 61.4 MBH		1.20	20		3,625	905		4,530	5,325
1080	24 KW, 81.8 MBH D3020-104		1.10	21.818		4,200	990		5,190	6,100
1120	36 KW, 123 MBH		1.10	21.818		4,650	990		5,640	6,575
1160	60 KW, 205 MBH		1	24		5,800	1,075		6,875	8,000
1220	112 KW, 382 MBH		.75	32		7,950	1,450		9,400	10,900
1240	148 KW, 505 MBH		.65	36.923		8,375	1,675		10,050	11,700
1260	168 KW, 573 MBH		.60	40		11,500	1,800		13,300	15,300
1280	222 KW, 758 MBH		.55	43.636		13,100	1,975		15,075	17,400
1300	296 KW, 1010 MBH		.45	53.333		13,200	2,425		15,625	18,100
1320	300 KW, 1023 MBH		.40	60		17,000	2,725		19,725	22,800
1340	370 KW, 1263 MBH		.35	68.571		18,300	3,100		21,400	24,800
1360	444 KW, 1515 MBH		.30	80		19,500	3,625		23,125	26,900
1380	518 KW, 1768 MBH	Q-21	.36	88.889		22,100	4,125		26,225	30,500
1400	592 KW, 2020 MBH		.34	94.118		25,000	4,350		29,350	34,000
1420	666 KW, 2273 MBH		.32	100		26,400	4,625		31,025	35,900
1460	740 KW, 2526 MBH		.28	114		27,600	5,300		32,900	38,300
1480	814 KW, 2778 MBH		.25	128		31,000	5,925		36,925	43,000
1500	962 KW, 3283 MBH		.22	145		31,800	6,725		38,525	45,000
1520	1036 KW, 3536 MBH		.20	160		33,400	7,400		40,800	47,800
1540	1110 KW, 3788 MBH		.19	168		35,000	7,800		42,800	50,000
1560	2070 KW, 7063 MBH		.18	177		49,000	8,225		57,225	66,500
1580	2250 KW, 7677 MBH		.17	188		57,000	8,725		65,725	76,000
1600	2,340 KW, 7984 MBH		.16	200		63,500	9,250		72,750	83,500
2000	Hot water, 7.5 KW, 25.6 MBH	Q-19	1.30	18.462		3,575	835		4,410	5,200
2020	15 KW, 51.2 MBH		1.30	18.462		3,700	835		4,535	5,300
2040	30 KW, 102 MBH		1.20	20		3,825	905		4,730	5,550
2060	45 KW, 164 MBH		1.20	20		4,275	905		5,180	6,050
2070	60 KW, 205 MBH		1.20	20		4,525	905		5,430	6,325
2080	75 KW, 256 MBH		1.10	21.818		4,850	990		5,840	6,825
2100	90 KW, 307 MBH		1.10	21.818		5,225	990		6,215	7,225
2120	105 KW, 358 MBH		1	24		5,600	1,075		6,675	7,775
2140	120 KW, 410 MBH		.90	26.667		5,875	1,200		7,075	8,250
2160	135 KW, 461 MBH		.75	32		6,375	1,450		7,825	9,175
2180	150 KW, 512 MBH		.65	36.923		6,625	1,675		8,300	9,775
2200	165 KW, 563 MBH		.60	40		6,850	1,800		8,650	10,300
2220	296 KW, 1010 MBH		.55	43.636		11,600	1,975		13,575	15,700

23 52 Heating Boilers

23 52 13 – Electric Boilers

23 52 13.10 Electric Boilers, ASME		Crew	Daily Output	Labor-Hours	Unit	2009 Bare Costs Material	Labor	Equipment	Total	Total Incl O&P
2280	370 KW, 1263 MBH	Q-19	.40	60	Ea.	12,800	2,725		15,525	18,200
2300	444 KW, 1515 MBH		.35	68.571		15,500	3,100		18,600	21,800
2340	518 KW, 1768 MBH	Q-21	.44	72.727		17,200	3,375		20,575	24,000
2360	592 KW, 2020 MBH		.43	74.419		19,000	3,450		22,450	26,100
2400	666 KW, 2273 MBH		.40	80		20,300	3,700		24,000	27,900
2420	740 KW, 2526 MBH		.39	82.051		20,800	3,800		24,600	28,500
2440	814 KW, 2778 MBH		.38	84.211		21,400	3,900		25,300	29,500
2460	888 KW, 3031 MBH		.37	86.486		22,600	4,000		26,600	30,800
2480	962 KW, 3283 MBH		.36	88.889		25,300	4,125		29,425	34,000
2500	1,036 KW, 3536 MBH		.34	94.118		26,900	4,350		31,250	36,100
2520	1110 KW, 3788 MBH		.33	96.970		29,700	4,500		34,200	39,400
2540	1440 KW, 4915 MBH		.32	100		34,200	4,625		38,825	44,500
2560	1560 KW, 5323 MBH		.31	103		38,400	4,775		43,175	49,400
2580	1680 KW, 5733 MBH		.30	106		41,000	4,950		45,950	52,500
2600	1800 KW, 6143 MBH		.29	110		43,100	5,100		48,200	55,000
2620	1980 KW, 6757 MBH		.28	114		46,100	5,300		51,400	58,500
2640	2100 KW, 7167 MBH		.27	118		51,500	5,475		56,975	64,500
2660	2220 KW, 7576 MBH		.26	123		53,000	5,700		58,700	67,000
2680	2,400 KW, 8191 MBH		.25	128		56,000	5,925		61,925	70,500
2700	2610 KW, 8905 MBH		.24	133		60,500	6,175		66,675	76,000
2720	2790 KW, 9519 MBH		.23	139		63,000	6,450		69,450	79,000
2740	2970 KW, 10133 MBH		.21	152		64,500	7,050		71,550	81,500
2760	3150 KW, 10748 MBH		.19	168		70,500	7,800		78,300	89,000
2780	3240 KW, 11055 MBH		.18	177		72,000	8,225		80,225	91,500
2800	3420 KW, 11669 MBH		.17	188		76,000	8,725		84,725	97,000
2820	3600 KW, 12,233 MBH		.16	200		79,000	9,250		88,250	101,000

23 52 16 – Condensing Boilers

23 52 16.24 Condensing Boilers		Crew	Daily Output	Labor-Hours	Unit	Material	Labor	Equipment	Total	Total Incl O&P
0010	**CONDENSING BOILERS**, Cast iron									
0020	Packaged with standard controls, circulator and trim									
0030	Intermittent (spark) pilot, natural or LP gas									
0040	Hot water, DOE MBH output, (AFUE)									
0100	42 MBH, (84.0%) G	Q-5	1.80	8.889	Ea.	1,300	395		1,695	2,050
0120	57 MBH, (84.3%) G		1.60	10		1,450	445		1,895	2,275
0140	85 MBH, (84.0%) G		1.40	11.429		1,600	510		2,110	2,500
0160	112 MBH, (83.7%) G		1.20	13.333		1,800	590		2,390	2,875
0180	140 MBH, (83.3%) G	Q-6	1.60	15		2,075	690		2,765	3,300
0200	167 MBH, (83.0%) G		1.40	17.143		2,325	790		3,115	3,750
0220	194 MBH, (82.7%) G		1.20	20		2,550	920		3,470	4,175

23 52 19 – Pulse Combustion Boilers

23 52 19.20 Pulse Type Combustion Boilers		Crew	Daily Output	Labor-Hours	Unit	Material	Labor	Equipment	Total	Total Incl O&P
0010	**PULSE TYPE COMBUSTION BOILERS**									
7990	Special feature gas fired boilers									
8000	Pulse combustion, standard controls / trim									
8050	88,000 BTU G	Q-5	1.40	11.429	Ea.	4,775	510		5,285	6,000
8080	134,000 BTU G	"	1.20	13.333	"	5,700	590		6,290	7,175

23 52 23 – Cast-Iron Boilers

23 52 23.20 Gas-Fired Boilers		Crew	Daily Output	Labor-Hours	Unit	Material	Labor	Equipment	Total	Total Incl O&P
0010	**GAS-FIRED BOILERS,** Natural or propane, standard controls, packaged. R235000-10									
1000	Cast iron, with insulated jacket									
2000	Steam, gross output, 81 MBH R235000-20	Q-7	1.40	22.857	Ea.	1,850	1,075		2,925	3,675

23 52 23.20 Gas-Fired Boilers			Crew	Daily Output	Labor-Hours	Unit	2009 Bare Costs Material	Labor	Equipment	Total	Total Incl O&P
2020	102 MBH		Q-7	1.30	24.615	Ea.	2,100	1,150		3,250	4,025
2040	122 MBH	R235000-30		1	32		2,250	1,500		3,750	4,725
2060	163 MBH			.90	35.556		2,700	1,675		4,375	5,475
2080	203 MBH	R235000-50		.90	35.556		2,950	1,675		4,625	5,750
2100	240 MBH			.85	37.647		2,950	1,775		4,725	5,900
2120	280 MBH	R235000-70		.80	40		3,950	1,875		5,825	7,150
2140	320 MBH			.70	45.714		4,200	2,150		6,350	7,825
2160	360 MBH	R235000-80		.63	51.200		4,675	2,400		7,075	8,750
2180	400 MBH			.56	56.838		4,950	2,675		7,625	9,450
2200	440 MBH			.51	62.500		5,275	2,950		8,225	10,200
2220	544 MBH			.45	71.588		8,125	3,375		11,500	14,000
2240	765 MBH			.43	74.419		9,825	3,500		13,325	16,100
2260	892 MBH			.38	84.211		11,400	3,950		15,350	18,500
2280	1275 MBH			.34	94.118		13,900	4,425		18,325	22,000
2300	1530 MBH			.32	100		15,200	4,700		19,900	23,800
2320	1,875 MBH			.30	106		17,800	5,025		22,825	27,000
2340	2170 MBH			.26	122		19,000	5,750		24,750	29,500
2360	2675 MBH			.20	163		21,800	7,675		29,475	35,500
2380	3060 MBH			.19	172		23,100	8,100		31,200	37,600
2400	3570 MBH			.18	181		25,600	8,550		34,150	41,000
2420	4207 MBH			.16	205		28,000	9,650		37,650	45,200
2440	4,720 MBH			.15	207		29,600	9,775		39,375	47,200
2460	5660 MBH			.15	220		65,000	10,400		75,400	87,000
2480	6,100 MBH			.13	246		74,000	11,600		85,600	99,000
2500	6390 MBH			.12	266		78,000	12,500		90,500	105,000
2520	6680 MBH			.11	290		79,500	13,700		93,200	108,000
2540	6,970 MBH			.10	320		83,000	15,000		98,000	114,000
3000	Hot water, gross output, 80 MBH			1.46	21.918		1,700	1,025		2,725	3,425
3020	100 MBH			1.35	23.704		1,950	1,125		3,075	3,825
3040	122 MBH			1.10	29.091		2,100	1,375		3,475	4,375
3060	163 MBH			1	32		2,550	1,500		4,050	5,050
3080	203 MBH			1	32		2,800	1,500		4,300	5,350
3100	240 MBH			.95	33.684		2,950	1,575		4,525	5,625
3120	280 MBH			.90	35.556		3,700	1,675		5,375	6,575
3140	320 MBH			.80	40		4,000	1,875		5,875	7,225
3160	360 MBH			.71	45.070		4,475	2,125		6,600	8,100
3180	400 MBH			.64	50		4,775	2,350		7,125	8,775
3200	440 MBH			.58	54.983		5,100	2,575		7,675	9,475
3220	544 MBH			.51	62.992		7,875	2,950		10,825	13,100
3240	765 MBH			.46	70.022		9,525	3,300		12,825	15,500
3260	1,088 MBH.			.40	80		11,900	3,750		15,650	18,800
3280	1,275 MBH			.36	89.888		13,600	4,225		17,825	21,300
3300	1,530 MBH			.31	104		14,900	4,925		19,825	23,800
3320	2,000 MBH			.26	125		21,000	5,875		26,875	31,900
3340	2,312 MBH			.22	148		21,000	6,975		27,975	33,600
3360	2,856 MBH			.20	160		23,100	7,525		30,625	36,800
3380	3,264 MBH			.18	179		24,300	8,450		32,750	39,500
3400	3,808 MBH			.16	195		26,800	9,175		35,975	43,300
3420	4,488 MBH			.15	210		29,100	9,900		39,000	46,900
3440	4,720 MBH			.15	220		62,500	10,400		72,900	84,000
3460	5,520 MBH			.14	228		80,000	10,700		90,700	104,000
3480	6,100 MBH			.13	250		82,000	11,800		93,800	107,500
3500	6,390 MBH			.11	285		86,500	13,400		99,900	115,500

23 52 23 – Cast-Iron Boilers

23 52 23.20 Gas-Fired Boilers		Crew	Daily Output	Labor-Hours	Unit	2009 Bare Costs Material	Labor	Equipment	Total	Total Incl O&P
3520	6,680 MBH	Q-7	.10	310	Ea.	86,500	14,600		101,100	117,500
3540	6,970 MBH		.09	359		89,500	16,900		106,400	124,000
7000	For tankless water heater, add					10%				
7050	For additional zone valves up to 312 MBH add					145			145	159

23 52 23.30 Gas/Oil Fired Boilers		Crew	Daily Output	Labor-Hours	Unit	Material	Labor	Equipment	Total	Total Incl O&P
0010	**GAS/OIL FIRED BOILERS,** Combination with burners and controls, packaged.									
1000	Cast iron with insulated jacket									
2000	Steam, gross output, 720 MBH	Q-7	.43	74.074	Ea.	13,400	3,475		16,875	20,000
2020	810 MBH		.38	83.990		13,400	3,950		17,350	20,700
2040	1,084 MBH		.34	93.023		14,800	4,375		19,175	22,800
2060	1,360 MBH		.33	98.160		16,400	4,625		21,025	25,000
2080	1,600 MBH		.30	107		17,800	5,025		22,825	27,200
2100	2,040 MBH		.25	130		23,100	6,125		29,225	34,600
2120	2,450 MBH		.21	156		25,300	7,350		32,650	38,900
2140	2,700 MBH		.19	165		26,600	7,800		34,400	41,000
2160	3,000 MBH		.18	175		28,100	8,275		36,375	43,300
2180	3,270 MBH		.17	183		29,500	8,650		38,150	45,400
2200	3,770 MBH		.17	191		60,500	9,000		69,500	80,000
2220	4,070 MBH		.16	200		63,500	9,400		72,900	84,000
2240	4,650 MBH		.15	210		67,000	9,900		76,900	89,000
2260	5,230 MBH		.14	223		74,500	10,500		85,000	98,000
2280	5,520 MBH		.14	235		81,000	11,100		92,100	105,500
2300	5,810 MBH		.13	248		82,000	11,700		93,700	107,500
2320	6,100 MBH		.12	260		82,500	12,200		94,700	109,500
2340	6,390 MBH		.11	296		87,000	13,900		100,900	116,500
2360	6,680 MBH		.10	320		91,500	15,000		106,500	123,000
2380	6,970 MBH		.09	372		92,000	17,500		109,500	128,000
2900	Hot water, gross output									
2910	200 MBH	Q-6	.62	39.024	Ea.	7,750	1,800		9,550	11,200
2920	300 MBH		.49	49.080		7,750	2,250		10,000	11,900
2930	400 MBH		.41	57.971		9,075	2,675		11,750	14,000
2940	500 MBH		.36	67.039		9,775	3,100		12,875	15,400
3000	584 MBH	Q-7	.44	72.072		13,200	3,400		16,600	19,600
3020	876 MBH		.41	79.012		14,300	3,725		18,025	21,300
3040	1,168 MBH		.31	103		16,000	4,875		20,875	25,000
3060	1,460 MBH		.28	113		18,700	5,325		24,025	28,500
3080	2,044 MBH		.26	122		24,700	5,750		30,450	35,700
3100	2,628 MBH		.21	150		28,900	7,075		35,975	42,400
3120	3,210 MBH		.18	174		35,700	8,225		43,925	51,500
3140	3,796 MBH		.17	186		39,900	8,750		48,650	57,000
3160	4,088 MBH		.16	195		45,400	9,175		54,575	63,500
3180	4,672 MBH		.16	203		49,500	9,575		59,075	69,000
3200	5,256 MBH		.15	217		54,500	10,200		64,700	75,500
3220	6,000 MBH, 179 BHP		.13	256		79,500	12,000		91,500	105,500
3240	7,130 MBH, 213 BHP		.08	385		83,000	18,100		101,100	118,500
3260	9,800 MBH, 286 BHP		.06	533		95,500	25,100		120,600	142,500
3280	10,900 MBH, 325.6 BHP		.05	592		116,000	27,900		143,900	169,500
3290	12,200 MBH, 364.5 BHP		.05	666		126,500	31,300		157,800	186,500
3300	13,500 MBH, 403.3 BHP		.04	727		140,000	34,200		174,200	205,500

23 52 23.40 Oil-Fired Boilers		Crew	Daily Output	Labor-Hours	Unit	Material	Labor	Equipment	Total	Total Incl O&P
0010	**OIL-FIRED BOILERS,** Standard controls, flame retention burner, packaged									
1000	Cast iron, with insulated flush jacket									

23 52 Heating Boilers

23 52 23 – Cast-Iron Boilers

23 52 23.40 Oil-Fired Boilers		Crew	Daily Output	Labor-Hours	Unit	2009 Bare Costs Material	Labor	Equipment	Total	Total Incl O&P
2000	Steam, gross output, 109 MBH	Q-7	1.20	26.667	Ea.	1,875	1,250		3,125	3,950
2020	144 MBH		1.10	29.091		2,125	1,375		3,500	4,375
2040	173 MBH		1	32		2,375	1,500		3,875	4,850
2060	207 MBH		.90	35.556		2,575	1,675		4,250	5,325
2080	236 MBH		.85	37.647		3,050	1,775		4,825	6,025
2100	300 MBH		.70	45.714		3,650	2,150		5,800	7,250
2120	480 MBH		.50	64		4,675	3,000		7,675	9,675
2140	665 MBH		.45	71.111		6,400	3,350		9,750	12,100
2160	794 MBH		.41	78.049		7,575	3,675		11,250	13,800
2180	1,084 MBH		.38	85.106		8,725	4,000		12,725	15,600
2200	1,360 MBH		.33	98.160		10,100	4,625		14,725	18,000
2220	1,600 MBH		.26	122		11,500	5,750		17,250	21,200
2240	2,175 MBH		.24	133		14,700	6,300		21,000	25,600
2260	2,480 MBH		.21	156		17,700	7,350		25,050	30,500
2280	3,000 MBH		.19	170		20,500	8,000		28,500	34,500
2300	3,550 MBH		.17	187		22,700	8,800		31,500	38,200
2320	3,820 MBH		.16	200		24,000	9,400		33,400	40,500
2340	4,360 MBH		.15	214		26,300	10,100		36,400	44,100
2360	4,940 MBH		.14	225		55,500	10,600		66,100	77,000
2380	5,520 MBH		.14	235		65,500	11,100		76,600	88,500
2400	6,100 MBH		.13	256		75,000	12,000		87,000	100,500
2420	6,390 MBH		.11	290		79,000	13,700		92,700	107,000
2440	6,680 MBH		.10	313		81,000	14,700		95,700	111,000
2460	6,970 MBH	↓	.09	363	↓	84,000	17,100		101,100	118,000
3000	Hot water, same price as steam									
4000	For tankless coil in smaller sizes, add				Ea.	15%				

23 52 23.60 Solid-Fuel Boilers		Crew	Daily Output	Labor-Hours	Unit	Material	Labor	Equipment	Total	Total Incl O&P
0010	**SOLID-FUEL BOILERS**									
3000	Stoker fired (coal) cast iron with flush jacket and									
3400	insulation, steam or water, gross output, 1280 MBH	Q-6	.36	66.667	Ea.	33,800	3,075		36,875	41,800
3420	1460 MBH		.30	80		37,100	3,675		40,775	46,300
3440	1640 MBH		.28	85.714		39,500	3,950		43,450	49,400
3460	1820 MBH		.26	92.308		42,300	4,250		46,550	53,000
3480	2000 MBH		.25	96		45,000	4,425		49,425	56,000
3500	2360 MBH		.23	104		51,000	4,800		55,800	63,000
3540	2725 MBH	↓	.20	120		56,500	5,525		62,025	71,000
3800	2950 MBH	Q-7	.16	200		94,500	9,400		103,900	118,000
3820	3210 MBH		.15	213		99,500	10,000		109,500	124,500
3840	3480 MBH		.14	228		104,000	10,700		114,700	130,500
3860	3745 MBH		.14	228		109,000	10,700		119,700	136,000
3880	4000 MBH		.13	246		113,500	11,600		125,100	142,000
3900	4200 MBH		.13	246		118,500	11,600		130,100	147,500
3920	4400 MBH		.12	266		123,000	12,500		135,500	154,000
3940	4600 MBH	↓	.12	266	↓	127,500	12,500		140,000	159,500

23 52 26 – Steel Boilers

23 52 26.40 Oil-Fired Boilers		Crew	Daily Output	Labor-Hours	Unit	Material	Labor	Equipment	Total	Total Incl O&P
0010	**OIL-FIRED BOILERS**, Standard controls, flame retention burner									
5000	Steel, with insulated flush jacket									
7000	Hot water, gross output, 103 MBH	Q-6	1.60	15	Ea.	1,675	690		2,365	2,875
7020	122 MBH		1.45	16.506		1,775	760		2,535	3,100
7040	137 MBH		1.36	17.595		1,900	810		2,710	3,300
7060	168 MBH	↓	1.30	18.405	↓	2,000	850		2,850	3,450

23 52 26 – Steel Boilers

23 52 26.40 Oil-Fired Boilers		Crew	Daily Output	Labor-Hours	Unit	2009 Bare Costs Material	Labor	Equipment	Total	Total Incl O&P
7080	225 MBH	Q-6	1.22	19.704	Ea.	2,625	910		3,535	4,250
7100	315 MBH		.96	25.105		5,425	1,150		6,575	7,700
7120	420 MBH		.70	34.483		5,700	1,600		7,300	8,650
7140	525 MBH		.57	42.403		6,850	1,950		8,800	10,500
7180	735 MBH		.48	50.104		8,425	2,300		10,725	12,800
7220	1,050 MBH		.37	65.753		13,400	3,025		16,425	19,300
7280	2,310 MBH		.21	114		21,200	5,300		26,500	31,300
7320	3,150 MBH		.13	184		30,200	8,500		38,700	46,100
7340	For tankless coil in steam or hot water, add					7%				

23 52 26.70 Packaged Water Tube Boilers		Crew	Daily Output	Labor-Hours	Unit	Material	Labor	Equipment	Total	Total Incl O&P
0010	**PACKAGED WATER TUBE BOILERS**									
2000	Packaged water tube, #2 oil, steam or hot water, gross output									
2010	200 MBH	Q-6	1.50	16	Ea.	4,450	735		5,185	6,000
2014	275 MBH		1.40	17.143		4,675	790		5,465	6,325
2018	360 MBH		1.10	21.818		4,900	1,000		5,900	6,900
2022	520 MBH		.65	36.923		5,250	1,700		6,950	8,325
2026	600 MBH		.60	40		5,400	1,850		7,250	8,700
2030	720 MBH		.55	43.636		7,325	2,000		9,325	11,100
2034	960 MBH		.48	50		8,925	2,300		11,225	13,300
2040	1200 MBH	Q-7	.50	64		18,200	3,000		21,200	24,500
2044	1440 MBH		.45	71.111		19,500	3,350		22,850	26,400
2060	1600 MBH		.40	80		22,900	3,750		26,650	30,900
2068	1920 MBH		.35	91.429		24,800	4,300		29,100	33,800
2072	2160 MBH		.33	96.970		27,500	4,550		32,050	37,200
2080	2400 MBH		.30	106		30,800	5,025		35,825	41,400
2100	3200 MBH		.25	128		35,200	6,025		41,225	47,700
2120	4800 MBH		.20	160		46,000	7,525		53,525	62,000
2200	Gas fired									
2204	200 MBH	Q-6	1.50	16	Ea.	5,650	735		6,385	7,325
2208	275 MBH		1.40	17.143		5,875	790		6,665	7,650
2212	360 MBH		1.10	21.818		6,100	1,000		7,100	8,225
2216	520 MBH		.65	36.923		6,450	1,700		8,150	9,650
2220	600 MBH		.60	40		6,600	1,850		8,450	10,000
2224	720 MBH		.55	43.636		7,800	2,000		9,800	11,600
2228	960 MBH		.48	50		9,825	2,300		12,125	14,300
2232	1220 MBH	Q-7	.50	64		14,300	3,000		17,300	20,200
2236	1440 MBH		.45	71.111		14,800	3,350		18,150	21,200
2240	1680 MBH		.40	80		19,800	3,750		23,550	27,400
2244	1920 MBH		.35	91.429		19,800	4,300		24,100	28,300
2248	2160 MBH		.33	96.970		20,200	4,550		24,750	29,100
2252	2400 MBH		.30	106		22,600	5,025		27,625	32,400

23 52 28 – Swimming Pool Boilers

23 52 28.10 Swimming Pool Heaters		Crew	Daily Output	Labor-Hours	Unit	Material	Labor	Equipment	Total	Total Incl O&P
0010	**SWIMMING POOL HEATERS**, Not including wiring, external									
0020	piping, base or pad.									
0160	Gas fired, input, 155 MBH	Q-6	1.50	16	Ea.	1,575	735		2,310	2,850
0200	199 MBH		1	24		1,700	1,100		2,800	3,525
0220	250 MBH		.70	34.286		1,850	1,575		3,425	4,400
0240	300 MBH		.60	40		1,925	1,850		3,775	4,900
0260	399 MBH		.50	48		2,175	2,200		4,375	5,700
0280	500 MBH		.40	60		7,225	2,775		10,000	12,100
0300	650 MBH		.35	68.571		7,675	3,150		10,825	13,200

23 52 28 – Swimming Pool Boilers

23 52 28.10 Swimming Pool Heaters		Crew	Daily Output	Labor-Hours	Unit	2009 Bare Costs Material	Labor	Equipment	Total	Total Incl O&P
0320	750 MBH	Q-6	.33	72.727	Ea.	8,400	3,350		11,750	14,300
0360	990 MBH		.22	109		11,300	5,025		16,325	20,000
0370	1,260 MBH		.21	114		13,200	5,275		18,475	22,400
0380	1,440 MBH		.19	126		14,100	5,825		19,925	24,200
0400	1,800 MBH		.14	171		15,700	7,900		23,600	29,200
0410	2,070 MBH		.13	184		18,500	8,500		27,000	33,200
1120	Oil fired, 315 MBH		.96	25		3,375	1,150		4,525	5,450
2000	Electric, 12 KW, 4,800 gallon pool	Q-19	3	8		1,950	360		2,310	2,700
2020	15 KW, 7,200 gallon pool		2.80	8.571		2,000	390		2,390	2,775
2040	24 KW, 9,600 gallon pool		2.40	10		2,275	455		2,730	3,175
2060	30 KW, 12,000 gallon pool		2	12		2,325	545		2,870	3,375
2080	36 KW, 14,400 gallon pool		1.60	15		2,675	680		3,355	3,975
2100	57 KW, 24,000 gallon pool		1.20	20		3,325	905		4,230	5,000
9000	To select pool heater: 12 BTUH x S.F. pool area									
9010	X temperature differential = required output									
9050	For electric, KW = gallons x 2.5 divided by 1000									
9100	For family home type pool, double the									
9110	Rated gallon capacity = 1/2° F rise per hour									

23 52 39 – Fire-Tube Boilers

23 52 39.13 Scotch Marine Boilers		Crew	Daily Output	Labor-Hours	Unit	Material	Labor	Equipment	Total	Total Incl O&P
0010	**SCOTCH MARINE BOILERS**									
1000	Packaged fire tube, #2 oil, gross output									
1006	15 PSI steam									
1010	1005 MBH, 30 HP	Q-6	.33	72.072	Ea.	34,200	3,325		37,525	42,600
1014	1675 MBH, 50 HP	″	.26	93.023		40,400	4,275		44,675	51,000
1020	3348 MBH, 100 HP	Q-7	.21	152		52,500	7,175		59,675	69,000
1030	5025 MBH, 150 HP		.17	192		66,500	9,050		75,550	86,500
1040	6696 MBH, 200 HP		.14	223		75,500	10,500		86,000	99,000
1050	8375 MBH, 250 HP		.14	231		84,500	10,900		95,400	109,500
1060	10,044 MBH, 300 HP		.13	251		103,500	11,800		115,300	132,000
1080	16,740 MBH, 500 HP		.08	380		140,500	17,900		158,400	181,500
1100	23,435 MBH, 700 HP		.07	484		172,500	22,800		195,300	223,500
1140	To fire #6, and gas, add		.42	76.190		15,300	3,575		18,875	22,300
1180	For duplex package feed system									
1200	To 3348 MBH boiler, add	Q-7	.54	59.259	Ea.	8,250	2,775		11,025	13,300
1220	To 6696 MBH boiler, add		.41	78.049		9,775	3,675		13,450	16,300
1240	To 10,044 MBH boiler, add		.38	84.211		11,200	3,950		15,150	18,300
1260	To 16,740 MBH boiler, add		.28	114		12,800	5,375		18,175	22,200
1280	To 23,435 MBH boiler, add		.25	128		15,100	6,025		21,125	25,600
1300	150 PSI steam									
1310	1005 MBH, 30 HP	Q-6	.33	72.072	Ea.	36,400	3,325		39,725	45,000
1320	1675 MBH, 50 HP	″	.26	93.023		45,400	4,275		49,675	56,500
1330	3350 MBH, 100 HP	Q-7	.21	152		62,000	7,175		69,175	79,000
1340	5025 MBH, 150 HP		.17	192		79,000	9,050		88,050	100,500
1350	6700 MBH, 200 HP		.14	223		87,500	10,500		98,000	112,500
1360	10,050 MBH, 300 HP		.13	251		130,000	11,800		141,800	160,500
1400	Packaged scotch marine, #6 oil									
1410	15 PSI steam									
1420	1675 MBH, 50 HP	Q-6	.25	97.166	Ea.	52,500	4,475		56,975	64,500
1430	3350 MBH, 100 HP	Q-7	.20	159		65,000	7,475		72,475	82,500
1440	5025 MBH, 150 HP		.16	202		81,500	9,525		91,025	104,000
1450	6700 MBH, 200 HP		.14	233		89,000	11,000		100,000	114,500

23 52 Heating Boilers

23 52 39 – Fire-Tube Boilers

23 52 39.13 Scotch Marine Boilers		Crew	Daily Output	Labor-Hours	Unit	2009 Bare Costs Material	Labor	Equipment	Total	Total Incl O&P
1460	8375 MBH, 250 HP	Q-7	.13	244	Ea.	114,000	11,500		125,500	142,000
1470	10,050 MBH, 300 HP		.12	262		134,000	12,300		146,300	165,500
1480	13,400 MBH, 400 HP		.09	351		153,500	16,500		170,000	193,500
1500	150 PSI steam									
1510	1675 MBH, 50 HP	Q-6	.25	97.166	Ea.	62,500	4,475		66,975	75,000
1520	3350 MBH, 100 HP	Q-7	.20	159		72,000	7,475		79,475	90,000
1530	5025 MBH, 150 HP		.16	202		84,000	9,525		93,525	107,000
1540	6700 MBH, 200 HP		.14	233		95,500	11,000		106,500	121,500
1550	8375 MBH, 250 HP		.13	244		122,000	11,500		133,500	151,500
1560	10,050 MBH, 300 HP		.12	262		136,500	12,300		148,800	169,000
1570	13,400 MBH, 400 HP		.09	351		160,500	16,500		177,000	202,000

23 52 84 – Boiler Blowdown

23 52 84.10 Boiler Blowdown Systems		Crew	Daily Output	Labor-Hours	Unit	Material	Labor	Equipment	Total	Total Incl O&P
0010	**BOILER BLOWDOWN SYSTEMS**									
1010	Boiler blowdown, auto/manual to 2000 MBH	Q-5	3.75	4.267	Ea.	3,125	190		3,315	3,700
1020	7300 MBH	″	3	5.333	″	4,350	237		4,587	5,150

23 52 88 – Burners

23 52 88.10 Replacement Type Burners		Crew	Daily Output	Labor-Hours	Unit	Material	Labor	Equipment	Total	Total Incl O&P
0010	**REPLACEMENT TYPE BURNERS**									
0990	Residential, conversion, gas fired, LP or natural									
1000	Gun type, atmospheric input 50 to 225 MBH	Q-1	2.50	6.400	Ea.	645	281		926	1,125
1020	100 to 400 MBH		2	8		1,075	350		1,425	1,700
1040	300 to 1000 MBH		1.70	9.412		3,300	415		3,715	4,250
2000	Commercial and industrial, gas/oil, input									
2050	400 MBH	Q-1	1.50	10.667	Ea.	3,400	470		3,870	4,450
2090	670 MBH		1.40	11.429		3,400	500		3,900	4,500
2140	1155 MBH		1.30	12.308		3,800	540		4,340	4,975
2200	1800 MBH		1.20	13.333		3,800	585		4,385	5,050
2260	3000 MBH		1.10	14.545		4,575	640		5,215	5,975
2320	4100 MBH		1	16		4,675	700		5,375	6,200
2500	Impinged jet, rectangular, burner only, 3-1/2″ WC,									
2520	Gas fired, input, 420 to 640 MBH	Q-1	2	8	Ea.	825	350		1,175	1,425
2540	560 to 860 MBH		2	8		1,175	350		1,525	1,825
2560	630 to 950 MBH		1.90	8.421		1,225	370		1,595	1,900
2580	950 to 1200 MBH		1.80	8.889		1,750	390		2,140	2,500
2600	1120 to 1700 MBH		1.80	8.889		2,350	390		2,740	3,150
2620	1700 to 1900 MBH		1.70	9.412		2,650	415		3,065	3,525
2640	1400 to 2100 MBH		1.60	10		2,925	440		3,365	3,875
2660	1600 to 2400 MBH		1.60	10		3,075	440		3,515	4,050
2680	1700 to 2500 MBH		1.50	10.667		3,525	470		3,995	4,575
2700	1880 to 2880 MBH		1.40	11.429		3,700	500		4,200	4,825
2720	2260 to 3680 MBH		1.40	11.429		4,325	500		4,825	5,500
2740	2590 to 4320 MBH		1.30	12.308		5,275	540		5,815	6,600
2760	3020 to 5040 MBH		1.20	13.333		6,150	585		6,735	7,650
2780	3450 to 5760 MBH		1.20	13.333		7,050	585		7,635	8,625
2850	Burner pilot		50	.320		21	14.05		35.05	44
2860	Thermocouple		50	.320		21.50	14.05		35.55	45
2870	Thermocouple & pilot bracket		45	.356		84	15.60		99.60	116
2900	For 7″ W.C. pressure, increase MBH 40%									
3000	Flame retention oil fired assembly, input									
3020	.50 to 2.25 GPH	Q-1	2.40	6.667	Ea.	289	293		582	760
3040	2.0 to 5.0 GPH		2	8		330	350		680	890

23 52 Heating Boilers

23 52 88 – Burners

23 52 88.10 Replacement Type Burners		Crew	Daily Output	Labor-Hours	Unit	Material	Labor	Equipment	Total	Total Incl O&P
						2009 Bare Costs				
3060	3.0 to 7.0 GPH	Q-1	1.80	8.889	Ea.	590	390		980	1,225
3080	6.0 to 12.0 GPH		1.60	10		935	440		1,375	1,675
4600	Gas safety, shut off valve, 3/4" threaded	1 Stpi	20	.400		178	19.75		197.75	226
4610	1" threaded		19	.421		172	21		193	220
4620	1-1/4" threaded		15	.533		194	26.50		220.50	253
4630	1-1/2" threaded		13	.615		209	30.50		239.50	276
4640	2" threaded		11	.727		235	36		271	315
4650	2-1/2" threaded	Q-1	15	1.067		269	47		316	365
4660	3" threaded		13	1.231		370	54		424	485
4670	4" flanged		3	5.333		2,625	234		2,859	3,250
4680	6" flanged	Q-2	3	8		5,875	365		6,240	7,025

23 53 Heating Boiler Feedwater Equipment

23 53 24 – Shot Chemical Feeder

23 53 24.10 Chemical Feeder

No.	Description	Crew	Daily Output	Labor-Hours	Unit	Material	Labor	Equipment	Total	Total Incl O&P
0010	**CHEMICAL FEEDER**									
0200	Shot chem feeder, by pass, in line mount, 125 PSIG, 1.7 gallon	2 Stpi	1.80	8.889	Ea.	298	440		738	990
0220	Floor mount, 175 PSIG, 5 gallon		1.24	12.903		350	635		985	1,350
0230	12 gallon		.93	17.204		540	850		1,390	1,875
0300	150 lb., ASME, 5 gallon		1.24	12.903		350	635		985	1,350
0320	10 gallon		.93	17.204		500	850		1,350	1,825
0400	300 lb., ASME, 5 gallon		1.24	12.903		350	635		985	1,350
0420	10 gallon		.93	17.204		500	850		1,350	1,825

23 54 Furnaces

23 54 13 – Electric-Resistance Furnaces

23 54 13.10 Electric Furnaces

No.	Description	Crew	Daily Output	Labor-Hours	Unit	Material	Labor	Equipment	Total	Total Incl O&P
0010	**ELECTRIC FURNACES**, Hot air, blowers, std. controls									
0011	not including gas, oil or flue piping									
1000	Electric, UL listed									
1020	10.2 MBH	Q-20	5	4	Ea.	360	174		534	660
1040	17.1 MBH		4.80	4.167		375	181		556	690
1060	27.3 MBH		4.60	4.348		455	189		644	790
1100	34.1 MBH		4.40	4.545		465	197		662	810
1120	51.6 MBH		4.20	4.762		575	207		782	950
1140	68.3 MBH		4	5		700	217		917	1,100
1160	85.3 MBH		3.80	5.263		765	228		993	1,200

23 54 16 – Fuel-Fired Furnaces

23 54 16.13 Gas-Fired Furnaces

No.	Description	Crew	Daily Output	Labor-Hours	Unit	Material	Labor	Equipment	Total	Total Incl O&P
0010	**GAS-FIRED FURNACES**									
3000	Gas, AGA certified, upflow, direct drive models									
3020	45 MBH input	Q-9	4	4	Ea.	575	170		745	895
3040	60 MBH input		3.80	4.211		615	179		794	950
3060	75 MBH input		3.60	4.444		685	189		874	1,050
3100	100 MBH input		3.20	5		695	212		907	1,075
3120	125 MBH input		3	5.333		700	227		927	1,125
3130	150 MBH input		2.80	5.714		705	243		948	1,150
3140	200 MBH input		2.60	6.154		2,425	261		2,686	3,050
3160	300 MBH input		2.30	6.957		2,525	296		2,821	3,250

23 54 Furnaces

23 54 16 – Fuel-Fired Furnaces

23 54 16.13 Gas-Fired Furnaces		Crew	Daily Output	Labor-Hours	Unit	Material	Labor	Equipment	Total	Total Incl O&P
						2009 Bare Costs				
3180	400 MBH input	Q-9	2	8	Ea.	2,825	340		3,165	3,625
3200	Gas fired wall furnace									
3204	Horizontal flow									
3208	7.7 MBH	Q-9	7	2.286	Ea.	460	97		557	655
3212	14 MBH		6.50	2.462		460	105		565	665
3216	24 MBH		5	3.200		485	136		621	745
3220	49 MBH		4	4		660	170		830	985
3224	65 MBH		3.60	4.444		875	189		1,064	1,250
3240	Up-flow									
3244	7.7 MBH	Q-9	7	2.286	Ea.	460	97		557	655
3248	14 MBH		6.50	2.462		460	105		565	665
3252	24 MBH		5	3.200		465	136		601	720
3254	35 MBH		4.50	3.556		615	151		766	910
3256	49 MBH		4	4		870	170		1,040	1,225
3260	65 MBH		3.60	4.444		965	189		1,154	1,375
3500	Gas furnace									
3510	Up-flow									
3514	1200 CFM, 45 MBH	Q-9	4	4	Ea.	1,125	170		1,295	1,500
3518	75 MBH		3.60	4.444		1,200	189		1,389	1,625
3522	100 MBH		3.20	5		1,275	212		1,487	1,725
3526	125 MBH		3	5.333		1,400	227		1,627	1,875
3530	150 MBH		2.80	5.714		1,425	243		1,668	1,925
3540	2000 CFM, 75 MBH		3.60	4.444		870	189		1,059	1,250
3544	100 MBH		3.20	5		875	212		1,087	1,275
3548	120 MBH		3	5.333		930	227		1,157	1,375
3552	150 MBH		2.80	5.714		1,000	243		1,243	1,475
3560	Horizontal flow									
3564	1200 CFM, 45 MBH	Q-9	4	4	Ea.	1,150	170		1,320	1,500
3568	60 MBH		3.80	4.211		1,200	179		1,379	1,600
3572	70 MBH		3.60	4.444		1,200	189		1,389	1,625
3576	2000 CFM, 90 MBH		3.30	4.848		1,250	206		1,456	1,700
3580	110 MBH		3.10	5.161		1,325	219		1,544	1,775

23 54 16.16 Oil-Fired Furnaces		Crew	Daily Output	Labor-Hours	Unit	Material	Labor	Equipment	Total	Total Incl O&P
0010	**OIL-FIRED FURNACES**									
6000	Oil, UL listed, atomizing gun type burner									
6020	56 MBH output	Q-9	3.60	4.444	Ea.	1,850	189		2,039	2,325
6030	84 MBH output		3.50	4.571		1,850	194		2,044	2,350
6040	95 MBH output		3.40	4.706		1,875	200		2,075	2,375
6060	134 MBH output		3.20	5		1,950	212		2,162	2,475
6080	151 MBH output		3	5.333		2,275	227		2,502	2,850
6100	200 MBH input		2.60	6.154		2,325	261		2,586	2,950
6120	300 MBH input		2.30	6.957		2,825	296		3,121	3,550
6140	400 MBH input		2	8		3,275	340		3,615	4,125
6200	Up flow									
6210	105 MBH	Q-9	3.40	4.706	Ea.	1,425	200		1,625	1,875
6220	140 MBH		3.30	4.848		1,725	206		1,931	2,225
6230	168 MBH		2.90	5.517		2,075	234		2,309	2,650
6260	Lo-boy style									
6270	105 MBH	Q-9	3.40	4.706	Ea.	1,575	200		1,775	2,025
6280	140 MBH		3.30	4.848		1,700	206		1,906	2,200
6290	189 MBH		2.80	5.714		2,300	243		2,543	2,900
6320	Down flow									

23 54 Furnaces

23 54 16 – Fuel-Fired Furnaces

23 54 16.16 Oil-Fired Furnaces

Code	23 54 16.16 Oil-Fired Furnaces		Crew	Daily Output	Labor-Hours	Unit	Material	Labor	Equipment	Total	Total Incl O&P
							2009 Bare Costs				
6330	105 MBH		Q-9	3.40	4.706	Ea.	1,500	200		1,700	1,950
6340	140 MBH			3.30	4.848		1,700	206		1,906	2,200
6350	168 MBH			2.90	5.517		2,150	234		2,384	2,725
6380	Horizontal										
6390	105 MBH		Q-9	3.40	4.706	Ea.	1,500	200		1,700	1,950
6400	140 MBH			3.30	4.848		1,700	206		1,906	2,200
6410	189 MBH			2.80	5.714		1,925	243		2,168	2,500
	23 54 16.21 Solid Fuel-Fired Furnaces										
0010	**SOLID FUEL-FIRED FURNACES**										
6020	Wood fired furnaces										
6030	Includes hot water coil, thermostat, and auto draft control										
6040	24″ long firebox	G	Q-9	4	4	Ea.	3,600	170		3,770	4,225
6050	30″ long firebox	G		3.60	4.444		4,200	189		4,389	4,900
6060	With fireplace glass doors	G		3.20	5		5,400	212		5,612	6,275
6200	Wood / oil fired furnaces, includes two thermostats										
6210	Includes hot water coil and auto draft control										
6240	24″ long firebox	G	Q-9	3.40	4.706	Ea.	4,400	200		4,600	5,125
6250	30″ long firebox	G		3	5.333		4,875	227		5,102	5,700
6260	With fireplace glass doors	G		2.80	5.714		5,400	243		5,643	6,325
6400	Wood / gas fired furnaces, includes two thermostats										
6410	Includes hot water coil and auto draft control										
6440	24″ long firebox	G	Q-9	2.80	5.714	Ea.	4,925	243		5,168	5,800
6450	30″ long firebox	G		2.40	6.667		5,525	283		5,808	6,500
6460	With fireplace glass doors	G		2	8		6,000	340		6,340	7,125
6600	Wood / oil /gas fired furnaces, optional accessories										
6610	Hot air plenum		Q-9	16	1	Ea.	81	42.50		123.50	155
6620	Safety heat dump			24	.667		72	28.50		100.50	123
6630	Auto air intake			18	.889		119	38		157	189
6640	Cold air return package			14	1.143		125	48.50		173.50	213
6650	Wood fork						39			39	43
6700	Wood fired outdoor furnace										
6740	24″ long firebox	G	Q-9	3.80	4.211	Ea.	3,775	179		3,954	4,425
6760	Wood fired outdoor furnace, optional accessories										
6770	Chimney section, stainless steel, 6″ ID x 3′ lg.	G	Q-9	36	.444	Ea.	95	18.90		113.90	134
6780	Chimney cap, stainless steel	G	″	40	.400	″	51	17		68	82
6800	Wood fired hot water furnace										
6820	Includes 200 gal. hot water storage, thermostat, and auto draft control										
6840	30″ long firebox	G	Q-9	2.10	7.619	Ea.	7,450	325		7,775	8,675
6850	Water to air heat exchanger										
6870	Includes mounting kit and blower relay										
6880	140 MBH, 18.75″ W x 18.75″ L		Q-9	7.50	2.133	Ea.	370	90.50		460.50	545
6890	200 MBH, 24″ W x 24″ L		″	7	2.286	″	525	97		622	730
6900	Water to water heat exchanger										
6940	100 MBH		Q-9	6.50	2.462	Ea.	370	105		475	570
6960	290 MBH		″	6	2.667	″	535	113		648	765
7000	Optional accessories										
7010	Large volume circulation pump, (2 Included)		Q-9	14	1.143	Ea.	242	48.50		290.50	340
7020	Air bleed fittings, (package)			24	.667		49	28.50		77.50	97.50
7030	Domestic water preheater			6	2.667		219	113		332	415
7040	Smoke pipe kit			4	4		62	170		232	330

23 54 24 – Furnace Components for Cooling

23 54 24.10 Furnace Components and Combinations		Crew	Daily Output	Labor-Hours	Unit	2009 Bare Costs Material	Labor	Equipment	Total	Total Incl O&P
0010	**FURNACE COMPONENTS AND COMBINATIONS**									
0080	Coils, A/C evaporator, for gas or oil furnaces									
0090	Add-on, with holding charge									
0100	Upflow									
0120	1-1/2 ton cooling	Q-5	4	4	Ea.	158	178		336	440
0130	2 ton cooling		3.70	4.324		188	192		380	495
0140	3 ton cooling		3.30	4.848		238	215		453	585
0150	4 ton cooling		3	5.333		325	237		562	715
0160	5 ton cooling		2.70	5.926		385	263		648	820
0300	Downflow									
0330	2-1/2 ton cooling	Q-5	3	5.333	Ea.	232	237		469	610
0340	3-1/2 ton cooling		2.60	6.154		276	273		549	715
0350	5 ton cooling		2.20	7.273		385	325		710	910
0600	Horizontal									
0630	2 ton cooling	Q-5	3.90	4.103	Ea.	156	182		338	445
0640	3 ton cooling		3.50	4.571		184	203		387	505
0650	4 ton cooling		3.20	5		224	222		446	580
0660	5 ton cooling		2.90	5.517		224	245		469	615
2000	Cased evaporator coils for air handlers									
2100	1-1/2 ton cooling	Q-5	4.40	3.636	Ea.	244	162		406	510
2110	2 ton cooling		4.10	3.902		248	173		421	530
2120	2-1/2 ton cooling		3.90	4.103		259	182		441	560
2130	3 ton cooling		3.70	4.324		286	192		478	605
2140	3-1/2 ton cooling		3.50	4.571		299	203		502	635
2150	4 ton cooling		3.20	5		360	222		582	730
2160	5 ton cooling		2.90	5.517		415	245		660	825
3010	Air handler, modular									
3100	With cased evaporator cooling coil									
3120	1-1/2 ton cooling	Q-5	3.80	4.211	Ea.	775	187		962	1,125
3130	2 ton cooling		3.50	4.571		865	203		1,068	1,250
3140	2-1/2 ton cooling		3.30	4.848		885	215		1,100	1,300
3150	3 ton cooling		3.10	5.161		945	229		1,174	1,375
3160	3-1/2 ton cooling		2.90	5.517		1,000	245		1,245	1,475
3170	4 ton cooling		2.50	6.400		1,125	284		1,409	1,675
3180	5 ton cooling		2.10	7.619		1,275	340		1,615	1,900
3500	With no cooling coil									
3520	1-1/2 ton coil size	Q-5	12	1.333	Ea.	575	59		634	720
3530	2 ton coil size		10	1.600		605	71		676	770
3540	2-1/2 ton coil size		10	1.600		605	71		676	770
3554	3 ton coil size		9	1.778		685	79		764	875
3560	3-1/2 ton coil size		9	1.778		745	79		824	940
3570	4 ton coil size		8.50	1.882		860	83.50		943.50	1,075
3580	5 ton coil size		8	2		1,000	89		1,089	1,225
4000	With heater									
4120	5 kW, 17.1 MBH	Q-5	16	1	Ea.	585	44.50		629.50	710
4130	7.5 kW, 25.6 MBH		15.60	1.026		590	45.50		635.50	720
4140	10 kW, 34.2 MBH		15.20	1.053		825	47		872	975
4150	12.5 KW, 42.7 MBH		14.80	1.081		910	48		958	1,075
4160	15 KW, 51.2 MBH		14.40	1.111		935	49.50		984.50	1,100
4170	25 KW, 85.4 MBH		14	1.143		1,050	51		1,101	1,225
4180	30 KW, 102 MBH		13	1.231		1,225	54.50		1,279.50	1,425

23 55 Fuel-Fired Heaters

23 55 13 – Fuel-Fired Duct Heaters

23 55 13.16 Gas-Fired Duct Heaters		Crew	Daily Output	Labor-Hours	Unit	Material	2009 Bare Costs Labor	Equipment	Total	Total Incl O&P
0010	**GAS-FIRED DUCT HEATERS**, Includes burner, controls, stainless steel									
0020	heat exchanger. Gas fired, electric ignition									
0030	Indoor installation									
0080	100 MBH output	Q-5	5	3.200	Ea.	1,875	142		2,017	2,300
0100	120 MBH output		4	4		2,125	178		2,303	2,625
0130	200 MBH output		2.70	5.926		2,850	263		3,113	3,525
0140	240 MBH output		2.30	6.957		2,975	310		3,285	3,725
0160	280 MBH output		2	8		3,350	355		3,705	4,225
0180	320 MBH output	↓	1.60	10		3,550	445		3,995	4,600
0300	For powered venter and adapter, add				↓	460			460	510
0502	For required flue pipe, see Div. 23 51 23.10									
1000	Outdoor installation, with vent cap									
1020	75 MBH output	Q-5	4	4	Ea.	2,775	178		2,953	3,325
1040	94 MBH output		4	4		3,125	178		3,303	3,725
1060	120 MBH output		4	4		3,525	178		3,703	4,150
1080	157 MBH output		3.50	4.571		4,100	203		4,303	4,800
1100	187 MBH output		3	5.333		5,050	237		5,287	5,900
1120	225 MBH output		2.50	6.400		5,225	284		5,509	6,175
1140	300 MBH output		1.80	8.889		6,575	395		6,970	7,850
1160	375 MBH output		1.60	10		6,575	445		7,020	7,900
1180	450 MBH output		1.40	11.429		6,875	510		7,385	8,325
1200	600 MBH output	↓	1	16		8,375	710		9,085	10,300
1300	Aluminized exchanger, subtract					15%				
1400	For powered venter, add					10%				
1500	For two stage gas valve, add				↓	253			253	278

23 55 33 – Fuel-Fired Unit Heaters

23 55 33.13 Oil-Fired Unit Heaters		Crew	Daily Output	Labor-Hours	Unit	Material	Labor	Equipment	Total	Total Incl O&P
0010	**OIL-FIRED UNIT HEATERS**, Cabinet, grilles, fan, ctrl, burner, no piping									
6000	Oil fired, suspension mounted, 94 MBH output	Q-5	4	4	Ea.	3,400	178		3,578	4,000
6040	140 MBH output		3	5.333		3,625	237		3,862	4,325
6060	184 MBH output	↓	3	5.333	↓	3,900	237		4,137	4,650

23 55 33.16 Gas-Fired Unit Heaters		Crew	Daily Output	Labor-Hours	Unit	Material	Labor	Equipment	Total	Total Incl O&P
0010	**GAS-FIRED UNIT HEATERS**, Cabinet, grilles, fan, ctrls., burner, no piping									
0022	thermostat, no piping. For flue see Div. 23 51 23.10									
1000	Gas fired, floor mounted									
1100	60 MBH output	Q-5	10	1.600	Ea.	805	71		876	990
1120	80 MBH output		9	1.778		815	79		894	1,025
1140	100 MBH output		8	2		890	89		979	1,125
1160	120 MBH output		7	2.286		1,050	102		1,152	1,300
1180	180 MBH output	↓	6	2.667	↓	1,275	118		1,393	1,575
1500	Rooftop mounted, gravity vent, stainless steel exchanger									
1520	75 MBH output	Q-6	4	6	Ea.	5,225	276		5,501	6,175
1540	95 MBH output		3.60	6.667		5,325	305		5,630	6,300
1560	120 MBH output		3.30	7.273		5,700	335		6,035	6,775
1580	159 MBH output		3	8		5,875	370		6,245	7,000
1600	190 MBH output		2.60	9.231		6,625	425		7,050	7,925
1620	225 MBH output		2.30	10.435		6,725	480		7,205	8,125
1640	300 MBH output		1.90	12.632		8,025	580		8,605	9,700
1660	375 MBH output		1.40	17.143		9,100	790		9,890	11,200
1680	450 MBH output		1.20	20		9,275	920		10,195	11,600
1700	600 MBH output		1	24		11,200	1,100		12,300	14,000
1720	750 MBH output	↓	.80	30	↓	19,100	1,375		20,475	23,100

23 55 Fuel-Fired Heaters

23 55 33 – Fuel-Fired Unit Heaters

23 55 33.16 Gas-Fired Unit Heaters		Crew	Daily Output	Labor-Hours	Unit	2009 Bare Costs Material	Labor	Equipment	Total	Total Incl O&P
1740	900 MBH output	Q-6	.60	40	Ea.	21,400	1,850		23,250	26,400
1760	1200 MBH output		.30	80		26,100	3,675		29,775	34,200
1800	For power vent, add					10%				
1900	For aluminized steel exchanger, subtract					10%				
2000	Suspension mounted, propeller fan, 20 MBH output	Q-5	8.50	1.882		985	83.50		1,068.50	1,200
2020	40 MBH output		7.50	2.133		1,075	95		1,170	1,325
2040	60 MBH output		7	2.286		1,200	102		1,302	1,475
2060	80 MBH output		6	2.667		1,225	118		1,343	1,525
2080	100 MBH output		5.50	2.909		1,350	129		1,479	1,700
2100	130 MBH output		5	3.200		1,500	142		1,642	1,875
2120	140 MBH output		4.50	3.556		1,550	158		1,708	1,925
2140	160 MBH output		4	4		1,600	178		1,778	2,025
2160	180 MBH output		3.50	4.571		1,700	203		1,903	2,175
2180	200 MBH output		3	5.333		1,800	237		2,037	2,325
2200	240 MBH output		2.70	5.926		2,075	263		2,338	2,675
2220	280 MBH output		2.30	6.957		2,425	310		2,735	3,150
2240	320 MBH output		2	8		2,625	355		2,980	3,425
2500	For powered venter and adapter, add					355			355	390
3000	Suspension mounted, blower type, 40 MBH output	Q-5	6.80	2.353		605	105		710	820
3020	60 MBH output		6.60	2.424		775	108		883	1,000
3040	84 MBH output		5.80	2.759		810	123		933	1,075
3060	104 MBH output		5.20	3.077		920	137		1,057	1,200
3080	140 MBH output		4.30	3.721		1,025	165		1,190	1,375
3100	180 MBH output		3.30	4.848		1,150	215		1,365	1,575
3120	240 MBH output		2.50	6.400		1,500	284		1,784	2,075
3140	280 MBH output		2	8		1,875	355		2,230	2,600
4000	Suspension mounted, sealed combustion system,									
4020	Aluminized steel exchanger, powered vent									
4040	100 MBH output	Q-5	5	3.200	Ea.	1,800	142		1,942	2,225
4060	120 MBH output		4.70	3.404		2,025	151		2,176	2,450
4080	160 MBH output		3.70	4.324		2,200	192		2,392	2,725
4100	200 MBH output		2.90	5.517		2,650	245		2,895	3,300
4120	240 MBH output		2.50	6.400		2,850	284		3,134	3,550
4140	320 MBH output		1.70	9.412		3,400	420		3,820	4,350
5000	Wall furnace, 17.5 MBH output		6	2.667		615	118		733	860
5020	24 MBH output		5	3.200		640	142		782	920
5040	35 MBH output		4	4		770	178		948	1,125

23 56 Solar Energy Heating Equipment

23 56 16 – Packaged Solar Heating Equipment

23 56 16.40 Solar Heating Systems		Crew	Daily Output	Labor-Hours	Unit	Material	Labor	Equipment	Total	Total Incl O&P
0010	**SOLAR HEATING SYSTEMS** R235616-60									
0020	System/Package prices, not including connecting									
0030	pipe, insulation, or special heating/plumbing fixtures									
0152	For solar ultraviolet pipe insulation see Div. 22 07 19.10									
0500	Hot water, standard package, low temperature									
0540	1 collector, circulator, fittings, 65 gal. tank G	Q-1	.50	32	Ea.	1,125	1,400		2,525	3,325
0580	2 collectors, circulator, fittings, 120 gal. tank G		.40	40		1,825	1,750		3,575	4,650
0620	3 collectors, circulator, fittings, 120 gal. tank G		.34	47.059		2,500	2,075		4,575	5,850
0700	Medium temperature package									
0720	1 collector, circulator, fittings, 80 gal. tank G	Q-1	.50	32	Ea.	1,850	1,400		3,250	4,125

23 56 Solar Energy Heating Equipment

23 56 16 – Packaged Solar Heating Equipment

23 56 16.40 Solar Heating Systems			Crew	Daily Output	Labor-Hours	Unit	Material	2009 Bare Costs Labor	Equipment	Total	Total Incl O&P
0740	2 collectors, circulator, fittings, 120 gal. tank	G	Q-1	.40	40	Ea.	2,625	1,750		4,375	5,525
0780	3 collectors, circulator, fittings, 120 gal. tank	G	↓	.30	53.333		3,400	2,350		5,750	7,250
0980	For each additional 120 gal. tank, add	G				↓	1,150			1,150	1,250

23 56 19 – Solar Heating Components

23 56 19.50 Solar Heating Ancillary			Crew	Daily Output	Labor-Hours	Unit	Material	Labor	Equipment	Total	Total Incl O&P
0010	**SOLAR HEATING ANCILLARY**										
2250	Controller, liquid temperature	G	1 Plum	5	1.600	Ea.	96	78		174	222
2300	Circulators, air										
2310	Blowers										
2330	100-300 S.F. system, 1/10 HP	G	Q-9	16	1	Ea.	229	42.50		271.50	320
2340	300-500 S.F. system, 1/5 HP	G		15	1.067		299	45.50		344.50	400
2350	Two speed, 100-300 S.F., 1/10 HP	G		14	1.143		136	48.50		184.50	225
2400	Reversible fan, 20" diameter, 2 speed	G		18	.889		108	38		146	177
2520	Space & DHW system, less duct work	G		.50	32		1,475	1,350		2,825	3,725
2550	Booster fan 6" diameter, 120 CFM	G		16	1		35	42.50		77.50	104
2570	6" diameter, 225 CFM	G		16	1		43	42.50		85.50	113
2580	8" diameter, 150 CFM	G		16	1		39.50	42.50		82	109
2590	8" diameter, 310 CFM	G		14	1.143		60.50	48.50		109	141
2600	8" diameter, 425 CFM	G		14	1.143		68	48.50		116.50	150
2650	Rheostat	G		32	.500		14.80	21		35.80	49
2660	Shutter/damper	G		12	1.333		49	56.50		105.50	141
2670	Shutter motor	G	↓	16	1		119	42.50		161.50	197
2800	Circulators, liquid, 1/25 HP, 5.3 GPM	G	Q-1	14	1.143		227	50		277	325
2820	1/20 HP, 17 GPM	G		12	1.333		154	58.50		212.50	258
2850	1/20 HP, 17 GPM, stainless steel	G		12	1.333		243	58.50		301.50	355
2870	1/12 HP, 30 GPM	G	↓	10	1.600	↓	330	70		400	470
3000	Collector panels, air with aluminum absorber plate										
3010	Wall or roof mount										
3040	Flat black, plastic glazing										
3080	4' x 8'	G	Q-9	6	2.667	Ea.	635	113		748	870
3100	4' x 10'	G		5	3.200	"	780	136		916	1,075
3200	Flush roof mount, 10' to 16' x 22" wide	G		96	.167	L.F.	550	7.10		557.10	615
3210	Manifold, by L.F. width of collectors	G	↓	160	.100	"	106	4.25		110.25	124
3300	Collector panels, liquid with copper absorber plate										
3320	Black chrome, tempered glass glazing										
3330	Alum. frame, 4' x 8', 5/32" single glazing	G	Q-1	9.50	1.684	Ea.	665	74		739	840
3390	Alum. frame, 4' x 10', 5/32" single glazing	G		6	2.667		805	117		922	1,050
3450	Flat black, alum. frame, 3.5' x 7.5'	G		9	1.778		560	78		638	735
3500	4' x 8'	G		5.50	2.909		690	128		818	945
3520	4' x 10'	G		10	1.600		830	70		900	1,025
3540	4' x 12.5'	G	↓	5	3.200	↓	1,025	140		1,165	1,325
3550	Liquid with fin tube absorber plate										
3560	Alum. frame 4' x 8' tempered glass	G	Q-1	10	1.600	Ea.	495	70		565	650
3580	Liquid with vacuum tubes, 4' x 6'-10"	G		9	1.778		850	78		928	1,050
3600	Liquid, full wetted, plastic, alum. frame, 3' x 10'	G		5	3.200		234	140		374	470
3650	Collector panel mounting, flat roof or ground rack	G		7	2.286	↓	217	100		317	390
3670	Roof clamps	G	↓	70	.229	Set	2.35	10.05		12.40	17.65
3700	Roof strap, teflon	G	1 Plum	205	.039	L.F.	19.95	1.90		21.85	25
3900	Differential controller with two sensors										
3930	Thermostat, hard wired	G	1 Plum	8	1	Ea.	79.50	49		128.50	160
3950	Line cord and receptacle	G		12	.667		345	32.50		377.50	430
4050	Pool valve system	G	↓	2.50	3.200	↓	264	156		420	525

23 56 Solar Energy Heating Equipment

23 56 19 – Solar Heating Components

23 56 19.50 Solar Heating Ancillary			Crew	Daily Output	Labor-Hours	Unit	2009 Bare Costs Material	Labor	Equipment	Total	Total Incl O&P
4070	With 12 VAC actuator	G	1 Plum	2	4	Ea.	276	195		471	600
4080	Pool pump system, 2" pipe size	G		6	1.333		172	65		237	288
4100	Five station with digital read-out	G		3	2.667		218	130		348	435
4150	Sensors										
4200	Brass plug, 1/2" MPT	G	1 Plum	32	.250	Ea.	17.95	12.20		30.15	38
4210	Brass plug, reversed	G		32	.250		24.50	12.20		36.70	45.50
4220	Freeze prevention	G		32	.250		22	12.20		34.20	42.50
4240	Screw attached	G		32	.250		8.35	12.20		20.55	27.50
4250	Brass, immersion	G		32	.250		26.50	12.20		38.70	47.50
4300	Heat exchanger										
4330	Fluid to air coil, up flow, 45 MBH	G	Q-1	4	4	Ea.	295	176		471	590
4380	70 MBH	G		3.50	4.571		335	201		536	665
4400	80 MBH	G		3	5.333		445	234		679	840
4580	Fluid to fluid package includes two circulating pumps										
4590	expansion tank, check valve, relief valve										
4600	controller, high temperature cutoff and sensors	G	Q-1	2.50	6.400	Ea.	695	281		976	1,175
4650	Heat transfer fluid										
4700	Propylene glycol, inhibited anti-freeze	G	1 Plum	28	.286	Gal.	13.85	13.95		27.80	36.50
4800	Solar storage tanks, knocked down										
4810	Air, galvanized steel clad, double wall, 4" fiberglass										
4820	insulation, 20 Mil PVC lining										
4870	4' high, 4' x 4', = 64 C.F./450 gallons	G	Q-9	2	8	Ea.	2,275	340		2,615	3,025
4880	4' x 8' = 128 C.F./900 gallons	G		1.50	10.667		3,225	455		3,680	4,250
4890	4' x 12' = 190 C.F./1300 gallons	G		1.30	12.308		4,300	525		4,825	5,525
4900	8' x 8' = 250 C.F./1700 gallons	G		1	16		4,800	680		5,480	6,325
5010	6'-3" high, 7' x 7' = 306 C.F./2000 gallons	G	Q-10	1.20	20		11,700	880		12,580	14,300
5020	7' x 10'-6" = 459 C.F./3000 gallons	G		.80	30		14,700	1,325		16,025	18,200
5030	7' x 14' = 613 C.F./4000 gallons	G		.60	40		17,200	1,750		18,950	21,700
5040	10'-6" x 10'-6" = 689 C.F./4500 gallons	G		.50	48		17,800	2,125		19,925	22,800
5050	10'-6" x 14' = 919 C.F./6000 gallons	G		.40	60		21,000	2,650		23,650	27,200
5060	14' x 14' = 1225 C.F./8000 gallons	G	Q-11	.40	80		24,500	3,600		28,100	32,400
5070	14' x 17'-6" = 1531 C.F./10,000 gallons	G		.30	106		27,600	4,800		32,400	37,700
5080	17'-6" x 17'-6" = 1914 C.F./12,500 gallons	G		.25	128		32,500	5,750		38,250	44,600
5090	17'-6" x 21' = 2297 C.F./15,000 gallons	G		.20	160		36,200	7,200		43,400	51,000
5100	21' x 21' = 2756 C.F./18,000 gallons	G		.18	177		41,500	8,000		49,500	58,000
5120	30 Mil reinforced Chemflex lining,										
5140	4' high, 4' x 4' = 64 C.F./450 gallons	G	Q-9	2	8	Ea.	2,525	340		2,865	3,300
5150	4' x 8' = 128 C.F./900 gallons	G		1.50	10.667		3,825	455		4,280	4,900
5160	4' x 12' = 190 C.F./1300 gallons	G		1.30	12.308		5,050	525		5,575	6,375
5170	8' x 8' = 250 C.F./1700 gallons	G		1	16		5,650	680		6,330	7,250
5190	6'-3" high, 7' x 7' = 306 C.F./2000 gallons	G	Q-10	1.20	20		12,500	880		13,380	15,100
5200	7' x 10'-6" = 459 C.F./3000 gallons	G		.80	30		15,200	1,325		16,525	18,800
5210	7' x 14' = 613 C.F./4000 gallons	G		.60	40		17,700	1,750		19,450	22,200
5220	10'-6" x 10'-6" = 689 C.F./4500 gallons	G		.50	48		18,300	2,125		20,425	23,400
5230	10'-6" x 14' = 919 C.F./6000 gallons	G		.40	60		21,500	2,650		24,150	27,700
5240	14' x 14' = 1225 C.F./8000 gallons	G	Q-11	.40	80		25,100	3,600		28,700	33,200
5250	14' x 17'-6" = 1531 C.F./10,000 gallons	G		.30	106		28,100	4,800		32,900	38,400
5260	17'-6" x 17'-6" = 1914 C.F./12,500 gallons	G		.25	128		33,100	5,750		38,850	45,300
5270	17'-6" x 21' = 2297 C.F./15,000 gallons	G		.20	160		37,300	7,200		44,500	52,000
5280	21' x 21' = 2756 C.F./18,000 gallons	G		.18	177		42,700	8,000		50,700	59,500
5290	30 Mil reinforced Hypalon lining, add						.02%				
7000	Solar control valves and vents										
7050	Air purger, 1" pipe size	G	1 Plum	12	.667	Ea.	43.50	32.50		76	96.50

23 56 Solar Energy Heating Equipment

23 56 19 – Solar Heating Components

23 56 19.50 Solar Heating Ancillary			Crew	Daily Output	Labor-Hours	Unit	2009 Bare Costs Material	Labor	Equipment	Total	Total Incl O&P
7070	Air eliminator, automatic 3/4" size	G	1 Plum	32	.250	Ea.	28.50	12.20		40.70	50
7090	Air vent, automatic, 1/8" fitting	G		32	.250		12.65	12.20		24.85	32.50
7100	Manual, 1/8" NPT	G		32	.250		2.78	12.20		14.98	21.50
7120	Backflow preventer, 1/2" pipe size	G		16	.500		58.50	24.50		83	101
7130	3/4" pipe size	G		16	.500		101	24.50		125.50	148
7150	Balancing valve, 3/4" pipe size	G		20	.400		40.50	19.50		60	74.50
7180	Draindown valve, 1/2" copper tube	G		9	.889		202	43.50		245.50	287
7200	Flow control valve, 1/2" pipe size	G		22	.364		102	17.75		119.75	139
7220	Expansion tank, up to 5 gal.	G		32	.250		61.50	12.20		73.70	86
7250	Hydronic controller (aquastat)	G		8	1		144	49		193	232
7400	Pressure gauge, 2" dial	G		32	.250		22	12.20		34.20	43
7450	Relief valve, temp. and pressure 3/4" pipe size	G		30	.267		15.15	13		28.15	36
7500	Solenoid valve, normally closed										
7520	Brass, 3/4" NPT, 24V	G	1 Plum	9	.889	Ea.	103	43.50		146.50	178
7530	1" NPT, 24V	G		9	.889		815	43.50		858.50	960
7750	Vacuum relief valve, 3/4" pipe size	G		32	.250		26.50	12.20		38.70	47.50
7800	Thermometers										
7820	Digital temperature monitoring, 4 locations	G	1 Plum	2.50	3.200	Ea.	128	156		284	375
7900	Upright, 1/2" NPT	G		8	1		24	49		73	99.50
7970	Remote probe, 2" dial	G		8	1		30.50	49		79.50	107
7990	Stem, 2" dial, 9" stem	G		16	.500		20	24.50		44.50	58.50
8250	Water storage tank with heat exchanger and electric element										
8270	66 gal. with 2" x 2 lb. density insulation	G	1 Plum	1.60	5	Ea.	925	244		1,169	1,400
8300	80 gal. with 2" x 2 lb. density insulation	G		1.60	5		1,050	244		1,294	1,525
8380	120 gal. with 2" x 2 lb. density insulation	G		1.40	5.714		1,175	279		1,454	1,700
8400	120 gal. with 2" x 2 lb. density insul., 40 S.F. heat coil	G		1.40	5.714		1,500	279		1,779	2,075
8500	Water storage module, plastic										
8600	Tubular, 12" diameter, 4' high	G	1 Carp	48	.167	Ea.	103	6.65		109.65	123
8610	12" diameter, 8' high	G		40	.200		159	8		167	187
8620	18" diameter, 5' high	G		38	.211		174	8.40		182.40	204
8630	18" diameter, 10' high	G		32	.250		229	10		239	268
8640	58" diameter, 5' high	G	2 Carp	32	.500		490	20		510	565
8650	Cap, 12" diameter	G					19			19	21
8660	18" diameter	G					24			24	26.50

23 57 Heat Exchangers for HVAC

23 57 16 – Steam-to-Water Heat Exchangers

23 57 16.10 Shell/Tube Type Steam-to-Water Heat Exch.		Crew	Daily Output	Labor-Hours	Unit	Material	Labor	Equipment	Total	Total Incl O&P
0010	**SHELL AND TUBE TYPE STEAM-TO-WATER HEAT EXCHANGERS**									
0016	Shell & tube type, 2 or 4 pass, 3/4" O.D. copper tubes,									
0020	C.I. heads, C.I. tube sheet, steel shell									
0100	Hot water 40° F to 180° F, by steam at 10 PSI									
0120	8 GPM	Q-5	6	2.667	Ea.	1,550	118		1,668	1,875
0140	10 GPM		5	3.200		2,350	142		2,492	2,800
0160	40 GPM		4	4		3,625	178		3,803	4,275
0180	64 GPM		2	8		5,550	355		5,905	6,625
0200	96 GPM		1	16		7,425	710		8,135	9,250
0220	120 GPM	Q-6	1.50	16		8,625	735		9,360	10,600
0240	168 GPM		1	24		12,000	1,100		13,100	14,900
0260	240 GPM		.80	30		18,700	1,375		20,075	22,700
0300	600 GPM		.70	34.286		49,000	1,575		50,575	56,500

23 57 Heat Exchangers for HVAC

23 57 16 – Steam-to-Water Heat Exchangers

23 57 16.10 Shell/Tube Type Steam-to-Water Heat Exch.		Crew	Daily Output	Labor-Hours	Unit	Material	2009 Bare Costs Labor	Equipment	Total	Total Incl O&P
0500	For bronze head and tube sheet, add					50%				

23 57 19 – Liquid-to-Liquid Heat Exchangers

23 57 19.13 Plate-Type, Liquid-to-Liquid Heat Exchangers

		Crew	Daily Output	Labor-Hours	Unit	Material	Labor	Equipment	Total	Total Incl O&P
0010	**PLATE-TYPE, LIQUID-TO-LIQUID HEAT EXCHANGERS**									
3000	Plate type,									
3100	400 GPM	Q-6	.80	30	Ea.	31,400	1,375		32,775	36,600
3120	800 GPM	″	.50	48		54,000	2,200		56,200	63,000
3140	1200 GPM	Q-7	.34	94.118		80,500	4,425		84,925	95,000
3160	1800 GPM	″	.24	133		106,500	6,275		112,775	126,500

23 57 19.16 Shell-Type, Liquid-to-Liquid Heat Exchangers

		Crew	Daily Output	Labor-Hours	Unit	Material	Labor	Equipment	Total	Total Incl O&P
0010	**SHELL-TYPE, LIQUID-TO-LIQUID HEAT EXCHANGERS**									
1000	Hot water 40° F to 140° F, by water at 200° F									
1020	7 GPM	Q-5	6	2.667	Ea.	1,900	118		2,018	2,275
1040	16 GPM		5	3.200		2,700	142		2,842	3,200
1060	34 GPM		4	4		4,100	178		4,278	4,775
1080	55 GPM		3	5.333		5,325	237		5,562	6,200
1100	74 GPM		1.50	10.667		7,400	475		7,875	8,850
1120	86 GPM		1.40	11.429		9,900	510		10,410	11,700
1140	112 GPM	Q-6	2	12		12,300	555		12,855	14,400
1160	126 GPM		1.80	13.333		15,400	615		16,015	17,800
1180	152 GPM		1	24		19,700	1,100		20,800	23,300

23 61 Refrigerant Compressors

23 61 15 – Rotary Refrigerant Compressors

23 61 15.10 Rotary Compressors

		Crew	Daily Output	Labor-Hours	Unit	Material	Labor	Equipment	Total	Total Incl O&P
0010	**ROTARY COMPRESSORS**									
0100	Refrigeration, hermetic, switches and protective devices									
0210	1.25 ton	1 Stpi	3	2.667	Ea.	221	132		353	440
0220	1.42 ton		3	2.667		223	132		355	440
0230	1.68 ton		2.80	2.857		243	141		384	480
0240	2.00 ton		2.60	3.077		250	152		402	505
0250	2.37 ton		2.50	3.200		282	158		440	545
0260	2.67 ton		2.40	3.333		296	165		461	570
0270	3.53 ton		2.30	3.478		325	172		497	615
0280	4.43 ton		2.20	3.636		415	179		594	725
0290	5.08 ton		2.10	3.810		455	188		643	780
0300	5.22 ton		2	4		640	197		837	1,000
0310	7.31 ton	Q-5	3	5.333		1,025	237		1,262	1,475
0320	9.95 ton		2.60	6.154		1,200	273		1,473	1,725
0330	11.6 ton		2.50	6.400		1,550	284		1,834	2,125
0340	15.25 ton		2.30	6.957		2,125	310		2,435	2,800
0350	17.7 ton		2.10	7.619		2,275	340		2,615	3,000

23 61 16 – Reciprocating Refrigerant Compressors

23 61 16.10 Reciprocating Compressors

		Crew	Daily Output	Labor-Hours	Unit	Material	Labor	Equipment	Total	Total Incl O&P
0010	**RECIPROCATING COMPRESSORS**									
0990	Refrigeration, recip. hermetic, switches & protective devices									
1000	10 ton	Q-5	1	16	Ea.	8,150	710		8,860	10,100
1100	20 ton	Q-6	.72	33.333		12,300	1,525		13,825	15,800
1200	30 ton		.64	37.500		14,700	1,725		16,425	18,700
1300	40 ton		.44	54.545		16,000	2,525		18,525	21,300

23 61 Refrigerant Compressors

23 61 16 – Reciprocating Refrigerant Compressors

23 61 16.10 Reciprocating Compressors		Crew	Daily Output	Labor-Hours	Unit	2009 Bare Costs Material	Labor	Equipment	Total	Total Incl O&P
1400	50 ton	Q-6	.20	120	Ea.	16,800	5,525		22,325	26,800
1500	75 ton	Q-7	.27	118		18,600	5,575		24,175	28,800
1600	130 ton	"	.21	152		21,200	7,175		28,375	34,100

23 61 19 – Scroll Refrigerant Compressors

23 61 19.10 Scroll Compressors		Crew	Daily Output	Labor-Hours	Unit	Material	Labor	Equipment	Total	Total Incl O&P
0010	**SCROLL COMPRESSORS**									
1800	Refrigeration, scroll type									
1810	1.9 ton	1 Stpi	2.70	2.963	Ea.	269	146		415	515
1820	2.35 ton		2.50	3.200		310	158		468	575
1830	2.82 ton		2.35	3.404		335	168		503	615
1840	3.3 ton		2.30	3.478		350	172		522	645
1850	3.83 ton		2.25	3.556		400	175		575	700
1860	4.1 ton		2.20	3.636		450	179		629	765
1870	4.8 ton		2.10	3.810		475	188		663	805
1880	5 ton		2.07	3.865		485	191		676	820

23 62 Packaged Compressor and Condenser Units

23 62 13 – Packaged Air-Cooled Refrigerant Compressor and Condenser Units

23 62 13.10 Packaged Air-Cooled Refrig. Condensing Units		Crew	Daily Output	Labor-Hours	Unit	Material	Labor	Equipment	Total	Total Incl O&P
0010	**PACKAGED AIR-COOLED REFRIGERANT CONDENSING UNITS**									
0020	Condensing unit									
0030	Air cooled, compressor, standard controls									
0050	1.5 ton	Q-5	2.50	6.400	Ea.	1,125	284		1,409	1,675
0100	2 ton		2.10	7.619		1,150	340		1,490	1,775
0200	2.5 ton		1.70	9.412		1,250	420		1,670	2,000
0300	3 ton		1.30	12.308		1,300	545		1,845	2,250
0350	3.5 ton		1.10	14.545		1,500	645		2,145	2,625
0400	4 ton		.90	17.778		1,700	790		2,490	3,025
0500	5 ton		.60	26.667		2,050	1,175		3,225	4,025
0550	7.5 ton		.55	29.091		3,350	1,300		4,650	5,650
0560	8.5 ton		.53	30.189		4,000	1,350		5,350	6,425
0600	10 ton		.50	32		4,350	1,425		5,775	6,900
0620	11 ton		.48	33.333		6,700	1,475		8,175	9,600
0650	15 ton		.40	40		7,150	1,775		8,925	10,600
0700	20 ton	Q-6	.40	60		9,950	2,775		12,725	15,200
0720	25 ton		.35	68.571		15,000	3,150		18,150	21,300
0750	30 ton		.30	80		17,100	3,675		20,775	24,300
0800	40 ton		.20	120		20,700	5,525		26,225	31,100
0840	50 ton		.18	133		24,700	6,150		30,850	36,300
0860	60 ton		.16	150		28,400	6,900		35,300	41,500
0900	75 ton		.14	171		32,500	7,900		40,400	47,500
1000	80 ton		.12	200		36,000	9,225		45,225	53,500
1100	100 ton		.09	266		50,500	12,300		62,800	74,000

23 62 23 – Packaged Water-Cooled Refrigerant Compressor and Condenser Units

23 62 23.10 Pckgd Water-Cooled Refrig Condens. Units		Crew	Daily Output	Labor-Hours	Unit	Material	Labor	Equipment	Total	Total Incl O&P
0010	**PACKAGED WATER-COOLED REFRIGERANT CONDENSING UNITS**									
2000	Water cooled, compressor, heat exchanger, controls									
2100	5 ton	Q-5	.70	22.857	Ea.	11,500	1,025		12,525	14,100
2110	10 ton		.60	26.667		13,300	1,175		14,475	16,400
2200	15 ton		.50	32		18,100	1,425		19,525	22,000

23 62 Packaged Compressor and Condenser Units

23 62 23 – Packaged Water-Cooled Refrigerant Compressor and Condenser Units

23 62 23.10 Pckgd Water-Cooled Refrig Condens. Units		Crew	Daily Output	Labor-Hours	Unit	2009 Bare Costs Material	Labor	Equipment	Total	Total Incl O&P
2300	20 ton	Q-6	.40	60	Ea.	20,200	2,775		22,975	26,400
2310	30 ton		.30	80		24,100	3,675		27,775	32,000
2400	40 ton		.20	120		27,400	5,525		32,925	38,400
2410	60 ton		.18	133		38,400	6,150		44,550	51,500
2420	80 ton		.14	171		44,300	7,900		52,200	60,500
2500	100 ton		.11	218		54,500	10,100		64,600	74,500
2510	120 ton		.10	240		87,500	11,100		98,600	112,500

23 63 Refrigerant Condensers

23 63 13 – Air-Cooled Refrigerant Condensers

23 63 13.10 Air-Cooled Refrig. Condensers

		Crew	Daily Output	Labor-Hours	Unit	Material	Labor	Equipment	Total	Total Incl O&P
0010	**AIR-COOLED REFRIG. CONDENSERS**, Ratings for 30° F TD, R-22.									
0080	Air cooled, belt drive, propeller fan									
0220	45 ton	Q-6	.70	34.286	Ea.	12,300	1,575		13,875	16,000
0240	50 ton		.69	34.985		12,700	1,600		14,300	16,400
0260	54 ton		.64	37.795		13,200	1,750		14,950	17,100
0280	59 ton		.58	41.308		14,200	1,900		16,100	18,500
0300	65 ton		.53	45.541		14,800	2,100		16,900	19,400
0320	73 ton		.47	51.173		16,700	2,350		19,050	22,000
0340	81 ton		.42	56.738		19,000	2,625		21,625	24,800
0360	86 ton		.40	60.302		19,800	2,775		22,575	26,000
0380	88 ton		.39	61.697		21,200	2,850		24,050	27,600
0400	101 ton	Q-7	.45	70.640		24,900	3,325		28,225	32,400
0500	159 ton		.31	102		37,200	4,825		42,025	48,200
0600	228 ton		.22	148		54,000	6,975		60,975	69,500
1500	May be specified single or multi-circuit									
1550	Air cooled, direct drive, propeller fan									
1590	1 ton	Q-5	3.80	4.211	Ea.	705	187		892	1,050
1600	1-1/2 ton		3.60	4.444		840	197		1,037	1,225
1620	2 ton		3.20	5		925	222		1,147	1,350
1630	3 ton		2.40	6.667		1,025	296		1,321	1,575
1640	5 ton		2	8		1,875	355		2,230	2,600
1650	8 ton		1.80	8.889		2,375	395		2,770	3,225
1660	10 ton		1.40	11.429		2,900	510		3,410	3,950
1670	12 ton		1.30	12.308		3,275	545		3,820	4,425
1680	14 ton		1.20	13.333		3,725	590		4,315	5,000
1690	16 ton		1.10	14.545		4,100	645		4,745	5,475
1700	21 ton		1	16		4,725	710		5,435	6,275
1720	26 ton		.84	19.002		5,300	845		6,145	7,100
1740	30 ton		.70	22.792		8,825	1,025		9,850	11,200
1760	41 ton	Q-6	.77	31.008		9,700	1,425		11,125	12,900
1780	52 ton		.66	36.419		13,800	1,675		15,475	17,700
1800	63 ton		.55	44.037		15,300	2,025		17,325	19,900
1820	76 ton		.45	52.980		17,300	2,450		19,750	22,800
1840	86 ton		.40	60		21,400	2,775		24,175	27,800
1860	97 ton		.35	67.989		24,400	3,125		27,525	31,500
1880	105 ton	Q-7	.44	73.563		28,100	3,450		31,550	36,100
1890	118 ton		.39	82.687		30,300	3,875		34,175	39,100
1900	126 ton		.36	88.154		35,200	4,150		39,350	44,900
1910	136 ton		.34	95.238		39,400	4,475		43,875	50,000
1920	142 ton		.32	99.379		40,400	4,675		45,075	51,500

23 63 Refrigerant Condensers

23 63 30 – Lubricants

23 63 30.10 Lubricant Oils		Crew	Daily Output	Labor-Hours	Unit	2009 Bare Costs Material	Labor	Equipment	Total	Total Incl O&P
0010	**LUBRICANT OILS**									
8000	Oils									
8100	Lubricating									
8120	Oil, lubricating				Oz.	.49			.49	.54
8500	Refrigeration									
8520	Oil, refrigeration				Gal.	21			21	23.50
8525	Oil, refrigeration				Qt.	5.30			5.30	5.85

23 63 33 – Evaporative Refrigerant Condensers

23 63 33.10 Evaporative Condensers

		Crew	Daily Output	Labor-Hours	Unit	Material	Labor	Equipment	Total	Total Incl O&P
0010	**EVAPORATIVE CONDENSERS**									
3400	Evaporative, copper coil, pump, fan motor									
3440	10 ton	Q-5	.54	29.630	Ea.	5,325	1,325		6,650	7,850
3460	15 ton		.50	32		5,550	1,425		6,975	8,225
3480	20 ton		.47	34.043		5,775	1,525		7,300	8,625
3500	25 ton		.45	35.556		6,175	1,575		7,750	9,150
3520	30 ton		.42	38.095		6,575	1,700		8,275	9,775
3540	40 ton	Q-6	.49	48.980		8,700	2,250		10,950	13,000
3560	50 ton		.39	61.538		10,900	2,825		13,725	16,300
3580	65 ton		.35	68.571		12,100	3,150		15,250	18,100
3600	80 ton		.33	72.727		13,100	3,350		16,450	19,400
3620	90 ton		.29	82.759		14,400	3,825		18,225	21,600
3640	100 ton	Q-7	.36	88.889		15,700	4,175		19,875	23,600
3660	110 ton		.33	96.970		17,400	4,550		21,950	26,000
3680	125 ton		.30	106		18,900	5,025		23,925	28,300
3700	135 ton		.28	114		20,200	5,375		25,575	30,400
3720	150 ton		.25	128		22,300	6,025		28,325	33,500
3740	165 ton		.23	139		24,500	6,550		31,050	36,700
3760	185 ton		.22	145		25,800	6,850		32,650	38,700
3860	For fan damper control, add	Q-5	2	8		555	355		910	1,150

23 64 Packaged Water Chillers

23 64 13 – Absorption Water Chillers

23 64 13.13 Direct-Fired Absorption Water Chillers

		Crew	Daily Output	Labor-Hours	Unit	Material	Labor	Equipment	Total	Total Incl O&P
0010	**DIRECT-FIRED ABSORPTION WATER CHILLERS**									
3000	Gas fired, air cooled									
3220	5 ton	Q-5	.60	26.667	Ea.	9,525	1,175		10,700	12,300
3270	10 ton	"	.40	40	"	25,500	1,775		27,275	30,800
4000	Water cooled, duplex									
4130	100 ton	Q-7	.13	246	Ea.	136,000	11,600		147,600	167,000
4140	200 ton		.11	283		133,000	13,300		146,300	166,500
4150	300 ton		.11	299		219,500	14,100		233,600	262,500
4160	400 ton		.10	316		285,500	14,900		300,400	336,500
4170	500 ton		.10	329		359,500	15,500		375,000	418,500
4180	600 ton		.09	340		424,500	16,000		440,500	490,500
4190	700 ton		.09	359		470,500	16,900		487,400	543,000
4200	800 ton		.08	380		568,000	17,900		585,900	652,000
4210	900 ton		.08	400		658,000	18,800		676,800	752,000
4220	1000 ton		.08	421		731,500	19,800		751,300	834,000

23 64 13.16 Indirect-Fired Absorption Water Chillers

		Crew	Daily Output	Labor-Hours	Unit	Material	Labor	Equipment	Total	Total Incl O&P
0010	**INDIRECT-FIRED ABSORPTION WATER CHILLERS**									

23 64 13 – Absorption Water Chillers

23 64 13.16 Indirect-Fired Absorption Water Chillers		Crew	Daily Output	Labor-Hours	Unit	2009 Bare Costs Material	Labor	Equipment	Total	Total Incl O&P
0020	Steam or hot water, water cooled									
0050	100 ton	Q-7	.13	240	Ea.	125,000	11,300		136,300	154,500
0100	148 ton		.12	258		191,000	12,100		203,100	228,000
0200	200 ton		.12	275		212,500	13,000		225,500	253,000
0240	250 ton		.11	290		233,500	13,700		247,200	277,500
0300	354 ton		.11	304		292,000	14,300		306,300	342,500
0400	420 ton		.10	323		319,000	15,200		334,200	374,000
0500	665 ton		.09	340		448,000	16,000		464,000	517,000
0600	750 ton		.09	363		523,500	17,100		540,600	601,000
0700	850 ton		.08	385		558,500	18,100		576,600	641,500
0800	955 ton		.08	410		597,500	19,300		616,800	686,000
0900	1125 ton		.08	421		689,500	19,800		709,300	788,000
1000	1250 ton		.07	444		741,500	20,900		762,400	847,500
1100	1465 ton		.07	463		879,500	21,800		901,300	999,500
1200	1660 ton		.07	477		1,034,500	22,500		1,057,000	1,171,500
2000	For two stage unit, add					80%	25%			

23 64 16 – Centrifugal Water Chillers

23 64 16.10 Centrifugal Type Water Chillers		Crew	Daily Output	Labor-Hours	Unit	Material	Labor	Equipment	Total	Total Incl O&P
0010	**CENTRIFUGAL TYPE WATER CHILLERS**, With standard controls									
0020	Centrifugal liquid chiller, water cooled									
0030	not including water tower									
0100	2000 ton, (twin 1000 ton units)	Q-7	.07	477	Ea.	840,000	22,500		862,500	957,500
0274	Centrifugal, packaged unit, water cooled, not incl. tower									
0280	400 ton	Q-7	.11	283	Ea.	133,000	13,300		146,300	166,000
0282	450 ton		.11	290		189,000	13,700		202,700	228,500
0290	500 ton		.11	296		210,000	13,900		223,900	252,000
0292	550 ton		.11	304		231,000	14,300		245,300	275,500
0300	600 ton		.10	310		252,000	14,600		266,600	299,500
0302	650 ton		.10	320		273,000	15,000		288,000	323,000
0304	700 ton		.10	326		294,000	15,400		309,400	346,500
0306	750 ton		.10	333		315,000	15,700		330,700	370,000
0310	800 ton		.09	340		336,000	16,000		352,000	393,500
0312	850 ton		.09	351		357,000	16,500		373,500	418,000
0316	900 ton		.09	359		378,000	16,900		394,900	441,500
0318	950 ton		.09	363		399,000	17,100		416,100	464,500
0320	1000 ton		.09	372		420,000	17,500		437,500	488,500
0324	1100 ton		.08	385		448,500	18,100		466,600	520,000
0330	1200 ton		.08	395		489,000	18,600		507,600	566,000
0340	1300 ton		.08	410		530,000	19,300		549,300	612,000
0350	1400 ton		.08	421		570,500	19,800		590,300	657,000
0360	1500 ton		.08	426		611,500	20,100		631,600	702,500
0370	1600 ton		.07	438		652,000	20,600		672,600	748,500
0380	1700 ton		.07	450		693,000	21,200		714,200	794,000
0390	1800 ton		.07	457		733,500	21,500		755,000	839,500
0400	1900 ton		.07	470		774,500	22,100		796,600	885,000
0430	2000 ton		.07	492		815,000	23,100		838,100	931,000
0440	2500 ton		.06	524		1,019,000	24,700		1,043,700	1,158,000

23 64 19 – Reciprocating Water Chillers

23 64 19.10 Reciprocating Type Water Chillers		Crew	Daily Output	Labor-Hours	Unit	Material	Labor	Equipment	Total	Total Incl O&P
0010	**RECIPROCATING TYPE WATER CHILLERS**, With standard controls.									
0494	Water chillers, integral air cooled condenser									
0538	60 ton cooling	Q-7	.28	115	Ea.	45,300	5,400		50,700	58,000

23 64 19.10 Reciprocating Type Water Chillers		Crew	Daily Output	Labor-Hours	Unit	2009 Bare Costs Material	Labor	Equipment	Total	Total Incl O&P
0540	65 ton cooling	Q-7	.27	118	Ea.	45,700	5,550		51,250	59,000
0546	70 ton cooling		.27	119		52,000	5,625		57,625	65,500
0552	80 ton cooling		.26	123		56,000	5,775		61,775	70,500
0554	90 ton cooling		.25	125		60,500	5,925		66,425	75,500
0600	100 ton cooling		.25	129		68,500	6,075		74,575	84,000
0620	110 ton cooling		.24	132		74,000	6,225		80,225	91,000
0630	130 ton cooling		.24	135		88,000	6,350		94,350	106,000
0640	150 ton cooling		.23	137		97,000	6,475		103,475	116,500
0650	175 ton cooling		.23	140		106,500	6,625		113,125	127,000
0654	190 ton cooling		.22	144		120,500	6,775		127,275	142,500
0660	210 ton cooling		.22	148		124,000	6,975		130,975	147,000
0662	240 ton cooling		.21	151		143,500	7,125		150,625	168,500
0664	270 ton cooling		.21	156		148,500	7,350		155,850	174,500
0666	300 ton cooling		.20	160		168,000	7,525		175,525	196,500
0668	330 ton cooling		.20	164		193,000	7,725		200,725	223,500
0670	360 ton cooling		.19	168		213,000	7,925		220,925	246,000
0672	390 ton cooling		.19	172		226,000	8,125		234,125	260,500
0674	420 ton cooling		.18	177		236,000	8,350		244,350	272,500
0980	Water cooled, multiple compressor, semi-hermetic, tower not incl.									
1000	15 ton cooling	Q-6	.36	65.934	Ea.	13,500	3,050		16,550	19,400
1020	20 ton cooling	Q-7	.41	78.049		14,800	3,675		18,475	21,800
1040	25 ton cooling		.36	89.888		16,600	4,225		20,825	24,700
1060	30 ton cooling		.31	101		17,500	4,800		22,300	26,400
1080	40 ton cooling		.30	108		20,300	5,075		25,375	29,900
1100	50 ton cooling		.28	113		31,300	5,350		36,650	42,400
1120	60 ton cooling		.25	125		34,500	5,925		40,425	46,900
1130	70 ton cooling		.23	139		46,600	6,550		53,150	61,500
1140	80 ton cooling		.21	151		48,800	7,125		55,925	64,000
1150	90 ton cooling		.19	164		52,000	7,750		59,750	69,000
1160	100 ton cooling		.18	179		56,000	8,450		64,450	74,000
1170	110 ton cooling		.17	190		59,000	8,950		67,950	78,500
1180	120 ton cooling		.16	196		63,500	9,225		72,725	83,500
1200	140 ton cooling		.16	202		74,500	9,525		84,025	96,500
1210	160 ton cooling		.15	210		79,000	9,900		88,900	102,000
1300	Water cooled, single compressor, semi-hermetic, tower not incl.									
1320	40 ton cooling,	Q-7	.30	108	Ea.	16,000	5,075		21,075	25,200
1340	50 ton cooling		.28	113		19,800	5,350		25,150	29,800
1360	60 ton cooling		.25	125		22,000	5,925		27,925	33,100
1451	Water cooled, dual compressors, semi-hermetic, tower not incl.									
1500	80 ton cooling	Q-7	.14	222	Ea.	27,400	10,400		37,800	45,800
1520	100 ton cooling		.14	228		37,000	10,700		47,700	56,500
1540	120 ton cooling		.14	231		44,800	10,900		55,700	65,500
4000	Packaged chiller, remote air cooled condensers not incl.									
4020	15 ton cooling	Q-7	.30	108	Ea.	13,500	5,075		18,575	22,400
4030	20 ton cooling		.28	115		14,800	5,450		20,250	24,500
4040	25 ton cooling		.25	125		16,700	5,925		22,625	27,200
4050	30 ton cooling		.24	133		17,500	6,300		23,800	28,800
4060	40 ton cooling		.22	144		20,300	6,775		27,075	32,500
4070	50 ton cooling		.21	153		25,000	7,225		32,225	38,400
4080	60 ton cooling		.20	164		29,100	7,725		36,825	43,600
4090	70 ton cooling		.18	173		46,600	8,175		54,775	64,000
4100	80 ton cooling		.17	183		48,800	8,650		57,450	66,500
4110	90 ton cooling		.17	193		52,000	9,125		61,125	71,000

23 64 Packaged Water Chillers

23 64 19 – Reciprocating Water Chillers

23 64 19.10 Reciprocating Type Water Chillers		Crew	Daily Output	Labor-Hours	Unit	2009 Bare Costs Material	Labor	Equipment	Total	Total Incl O&P
4120	100 ton cooling	Q-7	.16	203	Ea.	56,000	9,575		65,575	76,000
4130	110 ton cooling		.15	213		59,000	10,000		69,000	80,000
4140	120 ton cooling		.14	223		63,500	10,500		74,000	85,500
4150	140 ton cooling		.14	233		74,500	11,000		85,500	98,500

23 64 23 – Scroll Water Chillers

23 64 23.10 Scroll Water Chillers

		Crew	Daily Output	Labor-Hours	Unit	Material	Labor	Equipment	Total	Total Incl O&P
0010	**SCROLL WATER CHILLERS,** With standard controls.									
0490	Packaged w/integral air cooled condenser, 15 ton cool	Q-7	.37	86.486	Ea.	16,500	4,075		20,575	24,200
0500	20 ton cooling		.34	94.118		21,800	4,425		26,225	30,600
0510	25 ton cooling		.34	94.118		22,800	4,425		27,225	31,700
0515	30 ton cooling		.31	101		24,400	4,800		29,200	34,000
0517	35 ton cooling		.31	104		27,300	4,925		32,225	37,400
0520	40 ton cooling		.30	108		29,500	5,075		34,575	40,100
0528	45 ton cooling		.29	109		31,400	5,175		36,575	42,300
0536	50 ton cooling		.28	113		34,300	5,325		39,625	45,700
0680	Scroll water cooled, single compressor, hermetic, tower not incl.									
0700	2 ton cooling	Q-5	.57	28.070	Ea.	3,225	1,250		4,475	5,425
0710	5 ton cooling		.57	28.070		3,850	1,250		5,100	6,125
0720	6 ton cooling		.42	38.005		4,850	1,700		6,550	7,875
0740	8 ton cooling		.31	52.117		5,325	2,325		7,650	9,325
0760	10 ton cooling	Q-6	.36	67.039		6,125	3,100		9,225	11,400
0780	15 ton cooling	"	.33	72.727		9,625	3,350		12,975	15,600
0800	20 ton cooling	Q-7	.38	83.990		10,900	3,950		14,850	17,900
0820	30 ton cooling	"	.33	96.096		12,200	4,525		16,725	20,300

23 64 26 – Rotary-Screw Water Chillers

23 64 26.10 Rotary-Screw Type Water Chillers

		Crew	Daily Output	Labor-Hours	Unit	Material	Labor	Equipment	Total	Total Incl O&P
0010	**ROTARY-SCREW TYPE WATER CHILLERS,** With standard controls									
0110	Screw, liquid chiller, air cooled, insulated evaporator									
0120	130 ton	Q-7	.14	228	Ea.	79,500	10,700		90,200	103,500
0124	160 ton		.13	246		98,000	11,600		109,600	125,000
0128	180 ton		.13	250		110,000	11,800		121,800	138,500
0132	210 ton		.12	258		121,500	12,100		133,600	151,500
0136	270 ton		.12	266		138,500	12,500		151,000	171,500
0140	320 ton		.12	275		174,000	13,000		187,000	211,000
0200	Packaged unit, water cooled, not incl. tower									
0210	80 ton	Q-7	.14	223	Ea.	41,700	10,500		52,200	61,500
0220	100 ton		.14	230		52,000	10,800		62,800	73,000
0230	150 ton		.13	240		69,000	11,300		80,300	93,000
0240	200 ton		.13	251		82,000	11,800		93,800	108,000
0250	250 ton		.12	260		94,000	12,200		106,200	122,000
0260	300 ton		.12	266		108,500	12,500		121,000	138,500
0270	350 ton		.12	275		154,000	13,000		167,000	189,000
1450	Water cooled, tower not included									
1560	135 ton cooling, screw compressors	Q-7	.14	235	Ea.	55,000	11,100		66,100	77,000
1580	150 ton cooling, screw compressors		.13	240		63,500	11,300		74,800	87,000
1620	200 ton cooling, screw compressors		.13	250		87,500	11,800		99,300	114,000
1660	291 ton cooling, screw compressors		.12	260		91,500	12,200		103,700	119,500

23 64 33 – Direct Expansion Water Chillers

23 64 33.10 Direct Expansion Type Water Chillers

		Crew	Daily Output	Labor-Hours	Unit	Material	Labor	Equipment	Total	Total Incl O&P
0010	**DIRECT EXPANSION TYPE WATER CHILLERS,** With standard controls									
8000	Direct expansion, shell and tube type, for built up systems									

23 64 Packaged Water Chillers

23 64 33 – Direct Expansion Water Chillers

23 64 33.10 Direct Expansion Type Water Chillers		Crew	Daily Output	Labor-Hours	Unit	2009 Bare Costs Material	Labor	Equipment	Total	Total Incl O&P
8020	1 ton	Q-5	2	8	Ea.	6,350	355		6,705	7,525
8030	5 ton	↓	1.90	8.421	↓	10,600	375		10,975	12,200
8040	10 ton	↓	1.70	9.412	↓	13,100	420		13,520	15,000
8050	20 ton	↓	1.50	10.667	↓	14,900	475		15,375	17,100
8060	30 ton	↓	1	16	↓	19,000	710		19,710	22,000
8070	50 ton	↓	.90	17.778	↓	30,600	790		31,390	34,800
8080	100 ton	Q-6	.90	26.667	↓	54,500	1,225		55,725	62,000

23 65 Cooling Towers

23 65 13 – Forced-Draft Cooling Towers

23 65 13.10 Forced-Draft Type Cooling Towers		Crew	Daily Output	Labor-Hours	Unit	Material	Labor	Equipment	Total	Total Incl O&P
0010	**FORCED-DRAFT TYPE COOLING TOWERS**, Packaged units D3030–310									
0070	Galvanized steel									
0080	Induced draft, crossflow									
0100	Vertical, belt drive, 61 tons	Q-6	90	.267	TonAC	137	12.30		149.30	169
0150	100 ton	↓	100	.240	↓	150	11.05		161.05	182
0200	115 ton	↓	109	.220	↓	130	10.15		140.15	158
0250	131 ton	↓	120	.200	↓	169	9.20		178.20	200
0260	162 ton	↓	132	.182	↓	137	8.40		145.40	163
1000	For higher capacities, use multiples									
1500	Induced air, double flow									
1900	Vertical, gear drive, 167 ton	Q-6	126	.190	TonAC	133	8.80		141.80	159
2000	297 ton	↓	129	.186	↓	83	8.55		91.55	104
2100	582 ton	↓	132	.182	↓	67	8.40		75.40	86
2150	849 ton	↓	142	.169	↓	65	7.80		72.80	83
2200	1016 ton	↓	150	.160	↓	60	7.35		67.35	77
2500	Blow through, centrifugal type									
2510	50 ton	Q-6	2.28	10.526	Ea.	9,600	485		10,085	11,300
2520	75 ton	↓	1.52	15.789	↓	12,500	725		13,225	14,800
2524	100 ton	↓	1.22	19.672	↓	14,700	905		15,605	17,600
2528	125 ton	Q-7	1.31	24.427	↓	17,200	1,150		18,350	20,600
2532	200 ton	↓	.82	39.216	↓	22,500	1,850		24,350	27,600
2536	250 ton	↓	.65	49.231	↓	26,900	2,325		29,225	33,000
2540	300 ton	↓	.54	59.259	↓	30,900	2,775		33,675	38,200
2544	350 ton	↓	.47	68.085	↓	34,900	3,200		38,100	43,200
2548	400 ton	↓	.41	78.049	↓	36,300	3,675		39,975	45,500
2552	450 ton	↓	.36	88.154	↓	43,700	4,150		47,850	54,500
2556	500 ton	↓	.32	100	↓	46,600	4,700		51,300	58,500
2560	550 ton	↓	.30	107	↓	50,000	5,075		55,075	62,500
2564	600 ton	↓	.27	117	↓	52,500	5,525		58,025	65,500
2568	650 ton	↓	.25	127	↓	58,500	6,000		64,500	73,500
2572	700 ton	↓	.23	137	↓	68,500	6,450		74,950	85,000
2576	750 ton	↓	.22	146	↓	73,500	6,900		80,400	91,500
2580	800 ton	↓	.21	152	↓	80,000	7,175		87,175	93,500
2584	850 ton	↓	.20	156	↓	84,500	7,375		91,875	104,000
2588	900 ton	↓	.20	161	↓	89,500	7,600		97,100	110,000
2592	950 ton	↓	.19	170	↓	94,500	8,000		102,500	116,000
2596	1000 ton	↓	.18	180	↓	97,500	8,500		106,000	120,500
2700	Axial fan, induced draft									
2710	50 ton	Q-6	2.28	10.526	Ea.	8,325	485		8,810	9,875
2720	75 ton	↓	1.52	15.789	↓	9,800	725		10,525	11,900

23 65 Cooling Towers

23 65 13 – Forced-Draft Cooling Towers

23 65 13.10 Forced-Draft Type Cooling Towers		Crew	Daily Output	Labor-Hours	Unit	Material	Labor	Equipment	Total	Total Incl O&P
						2009 Bare Costs				
2724	100 ton	Q-6	1.22	19.672	Ea.	10,400	905		11,305	12,800
2728	125 ton	Q-7	1.31	24.427		12,800	1,150		13,950	15,700
2732	200 ton		.82	39.216		21,600	1,850		23,450	26,600
2736	250 ton		.65	49.231		24,400	2,325		26,725	30,300
2740	300 ton		.54	59.259		28,500	2,775		31,275	35,500
2744	350 ton		.47	68.085		30,500	3,200		33,700	38,400
2748	400 ton		.41	78.049		35,000	3,675		38,675	44,000
2752	450 ton		.36	88.154		38,800	4,150		42,950	48,900
2756	500 ton		.32	100		41,200	4,700		45,900	52,500
2760	550 ton		.30	107		46,100	5,075		51,175	58,000
2764	600 ton		.27	117		48,600	5,525		54,125	62,000
2768	650 ton		.25	127		53,000	6,000		59,000	67,500
2772	700 ton		.23	137		60,500	6,450		66,950	76,000
2776	750 ton		.22	146		63,000	6,900		69,900	79,500
2780	800 ton		.21	152		65,500	7,175		72,675	83,000
2784	850 ton		.20	156		66,000	7,375		73,375	83,500
2788	900 ton		.20	161		68,000	7,600		75,600	86,500
2792	950 ton		.19	170		70,500	8,000		78,500	89,500
2796	1000 ton	↓	.18	180	↓	76,000	8,500		84,500	97,000
3000	For higher capacities, use multiples									
3500	For pumps and piping, add	Q-6	38	.632	TonAC	68.50	29		97.50	119
4000	For absorption systems, add				"	75%	75%			
4100	Cooling water chemical feeder	Q-5	3	5.333	Ea.	292	237		529	675
5000	Fiberglass									
5010	Draw thru									
5100	60 ton	Q-6	1.50	16	Ea.	3,650	735		4,385	5,125
5120	125 ton		.99	24.242		7,475	1,125		8,600	9,875
5140	300 ton		.43	55.814		17,500	2,575		20,075	23,100
5160	600 ton		.22	109		31,400	5,025		36,425	42,100
5180	1000 ton	↓	.15	160	↓	54,000	7,375		61,375	70,500
6000	Stainless steel									
6010	Induced draft, crossflow, horizontal, belt drive									
6100	57 ton	Q-6	1.50	16	Ea.	20,400	735		21,135	23,600
6120	91 ton		.99	24.242		24,700	1,125		25,825	28,800
6140	111 ton		.43	55.814		37,100	2,575		39,675	44,700
6160	126 ton	↓	.22	109	↓	37,100	5,025		42,125	48,400
6170	Induced draft, crossflow, vertical, gear drive									
6172	167 ton	Q-6	.75	32	Ea.	47,100	1,475		48,575	54,000
6174	297 ton		.43	55.814		52,500	2,575		55,075	61,500
6176	582 ton		.23	104		83,000	4,800		87,800	98,000
6178	849 ton		.17	141		117,000	6,500		123,500	138,500
6180	1016 ton	↓	.15	160	↓	130,000	7,375		137,375	154,000

23 72 Air-to-Air Energy Recovery Equipment

23 72 13 – Heat-Wheel Air-to-Air Energy-Recovery Equipment

23 72 13.10	Heat-Wheel Air-to-air Energy Recovery Equip.		Crew	Daily Output	Labor-Hours	Unit	2009 Bare Costs Material	Labor	Equipment	Total	Total Incl O&P
0010	**HEAT-WHEEL AIR-TO-AIR ENERGY RECOVERY EQUIPMENT**										
0100	Air to air										
4000	Enthalpy recovery wheel										
4010	1000 max CFM	G	Q-9	1.20	13.333	Ea.	6,050	565		6,615	7,525
4020	2000 max CFM	G		1	16		7,075	680		7,755	8,825
4030	4000 max CFM	G		.80	20		8,200	850		9,050	10,300
4040	6000 max CFM	G		.70	22.857		9,575	970		10,545	12,000
4050	8000 max CFM	G	Q-10	1	24		10,600	1,050		11,650	13,200
4060	10,000 max CFM	G		.90	26.667		12,700	1,175		13,875	15,700
4070	20,000 max CFM	G		.80	30		22,900	1,325		24,225	27,200
4080	25,000 max CFM	G		.70	34.286		28,000	1,500		29,500	33,100
4090	30,000 max CFM	G		.50	48		31,100	2,125		33,225	37,500
4100	40,000 max CFM	G		.45	53.333		42,900	2,350		45,250	51,000
4110	50,000 max CFM	G		.40	60		49,900	2,650		52,550	59,000

23 72 16 – Heat-Pipe Air-To-Air Energy-Recovery Equipment

23 72 16.10 Heat Pipes

		Crew	Daily Output	Labor-Hours	Unit	Material	Labor	Equipment	Total	Total Incl O&P
0010	**HEAT PIPES**									
8000	Heat pipe type, glycol, 50% efficient									
8010	100 MBH, 1700 CFM	1 Stpi	.80	10	Ea.	3,550	495		4,045	4,675
8020	160 MBH, 2700 CFM		.60	13.333		4,775	660		5,435	6,225
8030	620 MBH, 4000 CFM		.40	20		7,225	985		8,210	9,425

23 73 Indoor Central-Station Air-Handling Units

23 73 13 – Modular Indoor Central-Station Air-Handling Units

23 73 13.10 Air Handling Units

		Crew	Daily Output	Labor-Hours	Unit	Material	Labor	Equipment	Total	Total Incl O&P
0010	**AIR HANDLING UNITS**, Built-Up									
0100	With cooling/heating coil section, filters, mixing box									
0880	Single zone, horizontal / vertical									
0890	Constant volume									
0900	1600 CFM	Q-5	1.20	13.333	Ea.	3,150	590		3,740	4,375
0906	2000 CFM		1.10	14.545		3,950	645		4,595	5,325
0910	3000 CFM		1	16		4,925	710		5,635	6,500
0916	4000 CFM		.95	16.842		6,575	750		7,325	8,350
0920	5000 CFM	Q-6	1.40	17.143		8,225	790		9,015	10,200
0926	6500 CFM		1.30	18.462		10,700	850		11,550	13,000
0930	7500 CFM		1.20	20		12,300	920		13,220	15,000
0936	9200 CFM		1.10	21.818		15,100	1,000		16,100	18,100
0940	11,500 CFM		1	24		16,600	1,100		17,700	20,000
0946	13,200 CFM		.90	26.667		19,100	1,225		20,325	22,900
0950	16,500 CFM		.80	30		23,900	1,375		25,275	28,300
0960	19,500 CFM		.70	34.286		28,200	1,575		29,775	33,400
0970	22,000 CFM		.60	40		31,800	1,850		33,650	37,800
0980	27,000 CFM		.50	48		35,500	2,200		37,700	42,300
0990	34,000 CFM		.40	60		44,700	2,775		47,475	53,500
1000	40,000 CFM		.30	80		48,700	3,675		52,375	59,000
1010	47,000 CFM		.20	120		62,000	5,525		67,525	76,500
1140	60,000 CFM	Q-7	.18	177		70,000	8,350		78,350	89,500
1160	75,000 CFM	"	.16	200		87,500	9,400		96,900	110,000
2000	Variable air volume									
2100	75,000 CFM	Q-7	.12	266	Ea.	89,000	12,500		101,500	117,000

23 73 13 – Modular Indoor Central-Station Air-Handling Units

23 73 13.10 Air Handling Units

	23 73 13.10 Air Handling Units	Crew	Daily Output	Labor-Hours	Unit	Material	2009 Bare Costs Labor	Equipment	Total	Total Incl O&P
2140	100,000 CFM	Q-7	.11	290	Ea.	123,500	13,700		137,200	156,500
2160	150,000 CFM		.10	320		188,000	15,000		203,000	229,000
2300	Multi-zone, horizontal or vertical, blow-thru fan									
2310	3000 CFM	Q-5	1	16	Ea.	5,750	710		6,460	7,400
2314	4000 CFM	"	.95	16.842		7,575	750		8,325	9,475
2318	5000 CFM	Q-6	1.40	17.143		9,450	790		10,240	11,600
2322	6500 CFM		1.30	18.462		12,300	850		13,150	14,800
2326	7500 CFM		1.20	20		14,200	920		15,120	17,000
2330	9200 CFM		1.10	21.818		17,400	1,000		18,400	20,700
2334	11,500 CFM		1	24		19,200	1,100		20,300	22,800
2336	13,200 CFM		.90	26.667		22,000	1,225		23,225	26,100
2340	16,500 CFM		.80	30		27,400	1,375		28,775	32,300
2344	19,500 CFM		.70	34.286		32,300	1,575		33,875	38,000
2348	22,000 CFM		.60	40		36,500	1,850		38,350	43,000
2352	27,000 CFM		.50	48		40,800	2,200		43,000	48,200
2356	34,000 CFM		.40	60		51,500	2,775		54,275	60,500
2360	40,000 CFM		.30	80		56,000	3,675		59,675	67,000
2364	47,000 CFM		.20	120		71,000	5,525		76,525	86,500

23 73 13.20 Air Handling Units

	23 73 13.20 Air Handling Units	Crew	Daily Output	Labor-Hours	Unit	Material	Labor	Equipment	Total	Total Incl O&P
0010	**AIR HANDLING UNITS**, Chilled water									
2300	Variable air volume, also incl. heating coil									
2340	5000 CFM	Q-6	.70	34.286	Ea.	10,400	1,575		11,975	13,800
2350	10,000 CFM		.50	48		13,000	2,200		15,200	17,600
2360	15,000 CFM		.40	60		18,600	2,775		21,375	24,700
2370	20,000 CFM		.29	82.759		24,100	3,825		27,925	32,300
2380	30,000 CFM		.20	120		40,900	5,525		46,425	53,500
3000	Packaged, 3000 CFM, 7.5 ton		1.70	14.118		2,450	650		3,100	3,675
3050	3200 CFM, 8 ton		1.60	15		2,800	690		3,490	4,100
3100	4000 CFM, 10 ton		1.40	17.143		2,950	790		3,740	4,425
3150	4400 CFM, 11 ton		1.35	17.778		3,500	820		4,320	5,075
3200	6000 CFM, 15 ton		1	24		3,850	1,100		4,950	5,875
3250	7000 CFM, 17.5 ton		.90	26.667		4,600	1,225		5,825	6,900
3300	10,000 CFM, 25 ton		.65	36.923		6,950	1,700		8,650	10,200
3400	12,000 CFM, 30 ton		.50	48		7,550	2,200		9,750	11,600
3500	For hot water heat coil, add					30%	5%			

23 73 39 – Indoor, Direct Gas-Fired Heating and Ventilating Units

23 73 39.10 Make-Up Air Unit

	23 73 39.10 Make-Up Air Unit	Crew	Daily Output	Labor-Hours	Unit	Material	Labor	Equipment	Total	Total Incl O&P
0010	**MAKE-UP AIR UNIT**									
0020	Indoor suspension, natural/LP gas, direct fired,									
0032	standard control. For flue see Div. 23 51 23.10									
0040	70° F temperature rise, MBH is input									
0100	2000 CFM, 168 MBH	Q-6	3	8	Ea.	9,050	370		9,420	10,500
0120	3000 CFM, 252 MBH		2	12		9,300	555		9,855	11,000
0140	4000 CFM, 336 MBH		1.80	13.333		9,450	615		10,065	11,300
0160	6000 CFM, 502 MBH		1.50	16		11,400	735		12,135	13,600
0180	8000 CFM, 670 MBH		1.40	17.143		12,500	790		13,290	14,900
0200	10,000 CFM, 838 MBH		1.20	20		14,000	920		14,920	16,800
0220	12,000 CFM, 1005 MBH		1	24		14,200	1,100		15,300	17,300
0240	14,000 CFM, 1180 MBH		.94	25.532		14,400	1,175		15,575	17,600
0260	18,000 CFM, 1340 MBH		.86	27.907		14,500	1,275		15,775	17,800
0280	20,000 CFM, 1675 MBH		.76	31.579		17,800	1,450		19,250	21,800
0300	24,000 CFM, 2007 MBH	Q-7	1	32		18,500	1,500		20,000	22,600

23 73 Indoor Central-Station Air-Handling Units

23 73 39 – Indoor, Direct Gas-Fired Heating and Ventilating Units

23 73 39.10 Make-Up Air Unit		Crew	Daily Output	Labor-Hours	Unit	Material	2009 Bare Costs Labor	Equipment	Total	Total Incl O&P
0320	30,000 CFM, 2510 MBH	Q-7	.96	33.333	Ea.	18,900	1,575		20,475	23,200
0340	35,000 CFM, 2930 MBH		.92	34.783		23,900	1,625		25,525	28,800
0360	40,000 CFM, 3350 MBH		.88	36.364		24,500	1,700		26,200	29,500
0380	45,000 CFM, 3770 MBH		.84	38.095		25,100	1,800		26,900	30,300
0400	50,000 CFM, 4180 MBH		.80	40		25,400	1,875		27,275	30,700
0420	55,000 CFM, 4600 MBH		.75	42.667		32,400	2,000		34,400	38,600
0440	60,000 CFM, 5020 MBH		.70	45.714		33,600	2,150		35,750	40,200
0460	65,000 CFM, 5435 MBH		.60	53.333		33,800	2,500		36,300	41,000
0480	75,000 CFM, 6275 MBH		.50	64		35,600	3,000		38,600	43,700
0600	For discharge louver assembly, add					10%				
0700	For filters, add					20%				
0800	For air shut-off damper section, add					10%				
0900	For vertical unit, add					10%				

23 74 Packaged Outdoor HVAC Equipment

23 74 13 – Packaged, Outdoor, Central-Station Air-Handling Units

23 74 13.10 Packaged, Outdoor Type, Central-Station Air-handling Units

		Crew	Daily Output	Labor-Hours	Unit	Material	Labor	Equipment	Total	Total Incl O&P
0010	**PACKAGED, OUTDOOR TYPE, CENTRAL-STATION AIR-HANDLING UNITS**									
3000	Rooftop									
3010	Constant volume									
3100	2000 CFM	Q-5	1	16	Ea.	5,250	710		5,960	6,850
3140	5000 CFM	Q-6	.79	30.380		9,400	1,400		10,800	12,500
3150	10,000 CFM		.52	46.154		11,900	2,125		14,025	16,300
3160	15,000 CFM		.39	61.538		18,000	2,825		20,825	24,100
3170	20,000 CFM		.30	80		22,400	3,675		26,075	30,100
3200	Variable air volume									
3240	5000 CFM	Q-6	.70	34.286	Ea.	11,100	1,575		12,675	14,600
3250	10,000 CFM		.50	48		13,900	2,200		16,100	18,600
3260	15,000 CFM		.40	60		20,000	2,775		22,775	26,200
3270	20,000 CFM		.29	82.759		25,900	3,825		29,725	34,200
3280	30,000 CFM		.20	120		43,900	5,525		49,425	56,500

23 74 23 – Packaged, Outdoor, Heating-Only Makeup-Air Units

23 74 23.16 Pkgd.,Ind.-Fired,Outdoor, Heat-Only Makeup-Air

		Crew	Daily Output	Labor-Hours	Unit	Material	Labor	Equipment	Total	Total Incl O&P
0010	**PACKAGED, IND.-FIRED, OUTDOOR, HEAT-ONLY MAKEUP-AIR UNITS**									
1000	Rooftop unit, natural gas, gravity vent, S.S. exchanger									
1010	70° F temperature rise, MBH is input									
1020	995 CFM, 75 MBH	Q-6	4	6	Ea.	6,075	276		6,351	7,100
1040	1245 CFM, 95 MBH		3.60	6.667		6,125	305		6,430	7,200
1060	1590 CFM, 120 MBH		3.30	7.273		6,600	335		6,935	7,775
1080	2090 CFM, 159 MBH		3	8		6,800	370		7,170	8,050
1100	2490 CFM, 190 MBH		2.60	9.231		7,800	425		8,225	9,225
1120	2985 CFM, 225 MBH		2.30	10.435		7,925	480		8,405	9,425
1140	3975 CFM, 300 MBH		1.90	12.632		9,375	580		9,955	11,200
1160	4970 CFM, 375 MBH		1.40	17.143		10,900	790		11,690	13,100
1180	5950 CFM, 450 MBH		1.20	20		11,400	920		12,320	14,000
1200	7940 CFM, 600 MBH		1	24		13,300	1,100		14,400	16,300
1220	9930 CFM, 750 MBH		.80	30		22,700	1,375		24,075	27,100
1240	11,900 CFM, 900 MBH		.60	40		25,400	1,850		27,250	30,700
1260	15,880 CFM, 1200 MBH		.30	80		30,500	3,675		34,175	39,000
1500	For power vent, add					8%				

23 74 Packaged Outdoor HVAC Equipment

23 74 23 – Packaged, Outdoor, Heating-Only Makeup-Air Units

23 74 23.16 Pkgd.,Ind.-Fired,Outdoor, Heat-Only Makeup-Air		Crew	Daily Output	Labor-Hours	Unit	Material	2009 Bare Costs Labor	Equipment	Total	Total Incl O&P
1600	For cleanable filters, add				Ea.	176			176	194
1700	For electric modulating gas control, add				"	1,900			1,900	2,075
1800	For aluminized steel exchanger, deduct					10%				
8000	Rooftop unit, electric, with curb, controls									
8010	33 kW	Q-5	4	4	Ea.	3,550	178		3,728	4,175
8020	100 kW		3.60	4.444		5,925	197		6,122	6,825
8030	150 kW		3.20	5		6,825	222		7,047	7,825
8040	250 kW		2.80	5.714		8,950	254		9,204	10,200
8050	400 kW		2.20	7.273		13,200	325		13,525	15,000

23 74 33 – Packaged, Outdoor, Heating and Cooling Makeup Air-Conditioners

23 74 33.10 Roof Top Air Conditioners

		Crew	Daily Output	Labor-Hours	Unit	Material	Labor	Equipment	Total	Total Incl O&P
0010	**ROOF TOP AIR CONDITIONERS**, Standard controls, curb, economizer.									
1000	Single zone, electric cool, gas heat									
1090	2 ton cooling, 55 MBH heating D3050–150	Q-5	.93	17.204	Ea.	3,100	765		3,865	4,550
1100	3 ton cooling, 60 MBH heating		.70	22.857		3,200	1,025		4,225	5,050
1120	4 ton cooling, 95 MBH heating D3050–155		.61	26.403		3,750	1,175		4,925	5,875
1140	5 ton cooling, 112 MBH heating R236000-10		.56	28.521		4,150	1,275		5,425	6,475
1145	6 ton cooling, 140 MBH heating R236000-20		.52	30.769		4,900	1,375		6,275	7,450
1150	7.5 ton cooling, 170 MBH heating		.50	32.258		6,275	1,425		7,700	9,050
1160	10 ton cooling, 200 MBH heating	Q-6	.67	35.982		8,300	1,650		9,950	11,600
1170	12.5 ton cooling, 230 MBH heating		.63	37.975		9,400	1,750		11,150	12,900
1180	15 ton cooling, 270 MBH heating		.57	42.032		12,500	1,925		14,425	16,700
1190	18 ton cooling, 330 MBH heating		.52	45.889		15,400	2,125		17,525	20,100
1200	20 ton cooling, 360 MBH heating	Q-7	.67	47.976		20,100	2,250		22,350	25,500
1210	25 ton cooling, 450 MBH heating		.56	57.554		23,700	2,700		26,400	30,100
1220	30 ton cooling, 540 MBH heating		.47	68.376		25,800	3,225		29,025	33,100
1240	40 ton cooling, 675 MBH heating		.35	91.168		32,900	4,275		37,175	42,600
1260	50 ton cooling, 810 MBH heating		.28	113		39,300	5,350		44,650	51,000
1265	60 ton cooling, 900 MBH heating		.23	136		46,900	6,425		53,325	61,000
1270	80 ton cooling, 1000 MBH heating		.20	160		76,000	7,525		83,525	95,000
1275	90 ton cooling, 1200 MBH heating		.17	188		79,000	8,850		87,850	100,500
1280	100 ton cooling, 1350 MBH heating		.14	228		88,000	10,700		98,700	113,000
1300	Electric cool, electric heat									
1310	5 ton	Q-5	.63	25.600	Ea.	3,450	1,125		4,575	5,500
1314	10 ton	Q-6	.74	32.432		6,250	1,500		7,750	9,125
1318	15 ton	"	.64	37.795		9,400	1,750		11,150	12,900
1322	20 ton	Q-7	.74	43.185		12,500	2,025		14,525	16,900
1326	25 ton		.62	51.696		15,700	2,425		18,125	20,900
1330	30 ton		.52	61.657		18,800	2,900		21,700	25,100
1334	40 ton		.39	82.051		25,000	3,850		28,850	33,300
1338	50 ton		.31	102		29,700	4,825		34,525	39,900
1342	60 ton		.26	123		35,700	5,775		41,475	48,000
1346	75 ton		.21	153		44,600	7,225		51,825	60,000
2000	Multizone, electric cool, gas heat, economizer									
2100	15 ton cooling, 360 MBH heating	Q-7	.61	52.545	Ea.	53,500	2,475		55,975	62,500
2120	20 ton cooling, 360 MBH heating		.53	60.038		57,500	2,825		60,325	67,500
2140	25 ton cooling, 450 MBH heating		.45	71.910		69,000	3,375		72,375	81,000
2160	28 ton cooling, 450 MBH heating		.41	79.012		78,500	3,725		82,225	92,000
2180	30 ton cooling, 540 MBH heating		.37	85.562		88,500	4,025		92,525	103,500
2200	40 ton cooling, 540 MBH heating		.28	113		101,000	5,350		106,350	119,000
2210	50 ton cooling, 540 MBH heating		.23	142		125,500	6,675		132,175	148,000
2220	70 ton cooling, 1500 MBH heating		.16	198		136,000	9,350		145,350	163,500

23 74 Packaged Outdoor HVAC Equipment

23 74 33 – Packaged, Outdoor, Heating and Cooling Makeup Air-Conditioners

23 74 33.10 Roof Top Air Conditioners		Crew	Daily Output	Labor-Hours	Unit	2009 Bare Costs Material	Labor	Equipment	Total	Total Incl O&P
2240	80 ton cooling, 1500 MBH heating	Q-7	.14	228	Ea.	155,500	10,700		166,200	187,000
2260	90 ton cooling, 1500 MBH heating		.13	256		163,000	12,000		175,000	197,500
2280	105 ton cooling, 1500 MBH heating		.11	290		179,500	13,700		193,200	218,000
2400	For hot water heat coil, deduct					5%				
2500	For steam heat coil, deduct					2%				
2600	For electric heat, deduct					3%	5%			
4000	Single zone, electric cool, gas heat, variable air volume									
4100	12.5 ton cooling, 230 MBH heating	Q-6	.55	43.716	Ea.	11,400	2,025		13,425	15,600
4120	18 ton cooling, 330 MBH heating	"	.46	52.747		17,700	2,425		20,125	23,100
4140	25 ton cooling, 450 MBH heating	Q-7	.48	66.116		27,100	3,100		30,200	34,500
4160	40 ton cooling, 675 MBH heating		.31	104		40,100	4,925		45,025	51,500
4180	60 ton cooling, 900 MBH heating		.20	157		58,500	7,400		65,900	75,500
4200	80 ton cooling, 1000 MBH heating		.15	209		88,000	9,825		97,825	111,500
5000	Single zone, electric cool only									
5050	3 ton cooling	Q-5	.88	18.203	Ea.	3,025	810		3,835	4,550
5060	4 ton cooling		.76	21.108		3,450	940		4,390	5,200
5070	5 ton cooling		.70	22.792		3,925	1,025		4,950	5,850
5080	6 ton cooling		.65	24.502		4,500	1,100		5,600	6,575
5090	7.5 ton cooling		.62	25.806		5,675	1,150		6,825	7,975
5100	8.5 ton cooling		.59	26.981		6,775	1,200		7,975	9,250
5110	10 ton cooling	Q-6	.83	28.812		7,075	1,325		8,400	9,775
5120	12 ton cooling		.80	30		8,900	1,375		10,275	11,900
5130	15 ton cooling		.71	33.613		12,200	1,550		13,750	15,700
5140	18 ton cooling		.65	36.697		14,500	1,700		16,200	18,400
5150	20 ton cooling	Q-7	.83	38.415		18,400	1,800		20,200	22,900
5160	25 ton cooling		.70	45.977		21,100	2,150		23,250	26,500
5170	30 ton cooling		.59	54.701		23,200	2,575		25,775	29,400
5180	40 ton cooling		.44	73.059		30,000	3,425		33,425	38,200
5400	For low heat, add					7%				
5410	For high heat, add					10%				
6000	Single zone electric cooling with variable volume distribution									
6020	20 ton cooling	Q-7	.72	44.199	Ea.	23,300	2,075		25,375	28,800
6030	25 ton cooling		.61	52.893		27,100	2,475		29,575	33,500
6040	30 ton cooling		.51	62.868		30,100	2,950		33,050	37,500
6050	40 ton cooling		.38	83.990		39,300	3,950		43,250	49,200
6060	50 ton cooling		.31	104		46,300	4,925		51,225	58,500
6070	60 ton cooling		.25	125		50,500	5,925		56,425	64,500
6200	For low heat, add					7%				
6210	For high heat, add					10%				
7000	Multizone, cool/heat, variable volume distribution									
7100	50 ton cooling	Q-7	.20	164	Ea.	93,500	7,725		101,225	114,000
7110	70 ton cooling		.14	228		130,500	10,700		141,200	159,500
7120	90 ton cooling		.11	296		150,000	13,900		163,900	186,000
7130	105 ton cooling		.10	333		168,000	15,700		183,700	208,000
7140	120 ton cooling		.08	380		192,000	17,900		209,900	238,000
7150	140 ton cooling		.07	444		205,500	20,900		226,400	257,500
7300	Penthouse unit, cool/heat, variable volume distribution									
7400	150 ton cooling	Q-7	.10	310	Ea.	336,000	14,600		350,600	391,500
7410	170 ton cooling		.10	320		359,000	15,000		374,000	417,500
7420	200 ton cooling		.10	326		422,500	15,400		437,900	487,500
7430	225 ton cooling		.09	340		462,000	16,000		478,000	532,500
7440	250 ton cooling		.09	359		476,500	16,900		493,400	550,000
7450	270 ton cooling		.09	363		515,000	17,100		532,100	592,000

23 74 Packaged Outdoor HVAC Equipment

23 74 33 – Packaged, Outdoor, Heating and Cooling Makeup Air-Conditioners

23 74 33.10 Roof Top Air Conditioners			Crew	Daily Output	Labor-Hours	Unit	Material	Labor	Equipment	Total	Total Incl O&P
							2009 Bare Costs				
7460	300 ton cooling		Q-7	.09	372	Ea.	572,000	17,500		589,500	656,000
9400	Sub assemblies for assembly systems										
9410	Duct work per ton roof top 1 zone units		Q-9	.96	16.667	Ton	284	710		994	1,425
9420	Ductwork per ton rooftop multizone units			.50	32		415	1,350		1,765	2,550
9430	Ductwork, VAV cooling/ton rooftop multizone, 1 zone/ton			.32	50		620	2,125		2,745	3,950
9440	Duct work per ton packaged water & air cooled units			1.26	12.698		107	540		647	950
9450	Duct work per ton split system remote condensing units		↓	1.32	12.121	↓	101	515		616	900

23 76 Evaporative Air-Cooling Equipment

23 76 13 – Direct Evaporative Air Coolers

23 76 13.10 Evaporative Coolers			Crew	Daily Output	Labor-Hours	Unit	Material	Labor	Equipment	Total	Total Incl O&P
0010	**EVAPORATIVE COOLERS**, (Swamp Coolers) ducted, not incl. duct.										
0100	Side discharge style, capacities at .25" S.P.										
0120	1785 CFM, 1/3 HP, 115 V	G	Q-9	5	3.200	Ea.	465	136		601	725
0140	2740 CFM, 1/3 HP, 115 V	G		4.50	3.556		570	151		721	860
0160	3235 CFM, 1/2 HP, 115 V	G		4	4		585	170		755	905
0180	3615 CFM, 1/2 HP, 230 V	G		3.60	4.444		595	189		784	945
0200	4215 CFM, 3/4 HP, 230 V	G		3.20	5		760	212		972	1,150
0220	5255 CFM, 1 HP, 115/230 V	G		3	5.333		1,250	227		1,477	1,725
0240	6090 CFM, 1 HP, 230/460 V	G		2.80	5.714		1,750	243		1,993	2,300
0260	8300 CFM, 1-1/2 HP, 230/460 V	G		2.60	6.154		1,750	261		2,011	2,325
0280	8360 CFM, 1-1/2 HP, 230/460 V	G		2.20	7.273		2,400	310		2,710	3,100
0300	9725 CFM, 2 HP, 230/460 V	G		1.80	8.889		2,400	380		2,780	3,225
0320	11,715 CFM, 3 HP, 230/460 V	G		1.40	11.429		2,450	485		2,935	3,450
0340	14,410 CFM, 5 HP, 230/460 V	G	↓	1	16	↓	2,575	680		3,255	3,900
0400	For two-speed motor, add						5%				
0500	For down discharge style, add						10%				
1000	High capacities at .4" S.P.										
1040	15,000 CFM, 5 HP, 230/460 V	G	Q-9	.54	29.630	Ea.	13,300	1,250		14,550	16,500
1080	20,000 CFM, 10 HP, 230/460 V	G	"	.48	33.333		16,300	1,425		17,725	20,200
1100	25,000 CFM, 20 HP, 230/460 V	G	Q-10	.54	44.444		18,200	1,950		20,150	23,000
1140	30,000 CFM, 20 HP, 230/460 V	G		.46	52.174		20,900	2,300		23,200	26,500
1160	35,000 CFM, 20 HP, 230/460 V	G		.44	54.545		22,600	2,400		25,000	28,600
1180	40,000 CFM, 30 HP, 230/460 V	G	↓	.42	57.143	↓	23,800	2,525		26,325	30,000

23 76 16 – Indirect Evaporative Air Coolers

23 76 16.10 Evaporators		Crew	Daily Output	Labor-Hours	Unit	Material	Labor	Equipment	Total	Total Incl O&P
0010	**EVAPORATORS**, DX coils, remote compressors not included.									
1000	Coolers, reach-in type, above freezing temperatures									
1300	Shallow depth, wall mount, 7 fins per inch, air defrost									
1310	600 BTUH, 8" fan, rust-proof core	Q-5	3.80	4.211	Ea.	380	187		567	700
1320	900 BTUH, 8" fan, rust-proof core		3.30	4.848		470	215		685	840
1330	1200 BTUH, 8" fan, rust-proof core		3	5.333		555	237		792	965
1340	1800 BTUH, 8" fan, rust-proof core		2.40	6.667		640	296		936	1,150
1350	2500 BTUH, 10" fan, rust-proof core		2.20	7.273		695	325		1,020	1,250
1360	3500 BTUH, 10" fan		2	8		815	355		1,170	1,425
1370	4500 BTUH, 10" fan	↓	1.90	8.421		860	375		1,235	1,500
1390	For pan drain, add				↓	45.50			45.50	50
1600	Undercounter refrigerators, ceiling or wall mount,									
1610	8 fins per inch, air defrost, rust-proof core									
1630	800 BTUH, one 6" fan	Q-5	5	3.200	Ea.	256	142		398	495

23 76 16.10 Evaporators		Crew	Daily Output	Labor-Hours	Unit	Material	2009 Bare Costs Labor	Equipment	Total	Total Incl O&P
1640	1300 BTUH, two 6" fans	Q-5	4	4	Ea.	310	178		488	605
1650	1700 BTUH, two 6" fans	↓	3.60	4.444	↓	345	197		542	675
2000	Coolers, reach-in and walk-in types, above freezing temperatures									
2600	Two-way discharge, ceiling mount, 150-4100 CFM,									
2610	above 34° F applications, air defrost									
2630	900 BTUH, 7 fins per inch, 8" fan	Q-5	4	4	Ea.	271	178		449	565
2660	2500 BTUH, 7 fins per inch, 10" fan		2.60	6.154		535	273		808	995
2690	5500 BTUH, 7 fins per inch, 12" fan		1.70	9.412		1,050	420		1,470	1,775
2720	8500 BTUH, 8 fins per inch, 16" fan		1.20	13.333		1,300	590		1,890	2,350
2750	15,000 BTUH, 7 fins per inch, 18" fan		1.10	14.545		1,825	645		2,470	2,975
2770	24,000 BTUH, 7 fins per inch, two 16" fans		1	16		2,750	710		3,460	4,075
2790	30,000 BTUH, 7 fins per inch, two 18" fans	↓	.90	17.778	↓	3,400	790		4,190	4,900
2850	Two-way discharge, low profile, ceiling mount,									
2860	8 fins per inch, 200-570 CFM, air defrost									
2880	800 BTUH, one 6" fan	Q-5	4	4	Ea.	289	178		467	585
2890	1300 BTUH, two 6" fans		3.50	4.571		340	203		543	680
2900	1800 BTUH, two 6" fans		3.30	4.848		360	215		575	720
2910	2700 BTUH, three 6" fans	↓	2.70	5.926	↓	425	263		688	860
3000	Coolers, walk-in type, above freezing temperatures									
3300	General use, ceiling mount, 108-2080 CFM									
3320	600 BTUH, 7 fins per inch, 6" fan	Q-5	5.20	3.077	Ea.	225	137		362	455
3340	1200 BTUH, 7 fins per inch, 8" fan		3.70	4.324		310	192		502	630
3360	1800 BTUH, 7 fins per inch, 10" fan		2.70	5.926		435	263		698	875
3380	3500 BTUH, 7 fins per inch, 12" fan		2.40	6.667		640	296		936	1,150
3400	5500 BTUH, 8 fins per inch, 12" fan		2	8		715	355		1,070	1,325
3430	8500 BTUH, 8 fins per inch, 16" fan		1.40	11.429		1,075	510		1,585	1,925
3460	15,000 BTUH, 7 fins per inch, two 16" fans	↓	1.10	14.545	↓	1,700	645		2,345	2,850
3640	Low velocity, high latent load, ceiling mount,									
3650	1050-2420 CFM, air defrost, 6 fins per inch									
3670	6700 BTUH, two 10" fans	Q-5	1.30	12.308	Ea.	1,500	545		2,045	2,475
3680	10,000 BTUH, three 10" fans		1	16		1,750	710		2,460	3,000
3690	13,500 BTUH, three 10" fans		1	16		2,100	710		2,810	3,375
3700	18,000 BTUH, four 10" fans		1	16		2,800	710		3,510	4,150
3710	26,500 BTUH, four 10" fans	↓	.80	20		3,450	890		4,340	5,100
3730	For electric defrost, add				↓	17%				
5000	Freezers and coolers, reach-in type, above 34° F									
5030	to sub-freezing temperature range, low latent load,									
5050	air defrost, 7 fins per inch									
5070	1200 BTUH, 8" fan, rustproof core	Q-5	4.40	3.636	Ea.	315	162		477	585
5080	1500 BTUH, 8" fan, rustproof core		4.20	3.810		325	169		494	615
5090	1800 BTUH, 8" fan, rustproof core		3.70	4.324		340	192		532	665
5100	2500 BTUH, 10" fan, rustproof core		2.90	5.517		390	245		635	795
5110	3500 BTUH, 10" fan, rustproof core		2.50	6.400		530	284		814	1,000
5120	4500 BTUH, 12" fan, rustproof core	↓	2.10	7.619	↓	705	340		1,045	1,275
6000	Freezers and coolers, walk-in type									
6050	1960-18,000 CFM, medium profile									
6060	Standard motor, 6 fins per inch, to -30° F, all aluminum									
6080	10,500 BTUH, air defrost, one 18" fan	Q-5	1.40	11.429	Ea.	1,275	510		1,785	2,175
6090	12,500 BTUH, air defrost, one 18" fan		1.40	11.429		1,400	510		1,910	2,275
6100	16,400 BTUH, air defrost, one 18" fan		1.30	12.308		1,500	545		2,045	2,475
6110	20,900 BTUH, air defrost, one 18" fans		1	16		1,750	710		2,460	3,000
6120	27,000 BTUH, air defrost, two 18" fans	↓	1	16		2,600	710		3,310	3,925
6130	32,900 BTUH, air defrost, two 20" fans	Q-6	1.30	18.462	↓	2,900	850		3,750	4,450

23 76 16 – Indirect Evaporative Air Coolers

23 76 16.10 Evaporators		Crew	Daily Output	Labor-Hours	Unit	Material	2009 Bare Costs Labor	Equipment	Total	Total Incl O&P
6140	39,000 BTUH, air defrost, two 20" fans	Q-6	1.25	19.200	Ea.	3,375	885		4,260	5,025
6150	44,100 BTUH, air defrost, three 20" fans		1	24		3,825	1,100		4,925	5,850
6155	53,000 BTUH, air defrost, three 20" fans		1	24		3,825	1,100		4,925	5,850
6160	66,200 BTUH, air defrost, four 20" fans		1	24		4,350	1,100		5,450	6,425
6170	78,000 BTUH, air defrost, four 20" fans		.80	30		5,400	1,375		6,775	8,025
6180	88,200 BTUH, air defrost, four 24" fans		.60	40		5,750	1,850		7,600	9,100
6190	110,000 BTUH, air defrost, four 24" fans		.50	48		6,525	2,200		8,725	10,500
6330	Hot gas defrost, standard motor units, add					35%				
6370	Electric defrost, 230V, standard motor, add					16%				
6410	460V, standard or high capacity motor, add					25%				
6800	Eight fins per inch, increases BTUH 75%									
6810	Standard capacity, add				Ea.	5%				
6830	Hot gas & electric defrost not recommended									
7000	800-4150 CFM, 12" fans									
7010	Four fins per inch									
7030	3400 BTUH, 1 fan	Q-5	2.80	5.714	Ea.	460	254		714	885
7040	4200 BTUH, 1 fan		2.40	6.667		470	296		766	960
7050	5300 BTUH, 1 fan		1.90	8.421		540	375		915	1,150
7060	6800 BTUH, 1 fan		1.70	9.412		595	420		1,015	1,275
7070	8400 BTUH, 2 fans		1.60	10		815	445		1,260	1,550
7080	10,500 BTUH, 2 fans		1.40	11.429		885	510		1,395	1,725
7090	13,000 BTUH, 3 fans		1.30	12.308		1,150	545		1,695	2,075
7100	17,000 BTUH, 4 fans		1.20	13.333		1,450	590		2,040	2,500
7110	21,500 BTUH, 5 fans		1	16		1,775	710		2,485	3,050
7160	Six fins per inch increases BTUH 35%, add					5%				
7180	Eight fins per inch increases BTUH 53%,									
7190	not recommended for below 35° F, add				Ea.	8%				
7210	For 12° F temperature differential, add					20%				
7230	For adjustable thermostat control, add					102			102	112
7240	For hot gas defrost, except 8 fin models									
7250	on applications below 35° F, add				Ea.	58%				
7260	For electric defrost, except 8 fin models									
7270	on applications below 35° F, add				Ea.	36%				
8000	Freezers, pass-thru door uprights, walk-in storage									
8300	Low temperature, thin profile, electric defrost									
8310	138-1800 CFM, 5 fins per inch									
8320	900 BTUH, one 6" fan	Q-5	3.30	4.848	Ea.	360	215		575	720
8340	1500 BTUH, two 6" fans		2.20	7.273		535	325		860	1,075
8360	2600 BTUH, three 6" fans		1.90	8.421		770	375		1,145	1,400
8370	3300 BTUH, four 6" fans		1.70	9.412		815	420		1,235	1,525
8380	4400 BTUH, five 6" fans		1.40	11.429		1,200	510		1,710	2,075
8400	7000 BTUH, three 10" fans		1.30	12.308		1,450	545		1,995	2,425
8410	8700 BTUH, four 10" fans		1.10	14.545		1,650	645		2,295	2,775
8450	For air defrost, deduct					176			176	194

23 81 Decentralized Unitary HVAC Equipment

23 81 13 – Packaged Terminal Air-Conditioners

23 81 13.10 Packaged Cabinet Type Air-Conditioners		Crew	Daily Output	Labor-Hours	Unit	Material	2009 Bare Costs Labor	Equipment	Total	Total Incl O&P
0010	**PACKAGED CABINET TYPE AIR-CONDITIONERS**, Cabinet, wall sleeve,									
0100	louver, electric heat, thermostat, manual changeover, 208 V									
0200	6,000 BTUH cooling, 8800 BTU heat	Q-5	6	2.667	Ea.	625	118		743	870
0220	9,000 BTUH cooling, 13,900 BTU heat		5	3.200		720	142		862	1,000
0240	12,000 BTUH cooling, 13,900 BTU heat		4	4		780	178		958	1,125
0260	15,000 BTUH cooling, 13,900 BTU heat		3	5.333		865	237		1,102	1,300
0280	18,000 BTUH cooling, 10 KW heat		2.40	6.667		1,575	296		1,871	2,200
0300	24,000 BTUH cooling, 10 KW heat		1.90	8.421		1,725	375		2,100	2,450
0320	30,000 BTUH cooling, 10 KW heat		1.40	11.429		1,875	510		2,385	2,825
0340	36,000 BTUH cooling, 10 KW heat		1.25	12.800		1,900	570		2,470	2,950
0360	42,000 BTUH cooling, 10 KW heat		1	16		2,400	710		3,110	3,700
0380	48,000 BTUH cooling, 10 KW heat		.90	17.778		2,600	790		3,390	4,025
0500	For hot water coil, increase heat by 10%, add					5%	10%			
1000	For steam, increase heat output by 30%, add					8%	10%			

23 81 19 – Self-Contained Air-Conditioners

23 81 19.10 Window Unit Air Conditioners		Crew	Daily Output	Labor-Hours	Unit	Material	Labor	Equipment	Total	Total Incl O&P
0010	**WINDOW UNIT AIR CONDITIONERS**									
4000	Portable/window, 15 amp 125V grounded receptacle required									
4060	5000 BTUH	1 Carp	8	1	Ea.	163	40		203	241
4340	6000 BTUH		8	1		265	40		305	355
4480	8000 BTUH		6	1.333		310	53.50		363.50	430
4500	10,000 BTUH		6	1.333		395	53.50		448.50	520
4520	12,000 BTUH	L-2	8	2		510	70.50		580.50	670
4600	Window/thru-the-wall, 15 amp 230V grounded receptacle required									
4780	17,000 BTUH	L-2	6	2.667	Ea.	695	93.50		788.50	910
4940	25,000 BTUH		4	4		880	141		1,021	1,200
4960	29,000 BTUH		4	4		950	141		1,091	1,275

23 81 19.20 Self-Contained Single Package		Crew	Daily Output	Labor-Hours	Unit	Material	Labor	Equipment	Total	Total Incl O&P
0010	**SELF-CONTAINED SINGLE PACKAGE**									
0100	Air cooled, for free blow or duct, not incl. remote condenser									
0110	Constant volume									
0200	3 ton cooling	Q-5	1	16	Ea.	3,200	710		3,910	4,600
0210	4 ton cooling	"	.80	20		3,500	890		4,390	5,175
0220	5 ton cooling	Q-6	1.20	20		3,825	920		4,745	5,575
0230	7.5 ton cooling	"	.90	26.667		4,975	1,225		6,200	7,325
0240	10 ton cooling	Q-7	1	32		6,600	1,500		8,100	9,500
0250	15 ton cooling		.95	33.684		9,200	1,575		10,775	12,500
0260	20 ton cooling		.90	35.556		13,300	1,675		14,975	17,100
0270	25 ton cooling		.85	37.647		15,400	1,775		17,175	19,600
0280	30 ton cooling		.80	40		19,100	1,875		20,975	23,800
0300	40 ton cooling		.60	53.333		35,200	2,500		37,700	42,500
0320	50 ton cooling		.50	64		44,100	3,000		47,100	53,000
0340	60 ton cooling	Q-8	.40	80		47,200	3,750	139	51,089	58,000
0380	With hot water heat, variable air volume									
0390	10 ton cooling	Q-7	.96	33.333	Ea.	9,275	1,575		10,850	12,600
0400	20 ton cooling		.86	37.209		18,000	1,750		19,750	22,400
0410	30 ton cooling		.76	42.105		25,900	1,975		27,875	31,400
0420	40 ton cooling		.58	55.172		45,800	2,600		48,400	54,500
0430	50 ton cooling		.48	66.667		56,500	3,125		59,625	66,500
0440	60 ton cooling	Q-8	.38	84.211		63,500	3,950	147	67,597	76,000
0490	For duct mounting no price change									
0500	For steam heating coils, add				Ea.	10%	10%			

23 81 Decentralized Unitary HVAC Equipment

23 81 19 – Self-Contained Air-Conditioners

23 81 19.20 Self-Contained Single Package		Crew	Daily Output	Labor-Hours	Unit	2009 Bare Costs Material	Labor	Equipment	Total	Total Incl O&P
0550	Packaged, with electric heat									
0552	Constant volume									
0560	5 ton cooling	Q-6	1.20	20	Ea.	7,050	920		7,970	9,125
0570	10 ton cooling	Q-7	1	32		11,500	1,500		13,000	15,000
0580	20 ton cooling		.90	35.556		20,300	1,675		21,975	24,800
0590	30 ton cooling		.80	40		30,900	1,875		32,775	36,800
0600	40 ton cooling		.60	53.333		47,900	2,500		50,400	56,500
0610	50 ton cooling		.50	64		62,000	3,000		65,000	72,500
0700	Variable air volume									
0710	10 ton cooling	Q-7	1	32	Ea.	13,600	1,500		15,100	17,200
0720	20 ton cooling		.90	35.556		23,800	1,675		25,475	28,600
0730	30 ton cooling		.80	40		35,800	1,875		37,675	42,200
0740	40 ton cooling		.60	53.333		55,000	2,500		57,500	64,500
0750	50 ton cooling		.50	64		69,500	3,000		72,500	81,000
0760	60 ton cooling	Q-8	.40	80		75,000	3,750	139	78,889	88,500
1000	Water cooled for free blow or duct, not including tower									
1010	Constant volume									
1100	3 ton cooling	Q-6	1	24	Ea.	3,075	1,100		4,175	5,050
1120	5 ton cooling		1	24		4,025	1,100		5,125	6,075
1130	7.5 ton cooling		.80	30		5,725	1,375		7,100	8,375
1140	10 ton cooling	Q-7	.90	35.556		7,900	1,675		9,575	11,200
1150	15 ton cooling		.85	37.647		11,400	1,775		13,175	15,200
1160	20 ton cooling		.80	40		25,800	1,875		27,675	31,100
1170	25 ton cooling		.75	42.667		30,400	2,000		32,400	36,400
1180	30 ton cooling		.70	45.714		34,200	2,150		36,350	40,800
1200	40 ton cooling		.40	80		43,300	3,750		47,050	53,500
1220	50 ton cooling		.30	106		54,000	5,025		59,025	66,500
1240	60 ton cooling	Q-8	.30	106		63,000	5,025	186	68,211	76,500
1300	For hot water or steam heat coils, add					12%	10%			
2000	Water cooled with electric heat, not including tower									
2010	Constant volume									
2100	5 ton cooling	Q-6	1.20	20	Ea.	7,250	920		8,170	9,350
2120	10 ton cooling	Q-7	1	32		12,800	1,500		14,300	16,400
2130	15 ton cooling	"	.90	35.556		18,400	1,675		20,075	22,800
2400	Variable air volume									
2440	10 ton cooling	Q-7	1	32	Ea.	14,900	1,500		16,400	18,600
2450	15 ton cooling	"	.90	35.556	"	21,900	1,675		23,575	26,600
2600	Water cooled, hot water coils, not including tower									
2610	Variable air volume									
2640	10 ton cooling	Q-7	1	32	Ea.	10,700	1,500		12,200	14,100
2650	15 ton cooling	"	.80	40	"	16,000	1,875		17,875	20,400

23 81 23 – Computer-Room Air-Conditioners

23 81 23.10 Computer Room Units		Crew	Daily Output	Labor-Hours	Unit	Material	Labor	Equipment	Total	Total Incl O&P
0010	**COMPUTER ROOM UNITS** D3050–185									
1000	Air cooled, includes remote condenser but not									
1020	interconnecting tubing or refrigerant									
1080	3 ton	Q-5	.50	32	Ea.	15,600	1,425		17,025	19,300
1120	5 ton		.45	35.556		16,700	1,575		18,275	20,800
1160	6 ton		.30	53.333		29,500	2,375		31,875	36,100
1200	8 ton		.27	59.259		31,400	2,625		34,025	38,500
1240	10 ton		.25	64		32,800	2,850		35,650	40,400
1280	15 ton		.22	72.727		36,100	3,225		39,325	44,600

23 81 Decentralized Unitary HVAC Equipment

23 81 23 – Computer-Room Air-Conditioners

23 81 23.10 Computer Room Units		Crew	Daily Output	Labor-Hours	Unit	2009 Bare Costs Material	Labor	Equipment	Total	Total Incl O&P
1290	18 ton	Q-5	.20	80	Ea.	40,200	3,550		43,750	49,500
1300	20 ton	Q-6	.26	92.308		43,200	4,250		47,450	54,000
1320	22 ton		.24	100		43,800	4,600		48,400	55,000
1360	30 ton		.21	114		54,000	5,275		59,275	67,500
2200	Chilled water, for connection to									
2220	existing chiller system of adequate capacity									
2260	5 ton	Q-5	.74	21.622	Ea.	12,000	960		12,960	14,700
2280	6 ton		.52	30.769		12,000	1,375		13,375	15,300
2300	8 ton		.50	32		12,100	1,425		13,525	15,400
2320	10 ton		.49	32.653		12,200	1,450		13,650	15,600
2330	12 ton		.49	32.990		12,500	1,475		13,975	16,000
2360	15 ton		.48	33.333		12,900	1,475		14,375	16,400
2400	20 ton		.46	34.783		13,600	1,550		15,150	17,300
2440	25 ton	Q-6	.63	38.095		14,500	1,750		16,250	18,600
2460	30 ton		.57	42.105		15,700	1,950		17,650	20,200
2480	35 ton		.52	46.154		15,700	2,125		17,825	20,500
2500	45 ton		.46	52.174		16,800	2,400		19,200	22,100
2520	50 ton		.42	57.143		19,400	2,625		22,025	25,300
2540	60 ton		.38	63.158		20,600	2,900		23,500	27,000
4000	Glycol system, complete except for interconnecting tubing									
4060	3 ton	Q-5	.40	40	Ea.	19,700	1,775		21,475	24,400
4100	5 ton		.38	42.105		21,600	1,875		23,475	26,500
4120	6 ton		.25	64		31,600	2,850		34,450	39,000
4140	8 ton		.23	69.565		34,700	3,100		37,800	42,800
4160	10 ton		.21	76.190		36,900	3,375		40,275	45,600
4200	15 ton	Q-6	.26	92.308		45,700	4,250		49,950	56,500
4240	20 ton		.24	100		50,500	4,600		55,100	62,500
4280	22 ton		.23	104		53,000	4,800		57,800	65,000
4430	30 ton		.22	109		65,500	5,025		70,525	79,500
8000	Water cooled system, not including condenser,									
8020	water supply or cooling tower									
8060	3 ton	Q-5	.62	25.806	Ea.	16,000	1,150		17,150	19,200
8100	5 ton		.54	29.630		17,400	1,325		18,725	21,100
8120	6 ton		.35	45.714		27,100	2,025		29,125	32,900
8140	8 ton		.33	48.485		28,500	2,150		30,650	34,500
8160	10 ton		.31	51.613		29,500	2,300		31,800	35,900
8200	15 ton		.27	59.259		34,500	2,625		37,125	42,000
8240	20 ton	Q-6	.38	63.158		37,200	2,900		40,100	45,300
8280	22 ton		.35	68.571		38,900	3,150		42,050	47,600
8300	30 ton		.30	80		48,500	3,675		52,175	59,000

23 81 26 – Split-System Air-Conditioners

23 81 26.10 Split Ductless Systems		Crew	Daily Output	Labor-Hours	Unit	Material	Labor	Equipment	Total	Total Incl O&P
0010	**SPLIT DUCTLESS SYSTEMS**									
0100	Cooling only, single zone									
0110	Wall mount									
0120	3/4 ton cooling	Q-5	2	8	Ea.	1,175	355		1,530	1,800
0130	1 ton cooling		1.80	8.889		1,300	395		1,695	2,025
0140	1-1/2 ton cooling		1.60	10		1,800	445		2,245	2,650
0150	2 ton cooling		1.40	11.429		2,125	510		2,635	3,100
1000	Ceiling mount									
1020	2 ton cooling	Q-5	1.40	11.429	Ea.	1,450	510		1,960	2,325
1030	3 ton cooling	"	1.20	13.333	"	3,900	590		4,490	5,200

23 81 Decentralized Unitary HVAC Equipment

23 81 26 – Split-System Air-Conditioners

23 81 26.10 Split Ductless Systems		Crew	Daily Output	Labor-Hours	Unit	Material	2009 Bare Costs Labor	Equipment	Total	Total Incl O&P
2000	T-Bar mount									
2010	2 ton cooling	Q-5	1.40	11.429	Ea.	3,400	510		3,910	4,500
2020	3 ton cooling		1.20	13.333		3,800	590		4,390	5,075
2030	3-1/2 ton cooling		1.10	14.545		4,200	645		4,845	5,600
3000	Multizone									
3010	Wall mount									
3020	2 @ 3/4 ton cooling	Q-5	1.80	8.889	Ea.	1,475	395		1,870	2,225
5000	Cooling / Heating									
5010	Wall mount									
5110	1 ton cooling	Q-5	1.70	9.412	Ea.	1,100	420		1,520	1,825
5120	1-1/2 ton cooling	"	1.50	10.667	"	1,750	475		2,225	2,625
5300	Ceiling mount									
5310	3 ton cooling	Q-5	1	16	Ea.	4,125	710		4,835	5,625
7000	Accessories for all split ductless systems									
7010	Add for ambient frost control	Q-5	8	2	Ea.	130	89		219	276
7020	Add for tube / wiring kit									
7030	15' kit	Q-5	32	.500	Ea.	38.50	22		60.50	76
7040	35' kit	"	24	.667	"	125	29.50		154.50	182

23 81 43 – Air-Source Unitary Heat Pumps

23 81 43.10 Air-Source Heat Pumps		Crew	Daily Output	Labor-Hours	Unit	Material	Labor	Equipment	Total	Total Incl O&P
0010	**AIR-SOURCE HEAT PUMPS**, Not including interconnecting tubing.									
1000	Air to air, split system, not including curbs, pads, fan coil and ductwork									
1010	For curbs/pads see Div. 23 91 00									
1012	Outside condensing unit only, for fan coil see Div. 23 82 19.10									
1015	1.5 ton cooling, 7 MBH heat @ 0° F	Q-5	2.40	6.667	Ea.	1,600	296		1,896	2,200
1020	2 ton cooling, 8.5 MBH heat @ 0° F		2	8		1,650	355		2,005	2,350
1030	2.5 ton cooling, 10 MBH heat @ 0° F		1.60	10		1,850	445		2,295	2,700
1040	3 ton cooling, 13 MBH heat @ 0° F		1.20	13.333		2,100	590		2,690	3,200
1050	3.5 ton cooling, 18 MBH heat @ 0° F		1	16		2,250	710		2,960	3,550
1054	4 ton cooling, 24 MBH heat @ 0° F		.80	20		2,450	890		3,340	4,025
1060	5 ton cooling, 27 MBH heat @ 0° F		.50	32		2,675	1,425		4,100	5,075
1080	7.5 ton cooling, 33 MBH heat @ 0° F		.45	35.556		5,050	1,575		6,625	7,925
1100	10 ton cooling, 50 MBH heat @ 0° F	Q-6	.64	37.500		7,050	1,725		8,775	10,400
1120	15 ton cooling, 64 MBH heat @ 0° F		.50	48		10,200	2,200		12,400	14,500
1130	20 ton cooling, 85 MBH heat @ 0° F		.35	68.571		15,300	3,150		18,450	21,600
1140	25 ton cooling, 119 MBH heat @ 0° F		.25	96		18,400	4,425		22,825	26,900
1500	Single package, not including curbs, pads, or plenums									
1502	1/2 ton cooling, supplementary heat not incl.	Q-5	8	2	Ea.	1,675	89		1,764	1,975
1504	3/4 ton cooling, supplementary heat not incl.		6	2.667		1,850	118		1,968	2,200
1506	1 ton cooling, supplementary heat not incl.		4	4		2,025	178		2,203	2,500
1510	1.5 ton cooling, 5 MBH heat @ 0° F		1.55	10.323		2,400	460		2,860	3,350
1520	2 ton cooling, 6.5 MBH heat @ 0° F		1.50	10.667		2,575	475		3,050	3,525
1540	2.5 ton cooling, 8 MBH heat @ 0° F		1.40	11.429		2,800	510		3,310	3,850
1560	3 ton cooling, 10 MBH heat @ 0° F		1.20	13.333		3,050	590		3,640	4,250
1570	3.5 ton cooling, 11 MBH heat @ 0° F		1	16		3,375	710		4,085	4,800
1580	4 ton cooling, 13 MBH heat @ 0° F		.96	16.667		3,650	740		4,390	5,125
1620	5 ton cooling, 27 MBH heat @ 0° F		.65	24.615		3,900	1,100		5,000	5,950
1640	7.5 ton cooling, 35 MBH heat @ 0° F		.40	40		6,450	1,775		8,225	9,775
1648	10 ton cooling, 45 MBH heat @ 0° F	Q-6	.40	60		8,350	2,775		11,125	13,300
1652	12 ton cooling, 50 MBH heat @ 0° F	"	.36	66.667		9,625	3,075		12,700	15,200
1696	Supplementary electric heat coil incl., except as noted									

23 81 Decentralized Unitary HVAC Equipment

23 81 46 – Water-Source Unitary Heat Pumps

23 81 46.10 Water Source Heat Pumps		Crew	Daily Output	Labor-Hours	Unit	2009 Bare Costs Material	Labor	Equipment	Total	Total Incl O&P
0010	**WATER SOURCE HEAT PUMPS**, Not incl. connecting tubing or water source									
2000	Water source to air, single package									
2100	1 ton cooling, 13 MBH heat @ 75° F	Q-5	2	8	Ea.	1,300	355		1,655	1,950
2120	1.5 ton cooling, 17 MBH heat @ 75° F		1.80	8.889		1,525	395		1,920	2,275
2140	2 ton cooling, 19 MBH heat @ 75° F		1.70	9.412		1,550	420		1,970	2,325
2160	2.5 ton cooling, 25 MBH heat @ 75° F		1.60	10		1,675	445		2,120	2,525
2180	3 ton cooling, 27 MBH heat @ 75° F		1.40	11.429		1,800	510		2,310	2,725
2190	3.5 ton cooling, 29 MBH heat @ 75° F		1.30	12.308		1,875	545		2,420	2,900
2200	4 ton cooling, 31 MBH heat @ 75° F		1.20	13.333		2,100	590		2,690	3,200
2220	5 ton cooling, 29 MBH heat @ 75° F		.90	17.778		2,450	790		3,240	3,875
2240	7.5 ton cooling, 35 MBH heat @ 75° F		.60	26.667		6,100	1,175		7,275	8,475
2250	8.5 ton cooling, 40 MBH heat @ 75° F		.58	27.586		6,475	1,225		7,700	8,975
2260	10 ton cooling, 50 MBH heat @ 75° F		.53	30.189		6,700	1,350		8,050	9,400
2280	15 ton cooling, 64 MBH heat @ 75° F	Q-6	.47	51.064		14,000	2,350		16,350	18,900
2300	20 ton cooling, 100 MBH heat @ 75° F		.41	58.537		15,300	2,700		18,000	20,900
2310	25 ton cooling, 100 MBH heat @ 75° F		.32	75		20,800	3,450		24,250	28,100
2320	30 ton cooling, 128 MBH heat @ 75° F		.24	102		22,800	4,700		27,500	32,200
2340	40 ton cooling, 200 MBH heat @ 75° F		.21	117		32,200	5,400		37,600	43,500
2360	50 ton cooling, 200 MBH heat @ 75° F		.15	160		36,400	7,375		43,775	51,000
3960	For supplementary heat coil, add					10%				
4000	For increase in capacity thru use									
4020	of solar collector, size boiler at 60%									

23 82 Convection Heating and Cooling Units

23 82 13 – Valance Heating and Cooling Units

23 82 13.16 Valance Units		Crew	Daily Output	Labor-Hours	Unit	Material	Labor	Equipment	Total	Total Incl O&P
0010	**VALANCE UNITS**									
6000	Valance units, complete with 1/2" cooling coil, enclosure									
6020	2 tube	Q-5	18	.889	L.F.	29.50	39.50		69	92
6040	3 tube		16	1		35	44.50		79.50	105
6060	4 tube		16	1		40	44.50		84.50	111
6080	5 tube		15	1.067		45.50	47.50		93	121
6100	6 tube		15	1.067		50.50	47.50		98	127
6120	8 tube		14	1.143		73	51		124	157
6200	For 3/4" cooling coil, add					10%				

23 82 16 – Air Coils

23 82 16.10 Flanged Coils		Crew	Daily Output	Labor-Hours	Unit	Material	Labor	Equipment	Total	Total Incl O&P
0010	**FLANGED COILS**									
0100	Basic water, DX, or condenser coils									
0110	Copper tubes, alum. fins, galv. end sheets									
0112	H is finned height, L is finned length									
0120	3/8" x .016 tube, .0065 aluminum fins									
0130	2 row, 8 fins per inch									
0140	4" H x 12" L	Q-5	48	.333	Ea.	570	14.80		584.80	645
0150	4" H x 24" L		38.38	.417		590	18.50		608.50	680
0160	4" H x 48" L		19.25	.831		625	37		662	745
0170	4" H x 72" L		12.80	1.250		810	55.50		865.50	975
0180	6" H x 12" L		48	.333		550	14.80		564.80	625
0190	6" H x 24" L		25.60	.625		575	28		603	670
0200	6" H x 48" L		12.80	1.250		620	55.50		675.50	770

23 82 16.10 Flanged Coils		Crew	Daily Output	Labor-Hours	Unit	Material	2009 Bare Costs Labor	Equipment	Total	Total Incl O&P
0210	6" H x 72" L	Q-5	8.53	1.876	Ea.	820	83.50		903.50	1,025
0220	10" H x 12" L		30.73	.521		585	23		608	680
0230	10" H x 24" L		15.33	1.044		620	46.50		666.50	755
0240	10" H x 48" L		7.69	2.081		695	92.50		787.50	900
0250	10" H x 72" L		5.12	3.125		920	139		1,059	1,225
0260	12" H x 12" L		25.60	.625		600	28		628	700
0270	12" H x 24" L		12.80	1.250		640	55.50		695.50	790
0280	12" H x 48" L		6.40	2.500		725	111		836	960
0290	12" H x 72" L		4.27	3.747		970	166		1,136	1,325
0300	16" H x 16" L		14.38	1.113		655	49.50		704.50	795
0310	16" H x 24" L		9.59	1.668		690	74		764	870
0320	16" H x 48" L		4.80	3.333		870	148		1,018	1,175
0330	16" H x 72" L		3.20	5		1,075	222		1,297	1,500
0340	20" H x 24" L		7.69	2.081		765	92.50		857.50	980
0350	20" H x 48" L		3.84	4.167		975	185		1,160	1,350
0360	20" H x 72" L		2.56	6.250		1,200	278		1,478	1,750
0370	24" H x 24" L		6.40	2.500		825	111		936	1,075
0380	24" H x 48" L		3.20	5		1,075	222		1,297	1,500
0390	24" H x 72" L		2.13	7.512		1,325	335		1,660	1,950
0400	30" H x 28" L		4.39	3.645		1,025	162		1,187	1,400
0410	30" H x 48" L		2.56	6.250		1,225	278		1,503	1,775
0420	30" H x 72" L		1.71	9.357		1,525	415		1,940	2,300
0430	30" H x 84" L		1.46	10.959		1,650	485		2,135	2,550
0440	34" H x 32" L		3.39	4.720		1,150	210		1,360	1,575
0450	34" H x 48" L		2.26	7.080		1,300	315		1,615	1,925
0460	34" H x 72" L		1.51	10.596		1,625	470		2,095	2,500
0470	34" H x 84" L		1.29	12.403		1,800	550		2,350	2,800
0600	For 10 fins per inch, add					1%				
0610	For 12 fins per inch, add					3%				
0620	For 4 row, add				Ea.	30%	50%			
0630	For 6 row, add					60%	100%			
0640	For 8 row, add					80%	150%			
1100	1/2" x .017 tube, .0065 aluminum fins									
1110	2 row, 8 fins per inch									
1120	5" H x 5" L	Q-5	40	.400	Ea.	565	17.75		582.75	645
1130	5" H x 20" L		37.10	.431		595	19.15		614.15	685
1140	5" H x 45" L		16.41	.975		650	43.50		693.50	780
1150	5" H x 90" L		8.21	1.949		950	86.50		1,036.50	1,175
1160	7.5" H x 5" L		38.50	.416		545	18.45		563.45	630
1170	7.5" H x 20" L		24.62	.650		590	29		619	690
1180	7.5" H x 45" L		10.94	1.463		660	65		725	825
1190	7.5" H x 90" L		5.46	2.930		1,000	130		1,130	1,300
1200	12.5" H x 10" L		29.49	.543		620	24		644	715
1210	12.5" H x 20" L		14.75	1.085		665	48		713	805
1220	12.5" H x 45" L		6.55	2.443		775	109		884	1,025
1230	12.5" H x 90" L		3.28	4.878		1,200	217		1,417	1,650
1240	15" H x 15" L		16.41	.975		665	43.50		708.50	800
1250	15" H x 30" L		8.21	1.949		745	86.50		831.50	950
1260	15" H x 45" L		5.46	2.930		820	130		950	1,100
1270	15" H x 90" L		2.73	5.861		1,300	260		1,560	1,825
1280	20" H x 20" L		9.22	1.735		760	77		837	950
1290	20" H x 45" L		4.10	3.902		1,000	173		1,173	1,350
1300	20" H x 90" L		2.05	7.805		1,475	345		1,820	2,150

23 82 16.10 Flanged Coils		Crew	Daily Output	Labor-Hours	Unit	Material	Labor	Equipment	Total	Total Incl O&P
						2009 Bare Costs				
1310	25" H x 25" L	Q-5	5.90	2.712	Ea.	905	120		1,025	1,175
1320	25" H x 45" L		3.28	4.878		1,150	217		1,367	1,600
1330	25" H x 90" L		1.64	9.756		1,700	435		2,135	2,525
1340	30" H x 30" L		4.10	3.902		1,100	173		1,273	1,475
1350	30" H x 45" L		2.73	5.861		1,275	260		1,535	1,800
1360	30" H x 90" L		1.37	11.679		1,900	520		2,420	2,875
1370	35" H x 35" L		3.01	5.316		1,275	236		1,511	1,750
1380	35" H x 45" L		2.34	6.838		1,400	305		1,705	1,975
1390	35" H x 90" L		1.17	13.675		2,100	610		2,710	3,200
1400	42.5" H x 45" L		1.93	8.290		1,625	370		1,995	2,325
1410	42.5" H x 75" L		1.16	13.793		2,150	615		2,765	3,300
1420	42.5" H x 105" L		.83	19.277		2,650	855		3,505	4,200
1600	For 10 fins per inch, add					2%				
1610	For 12 fins per inch, add					5%				
1620	For 4 row, add					35%	50%			
1630	For 6 row, add					65%	100%			
1640	For 8 row, add					90%	150%			
2010	5/8" x .020" tube, .0065 aluminum fins									
2020	2 row, 8 fins per inch									
2030	6" H x 6" L	Q-5	38	.421	Ea.	580	18.70		598.70	670
2040	6" H x 24" L		24.96	.641		630	28.50		658.50	735
2050	6" H x 48" L		12.48	1.282		690	57		747	845
2060	6" H x 96" L		6.24	2.564		1,025	114		1,139	1,300
2070	12" H x 12" L		24.96	.641		605	28.50		633.50	710
2080	12" H x 24" L		12.48	1.282		660	57		717	815
2090	12" H x 48" L		6.24	2.564		770	114		884	1,025
2100	12" H x 96" L		3.12	5.128		1,225	228		1,453	1,700
2110	18" H x 24" L		8.32	1.923		780	85.50		865.50	985
2120	18" H x 48" L		4.16	3.846		1,025	171		1,196	1,375
2130	18" H x 96" L		2.08	7.692		1,500	340		1,840	2,175
2140	24" H x 24" L		6.24	2.564		865	114		979	1,125
2150	24" H x 48" L		3.12	5.128		1,150	228		1,378	1,625
2160	24" H x 96" L		1.56	10.256		1,750	455		2,205	2,600
2170	30" H x 30" L		3.99	4.010		1,125	178		1,303	1,525
2180	30" H x 48" L		2.50	6.400		1,350	284		1,634	1,900
2190	30" H x 96" L		1.25	12.800		2,050	570		2,620	3,100
2200	30" H x 120" L		1	16		2,400	710		3,110	3,700
2210	36" H x 36" L		2.77	5.776		1,350	257		1,607	1,875
2220	36" H x 48" L		2.08	7.692		1,500	340		1,840	2,175
2230	36" H x 90" L		1.11	14.414		2,225	640		2,865	3,400
2240	36" H x 120" L		.83	19.277		2,700	855		3,555	4,250
2250	42" H x 48" L		1.78	8.989		1,675	400		2,075	2,425
2260	42" H x 90" L		.95	16.842		2,475	750		3,225	3,850
2270	42" H x 120" L		.71	22.535		3,025	1,000		4,025	4,825
2280	51" H x 48" L		1.47	10.884		1,950	485		2,435	2,875
2290	51" H x 96" L		.73	21.918		3,025	975		4,000	4,775
2300	51" H x 120" L		.59	27.119		3,550	1,200		4,750	5,700
2500	For 10 fins per inch, add					3%				
2510	For 12 fins per inch, add					6%				
2520	For 4 row, add				Ea.	40%	50%			
2530	For 6 row, add					80%	100%			
2540	For 8 row, add					110%	150%			
3000	Hot water booster coils									

23 82 16 – Air Coils

23 82 16.10 Flanged Coils		Crew	Daily Output	Labor-Hours	Unit	Material	Labor	Equipment	Total	Total Incl O&P
						2009 Bare Costs				
3010	Copper tubes, alum. fins, galv. end sheets									
3012	H is finned height, L is finned length									
3020	1/2" x .017" tube, .0065 Al fins									
3030	1 row, 10 fins per inch									
3040	5"H x 10"L	Q-5	40	.400	Ea.	265	17.75		282.75	320
3050	5"H x 25"L		33.18	.482		290	21.50		311.50	350
3060	10"H x 10"L		39	.410		292	18.25		310.25	350
3070	10"H x 20"L		20.73	.772		315	34.50		349.50	395
3080	10"H x 30"L		13.83	1.157		340	51.50		391.50	445
3090	15"H x 15"L		18.44	.868		330	38.50		368.50	425
3110	15"H x 20"L		13.83	1.157		345	51.50		396.50	455
3120	15"H x 30"L		9.22	1.735		380	77		457	535
3130	20"H x 20"L		10.37	1.543		380	68.50		448.50	520
3140	20"H x 30"L		6.91	2.315		420	103		523	620
3150	20"H x 40"L		5.18	3.089		465	137		602	715
3160	25"H x 25"L		6.64	2.410		435	107		542	640
3170	25"H x 35"L		4.74	3.376		555	150		705	835
3180	25"H x 45"L		3.69	4.336		625	193		818	980
3190	30"H x 30"L		4.61	3.471		570	154		724	860
3200	30"H x 40"L		3.46	4.624		655	205		860	1,025
3210	30"H x 50"L		2.76	5.797		730	258		988	1,175
3300	For 2 row, add					20%	50%			
3400	1/2" x .020" tube, .008 Al fins									
3410	2 row, 10 fins per inch									
3420	20" H x 27" L	Q-5	7.27	2.201	Ea.	560	98		658	760
3430	20" H x 36" L		5.44	2.941		635	131		766	895
3440	20" H x 45" L		4.35	3.678		790	163		953	1,100
3450	25" H x 45" L		3.49	4.585		910	204		1,114	1,300
3460	25" H x 54" L		2.90	5.517		1,075	245		1,320	1,575
3470	30" H x 54" L		2.42	6.612		1,225	294		1,519	1,800
3480	35" H x 54" L		2.07	7.729		1,375	345		1,720	2,025
3490	35" H x 63" L		1.78	8.989		1,500	400		1,900	2,250
3500	40" H x 72" L		1.36	11.765		1,850	525		2,375	2,800
3510	55" H x 63" L		1.13	14.159		2,175	630		2,805	3,350
3520	55" H x 72" L		.99	16.162		2,400	720		3,120	3,725
3530	55" H x 81" L		.88	18.182		2,650	810		3,460	4,125
3540	75" H x 81" L		.64	25		3,450	1,100		4,550	5,475
3550	75" H x 90" L		.58	27.586		3,775	1,225		5,000	6,000
3560	75" H x 108" L		.48	33.333		4,425	1,475		5,900	7,100
3800	5/8" x .020 tubing, .0065 aluminum fins									
3810	1 row, 10 fins per inch									
3820	6" H x 12" L	Q-5	38	.421	Ea.	296	18.70		314.70	355
3830	6" H x 24" L		28.80	.556		295	24.50		319.50	360
3840	6" H x 30" L		23.04	.694		305	31		336	380
3850	12" H x 12" L		28.80	.556		305	24.50		329.50	370
3860	12" H x 24" L		14.40	1.111		330	49.50		379.50	440
3870	12" H x 30" L		11.52	1.389		345	61.50		406.50	475
3880	18" H x 18" L		12.80	1.250		350	55.50		405.50	470
3890	18" H x 24" L		9.60	1.667		370	74		444	515
3900	18" H x 36" L		6.40	2.500		410	111		521	615
3910	24" H x 24" L		7.20	2.222		405	98.50		503.50	595
3920	24" H x 30" L		4.80	3.333		535	148		683	810
3930	24" H x 36" L		5.05	3.168		580	141		721	850

23 82 16.10 Flanged Coils		Crew	Daily Output	Labor-Hours	Unit	Material	Labor	Equipment	Total	Total Incl O&P
						2009 Bare Costs				
3940	30" H x 30" L	Q-5	4.61	3.471	Ea.	600	154		754	890
3950	30" H x 36" L		3.85	4.159		650	185		835	990
3960	30" H x 42" L		3.29	4.863		705	216		921	1,100
3970	36" H x 36" L		3.20	5		720	222		942	1,125
3980	36" H x 42" L		2.74	5.839		785	259		1,044	1,250
4100	For 2 row, add					30%	50%			
4400	Steam and hot water heating coils									
4410	Copper tubes, alum. fins. galv. end sheets									
4412	H is finned height, L is finned length									
4420	5/8" tubing, to 175 psig working pressure									
4430	1 row, 8 fins per inch									
4440	6"H x 6"L	Q-5	38	.421	Ea.	550	18.70		568.70	635
4450	6"H x 12"L		37	.432		565	19.20		584.20	650
4460	6"H x 24"L		24.48	.654		590	29		619	690
4470	6"H x 36"L		16.32	.980		610	43.50		653.50	735
4480	6"H x 48"L		12.24	1.307		635	58		693	785
4490	6"H x 84"L		6.99	2.289		1,225	102		1,327	1,500
4500	6" H x 108" L		5.44	2.941		1,375	131		1,506	1,700
4510	12" H x 12" L		24.48	.654		665	29		694	775
4520	12" H x 24" L		12.24	1.307		705	58		763	860
4530	12" H x 36" L		8.16	1.961		750	87		837	955
4540	12" H x 48" L		6.12	2.614		865	116		981	1,125
4550	12" H x 84" L		3.50	4.571		1,525	203		1,728	1,975
4560	12" H x 108" L		2.72	5.882		1,750	261		2,011	2,325
4570	18" H x18" L		10.88	1.471		795	65.50		860.50	975
4580	18" H x 24" L		8.16	1.961		825	87		912	1,025
4590	18" H x 48" L		4.08	3.922		1,025	174		1,199	1,375
4600	18" H x 84" L		2.33	6.867		1,825	305		2,130	2,475
4610	18" H x 108" L		1.81	8.840		2,125	395		2,520	2,925
4620	24" H x 24" L		6.12	2.614		1,000	116		1,116	1,275
4630	24" H x 48" L		3.06	5.229		1,200	232		1,432	1,675
4640	24" H x 84" L		1.75	9.143		2,150	405		2,555	2,950
4650	24" H x 108" L		1.36	11.765		2,500	525		3,025	3,525
4660	36" H x 36" L		2.72	5.882		1,400	261		1,661	1,925
4670	36" H x 48" L		2.04	7.843		1,525	350		1,875	2,200
4680	36" H x 84" L		1.17	13.675		2,750	610		3,360	3,925
4690	36" H x 108" L		.91	17.582		3,250	780		4,030	4,750
4700	48"H x 48"L		1.53	10.458		1,875	465		2,340	2,750
4710	48"H x 84"L		.87	18.391		3,400	815		4,215	4,975
4720	48"H x 108"L		.68	23.529		4,025	1,050		5,075	6,000
4730	48"H x 126"L		.58	27.586		4,300	1,225		5,525	6,575
4910	For 10 fins per inch, add					2%				
4920	For 12 fins per inch, add					5%				
4930	For 2 row, add					30%	50%			
4940	For 3 row, add					50%	100%			
4950	For 4 row, add					80%	123%			
4960	For 230 PSIG heavy duty, add					20%	30%			
6000	Steam, heavy duty, 200 PSIG									
6010	Copper tubes, alum. fins, galv. end sheets									
6012	H is finned height, L is finned length									
6020	1" x .035" tubing, .010 aluminum fins									
6030	1 row, 8 fins per inch									
6040	12"H x 12"L	Q-5	24.48	.654	Ea.	795	29		824	920

23 82 16.10 Flanged Coils		Crew	Daily Output	Labor-Hours	Unit	Material	2009 Bare Costs Labor	Equipment	Total	Total Incl O&P
6050	12"H x 24"L	Q-5	12.24	1.307	Ea.	900	58		958	1,075
6060	12"H x 36"L		8.16	1.961		1,000	87		1,087	1,225
6070	12"H x 48"L		6.12	2.614		1,200	116		1,316	1,475
6080	12"H x 84"L		3.50	4.571		1,675	203		1,878	2,125
6090	12"H x 108"L		2.72	5.882		1,975	261		2,236	2,575
6100	18"H x 24"L		8.16	1.961		1,125	87		1,212	1,350
6110	18"H x 36"L		5.44	2.941		1,350	131		1,481	1,675
6120	18"H x 48"L		4.08	3.922		1,525	174		1,699	1,925
6130	18"H x 84"L		2.33	6.867		2,150	305		2,455	2,800
6140	18"H x 108"L		1.81	8.840		2,575	395		2,970	3,450
6150	24"H x 24"L		6.12	2.614		1,450	116		1,566	1,775
6160	24"H x 36"L		4.08	3.922		1,625	174		1,799	2,025
6170	24"H x 48"L		3.06	5.229		1,825	232		2,057	2,375
6180	24"H x 84"L		1.75	9.143		2,625	405		3,030	3,500
6190	24"H x 108"L		1.36	11.765		3,175	525		3,700	4,275
6200	30"H x 36"L		3.26	4.908		1,875	218		2,093	2,400
6210	30"H x 48"L		2.45	6.531		2,150	290		2,440	2,800
6220	30"H x 84"L		1.40	11.429		3,100	510		3,610	4,175
6230	30"H x 108"L		1.09	14.679		3,775	650		4,425	5,125
6240	36"H x 36"L		2.72	5.882		2,150	261		2,411	2,775
6250	36"H x 48"L		2.04	7.843		2,475	350		2,825	3,250
6260	36"H x 84"L		1.17	13.675		3,600	610		4,210	4,850
6270	36"H x 108"L		.91	17.582		4,375	780		5,155	6,000
6280	36"H x 120"L		.82	19.512		4,700	865		5,565	6,475
6290	48"H x 48"L		1.53	10.458		4,525	465		4,990	5,675
6300	48"H x 84"L		.87	18.391		4,550	815		5,365	6,250
6310	48"H x 108"L		.68	23.529		5,600	1,050		6,650	7,725
6320	48"H x 120"L	↓	.61	26.230		6,025	1,175		7,200	8,375
6400	For 10 fins per inch, add					4%				
6410	For 12 fins per inch, add				↓	10%				

23 82 16.20 Duct Heaters		Crew	Daily Output	Labor-Hours	Unit	Material	Labor	Equipment	Total	Total Incl O&P
0010	**DUCT HEATERS**, Electric, 480 V, 3 Ph.									
0020	Finned tubular insert, 500 ° F									
0100	8" wide x 6" high, 4.0 kW	Q-20	16	1.250	Ea.	705	54		759	860
0120	12" high, 8.0 kW		15	1.333		1,175	58		1,233	1,375
0140	18" high, 12.0 kW		14	1.429		1,650	62		1,712	1,900
0160	24" high, 16.0 kW		13	1.538		2,100	66.50		2,166.50	2,425
0180	30" high, 20.0 kW		12	1.667		2,575	72.50		2,647.50	2,925
0300	12" wide x 6" high, 6.7 kW		15	1.333		750	58		808	915
0320	12" high, 13.3 kW		14	1.429		1,200	62		1,262	1,425
0340	18" high, 20.0 kW		13	1.538		1,700	66.50		1,766.50	1,975
0360	24" high, 26.7 kW		12	1.667		2,175	72.50		2,247.50	2,500
0380	30" high, 33.3 kW		11	1.818		2,650	79		2,729	3,050
0500	18" wide x 6" high, 13.3 kW		14	1.429		785	62		847	960
0520	12" high, 26.7 kW		13	1.538		1,375	66.50		1,441.50	1,625
0540	18" high, 40.0 kW		12	1.667		1,850	72.50		1,922.50	2,125
0560	24" high, 53.3 kW		11	1.818		2,450	79		2,529	2,825
0580	30" high, 66.7 kW		10	2		3,050	87		3,137	3,500
0700	24" wide x 6" high, 17.8 kW		13	1.538		865	66.50		931.50	1,050
0720	12" high, 35.6 kW		12	1.667		1,500	72.50		1,572.50	1,750
0740	18" high, 53.3 kW		11	1.818		2,050	79		2,129	2,375
0760	24" high, 71.1 kW	↓	10	2	↓	2,725	87		2,812	3,125

23 82 16.20 Duct Heaters		Crew	Daily Output	Labor-Hours	Unit	2009 Bare Costs Material	Labor	Equipment	Total	Total Incl O&P
0780	30" high, 88.9 kW	Q-20	9	2.222	Ea.	3,325	96.50		3,421.50	3,825
0900	30" wide x 6" high, 22.2 kW		12	1.667		920	72.50		992.50	1,100
0920	12" high, 44.4 kW		11	1.818		1,575	79		1,654	1,875
0940	18" high, 66.7 kW		10	2		2,175	87		2,262	2,525
0960	24" high, 88.9 kW		9	2.222		2,675	96.50		2,771.50	3,100
0980	30" high, 111.0 kW		8	2.500		3,525	108		3,633	4,050
1400	Note decreased kW available for									
1410	each duct size at same cost									
1420	See line 5000 for modifications and accessories									
2000	Finned tubular flange with insulated									
2020	terminal box, 500 ° F									
2100	12" wide x 36" high, 54 kW	Q-20	10	2	Ea.	3,475	87		3,562	3,950
2120	40" high, 60 kW		9	2.222		4,000	96.50		4,096.50	4,550
2200	24" wide x 36" high, 118.8 kW		9	2.222		4,150	96.50		4,246.50	4,700
2220	40" high, 132 kW		8	2.500		4,550	108		4,658	5,175
2400	36" wide x 8" high, 40 kW		11	1.818		1,750	79		1,829	2,050
2420	16" high, 80 kW		10	2		2,350	87		2,437	2,725
2440	24" high, 120 kW		9	2.222		3,075	96.50		3,171.50	3,550
2460	32" high, 160 kW		8	2.500		3,950	108		4,058	4,525
2480	36" high, 180 kW		7	2.857		4,875	124		4,999	5,550
2500	40" high, 200 kW		6	3.333		5,400	145		5,545	6,150
2600	40" wide x 8" high, 45 kW		11	1.818		1,850	79		1,929	2,150
2620	16" high, 90 kW		10	2		2,550	87		2,637	2,925
2640	24" high, 135 kW		9	2.222		3,225	96.50		3,321.50	3,700
2660	32" high, 180 kW		8	2.500		4,300	108		4,408	4,900
2680	36" high, 202.5 kW		7	2.857		4,950	124		5,074	5,650
2700	40" high, 225 kW		6	3.333		5,625	145		5,770	6,425
2800	48" wide x 8" high, 54.8 kW		10	2		1,925	87		2,012	2,250
2820	16" high, 109.8 kW		9	2.222		2,675	96.50		2,771.50	3,100
2840	24" high, 164.4 kW		8	2.500		3,400	108		3,508	3,925
2860	32" high, 219.2 kW		7	2.857		4,550	124		4,674	5,200
2880	36" high, 246.6 kW		6	3.333		5,225	145		5,370	5,975
2900	40" high, 274 kW		5	4		5,925	174		6,099	6,775
3000	56" wide x 8" high, 64 kW		9	2.222		2,200	96.50		2,296.50	2,575
3020	16" high, 128 kW		8	2.500		3,050	108		3,158	3,525
3040	24" high, 192 kW		7	2.857		3,700	124		3,824	4,275
3060	32" high, 256 kW		6	3.333		5,175	145		5,320	5,900
3080	36" high, 288kW		5	4		5,925	174		6,099	6,800
3100	40" high, 320kW		4	5		6,550	217		6,767	7,550
3200	64" wide x 8" high, 74kW		8	2.500		2,275	108		2,383	2,675
3220	16" high, 148kW		7	2.857		3,150	124		3,274	3,650
3240	24" high, 222kW		6	3.333		4,000	145		4,145	4,625
3260	32" high, 296kW		5	4		5,350	174		5,524	6,175
3280	36" high, 333kW		4	5		6,375	217		6,592	7,350
3300	40" high, 370kW		3	6.667		7,050	289		7,339	8,200
3800	Note decreased kW available for									
3820	each duct size at same cost									
5000	Duct heater modifications and accessories									
5120	T.C.O. limit auto or manual reset	Q-20	42	.476	Ea.	116	20.50		136.50	159
5140	Thermostat		28	.714		495	31		526	595
5160	Overheat thermocouple (removable)		7	2.857		700	124		824	960
5180	Fan interlock relay		18	1.111		168	48		216	259
5200	Air flow switch		20	1		146	43.50		189.50	227

23 82 Convection Heating and Cooling Units

23 82 16 – Air Coils

23 82 16.20 Duct Heaters		Crew	Daily Output	Labor-Hours	Unit	2009 Bare Costs Material	Labor	Equipment	Total	Total Incl O&P
5220	Split terminal box cover	Q-20	100	.200	Ea.	48	8.70		56.70	66.50
8000	To obtain BTU multiply kW by 3413									

23 82 19 – Fan Coil Units

23 82 19.10 Fan Coil Air Conditioning		Crew	Daily Output	Labor-Hours	Unit	Material	Labor	Equipment	Total	Total Incl O&P
0010	**FAN COIL AIR CONDITIONING** Cabinet mounted, filters, controls									
0100	Chilled water, 1/2 ton cooling	Q-5	8	2	Ea.	675	89		764	880
0110	3/4 ton cooling		7	2.286		700	102		802	920
0120	1 ton cooling		6	2.667		775	118		893	1,025
0140	1.5 ton cooling		5.50	2.909		875	129		1,004	1,150
0150	2 ton cooling		5.25	3.048		1,100	135		1,235	1,400
0160	2.5 ton cooling		5	3.200		1,450	142		1,592	1,825
0180	3 ton cooling		4	4		1,650	178		1,828	2,100
0190	7.5 ton cooling		2.70	5.926		1,950	263		2,213	2,550
0262	For hot water coil, add					40%	10%			
0300	Console, 2 pipe with electric heat									
0310	1/2 ton cooling	Q-5	8	2	Ea.	1,150	89		1,239	1,375
0315	3/4 ton cooling		7	2.286		1,250	102		1,352	1,525
0320	1 ton cooling		6	2.667		1,350	118		1,468	1,650
0330	1.5 ton cooling		5.50	2.909		1,650	129		1,779	2,025
0340	2 ton cooling		4.70	3.404		2,200	151		2,351	2,650
0345	2-1/2 ton cooling		4.70	3.404		3,175	151		3,326	3,725
0350	3 ton cooling		4	4		4,375	178		4,553	5,100
0360	4 ton cooling		4	4		4,550	178		4,728	5,275
0940	Direct expansion, for use w/air cooled condensing unit, 1.5 ton cooling		5	3.200		540	142		682	810
0950	2 ton cooling		4.80	3.333		575	148		723	855
0960	2.5 ton cooling		4.40	3.636		610	162		772	910
0970	3 ton cooling		3.80	4.211		750	187		937	1,100
0980	3.5 ton cooling		3.60	4.444		800	197		997	1,175
0990	4 ton cooling		3.40	4.706		900	209		1,109	1,300
1000	5 ton cooling		3	5.333		1,025	237		1,262	1,475
1020	7.5 ton cooling		3	5.333		2,350	237		2,587	2,925
1040	10 ton cooling	Q-6	2.60	9.231		2,750	425		3,175	3,675
1042	11 ton cooling		2.30	10.435		3,500	480		3,980	4,575
1050	15 ton cooling		1.60	15		3,650	690		4,340	5,050
1060	20 ton cooling		.70	34.286		4,850	1,575		6,425	7,700
1070	25 ton cooling		.65	36.923		6,100	1,700		7,800	9,250
1080	30 ton cooling		.60	40		6,850	1,850		8,700	10,300
1500	For hot water coil, add					40%	10%			
1512	For condensing unit add see Div. 23 62									
3000	Chilled water, horizontal unit, housing, 2 pipe, controls									
3100	1/2 ton cooling	Q-5	8	2	Ea.	1,050	89		1,139	1,300
3110	1 ton cooling		6	2.667		1,275	118		1,393	1,600
3120	1.5 ton cooling		5.50	2.909		1,525	129		1,654	1,900
3130	2 ton cooling		5.25	3.048		1,850	135		1,985	2,225
3140	3 ton cooling		4	4		2,125	178		2,303	2,625
3150	3.5 ton cooling		3.80	4.211		2,350	187		2,537	2,850
3160	4 ton cooling		3.80	4.211		2,350	187		2,537	2,850
3170	5 ton cooling		3.40	4.706		2,725	209		2,934	3,325
3180	6 ton cooling		3.40	4.706		2,725	209		2,934	3,325
3190	7 ton cooling		2.90	5.517		2,975	245		3,220	3,650
3200	8 ton cooling		2.90	5.517		2,975	245		3,220	3,650
3210	10 ton cooling	Q-6	2.80	8.571		3,400	395		3,795	4,325

23 82 19 – Fan Coil Units

23 82 19.10 Fan Coil Air Conditioning		Crew	Daily Output	Labor-Hours	Unit	Material	2009 Bare Costs Labor	Equipment	Total	Total Incl O&P
3240	For hot water coil, add					40%	10%			
4000	With electric heat, 2 pipe									
4100	1/2 ton cooling	Q-5	8	2	Ea.	1,300	89		1,389	1,550
4105	3/4 ton cooling		7	2.286		1,425	102		1,527	1,725
4110	1 ton cooling		6	2.667		1,575	118		1,693	1,900
4120	1.5 ton cooling		5.50	2.909		1,750	129		1,879	2,125
4130	2 ton cooling		5.25	3.048		2,200	135		2,335	2,625
4135	2.5 ton cooling		4.80	3.333		3,050	148		3,198	3,575
4140	3 ton cooling		4	4		4,175	178		4,353	4,875
4150	3.5 ton cooling		3.80	4.211		5,075	187		5,262	5,850
4160	4 ton cooling		3.60	4.444		5,875	197		6,072	6,750
4170	5 ton cooling		3.40	4.706		6,875	209		7,084	7,875
4180	6 ton cooling		3.20	5		6,875	222		7,097	7,875
4190	7 ton cooling		2.90	5.517		6,925	245		7,170	8,000
4200	8 ton cooling		2.20	7.273		7,850	325		8,175	9,100
4210	10 ton cooling	Q-6	2.80	8.571		8,325	395		8,720	9,750

23 82 19.20 Heating and Ventilating Units		Crew	Daily Output	Labor-Hours	Unit	Material	Labor	Equipment	Total	Total Incl O&P
0010	**HEATING AND VENTILATING UNITS**, Classroom units									
0020	Includes filter, heating/cooling coils, standard controls									
0080	750 CFM, 2 tons cooling	Q-6	2	12	Ea.	3,550	555		4,105	4,725
0100	1000 CFM, 2-1/2 tons cooling		1.60	15		4,200	690		4,890	5,650
0120	1250 CFM, 3 tons cooling		1.40	17.143		4,325	790		5,115	5,950
0140	1500 CFM, 4 tons cooling		.80	30		4,625	1,375		6,000	7,175
0160	2000 CFM 5 tons cooling		.50	48		5,450	2,200		7,650	9,325
0500	For electric heat, add					35%				
1000	For no cooling, deduct					25%	10%			

23 82 27 – Infrared Units

23 82 27.10 Infrared Type Heating Units		Crew	Daily Output	Labor-Hours	Unit	Material	Labor	Equipment	Total	Total Incl O&P
0010	**INFRARED TYPE HEATING UNITS**									
0020	Gas fired, unvented, electric ignition, 100% shutoff.									
0030	Piping and wiring not included									
0060	Input, 15 MBH	Q-5	7	2.286	Ea.	425	102		527	615
0100	30 MBH		6	2.667		430	118		548	655
0120	45 MBH		5	3.200		435	142		577	695
0140	50 MBH		4.50	3.556		450	158		608	730
0160	60 MBH		4	4		540	178		718	860
0180	75 MBH		3	5.333		545	237		782	950
0200	90 MBH		2.50	6.400		640	284		924	1,125
0220	105 MBH		2	8		645	355		1,000	1,250
0240	120 MBH		2	8		900	355		1,255	1,525
2000	Electric, single or three phase									
2050	6 kW, 20,478 BTU	1 Elec	2.30	3.478	Ea.	400	163		563	685
2100	13.5 KW, 40,956 BTU		2.20	3.636		605	171		776	920
2150	24 KW, 81,912 BTU		2	4		1,200	188		1,388	1,575
3000	Oil fired, pump, controls, fusible valve, oil supply tank									
3050	91,000 BTU	Q-5	2.50	6.400	Ea.	6,000	284		6,284	7,025
3080	105,000 BTU		2.25	7.111		6,375	315		6,690	7,500
3110	119,000 BTU		2	8		6,975	355		7,330	8,200

23 82 29 – Radiators

23 82 29.10 Hydronic Heating		Crew	Daily Output	Labor-Hours	Unit	Material	Labor	Equipment	Total	Total Incl O&P
0010	**HYDRONIC HEATING**, Terminal units, not incl. main supply pipe									

23 82 Convection Heating and Cooling Units

23 82 29 – Radiators

23 82 29.10 Hydronic Heating		Crew	Daily Output	Labor-Hours	Unit	Material	Labor	Equipment	Total	Total Incl O&P
						2009 Bare Costs				
1000	Radiation									
1100	Panel, baseboard, C.I., including supports, no covers	Q-5	46	.348	L.F.	32	15.45		47.45	58
3000	Radiators, cast iron									
3100	Free standing or wall hung, 6 tube, 25" high	Q-5	96	.167	Section	30	7.40		37.40	44
3150	4 tube 25" high	↓	96	.167	↓	27	7.40		34.40	40.50
3200	4 tube, 19" high	↓	96	.167	↓	23.50	7.40		30.90	37
3250	Adj. brackets, 2 per wall radiator up to 30 sections	1 Stpi	32	.250	Ea.	27	12.35		39.35	48.50
3500	Recessed, 20" high x 5" deep, without grille	Q-5	60	.267	Section	30.50	11.85		42.35	51.50
3600	For inlet grille, add				"	2.12			2.12	2.33
9500	To convert SFR to BTU rating: Hot water, 150 x SFR									
9510	Forced hot water, 180 x SFR; steam, 240 x SFR									

23 82 33 – Convectors

23 82 33.10 Convector Units

		Crew	Daily Output	Labor-Hours	Unit	Material	Labor	Equipment	Total	Total Incl O&P
0010	**CONVECTOR UNITS,** Terminal units, not incl. main supply pipe									
2204	Convector, multifin, 2 pipe w/ cabinet									
2210	17" H x 24" L	Q-5	10	1.600	Ea.	92	71		163	208
2214	17" H x 36" L	↓	8.60	1.860	↓	138	82.50		220.50	276
2218	17" H x 48" L	↓	7.40	2.162	↓	184	96		280	345
2222	21" H x 24" L	↓	9	1.778	↓	92	79		171	220
2226	21" H x 36" L	↓	8.20	1.951	↓	138	86.50		224.50	282
2228	21" H x 48" L	↓	6.80	2.353	↓	184	105		289	360
2240	For knob operated damper, add					140%				
2241	For metal trim strips, add	Q-5	64	.250	Ea.	18.30	11.10		29.40	36.50
2243	For snap-on inlet grille, add					10%	10%			
2245	For hinged access door, add	Q-5	64	.250	Ea.	34	11.10		45.10	53.50
2246	For air chamber, auto-venting, add	"	58	.276	"	5.65	12.25		17.90	24.50

23 82 36 – Finned-Tube Radiation Heaters

23 82 36.10 Finned Tube Radiation

		Crew	Daily Output	Labor-Hours	Unit	Material	Labor	Equipment	Total	Total Incl O&P
0010	**FINNED TUBE RADIATION**, Terminal units, not incl. main supply pipe									
1150	Fin tube, wall hung, 14" slope top cover, with damper									
1200	1-1/4" copper tube, 4-1/4" alum. fin	Q-5	38	.421	L.F.	39	18.70		57.70	70.50
1250	1-1/4" steel tube, 4-1/4" steel fin	↓	36	.444	↓	36.50	19.75		56.25	70
1255	2" steel tube, 4-1/4" steel fin	↓	32	.500	↓	39.50	22		61.50	77
1260	21", two tier, slope top, w/damper									
1262	1-1/4" copper tube, 4-1/4" alum. fin	Q-5	30	.533	L.F.	46	23.50		69.50	86
1264	1-1/4" steel tube, 4-1/4" steel fin	↓	28	.571	↓	44	25.50		69.50	86
1266	2" steel tube, 4-1/4" steel fin	↓	25	.640	↓	47	28.50		75.50	94
1270	10", flat top, with damper									
1272	1-1/4" copper tube, 4-1/4" alum. fin	Q-5	38	.421	L.F.	27.50	18.70		46.20	58
1274	1-1/4" steel tube, 4-1/4" steel fin	↓	36	.444	↓	36.50	19.75		56.25	70
1276	2" steel tube, 4-1/4" steel fin	↓	32	.500	↓	39.50	22		61.50	77
1280	17", two tier, flat top, w/damper									
1282	1-1/4" copper tube, 4-1/4" alum. fin	Q-5	30	.533	L.F.	46	23.50		69.50	86
1284	1-1/4" steel tube, 4-1/4" steel fin	↓	28	.571	↓	44	25.50		69.50	86
1286	2" steel tube, 4-1/4" steel fin	↓	25	.640	↓	46.50	28.50		75	94
1301	Rough in wall hung, steel fin w/supply & balance valves	1 Stpi	.91	8.791	Ea.	205	435		640	875
1310	Baseboard, pkgd, 1/2" copper tube, alum. fin, 7" high	Q-5	60	.267	L.F.	7.20	11.85		19.05	25.50
1320	3/4" copper tube, alum. fin, 7" high	↓	58	.276	↓	7.65	12.25		19.90	27
1340	1" copper tube, alum. fin, 8-7/8" high	↓	56	.286	↓	24.50	12.70		37.20	45.50
1360	1-1/4" copper tube, alum. fin, 8-7/8" high	↓	54	.296	↓	25	13.15		38.15	47.50
1380	1-1/4" IPS steel tube with steel fins	↓	52	.308	↓	24.50	13.65		38.15	47.50
1381	Rough in baseboard panel & fin tube, supply & balance valves	1 Stpi	1.06	7.547	Ea.	208	370		578	790

23 82 Convection Heating and Cooling Units

23 82 36 – Finned-Tube Radiation Heaters

23 82 36.10 Finned Tube Radiation		Crew	Daily Output	Labor-Hours	Unit	2009 Bare Costs Material	Labor	Equipment	Total	Total Incl O&P
1500	Note: fin tube may also require corners, caps, etc.									

23 82 39 – Unit Heaters

23 82 39.13 Cabinet Unit Heaters		Crew	Daily Output	Labor-Hours	Unit	2009 Bare Costs Material	Labor	Equipment	Total	Total Incl O&P
0010	**CABINET UNIT HEATERS**									
5400	Cabinet, horizontal, hot water, blower type									
5420	20 MBH	Q-5	10	1.600	Ea.	740	71		811	920
5430	60 MBH		7	2.286		1,300	102		1,402	1,575
5440	100 MBH		5	3.200		1,350	142		1,492	1,700
5450	120 MBH		4	4		1,400	178		1,578	1,825

23 82 39.16 Propeller Unit Heaters		Crew	Daily Output	Labor-Hours	Unit	2009 Bare Costs Material	Labor	Equipment	Total	Total Incl O&P
0010	**PROPELLER UNIT HEATERS**									
3950	Unit heaters, propeller, 115 V 2 psi steam, 60° F entering air									
4000	Horizontal, 12 MBH	Q-5	12	1.333	Ea.	315	59		374	435
4020	28.2 MBH		10	1.600		405	71		476	550
4040	36.5 MBH		8	2		475	89		564	660
4060	43.9 MBH		8	2		475	89		564	660
4080	56.5 MBH		7.50	2.133		510	95		605	700
4100	65.6 MBH		7	2.286		520	102		622	720
4120	87.6 MBH		6.50	2.462		560	109		669	780
4140	96.8 MBH		6	2.667		650	118		768	900
4160	133.3 MBH		5	3.200		745	142		887	1,025
4180	157.6 MBH		4	4		915	178		1,093	1,275
4200	197.7 MBH		3	5.333		1,050	237		1,287	1,500
4220	257.2 MBH		2.50	6.400		1,275	284		1,559	1,825
4240	286.9 MBH		2	8		1,425	355		1,780	2,100
4250	326.0 MBH		1.90	8.421		1,825	375		2,200	2,550
4260	364 MBH		1.80	8.889		2,000	395		2,395	2,775
4270	404 MBH		1.60	10		2,500	445		2,945	3,425
4280	Rough in unit heater, 72.7 MBH, piping, supports & controls	1 Stpi	.47	17.021	Se-	490	840		1,330	1,800
4300	For vertical diffuser, add				Ea	174			174	191
4310	Vertical flow, 40 MBH	Q-5	11	1.455		465	64.50		529.50	610
4314	58.5 MBH		8	2		485	89		574	665
4318	92 MBH		7	2.286		595	102		697	805
4322	109.7 MBH		6	2.667		730	118		848	980
4326	131.0 MBH		4	4		730	178		908	1,075
4330	160.0 MBH		3	5.333		775	237		1,012	1,200
4334	194.0 MBH		2.20	7.273		895	325		1,220	1,475
4338	212.0 MBH		2.10	7.619		1,100	340		1,440	1,725
4342	247.0 MBH		1.96	8.163		1,250	365		1,615	1,925
4346	297.0 MBH		1.80	8.889		1,400	395		1,795	2,125
4350	333.0 MBH	Q-6	1.90	12.632		1,525	580		2,105	2,550
4354	420 MBH,(460 V)		1.80	13.333		1,875	615		2,490	2,975
4358	500 MBH, (460 V)		1.71	14.035		2,475	645		3,120	3,700
4362	570 MBH, (460 V)		1.40	17.143		3,725	790		4,515	5,275
4366	620 MBH, (460 V)		1.30	18.462		4,275	850		5,125	5,975
4370	960 MBH, (460 V)		1.10	21.818		7,200	1,000		8,200	9,400
4400	Unit heaters, propeller, hot water, 115 V, 60 Deg. F entering air									
4404	Horizontal									
4408	6 MBH	Q-5	11	1.455	Ea.	315	64.50		379.50	440
4412	12 MBH		11	1.455		315	64.50		379.50	440
4416	16 MBH		10	1.600		350	71		421	490
4420	24 MBH		9	1.778		405	79		484	565

23 82 Convection Heating and Cooling Units

23 82 39 – Unit Heaters

23 82 39.16 Propeller Unit Heaters		Crew	Daily Output	Labor-Hours	Unit	Material	2009 Bare Costs Labor	Equipment	Total	Total Incl O&P
4424	29 MBH	Q-5	8	2	Ea.	475	89		564	655
4428	47 MBH		7	2.286		520	102		622	720
4432	63 MBH		6	2.667		545	118		663	780
4436	81 MBH		5.50	2.909		690	129		819	950
4440	90 MBH		5	3.200		745	142		887	1,025
4460	133 MBH		4	4		915	178		1,093	1,275
4464	139 MBH		3.50	4.571		1,050	203		1,253	1,450
4468	198 MBH		2.50	6.400		1,275	284		1,559	1,825
4472	224 MBH		2	8		1,425	355		1,780	2,100
4476	273 MBH	↓	1.50	10.667	↓	1,775	475		2,250	2,675
4500	Vertical									
4504	8 MBH	Q-5	13	1.231	Ea.	465	54.50		519.50	595
4508	12 MBH		13	1.231		465	54.50		519.50	595
4512	16 MBH		13	1.231		465	54.50		519.50	595
4516	30 MBH		12	1.333		465	59		524	605
4520	43 MBH		11	1.455		485	64.50		549.50	625
4524	57 MBH		8	2		595	89		684	790
4528	68 MBH		7	2.286		620	102		722	830
4540	105 MBH		6	2.667		730	118		848	980
4544	123 MBH		5	3.200		770	142		912	1,075
4550	140 MBH		5	3.200		895	142		1,037	1,200
4554	156 MBH		4	4		1,100	178		1,278	1,500
4560	210 MBH		4	4		1,400	178		1,578	1,800
4564	223 MBH		3	5.333		1,525	237		1,762	2,025
4568	257 MBH	↓	2	8	↓	1,525	355		1,880	2,200

23 83 Radiant Heating Units

23 83 16 – Radiant-Heating Hydronic Piping

23 83 16.10 Radiant Floor Heating

23 83 16.10 Radiant Floor Heating		Crew	Daily Output	Labor-Hours	Unit	Material	Labor	Equipment	Total	Total Incl O&P
0010	**RADIANT FLOOR HEATING**									
0100	Tubing, PEX (cross-linked polyethylene)									
0110	Oxygen barrier type for systems with ferrous materials									
0120	1/2"	Q-5	800	.020	L.F.	.94	.89		1.83	2.36
0130	3/4"		535	.030		1.15	1.33		2.48	3.26
0140	1"	↓	400	.040	↓	2.07	1.78		3.85	4.95
0200	Non barrier type for ferrous free systems									
0210	1/2"	Q-5	800	.020	L.F.	.51	.89		1.40	1.89
0220	3/4"		535	.030		.92	1.33		2.25	3
0230	1"	↓	400	.040	↓	1.60	1.78		3.38	4.43
1000	Manifolds									
1110	Brass									
1120	Valved									
1130	1", 2 circuit	Q-5	13	1.231	Ea.	47	54.50		101.50	134
1140	1", 3 circuit		11	1.455		63.50	64.50		128	167
1150	1", 4 circuit		9	1.778		87.50	79		166.50	216
1160	1-1/4", 3 circuit		10	1.600		73	71		144	187
1170	1-1/4", 4 circuit	↓	8	2	↓	88.50	89		177.50	231
1300	Valveless									
1310	1", 2 circuit	Q-5	14	1.143	Ea.	24.50	51		75.50	103
1320	1", 3 circuit		12	1.333		32	59		91	125
1330	1", 4 circuit	↓	10	1.600	↓	41.50	71		112.50	153

23 83 Radiant Heating Units

23 83 16 – Radiant-Heating Hydronic Piping

23 83 16.10 Radiant Floor Heating		Crew	Daily Output	Labor-Hours	Unit	2009 Bare Costs Material	Labor	Equipment	Total	Total Incl O&P
1340	1-1/4", 3 circuit	Q-5	11	1.455	Ea.	43.50	64.50		108	145
1350	1-1/4", 4 circuit	↓	9	1.778	↓	45	79		124	169
1400	Manifold connector kit									
1410	Mounting brackets, couplings, end cap, air vent, valve, plug									
1420	1"	Q-5	8	2	Ea.	81	89		170	222
1430	1-1/4"	"	7	2.286	"	95.50	102		197.50	257
1610	Copper, (cut to size)									
1620	1" x 1/2" x 72" Lg, 2" sweat drops, 24 circuit	Q-5	3.33	4.805	Ea.	84.50	213		297.50	415
1630	1-1/4" x 1/2" x 72" Lg, 2" sweat drops, 24 circuit		3.20	5		98	222		320	445
1640	1-1/2" x 1/2" x 72" Lg, 2" sweat drops, 24 circuit		3	5.333		120	237		357	485
1650	1-1/4" x 3/4" x 72" Lg, 2" sweat drops, 24 circuit		3.10	5.161		105	229		334	460
1660	1-1/2" x 3/4" x 72" Lg, 2" sweat drops, 24 circuit	↓	2.90	5.517	↓	127	245		372	510
1710	Copper, (modular)									
1720	1" x 1/2", 2" sweat drops, 2 circuit	Q-5	17.50	.914	Ea.	8.60	40.50		49.10	70.50
1730	1" x 1/2", 2" sweat drops, 3 circuit		14.50	1.103		12.05	49		61.05	87
1740	1" x 1/2", 2" sweat drops, 4 circuit		12.30	1.301		13.75	58		71.75	102
1750	1-1/4" x 1/2", 2" sweat drops, 2 circuit		17	.941		9.45	42		51.45	73.50
1760	1-1/4" x 1/2", 2" sweat drops, 3 circuit		14	1.143		13.75	51		64.75	91
1770	1-1/4" x 1/2", 2" sweat drops, 4 circuit	↓	11.80	1.356	↓	18.05	60		78.05	110
3000	Valves									
3110	Motorized zone valve operator, 24V	Q-5	30	.533	Ea.	49	23.50		72.50	89.50
3120	Motorized zone valve with operator complete, 24V									
3130	3/4"	Q-5	35	.457	Ea.	85	20.50		105.50	124
3140	1"		32	.500		96.50	22		118.50	140
3150	1-1/4"		29.60	.541		119	24		143	167
3200	Flow control/isolation valve, 1/2"	↓	32	.500	↓	27.50	22		49.50	64
3300	Thermostatic mixing valves									
3310	1/2"	Q-5	26.60	.602	Ea.	88.50	26.50		115	138
3320	3/4"		25	.640		107	28.50		135.50	161
3330	1"	↓	21.40	.748	↓	115	33		148	177
3400	3 Way mixing / diverting valve, manual, brass									
3410	1/2"	Q-5	21.30	.751	Ea.	117	33.50		150.50	179
3420	3/4"		20	.800		120	35.50		155.50	186
3430	1"	↓	17.80	.899	↓	139	40		179	213
3500	4 Way mixing / diverting valve, manual, brass									
3510	1/2"	Q-5	16	1	Ea.	122	44.50		166.50	201
3520	3/4"		15	1.067		136	47.50		183.50	220
3530	1"		13.30	1.203		152	53.50		205.50	247
3540	1-1/4"		11.40	1.404		171	62.50		233.50	282
3550	1-1/2"		10.60	1.509		224	67		291	350
3560	2"		10	1.600		242	71		313	375
3570	2-1/2"		8	2		665	89		754	870
3800	Motor control, 3 or 4 way, for valves up to 1-1/2"	↓	30	.533	↓	169	23.50		192.50	221
5000	Radiant floor heating, zone control panel									
5110	6 Zone control									
5120	For zone valves	Q-5	16	1	Ea.	158	44.50		202.50	241
5130	For circulators	"	16	1	"	225	44.50		269.50	315
7000	PEX tubing fittings, compression type									
7110	PEX tubing to brass valve conn., 1/2"	Q-5	54	.296	Ea.	2.72	13.15		15.87	22.50
7200	PEX x male NPT									
7210	1/2" x 1/2"	Q-5	54	.296	Ea.	5.30	13.15		18.45	25.50
7220	3/4" x 3/4"		46	.348		9.05	15.45		24.50	33
7230	1" x 1"	↓	40	.400	↓	21.50	17.75		39.25	50

23 83 Radiant Heating Units

23 83 16 – Radiant-Heating Hydronic Piping

23 83 16.10 Radiant Floor Heating		Crew	Daily Output	Labor-Hours	Unit	2009 Bare Costs Material	Labor	Equipment	Total	Total Incl O&P
7300	PEX coupling									
7310	1/2" x 1/2"	Q-5	50	.320	Ea.	6.45	14.20		20.65	28.50
7320	3/4" x 3/4"		42	.381		18.50	16.95		35.45	46
7330	1" x 1"		36	.444		25	19.75		44.75	57
7400	PEX x copper sweat									
7410	1/2" x 1/2"	Q-5	38	.421	Ea.	4.73	18.70		23.43	33
7420	3/4" x 3/4"		34	.471		9.90	21		30.90	42.50
7430	1" x 1"		29	.552		21.50	24.50		46	60.50

23 83 33 – Electric Radiant Heaters

23 83 33.10 Electric Heating		Crew	Daily Output	Labor-Hours	Unit	Material	Labor	Equipment	Total	Total Incl O&P
0010	**ELECTRIC HEATING**, not incl. conduit or feed wiring.									
1100	Rule of thumb: Baseboard units, including control	1 Elec	4.40	1.818	kW	92.50	85.50		178	229
1300	Baseboard heaters, 2' long, 375 watt		8	1	Ea.	40	47		87	114
1400	3' long, 500 watt		8	1		46	47		93	121
1600	4' long, 750 watt		6.70	1.194		54	56		110	143
1800	5' long, 935 watt		5.70	1.404		68	66		134	173
2000	6' long, 1125 watt		5	1.600		74	75		149	194
2200	7' long, 1310 watt		4.40	1.818		84	85.50		169.50	220
2400	8' long, 1500 watt		4	2		92	94		186	241
2600	9' long, 1680 watt		3.60	2.222		104	104		208	269
2800	10' long, 1875 watt		3.30	2.424		193	114		307	380
2950	Wall heaters with fan, 120 to 277 volt									
3160	Recessed, residential, 750 watt	1 Elec	6	1.333	Ea.	98.50	62.50		161	202
3170	1000 watt		6	1.333		101	62.50		163.50	205
3180	1250 watt		5	1.600		111	75		186	234
3190	1500 watt		4	2		111	94		205	262
3210	2000 watt		4	2		117	94		211	269
3230	2500 watt		3.50	2.286		252	107		359	440
3240	3000 watt		3	2.667		380	125		505	605
3250	4000 watt		2.70	2.963		385	139		524	625
3260	Commercial, 750 watt		6	1.333		176	62.50		238.50	287
3270	1000 watt		6	1.333		176	62.50		238.50	287
3280	1250 watt		5	1.600		176	75		251	305
3290	1500 watt		4	2		176	94		270	335
3300	2000 watt		4	2		176	94		270	335
3310	2500 watt		3.50	2.286		176	107		283	355
3320	3000 watt		3	2.667		320	125		445	535
3330	4000 watt		2.70	2.963		320	139		459	555
3600	Thermostats, integral		16	.500		20.50	23.50		44	57.50
3800	Line voltage, 1 pole		8	1		31	47		78	104
3810	2 pole		8	1		33.50	47		80.50	107
3820	Low voltage, 1 pole		8	1		27.50	47		74.50	100
4000	Heat trace system, 400 degree									
4020	115V, 2.5 watts per L.F.	1 Elec	530	.015	L.F.	6.70	.71		7.41	8.40
4030	5 watts per L.F.		530	.015		6.70	.71		7.41	8.40
4050	10 watts per L.F.		530	.015		6.70	.71		7.41	8.40
4060	208V, 5 watts per L.F.		530	.015		6.70	.71		7.41	8.40
4080	480V, 8 watts per L.F.		530	.015		6.70	.71		7.41	8.40
4200	Heater raceway									
4260	Heat transfer cement									
4280	1 gallon				Ea.	60.50			60.50	67
4300	5 gallon				"	249			249	274

23 83 33.10 Electric Heating		Crew	Daily Output	Labor-Hours	Unit	Material	2009 Bare Costs Labor	Equipment	Total	Total Incl O&P
4320	Snap band, clamp									
4340	3/4" pipe size	1 Elec	470	.017	Ea.		.80		.80	1.19
4360	1" pipe size		444	.018			.85		.85	1.26
4380	1-1/4" pipe size		400	.020			.94		.94	1.40
4400	1-1/2" pipe size		355	.023			1.06		1.06	1.58
4420	2" pipe size		320	.025			1.18		1.18	1.75
4440	3" pipe size		160	.050			2.35		2.35	3.50
4460	4" pipe size		100	.080			3.76		3.76	5.60
4480	Thermostat NEMA 3R, 22 amp, 0-150 Deg, 10' cap.		8	1		214	47		261	305
4500	Thermostat NEMA 4X, 25 amp, 40 Deg, 5-1/2' cap.		7	1.143		214	53.50		267.50	315
4520	Thermostat NEMA 4X, 22 amp, 25-325 Deg, 10' cap.		7	1.143		575	53.50		628.50	715
4540	Thermostat NEMA 4X, 22 amp, 15-140 Deg,		6	1.333		485	62.50		547.50	630
4580	Thermostat NEMA 4,7,9, 22 amp, 25-325 Deg, 10' cap.		3.60	2.222		710	104		814	935
4600	Thermostat NEMA 4,7,9, 22 amp, 15-140 Deg,		3	2.667		705	125		830	960
4720	Fiberglass application tape, 36 yard roll		11	.727		66.50	34		100.50	125
5000	Radiant heating ceiling panels, 2' x 4', 500 watt		16	.500		253	23.50		276.50	315
5050	750 watt		16	.500		279	23.50		302.50	340
5200	For recessed plaster frame, add		32	.250		62	11.75		73.75	86
5300	Infrared quartz heaters, 120 volts, 1000 watts		6.70	1.194		164	56		220	264
5350	1500 watt		5	1.600		164	75		239	292
5400	240 volts, 1500 watt		5	1.600		164	75		239	292
5450	2000 watt		4	2		164	94		258	320
5500	3000 watt		3	2.667		189	125		314	395
5550	4000 watt		2.60	3.077		189	145		334	425
5570	Modulating control	↓	.80	10	↓	89	470		559	800
5600	Unit heaters, heavy duty, with fan & mounting bracket									
5650	Single phase, 208-240-277 volt, 3 kW	1 Elec	3.20	2.500	Ea.	405	118		523	620
5750	5 kW		2.40	3.333		420	157		577	695
5800	7 kW		1.90	4.211		645	198		843	1,000
5850	10 kW		1.30	6.154		735	289		1,024	1,250
5950	15 kW		.90	8.889		1,200	420		1,620	1,925
6000	480 volt, 3 kW		3.30	2.424		400	114		514	610
6020	4 kW		3	2.667		455	125		580	685
6040	5 kW		2.60	3.077		465	145		610	725
6060	7 kW		2	4		690	188		878	1,050
6080	10 kW		1.40	5.714		755	269		1,024	1,225
6100	13 kW		1.10	7.273		1,200	340		1,540	1,800
6120	15 kW		1	8		1,200	375		1,575	1,850
6140	20 kW		.90	8.889		1,550	420		1,970	2,325
6300	3 phase, 208-240 volt, 5 kW		2.40	3.333		395	157		552	670
6320	7 kW		1.90	4.211		610	198		808	965
6340	10 kW		1.30	6.154		660	289		949	1,150
6360	15 kW		.90	8.889		1,100	420		1,520	1,850
6380	20 kW		.70	11.429		1,575	535		2,110	2,525
6400	25 kW		.50	16		1,850	750		2,600	3,150
6500	480 volt, 5 kW		2.60	3.077		550	145		695	820
6520	7 kW		2	4		695	188		883	1,050
6540	10 kW		1.40	5.714		735	269		1,004	1,200
6560	13 kW		1.10	7.273		1,200	340		1,540	1,800
6580	15 kW		1	8		1,200	375		1,575	1,850
6600	20 kW		.90	8.889		1,550	420		1,970	2,325
6620	25 kW		.60	13.333		1,850	625		2,475	2,950
6630	30 kW	↓	.70	11.429	↓	2,150	535		2,685	3,150

23 83 33 – Electric Radiant Heaters

23 83 33.10 Electric Heating		Crew	Daily Output	Labor-Hours	Unit	2009 Bare Costs Material	Labor	Equipment	Total	Total Incl O&P
6640	40 kW	1 Elec	.60	13.333	Ea.	2,725	625		3,350	3,925
6650	50 kW	↓	.50	16	↓	3,300	750		4,050	4,750
6800	Vertical discharge heaters, with fan									
6820	Single phase, 208-240-277 volt, 10 kW	1 Elec	1.30	6.154	Ea.	680	289		969	1,175
6840	15 kW		.90	8.889		1,150	420		1,570	1,875
6900	3 phase, 208-240 volt, 10 kW		1.30	6.154		655	289		944	1,150
6920	15 kW		.90	8.889		1,100	420		1,520	1,850
6940	20 kW		.70	11.429		1,575	535		2,110	2,525
6960	25 kW		.50	16		1,850	750		2,600	3,150
6980	30 kW		.40	20		2,150	940		3,090	3,750
7000	40 kW		.36	22.222		2,750	1,050		3,800	4,575
7020	50 kW		.32	25		3,300	1,175		4,475	5,375
7100	480 volt, 10 kW		1.40	5.714		735	269		1,004	1,200
7120	15 kW		1	8		1,200	375		1,575	1,850
7140	20 kW		.90	8.889		1,550	420		1,970	2,325
7160	25 kW		.60	13.333		1,850	625		2,475	2,950
7180	30 kW		.50	16		2,150	750		2,900	3,475
7200	40 kW		.40	20		2,725	940		3,665	4,400
7220	50 kW	↓	.35	22.857	↓	3,300	1,075		4,375	5,225
7900	Cabinet convector heaters, 240 volt									
7920	3′ long, 2000 watt	1 Elec	5.30	1.509	Ea.	1,775	71		1,846	2,075
7940	3000 watt		5.30	1.509		1,850	71		1,921	2,150
7960	4000 watt		5.30	1.509		1,925	71		1,996	2,200
7980	6000 watt		4.60	1.739		2,000	81.50		2,081.50	2,300
8000	8000 watt		4.60	1.739		2,075	81.50		2,156.50	2,400
8020	4′ long, 4000 watt		4.60	1.739		1,950	81.50		2,031.50	2,250
8040	6000 watt		4	2		2,025	94		2,119	2,375
8060	8000 watt		4	2		2,100	94		2,194	2,450
8080	10,000 watt	↓	4	2	↓	2,100	94		2,194	2,450
8100	Available also in 208 or 277 volt									
8200	Cabinet unit heaters, 120 to 277 volt, 1 pole,									
8220	wall mounted, 2 kW	1 Elec	4.60	1.739	Ea.	1,725	81.50		1,806.50	2,025
8230	3 kW		4.60	1.739		1,800	81.50		1,881.50	2,125
8240	4 kW		4.40	1.818		1,875	85.50		1,960.50	2,175
8250	5 kW		4.40	1.818		1,925	85.50		2,010.50	2,250
8260	6 kW		4.20	1.905		1,950	89.50		2,039.50	2,250
8270	8 kW		4	2		2,025	94		2,119	2,375
8280	10 kW		3.80	2.105		2,050	99		2,149	2,425
8290	12 kW		3.50	2.286		2,225	107		2,332	2,600
8300	13.5 kW		2.90	2.759		2,275	130		2,405	2,700
8310	16 kW		2.70	2.963		2,325	139		2,464	2,750
8320	20 kW		2.30	3.478		3,450	163		3,613	4,050
8330	24 kW		1.90	4.211		3,500	198		3,698	4,150
8350	Recessed, 2 kW		4.40	1.818		1,800	85.50		1,885.50	2,100
8370	3 kW		4.40	1.818		1,875	85.50		1,960.50	2,200
8380	4 kW		4.20	1.905		1,925	89.50		2,014.50	2,250
8390	5 kW		4.20	1.905		2,000	89.50		2,089.50	2,325
8400	6 kW		4	2		2,075	94		2,169	2,425
8410	8 kW		3.80	2.105		2,100	99		2,199	2,450
8420	10 kW		3.50	2.286		2,125	107		2,232	2,500
8430	12 kW		2.90	2.759		2,175	130		2,305	2,575
8440	13.5 kW		2.70	2.963		2,750	139		2,889	3,225
8450	16 kW	↓	2.30	3.478	↓	2,775	163		2,938	3,300

23 83 33.10 Electric Heating		Crew	Daily Output	Labor-Hours	Unit	Material	Labor	Equipment	Total	Total Incl O&P
8460	20 kW	1 Elec	1.90	4.211	Ea.	2,975	198		3,173	3,575
8470	24 kW		1.60	5		3,025	235		3,260	3,675
8490	Ceiling mounted, 2 kW		3.20	2.500		1,800	118		1,918	2,150
8510	3 kW		3.20	2.500		1,875	118		1,993	2,250
8520	4 kW		3	2.667		1,925	125		2,050	2,300
8530	5 kW		3	2.667		2,000	125		2,125	2,375
8540	6 kW		2.80	2.857		2,075	134		2,209	2,475
8550	8 kW		2.40	3.333		2,100	157		2,257	2,525
8560	10 kW		2.20	3.636		2,125	171		2,296	2,600
8570	12 kW		2	4		2,175	188		2,363	2,650
8580	13.5 kW		1.50	5.333		2,750	251		3,001	3,400
8590	16 kW		1.30	6.154		2,775	289		3,064	3,475
8600	20 kW		.90	8.889		2,975	420		3,395	3,900
8610	24 kW		.60	13.333		3,025	625		3,650	4,250
8630	208 to 480 V, 3 pole									
8650	Wall mounted, 2 kW	1 Elec	4.60	1.739	Ea.	1,925	81.50		2,006.50	2,250
8670	3 kW		4.60	1.739		2,025	81.50		2,106.50	2,350
8680	4 kW		4.40	1.818		2,075	85.50		2,160.50	2,400
8690	5 kW		4.40	1.818		2,125	85.50		2,210.50	2,475
8700	6 kW		4.20	1.905		2,150	89.50		2,239.50	2,475
8710	8 kW		4	2		2,225	94		2,319	2,600
8720	10 kW		3.80	2.105		2,250	99		2,349	2,625
8730	12 kW		3.50	2.286		2,300	107		2,407	2,675
8740	13.5 kW		2.90	2.759		2,350	130		2,480	2,775
8750	16 kW		2.70	2.963		2,375	139		2,514	2,825
8760	20 kW		2.30	3.478		3,275	163		3,438	3,850
8770	24 kW		1.90	4.211		3,325	198		3,523	3,950
8790	Recessed, 2 kW		4.40	1.818		2,000	85.50		2,085.50	2,325
8810	3 kW		4.40	1.818		2,075	85.50		2,160.50	2,425
8820	4 kW		4.20	1.905		2,125	89.50		2,214.50	2,450
8830	5 kW		4.20	1.905		2,200	89.50		2,289.50	2,550
8840	6 kW		4	2		2,225	94		2,319	2,575
8850	8 kW		3.80	2.105		2,275	99		2,374	2,675
8860	10 kW		3.50	2.286		2,325	107		2,432	2,725
8870	12 kW		2.90	2.759		2,375	130		2,505	2,800
8880	13.5 kW		2.70	2.963		2,400	139		2,539	2,850
8890	16 kW		2.30	3.478		2,450	163		2,613	2,950
8900	20 kW		1.90	4.211		3,350	198		3,548	3,975
8920	24 kW		1.60	5		3,400	235		3,635	4,075
8940	Ceiling mount, 2 kW		3.20	2.500		2,000	118		2,118	2,375
8950	3 kW		3.20	2.500		2,075	118		2,193	2,475
8960	4 kW		3	2.667		2,125	125		2,250	2,525
8970	5 kW		3	2.667		2,200	125		2,325	2,600
8980	6 kW		2.80	2.857		2,225	134		2,359	2,625
8990	8 kW		2.40	3.333		2,275	157		2,432	2,750
9000	10 kW		2.20	3.636		2,325	171		2,496	2,825
9020	13.5 kW		1.50	5.333		2,400	251		2,651	3,025
9030	16 kW		1.30	6.154		2,450	289		2,739	3,125
9040	20 kW		.90	8.889		3,350	420		3,770	4,300
9060	24 kW		.60	13.333		3,400	625		4,025	4,650

Note: Material, Labor, Equipment and Total columns are 2009 Bare Costs.

23 84 Humidity Control Equipment

23 84 13 – Humidifiers

23 84 13.10 Humidifier Units		Crew	Daily Output	Labor-Hours	Unit	2009 Bare Costs Material	Labor	Equipment	Total	Total Incl O&P
0010	**HUMIDIFIER UNITS**									
0520	Steam, room or duct, filter, regulators, auto. controls, 220 V									
0540	11 lb. per hour	Q-5	6	2.667	Ea.	2,275	118		2,393	2,675
0560	22 lb. per hour		5	3.200		2,525	142		2,667	3,000
0580	33 lb. per hour		4	4		2,575	178		2,753	3,125
0600	50 lb. per hour		4	4		3,175	178		3,353	3,750
0620	100 lb. per hour		3	5.333		3,800	237		4,037	4,525
0640	150 lb. per hour		2.50	6.400		5,025	284		5,309	5,950
0660	200 lb. per hour		2	8		6,200	355		6,555	7,350
0700	With blower									
0720	11 lb. per hour	Q-5	5.50	2.909	Ea.	3,175	129		3,304	3,700
0740	22 lb. per hour		4.75	3.368		3,425	150		3,575	3,975
0760	33 lb. per hour		3.75	4.267		3,475	190		3,665	4,100
0780	50 lb. per hour		3.50	4.571		4,200	203		4,403	4,900
0800	100 lb. per hour		2.75	5.818		4,725	259		4,984	5,600
0820	150 lb. per hour		2	8		7,000	355		7,355	8,225
0840	200 lb. per hour		1.50	10.667		8,225	475		8,700	9,725
1000	Steam for duct installation, manifold with pneum. controls									
1010	3 – 191 lb/hr, 24" x 24"	Q-5	5	3.200	Ea.	1,725	142		1,867	2,100
1020	3 – 191 lb/hr, 48" x 48"		4.60	3.478		3,375	155		3,530	3,950
1030	65 – 334 lb/hr, 36" x 36"		4.80	3.333		3,300	148		3,448	3,875
1040	65 – 334 lb/hr, 72" x 72"		4	4		3,575	178		3,753	4,200
1050	100 – 690 lb/hr, 48" x 48"		4	4		4,650	178		4,828	5,375
1060	100 – 690 lb/hr, 84" x 84"		3.60	4.444		4,800	197		4,997	5,575
1070	220 – 2000 lb/hr, 72" x 72"		3.80	4.211		4,850	187		5,037	5,625
1080	220 – 2000 lb/hr, 144" x 144"		3.40	4.706		5,675	209		5,884	6,575
1090	For electrically controlled unit, add					805			805	885
1092	For additional manifold, add					25%	15%			
1200	Electric steam, self generating, with manifold									
1210	1 – 96 lb/hr, 12" x 72"	Q-5	3.60	4.444	Ea.	7,575	197		7,772	8,625
1220	20 – 168 lb/hr, 12" x 72"		3.60	4.444		13,100	197		13,297	14,700
1290	For room mounting, add		16	1		217	44.50		261.50	305
5000	Furnace type, wheel bypass									
5020	10 GPD	1 Stpi	4	2	Ea.	192	98.50		290.50	360
5040	14 GPD		3.80	2.105		234	104		338	415
5060	19 GPD		3.60	2.222		272	110		382	465

23 84 16 – Dehumidifiers

23 84 16.10 Dehumidifier Units		Crew	Daily Output	Labor-Hours	Unit	Material	Labor	Equipment	Total	Total Incl O&P
0010	**DEHUMIDIFIER UNITS**									
6000	Self contained with filters and standard controls									
6040	1.5 lb/hr, 50 cfm	1 Plum	8	1	Ea.	5,300	49		5,349	5,900
6060	3 lb/hr, 150 cfm	Q-1	12	1.333		5,525	58.50		5,583.50	6,175
6065	6 lb/hr, 150 cfm		9	1.778		9,000	78		9,078	10,000
6070	16 to 20 lb/hr, 600 cfm		5	3.200		18,300	140		18,440	20,400
6080	30 to 40 lb/hr, 1125 cfm		4	4		28,600	176		28,776	31,700
6090	60 to 75 lb/hr, 2250 cfm		3	5.333		37,300	234		37,534	41,400
6100	120 to 155 lb/hr, 4500 cfm		2	8		66,000	350		66,350	73,000
6110	240 to 310 lb/hr, 9000 cfm		1.50	10.667		94,500	470		94,970	104,500
6120	400 to 515 lb/hr, 15,000 cfm	Q-2	1.60	15		128,000	685		128,685	142,000
6130	530 to 690 lb/hr, 20,000 cfm		1.40	17.143		139,000	780		139,780	153,500
6140	800 to 1030 lb/hr, 30,000 cfm		1.20	20		158,000	910		158,910	175,500
6150	1060 to 1375 lb/hr, 40,000 cfm		1	24		220,000	1,100		221,100	243,500

23 91 Prefabricated Equipment Supports

23 91 10 – Prefabricated Curbs, Pads and Stands

23 91 10.10 Prefabricated Pads and Stands		Crew	Daily Output	Labor-Hours	Unit	2009 Bare Costs Material	Labor	Equipment	Total	Total Incl O&P
0010	**PREFABRICATED PADS AND STANDS**									
6000	Pad, fiberglass reinforced concrete with polystyrene foam core									
6050	Condenser, 2″ thick, 20″ x 38″	1 Shee	8	1	Ea.	22	47		69	97
6070	24″ x 24″		16	.500		17.60	23.50		41.10	56
6090	24″ x 36″		12	.667		28	31.50		59.50	79.50
6110	24″ x 42″		8	1		32.50	47		79.50	109
6150	26″ x 36″	Q-9	8	2		31	85		116	165
6170	28″ x 38″	1 Shee	8	1		35	47		82	111
6190	30″ x 30″		12	.667		29.50	31.50		61	81
6220	30″ x 36″		8	1		35	47		82	111
6240	30″ x 40″	Q-9	8	2		38	85		123	173
6260	32″ x 32″		7	2.286		33	97		130	185
6280	36″ x 36″		8	2		40.50	85		125.50	176
6300	36″ x 40″		7	2.286		46	97		143	200
6320	36″ x 48″		7	2.286		54	97		151	209
6340	36″ x 54″		6	2.667		65.50	113		178.50	246
6360	36″ x 60″ x 3″		7	2.286		76	97		173	233
6400	41″ x 47″ x 3″		6	2.667		72	113		185	254
6490	26″ round		10	1.600		17.40	68		85.40	124
6550	30″ round		9	1.778		23	75.50		98.50	141
6600	36″ round		8	2		34.50	85		119.50	169
6700	For 3″ thick pad, add					20%	5%			
8800	Stand, corrosion-resistant plastic									
8850	Heat pump, 6″ high	1 Shee	32	.250	Ea.	8.15	11.80		19.95	27
8870	12″ high	″	32	.250	″	11	11.80		22.80	30.50

Division Notes

	CREW	DAILY OUTPUT	LABOR-HOURS	UNIT	2009 BARE COSTS MAT.	LABOR	EQUIP.	TOTAL	TOTAL INCL O&P

Estimating Tips

26 05 00 Common Work Results for Electrical

- Conduit should be taken off in three main categories—power distribution, branch power, and branch lighting—so the estimator can concentrate on systems and components, therefore making it easier to ensure all items have been accounted for.
- For cost modifications for elevated conduit installation, add the percentages to labor according to the height of installation, and only to the quantities exceeding the different height levels, not to the total conduit quantities.
- Remember that aluminum wiring of equal ampacity is larger in diameter than copper and may require larger conduit.
- If more than three wires at a time are being pulled, deduct percentages from the labor hours of that grouping of wires.
- When taking off grounding systems, identify separately the type and size of wire, and list each unique type of ground connection.
- The estimator should take the weights of materials into consideration when completing a takeoff. Topics to consider include: How will the materials be supported? What methods of support are available? How high will the support structure have to reach? Will the final support structure be able to withstand the total burden? Is the support material included or separate from the fixture, equipment, and material specified?
- Do not overlook the costs for equipment used in the installation. If scaffolding or highlifts are available in the field, contractors may use them in lieu of the proposed ladders and rolling staging.

26 20 00 Low-Voltage Electrical Transmission

- Supports and concrete pads may be shown on drawings for the larger equipment, or the support system may be only a piece of plywood for the back of a panelboard. In either case, it must be included in the costs.

26 40 00 Electrical and Cathodic Protection

- When taking off cathodic protections systems, identify the type and size of cable, and list each unique type of anode connection.

26 50 00 Lighting

- Fixtures should be taken off room by room, using the fixture schedule, specifications, and the ceiling plan. For large concentrations of lighting fixtures in the same area, deduct the percentages from labor hours.

Reference Numbers

Reference numbers are shown in shaded boxes at the beginning of some major classifications. These numbers refer to related items in the Reference Section. The reference information may be an estimating procedure, an alternate pricing method, or technical information.

Note: Not all subdivisions listed here necessarily appear in this publication.

Note: **Trade Service,** *in part, has been used as a reference source for some of the material prices used in Division 26.*

26 05 33.95 Cutting and Drilling		Crew	Daily Output	Labor-Hours	Unit	Material	Labor	Equipment	Total	Total Incl O&P
						2009 Bare Costs				
0010	**CUTTING AND DRILLING**									
0100	Hole drilling to 10' high, concrete wall									
0110	8" thick, 1/2" pipe size	R-31	12	.667	Ea.	4.71	31.50	4.41	40.62	56.50
0120	3/4" pipe size		12	.667		4.71	31.50	4.41	40.62	56.50
0130	1" pipe size		9.50	.842		9.50	39.50	5.55	54.55	75.50
0140	1-1/4" pipe size		9.50	.842		9.50	39.50	5.55	54.55	75.50
0150	1-1/2" pipe size		9.50	.842		9.50	39.50	5.55	54.55	75.50
0160	2" pipe size		4.40	1.818		10.15	85.50	12.05	107.70	151
0170	2-1/2" pipe size		4.40	1.818		10.15	85.50	12.05	107.70	151
0180	3" pipe size		4.40	1.818		10.15	85.50	12.05	107.70	151
0190	3-1/2" pipe size		3.30	2.424		11	114	16.05	141.05	200
0200	4" pipe size		3.30	2.424		11	114	16.05	141.05	200
0500	12" thick, 1/2" pipe size		9.40	.851		7.25	40	5.65	52.90	73.50
0520	3/4" pipe size		9.40	.851		7.25	40	5.65	52.90	73.50
0540	1" pipe size		7.30	1.096		13.75	51.50	7.25	72.50	99.50
0560	1-1/4" pipe size		7.30	1.096		13.75	51.50	7.25	72.50	99.50
0570	1-1/2" pipe size		7.30	1.096		13.75	51.50	7.25	72.50	99.50
0580	2" pipe size		3.60	2.222		15.70	104	14.70	134.40	188
0590	2-1/2" pipe size		3.60	2.222		15.70	104	14.70	134.40	188
0600	3" pipe size		3.60	2.222		15.70	104	14.70	134.40	188
0610	3-1/2" pipe size		2.80	2.857		18.25	134	18.90	171.15	241
0630	4" pipe size		2.50	3.200		18.25	150	21	189.25	268
0650	16" thick, 1/2" pipe size		7.60	1.053		9.75	49.50	6.95	66.20	92
0670	3/4" pipe size		7	1.143		9.75	53.50	7.55	70.80	99
0690	1" pipe size		6	1.333		18	62.50	8.85	89.35	123
0710	1-1/4" pipe size		5.50	1.455		18	68.50	9.65	96.15	132
0730	1-1/2" pipe size		5.50	1.455		18	68.50	9.65	96.15	132
0750	2" pipe size		3	2.667		21	125	17.65	163.65	230
0770	2-1/2" pipe size		2.70	2.963		21	139	19.60	179.60	252
0790	3" pipe size		2.50	3.200		21	150	21	192	271
0810	3-1/2" pipe size		2.30	3.478		25.50	163	23	211.50	297
0830	4" pipe size		2	4		25.50	188	26.50	240	335
0850	20" thick, 1/2" pipe size		6.40	1.250		12.25	59	8.30	79.55	110
0870	3/4" pipe size		6	1.333		12.25	62.50	8.85	83.60	117
0890	1" pipe size		5	1.600		22	75	10.60	107.60	148
0910	1-1/4" pipe size		4.80	1.667		22	78.50	11.05	111.55	154
0930	1-1/2" pipe size		4.60	1.739		22	81.50	11.50	115	159
0950	2" pipe size		2.70	2.963		26.50	139	19.60	185.10	258
0970	2-1/2" pipe size		2.40	3.333		26.50	157	22	205.50	287
0990	3" pipe size		2.20	3.636		26.50	171	24	221.50	310
1010	3-1/2" pipe size		2	4		32.50	188	26.50	247	345
1030	4" pipe size		1.70	4.706		32.50	221	31	284.50	400
1050	24" thick, 1/2" pipe size		5.50	1.455		14.80	68.50	9.65	92.95	129
1070	3/4" pipe size		5.10	1.569		14.80	73.50	10.40	98.70	138
1090	1" pipe size		4.30	1.860		26.50	87.50	12.30	126.30	173
1110	1-1/4" pipe size		4	2		26.50	94	13.25	133.75	184
1130	1-1/2" pipe size		4	2		26.50	94	13.25	133.75	184
1150	2" pipe size		2.40	3.333		32	157	22	211	293
1170	2-1/2" pipe size		2.20	3.636		32	171	24	227	315
1190	3" pipe size		2	4		32	188	26.50	246.50	345
1210	3-1/2" pipe size		1.80	4.444		40	209	29.50	278.50	385
1230	4" pipe size		1.50	5.333		40	251	35.50	326.50	460
1500	Brick wall, 8" thick, 1/2" pipe size		18	.444		4.71	21	2.94	28.65	39.50

26 05 33.95 Cutting and Drilling		Crew	Daily Output	Labor-Hours	Unit	2009 Bare Costs Material	Labor	Equipment	Total	Total Incl O&P
1520	3/4" pipe size	R-31	18	.444	Ea.	4.71	21	2.94	28.65	39.50
1540	1" pipe size		13.30	.602		9.50	28.50	3.98	41.98	57
1560	1-1/4" pipe size		13.30	.602		9.50	28.50	3.98	41.98	57
1580	1-1/2" pipe size		13.30	.602		9.50	28.50	3.98	41.98	57
1600	2" pipe size		5.70	1.404		10.15	66	9.30	85.45	119
1620	2-1/2" pipe size		5.70	1.404		10.15	66	9.30	85.45	119
1640	3" pipe size		5.70	1.404		10.15	66	9.30	85.45	119
1660	3-1/2" pipe size		4.40	1.818		11	85.50	12.05	108.55	152
1680	4" pipe size		4	2		11	94	13.25	118.25	167
1700	12" thick, 1/2" pipe size		14.50	.552		7.25	26	3.65	36.90	50.50
1720	3/4" pipe size		14.50	.552		7.25	26	3.65	36.90	50.50
1740	1" pipe size		11	.727		13.75	34	4.81	52.56	71.50
1760	1-1/4" pipe size		11	.727		13.75	34	4.81	52.56	71.50
1780	1-1/2" pipe size		11	.727		13.75	34	4.81	52.56	71.50
1800	2" pipe size		5	1.600		15.70	75	10.60	101.30	141
1820	2-1/2" pipe size		5	1.600		15.70	75	10.60	101.30	141
1840	3" pipe size		5	1.600		15.70	75	10.60	101.30	141
1860	3-1/2" pipe size		3.80	2.105		18.25	99	13.95	131.20	182
1880	4" pipe size		3.30	2.424		18.25	114	16.05	148.30	208
1900	16" thick, 1/2" pipe size		12.30	.650		9.75	30.50	4.31	44.56	61
1920	3/4" pipe size		12.30	.650		9.75	30.50	4.31	44.56	61
1940	1" pipe size		9.30	.860		18	40.50	5.70	64.20	86
1960	1-1/4" pipe size		9.30	.860		18	40.50	5.70	64.20	86
1980	1-1/2" pipe size		9.30	.860		18	40.50	5.70	64.20	86
2000	2" pipe size		4.40	1.818		21	85.50	12.05	118.55	164
2010	2-1/2" pipe size		4.40	1.818		21	85.50	12.05	118.55	164
2030	3" pipe size		4.40	1.818		21	85.50	12.05	118.55	164
2050	3-1/2" pipe size		3.30	2.424		25.50	114	16.05	155.55	216
2070	4" pipe size		3	2.667		25.50	125	17.65	168.15	234
2090	20" thick, 1/2" pipe size		10.70	.748		12.25	35	4.95	52.20	71.50
2110	3/4" pipe size		10.70	.748		12.25	35	4.95	52.20	71.50
2130	1" pipe size		8	1		22	47	6.60	75.60	102
2150	1-1/4" pipe size		8	1		22	47	6.60	75.60	102
2170	1-1/2" pipe size		8	1		22	47	6.60	75.60	102
2190	2" pipe size		4	2		26.50	94	13.25	133.75	184
2210	2-1/2" pipe size		4	2		26.50	94	13.25	133.75	184
2230	3" pipe size		4	2		26.50	94	13.25	133.75	184
2250	3-1/2" pipe size		3	2.667		32.50	125	17.65	175.15	242
2270	4" pipe size		2.70	2.963		32.50	139	19.60	191.10	265
2290	24" thick, 1/2" pipe size		9.40	.851		14.80	40	5.65	60.45	82
2310	3/4" pipe size		9.40	.851		14.80	40	5.65	60.45	82
2330	1" pipe size		7.10	1.127		26.50	53	7.45	86.95	116
2350	1-1/4" pipe size		7.10	1.127		26.50	53	7.45	86.95	116
2370	1-1/2" pipe size		7.10	1.127		26.50	53	7.45	86.95	116
2390	2" pipe size		3.60	2.222		32	104	14.70	150.70	207
2410	2-1/2" pipe size		3.60	2.222		32	104	14.70	150.70	207
2430	3" pipe size		3.60	2.222		32	104	14.70	150.70	207
2450	3-1/2" pipe size		2.80	2.857		40	134	18.90	192.90	265
2470	4" pipe size	↓	2.50	3.200	↓	40	150	21	211	292
3000	Knockouts to 8' high, metal boxes & enclosures									
3020	With hole saw, 1/2" pipe size	1 Elec	53	.151	Ea.		7.10		7.10	10.55
3040	3/4" pipe size		47	.170			8		8	11.90
3050	1" pipe size	↓	40	.200	↓		9.40		9.40	14

26 05 33 – Raceway and Boxes for Electrical Systems

26 05 33.95 Cutting and Drilling		Crew	Daily Output	Labor-Hours	Unit	2009 Bare Costs Material	Labor	Equipment	Total	Total Incl O&P
3060	1-1/4" pipe size	1 Elec	36	.222	Ea.		10.45		10.45	15.55
3070	1-1/2" pipe size		32	.250			11.75		11.75	17.50
3080	2" pipe size		27	.296			13.95		13.95	20.50
3090	2-1/2" pipe size		20	.400			18.80		18.80	28
4010	3" pipe size		16	.500			23.50		23.50	35
4030	3-1/2" pipe size		13	.615			29		29	43
4050	4" pipe size		11	.727			34		34	51
4070	With hand punch set, 1/2" pipe size		40	.200			9.40		9.40	14
4090	3/4" pipe size		32	.250			11.75		11.75	17.50
4110	1" pipe size		30	.267			12.55		12.55	18.65
4130	1-1/4" pipe size		28	.286			13.45		13.45	20
4150	1-1/2" pipe size		26	.308			14.45		14.45	21.50
4170	2" pipe size		20	.400			18.80		18.80	28
4190	2-1/2" pipe size		17	.471			22		22	33
4200	3" pipe size		15	.533			25		25	37.50
4220	3-1/2" pipe size		12	.667			31.50		31.50	46.50
4240	4" pipe size		10	.800			37.50		37.50	56
4260	With hydraulic punch, 1/2" pipe size		44	.182			8.55		8.55	12.70
4280	3/4" pipe size		38	.211			9.90		9.90	14.75
4300	1" pipe size		38	.211			9.90		9.90	14.75
4320	1-1/4" pipe size		38	.211			9.90		9.90	14.75
4340	1-1/2" pipe size		38	.211			9.90		9.90	14.75
4360	2" pipe size		32	.250			11.75		11.75	17.50
4380	2-1/2" pipe size		27	.296			13.95		13.95	20.50
4400	3" pipe size		23	.348			16.35		16.35	24.50
4420	3-1/2" pipe size		20	.400			18.80		18.80	28
4440	4" pipe size		18	.444			21		21	31

26 05 80 – Wiring Connections

26 05 80.10 Motor Connections

		Crew	Daily Output	Labor-Hours	Unit	Material	Labor	Equipment	Total	Total Incl O&P
0010	**MOTOR CONNECTIONS**									
0020	Flexible conduit and fittings, 115 volt, 1 phase, up to 1 HP motor	1 Elec	8	1	Ea.	9.10	47		56.10	80
0050	2 HP motor		6.50	1.231		9.35	58		67.35	96.50
0100	3 HP motor		5.50	1.455		14.55	68.50		83.05	118
0120	230 volt, 10 HP motor, 3 phase		4.20	1.905		15.60	89.50		105.10	150
0150	15 HP motor		3.30	2.424		27	114		141	200
0200	25 HP motor		2.70	2.963		31	139		170	241
0400	50 HP motor		2.20	3.636		79	171		250	340
0600	100 HP motor		1.50	5.333		194	251		445	590
1500	460 volt, 5 HP motor, 3 phase		8	1		9.95	47		56.95	81
1520	10 HP motor		8	1		9.95	47		56.95	81
1530	25 HP motor		6	1.333		17.50	62.50		80	113
1540	30 HP motor		6	1.333		17.50	62.50		80	113
1550	40 HP motor		5	1.600		27	75		102	142
1560	50 HP motor		5	1.600		29	75		104	144
1570	60 HP motor		3.80	2.105		32.50	99		131.50	183
1580	75 HP motor		3.50	2.286		45.50	107		152.50	211
1590	100 HP motor		2.50	3.200		68.50	150		218.50	300
1600	125 HP motor		2	4		90	188		278	380
1610	150 HP motor		1.80	4.444		95	209		304	415
1620	200 HP motor		1.50	5.333		150	251		401	540
2005	460 Volt, 5 HP motor, 3 Phase, w/sealtite		8	1		14.95	47		61.95	86.50
2010	10 HP motor		8	1		14.95	47		61.95	86.50

26 05 Common Work Results for Electrical

26 05 80 – Wiring Connections

26 05 80.10 Motor Connections		Crew	Daily Output	Labor-Hours	Unit	Material	2009 Bare Costs Labor	Equipment	Total	Total Incl O&P
2015	25 HP motor	1 Elec	6	1.333	Ea.	27	62.50		89.50	123
2020	30 HP motor		6	1.333		27	62.50		89.50	123
2025	40 HP motor		5	1.600		49.50	75		124.50	167
2030	50 HP motor		5	1.600		48.50	75		123.50	166
2035	60 HP motor		3.80	2.105		78	99		177	233
2040	75 HP motor		3.50	2.286		81.50	107		188.50	250
2045	100 HP motor		2.50	3.200		110	150		260	345
2055	150 HP motor		1.80	4.444		116	209		325	440
2060	200 HP motor		1.50	5.333		785	251		1,036	1,250

26 05 90 – Residential Applications

26 05 90.10 Residential Wiring		Crew	Daily Output	Labor-Hours	Unit	Material	2009 Bare Costs Labor	Equipment	Total	Total Incl O&P
0010	**RESIDENTIAL WIRING**									
0020	20' avg. runs and #14/2 wiring incl. unless otherwise noted									
1000	Service & panel, includes 24' SE-AL cable, service eye, meter,									
1010	Socket, panel board, main bkr., ground rod, 15 or 20 amp									
1020	1-pole circuit breakers, and misc. hardware									
1100	100 amp, with 10 branch breakers	1 Elec	1.19	6.723	Ea.	585	315		900	1,100
1110	With PVC conduit and wire		.92	8.696		660	410		1,070	1,350
1120	With RGS conduit and wire		.73	10.959		840	515		1,355	1,700
1150	150 amp, with 14 branch breakers		1.03	7.767		895	365		1,260	1,525
1170	With PVC conduit and wire		.82	9.756		1,050	460		1,510	1,825
1180	With RGS conduit and wire		.67	11.940		1,400	560		1,960	2,350
1200	200 amp, with 18 branch breakers	2 Elec	1.80	8.889		1,175	420		1,595	1,925
1220	With PVC conduit and wire		1.46	10.959		1,325	515		1,840	2,250
1230	With RGS conduit and wire		1.24	12.903		1,800	605		2,405	2,875
1800	Lightning surge suppressor for above services, add	1 Elec	32	.250		50.50	11.75		62.25	73
2000	Switch devices									
2100	Single pole, 15 amp, Ivory, with a 1-gang box, cover plate,									
2110	Type NM (Romex) cable	1 Elec	17.10	.468	Ea.	11.70	22		33.70	45.50
2120	Type MC (BX) cable		14.30	.559		28	26.50		54.50	70
2130	EMT & wire		5.71	1.401		34.50	66		100.50	136
2150	3-way, #14/3, type NM cable		14.55	.550		17.25	26		43.25	57.50
2170	Type MC cable		12.31	.650		39	30.50		69.50	88.50
2180	EMT & wire		5	1.600		38.50	75		113.50	155
2200	4-way, #14/3, type NM cable		14.55	.550		33.50	26		59.50	75.50
2220	Type MC cable		12.31	.650		55.50	30.50		86	107
2230	EMT & wire		5	1.600		55	75		130	173
2250	S.P., 20 amp, #12/2, type NM cable		13.33	.600		19.70	28		47.70	63.50
2270	Type MC cable		11.43	.700		33	33		66	85.50
2280	EMT & wire		4.85	1.649		44.50	77.50		122	164
2290	S.P. rotary dimmer, 600W, no wiring		17	.471		19.90	22		41.90	55
2300	S.P. rotary dimmer, 600W, type NM cable		14.55	.550		26.50	26		52.50	67.50
2320	Type MC cable		12.31	.650		43	30.50		73.50	92.50
2330	EMT & wire		5	1.600		51.50	75		126.50	169
2350	3-way rotary dimmer, type NM cable		13.33	.600		22.50	28		50.50	66.50
2370	Type MC cable		11.43	.700		39	33		72	92
2380	EMT & wire		4.85	1.649		47.50	77.50		125	167
2400	Interval timer wall switch, 20 amp, 1-30 min., #12/2									
2410	Type NM cable	1 Elec	14.55	.550	Ea.	53	26		79	96.50
2420	Type MC cable		12.31	.650		60	30.50		90.50	112
2430	EMT & wire		5	1.600		77.50	75		152.50	198
2500	Decorator style									

26 05 90.10 Residential Wiring		Crew	Daily Output	Labor-Hours	Unit	Material (2009 Bare Costs)	Labor (2009 Bare Costs)	Equipment (2009 Bare Costs)	Total (2009 Bare Costs)	Total Incl O&P
2510	S.P., 15 amp, type NM cable	1 Elec	17.10	.468	Ea.	16.15	22		38.15	50.50
2520	Type MC cable		14.30	.559		32.50	26.50		59	75
2530	EMT & wire		5.71	1.401		39	66		105	141
2550	3-way, #14/3, type NM cable		14.55	.550		21.50	26		47.50	62.50
2570	Type MC cable		12.31	.650		43.50	30.50		74	93
2580	EMT & wire		5	1.600		43	75		118	160
2600	4-way, #14/3, type NM cable		14.55	.550		38	26		64	80.50
2620	Type MC cable		12.31	.650		60	30.50		90.50	111
2630	EMT & wire		5	1.600		59.50	75		134.50	178
2650	S.P., 20 amp, #12/2, type NM cable		13.33	.600		24	28		52	68.50
2670	Type MC cable		11.43	.700		37.50	33		70.50	90.50
2680	EMT & wire		4.85	1.649		49	77.50		126.50	169
2700	S.P., slide dimmer, type NM cable		17.10	.468		30	22		52	66
2720	Type MC cable		14.30	.559		46.50	26.50		73	90.50
2730	EMT & wire		5.71	1.401		55	66		121	159
2750	S.P., touch dimmer, type NM cable		17.10	.468		27	22		49	62.50
2770	Type MC cable		14.30	.559		43.50	26.50		70	87
2780	EMT & wire		5.71	1.401		52	66		118	156
2800	3-way touch dimmer, type NM cable		13.33	.600		47	28		75	94
2820	Type MC cable		11.43	.700		63.50	33		96.50	119
2830	EMT & wire	↓	4.85	1.649	↓	72	77.50		149.50	194
3000	Combination devices									
3100	S.P. switch/15 amp recpt., Ivory, 1-gang box, plate									
3110	Type NM cable	1 Elec	11.43	.700	Ea.	22.50	33		55.50	74
3120	Type MC cable		10	.800		39	37.50		76.50	99
3130	EMT & wire		4.40	1.818		47.50	85.50		133	179
3150	S.P. switch/pilot light, type NM cable		11.43	.700		23.50	33		56.50	74.50
3170	Type MC cable		10	.800		40	37.50		77.50	100
3180	EMT & wire		4.43	1.806		48	85		133	179
3190	2-S.P. switches, 2-#14/2, no wiring		14	.571		7.80	27		34.80	48.50
3200	2-S.P. switches, 2-#14/2, type NM cables		10	.800		28	37.50		65.50	87
3220	Type MC cable		8.89	.900		53.50	42.50		96	122
3230	EMT & wire		4.10	1.951		52.50	91.50		144	194
3250	3-way switch/15 amp recpt., #14/3, type NM cable		10	.800		31.50	37.50		69	91
3270	Type MC cable		8.89	.900		53.50	42.50		96	122
3280	EMT & wire		4.10	1.951		53	91.50		144.50	195
3300	2-3 way switches, 2-#14/3, type NM cables		8.89	.900		44.50	42.50		87	112
3320	Type MC cable		8	1		80	47		127	158
3330	EMT & wire		4	2		62	94		156	208
3350	S.P. switch/20 amp recpt., #12/2, type NM cable		10	.800		36	37.50		73.50	95.50
3370	Type MC cable		8.89	.900		43	42.50		85.50	111
3380	EMT & wire	↓	4.10	1.951	↓	60.50	91.50		152	203
3400	Decorator style									
3410	S.P. switch/15 amp recpt., type NM cable	1 Elec	11.43	.700	Ea.	27	33		60	78.50
3420	Type MC cable		10	.800		43.50	37.50		81	104
3430	EMT & wire		4.40	1.818		52	85.50		137.50	184
3450	S.P. switch/pilot light, type NM cable		11.43	.700		28	33		61	79.50
3470	Type MC cable		10	.800		44.50	37.50		82	105
3480	EMT & wire		4.40	1.818		52.50	85.50		138	185
3500	2-S.P. switches, 2-#14/2, type NM cables		10	.800		32.50	37.50		70	92
3520	Type MC cable		8.89	.900		58	42.50		100.50	127
3530	EMT & wire		4.10	1.951		57	91.50		148.50	199
3550	3-way/15 amp recpt., #14/3, type NM cable	↓	10	.800	↓	36	37.50		73.50	95.50

26 05 90.10 Residential Wiring		Crew	Daily Output	Labor-Hours	Unit	Material	2009 Bare Costs Labor	Equipment	Total	Total Incl O&P
3570	Type MC cable	1 Elec	8.89	.900	Ea.	58	42.50		100.50	127
3580	EMT & wire		4.10	1.951		57.50	91.50		149	200
3650	2-3 way switches, 2-#14/3, type NM cables		8.89	.900		48.50	42.50		91	117
3670	Type MC cable		8	1		84.50	47		131.50	163
3680	EMT & wire		4	2		66.50	94		160.50	213
3700	S.P. switch/20 amp recpt., #12/2, type NM cable		10	.800		40.50	37.50		78	101
3720	Type MC cable		8.89	.900		47.50	42.50		90	116
3730	EMT & wire		4.10	1.951		65	91.50		156.50	208
4000	Receptacle devices									
4010	Duplex outlet, 15 amp recpt., Ivory, 1-gang box, plate									
4015	Type NM cable	1 Elec	14.55	.550	Ea.	10.20	26		36.20	49.50
4020	Type MC cable		12.31	.650		26.50	30.50		57	75
4030	EMT & wire		5.33	1.501		33	70.50		103.50	142
4050	With #12/2, type NM cable		12.31	.650		13.65	30.50		44.15	60.50
4070	Type MC cable		10.67	.750		27	35		62	82
4080	EMT & wire		4.71	1.699		38.50	80		118.50	162
4100	20 amp recpt., #12/2, type NM cable		12.31	.650		22.50	30.50		53	70
4120	Type MC cable		10.67	.750		35.50	35		70.50	91.50
4130	EMT & wire		4.71	1.699		47	80		127	171
4140	For GFI see Div. 26 05 90.10 line 4300 below									
4150	Decorator style, 15 amp recpt., type NM cable	1 Elec	14.55	.550	Ea.	14.65	26		40.65	54.50
4170	Type MC cable		12.31	.650		31	30.50		61.50	79.50
4180	EMT & wire		5.33	1.501		37.50	70.50		108	146
4200	With #12/2, type NM cable		12.31	.650		18.10	30.50		48.60	65.50
4220	Type MC cable		10.67	.750		31.50	35		66.50	87
4230	EMT & wire		4.71	1.699		43	80		123	166
4250	20 amp recpt. #12/2, type NM cable		12.31	.650		26.50	30.50		57	75
4270	Type MC cable		10.67	.750		40	35		75	96.50
4280	EMT & wire		4.71	1.699		51.50	80		131.50	176
4300	GFI, 15 amp recpt., type NM cable		12.31	.650		41.50	30.50		72	91
4320	Type MC cable		10.67	.750		58	35		93	116
4330	EMT & wire		4.71	1.699		64.50	80		144.50	190
4350	GFI with #12/2, type NM cable		10.67	.750		45	35		80	102
4370	Type MC cable		9.20	.870		58.50	41		99.50	125
4380	EMT & wire		4.21	1.900		69.50	89.50		159	210
4400	20 amp recpt., #12/2 type NM cable		10.67	.750		47	35		82	104
4420	Type MC cable		9.20	.870		60	41		101	127
4430	EMT & wire		4.21	1.900		71.50	89.50		161	212
4500	Weather-proof cover for above receptacles, add		32	.250		5.25	11.75		17	23.50
4550	Air conditioner outlet, 20 amp-240 volt recpt.									
4560	30′ of #12/2, 2 pole circuit breaker									
4570	Type NM cable	1 Elec	10	.800	Ea.	63	37.50		100.50	125
4580	Type MC cable		9	.889		79	42		121	149
4590	EMT & wire		4	2		87.50	94		181.50	236
4600	Decorator style, type NM cable		10	.800		67.50	37.50		105	130
4620	Type MC cable		9	.889		84	42		126	154
4630	EMT & wire		4	2		92	94		186	241
4650	Dryer outlet, 30 amp-240 volt recpt., 20′ of #10/3									
4660	2 pole circuit breaker									
4670	Type NM cable	1 Elec	6.41	1.248	Ea.	75	58.50		133.50	170
4680	Type MC cable		5.71	1.401		83	66		149	190
4690	EMT & wire		3.48	2.299		89	108		197	259
4700	Range outlet, 50 amp-240 volt recpt., 30′ of #8/3									

26 05 90.10 Residential Wiring		Crew	Daily Output	Labor-Hours	Unit	Material	Labor	Equipment	Total	Total Incl O&P
						2009 Bare Costs				
4710	Type NM cable	1 Elec	4.21	1.900	Ea.	110	89.50		199.50	253
4720	Type MC cable		4	2		195	94		289	355
4730	EMT & wire		2.96	2.703		121	127		248	320
4750	Central vacuum outlet, Type NM cable		6.40	1.250		69.50	59		128.50	164
4770	Type MC cable		5.71	1.401		94	66		160	201
4780	EMT & wire	↓	3.48	2.299	↓	96.50	108		204.50	267
4800	30 amp-110 volt locking recpt., #10/2 circ. bkr.									
4810	Type NM cable	1 Elec	6.20	1.290	Ea.	80.50	60.50		141	179
4830	EMT & wire	"	3.20	2.500	"	110	118		228	296
4900	Low voltage outlets									
4910	Telephone recpt., 20′ of 4/C phone wire	1 Elec	26	.308	Ea.	10.30	14.45		24.75	33
4920	TV recpt., 20′ of RG59U coax wire, F type connector	"	16	.500	"	18.35	23.50		41.85	55
4950	Door bell chime, transformer, 2 buttons, 60′ of bellwire									
4970	Economy model	1 Elec	11.50	.696	Ea.	75	32.50		107.50	131
4980	Custom model		11.50	.696		116	32.50		148.50	177
4990	Luxury model, 3 buttons	↓	9.50	.842	↓	310	39.50		349.50	400
6000	Lighting outlets									
6050	Wire only (for fixture), type NM cable	1 Elec	32	.250	Ea.	8.45	11.75		20.20	27
6070	Type MC cable		24	.333		18.55	15.65		34.20	44
6080	EMT & wire		10	.800		23.50	37.50		61	81.50
6100	Box (4″), and wire (for fixture), type NM cable		25	.320		17.45	15.05		32.50	41.50
6120	Type MC cable		20	.400		27.50	18.80		46.30	58.50
6130	EMT & wire	↓	11	.727	↓	32.50	34		66.50	86.50
6200	Fixtures (use with lines 6050 or 6100 above)									
6210	Canopy style, economy grade	1 Elec	40	.200	Ea.	29	9.40		38.40	46
6220	Custom grade		40	.200		51	9.40		60.40	70
6250	Dining room chandelier, economy grade		19	.421		82.50	19.80		102.30	121
6260	Custom grade		19	.421		243	19.80		262.80	297
6270	Luxury grade		15	.533		540	25		565	635
6310	Kitchen fixture (fluorescent), economy grade		30	.267		61.50	12.55		74.05	86
6320	Custom grade		25	.320		168	15.05		183.05	208
6350	Outdoor, wall mounted, economy grade		30	.267		29.50	12.55		42.05	51
6360	Custom grade		30	.267		108	12.55		120.55	138
6370	Luxury grade		25	.320		244	15.05		259.05	291
6410	Outdoor PAR floodlights, 1 lamp, 150 watt		20	.400		32	18.80		50.80	63
6420	2 lamp, 150 watt each		20	.400		52	18.80		70.80	85
6430	For infrared security sensor, add		32	.250		111	11.75		122.75	140
6450	Outdoor, quartz-halogen, 300 watt flood		20	.400		41	18.80		59.80	73
6600	Recessed downlight, round, pre-wired, 50 or 75 watt trim		30	.267		42	12.55		54.55	64.50
6610	With shower light trim		30	.267		52	12.55		64.55	75.50
6620	With wall washer trim		28	.286		62.50	13.45		75.95	89
6630	With eye-ball trim	↓	28	.286		62.50	13.45		75.95	89
6640	For direct contact with insulation, add					2.10			2.10	2.31
6700	Porcelain lamp holder	1 Elec	40	.200		4.10	9.40		13.50	18.50
6710	With pull switch		40	.200		4.59	9.40		13.99	19.05
6750	Fluorescent strip, 1-20 watt tube, wrap around diffuser, 24″		24	.333		54	15.65		69.65	83
6770	2-40 watt tubes, 48″		20	.400		80	18.80		98.80	116
6780	With residential ballast		20	.400		90	18.80		108.80	127
6800	Bathroom heat lamp, 1-250 watt		28	.286		46	13.45		59.45	70.50
6810	2-250 watt lamps	↓	28	.286	↓	71.50	13.45		84.95	98.50
6820	For timer switch, see Div. 26 05 90.10 line 2400									
6900	Outdoor post lamp, incl. post, fixture, 35′ of #14/2									
6910	Type NMC cable	1 Elec	3.50	2.286	Ea.	203	107		310	385

26 05 90.10 Residential Wiring		Crew	Daily Output	Labor-Hours	Unit	Material	2009 Bare Costs Labor	Equipment	Total	Total Incl O&P
6920	Photo-eye, add	1 Elec	27	.296	Ea.	32.50	13.95		46.45	56.50
6950	Clock dial time switch, 24 hr., w/enclosure, type NM cable		11.43	.700		63	33		96	118
6970	Type MC cable		11	.727		79.50	34		113.50	138
6980	EMT & wire		4.85	1.649		85.50	77.50		163	209
7000	Alarm systems									
7050	Smoke detectors, box, #14/3, type NM cable	1 Elec	14.55	.550	Ea.	37.50	26		63.50	79.50
7070	Type MC cable		12.31	.650		53	30.50		83.50	104
7080	EMT & wire		5	1.600		52.50	75		127.50	170
7090	For relay output to security system, add					12.90			12.90	14.20
8000	Residential equipment									
8050	Disposal hook-up, incl. switch, outlet box, 3′ of flex									
8060	20 amp-1 pole circ. bkr., and 25′ of #12/2									
8070	Type NM cable	1 Elec	10	.800	Ea.	33.50	37.50		71	93
8080	Type MC cable		8	1		48.50	47		95.50	124
8090	EMT & wire		5	1.600		61	75		136	180
8100	Trash compactor or dishwasher hook-up, incl. outlet box,									
8110	3′ of flex, 15 amp-1 pole circ. bkr., and 25′ of #14/2									
8120	Type NM cable	1 Elec	10	.800	Ea.	24.50	37.50		62	83
8130	Type MC cable		8	1		43.50	47		90.50	118
8140	EMT & wire		5	1.600		52.50	75		127.50	170
8150	Hot water sink dispensor hook-up, use line 8100									
8200	Vent/exhaust fan hook-up, type NM cable	1 Elec	32	.250	Ea.	8.45	11.75		20.20	27
8220	Type MC cable		24	.333		18.55	15.65		34.20	44
8230	EMT & wire		10	.800		23.50	37.50		61	81.50
8250	Bathroom vent fan, 50 CFM (use with above hook-up)									
8260	Economy model	1 Elec	15	.533	Ea.	24.50	25		49.50	64.50
8270	Low noise model		15	.533		32	25		57	72.50
8280	Custom model		12	.667		117	31.50		148.50	175
8300	Bathroom or kitchen vent fan, 110 CFM									
8310	Economy model	1 Elec	15	.533	Ea.	66.50	25		91.50	111
8320	Low noise model	″	15	.533	″	86	25		111	132
8350	Paddle fan, variable speed (w/o lights)									
8360	Economy model (AC motor)	1 Elec	10	.800	Ea.	105	37.50		142.50	172
8370	Custom model (AC motor)		10	.800		182	37.50		219.50	256
8380	Luxury model (DC motor)		8	1		360	47		407	465
8390	Remote speed switch for above, add		12	.667		28.50	31.50		60	78
8500	Whole house exhaust fan, ceiling mount, 36″, variable speed									
8510	Remote switch, incl. shutters, 20 amp-1 pole circ. bkr.									
8520	30′ of #12/2, type NM cable	1 Elec	4	2	Ea.	985	94		1,079	1,225
8530	Type MC cable		3.50	2.286		1,000	107		1,107	1,250
8540	EMT & wire		3	2.667		1,025	125		1,150	1,300
8600	Whirlpool tub hook-up, incl. timer switch, outlet box									
8610	3′ of flex, 20 amp-1 pole GFI circ. bkr.									
8620	30′ of #12/2, type NM cable	1 Elec	5	1.600	Ea.	127	75		202	257
8630	Type MC cable		4.20	1.905		135	89.50		224.50	282
8640	EMT & wire		3.40	2.353		146	111		257	325
8650	Hot water heater hook-up, incl. 1-2 pole circ. bkr., box;									
8660	3′ of flex, 20′ of #10/2, type NM cable	1 Elec	5	1.600	Ea.	36	75		111	152
8670	Type MC cable		4.20	1.905		58	89.50		147.50	197
8680	EMT & wire		3.40	2.353		55.50	111		166.50	226
9000	Heating/air conditioning									
9050	Furnace/boiler hook-up, incl. firestat, local on-off switch									
9060	Emergency switch, and 40′ of type NM cable	1 Elec	4	2	Ea.	57	94		151	203

26 05 Common Work Results for Electrical

26 05 90 – Residential Applications

26 05 90.10 Residential Wiring		Crew	Daily Output	Labor-Hours	Unit	Material	Labor	Equipment	Total	Total Incl O&P
						2009 Bare Costs				
9070	Type MC cable	1 Elec	3.50	2.286	Ea.	82.50	107		189.50	251
9080	EMT & wire	↓	1.50	5.333	↓	94	251		345	480
9100	Air conditioner hook-up, incl. local 60 amp disc. switch									
9110	3′ sealtite, 40 amp, 2 pole circuit breaker									
9130	40′ of #8/2, type NM cable	1 Elec	3.50	2.286	Ea.	229	107		336	410
9140	Type MC cable		3	2.667		360	125		485	580
9150	EMT & wire	↓	1.30	6.154	↓	268	289		557	725
9200	Heat pump hook-up, 1-40 & 1-100 amp 2 pole circ. bkr.									
9210	Local disconnect switch, 3′ sealtite									
9220	40′ of #8/2 & 30′ of #3/2									
9230	Type NM cable	1 Elec	1.30	6.154	Ea.	595	289		884	1,075
9240	Type MC cable		1.08	7.407		865	350		1,215	1,475
9250	EMT & wire	↓	.94	8.511	↓	675	400		1,075	1,350
9500	Thermostat hook-up, using low voltage wire									
9520	Heating only, 25′ of #18-3	1 Elec	24	.333	Ea.	7.70	15.65		23.35	32
9530	Heating/cooling, 25′ of #18-4	"	20	.400	"	9.15	18.80		27.95	38

26 24 Switchboards and Panelboards

26 24 19 – Motor-Control Centers

26 24 19.40 Motor Starters and Controls		Crew	Daily Output	Labor-Hours	Unit	Material	Labor	Equipment	Total	Total Incl O&P
0010	**MOTOR STARTERS AND CONTROLS**									
0050	Magnetic, FVNR, with enclosure and heaters, 480 volt									
0080	2 HP, size 00	1 Elec	3.50	2.286	Ea.	205	107		312	385
0100	5 HP, size 0		2.30	3.478		248	163		411	515
0200	10 HP, size 1	↓	1.60	5		278	235		513	655
0300	25 HP, size 2	2 Elec	2.20	7.273		525	340		865	1,075
0400	50 HP, size 3		1.80	8.889		855	420		1,275	1,550
0500	100 HP, size 4		1.20	13.333		1,900	625		2,525	3,000
0600	200 HP, size 5		.90	17.778		4,425	835		5,260	6,125
0610	400 HP, size 6	↓	.80	20		17,300	940		18,240	20,500
0620	NEMA 7, 5 HP, size 0	1 Elec	1.60	5		1,300	235		1,535	1,775
0630	10 HP, size 1	"	1.10	7.273		1,350	340		1,690	1,975
0640	25 HP, size 2	2 Elec	1.80	8.889		2,175	420		2,595	3,025
0650	50 HP, size 3		1.20	13.333		3,275	625		3,900	4,525
0660	100 HP, size 4		.90	17.778		5,275	835		6,110	7,050
0670	200 HP, size 5	↓	.50	32		12,600	1,500		14,100	16,100
0700	Combination, with motor circuit protectors, 5 HP, size 0	1 Elec	1.80	4.444		805	209		1,014	1,200
0800	10 HP, size 1	"	1.30	6.154		835	289		1,124	1,350
0900	25 HP, size 2	2 Elec	2	8		1,175	375		1,550	1,850
1000	50 HP, size 3		1.32	12.121		1,700	570		2,270	2,700
1200	100 HP, size 4	↓	.80	20		3,675	940		4,615	5,450
1220	NEMA 7, 5 HP, size 0	1 Elec	1.30	6.154		2,300	289		2,589	2,950
1230	10 HP, size 1	"	1	8		2,350	375		2,725	3,150
1240	25 HP, size 2	2 Elec	1.32	12.121		3,150	570		3,720	4,325
1250	50 HP, size 3		.80	20		5,200	940		6,140	7,125
1260	100 HP, size 4		.60	26.667		8,125	1,250		9,375	10,800
1270	200 HP, size 5	↓	.40	40		17,600	1,875		19,475	22,200
1400	Combination, with fused switch, 5 HP, size 0	1 Elec	1.80	4.444		615	209		824	985
1600	10 HP, size 1	"	1.30	6.154		660	289		949	1,150
1800	25 HP, size 2	2 Elec	2	8		1,075	375		1,450	1,725
2000	50 HP, size 3	↓	1.32	12.121	↓	1,800	570		2,370	2,825

26 24 19 – Motor-Control Centers

26 24 19.40 Motor Starters and Controls		Crew	Daily Output	Labor-Hours	Unit	2009 Bare Costs Material	Labor	Equipment	Total	Total Incl O&P
2200	100 HP, size 4	2 Elec	.80	20	Ea.	3,150	940		4,090	4,875
3500	Magnetic FVNR with NEMA 12, enclosure & heaters, 480 volt									
3600	5 HP, size 0	1 Elec	2.20	3.636	Ea.	234	171		405	510
3700	10 HP, size 1	"	1.50	5.333		350	251		601	760
3800	25 HP, size 2	2 Elec	2	8		660	375		1,035	1,275
3900	50 HP, size 3		1.60	10		1,025	470		1,495	1,825
4000	100 HP, size 4		1	16		2,425	750		3,175	3,800
4100	200 HP, size 5		.80	20		5,825	940		6,765	7,800
4200	Combination, with motor circuit protectors, 5 HP, size 0	1 Elec	1.70	4.706		775	221		996	1,175
4300	10 HP, size 1	"	1.20	6.667		805	315		1,120	1,350
4400	25 HP, size 2	2 Elec	1.80	8.889		1,200	420		1,620	1,950
4500	50 HP, size 3		1.20	13.333		1,950	625		2,575	3,075
4600	100 HP, size 4		.74	21.622		4,425	1,025		5,450	6,375
4700	Combination, with fused switch, 5 HP, size 0	1 Elec	1.70	4.706		750	221		971	1,150
4800	10 HP, size 1	"	1.20	6.667		780	315		1,095	1,325
4900	25 HP, size 2	2 Elec	1.80	8.889		1,200	420		1,620	1,925
5000	50 HP, size 3		1.20	13.333		1,900	625		2,525	3,025
5100	100 HP, size 4		.74	21.622		3,875	1,025		4,900	5,750
5200	Factory installed controls, adders to size 0 thru 5									
5300	Start-stop push button	1 Elec	32	.250	Ea.	49	11.75		60.75	71.50
5400	Hand-off-auto-selector switch		32	.250		49	11.75		60.75	71.50
5500	Pilot light		32	.250		92	11.75		103.75	117
5600	Start-stop-pilot		32	.250		141	11.75		152.75	173
5700	Auxiliary contact, NO or NC		32	.250		67.50	11.75		79.25	91.50
5800	NO-NC		32	.250		135	11.75		146.75	166

26 29 Low-Voltage Controllers

26 29 13 – Enclosed Controllers

26 29 13.20 Control Stations		Crew	Daily Output	Labor-Hours	Unit	Material	Labor	Equipment	Total	Total Incl O&P
0010	**CONTROL STATIONS**									
0050	NEMA 1, heavy duty, stop/start	1 Elec	8	1	Ea.	141	47		188	225
0100	Stop/start, pilot light		6.20	1.290		192	60.50		252.50	300
0200	Hand/off/automatic		6.20	1.290		104	60.50		164.50	205
0400	Stop/start/reverse		5.30	1.509		190	71		261	315
0500	NEMA 7, heavy duty, stop/start		6	1.333		320	62.50		382.50	450
0600	Stop/start, pilot light		4	2		390	94		484	570

26 29 23 – Variable-Frequency Motor Controllers

26 29 23.10 Variable Frequency Drives/Adj. Frequency Drives		Crew	Daily Output	Labor-Hours	Unit	Material	Labor	Equipment	Total	Total Incl O&P
0010	**VARIABLE FREQUENCY DRIVES/ADJ. FREQUENCY DRIVES**									
0100	Enclosed (NEMA 1), 460 volt, for 3 HP motor size G	1 Elec	.80	10	Ea.	1,575	470		2,045	2,425
0110	5 HP motor size G		.80	10		1,700	470		2,170	2,575
0120	7.5 HP motor size G		.67	11.940		2,025	560		2,585	3,050
0130	10 HP motor size G		.67	11.940		2,025	560		2,585	3,050
0140	15 HP motor size G	2 Elec	.89	17.978		2,325	845		3,170	3,825
0150	20 HP motor size G		.89	17.978		3,450	845		4,295	5,050
0160	25 HP motor size G		.67	23.881		4,025	1,125		5,150	6,100
0170	30 HP motor size G		.67	23.881		4,900	1,125		6,025	7,050
0180	40 HP motor size G		.67	23.881		7,175	1,125		8,300	9,575
0190	50 HP motor size G		.53	30.189		7,825	1,425		9,250	10,700
0200	60 HP motor size G	R-3	.56	35.714		8,575	1,650	231	10,456	12,200

26 29 Low-Voltage Controllers

26 29 23 – Variable-Frequency Motor Controllers

26 29 23.10 Variable Frequency Drives/Adj. Frequency Drives			Crew	Daily Output	Labor-Hours	Unit	2009 Bare Costs Material	Labor	Equipment	Total	Total Incl O&P
0210	75 HP motor size	G	R-3	.56	35.714	Ea.	11,700	1,650	231	13,581	15,600
0220	100 HP motor size	G		.50	40		12,000	1,850	258	14,108	16,100
0230	125 HP motor size	G		.50	40		13,200	1,850	258	15,308	17,600
0240	150 HP motor size	G		.50	40		16,900	1,850	258	19,008	21,600
0250	200 HP motor size	G		.42	47.619		21,000	2,200	310	23,510	26,700
1100	Custom-engineered, 460 volt, for 3 HP motor size	G	1 Elec	.56	14.286		2,600	670		3,270	3,850
1110	5 HP motor size	G		.56	14.286		2,725	670		3,395	4,000
1120	7.5 HP motor size	G		.47	17.021		3,275	800		4,075	4,800
1130	10 HP motor size	G		.47	17.021		3,275	800		4,075	4,800
1140	15 HP motor size	G	2 Elec	.62	25.806		3,575	1,225		4,800	5,750
1150	20 HP motor size	G		.62	25.806		4,900	1,225		6,125	7,175
1160	25 HP motor size	G		.47	34.043		5,325	1,600		6,925	8,225
1170	30 HP motor size	G		.47	34.043		7,250	1,600		8,850	10,400
1180	40 HP motor size	G		.47	34.043		8,075	1,600		9,675	11,300
1190	50 HP motor size	G		.37	43.243		8,075	2,025		10,100	11,900
1200	60 HP motor size	G	R-3	.39	51.282		12,100	2,375	330	14,805	17,200
1210	75 HP motor size	G		.39	51.282		14,000	2,375	330	16,705	19,200
1220	100 HP motor size	G		.35	57.143		14,500	2,650	370	17,520	20,400
1230	125 HP motor size	G		.35	57.143		15,300	2,650	370	18,320	21,300
1240	150 HP motor size	G		.35	57.143		17,200	2,650	370	20,220	23,300
1250	200 HP motor size	G		.29	68.966		23,800	3,200	445	27,445	31,400
2000	For complex & special design systems to meet specific										
2010	requirements, obtain quote from vendor.										

26 42 Cathodic Protection

26 42 16 – Passive Cathodic Protection for Underground Storage Tank

26 42 16.50 Cathodic Protection Wiring Methods		Crew	Daily Output	Labor-Hours	Unit	Material	Labor	Equipment	Total	Total Incl O&P
0010	**CATHODIC PROTECTION WIRING METHODS**									
1000	Anodes, magnesium type, 9 #	R-15	18.50	2.595	Ea.	47	119	17.75	183.75	249
1010	17 #		13	3.692		84	169	25.50	278.50	375
1020	32 #		10	4.800		153	220	33	406	530
1030	48 #		7.20	6.667		230	305	45.50	580.50	760
1100	Graphite type w/ epoxy cap, 3" x 60" (32 #)	R-22	8.40	4.438		82.50	177		259.50	360
1110	4" x 80" (68 #)		6	6.213		157	248		405	550
1120	6" x 72" (80 #)		5.20	7.169		1,275	286		1,561	1,825
1130	6" x 36" (45 #)		9.60	3.883		645	155		800	945
2000	Rectifiers, silicon type, air cooled, 28 V/10 A	R-19	3.50	5.714		2,025	269		2,294	2,625
2010	20 V/20 A		3.50	5.714		2,050	269		2,319	2,675
2100	Oil immersed, 28 V/10 A		3	6.667		3,425	315		3,740	4,250
2110	20 V/20 A		3	6.667		3,525	315		3,840	4,350
3000	Anode backfill, coke breeze	R-22	3850	.010	Lb.	.14	.39		.53	.73
4000	Cable, HMWPE, No. 8		2.40	15.533	M.L.F.	320	620		940	1,300
4010	No. 6		2.40	15.533		480	620		1,100	1,450
4020	No. 4		2.40	15.533		695	620		1,315	1,700
4030	No. 2		2.40	15.533		1,100	620		1,720	2,125
4040	No. 1		2.20	16.945		1,450	675		2,125	2,625
4050	No. 1/0		2.20	16.945		1,775	675		2,450	2,975
4060	No. 2/0		2.20	16.945		2,175	675		2,850	3,400
4070	No. 4/0		2	18.640		4,600	745		5,345	6,200
5000	Test station, 7 terminal box, flush curb type w/lockable cover	R-19	12	1.667	Ea.	70.50	78.50		149	195
5010	Reference cell, 2" dia PVC conduit, cplg, plug, set flush	"	4.80	4.167	"	148	196		344	455

Estimating Tips

- When estimating material costs for electronic safety and security systems, it is always prudent to obtain manufacturers' quotations for equipment prices and special installation requirements that affect the total cost.
- Fire alarm systems consist of control panels, annunciator panels, battery with rack, charger, and fire alarm actuating and indicating devices. Some fire alarm systems include speakers, telephone lines, door closer controls, and other components. Be careful not to overlook the costs related to installation for these items. Also be aware of costs for integrated automation instrumentation and terminal devices, control equipment, control wiring, and programming.
- Security equipment includes items such as CCTV, access control, and other detection and identification systems to perform alert and alarm functions. Be sure to consider the costs related to installation for this security equipment, such as for integrated automation instrumentation and terminal devices, control equipment, control wiring, and programming.

Reference Numbers

Reference numbers are shown in shaded boxes at the beginning of some major classifications. These numbers refer to related items in the Reference Section. The reference information may be an estimating procedure, an alternate pricing method, or technical information.

Note: Not all subdivisions listed here necessarily appear in this publication.

28 13 Access Control

28 13 53 – Security Access Detection

28 13 53.13 Security Access Metal Detectors		Crew	Daily Output	Labor-Hours	Unit	2009 Bare Costs Material	Labor	Equipment	Total	Total Incl O&P
0010	**SECURITY ACCESS METAL DETECTORS**									
0240	Metal detector, hand-held, wand type, unit only				Ea.	81.50			81.50	90
0250	Metal detector, walk through portal type, single zone	1 Elec	2	4		2,750	188		2,938	3,300
0260	Multi-zone	"	2	4		3,500	188		3,688	4,125

28 13 53.16 Security Access X-Ray Equipment		Crew	Daily Output	Labor-Hours	Unit	Material	Labor	Equipment	Total	Total Incl O&P
0010	**SECURITY ACCESS X-RAY EQUIPMENT**									
0290	X-ray machine, desk top, for mail/small packages/letters	1 Elec	4	2	Ea.	3,000	94		3,094	3,450
0300	Conveyor type, incl monitor, minimum		2	4		14,000	188		14,188	15,700
0310	Maximum		2	4		25,000	188		25,188	27,800
0320	X-ray machine, large unit, for airports, incl monitor, min	2 Elec	1	16		35,000	750		35,750	39,600
0330	Maximum	"	.50	32		60,000	1,500		61,500	68,500

28 13 53.23 Security Access Explosive Detection Equipment		Crew	Daily Output	Labor-Hours	Unit	Material	Labor	Equipment	Total	Total Incl O&P
0010	**SECURITY ACCESS EXPLOSIVE DETECTION EQUIPMENT**									
0270	Explosives detector, walk through portal type	1 Elec	2	4	Ea.	3,500	188		3,688	4,125
0280	Hand-held, battery operated				"				25,500	28,050

28 16 Intrusion Detection

28 16 16 – Intrusion Detection Systems Infrastructure

28 16 16.50 Intrusion Detection		Crew	Daily Output	Labor-Hours	Unit	Material	Labor	Equipment	Total	Total Incl O&P
0010	**INTRUSION DETECTION**, not including wires & conduits									
0100	Burglar alarm, battery operated, mechanical trigger	1 Elec	4	2	Ea.	272	94		366	440
0200	Electrical trigger		4	2		325	94		419	495
0400	For outside key control, add		8	1		77	47		124	155
0600	For remote signaling circuitry, add		8	1		122	47		169	205
0800	Card reader, flush type, standard		2.70	2.963		910	139		1,049	1,200
1000	Multi-code		2.70	2.963		1,175	139		1,314	1,500

28 23 Video Surveillance

28 23 23 – Video Surveillance Systems Infrastructure

28 23 23.50 Video Surveillance Equipments		Crew	Daily Output	Labor-Hours	Unit	Material	Labor	Equipment	Total	Total Incl O&P
0010	**VIDEO SURVEILLANCE EQUIPMENTS**									
0200	Video cameras, wireless, hidden in exit signs, clocks, etc, incl receiver	1 Elec	3	2.667	Ea.	149	125		274	350
0210	Accessories for VCR, single camera		3	2.667		570	125		695	810
0220	For multiple cameras		3	2.667		1,450	125		1,575	1,775
0230	Video cameras, wireless, for under vehicle searching, complete		2	4		9,975	188		10,163	11,300

28 31 Fire Detection and Alarm

28 31 23 – Fire Detection and Alarm Annunciation Panels and Fire Stations

28 31 23.50 Alarm Panels and Devices		Crew	Daily Output	Labor-Hours	Unit	Material	Labor	Equipment	Total	Total Incl O&P
0010	**ALARM PANELS AND DEVICES**, not including wires & conduits									
3594	Fire, alarm control panel									
3600	4 zone	2 Elec	2	8	Ea.	615	375		990	1,225
3800	8 zone		1	16		675	750		1,425	1,875
4000	12 zone		.67	23.988		2,300	1,125		3,425	4,200
4020	Alarm device	1 Elec	8	1		135	47		182	219
4050	Actuating device		8	1		325	47		372	425
4200	Battery and rack		4	2		805	94		899	1,025

28 31 Fire Detection and Alarm

28 31 23 – Fire Detection and Alarm Annunciation Panels and Fire Stations

28 31 23.50 Alarm Panels and Devices		Crew	Daily Output	Labor-Hours	Unit	2009 Bare Costs Material	Labor	Equipment	Total	Total Incl O&P
4400	Automatic charger	1 Elec	8	1	Ea.	495	47		542	615
4600	Signal bell		8	1		57.50	47		104.50	134
4800	Trouble buzzer or manual station		8	1		41	47		88	115
5600	Strobe and horn		5.30	1.509		105	71		176	222
5800	Fire alarm horn		6.70	1.194		40.50	56		96.50	128
6000	Door holder, electro-magnetic		4	2		86	94		180	235
6200	Combination holder and closer		3.20	2.500		475	118		593	700
6400	Code transmitter		4	2		765	94		859	980
6600	Drill switch		8	1		96	47		143	175
6800	Master box		2.70	2.963		3,425	139		3,564	3,975
7000	Break glass station		8	1		55.50	47		102.50	131
7800	Remote annunciator, 8 zone lamp		1.80	4.444		201	209		410	530
8000	12 zone lamp	2 Elec	2.60	6.154		345	289		634	810
8200	16 zone lamp	"	2.20	7.273		345	340		685	890

28 31 43 – Fire Detection Sensors

28 31 43.50 Fire and Heat Detectors		Crew	Daily Output	Labor-Hours	Unit	Material	Labor	Equipment	Total	Total Incl O&P
0010	**FIRE & HEAT DETECTORS**									
5000	Detector, rate of rise	1 Elec	8	1	Ea.	39	47		86	113
5100	Fixed temperature	"	8	1	"	32.50	47		79.50	106

28 31 46 – Smoke Detection Sensors

28 31 46.50 Smoke Detectors		Crew	Daily Output	Labor-Hours	Unit	Material	Labor	Equipment	Total	Total Incl O&P
0010	**SMOKE DETECTORS**									
5200	Smoke detector, ceiling type	1 Elec	6.20	1.290	Ea.	87.50	60.50		148	187
5400	Duct type	"	3.20	2.500	"	277	118		395	480

28 33 Fuel-Gas Detection and Alarm

28 33 33 – Fuel-Gas Detection Sensors

28 33 33.50 Tank Leak Detection Systems		Crew	Daily Output	Labor-Hours	Unit	Material	Labor	Equipment	Total	Total Incl O&P
0010	**TANK LEAK DETECTION SYSTEMS** Liquid and vapor									
0100	For hydrocarbons and hazardous liquids/vapors									
0120	Controller, data acquisition, incl. printer, modem, RS232 port									
0140	24 channel, for use with all probes				Ea.	9,500			9,500	10,500
0160	9 channel, for external monitoring				"	6,000			6,000	6,600
0200	Probes									
0210	Well monitoring									
0220	Liquid phase detection				Ea.	760			760	835
0230	Hydrocarbon vapor, fixed position					250			250	275
0240	Hydrocarbon vapor, float mounted					250			250	275
0250	Both liquid and vapor hydrocarbon					250			250	275
0300	Secondary containment, liquid phase									
0310	Pipe trench/manway sump				Ea.	250			250	275
0320	Double wall pipe and manual sump					250			250	275
0330	Double wall fiberglass annular space					250			250	275
0340	Double wall steel tank annular space					250			250	275
0500	Accessories									
0510	Modem, non-dedicated phone line				Ea.	273			273	300
0600	Monitoring, internal									
0610	Automatic tank gauge, incl. overfill				Ea.	1,200			1,200	1,325
0620	Product line				"	1,175			1,175	1,300
0700	Monitoring, special									

28 33 33 – Fuel-Gas Detection Sensors

28 33 33.50 Tank Leak Detection Systems		Crew	Daily Output	Labor-Hours	Unit	Material	Labor	Equipment	Total	Total Incl O&P
						2009 Bare Costs				
0710	Cathodic protection				Ea.	650			650	715
0720	Annular space chemical monitor				"	885			885	970

Estimating Tips

31 05 00 Common Work Results for Earthwork

- Estimating the actual cost of performing earthwork requires careful consideration of the variables involved. This includes items such as type of soil, whether water will be encountered, dewatering, whether banks need bracing, disposal of excavated earth, and length of haul to fill or spoil sites, etc. If the project has large quantities of cut or fill, consider raising or lowering the site to reduce costs, while paying close attention to the effect on site drainage and utilities.
- If the project has large quantities of fill, creating a borrow pit on the site can significantly lower the costs.
- It is very important to consider what time of year the project is scheduled for completion. Bad weather can create large cost overruns from dewatering, site repair, and lost productivity from cold weather.

Reference Numbers

Reference numbers are shown in shaded boxes at the beginning of some major classifications. These numbers refer to related items in the Reference Section. The reference information may be an estimating procedure, an alternate pricing method, or technical information.

Note: Not all subdivisions listed here necessarily appear in this publication.

Division 31 - Earthwork

31 05 Common Work Results for Earthwork

31 05 23 – Cement and Concrete for Earthwork

31 05 23.30 Plant Mixed Bituminous Concrete		Crew	Daily Output	Labor-Hours	Unit	2009 Bare Costs Material	Labor	Equipment	Total	Total Incl O&P
0010	**PLANT MIXED BITUMINOUS CONCRETE**									
0020	Asphaltic concrete plant mix (145 LB per C.F.)				Ton	57			57	62.50
0040	Asphaltic concrete less than 300 tons add trucking costs									
0050	See Div. 31 23 23.20 for hauling costs									
0200	All weather patching mix, hot				Ton	62			62	68
0250	Cold patch					62			62	68
0300	Berm mix					62			62	68

31 23 Excavation and Fill

31 23 16 – Excavation

31 23 16.13 Excavating, Trench		Crew	Daily Output	Labor-Hours	Unit	Material	Labor	Equipment	Total	Total Incl O&P
0010	**EXCAVATING, TRENCH** G									
0011	Or continuous footing									
0020	Common earth with no sheeting or dewatering included									
0050	1′ to 4′ deep, 3/8 C.Y. excavator	B-11C	150	.107	B.C.Y.		3.89	1.96	5.85	8.10
0060	1/2 C.Y. excavator	B-11M	200	.080			2.92	1.72	4.64	6.35
0090	4′ to 6′ deep, 1/2 C.Y. excavator	″	200	.080			2.92	1.72	4.64	6.35
0100	5/8 C.Y. excavator	B-12Q	250	.064			2.37	2.07	4.44	5.90
0110	3/4 C.Y. excavator	B-12F	300	.053			1.98	2	3.98	5.20
0300	1/2 C.Y. excavator, truck mounted	B-12J	200	.080			2.97	4.79	7.76	9.75
0500	6′ to 10′ deep, 3/4 C.Y. excavator	B-12F	225	.071			2.64	2.66	5.30	6.95
0600	1 C.Y. excavator, truck mounted	B-12K	400	.040			1.48	2.83	4.31	5.35
0900	10′ to 14′ deep, 3/4 C.Y. excavator	B-12F	200	.080			2.97	2.99	5.96	7.80
1000	1-1/2 C.Y. excavator	B-12B	540	.030			1.10	1.60	2.70	3.43
1300	14′ to 20′ deep, 1 C.Y. excavator	B-12A	320	.050			1.35	2.09	3.94	5.10
1340	20′ to 24′ deep, 1 C.Y. excavator	″	288	.056			2.06	2.32	4.38	5.70
1352	4′ to 6′ deep, 1/2 C.Y. excavator w/ trench box	B-13H	188	.085			3.16	5.70	8.86	11.05
1354	5/8 C.Y. excavator	″	235	.068			2.52	4.56	7.08	8.85
1356	3/4 C.Y. excavator	B-13G	282	.057			2.10	2.52	4.62	6
1362	6′ to 10′ deep, 3/4 C.Y. excavator w/trench box		212	.075			2.80	3.36	6.16	7.95
1374	10′ to 14′ deep, 3/4 C.Y. excavator w/trench box		188	.085			3.16	3.79	6.95	9
1376	1-1/2 C.Y. excavator/ trench box	B-13E	508	.032			1.17	1.93	3.10	3.90
1381	14′ to 20′ deep, 1 C.Y. excavator w/trench box	B-13D	301	.053			1.97	2.60	4.57	5.85
1386	20′ to 24′ deep, 1 C.Y. excavator w/trench box	″	271	.059			2.19	2.89	5.08	6.50
1400	By hand with pick and shovel 2′ to 6′ deep, light soil	1 Clab	8	1			31.50		31.50	49
1500	Heavy soil	″	4	2			63		63	98
1700	For tamping backfilled trenches, air tamp, add	A-1G	100	.080	E.C.Y.		2.53	.49	3.02	4.46
1900	Vibrating plate, add	B-18	180	.133	″		4.30	.22	4.52	6.90
2100	Trim sides and bottom for concrete pours, common earth		1500	.016	S.F.		.52	.03	.55	.83
2300	Hardpan		600	.040	″		1.29	.07	1.36	2.07
5020	Loam & Sandy clay with no sheeting or dewatering included									
5050	1′ to 4′ deep, 3/8 C.Y. tractor loader/backhoe	B-11C	162	.099	B.C.Y.		3.60	1.81	5.41	7.50
5060	1/2 C.Y. excavator	B-11M	216	.074			2.70	1.60	4.30	5.90
5080	4′ to 6′ deep, 1/2 C.Y. excavator	″	216	.074			2.70	1.60	4.30	5.90
5090	5/8 C.Y. excavator	B-12Q	276	.058			2.15	1.87	4.02	5.35
5100	3/4 C.Y. excavator	B-12F	324	.049			1.83	1.85	3.68	4.82
5130	1/2 C.Y. excavator, truck mounted	B-12J	216	.074			2.75	4.44	7.19	9.05
5140	6′ to 10′ deep, 3/4 C.Y. excavator	B-12F	243	.066			2.44	2.46	4.90	6.45
5160	1 C.Y. excavator, truck mounted	B-12K	432	.037			1.37	2.62	3.99	4.97
5190	10′ to 14′ deep, 3/4 C.Y. excavator	B-12F	216	.074			2.75	2.77	5.52	7.25
5210	1-1/2 C.Y. excavator	B-12B	583	.027			1.02	1.48	2.50	3.18

31 23 16.13 Excavating, Trench

	31 23 16.13 Excavating, Trench	Crew	Daily Output	Labor-Hours	Unit	Material	2009 Bare Costs Labor	Equipment	Total	Total Incl O&P
5250	14′ to 20′ deep, 1 C.Y. excavator	B-12A	346	.046	B.C.Y.		1.71	1.93	3.64	4.74
5300	20′ to 24′ deep, 1 C.Y. excavator	″	311	.051			1.91	2.15	4.06	5.30
5352	4′ to 6′ deep, 1/2 C.Y. excavator w/trench box	B-13H	205	.078			2.89	5.25	8.14	10.15
5354	5/8 C.Y. excavator w/trench box	″	257	.062			2.31	4.17	6.48	8.10
5356	3/4 C.Y. excavator w/trench box	B-13G	308	.052			1.93	2.31	4.24	5.45
5362	6′ to 10′ deep, 3/4 C.Y. excavator w/trench box		231	.069			2.57	3.08	5.65	7.30
5370	10′ to 14′ deep, 3/4 C.Y. excavator w/trench box		205	.078			2.89	3.47	6.36	8.25
5374	1-1/2 C.Y. excavator w/trench box	B-13E	554	.029			1.07	1.77	2.84	3.57
5382	14′ to 20′ deep, 1 C.Y. excavator w/ trench box	B-13D	329	.049			1.80	2.38	4.18	5.35
5392	20′ to 24′ deep, 1 C.Y. excavator w/ trench box	″	295	.054			2.01	2.65	4.66	6
6020	Sand & gravel with no sheeting or dewatering included									
6050	1′ to 4′ deep, 3/8 C.Y. excavator	B-11C	165	.097	B.C.Y.		3.54	1.78	5.32	7.35
6060	1/2 C.Y. excavator	B-11M	220	.073			2.65	1.57	4.22	5.75
6080	4′ to 6′ deep, 1/2 C.Y. excavator	″	220	.073			2.65	1.57	4.22	5.75
6090	5/8 C.Y. excavator	B-12Q	275	.058			2.16	1.88	4.04	5.35
6100	3/4 C.Y. excavator	B-12F	330	.048			1.80	1.81	3.61	4.74
6130	1/2 C.Y. excavator, truck mounted	B-12J	220	.073			2.70	4.36	7.06	8.90
6140	6′ to 10′ deep, 3/4 C.Y. excavator	B-12F	248	.065			2.39	2.41	4.80	6.30
6160	1 C.Y. excavator, truck mounted	B-12K	440	.036			1.35	2.57	3.92	4.87
6190	10′ to 14′ deep, 3/4 C.Y. excavator	B-12F	220	.073			2.70	2.72	5.42	7.10
6210	1-1/2 C.Y. excavator	B-12B	594	.027			1	1.46	2.46	3.12
6250	14′ to 20′ deep, 1 C.Y. excavator	B-12A	352	.045			1.69	1.90	3.59	4.66
6300	20′ to 24′ deep, 1 C.Y. excavator	″	317	.050			1.87	2.11	3.98	5.15
6352	4′ to 6′ deep, 1/2 C.Y. excavator w/ trench box	B-13H	209	.077			2.84	5.15	7.99	9.95
6354	5/8 C.Y. excavator w/ trench box	″	261	.061			2.27	4.11	6.38	8
6356	3/4 C.Y. excavator w/ trench box	B-13G	314	.051			1.89	2.27	4.16	5.35
6362	6′ to 10′ deep, 3/4 C.Y. excavator w/ trench box		236	.068			2.51	3.02	5.53	7.15
6370	10′ to 14′ deep, 3/4 C.Y. excavator w/trench box		209	.077			2.84	3.41	6.25	8.05
6374	1-1/2 C.Y. excavator w/trench box	B-13E	564	.028			1.05	1.74	2.79	3.51
6382	14′ to 20′ deep, 1 C.Y. excavator w/trench box	B-13D	334	.048			1.78	2.34	4.12	5.30
6392	20′ to 24′ deep, 1 C.Y. excavator w/ trench box	″	301	.053			1.97	2.60	4.57	5.85
7020	Dense hard clay with no sheeting or dewatering included									
7050	1′ to 4′ deep, 3/8 C.Y. excavator	B-11C	132	.121	B.C.Y.		4.42	2.23	6.65	9.20
7060	1/2 C.Y. excavator	B-11M	176	.091			3.32	1.96	5.28	7.20
7080	4′ to 6′ deep, 1/2 C.Y. excavator	″	176	.091			3.32	1.96	5.28	7.20
7090	5/8 C.Y. excavator	B-12Q	220	.073			2.70	2.35	5.05	6.70
7100	3/4 C.Y. excavator	B-12F	264	.061			2.25	2.27	4.52	5.90
7130	1/2 C.Y. excavator, truck mounted	B-12J	176	.091			3.37	5.45	8.82	11.15
7140	6′ to 10′ deep, 3/4 C.Y. excavator	B-12F	198	.081			3	3.02	6.02	7.90
7160	1 C.Y. excavator, truck mounted	B-12K	352	.045			1.69	3.21	4.90	6.10
7190	10′ to 14′ deep, 3/4 C.Y. excavator	B-12F	176	.091			3.37	3.40	6.77	8.90
7210	1-1/2 C.Y. excavator	B-12B	475	.034			1.25	1.82	3.07	3.90
7250	14′ to 20′ deep, 1 C.Y. excavator	B-12A	282	.057			2.10	2.37	4.47	5.80
7300	20′ to 24′ deep, 1 C.Y. excavator	″	254	.063			2.34	2.63	4.97	6.45

31 23 16.14 Excavating, Utility Trench

	31 23 16.14 Excavating, Utility Trench	Crew	Daily Output	Labor-Hours	Unit	Material	Labor	Equipment	Total	Total Incl O&P
0010	**EXCAVATING, UTILITY TRENCH**									
0011	Common earth									
0050	Trenching with chain trencher, 12 H.P., operator walking									
0100	4″ wide trench, 12″ deep	B-53	800	.010	L.F.		.39	.07	.46	.67
0150	18″ deep		750	.011			.42	.08	.50	.72
0200	24″ deep		700	.011			.45	.09	.54	.76
0300	6″ wide trench, 12″ deep		650	.012			.48	.09	.57	.82

31 23 Excavation and Fill

31 23 16 – Excavation

31 23 16.14 Excavating, Utility Trench

	31 23 16.14 Excavating, Utility Trench	Crew	Daily Output	Labor-Hours	Unit	Material	2009 Bare Costs Labor	Equipment	Total	Total Incl O&P
0350	18″ deep	B-53	600	.013	L.F.		.52	.10	.62	.89
0400	24″ deep		550	.015			.57	.11	.68	.97
0450	36″ deep		450	.018			.69	.13	.82	1.19
0600	8″ wide trench, 12″ deep		475	.017			.66	.13	.79	1.13
0650	18″ deep		400	.020			.78	.15	.93	1.33
0700	24″ deep		350	.023			.89	.17	1.06	1.53
0750	36″ deep		300	.027			1.04	.20	1.24	1.79
1000	Backfill by hand including compaction, add									
1050	4″ wide trench, 12″ deep	A-1G	800	.010	L.F.		.32	.06	.38	.56
1100	18″ deep		530	.015			.48	.09	.57	.84
1150	24″ deep		400	.020			.63	.12	.75	1.12
1300	6″ wide trench, 12″ deep		540	.015			.47	.09	.56	.83
1350	18″ deep		405	.020			.62	.12	.74	1.10
1400	24″ deep		270	.030			.94	.18	1.12	1.65
1450	36″ deep		180	.044			1.40	.27	1.67	2.48
1600	8″ wide trench, 12″ deep		400	.020			.63	.12	.75	1.12
1650	18″ deep		265	.030			.95	.19	1.14	1.69
1700	24″ deep		200	.040			1.26	.25	1.51	2.23
1750	36″ deep		135	.059			1.87	.37	2.24	3.30
2000	Chain trencher, 40 H.P. operator riding									
2050	6″ wide trench and backfill, 12″ deep	B-54	1200	.007	L.F.		.26	.25	.51	.67
2100	18″ deep		1000	.008			.31	.30	.61	.80
2150	24″ deep		975	.008			.32	.31	.63	.82
2200	36″ deep		900	.009			.35	.33	.68	.89
2250	48″ deep		750	.011			.42	.40	.82	1.07
2300	60″ deep		650	.012			.48	.46	.94	1.23
2400	8″ wide trench and backfill, 12″ deep		1000	.008			.31	.30	.61	.80
2450	18″ deep		950	.008			.33	.32	.65	.84
2500	24″ deep		900	.009			.35	.33	.68	.89
2550	36″ deep		800	.010			.39	.38	.77	1
2600	48″ deep		650	.012			.48	.46	.94	1.23
2700	12″ wide trench and backfill, 12″ deep		975	.008			.32	.31	.63	.82
2750	18″ deep		860	.009			.36	.35	.71	.93
2800	24″ deep		800	.010			.39	.38	.77	1
2850	36″ deep		725	.011			.43	.42	.85	1.11
3000	16″ wide trench and backfill, 12″ deep		835	.010			.37	.36	.73	.96
3050	18″ deep		750	.011			.42	.40	.82	1.07
3100	24″ deep		700	.011			.45	.43	.88	1.14
3200	Compaction with vibratory plate, add								50%	50%
5100	Hand excavate and trim for pipe bells after trench excavation									
5200	8″ pipe	1 Clab	155	.052	L.F.		1.63		1.63	2.53
5300	18″ pipe	″	130	.062	″		1.94		1.94	3.02

31 23 16.16 Structural Excavation for Minor Structures

	31 23 16.16 Structural Excavation for Minor Structures	Crew	Daily Output	Labor-Hours	Unit	Material	2009 Bare Costs Labor	Equipment	Total	Total Incl O&P
0010	**STRUCTURAL EXCAVATION FOR MINOR STRUCTURES**									
0015	Hand, pits to 6′ deep, sandy soil	1 Clab	8	1	B.C.Y.		31.50		31.50	49
0100	Heavy soil or clay		4	2			63		63	98
0300	Pits 6′ to 12′ deep, sandy soil		5	1.600			50.50		50.50	78.50
0500	Heavy soil or clay		3	2.667			84.50		84.50	131
0700	Pits 12′ to 18′ deep, sandy soil		4	2			63		63	98
0900	Heavy soil or clay		2	4			126		126	196
1500	For wet or muck hand excavation, add to above				%				50%	50%

31 23 Excavation and Fill

31 23 16 – Excavation

31 23 16.30 Drilling and Blasting Rock		Crew	Daily Output	Labor-Hours	Unit	Material	2009 Bare Costs Labor	Equipment	Total	Total Incl O&P
0010	**DRILLING AND BLASTING ROCK**									
0020	Rock, open face, under 1500 C.Y.	B-47	225	.107	B.C.Y.	2.75	3.71	5.85	12.31	15.20
0100	Over 1500 C.Y.		300	.080		2.75	2.78	4.39	9.92	12.10
0200	Areas where blasting mats are required, under 1500 C.Y.		175	.137		2.75	4.77	7.55	15.07	18.65
0250	Over 1500 C.Y.		250	.096		2.75	3.34	5.25	11.34	13.95
2200	Trenches, up to 1500 C.Y.		22	1.091		8	38	60	106	133
2300	Over 1500 C.Y.		26	.923		8	32	50.50	90.50	113

31 23 19 – Dewatering

31 23 19.20 Dewatering Systems

Line	Description	Crew	Daily Output	Labor-Hours	Unit	Material	Labor	Equipment	Total	Total Incl O&P
0010	**DEWATERING SYSTEMS**									
0020	Excavate drainage trench, 2′ wide, 2′ deep	B-11C	90	.178	C.Y.		6.50	3.26	9.76	13.50
0100	2′ wide, 3′ deep, with backhoe loader	"	135	.119			4.32	2.18	6.50	9
0200	Excavate sump pits by hand, light soil	1 Clab	7.10	1.127			35.50		35.50	55
0300	Heavy soil	"	3.50	2.286			72		72	112
0500	Pumping 8 hr., attended 2 hrs. per day, including 20 L.F.									
0550	of suction hose & 100 L.F. discharge hose									
0600	2″ diaphragm pump used for 8 hours	B-10H	4	3	Day		114	16.85	130.85	192
0650	4″ diaphragm pump used for 8 hours	B-10I	4	3			114	23.50	137.50	199
0800	8 hrs. attended, 2″ diaphragm pump	B-10H	1	12			455	67.50	522.50	770
0900	3″ centrifugal pump	B-10J	1	12			455	76	531	780
1000	4″ diaphragm pump	B-10I	1	12			455	94.50	549.50	800
1100	6″ centrifugal pump	B-10K	1	12			455	335	790	1,050
1300	CMP, incl. excavation 3′ deep, 12″ diameter	B-6	115	.209	L.F.	13.65	7.10	2.55	23.30	28.50
1400	18″ diameter		100	.240	"	17	8.20	2.94	28.14	34.50
1600	Sump hole construction, incl. excavation and gravel, pit		1250	.019	C.F.	.96	.65	.24	1.85	2.31
1700	With 12″ gravel collar, 12″ pipe, corrugated, 16 ga.		70	.343	L.F.	22.50	11.70	4.20	38.40	47.50
1800	15″ pipe, corrugated, 16 ga.		55	.436		29	14.85	5.35	49.20	60.50
1900	18″ pipe, corrugated, 16 ga.		50	.480		33.50	16.35	5.90	55.75	68.50
2000	24″ pipe, corrugated, 14 ga.		40	.600		40.50	20.50	7.35	68.35	84
2200	Wood lining, up to 4′ x 4′, add		300	.080	SFCA	16.65	2.73	.98	20.36	23.50
9950	See Div. 31 23 19.40 for wellpoints									
9960	See Div. 31 23 19.30 for deep well systems									

31 23 19.30 Wells

Line	Description	Crew	Daily Output	Labor-Hours	Unit	Material	Labor	Equipment	Total	Total Incl O&P
0010	**WELLS**									
0011	For dewatering 10′ to 20′ deep, 2′ diameter									
0020	with steel casing, minimum	B-6	165	.145	V.L.F.	47	4.96	1.78	53.74	61
0050	Average		98	.245		47	8.35	3	58.35	67.50
0100	Maximum		49	.490		47	16.70	6	69.70	83.50
0300	For dewatering pumps see 01 54 33 in Reference Section									
0500	For domestic water wells, see Div. 33 21 13.10									

31 23 19.40 Wellpoints

Line	Description	Crew	Daily Output	Labor-Hours	Unit	Material	Labor	Equipment	Total	Total Incl O&P
0010	**WELLPOINTS** R312319-90									
0011	For equipment rental, see 01 54 33 in Reference Section									
0100	Installation and removal of single stage system									
0110	Labor only, .75 labor-hours per L.F., minimum	1 Clab	10.70	.748	LF Hdr		23.50		23.50	36.50
0200	2.0 labor-hours per L.F., maximum	"	4	2	"		63		63	98
0400	Pump operation, 4 @ 6 hr. shifts									
0410	Per 24 hour day	4 Eqlt	1.27	25.197	Day		985		985	1,475
0500	Per 168 hour week, 160 hr. straight, 8 hr. double time		.18	177	Week		6,950		6,950	10,400
0550	Per 4.3 week month		.04	800	Month		31,200		31,200	47,000
0600	Complete installation, operation, equipment rental, fuel &									

31 23 19 – Dewatering

31 23 19.40 Wellpoints		Crew	Daily Output	Labor-Hours	Unit	2009 Bare Costs Material	Labor	Equipment	Total	Total Incl O&P
0610	removal of system with 2" wellpoints 5' O.C.									
0700	100' long header, 6" diameter, first month	4 Eqlt	3.23	9.907	LF Hdr	157	385		542	755
0800	Thereafter, per month		4.13	7.748		125	305		430	595
1000	200' long header, 8" diameter, first month		6	5.333		142	208		350	470
1100	Thereafter, per month		8.39	3.814		70.50	149		219.50	300
1300	500' long header, 8" diameter, first month		10.63	3.010		55	118		173	238
1400	Thereafter, per month		20.91	1.530		39	60		99	133
1600	1,000' long header, 10" diameter, first month		11.62	2.754		47	108		155	214
1700	Thereafter, per month		41.81	.765		23.50	30		53.50	71
1900	Note: above figures include pumping 168 hrs. per week									
1910	and include the pump operator and one stand-by pump.									

31 23 23 – Fill

31 23 23.13 Backfill

		Crew	Daily Output	Labor-Hours	Unit	Material	Labor	Equipment	Total	Total Incl O&P
0010	**BACKFILL** G1030-805									
0015	By hand, no compaction, light soil	1 Clab	14	.571	L.C.Y.		18.05		18.05	28
0100	Heavy soil		11	.727	"		23		23	35.50
0300	Compaction in 6" layers hand tamp, add to above		20.60	.388	E.C.Y.		12.25		12.25	19.05
0400	Roller compaction operator walking, add	B-10A	100	.120			4.57	1.45	6.02	8.55
0500	Air tamp, add	B-9D	190	.211			6.75	1.17	7.92	11.75
0600	Vibrating plate, add	A-1D	60	.133			4.21	.53	4.74	7.15
0800	Compaction in 12" layers, hand tamp, add to above	1 Clab	34	.235			7.45		7.45	11.55
0900	Roller compaction operator walking, add	B-10A	150	.080			3.05	.97	4.02	5.70
1000	Air tamp, add	B-9	285	.140			4.49	.67	5.16	7.70
1100	Vibrating plate, add	A-1E	90	.089			2.81	.44	3.25	4.84
1300	Dozer backfilling, bulk, up to 300' haul, no compaction	B-10B	1200	.010	L.C.Y.		.38	.90	1.28	1.57
1400	Air tamped, add	B-11B	80	.200	E.C.Y.		7.05	3.10	10.15	14.15
1600	Compacting backfill, 6" to 12" lifts, vibrating roller	B-10C	800	.015			.57	1.87	2.44	2.92
1700	Sheepsfoot roller	B-10D	750	.016			.61	2.06	2.67	3.19
1900	Dozer backfilling, trench, up to 300' haul, no compaction	B-10B	900	.013	L.C.Y.		.51	1.20	1.71	2.09
2000	Air tamped, add	B-11B	80	.200	E.C.Y.		7.05	3.10	10.15	14.15
2200	Compacting backfill, 6" to 12" lifts, vibrating roller	B-10C	700	.017			.65	2.13	2.78	3.34
2300	Sheepsfoot roller	B-10D	650	.018			.70	2.38	3.08	3.69

31 23 23.16 Fill By Borrow and Utility Bedding

		Crew	Daily Output	Labor-Hours	Unit	Material	Labor	Equipment	Total	Total Incl O&P
0010	**FILL BY BORROW AND UTILITY BEDDING**									
0049	Utility bedding, for pipe & conduit, not incl. compaction									
0050	Crushed or screened bank run gravel	B-6	150	.160	L.C.Y.	31	5.45	1.96	38.41	44.50
0100	Crushed stone 3/4" to 1/2"		150	.160		43.50	5.45	1.96	50.91	58.50
0200	Sand, dead or bank		150	.160		10.50	5.45	1.96	17.91	22
0500	Compacting bedding in trench	A-1D	90	.089	E.C.Y.		2.81	.35	3.16	4.75
0600	If material source exceeds 2 miles, add for extra mileage.									
0610	See Div. 31 23 23 .20 for hauling mileage add.									

31 23 23.17 General Fill

		Crew	Daily Output	Labor-Hours	Unit	Material	Labor	Equipment	Total	Total Incl O&P
0010	**GENERAL FILL**									
0011	Spread dumped material, no compaction									
0020	By dozer, no compaction	B-10B	1000	.012	L.C.Y.		.46	1.08	1.54	1.88
0100	By hand	1 Clab	12	.667	"		21		21	32.50
0500	Gravel fill, compacted, under floor slabs, 4" deep	B-37	10000	.005	S.F.	.25	.16	.01	.42	.55
0600	6" deep		8600	.006		.38	.19	.02	.59	.73
0700	9" deep		7200	.007		.63	.22	.02	.87	1.05
0800	12" deep		6000	.008		.88	.27	.02	1.17	1.41
1000	Alternate pricing method, 4" deep		120	.400	E.C.Y.	18.90	13.25	1.15	33.30	43
1100	6" deep		160	.300		18.90	9.95	.86	29.71	37.50

31 23 23.17 General Fill		Crew	Daily Output	Labor-Hours	Unit	2009 Bare Costs Material	Labor	Equipment	Total	Total Incl O&P
1200	9" deep	B-37	200	.240	E.C.Y.	18.90	7.95	.69	27.54	34
1300	12" deep	↓	220	.218	↓	18.90	7.25	.63	26.78	33
1400	Granular fill				L.C.Y.	12.60			12.60	13.85

31 23 23.20 Hauling

Line	Description	Crew	Daily Output	Labor-Hours	Unit	Material	Labor	Equipment	Total	Total Incl O&P
0010	**HAULING**									
0011	Excavated or borrow, loose cubic yards									
0012	no loading equipment, including hauling,waiting, loading/dumping									
0013	time per cycle (wait, load, travel, unload or dump & return)									
0014	8 CY truck, 15 MPH ave, cycle 0.5 miles, 10 min. wait/Ld./Uld.	B-34A	320	.025	L.C.Y.		.80	1.02	1.82	2.35
0016	cycle 1 mile		272	.029			.94	1.20	2.14	2.77
0018	cycle 2 miles		208	.038			1.23	1.57	2.80	3.62
0020	cycle 4 miles		144	.056			1.78	2.27	4.05	5.20
0022	cycle 6 miles		112	.071			2.28	2.91	5.19	6.70
0024	cycle 8 miles		88	.091			2.90	3.71	6.61	8.55
0026	20 MPH ave, cycle 0.5 mile		336	.024			.76	.97	1.73	2.24
0028	cycle 1 mile		296	.027			.86	1.10	1.96	2.54
0030	cycle 2 miles		240	.033			1.06	1.36	2.42	3.14
0032	cycle 4 miles		176	.045			1.45	1.85	3.30	4.27
0034	cycle 6 miles		136	.059			1.88	2.40	4.28	5.55
0036	cycle 8 miles		112	.071			2.28	2.91	5.19	6.70
0044	25 MPH ave, cycle 4 miles		192	.042			1.33	1.70	3.03	3.92
0046	cycle 6 miles		160	.050			1.60	2.04	3.64	4.70
0048	cycle 8 miles		128	.063			2	2.55	4.55	5.90
0050	30 MPH ave, cycle 4 miles		216	.037			1.18	1.51	2.69	3.48
0052	cycle 6 miles		176	.045			1.45	1.85	3.30	4.27
0054	cycle 8 miles		144	.056			1.78	2.27	4.05	5.20
0114	15 MPH ave, cycle 0.5 mile, 15 min. wait/Ld./Uld.		224	.036			1.14	1.46	2.60	3.36
0116	cycle 1 mile		200	.040			1.28	1.63	2.91	3.77
0118	cycle 2 miles		168	.048			1.52	1.94	3.46	4.48
0120	cycle 4 miles		120	.067			2.13	2.72	4.85	6.25
0122	cycle 6 miles		96	.083			2.66	3.40	6.06	7.85
0124	cycle 8 miles		80	.100			3.20	4.08	7.28	9.40
0126	20 MPH ave, cycle 0.5 mile		232	.034			1.10	1.41	2.51	3.24
0128	cycle 1 mile		208	.038			1.23	1.57	2.80	3.62
0130	cycle 2 miles		184	.043			1.39	1.77	3.16	4.09
0132	cycle 4 miles		144	.056			1.78	2.27	4.05	5.20
0134	cycle 6 miles		112	.071			2.28	2.91	5.19	6.70
0136	cycle 8 miles		96	.083			2.66	3.40	6.06	7.85
0144	25 MPH ave, cycle 4 miles		152	.053			1.68	2.15	3.83	4.95
0146	cycle 6 miles		128	.063			2	2.55	4.55	5.90
0148	cycle 8 miles		112	.071			2.28	2.91	5.19	6.70
0150	30 MPH ave, cycle 4 miles		168	.048			1.52	1.94	3.46	4.48
0152	cycle 6 miles		144	.056			1.78	2.27	4.05	5.20
0154	cycle 8 miles		120	.067			2.13	2.72	4.85	6.25
0214	15 MPH ave, cycle 0.5 mile, 20 min wait/Ld./Uld.		176	.045			1.45	1.85	3.30	4.27
0216	cycle 1 mile		160	.050			1.60	2.04	3.64	4.70
0218	cycle 2 miles		136	.059			1.88	2.40	4.28	5.55
0220	cycle 4 miles		104	.077			2.46	3.14	5.60	7.25
0222	cycle 6 miles		88	.091			2.90	3.71	6.61	8.55
0224	cycle 8 miles		72	.111			3.55	4.53	8.08	10.45
0226	20 MPH ave, cycle 0.5 mile		176	.045			1.45	1.85	3.30	4.27
0228	cycle 1 mile	↓	168	.048	↓		1.52	1.94	3.46	4.48

31 23 23.20 Hauling		Crew	Daily Output	Labor-Hours	Unit	Material	2009 Bare Costs Labor	Equipment	Total	Total Incl O&P
0230	cycle 2 miles	B-34A	144	.056	L.C.Y.		1.78	2.27	4.05	5.20
0232	cycle 4 miles		120	.067			2.13	2.72	4.85	6.25
0234	cycle 6 miles		96	.083			2.66	3.40	6.06	7.85
0236	cycle 8 miles		88	.091			2.90	3.71	6.61	8.55
0244	25 MPH ave, cycle 4 miles		128	.063			2	2.55	4.55	5.90
0246	cycle 6 miles		112	.071			2.28	2.91	5.19	6.70
0248	cycle 8 miles		96	.083			2.66	3.40	6.06	7.85
0250	30 MPH ave, cycle 4 miles		136	.059			1.88	2.40	4.28	5.55
0252	cycle 6 miles		120	.067			2.13	2.72	4.85	6.25
0254	cycle 8 miles		104	.077			2.46	3.14	5.60	7.25
0314	15 MPH ave, cycle 0.5 mile, 25 min wait/Ld./Uld		144	.056			1.78	2.27	4.05	5.20
0316	cycle 1 mile		128	.063			2	2.55	4.55	5.90
0318	cycle 2 miles		112	.071			2.28	2.91	5.19	6.70
0320	cycle 4 miles		96	.083			2.66	3.40	6.06	7.85
0322	cycle 6 miles		80	.100			3.20	4.08	7.28	9.40
0324	cycle 8 miles		64	.125			3.99	5.10	9.09	11.75
0326	20 MPH ave, cycle 0.5 mile		144	.056			1.78	2.27	4.05	5.20
0328	cycle 1 mile		136	.059			1.88	2.40	4.28	5.55
0330	cycle 2 miles		120	.067			2.13	2.72	4.85	6.25
0332	cycle 4 miles		104	.077			2.46	3.14	5.60	7.25
0334	cycle 6 miles		88	.091			2.90	3.71	6.61	8.55
0336	cycle 8 miles		80	.100			3.20	4.08	7.28	9.40
0344	25 MPH ave, cycle 4 miles		112	.071			2.28	2.91	5.19	6.70
0346	cycle 6 miles		96	.083			2.66	3.40	6.06	7.85
0348	cycle 8 miles		88	.091			2.90	3.71	6.61	8.55
0350	30 MPH ave, cycle 4 miles		112	.071			2.28	2.91	5.19	6.70
0352	cycle 6 miles		104	.077			2.46	3.14	5.60	7.25
0354	cycle 8 miles		96	.083			2.66	3.40	6.06	7.85
0414	15 MPH ave, cycle 0.5 mile, 30 min wait/Ld./Uld		120	.067			2.13	2.72	4.85	6.25
0416	cycle 1 mile		112	.071			2.28	2.91	5.19	6.70
0418	cycle 2 miles		96	.083			2.66	3.40	6.06	7.85
0420	cycle 4 miles		80	.100			3.20	4.08	7.28	9.40
0422	cycle 6 miles		72	.111			3.55	4.53	8.08	10.45
0424	cycle 8 miles		64	.125			3.99	5.10	9.09	11.75
0426	20 MPH ave, cycle 0.5 mile		120	.067			2.13	2.72	4.85	6.25
0428	cycle 1 mile		112	.071			2.28	2.91	5.19	6.70
0430	cycle 2 miles		104	.077			2.46	3.14	5.60	7.25
0432	cycle 4 miles		88	.091			2.90	3.71	6.61	8.55
0434	cycle 6 miles		80	.100			3.20	4.08	7.28	9.40
0436	cycle 8 miles		72	.111			3.55	4.53	8.08	10.45
0444	25 MPH ave, cycle 4 miles		96	.083			2.66	3.40	6.06	7.85
0446	cycle 6 miles		88	.091			2.90	3.71	6.61	8.55
0448	cycle 8 miles		80	.100			3.20	4.08	7.28	9.40
0450	30 MPH ave, cycle 4 miles		96	.083			2.66	3.40	6.06	7.85
0452	cycle 6 miles		88	.091			2.90	3.71	6.61	8.55
0454	cycle 8 miles		80	.100			3.20	4.08	7.28	9.40
0514	15 MPH ave, cycle 0.5 mile, 35 min wait/Ld./Uld		104	.077			2.46	3.14	5.60	7.25
0516	cycle 1 mile		96	.083			2.66	3.40	6.06	7.85
0518	cycle 2 miles		88	.091			2.90	3.71	6.61	8.55
0520	cycle 4 miles		72	.111			3.55	4.53	8.08	10.45
0522	cycle 6 miles		64	.125			3.99	5.10	9.09	11.75
0524	cycle 8 miles		56	.143			4.56	5.85	10.41	13.40
0526	20 MPH ave, cycle 0.5 mile	↓	104	.077	↓		2.46	3.14	5.60	7.25

31 23 23.20 Hauling		Crew	Daily Output	Labor-Hours	Unit	Material	2009 Bare Costs Labor	Equipment	Total	Total Incl O&P
0528	cycle 1 mile	B-34A	96	.083	L.C.Y.		2.66	3.40	6.06	7.85
0530	cycle 2 miles		96	.083			2.66	3.40	6.06	7.85
0532	cycle 4 miles		80	.100			3.20	4.08	7.28	9.40
0534	cycle 6 miles		72	.111			3.55	4.53	8.08	10.45
0536	cycle 8 miles		64	.125			3.99	5.10	9.09	11.75
0544	25 MPH ave, cycle 4 miles		88	.091			2.90	3.71	6.61	8.55
0546	cycle 6 miles		80	.100			3.20	4.08	7.28	9.40
0548	cycle 8 miles		72	.111			3.55	4.53	8.08	10.45
0550	30 MPH ave, cycle 4 miles		88	.091			2.90	3.71	6.61	8.55
0552	cycle 6 miles		80	.100			3.20	4.08	7.28	9.40
0554	cycle 8 miles		72	.111			3.55	4.53	8.08	10.45
1014	12 CY truck, cycle 0.5 mile, 15 MPH ave, 15 min. wait/Ld./Uld.	B-34B	336	.024			.76	1.59	2.35	2.92
1016	cycle 1 mile		300	.027			.85	1.78	2.63	3.26
1018	cycle 2 miles		252	.032			1.01	2.12	3.13	3.89
1020	cycle 4 miles		180	.044			1.42	2.96	4.38	5.45
1022	cycle 6 miles		144	.056			1.78	3.70	5.48	6.80
1024	cycle 8 miles		120	.067			2.13	4.44	6.57	8.15
1025	cycle 10 miles		96	.083			2.66	5.55	8.21	10.20
1026	20 MPH ave, cycle 0.5 mile		348	.023			.73	1.53	2.26	2.81
1028	cycle 1 mile		312	.026			.82	1.71	2.53	3.14
1030	cycle 2 miles		276	.029			.93	1.93	2.86	3.54
1032	cycle 4 miles		216	.037			1.18	2.47	3.65	4.53
1034	cycle 6 miles		168	.048			1.52	3.17	4.69	5.85
1036	cycle 8 miles		144	.056			1.78	3.70	5.48	6.80
1038	cycle 10 miles		120	.067			2.13	4.44	6.57	8.15
1040	25 MPH ave, cycle 4 miles		228	.035			1.12	2.34	3.46	4.29
1042	cycle 6 miles		192	.042			1.33	2.78	4.11	5.10
1044	cycle 8 miles		168	.048			1.52	3.17	4.69	5.85
1046	cycle 10 miles		144	.056			1.78	3.70	5.48	6.80
1050	30 MPH ave, cycle 4 miles		252	.032			1.01	2.12	3.13	3.89
1052	cycle 6 miles		216	.037			1.18	2.47	3.65	4.53
1054	cycle 8 miles		180	.044			1.42	2.96	4.38	5.45
1056	cycle 10 miles		156	.051			1.64	3.42	5.06	6.30
1060	35 MPH ave, cycle 4 miles		264	.030			.97	2.02	2.99	3.71
1062	cycle 6 miles		228	.035			1.12	2.34	3.46	4.29
1064	cycle 8 miles		204	.039			1.25	2.61	3.86	4.80
1066	cycle 10 miles		180	.044			1.42	2.96	4.38	5.45
1068	cycle 20 miles		120	.067			2.13	4.44	6.57	8.15
1069	cycle 30 miles		84	.095			3.04	6.35	9.39	11.70
1070	cycle 40 miles		72	.111			3.55	7.40	10.95	13.60
1072	40 MPH ave, cycle 6 miles		240	.033			1.06	2.22	3.28	4.08
1074	cycle 8 miles		216	.037			1.18	2.47	3.65	4.53
1076	cycle 10 miles		192	.042			1.33	2.78	4.11	5.10
1078	cycle 20 miles		120	.067			2.13	4.44	6.57	8.15
1080	cycle 30 miles		96	.083			2.66	5.55	8.21	10.20
1082	cycle 40 miles		72	.111			3.55	7.40	10.95	13.60
1084	cycle 50 miles		60	.133			4.26	8.90	13.16	16.30
1094	45 MPH ave, cycle 8 miles		216	.037			1.18	2.47	3.65	4.53
1096	cycle 10 miles		204	.039			1.25	2.61	3.86	4.80
1098	cycle 20 miles		132	.061			1.94	4.04	5.98	7.40
1100	cycle 30 miles		108	.074			2.37	4.94	7.31	9.10
1102	cycle 40 miles		84	.095			3.04	6.35	9.39	11.70
1104	cycle 50 miles		72	.111			3.55	7.40	10.95	13.60

31 23 23.20 Hauling		Crew	Daily Output	Labor-Hours	Unit	Material	2009 Bare Costs Labor	Equipment	Total	Total Incl O&P
1106	50 MPH ave, cycle 10 miles	B-34B	216	.037	L.C.Y.		1.18	2.47	3.65	4.53
1108	cycle 20 miles		144	.056			1.78	3.70	5.48	6.80
1110	cycle 30 miles		108	.074			2.37	4.94	7.31	9.10
1112	cycle 40 miles		84	.095			3.04	6.35	9.39	11.70
1114	cycle 50 miles		72	.111			3.55	7.40	10.95	13.60
1214	15 MPH ave, cycle 0.5 mile, 20 min. wait/Ld./Uld		264	.030			.97	2.02	2.99	3.71
1216	cycle 1 mile		240	.033			1.06	2.22	3.28	4.08
1218	cycle 2 miles		204	.039			1.25	2.61	3.86	4.80
1220	cycle 4 miles		156	.051			1.64	3.42	5.06	6.30
1222	cycle 6 miles		132	.061			1.94	4.04	5.98	7.40
1224	cycle 8 miles		108	.074			2.37	4.94	7.31	9.10
1225	cycle 10 miles		96	.083			2.66	5.55	8.21	10.20
1226	20 MPH ave, cycle 0.5 mile		264	.030			.97	2.02	2.99	3.71
1228	cycle 1 mile		252	.032			1.01	2.12	3.13	3.89
1230	cycle 2 miles		216	.037			1.18	2.47	3.65	4.53
1232	cycle 4 miles		180	.044			1.42	2.96	4.38	5.45
1234	cycle 6 miles		144	.056			1.78	3.70	5.48	6.80
1236	cycle 8 miles		132	.061			1.94	4.04	5.98	7.40
1238	cycle 10 miles		108	.074			2.37	4.94	7.31	9.10
1240	25 MPH ave, cycle 4 miles		192	.042			1.33	2.78	4.11	5.10
1242	cycle 6 miles		168	.048			1.52	3.17	4.69	5.85
1244	cycle 8 miles		144	.056			1.78	3.70	5.48	6.80
1246	cycle 10 miles		132	.061			1.94	4.04	5.98	7.40
1250	30 MPH ave, cycle 4 miles		204	.039			1.25	2.61	3.86	4.80
1252	cycle 6 miles		180	.044			1.42	2.96	4.38	5.45
1254	cycle 8 miles		156	.051			1.64	3.42	5.06	6.30
1256	cycle 10 miles		144	.056			1.78	3.70	5.48	6.80
1260	35 MPH ave, cycle 4 miles		216	.037			1.18	2.47	3.65	4.53
1262	cycle 6 miles		192	.042			1.33	2.78	4.11	5.10
1264	cycle 8 miles		168	.048			1.52	3.17	4.69	5.85
1266	cycle 10 miles		156	.051			1.64	3.42	5.06	6.30
1268	cycle 20 miles		108	.074			2.37	4.94	7.31	9.10
1269	cycle 30 miles		72	.111			3.55	7.40	10.95	13.60
1270	cycle 40 miles		60	.133			4.26	8.90	13.16	16.30
1272	40 MPH ave, cycle 6 miles		192	.042			1.33	2.78	4.11	5.10
1274	cycle 8 miles		180	.044			1.42	2.96	4.38	5.45
1276	cycle 10 miles		156	.051			1.64	3.42	5.06	6.30
1278	cycle 20 miles		108	.074			2.37	4.94	7.31	9.10
1280	cycle 30 miles		84	.095			3.04	6.35	9.39	11.70
1282	cycle 40 miles		72	.111			3.55	7.40	10.95	13.60
1284	cycle 50 miles		60	.133			4.26	8.90	13.16	16.30
1294	45 MPH ave, cycle 8 miles		180	.044			1.42	2.96	4.38	5.45
1296	cycle 10 miles		168	.048			1.52	3.17	4.69	5.85
1298	cycle 20 miles		120	.067			2.13	4.44	6.57	8.15
1300	cycle 30 miles		96	.083			2.66	5.55	8.21	10.20
1302	cycle 40 miles		72	.111			3.55	7.40	10.95	13.60
1304	cycle 50 miles		60	.133			4.26	8.90	13.16	16.30
1306	50 MPH ave, cycle 10 miles		180	.044			1.42	2.96	4.38	5.45
1308	cycle 20 miles		132	.061			1.94	4.04	5.98	7.40
1310	cycle 30 miles		96	.083			2.66	5.55	8.21	10.20
1312	cycle 40 miles		84	.095			3.04	6.35	9.39	11.70
1314	cycle 50 miles		72	.111			3.55	7.40	10.95	13.60
1414	15 MPH ave, cycle 0.5 mile, 25 min. wait/Ld./Uld	↓	204	.039	↓		1.25	2.61	3.86	4.80

31 23 23.20 Hauling		Crew	Daily Output	Labor-Hours	Unit	Material	2009 Bare Costs Labor	Equipment	Total	Total Incl O&P
1416	cycle 1 mile	B-34B	192	.042	L.C.Y.		1.33	2.78	4.11	5.10
1418	cycle 2 miles		168	.048			1.52	3.17	4.69	5.85
1420	cycle 4 miles		132	.061			1.94	4.04	5.98	7.40
1422	cycle 6 miles		120	.067			2.13	4.44	6.57	8.15
1424	cycle 8 miles		96	.083			2.66	5.55	8.21	10.20
1425	cycle 10 miles		84	.095			3.04	6.35	9.39	11.70
1426	20 MPH ave, cycle 0.5 mile		216	.037			1.18	2.47	3.65	4.53
1428	cycle 1 mile		204	.039			1.25	2.61	3.86	4.80
1430	cycle 2 miles		180	.044			1.42	2.96	4.38	5.45
1432	cycle 4 miles		156	.051			1.64	3.42	5.06	6.30
1434	cycle 6 miles		132	.061			1.94	4.04	5.98	7.40
1436	cycle 8 miles		120	.067			2.13	4.44	6.57	8.15
1438	cycle 10 miles		96	.083			2.66	5.55	8.21	10.20
1440	25 MPH ave, cycle 4 miles		168	.048			1.52	3.17	4.69	5.85
1442	cycle 6 miles		144	.056			1.78	3.70	5.48	6.80
1444	cycle 8 miles		132	.061			1.94	4.04	5.98	7.40
1446	cycle 10 miles		108	.074			2.37	4.94	7.31	9.10
1450	30 MPH ave, cycle 4 miles		168	.048			1.52	3.17	4.69	5.85
1452	cycle 6 miles		156	.051			1.64	3.42	5.06	6.30
1454	cycle 8 miles		132	.061			1.94	4.04	5.98	7.40
1456	cycle 10 miles		120	.067			2.13	4.44	6.57	8.15
1460	35 MPH ave, cycle 4 miles		180	.044			1.42	2.96	4.38	5.45
1462	cycle 6 miles		156	.051			1.64	3.42	5.06	6.30
1464	cycle 8 miles		144	.056			1.78	3.70	5.48	6.80
1466	cycle 10 miles		132	.061			1.94	4.04	5.98	7.40
1468	cycle 20 miles		96	.083			2.66	5.55	8.21	10.20
1469	cycle 30 miles		72	.111			3.55	7.40	10.95	13.60
1470	cycle 40 miles		60	.133			4.26	8.90	13.16	16.30
1472	40 MPH ave, cycle 6 miles		168	.048			1.52	3.17	4.69	5.85
1474	cycle 8 miles		156	.051			1.64	3.42	5.06	6.30
1476	cycle 10 miles		144	.056			1.78	3.70	5.48	6.80
1478	cycle 20 miles		96	.083			2.66	5.55	8.21	10.20
1480	cycle 30 miles		84	.095			3.04	6.35	9.39	11.70
1482	cycle 40 miles		60	.133			4.26	8.90	13.16	16.30
1484	cycle 50 miles		60	.133			4.26	8.90	13.16	16.30
1494	45 MPH ave, cycle 8 miles		156	.051			1.64	3.42	5.06	6.30
1496	cycle 10 miles		144	.056			1.78	3.70	5.48	6.80
1498	cycle 20 miles		108	.074			2.37	4.94	7.31	9.10
1500	cycle 30 miles		84	.095			3.04	6.35	9.39	11.70
1502	cycle 40 miles		72	.111			3.55	7.40	10.95	13.60
1504	cycle 50 miles		60	.133			4.26	8.90	13.16	16.30
1506	50 MPH ave, cycle 10 miles		156	.051			1.64	3.42	5.06	6.30
1508	cycle 20 miles		120	.067			2.13	4.44	6.57	8.15
1510	cycle 30 miles		96	.083			2.66	5.55	8.21	10.20
1512	cycle 40 miles		72	.111			3.55	7.40	10.95	13.60
1514	cycle 50 miles		60	.133			4.26	8.90	13.16	16.30
1614	15 MPH, cycle 0.5 mile, 30 min. wait/Ld./Uld		180	.044			1.42	2.96	4.38	5.45
1616	cycle 1 mile		168	.048			1.52	3.17	4.69	5.85
1618	cycle 2 miles		144	.056			1.78	3.70	5.48	6.80
1620	cycle 4 miles		120	.067			2.13	4.44	6.57	8.15
1622	cycle 6 miles		108	.074			2.37	4.94	7.31	9.10
1624	cycle 8 miles		84	.095			3.04	6.35	9.39	11.70
1625	cycle 10 miles		84	.095			3.04	6.35	9.39	11.70

31 23 23.20 Hauling		Crew	Daily Output	Labor-Hours	Unit	Material	2009 Bare Costs Labor	Equipment	Total	Total Incl O&P
1626	20 MPH ave, cycle 0.5 mile	B-34B	180	.044	L.C.Y.		1.42	2.96	4.38	5.45
1628	cycle 1 mile		168	.048			1.52	3.17	4.69	5.85
1630	cycle 2 miles		156	.051			1.64	3.42	5.06	6.30
1632	cycle 4 miles		132	.061			1.94	4.04	5.98	7.40
1634	cycle 6 miles		120	.067			2.13	4.44	6.57	8.15
1636	cycle 8 miles		108	.074			2.37	4.94	7.31	9.10
1638	cycle 10 miles		96	.083			2.66	5.55	8.21	10.20
1640	25 MPH ave, cycle 4 miles		144	.056			1.78	3.70	5.48	6.80
1642	cycle 6 miles		132	.061			1.94	4.04	5.98	7.40
1644	cycle 8 miles		108	.074			2.37	4.94	7.31	9.10
1646	cycle 10 miles		108	.074			2.37	4.94	7.31	9.10
1650	30 MPH ave, cycle 4 miles		144	.056			1.78	3.70	5.48	6.80
1652	cycle 6 miles		132	.061			1.94	4.04	5.98	7.40
1654	cycle 8 miles		120	.067			2.13	4.44	6.57	8.15
1656	cycle 10 miles		108	.074			2.37	4.94	7.31	9.10
1660	35 MPH ave, cycle 4 miles		156	.051			1.64	3.42	5.06	6.30
1662	cycle 6 miles		144	.056			1.78	3.70	5.48	6.80
1664	cycle 8 miles		132	.061			1.94	4.04	5.98	7.40
1666	cycle 10 miles		120	.067			2.13	4.44	6.57	8.15
1668	cycle 20 miles		84	.095			3.04	6.35	9.39	11.70
1669	cycle 30 miles		72	.111			3.55	7.40	10.95	13.60
1670	cycle 40 miles		60	.133			4.26	8.90	13.16	16.30
1672	40 MPH, cycle 6 miles		144	.056			1.78	3.70	5.48	6.80
1674	cycle 8 miles		132	.061			1.94	4.04	5.98	7.40
1676	cycle 10 miles		120	.067			2.13	4.44	6.57	8.15
1678	cycle 20 miles		96	.083			2.66	5.55	8.21	10.20
1680	cycle 30 miles		72	.111			3.55	7.40	10.95	13.60
1682	cycle 40 miles		60	.133			4.26	8.90	13.16	16.30
1684	cycle 50 miles		48	.167			5.35	11.10	16.45	20.50
1694	45 MPH ave, cycle 8 miles		144	.056			1.78	3.70	5.48	6.80
1696	cycle 10 miles		132	.061			1.94	4.04	5.98	7.40
1698	cycle 20 miles		96	.083			2.66	5.55	8.21	10.20
1700	cycle 30 miles		84	.095			3.04	6.35	9.39	11.70
1702	cycle 40 miles		60	.133			4.26	8.90	13.16	16.30
1704	cycle 50 miles		60	.133			4.26	8.90	13.16	16.30
1706	50 MPH ave, cycle 10 miles		132	.061			1.94	4.04	5.98	7.40
1708	cycle 20 miles		108	.074			2.37	4.94	7.31	9.10
1710	cycle 30 miles		84	.095			3.04	6.35	9.39	11.70
1712	cycle 40 miles		72	.111			3.55	7.40	10.95	13.60
1714	cycle 50 miles		60	.133			4.26	8.90	13.16	16.30
2000	Hauling, 8 CY truck, small project cost per hour	B-34A	8	1	Hr.		32	41	73	94
2100	12 CY Truck	B-34B	8	1			32	66.50	98.50	123
2150	16.5 CY Truck	B-34C	8	1			32	74.50	106.50	131
2200	20 CY Truck	B-34D	8	1			32	76	108	133
2300	Grading at dump, or embankment if required, by dozer	B-10B	1000	.012	L.C.Y.		.46	1.08	1.54	1.88
2310	Spotter at fill or cut, if required	1 Clab	8	1	Hr.		31.50		31.50	49
3014	16.5 CY truck, 15 min. wait/Ld./Uld., 15 MPH, cycle 0.5 mile	B-34C	462	.017	L.C.Y.		.55	1.29	1.84	2.27
3016	cycle 1 mile		413	.019			.62	1.44	2.06	2.53
3018	cycle 2 miles		347	.023			.74	1.71	2.45	3.01
3020	cycle 4 miles		248	.032			1.03	2.40	3.43	4.23
3022	cycle 6 miles		198	.040			1.29	3	4.29	5.30
3024	cycle 8 miles		165	.048			1.55	3.60	5.15	6.35
3025	cycle 10 miles		132	.061			1.94	4.50	6.44	7.95

31 23 23.20 Hauling		Crew	Daily Output	Labor-Hours	Unit	Material	2009 Bare Costs Labor	Equipment	Total	Total Incl O&P
3026	20 MPH ave, cycle 0.5 mile	B-34C	479	.017	L.C.Y.		.53	1.24	1.77	2.18
3028	cycle 1 mile		429	.019			.60	1.39	1.99	2.44
3030	cycle 2 miles		380	.021			.67	1.56	2.23	2.75
3032	cycle 4 miles		281	.028			.91	2.12	3.03	3.73
3034	cycle 6 miles		231	.035			1.11	2.57	3.68	4.53
3036	cycle 8 miles		198	.040			1.29	3	4.29	5.30
3038	cycle 10 miles		165	.048			1.55	3.60	5.15	6.35
3040	25 MPH ave, cycle 4 miles		314	.025			.81	1.89	2.70	3.33
3042	cycle 6 miles		264	.030			.97	2.25	3.22	3.97
3044	cycle 8 miles		231	.035			1.11	2.57	3.68	4.53
3046	cycle 10 miles		198	.040			1.29	3	4.29	5.30
3050	30 MPH ave, cycle 4 miles		347	.023			.74	1.71	2.45	3.01
3052	cycle 6 miles		281	.028			.91	2.12	3.03	3.73
3054	cycle 8 miles		248	.032			1.03	2.40	3.43	4.23
3056	cycle 10 miles		215	.037			1.19	2.76	3.95	4.87
3060	35 MPH ave, cycle 4 miles		363	.022			.70	1.64	2.34	2.88
3062	cycle 6 miles		314	.025			.81	1.89	2.70	3.33
3064	cycle 8 miles		264	.030			.97	2.25	3.22	3.97
3066	cycle 10 miles		248	.032			1.03	2.40	3.43	4.23
3068	cycle 20 miles		149	.054			1.72	3.99	5.71	7.05
3070	cycle 30 miles		116	.069			2.20	5.10	7.30	9.05
3072	cycle 40 miles		83	.096			3.08	7.15	10.23	12.65
3074	40 MPH ave, cycle 6 miles		330	.024			.77	1.80	2.57	3.17
3076	cycle 8 miles		281	.028			.91	2.12	3.03	3.73
3078	cycle 10 miles		264	.030			.97	2.25	3.22	3.97
3080	cycle 20 miles		165	.048			1.55	3.60	5.15	6.35
3082	cycle 30 miles		132	.061			1.94	4.50	6.44	7.95
3084	cycle 40 miles		99	.081			2.58	6	8.58	10.55
3086	cycle 50 miles		83	.096			3.08	7.15	10.23	12.65
3094	45 MPH ave, cycle 8 miles		297	.027			.86	2	2.86	3.52
3096	cycle 10 miles		281	.028			.91	2.12	3.03	3.73
3098	cycle 20 miles		182	.044			1.40	3.27	4.67	5.75
3100	cycle 30 miles		132	.061			1.94	4.50	6.44	7.95
3102	cycle 40 miles		116	.069			2.20	5.10	7.30	9.05
3104	cycle 50 miles		99	.081			2.58	6	8.58	10.55
3106	50 MPH ave, cycle 10 miles		281	.028			.91	2.12	3.03	3.73
3108	cycle 20 miles		198	.040			1.29	3	4.29	5.30
3110	cycle 30 miles		149	.054			1.72	3.99	5.71	7.05
3112	cycle 40 miles		116	.069			2.20	5.10	7.30	9.05
3114	cycle 50 miles		99	.081			2.58	6	8.58	10.55
3214	20 min. wait/Ld./Uld., 15 MPH, cycle 0.5 mile		363	.022			.70	1.64	2.34	2.88
3216	cycle 1 mile		330	.024			.77	1.80	2.57	3.17
3218	cycle 2 miles		281	.028			.91	2.12	3.03	3.73
3220	cycle 4 miles		215	.037			1.19	2.76	3.95	4.87
3222	cycle 6 miles		182	.044			1.40	3.27	4.67	5.75
3224	cycle 8 miles		149	.054			1.72	3.99	5.71	7.05
3225	cycle 10 miles		132	.061			1.94	4.50	6.44	7.95
3226	20 MPH ave, cycle 0.5 mile		363	.022			.70	1.64	2.34	2.88
3228	cycle 1 mile		347	.023			.74	1.71	2.45	3.01
3230	cycle 2 miles		297	.027			.86	2	2.86	3.52
3232	cycle 4 miles		248	.032			1.03	2.40	3.43	4.23
3234	cycle 6 miles		198	.040			1.29	3	4.29	5.30
3236	cycle 8 miles	↓	182	.044	↓		1.40	3.27	4.67	5.75

31 23 23.20 Hauling		Crew	Daily Output	Labor-Hours	Unit	Material	2009 Bare Costs Labor	Equipment	Total	Total Incl O&P
3238	cycle 10 miles	B-34C	149	.054	L.C.Y.		1.72	3.99	5.71	7.05
3240	25 MPH ave, cycle 4 miles		264	.030			.97	2.25	3.22	3.97
3242	cycle 6 miles		231	.035			1.11	2.57	3.68	4.53
3244	cycle 8 miles		198	.040			1.29	3	4.29	5.30
3246	cycle 10 miles		182	.044			1.40	3.27	4.67	5.75
3250	30 MPH ave, cycle 4 miles		281	.028			.91	2.12	3.03	3.73
3252	cycle 6 miles		248	.032			1.03	2.40	3.43	4.23
3254	cycle 8 miles		215	.037			1.19	2.76	3.95	4.87
3256	cycle 10 miles		198	.040			1.29	3	4.29	5.30
3260	35 MPH ave, cycle 4 miles		297	.027			.86	2	2.86	3.52
3262	cycle 6 miles		264	.030			.97	2.25	3.22	3.97
3264	cycle 8 miles		231	.035			1.11	2.57	3.68	4.53
3266	cycle 10 miles		215	.037			1.19	2.76	3.95	4.87
3268	cycle 20 miles		149	.054			1.72	3.99	5.71	7.05
3270	cycle 30 miles		99	.081			2.58	6	8.58	10.55
3272	cycle 40 miles		83	.096			3.08	7.15	10.23	12.65
3274	40 MPH ave, cycle 6 miles		264	.030			.97	2.25	3.22	3.97
3276	cycle 8 miles		248	.032			1.03	2.40	3.43	4.23
3278	cycle 10 miles		215	.037			1.19	2.76	3.95	4.87
3280	cycle 20 miles		149	.054			1.72	3.99	5.71	7.05
3282	cycle 30 miles		116	.069			2.20	5.10	7.30	9.05
3284	cycle 40 miles		99	.081			2.58	6	8.58	10.55
3286	cycle 50 miles		83	.096			3.08	7.15	10.23	12.65
3294	45 MPH ave, cycle 8 miles		248	.032			1.03	2.40	3.43	4.23
3296	cycle 10 miles		231	.035			1.11	2.57	3.68	4.53
3298	cycle 20 miles		165	.048			1.55	3.60	5.15	6.35
3300	cycle 30 miles		132	.061			1.94	4.50	6.44	7.95
3302	cycle 40 miles		99	.081			2.58	6	8.58	10.55
3304	cycle 50 miles		83	.096			3.08	7.15	10.23	12.65
3306	50 MPH ave, cycle 10 miles		248	.032			1.03	2.40	3.43	4.23
3308	cycle 20 miles		182	.044			1.40	3.27	4.67	5.75
3310	cycle 30 miles		132	.061			1.94	4.50	6.44	7.95
3312	cycle 40 miles		116	.069			2.20	5.10	7.30	9.05
3314	cycle 50 miles		99	.081			2.58	6	8.58	10.55
3414	25 min. wait/Ld./Uld., 15 MPH, cycle 0.5 mile		281	.028			.91	2.12	3.03	3.73
3416	cycle 1 mile		264	.030			.97	2.25	3.22	3.97
3418	cycle 2 miles		231	.035			1.11	2.57	3.68	4.53
3420	cycle 4 miles		182	.044			1.40	3.27	4.67	5.75
3422	cycle 6 miles		165	.048			1.55	3.60	5.15	6.35
3424	cycle 8 miles		132	.061			1.94	4.50	6.44	7.95
3425	cycle 10 miles		116	.069			2.20	5.10	7.30	9.05
3426	20 MPH ave, cycle 0.5 mile		297	.027			.86	2	2.86	3.52
3428	cycle 1 mile		281	.028			.91	2.12	3.03	3.73
3430	cycle 2 miles		248	.032			1.03	2.40	3.43	4.23
3432	cycle 4 miles		215	.037			1.19	2.76	3.95	4.87
3434	cycle 6 miles		182	.044			1.40	3.27	4.67	5.75
3436	cycle 8 miles		165	.048			1.55	3.60	5.15	6.35
3438	cycle 10 miles		132	.061			1.94	4.50	6.44	7.95
3440	25 MPH ave, cycle 4 miles		231	.035			1.11	2.57	3.68	4.53
3442	cycle 6 miles		198	.040			1.29	3	4.29	5.30
3444	cycle 8 miles		182	.044			1.40	3.27	4.67	5.75
3446	cycle 10 miles		165	.048			1.55	3.60	5.15	6.35
3450	30 MPH ave, cycle 4 miles		231	.035			1.11	2.57	3.68	4.53

31 23 23.20 Hauling		Crew	Daily Output	Labor-Hours	Unit	Material	2009 Bare Costs Labor	Equipment	Total	Total Incl O&P
3452	cycle 6 miles	B-34C	215	.037	L.C.Y.		1.19	2.76	3.95	4.87
3454	cycle 8 miles		182	.044			1.40	3.27	4.67	5.75
3456	cycle 10 miles		165	.048			1.55	3.60	5.15	6.35
3460	35 MPH ave, cycle 4 miles		248	.032			1.03	2.40	3.43	4.23
3462	cycle 6 miles		215	.037			1.19	2.76	3.95	4.87
3464	cycle 8 miles		198	.040			1.29	3	4.29	5.30
3466	cycle 10 miles		182	.044			1.40	3.27	4.67	5.75
3468	cycle 20 miles		132	.061			1.94	4.50	6.44	7.95
3470	cycle 30 miles		99	.081			2.58	6	8.58	10.55
3472	cycle 40 miles		83	.096			3.08	7.15	10.23	12.65
3474	40 MPH, cycle 6 miles		231	.035			1.11	2.57	3.68	4.53
3476	cycle 8 miles		215	.037			1.19	2.76	3.95	4.87
3478	cycle 10 miles		198	.040			1.29	3	4.29	5.30
3480	cycle 20 miles		132	.061			1.94	4.50	6.44	7.95
3482	cycle 30 miles		116	.069			2.20	5.10	7.30	9.05
3484	cycle 40 miles		83	.096			3.08	7.15	10.23	12.65
3486	cycle 50 miles		83	.096			3.08	7.15	10.23	12.65
3494	45 MPH ave, cycle 8 miles		215	.037			1.19	2.76	3.95	4.87
3496	cycle 10 miles		198	.040			1.29	3	4.29	5.30
3498	cycle 20 miles		149	.054			1.72	3.99	5.71	7.05
3500	cycle 30 miles		116	.069			2.20	5.10	7.30	9.05
3502	cycle 40 miles		99	.081			2.58	6	8.58	10.55
3504	cycle 50 miles		83	.096			3.08	7.15	10.23	12.65
3506	50 MPH ave, cycle 10 miles		215	.037			1.19	2.76	3.95	4.87
3508	cycle 20 miles		165	.048			1.55	3.60	5.15	6.35
3510	cycle 30 miles		132	.061			1.94	4.50	6.44	7.95
3512	cycle 40 miles		99	.081			2.58	6	8.58	10.55
3514	cycle 50 miles		83	.096			3.08	7.15	10.23	12.65
3614	30 min. wait/Ld./Uld., 15 MPH, cycle 0.5 mile		248	.032			1.03	2.40	3.43	4.23
3616	cycle 1 mile		231	.035			1.11	2.57	3.68	4.53
3618	cycle 2 miles		198	.040			1.29	3	4.29	5.30
3620	cycle 4 miles		165	.048			1.55	3.60	5.15	6.35
3622	cycle 6 miles		149	.054			1.72	3.99	5.71	7.05
3624	cycle 8 miles		116	.069			2.20	5.10	7.30	9.05
3625	cycle 10 miles		116	.069			2.20	5.10	7.30	9.05
3626	20 MPH ave, cycle 0.5 mile		248	.032			1.03	2.40	3.43	4.23
3628	cycle 1 mile		231	.035			1.11	2.57	3.68	4.53
3630	cycle 2 miles		215	.037			1.19	2.76	3.95	4.87
3632	cycle 4 miles		182	.044			1.40	3.27	4.67	5.75
3634	cycle 6 miles		165	.048			1.55	3.60	5.15	6.35
3636	cycle 8 miles		149	.054			1.72	3.99	5.71	7.05
3638	cycle 10 miles		132	.061			1.94	4.50	6.44	7.95
3640	25 MPH ave, cycle 4 miles		198	.040			1.29	3	4.29	5.30
3642	cycle 6 miles		182	.044			1.40	3.27	4.67	5.75
3644	cycle 8 miles		149	.054			1.72	3.99	5.71	7.05
3646	cycle 10 miles		149	.054			1.72	3.99	5.71	7.05
3650	30 MPH ave, cycle 4 miles		198	.040			1.29	3	4.29	5.30
3652	cycle 6 miles		182	.044			1.40	3.27	4.67	5.75
3654	cycle 8 miles		165	.048			1.55	3.60	5.15	6.35
3656	cycle 10 miles		149	.054			1.72	3.99	5.71	7.05
3660	35 MPH ave, cycle 4 miles		215	.037			1.19	2.76	3.95	4.87
3662	cycle 6 miles		198	.040			1.29	3	4.29	5.30
3664	cycle 8 miles		182	.044			1.40	3.27	4.67	5.75

31 23 23.20 Hauling		Crew	Daily Output	Labor-Hours	Unit	Material	2009 Bare Costs Labor	Equipment	Total	Total Incl O&P
3666	cycle 10 miles	B-34C	165	.048	L.C.Y.		1.55	3.60	5.15	6.35
3668	cycle 20 miles		116	.069			2.20	5.10	7.30	9.05
3670	cycle 30 miles		99	.081			2.58	6	8.58	10.55
3672	cycle 40 miles		83	.096			3.08	7.15	10.23	12.65
3674	40 MPH, cycle 6 miles		198	.040			1.29	3	4.29	5.30
3676	cycle 8 miles		182	.044			1.40	3.27	4.67	5.75
3678	cycle 10 miles		165	.048			1.55	3.60	5.15	6.35
3680	cycle 20 miles		132	.061			1.94	4.50	6.44	7.95
3682	cycle 30 miles		99	.081			2.58	6	8.58	10.55
3684	cycle 40 miles		83	.096			3.08	7.15	10.23	12.65
3686	cycle 50 miles		66	.121			3.87	9	12.87	15.85
3694	45 MPH ave, cycle 8 miles		198	.040			1.29	3	4.29	5.30
3696	cycle 10 miles		182	.044			1.40	3.27	4.67	5.75
3698	cycle 20 miles		132	.061			1.94	4.50	6.44	7.95
3700	cycle 30 miles		116	.069			2.20	5.10	7.30	9.05
3702	cycle 40 miles		83	.096			3.08	7.15	10.23	12.65
3704	cycle 50 miles		83	.096			3.08	7.15	10.23	12.65
3706	50 MPH ave, cycle 10 miles		182	.044			1.40	3.27	4.67	5.75
3708	cycle 20 miles		149	.054			1.72	3.99	5.71	7.05
3710	cycle 30 miles		116	.069			2.20	5.10	7.30	9.05
3712	cycle 40 miles		99	.081			2.58	6	8.58	10.55
3714	cycle 50 miles		83	.096			3.08	7.15	10.23	12.65
4014	20 CY truck, 15 min. wait/Ld./Uld., 15 MPH, cycle 0.5 mile	B-34D	560	.014			.46	1.09	1.55	1.90
4016	cycle 1 mile		500	.016			.51	1.22	1.73	2.13
4018	cycle 2 miles		420	.019			.61	1.45	2.06	2.54
4020	cycle 4 miles		300	.027			.85	2.03	2.88	3.54
4022	cycle 6 miles		240	.033			1.06	2.54	3.60	4.43
4024	cycle 8 miles		200	.040			1.28	3.05	4.33	5.30
4025	cycle 10 miles		160	.050			1.60	3.81	5.41	6.65
4026	20 MPH ave, cycle 0.5 mile		580	.014			.44	1.05	1.49	1.83
4028	cycle 1 mile		520	.015			.49	1.17	1.66	2.05
4030	cycle 2 miles		460	.017			.56	1.32	1.88	2.31
4032	cycle 4 miles		340	.024			.75	1.79	2.54	3.13
4034	cycle 6 miles		280	.029			.91	2.18	3.09	3.79
4036	cycle 8 miles		240	.033			1.06	2.54	3.60	4.43
4038	cycle 10 miles		200	.040			1.28	3.05	4.33	5.30
4040	25 MPH ave, cycle 4 miles		380	.021			.67	1.60	2.27	2.79
4042	cycle 6 miles		320	.025			.80	1.90	2.70	3.32
4044	cycle 8 miles		280	.029			.91	2.18	3.09	3.79
4046	cycle 10 miles		240	.033			1.06	2.54	3.60	4.43
4050	30 MPH ave, cycle 4 miles		420	.019			.61	1.45	2.06	2.54
4052	cycle 6 miles		340	.024			.75	1.79	2.54	3.13
4054	cycle 8 miles		300	.027			.85	2.03	2.88	3.54
4056	cycle 10 miles		260	.031			.98	2.34	3.32	4.09
4060	35 MPH ave, cycle 4 miles		440	.018			.58	1.38	1.96	2.41
4062	cycle 6 miles		380	.021			.67	1.60	2.27	2.79
4064	cycle 8 miles		320	.025			.80	1.90	2.70	3.32
4066	cycle 10 miles		300	.027			.85	2.03	2.88	3.54
4068	cycle 20 miles		180	.044			1.42	3.38	4.80	5.90
4070	cycle 30 miles		140	.057			1.83	4.35	6.18	7.60
4072	cycle 40 miles		100	.080			2.56	6.10	8.66	10.65
4074	40 MPH ave, cycle 6 miles		400	.020			.64	1.52	2.16	2.65
4076	cycle 8 miles		340	.024			.75	1.79	2.54	3.13

31 23 23.20 Hauling		Crew	Daily Output	Labor-Hours	Unit	Material	2009 Bare Costs Labor	Equipment	Total	Total Incl O&P
4078	cycle 10 miles	B-34D	320	.025	L.C.Y.		.80	1.90	2.70	3.32
4080	cycle 20 miles		200	.040			1.28	3.05	4.33	5.30
4082	cycle 30 miles		160	.050			1.60	3.81	5.41	6.65
4084	cycle 40 miles		120	.067			2.13	5.10	7.23	8.90
4086	cycle 50 miles		100	.080			2.56	6.10	8.66	10.65
4094	45 MPH ave, cycle 8 miles		360	.022			.71	1.69	2.40	2.95
4096	cycle 10 miles		340	.024			.75	1.79	2.54	3.13
4098	cycle 20 miles		220	.036			1.16	2.77	3.93	4.83
4100	cycle 30 miles		160	.050			1.60	3.81	5.41	6.65
4102	cycle 40 miles		140	.057			1.83	4.35	6.18	7.60
4104	cycle 50 miles		120	.067			2.13	5.10	7.23	8.90
4106	50 MPH ave, cycle 10 miles		340	.024			.75	1.79	2.54	3.13
4108	cycle 20 miles		240	.033			1.06	2.54	3.60	4.43
4110	cycle 30 miles		180	.044			1.42	3.38	4.80	5.90
4112	cycle 40 miles		140	.057			1.83	4.35	6.18	7.60
4114	cycle 50 miles		120	.067			2.13	5.10	7.23	8.90
4214	20 min. wait/Ld./Uld., 15 MPH, cycle 0.5 mile		440	.018			.58	1.38	1.96	2.41
4216	cycle 1 mile		400	.020			.64	1.52	2.16	2.65
4218	cycle 2 miles		340	.024			.75	1.79	2.54	3.13
4220	cycle 4 miles		260	.031			.98	2.34	3.32	4.09
4222	cycle 6 miles		220	.036			1.16	2.77	3.93	4.83
4224	cycle 8 miles		180	.044			1.42	3.38	4.80	5.90
4225	cycle 10 miles		160	.050			1.60	3.81	5.41	6.65
4226	20 MPH ave, cycle 0.5 mile		440	.018			.58	1.38	1.96	2.41
4228	cycle 1 mile		420	.019			.61	1.45	2.06	2.54
4230	cycle 2 miles		360	.022			.71	1.69	2.40	2.95
4232	cycle 4 miles		300	.027			.85	2.03	2.88	3.54
4234	cycle 6 miles		240	.033			1.06	2.54	3.60	4.43
4236	cycle 8 miles		220	.036			1.16	2.77	3.93	4.83
4238	cycle 10 miles		180	.044			1.42	3.38	4.80	5.90
4240	25 MPH ave, cycle 4 miles		320	.025			.80	1.90	2.70	3.32
4242	cycle 6 miles		280	.029			.91	2.18	3.09	3.79
4244	cycle 8 miles		240	.033			1.06	2.54	3.60	4.43
4246	cycle 10 miles		220	.036			1.16	2.77	3.93	4.83
4250	30 MPH ave, cycle 4 miles		340	.024			.75	1.79	2.54	3.13
4252	cycle 6 miles		300	.027			.85	2.03	2.88	3.54
4254	cycle 8 miles		260	.031			.98	2.34	3.32	4.09
4256	cycle 10 miles		240	.033			1.06	2.54	3.60	4.43
4260	35 MPH ave, cycle 4 miles		360	.022			.71	1.69	2.40	2.95
4262	cycle 6 miles		320	.025			.80	1.90	2.70	3.32
4264	cycle 8 miles		280	.029			.91	2.18	3.09	3.79
4266	cycle 10 miles		260	.031			.98	2.34	3.32	4.09
4268	cycle 20 miles		180	.044			1.42	3.38	4.80	5.90
4270	cycle 30 miles		120	.067			2.13	5.10	7.23	8.90
4272	cycle 40 miles		100	.080			2.56	6.10	8.66	10.65
4274	40 MPH ave, cycle 6 miles		320	.025			.80	1.90	2.70	3.32
4276	cycle 8 miles		300	.027			.85	2.03	2.88	3.54
4278	cycle 10 miles		260	.031			.98	2.34	3.32	4.09
4280	cycle 20 miles		180	.044			1.42	3.38	4.80	5.90
4282	cycle 30 miles		140	.057			1.83	4.35	6.18	7.60
4284	cycle 40 miles		120	.067			2.13	5.10	7.23	8.90
4286	cycle 50 miles		100	.080			2.56	6.10	8.66	10.65
4294	45 MPH ave, cycle 8 miles		300	.027			.85	2.03	2.88	3.54

31 23 23.20 Hauling		Crew	Daily Output	Labor-Hours	Unit	Material	2009 Bare Costs Labor	Equipment	Total	Total Incl O&P
4296	cycle 10 miles	B-34D	280	.029	L.C.Y.		.91	2.18	3.09	3.79
4298	cycle 20 miles		200	.040			1.28	3.05	4.33	5.30
4300	cycle 30 miles		160	.050			1.60	3.81	5.41	6.65
4302	cycle 40 miles		120	.067			2.13	5.10	7.23	8.90
4304	cycle 50 miles		100	.080			2.56	6.10	8.66	10.65
4306	50 MPH ave, cycle 10 miles		300	.027			.85	2.03	2.88	3.54
4308	cycle 20 miles		220	.036			1.16	2.77	3.93	4.83
4310	cycle 30 miles		180	.044			1.42	3.38	4.80	5.90
4312	cycle 40 miles		140	.057			1.83	4.35	6.18	7.60
4314	cycle 50 miles		120	.067			2.13	5.10	7.23	8.90
4414	25 min. wait/Ld./Uld., 15 MPH, cycle 0.5 mile		340	.024			.75	1.79	2.54	3.13
4416	cycle 1 mile		320	.025			.80	1.90	2.70	3.32
4418	cycle 2 miles		280	.029			.91	2.18	3.09	3.79
4420	cycle 4 miles		220	.036			1.16	2.77	3.93	4.83
4422	cycle 6 miles		200	.040			1.28	3.05	4.33	5.30
4424	cycle 8 miles		160	.050			1.60	3.81	5.41	6.65
4425	cycle 10 miles		140	.057			1.83	4.35	6.18	7.60
4426	20 MPH ave, cycle 0.5 mile		360	.022			.71	1.69	2.40	2.95
4428	cycle 1 mile		340	.024			.75	1.79	2.54	3.13
4430	cycle 2 miles		300	.027			.85	2.03	2.88	3.54
4432	cycle 4 miles		260	.031			.98	2.34	3.32	4.09
4434	cycle 6 miles		220	.036			1.16	2.77	3.93	4.83
4436	cycle 8 miles		200	.040			1.28	3.05	4.33	5.30
4438	cycle 10 miles		160	.050			1.60	3.81	5.41	6.65
4440	25 MPH ave, cycle 4 miles		280	.029			.91	2.18	3.09	3.79
4442	cycle 6 miles		240	.033			1.06	2.54	3.60	4.43
4444	cycle 8 miles		220	.036			1.16	2.77	3.93	4.83
4446	cycle 10 miles		200	.040			1.28	3.05	4.33	5.30
4450	30 MPH ave, cycle 4 miles		280	.029			.91	2.18	3.09	3.79
4452	cycle 6 miles		260	.031			.98	2.34	3.32	4.09
4454	cycle 8 miles		220	.036			1.16	2.77	3.93	4.83
4456	cycle 10 miles		200	.040			1.28	3.05	4.33	5.30
4460	35 MPH ave, cycle 4 miles		300	.027			.85	2.03	2.88	3.54
4462	cycle 6 miles		260	.031			.98	2.34	3.32	4.09
4464	cycle 8 miles		240	.033			1.06	2.54	3.60	4.43
4466	cycle 10 miles		220	.036			1.16	2.77	3.93	4.83
4468	cycle 20 miles		160	.050			1.60	3.81	5.41	6.65
4470	cycle 30 miles		120	.067			2.13	5.10	7.23	8.90
4472	cycle 40 miles		100	.080			2.56	6.10	8.66	10.65
4474	40 MPH, cycle 6 miles		280	.029			.91	2.18	3.09	3.79
4476	cycle 8 miles		260	.031			.98	2.34	3.32	4.09
4478	cycle 10 miles		240	.033			1.06	2.54	3.60	4.43
4480	cycle 20 miles		160	.050			1.60	3.81	5.41	6.65
4482	cycle 30 miles		140	.057			1.83	4.35	6.18	7.60
4484	cycle 40 miles		100	.080			2.56	6.10	8.66	10.65
4486	cycle 50 miles		100	.080			2.56	6.10	8.66	10.65
4494	45 MPH ave, cycle 8 miles		260	.031			.98	2.34	3.32	4.09
4496	cycle 10 miles		240	.033			1.06	2.54	3.60	4.43
4498	cycle 20 miles		180	.044			1.42	3.38	4.80	5.90
4500	cycle 30 miles		140	.057			1.83	4.35	6.18	7.60
4502	cycle 40 miles		120	.067			2.13	5.10	7.23	8.90
4504	cycle 50 miles		100	.080			2.56	6.10	8.66	10.65
4506	50 MPH ave, cycle 10 miles		260	.031			.98	2.34	3.32	4.09

31 23 23.20 Hauling		Crew	Daily Output	Labor-Hours	Unit	Material	2009 Bare Costs Labor	Equipment	Total	Total Incl O&P
4508	cycle 20 miles	B-34D	200	.040	L.C.Y.		1.28	3.05	4.33	5.30
4510	cycle 30 miles		160	.050			1.60	3.81	5.41	6.65
4512	cycle 40 miles		120	.067			2.13	5.10	7.23	8.90
4514	cycle 50 miles		100	.080			2.56	6.10	8.66	10.65
4614	30 min. wait/Ld./Uld., 15 MPH, cycle 0.5 mile		300	.027			.85	2.03	2.88	3.54
4616	cycle 1 mile		280	.029			.91	2.18	3.09	3.79
4618	cycle 2 miles		240	.033			1.06	2.54	3.60	4.43
4620	cycle 4 miles		200	.040			1.28	3.05	4.33	5.30
4622	cycle 6 miles		180	.044			1.42	3.38	4.80	5.90
4624	cycle 8 miles		140	.057			1.83	4.35	6.18	7.60
4625	cycle 10 miles		140	.057			1.83	4.35	6.18	7.60
4626	20 MPH ave, cycle 0.5 mile		300	.027			.85	2.03	2.88	3.54
4628	cycle 1 mile		280	.029			.91	2.18	3.09	3.79
4630	cycle 2 miles		260	.031			.98	2.34	3.32	4.09
4632	cycle 4 miles		220	.036			1.16	2.77	3.93	4.83
4634	cycle 6 miles		200	.040			1.28	3.05	4.33	5.30
4636	cycle 8 miles		180	.044			1.42	3.38	4.80	5.90
4638	cycle 10 miles		160	.050			1.60	3.81	5.41	6.65
4640	25 MPH ave, cycle 4 miles		240	.033			1.06	2.54	3.60	4.43
4642	cycle 6 miles		220	.036			1.16	2.77	3.93	4.83
4644	cycle 8 miles		180	.044			1.42	3.38	4.80	5.90
4646	cycle 10 miles		180	.044			1.42	3.38	4.80	5.90
4650	30 MPH ave, cycle 4 miles		240	.033			1.06	2.54	3.60	4.43
4652	cycle 6 miles		220	.036			1.16	2.77	3.93	4.83
4654	cycle 8 miles		200	.040			1.28	3.05	4.33	5.30
4656	cycle 10 miles		180	.044			1.42	3.38	4.80	5.90
4660	35 MPH ave, cycle 4 miles		260	.031			.98	2.34	3.32	4.09
4662	cycle 6 miles		240	.033			1.06	2.54	3.60	4.43
4664	cycle 8 miles		220	.036			1.16	2.77	3.93	4.83
4666	cycle 10 miles		200	.040			1.28	3.05	4.33	5.30
4668	cycle 20 miles		140	.057			1.83	4.35	6.18	7.60
4670	cycle 30 miles		120	.067			2.13	5.10	7.23	8.90
4672	cycle 40 miles		100	.080			2.56	6.10	8.66	10.65
4674	40 MPH, cycle 6 miles		240	.033			1.06	2.54	3.60	4.43
4676	cycle 8 miles		220	.036			1.16	2.77	3.93	4.83
4678	cycle 10 miles		200	.040			1.28	3.05	4.33	5.30
4680	cycle 20 miles		160	.050			1.60	3.81	5.41	6.65
4682	cycle 30 miles		120	.067			2.13	5.10	7.23	8.90
4684	cycle 40 miles		100	.080			2.56	6.10	8.66	10.65
4686	cycle 50 miles		80	.100			3.20	7.60	10.80	13.25
4694	45 MPH ave, cycle 8 miles		220	.036			1.16	2.77	3.93	4.83
4696	cycle 10 miles		220	.036			1.16	2.77	3.93	4.83
4698	cycle 20 miles		160	.050			1.60	3.81	5.41	6.65
4700	cycle 30 miles		140	.057			1.83	4.35	6.18	7.60
4702	cycle 40 miles		100	.080			2.56	6.10	8.66	10.65
4704	cycle 50 miles		100	.080			2.56	6.10	8.66	10.65
4706	50 MPH ave, cycle 10 miles		220	.036			1.16	2.77	3.93	4.83
4708	cycle 20 miles		180	.044			1.42	3.38	4.80	5.90
4710	cycle 30 miles		140	.057			1.83	4.35	6.18	7.60
4712	cycle 40 miles		120	.067			2.13	5.10	7.23	8.90
4714	cycle 50 miles		100	.080			2.56	6.10	8.66	10.65
5000	22 CY off-road, 15 min. wait/Ld./Uld., 5 MPH, cycle 2000 ft	B-34F	528	.015			.48	2.17	2.65	3.12
5010	cycle 3000 ft		484	.017			.53	2.36	2.89	3.41

31 23 23.20 Hauling		Crew	Daily Output	Labor-Hours	Unit	Material	2009 Bare Costs Labor	Equipment	Total	Total Incl O&P
5020	cycle 4000 ft	B-34F	440	.018	L.C.Y.		.58	2.60	3.18	3.75
5030	cycle 0.5 mile		506	.016			.51	2.26	2.77	3.27
5040	cycle 1 mile		374	.021			.68	3.06	3.74	4.41
5050	cycle 2 miles		264	.030			.97	4.33	5.30	6.25
5060	10 MPH, cycle 2000 ft		594	.013			.43	1.93	2.36	2.78
5070	cycle 3000 ft		572	.014			.45	2	2.45	2.89
5080	cycle 4000 ft		528	.015			.48	2.17	2.65	3.12
5090	cycle 0.5 mile		572	.014			.45	2	2.45	2.89
5100	cycle 1 mile		506	.016			.51	2.26	2.77	3.27
5110	cycle 2 miles		374	.021			.68	3.06	3.74	4.41
5120	cycle 4 miles		264	.030			.97	4.33	5.30	6.25
5130	15 MPH, cycle 2000 ft		638	.013			.40	1.79	2.19	2.59
5140	cycle 3000 ft		594	.013			.43	1.93	2.36	2.78
5150	cycle 4000 ft		572	.014			.45	2	2.45	2.89
5160	cycle 0.5 mile		616	.013			.42	1.86	2.28	2.68
5170	cycle 1 mile		550	.015			.46	2.08	2.54	3.01
5180	cycle 2 miles		462	.017			.55	2.48	3.03	3.57
5190	cycle 4 miles		330	.024			.77	3.47	4.24	5
5200	20 MPH, cycle 2 miles		506	.016			.51	2.26	2.77	3.27
5210	cycle 4 miles		374	.021			.68	3.06	3.74	4.41
5220	25 MPH, cycle 2 miles		528	.015			.48	2.17	2.65	3.12
5230	cycle 4 miles		418	.019			.61	2.74	3.35	3.95
5300	20 min. wait/Ld./Uld., 5 MPH, cycle 2000 ft		418	.019			.61	2.74	3.35	3.95
5310	cycle 3000 ft		396	.020			.65	2.89	3.54	4.17
5320	cycle 4000 ft		352	.023			.73	3.25	3.98	4.70
5330	cycle 0.5 mile		396	.020			.65	2.89	3.54	4.17
5340	cycle 1 mile		330	.024			.77	3.47	4.24	5
5350	cycle 2 miles		242	.033			1.06	4.73	5.79	6.80
5360	10 MPH, cycle 2000 ft		462	.017			.55	2.48	3.03	3.57
5370	cycle 3000 ft		440	.018			.58	2.60	3.18	3.75
5380	cycle 4000 ft		418	.019			.61	2.74	3.35	3.95
5390	cycle 0.5 mile		462	.017			.55	2.48	3.03	3.57
5400	cycle 1 mile		396	.020			.65	2.89	3.54	4.17
5410	cycle 2 miles		330	.024			.77	3.47	4.24	5
5420	cycle 4 miles		242	.033			1.06	4.73	5.79	6.80
5430	15 MPH, cycle 2000 ft		484	.017			.53	2.36	2.89	3.41
5440	cycle 3000 ft		462	.017			.55	2.48	3.03	3.57
5450	cycle 4000 ft		462	.017			.55	2.48	3.03	3.57
5460	cycle 0.5 mile		484	.017			.53	2.36	2.89	3.41
5470	cycle 1 mile		440	.018			.58	2.60	3.18	3.75
5480	cycle 2 miles		374	.021			.68	3.06	3.74	4.41
5490	cycle 4 miles		286	.028			.89	4	4.89	5.75
5500	20 MPH, cycle 2 miles		396	.020			.65	2.89	3.54	4.17
5510	cycle 4 miles		330	.024			.77	3.47	4.24	5
5520	25 MPH, cycle 2 miles		418	.019			.61	2.74	3.35	3.95
5530	cycle 4 miles		352	.023			.73	3.25	3.98	4.70
5600	25 min. wait/Ld./Uld., 5 MPH, cycle 2000 ft		352	.023			.73	3.25	3.98	4.70
5610	cycle 3000 ft		330	.024			.77	3.47	4.24	5
5620	cycle 4000 ft		308	.026			.83	3.71	4.54	5.35
5630	cycle 0.5 mile		330	.024			.77	3.47	4.24	5
5640	cycle 1 mile		286	.028			.89	4	4.89	5.75
5650	cycle 2 miles		220	.036			1.16	5.20	6.36	7.50
5660	10 MPH, cycle 2000 ft		374	.021			.68	3.06	3.74	4.41

31 23 23.20 Hauling		Crew	Daily Output	Labor-Hours	Unit	Material	2009 Bare Costs Labor	Equipment	Total	Total Incl O&P
5670	cycle 3000 ft	B-34F	374	.021	L.C.Y.		.68	3.06	3.74	4.41
5680	cycle 4000 ft		352	.023			.73	3.25	3.98	4.70
5690	cycle 0.5 mile		374	.021			.68	3.06	3.74	4.41
5700	cycle 1 mile		330	.024			.77	3.47	4.24	5
5710	cycle 2 miles		286	.028			.89	4	4.89	5.75
5720	cycle 4 miles		220	.036			1.16	5.20	6.36	7.50
5730	15 MPH, cycle 2000 ft		396	.020			.65	2.89	3.54	4.17
5740	cycle 3000 ft		374	.021			.68	3.06	3.74	4.41
5750	cycle 4000 ft		374	.021			.68	3.06	3.74	4.41
5760	cycle 0.5 mile		374	.021			.68	3.06	3.74	4.41
5770	cycle 1 mile		352	.023			.73	3.25	3.98	4.70
5780	cycle 2 miles		308	.026			.83	3.71	4.54	5.35
5790	cycle 4 miles		242	.033			1.06	4.73	5.79	6.80
5800	20 MPH, cycle 2 miles		330	.024			.77	3.47	4.24	5
5810	cycle 4 miles		286	.028			.89	4	4.89	5.75
5820	25 MPH, cycle 2 miles		352	.023			.73	3.25	3.98	4.70
5830	cycle 4 miles		308	.026			.83	3.71	4.54	5.35
6000	34 CY off-road, 15 min. wait/Ld./Uld., 5 MPH, cycle 2000 ft	B-34G	816	.010			.31	1.94	2.25	2.61
6010	cycle 3000 ft		748	.011			.34	2.12	2.46	2.86
6020	cycle 4000 ft		680	.012			.38	2.33	2.71	3.14
6030	cycle 0.5 mile		782	.010			.33	2.02	2.35	2.73
6040	cycle 1 mile		578	.014			.44	2.74	3.18	3.69
6050	cycle 2 miles		408	.020			.63	3.88	4.51	5.25
6060	10 MPH, cycle 2000 ft		918	.009			.28	1.72	2	2.32
6070	cycle 3000 ft		884	.009			.29	1.79	2.08	2.41
6080	cycle 4000 ft		816	.010			.31	1.94	2.25	2.61
6090	cycle 0.5 mile		884	.009			.29	1.79	2.08	2.41
6100	cycle 1 mile		782	.010			.33	2.02	2.35	2.73
6110	cycle 2 miles		578	.014			.44	2.74	3.18	3.69
6120	cycle 4 miles		408	.020			.63	3.88	4.51	5.25
6130	15 MPH, cycle 2000 ft		986	.008			.26	1.60	1.86	2.16
6140	cycle 3000 ft		918	.009			.28	1.72	2	2.32
6150	cycle 4000 ft		884	.009			.29	1.79	2.08	2.41
6160	cycle 0.5 mile		952	.008			.27	1.66	1.93	2.24
6170	cycle 1 mile		850	.009			.30	1.86	2.16	2.51
6180	cycle 2 miles		714	.011			.36	2.21	2.57	2.99
6190	cycle 4 miles		510	.016			.50	3.10	3.60	4.18
6200	20 MPH, cycle 2 miles		782	.010			.33	2.02	2.35	2.73
6210	cycle 4 miles		578	.014			.44	2.74	3.18	3.69
6220	25 MPH, cycle 2 miles		816	.010			.31	1.94	2.25	2.61
6230	cycle 4 miles		646	.012			.40	2.45	2.85	3.30
6300	20 min. wait/Ld./Uld., 5 MPH, cycle 2000 ft		646	.012			.40	2.45	2.85	3.30
6310	cycle 3000 ft		612	.013			.42	2.58	3	3.48
6320	cycle 4000 ft		544	.015			.47	2.91	3.38	3.92
6330	cycle 0.5 mile		612	.013			.42	2.58	3	3.48
6340	cycle 1 mile		510	.016			.50	3.10	3.60	4.18
6350	cycle 2 miles		374	.021			.68	4.23	4.91	5.70
6360	10 MPH, cycle 2000 ft		714	.011			.36	2.21	2.57	2.99
6370	cycle 3000 ft		680	.012			.38	2.33	2.71	3.14
6380	cycle 4000 ft		646	.012			.40	2.45	2.85	3.30
6390	cycle 0.5 mile		714	.011			.36	2.21	2.57	2.99
6400	cycle 1 mile		612	.013			.42	2.58	3	3.48
6410	cycle 2 miles		510	.016			.50	3.10	3.60	4.18

31 23 23.20 Hauling		Crew	Daily Output	Labor-Hours	Unit	Material	2009 Bare Costs Labor	Equipment	Total	Total Incl O&P
6420	cycle 4 miles	B-34G	374	.021	L.C.Y.		.68	4.23	4.91	5.70
6430	15 MPH, cycle 2000 ft		748	.011			.34	2.12	2.46	2.86
6440	cycle 3000 ft		714	.011			.36	2.21	2.57	2.99
6450	cycle 4000 ft		714	.011			.36	2.21	2.57	2.99
6460	cycle 0.5 mile		748	.011			.34	2.12	2.46	2.86
6470	cycle 1 mile		680	.012			.38	2.33	2.71	3.14
6480	cycle 2 miles		578	.014			.44	2.74	3.18	3.69
6490	cycle 4 miles		442	.018			.58	3.58	4.16	4.83
6500	20 MPH, cycle 2 miles		612	.013			.42	2.58	3	3.48
6510	cycle 4 miles		510	.016			.50	3.10	3.60	4.18
6520	25 MPH, cycle 2 miles		646	.012			.40	2.45	2.85	3.30
6530	cycle 4 miles		544	.015			.47	2.91	3.38	3.92
6600	25 min. wait/Ld./Uld., 5 MPH, cycle 2000 ft		544	.015			.47	2.91	3.38	3.92
6610	cycle 3000 ft		510	.016			.50	3.10	3.60	4.18
6620	cycle 4000 ft		476	.017			.54	3.32	3.86	4.49
6630	cycle 0.5 mile		510	.016			.50	3.10	3.60	4.18
6640	cycle 1 mile		442	.018			.58	3.58	4.16	4.83
6650	cycle 2 miles		340	.024			.75	4.65	5.40	6.25
6660	10 MPH, cycle 2000 ft		578	.014			.44	2.74	3.18	3.69
6670	cycle 3000 ft		578	.014			.44	2.74	3.18	3.69
6680	cycle 4000 ft		544	.015			.47	2.91	3.38	3.92
6690	cycle 0.5 mile		578	.014			.44	2.74	3.18	3.69
6700	cycle 1 mile		510	.016			.50	3.10	3.60	4.18
6710	cycle 2 miles		442	.018			.58	3.58	4.16	4.83
6720	cycle 4 miles		340	.024			.75	4.65	5.40	6.25
6730	15 MPH, cycle 2000 ft		612	.013			.42	2.58	3	3.48
6740	cycle 3000 ft		578	.014			.44	2.74	3.18	3.69
6750	cycle 4000 ft		578	.014			.44	2.74	3.18	3.69
6760	cycle 0.5 mile		612	.013			.42	2.58	3	3.48
6770	cycle 1 mile		544	.015			.47	2.91	3.38	3.92
6780	cycle 2 miles		476	.017			.54	3.32	3.86	4.49
6790	cycle 4 miles		374	.021			.68	4.23	4.91	5.70
6800	20 MPH, cycle 2 miles		510	.016			.50	3.10	3.60	4.18
6810	cycle 4 miles		442	.018			.58	3.58	4.16	4.83
6820	25 MPH, cycle 2 miles		544	.015			.47	2.91	3.38	3.92
6830	cycle 4 miles		476	.017			.54	3.32	3.86	4.49
7000	42 CY off-road, 20 min. wait/Ld./Uld., 5 MPH, cycle 2000 ft	B-34H	798	.010			.32	2	2.32	2.69
7010	cycle 3000 ft		756	.011			.34	2.11	2.45	2.84
7020	cycle 4000 ft		672	.012			.38	2.37	2.75	3.19
7030	cycle 0.5 mile		756	.011			.34	2.11	2.45	2.84
7040	cycle 1 mile		630	.013			.41	2.53	2.94	3.40
7050	cycle 2 miles		462	.017			.55	3.45	4	4.65
7060	10 MPH, cycle 2000 ft		882	.009			.29	1.81	2.10	2.44
7070	cycle 3000 ft		840	.010			.30	1.90	2.20	2.56
7080	cycle 4000 ft		798	.010			.32	2	2.32	2.69
7090	cycle 0.5 mile		882	.009			.29	1.81	2.10	2.44
7100	cycle 1 mile		798	.010			.32	2	2.32	2.69
7110	cycle 2 miles		630	.013			.41	2.53	2.94	3.40
7120	cycle 4 miles		462	.017			.55	3.45	4	4.65
7130	15 MPH, cycle 2000 ft		924	.009			.28	1.73	2.01	2.33
7140	cycle 3000 ft		882	.009			.29	1.81	2.10	2.44
7150	cycle 4000 ft		882	.009			.29	1.81	2.10	2.44
7160	cycle 0.5 mile		882	.009			.29	1.81	2.10	2.44

31 23 23.20 Hauling		Crew	Daily Output	Labor-Hours	Unit	Material	2009 Bare Costs Labor	Equipment	Total	Total Incl O&P
7170	cycle 1 mile	B-34H	840	.010	L.C.Y.		.30	1.90	2.20	2.56
7180	cycle 2 miles		714	.011			.36	2.23	2.59	3
7190	cycle 4 miles		546	.015			.47	2.92	3.39	3.93
7200	20 MPH, cycle 2 miles		756	.011			.34	2.11	2.45	2.84
7210	cycle 4 miles		630	.013			.41	2.53	2.94	3.40
7220	25 MPH, cycle 2 miles		798	.010			.32	2	2.32	2.69
7230	cycle 4 miles		672	.012			.38	2.37	2.75	3.19
7300	25 min. wait/Ld./Uld., 5 MPH, cycle 2000 ft		672	.012			.38	2.37	2.75	3.19
7310	cycle 3000 ft		630	.013			.41	2.53	2.94	3.40
7320	cycle 4000 ft		588	.014			.43	2.71	3.14	3.65
7330	cycle 0.5 mile		630	.013			.41	2.53	2.94	3.40
7340	cycle 1 mile		546	.015			.47	2.92	3.39	3.93
7350	cycle 2 miles		378	.021			.68	4.22	4.90	5.70
7360	10 MPH, cycle 2000 ft		714	.011			.36	2.23	2.59	3
7370	cycle 3000 ft		714	.011			.36	2.23	2.59	3
7380	cycle 4000 ft		672	.012			.38	2.37	2.75	3.19
7390	cycle 0.5 mile		714	.011			.36	2.23	2.59	3
7400	cycle 1 mile		630	.013			.41	2.53	2.94	3.40
7410	cycle 2 miles		546	.015			.47	2.92	3.39	3.93
7420	cycle 4 miles		378	.021			.68	4.22	4.90	5.70
7430	15 MPH, cycle 2000 ft		756	.011			.34	2.11	2.45	2.84
7440	cycle 3000 ft		714	.011			.36	2.23	2.59	3
7450	cycle 4000 ft		714	.011			.36	2.23	2.59	3
7460	cycle 0.5 mile		714	.011			.36	2.23	2.59	3
7470	cycle 1 mile		672	.012			.38	2.37	2.75	3.19
7480	cycle 2 miles		588	.014			.43	2.71	3.14	3.65
7490	cycle 4 miles		462	.017			.55	3.45	4	4.65
7500	20 MPH, cycle 2 miles		630	.013			.41	2.53	2.94	3.40
7510	cycle 4 miles		546	.015			.47	2.92	3.39	3.93
7520	25 MPH, cycle 2 miles		672	.012			.38	2.37	2.75	3.19
7530	cycle 4 miles		588	.014			.43	2.71	3.14	3.65
8000	60 CY off-road, 20 min. wait/Ld./Uld., 5 MPH, cycle 2000 ft	B-34J	1140	.007			.22	1.80	2.02	2.33
8010	cycle 3000 ft		1080	.007			.24	1.90	2.14	2.45
8020	cycle 4000 ft		960	.008			.27	2.13	2.40	2.75
8030	cycle 0.5 mile		1080	.007			.24	1.90	2.14	2.45
8040	cycle 1 mile		900	.009			.28	2.27	2.55	2.94
8050	cycle 2 miles		660	.012			.39	3.10	3.49	4.01
8060	10 MPH, cycle 2000 ft		1260	.006			.20	1.62	1.82	2.10
8070	cycle 3000 ft		1200	.007			.21	1.71	1.92	2.21
8080	cycle 4000 ft		1140	.007			.22	1.80	2.02	2.33
8090	cycle 0.5 mile		1260	.006			.20	1.62	1.82	2.10
8100	cycle 1 mile		1080	.007			.24	1.90	2.14	2.45
8110	cycle 2 miles		900	.009			.28	2.27	2.55	2.94
8120	cycle 4 miles		660	.012			.39	3.10	3.49	4.01
8130	15 MPH, cycle 2000 ft		1320	.006			.19	1.55	1.74	2.01
8140	cycle 3000 ft		1260	.006			.20	1.62	1.82	2.10
8150	cycle 4000 ft		1260	.006			.20	1.62	1.82	2.10
8160	cycle 0.5 mile		1320	.006			.19	1.55	1.74	2.01
8170	cycle 1 mile		1200	.007			.21	1.71	1.92	2.21
8180	cycle 2 miles		1020	.008			.25	2.01	2.26	2.60
8190	cycle 4 miles		780	.010			.33	2.63	2.96	3.39
8200	20 MPH, cycle 2 miles		1080	.007			.24	1.90	2.14	2.45
8210	cycle 4 miles		900	.009			.28	2.27	2.55	2.94

31 23 Excavation and Fill

31 23 23 – Fill

31 23 23.20 Hauling

	31 23 23.20 Hauling	Crew	Daily Output	Labor-Hours	Unit	Material (2009 Bare Costs)	Labor (2009 Bare Costs)	Equipment (2009 Bare Costs)	Total (2009 Bare Costs)	Total Incl O&P
8220	25 MPH, cycle 2 miles	B-34J	1140	.007	L.C.Y.		.22	1.80	2.02	2.33
8230	cycle 4 miles		960	.008			.27	2.13	2.40	2.75
8300	25 min. wait/Ld./Uld., 5 MPH, cycle 2000 ft		960	.008			.27	2.13	2.40	2.75
8310	cycle 3000 ft		900	.009			.28	2.27	2.55	2.94
8320	cycle 4000 ft		840	.010			.30	2.44	2.74	3.15
8330	cycle 0.5 mile		900	.009			.28	2.27	2.55	2.94
8340	cycle 1 mile		780	.010			.33	2.63	2.96	3.39
8350	cycle 2 miles		600	.013			.43	3.41	3.84	4.41
8360	10 MPH, cycle 2000 ft		1020	.008			.25	2.01	2.26	2.60
8370	cycle 3000 ft		1020	.008			.25	2.01	2.26	2.60
8380	cycle 4000 ft		960	.008			.27	2.13	2.40	2.75
8390	cycle 0.5 mile		1020	.008			.25	2.01	2.26	2.60
8400	cycle 1 mile		900	.009			.28	2.27	2.55	2.94
8410	cycle 2 miles		780	.010			.33	2.63	2.96	3.39
8420	cycle 4 miles		600	.013			.43	3.41	3.84	4.41
8430	15 MPH, cycle 2000 ft		1080	.007			.24	1.90	2.14	2.45
8440	cycle 3000 ft		1020	.008			.25	2.01	2.26	2.60
8450	cycle 4000 ft		1020	.008			.25	2.01	2.26	2.60
8460	cycle 0.5 mile		1080	.007			.24	1.90	2.14	2.45
8470	cycle 1 mile		960	.008			.27	2.13	2.40	2.75
8480	cycle 2 miles		840	.010			.30	2.44	2.74	3.15
8490	cycle 4 miles		660	.012			.39	3.10	3.49	4.01
8500	20 MPH, cycle 2 miles		900	.009			.28	2.27	2.55	2.94
8510	cycle 4 miles		780	.010			.33	2.63	2.96	3.39
8520	25 MPH, cycle 2 miles		960	.008			.27	2.13	2.40	2.75
8530	cycle 4 miles	↓	840	.010	↓		.30	2.44	2.74	3.15

31 41 Shoring

31 41 16 – Sheet Piling

31 41 16.10 Sheet Piling Systems

	31 41 16.10 Sheet Piling Systems	Crew	Daily Output	Labor-Hours	Unit	Material (2009 Bare Costs)	Labor (2009 Bare Costs)	Equipment (2009 Bare Costs)	Total (2009 Bare Costs)	Total Incl O&P
0010	**SHEET PILING SYSTEMS**									
0020	Sheet piling steel, not incl. wales, 22 psf, 15′ excav., left in place	B-40	10.81	5.920	Ton	1,225	234	305	1,764	2,050
0100	Drive, extract & salvage R314116-45		6	10.667		505	420	550	1,475	1,825
0300	20′ deep excavation, 27 psf, left in place		12.95	4.942		1,225	195	256	1,676	1,950
0400	Drive, extract & salvage		6.55	9.771	↓	505	385	505	1,395	1,725
1200	15′ deep excavation, 22 psf, left in place		983	.065	S.F.	14.35	2.58	3.37	20.30	23.50
1300	Drive, extract & salvage		545	.117		5.65	4.64	6.10	16.39	20
1500	20′ deep excavation, 27 psf, left in place		960	.067		18.05	2.64	3.45	24.14	28
1600	Drive, extract & salvage	↓	485	.132	↓	7.35	5.20	6.85	19.40	24
2100	Rent steel sheet piling and wales, first month				Ton	263			263	289
2200	Per added month					26.50			26.50	29
2300	Rental piling left in place, add to rental					980			980	1,075
2500	Wales, connections & struts, 2/3 salvage				↓	275			275	300
3900	Wood, solid sheeting, incl. wales, braces and spacers, R314116-40									
3910	drive, extract & salvage, 8′ deep excavation	B-31	330	.121	S.F.	1.47	4.08	.55	6.10	8.55
4000	10′ deep, 50 S.F./hr. in & 150 S.F./hr. out		300	.133		1.52	4.49	.60	6.61	9.30
4100	12′ deep, 45 S.F./hr. in & 135 S.F./hr. out		270	.148		1.56	4.99	.67	7.22	10.20
4200	14′ deep, 42 S.F./hr. in & 126 S.F./hr. out		250	.160		1.61	5.40	.72	7.73	10.90
4300	16′ deep, 40 S.F./hr. in & 120 S.F./hr. out		240	.167		1.66	5.60	.75	8.01	11.35
4400	18′ deep, 38 S.F./hr. in & 114 S.F./hr. out		230	.174		1.71	5.85	.79	8.35	11.85
4520	Left in place, 8′ deep, 55 S.F./hr.	↓	440	.091	↓	2.65	3.06	.41	6.12	8.10

31 41 Shoring

31 41 16 – Sheet Piling

31 41 16.10 Sheet Piling Systems		Crew	Daily Output	Labor-Hours	Unit	Material	Labor	Equipment	Total	Total Incl O&P
						2009 Bare Costs				
4540	10′ deep, 50 S.F./hr.	B-31	400	.100	S.F.	2.79	3.37	.45	6.61	8.75
4560	12′ deep, 45 S.F./hr.		360	.111		2.95	3.74	.50	7.19	9.60
4565	14′ deep, 42 S.F./hr.		335	.119		3.12	4.02	.54	7.68	10.25
4570	16′ deep, 40 S.F./hr.		320	.125		3.32	4.21	.57	8.10	10.80
4700	Alternate pricing, left in place, 8′ deep		1.76	22.727	M.B.F.	595	765	103	1,463	1,950
4800	Drive, extract and salvage, 8′ deep		1.32	30.303	"	530	1,025	137	1,692	2,300

31 52 Cofferdams

31 52 16 – Timber Cofferdams

31 52 16.10 Cofferdams

		Crew	Daily Output	Labor-Hours	Unit	Material	Labor	Equipment	Total	Total Incl O&P
0010	**COFFERDAMS**									
0011	Incl. mobilization and temporary sheeting									
0080	Soldier beams & lagging H piles with 3" wood sheeting									
0090	horizontal between piles, including removal of wales & braces									
0100	No hydrostatic head, 15′ deep, 1 line of braces, minimum	B-50	545	.206	S.F.	6.95	7.75	4.05	18.75	24.50
0200	Maximum		495	.226		7.70	8.55	4.46	20.71	27
1300	No hydrostatic head, left in place, 15′ dp., 1 line of braces, min.		635	.176		9.25	6.65	3.48	19.38	24.50
1400	Maximum		575	.195		9.90	7.35	3.84	21.09	27
2350	Lagging only, 3" thick wood between piles 8′ O.C., minimum	B-46	400	.120		1.54	4.25	.09	5.88	8.55
2370	Maximum		250	.192		2.31	6.80	.15	9.26	13.50
2400	Open sheeting no bracing, for trenches to 10′ deep, min.		1736	.028		.69	.98	.02	1.69	2.34
2450	Maximum		1510	.032		.77	1.12	.03	1.92	2.67
2500	Tie-back method, add to open sheeting, add, minimum								20%	20%
2550	Maximum								60%	60%
2700	Tie-backs only, based on tie-backs total length, minimum	B-46	86.80	.553	L.F.	13.30	19.55	.44	33.29	46
2750	Maximum		38.50	1.247	"	23.50	44	.98	68.48	96.50
3500	Tie-backs only, typical average, 25′ long		2	24	Ea.	585	850	18.95	1,453.95	2,025
3600	35′ long		1.58	30.380	"	780	1,075	24	1,879	2,575

Division Notes

I	CREW	DAILY OUTPUT	LABOR-HOURS	UNIT	2009 BARE COSTS MAT.	LABOR	EQUIP.	TOTAL	TOTAL INCL O&P

Estimating Tips

32 01 00 Operations and Maintenance of Exterior Improvements

- Recycling of asphalt pavement is becoming very popular and is an alternative to removal and replacement of asphalt pavement. It can be a good value engineering proposal if removed pavement can be recycled either at the site or another site that is reasonably close to the project site.

32 10 00 Bases, Ballasts, and Paving

- When estimating paving, keep in mind the project schedule. If an asphaltic paving project is in a colder climate and runs through to the spring, consider placing the base course in the autumn and then topping it in the spring, just prior to completion. This could save considerable costs in spring repair. Keep in mind that prices for asphalt and concrete are generally higher in the cold seasons.

32 90 00 Planting

- The timing of planting and guarantee specifications often dictate the costs for establishing tree and shrub growth and a stand of grass or ground cover. Establish the work performance schedule to coincide with the local planting season. Maintenance and growth guarantees can add from 20%–100% to the total landscaping cost. The cost to replace trees and shrubs can be as high as 5% of the total cost, depending on the planting zone, soil conditions, and time of year.

Reference Numbers

Reference numbers are shown in shaded boxes at the beginning of some major classifications. These numbers refer to related items in the Reference Section. The reference information may be an estimating procedure, an alternate pricing method, or technical information.

Note: Not all subdivisions listed here necessarily appear in this publication.

32 12 16.13 Plant-Mix Asphalt Paving		Crew	Daily Output	Labor-Hours	Unit	2009 Bare Costs Material	Labor	Equipment	Total	Total Incl O&P
0010	**PLANT-MIX ASPHALT PAVING**									
0020	And large paved areas with no hauling included									
0025	See Div. 31 23 23.20 for hauling costs									
0080	Binder course, 1-1/2" thick	B-25	7725	.011	S.Y.	4.66	.39	.32	5.37	6.10
0120	2" thick		6345	.014		6.20	.48	.39	7.07	7.95
0130	2-1/2" thick		5620	.016		7.75	.54	.44	8.73	9.80
0160	3" thick		4905	.018		9.30	.62	.51	10.43	11.75
0170	3-1/2 " thick		4520	.019		10.85	.67	.55	12.07	13.60
0200	4" thick	↓	4140	.021		12.40	.73	.60	13.73	15.45
0300	Wearing course, 1" thick	B-25B	10575	.009		3.08	.32	.26	3.66	4.15
0340	1-1/2" thick		7725	.012		5.15	.44	.35	5.94	6.75
0380	2" thick		6345	.015		6.95	.53	.43	7.91	8.95
0420	2-1/2" thick		5480	.018		8.55	.61	.49	9.65	10.90
0460	3" thick		4900	.020		10.20	.69	.55	11.44	12.90
0470	3-1/2" thick		4520	.021		12	.74	.60	13.34	14.95
0480	4 " thick	↓	4140	.023		13.70	.81	.65	15.16	17
0500	Open graded friction course	B-25C	5000	.010	↓	3.29	.34	.43	4.06	4.61
0800	Alternate method of figuring paving costs									
0810	Binder course, 1-1/2" thick	B-25	630	.140	Ton	57	4.81	3.93	65.74	74
0811	2" thick		690	.128		57	4.39	3.59	64.98	73
0812	3" thick		800	.110		57	3.79	3.10	63.89	71.50
0813	4" thick	↓	900	.098		57	3.37	2.75	63.12	70.50
0850	Wearing course, 1" thick	B-25B	575	.167		62	5.85	4.71	72.56	82
0851	1-1/2" thick		630	.152		62	5.35	4.30	71.65	81
0852	2" thick		690	.139		62	4.87	3.93	70.80	80
0853	2-1/2" thick		765	.125		62	4.39	3.54	69.93	78.50
0854	3" thick	↓	800	.120	↓	62	4.20	3.39	69.59	78
1000	Pavement replacement over trench, 2" thick	B-37	90	.533	S.Y.	6.40	17.70	1.53	25.63	36
1050	4" thick		70	.686		12.65	23	1.97	37.62	51
1080	6" thick	↓	55	.873	↓	20	29	2.50	51.50	69.50

32 12 16.14 Asphaltic Concrete Paving, Parking Lots and Driveways		Crew	Daily Output	Labor-Hours	Unit	Material	Labor	Equipment	Total	Total Incl O&P
0011	**ASPHALTIC CONCRETE PAVING, PARKING LOTS & DRIVEWAYS**									
0020	6" stone base, 2" binder course, 1" topping	B-25C	9000	.005	S.F.	2.07	.19	.24	2.50	2.83
0300	Binder course, 1-1/2" thick		35000	.001		.52	.05	.06	.63	.71
0400	2" thick		25000	.002		.67	.07	.09	.83	.94
0500	3" thick		15000	.003		1.04	.11	.14	1.29	1.47
0600	4" thick		10800	.004		1.36	.16	.20	1.72	1.95
0800	Sand finish course, 3/4" thick		41000	.001		.30	.04	.05	.39	.44
0900	1" thick	↓	34000	.001		.36	.05	.06	.47	.55
1000	Fill pot holes, hot mix, 2" thick	B-16	4200	.008		.77	.25	.13	1.15	1.36
1100	4" thick		3500	.009		1.12	.29	.15	1.56	1.86
1120	6" thick	↓	3100	.010		1.50	.33	.17	2	2.35
1140	Cold patch, 2" thick	B-51	3000	.016		.78	.51	.06	1.35	1.71
1160	4" thick		2700	.018		1.48	.57	.07	2.12	2.58
1180	6" thick	↓	1900	.025	↓	2.30	.80	.10	3.20	3.89

32 84 Planting Irrigation

32 84 23 – Underground Sprinklers

32 84 23.10 Sprinkler Irrigation System		Crew	Daily Output	Labor-Hours	Unit	2009 Bare Costs Material	Labor	Equipment	Total	Total Incl O&P
0010	**SPRINKLER IRRIGATION SYSTEM**									
0011	For lawns									
0100	Golf course with fully automatic system	C-17	.05	1600	9 holes	97,500	66,000		163,500	209,500
0200	24′ diam. head at 15′ O.C incl. piping, auto oper., minimum	B-20	70	.343	Head	22	12.10		34.10	43
0300	Maximum		40	.600		50.50	21		71.50	88.50
0500	60′ diam. head at 40′ O.C. incl. piping, auto oper., minimum		28	.857		66.50	30.50		97	120
0600	Maximum		23	1.043		186	37		223	261
0800	Residential system, custom, 1″ supply		2000	.012	S.F.	.32	.42		.74	1.01
0900	1-1/2″ supply		1800	.013	″	.39	.47		.86	1.16

Division Notes

		CREW	DAILY OUTPUT	LABOR-HOURS	UNIT	2009 BARE COSTS MAT.	LABOR	EQUIP.	TOTAL	TOTAL INCL O&P

Estimating Tips

33 10 00 Water Utilities
33 30 00 Sanitary Sewerage Utilities
33 40 00 Storm Drainage Utilities

- Never assume that the water, sewer, and drainage lines will go in at the early stages of the project. Consider the site access needs before dividing the site in half with open trenches, loose pipe, and machinery obstructions. Always inspect the site to establish that the site drawings are complete. Check off all existing utilities on your drawings as you locate them. If you find any discrepancies, mark up the site plan for further research. Differing site conditions can be very costly if discovered later in the project.
- See also Section 33 01 00 for restoration of pipe where removal/replacement may be undesirable. Use of new types of piping materials can reduce the overall project cost. Owners/design engineers should consider the installing contractor as a valuable source of current information on utility products and local conditions that could lead to significant cost savings.

Reference Numbers

Reference numbers are shown in shaded boxes at the beginning of some major classifications. These numbers refer to related items in the Reference Section. The reference information may be an estimating procedure, an alternate pricing method, or technical information.

Note: Not all subdivisions listed here necessarily appear in this publication.

Note: **Trade Service,** *in part, has been used as a reference source for some of the material prices used in Division 33.*

33 01 10 – Operation and Maintenance of Water Utilities

33 01 10.10 Corrosion Resistance		Crew	Daily Output	Labor-Hours	Unit	Material (2009 Bare Costs)	Labor	Equipment	Total	Total Incl O&P
0010	**CORROSION RESISTANCE**									
0012	Wrap & coat, add to pipe, 4" diameter				L.F.	2			2	2.20
0020	5" diameter					2.08			2.08	2.29
0040	6" diameter					2.08			2.08	2.29
0060	8" diameter					3.22			3.22	3.54
0080	10" diameter					4.81			4.81	5.30
0100	12" diameter					4.81			4.81	5.30
0120	14" diameter					7.05			7.05	7.75
0140	16" diameter					7.05			7.05	7.75
0160	18" diameter					7.15			7.15	7.85
0180	20" diameter					7.45			7.45	8.20
0200	24" diameter					9.10			9.10	10.05
0220	Small diameter pipe, 1" diameter, add					1.50			1.50	1.65
0240	2" diameter					1.65			1.65	1.82
0260	2-1/2" diameter					1.82			1.82	2
0280	3" diameter					1.82			1.82	2
0300	Fittings, field covered, add				S.F.	9.50			9.50	10.45
0500	Coating, bituminous, per diameter inch, 1 coat, add				L.F.	.37			.37	.41
0540	3 coat					.57			.57	.63
0560	Coal tar epoxy, per diameter inch, 1 coat, add					.21			.21	.23
0600	3 coat					.46			.46	.51
1000	Polyethylene H.D. extruded, .025" thk., 1/2" diameter add					.33			.33	.36
1020	3/4" diameter					.35			.35	.39
1040	1" diameter					.39			.39	.43
1060	1-1/4" diameter					.48			.48	.53
1080	1-1/2" diameter					.53			.53	.58
1100	.030" thk., 2" diameter					.57			.57	.63
1120	2-1/2" diameter					.67			.67	.74
1140	.035" thk., 3" diameter					.84			.84	.92
1160	3-1/2" diameter					.95			.95	1.05
1180	4" diameter					1.06			1.06	1.17
1200	.040" thk, 5" diameter					1.46			1.46	1.61
1220	6" diameter					1.52			1.52	1.67
1240	8" diameter					1.98			1.98	2.18
1260	10" diameter					2.43			2.43	2.67
1280	12" diameter					2.91			2.91	3.20
1300	.060" thk., 14" diameter					4			4	4.40
1320	16" diameter					4.68			4.68	5.15
1340	18" diameter					5.55			5.55	6.10
1360	20" diameter					6.25			6.25	6.85
1380	Fittings, field wrapped, add				S.F.	6.45			6.45	7.10

33 01 10.20 Pipe Repair		Crew	Daily Output	Labor-Hours	Unit	Material	Labor	Equipment	Total	Total Incl O&P
0010	**PIPE REPAIR**									
0020	Not including excavation or backfill									
0100	Clamp, stainless steel, lightweight, for steel pipe									
0110	3" long, 1/2" diameter pipe	1 Plum	34	.235	Ea.	13.80	11.45		25.25	32.50
0120	3/4" diameter pipe		32	.250		14.20	12.20		26.40	34
0130	1" diameter pipe		30	.267		15.25	13		28.25	36.50
0140	1-1/4" diameter pipe		28	.286		16.15	13.95		30.10	39
0150	1-1/2" diameter pipe		26	.308		16.70	15		31.70	41
0160	2" diameter pipe		24	.333		18.25	16.25		34.50	44.50
0170	2-1/2" diameter pipe		23	.348		19.85	16.95		36.80	47.50

33 01 10 – Operation and Maintenance of Water Utilities

33 01 10.20 Pipe Repair		Crew	Daily Output	Labor-Hours	Unit	Material	Labor	Equipment	Total	Total Incl O&P
						2009 Bare Costs				
0180	3″ diameter pipe	1 Plum	22	.364	Ea.	22	17.75		39.75	50.50
0190	3-1/2″ diameter pipe	↓	21	.381		23.50	18.55		42.05	54
0200	4″ diameter pipe	B-20	56	.429		24.50	15.15		39.65	50.50
0210	5″ diameter pipe		53	.453		28	16		44	56
0220	6″ diameter pipe		48	.500		31.50	17.70		49.20	62
0230	8″ diameter pipe		30	.800		36.50	28.50		65	84
0240	10″ diameter pipe		28	.857		94	30.50		124.50	150
0250	12″ diameter pipe		24	1		100	35.50		135.50	165
0260	14″ diameter pipe		22	1.091		104	38.50		142.50	175
0270	16″ diameter pipe		20	1.200		113	42.50		155.50	190
0280	18″ diameter pipe		18	1.333		119	47		166	204
0290	20″ diameter pipe		16	1.500		131	53		184	227
0300	24″ diameter pipe	↓	14	1.714		146	60.50		206.50	255
0360	For 6″ long, add					100%	40%			
0370	For 9″ long, add					200%	100%			
0380	For 12″ long, add					300%	150%			
0390	For 18″ long, add				↓	500%	200%			
0400	Pipe Freezing for live repairs of systems 3/8 inch to 6 inch									
0410	Note: Pipe Freezing can also be used to install a valve into a live system									
0420	Pipe Freezing each side 3/8 inch	2 Skwk	8	2	Ea.	495	81.50		576.50	670
0425	Pipe Freezing each side 3/8 inch, second location same kit		8	2		19.95	81.50		101.45	149
0430	Pipe Freezing each side 3/4 inch		8	2		495	81.50		576.50	670
0435	Pipe Freezing each side 3/4 inch, second location same kit		8	2		19.95	81.50		101.45	149
0440	Pipe Freezing each side 1 1/2 inch		6	2.667		495	109		604	715
0445	Pipe Freezing each side 1 1/2 inch, second location same kit		6	2.667		19.95	109		128.95	191
0450	Pipe Freezing each side 2 inch		6	2.667		735	109		844	975
0455	Pipe Freezing each side 2 inch, second location same kit		6	2.667		19.95	109		128.95	191
0460	Pipe Freezing each side 2 1/2 inch-3 inch		6	2.667		1,025	109		1,134	1,300
0465	Pipe Freeze each side 2 1/2-3 inch, second location same kit		6	2.667		40	109		149	213
0470	Pipe Freezing each side 4 inch		4	4		1,675	163		1,838	2,075
0475	Pipe Freezing each side 4 inch, second location same kit		4	4		65	163		228	325
0480	Pipe Freezing each side 5-6 inch		4	4		4,025	163		4,188	4,675
0485	Pipe Freezing each side 5-6 inch, second location same kit	↓	4	4		195	163		358	470
0490	Pipe Frz extra 20 lb CO_2 cylinders (3/8 to 2″ - 1 ea, 3″ -2 ea)					218			218	240
0500	Pipe Freezing extra 50 lb CO_2 cylinders (4″ - 2 ea, 5-6″ -6 ea)				↓	490			490	535
1000	Clamp, stainless steel, with threaded service tap									
1040	Full seal for iron, steel, PVC pipe									
1100	6″ long, 2″ diameter pipe	1 Plum	17	.471	Ea.	177	23		200	230
1110	2-1/2″ diameter pipe		16	.500		188	24.50		212.50	244
1120	3″ diameter pipe		15.60	.513		200	25		225	258
1130	3-1/2″ diameter pipe	↓	15	.533		209	26		235	269
1140	4″ diameter pipe	B-20	40	.600		230	21		251	285
1150	6″ diameter pipe		34	.706		265	25		290	330
1160	8″ diameter pipe		21	1.143		330	40.50		370.50	430
1170	10″ diameter pipe		20	1.200		365	42.50		407.50	465
1180	12″ diameter pipe	↓	17	1.412		415	50		465	540
1205	For 9″ long, add					20%	45%			
1210	For 12″ long, add					40%	80%			
1220	For 18″ long, add				↓	70%	110%			
1600	Clamp, stainless steel, single section									
1640	Full seal for iron, steel, PVC pipe									
1700	6″ long, 2″ diameter pipe	1 Plum	17	.471	Ea.	92	23		115	136
1710	2-1/2″ diameter pipe	↓	16	.500	↓	100	24.50		124.50	147

33 01 Operation and Maintenance of Utilities

33 01 10 – Operation and Maintenance of Water Utilities

33 01 10.20 Pipe Repair		Crew	Daily Output	Labor-Hours	Unit	2009 Bare Costs Material	Labor	Equipment	Total	Total Incl O&P
1720	3" diameter pipe	1 Plum	15.60	.513	Ea.	104	25		129	153
1730	3-1/2" diameter pipe		15	.533		115	26		141	165
1740	4" diameter pipe	B-20	40	.600		125	21		146	171
1750	6" diameter pipe		34	.706		156	25		181	211
1760	8" diameter pipe		21	1.143		209	40.50		249.50	293
1770	10" diameter pipe		20	1.200		240	42.50		282.50	330
1780	12" diameter pipe		17	1.412		261	50		311	365
1800	For 9" long, add					40%	45%			
1810	For 12" long, add					60%	80%			
1820	For 18" long, add					120%	110%			
2000	Clamp, stainless steel, two section									
2040	Full seal, for iron, steel, PVC pipe									
2100	6" long, 4" diameter pipe	B-20	24	1	Ea.	177	35.50		212.50	250
2110	6" diameter pipe		20	1.200		209	42.50		251.50	296
2120	8" diameter pipe		13	1.846		230	65.50		295.50	355
2130	10" diameter pipe		12	2		335	70.50		405.50	475
2140	12" diameter pipe		10	2.400		365	85		450	530
2200	9" long, 4" diameter pipe		16	1.500		230	53		283	335
2210	6" diameter pipe		13	1.846		261	65.50		326.50	390
2220	8" diameter pipe		9	2.667		271	94.50		365.50	445
2230	10" diameter pipe		8	3		375	106		481	580
2240	12" diameter pipe		7	3.429		415	121		536	650
2250	14" diameter pipe		6.40	3.750		625	133		758	895
2260	16" diameter pipe		6	4		655	141		796	945
2270	18" diameter pipe		5	4.800		680	170		850	1,000
2280	20" diameter pipe		4.60	5.217		730	184		914	1,100
2290	24" diameter pipe		4	6		1,250	212		1,462	1,700
2320	For 12" long, add to 9"					15%	25%			
2330	For 18" long, add to 9"					70%	55%			
8000	For internal cleaning and inspection, see Div. 33 01 30.16									
8100	For pipe testing, see Div. 23 05 93.50									

33 01 30 – Operation and Maintenance of Sewer Utilities

33 01 30.16 TV Inspection of Sewer Pipelines		Crew	Daily Output	Labor-Hours	Unit	Material	Labor	Equipment	Total	Total Incl O&P
0010	**TV INSPECTION OF SEWER PIPELINES**									
0100	Pipe internal cleaning & inspection, cleaning, pressure pipe systems									
0120	Pig method, lengths 1000' to 10,000'									
0140	4" diameter thru 24" diameter, minimum				L.F.				3.50	4.05
0160	Maximum				"				8.50	9.80
6000	Sewage/sanitary systems									
6100	Power rodder with header & cutters									
6110	Mobilization charge, minimum				Total				425	490
6120	Mobilization charge, maximum				"				1,000	1,150
6140	Cleaning 4-12" diameter				L.F.				3.50	4.05
6190	14-24" diameter								4	4.60
6240	30" diameter								6	6.90
6250	36" diameter								7	8.05
6260	48" diameter								8	9.20
6270	60" diameter								9	10.35
6280	72" diameter								10	11.50
9000	Inspection, television camera with video									
9060	up to 500 linear feet				Total				750	865

33 01 Operation and Maintenance of Utilities

33 01 30 – Operation and Maintenance of Sewer Utilities

33 01 30.71 Pipebursting

Line	33 01 30.71 Pipebursting	Crew	Daily Output	Labor-Hours	Unit	2009 Bare Costs Material	Labor	Equipment	Total	Total Incl O&P
0010	**PIPEBURSTING**									
0011	300′ runs, replace with HDPE pipe									
0020	Not including excavation, backfill, shoring, or dewatering									
0100	6″ to 15″ diameter, minimum				L.F.				100	110
0200	Maximum								200	220
0300	18″ to 36″ diameter, minimum								225	248
0400	Maximum								400	440
0500	Mobilize and demobilize, minimum				Job				3,000	3,300
0600	Maximum				″				30,000	33,000

33 01 30.74 HDPE Pipe Lining

Line	33 01 30.74 HDPE Pipe Lining	Crew	Daily Output	Labor-Hours	Unit	Material	Labor	Equipment	Total	Total Incl O&P
0010	**HDPE PIPE LINING**, excludes cleaning and video inspection									
0020	Pipe relined with one pipe size smaller than original (4″ for 6 ″)									
0100	6″ diameter, original size	B-6B	600	.080	L.F.	3.61	2.58	1.09	7.28	9.15
0150	8″ diameter, original size		600	.080		10.60	2.58	1.09	14.27	16.85
0200	10″ diameter, original size		600	.080		12.20	2.58	1.09	15.87	18.65
0250	12″ diameter, original size		400	.120		18.90	3.87	1.63	24.40	29
0300	14″ diameter, original size		400	.120		26.50	3.87	1.63	32	37.50
0350	16″ diameter, original size		300	.160		32.50	5.15	2.17	39.82	46
0400	18″ diameter, original size		300	.160		42	5.15	2.17	49.32	56.50
1000	Pipe HDPE lining, make service line taps	B-6	4	6	Ea.	116	204	73.50	393.50	525

33 05 Common Work Results for Utilities

33 05 23 – Trenchless Utility Installation

33 05 23.19 Microtunneling

Line	33 05 23.19 Microtunneling	Crew	Daily Output	Labor-Hours	Unit	Material	Labor	Equipment	Total	Total Incl O&P
0010	**MICROTUNNELING**									
0011	Not including excavation, backfill, shoring,									
0020	or dewatering, average 50′/day, slurry method									
0100	24″ to 48″ outside diameter, minimum				L.F.				800	880
0110	Adverse conditions, add				%				50%	50%
1000	Rent microtunneling machine, average monthly lease				Month				90,000	95,000
1010	Operating technician				Day				560	620
1100	Mobilization and demobilization, minimum				Job				40,000	44,000
1110	Maximum				″				400,000	440,000

33 05 23.20 Horizontal Boring

Line	33 05 23.20 Horizontal Boring	Crew	Daily Output	Labor-Hours	Unit	Material	Labor	Equipment	Total	Total Incl O&P
0010	**HORIZONTAL BORING**									
0011	Casing only, 100′ minimum,									
0020	not incl. jacking pits or dewatering									
0100	Roadwork, 1/2″ thick wall, 24″ diameter casing	B-42	20	3.200	L.F.	119	114	68.50	301.50	385
0200	36″ diameter		16	4		189	142	85.50	416.50	525
0300	48″ diameter		15	4.267		278	152	91	521	645
0500	Railroad work, 24″ diameter		15	4.267		119	152	91	362	470
0600	36″ diameter		14	4.571		189	162	97.50	448.50	570
0700	48″ diameter		12	5.333		278	189	114	581	730
0900	For ledge, add									20%
1000	Small diameter boring, 3″, sandy soil	B-82	900	.018		29	.63	.09	29.72	32.50
1040	Rocky soil	″	500	.032		29	1.13	.16	30.29	33.50
1100	Prepare jacking pits, incl. mobilization & demobilization, minimum				Ea.				3,000	3,450
1101	Maximum				″				20,000	23,000

33 05 Common Work Results for Utilities

33 05 26 – Utility Line Signs, Markers, and Flags

33 05 26.10 Utility Accessories		Crew	Daily Output	Labor-Hours	Unit	Material	Labor	Equipment	Total	Total Incl O&P
						2009 Bare Costs				
0010	**UTILITY ACCESSORIES** G1030-805									
0400	Underground tape, detectable, reinforced, alum. foil core, 2"	1 Clab	150	.053	C.L.F.	2.10	1.69		3.79	4.92
0500	6"	"	140	.057	"	5.25	1.81		7.06	8.60

33 11 Water Utility Distribution Piping

33 11 13 – Public Water Utility Distribution Piping

33 11 13.15 Water Supply, Ductile Iron Pipe

Line	Description	Crew	Daily Output	Labor-Hours	Unit	Material	Labor	Equipment	Total	Total Incl O&P
0010	**WATER SUPPLY, DUCTILE IRON PIPE** R331113-80									
0020	Not including excavation or backfill									
2000	Pipe, class 50 water piping, 18' lengths									
2020	Mechanical joint, 4" diameter	B-21A	200	.200	L.F.	15.35	7.80	3.01	26.16	32
2040	6" diameter		160	.250		17.80	9.80	3.76	31.36	38.50
2060	8" diameter		133.33	.300		19.70	11.75	4.52	35.97	44.50
2080	10" diameter		114.29	.350		26.50	13.70	5.25	45.45	56.50
2100	12" diameter		105.26	.380		33	14.85	5.70	53.55	65
2120	14" diameter		100	.400		42	15.65	6	63.65	76
2140	16" diameter		72.73	.550		45.50	21.50	8.30	75.30	91.50
2160	18" diameter		68.97	.580		57.50	22.50	8.75	88.75	107
2170	20" diameter		57.14	.700		67	27.50	10.55	105.05	127
2180	24" diameter		47.06	.850		86	33	12.80	131.80	159
3000	Tyton, push-on joint, 4" diameter		400	.100		11.50	3.91	1.51	16.92	20.50
3020	6" diameter		333.33	.120		14.70	4.69	1.81	21.20	25
3040	8" diameter		200	.200		20.50	7.80	3.01	31.31	37.50
3060	10" diameter		181.82	.220		29	8.60	3.31	40.91	48
3080	12" diameter		160	.250		30.50	9.80	3.76	44.06	52.50
3100	14" diameter		133.33	.300		33.50	11.75	4.52	49.77	59.50
3120	16" diameter		114.29	.350		54	13.70	5.25	72.95	86.50
3140	18" diameter		100	.400		60	15.65	6	81.65	96
3160	20" diameter		88.89	.450		66	17.60	6.75	90.35	106
3180	24" diameter		76.92	.520		78	20.50	7.85	106.35	126
8000	Fittings, mechanical joint									
8006	90° bend, 4" diameter	B-20A	16	2	Ea.	213	76.50		289.50	350
8020	6" diameter		12.80	2.500		292	95.50		387.50	465
8040	8" diameter		10.67	2.999		425	115		540	640
8060	10" diameter	B-21A	11.43	3.500		465	137	52.50	654.50	775
8080	12" diameter		10.53	3.799		840	149	57	1,046	1,225
8100	14" diameter		10	4		1,275	156	60	1,491	1,700
8120	16" diameter		7.27	5.502		1,900	215	83	2,198	2,525
8140	18" diameter		6.90	5.797		3,150	227	87.50	3,464.50	3,925
8160	20" diameter		5.71	7.005		4,250	274	105	4,629	5,200
8180	24" diameter		4.70	8.511		4,675	335	128	5,138	5,800
8200	Wye or tee, 4" diameter	B-20A	10.67	2.999		325	115		440	530
8220	6" diameter		8.53	3.751		435	143		578	695
8240	8" diameter		7.11	4.501		620	172		792	940
8260	10" diameter	B-21A	7.62	5.249		1,275	205	79	1,559	1,800
8280	12" diameter		7.02	5.698		1,625	223	86	1,934	2,200
8300	14" diameter		6.67	5.997		1,800	234	90.50	2,124.50	2,450
8320	16" diameter		4.85	8.247		1,875	320	124	2,319	2,700
8340	18" diameter		4.60	8.696		4,900	340	131	5,371	6,050
8360	20" diameter		3.81	10.499		7,675	410	158	8,243	9,250

33 11 13 – Public Water Utility Distribution Piping

33 11 13.15 Water Supply, Ductile Iron Pipe		Crew	Daily Output	Labor-Hours	Unit	2009 Bare Costs Material	Labor	Equipment	Total	Total Incl O&P
8380	24" diameter	B-21A	3.14	12.739	Ea.	9,100	500	192	9,792	11,000
8398	45° bends, 4" diameter	B-20A	16	2		167	76.50		243.50	300
8400	6" diameter	"	12.80	2.500		231	95.50		326.50	400
8410	12" diameter	B-21A	10.53	3.799		505	149	57	711	845
8420	16" diameter		7.27	5.502		1,200	215	83	1,498	1,750
8430	20" diameter		5.71	7.005		2,650	274	105	3,029	3,425
8440	24" diameter		4.70	8.511		3,650	335	128	4,113	4,650
8450	Decreaser, 6" x 4" diameter	B-20A	14.22	2.250		179	86		265	330
8460	8" x 6" diameter	"	11.64	2.749		229	105		334	410
8470	10" x 6 " diameter	B-21A	13.33	3.001		290	117	45	452	550
8480	12" x 6" diameter		12.70	3.150		355	123	47.50	525.50	630
8490	16" x 6" diameter		10	4		820	156	60	1,036	1,200
8500	20" x 6" diameter		8.42	4.751		2,025	186	71.50	2,282.50	2,575
8552	For water utility valves see 33 12 16									
8600	12" diameter	B-21	3	9.333	Ea.	2,700	340	43	3,083	3,550
8610	14" diameter		2	14		3,325	510	64.50	3,899.50	4,500
8620	16" diameter		2	14		5,125	510	64.50	5,699.50	6,475
8700	Joint restraint, ductile iron mechanical joints									
8710	4" diameter	B-20A	32	1	Ea.	33	38		71	94
8720	6" diameter		25.60	1.250		39.50	48		87.50	117
8730	8" diameter		21.33	1.500		57	57.50		114.50	150
8740	10" diameter		18.28	1.751		83	67		150	194
8750	12" diameter		16.84	1.900		116	72.50		188.50	239
8760	14" diameter		16	2		159	76.50		235.50	291
8770	16" diameter		11.64	2.749		210	105		315	390
8780	18" diameter		11.03	2.901		292	111		403	490
8785	20" diameter		9.14	3.501		360	134		494	600
8790	24" diameter		7.53	4.250		435	163		598	725
9600	Steel sleeve and tap, 4" diameter	B-20	3	8		610	283		893	1,100
9620	6" diameter		2	12		650	425		1,075	1,375
9630	8" diameter		2	12		890	425		1,315	1,625

33 11 13.25 Water Supply, Polyvinyl Chloride Pipe		Crew	Daily Output	Labor-Hours	Unit	Material	Labor	Equipment	Total	Total Incl O&P
0010	**WATER SUPPLY, POLYVINYL CHLORIDE PIPE** R331113-80									
0020	Not including excavation or backfill, unless specified									
2100	AWWA Class 150, S.D.R. 18, 1-1/2" diameter	Q-1A	750	.013	L.F.	.34	.66		1	1.35
2120	2" diameter		686	.015		.51	.72		1.23	1.64
2140	2-1/2" diameter		500	.020		.87	.98		1.85	2.44
2160	3" diameter	B-20	430	.056		1.09	1.97		3.06	4.26
2180	4" diameter		375	.064		1.65	2.26		3.91	5.35
2200	6" diameter		316	.076		3.30	2.68		5.98	7.80
2210	8" diameter		260	.092		5.70	3.26		8.96	11.30
3010	AWWA C905, PR 100, DR 25									
3030	14" diameter	B-20A	213	.150	L.F.	19	5.75		24.75	30
3040	16" diameter		200	.160		31.50	6.10		37.60	44
3050	18" diameter		160	.200		38.50	7.65		46.15	54
3060	20" diameter		133	.241		53	9.20		62.20	72.50
3070	24" diameter		107	.299		68	11.45		79.45	92.50
3080	30" diameter		80	.400		110	15.30		125.30	145
3090	36" diameter		80	.400		170	15.30		185.30	211
3100	42" diameter		60	.533		230	20.50		250.50	284
3200	48" diameter		60	.533		300	20.50		320.50	360
4520	Pressure pipe Class 150, SDR 18, AWWA C900, 4" diameter		380	.084		2.32	3.22		5.54	7.45

33 11 Water Utility Distribution Piping

33 11 13 – Public Water Utility Distribution Piping

33 11 13.25 Water Supply, Polyvinyl Chloride Pipe		Crew	Daily Output	Labor-Hours	Unit	2009 Bare Costs Material	Labor	Equipment	Total	Total Incl O&P
4530	6″ diameter	B-20A	316	.101	L.F.	4.59	3.87		8.46	10.95
4540	8″ diameter		264	.121		8.05	4.64		12.69	15.90
4550	10″ diameter		220	.145		12.20	5.55		17.75	22
4560	12″ diameter		186	.172		17.25	6.60		23.85	29
8000	Fittings with rubber gasket									
8003	Class 150, D.R. 18									
8006	90° Bend , 4″ diameter	B-20	100	.240	Ea.	51	8.50		59.50	69
8020	6″ diameter		90	.267		92	9.45		101.45	116
8040	8″ diameter		80	.300		174	10.60		184.60	207
8060	10″ diameter		50	.480		265	16.95		281.95	320
8080	12″ diameter		30	.800		405	28.50		433.50	490
8100	Tee, 4″ diameter		90	.267		54	9.45		63.45	74
8120	6″ diameter		80	.300		145	10.60		155.60	176
8140	8″ diameter		70	.343		203	12.10		215.10	242
8160	10″ diameter		40	.600		430	21		451	505
8180	12″ diameter		20	1.200		600	42.50		642.50	725
8200	45° Bend, 4″ diameter		100	.240		50.50	8.50		59	68.50
8220	6″ diameter		90	.267		91	9.45		100.45	115
8240	8″ diameter		50	.480		171	16.95		187.95	215
8260	10″ diameter		50	.480		270	16.95		286.95	325
8280	12″ diameter		30	.800		360	28.50		388.50	440
8300	Reducing tee 6″x4″		100	.240		119	8.50		127.50	144
8320	8″ x 6″		90	.267		287	9.45		296.45	330
8330	10″ x 6″		90	.267		330	9.45		339.45	380
8340	10″x8″		90	.267		355	9.45		364.45	405
8350	12″ x 6″		90	.267		445	9.45		454.45	505
8360	12″ x 8″		90	.267		495	9.45		504.45	560
8400	Tapped service tee (threaded type) 6″ x 6″ x 3/4″		100	.240		77.50	8.50		86	98
8430	6″ x 6″ x 1″		90	.267		77.50	9.45		86.95	99.50
8440	6″ x 6″ x 1 1/2″		90	.267		77.50	9.45		86.95	99.50
8450	6″ x 6″ x 2″		90	.267		77.50	9.45		86.95	99.50
8460	8″ x 8″ x 3/4″		90	.267		122	9.45		131.45	149
8470	8″ x 8″ x 1″		90	.267		122	9.45		131.45	149
8480	8″ x 8″ x 1 1/2″		90	.267		122	9.45		131.45	149
8490	8″ x 8″ x 2″		90	.267		122	9.45		131.45	149
8500	Repair coupling 4″		100	.240		111	8.50		119.50	135
8520	6″ diameter		90	.267		135	9.45		144.45	164
8540	8″ diameter		50	.480		153	16.95		169.95	195
8560	10″ diameter		50	.480		178	16.95		194.95	223
8580	12″ diameter		50	.480		218	16.95		234.95	267
8600	Plug end 4″		100	.240		38	8.50		46.50	54.50
8620	6″ diameter		90	.267		73	9.45		82.45	94.50
8640	8″ diameter		50	.480		78.50	16.95		95.45	113
8660	10″ diameter		50	.480		125	16.95		141.95	165
8680	12″ diameter		50	.480		155	16.95		171.95	197
8700	PVC pipe, joint restraint									
8710	4″ diameter	B-20A	32	1	Ea.	34.50	38		72.50	96
8720	6″ diameter		25.60	1.250		42	48		90	119
8730	8″ diameter		21.33	1.500		62	57.50		119.50	156
8740	10″ diameter		18.28	1.751		117	67		184	231
8750	12″ diameter		16.84	1.900		123	72.50		195.50	246
8760	14″ diameter		16	2		196	76.50		272.50	330
8770	16″ diameter		11.64	2.749		264	105		369	450

33 11 Water Utility Distribution Piping

33 11 13 – Public Water Utility Distribution Piping

33 11 13.25 Water Supply, Polyvinyl Chloride Pipe		Crew	Daily Output	Labor-Hours	Unit	2009 Bare Costs Material	Labor	Equipment	Total	Total Incl O&P
8780	18″ diameter	B-20A	11.03	2.901	Ea.	325	111		436	530
8785	20″ diameter		9.14	3.501		405	134		539	650
8790	24″ diameter		7.53	4.250		470	163		633	760
33 11 13.35 Water Supply, HDPE										
0010	**WATER SUPPLY, HDPE**									
0011	Butt fusion joints, SDR 21,40′ lengths not including excavation or backfill									
0100	4″ diameter	B-22A	400	.095	L.F.	3.61	3.39	1.08	8.08	10.40
0200	6″ diameter		380	.100		10.60	3.57	1.14	15.31	18.40
0300	8″ diameter		320	.119		12.20	4.24	1.35	17.79	21.50
0400	10″ diameter		300	.127		18.90	4.52	1.45	24.87	29.50
0500	12″ diameter		260	.146		26.50	5.20	1.67	33.37	39.50
0600	14″ diameter		220	.173		32.50	6.15	1.97	40.62	47
0700	16″ diameter		180	.211		42	7.55	2.41	51.96	60.50
0800	18″ diameter		140	.271		53	9.70	3.10	65.80	77
0900	24″ diameter		100	.380		94.50	13.55	4.34	112.39	130
1000	Fittings									
1100	Elbows, 90 degrees									
1200	4″ diameter	B-22B	32	.500	Ea.	33.50	18.10	4.59	56.19	69.50
1300	6″ diameter		28	.571		103	20.50	5.25	128.75	151
1400	8″ diameter		24	.667		217	24	6.10	247.10	283
1500	10″ diameter		18	.889		350	32	8.15	390.15	445
1600	12″ diameter		12	1.333		525	48.50	12.25	585.75	670
1700	14″ diameter		9	1.778		630	64.50	16.30	710.80	810
1800	16″ diameter		6	2.667		760	96.50	24.50	881	1,025
1900	18″ diameter		4	4		1,050	145	36.50	1,231.50	1,425
2000	24″ diameter		3	5.333		2,100	193	49	2,342	2,675
2100	Tees									
2200	4″ diameter	B-22B	30	.533	Ea.	41	19.30	4.90	65.20	81
2300	6″ diameter		26	.615		122	22.50	5.65	150.15	175
2400	8″ diameter		22	.727		305	26.50	6.70	338.20	385
2500	10″ diameter		15	1.067		400	38.50	9.80	448.30	510
2600	12″ diameter		10	1.600		560	58	14.70	632.70	720
2700	14″ diameter		8	2		660	72.50	18.35	750.85	855
2800	16″ diameter		6	2.667		780	96.50	24.50	901	1,025
2900	18″ diameter		4	4		1,025	145	36.50	1,206.50	1,425
3000	24″ diameter		2	8		2,150	290	73.50	2,513.50	2,875

33 12 Water Utility Distribution Equipment

33 12 13 – Water Service Connections

33 12 13.15 Tapping, Crosses and Sleeves		Crew	Daily Output	Labor-Hours	Unit	Material	Labor	Equipment	Total	Total Incl O&P
0010	**TAPPING, CROSSES AND SLEEVES**									
4000	Drill and tap pressurized main (labor only)									
4100	6″ main, 1″ to 2″ service	Q-1	3	5.333	Ea.		234		234	350
4150	8″ main, 1″ to 2″ service	″	2.75	5.818	″		255		255	385
4500	Tap and insert gate valve									
4600	8″ main, 4″ branch	B-21	3.20	8.750	Ea.		320	40.50	360.50	535
4650	6″ branch		2.70	10.370			375	48	423	635
4700	10″ Main, 4″ branch		2.70	10.370			375	48	423	635
4750	6″ branch		2.35	11.915			435	55	490	730
4800	12″ main, 6″ branch		2.35	11.915			435	55	490	730
4850	8″ branch		2.35	11.915			435	55	490	730

33 12 Water Utility Distribution Equipment

33 12 13 – Water Service Connections

33 12 13.15 Tapping, Crosses and Sleeves		Crew	Daily Output	Labor-Hours	Unit	2009 Bare Costs Material	Labor	Equipment	Total	Total Incl O&P
7020	Crosses, 4" x 4"	B-21	37	.757	Ea.	585	27.50	3.49	615.99	685
7030	6" x 4"		25	1.120		695	41	5.15	741.15	830
7040	6" x 6"		25	1.120		695	41	5.15	741.15	830
7060	8" x 6"		21	1.333		850	48.50	6.15	904.65	1,025
7080	8" x 8"		21	1.333		920	48.50	6.15	974.65	1,100
7100	10" x 6"		21	1.333		1,675	48.50	6.15	1,729.65	1,925
7120	10" x 10"		21	1.333		1,825	48.50	6.15	1,879.65	2,075
7140	12" x 6"		18	1.556		1,675	56.50	7.15	1,738.65	1,950
7160	12" x 12"		18	1.556		2,125	56.50	7.15	2,188.65	2,450
7180	14" x 6"		16	1.750		4,225	63.50	8.05	4,296.55	4,750
7200	14" x 14"		16	1.750		4,475	63.50	8.05	4,546.55	5,000
7220	16" x 6"		14	2		4,525	73	9.20	4,607.20	5,100
7240	16" x 10"		14	2		4,575	73	9.20	4,657.20	5,175
7260	16" x 16"		14	2		4,825	73	9.20	4,907.20	5,425
7280	18" x 6"		10	2.800		6,775	102	12.90	6,889.90	7,625
7300	18" x 12"		10	2.800		6,800	102	12.90	6,914.90	7,675
7320	18" x 18"		10	2.800		6,225	102	12.90	6,339.90	7,025
7340	20" x 6"		8	3.500		5,475	127	16.15	5,618.15	6,250
7360	20" x 12"		8	3.500		5,850	127	16.15	5,993.15	6,675
7380	20" x 20"		8	3.500		8,575	127	16.15	8,718.15	9,650
7400	24" x 6"		6	4.667		7,075	170	21.50	7,266.50	8,050
7420	24" x 12"		6	4.667		7,175	170	21.50	7,366.50	8,150
7440	24" x 18"		6	4.667		10,700	170	21.50	10,891.50	12,100
7460	24" x 24"		6	4.667		11,100	170	21.50	11,291.50	12,500
7600	Cut-in sleeves with rubber gaskets, 4"		18	1.556		210	56.50	7.15	273.65	325
7620	6"		12	2.333		268	85	10.75	363.75	440
7640	8"		10	2.800		365	102	12.90	479.90	570
7660	10"		10	2.800		505	102	12.90	619.90	725
7680	12"		9	3.111		600	113	14.35	727.35	850
7800	Cut-in valves with rubber gaskets, 4"		18	1.556		430	56.50	7.15	493.65	565
7820	6"		12	2.333		580	85	10.75	675.75	785
7840	8"		10	2.800		805	102	12.90	919.90	1,050
7860	10"		10	2.800		960	102	12.90	1,074.90	1,225
7880	12"		9	3.111		1,200	113	14.35	1,327.35	1,500
7900	Tapping Valve 4 inch, MJ, ductile iron		18	1.556		565	56.50	7.15	628.65	715
7920	Tapping Valve 6 inch, MJ, ductile iron		12	2.333		745	85	10.75	840.75	965
8000	Sleeves with rubber gaskets, 4" x 4"		37	.757		655	27.50	3.49	685.99	770
8010	6" x 4"		25	1.120		710	41	5.15	756.15	850
8020	6" x 6"		25	1.120		770	41	5.15	816.15	915
8030	8" x 4"		21	1.333		760	48.50	6.15	814.65	915
8040	8" x 6"		21	1.333		750	48.50	6.15	804.65	905
8060	8" x 8"		21	1.333		985	48.50	6.15	1,039.65	1,150
8070	10" x 4"		21	1.333		780	48.50	6.15	834.65	940
8080	10" x 6"		21	1.333		850	48.50	6.15	904.65	1,025
8090	10" x 8"		21	1.333		1,000	48.50	6.15	1,054.65	1,175
8100	10" x 10"		21	1.333		1,500	48.50	6.15	1,554.65	1,725
8110	12" x 4"		18	1.556		850	56.50	7.15	913.65	1,025
8120	12" x 6"		18	1.556		880	56.50	7.15	943.65	1,075
8130	12" x 8"		18	1.556		1,200	56.50	7.15	1,263.65	1,400
8135	12" x 10"		18	1.556		1,000	56.50	7.15	1,063.65	1,200
8140	12" x 12"		18	1.556		1,800	56.50	7.15	1,863.65	2,100
8160	14" x 6"		16	1.750		1,150	63.50	8.05	1,221.55	1,350
8180	14" x 14"	B-21	16	1.750	Ea.	2,225	63.50	8.05	2,296.55	2,550

33 12 13 – Water Service Connections

33 12 13.15 Tapping, Crosses and Sleeves		Crew	Daily Output	Labor-Hours	Unit	2009 Bare Costs Material	Labor	Equipment	Total	Total Incl O&P
8200	16″ x 6″	B-21	14	2	Ea.	1,675	73	9.20	1,757.20	1,975
8220	16″ x 10″		14	2		2,725	73	9.20	2,807.20	3,100
8240	16″ x 16″		14	2		2,425	73	9.20	2,507.20	2,800
8260	18″ x 6″		10	2.800		975	102	12.90	1,089.90	1,250
8280	18″ x 12″		10	2.800		1,725	102	12.90	1,839.90	2,075
8300	18″ x 18″		10	2.800		2,675	102	12.90	2,789.90	3,125
8320	20″ x 6″		8	3.500		2,000	127	16.15	2,143.15	2,425
8340	20″ x 12″		8	3.500		3,425	127	16.15	3,568.15	3,975
8360	20″ x 20″		8	3.500		2,900	127	16.15	3,043.15	3,425
8380	24″ x 6″		6	4.667		2,100	170	21.50	2,291.50	2,575
8400	24″ x 12″		6	4.667		3,550	170	21.50	3,741.50	4,175
8420	24″ x 18″		6	4.667		3,050	170	21.50	3,241.50	3,650
8440	24″ x 24″	↓	6	4.667		4,675	170	21.50	4,866.50	5,400
8800	Curb box, 6′ long	B-20	20	1.200		282	42.50		324.50	375
8820	8′ long		18	1.333		355	47		402	465
8830	Valve box w/ lid		14	1.714		148	60.50		208.50	257
8840	Valve box and large base w/ lid	↓	14	1.714	↓	249	60.50		309.50	370

33 12 16 – Water Utility Distribution Valves

33 12 16.10 Valves		Crew	Daily Output	Labor-Hours	Unit	Material	Labor	Equipment	Total	Total Incl O&P
0010	**VALVES**, water distribution									
0011	See Div. 22 05 23.20 and 22 05 23.60									
3000	Butterfly valves with boxes, cast iron, mech. jt.									
3100	4″ diameter	B-6	6	4	Ea.	1,000	136	49	1,185	1,375
3140	6″ diameter		6	4		1,450	136	49	1,635	1,850
3180	8″ diameter		6	4		1,650	136	49	1,835	2,100
3300	10″ diameter		6	4		2,100	136	49	2,285	2,575
3340	12″ diameter		6	4		2,850	136	49	3,035	3,425
3400	14″ diameter		4	6		3,575	204	73.50	3,852.50	4,325
3440	16″ diameter		4	6		5,375	204	73.50	5,652.50	6,300
3460	18″ diameter		4	6		5,975	204	73.50	6,252.50	6,975
3480	20″ diameter		4	6		6,825	204	73.50	7,102.50	7,925
3500	24″ diameter	↓	4	6	↓	7,425	204	73.50	7,702.50	3,575
3600	With lever operator									
3610	4″ diameter	B-6	6	4	Ea.	860	136	49	1,045	1,200
3614	6″ diameter		6	4		1,300	136	49	1,485	1,700
3616	8″ diameter		6	4		1,525	136	49	1,710	1,950
3618	10″ diameter		6	4		1,950	136	49	2,135	2,425
3620	12″ diameter		6	4		2,700	136	49	2,885	3,250
3622	14″ diameter		4	6		3,325	204	73.50	3,602.50	4,050
3624	16″ diameter		4	6		5,125	204	73.50	5,402.50	6,025
3626	18″ diameter		4	6		5,725	204	73.50	6,002.50	6,700
3628	20″ diameter		4	6		6,575	204	73.50	6,852.50	7,650
3630	24″ diameter	↓	4	6	↓	7,175	204	73.50	7,452.50	8,300
3700	Check valves, flanged									
3710	4″ diameter	B-6	6	4	Ea.	505	136	49	690	820
3714	6″ diameter		6	4		905	136	49	1,090	1,250
3716	8″ diameter		6	4		1,700	136	49	1,885	2,150
3718	10″ diameter		6	4		2,675	136	49	2,860	3,225
3720	12″ diameter		6	4		3,900	136	49	4,085	4,550
3722	14″ diameter		4	6		6,550	204	73.50	6,827.50	7,600
3724	16″ diameter		4	6		6,050	204	73.50	6,327.50	7,050
3726	18″ diameter	↓	4	6	↓	11,400	204	73.50	11,677.50	13,000

33 12 16 – Water Utility Distribution Valves

33 12 16.10 Valves		Crew	Daily Output	Labor-Hours	Unit	Material	Labor	Equipment	Total	Total Incl O&P
						2009 Bare Costs				
3728	20″ diameter	B-6	4	6	Ea.	14,400	204	73.50	14,677.50	16,200
3730	24″ diameter	↓	4	6	↓	19,300	204	73.50	19,577.50	21,700
3800	Gate valves, C.I., 250 PSI, mechanical joint, w/boxes									
3810	4″ diameter	B-6	6	4	Ea.	600	136	49	785	925
3814	6″ diameter		6	4		685	136	49	870	1,025
3816	8″ diameter		6	4		995	136	49	1,180	1,375
3818	10″ diameter		6	4		1,600	136	49	1,785	2,025
3820	12″ diameter		6	4		2,175	136	49	2,360	2,650
3822	14″ diameter		4	6		5,200	204	73.50	5,477.50	6,100
3824	16″ diameter		4	6		8,875	204	73.50	9,152.50	10,100
3826	18″ diameter		4	6		11,700	204	73.50	11,977.50	13,300
3828	20″ diameter		4	6		14,600	204	73.50	14,877.50	16,500
3830	24″ diameter		4	6		19,600	204	73.50	19,877.50	21,900
3831	30″ diameter		4	6		43,500	204	73.50	43,777.50	48,300
3832	36″ diameter	↓	4	6	↓	73,000	204	73.50	73,277.50	81,000

33 12 16.20 Valves		Crew	Daily Output	Labor-Hours	Unit	Material	Labor	Equipment	Total	Total Incl O&P
0010	**VALVES**									
0011	Special trim or use									
9000	Valves, gate valve, N.R.S. post type, 4″ diameter	B-6	6	4	Ea.	600	136	49	785	925
9020	6″ diameter		6	4		685	136	49	870	1,025
9040	8″ diameter		6	4		995	136	49	1,180	1,375
9060	10″ diameter		6	4		1,600	136	49	1,785	2,025
9080	12″ diameter		6	4		2,175	136	49	2,360	2,650
9100	14″ diameter		6	4		5,100	136	49	5,285	5,875
9120	O.S.&Y., 4″ diameter		6	4		525	136	49	710	840
9140	6″ diameter		6	4		860	136	49	1,045	1,225
9160	8″ diameter		6	4		1,375	136	49	1,560	1,775
9180	10″ diameter		6	4		2,100	136	49	2,285	2,600
9200	12″ diameter		6	4		2,800	136	49	2,985	3,350
9220	14″ diameter		4	6		6,625	204	73.50	6,902.50	7,700
9400	Check valves, rubber disc, 2-1/2″ diameter		6	4		340	136	49	525	640
9420	3″ diameter		6	4		345	136	49	530	645
9440	4″ diameter		6	4		505	136	49	690	820
9480	6″ diameter		6	4		905	136	49	1,090	1,250
9500	8″ diameter		6	4		1,700	136	49	1,885	2,150
9520	10″ diameter		6	4		2,675	136	49	2,860	3,225
9540	12″ diameter		6	4		3,900	136	49	4,085	4,550
9542	14″ diameter		4	6		6,550	204	73.50	6,827.50	7,600
9700	Detector check valves, reducing, 4″ diameter		6	4		1,275	136	49	1,460	1,675
9720	6″ diameter		6	4		1,650	136	49	1,835	2,100
9740	8″ diameter		6	4		2,525	136	49	2,710	3,050
9760	10″ diameter		6	4		4,400	136	49	4,585	5,125
9800	Galvanized, 4″ diameter		6	4		1,575	136	49	1,760	2,025
9820	6″ diameter		6	4		2,300	136	49	2,485	2,800
9840	8″ diameter		6	4		3,675	136	49	3,860	4,325
9860	10″ diameter	↓	6	4	↓	7,300	136	49	7,485	8,300

33 16 13 – Aboveground Water Utility Storage Tanks

33 16 13.13 Steel Water Storage Tanks		Crew	Daily Output	Labor-Hours	Unit	Material (2009 Bare Costs)	Labor (2009 Bare Costs)	Equipment (2009 Bare Costs)	Total (2009 Bare Costs)	Total Incl O&P
0010	**STEEL WATER STORAGE TANKS**									
0910	Steel, ground level, ht/diam. less than 1, not incl. fdn., 100,000 gallons				Ea.				173,000	190,000
1000	250,000 gallons								184,000	202,000
1200	500,000 gallons								280,000	308,000
1250	750,000 gallons								334,000	367,000
1300	1,000,000 gallons								509,000	560,000
1500	2,000,000 gallons								822,000	904,000
1600	4,000,000 gallons								1,277,000	1,405,000
1800	6,000,000 gallons								1,772,000	1,949,000
1850	8,000,000 gallons								2,320,000	2,550,000
1910	10,000,000 gallons				↓				3,460,000	3,805,000
2100	Steel standpipes, hgt/diam. more than 1, 100' to overflow, no fdn									
2200	500,000 gallons				Ea.				367,000	404,000
2400	750,000 gallons								448,000	493,000
2500	1,000,000 gallons								545,000	600,000
2700	1,500,000 gallons								762,000	838,000
2800	2,000,000 gallons				↓				936,000	1,030,000

33 16 13.16 Prestressed Conc. Aboveground Water Utility Storage Tanks

		Crew	Daily Output	Labor-Hours	Unit	Material	Labor	Equipment	Total	Total Incl O&P
0010	**PRESTRESSED CONC. ABOVEGROUND WATER UTILITY STORAGE TANKS**									
0020	Not including pipe or pumps, 250,000 gallons				Ea.				270,000	297,000
0100	500,000 gallons								400,000	440,000
0300	1,000,000 gallons								580,000	638,000
0400	2,000,000 gallons								880,000	968,000
0600	4,000,000 gallons								1,400,000	1,540,000
0700	6,000,000 gallons								1,860,000	2,046,000
0750	8,000,000 gallons								2,400,000	2,640,000
0800	10,000,000 gallons				↓				2,900,000	3,190,000

33 16 13.23 Plastic-Coated Fabric Pillow Water Tanks

		Crew	Daily Output	Labor-Hours	Unit	Material	Labor	Equipment	Total	Total Incl O&P
0010	**PLASTIC-COATED FABRIC PILLOW WATER TANKS**									
7000	Water tanks, vinyl coated fabric pillow tanks, freestanding, 5,000 gallons	4 Clab	4	8	Ea.	9,500	253		9,753	10,900
7100	Supporting embankment not included, 25,000 gallons	6 Clab	2	24		14,900	760		15,660	17,600
7200	50,000 gallons	8 Clab	1.50	42.667		27,600	1,350		28,950	32,500
7300	100,000 gallons	9 Clab	.90	80		60,000	2,525		62,525	70,000
7400	150,000 gallons		.50	144		82,000	4,550		86,550	97,000
7500	200,000 gallons		.40	180		100,500	5,700		106,200	119,500
7600	250,000 gallons	↓	.30	240	↓	143,000	7,575		150,575	169,500

33 16 19 – Elevated Water Utility Storage Tanks

33 16 19.50 Elevated Water Storage Tanks

		Crew	Daily Output	Labor-Hours	Unit	Material	Labor	Equipment	Total	Total Incl O&P
0010	**ELEVATED WATER STORAGE TANKS**									
0011	Not incl pipe, pumps or foundation									
3000	Elevated water tanks, 100' to bottom capacity line, incl painting									
3010	50,000 gallons				Ea.				222,000	242,000
3300	100,000 gallons								306,000	337,000
3400	250,000 gallons								422,000	464,000
3600	500,000 gallons								681,000	749,000
3700	750,000 gallons								940,000	1,034,000
3900	1,000,000 gallons				↓				944,000	1,038,000

33 21 13 – Public Water Supply Wells

33 21 13.10 Wells and Accessories		Crew	Daily Output	Labor-Hours	Unit	2009 Bare Costs Material	Labor	Equipment	Total	Total Incl O&P
0010	**WELLS & ACCESSORIES**									
0011	Domestic									
0100	Drilled, 4″ to 6″ diameter	B-23	120	.333	L.F.		10.65	23	33.65	42
0200	8″ diameter	″	95.20	.420	″		13.45	29	42.45	53
0400	Gravel pack well, 40′ deep, incl. gravel & casing, complete									
0500	24″ diameter casing x 18″ diameter screen	B-23	.13	307	Total	30,600	9,850	21,400	61,850	72,500
0600	36″ diameter casing x 18″ diameter screen		.12	333	″	32,900	10,700	23,100	66,700	78,000
0800	Observation wells, 1-1/4″ riser pipe		163	.245	V.L.F.	16.85	7.85	17.05	41.75	49.50
0900	For flush Buffalo roadway box, add	1 Skwk	16.60	.482	Ea.	46	19.70		65.70	81
1200	Test well, 2-1/2″ diameter, up to 50′ deep (15 to 50 GPM)	B-23	1.51	26.490	″	690	850	1,850	3,390	4,100
1300	Over 50′ deep, add	″	121.80	.328	L.F.	18.35	10.50	23	51.85	61.50
1500	Pumps, installed in wells to 100′ deep, 4″ submersible									
1510	1/2 H.P.	Q-1	3.22	4.969	Ea.	460	218		678	830
1520	3/4 H.P.		2.66	6.015		565	264		829	1,025
1600	1 H.P.		2.29	6.987		615	305		920	1,150
1700	1-1/2 H.P.	Q-22	1.60	10		800	440	480	1,720	2,075
1800	2 H.P.		1.33	12.030		1,200	530	580	2,310	2,725
1900	3 H.P.		1.14	14.035		1,975	615	675	3,265	3,850
2000	5 H.P.		1.14	14.035		2,775	615	675	4,065	4,725
2050	Remove and install motor only, 4 H.P.		1.14	14.035		2,375	615	675	3,665	4,275
3000	Pump, 6″ submersible, 25′ to 150′ deep, 25 H.P., 249 to 297 GPM		.89	17.978		5,425	790	865	7,080	8,100
3100	25′ to 500′ deep, 30 H.P., 100 to 300 GPM		.73	21.918		6,300	960	1,050	8,310	9,525
8110	Well screen assembly, stainless steel, 2″ diameter	B-23A	273	.088	L.F.	78	3.12	9.70	90.82	101
8120	3″ diameter		253	.095		108	3.37	10.45	121.82	136
8130	4″ diameter		200	.120		124	4.26	13.25	141.51	157
8140	5″ diameter		168	.143		144	5.05	15.75	164.80	184
8150	6″ diameter		126	.190		173	6.75	21	200.75	223
8160	8″ diameter		98.50	.244		225	8.65	27	260.65	290
8170	10″ diameter		73	.329		283	11.70	36.50	331.20	370
8180	12″ diameter		62.50	.384		335	13.65	42.50	391.15	435
8190	14″ diameter		54.30	.442		380	15.70	49	444.70	495
8200	16″ diameter		48.30	.497		415	17.65	55	487.65	550
8210	18″ diameter		39.20	.612		520	22	67.50	609.50	680
8220	20″ diameter		31.20	.769		590	27.50	85	702.50	785
8230	24″ diameter		23.80	1.008		730	36	111	877	980
8240	26″ diameter		21	1.143		820	40.50	126	986.50	1,100
8300	Slotted PVC, 1-1/4″ diameter		521	.046		1.93	1.64	5.10	8.67	10.25
8310	1-1/2″ diameter		488	.049		2.79	1.75	5.45	9.99	11.70
8320	2″ diameter		273	.088		3.87	3.12	9.70	16.69	19.75
8330	3″ diameter		253	.095		4.11	3.37	10.45	17.93	21
8340	4″ diameter		200	.120		4.74	4.26	13.25	22.25	26.50
8350	5″ diameter		168	.143		4.98	5.05	15.75	25.78	30.50
8360	6″ diameter		126	.190		6.90	6.75	21	34.65	41
8370	8″ diameter		98.50	.244		10.45	8.65	27	46.10	54.50
8400	Artificial gravel pack, 2″ screen, 6″ casing	B-23B	174	.138		3.59	4.90	17	25.49	30
8405	8″ casing		111	.216		4.91	7.70	26.50	39.11	46.50
8410	10″ casing		74.50	.322		6.15	11.45	39.50	57.10	68
8415	12″ casing		60	.400		8.35	14.20	49.50	72.05	85.50
8420	14″ casing		50.20	.478		10.80	17	59	86.80	103
8425	16″ casing		40.70	.590		14.85	21	72.50	108.35	128
8430	18″ casing		36	.667		17.25	23.50	82	122.75	146
8435	20″ casing		29.50	.814		19.90	29	100	148.90	177
8440	24″ casing		25.70	.934		22	33	115	170	203

33 21 Water Supply Wells

33 21 13 – Public Water Supply Wells

33 21 13.10 Wells and Accessories		Crew	Daily Output	Labor-Hours	Unit	Material	Labor	Equipment	Total	Total Incl O&P
						2009 Bare Costs				
8445	26″ casing	B-23B	24.60	.976	L.F.	24.50	34.50	120	179	212
8450	30″ casing		20	1.200		28.50	42.50	148	219	260
8455	36″ casing		16.40	1.463		30.50	52	180	262.50	310
8500	Develop well		8	3	Hr.	269	107	370	746	865
8550	Pump test well		8	3		71.50	107	370	548.50	645
8560	Standby well	B-23A	8	3		69.50	107	330	506.50	605
8570	Standby, drill rig		8	3			107	330	437	530
8580	Surface seal well, concrete filled		1	24	Ea.	740	850	2,650	4,240	5,050
8590	Well test pump, install & remove	B-23	1	40			1,275	2,775	4,050	5,025
8600	Well sterilization, chlorine	2 Clab	1	16		500	505		1,005	1,325
8610	Well water pressure switch	1 Clab	12	.667		66	21		87	106
9950	See Div. 31 23 19.40 for wellpoints									
9960	See Div. 31 23 19.30 for drainage wells									

33 21 13.20 Water Supply Wells, Pumps

		Crew	Daily Output	Labor-Hours	Unit	Material	Labor	Equipment	Total	Total Incl O&P
0010	**WATER SUPPLY WELLS, PUMPS**									
0011	With pressure control									
1000	Deep well, jet, 42 gal. galvanized tank									
1040	3/4 HP	1 Plum	.80	10	Ea.	685	490		1,175	1,475
3000	Shallow well, jet, 30 gal. galvanized tank									
3040	1/2 HP	1 Plum	2	4	Ea.	495	195		690	840

33 31 Sanitary Utility Sewerage Piping

33 31 13 – Public Sanitary Utility Sewerage Piping

33 31 13.15 Sewage Collection, Concrete Pipe

		Crew	Daily Output	Labor-Hours	Unit	Material	Labor	Equipment	Total	Total Incl O&P
0010	**SEWAGE COLLECTION, CONCRETE PIPE**									
0020	See Div. 33 41 13.60 for sewage/drainage collection, concrete pipe									

33 31 13.25 Sewage Collection, Polyvinyl Chloride Pipe

		Crew	Daily Output	Labor-Hours	Unit	Material	Labor	Equipment	Total	Total Incl O&P
0010	**SEWAGE COLLECTION, POLYVINYL CHLORIDE PIPE**									
0020	Not including excavation or backfill									
2000	20′ lengths, S.D.R. 35, B&S, 4″ diameter	B-20	375	.064	L.F.	1.66	2.26		3.92	5.35
2040	6″ diameter		350	.069		3.40	2.42		5.82	7.50
2080	13′ lengths , S.D.R. 35, B&S, 8″ diameter		335	.072		7.05	2.53		9.58	11.65
2120	10″ diameter	B-21	330	.085		11.15	3.09	.39	14.63	17.45
2160	12″ diameter		320	.088		12.70	3.18	.40	16.28	19.30
2200	15″ diameter		240	.117		12.10	4.24	.54	16.88	20.50
4000	Piping, DWV PVC, no exc/bkfill, 10′ L, Sch 40, 4″ diameter	B-20	375	.064		3.84	2.26		6.10	7.75
4010	6″ diameter		350	.069		7.65	2.42		10.07	12.15
4020	8″ diameter		335	.072		13.30	2.53		15.83	18.55

33 41 Storm Utility Drainage Piping

33 41 13 – Public Storm Utility Drainage Piping

33 41 13.40 Piping, Storm Drainage, Corrugated Metal		Crew	Daily Output	Labor-Hours	Unit	Material	2009 Bare Costs Labor	Equipment	Total	Total Incl O&P
0010	**PIPING, STORM DRAINAGE, CORRUGATED METAL**									
0020	Not including excavation or backfill									
2000	Corrugated metal pipe, galvanized									
2020	Bituminous coated with paved invert, 20′ lengths									
2040	8″ diameter, 16 ga.	B-14	330	.145	L.F.	12.65	4.83	.89	18.37	22.50
2060	10″ diameter, 16 ga.		260	.185		15.20	6.15	1.13	22.48	27.50
2080	12″ diameter, 16 ga.		210	.229		18.20	7.60	1.40	27.20	33
2100	15″ diameter, 16 ga.		200	.240		22	7.95	1.47	31.42	38.50
2120	18″ diameter, 16 ga.		190	.253		28.50	8.40	1.55	38.45	45.50
2140	24″ diameter, 14 ga.		160	.300		34.50	9.95	1.84	46.29	55.50
2160	30″ diameter, 14 ga.	B-13	120	.467		46	15.95	6.60	68.55	82.50
2180	36″ diameter, 12 ga.		120	.467		67.50	15.95	6.60	90.05	106
2200	48″ diameter, 12 ga.		100	.560		102	19.15	7.90	129.05	151
2220	60″ diameter, 10 ga.	B-13B	75	.747		132	25.50	15	172.50	202
2240	72″ diameter, 8 ga.	″	45	1.244		198	42.50	25	265.50	310
2500	Galvanized, uncoated, 20′ lengths									
2520	8″ diameter, 16 ga.	B-14	355	.135	L.F.	10.35	4.49	.83	15.67	19.20
2540	10″ diameter, 16 ga.		280	.171		11.35	5.70	1.05	18.10	22.50
2560	12″ diameter, 16 ga.		220	.218		12.95	7.25	1.34	21.54	27
2580	15″ diameter, 16 ga.		220	.218		16.50	7.25	1.34	25.09	31
2600	18″ diameter, 16 ga.		205	.234		22	7.75	1.43	31.18	37.50
2620	24″ diameter, 14 ga.		175	.274		31.50	9.10	1.68	42.28	51
2640	30″ diameter, 14 ga.	B-13	130	.431		40.50	14.75	6.05	61.30	73.50
2660	36″ diameter, 12 ga.		130	.431		66	14.75	6.05	86.80	102
2680	48″ diameter, 12 ga.		110	.509		88.50	17.40	7.20	113.10	131
2690	60″ diameter, 10 ga.	B-13B	78	.718		139	24.50	14.40	177.90	206
2780	End sections, 8″ diameter	B-14	24	2	Ea.	101	66.50	12.25	179.75	226
2785	10″ diameter		22	2.182		104	72.50	13.35	189.85	241
2790	12″ diameter		35	1.371		117	45.50	8.40	170.90	208
2800	18″ diameter		30	1.600		139	53	9.80	201.80	246
2810	24″ diameter	B-13	25	2.240		202	76.50	31.50	310	375
2820	30″ diameter		25	2.240		385	76.50	31.50	493	580
2825	36″ diameter		20	2.800		510	95.50	39.50	645	750
2830	48″ diameter		10	5.600		1,050	191	79	1,320	1,525
2835	60″ diameter	B-13B	5	11.200		1,375	385	225	1,985	2,375
2840	72″ diameter	″	4	14		2,475	480	281	3,236	3,775

33 41 13.60 Sewage/Drainage Collection, Concrete Pipe		Crew	Daily Output	Labor-Hours	Unit	Material	Labor	Equipment	Total	Total Incl O&P
0010	**SEWAGE/DRAINAGE COLLECTION, CONCRETE PIPE**									
0020	Not including excavation or backfill									
1000	Non-reinforced pipe, extra strength, B&S or T&G joints									
1010	6″ diameter	B-14	265.04	.181	L.F.	6.05	6	1.11	13.16	17.15
1020	8″ diameter		224	.214		6.70	7.10	1.31	15.11	19.75
1030	10″ diameter		216	.222		7.40	7.35	1.36	16.11	21
1040	12″ diameter		200	.240		9.10	7.95	1.47	18.52	24
1050	15″ diameter		180	.267		10.60	8.85	1.63	21.08	27
1060	18″ diameter		144	.333		13.05	11.05	2.04	26.14	33.50
1070	21″ diameter		112	.429		16.10	14.20	2.62	32.92	42.50
1080	24″ diameter		100	.480		19.75	15.95	2.94	38.64	49
2000	Reinforced culvert, class 3, no gaskets									
2010	12″ diameter	B-14	150	.320	L.F.	10.20	10.60	1.96	22.76	29.50
2020	15″ diameter		150	.320		14.95	10.60	1.96	27.51	35
2030	18″ diameter		132	.364		17.45	12.05	2.23	31.73	40.50

33 41 Storm Utility Drainage Piping

33 41 13 – Public Storm Utility Drainage Piping

33 41 13.60 Sewage/Drainage Collection, Concrete Pipe		Crew	Daily Output	Labor-Hours	Unit	Material	Labor	Equipment	Total	Total Incl O&P
						2009 Bare Costs				
2035	21" diameter	B-14	120	.400	L.F.	21.50	13.25	2.45	37.20	47
2040	24" diameter	↓	100	.480		29	15.95	2.94	47.89	59
2045	27" diameter	B-13	92	.609		36.50	21	8.60	66.10	81.50
2050	30" diameter		88	.636		42	22	8.95	72.95	89.50
2060	36" diameter	↓	72	.778		53.50	26.50	10.95	90.95	112
2070	42" diameter	B-13B	72	.778		77.50	26.50	15.65	119.65	144
2080	48" diameter		64	.875		113	30	17.60	160.60	189
2090	60" diameter		48	1.167		202	40	23.50	265.50	310
2100	72" diameter		40	1.400		225	48	28	301	350
2120	84" diameter		32	1.750		255	60	35	350	410
2140	96" diameter	↓	24	2.333		305	80	47	432	510
2200	With gaskets, class 3, 12" diameter	B-21	168	.167		14.15	6.05	.77	20.97	26
2220	15" diameter		160	.175		17.55	6.35	.81	24.71	30
2230	18" diameter		152	.184		23.50	6.70	.85	31.05	37.50
2240	24" diameter	↓	136	.206		36	7.50	.95	44.45	52
2260	30" diameter	B-13	88	.636		50.50	22	8.95	81.45	99
2270	36" diameter	"	72	.778		64	26.50	10.95	101.45	124
2290	48" diameter	B-13B	64	.875		129	30	17.60	176.60	207
2310	72" diameter	"	40	1.400		251	48	28	327	380
2330	Flared ends, 6'-1" long, 12" diameter	B-21	190	.147		51.50	5.35	.68	57.53	65.50
2340	15" diameter		155	.181		58.50	6.55	.83	65.88	75.50
2400	6'-2" long, 18" diameter		122	.230		61.50	8.35	1.06	70.91	87.50
2420	24" diameter	↓	88	.318		70.50	11.60	1.47	83.57	97
2440	36" diameter	B-13	60	.933		128	32	13.15	173.15	205
3080	Radius pipe, add to pipe prices, 12" to 60" diameter					50%				
3090	Over 60" diameter, add				↓	20%				
3500	Reinforced elliptical, 8' lengths, C507 class 3									
3520	14" x 23" inside, round equivalent 18" diameter	B-21	82	.341	L.F.	49.50	12.40	1.57	63.47	75.50
3530	24" x 38" inside, round equivalent 30" diameter	B-13	58	.966		73.50	33	13.60	120.10	147
3540	29" x 45" inside, round equivalent 36" diameter		52	1.077		122	37	15.20	174.20	207
3550	38" x 60" inside, round equivalent 48" diameter		38	1.474		209	50.50	21	280.50	330
3560	48" x 76" inside, round equivalent 60" diameter		26	2.154		294	73.50	30.50	398	470
3570	58" x 91" inside, round equivalent 72" diameter	↓	22	2.545	↓	410	87	36	533	625
3780	Concrete slotted pipe, class 4 mortar joint									
3800	12" diameter	B-21	168	.167	L.F.	18.80	6.05	.77	25.62	30.50
3840	18" diameter	"	152	.184	"	29	6.70	.85	36.55	43.50
3900	Class 4 O-ring									
3940	12" diameter	B-21	168	.167	L.F.	19.70	6.05	.77	26.52	31.50
3960	18" diameter	"	152	.184	"	26.50	6.70	.85	34.05	40.50

33 42 Culverts

33 42 16 – Concrete Culverts

33 42 16.15 Oval Arch Culverts

		Crew	Daily Output	Labor-Hours	Unit	Material	Labor	Equipment	Total	Total Incl O&P
0010	**OVAL ARCH CULVERTS**									
3000	Corrugated galvanized or aluminum, coated & paved									
3020	17" x 13", 16 ga., 15" equivalent	B-14	200	.240	L.F.	39.50	7.95	1.47	48.92	57.50
3040	21" x 15", 16 ga., 18" equivalent		150	.320		51	10.60	1.96	63.56	74.50
3060	28" x 20", 14 ga., 24" equivalent		125	.384		73	12.75	2.35	88.10	103
3080	35" x 24", 14 ga., 30" equivalent	↓	100	.480		89	15.95	2.94	107.89	126
3100	42" x 29", 12 ga., 36" equivalent	B-13	100	.560		133	19.15	7.90	160.05	184
3120	49" x 33", 12 ga., 42" equivalent	↓	90	.622	↓	160	21.50	8.75	190.25	218

33 42 Culverts

33 42 16 – Concrete Culverts

33 42 16.15 Oval Arch Culverts		Crew	Daily Output	Labor-Hours	Unit	2009 Bare Costs Material	Labor	Equipment	Total	Total Incl O&P
3140	57" x 38", 12 ga., 48" equivalent	B-13	75	.747	L.F.	178	25.50	10.55	214.05	246
3160	Steel, plain oval arch culverts, plain									
3180	17" x 13", 16 ga., 15" equivalent	B-14	225	.213	L.F.	21.50	7.10	1.31	29.91	36
3200	21" x 15", 16 ga., 18" equivalent		175	.274		25.50	9.10	1.68	36.28	44
3220	28" x 20", 14 ga., 24" equivalent		150	.320		40.50	10.60	1.96	53.06	63.50
3240	35" x 24", 14 ga., 30" equivalent	B-13	108	.519		51	17.75	7.30	76.05	91
3260	42" x 29", 12 ga., 36" equivalent		108	.519		84.50	17.75	7.30	109.55	128
3280	49" x 33", 12 ga., 42" equivalent		92	.609		99.50	21	8.60	129.10	151
3300	57" x 38", 12 ga., 48" equivalent		75	.747		88	25.50	10.55	124.05	147
3320	End sections, 17" x 13"		22	2.545	Ea.	117	87	36	240	305
3340	42" x 29"		17	3.294	"	480	113	46.50	639.50	755
3360	Multi-plate arch, steel	B-20	1690	.014	Lb.	1.34	.50		1.84	2.25

33 44 Storm Utility Water Drains

33 44 13 – Utility Area Drains

33 44 13.13 Catchbasins		Crew	Daily Output	Labor-Hours	Unit	Material	Labor	Equipment	Total	Total Incl O&P
0010	**CATCHBASINS**									
0011	Not including footing & excavation									
1600	Frames & covers, C.I., 24" square, 500 lb.	B-6	7.80	3.077	Ea.	298	105	37.50	440.50	535
1700	26" D shape, 600 lb.		7	3.429		510	117	42	669	790
1800	Light traffic, 18" diameter, 100 lb.		10	2.400		167	82	29.50	278.50	340
1900	24" diameter, 300 lb.		8.70	2.759		259	94	34	387	465
2000	36" diameter, 900 lb.		5.80	4.138		520	141	50.50	711.50	840
2100	Heavy traffic, 24" diameter, 400 lb.		7.80	3.077		249	105	37.50	391.50	475
2200	36" diameter, 1150 lb.		3	8		825	273	98	1,196	1,450
2300	Mass. State standard, 26" diameter, 475 lb.		7	3.429		625	117	42	784	910
2400	30" diameter, 620 lb.		7	3.429		395	117	42	554	660
2500	Watertight, 24" diameter, 350 lb.		7.80	3.077		425	105	37.50	567.50	670
2600	26" diameter, 500 lb.		7	3.429		400	117	42	559	665
2700	32" diameter, 575 lb.		6	4		895	136	49	1,080	1,250
2800	3 piece cover & frame, 10" deep,									
2900	1200 lbs., for heavy equipment	B-6	3	8	Ea.	1,275	273	98	1,646	1,925
3000	Raised for paving 1-1/4" to 2" high									
3100	4 piece expansion ring									
3200	20" to 26" diameter	1 Clab	3	2.667	Ea.	140	84.50		224.50	285
3300	30" to 36" diameter	"	3	2.667	"	195	84.50		279.50	345
3320	Frames and covers, existing, raised for paving, 2", including									
3340	row of brick, concrete collar, up to 12" wide frame	B-6	18	1.333	Ea.	43.50	45.50	16.30	105.30	135
3360	20" to 26" wide frame		11	2.182		66	74.50	26.50	167	216
3380	30" to 36" wide frame		9	2.667		82	91	32.50	205.50	265
3400	Inverts, single channel brick	D-1	3	5.333		92	194		286	395
3500	Concrete		5	3.200		72	116		188	255
3600	Triple channel, brick		2	8		139	291		430	595
3700	Concrete		3	5.333		123	194		317	430

33 46 16.20 Piping, Subdrainage, Concrete

	33 46 16.20 Piping, Subdrainage, Concrete	Crew	Daily Output	Labor-Hours	Unit	2009 Bare Costs Material	Labor	Equipment	Total	Total Incl O&P
0010	**PIPING, SUBDRAINAGE, CONCRETE**									
0021	Not including excavation and backfill									
3000	Porous wall concrete underdrain, std. strength, 4″ diameter	B-20	335	.072	L.F.	3.30	2.53		5.83	7.55
3020	6″ diameter	″	315	.076		4.29	2.69		6.98	8.90
3040	8″ diameter	B-21	310	.090		5.30	3.29	.42	9.01	11.35
3100	18″ diameter	″	165	.170		17	6.15	.78	23.93	29
4000	Extra strength, 6″ diameter	B-20	315	.076		3.97	2.69		6.66	8.55
4020	8″ diameter	B-21	310	.090		5.95	3.29	.42	9.66	12.05
4040	10″ diameter		295	.095		11.90	3.45	.44	15.79	18.90
4060	12″ diameter		285	.098		12.90	3.57	.45	16.92	20
4080	15″ diameter		230	.122		14.30	4.43	.56	19.29	23
4100	18″ diameter		165	.170		21	6.15	.78	27.93	33.50

33 46 16.25 Piping, Subdrainage, Corrugated Metal

	33 46 16.25 Piping, Subdrainage, Corrugated Metal	Crew	Daily Output	Labor-Hours	Unit	Material	Labor	Equipment	Total	Total Incl O&P
0010	**PIPING, SUBDRAINAGE, CORRUGATED METAL**									
0021	Not including excavation and backfill									
2010	Aluminum, perforated									
2020	6″ diameter, 18 ga.	B-14	380	.126	L.F.	4.70	4.19	.77	9.66	12.45
2200	8″ diameter, 16 ga.		370	.130		7.15	4.30	.79	12.24	15.35
2220	10″ diameter, 16 ga.		360	.133		8.90	4.42	.82	14.14	17.50
2240	12″ diameter, 16 ga.		285	.168		10	5.60	1.03	16.63	20.50
2260	18″ diameter, 16 ga.		205	.234		14.95	7.75	1.43	24.13	30
3000	Uncoated galvanized, perforated									
3020	6″ diameter, 18 ga.	B-20	380	.063	L.F.	8.25	2.23		10.48	12.50
3200	8″ diameter, 16 ga.	″	370	.065		11.35	2.29		13.64	16
3220	10″ diameter, 16 ga.	B-21	360	.078		17	2.83	.36	20.19	23.50
3240	12″ diameter, 16 ga.		285	.098		17.80	3.57	.45	21.82	25.50
3260	18″ diameter, 16 ga.		205	.137		27	4.97	.63	32.60	38.50
4000	Steel, perforated, asphalt coated									
4020	6″ diameter 18 ga.	B-20	380	.063	L.F.	6.60	2.23		8.83	10.70
4030	8″ diameter 18 ga.	″	370	.065		10.30	2.29		12.59	14.90
4040	10″ diameter 16 ga.	B-21	360	.078		11.85	2.83	.36	15.04	17.75
4050	12″ diameter 16 ga.		285	.098		13.60	3.57	.45	17.62	21
4060	18″ diameter 16 ga.		205	.137		18.50	4.97	.63	24.10	29

33 46 16.30 Piping, Subdrainage, Plastic

	33 46 16.30 Piping, Subdrainage, Plastic	Crew	Daily Output	Labor-Hours	Unit	Material	Labor	Equipment	Total	Total Incl O&P
0010	**PIPING, SUBDRAINAGE, PLASTIC**									
0020	Not including excavation and backfill									
2100	Perforated PVC, 4″ diameter	B-14	314	.153	L.F.	1.09	5.05	.94	7.08	10.05
2110	6″ diameter		300	.160		2.02	5.30	.98	8.30	11.50
2120	8″ diameter		290	.166		2.29	5.50	1.01	8.80	12.10
2130	10″ diameter		280	.171		3.54	5.70	1.05	10.29	13.80
2140	12″ diameter		270	.178		4.91	5.90	1.09	11.90	15.70

33 49 Storm Drainage Structures

33 49 13 – Storm Drainage Manholes, Frames, and Covers

33 49 13.10 Storm Drainage Manholes, Frames and Covers		Crew	Daily Output	Labor-Hours	Unit	2009 Bare Costs Material	Labor	Equipment	Total	Total Incl O&P
0010	**STORM DRAINAGE MANHOLES, FRAMES & COVERS**									
0020	Excludes footing, excavation, backfill (See line items for frame & cover)									
0050	Brick, 4′ inside diameter, 4′ deep	D-1	1	16	Ea.	400	580		980	1,325
0100	6′ deep		.70	22.857		555	830		1,385	1,850
0150	8′ deep		.50	32		710	1,175		1,885	2,550
0200	For depths over 8′, add		4	4	V.L.F.	162	145		307	400
0400	Concrete blocks (radial), 4′ I.D., 4′ deep		1.50	10.667	Ea.	350	390		740	970
0500	6′ deep		1	16		460	580		1,040	1,375
0600	8′ deep		.70	22.857		575	830		1,405	1,875
0700	For depths over 8′, add		5.50	2.909	V.L.F.	59	106		165	225
0800	Concrete, cast in place, 4′ x 4′, 8″ thick, 4′ deep	C-14H	2	24	Ea.	560	945	12.95	1,517.95	2,100
0900	6′ deep		1.50	32		810	1,250	17.30	2,077.30	2,850
1000	8′ deep		1	48		1,150	1,900	26	3,076	4,225
1100	For depths over 8′, add		8	6	V.L.F.	132	236	3.24	371.24	515
1110	Precast, 4′ I.D., 4′ deep	B-22	4.10	7.317	Ea.	880	269	47.50	1,196.50	1,425
1120	6′ deep		3	10		1,100	370	64.50	1,534.50	1,850
1130	8′ deep		2	15		1,325	550	97	1,972	2,400
1140	For depths over 8′, add		16	1.875	V.L.F.	181	69	12.10	262.10	320
1150	5′ I.D., 4′ deep	B-6	3	8	Ea.	910	273	98	1,281	1,525
1160	6′ deep		2	12		1,225	410	147	1,782	2,125
1170	8′ deep		1.50	16		1,550	545	196	2,291	2,750
1180	For depths over 8′, add		12	2	V.L.F.	202	68	24.50	294.50	355
1190	6′ I.D., 4′ deep		2	12	Ea.	1,500	410	147	2,057	2,400
1200	6′ deep		1.50	16		1,925	545	196	2,666	3,175
1210	8′ deep		1	24		2,375	820	294	3,489	4,200
1220	For depths over 8′, add		8	3	V.L.F.	310	102	36.50	448.50	545
1250	Slab tops, precast, 8″ thick									
1300	4′ diameter manhole	B-6	8	3	Ea.	209	102	36.50	347.50	430
1400	5′ diameter manhole		7.50	3.200		415	109	39	563	665
1500	6′ diameter manhole		7	3.429		595	117	42	754	880
3800	Steps, heavyweight cast iron, 7″ x 9″	1 Bric	40	.200		12.95	8.10		21.05	26.50
3900	8″ x 9″		40	.200		19.45	8.10		27.55	34
3928	12″ x 10-1/2″		40	.200		22	8.10		30.10	37
4000	Standard sizes, galvanized steel		40	.200		18.75	8.10		26.85	33
4100	Aluminum		40	.200		25.50	8.10		33.60	40.50

33 51 Natural-Gas Distribution

33 51 13 – Natural-Gas Piping

33 51 13.10 Piping, Gas Service and Distribution, P.E.		Crew	Daily Output	Labor-Hours	Unit	Material	Labor	Equipment	Total	Total Incl O&P
0010	**PIPING, GAS SERVICE AND DISTRIBUTION, POLYETHYLENE**									
0020	Not including excavation or backfill									
1000	60 psi coils, compression coupling @ 100′, 1/2″ diameter, SDR 11	B-20A	608	.053	L.F.	1.61	2.01		3.62	4.83
1040	1-1/4″ diameter, SDR 11		544	.059		2.49	2.25		4.74	6.15
1100	2″ diameter, SDR 11		488	.066		3.09	2.51		5.60	7.20
1160	3″ diameter, SDR 11		408	.078		6.45	3		9.45	11.65
1500	60 PSI 40′ joints with coupling, 3″ diameter, SDR 11	B-21A	408	.098		6.45	3.83	1.48	11.76	14.50
1540	4″ diameter, SDR 11		352	.114		14.75	4.44	1.71	20.90	25
1600	6″ diameter, SDR 11		328	.122		46	4.77	1.84	52.61	60
1640	8″ diameter, SDR 11		272	.147		62.50	5.75	2.21	70.46	80

33 52 Liquid Fuel Distribution

33 52 16 – Gasoline Distribution

33 52 16.13 Gasoline Piping		Crew	Daily Output	Labor-Hours	Unit	2009 Bare Costs Material	Labor	Equipment	Total	Total Incl O&P
0010	**GASOLINE PIPING**									
0020	Primary containment pipe, fiberglass-reinforced									
0030	Plastic pipe 15′ & 30′ lengths									
0040	2″ diameter	Q-6	425	.056	L.F.	4.94	2.60		7.54	9.35
0050	3″ diameter		400	.060		6.45	2.76		9.21	11.25
0060	4″ diameter	↓	375	.064	↓	8.35	2.95		11.30	13.55
0100	Fittings									
0110	Elbows, 90° & 45°, bell-ends, 2″	Q-6	24	1	Ea.	50	46		96	124
0120	3″ diameter		22	1.091		52.50	50.50		103	133
0130	4″ diameter		20	1.200		69.50	55.50		125	159
0200	Tees, bell ends, 2″		21	1.143		60.50	52.50		113	146
0210	3″ diameter		18	1.333		61	61.50		122.50	159
0220	4″ diameter		15	1.600		83.50	73.50		157	203
0230	Flanges bell ends, 2″		24	1		20	46		66	91
0240	3″ diameter		22	1.091		25	50.50		75.50	103
0250	4″ diameter		20	1.200		34.50	55.50		90	121
0260	Sleeve couplings, 2″		21	1.143		12.75	52.50		65.25	93
0270	3″ diameter		18	1.333		18.05	61.50		79.55	112
0280	4″ diameter		15	1.600		25	73.50		98.50	139
0290	Threaded adapters 2″		21	1.143		16.75	52.50		69.25	97.50
0300	3″ diameter		18	1.333		29.50	61.50		91	125
0310	4″ diameter		15	1.600		40	73.50		113.50	155
0320	Reducers, 2″		27	.889		22.50	41		63.50	86
0330	3″ diameter		22	1.091		26	50.50		76.50	104
0340	4″ diameter	↓	20	1.200	↓	33.50	55.50		89	120
1010	Gas station product line for secondary containment (double wall)									
1100	Fiberglass reinforced plastic pipe 25′ lengths									
1120	Pipe, plain end, 3″ diameter	Q-6	375	.064	L.F.	8.50	2.95		11.45	13.75
1130	4″ diameter		350	.069		13.90	3.16		17.06	20
1140	5″ diameter		325	.074		18.05	3.40		21.45	25
1150	6″ diameter	↓	300	.080	↓	18.60	3.69		22.29	26
1200	Fittings									
1230	Elbows, 90° & 45°, 3″ diameter	Q-6	18	1.333	Ea.	61	61.50		122.50	159
1240	4″ diameter		16	1.500		106	69		175	221
1250	5″ diameter		14	1.714		247	79		326	390
1260	6″ diameter		12	2		250	92		342	415
1270	Tees, 3″ diameter		15	1.600		90.50	73.50		164	211
1280	4″ diameter		12	2		133	92		225	284
1290	5″ diameter		9	2.667		269	123		392	480
1300	6″ diameter		6	4		281	184		465	585
1310	Couplings, 3″ diameter		18	1.333		43	61.50		104.50	139
1320	4″ diameter		16	1.500		111	69		180	226
1330	5″ diameter		14	1.714		230	79		309	370
1340	6″ diameter		12	2		238	92		330	400
1350	Cross-over nipples, 3″ diameter		18	1.333		9.80	61.50		71.30	103
1360	4″ diameter		16	1.500		11.50	69		80.50	117
1370	5″ diameter		14	1.714		17.05	79		96.05	138
1380	6″ diameter		12	2		17.85	92		109.85	158
1400	Telescoping, reducers, concentric 4″ x 3″		18	1.333		32.50	61.50		94	128
1410	5″ x 4″		17	1.412		85	65		150	191
1420	6″ x 5″	↓	16	1.500	↓	204	69		273	330

33 61 Hydronic Energy Distribution

33 61 13 – Underground Hydronic Energy Distribution

33 61 13.20 Pipe Conduit, Prefabricated/Preinsulated		Crew	Daily Output	Labor-Hours	Unit	2009 Bare Costs Material	Labor	Equipment	Total	Total Incl O&P
0010	**PIPE CONDUIT, PREFABRICATED/PREINSULATED**									
0020	Does not include trenching, fittings or crane.									
0300	For cathodic protection, add 12 to 14%									
0310	of total built-up price (casing plus service pipe)									
0580	Polyurethane insulated system, 250° F. max. temp.									
0620	Black steel service pipe, standard wt., 1/2" insulation									
0660	3/4" diam. pipe size	Q-17	54	.296	L.F.	50	13.15	1.03	64.18	76
0670	1" diam. pipe size		50	.320		55	14.20	1.11	70.31	83
0680	1-1/4" diam. pipe size		47	.340		61	15.15	1.18	77.33	91.50
0690	1-1/2" diam. pipe size		45	.356		66.50	15.80	1.24	83.54	98
0700	2" diam. pipe size		42	.381		69	16.95	1.33	87.28	103
0710	2-1/2" diam. pipe size		34	.471		70	21	1.64	92.64	110
0720	3" diam. pipe size		28	.571		81	25.50	1.99	108.49	129
0730	4" diam. pipe size		22	.727		102	32.50	2.53	137.03	163
0740	5" diam. pipe size	↓	18	.889		130	39.50	3.09	172.59	206
0750	6" diam. pipe size	Q-18	23	1.043		152	48	2.42	202.42	242
0760	8" diam. pipe size		19	1.263		222	58	2.93	282.93	335
0770	10" diam. pipe size		16	1.500		293	69	3.48	365.48	430
0780	12" diam. pipe size		13	1.846		365	85	4.28	454.28	535
0790	14" diam. pipe size		11	2.182		405	101	5.05	511.05	600
0800	16" diam. pipe size		10	2.400		465	111	5.55	581.55	680
0810	18" diam. pipe size		8	3		535	138	6.95	679.95	805
0820	20" diam. pipe size		7	3.429		595	158	7.95	760.95	900
0830	24" diam. pipe size	↓	6	4		730	184	9.30	923.30	1,075
0900	For 1" thick insulation, add					10%				
0940	For 1-1/2" thick insulation, add					13%				
0980	For 2" thick insulation, add				↓	20%				
1500	Gland seal for system, 3/4" diam. pipe size	Q-17	32	.500	Ea.	675	22	1.74	698.74	780
1510	1" diam. pipe size		32	.500		675	22	1.74	698.74	780
1540	1-1/4" diam. pipe size		30	.533		725	23.50	1.86	750.36	835
1550	1-1/2" diam. pipe size		30	.533		725	23.50	1.86	750.36	835
1560	2" diam. pipe size		28	.571		860	25.50	1.99	887.49	985
1570	2-1/2" diam. pipe size		26	.615		925	27.50	2.14	954.64	1,075
1580	3" diam. pipe size		24	.667		995	29.50	2.32	1,026.82	1,150
1590	4" diam. pipe size		22	.727		1,175	32.50	2.53	1,210.03	1,350
1600	5" diam. pipe size	↓	19	.842		1,450	37.50	2.93	1,490.43	1,650
1610	6" diam. pipe size	Q-18	26	.923		1,525	42.50	2.14	1,569.64	1,775
1620	8" diam. pipe size		25	.960		1,775	44	2.23	1,821.23	2,025
1630	10" diam. pipe size		23	1.043		2,175	48	2.42	2,225.42	2,450
1640	12" diam. pipe size		21	1.143		2,400	52.50	2.65	2,455.15	2,700
1650	14" diam. pipe size		19	1.263		2,675	58	2.93	2,735.93	3,050
1660	16" diam. pipe size		18	1.333		3,150	61.50	3.09	3,214.59	3,550
1670	18" diam. pipe size		16	1.500		3,375	69	3.48	3,447.48	3,800
1680	20" diam. pipe size		14	1.714		3,825	79	3.98	3,907.98	4,325
1690	24" diam. pipe size	↓	12	2	↓	4,250	92	4.64	4,346.64	4,825
2000	Elbow, 45° for system									
2020	3/4" diam. pipe size	Q-17	14	1.143	Ea.	425	51	3.98	479.98	550
2040	1" diam. pipe size		13	1.231		440	54.50	4.28	498.78	565
2050	1-1/4" diam. pipe size		11	1.455		495	64.50	5.05	564.55	650
2060	1-1/2" diam. pipe size		9	1.778		520	79	6.20	605.20	695
2070	2" diam. pipe size		6	2.667		545	118	9.30	672.30	790
2080	2-1/2" diam. pipe size		4	4		585	178	13.90	776.90	925
2090	3" diam. pipe size	↓	3.50	4.571	↓	680	203	15.90	898.90	1,075

33 61 Hydronic Energy Distribution

33 61 13 – Underground Hydronic Energy Distribution

33 61 13.20 Pipe Conduit, Prefabricated/Preinsulated		Crew	Daily Output	Labor-Hours	Unit	Material	2009 Bare Costs Labor	Equipment	Total	Total Incl O&P
2100	4" diam. pipe size	Q-17	3	5.333	Ea.	790	237	18.55	1,045.55	1,250
2110	5" diam. pipe size	↓	2.80	5.714		1,025	254	19.90	1,298.90	1,525
2120	6" diam. pipe size	Q-18	4	6		1,150	276	13.90	1,439.90	1,700
2130	8" diam. pipe size		3	8		1,675	370	18.55	2,063.55	2,425
2140	10" diam. pipe size		2.40	10		2,125	460	23	2,608	3,075
2150	12" diam. pipe size		2	12		2,800	555	28	3,383	3,925
2160	14" diam. pipe size		1.80	13.333		3,475	615	31	4,121	4,775
2170	16" diam. pipe size		1.60	15		4,150	690	35	4,875	5,625
2180	18" diam. pipe size		1.30	18.462		5,200	850	43	6,093	7,050
2190	20" diam. pipe size		1	24		6,550	1,100	55.50	7,705.50	8,900
2200	24" diam. pipe size	↓	.70	34.286		8,225	1,575	79.50	9,879.50	11,500
2260	For elbow, 90°, add					25%				
2300	For tee, straight, add					85%	30%			
2340	For tee, reducing, add					170%	30%			
2380	For weldolet, straight, add				↓	50%				
2400	Polyurethane insulation, 1"									
2410	FRP carrier and casing									
2420	4"	Q-5	18	.889	L.F.	54	39.50		93.50	119
2422	6"	Q-6	23	1.043		94	48		142	175
2424	8"		19	1.263		153	58		211	256
2426	10"		16	1.500		206	69		275	330
2428	12"	↓	13	1.846	↓	269	85		354	425
2430	FRP carrier and PVC casing									
2440	4"	Q-5	18	.889	L.F.	21.50	39.50		61	83
2444	8"	Q-6	19	1.263		44.50	58		102.50	137
2446	10"		16	1.500		64	69		133	175
2448	12"	↓	13	1.846	↓	82	85		167	218
2450	PVC carrier and casing									
2460	4"	Q-1	36	.444	L.F.	10.55	19.50		30.05	41
2462	6"	"	29	.552		14.75	24		38.75	53
2464	8"	Q-2	36	.667		20.50	30.50		51	68
2466	10"		32	.750		28	34		62	82
2468	12"	↓	31	.774	↓	32.50	35		67.50	89

33 63 Steam Energy Distribution

33 63 13 – Underground Steam and Condensate Distribution Piping

33 63 13.10 Calcium Silicate Insulated System		Crew	Daily Output	Labor-Hours	Unit	Material	Labor	Equipment	Total	Total Incl O&P
0010	**CALCIUM SILICATE INSULATED SYSTEM**									
0011	High temp. (1200 degrees F)									
2840	Steel casing with protective exterior coating									
2850	6-5/8" diameter	Q-18	52	.462	L.F.	96	21.50	1.07	118.57	138
2860	8-5/8" diameter		50	.480		105	22	1.11	128.11	149
2870	10-3/4" diameter		47	.511		123	23.50	1.18	147.68	172
2880	12-3/4" diameter		44	.545		134	25	1.27	160.27	186
2890	14" diameter		41	.585		150	27	1.36	178.36	208
2900	16" diameter		39	.615		162	28.50	1.43	191.93	222
2910	18" diameter		36	.667		178	30.50	1.55	210.05	244
2920	20" diameter		34	.706		201	32.50	1.64	235.14	272
2930	22" diameter		32	.750		279	34.50	1.74	315.24	360
2940	24" diameter		29	.828		320	38	1.92	359.92	410
2950	26" diameter	↓	26	.923	↓	360	42.50	2.14	404.64	465

33 63 13.10 Calcium Silicate Insulated System		Crew	Daily Output	Labor-Hours	Unit	2009 Bare Costs Material	Labor	Equipment	Total	Total Incl O&P
2960	28″ diameter	Q-18	23	1.043	L.F.	450	48	2.42	500.42	570
2970	30″ diameter		21	1.143		480	52.50	2.65	535.15	605
2980	32″ diameter		19	1.263		540	58	2.93	600.93	685
2990	34″ diameter		18	1.333		545	61.50	3.09	609.59	695
3000	36″ diameter		16	1.500		585	69	3.48	657.48	755
3040	For multi-pipe casings, add					10%				
3060	For oversize casings, add					2%				
3400	Steel casing gland seal, single pipe									
3420	6-5/8″ diameter	Q-18	25	.960	Ea.	1,325	44	2.23	1,371.23	1,550
3440	8-5/8″ diameter		23	1.043		1,550	48	2.42	1,600.42	1,800
3450	10-3/4″ diameter		21	1.143		1,750	52.50	2.65	1,805.15	2,000
3460	12-3/4″ diameter		19	1.263		2,100	58	2.93	2,160.93	2,400
3470	14″ diameter		17	1.412		2,275	65	3.28	2,343.28	2,625
3480	16″ diameter		16	1.500		2,675	69	3.48	2,747.48	3,050
3490	18″ diameter		15	1.600		2,975	73.50	3.71	3,052.21	3,400
3500	20″ diameter		13	1.846		3,300	85	4.28	3,389.28	3,750
3510	22″ diameter		12	2		3,700	92	4.64	3,796.64	4,225
3520	24″ diameter		11	2.182		4,150	101	5.05	4,256.05	4,725
3530	26″ diameter		10	2.400		4,725	111	5.55	4,841.55	5,375
3540	28″ diameter		9.50	2.526		5,350	116	5.85	5,471.85	6,050
3550	30″ diameter		9	2.667		5,450	123	6.20	5,579.20	6,200
3560	32″ diameter		8.50	2.824		6,125	130	6.55	6,261.55	6,950
3570	34″ diameter		8	3		6,675	138	6.95	6,819.95	7,575
3580	36″ diameter		7	3.429		7,100	158	7.95	7,265.95	8,075
3620	For multi-pipe casings, add					5%				
4000	Steel casing anchors, single pipe									
4020	6-5/8″ diameter	Q-18	8	3	Ea.	1,200	138	6.95	1,344.95	1,550
4040	8-5/8″ diameter		7.50	3.200		1,250	147	7.40	1,404.40	1,600
4050	10-3/4″ diameter		7	3.429		1,650	158	7.95	1,815.95	2,050
4060	12-3/4″ diameter		6.50	3.692		1,750	170	8.55	1,928.55	2,200
4070	14″ diameter		6	4		2,050	184	9.30	2,243.30	2,550
4080	16″ diameter		5.50	4.364		2,400	201	10.10	2,611.10	2,925
4090	18″ diameter		5	4.800		2,700	221	11.15	2,932.15	3,325
4100	20″ diameter		4.50	5.333		2,975	246	12.35	3,233.35	3,650
4110	22″ diameter		4	6		3,325	276	13.90	3,614.90	4,075
4120	24″ diameter		3.50	6.857		3,625	315	15.90	3,955.90	4,475
4130	26″ diameter		3	8		4,100	370	18.55	4,488.55	5,075
4140	28″ diameter		2.50	9.600		4,475	440	22.50	4,937.50	5,625
4150	30″ diameter		2	12		4,775	555	28	5,358	6,100
4160	32″ diameter		1.50	16		5,700	735	37	6,472	7,425
4170	34″ diameter		1	24		6,400	1,100	55.50	7,555.50	8,750
4180	36″ diameter		1	24		6,975	1,100	55.50	8,130.50	9,375
4220	For multi-pipe, add					5%	20%			
4800	Steel casing elbow									
4820	6-5/8″ diameter	Q-18	15	1.600	Ea.	1,650	73.50	3.71	1,727.21	1,925
4830	8-5/8″ diameter		15	1.600		1,750	73.50	3.71	1,827.21	2,050
4850	10-3/4″ diameter		14	1.714		2,125	79	3.98	2,207.98	2,450
4860	12-3/4″ diameter		13	1.846		2,500	85	4.28	2,589.28	2,875
4870	14″ diameter		12	2		2,675	92	4.64	2,771.64	3,100
4880	16″ diameter		11	2.182		2,925	101	5.05	3,031.05	3,375
4890	18″ diameter		10	2.400		3,350	111	5.55	3,466.55	3,850
4900	20″ diameter		9	2.667		3,575	123	6.20	3,704.20	4,125
4910	22″ diameter		8	3		3,850	138	6.95	3,994.95	4,450

33 63 13 – Underground Steam and Condensate Distribution Piping

33 63 13.10 Calcium Silicate Insulated System		Crew	Daily Output	Labor-Hours	Unit	2009 Bare Costs Material	Labor	Equipment	Total	Total Incl O&P
4920	24" diameter	Q-18	7	3.429	Ea.	4,225	158	7.95	4,390.95	4,900
4930	26" diameter		6	4		4,625	184	9.30	4,818.30	5,375
4940	28" diameter		5	4.800		5,000	221	11.15	5,232.15	5,825
4950	30" diameter		4	6		5,025	276	13.90	5,314.90	5,950
4960	32" diameter		3	8		5,700	370	18.55	6,088.55	6,850
4970	34" diameter		2	12		6,275	555	28	6,858	7,750
4980	36" diameter		2	12		6,675	555	28	7,258	8,200
5500	Black steel service pipe, std. wt., 1" thick insulation									
5510	3/4" diameter pipe size	Q-17	54	.296	L.F.	42	13.15	1.03	56.18	67
5540	1" diameter pipe size		50	.320		43.50	14.20	1.11	58.81	70.50
5550	1-1/4" diameter pipe size		47	.340		49	15.15	1.18	65.33	77.50
5560	1-1/2" diameter pipe size		45	.356		53	15.80	1.24	70.04	83.50
5570	2" diameter pipe size		42	.381		59	16.95	1.33	77.28	92
5580	2-1/2" diameter pipe size		34	.471		62	21	1.64	84.64	102
5590	3" diameter pipe size		28	.571		71	25.50	1.99	98.49	118
5600	4" diameter pipe size		22	.727		90.50	32.50	2.53	125.53	151
5610	5" diameter pipe size		18	.889		123	39.50	3.09	165.59	198
5620	6" diameter pipe size	Q-18	23	1.043		134	48	2.42	184.42	222
6000	Black steel service pipe, std. wt., 1-1/2" thick insul.									
6010	3/4" diameter pipe size	Q-17	54	.296	L.F.	42.50	13.15	1.03	56.68	67.50
6040	1" diameter pipe size		50	.320		47.50	14.20	1.11	62.81	74.50
6050	1-1/4" diameter pipe size		47	.340		53	15.15	1.18	69.33	82.50
6060	1-1/2" diameter pipe size		45	.356		58	15.80	1.24	75.04	89
6070	2" diameter pipe size		42	.381		63	16.95	1.33	81.28	96.50
6080	2-1/2" diameter pipe size		34	.471		67	21	1.64	89.64	107
6090	3" diameter pipe size		28	.571		75.50	25.50	1.99	102.99	123
6100	4" diameter pipe size		22	.727		96	32.50	2.53	131.03	157
6110	5" diameter pipe size		18	.889		123	39.50	3.09	165.59	198
6120	6" diameter pipe size	Q-18	23	1.043		139	48	2.42	189.42	228
6130	8" diameter pipe size		19	1.263		201	58	2.93	261.93	310
6140	10" diameter pipe size		16	1.500		262	69	3.48	334.48	395
6150	12" diameter pipe size		13	1.846		310	85	4.28	399.28	480
6190	For 2" thick insulation, add					15%				
6220	For 2-1/2" thick insulation, add					25%				
6260	For 3" thick insulation, add					30%				
6800	Black steel service pipe, ex. hvy. wt., 1" thick insul.									
6820	3/4" diameter pipe size	Q-17	50	.320	L.F.	44	14.20	1.11	59.31	71
6840	1" diameter pipe size		47	.340		47	15.15	1.18	63.33	75.50
6850	1-1/4" diameter pipe size		44	.364		54	16.15	1.27	71.42	85.50
6860	1-1/2" diameter pipe size		42	.381		57.50	16.95	1.33	75.78	90
6870	2" diameter pipe size		40	.400		61	17.75	1.39	80.14	95
6880	2-1/2" diameter pipe size		31	.516		75.50	23	1.80	100.30	119
6890	3" diameter pipe size		27	.593		84.50	26.50	2.06	113.06	135
6900	4" diameter pipe size		21	.762		112	34	2.65	148.65	177
6910	5" diameter pipe size		17	.941		156	42	3.28	201.28	239
6920	6" diameter pipe size	Q-18	22	1.091		144	50.50	2.53	197.03	236
7400	Black steel service pipe, ex. hvy. wt., 1-1/2" thick insul.									
7420	3/4" diameter pipe size	Q-17	50	.320	L.F.	36.50	14.20	1.11	51.81	63
7440	1" diameter pipe size		47	.340		42	15.15	1.18	58.33	70
7450	1-1/4" diameter pipe size		44	.364		48.50	16.15	1.27	65.92	79.50
7460	1-1/2" diameter pipe size		42	.381		53	16.95	1.33	71.28	85.50
7470	2" diameter pipe size		40	.400		53.50	17.75	1.39	72.64	87
7480	2-1/2" diameter pipe size		31	.516		63.50	23	1.80	88.30	106

33 63 13 – Underground Steam and Condensate Distribution Piping

33 63 13.10 Calcium Silicate Insulated System		Crew	Daily Output	Labor-Hours	Unit	2009 Bare Costs Material	Labor	Equipment	Total	Total Incl O&P
7490	3" diameter pipe size	Q-17	27	.593	L.F.	75	26.50	2.06	103.56	124
7500	4" diameter pipe size		21	.762		97.50	34	2.65	134.15	161
7510	5" diameter pipe size		17	.941		135	42	3.28	180.28	215
7520	6" diameter pipe size	Q-18	22	1.091		149	50.50	2.53	202.03	241
7530	8" diameter pipe size		18	1.333		223	61.50	3.09	287.59	340
7540	10" diameter pipe size		15	1.600		265	73.50	3.71	342.21	405
7550	12" diameter pipe size		13	1.846		325	85	4.28	414.28	495
7590	For 2" thick insulation, add					13%				
7640	For 2-1/2" thick insulation, add					18%				
7680	For 3" thick insulation, add					24%				

33 63 13.20 Combined Steam Pipe With Condensate Return		Crew	Daily Output	Labor-Hours	Unit	Material	Labor	Equipment	Total	Total Incl O&P
9000	**COMBINED STEAM PIPE WITH CONDENSATE RETURN**									
9010	8" and 4" in 24" case	Q-18	100	.240	L.F.	470	11.05	.56	481.61	530
9020	6" and 3" in 20" case		104	.231		320	10.65	.54	331.19	370
9030	3" and 1-1/2" in 16" case		110	.218		240	10.05	.51	250.56	280
9040	2" and 1-1/4" in 12-3/4" case		114	.211		209	9.70	.49	219.19	245
9050	1-1/2" and 1-1/4" in 10-3/4" case		116	.207		183	9.55	.48	193.03	216
9100	Steam pipe only (no return)									
9110	6" in 18" case	Q-18	108	.222	L.F.	264	10.25	.52	274.77	305
9120	4" in 14" case		112	.214		209	9.85	.50	219.35	245
9130	3" in 14" case		114	.211		193	9.70	.49	203.19	227
9140	2-1/2" in 12-3/4" case		118	.203		180	9.35	.47	189.82	213
9150	2" in 12-3/4" case		122	.197		172	9.05	.46	181.51	203

Division 44 - Pollution Control Equipment

Estimating Tips

There are four basic types of pollution: Air, Noise, Water, and Solid.

These systems may be interrelated and care must be taken that the complete systems are estimated. For example, Air Pollution Equipment may include dust and air-entrained particles that have to be collected. The vacuum systems could be noisy, requiring silencers to reduce noise pollution, and the collected solids have to be disposed of to prevent Solid Pollution.

Water Treatment may be accomplished as comparatively small packaged plants providing Chemical, Biological or Thermal Treatment. These are usually priced by GPH or GPD.

Large plants and municipal treatment facilities are sized MGD. These facilities are usually priced in two steps. First, the large concrete structures used as settling ponds, etc.; and second, the equipment required. Storm water does not need as much treatment as sanitary wastewater.

The total cost of the system is greatly affected by the level of purification required by specifications and codes before release.

An additional cost may be encountered if there is a possibility that the water being treated may contain petroleum products, which have to be isolated and disposed of in an environmentally acceptable manner.

Reference Numbers

Reference numbers are shown in shaded boxes at the beginning of some major classifications. These numbers refer to related items in the Reference Section. The reference information may be an estimating procedure, an alternate pricing method, or technical information.

Note: Not all subdivisions listed here necessarily appear in this publication.

44 11 16 – Industrial Dust Collectors

44 11 16.10 Dust Collection Systems		Crew	Daily Output	Labor-Hours	Unit	2009 Bare Costs Material	Labor	Equipment	Total	Total Incl O&P
0010	**DUST COLLECTION SYSTEMS** Commercial / industrial									
0120	Central vacuum units									
0130	Includes stand, filters and motorized shaker									
0200	500 CFM, 10″ inlet, 2 HP	Q-20	2.40	8.333	Ea.	3,725	360		4,085	4,650
0220	1000 CFM, 10″ inlet, 3 HP		2.20	9.091		3,900	395		4,295	4,900
0240	1500 CFM, 10″ inlet, 5 HP		2	10		4,100	435		4,535	5,175
0260	3000 CFM, 13″ inlet, 10 HP		1.50	13.333		10,400	580		10,980	12,400
0280	5000 CFM, 16″ inlet, 2 @ 10 HP		1	20		11,000	870		11,870	13,400
1000	Vacuum tubing, galvanized									
1100	2-1/8″ OD, 16 ga.	Q-9	440	.036	L.F.	2.45	1.54		3.99	5.10
1110	2-1/2″ OD, 16 ga.		420	.038		2.82	1.62		4.44	5.60
1120	3″ OD, 16 ga.		400	.040		3.23	1.70		4.93	6.15
1130	3-1/2″ OD, 16 ga.		380	.042		4.67	1.79		6.46	7.90
1140	4″ OD, 16 ga.		360	.044		5.25	1.89		7.14	8.65
1150	5″ OD, 14 ga.		320	.050		9.40	2.12		11.52	13.60
1160	6″ OD, 14 ga.		280	.057		10.90	2.43		13.33	15.70
1170	8″ OD, 14 ga.		200	.080		16.40	3.40		19.80	23.50
1180	10″ OD, 12 ga.		160	.100		16.40	4.25		20.65	24.50
1190	12″ OD, 12 ga.		120	.133		28.50	5.65		34.15	40
1200	14″ OD, 12 ga.		80	.200		34.50	8.50		43	51
1940	Hose, flexible wire reinforced rubber									
1956	3″ dia.	Q-9	400	.040	L.F.	7.05	1.70		8.75	10.35
1960	4″ dia.		360	.044		7.95	1.89		9.84	11.65
1970	5″ dia.		320	.050		9.10	2.12		11.22	13.25
1980	6″ dia.		280	.057		9.80	2.43		12.23	14.55
2000	90° Elbow, slip fit									
2110	2-1/8″ dia.	Q-9	70	.229	Ea.	10.25	9.70		19.95	26
2120	2-1/2″ dia.		65	.246		14.30	10.45		24.75	32
2130	3″ dia.		60	.267		19.40	11.35		30.75	39
2140	3-1/2″ dia.		55	.291		24	12.35		36.35	45.50
2150	4″ dia.		50	.320		30	13.60		43.60	53.50
2160	5″ dia.		45	.356		56.50	15.10		71.60	85.50
2170	6″ dia.		40	.400		77.50	17		94.50	112
2180	8″ dia.		30	.533		148	22.50		170.50	197
2400	45° Elbow, slip fit									
2410	2-1/8″ dia.	Q-9	70	.229	Ea.	8.95	9.70		18.65	25
2420	2-1/2″ dia.		65	.246		13.20	10.45		23.65	30.50
2430	3″ dia.		60	.267		15.95	11.35		27.30	35
2440	3-1/2″ dia.		55	.291		20	12.35		32.35	41
2450	4″ dia.		50	.320		26	13.60		39.60	49.50
2460	5″ dia.		45	.356		44	15.10		59.10	71.50
2470	6″ dia.		40	.400		59	17		76	91
2480	8″ dia.		35	.457		116	19.40		135.40	158
2800	90° TY, slip fit thru 6″ dia									
2810	2-1/8″ dia.	Q-9	42	.381	Ea.	20	16.20		36.20	47
2820	2-1/2″ dia.		39	.410		26	17.45		43.45	55.50
2830	3″ dia.		36	.444		34.50	18.90		53.40	67
2840	3-1/2″ dia.		33	.485		44	20.50		64.50	79.50
2850	4″ dia.		30	.533		63	22.50		85.50	104
2860	5″ dia.		27	.593		120	25		145	171
2870	6″ dia.		24	.667		163	28.50		191.50	223
2880	8″ dia., butt end					345			345	380
2890	10″ dia., butt end					625			625	685

44 11 16 – Industrial Dust Collectors

44 11 16.10 Dust Collection Systems		Crew	Daily Output	Labor-Hours	Unit	2009 Bare Costs Material	Labor	Equipment	Total	Total Incl O&P
2900	12″ dia., butt end				Ea.	625			625	685
2910	14″ dia., butt end					1,125			1,125	1,250
2920	6″ x 4″ dia., butt end					99.50			99.50	109
2930	8″ x 4″ dia., butt end					188			188	207
2940	10″ x 4″ dia., butt end					247			247	271
2950	12″ x 4″ dia., butt end				↓	283			283	310
3100	90° Elbow, butt end segmented									
3110	8″ dia., butt end, segmented				Ea.	148			148	162
3120	10″ dia., butt end, segmented					365			365	405
3130	12″ dia., butt end, segmented					465			465	510
3140	14″ dia., butt end, segmented				↓	580			580	640
3200	45° Elbow, butt end segmented									
3210	8″ dia., butt end, segmented				Ea.	116			116	128
3220	10″ dia., butt end, segmented					265			265	292
3230	12″ dia., butt end, segmented					272			272	299
3240	14″ dia., butt end, segmented				↓	335			335	365
3400	All butt end fittings require one coupling per joint.									
3410	Labor for fitting included with couplings.									
3460	Compression coupling, galvanized, neoprene gasket									
3470	2-1/8″ dia.	Q-9	44	.364	Ea.	11.85	15.45		27.30	37
3480	2-1/2″ dia.		44	.364		11.85	15.45		27.30	37
3490	3″ dia.		38	.421		17.10	17.90		35	46.50
3500	3-1/2″ dia.		35	.457		19.40	19.40		38.80	51.50
3510	4″ dia.		33	.485		21	20.50		41.50	54.50
3520	5″ dia.		29	.552		24	23.50		47.50	62.50
3530	6″ dia.		26	.615		24	26		50	66.50
3540	8″ dia.		22	.727		50	31		81	103
3550	10″ dia.		20	.800		67.50	34		101.50	127
3560	12″ dia.		18	.889		83.50	38		121.50	150
3570	14″ dia.	↓	16	1	↓	107	42.50		149.50	184
3800	Air gate valves, galvanized									
3810	2-1/8″ dia.	Q-9	30	.533	Ea.	173	22.50		195.50	225
3820	2-1/2″ dia.		28	.571		191	24.50		215.50	248
3830	3″ dia.		26	.615		209	26		235	270
3840	4″ dia.		23	.696		263	29.50		292.50	335
3850	6″ dia.	↓	18	.889	↓	355	38		393	450

44 11 26 – Electrostatic Precipitators

44 11 26.20 Electrostatic Precipitators		Crew	Daily Output	Labor-Hours	Unit	Material	Labor	Equipment	Total	Total Incl O&P
0010	**ELECTROSTATIC PRECIPITATORS**									
4100	Smoke pollution control									
4110	Exhaust air volume									
4120	1 to 2.4 MCFM	Q-10	2.15	11.163	MCFM	7,725	490		8,215	9,250
4130	2.4 to 4.8 MCFM		2.70	8.889		8,800	390		9,190	10,300
4140	4.8 to 7.2 MCFM		3.60	6.667		8,950	294		9,244	10,300
4150	7.2 to 12 MCFM	↓	5.40	4.444	↓	9,375	196		9,571	10,600
4400	Electrostatic precipitators with built-in exhaust fans									
4410	Smoke pollution control									
4420	Exhaust air volume									
4440	1 to 2.4 MCFM	Q-10	2.15	11.163	MCFM	9,300	490		9,790	11,000
4450	2.4 to 4.8 MCFM		2.70	8.889		7,375	390		7,765	8,700
4460	4.8 to 7.2 MCFM		3.60	6.667		6,325	294		6,619	7,400
4470	7.2 to 12 MCFM	↓	5.40	4.444	↓	5,450	196		5,646	6,300

44 42 General Water Treatment Equipment

44 42 56 – Water Treatment Pumps

44 42 56.10 Pumps, Pneumatic Ejector		Crew	Daily Output	Labor-Hours	Unit	Material	Labor	Equipment	Total	Total Incl O&P
						2009 Bare Costs				
0010	**PUMPS, PNEUMATIC EJECTOR**									
0020	With steel receiver, level controls, inlet/outlet gate/check valves									
0030	Cross connect. not incl. compressor, fittings or piping									
0040	Duplex									
0050	30 GPM	Q-2	1.70	14.118	Ea.	20,500	640		21,140	23,500
0060	50 GPM		1.56	15.385		24,000	700		24,700	27,500
0070	100 GPM		1.30	18.462		33,200	840		34,040	37,800
0080	150 GPM		1.10	21.818		61,000	995		61,995	68,500
0090	200 GPM	↓	.85	28.235		86,000	1,275		87,275	96,500
0100	250 GPM	Q-3	.91	35.165		98,000	1,625		99,625	110,500
0110	300 GPM	Q-3	.57	56.140	↓	133,000	2,600		135,600	150,500

Assemblies Section

Table of Contents

How to Use the Assemblies Cost Tables

The following is a detailed explanation of a sample Assemblies Cost Table. Most Assembly Tables are separated into three parts: 1) an illustration of the system to be estimated; 2) the components and related costs of a typical system; and 3) the costs for similar systems with dimensional and/or size variations. For costs of the components that comprise these systems, or assemblies, refer to the Unit Price Section. Next to each bold number below is the described item with the appropriate component of the sample entry following in parentheses. In most cases, if the work is to be subcontracted, the general contractor will need to add an additional markup (RSMeans suggests using 10%) to the "Total" figures.

1 *System/Line Numbers (D3010 510 1760)*

Each Assemblies Cost Line has been assigned a unique identification number based on the UniFormat classification sytem.

UNIFORMAT II Major Group

D3010 510 1760

UNIFORMAT II Level 3 (D3010)

Means Major Classification (510)

Means Individual Line Number (1760)

D30 HVAC

D3010 Energy Supply 1

2

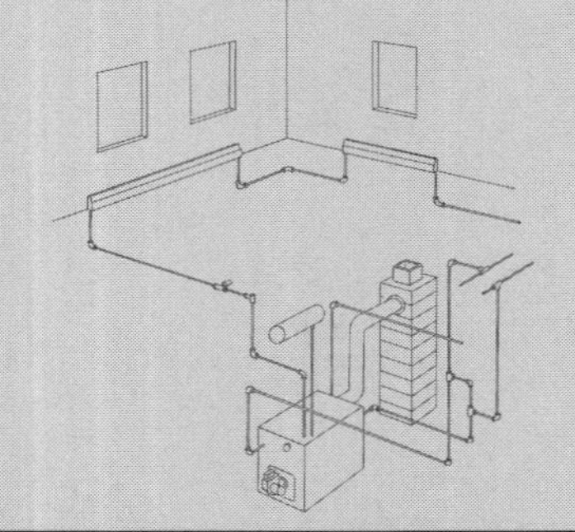

Basis for Heat Loss Estimate, Apartment Type Structures:

1. Masonry walls and flat roof are insulated. U factor is assumed at .08.
2. Window glass area taken as BOCA minimum, 1/10th of floor area. Double insulating glass with 1/4" air space, U = .65.
3. Infiltration = 0.3 C.F. per hour per S.F. of net wall.
4. Concrete floor loss is 2 BTUH per S.F.
5. Temperature difference taken as 70° F.
6. Ventilating or makeup air has not been included and must be added if desired. Air shafts are not used.

System Components 3	QUANTITY 4	UNIT 5	COST EACH 6 MAT. 7	COST EACH INST. 8	COST EACH TOTAL 9
SYSTEM D3010 510 1760 HEATING SYSTEM, FIN TUBE RADIATION, FORCED HOT WATER 1,000 S.F. AREA, 10,000 C.F. VOLUME					
Boiler, oil fired, CI, burner, ctrls/insul/breech/pipe/ftng/valves, 109 MBH	1.000	Ea.	3,631.25	3,281.25	6,912.50
Circulating pump, CI flange connection, 1/12 HP	1.000	Ea.	315	176	491
Expansion tank, painted steel, ASME 18 Gal capacity	1.000	Ea.	520	76	596
Storage tank, steel, above ground, 275 Gal capacity w/supports	1.000	Ea.	420	213	633
Copper tubing type L, solder joint, hanger 10' OC, 3/4" diam	100.000	L.F.	660	770	1,430
Radiation, 3/4" copper tube w/alum fin baseboard pkg, 7" high	30.000	L.F.	252	552	804
Pipe covering, calcium silicate w/cover, 1" wall, 3/4" diam	100.000	L.F.	282	585	867
TOTAL			6,080.25	5,653.25	11,733.50
COST PER S.F.			6.08	5.65	11.73

D3010 510	Apartment Building Heating - Fin Tube Radiation		COST PER S.F. MAT.	COST PER S.F. INST.	COST PER S.F. TOTAL
1740	Heating systems, fin tube radiation, forced hot water				
1760	1,000 S.F. area, 10,000 C.F. volume		6.08	5.65	11.73
1800	10,000 S.F. area, 100,000 C.F. volume	R235000 -10	2.84	3.42	6.26
1840	20,000 S.F. area, 200,000 C.F. volume		3.31	3.83	7.14
1880	30,000 S.F. area, 300,000 C.F. volume	R235000 -20	3.19	3.72	6.91
1890					

2

Illustration

At the top of most assembly pages are an illustration, a brief description, and the design criteria used to develop the cost.

3

System Components

The components of a typical system are listed separately to show what has been included in the development of the total system price. The table below contains prices for other similar systems with dimensional and/or size variations.

4

Quantity

This is the number of line item units required for one system unit. For example, we assume that it will take 30 linear feet of radiation for the 1,000 S.F. area shown.

5

Unit of Measure for Each Item

The abbreviated designation indicates the unit of measure, as defined by industry standards, upon which the price of the component is based. For example, baseboard radiation is priced by the linear foot. For a complete listing of abbreviations, see the Reference Section.

6

Unit of Measure for Each System (Each)

Costs shown in the three right-hand columns have been adjusted by the component quantity and unit of measure for the entire system. In this example, "Cost Each" is the unit of measure for this system, or assembly.

7

Materials (6,080.25)

This column contains the Materials Cost of each component. These cost figures are bare costs plus 10% for profit.

8

Installation (5,653.25)

Installation includes labor and equipment plus the installing contractor's overhead and profit. Equipment costs are the bare rental costs plus 10% for profit. The labor overhead and profit is defined on the inside back cover of this book.

9

Total (11,733.50)

The figure in this column is the sum of the material and installation costs.

Material Cost	+	Installation Cost	=	Total
$6,080.25	+	$5,653.25	=	$11,733.50

A Substructure

A2020 Basement Walls

A2020 220	Subdrainage Piping	COST PER L.F.		
		MAT.	INST.	TOTAL
2000	Piping, excavation & backfill excluded, PVC, perforated			
2110	3″ diameter	1.83	3.51	5.34
2130	4″ diameter	1.83	3.51	5.34
2140	5″ diameter	3.74	3.76	7.50
2150	6″ diameter	3.74	3.76	7.50
3000	Metal alum. or steel, perforated asphalt coated			
3150	6″ diameter	5.15	7.30	12.45
3160	8″ diameter	7.85	7.50	15.35
3170	10″ diameter	9.80	7.70	17.50
3180	12″ diameter	11	9.75	20.75
3220	18″ diameter	16.45	13.55	30
4000	Porous wall concrete			
4130	4″ diameter	3.63	3.92	7.55
4150	6″ diameter	4.72	4.17	8.89
4160	8″ diameter	5.85	5.50	11.35
4180	12″ diameter	12.35	6	18.35
4200	15″ diameter	14.15	7.45	21.60
4220	18″ diameter	18.70	10.35	29.05

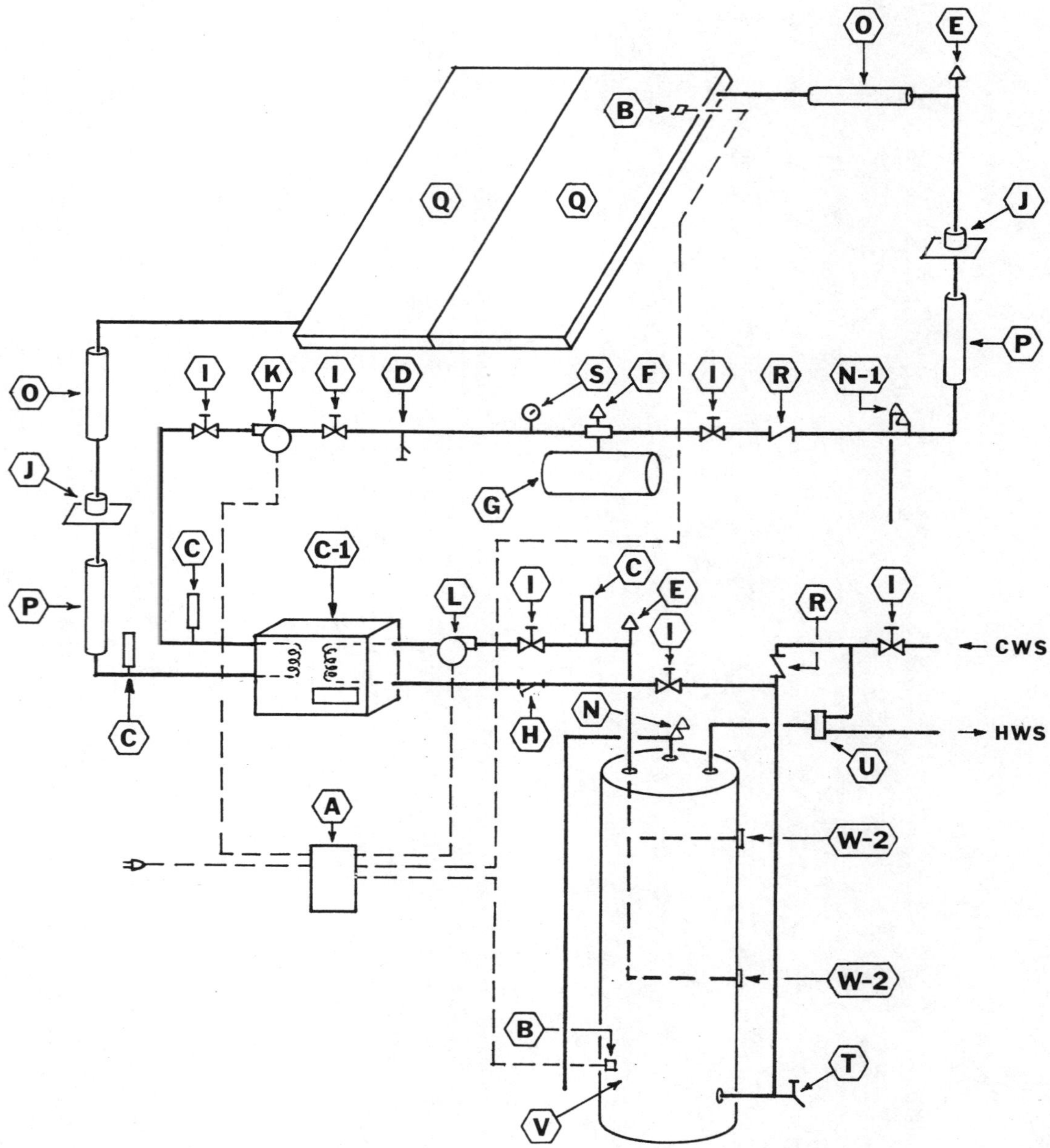

In this closed-loop indirect collection system, fluid with a low freezing temperature, such as propylene glycol, transports heat from the collectors to water storage. The transfer fluid is contained in a closedloop consisting of collectors, supply and return piping, and a remote heat exchanger. The heat exchanger transfers heat energy from the fluid in the collector loop to potable water circulated in a storage loop. A typical two-or-three panel system contains 5 to 6 gallons of heat transfer fluid.

When the collectors become approximately 20° F warmer than the storage temperature, a controller activates the circulator on the collector and storage loops. The circulators will move the fluid and potable water through the heat exchanger until heat collection no longer occurs. At that point, the system shuts down. Since the heat transfer medium is a fluid with a very low freezing temperature, there is no need for it to be drained from the system between periods of collection.

D2020 Domestic Water Distribution

System Components

		QUANTITY	UNIT	COST EACH MAT.	COST EACH INST.	COST EACH TOTAL
	SYSTEM D2020 265 2760					
	SOLAR, CLOSED LOOP, ADD-ON HOT WATER SYS., EXTERNAL HEAT EXCHANGER					
	3/4" TUBING, TWO 3'X7' BLACK CHROME COLLECTORS					
A,B,G,L,K,M	Heat exchanger fluid-fluid pkg incl 2 circulators, expansion tank, Check valve, relief valve, controller, hi temp cutoff, & 2 sensors	1.000	Ea.	765	420	1,185
C	Thermometer, 2" dial	3.000	Ea.	66	109.50	175.50
D, T	Fill & drain valve, brass, 3/4" connection	1.000	Ea.	8.85	24.50	33.35
E	Air vent, manual, 1/8" fitting	2.000	Ea.	6.12	36.60	42.72
F	Air purger	1.000	Ea.	47.50	49	96.50
H	Strainer, Y type, bronze body, 3/4" IPS	1.000	Ea.	24	31	55
I	Valve, gate, bronze, NRS, soldered 3/4" diam	6.000	Ea.	234	177	411
J	Neoprene vent flashing	2.000	Ea.	23.80	59	82.80
N-1, N	Relief valve temp & press, 150 psi 210°F self-closing 3/4" IPS	1.000	Ea.	16.70	19.50	36.20
O	Pipe covering, urethane, ultraviolet cover, 1" wall 3/4" diam	20.000	L.F.	52.40	108	160.40
P	Pipe covering, fiberglass, all service jacket, 1" wall, 3/4" diam	50.000	L.F.	53.50	216.50	270
Q	Collector panel solar energy blk chrome on copper, 1/8" temp glass 3'x7'	2.000	Ea.	1,460	222	1,682
	Roof clamps for solar energy collector panels	2.000	Set	5.18	30.10	35.28
R	Valve, swing check, bronze, regrinding disc, 3/4" diam	2.000	Ea.	140	59	199
S	Pressure gauge, 60 psi, 2" dial	1.000	Ea.	24.50	18.30	42.80
U	Valve, water tempering, bronze, sweat connections, 3/4" diam	1.000	Ea.	95.50	29.50	125
W-2, V	Tank water storage w/heating element, drain, relief valve, existing	1.000	Ea.			
	Copper tubing type L, solder joint, hanger 10' OC 3/4" diam	20.000	L.F.	132	154	286
	Copper tubing, type M, solder joint, hanger 10' OC 3/4" diam	70.000	L.F.	339.50	525	864.50
	Sensor wire, #22-2 conductor multistranded	.500	C.L.F.	8.45	28	36.45
	Solar energy heat transfer fluid, propylene glycol anti-freeze	6.000	Gal.	91.50	126	217.50
	Wrought copper fittings & solder, 3/4" diam	76.000	Ea.	231.04	2,356	2,587.04
	TOTAL			3,825.54	4,798.50	8,624.04

D2020 265	Solar, Closed Loop, Add-On Hot Water Systems		COST EACH MAT.	COST EACH INST.	COST EACH TOTAL
2550	Solar, closed loop, add-on hot water system, external heat exchanger				
2570	3/8" tubing, 3 ea 4' x 4'-4" vacuum tube collectors		5,000	4,475	9,475
2580	1/2" tubing, 4 ea 4 x 4'-4" vacuum tube collectors	R235616-60	5,700	4,850	10,550
2600	2 ea 3'x7' black chrome collectors		3,425	4,575	8,000
2620	3 ea 3'x7' black chrome collectors		4,150	4,700	8,850
2640	2 ea 3'x7' flat black collectors		3,200	4,600	7,800
2660	3 ea 3'x7' flat black collectors		3,825	4,725	8,550
2700	3/4" tubing, 3 ea 3'x7' black chrome collectors		4,550	4,925	9,475
2720	3 ea 3'x7' flat black absorber plate collectors		4,225	4,950	9,175
2740	2 ea 4'x9' flat black w/plastic glazing collectors		3,875	4,950	8,825
2760	2 ea 3'x7' black chrome collectors		3,825	4,800	8,625
2780	1" tubing, 4 ea 2'x9' plastic absorber & glazing collectors		5,325	5,550	10,875
2800	4 ea 3'x7' black chrome absorber collectors		6,050	5,575	11,625
2820	4 ea 3'x7' flat black absorber collectors		5,625	5,600	11,225

D2020 Domestic Water Distribution

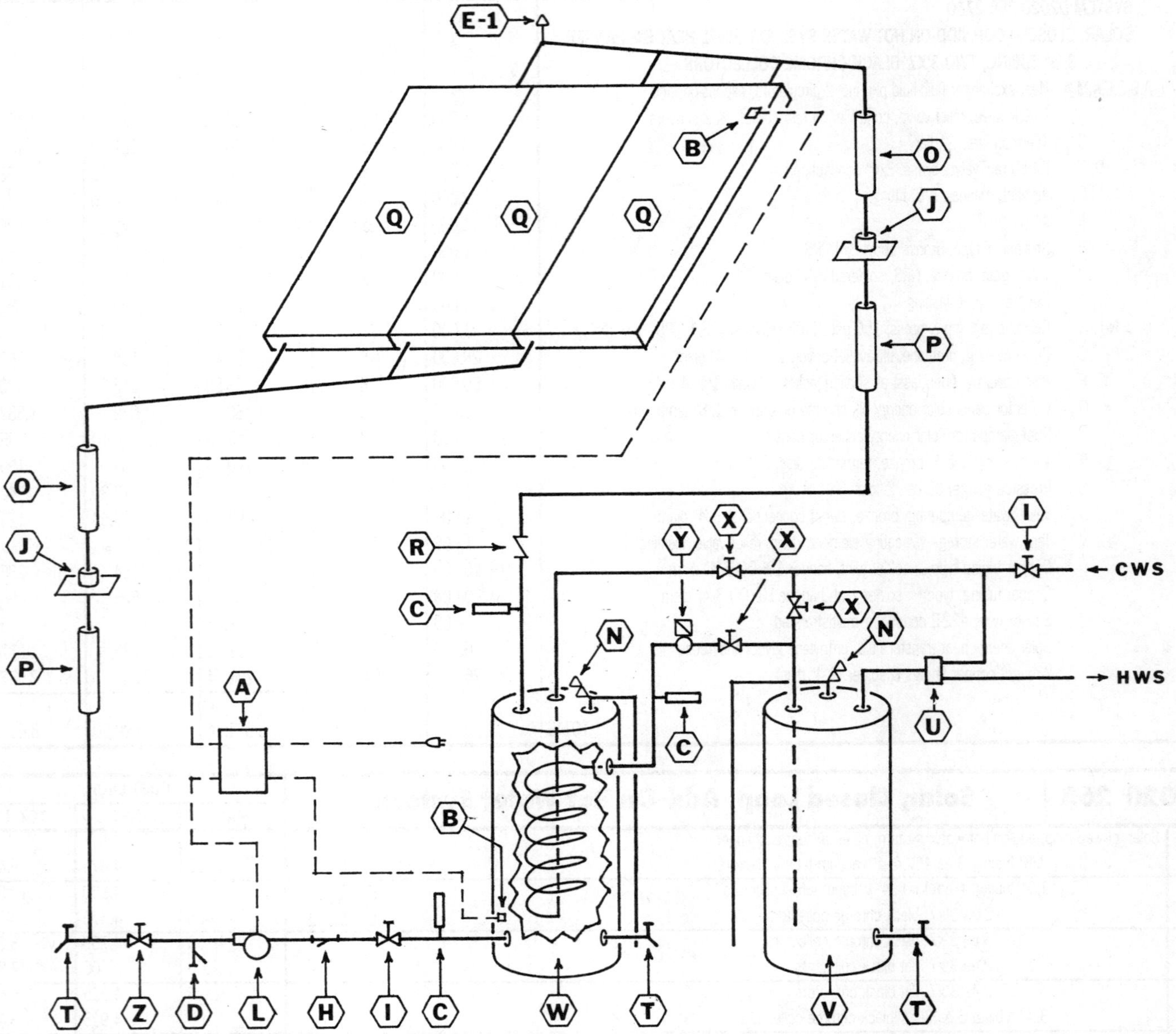

In the drainback indirect-collection system, the heat transfer fluid is distilled water contained in a loop consisting of collectors, supply and return piping, and an unpressurized holding tank. A large heat exchanger containing incoming potable water is immersed in the holding tank. When a controller activates solar collection, the distilled water is pumped through the collectors and heated and pumped back down to the holding tank. When the temperature differential between the water in the collectors and water in storage is such that collection no longer occurs, the pump turns off and gravity causes the distilled water in the collector loop to drain back to the holding tank. All the loop piping is pitched so that the water can drain out of the collectors and piping and not freeze there. As hot water is needed in the home, incoming water first flows through the holding tank with the immersed heat exchanger and is warmed and then flows through a conventional heater for any supplemental heating that is necessary.

D2020 Domestic Water Distribution

System Components		QUANTITY	UNIT	COST EACH MAT.	COST EACH INST.	COST EACH TOTAL
	SYSTEM D2020 270 2760					
	SOLAR, DRAINBACK, ADD ON, HOT WATER, IMMERSED HEAT EXCHANGER					
	3/4″ TUBING, THREE EA 3′X7′ BLACK CHROME COLLECTOR					
A, B	Differential controller 2 sensors, thermostat, solar energy system	1.000	Ea.	380	49	429
C	Thermometer 2″ dial	3.000	Ea.	66	109.50	175.50
D, T	Fill & drain valve, brass, 3/4″ connection	1.000	Ea.	8.85	24.50	33.35
E-1	Automatic air vent 1/8″ fitting	1.000	Ea.	13.95	18.30	32.25
H	Strainer, Y type, bronze body, 3/4″ IPS	1.000	Ea.	24	31	55
I	Valve, gate, bronze, NRS, soldered 3/4″ diam	2.000	Ea.	78	59	137
J	Neoprene vent flashing	2.000	Ea.	23.80	59	82.80
L	Circulator, solar heated liquid, 1/20 HP	1.000	Ea.	170	88	258
N	Relief valve temp. & press. 150 psi 210°F self-closing 3/4″ IPS	1.000	Ea.	16.70	19.50	36.20
O	Pipe covering, urethane, ultraviolet cover, 1″ wall, 3/4″ diam	20.000	L.F.	52.40	108	160.40
P	Pipe covering, fiberglass, all service jacket, 1″ wall, 3/4″ diam	50.000	L.F.	53.50	216.50	270
Q	Collector panel solar energy blk chrome on copper, 1/8″ temp glas 3′x7′	3.000	Ea.	2,190	333	2,523
	Roof clamps for solar energy collector panels	3.000	Set	7.77	45.15	52.92
R	Valve, swing check, bronze, regrinding disc, 3/4″ diam	1.000	Ea.	70	29.50	99.50
U	Valve, water tempering, bronze sweat connections, 3/4″ diam	1.000	Ea.	95.50	29.50	125
V	Tank, water storage w/heating element, drain, relief valve, existing	1.000	Ea.			
W	Tank, water storage immersed heat exchr elec 2″x1/2# insul 120 gal	1.000	Ea.	1,275	420	1,695
X	Valve, globe, bronze, rising stem, 3/4″ diam, soldered	3.000	Ea.	297	88.50	385.50
Y	Flow control valve	1.000	Ea.	112	26.50	138.50
Z	Valve, ball, bronze, solder 3/4″ diam, solar loop flow control	1.000	Ea.	18.65	29.50	48.15
	Copper tubing, type L, solder joint, hanger 10′ OC 3/4″ diam	20.000	L.F.	132	154	286
	Copper tubing, type M, solder joint, hanger 10′ OC 3/4″ diam	70.000	L.F.	339.50	525	864.50
	Sensor wire, #22-2 conductor, multistranded	.500	C.L.F.	8.45	28	36.45
	Wrought copper fittings & solder, 3/4″ diam	76.000	Ea.	231.04	2,356	2,587.04
	TOTAL			5,664.11	4,846.95	10,511.06

D2020 270	Solar, Drainback, Hot Water Systems	COST EACH MAT.	COST EACH INST.	COST EACH TOTAL
2550	Solar, drainback, hot water, immersed heat exchanger R235616-60			
2560	3/8″ tubing, 3 ea. 4′ x 4′-4″ vacuum tube collectors	5,900	4,350	10,250
2580	1/2″ tubing, 4 ea. 4′ x 4′-4″ vacuum tube collectors, 80 gal tank	6,650	4,725	11,375
2600	120 gal tank	6,775	4,775	11,550
2640	2 ea. 3′x7′ blk chrome collectors, 80 gal tank	4,350	4,450	8,800
2660	3 ea. 3′x7′ blk chrome collectors, 120 gal tank	5,225	4,625	9,850
2700	2 ea. 3′x7′ flat blk collectors, 120 gal tank	4,275	4,525	8,800
2720	3 ea. 3′x7′ flat blk collectors, 120 gal tank	4,875	4,650	9,525
2760	3/4″ tubing, 3 ea. 3′x7′ black chrome collectors, 120 gal tank	5,675	4,850	10,525
2780	3 ea. 3′x7′ flat black absorber collectors, 120 gal tank	5,325	4,875	10,200
2800	2 ea. 4′x9′ flat blk w/plastic glazing collectors 120 gal tank	4,975	4,875	9,850
2840	1″ tubing, 4 ea. 2′x9′ plastic absorber & glazing collectors, 120 gal tank	6,500	5,475	11,975
2860	4 ea. 3′x7′ black chrome absorber collectors, 120 gal tank	7,250	5,500	12,750
2880	4 ea. 3′x7′ flat black absorber collectors, 120 gal tank	6,800	5,525	12,325

D2020 Domestic Water Distribution

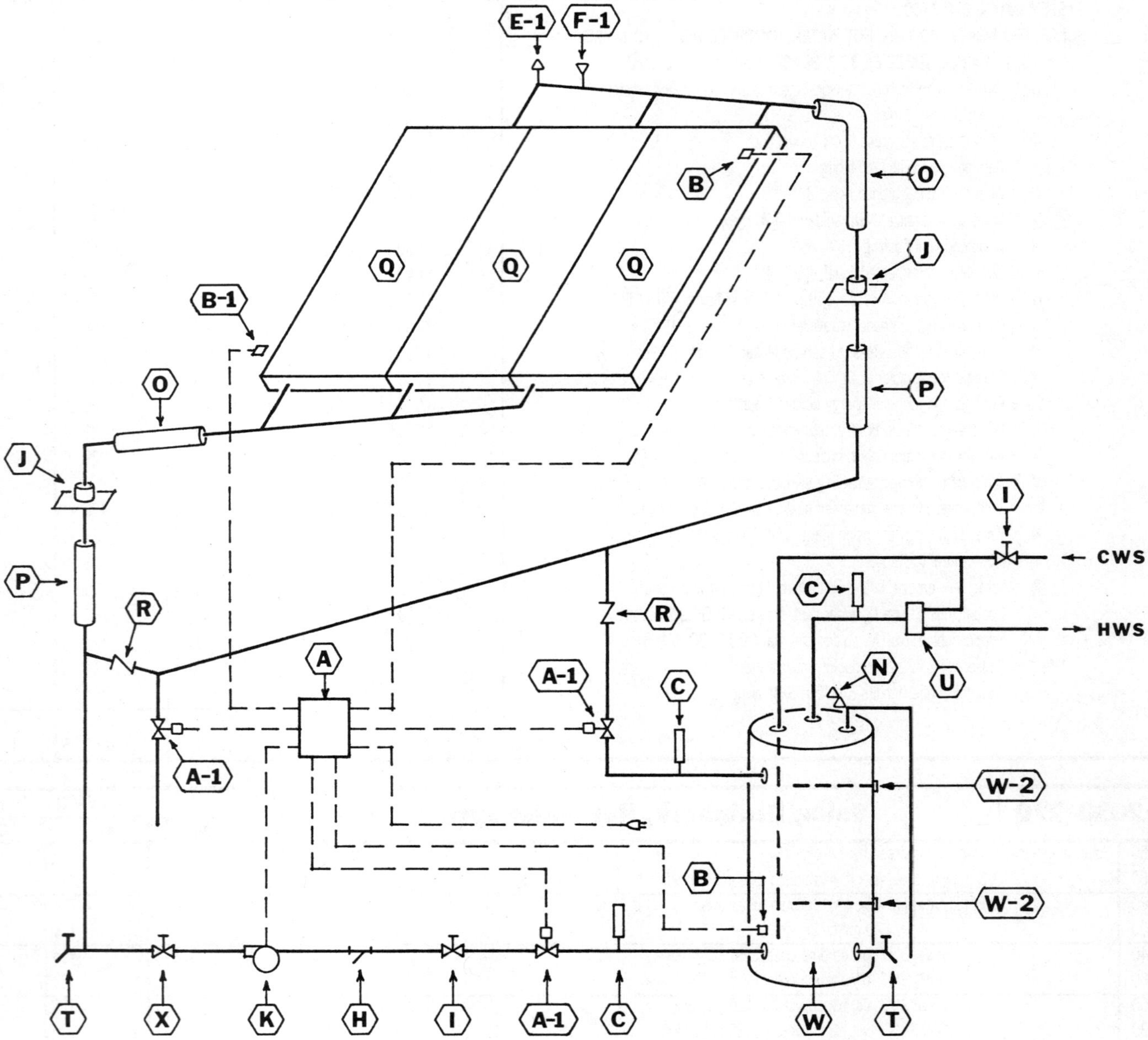

In the draindown direct-collection system, incoming domestic water is heated in the collectors. When the controller activates solar collection, domestic water is first heated as it flows through the collectors and is then pumped to storage. When conditions are no longer suitable for heat collection, the pump shuts off and the water in the loop drains down and out of the system by means of solenoid valves and properly pitched piping.

D2020 Domestic Water Distribution

System Components		QUANTITY	UNIT	COST EACH MAT.	COST EACH INST.	COST EACH TOTAL
	SYSTEM D2020 275 2760					
	SOLAR, DRAINDOWN, HOT WATER, DIRECT COLLECTION					
	3/4" TUBING, THREE 3'X7' BLACK CHROME COLLECTORS					
A, B	Differential controller, 2 sensors, thermostat, solar energy system	1.000	Ea.	380	49	429
A-1	Solenoid valve, solar heating loop, brass, 3/4" diam, 24 volts	3.000	Ea.	339	195	534
B-1	Solar energy sensor, freeze prevention	1.000	Ea.	24	18.30	42.30
C	Thermometer, 2" dial	3.000	Ea.	66	109.50	175.50
E-1	Vacuum relief valve, 3/4" diam	1.000	Ea.	29	18.30	47.30
F-1	Air vent, automatic, 1/8" fitting	1.000	Ea.	13.95	18.30	32.25
H	Strainer, Y type, bronze body, 3/4" IPS	1.000	Ea.	24	31	55
I	Valve, gate, bronze, NRS, soldered, 3/4" diam	2.000	Ea.	78	59	137
J	Vent flashing neoprene	2.000	Ea.	23.80	59	82.80
K	Circulator, solar heated liquid, 1/25 HP	1.000	Ea.	250	75.50	325.50
N	Relief valve temp & press 150 psi 210° F self-closing 3/4" IPS	1.000	Ea.	16.70	19.50	36.20
O	Pipe covering, urethane, ultraviolet cover, 1" wall, 3/4" diam	20.000	L.F.	52.40	108	160.40
P	Pipe covering, fiberglass, all service jacket, 1" wall, 3/4" diam	50.000	L.F.	53.50	216.50	270
	Roof clamps for solar energy collector panels	3.000	Set	7.77	45.15	52.92
Q	Collector panel solar energy blk chrome on copper, 1/8" temp glass 3'x7'	3.000	Ea.	2,190	333	2,523
R	Valve, swing check, bronze, regrinding disc, 3/4" diam, soldered	2.000	Ea.	140	59	199
T	Drain valve, brass, 3/4" connection	2.000	Ea.	17.70	49	66.70
U	Valve, water tempering, bronze, sweat connections, 3/4" diam	1.000	Ea.	95.50	29.50	125
W-2, W	Tank, water storage elec elem 2"x1/2# insul 120 gal	1.000	Ea.	1,275	420	1,695
X	Valve, globe, bronze, rising stem, 3/4" diam, soldered	1.000	Ea.	99	29.50	128.50
	Copper tubing, type L, solder joints, hangers 10' OC 3/4" diam	20.000	L.F.	132	154	286
	Copper tubing, type M, solder joints, hangers 10' OC 3/4" diam	70.000	L.F.	339.50	525	864.50
	Sensor wire, #22-2 conductor, multistranded	.500	C.L.F.	8.45	28	36.45
	Wrought copper fittings & solder, 3/4" diam	76.000	Ea.	231.04	2,356	2,587.04
	TOTAL			5,886.31	5,005.05	10,891.36

D2020 275	Solar, Draindown, Hot Water Systems		COST EACH MAT.	COST EACH INST.	COST EACH TOTAL
2550	Solar, draindown, hot water				
2560	3/8" tubing, 3 ea 4' x 4'-4" vacuum tube collectors, 80 gal tank		5,975	4,775	10,750
2580	1/2" tubing, 4 ea 4' x 4'-4" vacuum tube collectors, 80 gal tank	R235616 -60	6,900	4,900	11,800
2600	120 gal tank		7,025	4,950	11,975
2640	2 ea 3'x7' black chrome collectors, 80 gal tank		4,625	4,625	9,250
2660	3 ea 3'x7' black chrome collectors, 120 gal tank		5,475	4,800	10,275
2700	2 ea 3'x7' flat black collectors, 120 gal tank		4,525	4,675	9,200
2720	3 ea 3'x7' flat black collectors, 120 gal tank		5,100	4,800	9,900
2760	3/4" tubing, 3 ea 3'x7' black chrome collectors, 120 gal tank		5,875	5,000	10,875
2780	3 ea 3'x7' flat collectors, 120 gal tank		5,550	5,025	10,575
2800	2 ea 4'x9' flat black & plastic glazing collectors, 120 gal tank		5,200	5,050	10,250
2840	1" tubing, 4 ea 2'x9' plastic absorber & glazing collectors, 120 gal tank		8,975	5,625	14,600
2860	4 ea 3'x7' black chrome absorber collectors, 120 gal tank		9,725	5,650	15,375
2880	4 ea 3'x7' flat black absorber collectors, 120 gal tank		9,275	5,675	14,950

D2020 Domestic Water Distribution

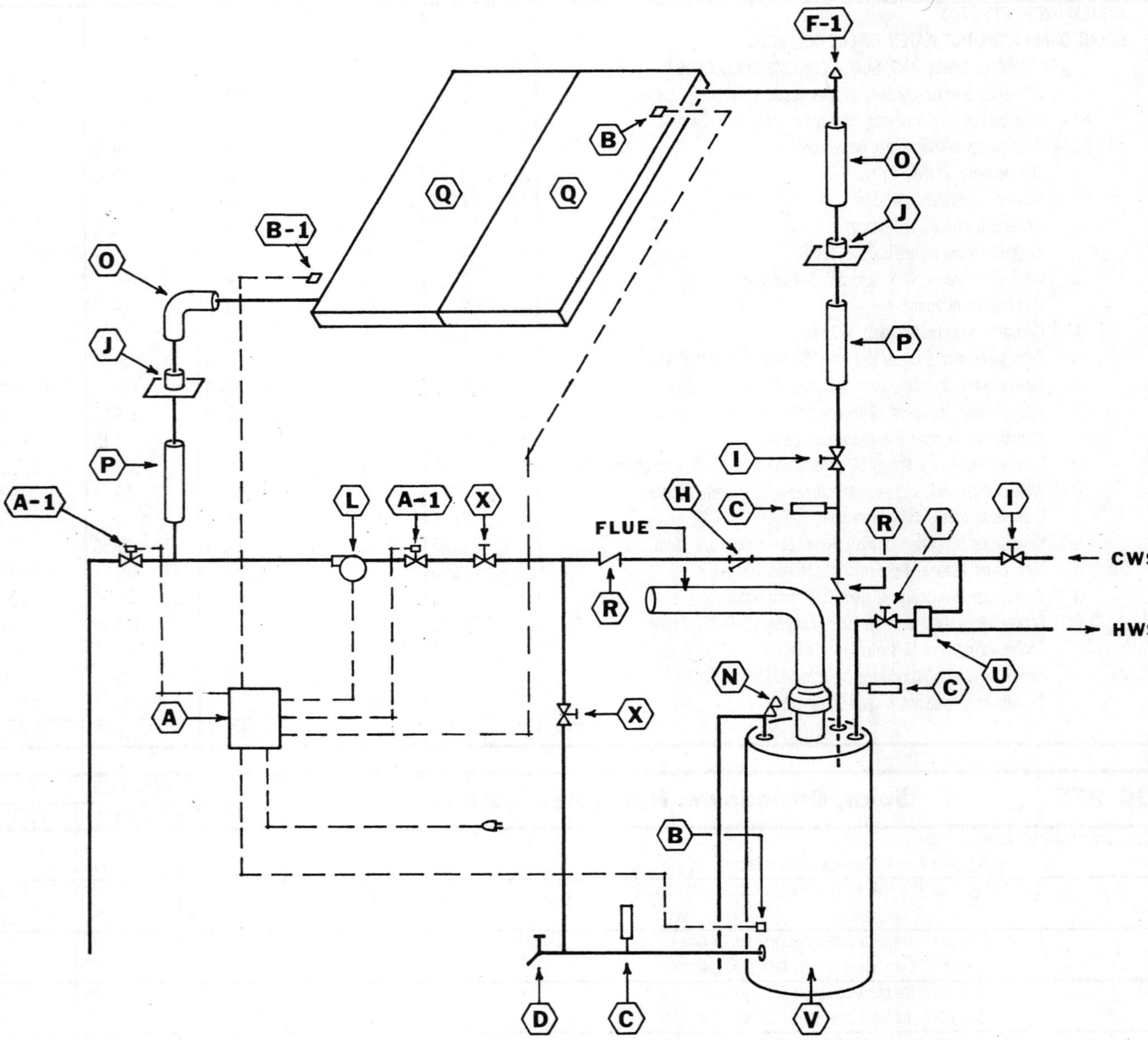

In the recirculation system, (a direct-collection system), incoming domestic water is heated in the collectors. When the controller activates solar collection, domestic water is heated as it flows through the collectors and then it flows back to storage. When conditions are not suitable for heat collection, the pump shuts off and the flow of the water stops. In this type of system, water remains in the collector loop at all times. A "frost sensor" at the collector activates the circulation of warm water from storage through the collectors when protection from freezing is required.

D2020 Domestic Water Distribution

System Components	QUANTITY	UNIT	COST EACH MAT.	COST EACH INST.	COST EACH TOTAL
SYSTEM D2020 280 2820					
SOLAR, RECIRCULATION, HOT WATER					
3/4" TUBING, TWO 3'X7' BLACK CHROME COLLECTORS					
A, B Differential controller 2 sensors, thermostat, for solar energy system	1.000	Ea.	380	49	429
A-1 Solenoid valve, solar heating loop, brass, 3/4" IPS, 24 volts	2.000	Ea.	226	130	356
B-1 Solar energy sensor freeze prevention	1.000	Ea.	24	18.30	42.30
C Thermometer, 2" dial	3.000	Ea.	66	109.50	175.50
D Drain valve, brass, 3/4" connection	1.000	Ea.	8.85	24.50	33.35
F-1 Air vent, automatic, 1/8" fitting	1.000	Ea.	13.95	18.30	32.25
H Strainer, Y type, bronze body, 3/4" IPS	1.000	Ea.	24	31	55
I Valve, gate, bronze, 125 lb, soldered 3/4" diam	3.000	Ea.	117	88.50	205.50
J Vent flashing, neoprene	2.000	Ea.	23.80	59	82.80
L Circulator, solar heated liquid, 1/20 HP	1.000	Ea.	170	88	258
N Relief valve, temp & press 150 psi 210° F self-closing 3/4" IPS	2.000	Ea.	33.40	39	72.40
O Pipe covering, urethane, ultraviolet cover, 1" wall, 3/4" diam	20.000	L.F.	52.40	108	160.40
P Pipe covering, fiberglass, all service jacket, 1" wall 3/4" diam	50.000	L.F.	53.50	216.50	270
Q Collector panel solar energy blk chrome on copper, 1/8" temp glass 3'x7'	2.000	Ea.	1,460	222	1,682
Roof clamps for solar energy collector panels	2.000	Set	5.18	30.10	35.28
R Valve, swing check, bronze, 125 lb, regrinding disc, soldered 3/4" diam	2.000	Ea.	140	59	199
U Valve, water tempering, bronze, sweat connections, 3/4" diam	1.000	Ea.	95.50	29.50	125
V Tank, water storage, w/heating element, drain, relief valve, existing	1.000	Ea.			
X Valve, globe, bronze, 125 lb, 3/4" diam	2.000	Ea.	198	59	257
Copper tubing, type L, solder joints, hangers 10' OC 3/4" diam	20.000	L.F.	132	154	286
Copper tubing, type M, solder joints, hangers 10' OC 3/4" diam	70.000	L.F.	339.50	525	364.50
Wrought copper fittings & solder, 3/4" diam	76.000	Ea.	231.04	2,356	2,587.04
Sensor wire, #22-2 conductor, multistranded	.500	C.L.F.	8.45	28	36.45
TOTAL			3,802.57	4,442.20	8,244.77

D2020 280	Solar Recirculation, Domestic Hot Water Systems	COST EACH MAT.	COST EACH INST.	COST EACH TOTAL
2550	Solar recirculation, hot water			
2560	3/8" tubing, 3 ea. 4' x 4'-4" vacuum tube collectors	4,900	4,125	9,025
2580	1/2" tubing, 4 ea. 4' x 4'-4" vacuum tube collectors	5,650	4,500	10,150
2640	2 ea. 3'x7' black chrome collectors (R235616-60)	3,350	4,225	7,575
2660	3 ea. 3'x7' black chrome collectors	4,100	4,350	8,450
2700	2 ea. 3'x7' flat black collectors	3,150	4,250	7,400
2720	3 ea. 3'x7' flat black collectors	3,775	4,375	8,150
2760	3/4" tubing, 3 ea. 3'x7' black chrome collectors	4,525	4,575	9,100
2780	3 ea. 3'x7' flat black absorber plate collectors	4,200	4,575	8,775
2800	2 ea. 4'x9' flat black w/plastic glazing collectors	3,850	4,600	8,450
2820	2 ea. 3'x7' black chrome collectors	3,800	4,450	8,250
2840	1" tubing, 4 ea. 2'x9' black plastic absorber & glazing collectors	6,925	5,200	12,125
2860	4 ea. 3'x7' black chrome absorber collectors	7,650	5,225	12,875
2880	4 ea. 3'x7' flat black absorber collectors	7,225	5,250	12,475

D2020 Domestic Water Distribution

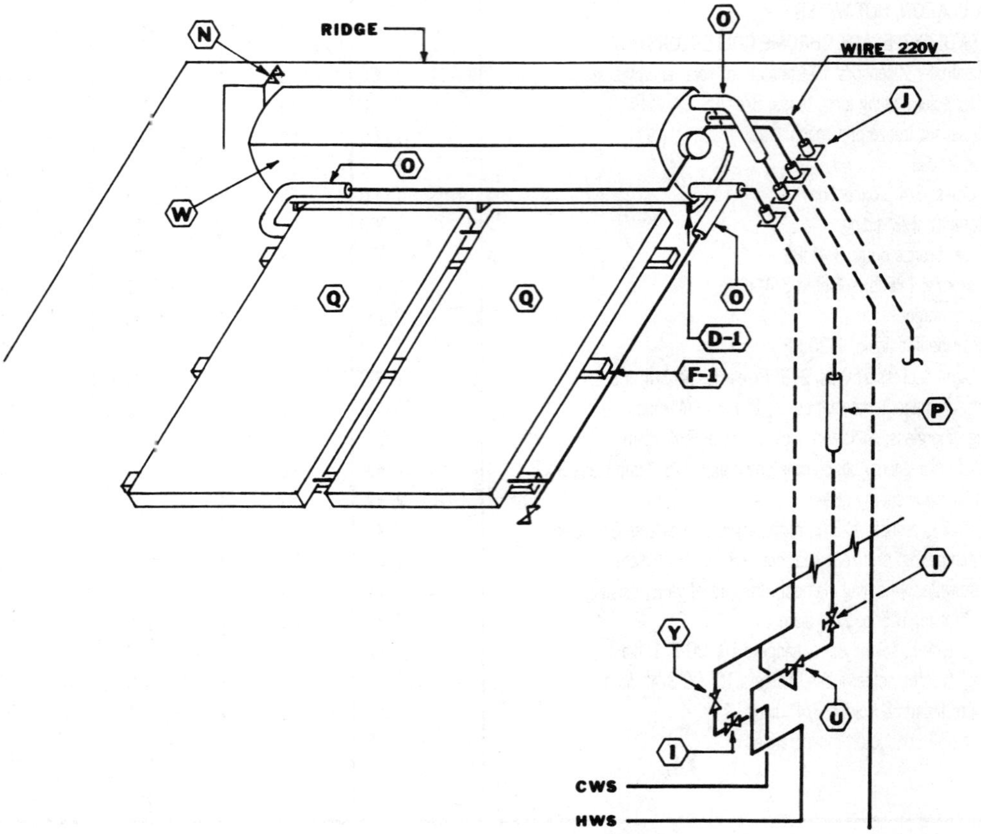

The thermosyphon domestic hot water system, a direct collection system, operates under city water pressure and does not require pumps for system operation. An insulated water storage tank is located above the collectors. As the sun heats the collectors, warm water in them rises by means of natural convection; the colder water in the storage tank flows into the collectors by means of gravity. As long as the sun is shining the water continues to flow through the collectors and to become warmer.

To prevent freezing, the system must be drained or the collectors covered with an insulated lid when the temperature drops below 32°F.

D2020 Domestic Water Distribution

	System Components	QUANTITY	UNIT	COST EACH MAT.	COST EACH INST.	COST EACH TOTAL
	SYSTEM D2020 285 0960					
	SOLAR, THERMOSYPHON, WATER HEATER					
	3/4″ TUBING, TWO 3′X7′ BLACK CHROME COLLECTORS					
D-1	Framing lumber, fir, 2″ x 6″ x 8′, tank cradle	.008	M.B.F.	4.60	6.36	10.96
F-1	Framing lumber, fir, 2″ x 4″ x 24′, sleepers	.016	M.B.F.	7.68	19.20	26.88
I	Valve, gate, bronze, 125 lb, soldered 1/2″ diam	2.000	Ea.	68	49	117
J	Vent flashing, neoprene	4.000	Ea.	47.60	118	165.60
O	Pipe covering, urethane, ultraviolet cover, 1″ wall, 1/2″ diam	40.000	L.F.	90.80	212	302.80
P	Pipe covering fiberglass all service jacket 1″ wall 1/2″ diam	160.000	L.F.	156.80	664	820.80
Q	Collector panel solar, blk chrome on copper, 3/16″ temp glass 3′-6″ x 7.5′	2.000	Ea.	1,240	234	1,474
Y	Flow control valve, globe, bronze, 125#, soldered, 1/2″ diam	1.000	Ea.	73.50	24.50	98
U	Valve, water tempering, bronze, sweat connections, 1/2″ diam	1.000	Ea.	78	24.50	102.50
W	Tank, water storage, solar energy system, 80 Gal, 2″ x 1/2 lb insul	1.000	Ea.	1,150	365	1,515
	Copper tubing type L, solder joints, hangers 10′ OC 1/2″ diam	150.000	L.F.	631.50	1,087.50	1,719
	Copper tubing type M, solder joints, hangers 10′ OC 1/2″ diam	50.000	L.F.	153.50	347.50	501
	Sensor wire, #22-2 gauge multistranded	.500	C.L.F.	8.45	28	36.45
	Wrought copper fittings & solder, 1/2″ diam	75.000	Ea.	101.25	2,212.50	2,313.75
	TOTAL			3,811.68	5,392.06	9,203.74

D2020 285	Thermosyphon, Hot Water		MAT.	INST.	TOTAL
0960	Solar, thermosyphon, hot water, two collector system	R235616-60	3,800	5,400	9,200
0970					

D2020 Domestic Water Distribution

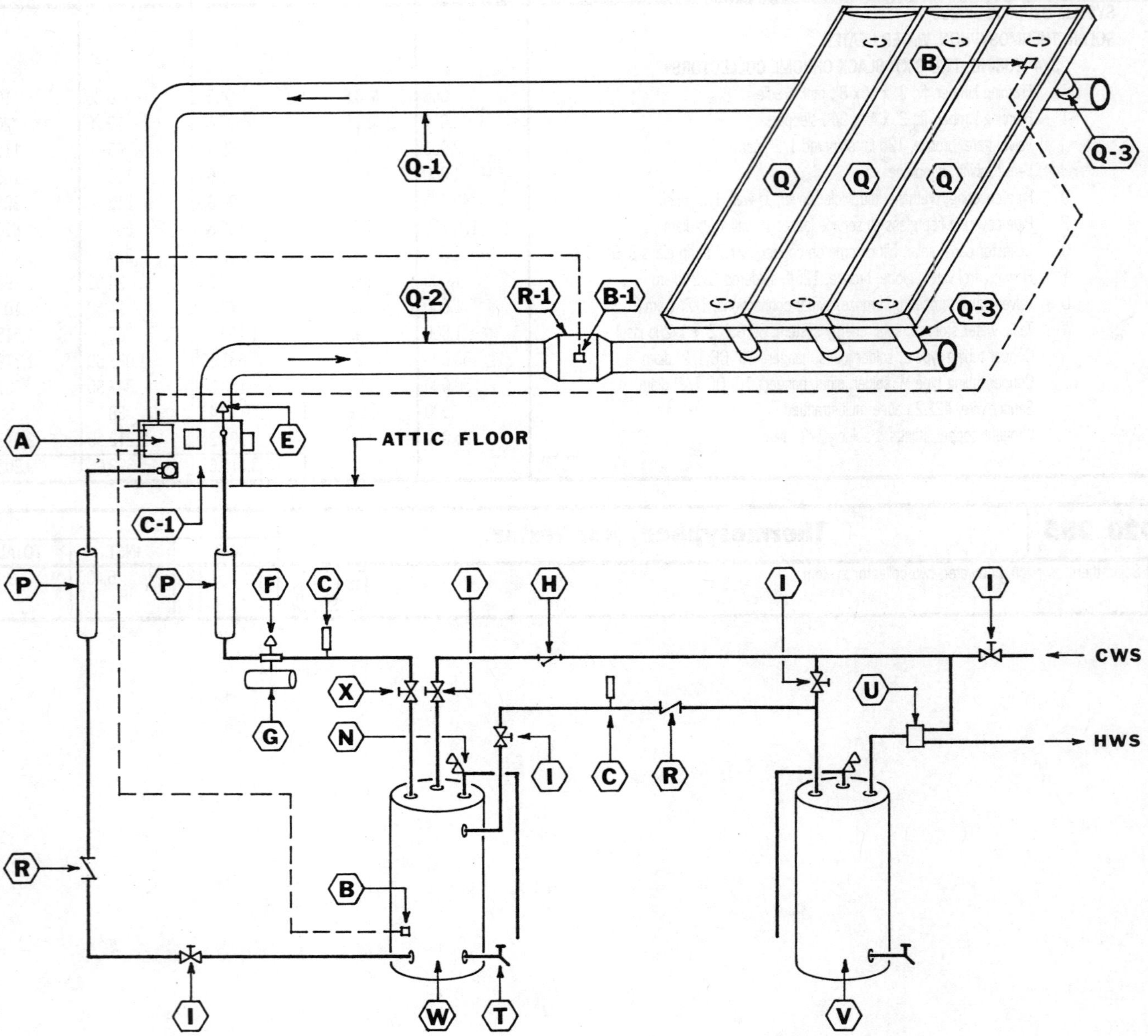

This domestic hot water pre-heat system includes heat exchanger with a circulating pump, blower, air-to-water, coil and controls, mounted in the upper collector manifold. Heat from the hot air coming out of the collectors is transferred through the heat exchanger. For each degree of DHW preheating gained, one degree less heating is needed from the fuel fired water heater. The system is simple, inexpensive to operate and can provide a substantial portion of DHW requirements for modest additional cost.

D2020 Domestic Water Distribution

System Components

System Components	QUANTITY	UNIT	COST EACH MAT.	COST EACH INST.	COST EACH TOTAL
SYSTEM D2020 290 2560					
SOLAR, HOT WATER, AIR TO WATER HEAT EXCHANGE					
THREE COLLECTORS, OPTICAL BLACK ON ALUMINUM, 10'X2',80GAL TANK					
A, B Differential controller, 2 sensors, thermostat, solar energy system	1.000	Ea.	380	49	429
C Thermometer, 2" dial	2.000	Ea.	44	73	117
C-1 Heat exchanger, air to fluid, up flow 70 MBH	1.000	Ea.	365	300	665
E Air vent, manual, for solar energy system 1/8" fitting	1.000	Ea.	3.06	18.30	21.36
F Air purger	1.000	Ea.	47.50	49	96.50
G Expansion tank	1.000	Ea.	67.50	18.30	85.80
H Strainer, Y type, bronze body, 1/2" IPS	1.000	Ea.	21	29.50	50.50
I Valve, gate, bronze, 125 lb, NRS, soldered 1/2" diam	5.000	Ea.	170	122.50	292.50
N Relief valve, temp & pressure solar 150 psi 210° F self-closing	1.000	Ea.	16.70	19.50	36.20
P Pipe covering, fiberglass, all service jacket, 1" wall, 1/2" diam	60.000	L.F.	58.80	249	307.80
Q Collector panel solar energy, air, black on alum plate, flush mount, 10'x2'	30.000	L.F.	18,150	327	18,477
B-1, R-1 Shutter damper for solar heater circulator	1.000	Ea.	54	87	141
B-1, R-1 Shutter motor for solar heater circulator	1.000	Ea.	131	65.50	196.50
R Backflow preventer for solar energy system, 1/2" diam	2.000	Ea.	128	73	201
T Drain valve, brass, 3/4" connection	1.000	Ea.	8.85	24.50	33.35
U Valve, water tempering, bronze, sweat connections, 1/2" diam	1.000	Ea.	78	24.50	102.50
W Tank, water storage, solar energy system, 80 Gal, 2" x 2 lb insul	1.000	Ea.	1,150	365	1,515
V Tank, water storage, w/heating element, drain, relief valve, existing	1.000	System			
X Valve, globe, bronze, 125 lb, rising stem, 1/2" diam	1.000	Ea.	73.50	24.50	98
Copper tubing type M, solder joints, hangers 10' OC 1/2" diam	50.000	L.F.	153.50	347.50	501
Copper tubing type L, solder joints, hangers 10' OC 1/2" diam	10.000	L.F.	42.10	72.50	114.60
Wrought copper fittings & solder, 1/2" diam	10.000	Ea.	13.50	295	308.50
Sensor wire, #22-2 conductor multistranded	.500	C.L.F.	8.45	28	36.45
Q-1, Q-2 Ductwork, fiberglass, aluminized jacket, 1-1/2" thick, 8" diam	32.000	S.F.	177.60	185.60	363.20
Q-3 Manifold for flush mount solar energy collector panels, air	6.000	L.F.	702	39.30	741.30
TOTAL			22,044.06	2,887	24,931.06

D2020 290	Solar Hot Water, Air To Water Heat Exchange	COST EACH MAT.	COST EACH INST.	COST EACH TOTAL
2550	Solar hot water, air to water heat exchange			
2560	Three collectors, optical black on aluminum, 10' x 2', 80 Gal tank	22,000	2,875	24,875
2580	Four collectors, optical black on aluminum, 10' x 2', 80 Gal tank R235616-60	28,400	3,050	31,450
2600	Four collectors, optical black on aluminum, 10' x 2', 120 Gal tank	28,500	3,100	31,600

D2020 Domestic Water Distribution

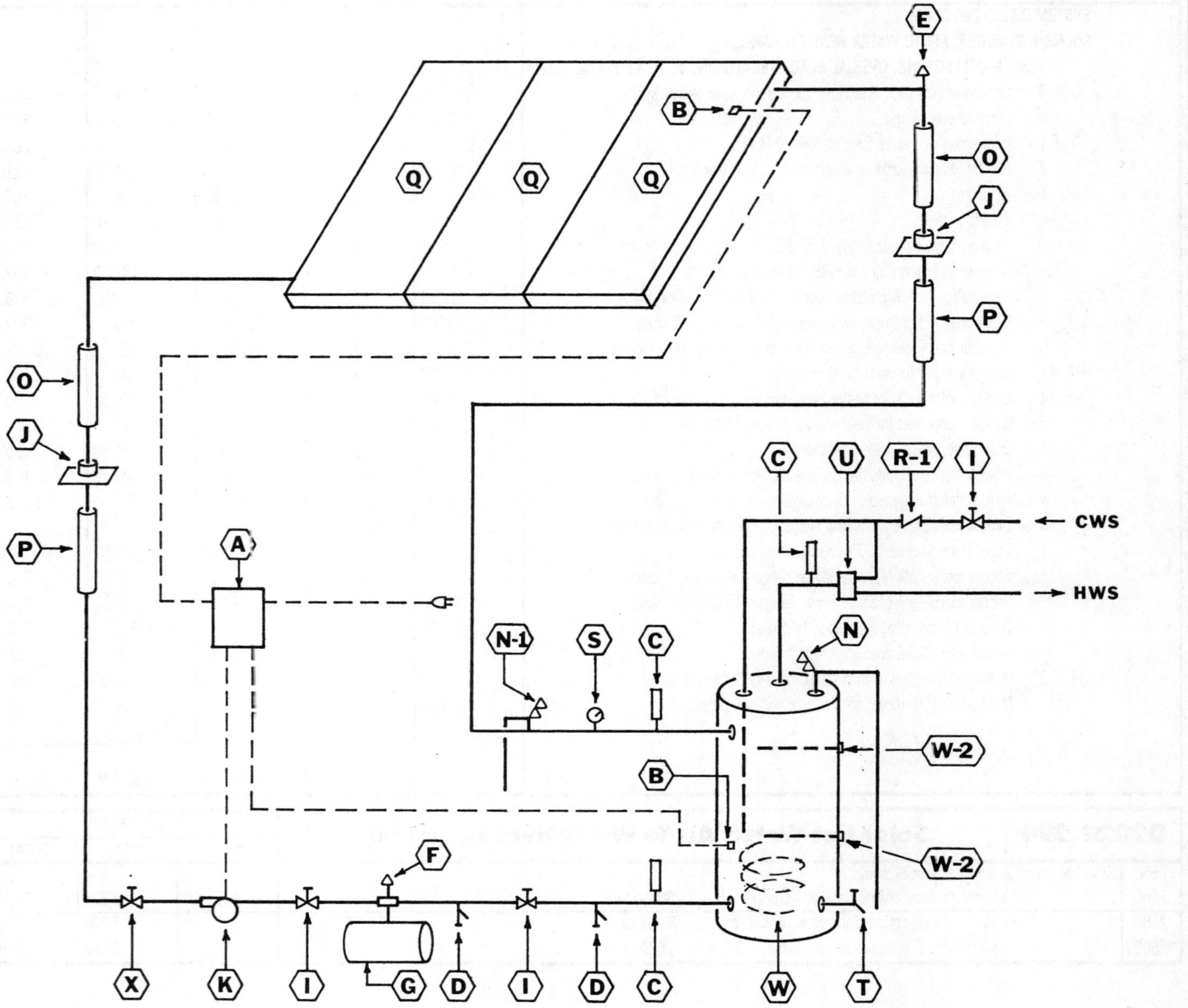

In this closed-loop indirect collection system, fluid with a low freezing temperature, such as propylene glycol, transports heat from the collectors to water storage. The transfer fluid is contained in a closed-loop consisting of collectors, supply and return piping, and a heat exchanger immersed in the storage tank. A typical two-or-three panel system contains 5 to 6 gallons of heat transfer fluid.

When the collectors become approximately 20° F warmer than the storage temperature, a controller activates the circulator. The circulator moves the fluid continuously through the collectors until the temperature difference between the collectors and storage is such that heat collection no longer occurs; at that point, the circulator shuts off. Since the heat transfer fluid has a very low freezing temperature, there is no need for it to be drained from the collectors between periods of collection.

D2020 Domestic Water Distribution

System Components

	System Components	QUANTITY	UNIT	COST EACH MAT.	COST EACH INST.	COST EACH TOTAL
	SYSTEM D2020 295 2760					
	SOLAR, CLOSED LOOP, HOT WATER SYSTEM, IMMERSED HEAT EXCHANGER					
	3/4" TUBING, THREE 3' X 7' BLACK CHROME COLLECTORS					
A, B	Differential controller, 2 sensors, thermostat, solar energy system	1.000	Ea.	380	49	429
C	Thermometer 2" dial	3.000	Ea.	66	109.50	175.50
D, T	Fill & drain valves, brass, 3/4" connection	3.000	Ea.	26.55	73.50	100.05
E	Air vent, manual, 1/8" fitting	1.000	Ea.	3.06	18.30	21.36
F	Air purger	1.000	Ea.	47.50	49	96.50
G	Expansion tank	1.000	Ea.	67.50	18.30	85.80
I	Valve, gate, bronze, NRS, soldered 3/4" diam	3.000	Ea.	117	88.50	205.50
J	Neoprene vent flashing	2.000	Ea.	23.80	59	82.80
K	Circulator, solar heated liquid, 1/25 HP	1.000	Ea.	250	75.50	325.50
N-1, N	Relief valve, temp & press 150 psi 210°F self-closing 3/4" IPS	2.000	Ea.	33.40	39	72.40
O	Pipe covering, urethane ultraviolet cover, 1" wall, 3/4" diam	20.000	L.F.	52.40	108	160.40
P	Pipe covering, fiberglass, all service jacket, 1" wall, 3/4" diam	50.000	L.F.	53.50	216.50	270
	Roof clamps for solar energy collector panel	3.000	Set	7.77	45.15	52.92
Q	Collector panel solar blk chrome on copper, 1/8" temp glass, 3'x7'	3.000	Ea.	2,190	333	2,523
R-1	Valve, swing check, bronze, regrinding disc, 3/4" diam, soldered	1.000	Ea.	70	29.50	99.50
S	Pressure gauge, 60 psi, 2-1/2" dial	1.000	Ea.	24.50	18.30	42.80
U	Valve, water tempering, bronze, sweat connections, 3/4" diam	1.000	Ea.	95.50	29.50	125
W-2, W	Tank, water storage immersed heat exchr elec elem 2"x2# insul 120 Gal	1.000	Ea.	1,275	420	1,695
X	Valve, globe, bronze, rising stem, 3/4" diam, soldered	1.000	Ea.	99	29.50	128.50
	Copper tubing type L, solder joint, hanger 10' OC 3/4" diam	20.000	L.F.	132	154	286
	Copper tubing, type M, solder joint, hanger 10' OC 3/4" diam	70.000	L.F.	339.50	525	864.50
	Sensor wire, #22-2 conductor multistranded	.500	C.L.F.	8.45	28	36.45
	Solar energy heat transfer fluid, propylene glycol, anti-freeze	6.000	Gal.	91.50	126	217.50
	Wrought copper fittings & solder, 3/4" diam	76.000	Ea.	231.04	2,356	2,587.04
	TOTAL			5,684.97	4,998.05	10,683.02

D2020 295	Solar, Closed Loop, Hot Water Systems	COST EACH MAT.	COST EACH INST.	COST EACH TOTAL
2550	Solar, closed loop, hot water system, immersed heat exchanger			
2560	3/8" tubing, 3 ea. 4' x 4'-4" vacuum tube collectors, 80 gal. tank	5,975	4,500	10,475
2580	1/2" tubing, 4 ea. 4' x 4'-4" vacuum tube collectors, 80 gal. tank	6,725	4,900	11,625
2600	120 gal. tank	6,850	4,950	11,800
2640	2 ea. 3'x7' black chrome collectors, 80 gal. tank (R235616-60)	4,500	4,675	9,175
2660	120 gal. tank	4,550	4,675	9,225
2700	2 ea. 3'x7' flat black collectors, 120 gal. tank	4,350	4,675	9,025
2720	3 ea. 3'x7' flat black collectors, 120 gal. tank	4,975	4,825	9,800
2760	3/4" tubing, 3 ea. 3'x7' black chrome collectors, 120 gal. tank	5,675	5,000	10,675
2780	3 ea. 3'x7' flat black collectors, 120 gal. tank	5,350	5,025	10,375
2800	2 ea. 4'x9' flat black w/plastic glazing collectors 120 gal. tank	5,000	5,025	10,025
2840	1" tubing, 4 ea. 2'x9' plastic absorber & glazing collectors 120 gal. tank	6,350	5,600	11,950
2860	4 ea. 3'x7' black chrome collectors, 120 gal. tank	7,100	5,625	12,725
2880	4 ea. 3'x7' flat black absorber collectors, 120 gal. tank	6,650	5,650	12,300

D3010 Energy Supply

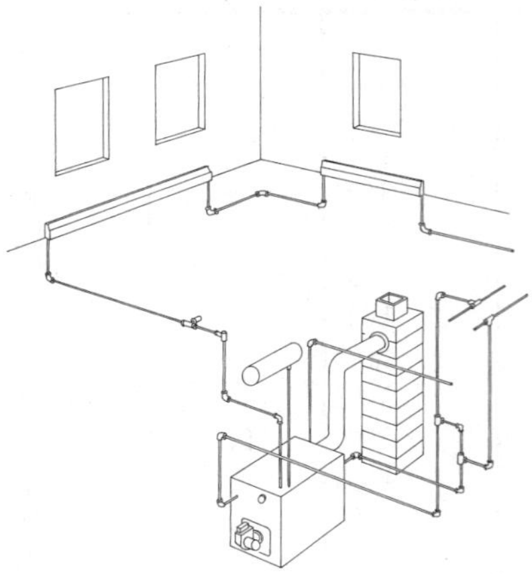

Basis for Heat Loss Estimate, Apartment Type Structures:

1. Masonry walls and flat roof are insulated. U factor is assumed at .08.
2. Window glass area taken as BOCA minimum, 1/10th of floor area. Double insulating glass with 1/4″ air space, U = .65.
3. Infiltration = 0.3 C.F. per hour per S.F. of net wall.
4. Concrete floor loss is 2 BTUH per S.F.
5. Temperature difference taken as 70° F.
6. Ventilating or makeup air has not been included and must be added if desired. Air shafts are not used.

System Components	QUANTITY	UNIT	COST EACH MAT.	COST EACH INST.	COST EACH TOTAL
SYSTEM D3010 510 1760					
HEATING SYSTEM, FIN TUBE RADIATION, FORCED HOT WATER					
1,000 S.F. AREA, 10,000 C.F. VOLUME					
Boiler, oil fired, CI, burner, ctrls/insul/breech/pipe/ftng/valves, 109 MBH	1.000	Ea.	3,631.25	3,281.25	6,912.50
Circulating pump, CI flange connection, 1/12 HP	1.000	Ea.	315	176	491
Expansion tank, painted steel, ASME 18 Gal capacity	1.000	Ea.	520	76	596
Storage tank, steel, above ground, 275 Gal capacity w/supports	1.000	Ea.	420	213	633
Copper tubing type L, solder joint, hanger 10′ OC, 3/4″ diam	100.000	L.F.	660	770	1,430
Radiation, 3/4″ copper tube w/alum fin baseboard pkg, 7″ high	30.000	L.F.	252	552	804
Pipe covering, calcium silicate w/cover, 1″ wall, 3/4″ diam	100.000	L.F.	282	585	867
TOTAL			6,080.25	5,653.25	11,733.50
COST PER S.F.			6.08	5.65	11.73

D3010 510	Apartment Building Heating - Fin Tube Radiation		COST PER S.F. MAT.	COST PER S.F. INST.	COST PER S.F. TOTAL
1740	Heating systems, fin tube radiation, forced hot water				
1760	1,000 S.F. area, 10,000 C.F. volume		6.08	5.65	11.73
1800	10,000 S.F. area, 100,000 C.F. volume	R235000 -10	2.84	3.42	6.26
1840	20,000 S.F. area, 200,000 C.F. volume		3.31	3.83	7.14
1880	30,000 S.F. area, 300,000 C.F. volume	R235000 -20	3.19	3.72	6.91
1890					

D3010 Energy Supply

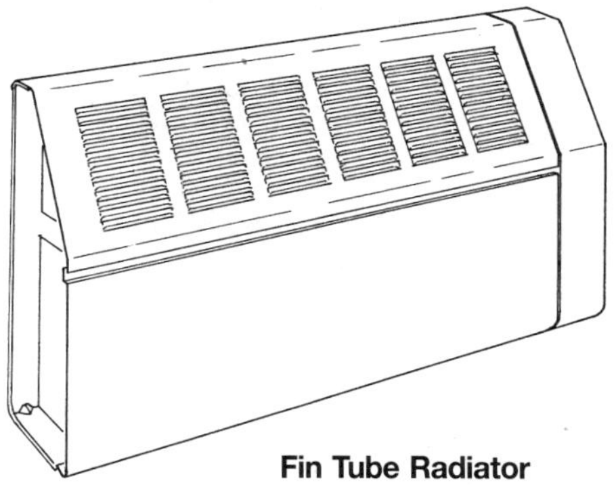

Fin Tube Radiator

Basis for Heat Loss Estimate, Factory or Commercial Type Structures:

1. Walls and flat roof are of lightly insulated concrete block or metal. U factor is assumed at .17.
2. Windows and doors are figured at 1-1/2 S.F. per linear foot of building perimeter. The U factor for flat single glass is 1.13.
3. Infiltration is approximately 5 C.F. per hour per S.F. of wall.
4. Concrete floor loss is 2 BTUH per S.F. of floor.
5. Temperature difference is assumed at 70° F.
6. Ventilation or makeup air has not been included and must be added if desired.

System Components	QUANTITY	UNIT	COST EACH		
			MAT.	INST.	TOTAL
SYSTEM D3010 520 1960					
HEATING SYSTEM, FIN TUBE RADIATION, FORCED HOT WATER					
1,000 S.F. BLDG., ONE FLOOR					
Boiler, oil fired, CI, burner/ctrls/insul/breech/pipe/ftngs/valves, 109 MBH	1.000	Ea.	3,631.25	3,281.25	6,912.50
Expansion tank, painted steel, ASME 18 Gal capacity	1.000	Ea.	2,125	89	2,214
Storage tank, steel, above ground, 550 Gal capacity w/supports	1.000	Ea.	2,525	395	2,920
Circulating pump, CI flanged, 1/8 HP	1.000	Ea.	525	176	701
Pipe, steel, black, sch. 40, threaded, cplg & hngr 10'OC, 1-1/2" diam.	260.000	L.F.	1,651	3,419	5,070
Pipe covering, calcium silicate w/cover , 1" wall, 1-1/2" diam	260.000	L.F.	743.60	1,573	2,316.60
Radiation, steel 1-1/4" tube & 4-1/4" fin w/cover & damper, wall hung	42.000	L.F.	1,701	1,239	2,940
Rough in, steel fin tube radiation w/supply & balance valves	4.000	Set	900	2,600	3,500
TOTAL			13,801.85	12,772.25	26,574.10
COST PER S.F.			13.80	12.77	26.57

D3010 520	Commercial Building Heating - Fin Tube Radiation		COST PER S.F.		
			MAT.	INST.	TOTAL
1940	Heating systems, fin tube radiation, forced hot water				
1960	1,000 S.F. bldg, one floor		13.80	12.80	26.60
2000	10,000 S.F., 100,000 C.F., total two floors	R235000 -10	3.87	4.86	8.73
2040	100,000 S.F., 1,000,000 C.F., total three floors		1.61	2.16	3.77
2080	1,000,000 S.F., 10,000,000 C.F., total five floors	R235000 -20	.82	1.15	1.97
2090					

D3010 Energy Supply

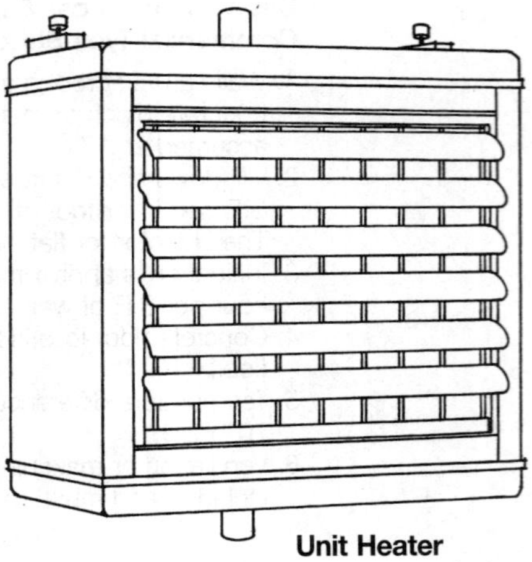

Unit Heater

Basis for Heat Loss Estimate, Factory or Commercial Type Structures:

1. Walls and flat roof are of lightly insulated concrete block or metal. U factor is assumed at .17.
2. Windows and doors are figured at 1-1/2 S.F. per linear foot of building perimeter. The U factor for flat single glass is 1.13.
3. Infiltration is approximately 5 C.F. per hour per S.F. of wall.
4. Concrete floor loss is 2 BTUH per S.F. of floor.
5. Temperature difference is assumed at 70° F.
6. Ventilation or makeup air has not been included and must be added if desired.

System Components	QUANTITY	UNIT	COST EACH MAT.	COST EACH INST.	COST EACH TOTAL
SYSTEM D3010 530 1880					
HEATING SYSTEM, TERMINAL UNIT HEATERS, FORCED HOT WATER					
1,000 S.F. BLDG., ONE FLOOR					
Boiler oil fired, CI, burner/ctrls/insul/breech/pipe/ftngs/valves, 109 MBH	1.000	Ea.	3,631.25	3,281.25	6,912.50
Expansion tank, painted steel, ASME 18 Gal capacity	1.000	Ea.	2,125	89	2,214
Storage tank, steel, above ground, 550 Gal capacity w/ supports	1.000	Ea.	2,525	395	2,920
Circulating pump, CI, flanged, 1/8 HP	1.000	Ea.	525	176	701
Pipe, steel, black, schedule 40, threaded, cplg & hngr 10′ OC 1-1/2″ diam	260.000	L.F.	1,651	3,419	5,070
Pipe covering, calcium silicate w/ cover, 1″ wall, 1-1/2″ diam	260.000	L.F.	743.60	1,573	2,316.60
Unit heater, 1 speed propeller, horizontal, 200° EWT, 26.9 MBH	2.000	Ea.	1,040	266	1,306
Unit heater piping hookup with controls	2.000	Set	1,080	2,500	3,580
TOTAL			13,320.85	11,699.25	25,020.10
COST PER S.F.			13.32	11.70	25.02

D3010 530	Commercial Bldg. Heating - Terminal Unit Heaters		COST PER S.F. MAT.	COST PER S.F. INST.	COST PER S.F. TOTAL
1860	Heating systems, terminal unit heaters, forced hot water				
1880	1,000 S.F. bldg., one floor		13.30	11.70	25
1920	10,000 S.F. bldg., 100,000 C.F. total two floors	R235000 -10	3.44	4.02	7.46
1960	100,000 S.F. bldg., 1,000,000 C.F. total three floors		1.60	1.98	3.58
2000	1,000,000 S.F. bldg., 10,000,000 C.F. total five floors	R235000 -20	1.03	1.28	2.31
2010					

D3010 Energy Supply

Many styles of active solar energy systems exist. Those shown on the following page represent the majority of systems now being installed in different regions of the country.

The five active domestic hot water (DHW) systems that follow are typically specified as two- or three-panel systems with additional variations being the type of glazing and size of the storage tanks. Various combinations have been costed for the user's evaluation and comparison. The basic specifications from which the following systems were developed satisfy the construction detail requirements specified in the HUD Intermediate Minimum Property Standards (IMPS). If these standards are not complied with, the renewable energy system's costs could be significantly lower than shown.

To develop the system's specifications and costs it was necessary to make a number of assumptions about the systems. Certain systems are more appropriate to one climatic region than another or the systems may require modifications to be usable in particular locations. Specific instances in which the systems are not appropriate throughout the country as specified, or in which modification will be needed include the following:

- The freeze protection mechanisms provided in the DHW systems vary greatly. In harsh climates, where freeze is a major concern, a closed-loop indirect collection system may be more appropriate than a direct collection system.
- The thermosyphon water heater system described cannot be used when temperatures drop below 32° F.
- In warm climates it may be necessary to modify the systems installed to prevent overheating.

For each renewable resource (solar) system, a schematic diagram and descriptive summary of the system is provided along with a list of all the components priced as part of the system. The costs were developed based on these specifications.

Considerations affecting costs that may increase or decrease beyond the estimates presented here include the following:

- Special structural qualities (allowance for earthquake, future expansion, high winds, and unusual spans or shapes).
- Isolated building site or rough terrain that would affect the transportation of personnel, material, or equipment.
- Unusual climatic conditions during the construction process.
- Substitution of other materials or system components for those used in the system specifications.

D3010 Energy Supply

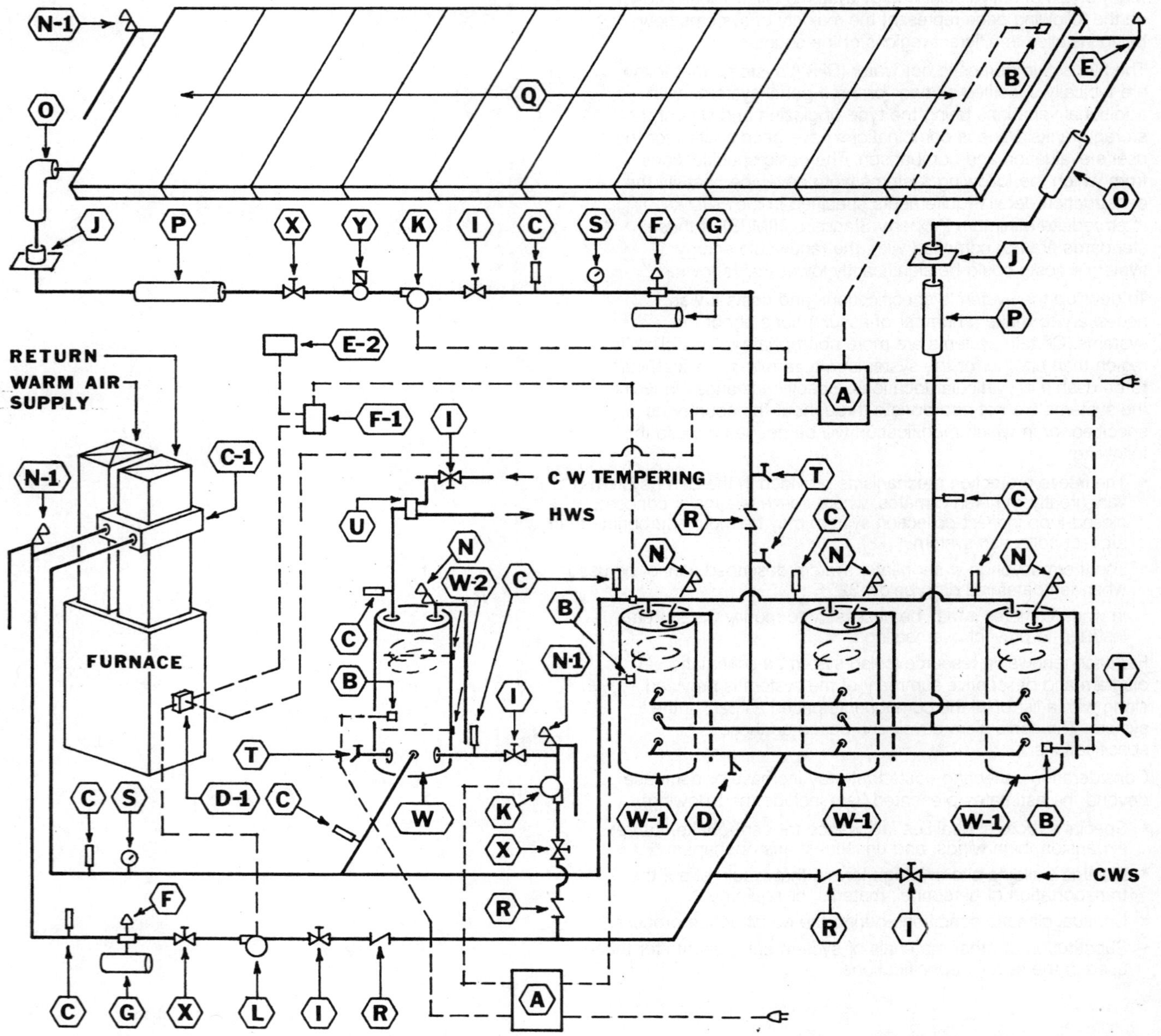

In this closed-loop indirect collection system, fluid with a low freezing temperature, propylene glycol, transports heat from the collectors to water storage. The transfer fluid is contained in a closed loop consisting of collectors, supply and return piping, and a heat exchanger immersed in the storage tank.

When the collectors become approximately 20° F warmer than the storage temperature, the controller activates the circulator. The circulator moves the fluid continuously until the temperature difference between fluid in the collectors and storage is such that the collection will no longer occur and then the circulator turns off. Since the heat transfer fluid has a very low freezing temperature, there is no need for it to be drained from the collectors between periods of collection.

D3010 Energy Supply

System Components

System Components		QUANTITY	UNIT	COST EACH MAT.	COST EACH INST.	COST EACH TOTAL
	SYSTEM D3010 650 2750					
	SOLAR, CLOSED LOOP, SPACE/HOT WATER					
	1″ TUBING, TEN 3′X7′ BLK CHROME ON COPPER ABSORBER COLLECTORS					
A, B	Differential controller 2 sensors, thermostat, solar energy system	2.000	Ea.	760	98	858
C	Thermometer, 2″ dial	10.000	Ea.	220	365	585
C-1	Heat exchanger, solar energy system, fluid to air, up flow, 80 MBH	1.000	Ea.	490	350	840
D, T	Fill & drain valves, brass, 3/4″ connection	5.000	Ea.	44.25	122.50	166.75
D-1	Fan center	1.000	Ea.	135	124	259
E	Air vent, manual, 1/8″ fitting	1.000	Ea.	3.06	18.30	21.36
E-2	Thermostat, 2 stage for sensing room temperature	1.000	Ea.	229	72.50	301.50
F	Air purger	2.000	Ea.	95	98	193
F-1	Controller, liquid temperature, solar energy system	1.000	Ea.	105	117	222
G	Expansion tank	2.000	Ea.	135	36.60	171.60
I	Valve, gate, bronze, 125 lb, soldered, 1″ diam	5.000	Ea.	277.50	155	432.50
J	Vent flashing, neoprene	2.000	Ea.	23.80	59	82.80
K	Circulator, solar heated liquid, 1/25 HP	2.000	Ea.	500	151	651
L	Circulator, solar heated liquid, 1/20 HP	1.000	Ea.	170	88	258
N	Relief valve, temp & pressure 150 psi 210° F self-closing	4.000	Ea.	66.80	78	144.80
N-1	Relief valve, pressure poppet, bronze, 30 psi, 3/4″ IPS	3.000	Ea.	115.50	63	178.50
O	Pipe covering, urethane, ultraviolet cover, 1″ wall, 1″ diam	50.000	L.F.	131	270	401
P	Pipe covering, fiberglass, all service jacket, 1″ wall, 1″ diam	60.000	L.F.	64.20	259.80	324
Q	Collector panel solar energy blk chrome on copper 1/8″ temp glass 3′x7′	10.000	Ea.	7,300	1,110	8,410
	Roof clamps for solar energy collector panel	10.000	Set	25.90	150.50	176.40
R	Valve, swing check, bronze, 125 lb, regrinding disc, 3/4″ & 1″ diam	4.000	Ea.	404	124	528
S	Pressure gage, 0-60 psi, for solar energy system	2.000	Ea.	49	36.60	85.60
U	Valve, water tempering, bronze, sweat connections, 3/4″ diam	1.000	Ea.	95.50	29.50	125
W-2, W-1, W	Tank, water storage immersed heat xchr elec elem 2″x1/2# ins 120 gal	4.000	Ea.	5,100	1,680	6,780
X	Valve, globe, bronze, 125 lb, rising stem, 1″ diam	3.000	Ea.	465	93	558
Y	Valve, flow control	1.000	Ea.	112	26.50	138.50
	Copper tubing type M, solder joint, hanger 10′ OC, 1″ diam	110.000	L.F.	836	918.50	1,754.50
	Copper tubing, type L, solder joint, hanger 10′ OC 3/4″ diam	20.000	L.F.	132	154	286
	Wrought copper fittings & solder, 3/4″ & 1″ diam	121.000	Ea.	901.45	4,416.50	5,317.95
	Sensor, wire, #22-2 conductor, multistranded	.700	C.L.F.	11.83	39.20	51.03
	Ductwork, galvanized steel, for heat exchanger	8.000	Lb.	9.84	55.20	65.04
	Solar energy heat transfer fluid propylene glycol, anti-freeze	25.000	Gal.	381.25	525	906.25
	TOTAL			19,388.88	11,884.20	31,273.08

D3010 650 Solar, Closed Loop, Space/Hot Water Systems

			COST EACH MAT.	COST EACH INST.	COST EACH TOTAL
2540	Solar, closed loop, space/hot water				
2550	1/2″ tubing, 12 ea. 4′x4′4″ vacuum tube collectors		21,500	11,100	32,600
2600	3/4″ tubing, 12 ea. 4′x4′4″ vacuum tube collectors	R235616 -60	22,400	11,400	33,800
2650	10 ea. 3′ x 7′ black chrome absorber collectors		18,500	11,100	29,600
2700	10 ea. 3′ x 7′ flat black absorber collectors		17,400	11,200	28,600
2750	1″ tubing, 10 ea. 3′ x 7′ black chrome absorber collectors		19,400	11,900	31,300
2800	10 ea. 3′ x 7′ flat black absorber collectors		18,400	12,000	30,400
2850	6 ea. 4′ x 9′ flat black w/plastic glazing collectors		16,700	11,900	28,600
2900	12 ea. 2′ x 9′ plastic absorber and glazing collectors		18,700	12,100	30,800

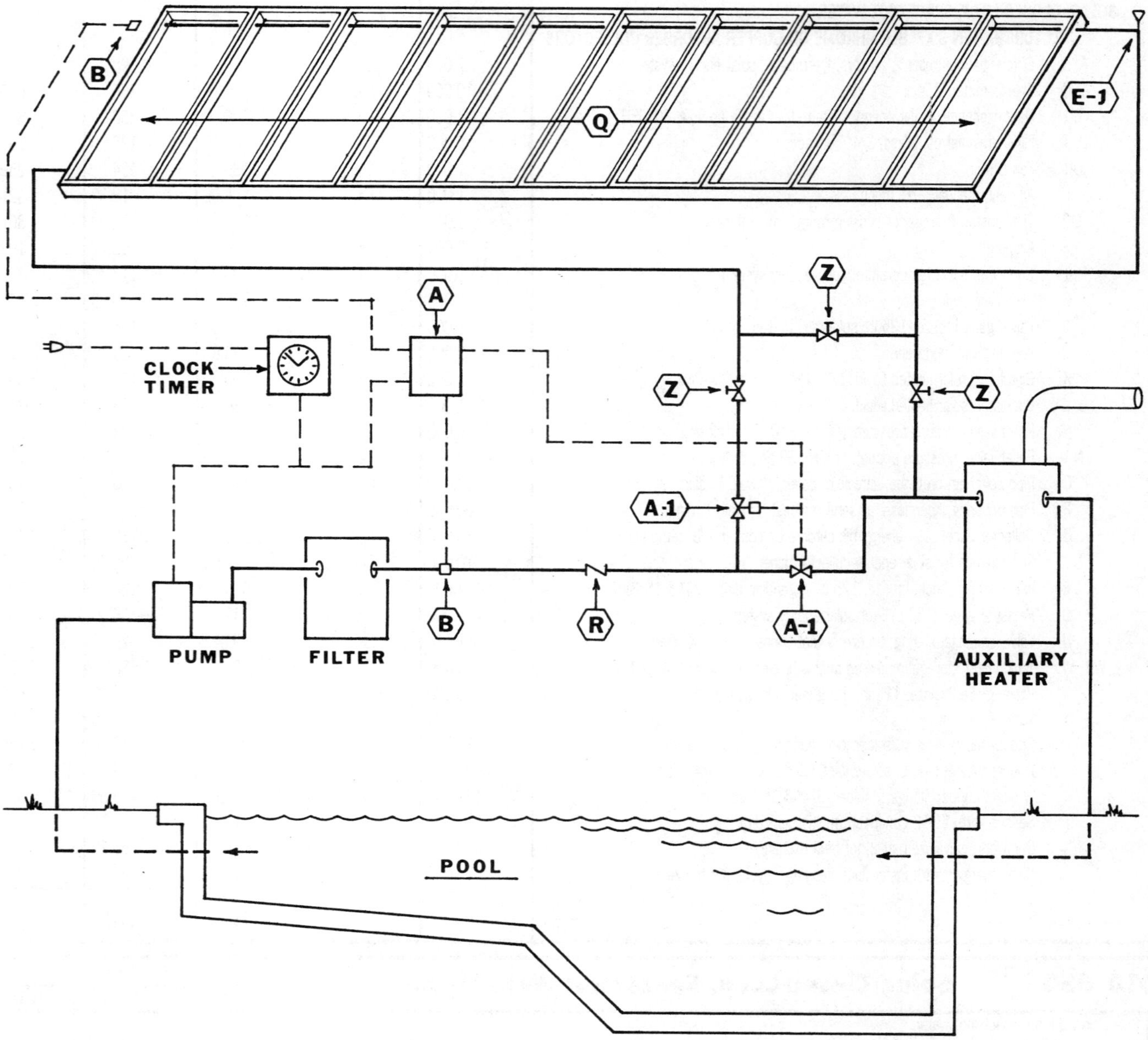

This draindown pool system uses a differential thermostat similar to those used in solar domestic hot water and space heating applications. To heat the pool, the pool water passes through the conventional pump-filter loop and then flows through the collectors. When collection is not possible, or when the pool temperature is reached, all water drains from the solar loop back to the pool through the existing piping. The modes are controlled by solenoid valves or other automatic valves in conjunction with a vacuum breaker relief valve, which facilitates draindown.

D3010 Energy Supply

System Components

System Components	QUANTITY	UNIT	COST EACH MAT.	COST EACH INST.	COST EACH TOTAL
SYSTEM D3010 660 2640					
SOLAR SWIMMING POOL HEATER, ROOF MOUNTED COLLECTORS					
TEN 4' X 10' FULLY WETTED UNGLAZED PLASTIC ABSORBERS					
A Differential thermostat/controller, 110V, adj pool pump system	1.000	Ea.	305	293	598
A-1 Solenoid valve, PVC, normally 1 open 1 closed (included)	2.000	Ea.			
B Sensor, thermistor type (included)	2.000	Ea.			
E-1 Valve, vacuum relief	1.000	Ea.	29	18.30	47.30
Q Collector panel, solar energy, plastic, liquid full wetted, 4' x 10'	10.000	Ea.	2,570	2,110	4,680
R Valve, ball check, PVC, socket, 1-1/2" diam	1.000	Ea.	83	29.50	112.50
Z Valve, ball, PVC, socket, 1-1/2" diam	3.000	Ea.	156	88.50	244.50
Pipe, PVC, sch 40, 1-1/2" diam	80.000	L.F.	240	1,300	1,540
Pipe fittings, PVC sch 40, socket joint, 1-1/2" diam	10.000	Ea.	17.10	295	312.10
Sensor wire, #22-2 conductor, multistranded	.500	C.L.F.	8.45	28	36.45
Roof clamps for solar energy collector panels	10.000	Set	25.90	150.50	176.40
Roof strap, teflon for solar energy collector panels	26.000	L.F.	572	74.10	646.10
TOTAL			4,006.45	4,386.90	8,393.35

D3010 660 Solar Swimming Pool Heater Systems

D3010 660	Solar Swimming Pool Heater Systems	COST EACH MAT.	COST EACH INST.	COST EACH TOTAL
2530	Solar swimming pool heater systems, roof mounted collectors			
2540	10 ea. 3'x7' black chrome absorber, 1/8" temp. glass	8,725	3,375	12,100
2560	10 ea. 4'x8' black chrome absorber, 3/16" temp. glass R235616-60	10,300	4,025	14,325
2580	10 ea. 3'8"x6' flat black absorber, 3/16" temp. glass	7,625	3,450	11,075
2600	10 ea. 4'x9' flat black absorber, plastic glazing	8,975	4,200	13,175
2620	10 ea. 2'x9' rubber absorber, plastic glazing	12,400	4,525	16,925
2640	10 ea. 4'x10' fully wetted unglazed plastic absorber	4,000	4,375	8,375
2660	Ground mounted collectors			
2680	10 ea. 3'x7' black chrome absorber, 1/8" temp. glass	8,825	3,900	12,725
2700	10 ea. 4'x8' black chrome absorber, 3/16" temp glass	10,400	4,550	14,950
2720	10 ea. 3'8"x6' flat blk absorber, 3/16" temp. glass	7,725	3,950	11,675
2740	10 ea. 4'x9' flat blk absorber, plastic glazing	9,075	4,700	13,775
2760	10 ea. 2'x9' rubber absorber, plastic glazing	12,400	4,875	17,275
2780	10 ea. 4'x10' fully wetted unglazed plastic absorber	4,100	4,900	9,000

D3010 Energy Supply

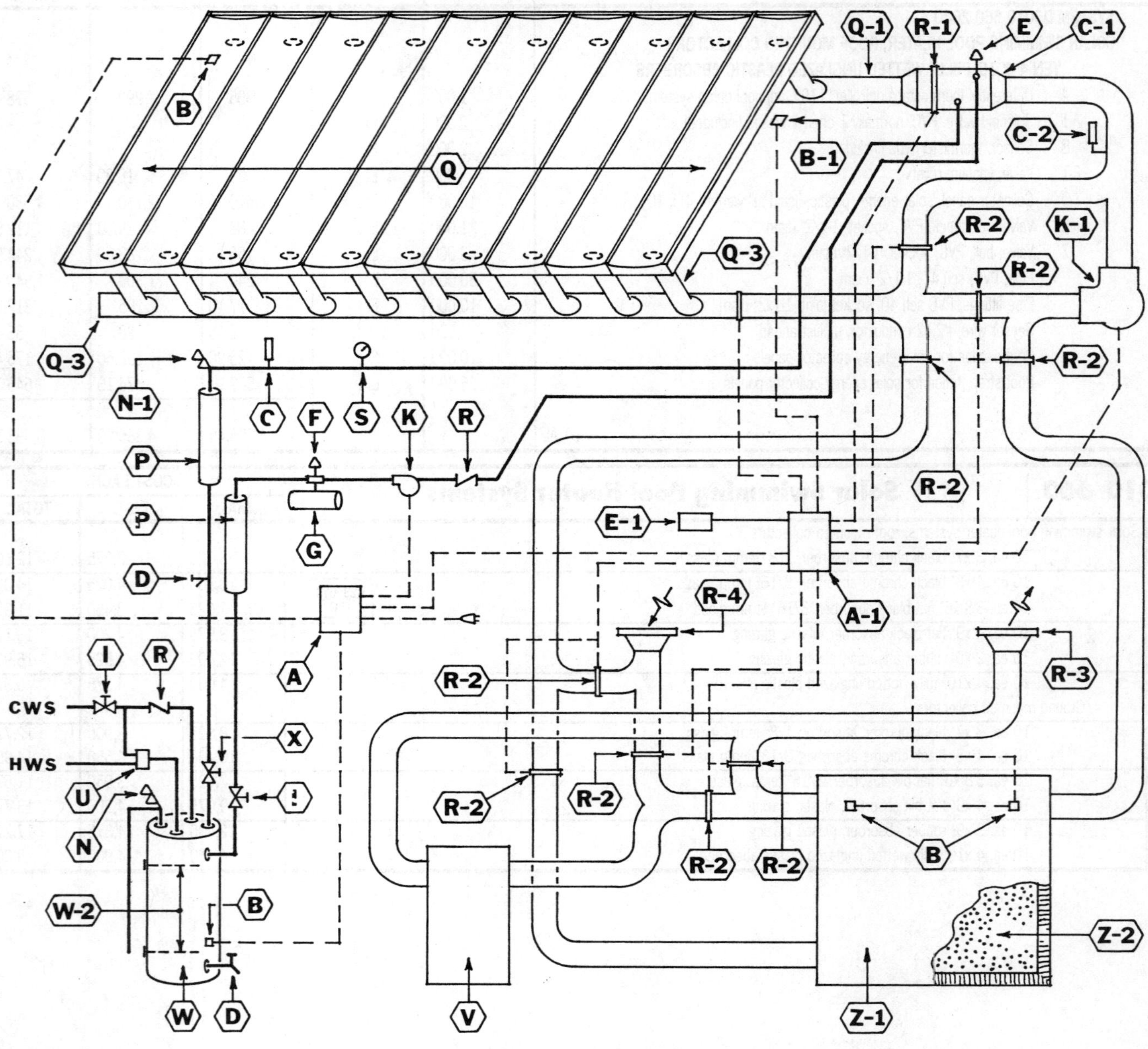

The complete Solar Air Heating System provides maximum savings of conventional fuel with both space heating and year-round domestic hot water heating. It allows for the home air conditioning to operate simultaneously and independently from the solar domestic water heating in summer. The system's modes of operation are:

Mode 1: The building is heated directly from the collectors with air circulated by the Solar Air Mover.

Mode 2: When heat is not needed in the building, the dampers change within the air mover to circulate the air from the collectors to the rock storage bin.

Mode 3: When heat is not available from the collector array and is available in rock storage, the air mover draws heated air from rock storage and directs it into the building. When heat is not available from the collectors or the rock storage bin, the auxiliary heating unit will provide heat for the building. The size of the collector array is typically 25% the size of the main floor area.

D3010 Energy Supply

System Components

System Components		QUANTITY	UNIT	COST EACH MAT.	COST EACH INST.	COST EACH TOTAL
	SYSTEM D3010 675 1210					
	SOLAR, SPACE/HOT WATER, AIR TO WATER HEAT EXCHANGE					
A, B	Differential controller 2 sensors thermos., solar energy sys liquid loop	1.000	Ea.	380	49	429
A-1, B	Differential controller 2 sensors 6 station solar energy sys air loop	1.000	Ea.	239	195	434
B-1	Solar energy sensor, freeze prevention	1.000	Ea.	24	18.30	42.30
C	Thermometer for solar energy system, 2″ dial	2.000	Ea.	44	73	117
C-1	Heat exchanger, solar energy system, air to fluid, up flow, 70 MBH	1.000	Ea.	365	300	665
D	Drain valve, brass, 3/4″ connection	2.000	Ea.	17.70	49	66.70
E	Air vent, manual, for solar energy system 1/8″ fitting	1.000	Ea.	3.06	18.30	21.36
E-1	Thermostat, 2 stage for sensing room temperature	1.000	Ea.	229	72.50	301.50
F	Air purger	1.000	Ea.	47.50	49	96.50
G	Expansion tank, for solar energy system	1.000	Ea.	67.50	18.30	85.80
I	Valve, gate, bronze, 125 lb, NRS, soldered 3/4″ diam	2.000	Ea.	78	59	137
K	Circulator, solar heated liquid, 1/25 HP	1.000	Ea.	250	75.50	325.50
N	Relief valve temp & press 150 psi 210° F self-closing, 3/4″ IPS	1.000	Ea.	16.70	19.50	36.20
N-1	Relief valve, pressure, poppet, bronze, 30 psi, 3/4″ IPS	1.000	Ea.	38.50	21	59.50
P	Pipe covering, fiberglass, all service jacket, 1″ wall, 3/4″ diam	60.000	L.F.	64.20	259.80	324
Q-3	Manifold for flush mount solar energy collector panels	20.000	L.F.	2,340	131	2,471
Q	Collector panel solar energy, air, black on alum. plate, flush mount 10′x2′	100.000	L.F.	60,500	1,090	61,590
R	Valve, swing check, bronze, 125 lb, regrinding disc, 3/4″ diam	2.000	Ea.	140	59	199
S	Pressure gage, 2″ dial, for solar energy system	1.000	Ea.	24.50	18.30	42.80
U	Valve, water tempering, bronze, sweat connections, 3/4″ diam	1.000	Ea.	95.50	29.50	125
W-2, W	Tank, water storage, solar, elec element 2″x1/2# insul, 80 Gal	1.000	Ea.	1,150	365	1,515
X	Valve, globe, bronze, 125 lb, soldered, 3/4″ diam	1.000	Ea.	99	29.50	128.50
	Copper tubing type L, solder joints, hangers 10′ OC 3/4″ diam	10.000	L.F.	66	77	143
	Copper tubing type M, solder joints, hangers 10′ OC 3/4″ diam	60.000	L.F.	291	450	741
	Wrought copper fittings & solder, 3/4″ diam	26.000	Ea.	79.04	806	885.04
	Sensor wire, #22-2 conductor multistranded	1.200	C.L.F.	20.28	67.20	87.48
Q-1	Duct work, rigid fiberglass, rectangular	400.000	S.F.	320	1,860	2,180
	Ductwork, transition pieces	8.000	Ea.	228	348	576
R-2	Shutter/damper for solar heater circulator	9.000	Ea.	486	783	1,269
K-1	Shutter motor for solar heater circulator blower	9.000	Ea.	1,179	589.50	1,768.50
K-1	Fan, solar energy heated air circulator, space & DHW system	1.000	Ea.	1,625	2,100	3,725
Z-1	Tank, solar energy air storage, 6′-3″H 7′x7′ = 306 CF/2000 Gal	1.000	Ea.	12,900	1,350	14,250
Z-2	Crushed stone 1-1/2″	11.000	C.Y.	528	70.51	598.51
C-2	Themometer, remote probe, 2″ dial	1.000	Ea.	33.50	73	106.50
R-1	Solenoid valve	1.000	Ea.	113	65	178
V, R-4, R-3	Furnace, supply diffusers, return grilles, existing	1.000	Ea.	12,900	1,350	14,250
	TOTAL			96,981.98	12,988.71	109,970.69

D3010 675	Air To Water Heat Exchange		COST EACH MAT.	COST EACH INST.	COST EACH TOTAL
1210	Solar, air to water heat exchange, for space / hot water heating	R235616 -60	97,000	12,900	109,900
1220					

D3020 Heat Generating Systems

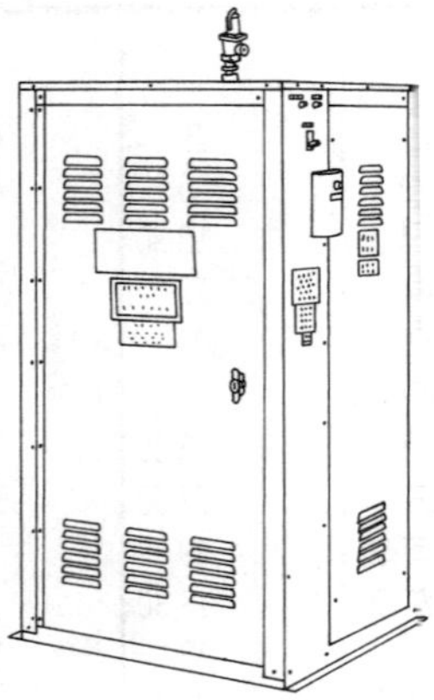

Boiler

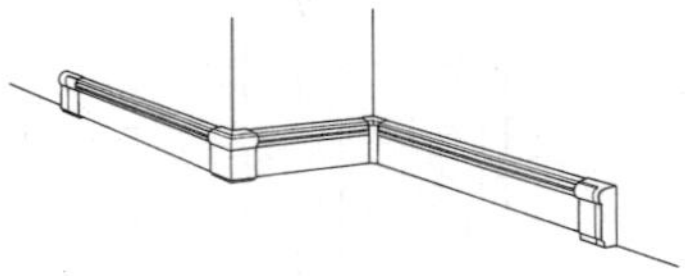

Baseboard Radiation

Small Electric Boiler System Considerations:

1. Terminal units are fin tube baseboard radiation rated at 720 BTU/hr with 200° water temperature or 820 BTU/hr steam.
2. Primary use being for residential or smaller supplementary areas, the floor levels are based on 7-1/2′ ceiling heights.
3. All distribution piping is copper for boilers through 205 MBH. All piping for larger systems is steel pipe.

System Components	QUANTITY	UNIT	COST EACH MAT.	COST EACH INST.	COST EACH TOTAL
SYSTEM D3020 102 1120 SMALL HEATING SYSTEM, HYDRONIC, ELECTRIC BOILER 1,480 S.F., 61 MBH, STEAM, 1 FLOOR					
Boiler, electric steam, std cntrls, trim, ftngs and valves, 18 KW, 61.4 MBH	1.000	Ea.	4,372.50	1,485	5,857.50
Copper tubing type L, solder joint, hanger 10′OC, 1-1/4″ diam	160.000	L.F.	2,200	1,616	3,816
Radiation, 3/4″ copper tube w/alum fin baseboard pkg 7″ high	60.000	L.F.	504	1,104	1,608
Rough in baseboard panel or fin tube with valves & traps	10.000	Set	2,290	5,600	7,890
Pipe covering, calcium silicate w/cover 1″ wall 1-1/4″ diam	160.000	L.F.	451.20	968	1,419.20
Low water cut-off, quick hookup, in gage glass tappings	1.000	Ea.	218	37	255
TOTAL			10,035.70	10,810	20,845.70
COST PER S.F.			6.78	7.30	14.08

D3020 102	Small Heating Systems, Hydronic, Electric Boilers		COST PER S.F. MAT.	COST PER S.F. INST.	COST PER S.F. TOTAL
1100	Small heating systems, hydronic, electric boilers				
1120	Steam, 1 floor, 1480 S.F., 61 M.B.H.		6.76	7.30	14.06
1160	3,000 S.F., 123 M.B.H.	R235000-10	5.25	6.40	11.65
1200	5,000 S.F., 205 M.B.H.		4.60	5.90	10.50
1240	2 floors, 12,400 S.F., 512 M.B.H.	R235000-20	3.38	5.85	9.23
1280	3 floors, 24,300 S.F., 1023 M.B.H.		3.53	5.80	9.33
1320	34,750 S.F., 1,433 M.B.H.	R235000-30	3.25	5.65	8.90
1360	Hot water, 1 floor, 1,000 S.F., 41 M.B.H.		9.75	4.07	13.82
1400	2,500 S.F., 103 M.B.H.		7.70	7.30	15
1440	2 floors, 4,850 S.F., 205 M.B.H.		8.10	8.75	16.85
1480	3 floors, 9,700 S.F., 410 M.B.H.		9.15	9.05	18.20

D3020 Heat Generating Systems

Boiler

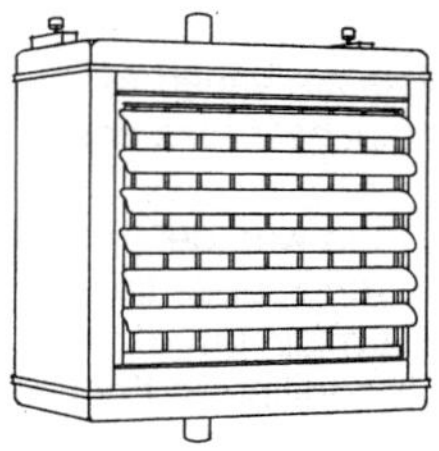

Unit Heater

Large Electric Boiler System Considerations:

1. Terminal units are all unit heaters of the same size. Quantities are varied to accommodate total requirements.
2. All air is circulated through the heaters a minimum of three times per hour.
3. As the capacities are adequate for commercial use, floor levels are based on 10′ ceiling heights.
4. All distribution piping is black steel pipe.

System Components	QUANTITY	UNIT	COST EACH MAT.	COST EACH INST.	COST EACH TOTAL
SYSTEM D3020 104 1240					
LARGE HEATING SYSTEM, HYDRONIC, ELECTRIC BOILER					
9,280 S.F., 135 KW, 461 MBH, 1 FLOOR					
Boiler, electric hot water, std ctrls, trim, ftngs, valves, 135 KW, 461 MBH	1.000	Ea.	10,500	3,262.50	13,762.50
Expansion tank, painted steel, 60 Gal capacity ASME	1.000	Ea.	3,325	178	3,503
Circulating pump, CI, close cpld, 50 GPM, 2 HP, 2″ pipe conn	1.000	Ea.	2,100	350	2,450
Unit heater, 1 speed propeller, horizontal, 200° EWT, 72.7 MBH	7.000	Ea.	5,285	1,358	6,643
Unit heater piping hookup with controls	7.000	Set	3,780	8,750	12,530
Pipe, steel, black, schedule 40, welded, 2-1/2″ diam	380.000	L.F.	4,104	9,044	13,148
Pipe covering, calcium silicate w/cover, 1″ wall, 2-1/2″ diam	380.000	L.F.	1,326.20	2,356	3,682.20
TOTAL			30,420.20	25,298.50	55,718.70
COST PER S.F.			3.28	2.73	6.01

D3020 104	Large Heating Systems, Hydronic, Electric Boilers		COST PER S.F. MAT.	COST PER S.F. INST.	COST PER S.F. TOTAL
1230	Large heating systems, hydronic, electric boilers				
1240	9,280 S.F., 135 K.W., 461 M.B.H., 1 floor		3.28	2.73	6.01
1280	14,900 S.F., 240 K.W., 820 M.B.H., 2 floors	R235000-10	4.11	4.43	8.54
1320	18,600 S.F., 296 K.W., 1,010 M.B.H., 3 floors		4.07	4.81	8.88
1360	26,100 S.F., 420 K.W., 1,432 M.B.H., 4 floors	R235000-20	4.02	4.74	8.76
1400	39,100 S.F., 666 K.W., 2,273 M.B.H., 4 floors		3.45	3.95	7.40
1440	57,700 S.F., 900 K.W., 3,071 M.B.H., 5 floors	R235000-30	3.58	3.87	7.45
1480	111,700 S.F., 1,800 K.W., 6,148 M.B.H., 6 floors		3.23	3.33	6.56
1520	149,000 S.F., 2,400 K.W., 8,191 M.B.H., 8 floors		3.22	3.33	6.55
1560	223,300 S.F., 3,600 K.W., 12,283 M.B.H., 14 floors		3.46	3.78	7.24

D3020 Heat Generating Systems

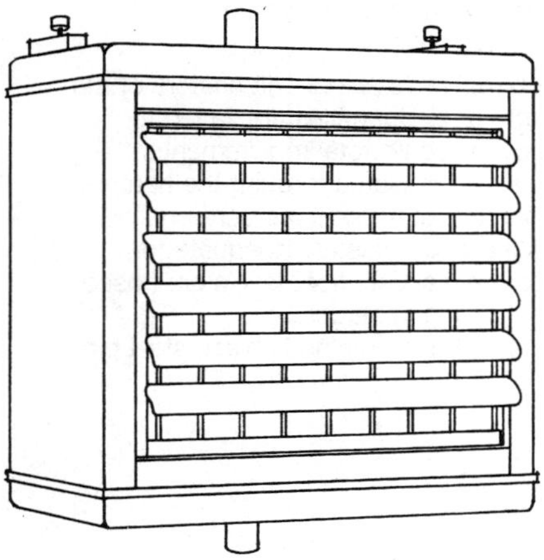

Unit Heater

Fossil Fuel Boiler System Considerations:

1. Terminal units are horizontal unit heaters. Quantities are varied to accommodate total heat loss per building.
2. Unit heater selection was determined by their capacity to circulate the building volume a minimum of three times per hour in addition to the BTU output.
3. Systems shown are forced hot water. Steam boilers cost slightly more than hot water boilers. However, this is compensated for by the smaller size or fewer terminal units required with steam.
4. Floor levels are based on 10′ story heights.
5. MBH requirements are gross boiler output.

System Components	QUANTITY	UNIT	COST EACH MAT.	INST.	TOTAL
SYSTEM D3020 108 1280					
HEATING SYSTEM, HYDRONIC, FOSSIL FUEL, TERMINAL UNIT HEATERS					
CAST IRON BOILER, GAS, 80 MBH, 1,070 S.F. BUILDING					
Boiler, gas, hot water, CI, burner, controls, insulation, breeching, 80 MBH	1.000	Ea.	1,968.75	1,627.50	3,596.25
Pipe, steel, black, schedule 40, threaded, cplg & hngr 10′OC, 2″ diam	200.000	L.F.	1,690	3,290	4,980
Unit heater, 1 speed propeller, horizontal, 200° EWT, 72.7 MBH	2.000	Ea.	1,510	388	1,898
Unit heater piping hookup with controls	2.000	Set	1,080	2,500	3,580
Expansion tank, painted steel, ASME, 18 Gal capacity	1.000	Ea.	2,125	89	2,214
Circulating pump, CI, flange connection, 1/12 HP	1.000	Ea.	315	176	491
Pipe covering, calcium silicate w/cover, 1″ wall, 2″ diam	200.000	L.F.	656	1,240	1,896
TOTAL			9,344.75	9,310.50	18,655.25
COST PER S.F.			8.73	8.70	17.43

D3020 108	Heating Systems, Unit Heaters		COST PER S.F. MAT.	INST.	TOTAL
1260	Heating systems, hydronic, fossil fuel, terminal unit heaters,				
1280	Cast iron boiler, gas, 80 M.B.H., 1,070 S.F. bldg.		8.74	8.69	17.43
1320	163 M.B.H., 2,140 S.F. bldg.	R235000 -10	5.80	5.90	11.70
1360	544 M.B.H., 7,250 S.F. bldg.		4.23	4.18	8.41
1400	1,088 M.B.H., 14,500 S.F. bldg.	R235000 -20	3.59	3.96	7.55
1440	3,264 M.B.H., 43,500 S.F. bldg.		3	3.01	6.01
1480	5,032 M.B.H., 67,100 S.F. bldg.	R235000 -30	3.45	3.06	6.51
1520	Oil, 109 M.B.H., 1,420 S.F. bldg.		9	7.75	16.75
1560	235 M.B.H., 3,150 S.F. bldg.		6	5.50	11.50
1600	940 M.B.H., 12,500 S.F. bldg.		4.36	3.62	7.98
1640	1,600 M.B.H., 21,300 S.F. bldg.		3.97	3.48	7.45
1680	2,480 M.B.H., 33,100 S.F. bldg.		3.95	3.13	7.08
1720	3,350 M.B.H., 44,500 S.F. bldg.		3.63	3.20	6.83
1800	300 M.B.H., 4,000 S.F. bldg.		6.10	3.94	10.04
1840	2,360 M.B.H., 31,500 S.F. bldg.		4.45	3.19	7.64

D3020 Heat Generating Systems

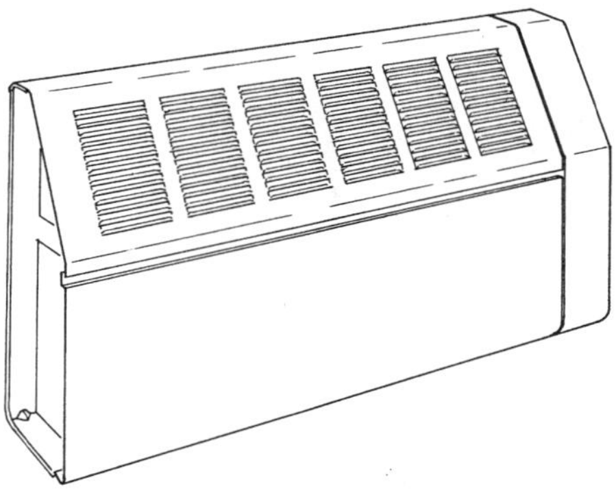

Fin Tube Radiator

Fossil Fuel Boiler System Considerations:

1. Terminal units are commercial steel fin tube radiation. Quantities are varied to accommodate total heat loss per building.
2. Systems shown are forced hot water. Steam boilers cost slightly more than hot water boilers. However, this is compensated for by the smaller size or fewer terminal units required with steam.
3. Floor levels are based on 10′ story heights.
4. MBH requirements are gross boiler output.

System Components	QUANTITY	UNIT	COST EACH MAT.	COST EACH INST.	COST EACH TOTAL
SYSTEM D3020 110 3240					
HEATING SYSTEM, HYDRONIC, FOSSIL FUEL, FIN TUBE RADIATION					
CAST IRON BOILER, GAS, 80 MBH, 1,070 S.F. BLDG.					
Boiler, gas, hot water, CI, burner, controls, insulation, breeching, 80 MBH	1.000	Ea.	1,968.75	1,627.50	3,596.25
Pipe, steel, black, schedule 40, threaded, 2″ diam	200.000	L.F.	1,690	3,290	4,980
Radiation,steel 1-1/4″ tube & 4-1/4″ fin w/cover & damper, wall hung	80.000	L.F.	3,240	2,360	5,600
Rough-in, wall hung steel fin radiation with supply & balance valves	8.000	Set	1,800	5,200	7,000
Expansion tank, painted steel, ASME, 18 Gal capacity	1.000	Ea.	2,125	89	2,214
Circulating pump, CI flanged, 1/12 HP	1.000	Ea.	315	176	491
Pipe covering, calcium silicate w/cover, 1″ wall, 2″ diam	200.000	L.F.	656	1,240	1,896
TOTAL			11,794.75	13,982.50	25,777.25
COST PER S.F.			11.02	13.07	24.09

D3020 110	Heating System, Fin Tube Radiation		COST PER S.F. MAT.	COST PER S.F. INST.	COST PER S.F. TOTAL
3230	Heating systems, hydronic, fossil fuel, fin tube radiation				
3240	Cast iron boiler, gas, 80 MBH, 1,070 S.F. bldg.		11.03	13.08	24.11
3280	169 M.B.H., 2,140 S.F. bldg.	R235000-10	6.90	8.25	15.15
3320	544 M.B.H., 7,250 S.F. bldg.		5.85	7.10	12.95
3360	1,088 M.B.H., 14,500 S.F. bldg.	R235000-20	5.30	6.95	12.25
3400	3,264 M.B.H., 43,500 S.F. bldg.		4.82	6.15	10.97
3440	5,032 M.B.H., 67,100 S.F. bldg.	R235000-30	5.30	6.20	11.50
3480	Oil, 109 M.B.H., 1,420 S.F. bldg.		12.80	14.25	27.05
3520	235 M.B.H., 3,150 S.F. bldg.		7.60	8.35	15.95
3560	940 M.B.H., 12,500 S.F. bldg.		6.10	6.65	12.75
3600	1,600 M.B.H., 21,300 S.F. bldg.		5.80	6.60	12.40
3640	2,480 M.B.H., 33,100 S.F. bldg.		5.80	6.30	12.10
3680	3,350 M.B.H., 44,500 S.F. bldg.		5.45	6.35	11.80
3720	Coal, 148 M.B.H., 1,975 S.F. bldg.		9.65	7.90	17.55
3760	300 M.B.H., 4,000 S.F. bldg.		7.65	6.75	14.40
3800	2,360 M.B.H., 31,500 S.F. bldg.		6.25	6.25	12.50
4080	Steel boiler, oil, 97 M.B.H., 1,300 S.F. bldg.		11.55	11.75	23.30
4120	315 M.B.H., 4,550 S.F. bldg.		5.20	5.95	11.15
4160	525 M.B.H., 7,000 S.F. bldg.		7.15	6.85	14
4200	1,050 M.B.H., 14,000 S.F. bldg.		6.25	6.85	13.10
4240	2,310 M.B.H., 30,800 S.F. bldg.		5.75	6.30	12.05
4280	3,150 M.B.H., 42,000 S.F. bldg.		5.75	6.40	12.15

D3020 Heat Generating Systems

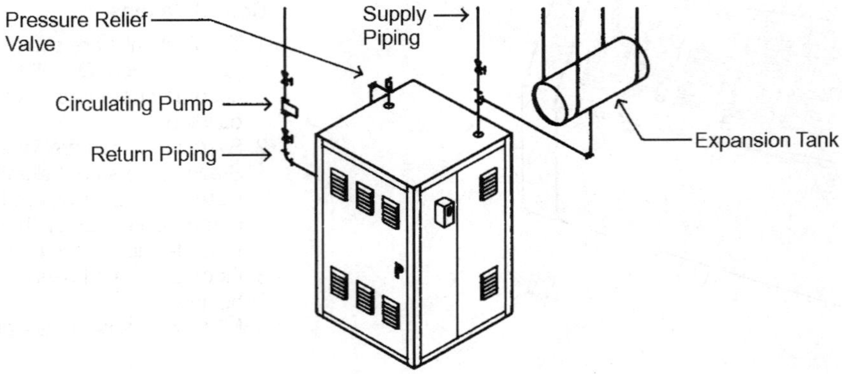

Electric Boiler, Hot Water

System Components	QUANTITY	UNIT	COST EACH MAT.	COST EACH INST.	COST EACH TOTAL
SYSTEM D3020 126 1010					
BOILER, ELECTRIC, HOT WATER, 15 kW, 52 MBH					
Boilers, electric, ASME, hot water, 15 KW, 52 MBH	1.000	Ea.	4,050	1,250	5,300
Pipe, black steel, Sch 40, threaded, W/coupling & hangers, 10′ OC, 3/4″ dia	10.000	L.F.	33.35	110.40	143.75
Pipe, black steel, Sch 40, threaded, W/coupling & hangers, 10′ OC, 1″ dia	20.000	L.F.	91.59	237.58	329.17
Elbow, 90°, black, straight, 3/4″ dia.	6.000	Ea.	22.26	252	274.26
Elbow, 90°, black, straight, 1″ dia.	6.000	Ea.	28.20	270	298.20
Tee, black, straight, 3/4″ dia.	1.000	Ea.	4.30	65	69.30
Tee, black, straight, 1″ dia.	2.000	Ea.	14.70	146	160.70
Tee, black, reducing, 1″ dia.	2.000	Ea.	23.60	146	169.60
Union, black with brass seat, 3/4″ dia.	1.000	Ea.	11.35	45	56.35
Union, black with brass seat, 1″ dia.	2.000	Ea.	29.70	98	127.70
Pipe nipples, 3/4″ diam.	3.000	Ea.	4.35	14.40	18.75
Pipe nipples, 1″ diam.	3.000	Ea.	6.39	16.58	22.97
Valves, bronze, gate, N.R.S., threaded, class 150, 3/4″ size	2.000	Ea.	124	59	183
Valves, bronze, gate, N.R.S., threaded, class 150, 1″ size	2.000	Ea.	157	62	219
Thermometer, stem type, 9″ case, 8″ stem, 3/4″ NPT	1.000	Ea.	240	42	282
Tank, steel, liquid expansion, ASME, painted, 15 gallon capacity	1.000	Ea.	485	63	548
Pump, circulating, bronze, flange connection, 3/4″ to 1-1/2″ size, 1/8 HP	1.000	Ea.	825	176	1,001
Insulation, fiberglass pipe covering, 1″ wall, 1″ IPS	20.000	L.F.	22.80	90.40	113.20
TOTAL			6,173.59	3,143.36	9,316.95

D3020 126	Electric Boiler, Hot Water	COST EACH MAT.	COST EACH INST.	COST EACH TOTAL
1010	Boiler, electric, hot water, 15 kW, 52 MBH	6,175	3,150	9,325
1020	30 kW, 103 MBH	6,950	3,300	10,250
1030	60 kW, 205 MBH	7,100	3,425	10,525
1040	120 kW, 410 MBH	9,450	4,100	13,550
1050	150 kW, 510 MBH	14,500	7,900	22,400
1060	210 kW, 716 MBH	17,000	8,675	25,675
1070	296 kW, 1010 MBH	22,500	8,925	31,425
1080	444 kW, 1515 MBH	27,400	12,000	39,400
1090	666 kW, 2273 MBH	32,800	13,000	45,800
1100	900 kW, 3071 MBH	40,700	15,200	55,900
1110	1320 kW, 4505 MBH	48,600	15,900	64,500
1120	2100 kW, 7167 MBH	73,500	19,300	92,800
1130	2610 kW, 8905 MBH	91,000	23,200	114,200
1140	3600 kW, 12,283 MBH	111,500	27,900	139,400

D3020 Heat Generating Systems

D3020 128	Electric Boiler, Steam	COST EACH		
		MAT.	INST.	TOTAL
1010	Boiler, electric, steam, 18 kW, 61.4 MBH	4,850	3,000	7,850
1020	36 kW, 123 MBH	6,100	3,200	9,300
1030	60 kW, 205 MBH	12,500	6,475	18,975
1040	120 kW, 409 MBH	14,800	7,150	21,950
1050	150 kW, 512 MBH	17,800	10,800	28,600
1060	210 kW, 716 MBH	23,700	11,600	35,300
1070	300 kW, 1023 MBH	28,000	12,700	40,700
1080	510 kW, 1740 MBH	35,000	16,400	51,400
1090	720 kW, 2456 MBH	39,700	17,200	56,900
1100	1080 kW, 3685 MBH	46,700	20,600	67,300
1110	1260 kW, 4300 MBH	49,400	21,900	71,300
1120	1620 kW, 5527 MBH	52,000	24,700	76,700
1130	2070 kW, 7063 MBH	69,000	25,900	94,900
1140	2340 kW, 7984 MBH	90,000	29,500	119,500

D3020 Heat Generating Systems

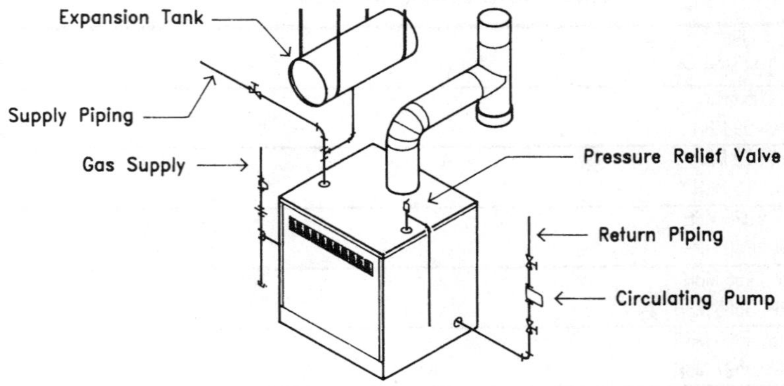

Cast Iron Boiler, Hot Water, Gas Fired

System Components	QUANTITY	UNIT	COST EACH MAT.	COST EACH INST.	COST EACH TOTAL
SYSTEM D3020 130 1010					
BOILER, CAST IRON, GAS, HOT WATER, 100 MBH					
Boilers, gas fired, std controls, CI, insulated, HW, gross output 100 MBH	1.000	Ea.	2,150	1,675	3,825
Pipe, black steel, Sch 40, threaded, W/coupling & hangers, 10′ OC, 3/4″ dia	20.000	L.F.	65.25	216	281.25
Pipe, black steel, Sch 40, threaded, W/coupling & hangers, 10′ OC, 1″ dia.	20.000	L.F.	91.59	237.58	329.17
Elbow, 90°, black, straight, 3/4″ dia.	9.000	Ea.	33.39	378	411.39
Elbow, 90°, black, straight, 1″ dia.	6.000	Ea.	28.20	270	298.20
Tee, black, straight, 3/4″ dia.	2.000	Ea.	8.60	130	138.60
Tee, black, straight, 1″ dia.	2.000	Ea.	14.70	146	160.70
Tee, black, reducing, 1″ dia.	2.000	Ea.	23.60	146	169.60
Pipe cap, black, 3/4″ dia.	1.000	Ea.	3.07	18.30	21.37
Union, black with brass seat, 3/4″ dia.	2.000	Ea.	22.70	90	112.70
Union, black with brass seat, 1″ dia.	2.000	Ea.	29.70	98	127.70
Pipe nipples, black, 3/4″ dia.	5.000	Ea.	7.25	24	31.25
Pipe nipples, black, 1″ dia.	3.000	Ea.	6.39	16.58	22.97
Valves, bronze, gate, N.R.S., threaded, class 150, 3/4″ size	2.000	Ea.	124	59	183
Valves, bronze, gate, N.R.S., threaded, class 150, 1″ size	2.000	Ea.	157	62	219
Gas cock, brass, 3/4″ size	1.000	Ea.	15.05	26.50	41.55
Thermometer, stem type, 9″ case, 8″ stem, 3/4″ NPT	2.000	Ea.	240	42	282
Tank, steel, liquid expansion, ASME, painted, 15 gallon capacity	1.000	Ea.	485	63	548
Pump, circulating, bronze, flange connection, 3/4″ to 1-1/2″ size, 1/8 HP	1.000	Ea.	825	176	1,001
Vent chimney, all fuel, pressure tight, double wall, SS, 6″ dia.	20.000	L.F.	1,040	349	1,389
Vent chimney, elbow, 90° fixed, 6″ dia.	2.000	Ea.	586	70	656
Vent chimney, Tee, 6″ dia.	2.000	Ea.	386	87	473
Vent chimney, ventilated roof thimble, 6″ dia.	1.000	Ea.	247	40	287
Vent chimney, adjustable roof flashing, 6″ dia.	1.000	Ea.	69	35	104
Vent chimney, stack cap, 6″ diameter	1.000	Ea.	218	22.50	240.50
Insulation, fiberglass pipe covering, 1″ wall, 1″ IPS	20.000	L.F.	22.80	90.40	113.20
TOTAL			6,899.29	4,567.86	11,467.15

D3020 130	Boiler, Cast Iron, Hot Water, Gas	COST EACH MAT.	COST EACH INST.	COST EACH TOTAL
1010	Boiler, cast iron, gas, hot water, 100 MBH	6,900	4,575	11,475
1020	200 MBH	8,000	5,225	13,225
1030	320 MBH	8,650	5,675	14,325
1040	440 MBH	11,700	7,325	19,025
1050	544 MBH	19,300	11,500	30,800
1060	765 MBH	24,000	12,300	36,300

D3020 Heat Generating Systems

D3020 130	Boiler, Cast Iron, Hot Water, Gas	COST EACH		
		MAT.	INST.	TOTAL
1070	1088 MBH	26,700	13,200	39,900
1080	1530 MBH	31,300	17,400	48,700
1090	2312 MBH	38,000	20,500	58,500
1100	2856 MBH	45,600	22,100	67,700
1110	3808 MBH	50,500	25,800	76,300
1120	4720 MBH	90,500	29,000	119,500
1130	6100 MBH	113,000	32,300	145,300
1140	6970 MBH	121,000	39,900	160,900

D3020 134	Boiler, Cast Iron, Steam, Gas	COST EACH		
		MAT.	INST.	TOTAL
1010	Boiler, cast iron, gas, steam, 100 MBH	5,875	4,400	10,275
1020	200 MBH	11,900	8,300	20,200
1030	320 MBH	13,700	9,325	23,025
1040	544 MBH	20,600	14,600	35,200
1050	765 MBH	23,600	15,200	38,800
1060	1275 MBH	28,200	16,700	44,900
1070	2675 MBH	39,100	24,300	63,400
1080	3570 MBH	46,000	27,300	73,300
1090	4720 MBH	50,500	29,200	79,700
1100	6100 MBH	102,000	33,900	135,900
1110	6970 MBH	112,000	39,700	151,700

D3020 136	Boiler, Cast Iron, Hot Water, Gas/Oil	COST EACH		
		MAT.	INST.	TOTAL
1010	Boiler, cast iron, gas & oil, hot water, 584 MBH	25,400	12,700	38,100
1020	876 MBH	29,500	13,700	43,200
1030	1168 MBH	31,500	15,400	46,900
1040	1460 MBH	35,600	18,500	54,100
1050	2044 MBH	42,200	19,200	61,400
1060	2628 MBH	53,000	23,200	76,200
1070	3210 MBH	60,500	24,900	85,400
1080	3796 MBH	65,500	25,700	91,200
1090	4672 MBH	76,000	27,000	103,000
1100	5256 MBH	91,500	32,800	124,300
1110	6000 MBH	110,500	33,300	143,800
1120	9800 MBH	137,500	56,500	194,000
1130	12,200 MBH	172,000	66,000	238,000
1140	13,500 MBH	186,500	70,500	257,000

D3020 138	Boiler, Cast Iron, Steam, Gas/Oil	COST EACH		
		MAT.	INST.	TOTAL
1010	Boiler, cast iron, gas & oil, steam, 810 MBH	27,900	16,500	44,400
1020	1360 MBH	31,300	17,600	48,900
1030	2040 MBH	40,900	22,600	63,500
1040	2700 MBH	44,800	25,100	69,900
1050	3270 MBH	50,500	28,100	78,600
1060	3770 MBH	84,500	28,600	113,100
1070	4650 MBH	95,000	32,600	127,600
1080	5520 MBH	110,000	34,300	144,300
1090	6100 MBH	112,000	36,100	148,100
1100	6970 MBH	122,500	44,000	166,500

D3020 Heat Generating Systems

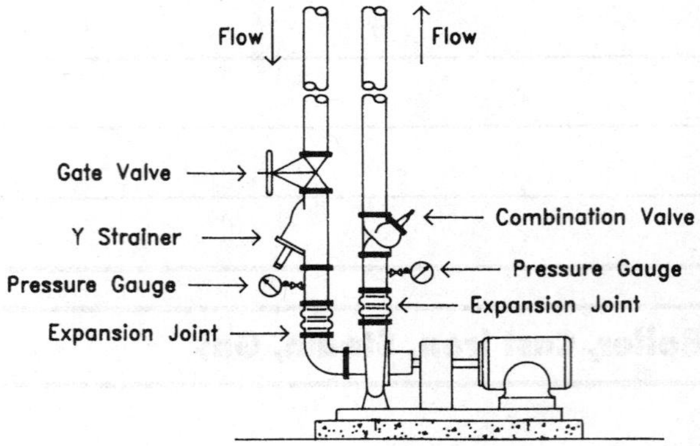

Base Mounted End–Suction Pump

System Components	QUANTITY	UNIT	COST EACH MAT.	COST EACH INST.	COST EACH TOTAL
SYSTEM D3020 330 1010					
PUMP, BASE MTD. WITH MOTOR, END-SUCTION, 2-1/2" SIZE, 3 HP, TO 150 GPM					
Pump, circulating, CI, base mounted, 2-1/2" size, 3 HP, to 150 GPM	1.000	Ea.	4,075	585	4,660
Pipe, black steel, Sch. 40, on yoke & roll hangers, 10' O.C., 2-1/2" dia	12.000	L.F.	129.60	285.60	415.20
Elbow, 90°, weld joint, steel, 2-1/2" pipe size	1.000	Ea.	18.65	139.65	158.30
Flange, weld neck, 150 LB, 2-1/2" pipe size	8.000	Ea.	204	558.64	762.64
Valve, iron body, gate, 125 lb., N.R.S., flanged, 2-1/2" size	1.000	Ea.	560	211	771
Strainer, Y type, iron body, flanged, 125 lb., 2-1/2" pipe size	1.000	Ea.	141	213	354
Multipurpose valve, CI body, 2" size	1.000	Ea.	415	74	489
Expansion joint, flanged spool, 6" F to F, 2-1/2" dia	2.000	Ea.	572	172	744
T-O-L, weld joint, socket, 1/4" pipe size, nozzle	2.000	Ea.	13.40	97.32	110.72
T-O-L, weld joint, socket, 1/2" pipe size, nozzle	2.000	Ea.	13.40	101.58	114.98
Control gauges, pressure or vacuum, 3-1/2" diameter dial	2.000	Ea.	79	37	116
Pressure/temperature relief plug, 316 SS, 3/4" OD, 7-1/2" insertion	2.000	Ea.	150	37	187
Insulation, fiberglass pipe covering, 1-1/2" wall, 2-1/2" IPS	12.000	L.F.	31.92	66.60	98.52
Pump control system	1.000	Ea.	1,200	540	1,740
Pump balancing	1.000	Ea.		261	261
TOTAL			7,602.97	3,379.39	10,982.36

D3020 330	Circulating Pump Systems, End Suction	COST EACH MAT.	COST EACH INST.	COST EACH TOTAL
1010	Pump, base mtd with motor, end-suction, 2-1/2" size, 3 HP, to 150 GPM	7,600	3,400	11,000
1020	3" size, 5 HP, to 225 GPM	8,300	3,875	12,175
1030	4" size, 7-1/2 HP, to 350 GPM	9,475	4,650	14,125
1040	5" size, 15 HP, to 1000 GPM	12,300	7,075	19,375
1050	6" size, 25 HP, to 1550 GPM	17,100	8,575	25,675

D3020 340	Circulating Pump Systems, Double Suction	COST EACH MAT.	COST EACH INST.	COST EACH TOTAL
1010	Pump, base mtd w/ motor, double suction, 6" size, 50 HP, to 1200 GPM	17,300	12,300	29,600
1020	8" size, 75 HP, to 2500 GPM	32,700	16,700	49,400
1030	8" size, 100 HP, to 3000 GPM	47,100	18,700	65,800
1040	10" size, 150 HP, to 4000 GPM	69,500	21,200	90,700

D3030 Cooling Generating Systems

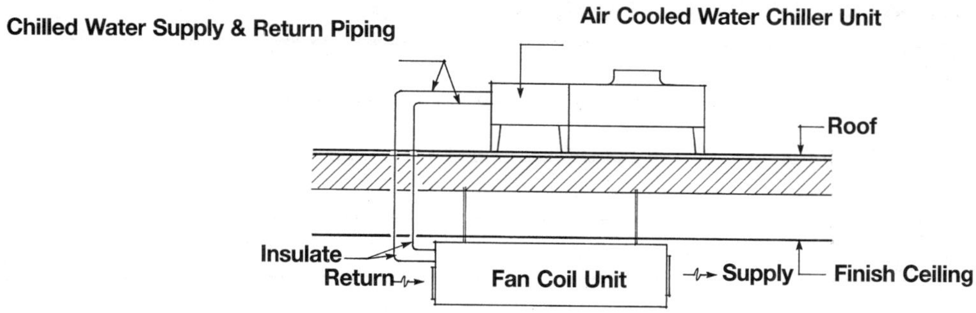

Design Assumptions: The chilled water, air cooled systems priced, utilize reciprocating hermetic compressors and propeller-type condenser fans. Piping with pumps and expansion tanks is included based on a two pipe system. No ducting is included and the fan-coil units are cooling only. Water treatment and balancing are not included. Chilled water piping is insulated. Area distribution is through the use of multiple fan coil units. Fewer but larger fan coil units with duct distribution would be approximately the same S.F. cost.

System Components	QUANTITY	UNIT	COST EACH MAT.	COST EACH INST.	COST EACH TOTAL
SYSTEM D3030 110 1200					
PACKAGED CHILLER, AIR COOLED, WITH FAN COIL UNIT					
APARTMENT CORRIDORS, 3,000 S.F., 5.50 TON					
Fan coil air conditioning unit, cabinet mounted & filters chilled water	1.000	Ea.	3,345.23	489.41	3,834.64
Water chiller,air conditioning unit, reciprocating, air cooled,	1.000	Ea.	6,572.50	1,828.75	8,401.25
Chilled water unit coil connections	1.000	Ea.	980	1,325	2,305
Chilled water distribution piping	440.000	L.F.	10,340	18,040	28,380
TOTAL			21,237.73	21,683.16	42,920.89
COST PER S.F.			7.08	7.23	14.31
*Cooling requirements would lead to choosing a water cooled unit					

D3030 110	Chilled Water, Air Cooled Condenser Systems	COST PER S.F. MAT.	COST PER S.F. INST.	COST PER S.F. TOTAL
1180	Packaged chiller, air cooled, with fan coil unit			
1200	Apartment corridors, 3,000 S.F., 5.50 ton	7.07	7.23	14.30
1240	6,000 S.F., 11.00 ton	5.85	5.80	11.65
1280	10,000 S.F., 18.33 ton	5	4.40	9.40
1320	20,000 S.F., 36.66 ton	3.73	3.18	6.91
1360	40,000 S.F., 73.33 ton	4.16	3.20	7.36
1440	Banks and libraries, 3,000 S.F., 12.50 ton	10.60	8.50	19.10
1480	6,000 S.F., 25.00 ton	9.60	6.95	16.55
1520	10,000 S.F., 41.66 ton	7.15	4.83	11.98
1560	20,000 S.F., 83.33 ton	7.30	4.39	11.69
1600	40,000 S.F., 167 ton*			
1680	Bars and taverns, 3,000 S.F., 33.25 ton	17.30	10.10	27.40
1720	6,000 S.F., 66.50 ton	15.95	7.80	23.75
1760	10,000 S.F., 110.83 ton	12.75	2.50	15.25
1800	20,000 S.F., 220 ton*			
1840	40,000 S.F., 440 ton*			
1920	Bowling alleys, 3,000 S.F., 17.00 ton	13.15	9.55	22.70
1960	6,000 S.F., 34.00 ton	9.70	6.75	16.45
2000	10,000 S.F., 56.66 ton	8.60	4.88	13.48
2040	20,000 S.F., 113.33 ton	8.40	4.44	12.84
2080	40,000 S.F., 227 ton*			
2160	Department stores, 3,000 S.F., 8.75 ton	9.55	8.05	17.60
2200	6,000 S.F., 17.50 ton	7.50	6.25	13.75
2240	10,000 S.F., 29.17 ton	5.75	4.50	10.25
2280	20,000 S.F., 58.33 ton	4.95	3.33	8.28
2320	40,000 S.F., 116.66 ton	5.45	3.31	8.76

D3030 Cooling Generating Systems

D3030 110	Chilled Water, Air Cooled Condenser Systems	COST PER S.F.		
		MAT.	INST.	TOTAL
2400	Drug stores, 3,000 S.F., 20.00 ton	14.85	9.90	24.75
2440	6,000 S.F., 40.00 ton	11.35	7.35	18.70
2480	10,000 S.F., 66.66 ton	11.25	6.30	17.55
2520	20,000 S.F., 133.33 ton	10.40	4.68	15.08
2560	40,000 S.F., 267 ton*			
2640	Factories, 2,000 S.F., 10.00 ton	9.30	8.15	17.45
2680	6,000 S.F., 20.00 ton	8.35	6.60	14.95
2720	10,000 S.F., 33.33 ton	6.20	4.61	10.81
2760	20,000 S.F., 66.66 ton	6.35	4.17	10.52
2800	40,000 S.F., 133.33 ton	5.95	3.40	9.35
2880	Food supermarkets, 3,000 S.F., 8.50 ton	9.40	8	17.40
2920	6,000 S.F., 17.00 ton	7.35	6.20	13.55
2960	10,000 S.F., 28.33 ton	5.50	4.38	9.88
3000	20,000 S.F., 56.66 ton	4.78	3.26	8.04
3040	40,000 S.F., 113.33 ton	5.20	3.32	8.52
3120	Medical centers, 3,000 S.F., 7.00 ton	8.50	7.80	16.30
3160	6,000 S.F., 14.00 ton	6.60	6	12.60
3200	10,000 S.F., 23.33 ton	5.80	4.58	10.38
3240	20,000 S.F., 46.66 ton	4.39	3.36	7.75
3280	40,000 S.F., 93.33 ton	4.68	3.25	7.93
3360	Offices, 3,000 S.F., 9.50 ton	8.65	7.95	16.60
3400	6,000 S.F., 19.00 ton	8.10	6.50	14.60
3440	10,000 S.F., 31.66 ton	6	4.56	10.56
3480	20,000 S.F., 63.33 ton	6.30	4.24	10.54
3520	40,000 S.F., 126.66 ton	5.85	3.43	9.28
3600	Restaurants, 3,000 S.F., 15.00 ton	11.85	8.85	20.70
3640	6,000 S.F., 30.00 ton	9.20	6.75	15.95
3680	10,000 S.F., 50.00 ton	8.20	4.96	13.16
3720	20,000 S.F., 100.00 ton	8.45	4.65	13.10
3760	40,000 S.F., 200 ton*			
3840	Schools and colleges, 3,000 S.F., 11.50 ton	10.10	8.35	18.45
3880	6,000 S.F., 23.00 ton	9.10	6.80	15.90
3920	10,000 S.F., 38.33 ton	6.75	4.74	11.49
3960	20,000 S.F., 76.66 ton	7.10	4.43	11.53
4000	40,000 S.F., 153 ton*			

D3030 Cooling Generating Systems

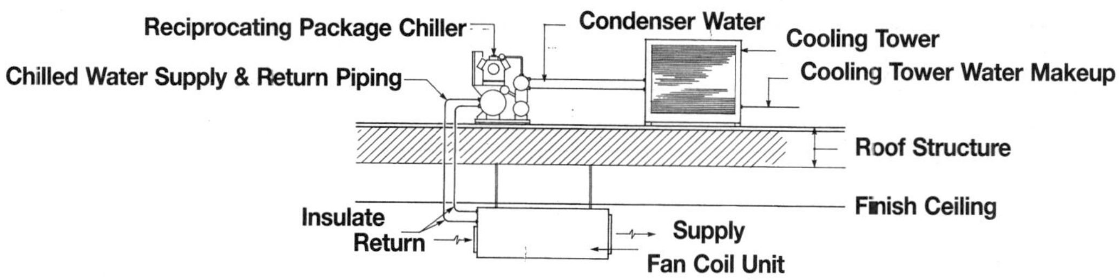

General: Water cooled chillers are available in the same sizes as air cooled units. They are also available in larger capacities.

Design Assumptions: The chilled water systems with water cooled condenser include reciprocating hermetic compressors, water cooling tower, pumps, piping and expansion tanks and are based on a two pipe system. Chilled water piping is insulated. No ducts are included and fan-coil units are cooling only. Area distribution is through use of multiple fan coil units. Fewer but larger fan coil units with duct distribution would be approximately the same S.F. cost. Water treatment and balancing are not included.

System Components	QUANTITY	UNIT	COST EACH MAT.	COST EACH INST.	COST EACH TOTAL
SYSTEM D3030 115 1320					
PACKAGED CHILLER, WATER COOLED, WITH FAN COIL UNIT					
APARTMENT CORRIDORS, 4,000 S.F., 7.33 TON					
Fan coil air conditioner unit, cabinet mounted & filters, chilled water	2.000	Ea.	4,458.48	652.28	5,110.76
Water chiller, water cooled, 1 compressor, hermetic scroll,	1.000	Ea.	5,358.60	3,183.10	8,541.70
Cooling tower, draw thru single flow, belt drive	1.000	Ea.	1,106.83	135.24	1,242.07
Cooling tower pumps & piping	1.000	System	553.42	318.86	872.28
Chilled water unit coil connections	2.000	Ea.	1,960	2,650	4,610
Chilled water distribution piping	520.000	L.F.	12,220	21,320	33,540
TOTAL			25,657.33	28,259.48	53,916.81
COST PER S.F.			6.41	7.06	13.47
*Cooling requirements would lead to choosing a water cooled unit					

D3030 115	Chilled Water, Cooling Tower Systems	COST PER S.F. MAT.	COST PER S.F. INST.	COST PER S.F. TOTAL
1300	Packaged chiller, water cooled, with fan coil unit			
1320	Apartment corridors, 4,000 S.F., 7.33 ton	6.43	7.08	13.51
1360	6,000 S.F., 11.00 ton (R236000 -20)	5.40	6.10	11.50
1400	10,000 S.F., 18.33 ton	4.88	4.51	9.39
1440	20,000 S.F., 26.66 ton	3.86	3.37	7.23
1480	40,000 S.F., 73.33 ton	3.85	3.44	7.29
1520	60,000 S.F., 110.00 ton	4.06	3.54	7.60
1600	Banks and libraries, 4,000 S.F., 16.66 ton	10.05	7.75	17.80
1640	6,000 S.F., 25.00 ton	8.65	6.85	15.50
1680	10,000 S.F., 41.66 ton	7	5.10	12.10
1720	20,000 S.F., 83.33 ton	6.50	4.89	11.39
1760	40,000 S.F., 166.66 ton	7.20	5.90	13.10
1800	60,000 S.F., 250.00 ton	7.30	6.30	13.60
1880	Bars and taverns, 4,000 S.F., 44.33 ton	16.25	9.80	26.05
1920	6,000 S.F., 66.50 ton	18.60	10.15	28.75
1960	10,000 S.F., 110.83 ton	14.75	8.05	22.80
2000	20,000 S.F., 221.66 ton	17.80	8.15	25.95
2040	40,000 S.F., 440 ton*			
2080	60,000 S.F., 660 ton*			
2160	Bowling alleys, 4,000 S.F., 22.66 ton	11.75	8.55	20.30
2200	6,000 S.F., 34.00 ton	10.30	7.55	17.85
2240	10,000 S.F., 56.66 ton	9.55	5.50	15.05
2280	20,000 S.F., 113.33 ton	8.25	5.15	13.40
2320	40,000 S.F., 226.66 ton	10.05	5.90	15.95

D3030 Cooling Generating Systems

D3030 115	Chilled Water, Cooling Tower Systems	COST PER S.F.		
		MAT.	INST.	TOTAL
2360	60,000 S.F., 340 ton			
2440	Department stores, 4,000 S.F., 11.66 ton	7.25	7.65	14.90
2480	6,000 S.F., 17.50 ton	7.55	6.30	13.85
2520	10,000 S.F., 29.17 ton	5.75	4.72	10.47
2560	20,000 S.F., 58.33 ton	5.10	3.51	8.61
2600	40,000 S.F., 116.66 ton	4.94	3.63	8.57
2640	60,000 S.F., 175.00 ton	6.55	5.75	12.30
2720	Drug stores, 4,000 S.F., 26.66 ton	12.55	8.80	21.35
2760	6,000 S.F., 40.00 ton	11	7.55	18.55
2800	10,000 S.F., 66.66 ton	11.80	6.85	18.65
2840	20,000 S.F., 133.33 ton	9.50	5.40	14.90
2880	40,000 S.F., 266.67 ton	9.90	6.70	16.60
2920	60,000 S.F., 400 ton*			
3000	Factories, 4,000 S.F., 13.33 ton	8.80	7.40	16.20
3040	6,000 S.F., 20.00 ton	7.70	6.40	14.10
3080	10,000 S.F., 33.33 ton	6.20	4.87	11.07
3120	20,000 S.F., 66.66 ton	6.55	4.38	10.93
3160	40,000 S.F., 133.33 ton	5.55	3.75	9.30
3200	60,000 S.F., 200.00 ton	6.60	6	12.60
3280	Food supermarkets, 4,000 S.F., 11.33 ton	7.10	7.60	14.70
3320	6,000 S.F., 17.00 ton	6.90	6.20	13.10
3360	10,000 S.F., 28.33 ton	5.65	4.69	10.34
3400	20,000 S.F., 56.66 ton	5.15	3.52	8.67
3440	40,000 S.F., 113.33 ton	4.89	3.64	8.53
3480	60,000 S.F., 170.00 ton	6.50	5.70	12.20
3560	Medical centers, 4.000 S.F., 9.33 ton	6.25	7	13.25
3600	6,000 S.F., 14.00 ton	6.65	6.05	12.70
3640	10,000 S.F., 23.33 ton	5.25	4.53	9.78
3680	20,000 S.F., 46.66 ton	4.63	3.43	8.06
3720	40,000 S.F., 93.33 ton	4.41	3.56	7.97
3760	60,000 S.F., 140.00 ton	5.40	5.80	11.20
3840	Offices, 4,000 S.F., 12.66 ton	8.55	7.30	15.85
3880	6,000 S.F., 19.00 ton	7.65	6.55	14.20
3920	10,000 S.F., 31.66 ton	6.20	4.92	11.12
3960	20,000 S.F., 63.33 ton	6.40	4.40	10.80
4000	40,000 S.F., 126.66 ton	5.90	5.65	11.55
4040	60,000 S.F., 190.00 ton	6.40	5.95	12.35
4120	Restaurants, 4,000 S.F., 20.00 ton	10.60	7.95	18.55
4160	6,000 S.F., 30.00 ton	9.35	7.05	16.40
4200	10,000 S.F., 50.00 ton	8.80	5.30	14.10
4240	20,000 S.F., 100.00 ton	7.70	5.15	12.85
4280	40,000 S.F., 200.00 ton	7.75	5.70	13.45
4320	60,000 S.F., 300.00 ton	8.20	6.50	14.70
4400	Schools and colleges, 4,000 S.F., 15.33 ton	9.55	7.60	17.15
4440	6,000 S.F., 23.00 ton	8.20	6.70	14.90
4480	10,000 S.F., 38.33 ton	6.65	4.98	11.63
4520	20,000 S.F., 76.66 ton	6.30	4.82	11.12
4560	40,000 S.F., 153.33 ton	6.80	5.75	12.55
4600	60,000 S.F., 230.00 ton	6.80	6.05	12.85

D3030 Cooling Generating Systems

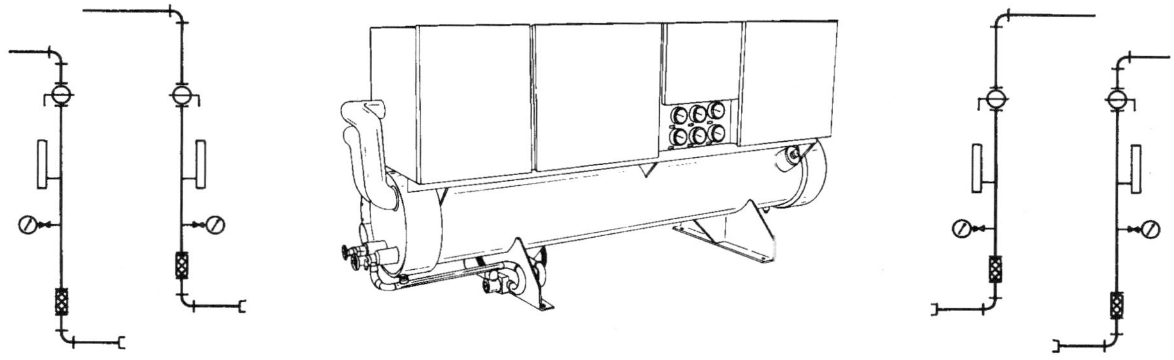

Chilled Water Piping — Reciprocating Chiller, Water Cooled — Condenser Water Piping

System Components

System Components	QUANTITY	UNIT	COST EACH MAT.	COST EACH INST.	COST EACH TOTAL
SYSTEM D3030 130 1010					
CHILLER, RECIPROCATING, WATER COOLED, STD. CONTROLS, 60 TON					
Water chiller, reciprocating, water cooled, 60 ton cooling	1.000	Ea.	38,000	8,900	46,900
Pipe, black steel, welded, Sch. 40, on yoke & roll hgrs, 10′ O.C., 4″ dia	40.000	L.F.	796	1,206.40	2,002.40
Elbow, 90°, weld joint, butt, 4″ pipe size	8.000	Ea.	300	1,786	2,086
Flange, welding neck, 150 lb., 4″ pipe size	16.000	Ea.	608	1,778.40	2,386.40
Valves, iron body, butterfly, lug type, lever actuator, 4″ size	4.000	Ea.	656	844	1,500
Expansion joints, bellows type, flange spool, 6″ F to F, 4″ dia	4.000	Ea.	1,380	508	1,888
T-O-L socket, weld joint, 1/4″ pipe size, nozzle	4.000	Ea.	26.80	194.64	221.44
T-O-L socket, weld joint, 3/4″ pipe size, nozzle	4.000	Ea.	30.80	211.68	242.48
Thermometers, stem type, 9″ case, 8″ stem, 3/4″ NPT	4.000	Ea.	480	84	564
Gauges, pressure or vacuum, 3-1/2″ diameter dial	4.000	Ea.	158	74	232
Insulation, fiberglass pipe covering, 1-1/2″ wall, 4″ IPS	20.000	L.F.	63.40	142	205.40
Chiller balancing	1.000	Ea.		505	505
TOTAL			42,499	16,234.12	58,733.12

D3030 130 Reciprocating Chiller, Water Cooled

D3030 130	Reciprocating Chiller, Water Cooled	COST EACH MAT.	COST EACH INST.	COST EACH TOTAL
1010	Chiller, reciprocating, water cooled, std. controls, 60 Ton	42,500	16,200	58,700
1020	100 Ton	45,900	24,400	70,300
1030	150 Ton	76,500	26,900	103,400
1040	200 Ton	103,000	28,300	131,300

D3030 135 Reciprocating Chiller, Air Cooled

D3030 135	Reciprocating Chiller, Air Cooled	COST EACH MAT.	COST EACH INST.	COST EACH TOTAL
1010	Chiller, reciprocating, air cooled, std. controls, 20 Ton	25,600	9,550	35,150
1020	30 Ton	28,500	10,100	38,600
1030	40 Ton	34,500	10,800	45,300
1040	60 Ton	52,000	12,100	64,100
1050	70 Ton	59,500	12,400	71,900
1060	80 Ton	64,500	12,600	77,100
1070	90 Ton	69,000	12,800	81,800
1080	100 Ton	77,500	13,000	90,500
1090	110 Ton	84,500	14,200	98,700
1100	130 Ton	99,500	14,400	113,900

D3030 Cooling Generating Systems

D3030 135	Reciprocating Chiller, Air Cooled	COST EACH		
		MAT.	INST.	TOTAL
1110	150 Ton	110,000	14,600	124,600
1120	175 Ton	120,000	14,800	134,800
1130	210 Ton	140,000	16,100	156,100

D3030 Cooling Generating Systems

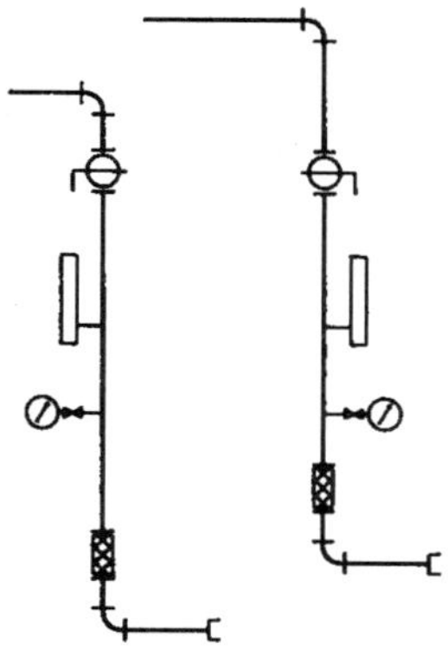

Chilled Water Piping

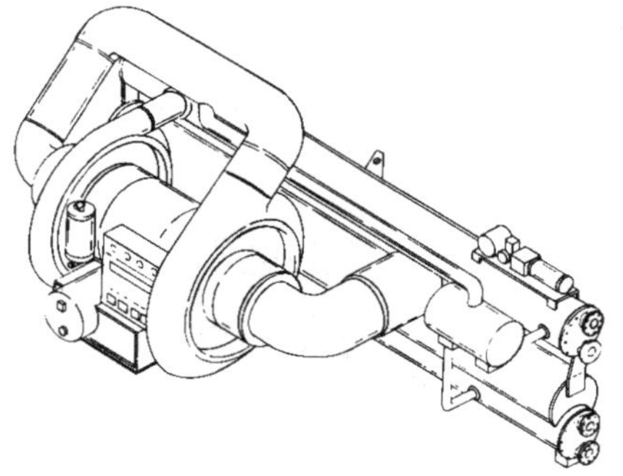

Centrifugal Chiller, Water Cooled, Hermetic

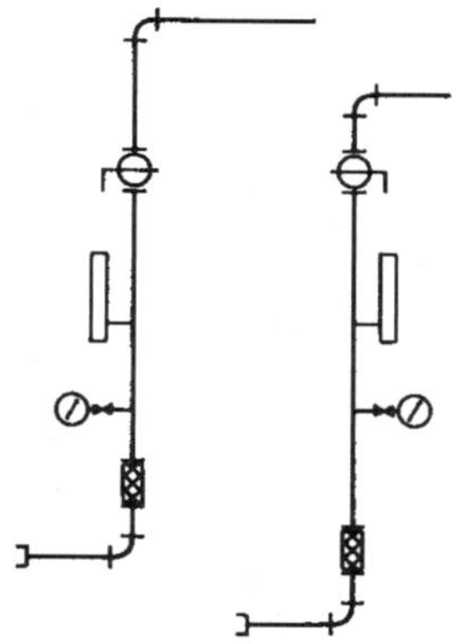

Condenser Water Piping

System Components	QUANTITY	UNIT	COST EACH		
			MAT.	INST.	TOTAL
SYSTEM D3030 140 1010					
CHILLER, CENTRIF., WATER COOLED, PKGD. HERMETIC, STD. CONTROLS, 200 TON					
Water chiller, screw, water cooled, not incl tower, 200 ton	1.000	Ea.	90,000	17,800	107,800
Pipe, black steel, welded, Sch. 40, on yoke & roll hgrs, 10′ o.c., 6″ dia	40.000	L.F.	1,360	1,888	3,248
Elbow, 90°, weld joint, 6″ pipe size	8.000	Ea.	656	2,738	3,394
Flange, welding neck, 150 lb., 6″ pipe size	16.000	Ea.	960	2,721.60	3,681.60
Valves, iron body, butterfly, lug type, lever actuator, 6″ size	4.000	Ea.	1,100	1,320	2,420
Expansion joints, bellows type, flange spool, 6″ F to F, 6″ dia	4.000	Ea.	1,740	628	2,368
T-O-L socket, weld jiont, 1/4″ pipe size, nozzle	4.000	Ea.	26.80	194.64	221.44
T-O-L socket, weld joint, 3/4″ pipe size, nozzle	4.000	Ea.	30.80	211.68	242.48
Thermometers, stem type, 9″ case, 8″ stem, 3/4″ NPT	4.000	Ea.	480	84	564
Gauges, pressure or vacuum, 3-1/2″ diameter dial	4.000	Ea.	158	74	232
Insulation, fiberglass pipe covering, 1-1/2″ wall, 6″ IPS	40.000	L.F.	150	362	512
Chiller balancing	1.000	Ea.		505	505
TOTAL			96,661.60	28,526.92	125,188.52

D3030 140	Centrifugal Chiller, Water Cooled	COST EACH		
		MAT.	INST.	TOTAL
1010	Chiller, centrif., water cooled, pkgd. hermetic, std controls, 200 Ton	96,500	28,500	125,000
1020	400 Ton	158,500	39,000	197,500
1030	1,000 Ton	491,000	59,000	550,000
1040	1,500 Ton	717,500	72,000	789,500

D3030 Cooling Generating Systems

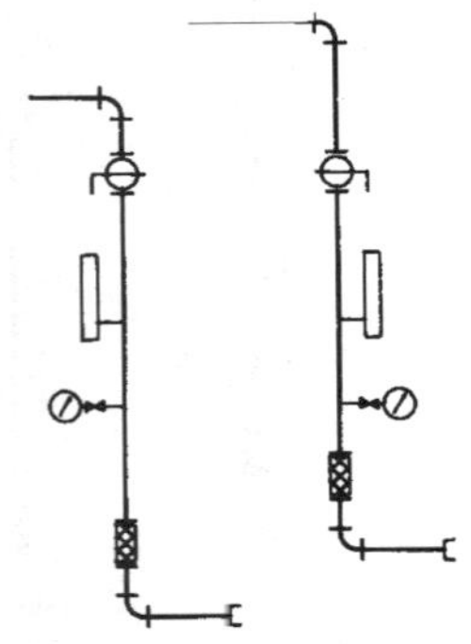

Chilled Water Piping
(Condenser Water Piping Similar)

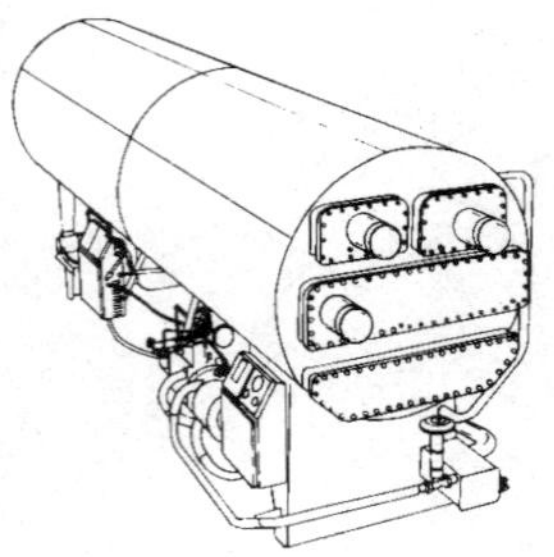

Gas Absorption Chiller, Water Cooled

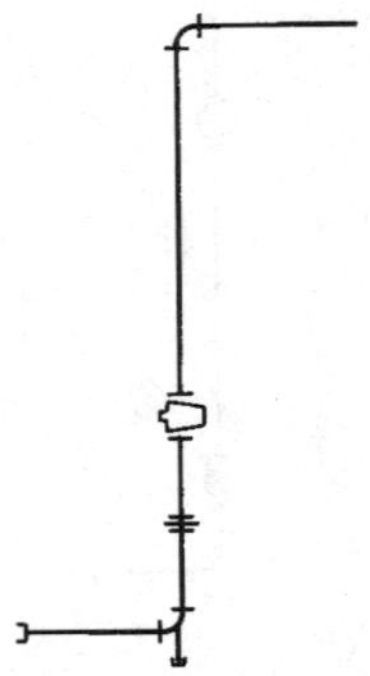

Gas Piping

System Components	QUANTITY	UNIT	COST EACH MAT.	COST EACH INST.	COST EACH TOTAL
SYSTEM D3030 145 1010					
CHILLER, GAS FIRED ABSORPTION, WATER COOLED, STD. CONTROLS, 800 TON					
Absorption water chiller, gas fired, water cooled, 800 ton	1.000	Ea.	625,000	26,900	651,900
Pipe, black steel weld, Sch. 40, on yoke & roll hangers, 10′ O.C., 10″ dia	20.000	L.F.	1,480	1,421	2,901
Pipe, black steel welded, Sch. 40, 14″ dia.	44.000	L.F.	4,400	4,975.52	9,375.52
Pipe, black steel welded, Sch. 40, on yoke & roll hangers, 10′ O.C., 4″ dia	20.000	L.F.	398	603.20	1,001.20
Elbow, 90°, weld joint, 10″ pipe size	4.000	Ea.	1,240	2,262	3,502
Elbow, 90°, weld joint, 14″ pipe size	6.000	Ea.	4,110	5,103	9,213
Elbow, 90°, weld joint, 4″ pipe size	3.000	Ea.	112.50	669.75	782.25
Flange, welding neck, 150 lb., 10″ pipe size	8.000	Ea.	1,256	2,265.60	3,521.60
Flange, welding neck, 150 lb., 14″ pipe size	14.000	Ea.	4,270	5,300.40	9,570.40
Flange, welding neck, 150 lb., 4″ pipe size	3.000	Ea.	114	333.45	447.45
Valves, iron body, butterfly, lug type, gear operated, 10″ size	2.000	Ea.	1,570	820	2,390
Valves, iron body, butterfly, lug type, gear operated, 14″ size	5.000	Ea.	10,875	3,550	14,425
Valves, semi-steel, lubricated plug valve, flanged, 4″ pipe size	1.000	Ea.	420	350	770
Expansion joints, bellows type, flange spool, 6″ F to F, 10″ dia	2.000	Ea.	1,400	426	1,826
Expansion joints, bellows type, flange spool, 10″ F to F, 14″ dia	2.000	Ea.	2,400	562	2,962
T-O-L socket, weld joint, 1/4″ pipe size, nozzle	4.000	Ea.	26.80	194.64	221.44
T-O-L socket, weld joint, 3/4″ pipe size, nozzle	4.000	Ea.	30.80	211.68	242.48
Thermometers, stem type, 9″ case, 8″ stem, 3/4″ NPT	4.000	Ea.	480	84	564
Gauges, pressure or vacuum, 3-1/2″ diameter dial	4.000	Ea.	158	74	232
Insulation, fiberglass pipe covering, 2″ wall, 10″ IPS	20.000	L.F.	167	284	451
Vent chimney, all fuel, pressure tight, double wall, SS, 12″ dia	20.000	L.F.	1,610	480	2,090
Chiller balancing	1.000	Ea.		505	505
TOTAL			661,518.10	57,375.24	718,893.34

D3030 145	Gas Absorption Chiller, Water Cooled	MAT.	INST.	TOTAL
1010	Chiller, gas fired absorption, water cooled, std. ctrls, 800 Ton	661,500	57,500	719,000
1020	1000 Ton	864,500	74,000	938,500

D3030 150	Steam Absorption Chiller, Water Cooled	COST EACH MAT.	COST EACH INST.	COST EACH TOTAL
1010	Chiller, steam absorption, water cooled, std. ctrls, 750 Ton	624,500	67,500	692,000
1020	955 Ton	727,000	81,000	808,000
1030	1465 Ton	1,052,500	93,000	1,145,500
1040	1660 Ton	1,269,000	105,500	1,374,500

D3030 Cooling Generating Systems

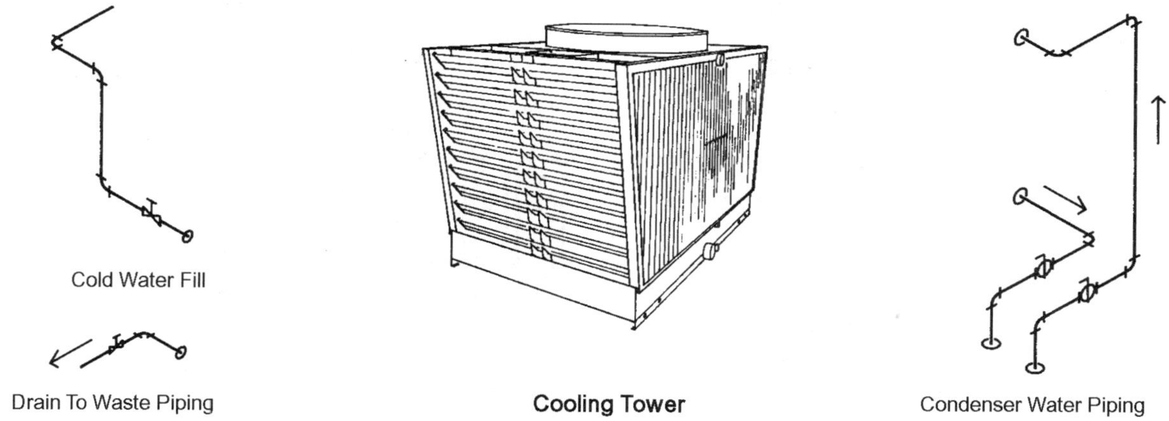

System Components	QUANTITY	UNIT	COST EACH MAT.	COST EACH INST.	COST EACH TOTAL
SYSTEM D3030 310 1010					
COOLING TOWER, GALVANIZED STEEL, PACKAGED UNIT, DRAW THRU, 60 TON					
Cooling towers, draw thru, single flow, belt drive, 60 tons	1.000	Ea.	9,060	1,107	10,167
Pipe, black steel, welded, Sch. 40, on yoke & roll hangers, 10′ OC, 4″ dia.	40.000	L.F.	796	1,206.40	2,002.40
Tubing, copper, Type L, couplings & hangers 10′ OC, 1-1/2″ dia.	20.000	L.F.	354	225	579
Elbow, 90°, copper, cu x cu, 1-1/2″ dia.	3.000	Ea.	53.25	135	188.25
Elbow, 90°, weld joint, 4″ pipe size	6.000	Ea.	225	1,339.50	1,564.50
Flange, welding neck, 150 lb., 4″ pipe size	6.000	Ea.	228	666.90	894.90
Valves, bronze, gate, N.R.S., soldered, 125 psi, 1-1/2″ size	1.000	Ea.	94.50	45	139.50
Valves, iron body, butterfly, lug type, lever actuator, 4″ size	2.000	Ea.	328	422	750
Electric heat trace system, 400 degree, 115v, 5 watts per L.F.	160.000	L.F.	1,176	169.60	1,345.60
Insulation, fiberglass pipe covering, , 1-1/2″ wall, 4″ IPS	40.000	L.F.	126.80	284	410.80
Insulation, fiberglass pipe covering, 1″ wall, 1-1/2″ IPS	20.000	L.F.	26.60	94.80	121.40
Cooling tower condenser control system	1.000	Ea.	5,175	2,425	7,600
Cooling tower balancing	1.000	Ea.		385	385
TOTAL			17,643.15	8,505.20	26,148.35

D3030 310	Galvanized Draw Through Cooling Tower	COST EACH MAT.	COST EACH INST.	COST EACH TOTAL
1010	Cooling tower, galvanized steel, packaged unit, draw thru, 60 Ton	17,600	8,500	26,100
1020	110 Ton	25,100	10,000	35,100
1030	300 Ton	38,800	15,000	53,800
1040	600 Ton	61,500	22,600	84,100
1050	1000 Ton	92,500	33,200	125,700

D3030 320	Fiberglass Draw Through Cooling Tower	COST EACH MAT.	COST EACH INST.	COST EACH TOTAL
1010	Cooling tower, fiberglass, packaged unit, draw thru, 60 Ton	12,600	8,500	21,100
1020	125 Ton	17,600	10,000	27,600
1030	300 Ton	30,700	15,000	45,700
1040	600 Ton	52,000	22,600	74,600
1050	1000 Ton	86,000	33,300	119,300

D3030 330	Stainless Steel Draw Through Cooling Tower	COST EACH MAT.	COST EACH INST.	COST EACH TOTAL
1010	Cooling tower, stainless steel, packaged unit, draw thru, 60 Ton	31,100	8,500	39,600
1020	110 Ton	36,500	10,000	46,500
1030	300 Ton	69,000	15,000	84,000
1040	600 Ton	108,500	22,300	130,800
1050	1000 Ton	169,500	33,300	202,800

D3040 Distribution Systems

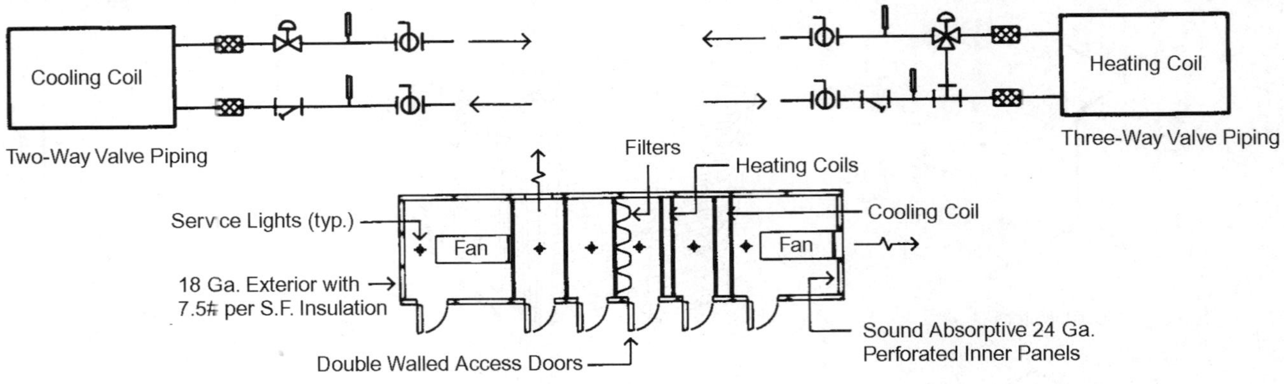

Built-Up Air Handling Unit

System Components	QUANTITY	UNIT	COST EACH MAT.	COST EACH INST.	COST EACH TOTAL
SYSTEM D3040 106 1010					
AHU, FIELD FAB., BLT. UP, COOL/HEAT COILS, FLTRS., CONST. VOL., 40,000 CFM					
AHU, Built-up, cool/heat coils, filters, mix box, constant volume	1.000	Ea.	53,500	5,525	59,025
Pipe, black steel, welded, Sch. 40, on yoke & roll hgrs, 10′ OC, 3″ dia.	26.000	L.F.	444.60	674.18	1,118.78
Pipe, black steel, welded, Sch. 40, on yoke & roll hgrs, 10′ OC, 4″ dia	20.000	L.F.	398	603.20	1,001.20
Elbow, 90°, weld joint, 3″ pipe size	5.000	Ea.	112.50	798.75	911.25
Elbow, 90°, weld joint, 4″ pipe size	4.000	Ea.	150	893	1,043
Tee, welded, reducing on outlet, 3″ by 2-1/2″ pipe	3.000	Ea.	241.50	834.90	1,076.40
Tee, welded, reducing on outlet, 4″ by 3″ pipe	2.000	Ea.	168	741	909
Reducer, eccentric, weld joint, 3″ pipe size	4.000	Ea.	736	558.60	1,294.60
Reducer, eccentric, weld joint, 4″ pipe size	4.000	Ea.	804	744.80	1,548.80
Flange, weld neck, 150 LB, 2-1/2″ pipe size	3.000	Ea.	76.50	209.49	285.99
Flange, weld neck, 150 lb., 3″ pipe size	14.000	Ea.	385	1,118.32	1,503.32
Flange, weld neck, 150 lb., 4″ pipe size	10.000	Ea.	380	1,111.50	1,491.50
Valves, iron body, butterfly, lug type, lever actuator, 3″ size	2.000	Ea.	268	264	532
Valves, iron body, butterfly, lug type, lever actuator, 4″ size	2.000	Ea.	328	422	750
Strainer, Y type iron body, flanged, 125 lb., 3″ pipe size	1.000	Ea.	157	237	394
Strainer, Y type iron body, flanged, 125 lb., 4″ pipe size	1.000	Ea.	199	355	554
Valve, electric motor actuated, iron body, 3 way, flanged, 2-1/2″ pipe	1.000	Ea.	1,225	288	1,513
Valve, electric motor actuated, iron body, two way, flanged, 3″ pipe size	1.000	Ea.	1,275	288	1,563
Expansion joints, bellows type, flange spool, 6″ F to F, 3″ dia	2.000	Ea.	640	187	827
Expansion joints, bellows type, flange spool, 6″ F to F, 4″ dia	2.000	Ea.	690	254	944
T-O-L socket, weld joint, 1/4″ pipe size, nozzle	4.000	Ea.	26.80	194.64	221.44
T-O-L socket, weld joint, 1/2″ pipe size, nozzle	12.000	Ea.	80.40	609.48	689.88
Thermometers, stem type, 9″ case, 8″ stem, 3/4″ NPT	4.000	Ea.	480	84	564
Pressure/temperature relief plug, 316 SS, 3/4″OD, 7-1/2″ insertion	4.000	Ea.	300	74	374
Insulation, fiberglass pipe covering, 1-1/2″ wall, 3″ IPS	26.000	L.F.	72.54	152.10	224.64
Insulation, fiberglass pipe covering, 1-1/2″ wall, 4″ IPS	20.000	L.F.	63.40	142	205.40
Fan coil control system	1.000	Ea.	1,175	540	1,715
Re-heat coil balancing	2.000	Ea.		220	220
Air conditioner balancing	1.000	Ea.		700	700
TOTAL			64,376.24	18,823.96	83,200.20

D3040 106	Field Fabricated Air Handling Unit	COST EACH MAT.	COST EACH INST.	COST EACH TOTAL
1010	AHU, Field fab, blt up, cool/heat coils, fltrs, const vol, 40,000 CFM	64,500	18,800	83,300
1020	60,000 CFM	92,500	29,100	121,600
1030	75,000 CFM	111,500	30,700	142,200

D3040 Distribution Systems

D3040 108	Field Fabricated VAV Air Handling Unit	COST EACH		
		MAT.	INST.	TOTAL
1010	AHU, Field fab, blt up, cool/heat coils, fltrs, VAV, 75,000 CFM	113,500	35,200	148,700
1020	100,000 CFM	172,500	46,700	219,200
1030	150,000 CFM	283,000	75,000	358,000

D3040 Distribution Systems

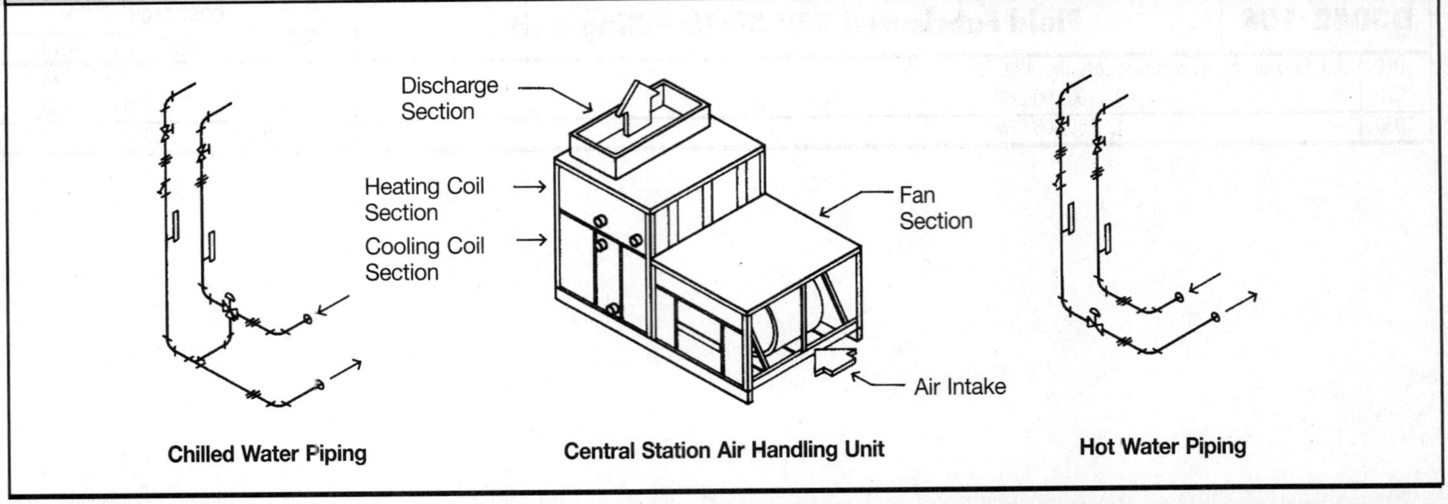

System Components	QUANTITY	UNIT	COST EACH MAT.	COST EACH INST.	COST EACH TOTAL
SYSTEM D3040 110 1010					
AHU, CENTRAL STATION, COOL/HEAT COILS, CONST. VOL., FLTRS., 2,000 CFM					
Central station air handling unit, chilled water, 1900 CFM	1.000	Ea.	5,117.50	1,115.50	6,233
Pipe, black steel, Sch 40, threaded, W/cplgs & hangers, 10′ OC, 1″ dia.	26.000	L.F.	138.45	359.13	497.58
Pipe, black steel, Sch 40, threaded, W/cplg & hangers, 10′ OC, 1-1/4″ dia.	20.000	L.F.	140.40	308.10	448.50
Elbow, 90°, black, straight, 1″ dia.	4.000	Ea.	18.80	180	198.80
Elbow, 90°, black, straight, 1-1/4″ dia.	4.000	Ea.	31	192	223
Tee, black, reducing, 1″ dia.	8.000	Ea.	94.40	584	678.40
Tee, black, reducing, 1-1/4″ dia.	8.000	Ea.	160	604	764
Reducer, concentric, black, 1″ dia.	4.000	Ea.	32	156	188
Reducer, concentric, black, 1-1/4″ dia.	4.000	Ea.	38.20	162	200.20
Union, black with brass seat, 1″ dia.	4.000	Ea.	59.40	196	255.40
Union, black with brass seat, 1-1/4″ dia.	2.000	Ea.	43	100	143
Pipe nipple, black, 1″ dia	13.000	Ea.	27.69	71.83	99.52
Pipe nipple, black, 1-1/4″ dia	12.000	Ea.	32.40	71.10	103.50
Valves, bronze, gate, N.R.S., threaded, class 150, 1″ size	2.000	Ea.	157	62	219
Valves, bronze, gate, N.R.S., threaded, class 150, 1-1/4″ size	2.000	Ea.	210	78	288
Strainer, Y type, bronze body, screwed, 150 lb., 1″ pipe size	1.000	Ea.	25	35	60
Strainer, Y type, bronze body, screwed, 150 lb., 1-1/4″ pipe size	1.000	Ea.	57	39.50	96.50
Valve, electric motor actuated, brass, 3 way, screwed, 3/4″ pipe	1.000	Ea.	325	32	357
Valve, electric motor actuated, brass, 2 way, screwed, 1″ pipe size	1.000	Ea.	330	31	361
Thermometers, stem type, 9″ case, 8″ stem, 3/4″ NPT	4.000	Ea.	480	84	564
Pressure/temperature relief plug, 316 SS, 3/4″OD, 7-1/2″ insertion	4.000	Ea.	300	74	374
Insulation, fiberglass pipe covering, 1″ wall, 1″ IPS	26.000	L.F.	29.64	117.52	147.16
Insulation, fiberglass pipe covering, 1″ wall, 1-1/4″ IPS	20.000	L.F.	24.60	94.80	119.40
Fan coil control system	1.000	Ea.	1,175	540	1,715
Re-heat coil balancing	2.000	Ea.		220	220
Multizone A/C balancing	1.000	Ea.		525	525
TOTAL			9,046.48	6,032.48	15,078.96

D3040 110	Central Station Air Handling Unit	COST EACH MAT.	COST EACH INST.	COST EACH TOTAL
1010	AHU, Central station, cool/heat coils, constant vol, fltrs, 2,000 CFM	9,050	6,025	15,075
1020	5,000 CFM	14,300	7,875	22,175
1030	10,000 CFM	19,200	11,100	30,300
1040	15,000 CFM	28,000	15,000	43,000
1050	20,000 CFM	32,700	17,500	50,200

D3040 Distribution Systems

D3040 112	Central Station Air Handling Unit, VAV	COST EACH		
		MAT.	INST.	TOTAL
1010	AHU, Central station, cool/heat coils, VAV, fltrs, 5,000 CFM	16,500	7,875	24,375
1020	10,000 CFM	21,900	10,900	32,800
1030	15,000 CFM	31,000	14,100	45,100
1040	20,000 CFM	37,500	17,100	54,600
1050	30,000 CFM	56,000	19,700	75,700

D3040 Distribution Systems

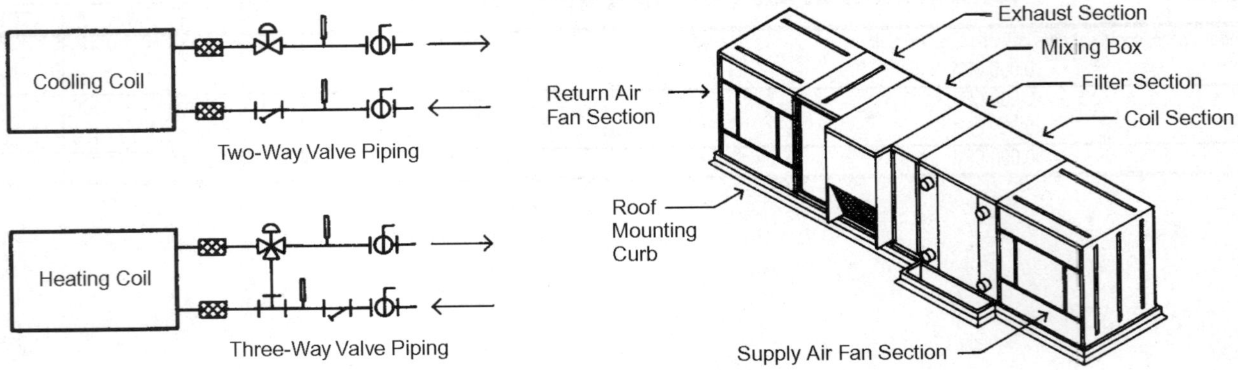

Rooftop Air Handling Unit

System Components	QUANTITY	UNIT	COST EACH MAT.	INST.	TOTAL
SYSTEM D3040 114 1010					
AHU, ROOFTOP, COOL/HEAT COILS, FLTRS., CONST. VOL., 2,000 CFM					
AHU, Built-up, cool/heat coils, filter, mix box, rooftop, constant volume	1.000	Ea.	5,775	1,075	6,850
Pipe, black steel, Sch 40, threaded, W/cplgs & hangers, 10′ OC, 1″ dia.	26.000	L.F.	138.45	359.13	497.58
Pipe, black steel, Sch 40, threaded, W/cplg & hangers, 10′ OC, 1-1/4″ dia.	20.000	L.F.	140.40	308.10	448.50
Elbow, 90°, black, straight, 1″ dia.	4.000	Ea.	18.80	180	198.80
Elbow, 90°, black, straight, 1-1/4″ dia.	4.000	Ea.	31	192	223
Tee, black, reducing, 1″ dia.	8.000	Ea.	94.40	584	678.40
Tee, black, reducing, 1-1/4″ dia.	8.000	Ea.	160	604	764
Reducer, black, concentric, 1″ dia.	4.000	Ea.	32	156	188
Reducer, black, concentric, 1-1/4″ dia.	4.000	Ea.	38.20	162	200.20
Union, black with brass seat, 1″ dia.	4.000	Ea.	59.40	196	255.40
Union, black with brass seat, 1-1/4″ dia.	2.000	Ea.	43	100	143
Pipe nipples, 1″ dia	13.000	Ea.	27.69	71.83	99.52
Pipe nipples, 1-1/4″ dia	12.000	Ea.	32.40	71.10	103.50
Valves, bronze, gate, N.R.S., threaded, class 150, 1″ size	2.000	Ea.	157	62	219
Valves, bronze, gate, N.R.S., threaded, class 150, 1-1/4″ size	2.000	Ea.	210	78	288
Strainer, Y type, bronze body, screwed, 150 lb., 1″ pipe size	1.000	Ea.	25	35	60
Strainer, Y type, bronze body, screwed, 150 lb., 1-1/4″ pipe size	1.000	Ea.	57	39.50	96.50
Valve, electric motor actuated, brass, 3 way, screwed, 3/4″ pipe	1.000	Ea.	325	32	357
Valve, electric motor actuated, brass, 2 way, screwed, 1″ pipe size	1.000	Ea.	330	31	361
Thermometers, stem type, 9″ case, 8″ stem, 3/4″ NPT	4.000	Ea.	480	84	564
Pressure/temperature relief plug, 316 SS, 3/4″OD, 7-1/2″ insertion	4.000	Ea.	300	74	374
Insulation, fiberglass pipe covering, 1″ wall, 1″ IPS	26.000	L.F.	29.64	117.52	147.16
Insulation, fiberglass pipe covering, 1″ wall, 1-1/4″ IPS	20.000	L.F.	24.60	94.80	119.40
Insulation pipe covering .016″ aluminum jacket finish, add	45.000	S.F.	61.20	224.10	285.30
Fan coil control system	1.000	Ea.	1,175	540	1,715
Re-heat coil balancing	2.000	Ea.		220	220
Rooftop unit heat/cool balancing	1.000	Ea.		405	405
TOTAL			9,765.18	6,096.08	15,861.26

D3040 114	Rooftop Air Handling Unit	COST EACH MAT.	INST.	TOTAL
1010	AHU, Rooftop, cool/heat coils, constant vol., fltrs, 2,000 CFM	9,775	6,100	15,875
1020	5,000 CFM	15,600	7,750	23,350
1030	10,000 CFM	20,800	10,900	31,700
1040	15,000 CFM	30,400	14,500	44,900
1050	20,000 CFM	35,600	17,200	52,800

D3040 Distribution Systems

D3040 116	Rooftop Air Handling Unit, VAV	COST EACH		
		MAT.	INST.	TOTAL
1010	AHU, Rooftop, cool/heat coils, VAV, fltrs, 5,000 CFM	17,400	8,025	25,425
1020	10,000 CFM	23,000	11,000	34,000
1030	15,000 CFM	32,600	14,400	47,000
1040	20,000 CFM	39,500	17,400	56,900
1050	30,000 CFM	59,000	20,000	79,000

D3040 Distribution Systems

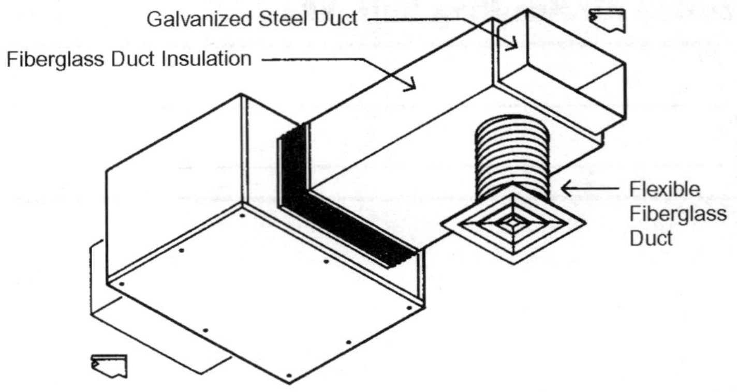

Horizontal Fan Coil Air Conditioning System

System Components	QUANTITY	UNIT	COST EACH MAT.	COST EACH INST.	COST EACH TOTAL
SYSTEM D3040 124 1010					
FAN COIL A/C SYSTEM, HORIZONTAL, W/HOUSING, CONTROLS, 2 PIPE, 1/2 TON					
Fan coil A/C, horizontal housing, filters, chilled water, 1/2 ton cooling	1.000	Ea.	1,175	133	1,308
Pipe, black steel, Sch 40, threaded, W/cplgs & hangers, 10′ OC, 3/4″ dia.	20.000	L.F.	62.35	206.40	268.75
Elbow, 90°, black, straight, 3/4″ dia.	6.000	Ea.	22.26	252	274.26
Tee, black, straight, 3/4″ dia.	2.000	Ea.	8.60	130	138.60
Union, black with brass seat, 3/4″ dia.	2.000	Ea.	22.70	90	112.70
Pipe nipples, 3/4″ diam	3.000	Ea.	4.35	14.40	18.75
Valves, bronze, gate, N.R.S., threaded, class 150, 3/4″ size	2.000	Ea.	124	59	183
Circuit setter, balance valve, bronze body, threaded, 3/4″ pipe size	1.000	Ea.	68.50	29.50	98
Insulation, fiberglass pipe covering, 1″ wall, 3/4″ IPS	20.000	L.F.	21.40	86.60	108
Ductwork, 12″ x 8″ fabricated, galvanized steel, 12 LF	55.000	Lb.	67.65	379.50	447.15
Insulation, ductwork, blanket type, fiberglass, 1″ thk, 1-1/2 LB density	40.000	S.F.	30	113.60	143.60
Diffusers, aluminum, OB damper, ceiling, perf, 24″x24″ panel size, 6″x6″	2.000	Ea.	123	73	196
Ductwork, flexible, fiberglass fabric, insulated, 1″thk, PE jacket, 6″ dia	16.000	L.F.	40.64	64.32	104.96
Round volume control damper 6″ dia.	2.000	Ea.	51	53	104
Fan coil unit balancing	1.000	Ea.		75.50	75.50
Re-heat coil balancing	1.000	Ea.		110	110
Diffuser/register, high, balancing	2.000	Ea.		210	210
TOTAL			1,821.45	2,079.82	3,901.27

D3040 118	Fan Coil A/C Unit, Two Pipe	COST EACH MAT.	COST EACH INST.	COST EACH TOTAL
1010	Fan coil A/C system, cabinet mounted, controls, 2 pipe, 1/2 Ton	1,075	1,100	2,175
1020	1 Ton	1,300	1,250	2,550
1030	1-1/2 Ton	1,400	1,250	2,650
1040	2 Ton	2,075	1,525	3,600
1050	3 Ton	2,700	1,600	4,300

D3040 120	Fan Coil A/C Unit, Two Pipe, Electric Heat	COST EACH MAT.	COST EACH INST.	COST EACH TOTAL
1010	Fan coil A/C system, cabinet mntd, elect. ht, controls, 2 pipe, 1/2 Ton	1,575	1,100	2,675
1020	1 Ton	1,925	1,250	3,175
1030	1-1/2 Ton	2,275	1,250	3,525
1040	2 Ton	3,300	1,550	4,850
1050	3 Ton	5,700	1,600	7,300

D3040 Distribution Systems

D3040 122	Fan Coil A/C Unit, Four Pipe	COST EACH		
		MAT.	INST.	TOTAL
1010	Fan coil A/C system, cabinet mounted, controls, 4 pipe, 1/2 Ton	1,725	2,150	3,875
1020	1 Ton	1,975	2,275	4,250
1030	1-1/2 Ton	2,125	2,300	4,425
1040	2 Ton	2,625	2,400	5,025
1050	3 Ton	3,600	2,575	6,175

D3040 124	Fan Coil A/C, Horizontal, Duct Mount, 2 Pipe	COST EACH		
		MAT.	INST.	TOTAL
1010	Fan coil A/C system, horizontal w/housing, controls, 2 pipe, 1/2 Ton	1,825	2,075	3,900
1020	1 Ton	2,375	3,025	5,400
1030	1-1/2 Ton	2,950	3,925	6,875
1040	2 Ton	3,800	4,725	8,525
1050	3 Ton	4,575	6,275	10,850
1060	3-1/2 Ton	4,800	6,500	11,300
1070	4 Ton	4,975	7,375	12,350
1080	5 Ton	5,925	9,375	15,300
1090	6 Ton	6,050	10,100	16,150
1100	7 Ton	6,225	10,400	16,625
1110	8 Ton	6,225	10,700	16,925
1120	10 Ton	6,925	11,100	18,025

D3040 126	Fan Coil A/C, Horiz., Duct Mount, 2 Pipe, Elec. Ht.	COST EACH		
		MAT.	INST.	TOTAL
1010	Fan coil A/C system, horiz. hsng, elect. ht, ctrls, 2 pipe, 1/2 Ton	2,075	2,075	4,150
1020	1 Ton	2,675	3,025	5,700
1030	1-1/2 Ton	3,175	3,925	7,100
1040	2 Ton	4,200	4,725	8,925
1050	3 Ton	6,825	6,475	13,300
1060	3-1/2 Ton	7,800	6,500	14,300
1070	4 Ton	8,850	7,375	16,225
1080	5 Ton	10,500	9,375	19,875
1090	6 Ton	10,600	10,100	20,700
1100	7 Ton	10,600	10,000	20,600
1110	8 Ton	11,600	10,100	21,700
1120	10 Ton	12,400	10,400	22,800

D3040 128	Fan Coil A/C, Horizontal, Duct Mount, 4 Pipe	COST EACH		
		MAT.	INST.	TOTAL
1010	Fan coil A/C system, horiz. w/cabinet, controls, 4 pipe, 1/2 Ton	2,625	3,100	5,725
1020	1 Ton	3,300	4,175	7,475
1030	1-1/2 Ton	3,950	4,975	8,925
1040	2 Ton	4,650	5,600	10,250
1050	3 Ton	5,675	7,450	13,125
1060	3.5 Ton	6,475	9,125	15,600
1070	4 Ton	6,475	9,225	15,700
1080	5 Ton	7,875	12,700	20,575
1090	6 Ton	8,525	15,000	23,525
1100	7 Ton	9,700	16,900	26,600
1110	8 Ton	9,875	17,800	27,675
1120	10 Ton	11,900	21,800	33,700

D3040 Distribution Systems

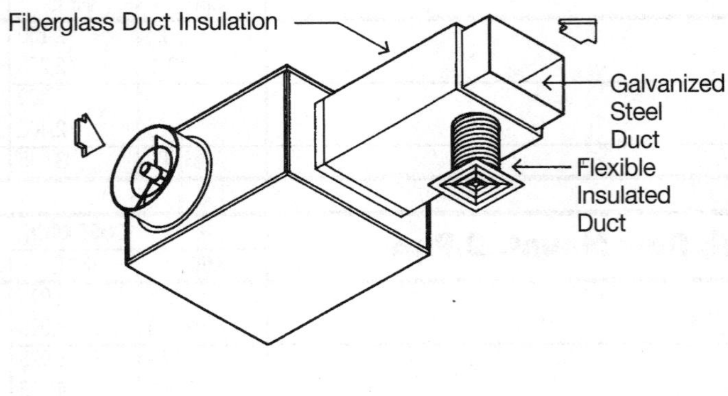

VAV Terminal

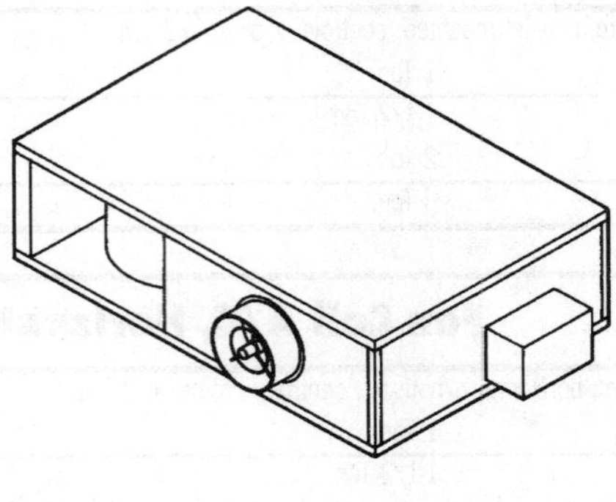
Fan Powered VAV Terminal

System Components	QUANTITY	UNIT	COST EACH MAT.	COST EACH INST.	COST EACH TOTAL
SYSTEM D3040 132 1010					
VAV TERMINAL, COOLING ONLY, WITH ACTUATOR / CONTROLS, 200 CFM					
Mixing box variable volume 300 to 600 CFM, cool only	1.000	Ea.	484	104.50	588.50
Ductwork, 12″ x 8″ fabricated, galvanized steel, 12 LF	55.000	Lb.	67.65	379.50	447.15
Insulation, ductwork, blanket type, fiberglass, 1″ thk, 1-1/2 LB density	40.000	S.F.	30	113.60	143.60
Diffusers, aluminum, OB damper, ceiling, perf, 24″x24″ panel size, 6″x6″	2.000	Ea.	123	73	196
Ductwork, flexible, fiberglass fabric, insulated, 1″thk, PE jacket, 6″ dia	16.000	L.F.	40.64	64.32	104.96
Round volume control damper 6″ dia.	2.000	Ea.	51	53	104
VAV box balancing	1.000	Ea.		69.50	69.50
Diffuser/register, high, balancing	2.000	Ea.		210	210
TOTAL			796.29	1,067.42	1,863.71

D3040 132	VAV Terminal, Cooling Only	COST EACH MAT.	COST EACH INST.	COST EACH TOTAL
1010	VAV Terminal, cooling only, with actuator / controls, 200 CFM	795	1,075	1,870
1020	400 CFM	1,000	1,875	2,875
1030	600 CFM	1,300	2,800	4,100
1040	800 CFM	1,425	3,400	4,825
1050	1000 CFM	1,625	3,925	5,550
1060	1250 CFM	1,875	5,000	6,875
1070	1500 CFM	2,075	5,900	7,975
1080	2000 CFM	2,750	8,400	11,150

D3040 134	VAV Terminal, Hot Water Reheat	COST EACH MAT.	COST EACH INST.	COST EACH TOTAL
1010	VAV Terminal, cooling, HW reheat, with actuator / controls, 200 CFM	1,725	1,975	3,700
1020	400 CFM	1,950	2,700	4,650
1030	600 CFM	2,225	3,700	5,925
1040	800 CFM	2,375	4,325	6,700
1050	1000 CFM	2,550	4,850	7,400
1060	1250 CFM	2,875	5,975	8,850
1070	1500 CFM	3,225	7,100	10,325
1080	2000 CFM	3,950	9,625	13,575

D3040 136	Fan Powered VAV Terminal, Cooling Only	COST EACH MAT.	COST EACH INST.	COST EACH TOTAL
1010	VAV Terminal, cooling, fan powrd, with actuator / controls, 200 CFM	1,350	1,075	2,425
1020	400 CFM	1,600	1,750	3,350

D3040 Distribution Systems

D3040 136	Fan Powered VAV Terminal, Cooling Only	COST EACH		
		MAT.	INST.	TOTAL
1030	600 CFM	1,900	2,775	4,675
1040	800 CFM	2,075	3,350	5,425
1050	1000 CFM	2,225	3,875	6,100
1060	1250 CFM	2,600	5,075	7,675
1070	1500 CFM	2,775	5,950	8,725
1080	2000 CFM	3,575	8,650	12,225

D3040 138	Fan Powered VAV Terminal, Hot Water Reheat	COST EACH		
		MAT.	INST.	TOTAL
1010	VAV Terminal, cool, HW reht, fan powrd, with actuator/ctrls, 200 CFM	2,225	1,975	4,200
1020	400 CFM	2,475	2,725	5,200
1030	600 CFM	2,750	3,700	6,450
1040	800 CFM	2,900	4,275	7,175
1050	1000 CFM	3,075	4,800	7,875
1060	1250 CFM	3,575	6,025	9,600
1070	1500 CFM	3,950	7,150	11,100
1080	2000 CFM	4,725	9,900	14,625

D3040 Distribution Systems

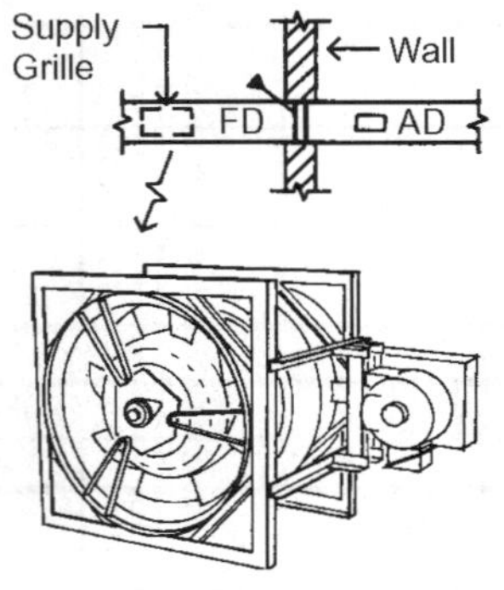

Axial Flow Centrifugal Fan
Belt Drive

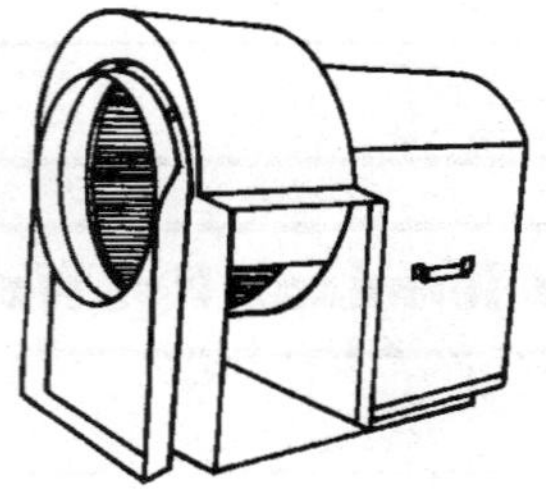

Belt Drive Utility Set

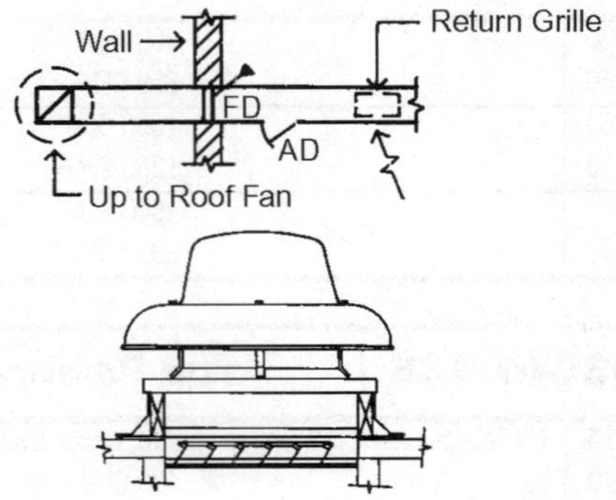

Centrifugal Roof Exhaust Fan
Direct Drive

System Components	QUANTITY	UNIT	COST EACH		
			MAT.	INST.	TOTAL
SYSTEM D3040 220 1010					
FAN SYSTEM, IN-LINE CENTRIFUGAL, 500 CFM					
Fans, in-line centrifugal, supply/exhaust, 500 CFM, 10″ dia conn	1.000	Ea.	1,225	440	1,665
Louver, alum., W/screen, damper, mill fin, fxd blade, cont line, stormproof	4.050	S.F.	186.30	85.05	271.35
Motor for louver damper, electric or pneumatic	1.000	Ea.	530	41.50	571.50
Ductwork, 12″ x 8″ fabricated, galvanized steel, 60 LF	200.000	Lb.	283.40	1,729	2,012.40
Grilles, aluminum, air supply, single deflection, adjustable, 12″ x 6″	1.000	Ea.	76.50	76.50	153
Fire damper, curtain type, vertical, 12″ x 8″	1.000	Ea.	26.50	26.50	53
Duct access door, insulated, 10″ x 10″	1.000	Ea.	19.35	53	72.35
Duct access door, insulated, 16″ x 12″	1.000	Ea.	29.50	64.50	94
Pressure controller/switch, air flow controller	1.000	Ea.	180	49.50	229.50
Fan control switch and mounting base	1.000	Ea.	107	37	144
Fan balancing	1.000	Ea.		350	350
Diffuser/register balancing	3.000	Ea.		210	210
TOTAL			2,663.55	3,162.55	5,826.10

D3040 220	Centrifugal In-Line Fan Systems	COST EACH		
		MAT.	INST.	TOTAL
1010	Fan system, in-line centrifugal, 500 CFM	2,675	3,175	5,850
1020	1300 CFM	3,275	5,475	8,750
1030	1500 CFM	3,725	6,350	10,075
1040	2500 CFM	6,475	19,300	25,775
1050	3500 CFM	8,275	26,000	34,275
1060	5000 CFM	11,200	38,600	49,800
1070	7500 CFM	12,700	39,400	52,100
1080	10,000 CFM	15,300	46,200	61,500

D3040 230	Utility Set Fan Systems	COST EACH		
		MAT.	INST.	TOTAL
1010	Utility fan set system, belt drive, 2000 CFM	2,900	10,200	13,100
1020	3500 CFM	5,075	18,000	23,075
1030	5000 CFM	6,625	25,100	31,725
1040	7500 CFM	8,800	29,100	37,900
1050	10,000 CFM	11,600	38,200	49,800
1060	15,000 CFM	12,200	39,000	51,200
1070	20,000 CFM	14,000	44,200	58,200

D3040 Distribution Systems

D3040 240	Roof Exhaust Fan Systems	COST EACH		
		MAT.	INST.	TOTAL
1010	Roof vent. system, centrifugal, alum., galv curb, BDD, 500 CFM	1,025	2,075	3,100
1020	800 CFM	1,350	4,000	5,350
1030	1500 CFM	1,825	5,950	7,775
1040	2750 CFM	3,675	13,400	17,075
1050	3500 CFM	4,625	17,200	21,825
1060	5000 CFM	6,925	28,400	35,325
1070	8500 CFM	8,850	36,500	45,350
1080	13,800 CFM	12,600	53,000	65,600

D3040 Controls & Instrumentation

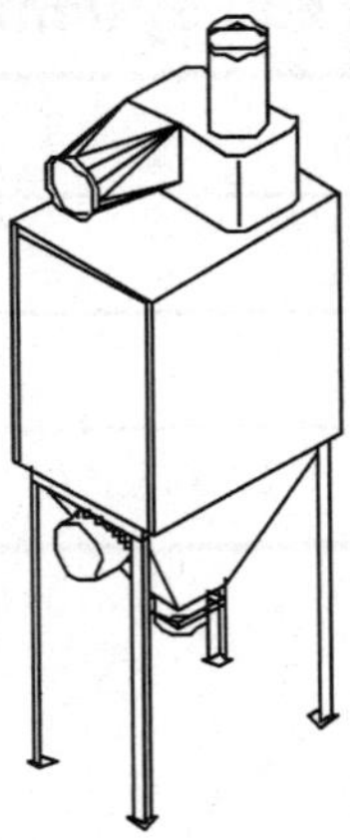

Commercial/Industrial Dust Collection System

System Components	QUANTITY	UNIT	COST EACH MAT.	COST EACH INST.	COST EACH TOTAL
SYSTEM D3040 250 1010					
COMMERCIAL/INDUSTRIAL VACUUM DUST COLLECTION SYSTEM, 500 CFM					
Dust collection central vac unit inc. stand, filter & shaker, 500 CFM 2 HP	1.000	Ea.	4,100	550	4,650
Galvanized tubing, 16 ga., 4″ OD	80.000	L.F.	460	232.80	692.80
Galvanized tubing, 90° ell, 4″ dia.	3.000	Ea.	97.50	63	160.50
Galvanized tubing, 45° ell, 4″ dia.	4.000	Ea.	114	84	198
Galvanized tubing, TY slip fit, 4″ dia.	4.000	Ea.	276	140	416
Flexible rubber hose, 4″ dia.	32.000	L.F.	280	93.12	373.12
Galvanized air gate valve, 4″ dia.	4.000	Ea.	1,156	182	1,338
Galvanized compression coupling, 4″ dia.	36.000	Ea.	828	1,134	1,962
Pipe hangers, steel, 4″ pipe size	16.000	Ea.	50.72	153.60	204.32
TOTAL			7,362.22	2,632.52	9,994.74

D3040 250	Commercial/Industrial Vacuum Dust Collection	COST EACH MAT.	COST EACH INST.	COST EACH TOTAL
1010	Commercial / industrial, vacuum dust collection system, 500 CFM	7,350	2,625	9,975
1020	1000 CFM	9,650	3,000	12,650
1030	1500 CFM	12,100	3,325	15,425
1040	3000 CFM	23,600	4,650	28,250
1050	5000 CFM	31,800	6,975	38,775

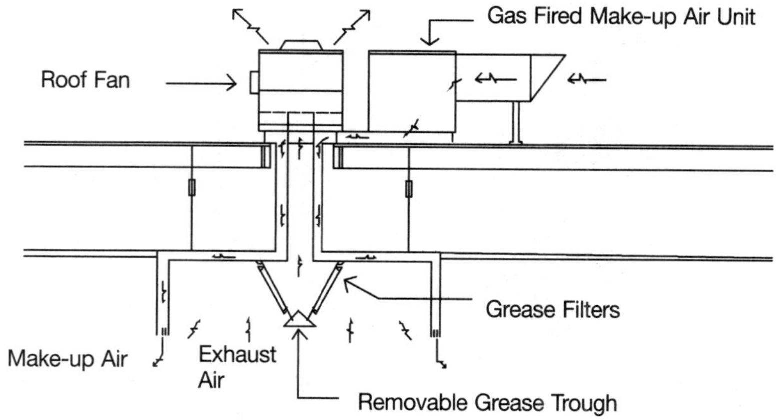

Commercial Kitchen Exhaust/Make-up Air Rooftop System

System Components	QUANTITY	UNIT	COST EACH MAT.	COST EACH INST.	COST EACH TOTAL
SYSTEM D3040 260 1010					
COMMERCIAL KITCHEN EXHAUST/MAKE-UP AIR SYSTEM, ROOFTOP, GAS, 2000 CFM					
Make-up air rooftop unit, gas, 2090 CFM, 159 MBH	1.000	Ea.	8,100	599.40	8,699.40
Pipe, black steel, Sch 40, thrded, nipples, W/cplngs/hngrs, 10′ OC, 1″ dia.	20.000	L.F.	93.72	243.10	336.82
Elbow, 90°, black steel, straight, 3/4″ dia.	2.000	Ea.	7.42	84	91.42
Elbow, 90°, black steel, straight, 1″ dia.	3.000	Ea.	14.10	135	149.10
Tee, black steel, reducing, 1″ dia.	1.000	Ea.	11.80	73	84.80
Union, black with brass seat, 1″ dia.	1.000	Ea.	14.85	49	63.85
Pipe nipples, black, 3/4″ dia	2.000	Ea.	5.80	19.20	25
Pipe cap, malleable iron, black, 1″ dia.	1.000	Ea.	3.71	19.50	23.21
Gas cock, brass, 1″ size	1.000	Ea.	19.70	31	50.70
Ductwork, 20″ x 14″ fabricated, galv. steel, 30 LF	220.000	Lb.	235.40	1,408	1,643.40
Ductwork, 18″ x 10″ fabricated, galv. steel, 12 LF	72.000	Lb.	77.04	460.80	537.84
Ductwork, 30″ x 10″ fabricated, galv. steel, 30 LF	260.000	Lb.	278.20	1,664	1,942.20
Ductwork, 16″ x 10″ fabricated, galv. steel, 12 LF	70.000	Lb.	74.90	448	522.90
Kitchen ventilation, island style, water wash, Stainless Steel	5.000	L.F.	11,500	1,145	12,645
Control system, Make-up Air Unit / Exhauster	1.000	Ea.	2,575	2,100	4,675
Rooftop unit heat/cool balancing	1.000	Ea.		405	405
Roof fan balancing	1.000	Ea.		232	232
TOTAL			23,011.64	9,116	32,127.64

D3040 260	Kitchen Exhaust/Make-Up Air	COST EACH MAT.	COST EACH INST.	COST EACH TOTAL
1010	Commercial kitchen exhaust/make-up air system, rooftop, gas, 2000 CFM	23,000	9,125	32,125
1020	3000 CFM	24,900	12,400	37,300
1030	5000 CFM	31,000	15,100	46,100
1040	8000 CFM	48,600	22,400	71,000
1050	12,000 CFM	73,000	27,200	100,200
1060	16,000 CFM	90,500	43,100	133,600

D3040 Distribution Systems

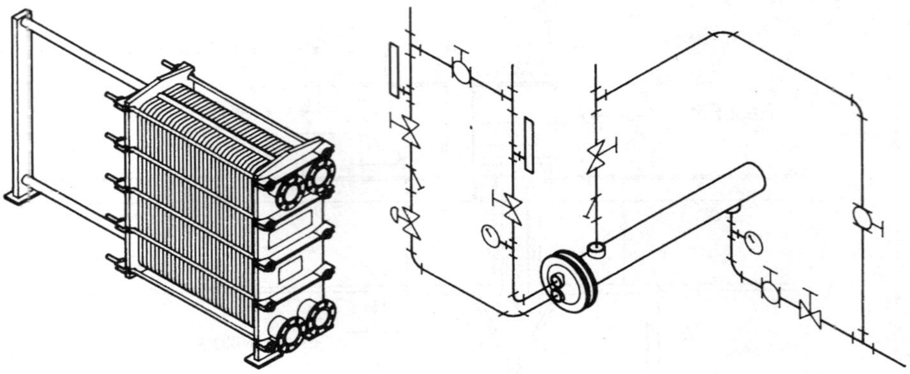

Plate Heat Exchanger Shell and Tube Heat Exchanger

System Components	QUANTITY	UNIT	COST EACH MAT.	INST.	TOTAL
SYSTEM D3040 620 1010					
SHELL & TUBE HEAT EXCHANGER, 40 GPM					
Heat exchanger, 4 pass, 3/4" O.D. copper tubes, by steam at 10 psi, 40 GPM	1.000	Ea.	4,000	267	4,267
Pipe, black steel, Sch 40, threaded, W/couplings & hangers, 10' OC, 2" dia.	40.000	L.F.	426.73	830.73	1,257.46
Elbow, 90°, straight, 2" dia.	16.000	Ea.	293.60	936	1,229.60
Reducer, black steel, concentric, 2" dia.	2.000	Ea.	39.60	100	139.60
Valves, bronze, gate, rising stem, threaded, class 150, 2" size	6.000	Ea.	1,038	318	1,356
Valves, bronze, globe, class 150, rising stem, threaded, 2" size	2.000	Ea.	900	106	1,006
Strainers, Y type, bronze body, screwed, 150 lb., 2" pipe size	2.000	Ea.	145	91	236
Valve, electric motor actuated, brass, 2 way, screwed, 1-1/2" pipe size	1.000	Ea.	435	45.50	480.50
Union, black with brass seat, 2" dia.	8.000	Ea.	248	496	744
Tee, black, straight, 2" dia.	9.000	Ea.	193.50	864	1,057.50
Tee, black, reducing run and outlet, 2" dia.	4.000	Ea.	100	384	484
Pipe nipple, black, 2" dia.	21.000	Ea.	4.23	8.23	12.46
Thermometers, stem type, 9" case, 8" stem, 3/4" NPT	2.000	Ea.	240	42	282
Gauges, pressure or vacuum, 3-1/2" diameter dial	2.000	Ea.	79	37	116
Insulation, fiberglass pipe covering, 1" wall, 2" IPS	40.000	L.F.	57.60	199.20	256.80
Heat exchanger control system	1.000	Ea.	2,500	1,900	4,400
Coil balancing	1.000	Ea.		110	110
TOTAL			10,700.26	6,734.66	17,434.92

D3040 610	Heat Exchanger, Plate Type	COST EACH MAT.	INST.	TOTAL
1010	Plate heat exchanger, 400 GPM	46,700	13,600	60,300
1020	800 GPM	75,000	17,300	92,300
1030	1200 GPM	113,000	23,400	136,400
1040	1800 GPM	152,000	29,200	181,200

D3040 620	Heat Exchanger, Shell & Tube	COST EACH MAT.	INST.	TOTAL
1010	Shell & tube heat exchanger, 40 GPM	10,700	6,725	17,425
1020	96 GPM	19,700	12,000	31,700
1030	240 GPM	35,400	16,200	51,600
1040	600 GPM	77,500	22,800	100,300

D3050 Terminal & Package Units

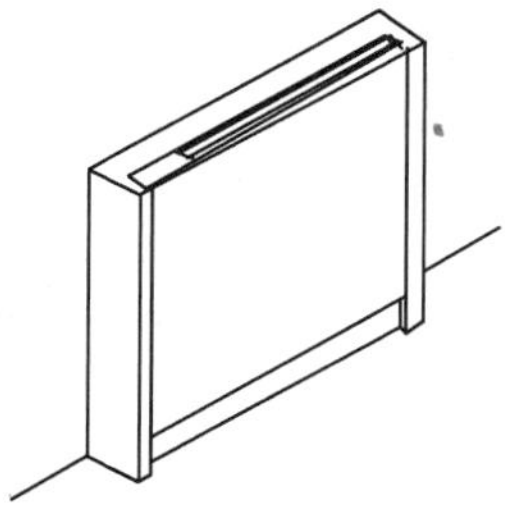

Cabinet Unit Heater

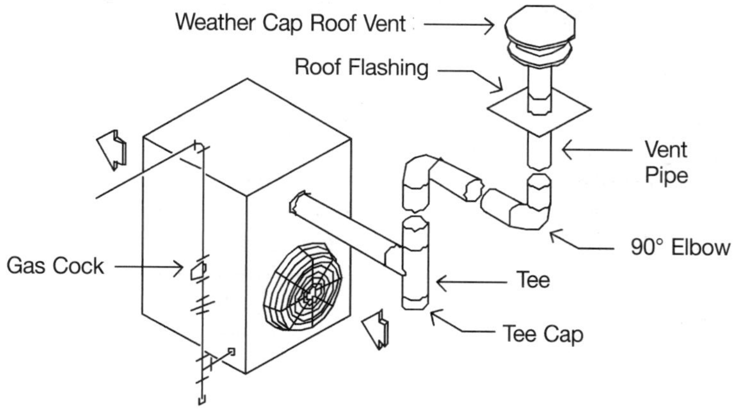

Gas Unit Heater

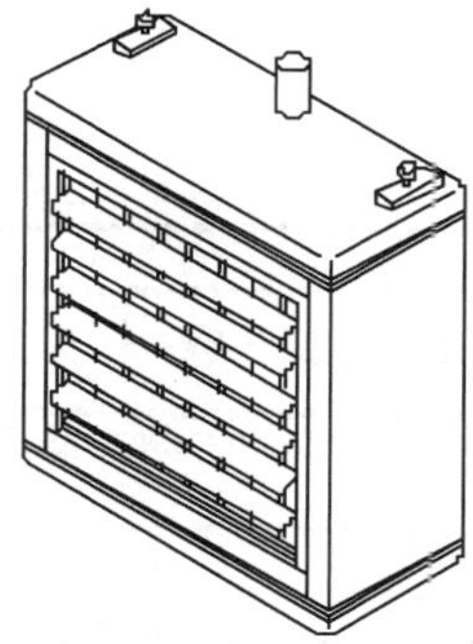

Hydronic Unit Heater

System Components

System Components	QUANTITY	UNIT	COST EACH MAT.	COST EACH INST.	COST EACH TOTAL
SYSTEM D3050 120 1010					
SPACE HEATER, SUSPENDED, GAS FIRED, PROPELLER FAN, 20 MBH					
Space heater, propeller fan, 20 MBH output	1.000	Ea.	1,075	126	1,201
Pipe, black steel, Sch 40, threaded, W/coupling & hangers, 10′ OC, 3/4″ dia	20.000	L.F.	60.90	201.60	262.50
Elbow, 90°, black steel, straight, 1/2″ dia.	3.000	Ea.	9.69	117	126.69
Elbow, 90°, black steel, straight, 3/4″ dia.	3.000	Ea.	11.13	126	137.13
Tee, black steel, reducing, 3/4″ dia.	1.000	Ea.	7.40	65	72.40
Union, black with brass seat, 3/4″ dia.	1.000	Ea.	11.35	45	56.35
Pipe nipples, black, 1/2″ dia	2.000	Ea.	4.92	18.60	23.52
Pipe nipples, black, 3/4″ dia	2.000	Ea.	2.90	9.60	12.50
Pipe cap, black, 3/4″ dia.	1.000	Ea.	3.07	18.30	21.37
Gas cock, brass, 3/4″ size	1.000	Ea.	15.05	26.50	41.55
Thermostat, 1 set back, electric, timed	1.000	Ea.	104	72.50	176.50
Wiring, thermostat hook-up	1.000	Ea.	8.45	23.50	31.95
Vent chimney, prefab metal, U.L. listed, gas, double wall, galv. st, 4″ dia	12.000	L.F.	88.20	184.80	273
Vent chimney, gas, double wall, galv. steel, elbow 90°, 4″ dia.	2.000	Ea.	59	62	121
Vent chimney, gas, double wall, galv. steel, Tee, 4″ dia.	1.000	Ea.	37.50	40	77.50
Vent chimney, gas, double wall, galv. steel, T cap, 4″ dia.	1.000	Ea.	2.89	25	27.89
Vent chimney, gas, double wall, galv. steel, roof flashing, 4″ dia.	1.000	Ea.	31	31	62
Vent chimney, gas, double wall, galv. steel, top, 4″ dia.	1.000	Ea.	17.80	24	41.80
TOTAL			1,550.25	1,216.40	2,766.65

D3050 120 Unit Heaters, Gas

D3050 120	Unit Heaters, Gas	COST EACH MAT.	COST EACH INST.	COST EACH TOTAL
1010	Space heater, suspended, gas fired, propeller fan, 20 MBH	1,550	1,225	2,775
1020	60 MBH	1,825	1,250	3,075
1030	100 MBH	2,050	1,350	3,400
1040	160 MBH	2,350	1,475	3,825
1050	200 MBH	2,775	1,625	4,400
1060	280 MBH	3,475	1,725	5,200
1070	320 MBH	4,550	1,975	6,525

D3050 Terminal & Package Units

D3050 130	Unit Heaters, Hydronic	COST EACH		
		MAT.	INST.	TOTAL
1010	Space heater, suspended, horiz. mount, HW, prop. fan, 20 MBH	1,700	1,300	3,000
1020	60 MBH	2,150	1,450	3,600
1030	100 MBH	2,575	1,600	4,175
1040	150 MBH	2,975	1,750	4,725
1050	200 MBH	3,425	1,975	5,400
1060	300 MBH	4,075	2,325	6,400

D3050 140	Cabinet Unit Heaters, Hydronic	COST EACH		
		MAT.	INST.	TOTAL
1010	Unit heater, cabinet type, horizontal blower, hot water, 20 MBH	1,975	1,275	3,250
1020	60 MBH	2,775	1,400	4,175
1030	100 MBH	3,050	1,550	4,600
1040	120 MBH	3,125	1,600	4,725

D3050 Terminal & Package Units

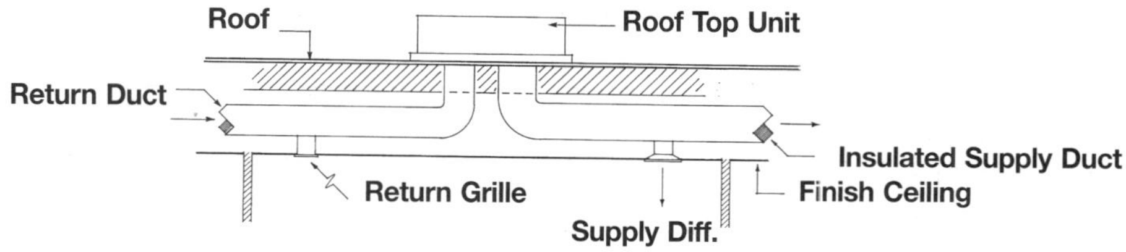

System Description: Rooftop single zone units are electric cooling and gas heat. Duct systems are low velocity, galvanized steel supply and return. Price variations between sizes are due to several factors. Jumps in the cost of the rooftop unit occur when the manufacturer shifts from the largest capacity unit on a small frame to the smallest capacity on the next larger frame, or changes from one compressor to two. As the unit capacity increases for larger areas the duct distribution grows in proportion. For most applications there is a tradeoff point where it is less expensive and more efficient to utilize smaller units with short simple distribution systems. Larger units also require larger initial supply and return ducts which can create a space problem. Supplemental heat may be desired in colder locations. The table below is based on one unit supplying the area listed. The 10,000 S.F. unit for bars and taverns is not listed because a nominal 110 ton unit would be required and this is above the normal single zone rooftop capacity.

System Components	QUANTITY	UNIT	COST EACH MAT.	COST EACH INST.	COST EACH TOTAL
SYSTEM D3050 150 1280					
ROOFTOP, SINGLE ZONE, AIR CONDITIONER					
APARTMENT CORRIDORS, 500 S.F., .92 TON					
Rooftop air conditioner, 1 zone, electric cool, standard controls, curb	1.000	Ea.	1,621.50	701.50	2,323
Ductwork package for rooftop single zone units	1.000	System	289.80	1,012	1,301.80
TOTAL			1,911.30	1,713.50	3,624.80
COST PER S.F.			3.82	3.43	7.25
*Size would suggest multiple units					

D3050 150	Rooftop Single Zone Unit Systems	COST PER S.F. MAT.	COST PER S.F. INST.	COST PER S.F. TOTAL
1260	Rooftop, single zone, air conditioner			
1280	Apartment corridors, 500 S.F., .92 ton	3.80	3.45	7.25
1320	1,000 S.F., 1.83 ton	3.81	3.41	7.22
1360	1500 S.F., 2.75 ton R236000-20	2.47	2.82	5.29
1400	3,000 S.F., 5.50 ton	2.26	2.72	4.98
1440	5,000 S.F., 9.17 ton	2.25	2.47	4.72
1480	10,000 S.F., 18.33 ton	2.61	2.33	4.94
1560	Banks or libraries, 500 S.F., 2.08 ton	8.65	7.75	16.40
1600	1,000 S.F., 4.17 ton	5.60	6.40	12
1640	1,500 S.F., 6.25 ton	5.10	6.15	11.25
1680	3,000 S.F., 12.50 ton	5.10	5.60	10.70
1720	5,000 S.F., 20.80 ton	5.90	5.30	11.20
1760	10,000 S.F., 41.67 ton	5.10	5.25	10.35
1840	Bars and taverns, 500 S.F. 5.54 ton	11.90	10.30	22.20
1880	1,000 S.F., 11.08 ton	11.85	8.85	20.70
1920	1,500 S.F., 16.62 ton	11.95	8.25	20.20
1960	3,000 S.F., 33.25 ton	12.20	7.90	20.10
2000	5,000 S.F., 55.42 ton	11.30	7.90	19.20
2040	10,000 S.F., 110.83 ton*			
2080	Bowling alleys, 500 S.F., 2.83 ton	7.60	8.70	16.30
2120	1,000 S.F., 5.67 ton	6.95	8.40	15.35
2160	1,500 S.F., 8.50 ton	6.95	7.65	14.60
2200	3,000 S.F., 17.00 ton	7	7.35	14.35
2240	5,000 S.F., 28.33 ton	7.10	7.15	14.25
2280	10,000 S.F., 56.67 ton	6.70	7.15	13.85
2360	Department stores, 500 S.F., 1.46 ton	6.05	5.45	11.50

D3050 Terminal & Package Units

D3050 150	Rooftop Single Zone Unit Systems	COST PER S.F. MAT.	COST PER S.F. INST.	COST PER S.F. TOTAL
2400	1,000 S.F., 2.92 ton	3.93	4.49	8.42
2440	1,500 S.F., 4.37 ton	3.59	4.31	7.90
2480	3,000 S.F., 8.75 ton	3.58	3.93	7.51
2520	5,000 S.F., 14.58 ton	3.60	3.77	7.37
2560	10,000 S.F., 29.17 ton	3.67	3.68	7.35
2640	Drug stores, 500 S.F., 3.33 ton	8.95	10.25	19.20
2680	1,000 S.F., 6.67 ton	8.20	9.85	18.05
2720	1,500 S.F., 10.00 ton	8.20	9	17.20
2760	3,000 S.F., 20.00 ton	9.45	8.45	17.90
2800	5,000 S.F., 33.33 ton	8.40	8.40	16.80
2840	10,000 S.F., 66.67 ton	7.85	8.40	16.25
2920	Factories, 500 S.F., 1.67 ton	6.90	6.20	13.10
2960	1,000 S.F., 3.33 ton	4.49	5.10	9.59
3000	1,500 S.F., 5.00 ton	4.10	4.94	9.04
3040	3,000 S.F., 10.00 ton	4.09	4.50	8.59
3080	5,000 S.F., 16.67 ton	4.12	4.31	8.43
3120	10,000 S.F., 33.33 ton	4.19	4.21	8.40
3200	Food supermarkets, 500 S.F., 1.42 ton	5.90	5.30	11.20
3240	1,000 S.F., 2.83 ton	3.78	4.34	8.12
3280	1,500 S.F., 4.25 ton	3.48	4.20	7.68
3320	3,000 S.F., 8.50 ton	3.48	3.82	7.30
3360	5,000 S.F., 14.17 ton	3.50	3.67	7.17
3400	10,000 S.F., 28.33 ton	3.56	3.58	7.14
3480	Medical centers, 500 S.F., 1.17 ton	4.85	4.35	9.20
3520	1,000 S.F., 2.33 ton	4.84	4.34	9.18
3560	1,500 S.F., 3.50 ton	3.15	3.59	6.74
3600	3,000 S.F., 7.00 ton	2.88	3.46	6.34
3640	5,000 S.F., 11.67 ton	2.87	3.15	6.02
3680	10,000 S.F., 23.33 ton	3.31	2.96	6.27
3760	Offices, 500 S.F., 1.58 ton	6.60	5.90	12.50
3800	1,000 S.F., 3.17 ton	4.27	4.88	9.15
3840	1,500 S.F., 4.75 ton	3.90	4.68	8.58
3880	3,000 S.F., 9.50 ton	3.89	4.26	8.15
3920	5,000 S.F., 15.83 ton	3.91	4.09	8
3960	10,000 S.F., 31.67 ton	3.99	3.99	7.98
4000	Restaurants, 500 S.F., 2.50 ton	10.40	9.30	19.70
4040	1,000 S.F., 5.00 ton	6.15	7.40	13.55
4080	1,500 S.F., 7.50 ton	6.15	6.75	12.90
4120	3,000 S.F., 15.00 ton	6.20	6.45	12.65
4160	5,000 S.F., 25.00 ton	7.10	6.35	13.45
4200	10,000 S.F., 50.00 ton	5.90	6.30	12.20
4240	Schools and colleges, 500 S.F., 1.92 ton	7.95	7.15	15.10
4280	1,000 S.F., 3.83 ton	5.15	5.90	11.05
4320	1,500 S.F., 5.75 ton	4.72	5.70	10.42
4360	3,000 S.F., 11.50 ton	4.71	5.15	9.86
4400	5,000 S.F., 19.17 ton	5.45	4.87	10.32
4440	10,000 S.F., 38.33 ton	4.68	4.84	9.52
5000				
5100				
5200				
9000	Components of ductwork packages for above systems, per ton of cooling:			
9010	Ductwork; galvanized steel, 120 pounds			
9020	Insulation; fiberglass 2" thick, FRK faced, 52 SF			
9030	Diffusers; aluminum, 24" x 12", one			
9040	Registers; aluminum, return with dampers, one			

D3050 Terminal & Package Units

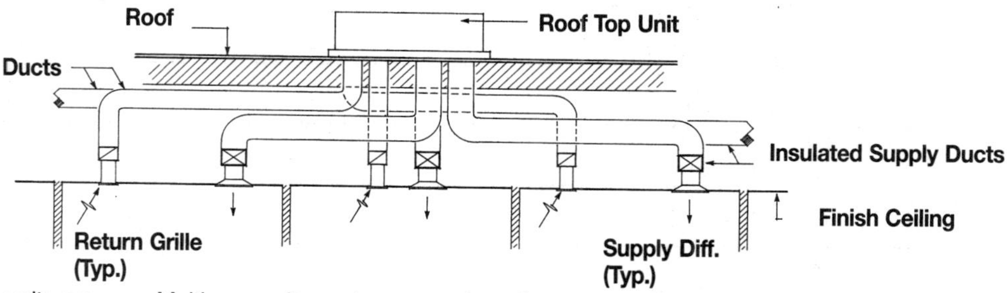

System Description: Rooftop units are multizone with up to 12 zones, and include electric cooling, gas heat, thermostats, filters, supply and return fans complete. Duct systems are low velocity, galvanized steel supply and return with insulated supplies.

Multizone units cost more per ton of cooling than single zone. However, they offer flexibility where load conditions are varied due to heat generating areas or exposure to radiational heating. For example, perimeter offices on the "sunny side" may require cooling at the same time "shady side" or central offices may require heating. It is possible to accomplish similar results using duct heaters in branches of the single zone unit. However, heater location could be a problem and total system operating energy efficiency could be lower.

System Components	QUANTITY	UNIT	COST EACH MAT.	COST EACH INST.	COST EACH TOTAL
SYSTEM D3050 155 1280					
ROOFTOP, MULTIZONE, AIR CONDITIONER					
APARTMENT CORRIDORS, 3,000 S.F., 5.50 TON					
Rooftop multizone unit, standard controls, curb	1.000	Ea.	25,960	1,628	27 588
Ductwork package for rooftop multizone units	1.000	System	2,530	11,550	14 080
TOTAL			28,490	13,178	41 668
COST PER S.F.			9.50	4.39	13.89
Note A: Small single zone unit recommended					
Note B: A combination of multizone units recommended					

D3050 155	Rooftop Multizone Unit Systems		COST PER S.F. MAT.	COST PER S.F. INST.	COST PER S.F. TOTAL
1240	Rooftop, multizone, air conditioner				
1260	Apartment corridors, 1,500 S.F., 2.75 ton, See Note A				
1280	3,000 S.F., 5.50 ton	R236000 -20	9.50	4.40	13.90
1320	10,000 S.F., 18.30 ton		6.65	4.23	10.88
1360	15,000 S.F., 27.50 ton		6.50	4.21	10.71
1400	20,000 S.F., 36.70 ton		6.80	4.22	11.02
1440	25,000 S.F., 45.80 ton		6.35	4.25	10.60
1520	Banks or libraries, 1,500 S.F., 6.25 ton		21.50	10	31.50
1560	3,000 S.F., 12.50 ton		18.30	9.80	28.10
1600	10,000 S.F., 41.67 ton		14.40	9.65	24.05
1640	15,000 S.F., 62.50 ton		10.80	9.60	20.40
1680	20,000 S.F., 83.33 ton		10.85	9.60	20.45
1720	25,000 S.F., 104.00 ton		9.75	9.55	19.30
1800	Bars and taverns, 1,500 S.F., 16.62 ton		46	14.35	60.35
1840	3,000 S.F., 33.24 ton		38.50	13.85	52.35
1880	10,000 S.F., 110.83 ton		23.50	13.80	37.30
1920	15,000 S.F., 165 ton, See Note B				
1960	20,000 S.F., 220 ton, See Note B				
2000	25,000 S.F., 275 ton, See Note B				
2080	Bowling alleys, 1,500 S.F., 8.50 ton		29.50	13.60	43.10
2120	3,000 S.F., 17.00 ton		25	13.30	38.30
2160	10,000 S.F., 56.70 ton		19.60	13.15	32.75
2200	15,000 S.F., 85.00 ton		14.75	13.05	27.80
2240	20,000 S.F., 113.00 ton		13.25	12.95	26.20
2280	25,000 S.F., 140.00 ton see Note B				
2360	Department stores, 1,500 S.F., 4.37 ton, See Note A				

D3050 Terminal & Package Units

D3050 155	Rooftop Multizone Unit Systems	COST PER S.F.		
		MAT.	INST.	TOTAL
2400	3,000 S.F., 8.75 ton	15.10	7	22.10
2440	10,000 S.F., 29.17 ton	10.80	6.70	17.50
2480	15,000 S.F., 43.75 ton	10.10	6.75	16.85
2520	20,000 S.F., 58.33 ton	7.55	6.70	14.25
2560	25,000 S.F., 72.92 ton	7.55	6.70	14.25
2640	Drug stores, 1,500 S.F., 10.00 ton	34.50	15.95	50.45
2680	3,000 S.F., 20.00 ton	24	15.40	39.40
2720	10,000 S.F., 66.66 ton	17.30	15.35	32.65
2760	15,000 S.F., 100.00 ton	15.60	15.30	30.90
2800	20,000 S.F., 135 ton, See Note B			
2840	25,000 S.F., 165 ton, See Note B			
2920	Factories, 1,500 S.F., 5 ton, See Note A			
2960	3,000 S.F., 10.00 ton	17.25	8	25.25
3000	10,000 S.F., 33.33 ton	12.35	7.65	20
3040	15,000 S.F., 50.00 ton	11.55	7.70	19.25
3080	20,000 S.F., 66.66 ton	8.65	7.65	16.30
3120	25,000 S.F., 83.33 ton	8.65	7.65	16.30
3200	Food supermarkets, 1,500 S.F., 4.25 ton, See Note A			
3240	3,000 S.F., 8.50 ton	14.65	6.80	21.45
3280	10,000 S.F., 28.33 ton	10.05	6.50	16.55
3320	15,000 S.F., 42.50 ton	9.80	6.55	16.35
3360	20,000 S.F., 56.67 ton	7.35	6.50	13.85
3400	25,000 S.F., 70.83 ton	7.35	6.50	13.85
3480	Medical centers 1,500 S.F., 3.5 ton, See Note A			
3520	3,000 S.F., 7.00 ton	12.10	5.60	17.70
3560	10,000 S.F., 23.33 ton	8.15	5.35	13.50
3600	15,000 S.F., 35.00 ton	8.65	5.35	14
3640	20,000 S.F., 46.66 ton	8.05	5.40	13.45
3680	25,000 S.F., 58.33 ton	6.05	5.35	11.40
3760	Offices, 1,500 S.F., 4.75 ton, See Note A			
3800	3,000 S.F., 9.50 ton	16.40	7.60	24
3840	10,000 S.F., 31.66 ton	11.75	7.30	19.05
3880	15,000 S.F., 47.50 ton	10.95	7.35	18.30
3920	20,000 S.F., 63.33 ton	8.20	7.30	15.50
3960	25,000 S.F., 79.16 ton	8.20	7.30	15.50
4000	Restaurants, 1,500 S.F., 7.50 ton	26	12	38
4040	3,000 S.F., 15.00 ton	22	11.75	33.75
4080	10,000 S.F., 50.00 ton	17.30	11.60	28.90
4120	15,000 S.F., 75.00 ton	12.95	11.50	24.45
4160	20,000 S.F., 100.00 ton	11.70	11.50	23.20
4200	25,000 S.F., 125 ton, See Note B			
4240	Schools and colleges, 1,500 S.F., 5.75 ton	19.85	9.20	29.05
4280	3,000 S.F., 11.50 ton	16.80	9	25.80
4320	10,000 S.F., 38.33 ton	13.25	8.90	22.15
4360	15,000 S.F., 57.50 ton	9.95	8.80	18.75
4400	20,000 S.F.,76.66 ton	9.95	8.80	18.75
4440	25,000 S.F., 95.83 ton	9.40	8.80	18.20
4450				
9000	Components of ductwork packages for above systems, per ton of cooling:			
9010	Ductwork; galvanized steel, 240 pounds			
9020	Insulation; fiberglass, 2" thick, FRK faced, 104 SF			
9030	Diffusers; aluminum, 24" x 12", two			
9040	Registers; aluminum, return with dampers, one			

D3050 Terminal & Package Units

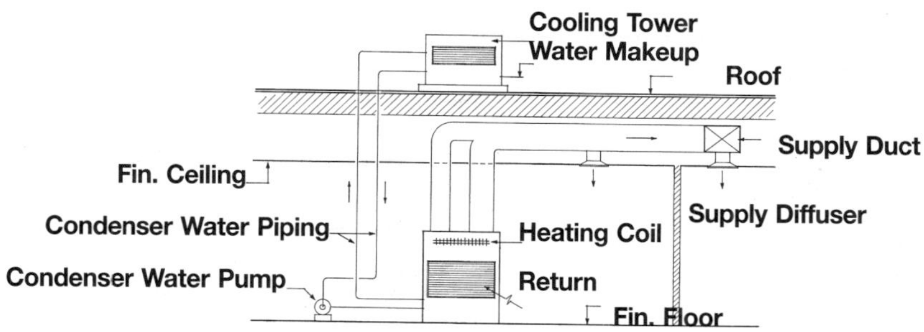

System Description: Self-contained, single package water cooled units include cooling tower, pump, piping allowance. Systems for 1000 S.F. and up include duct and diffusers to provide for even distribution of air. Smaller units distribute air through a supply air plenum, which is integral with the unit.

Returns are not ducted and supplies are not insulated.

Hot water or steam heating coils are included but piping to boiler and the boiler itself is not included.

Where local codes or conditions permit single pass cooling for the smaller units, deduct 10%.

System Components	QUANTITY	UNIT	COST EACH		
			MAT.	INST.	TOTAL
SYSTEM D3050 160 1300					
SELF-CONTAINED, WATER COOLED UNIT					
APARTMENT CORRIDORS, 500 S.F., .92 TON					
Self-contained, water cooled, single package air conditioner unit	1.000	Ea.	1,142.40	554.40	1,696.80
Ductwork package for water or air cooled packaged units	1.000	System	108.56	763.60	872.16
Cooling tower, draw thru single flow, belt drive	1.000	Ea.	138.92	16.97	155.89
Cooling tower pumps & piping	1.000	System	69.46	40.02	109.48
TOTAL			1,459.34	1,374.99	2,834.33
COST PER S.F.			2.92	2.75	5.67

D3050 160	Self-contained, Water Cooled Unit Systems	COST PER S.F.		
		MAT.	INST.	TOTAL
1280	Self-contained, water cooled unit	2.92	2.75	5.67
1300	Apartment corridors, 500 S.F., .92 ton	2.90	2.75	5.65
1320	1,000 S.F., 1.83 ton R236000-20	2.92	2.74	5.66
1360	3,000 S.F., 5.50 ton	2.42	2.30	4.72
1400	5,000 S.F., 9.17 ton	2.40	2.13	4.53
1440	10,000 S.F., 18.33 ton	3.50	1.92	5.42
1520	Banks or libraries, 500 S.F., 2.08 ton	6.10	2.77	8.87
1560	1,000 S.F., 4.17 ton	5.50	5.25	10.75
1600	3,000 S.F., 12.50 ton	5.40	4.87	10.27
1640	5,000 S.F., 20.80 ton	7.90	4.36	12.26
1680	10,000 S.F., 41.66 ton	6.90	4.37	11.27
1760	Bars and taverns, 500 S.F., 5.54 ton	13.20	4.67	17.87
1800	1,000 S.F., 11.08 ton	13.70	8.30	22
1840	3,000 S.F., 33.25 ton	18.35	6.60	24.95
1880	5,000 S.F., 55.42 ton	17.20	6.85	24.05
1920	10,000 S.F., 110.00 ton	16.95	6.80	23.75
2000	Bowling alleys, 500 S.F., 2.83 ton	8.35	3.78	12.13
2040	1,000 S.F., 5.66 ton	7.45	7.10	14.55
2080	3,000 S.F., 17.00 ton	10.80	5.95	16.75
2120	5,000 S.F., 28.33 ton	9.75	5.70	15.45
2160	10,000 S.F., 56.66 ton	9.10	5.85	14.95
2200	Department stores, 500 S.F., 1.46 ton	4.30	1.95	6.25
2240	1,000 S.F., 2.92 ton	3.84	3.66	7.50
2280	3,000 S.F., 8.75 ton	3.79	3.40	7.19
2320	5,000 S.F., 14.58 ton	5.55	3.05	8.60
2360	10,000 S.F., 29.17 ton	5	2.94	7.94

D3050 Terminal & Package Units

D3050 160	Self-contained, Water Cooled Unit Systems	COST PER S.F.		
		MAT.	INST.	TOTAL
2440	Drug stores, 500 S.F., 3.33 ton	9.80	4.44	14.24
2480	1,000 S.F., 6.66 ton	8.80	8.35	17.15
2520	3,000 S.F., 20.00 ton	12.70	7	19.70
2560	5,000 S.F., 33.33 ton	13.30	6.90	20.20
2600	10,000 S.F., 66.66 ton	10.75	6.85	17.60
2680	Factories, 500 S.F., 1.66 ton	4.90	2.21	7.11
2720	1,000 S.F. 3.37 ton	4.44	4.23	8.67
2760	3,000 S.F., 10.00 ton	4.33	3.90	8.23
2800	5,000 S.F., 16.66 ton	6.35	3.49	9.84
2840	10,000 S.F., 33.33 ton	5.75	3.36	9.11
2920	Food supermarkets, 500 S.F., 1.42 ton	4.17	1.88	6.05
2960	1,000 S.F., 2.83 ton	4.50	4.23	8.73
3000	3,000 S.F., 8.50 ton	3.73	3.55	7.28
3040	5,000 S.F., 14.17 ton	3.68	3.30	6.98
3080	10,000 S.F., 28.33 ton	4.88	2.86	7.74
3160	Medical centers, 500 S.F., 1.17 ton	3.43	1.55	4.98
3200	1,000 S.F., 2.33 ton	3.72	3.49	7.21
3240	3,000 S.F., 7.00 ton	3.08	2.93	6.01
3280	5,000 S.F., 11.66 ton	3.04	2.72	5.76
3320	10,000 S.F., 23.33 ton	4.44	2.44	6.88
3400	Offices, 500 S.F., 1.58 ton	4.66	2.11	6.77
3440	1,000 S.F., 3.17 ton	5.05	4.74	9.79
3480	3,000 S.F., 9.50 ton	4.12	3.70	7.82
3520	5,000 S.F., 15.83 ton	6	3.32	9.32
3560	10,000 S.F., 31.67 ton	5.45	3.20	8.65
3640	Restaurants, 500 S.F., 2.50 ton	7.35	3.33	10.68
3680	1,000 S.F., 5.00 ton	6.60	6.30	12.90
3720	3,000 S.F., 15.00 ton	6.50	5.85	12.35
3760	5,000 S.F., 25.00 ton	9.50	5.25	14.75
3800	10,000 S.F., 50.00 ton	5.85	4.81	10.66
3880	Schools and colleges, 500 S.F., 1.92 ton	5.65	2.56	8.21
3920	1,000 S.F., 3.83 ton	5.05	4.81	9.86
3960	3,000 S.F., 11.50 ton	4.99	4.47	9.46
4000	5,000 S.F., 19.17 ton	7.30	4.02	11.32
4040	10,000 S.F., 38.33 ton	6.35	4.02	10.37
4050				
9000	Components of ductwork packages for above systems, per ton of cooling:			
9010	Ductwork; galvanized steel, 108 pounds			
9020	Diffusers, aluminum, 24" x 12", two			

D3050 Terminal & Package Units

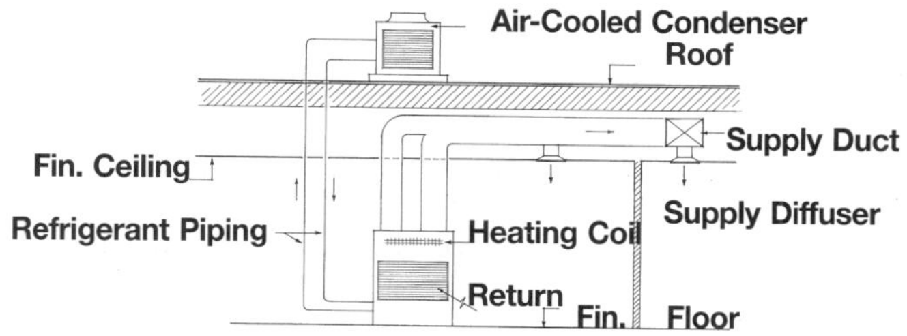

System Description: Self-contained air cooled units with remote air cooled condenser and interconnecting tubing. Systems for 1000 S.F. and up include duct and diffusers. Smaller units distribute air directly.

Returns are not ducted and supplies are not insulated.

Potential savings may be realized by using a single zone rooftop system or through-the-wall unit, especially in the smaller capacities, if the application permits.

Hot water or steam heating coils are included but piping to boiler and the boiler itself is not included.

Condenserless models are available for 15% less where remote refrigerant source is available.

System Components	QUANTITY	UNIT	COST EACH		
			MAT.	INST.	TOTAL
SYSTEM D3050 165 1320					
SELF-CONTAINED, AIR COOLED UNIT					
APARTMENT CORRIDORS, 500 S.F., .92 TON					
Air cooled, package unit	1.000	Ea.	1,198.50	365.50	1,564
Ductwork package for water or air cooled packaged units	1.000	System	108.56	763.60	872.16
Refrigerant piping	1.000	System	506	492.20	998.20
Air cooled condenser, direct drive, propeller fan	1.000	Ea.	348.75	137.95	486.70
TOTAL			2,161.81	1,759.25	3,921.06
COST PER S.F.			4.32	3.52	7.84

D3050 165	Self-contained, Air Cooled Unit Systems		COST PER S.F.		
			MAT.	INST.	TOTAL
1300	Self-contained, air cooled unit				
1320	Apartment corridors, 500 S.F., .92 ton		4.30	3.50	7.80
1360	1,000 S.F., 1.83 ton	R236000-20	4.28	3.49	7.77
1400	3,000 S.F., 5.50 ton		3.68	3.25	6.93
1440	5,000 S.F., 9.17 ton		3.28	3.09	6.37
1480	10,000 S.F., 18.33 ton		3.15	2.84	5.99
1560	Banks or libraries, 500 S.F., 2.08 ton		9.20	4.47	13.67
1600	1,000 S.F., 4.17 ton		8.30	7.40	15.70
1640	3,000 S.F., 12.50 ton		7.45	7.05	14.50
1680	5,000 S.F., 20.80 ton		7.20	6.45	13.65
1720	10,000 S.F., 41.66 ton		8.30	6.35	14.65
1800	Bars and taverns, 500 S.F., 5.54 ton		19.30	9.75	29.05
1840	1,000 S.F., 11.08 ton		19.15	14.10	33.25
1880	3,000 S.F., 33.25 ton		18.90	12.25	31.15
1920	5,000 S.F., 55.42 ton		20.50	12.25	32.75
1960	10,000 S.F., 110.00 ton		20.50	12.20	32.70
2040	Bowling alleys, 500 S.F., 2.83 ton		12.55	6.10	18.65
2080	1,000 S.F., 5.66 ton		11.35	10.05	21.40
2120	3,000 S.F., 17.00 ton		9.75	8.80	18.55
2160	5,000 S.F., 28.33 ton		10	8.60	18.60
2200	10,000 S.F., 56.66 ton		10.70	8.60	19.30
2240	Department stores, 500 S.F., 1.46 ton		6.45	3.14	9.59
2280	1,000 S.F., 2.92 ton		5.85	5.15	11
2320	3,000 S.F., 8.75 ton		5.20	4.92	10.12

D3050 Terminal & Package Units

D3050 165	Self-contained, Air Cooled Unit Systems	COST PER S.F.		
		MAT.	INST.	TOTAL
2360	5,000 S.F., 14.58 ton	5.20	4.92	10.12
2400	10,000 S.F., 29.17 ton	5.10	4.43	9.53
2480	Drug stores, 500 S.F., 3.33 ton	14.75	7.20	21.95
2520	1,000 S.F., 6.66 ton	13.35	11.80	25.15
2560	3,000 S.F., 20.00 ton	11.55	10.40	21.95
2600	5,000 S.F., 33.33 ton	11.75	10.15	21.90
2640	10,000 S.F., 66.66 ton	12.65	10.15	22.80
2720	Factories, 500 S.F., 1.66 ton	7.45	3.61	11.06
2760	1,000 S.F., 3.33 ton	6.70	5.95	12.65
2800	3,000 S.F., 10.00 ton	5.95	5.65	11.60
2840	5,000 S.F., 16.66 ton	5.70	5.20	10.90
2880	10,000 S.F., 33.33 ton	5.85	5.05	10.90
2960	Food supermarkets, 500 S.F., 1.42 ton	6.30	3.06	9.36
3000	1,000 S.F., 2.83 ton	6.60	5.40	12
3040	3,000 S.F., 8.50 ton	5.70	5.05	10.75
3080	5,000 S.F., 14.17 ton	5.05	4.79	9.84
3120	10,000 S.F., 28.33 ton	4.99	4.30	9.29
3200	Medical centers, 500 S.F., 1.17 ton	5.20	2.52	7.72
3240	1,000 S.F., 2.33 ton	5.45	4.46	9.91
3280	3,000 S.F., 7.00 ton	4.69	4.15	8.84
3320	5,000 S.F., 16.66 ton	4.18	3.95	8.13
3360	10,000 S.F., 23.33 ton	4.02	3.63	7.65
3440	Offices, 500 S.F., 1.58 ton	7	3.41	10.41
3480	1,000 S.F., 3.16 ton	7.40	6.05	13.45
3520	3,000 S.F., 9.50 ton	5.65	5.35	11
3560	5,000 S.F., 15.83 ton	5.45	4.93	10.38
3600	10,000 S.F., 31.66 ton	5.60	4.81	10.41
3680	Restaurants, 500 S.F., 2.50 ton	11.05	5.40	16.45
3720	1,000 S.F., 5.00 ton	10.05	8.90	18.95
3760	3,000 S.F., 15.00 ton	8.95	8.45	17.40
3800	5,000 S.F., 25.00 ton	8.75	7.60	16.35
3840	10,000 S.F., 50.00 ton	10.15	7.55	17.70
3920	Schools and colleges, 500 S.F., 1.92 ton	8.50	4.13	12.63
3960	1,000 S.F., 3.83 ton	7.75	6.80	14.55
4000	3,000 S.F., 11.50 ton	6.85	6.45	13.30
4040	5,000 S.F., 19.17 ton	6.60	5.95	12.55
4080	10,000 S.F., 38.33 ton	7.65	5.85	13.50
4090				
9000	Components of ductwork packages for above systems, per ton of cooling:			
9010	Ductwork; galvanized steel, 102 pounds			
9020	Diffusers; aluminum, 24″ x 12″, two			

D3050 Terminal & Package Units

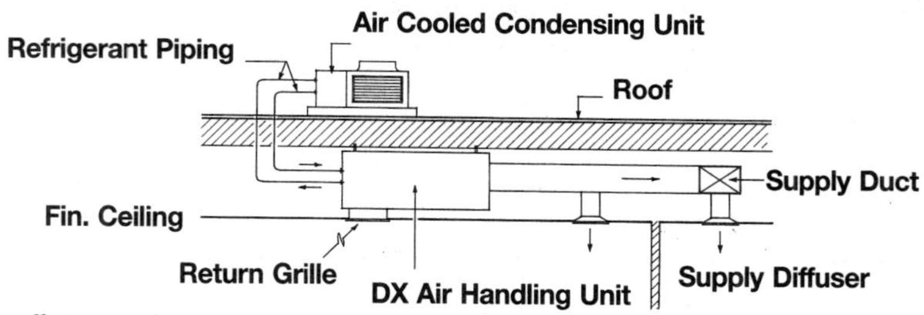

General: Split systems offer several important advantages which should be evaluated when a selection is to be made. They provide a greater degree of flexibility in component selection which permits an accurate match-up of the proper equipment size and type with the particular needs of the building. This allows for maximum use of modern energy saving concepts in heating and cooling. Outdoor installation of the air cooled condensing unit allows space savings in the building and also isolates the equipment operating sounds from building occupants.

Design Assumptions: The systems below are comprised of a direct expansion air handling unit and air cooled condensing unit with interconnecting copper tubing. Ducts and diffusers are also included for distribution of air. Systems are priced for cooling only. Heat can be added as desired either by putting hot water/steam coils into the air unit or into the duct supplying the particular area of need. Gas fired duct furnaces are also available. Refrigerant liquid line is insulated.

System Components	QUANTITY	UNIT	COST EACH MAT.	COST EACH INST.	COST EACH TOTAL
SYSTEM D3050 170 1280					
SPLIT SYSTEM, AIR COOLED CONDENSING UNIT					
APARTMENT CORRIDORS, 1,000 S.F., 1.80 TON					
Fan coil AC unit, cabinet mntd. & filters direct expansion air cool	1.000	Ea.	411.75	129.93	541.68
Ductwork package, for split system, remote condensing unit	1.000	System	101.57	722.85	824.42
Refrigeration piping	1.000	System	406.26	814.35	1,220.61
Condensing unit, air cooled, incls. compressor & standard controls	1.000	Ea.	1,525	518.50	2,043.50
TOTAL			2,444.58	2,185.63	4,630.21
COST PER S.F.			2.44	2.19	4.63
*Cooling requirements would lead to choosing a water cooled unit					

D3050 170	Split Systems With Air Cooled Condensing Units	COST PER S.F. MAT.	COST PER S.F. INST.	COST PER S.F. TOTAL
1260	Split system, air cooled condensing unit			
1280	Apartment corridors, 1,000 S.F., 1.83 ton	2.45	2.18	4.63
1320	2,000 S.F., 3.66 ton	1.77	2.20	3.97
1360	5,000 S.F., 9.17 ton (R236000 -20)	2.04	2.78	4.82
1400	10,000 S.F., 18.33 ton	2.32	3.03	5.35
1440	20,000 S.F., 36.66 ton	2.58	3.08	5.66
1520	Banks and libraries, 1,000 S.F., 4.17 ton	4.02	5	9.02
1560	2,000 S.F., 8.33 ton	4.63	6.30	10.93
1600	5,000 S.F., 20.80 ton	5.25	6.85	12.10
1640	10,000 S.F., 41.66 ton	5.85	6.95	12.80
1680	20,000 S.F., 83.32 ton	7.05	7.25	14.30
1760	Bars and taverns, 1,000 S.F., 11.08 ton	10.50	9.90	20.40
1800	2,000 S.F., 22.16 ton	13.80	11.85	25.65
1840	5,000 S.F., 55.42 ton	12.45	11.10	23.55
1880	10,000 S.F., 110.84 ton	16.05	11.55	27.60
1920	20,000 S.F., 220 ton*			
2000	Bowling alleys, 1,000 S.F., 5.66 ton	5.70	9.40	15.10
2040	2,000 S.F., 11.33 ton	6.30	8.55	14.85
2080	5,000 S.F., 28.33 ton	7.20	9.35	16.55
2120	10,000 S.F., 56.66 ton	8	9.50	17.50
2160	20,000 S.F., 113.32 ton	11.10	10.30	21.40
2320	Department stores, 1,000 S.F., 2.92 ton	2.86	3.46	6.32
2360	2,000 S.F., 5.83 ton	2.94	4.85	7.79
2400	5,000 S.F., 14.58 ton	3.24	4.41	7.65
2440	10,000 S.F., 29.17 ton	3.69	4.82	8.51
2480	20,000 S.F., 58.33 ton	4.10	4.89	8.99

D3050 Terminal & Package Units

D3050 170	Split Systems With Air Cooled Condensing Units	COST PER S.F.		
		MAT.	INST.	TOTAL
2560	Drug stores, 1,000 S.F., 6.66 ton	6.70	11.05	17.75
2600	2,000 S.F., 13.32 ton	7.40	10.05	17.45
2640	5,000 S.F., 33.33 ton	9.40	11.15	20.55
2680	10,000 S.F., 66.66 ton	10.20	11.70	21.90
2720	20,000 S.F., 133.32 ton*			
2800	Factories, 1,000 S.F., 3.33 ton	3.25	3.94	7.19
2840	2,000 S.F., 6.66 ton	3.36	5.55	8.91
2880	5,000 S.F., 16.66 ton	4.22	5.50	9.72
2920	10,000 S.F., 33.33 ton	4.69	5.60	10.29
2960	20,000 S.F., 66.66 ton	5.10	5.85	10.95
3040	Food supermarkets, 1,000 S.F., 2.83 ton	2.78	3.35	6.13
3080	2,000 S.F., 5.66 ton	2.85	4.71	7.56
3120	5,000 S.F., 14.66 ton	3.15	4.28	7.43
3160	10,000 S.F., 28.33 ton	3.58	4.69	8.27
3200	20,000 S.F., 56.66 ton	3.98	4.76	8.74
3280	Medical centers 1,000 S.F., 2.33 ton	2.46	2.72	5.18
3320	2,000 S.F., 4.66 ton	2.35	3.88	6.23
3360	5,000 S.F., 11.66 ton	2.60	3.53	6.13
3400	10,000 S.F., 23.33 ton	2.95	3.85	6.80
3440	20,000 S.F., 46.66 ton	3.28	3.91	7.19
3520	Offices, 1,000 S.F., 3.17 ton	3.09	3.75	6.84
3560	2,000 S.F., 6.33 ton	3.18	5.25	8.43
3600	5,000 S.F., 15.83 ton	3.52	4.78	8.30
3640	10,000 S.F., 31.66 ton	4.01	5.25	9.26
3680	20,000 S.F., 63.32 ton	4.85	5.55	10.40
3760	Restaurants, 1,000 S.F., 5.00 ton	5.05	8.30	13.35
3800	2,000 S.F., 10.00 ton	5.55	7.55	13.10
3840	5,000 S.F., 25.00 ton	6.35	8.25	14.60
3880	10,000 S.F., 50.00 ton	7.05	8.40	15.45
3920	20,000 S.F., 100.00 ton	9.80	9.10	18.90
4000	Schools and colleges, 1,000 S.F., 3.83 ton	3.69	4.62	8.31
4040	2,000 S.F., 7.66 ton	4.27	5.80	10.07
4080	5,000 S.F., 19.17 ton	4.86	6.35	11.21
4120	10,000 S.F., 38.33 ton	5.40	6.45	11.85
4160	20,000 S.F., 76.66 ton	5.90	6.70	12.60
5000				
5100				
5200				
9000	Components of ductwork packages for above systems, per ton of cooling:			
9010	Ductwork; galvanized steel, 102 pounds			
9020	Diffusers; aluminum, 24" x 12", two			

D3050 Terminal & Package Units

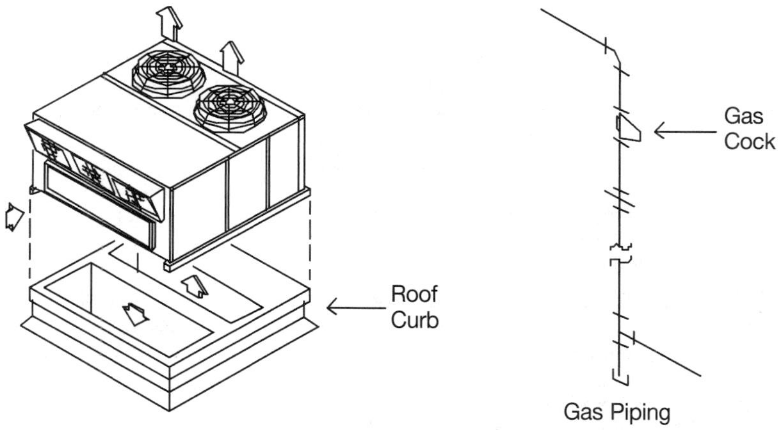

Rooftop Air Conditioner Unit

System Components	QUANTITY	UNIT	COST EACH MAT.	COST EACH INST.	COST EACH TOTAL
SYSTEM D3050 175 1010					
A/C, ROOFTOP, DX COOL, GAS HEAT, CURB, ECONOMIZER, FILTERS., 5 TON					
Roof top A/C, curb, economizer, sgl zone, elec cool, gas ht, 5 ton, 112 MBH	1.000	Ea.	4,575	1,900	6,475
Pipe, black steel, Sch 40, threaded, W/cplgs & hangers, 10′ OC, 1″ dia	20.000	L.F.	89.46	232.05	321.51
Elbow, 90°, black, straight, 3/4″ dia.	3.000	Ea.	11.13	126	137.13
Elbow, 90°, black, straight, 1″ dia.	3.000	Ea.	14.10	135	149.10
Tee, black, reducing, 1″ dia.	1.000	Ea.	11.80	73	84.80
Union, black with brass seat, 1″ dia.	1.000	Ea.	14.85	49	63.85
Pipe nipple, black, 3/4″ dia	2.000	Ea.	5.80	19.20	25
Pipe nipple, black, 1″ dia	2.000	Ea.	4.26	11.05	15.31
Cap, black, 1″ dia.	1.000	Ea.	3.71	19.50	23.21
Gas cock, brass, 1″ size	1.000	Ea.	19.70	31	50.70
Control system, pneumatic, Rooftop A/C unit	1.000	Ea.	2,575	2,100	4,675
Rooftop unit heat/cool balancing	1.000	Ea.		405	405
TOTAL			7,324.81	5,100.80	12,425.61

D3050 175	Rooftop Air Conditioner, Const. Volume	COST EACH MAT.	COST EACH INST.	COST EACH TOTAL
1010	A/C, Rooftop, DX cool, gas heat, curb, economizer, fltrs, 5 Ton	7,325	5,100	12,425
1020	7-1/2 Ton	9,650	5,300	14,950
1030	12-1/2 Ton	13,100	5,825	18,925
1040	18 Ton	19,700	6,375	26,075
1050	25 Ton	28,800	7,250	36,050
1060	40 Ton	39,100	9,700	48,800

D3050 180	Rooftop Air Conditioner, Variable Air Volume	COST EACH MAT.	COST EACH INST.	COST EACH TOTAL
1010	A/C, Rooftop, DX cool, gas heat, curb, ecmizr, fltrs, VAV, 12-1/2 Ton	15,400	6,225	21,625
1020	18 Ton	22,200	6,850	29,050
1030	25 Ton	32,600	7,875	40,475
1040	40 Ton	47,000	10,700	57,700
1050	60 Ton	67,500	14,500	82,000
1060	80 Ton	99,500	18,200	117,700

D3050 Terminal & Package Units

Computer rooms impose special requirements on air conditioning systems. A prime requirement is reliability, due to the potential monetary loss that could be incurred by a system failure. A second basic requirement is the tolerance of control with which temperature and humidity are regulated, and dust eliminated. As the air conditioning system reliability is so vital, the additional cost of reserve capacity and redundant components is often justified.

System Descriptions: Computer areas may be environmentally controlled by one of three methods as follows:

1. **Self-contained Units**
   These are units built to higher standards of performance and reliability. They usually contain alarms and controls to indicate component operation failure, filter change, etc. It should be remembered that these units in the room will occupy space that is relatively expensive to build and that all alterations and service of the equipment will also have to be accomplished within the computer area.
2. **Decentralized Air Handling Units** In operation these are similar to the self-contained units except that their cooling capability comes from remotely located refrigeration equipment as refrigerant or chilled water. As no compressors or refrigerating equipment are required in the air units, they are smaller and require less service than self-contained units. An added plus for this type of system occurs if some of the computer components themselves also require chilled water for cooling.
3. **Central System Supply** Cooling is obtained from a central source which, since it is not located within the computer room, may have excess capacity and permit greater flexibility without interfering with the computer components. System performance criteria must still be met.

Note: The costs shown below do not include an allowance for ductwork or piping.

D3050 185	Computer Room Cooling Units	COST EACH		
		MAT.	INST.	TOTAL
0560	Computer room unit, air cooled, includes remote condenser			
0580	3 ton	17,200	2,125	19,325
0600	5 ton	18,400	2,375	20,775
0620	8 ton	34,500	3,950	38,450
0640	10 ton	36,100	4,275	40,375
0660	15 ton	39,700	4,850	44,550
0680	20 ton	48,200	6,925	55,125
0700	23 ton	59,500	7,900	67,400
0800	Chilled water, for connection to existing chiller system			
0820	5 ton	13,200	1,450	14,650
0840	8 ton	13,300	2,125	15,425
0860	10 ton	13,400	2,175	15,575
0880	15 ton	14,200	2,225	16,425
0900	20 ton	15,000	2,325	17,325
0920	23 ton	16,000	2,625	18,625
1000	Glycol system, complete except for interconnecting tubing			
1020	3 ton	21,700	2,675	24,375
1040	5 ton	23,700	2,800	26,500
1060	8 ton	38,100	4,650	42,750
1080	10 ton	40,500	5,075	45,575
1100	15 ton	50,000	6,375	56,375
1120	20 ton	55,500	6,925	62,425
1140	23 ton	58,000	7,225	65,225
1240	Water cooled, not including condenser water supply or cooling tower			
1260	3 ton	17,500	1,725	19,225
1280	5 ton	19,100	1,975	21,075
1300	8 ton	31,300	3,225	34,525
1320	15 ton	38,000	3,950	41,950
1340	20 ton	40,900	4,375	45,275
1360	23 ton	42,800	4,750	47,550

D3050 201	Packaged A/C, Elec. Ht., Const. Volume	COST EACH		
		MAT.	INST.	TOTAL
1010	A/C Packaged, DX, air cooled, electric heat, constant vol., 5 Ton	8,100	3,650	11,750
1020	10 Ton	13,100	5,050	18,150
1030	20 Ton	23,000	6,625	29,625
1040	30 Ton	34,800	7,375	42,175

D3050 Terminal & Package Units

D3050 201	Packaged A/C, Elec. Ht., Const. Volume	COST EACH		
		MAT.	INST.	TOTAL
1050	40 Ton	53,500	8,475	61,975
1060	50 Ton	69,000	9,500	78,500

D3050 202	Packaged Air Conditioner, Elect. Heat, VAV	COST EACH		
		MAT.	INST.	TOTAL
1010	A/C Packaged, DX, air cooled, electric heat, VAV, 10 Ton	15,500	6,300	21,800
1020	20 Ton	27,000	8,325	35,325
1030	30 Ton	40,300	9,075	49,375
1040	40 Ton	61,500	10,400	71,900
1050	50 Ton	77,500	11,600	89,100
1060	60 Ton	83,500	13,600	97,100

D3050 203	Packaged A/C, Hot Wtr. Heat, Const. Volume	COST EACH		
		MAT.	INST.	TOTAL
1010	A/C Packaged, DX, air cooled, H/W heat, constant vol., 5 Ton	6,400	5,400	11,800
1020	10 Ton	9,925	7,075	17,000
1030	20 Ton	19,100	9,525	28,625
1040	30 Ton	27,400	11,800	39,200
1050	40 Ton	47,200	13,300	60,500
1060	50 Ton	59,000	14,500	73,500

D3050 204	Packaged A/C, Hot Wtr. Heat, VAV	COST EACH		
		MAT.	INST.	TOTAL
1010	A/C Packaged, DX, air cooled, H/W heat, VAV, 10 Ton	12,300	8,200	20,500
1020	20 Ton	23,100	10,800	33,900
1030	30 Ton	32,900	13,400	46,300
1040	40 Ton	55,500	14,500	70,000
1050	50 Ton	67,500	16,800	84,300
1060	60 Ton	75,500	18,900	94,400

D3050 Terminal & Package Units

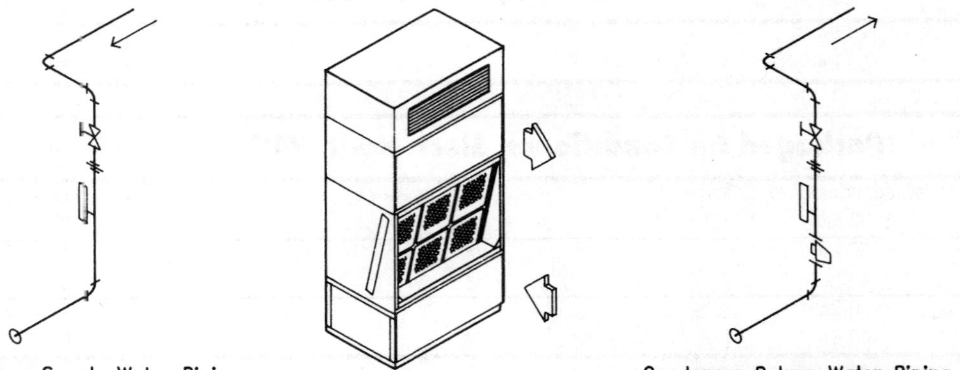

Self-Contained Air Conditioner Unit, Water Cooled

System Components	QUANTITY	UNIT	COST EACH MAT.	COST EACH INST.	COST EACH TOTAL
SYSTEM D3050 210 1010					
A/C, SELF CONTAINED, SINGLE PKG., WATER COOLED, ELECT. HEAT, 5 TON					
Self-contained, water cooled, elect. heat, not inc tower, 5 ton, const vol	1.000	Ea.	7,975	1,375	9,350
Pipe, black steel Sch 40, threaded, W/cplg & hangers, 10′ OC, 1-1/4″ dia.	20.000	L.F.	116.10	254.78	370.88
Elbow, 90°, black, straight, 1-1/4″ dia.	6.000	Ea.	46.50	288	334.50
Tee, black, straight, 1-1/4″ dia.	2.000	Ea.	23.80	151	174.80
Tee, black, reducing, 1-1/4″ dia.	2.000	Ea.	40	151	191
Thermometers, stem type, 9″ case, 8″ stem, 3/4″ NPT	2.000	Ea.	240	42	282
Union, black with brass seat, 1-1/4″ dia.	2.000	Ea.	43	100	143
Pipe nipple, black, 1-1/4″ dia	3.000	Ea.	8.10	17.78	25.88
Valves, bronze, gate, N.R.S., threaded, class 150, 1-1/4″ size	2.000	Ea.	210	78	288
Circuit setter, bal valve, bronze body, threaded, 1-1/4″ pipe size	1.000	Ea.	131	39.50	170.50
Insulation, fiberglass pipe covering, 1″ wall, 1-1/4″ IPS	20.000	L.F.	24.60	94.80	119.40
Control system, pneumatic, A/C Unit with heat	1.000	Ea.	2,575	2,100	4,675
Re-heat coil balancing	1.000	Ea.		110	110
Rooftop unit heat/cool balancing	1.000	Ea.		405	405
TOTAL			11,433.10	5,206.86	16,639.96

D3050 210	AC Unit, Package, Elec. Ht., Water Cooled	COST EACH MAT.	COST EACH INST.	COST EACH TOTAL
1010	A/C, Self contained, single pkg., water cooled, elect. heat, 5 Ton	11,400	5,200	16,600
1020	10 Ton	18,000	6,400	24,400
1030	20 Ton	24,500	8,150	32,650
1040	30 Ton	38,400	8,875	47,275
1050	40 Ton	57,000	9,825	66,825
1060	50 Ton	73,000	11,800	84,800

D3050 215	AC Unit, Package, Elec. Ht., Water Cooled, VAV	COST EACH MAT.	COST EACH INST.	COST EACH TOTAL
1010	A/C, Self contained, single pkg., water cooled, elect. ht, VAV, 10 Ton	20,200	6,400	26,600
1020	20 Ton	28,300	8,150	36,450
1030	30 Ton	43,800	8,875	52,675
1040	40 Ton	65,000	9,825	74,825
1050	50 Ton	81,500	11,800	93,300
1060	60 Ton	88,000	14,300	102,300

D3050 Terminal & Package Units

D3050 220	AC Unit, Package, Hot Water Coil, Water Cooled	COST EACH		
		MAT.	INST.	TOTAL
1010	A/C, Self contn'd, single pkg., water cool, H/W ht, const. vol, 5 Ton	9,600	7,225	16,825
1020	10 Ton	14,600	8,600	23,200
1030	20 Ton	37,700	10,700	48,400
1040	30 Ton	49,500	13,600	63,100
1050	40 Ton	61,000	16,600	77,600
1060	50 Ton	75,000	20,400	95,400

D3050 225	AC Unit, Package, HW Coil, Water Cooled, VAV	COST EACH		
		MAT.	INST.	TOTAL
1010	A/C, Self contn'd, single pkg., water cool, H/W ht, VAV, 10 Ton	16,800	8,100	24,900
1020	20 Ton	24,200	10,500	34,700
1030	30 Ton	36,600	13,100	49,700
1040	40 Ton	59,000	14,200	73,200
1050	50 Ton	72,000	16,800	88,800
1060	60 Ton	80,000	18,300	98,300

D3050 Terminal & Package Units

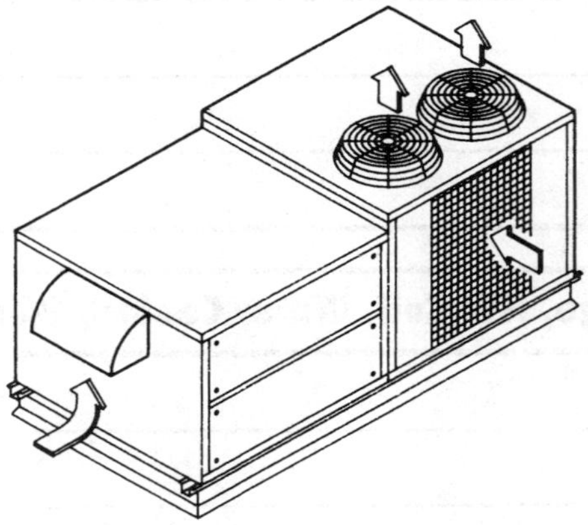

Heat Pump, Air/Air, Rooftop w/ Economizer

System Components	QUANTITY	UNIT	COST EACH MAT.	COST EACH INST.	COST EACH TOTAL
SYSTEM D3050 245 1010					
HEAT PUMP, ROOF TOP, AIR/AIR, CURB., ECONOMIZER., SUP. ELECT. HEAT, 10 TON					
Heat pump, air to air, rooftop, curb, economizer, elect ht, 10 ton cooling	1.000	Ea.	52,500	4,150	56,650
Ductwork, 32" x 12" fabricated, galvanized steel, 16 LF	176.000	Lb.	216.48	1,214.40	1,430.88
Insulation, ductwork, blanket type, fiberglass, 1" thk, 1-1/2 LB density	117.000	S.F.	87.75	332.28	420.03
Rooftop unit heat/cool balancing	1.000	Ea.		405	405
TOTAL			52,804.23	6,101.68	58,905.91

D3050 230	Heat Pump, Water Source, Central Station	COST EACH MAT.	COST EACH INST.	COST EACH TOTAL
1010	Heat pump, central station, water source, constant vol., 5 Ton	3,575	2,900	6,475
1020	10 Ton	8,700	4,075	12,775
1030	20 Ton	18,400	7,600	26,000
1040	30 Ton	26,900	11,100	38,000
1050	40 Ton	37,200	12,100	49,300
1060	50 Ton	42,200	16,200	58,400

D3050 235	Heat Pump, Water Source, Console	COST EACH MAT.	COST EACH INST.	COST EACH TOTAL
1020	Heat pump, console, water source, 1 Ton	1,875	1,600	3,475
1030	1-1/2 Ton	2,125	1,650	3,775
1040	2 Ton	2,575	1,950	4,525
1050	3 Ton	2,850	2,075	4,925
1060	3-1/2 Ton	2,950	2,150	5,100

D3050 240	Heat Pump, Horizontal, Duct Mounted	COST EACH MAT.	COST EACH INST.	COST EACH TOTAL
1020	Heat pump, horizontal, ducted, water source, 1 Ton	2,250	3,425	5,675
1030	1-1/2 Ton	2,700	4,325	7,025
1040	2 Ton	3,225	5,175	8,400
1050	3 Ton	3,850	6,775	10,625
1060	3-1/2 Ton	3,975	7,000	10,975

D3050 245	Heat Pump, Roof Top, Air/Air	COST EACH MAT.	COST EACH INST.	COST EACH TOTAL
1010	Heat pump, roof top, air/air, curb, economizer, sup. elect heat, 10 Ton	53,000	6,100	59,100
1020	20 Ton	75,500	10,400	85,900

D3050 Terminal & Package Units

D3050 245	Heat Pump, Roof Top, Air/Air	COST EACH		
		MAT.	INST.	TOTAL
1030	30 Ton	102,000	14,500	116 500
1040	40 Ton	113,000	16,700	129 700
1050	50 Ton	132,000	18,800	150 800
1060	60 Ton	139,500	21,400	160 900

D3050 Terminal & Package Units

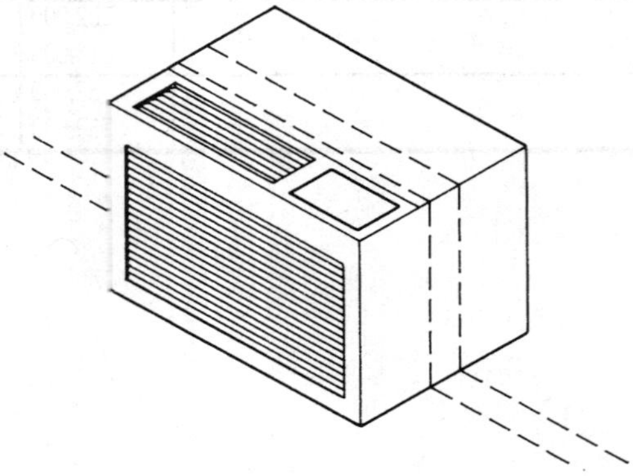

Thru–Wall Heat Pump

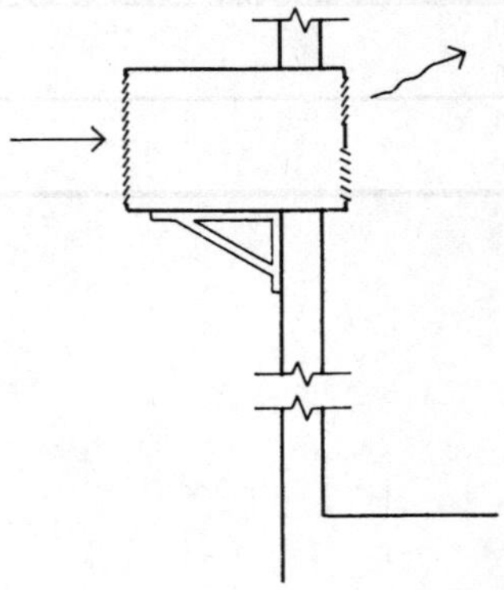

Thru–Wall Air Conditioner

System Components	QUANTITY	UNIT	COST EACH MAT.	COST EACH INST.	COST EACH TOTAL
SYSTEM D3050 255 1010					
A/C UNIT, THRU-THE-WALL, SUPPLEMENTAL ELECT. HEAT, CABINET, LOUVER, 1/2 TON					
A/C unit, thru-wall, electric heat., cabinet, louver, 1/2 ton	1.000	Ea.	690	178	868
TOTAL			690	178	868

D3050 255	Thru-Wall A/C Unit	COST EACH MAT.	COST EACH INST.	COST EACH TOTAL
1010	A/C Unit, thru-the-wall, supplemental elect. heat, cabinet, louver, 1/2 Ton	690	178	868
1020	3/4 Ton	790	213	1,003
1030	1 Ton	860	267	1,127
1040	1-1/2 Ton	1,750	445	2,195
1050	2 Ton	1,900	560	2,460

D3050 260	Thru-Wall Heat Pump	COST EACH MAT.	COST EACH INST.	COST EACH TOTAL
1010	Heat pump, thru-the-wall, cabinet, louver, 1/2 Ton	1,850	133	1,983
1020	3/4 Ton	2,025	178	2,203
1030	1 Ton	2,225	267	2,492
1040	Supplemental. elect. heat, 1-1/2 Ton	2,650	690	3,340
1050	2 Ton	2,825	710	3,535

D3090 Fume Exhaust Systems

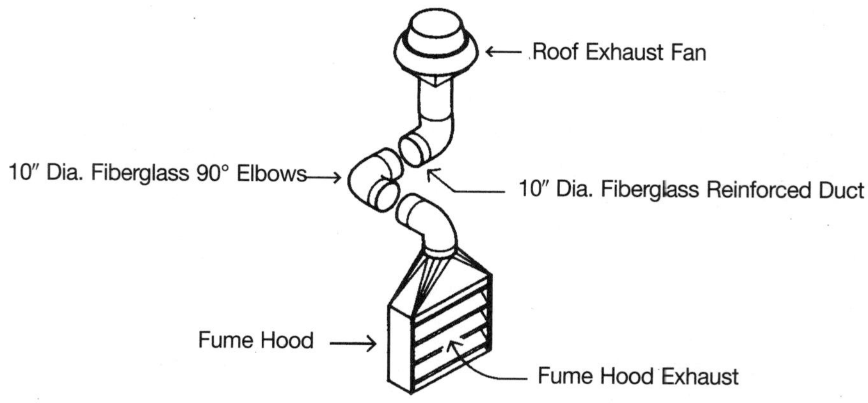

System Components	QUANTITY	UNIT	COST EACH MAT.	COST EACH INST.	COST EACH TOTAL
SYSTEM D3090 310 1010					
FUME HOOD EXHAUST SYSTEM, 3′ LONG, 1000 CFM		Ea.			
Fume hood	3.000	L.F.	5,550	1,245	6,795
Ductwork, plastic, FM approved for acid fume & smoke, straight, 10″ dia	60.000	L.F.	1,320	738	2,058
Ductwork, plastic, FM approved for acid fume & smoke, elbow 90°, 10″ dia	3.000	Ea.	342	402	744
Fan, corrosive fume resistant plastic, 1630 CFM, 1/2 HP	1.000	Ea.	4,750	330	5,080
Fan control switch & mounting base	1.000	Ea.	107	37	144
Roof fan balancing	1.000	Ea.		232	232
TOTAL			12,069	2,984	15,053

D3090 310	Fume Hood Exhaust Systems	COST EACH MAT.	COST EACH INST.	COST EACH TOTAL
1010	Fume hood exhaust system, 3′ long, 1000 CFM	12,100	2,975	15,075
1020	4′ long, 2000 CFM	14,400	3,775	18,175
1030	6′ long, 3500 CFM	19,900	5,650	25,550
1040	6′ long, 5000 CFM	27,300	6,625	33,925
1050	10′ long, 8000 CFM	36,900	9,075	45,975

D3090 Other HVAC Systems/Equip

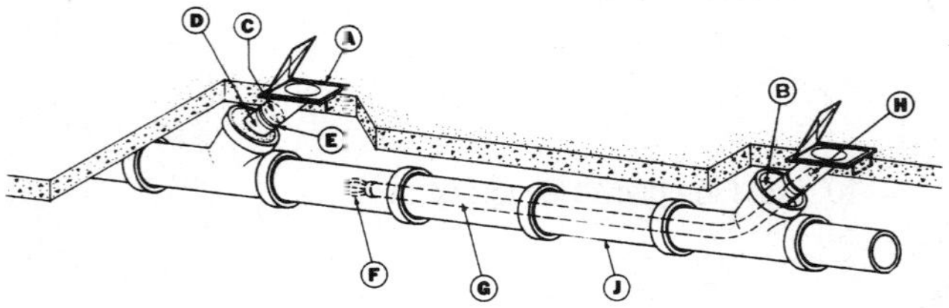

Vitrified Clay Garage Exhaust System

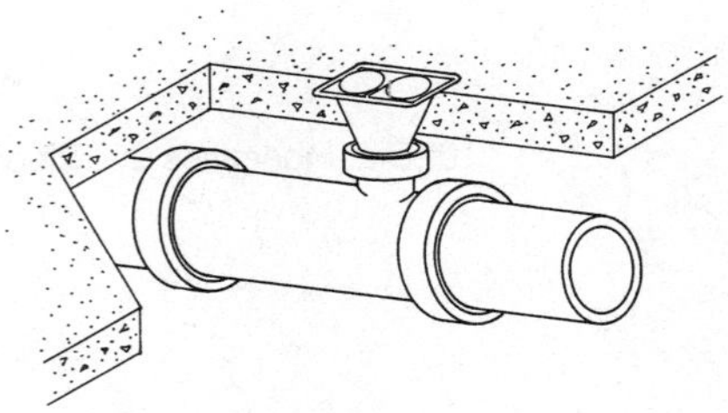

Dual Exhaust System

System Components		QUANTITY	UNIT	COST EACH MAT.	COST EACH INST.	COST EACH TOTAL
SYSTEM D3090 320 1040						
GARAGE, EXHAUST, SINGLE OUTLET, 3″ EXHAUST, CARS & LIGHT TRUCKS						
A	Outlet top assy for engine exhaust system with adapters and ftngs, 3″ diam	1.000	Ea.	239	36.50	275.50
F	Bullnose (guide) for engine exhaust system, 3″ diam	1.000	Ea.	26.50		26.50
G	Galvanized flexible tubing for engine exhaust system, 3″ diam	8.000	L.F.	56.80		56.80
H	Adapter for metal tubing end of engine exhaust system, 3″ tail pipe	1.000	Ea.	53		53
J	Pipe, sewer, vitrified clay, premium joint, 8″ diam.	18.000	L.F.	83.88	154.08	237.96
	24″ deep	18.000	L.F.		27.54	27.54
	24″ deep	18.000	L.F.		40.14	40.14
	Stand for blower, concrete over polystyrene core, 6″ high	1.000	Ea.	9	18.15	27.15
	AC&V duct spiral reducer 10″x8″	1.000	Ea.	12	32.50	44.50
	AC&V duct spiral reducer 12″x10″	1.000	Ea.	17.15	43.50	60.65
	AC&V duct spiral preformed 45° elbow 8″ diam	1.000	Ea.	5.60	37.50	43.10
	AC&V utility fan, belt drive, 3 phase, 2000 CFM, 1 HP	2.000	Ea.	2,000	576	2,576
	Safety switch, heavy duty fused, 240V, 3 pole, 30 amp	1.000	Ea.	201	175	376
	TOTAL			2,703.93	1,140.91	3,844.84

D3090 320	Garage Exhaust Systems	COST PER BAY MAT.	COST PER BAY INST.	COST PER BAY TOTAL
1040	Garage, single exhaust, 3″ outlet, cars & light trucks, one bay	2,700	1,150	3,850
1060	Additional bays up to seven bays	420	160	580
1500	4″ outlet, trucks, one bay	2,725	1,150	3,875
1520	Additional bays up to six bays	435	160	595
1600	5″ outlet, diesel trucks, one bay	3,050	1,150	4,200
1650	Additional single bays up to six	770	184	954
1700	Two adjoining bays	3,050	1,150	4,200
2000	Dual exhaust, 3″ outlets, pair of adjoining bays	3,125	1,250	4,375
2100	Additional pairs of adjoining bays	795	184	979

D3090 Other HVAC Systems/Equip

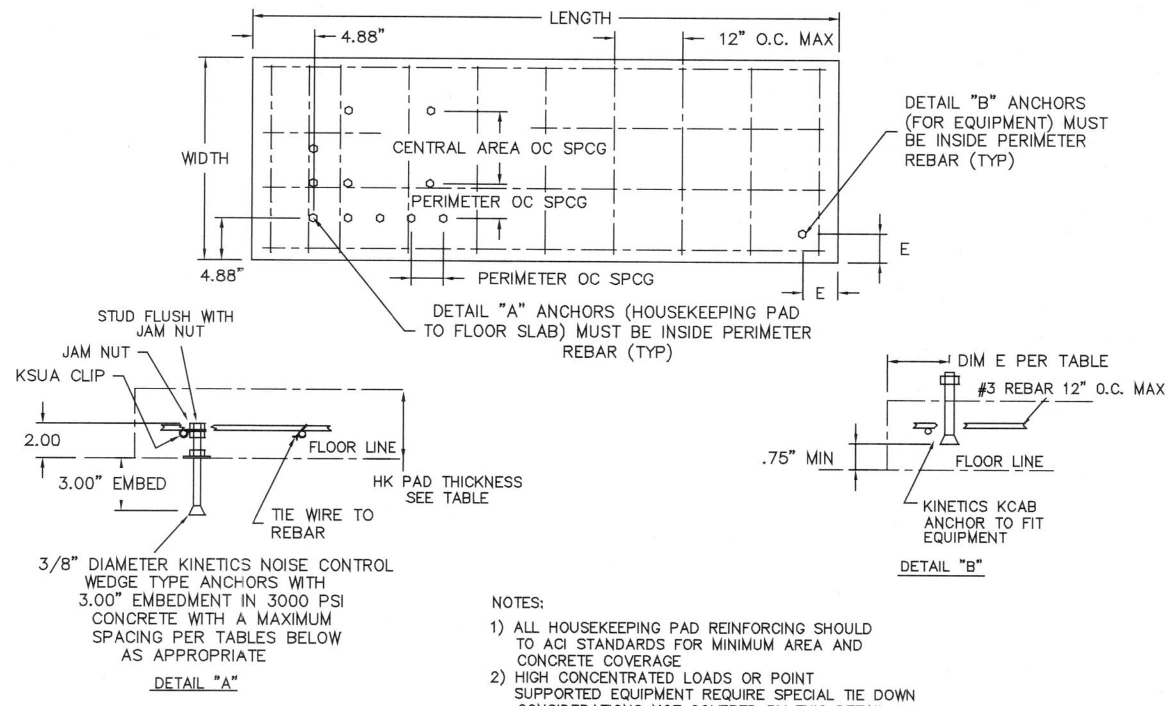

NOTES:

1) ALL HOUSEKEEPING PAD REINFORCING SHOULD TO ACI STANDARDS FOR MINIMUM AREA AND CONCRETE COVERAGE
2) HIGH CONCENTRATED LOADS OR POINT SUPPORTED EQUIPMENT REQUIRE SPECIAL TIE DOWN CONSIDERATIONS NOT COVERED BY THIS DETAIL
3) MINIMUM OF (4) HK PAD TO FLOOR ANCHOR BOLTS REQUIRED FOR EACH HOUSEKEEPING PAD.

System Components	QUANTITY	UNIT	COST EACH MAT.	INST.	TOTAL
SYSTEM D3090 750 1000					
SEISMIC PAD FOR 20-38 SELF-CONTAINED UNIT					
Form equipment pad	28.000	L.F.	25.76	103.04	128.80
Reinforcing steel for equipment pad	.030	Ton	53.25	39	92.25
Concrete 3000 PSI	.866	C.Y.	96.13		96.13
Place concrete pad	.866	C.Y.		27.88	27.88
Finish concrete equipment pad	47.000	S.F.		32.90	32.90
Drill for 3/8 inch slab anchors	10.000	Ea.	.60	78.50	79.10
Exisiting slab anchors 3/8 inch	10.000	Ea.	6.30	35.40	41.70
Equipment anchors 3/8 inch for plate/foot	8.000	Ea.	11.12	122	133.12
TOTAL			193.16	438.72	631.88

D3090 750	Seismic Concrete Equipment Pads	COST PER EACH MAT.	INST.	TOTAL
0990	Seismic pad for self-contained AC units for plate/foot attachment			
1000	Seismic pad for 20-38 tons Self-contained AC unit without economizer .25g	193	440	633
1010	0.5 g seismic acceleration	193	440	633
1020	0.75 g seismic acceleration	193	440	633
1030	1.0 g seismic acceleration	193	440	633
1040	1.5 g seismic acceleration	196	485	681
1050	2.0 g seismic acceleration	207	650	857
1060	2.5 g seismic acceleration	215	785	1,000
1100	Seismic pad for 40-80 tons Self-contained AC unit without economizer .25g	335	645	980
1110	0.5 g seismic acceleration	335	645	980
1120	0.75 g seismic acceleration	335	645	980
1130	1.0 g seismic acceleration	335	645	980
1140	1.5 g seismic acceleration	365	850	1,215
1150	2.0 g seismic acceleration	420	1,275	1,695

D3090 Other HVAC Systems/Equip

D3090 750	Seismic Concrete Equipment Pads	COST PER EACH		
		MAT.	INST.	TOTAL
1160	2.5 g seismic acceleration	485	1,775	2,260
1200	Seismic pad for 80-110 tons Self contained AC unit without economizer .25g	370	690	1,060
1210	0.5 g seismic acceleration	370	690	1,060
1220	0.75 g seismic acceleration	370	690	1,060
1230	1.0 g seismic acceleration	370	690	1,060
1240	1.5 g seismic acceleration	400	910	1,310
1250	2.0 g seismic acceleration	465	1,375	1,840
1260	2.5 g seismic acceleration	535	1,900	2,435
1990	Seismic pad for self-contained AC units for isolator/snubber bolts			
2000	Seismic pad for 20-38 tons Self contained AC unit without economizer .25g	204	560	764
2010	0.5g seismic acceleration	204	560	764
2020	0.75g seismic acceleration	204	560	764
2030	1.0 g seismic acceleration	204	560	764
2040	1.5 g seismic acceleration	207	605	812
2050	2.0 g seismic acceleration	221	805	1,026
2060	2.5 g seismic acceleration	229	940	1,169
2100	Seismic pad for 40-80 tons Self contained AC unit without economizer .25g	350	795	1,145
2110	0.5 g seismic acceleration	350	795	1,145
2120	0.75 g seismic acceleration	350	795	1,145
2130	1.0 g seismic acceleration	350	795	1,145
2140	1.5 g seismic acceleration	375	1,000	1,375
2150	2.0 g seismic acceleration	440	1,500	1,940
2160	2.5 g seismic acceleration	505	2,000	2,505
2200	Seismic pad for 80-110 tons Self contained AC unit without economizer .25g	385	840	1,225
2210	0.5 g seismic acceleration	385	840	1,225
2220	0.75 g seismic acceleration	385	840	1,225
2230	1.0 g seismic acceleration	385	840	1,225
2240	1.5 g seismic acceleration	415	1,050	1,465
2250	2.0 g seismic acceleration	485	1,600	2,085
2260	2.5 g seismic acceleration	555	2,125	2,680
2990	Seismic pad for self-contained AC economizer units for plate/foot attachment			
3000	Seismic pad for 20-38 tons Self contained AC unit economizer .25 g-2.0 g	118	225	343
3010	2.5 g seismic acceleration	124	315	439
3100	Seismic pad for 40-80 tons Self contained AC unit economizer .25-1.5 g	168	330	498
3110	2.0 g seismic acceleration	173	375	548
3120	2.5 g seismic acceleration	181	430	611
3190	Seismic pad for self-contained AC unit economizer for isolator/snubber bolts			
3200	Seismic pad for 20-38 tons Self contained AC unit economizer .25 g-2.0 g	122	271	393
3210	2.5 g seismic acceleration	131	390	521
3300	Seismic pad for 40-80 tons Self contained AC unit economizer .25-1.5 g	176	425	601
3310	2.0 g seismic acceleration	182	465	647
3320	2.5 g seismic acceleration	189	520	709

G Building Sitework

G1030 Site Earthwork

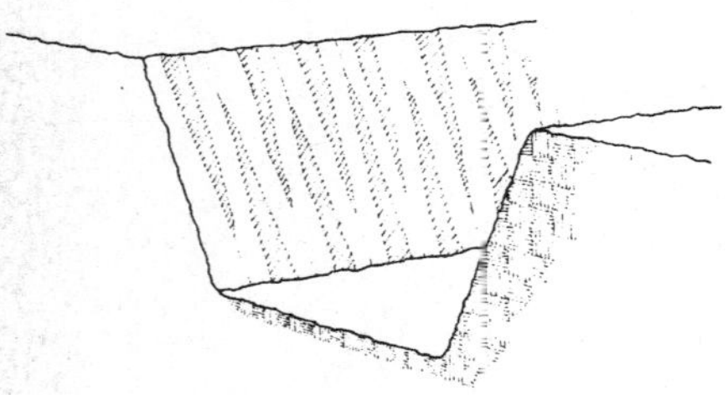

Trenching Systems are shown on a cost per linear foot basis. The systems include: excavation; backfill and removal of spoil; and compaction for various depths and trench bottom widths. The backfill has been reduced to accomodate a pipe of suitable diameter and bedding.

The slope for trench sides varies from none to 1:1.

The Expanded System Listing shows Trenching Systems that range from 2' to 12' in width. Depths range from 2' to 25'.

System Components	QUANTITY	UNIT	COST PER L.F. EQUIP.	COST PER L.F. LABOR	COST PER L.F. TOTAL
SYSTEM G1030 805 1310					
TRENCHING COMMON EARTH, NO SLOPE, 2′ WIDE, 2′ DP, 3/8 C.Y. BUCKET					
Excavation, trench, hyd. backhoe, track mtd., 3/8 C.Y. bucket	.148	B.C.Y.	.32	.88	1.20
Backfill and load spoil, from stockpile	.153	L.C.Y.	.10	.26	.36
Compaction by vibrating plate, 6″ lifts, 4 passes	.118	E.C.Y.	.03	.33	.36
Remove excess spoil, 8 C.Y. dump truck, 2 mile roundtrip	.040	L.C.Y.	.12	.13	.25
TOTAL			.57	1.60	2.17

G1030 805	Trenching Common Earth	COST PER L.F. EQUIP.	COST PER L.F. LABOR	COST PER L.F. TOTAL
1310	Trenching, common earth, no slope, 2′ wide, 2′ deep, 3/8 C.Y. bucket	.57	1.60	2.17
1320	3′ deep, 3/8 C.Y. bucket	.82	2.42	3.24
1330	4′ deep, 3/8 C.Y. bucket	1.06	3.23	4.29
1340	6′ deep, 3/8 C.Y. bucket	1.42	4.20	5.62
1350	8′ deep, 1/2 C.Y. bucket	1.88	5.60	7.48
1360	10′ deep, 1 C.Y. bucket	3.07	6.65	9.72
1400	4′ wide, 2′ deep, 3/8 C.Y. bucket	1.29	3.16	4.45
1410	3′ deep, 3/8 C.Y. bucket	1.76	4.80	6.56
1420	4′ deep, 1/2 C.Y. bucket	2.15	5.40	7.55
1430	6′ deep, 1/2 C.Y. bucket	3.44	8.60	12.04
1440	8′ deep, 1/2 C.Y. bucket	5.70	11.05	16.75
1450	10′ deep, 1 C.Y. bucket	6.95	13.75	20.70
1460	12′ deep, 1 C.Y. bucket	8.95	17.55	26.50
1470	15′ deep, 1-1/2 C.Y. bucket	7.70	15.65	23.35
1480	18′ deep, 2-1/2 C.Y. bucket	10.60	22	32.60
1520	6′ wide, 6′ deep, 5/8 C.Y. bucket w/trench box	7.65	12.70	20.35
1530	8′ deep, 3/4 C.Y. bucket	10.15	16.75	26.90
1540	10′ deep, 1 C.Y. bucket	9.75	17.35	27.10
1550	12′ deep, 1-1/2 C.Y. bucket	10.45	18.65	29.10
1560	16′ deep, 2-1/2 C.Y. bucket	14.75	23	37.75
1570	20′ deep, 3-1/2 C.Y. bucket	18.85	28	46.85
1580	24′ deep, 3-1/2 C.Y. bucket	22.50	33.50	56
1640	8′ wide, 12′ deep, 1-1/2 C.Y. bucket w/trench box	14.60	23.50	38.10
1650	15′ deep, 1-1/2 C.Y. bucket	19.10	31	50.10
1660	18′ deep, 2-1/2 C.Y. bucket	21.50	30.50	52
1680	24′ deep, 3-1/2 C.Y. bucket	30.50	43	73.50
1730	10′ wide, 20′ deep, 3-1/2 C.Y. bucket w/trench box	24.50	41	65.50
1740	24′ deep, 3-1/2 C.Y. bucket	37	49.50	86.50
1780	12′ wide, 20′ deep, 3-1/2 C.Y. bucket w/trench box	39	52	91
1790	25′ deep, bucket	48	66.50	114.50
1800	1/2 to 1 slope, 2′ wide, 2′ deep, 3/8 C.Y. bucket	.82	2.42	3.24
1810	3′ deep, 3/8 C.Y. bucket	1.36	4.26	5.62
1820	4′ deep, 3/8 C.Y. bucket	2.01	6.50	8.51
1840	6′ deep, 3/8 C.Y. bucket	3.43	10.55	13.98

G1030 Site Earthwork

G1030 805	Trenching Common Earth	COST PER L.F.		
		EQUIP.	LABOR	TOTAL
1860	8′ deep, 1/2 C.Y. bucket	5.45	16.80	22.25
1880	10′ deep, 1 C.Y. bucket	10.55	23.50	34.05
2300	4′ wide, 2′ deep, 3/8 C.Y. bucket	1.53	3.98	5.51
2310	3′ deep, 3/8 C.Y. bucket	2.31	6.65	8.96
2320	4′ deep, 1/2 C.Y. bucket	3.04	8.20	11.24
2340	6′ deep, 1/2 C.Y. bucket	5.75	15.25	21
2360	8′ deep, 1/2 C.Y. bucket	11.05	22.50	33.55
2380	10′ deep, 1 C.Y. bucket	15.40	31.50	46.90
2400	12′ deep, 1 C.Y. bucket	19.75	42.50	62.25
2430	15′ deep, 1-1/2 C.Y. bucket	22	45.50	67.50
2460	18′ deep, 2-1/2 C.Y. bucket	39.50	70	109.50
2840	6′ wide, 6′ deep, 5/8 C.Y. bucket w/trench box	11.30	18.90	30.20
2860	8′ deep, 3/4 C.Y. bucket	16.40	28.50	44.90
2880	10′ deep, 1 C.Y. bucket	15.70	29	44.70
2900	12′ deep, 1-1/2 C.Y. bucket	19.90	38	57.90
2940	16′ deep, 2-1/2 C.Y. bucket	33.50	54.50	88
2980	20′ deep, 3-1/2 C.Y. bucket	47	75	122
3020	24′ deep, 3-1/2 C.Y. bucket	67	103	170
3100	8′ wide, 12′ deep, 1-1/2 C.Y. bucket w/trench box	24.50	43	67.50
3120	15′ deep, 1-1/2 C.Y. bucket	36	62.50	98.50
3140	18′ deep, 2-1/2 C.Y. bucket	46.50	73	119.50
3180	24′ deep, 3-1/2 C.Y. bucket	75	112	187
3270	10′ wide, 20′ deep, 3-1/2 C.Y. bucket w/trench box	47.50	86	133.50
3280	24′ deep, 3-1/2 C.Y. bucket	82.50	122	204.50
3370	12′ wide, 20′ deep, 3-1/2 C.Y. bucket w/ trench box	69.50	100	169.50
3380	25′ deep, 3-1/2 C.Y. bucket	96.50	143	239.50
3500	1 to 1 slope, 2′ wide, 2′ deep, 3/8 C.Y. bucket	1.06	3.24	4.30
3520	3′ deep, 3/8 C.Y. bucket	2.98	7.25	10.23
3540	4′ deep, 3/8 C.Y. bucket	2.98	9.75	12.73
3560	6′ deep, 3/8 C.Y. bucket	3.43	10.55	13.98
3580	8′ deep, 1/2 C.Y. bucket	6.80	21	27.80
3600	10′ deep, 1 C.Y. bucket	18	40	58
3800	4′ wide, 2′ deep, 3/8 C.Y. bucket	1.76	4.80	6.56
3820	3′ deep, 3/8 C.Y. bucket	2.86	8.45	11.31
3840	4′ deep, 1/2 C.Y. bucket	3.93	11	14.93
3860	6′ deep, 1/2 C.Y. bucket	8.10	22	30.10
3880	8′ deep, 1/2 C.Y. bucket	16.50	33.50	50
3900	10′ deep, 1 C.Y. bucket	24	49	73
3920	12′ deep, 1 C.Y. bucket	35	71	106
3940	15′ deep, 1-1/2 C.Y. bucket	36	76	112
3960	18′ deep, 2-1/2 C.Y. bucket	54.50	96.50	151
4030	6′ wide, 6′ deep, 5/8 C.Y. bucket w/trench box	14.85	25.50	40.35
4040	8′ deep, 3/4 C.Y. bucket	21.50	35	56.50
4050	10′ deep, 1 C.Y. bucket	22.50	42.50	65
4060	12′ deep, 1-1/2 C.Y. bucket	30.50	57.50	88
4070	16′ deep, 2-1/2 C.Y. bucket	52.50	86	138.50
4080	20′ deep, 3-1/2 C.Y. bucket	76	122	198
4090	24′ deep, 3-1/2 C.Y. bucket	111	172	283
4500	8′ wide, 12′ deep, 1-1/2 C.Y. bucket w/trench box	34.50	62.50	97
4550	15′ deep, 1-1/2 C.Y. bucket	52.50	94	146.50
4600	18′ deep, 2-1/2 C.Y. bucket	70.50	113	183.50
4650	24′ deep, 3-1/2 C.Y. bucket	119	181	300
4800	10′ wide, 20′ deep, 3-1/2 C.Y. bucket w/trench box	70	131	201
4850	24′ deep, 3-1/2 C.Y. bucket	127	191	318
4950	12′ wide, 20′ deep, 3-1/2 C.Y. bucket w/ trench box	100	148	248
4980	25′ deep, 3-1/2 C.Y. bucket	144	216	360

G1030 Site Earthwork

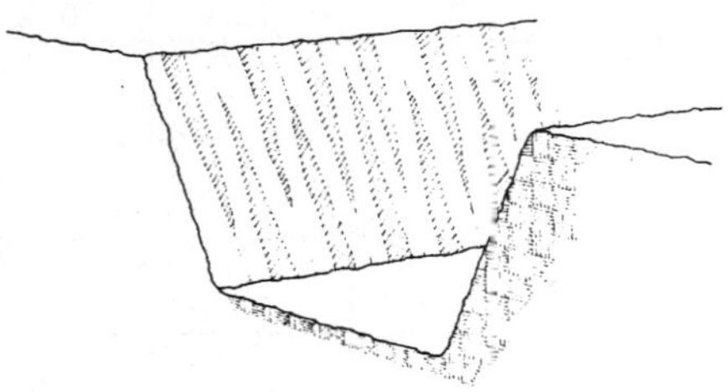

Trenching Systems are shown on a cost per linear foot basis. The systems include: excavation; backfill and removal of spoil; and compaction for various depths and trench bottom widths. The backfill has been reduced to accomodate a pipe of suitable diameter and bedding.

The slope for trench sides varies from none to 1:1.

The Expanded System Listing shows Trenching Systems that range from 2' to 12' in width. Depths range from 2' to 25'.

System Components	QUANTITY	UNIT	COST PER L.F. EQUIP.	LABOR	TOTAL
SYSTEM G1030 806 1310					
TRENCHING LOAM & SANDY CLAY, NO SLOPE, 2′ WIDE, 2′ DP, 3/8 C.Y. BUCKET					
Excavation, trench, hyd. backhoe, track mtd., 3/8 C.Y. bucket	.148	B.C.Y.	.30	.81	1.11
Backfill and load spoil, from stockpile	.165	L.C.Y.	.11	.29	.40
Compaction by vibrating plate 18″ wide, 6″ lifts, 4 passes	.118	E.C.Y.	.03	.33	.36
Remove excess spoil, 8 C.Y. dump truck, 2 mile roundtrip	.042	L.C.Y.	.13	.14	.27
TOTAL			.57	1.60	2.17

G1030 806	Trenching Loam & Sandy Clay	COST PER L.F. EQUIP.	LABOR	TOTAL
1310	Trenching, loam & sandy clay, no slope, 2′ wide, 2′ deep, 3/8 C.Y. bucket	.57	1.57	2.14
1320	3′ deep, 3/8 C.Y. bucket	.89	2.59	3.48
1330	4′ deep, 3/8 C.Y. bucket	1.04	3.16	4.20
1340	6′ deep, 3/8 C.Y. bucket	1.52	3.75	5.25
1350	8′ deep, 1/2 C.Y. bucket	2.02	4.97	7
1360	10′ deep, 1 C.Y. bucket	2.21	5.35	7.55
1400	4′ wide, 2′ deep, 3/8 C.Y. bucket	1.29	3.10	4.39
1410	3′ deep, 3/8 C.Y. bucket	1.75	4.69	6.45
1420	4′ deep, 1/2 C.Y. bucket	2.13	5.30	7.45
1430	6′ deep, 1/2 C.Y. bucket	3.66	7.70	11.35
1440	8′ deep, 1/2 C.Y. bucket	5.55	10.90	16.45
1450	10′ deep, 1 C.Y. bucket	5.25	11.15	16.40
1460	12′ deep, 1 C.Y. bucket	6.60	13.90	20.50
1470	15′ deep, 1-1/2 C.Y. bucket	8.05	16.15	24
1480	18′ deep, 2-1/2 C.Y. bucket	9.85	17.45	27.50
1520	6′ wide, 6′ deep, 5/8 C.Y. bucket w/trench box	7.35	12.50	19.85
1530	8′ deep, 3/4 C.Y. bucket	9.75	16.40	26
1540	10′ deep, 1 C.Y. bucket	8.90	16.65	25.50
1550	12′ deep, 1-1/2 C.Y. bucket	10.15	18.70	29
1560	16′ deep, 2-1/2 C.Y. bucket	14.30	23	37.50
1570	20′ deep, 3-1/2 C.Y. bucket	17.25	28	45.50
1580	24′ deep, 3-1/2 C.Y. bucket	21.50	34	55.50
1640	8′ wide, 12′ deep, 1-1/4 C.Y. bucket w/trench box	14.35	23.50	38
1650	15′ deep, 1-1/2 C.Y. bucket	17.55	30	47.50
1660	18′ deep, 2-1/2 C.Y. bucket	21.50	33	54.50
1680	24′ deep, 3-1/2 C.Y. bucket	29.50	44	73.50
1730	10′ wide, 20′ deep, 3-1/2 C.Y. bucket w/trench box	29.50	44	73.50
1740	24′ deep, 3-1/2 C.Y. bucket	37	54.50	91.50
1780	12′ wide, 20′ deep, 3-1/2 C.Y. bucket w/trench box	36	52	88
1790	25′ deep, bucket	46.50	67.50	114
1800	1/2:1 slope, 2′ wide, 2′ deep, 3/8 C.Y. BK	.80	2.37	3.17
1810	3′ deep, 3/8 C.Y. bucket	1.33	4.16	5.50
1820	4′ deep, 3/8 C.Y. bucket	1.98	6.35	8.35
1840	6′ deep, 3/8 C.Y. bucket	3.69	9.40	13.10
1860	8′ deep, 1/2 C.Y. bucket	5.85	15.05	21
1880	10′ deep, 1 C.Y. bucket	7.55	18.90	26.50

G1030 Site Earthwork

G1030 806	Trenching Loam & Sandy Clay	COST PER L.F.		
		EQUIP.	LABOR	TOTAL
2300	4′ wide, 2′ deep, 3/8 C.Y. bucket	1.52	3.90	5.40
2310	3′ deep, 3/8 C.Y. bucket	2.28	6.50	8.80
2320	4′ deep, 1/2 C.Y. bucket	2.99	8.05	11.05
2340	6′ deep, 1/2 C.Y. bucket	6.15	13.70	19.85
2360	8′ deep, 1/2 C.Y. bucket	10.75	22	33
2380	10′ deep, 1 C.Y. bucket	11.50	25.50	37
2400	12′ deep, 1 C.Y. bucket	19.30	42.50	62
2430	15′ deep, 1-1/2 C.Y. bucket	23	47.50	70.50
2460	18′ deep, 2-1/2 C.Y. bucket	39	71	110
2840	6′ wide, 6′ deep, 5/8 C.Y. bucket w/trench box	10.75	18.95	29.50
2860	8′ deep, 3/4 C.Y. bucket	15.75	28	44
2880	10′ deep, 1 C.Y. bucket	15.90	31.50	47.50
2900	12′ deep, 1-1/2 C.Y. bucket	19.80	38.50	58.50
2940	16′ deep, 2-1/2 C.Y. bucket	32.50	55	87.50
2980	20′ deep, 3-1/2 C.Y. bucket	45.50	76	122
3020	24′ deep, 3-1/2 C.Y. bucket	64.50	104	169
3100	8′ wide, 12′ deep, 1-1/2 C.Y. bucket w/trench box	24	43	67
3120	15′ deep, 1-1/2 C.Y. bucket	32.50	61	93.50
3140	18′ deep, 2-1/2 C.Y. bucket	45	74	119
3180	24′ deep, 3-1/2 C.Y. bucket	72	114	186
3270	10′ wide, 20′ deep, 3-1/2 C.Y. bucket w/trench box	58	92	150
3280	24′ deep, 3-1/2 C.Y. bucket	80	124	204
3320	12′ wide, 20′ deep, 3-1/2 C.Y. bucket w/trench box	64	100	164
3380	25′ deep, 3-1/2 C.Y. bucket w/trench box	87.50	134	222
3500	1:1 slope, 2′ wide, 2′ deep, 3/8 C.Y. bucket	1.04	3.16	4.20
3520	3′ deep, 3/8 C.Y. bucket	1.86	5.95	7.80
3540	4′ deep, 3/8 C.Y. bucket	2.91	9.50	12.40
3560	6′ deep, 3/8 C.Y. bucket	3.69	9.40	13.10
3580	8′ deep, 1/2 C.Y. bucket	9.70	25	34.50
3600	10′ deep, 1 C.Y. bucket	12.85	32.50	45.50
3800	2′ deep, 3/8 C.Y. bucket	1.75	4.69	6.45
3820	4′ wide, 3′ deep, 3/8 C.Y. bucket	2.82	8.25	11.05
3840	4′ deep, 1/2 C.Y. bucket	3.88	10.80	14.70
3860	6′ deep, 1/2 C.Y. bucket	8.65	19.65	28.50
3880	8′ deep, 1/2 C.Y. bucket	15.95	33.50	49.50
3900	10′ deep, 1 C.Y. bucket	17.75	40	58
3920	12′ deep, 1 C.Y. bucket	25.50	57	82.50
3940	15′ deep, 1-1/2 C.Y. bucket	37.50	78.50	116
3960	18′ deep, 2-1/2 C.Y. bucket	53.50	98	152
4030	6′ wide, 6′ deep, 5/8 C.Y. bucket w/trench box	14.15	25.50	39.50
4040	8′ deep, 3/4 C.Y. bucket	22	39	61
4050	10′ deep, 1 C.Y. bucket	23	46	69
4060	12′ deep, 1-1/2 C.Y. bucket	29.50	58	87.50
4070	16′ deep, 2-1/2 C.Y. bucket	51	87	138
4080	20′ deep, 3-1/2 C.Y. bucket	73.50	124	198
4090	24′ deep, 3-1/2 C.Y. bucket	107	174	281
4500	8′ wide, 12′ deep, 1-1/4 C.Y. bucket w/trench box	33.50	63	96.50
4550	15′ deep, 1-1/2 C.Y. bucket	48	91.50	140
4600	18′ deep, 2-1/2 C.Y. bucket	68	114	182
4650	24′ deep, 3-1/2 C.Y. bucket	115	184	299
4800	10′ wide, 20′ deep, 3-1/2 C.Y. bucket w/trench box	86	140	226
4850	24′ deep, 3-1/2 C.Y. bucket	123	194	315
4950	12′ wide, 20′ deep, 3-1/2 C.Y. bucket w/trench box	92	148	240
4980	25′ deep, bucket	139	219	360

G1030 Site Earthwork

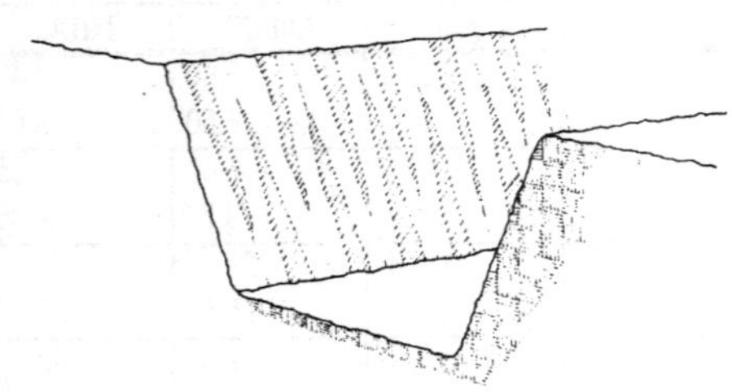

Trenching Systems are shown on a cost per linear foot basis. The systems include: excavation; backfill and removal of spoil; and compaction for various depths and trench bottom widths. The backfill has been reduced to accomodate a pipe of suitable diameter and bedding.

The slope for trench sides varies from none to 1:1.

The Expanded System Listing shows Trenching Systems that range from 2' to 12' in width. Depths range from 2' to 25'.

System Components	QUANTITY	UNIT	COST PER L.F. EQUIP.	COST PER L.F. LABOR	COST PER L.F. TOTAL
SYSTEM G1030 807 1310					
TRENCHING SAND & GRAVEL, NO SLOPE, 2′ WIDE, 2′ DEEP, 3/8 C.Y. BUCKET					
Excavation, trench, hyd. backhoe, track mtd., 3/8 C.Y. bucket	.148	B.C.Y.	.29	.80	1.09
Backfill and load spoil, from stockpile	.140	L.C.Y.	.09	.24	.33
Compaction by vibrating plate 18″ wide, 6″ lifts, 4 passes	.118	E.C.Y.	.03	.33	.36
Remove excess spoil, 8 C.Y. dump truck, 2 mile roundtrip	.035	L.C.Y.	.11	.12	.23
TOTAL			.52	1.49	2.01

G1030 807	Trenching Sand & Gravel	COST PER L.F. EQUIP.	COST PER L.F. LABOR	COST PER L.F. TOTAL
1310	Trenching, sand & gravel, no slope, 2′ wide, 2′ deep, 3/8 C.Y. bucket	.52	1.49	2.01
1320	3′ deep, 3/8 C.Y. bucket	.84	2.49	3.33
1330	4′ deep, 3/8 C.Y. bucket	.97	3	3.97
1340	6′ deep, 3/8 C.Y. bucket	1.45	3.59	5.05
1350	8′ deep, 1/2 C.Y. bucket	1.93	4.76	6.70
1360	10′ deep, 1 C.Y. bucket	2.06	5	7.05
1400	4′ wide, 2′ deep, 3/8 C.Y. bucket	1.18	2.92	4.10
1410	3′ deep, 3/8 C.Y. bucket	1.62	4.44	6.05
1420	4′ deep, 1/2 C.Y. bucket	1.97	5	6.95
1430	6′ deep, 1/2 C.Y. bucket	3.48	7.35	10.85
1440	8′ deep, 1/2 C.Y. bucket	5.25	10.35	15.60
1450	10′ deep, 1 C.Y. bucket	4.93	10.55	15.50
1460	12′ deep, 1 C.Y. bucket	6.20	13.15	19.35
1470	15′ deep, 1-1/2 C.Y. bucket	7.60	15.25	23
1480	18′ deep, 2-1/2 C.Y. bucket	9.30	16.40	25.50
1520	6′ wide, 6′ deep, 5/8 C.Y. bucket w/trench box	7	11.80	18.80
1530	8′ deep, 3/4 C.Y. bucket	9.25	15.60	25
1540	10′ deep, 1 C.Y. bucket	8.40	15.75	24
1550	12′ deep, 1-1/2 C.Y. bucket	9.60	17.60	27
1560	16′ deep, 2 C.Y. bucket	13.70	22	35.50
1570	20′ deep, 3-1/2 C.Y. bucket	16.40	26	42.50
1580	24′ deep, 3-1/2 C.Y. bucket	20.50	32	52.50
1640	8′ wide, 12′ deep, 1-1/2 C.Y. bucket w/trench box	13.50	22.50	36
1650	15′ deep, 1-1/2 C.Y. bucket	16.45	28.50	45
1660	18′ deep, 2-1/2 C.Y. bucket	20.50	31	51.50
1680	24′ deep, 3-1/2 C.Y. bucket	28	41	69
1730	10′ wide, 20′ deep, 3-1/2 C.Y. bucket w/trench box	28	41	69
1740	24′ deep, 3-1/2 C.Y. bucket	35	51	86
1780	12′ wide, 20′ deep, 3-1/2 C.Y. bucket w/trench box	33.50	48.50	82
1790	25′ deep, bucket	44	63	107
1800	1/2:1 slope, 2′ wide, 2′ deep, 3/8 CY bk	.75	2.25	3
1810	3′ deep, 3/8 C.Y. bucket	1.24	3.96	5.20
1820	4′ deep, 3/8 C.Y. bucket	1.85	6.05	7.90
1840	6′ deep, 3/8 C.Y. bucket	3.52	9	12.50
1860	8′ deep, 1/2 C.Y. bucket	5.60	14.40	20
1880	10′ deep, 1 C.Y. bucket	7.05	17.80	25

G1030 Site Earthwork

G1030 807	Trenching Sand & Gravel	COST PER L.F.		
		EQUIP.	LABOR	TOTAL
2300	4' wide, 2' deep, 3/8 C.Y. bucket	1.40	3.68	5.10
2310	3' deep, 3/8 C.Y. bucket	2.12	6.15	8.25
2320	4' deep, 1/2 C.Y. bucket	2.78	7.65	10.45
2340	6' deep, 1/2 C.Y. bucket	5.90	13.10	19
2360	8' deep, 1/2 C.Y. bucket	10.25	21	31.50
2380	10' deep, 1 C.Y. bucket	10.90	24	35
2400	12' deep, 1 C.Y. bucket	18.30	40.50	59
2430	15' deep, 1-1/2 C.Y. bucket	21.50	44.50	66
2460	18' deep, 2-1/2 C.Y. bucket	37	67	104
2840	6' wide, 6' deep, 5/8 C.Y. bucket w/trench box	10.25	17.95	28
2860	8' deep, 3/4 C.Y. bucket	15	26.50	41.50
2880	10' deep, 1 C.Y. bucket	15.10	29.50	44.50
2900	12' deep, 1-1/2 C.Y. bucket	18.80	36	55
2940	16' deep, 2 C.Y. bucket	31.50	52	83.50
2980	20' deep, 3-1/2 C.Y. bucket	43	71	114
3020	24' deep, 3-1/2 C.Y. bucket	61.50	97.50	159
3100	8' wide, 12' deep, 1-1/4 C.Y. bucket w/trench box	23	41	64
3120	15' deep, 1-1/2 C.Y. bucket	31	57.50	88.50
3140	18' deep, 2-1/2 C.Y. bucket	42.50	69.50	112
3180	24' deep, 3-1/2 C.Y. bucket	68.50	107	176
3270	10' wide, 20' deep, 3-1/2 C.Y. bucket w/trench box	55	86.50	142
3280	24' deep, 3-1/2 C.Y. bucket	76	117	193
3370	12' wide, 20' deep, 3-1/2 C.Y. bucket w/trench box	61	96	157
3380	25' deep, bucket	88	134	222
3500	1:1 slope, 2' wide, 2' deep, 3/8 CY bk	1.64	3.68	5.30
3520	3' deep, 3/8 C.Y. bucket	1.74	5.65	7.40
3540	4' deep, 3/8 C.Y. bucket	2.73	9.05	11.80
3560	6' deep, 3/8 C.Y. bucket	3.52	9	12.50
3580	8' deep, 1/2 C.Y. bucket	9.25	24	33.50
3600	10' deep, 1 C.Y. bucket	12.05	30.50	42.50
3800	4' wide, 2' deep, 3/8 C.Y. bucket	1.62	4.44	6.05
3820	3' deep, 3/8 C.Y. bucket	2.62	7.85	10.45
3840	4' deep, 1/2 C.Y. bucket	3.60	10.25	13.85
3860	6' deep, 1/2 C.Y. bucket	8.30	18.85	27
3880	8' deep, 1/2 C.Y. bucket	15.20	31.50	46.50
3900	10' deep, 1 C.Y. bucket	16.80	38	55
3920	12' deep, 1 C.Y. bucket	24.50	54	78.50
3940	15' deep, 1-1/2 C.Y. bucket	35.50	74	110
3960	18' deep, 2-1/2 C.Y. bucket	50.50	92	143
4030	6' wide, 6' deep, 5/8 C.Y. bucket w/trench box	13.45	24	37.50
4040	8' deep, 3/4 C.Y. bucket	21	37.50	58.50
4050	10' deep, 1 C.Y. bucket	21.50	43.50	65
4060	12' deep, 1-1/2 C.Y. bucket	28	54.50	82.50
4070	16' deep, 2 C.Y. bucket	49	82.50	132
4080	20' deep, 3-1/2 C.Y. bucket	70	116	186
4090	24' deep, 3-1/2 C.Y. bucket	102	163	265
4500	8' wide, 12' deep, 1-1/2 C.Y. bucket w/trench box	32	59.50	91.50
4550	15' deep, 1-1/2 C.Y. bucket	45	86	131
4600	18' deep, 2-1/2 C.Y. bucket	64.50	108	173
4650	24' deep, 3-1/2 C.Y. bucket	109	172	281
4800	10' wide, 20' deep, 3-1/2 C.Y. bucket w/trench box	81.50	131	213
4850	24' deep, 3-1/2 C.Y. bucket	117	182	299
4950	12' wide, 20' deep, 3-1/2 C.Y. bucket w/trench box	87.50	139	227
4980	25' deep, bucket	132	206	340

G1030 Site Earthwork

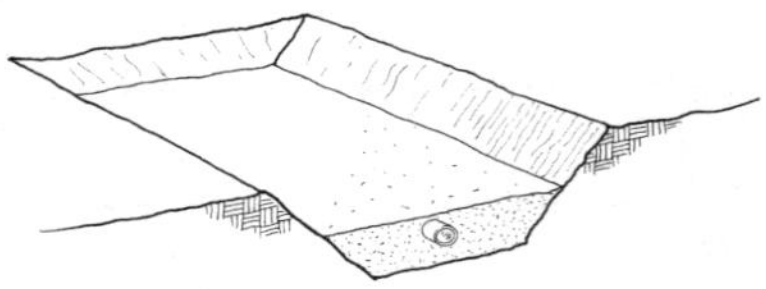

The Pipe Bedding System is shown for various pipe diameters. Compacted bank sand is used for pipe bedding and to fill 12″ over the pipe. No backfill is included. Various side slopes are shown to accommodate different soil conditions. Pipe sizes vary from 6″ to 84″ diameter.

System Components	QUANTITY	UNIT	COST PER L.F. MAT.	COST PER L.F. INST.	COST PER L.F. TOTAL
SYSTEM G1030 815 1440					
PIPE BEDDING, SIDE SLOPE 0 TO 1, 1′ WIDE, PIPE SIZE 6″ DIAMETER					
Borrow, bank sand, 2 mile haul, machine spread	.067	C.Y.	.77	.43	1.20
Compaction, vibrating plate	.067	C.Y.		.14	.14
TOTAL			.77	.57	1.34

G1030 815	Pipe Bedding	COST PER L.F. MAT.	COST PER L.F. INST.	COST PER L.F. TOTAL
1440	Pipe bedding, side slope 0 to 1, 1′ wide, pipe size 6″ diameter	.77	.57	1.34
1460	2′ wide, pipe size 8″ diameter	1.71	1.27	2.98
1480	Pipe size 10″ diameter	1.76	1.31	3.07
1500	Pipe size 12″ diameter	1.80	1.34	3.14
1520	3′ wide, pipe size 14″ diameter	2.97	2.18	5.15
1540	Pipe size 15″ diameter	3	2.21	5.21
1560	Pipe size 16″ diameter	3.04	2.25	5.29
1580	Pipe size 18″ diameter	3.11	2.30	5.41
1600	4′ wide, pipe size 20″ diameter	4.48	3.32	7.80
1620	Pipe size 21″ diameter	4.53	3.35	7.88
1640	Pipe size 24″ diameter	4.64	3.43	8.07
1660	Pipe size 30″ diameter	4.75	3.52	8.27
1680	6′ wide, pipe size 32″ diameter	8.30	6.15	14.45
1700	Pipe size 36″ diameter	8.50	6.30	14.80
1720	7′ wide, pipe size 48″ diameter	11.10	8.20	19.30
1740	8′ wide, pipe size 60″ diameter	13.85	10.20	24.05
1760	10′ wide, pipe size 72″ diameter	20	14.75	34.75
1780	12′ wide, pipe size 84″ diameter	27	20	47
2140	Side slope 1/2 to 1, 1′ wide, pipe size 6″ diameter	1.63	1.21	2.84
2160	2′ wide, pipe size 8″ diameter	2.70	2	4.70
2180	Pipe size 10″ diameter	2.93	2.17	5.10
2200	Pipe size 12″ diameter	3.13	2.32	5.45
2220	3′ wide, pipe size 14″ diameter	4.49	3.32	7.81
2240	Pipe size 15″ diameter	4.61	3.41	8.02
2260	Pipe size 16″ diameter	4.75	3.52	8.27
2280	Pipe size 18″ diameter	5	3.70	8.70
2300	4′ wide, pipe size 20″ diameter	6.65	4.91	11.56
2320	Pipe size 21″ diameter	6.80	5	11.80
2340	Pipe size 24″ diameter	7.30	5.40	12.70
2360	Pipe size 30″ diameter	8.15	6.05	14.20
2380	6′ wide, pipe size 32″ diameter	12	8.90	20.90
2400	Pipe size 36″ diameter	12.85	9.50	22.35
2420	7′ wide, pipe size 48″ diameter	17.60	13	30.60
2440	8′ wide, pipe size 60″ diameter	23	16.90	39.90
2460	10′ wide, pipe size 72″ diameter	32	23.50	55.50
2480	12′ wide, pipe size 84″ diameter	42.50	31.50	74
2620	Side slope 1 to 1, 1′ wide, pipe size 6″ diameter	2.48	1.84	4.32
2640	2′ wide,pipe size 8″ diameter	3.73	2.75	6.48

G1030 Site Earthwork

G1030 815	Pipe Bedding	COST PER L.F.		
		MAT.	INST.	TOTAL
2660	Pipe size 10" diameter	4.09	3.01	7.10
2680	Pipe size 12" diameter	4.49	3.32	7.81
2700	3' wide, pipe size 14" diameter	6	4.45	10.45
2720	Pipe size 15" diameter	6.25	4.60	10.85
2740	Pipe size 15" diameter	6.45	4.78	11.23
2760	Pipe size 18" diameter	6.95	5.15	12.10
2780	4' wide, pipe size 20" diameter	8.80	6.50	15.30
2800	Pipe size 21" diameter	9.05	6.70	15.75
2820	Pipe size 24" diameter	9.90	7.30	17.20
2840	Pipe size 30" diameter	11.55	8.55	20.10
2860	6' wide, pipe size 32" diameter	15.75	11.65	27.40
2880	Pipe size 36" diameter	17.20	12.70	29.90
2900	7' wide, pipe size 48" diameter	24	17.80	41.80
2920	8' wide, pipe size 60" diameter	32	23.50	55.50
2940	10' wide, pipe size 72" diameter	44	32.50	76.50
2960	12' wide, pipe size 84" diameter	58	43	101

G3010 Water Supply

G3010 110	Water Distribution Piping	COST PER L.F.		
		MAT.	INST.	TOTAL
2000	Piping, excav. & backfill excl., ductile iron class 250, mech. joint			
2130	4" diameter	16.85	15.15	32
2150	6" diameter	19.55	19	38.55
2160	8" diameter	21.50	23	44.50
2170	10" diameter	29.50	27	56.50
2180	12" diameter	36	29	65
2210	16" diameter	50	41.50	91.50
2220	18" diameter	63	44	107
3000	Tyton joint			
3130	4" diameter	12.65	7.60	20.25
3150	6" diameter	16.15	9.10	25.25
3160	8" diameter	22.50	15.15	37.65
3170	10" diameter	31.50	16.70	48.20
3180	12" diameter	33.50	19	52.50
3210	16" diameter	59.50	27	86.50
3220	18" diameter	66	30	96
3230	20" diameter	72.50	34	106.50
3250	24" diameter	86	39.50	125.50
4000	Copper tubing, type K			
4050	3/4" diameter	9.65	2.63	12.28
4060	1" diameter	12.60	3.29	15.89
4080	1-1/2" diameter	20.50	3.98	24.48
4090	2" diameter	31.50	4.58	36.08
4110	3" diameter	64.50	7.85	72.35
4130	4" diameter	106	11.10	117.10
4150	6" diameter	269	20.50	289.50
5000	Polyvinyl chloride class 160, S.D.R. 26			
5130	1-1/2" diameter	.37	.98	1.35
5150	2" diameter	.56	1.08	1.64
5160	4" diameter	1.82	3.51	5.33
5170	6" diameter	3.63	4.16	7.79
5180	8" diameter	6.25	5.05	11.30
6000	Polyethylene 160 psi, S.D.R. 7			
6050	3/4" diameter	.52	1.41	1.93
6060	1" diameter	.81	1.52	2.33
6080	1-1/2" diameter	1.76	1.64	3.40
6090	2" diameter	3.04	2.02	5.06

G3020 Sanitary Sewer

G3020 110	Drainage & Sewage Piping	COST PER L.F.		
		MAT.	INST.	TOTAL
2000	Piping, excavation & backfill excluded, PVC, plain			
2130	4″ diameter	1.83	3.51	5.34
2150	6″ diameter	3.74	3.76	7.50
2160	8″ diameter	7.75	3.92	11.67
2900	Box culvert, precast, 8′ long			
3000	6′ x 3′	315	28.50	343.50
3020	6′ x 7′	475	32	507
3040	8′ x 3′	430	30	460
3060	8′ x 8′	585	40	625
3080	10′ x 3′	640	36	676
3100	10′ x 8′	730	50	780
3120	12′ x 3′	625	40	665
3140	12′ x 8′	1,075	59.50	1,134.50
4000	Concrete, nonreinforced			
4150	6″ diameter	6.70	10.45	17.15
4160	8″ diameter	7.35	12.40	19.75
4170	10″ diameter	8.15	12.85	21
4180	12″ diameter	10	13.85	23.85
4200	15″ diameter	11.70	15.45	27.15
4220	18″ diameter	14.35	19.30	33.65
4250	24″ diameter	21.50	27.50	49
4400	Reinforced, no gasket			
4580	12″ diameter	11.20	18.50	29.70
4600	15″ diameter	16.45	18.50	34.95
4620	18″ diameter	19.20	21	40.20
4650	24″ diameter	31.50	27.50	59
4670	30″ diameter	46	43.50	89.50
4680	36″ diameter	59	53	112
4690	42″ diameter	85.50	58	143.50
4700	48″ diameter	124	65.50	189.50
4720	60″ diameter	222	87	309
4730	72″ diameter	247	105	352
4740	84″ diameter	280	131	411
4800	With gasket			
4980	12″ diameter	15.55	10.20	25.75
5000	15″ diameter	19.30	10.70	30
5020	18″ diameter	26	11.30	37.30
5050	24″ diameter	39.50	12.60	52.10
5070	30″ diameter	55.50	43.50	99
5080	36″ diameter	70.50	53	123.50
5090	42″ diameter	114	52.50	166.50
5100	48″ diameter	142	65.50	207.50
5120	60″ diameter	198	84	282
5130	72″ diameter	276	105	381
5140	84″ diameter	335	174	509
5700	Corrugated metal, alum. or galv. bit. coated			
5760	8″ diameter	13.95	8.45	22.40
5770	10″ diameter	16.70	10.70	27.40
5780	12″ diameter	20	13.25	33.25
5800	15″ diameter	24.50	13.85	38.35
5820	18″ diameter	31	14.60	45.60
5850	24″ diameter	38	17.35	55.35
5870	30″ diameter	50.50	32	82.50
5880	36″ diameter	74	32	106
5900	48″ diameter	113	38	151
5920	60″ diameter	146	55.50	201.50
5930	72″ diameter	217	93	310
6000	Plain			

G3020 Sanitary Sewer

G3020 110	Drainage & Sewage Piping	COST PER L.F.		
		MAT.	INST.	TOTAL
6060	8″ diameter	11.40	7.80	19.20
6070	10″ diameter	12.50	9.90	22.40
6080	12″ diameter	14.25	12.60	26.85
6100	15″ diameter	18.15	12.60	30.75
6120	18″ diameter	24	13.55	37.55
6140	24″ diameter	35	15.85	50.85
6170	30″ diameter	44.50	29	73.50
6180	36″ diameter	72.50	29	101.50
6200	48″ diameter	97	34.50	131.50
6220	60″ diameter	153	53.50	206.50
6230	72″ diameter	214	130	344
6300	Steel or alum. oval arch, coated & paved invert			
6400	15″ equivalent diameter	43.50	13.85	57.35
6420	18″ equivalent diameter	56	18.50	74.50
6450	24″ equivalent diameter	80.50	22	102.50
6470	30″ equivalent diameter	98	27.50	125.50
6480	36″ equivalent diameter	146	38	184
6490	42″ equivalent diameter	176	42	218
6500	48″ equivalent diameter	195	50.50	245.50
6600	Plain			
6700	15″ equivalent diameter	23.50	12.35	35.85
6720	18″ equivalent diameter	28	15.85	43.85
6750	24″ equivalent diameter	45	18.50	63.50
6770	30″ equivalent diameter	56	35	91
6780	36″ equivalent diameter	93	35	128
6790	42″ equivalent diameter	110	41.50	151.50
6800	48″ equivalent diameter	96.50	50.50	147
8000	Polyvinyl chloride SDR 35			
8130	4″ diameter	1.83	3.51	5.34
8150	6″ diameter	3.74	3.76	7.50
8160	8″ diameter	7.75	3.92	11.67
8170	10″ diameter	12.25	5.20	17.45
8180	12″ diameter	13.95	5.35	19.30
8200	15″ diameter	13.30	7.15	20.45

G3030 Storm Sewer

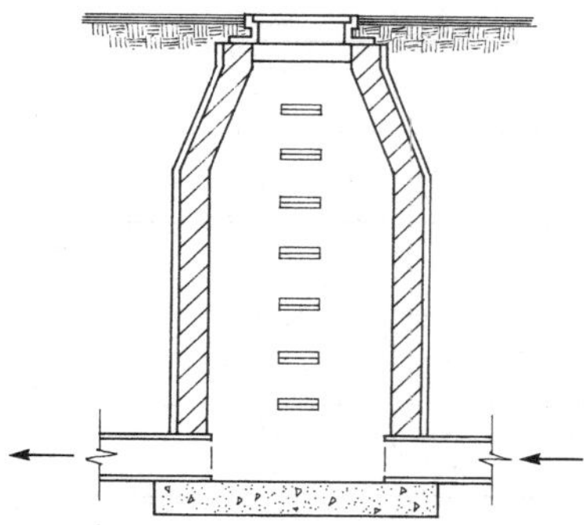

Manhole

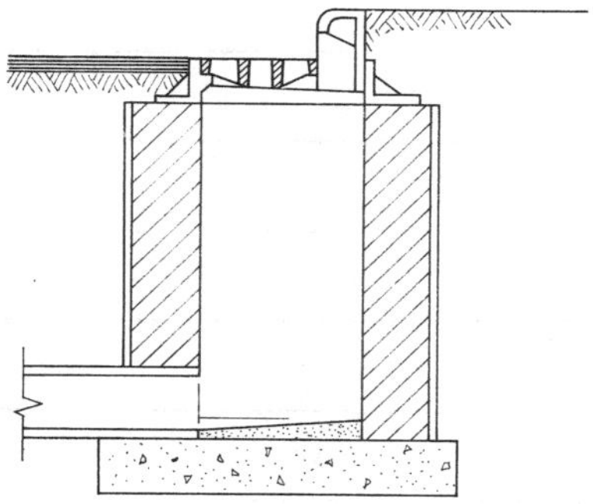

Catch Basin

The Manhole and Catch Basin System includes: excavation with a backhoe; a formed concrete footing; frame and cover; cast iron steps and compacted backfill.

The Expanded System Listing shows manholes that have a 4′, 5′, and 6′ inside diameter riser. Depths range from 4′ to 14′. Construction material shown is either concrete, concrete block, precast concrete, or brick.

System Components	QUANTITY	UNIT	COST PER EACH MAT.	INST.	TOTAL
SYSTEM G3030 210 1920					
MANHOLE/CATCH BASIN, BRICK, 4′ I.D. RISER, 4′ DEEP					
Excavation, hydraulic backhoe, 3/8 C.Y. bucket	14.815	C.Y.		102.97	102.97
Trim sides and bottom of excavation	64.000	S.F.		53.12	53.12
Forms in place, manhole base, 4 uses	20.000	SFCA	15.40	90.80	106.20
Reinforcing in place footings, #4 to #7	.019	Ton	30.88	20.90	51.78
Concrete, 3000 psi	.925	C.Y.	102.68		102.68
Place and vibrate concrete, footing, direct chute	.925	C.Y.		41.66	41.66
Catch basin or MH, brick, 4′ ID, 4′ deep	1.000	Ea.	440	880	1,320
Catch basin or MH steps; heavy galvanized cast iron	1.000	Ea.	14.25	12.30	26.55
Catch basin or MH frame and cover	1.000	Ea.	274	202.50	476.50
Fill, granular	12.954	C.Y.	179.41		179.41
Backfill, spread with wheeled front end loader	12.954	C.Y.		27.07	27.07
Air tamp, add	12.954	C.Y.		99.62	99.62
TOTAL			1,056.62	1,530.94	2,587.56

G3030 210	Manholes & Catch Basins	COST PER EACH MAT.	INST.	TOTAL
1920	Manhole/catch basin, brick, 4′ I.D. riser, 4′ deep	1,050	1,550	2,600
1940	6′ deep	1,400	2,125	3,525
1960	8′ deep	1,825	2,925	4,750
1980	10′ deep	2,325	3,625	5,950
3000	12′ deep	3,000	3,925	6,925
3020	14′ deep	3,775	5,500	9,275
3200	Block, 4′ I.D. riser, 4′ deep	995	1,250	2,245
3220	6′ deep	1,300	1,750	3,050
3240	8′ deep	1,675	2,400	4,075
3260	10′ deep	1,950	2,975	4,925
3280	12′ deep	2,400	3,775	6,175
3300	14′ deep	2,925	4,600	7,525
4620	Concrete, cast-in-place, 4′ I.D. riser, 4′ deep	1,225	2,125	3,350
4640	6′ deep	1,700	2,850	4,550
4660	8′ deep	2,300	4,100	6,400
4680	10′ deep	2,750	5,100	7,850
4700	12′ deep	3,375	6,325	9,700
4720	14′ deep	4,050	7,550	11,600
5820	Concrete, precast, 4′ I.D. riser, 4′ deep	1,575	1,125	2,700
5840	6′ deep	2,025	1,500	3,525

G3030 Storm Sewer

G3030 210	Manholes & Catch Basins	COST PER EACH		
		MAT.	INST.	TOTAL
5860	8′ deep	2,475	2,100	4,575
5880	10′ deep	3,025	2,575	5,600
5900	12′ deep	3,750	3,200	6,950
5920	14′ deep	4,550	4,075	8,625
6000	5′ I.D. riser, 4 deep	1,675	1,250	2,925
6020	6′ deep	2,225	1,750	3,975
6040	8′ deep	2,825	2,325	5,150
6060	10′ deep	3,575	2,975	6,550
6080	12′ deep	4,400	3,775	8,175
6100	14′ deep	5,300	4,625	9,925
6200	6′ I.D. riser, 4 deep	2,425	1,650	4,075
6220	6′ deep	3,150	2,175	5,325
6240	8′ deep	3,925	3,075	7,000
6260	10′ deep	4,975	3,875	8,850
6280	12′ deep	6,075	4,900	10,975
6300	14′ deep	7,275	5,900	13,175

G3060 Fuel Distribution

G3060 110	Gas Service Piping	COST PER L.F.		
		MAT.	INST.	TOTAL
2000	Piping, excavation & backfill excluded, polyethylene			
2070	1-1/4" diam, SDR 10	2.74	3.42	6.16
2090	2" diam, SDR 11	3.40	3.82	7.22
2110	3" diam, SDR 11.5	7.10	4.56	11.66
2130	4" diam,SDR 11	16.20	8.65	24.85
2150	6" diam, SDR 21	50.50	9.25	59.75
2160	8" diam, SDR 21	69	11.20	80.20
3000	Steel, schedule 40, plain end, tarred & wrapped			
3060	1" diameter	8.30	7.65	15.95
3090	2" diameter	13	8.15	21.15
3110	3" diameter	21.50	8.85	30.35
3130	4" diameter	28	14	42
3140	5" diameter	40.50	16.20	56.70
3150	6" diameter	50	19.85	69.85
3160	8" diameter	79	25.50	104.50

G3060 Fuel Distribution

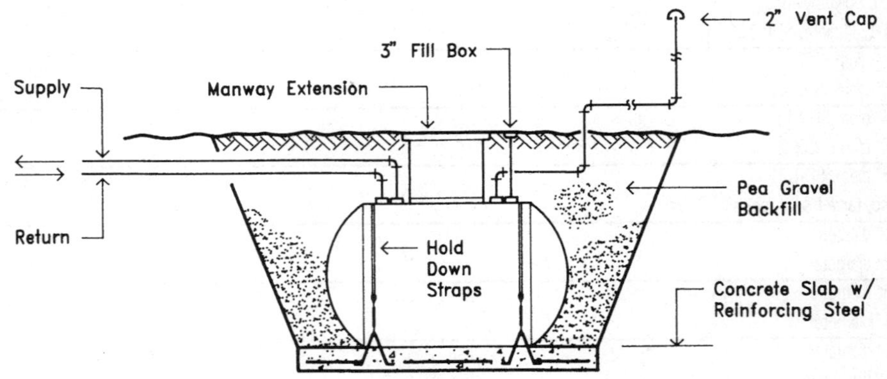

Fiberglass Underground Fuel Storage Tank, Double Wall

System Components	QUANTITY	UNIT	COST EACH MAT.	COST EACH INST.	COST EACH TOTAL
SYSTEM G3060 310 1010					
STORAGE TANK, FUEL, UNDERGROUND, DOUBLE WALL FIBERGLASS, 550 GAL.		Ea.			
Excavating trench, no sheeting or dewatering, 6'-10 ' D, 3/4 CY hyd backhoe	18.000	Ea.		125.10	125.10
Corrosion resistance wrap & coat, small diam pipe, 1" diam, add	120.000	Ea.	198		198
Fiberglass-reinforced elbows 90 & 45 degree, 2 inch	7.000	Ea.	385	483	868
Fiberglass-reinforced threaded adapters, 2 inch	2.000	Ea.	36.80	158	194.80
Remote tank gauging system, 30', 5" pointer travel	1.000	Ea.	4,850	296	5,146
Corrosion resistance wrap & coat, small diam pipe. 4 " diam, add	30.000	Ea.	66		66
Stone back-fill, hand spread, pea gravel	9.000	Ea.	675	387	1,062
Reinforcing in hold down pad, #3 to #7	120.000	Ea.	102	67.20	169.20
Tank leak detector system, 8 channel, external monitoring	1.000	Ea.	6,600		6,600
Tubbing copper, type L, 3/4" diam	120.000	Ea.	792	924	1,716
Elbow 90 degree ,wrought cu x cu 3/4 " diam	4.000	Ea.	12.16	124	136.16
Pipe, black steel welded, sch 40, 3" diam	30.000	Ea.	513	777.90	1,290.90
Elbow 90 degree, steel welded butt joint, 3 " diam	3.000	Ea.	67.50	479.25	546.75
Fiberglass-reinforced plastic pipe, 2 inch	156.000	Ea.	850.20	608.40	1,458.60
Fiberglass-reinforced plastic pipe, 3 inch	36.000	Ea.	336.60	159.12	495.72
Fiberglass-reinforced elbows 90 & 45 degree, 3 inch	3.000	Ea.	201	276	477
Tank leak detection system, probes, well monitoring, liquid phase detection	2.000	Ea.	1,670		1,670
Double wall fiberglass annular space	1.000	Ea.	275		275
Flange, weld neck, 150 lb, 3 " pipe size	1.000	Ea.	27.50	79.88	107.38
Vent protector / breather, 2" dia.	1.000	Ea.	23	18.50	41.50
Corrosion resistance wrap & coat, small diam pipe, 2" diam, add	36.000	Ea.	65.52		65.52
Concrete hold down pad, 6' x 5' x 8" thick	30.000	Ea.	88.20	39	127.20
600 gallon capacity	1.000	Ea.	7,300	440	7,740
Tank, for hold-downs 500-4000 gal, add	1.000	Ea.	460	141	601
Foot valve, single poppet, 3/4" dia	1.000	Ea.	62.50	33	95.50
Fuel fill box, locking inner cover, 3" dia.	1.000	Ea.	130	118	248
Fuel oil specialties, valve,ball chk, globe type fusible, 3/4" dia	2.000	Ea.	134	59	193
TOTAL			25,920.98	5,793.35	31,714.33

G3060 310	Fiberglass Fuel Tank, Double Wall	COST EACH MAT.	COST EACH INST.	COST EACH TOTAL
1010	Storage tank, fuel, underground, double wall fiberglass, 550 Gal.	25,900	5,800	31,700
1020	2500 Gal.	36,100	8,050	44,150
1030	4000 Gal.	43,200	9,950	53,150
1040	6000 Gal.	44,600	11,300	55,900

G3060 Fuel Distribution

G3060 310	Fiberglass Fuel Tank, Double Wall	COST EACH		
		MAT.	INST.	TOTAL
1050	8000 Gal.	54,000	13,500	67,500
1060	10,000 Gal.	60,500	15,600	76,100
1070	15,000 Gal.	78,000	18,000	96,000
1080	20,000 Gal.	96,000	21,800	117,800
1090	25,000 Gal.	117,500	25,400	142,900
1100	30,000 Gal.	133,000	26,900	159,900

G3060 Fuel Distribution

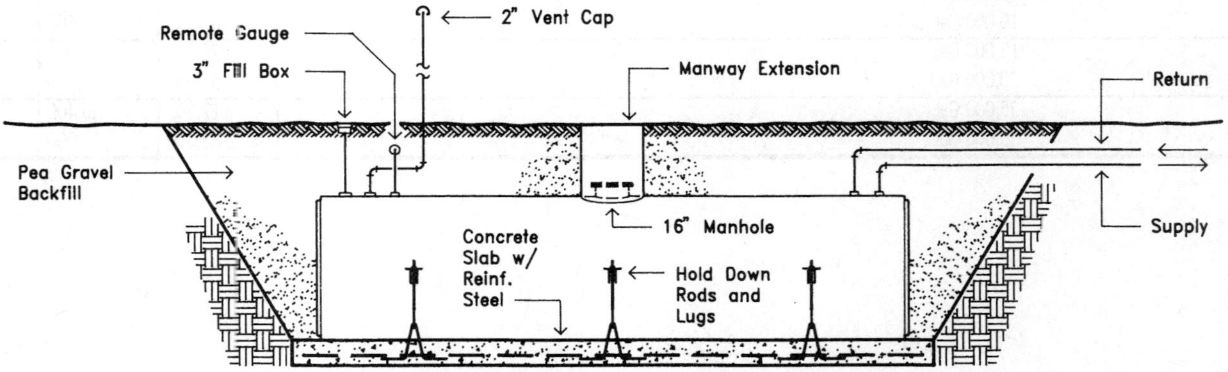

Steel Underground Fuel Storage Tank, Double Wall

System Components	QUANTITY	UNIT	COST EACH MAT.	COST EACH INST.	COST EACH TOTAL
SYSTEM G3060 320 1010					
STORAGE TANK, FUEL, UNDERGROUND, DOUBLE WALL STEEL, 550 GAL.		Ea.			
Fiberglass-reinforced plastic pipe, 2 inch	156.000	Ea.	850.20	608.40	1,458.60
Fiberglass-reinforced plastic pipe, 3 inch	36.000	Ea.	336.60	159.12	495.72
Fiberglass-reinforced elbows 90 & 45 degree, 3 inch	3.000	Ea.	201	276	477
Tank leak detection system, probes, well monitoring, liquid phase detection	2.000	Ea.	1,670		1,670
Flange, weld neck, 150 lb, 3″ pipe size	1.000	Ea.	27.50	79.88	107.38
Vent protector / breather, 2″ dia.	1.000	Ea.	23	18.50	41.50
Concrete hold down pad, 6′ x 5′ x 8″ thick	30.000	Ea.	88.20	39	127.20
Tanks, for hold-downs, 500-2000 gal, add	1.000	Ea.	166	141	307
Double wall steel tank annular space	1.000	Ea.	275		275
Foot valve, single poppet, 3/4″ dia	1.000	Ea.	62.50	33	95.50
Fuel fill box, locking inner cover, 3″ dia.	1.000	Ea.	130	118	248
Fuel oil specialties, valve, ball chk, globe type fusible, 3/4″ dia	2.000	Ea.	134	59	193
Stone back-fill, hand spread, pea gravel	9.000	Ea.	675	387	1,062
Reinforcing in hold down pad, #3 to #7	120.000	Ea.	102	67.20	169.20
Tank, for manways, add	1.000	Ea.	1,650		1,650
Tank leak detector system, 8 channel, external monitoring	1.000	Ea.	6,600		6,600
Tubbing copper, type L, 3/4″ diam	120.000	Ea.	792	924	1,716
Elbow 90 degree, wrought cu x cu 3/4″ diam	8.000	Ea.	24.32	248	272.32
Pipe, black steel welded, sch 40, 3″ diam	30.000	Ea.	513	777.90	1,290.90
Elbow 90 degree, steel welded butt joint, 3″ diam	3.000	Ea.	67.50	479.25	546.75
Excavating trench, no sheeting or dewatering, 6′-10′ D, 3/4 CY hyd backhoe	18.000	Ea.		125.10	125.10
Fiberglass-reinforced elbows 90 & 45 degree, 2 inch	7.000	Ea.	385	483	868
Fiberglass-reinforced threaded adapters, 2 inch	2.000	Ea.	36.80	158	194.80
500 gallon capacity	1.000	Ea.	4,300	445	4,745
Remote tank gauging system, 30′, 5″ pointer travel	1.000	Ea.	4,850	296	5,146
TOTAL			23,959.62	5,922.35	29,881.97

G3060 320	Steel Fuel Tank, Double Wall	COST EACH MAT.	COST EACH INST.	COST EACH TOTAL
1010	Storage tank, fuel, underground, double wall steel, 500 Gal.	24,000	5,925	29,925
1020	2000 Gal.	30,700	7,775	38,475
1030	4000 Gal.	40,100	10,100	50,200
1040	6000 Gal.	45,900	11,200	57,100
1050	8000 Gal.	53,000	13,700	66,700
1060	10,000 Gal.	60,500	15,800	76,300

G3060 Fuel Distribution

G3060 320	Steel Fuel Tank, Double Wall	COST EACH		
		MAT.	INST.	TOTAL
1070	15,000 Gal.	71,000	18,500	89,500
1080	20,000 Gal.	90,500	22,100	112,600
1090	25,000 Gal.	118,000	25,300	143,300
1100	30,000 Gal.	134,500	27,000	161,500
1110	40,000 Gal.	165,500	34,000	199,500
1120	50,000 Gal.	198,000	40,800	238,800

G3060 325	Steel Tank, Above Ground, Single Wall	COST EACH		
		MAT.	INST.	TOTAL
1010	Storage tank, fuel, above ground, single wall steel, 550 Gal.	11,400	4,150	15,550
1020	2000 Gal.	14,300	4,250	13,550
1030	5000 Gal.	17,600	6,500	24,100
1040	10,000 Gal.	30,900	6,925	37,825
1050	15,000 Gal.	33,700	7,125	40,825
1060	20,000 Gal.	39,500	7,350	46,850
1070	25,000 Gal.	45,700	7,525	53,225
1080	30,000 Gal.	50,500	7,850	58,350

G3060 330	Steel Tank, Above Ground, Double Wall	COST EACH		
		MAT.	INST.	TOTAL
1010	Storage tank, fuel, above ground, double wall steel, 500 Gal.	18,300	4,200	22,500
1020	2000 Gal.	24,800	4,300	29,100
1030	4000 Gal.	32,200	4,375	36,575
1040	6000 Gal.	36,800	6,750	43,550
1050	8000 Gal.	42,500	6,925	49,425
1060	10,000 Gal.	45,500	7,050	52,550
1070	15,000 Gal.	59,500	7,300	66,800
1080	20,000 Gal.	65,500	7,525	73,025
1090	25,000 Gal.	75,500	7,775	83,275
1100	30,000 Gal.	81,500	8,050	89,550

Reference Section

All the reference information is in one section, making it easy to find what you need to know . . . and easy to use the book on a daily basis. This section is visually identified by a vertical gray bar on the page edges.

In this Reference Section, we've included Equipment Rental Costs, a listing of rental and operating costs; Crew Listings, a full listing of all crews and equipment, and their costs; Historical Cost Indexes for cost comparisons over time; City Cost Indexes and Location Factors for adjusting costs to the region you are in; Reference Tables, where you will find explanations, estimating information and procedures, or technical data; Change Orders, information on pricing changes to contract documents; Square Foot Costs that allow you to make a rough estimate for the overall cost of a project; and an explanation of all the Abbreviations in the book.

Table of Contents

Estimating Tips

- This section contains the average costs to rent and operate hundreds of pieces of construction equipment. This is useful information when estimating the time and material requirements of any particular operation in order to establish a unit or total cost. Equipment costs include not only rental, but also operating costs for equipment under normal use.

Rental Costs

- Equipment rental rates are obtained from industry sources throughout North America–contractors, suppliers, dealers, manufacturers, and distributors.
- Rental rates vary throughout the country, with larger cities generally having lower rates. Lease plans for new equipment are available for periods in excess of six months, with a percentage of payments applying toward purchase.
- Monthly rental rates vary from 2% to 5% of the purchase price of the equipment depending on the anticipated life of the equipment and its wearing parts.
- Weekly rental rates are about 1/3 the monthly rates, and daily rental rates are about 1/3 the weekly rate.

Operating Costs

- The operating costs include parts and labor for routine servicing, such as repair and replacement of pumps, filters and worn lines. Normal operating expendables, such as fuel, lubricants, tires and electricity (where applicable), are also included.
- Extraordinary operating expendables with highly variable wear patterns, such as diamond bits and blades, are excluded. These costs can be found as material costs in the Unit Price section.
- The hourly operating costs listed do not include the operator's wages.

Equipment Cost/Day

- Any power equipment required by a crew is shown in the Crew Listings with a daily cost.
- The daily cost of equipment needed by a crew is based on dividing the weekly rental rate by 5 (number of working days in the week), and then adding the hourly operating cost times 8 (the number of hours in a day). This "Equipment Cost/Day" is shown in the far right column of the Equipment Rental pages.
- If equipment is needed for only one or two days, it is best to develop your own cost by including components for daily rent and hourly operating cost. This is important when the listed Crew for a task does not contain the equipment needed, such as a crane for lifting mechanical heating/cooling equipment up onto a roof.
- If the quantity of work is less than the crew's Daily Output shown for a Unit Price line item that includes a bare unit equipment cost, it is recommended to estimate one day's rental cost and operating cost for equipment shown in the Crew Listing for that line item.

Mobilization/ Demobilization

- The cost to move construction equipment from an equipment yard or rental company to the jobsite and back again is not included in equipment rental costs listed in the Reference section, nor in the bare equipment cost of any Unit Price line item, nor in any equipment costs shown in the Crew listings.
- Mobilization (to the site) and demobilization (from the site) costs can be found in the Unit Price section.
- If a piece of equipment is already at the jobsite, it is not appropriate to utilize mobil./ demob. costs again in an estimate.

01 54 33 | Equipment Rental

				UNIT	HOURLY OPER. COST	RENT PER DAY	RENT PER WEEK	RENT PER MONTH	EQUIPMENT COST/DAY
10	0010	**CONCRETE EQUIPMENT RENTAL** without operators	R015433 -10						
	0200	Bucket, concrete lightweight, 1/2 C.Y.		Ea.	.65	20	60	180	17.20
	0300	1 C.Y.			.70	24.50	73	219	20.20
	0400	1-1/2 C.Y.			.90	33	99	297	27
	0500	2 C.Y.			1.00	40	120	360	32
	0580	8 C.Y.			5.50	262	785	2,350	201
	0600	Cart, concrete, self propelled, operator walking, 10 C.F.			2.65	56.50	170	510	55.20
	0700	Operator riding, 18 C.F.			4.30	86.50	260	780	86.40
	0800	Conveyer for concrete, portable, gas, 16" wide, 26' long			10.00	123	370	1,100	154
	0900	46' long			10.40	150	450	1,350	173.20
	1000	56' long			10.50	158	475	1,425	179
	1100	Core drill, electric, 2-1/2 H.P., 1" to 8" bit diameter			1.72	65.50	196	590	52.95
	1150	11 HP, 8" to 18" cores			5.15	115	345	1,025	110.20
	1200	Finisher, concrete floor, gas, riding trowel, 96" wide			9.95	145	435	1,300	166.60
	1300	Gas, walk-behind, 3 blade, 36" trowel			1.65	19.65	59	177	25
	1400	4 blade, 48" trowel			3.20	28	84	252	42.40
	1500	Float, hand-operated (Bull float) 48" wide			.08	13.65	41	123	8.85
	1570	Curb builder, 14 H.P., gas, single screw			11.85	243	730	2,200	240.80
	1590	Double screw			12.55	285	855	2,575	271.40
	1600	Grinder, concrete and terrazzo, electric, floor			2.45	103	310	930	81.60
	1700	Wall grinder			1.23	51.50	155	465	40.85
	1800	Mixer, powered, mortar and concrete, gas, 6 C.F., 18 H.P.			6.95	118	355	1,075	126.60
	1900	10 C.F., 25 H.P.			8.65	143	430	1,300	155.20
	2000	16 C.F.			8.95	167	500	1,500	171.60
	2100	Concrete, stationary, tilt drum, 2 C.Y.			6.10	232	695	2,075	187.80
	2120	Pump, concrete, truck mounted 4" line 80' boom			22.85	925	2,775	8,325	737.80
	2140	5" line, 110' boom			29.75	1,225	3,695	11,100	977
	2160	Mud jack, 50 C.F. per hr.			6.10	130	390	1,175	126.80
	2180	225 C.F. per hr.			7.95	147	440	1,325	151.60
	2190	Shotcrete pump rig, 12 CY/hr			12.60	228	685	2,050	237.80
	2600	Saw, concrete, manual, gas, 18 H.P.			5.15	36.50	110	330	63.20
	2650	Self-propelled, gas, 30 H.P.			10.25	98.50	295	885	141
	2700	Vibrators, concrete, electric, 60 cycle, 2 H.P.			.46	8.35	25	75	8.70
	2800	3 H.P.			.64	10.35	31	93	11.30
	2900	Gas engine, 5 H.P.			1.55	14.35	43	129	21
	3000	8 H.P.			2.10	15.35	46	138	26
	3050	Vibrating screed, gas engine, 8 H.P.			3.32	73	219	655	70.35
	3120	Concrete transit mixer, 6 x 4, 250 H.P., 8 C.Y., rear discharge			47.15	575	1,720	5,150	721.20
	3200	Front discharge			55.50	710	2,125	6,375	869
	3300	6 x 6, 285 H.P., 12 C.Y., rear discharge			54.60	665	2,000	6,000	836.80
	3400	Front discharge		↓	57.05	715	2,150	6,450	886.40
20	0010	**EARTHWORK EQUIPMENT RENTAL** without operators	R015433 -10						
	0040	Aggregate spreader, push type 8' to 12' wide		Ea.	2.45	25.50	77	231	35
	0045	Tailgate type, 8' wide			2.35	33.50	100	300	38.80
	0055	Earth auger, truck-mounted, for fence & sign posts, utility poles			13.50	505	1,520	4,550	412
	0060	For borings and monitoring wells			38.20	670	2,005	6,025	706.60
	0070	Portable, trailer mounted			2.65	26	78	234	36.80
	0075	Truck-mounted, for caissons, water wells			85.35	3,075	9,250	27,800	2,533
	0080	Horizontal boring machine, 12" to 36" diameter, 45 H.P.			20.30	202	605	1,825	283.40
	0090	12" to 48" diameter, 65 H.P.			28.65	360	1,075	3,225	444.20
	0095	Auger, for fence posts, gas engine, hand held			.50	5.65	17	51	7.40
	0100	Excavator, diesel hydraulic, crawler mounted, 1/2 C.Y. cap.			20.95	380	1,135	3,400	394.60
	0120	5/8 C.Y. capacity			25.40	525	1,570	4,700	517.20
	0140	3/4 C.Y. capacity			29.70	600	1,805	5,425	598.60
	0150	1 C.Y. capacity			36.90	625	1,870	5,600	669.20
	0200	1-1/2 C.Y. capacity			45.10	840	2,525	7,575	865.80
	0300	2 C.Y. capacity		↓	61.75	1,150	3,440	10,300	1,182

01 54 33 | Equipment Rental

			UNIT	HOURLY OPER. COST	RENT PER DAY	RENT PER WEEK	RENT PER MONTH	EQUIPMENT COST/DAY	
20	0320	2-1/2 C.Y. capacity	Ea.	83.30	1,600	4,765	14,300	1,619	20
	0325	3-1/2 C.Y. capacity		111.30	2,125	6,360	19,100	2,162	
	0330	4-1/2 C.Y. capacity		134.65	2,600	7,790	23,400	2,635	
	0335	6 C.Y. capacity		168.45	2,875	8,660	26,000	3,080	
	0340	7 C.Y. capacity		173.20	3,125	9,350	28,100	3,256	
	0342	Excavator attachments, bucket thumbs		2.70	215	645	1,925	150.60	
	0345	Grapples		2.45	188	565	1,700	132.60	
	0347	Hydraulic hammer for boom mounting, 5000 ft-lb		12.10	405	1,220	3,650	340.80	
	0349	11,000 ft-lb		20.40	730	2,190	6,575	601.20	
	0350	Gradall type, truck mounted, 3 ton @ 15' radius, 5/8 C.Y.		49.60	935	2,810	8,425	958.80	
	0370	1 C.Y. capacity		55.10	1,150	3,445	10,300	1,130	
	0400	Backhoe-loader, 40 to 45 H.P., 5/8 C.Y. capacity		11.90	227	680	2,050	231.20	
	0450	45 H.P. to 60 H.P., 3/4 C.Y. capacity		15.85	278	835	2,500	293.80	
	0460	80 H.P., 1-1/4 C.Y. capacity		19.00	320	965	2,900	345	
	0470	112 H.P., 1-1/2 C.Y. capacity		25.45	465	1,400	4,200	483.60	
	0482	Backhoe-loader attachment, compactor, 20,000 lb		4.70	127	380	1,150	113.60	
	0485	Hydraulic hammer, 750 ft-lbs		2.80	90	270	810	76.40	
	0486	Hydraulic hammer, 1200 ft-lbs		5.25	185	555	1,675	153	
	0500	Brush chipper, gas engine, 6" cutter head, 35 H.P.		8.95	108	325	975	136.60	
	0550	12" cutter head, 130 H.P.		14.95	177	530	1,600	225.60	
	0600	15" cutter head, 165 H.P.		22.45	217	650	1,950	309.60	
	0750	Bucket, clamshell, general purpose, 3/8 C.Y.		1.10	36.50	110	330	30.80	
	0800	1/2 C.Y.		1.20	45	135	405	36.60	
	0850	3/4 C.Y.		1.35	55	165	495	43.80	
	0900	1 C.Y.		1.40	58.50	175	525	46.20	
	0950	1-1/2 C.Y.		2.25	80	240	720	66	
	1000	2 C.Y.		2.40	88.50	265	795	72.20	
	1010	Bucket, dragline, medium duty, 1/2 C.Y.		.65	23.50	70	210	19.20	
	1020	3/4 C.Y.		.65	24.50	74	222	20	
	1030	1 C.Y.		.70	26.50	80	240	21.60	
	1040	1-1/2 C.Y.		1.10	40	120	360	32.80	
	1050	2 C.Y.		1.15	45	135	405	36.20	
	1070	3 C.Y.		1.65	55	165	495	46.20	
	1200	Compactor, manually guided 2-drum vibratory smooth roller, 7.5 H.P.		5.65	167	500	1,500	145.20	
	1250	Rammer/tamper, gas, 8"		2.30	41.50	125	375	43.40	
	1260	15"		2.55	48.50	145	435	49.40	
	1300	Vibratory plate, gas, 18" plate, 3000 lb. blow		2.25	23	69	207	31.80	
	1350	21" plate, 5000 lb. blow		2.80	28	84	252	39.20	
	1370	Curb builder/extruder, 14 H.P., gas, single screw		11.85	243	730	2,200	240.80	
	1390	Double screw		12.55	285	855	2,575	271.40	
	1500	Disc harrow attachment, for tractor		.39	65.50	197	590	42.50	
	1810	Feller buncher, shearing & accumulating trees, 100 H.P.		28.55	555	1,665	5,000	561.40	
	1860	Grader, self-propelled, 25,000 lb.		24.85	450	1,345	4,025	467.80	
	1910	30,000 lb.		29.65	520	1,565	4,700	550.20	
	1920	40,000 lb.		45.50	995	2,985	8,950	961	
	1930	55,000 lb.		59.10	1,400	4,215	12,600	1,316	
	1950	Hammer, pavement breaker, self-propelled, diesel, 1000 to 1250 lb		23.55	340	1,025	3,075	393.40	
	2000	1300 to 1500 lb.		35.33	685	2,050	6,150	692.65	
	2050	Pile driving hammer, steam or air, 4150 ft.-lb. @ 225 BPM		9.35	470	1,405	4,225	355.80	
	2100	8750 ft.-lb. @ 145 BPM		11.35	650	1,955	5,875	481.80	
	2150	15,000 ft.-lb. @ 60 BPM		12.95	790	2,375	7,125	578.60	
	2200	24,450 ft.-lb. @ 111 BPM		13.95	885	2,655	7,975	642.60	
	2250	Leads, 60' high for pile driving hammers up to 20,000 ft.-lb.		2.85	78	234	700	69.60	
	2300	90' high for hammers over 20,000 ft.-lb.		4.35	140	420	1,250	118.80	
	2350	Diesel type hammer, 22,400 ft.-lb.		25.75	640	1,920	5,750	590	
	2400	41,300 ft.-lb.		32.70	695	2,080	6,250	677.60	
	2450	141,000 ft.-lb.		49.25	1,200	3,570	10,700	1,108	
20	2500	Vib. elec. hammer/extractor, 200 KW diesel generator, 34 H.P.	↓	41.05	665	2,000	6,000	728.40	

01 54 33 | Equipment Rental

		Description	Unit	Hourly Oper. Cost	Rent Per Day	Rent Per Week	Rent Per Month	Equipment Cost/Day
20	2550	80 H.P.	Ea.	73.75	980	2,940	8,825	1,178
	2600	150 H.P.		139.15	1,900	5,720	17,200	2,257
	2700	Extractor, steam or air, 700 ft.-lb.		17.15	510	1,530	4,600	443.20
	2750	1000 ft.-lb.		19.55	625	1,875	5,625	531.40
	2800	Log chipper, up to 22" diam, 600 H.P.		40.05	445	1,340	4,025	588.40
	2850	Logger, for skidding & stacking logs, 150 H.P.		44.95	895	2,690	8,075	897.60
	2860	Mulcher, diesel powered, trailer mounted		17.60	218	655	1,975	271.80
	2900	Rake, spring tooth, with tractor		11.34	278	835	2,500	257.70
	3000	Roller, vibratory, tandem, smooth drum, 20 H.P.		7.35	132	395	1,175	137.80
	3050	35 H.P.		9.50	262	785	2,350	233
	3100	Towed type vibratory compactor, smooth drum, 50 H.P.		22.75	380	1,145	3,425	411
	3150	Sheepsfoot, 50 H.P.		25.05	440	1,325	3,975	465.40
	3170	Landfill compactor, 220 HP		69.10	1,450	4,315	12,900	1,416
	3200	Pneumatic tire roller, 80 H.P.		13.10	340	1,015	3,050	307.80
	3250	120 H.P.		19.90	600	1,805	5,425	520.20
	3300	Sheepsfoot vibratory roller, 240 H.P.		56.65	1,150	3,445	10,300	1,142
	3320	340 H.P.		75.60	1,425	4,240	12,700	1,453
	3350	Smooth drum vibratory roller, 75 H.P.		19.70	520	1,560	4,675	469.60
	3400	125 H.P.		25.70	650	1,945	5,825	594.60
	3410	Rotary mower, brush, 60", with tractor		15.30	252	755	2,275	273.40
	3420	Rototiller, walk-behind, gas, 5 H.P.		2.43	62.50	188	565	57.05
	3422	8 H.P.		3.89	100	300	900	91.10
	3440	Scrapers, towed type, 7 CY capacity		4.90	105	315	945	102.20
	3450	10 C.Y. capacity		6.20	173	520	1,550	153.60
	3500	15 C.Y. capacity		6.85	190	570	1,700	168.80
	3525	Self-propelled, single engine, 14 C.Y. capacity		83.90	1,400	4,190	12,600	1,509
	3550	Dual engine, 21 C.Y. capacity		119.25	1,625	4,850	14,600	1,924
	3600	31 C.Y. capacity		165.60	2,600	7,765	23,300	2,878
	3640	44 C.Y. capacity		195.70	3,150	9,430	28,300	3,452
	3650	Elevating type, single engine, 11 CY capacity		56.70	1,075	3,205	9,625	1,095
	3700	22 C.Y. capacity		105.60	2,125	6,410	19,200	2,127
	3710	Screening plant 110 H.P. w/ 5' x 10' screen		30.75	410	1,230	3,700	492
	3720	5' x 16' screen		32.90	515	1,540	4,625	571.20
	3850	Shovel, crawler-mounted, front-loading, 7 C.Y. capacity		171.90	2,850	8,575	25,700	3,090
	3855	12 C.Y. capacity		244.40	3,650	10,965	32,900	4,148
	3860	Shovel/backhoe bucket, 1/2 C.Y.		2.10	60	180	540	52.80
	3870	3/4 C.Y.		2.15	68.50	205	615	58.20
	3880	1 C.Y.		2.25	76.50	230	690	64
	3890	1-1/2 C.Y.		2.40	91.50	275	825	74.20
	3910	3 C.Y.		2.70	127	380	1,150	97.60
	3950	Stump chipper, 18" deep, 30 H.P.		8.99	215	646	1,950	201.10
	4110	Tractor, crawler, with bulldozer, torque converter, diesel 80 H.P.		21.90	375	1,120	3,350	399.20
	4150	105 H.P.		31.70	580	1,735	5,200	600.60
	4200	140 H.P.		36.50	650	1,950	5,850	682
	4260	200 H.P.		55.85	1,050	3,175	9,525	1,082
	4310	300 H.P.		71.45	1,425	4,255	12,800	1,423
	4360	410 H.P.		95.30	1,800	5,435	16,300	1,849
	4370	500 H.P.		127.20	2,400	7,230	21,700	2,464
	4380	700 H.P.		204.15	4,325	12,995	39,000	4,232
	4400	Loader, crawler, torque conv., diesel, 1-1/2 C.Y., 80 H.P.		20.00	345	1,035	3,100	367
	4450	1-1/2 to 1-3/4 C.Y., 95 H.P.		23.75	455	1,365	4,100	463
	4510	1-3/4 to 2-1/4 C.Y., 130 H.P.		35.05	810	2,435	7,300	767.40
	4530	2-1/2 to 3-1/4 C.Y., 190 H.P.		47.75	1,025	3,040	9,125	990
	4560	3-1/2 to 5 C.Y., 275 H.P.		68.30	1,500	4,520	13,600	1,450
	4610	Front end loader, 4WD, articulated frame, 1 to 1-1/4 C.Y., 70 H.P.		14.25	208	625	1,875	239
	4620	1-1/2 to 1-3/4 C.Y., 95 H.P.		18.90	295	885	2,650	328.20
	4650	1-3/4 to 2 C.Y., 130 H.P.		21.60	345	1,035	3,100	379.80
	4710	2-1/2 to 3-1/2 C.Y., 145 H.P.		22.85	355	1,065	3,200	395.80

01 54 33 | Equipment Rental

		01 54 33 | Equipment Rental	UNIT	HOURLY OPER. COST	RENT PER DAY	RENT PER WEEK	RENT PER MONTH	EQUIPMENT COST/DAY	
20	4730	3 to 4-1/2 C.Y., 185 H.P.	Ea.	29.75	505	1,515	4,550	541	20
	4760	5-1/4 to 5-3/4 C.Y., 270 H.P.		48.05	785	2,350	7,050	854.40	
	4810	7 to 9 C.Y., 475 H.P.		81.85	1,400	4,210	12,600	1,497	
	4870	9 - 11 C.Y., 620 H.P.		112.20	2,200	6,635	19,900	2,225	
	4880	Skid steer loader, wheeled, 10 C.F., 30 H.P. gas		8.15	140	420	1,250	149.20	
	4890	1 C.Y., 78 H.P., diesel		14.95	233	700	2,100	259.60	
	4892	Skid-steer attachment, auger		.54	89.50	268	805	57.90	
	4893	Backhoe		.67	112	336	1,000	72.55	
	4894	Broom		.70	117	352	1,050	76	
	4895	Forks		.24	39.50	118	355	25.50	
	4896	Grapple		.57	95.50	287	860	61.95	
	4897	Concrete hammer		1.06	176	528	1,575	114.10	
	4898	Tree spade		.80	133	398	1,200	86	
	4899	Trencher		.77	128	385	1,150	83.15	
	4900	Trencher, chain, boom type, gas, operator walking, 12 H.P.		3.95	46.50	140	420	59.60	
	4910	Operator riding, 40 H.P.		15.50	295	885	2,650	301	
	5000	Wheel type, diesel, 4' deep, 12" wide		65.30	810	2,435	7,300	1,009	
	5100	6' deep, 20" wide		77.15	1,900	5,680	17,000	1,753	
	5150	Chain type, diesel, 5' deep, 8" wide		28.90	580	1,745	5,225	580.20	
	5200	Diesel, 8' deep, 16" wide		74.70	1,975	5,895	17,700	1,777	
	5210	Tree spade, self-propelled		16.78	267	800	2,400	294.25	
	5250	Truck, dump, 2-axle, 12 ton, 8 CY payload, 220 H.P.		24.55	217	650	1,950	326.40	
	5300	Three axle dump, 16 ton, 12 CY payload, 400 H.P.		43.25	310	935	2,800	533	
	5350	Dump trailer only, rear dump, 16-1/2 C.Y.		4.70	130	390	1,175	115.60	
	5400	20 C.Y.		5.15	148	445	1,325	130.20	
	5450	Flatbed, single axle, 1-1/2 ton rating		19.30	66.50	200	600	194.40	
	5500	3 ton rating		23.20	95	285	855	242.60	
	5550	Off highway rear dump, 25 ton capacity		59.95	1,325	4,000	12,000	1,280	
	5600	35 ton capacity		55.80	1,175	3,490	10,500	1,144	
	5610	50 ton capacity		76.40	1,625	4,855	14,600	1,582	
	5620	65 ton capacity		78.80	1,600	4,820	14,500	1,594	
	5630	100 ton capacity		101.35	2,050	6,180	18,500	2,047	
	6000	Vibratory plow, 25 H.P., walking	↓	6.65	61.50	185	555	90.20	
40	0010	**GENERAL EQUIPMENT RENTAL** without operators R015433 -10							40
	0150	Aerial lift, scissor type, to 15' high, 1000 lb. cap., electric	Ea.	2.70	56.50	170	510	55.60	
	0160	To 25' high, 2000 lb. capacity		3.15	80	240	720	73.20	
	0170	Telescoping boom to 40' high, 500 lb. capacity, gas		17.10	320	960	2,875	328.80	
	0180	To 45' high, 500 lb. capacity		18.10	365	1,100	3,300	364.80	
	0190	To 60' high, 600 lb. capacity		20.45	505	1,510	4,525	465.60	
	0195	Air compressor, portable, 6.5 CFM, electric		.47	12.65	38	114	11.35	
	0196	Gasoline		.96	19	57	171	19.10	
	0200	Towed type, gas engine, 60 C.F.M.		11.70	48.50	145	435	122.60	
	0300	160 C.F.M.		13.60	50	150	450	138.80	
	0400	Diesel engine, rotary screw, 250 C.F.M.		12.80	100	300	900	162.40	
	0500	365 C.F.M.		17.15	123	370	1,100	211.20	
	0550	450 C.F.M.		21.75	153	460	1,375	266	
	0600	600 C.F.M.		38.05	217	650	1,950	434.40	
	0700	750 C.F.M.	↓	38.20	223	670	2,000	439.60	
	0800	For silenced models, small sizes, add to rent		3%	5%	5%	5%		
	0900	Large sizes, add to rent		5%	7%	7%	7%		
	0930	Air tools, breaker, pavement, 60 lb.	Ea.	.40	8.35	25	75	8.20	
	0940	80 lb.		.40	8.35	25	75	8.20	
	0950	Drills, hand (jackhammer) 65 lb.		.50	16	48	144	13.60	
	0960	Track or wagon, swing boom, 4" drifter		49.20	760	2,285	6,850	850.60	
	0970	5" drifter		60.00	825	2,470	7,400	974	
	0975	Track mounted quarry drill, 6" diameter drill		82.25	1,250	3,780	11,300	1,414	
	0980	Dust control per drill	↓	.99	19	57	171	19.30	

	01 54 33 | Equipment Rental		UNIT	HOURLY OPER. COST	RENT PER DAY	RENT PER WEEK	RENT PER MONTH	EQUIPMENT COST/DAY	
40	0990	Hammer, chipping, 12 lb.	Ea.	.45	22.50	67	201	17	40
	1000	Hose, air with couplings, 50' long, 3/4" diameter		.03	4.33	13	39	2.85	
	1100	1" diameter		.04	6.35	19	57	4.10	
	1200	1-1/2" diameter		.06	9.65	29	87	6.30	
	1300	2" diameter		.13	21.50	65	195	14.05	
	1400	2-1/2" diameter		.12	19.65	59	177	12.75	
	1410	3" diameter		.15	25	75	225	16.20	
	1450	Drill, steel, 7/8" x 2'		.05	8.35	25	75	5.40	
	1460	7/8" x 6'		.06	9.35	28	84	6.10	
	1520	Moil points		.02	3.67	11	33	2.35	
	1525	Pneumatic nailer w/accessories		.45	30	90	270	21.60	
	1530	Sheeting driver for 60 lb. breaker		.04	6	18	54	3.90	
	1540	For 90 lb. breaker		.12	8	24	72	5.75	
	1550	Spade, 25 lb.		.40	6	18	54	6.80	
	1560	Tamper, single, 35 lb.		.50	33.50	100	300	24	
	1570	Triple, 140 lb.		.75	50	150	450	36	
	1580	Wrenches, impact, air powered, up to 3/4" bolt		.30	8	24	72	7.20	
	1590	Up to 1-1/4" bolt		.45	21	63	189	16.20	
	1600	Barricades, barrels, reflectorized, 1 to 50 barrels		.03	4.60	13.80	41.50	3	
	1610	100 to 200 barrels		.02	3.53	10.60	32	2.30	
	1620	Barrels with flashers 1 to 50 barrels		.03	5.25	15.80	47.50	3.40	
	1630	100 to 200 barrels		.03	4.20	12.60	38	2.75	
	1640	Barrels with steady burn type C lights		.04	7	21	63	4.50	
	1650	Illuminated board, trailer mounted, with generator		.70	120	360	1,075	77.60	
	1670	Portable barricade, stock, with flashers, 1 to 6 units		.03	5.25	15.80	47.50	3.40	
	1680	25 to 50 units		.03	4.90	14.70	44	3.20	
	1690	Butt fusion machine, electric		20.85	55	165	495	199.80	
	1695	Electro fusion machine		15.35	40	120	360	146.80	
	1700	Carts, brick, hand powered, 1000 lb. capacity		.38	63	189	565	40.85	
	1800	Gas engine, 1500 lb., 7-1/2' lift		4.73	119	357	1,075	109.25	
	1822	Dehumidifier, medium, 6 lb/hr, 150 CFM		.85	51.50	155	465	37.80	
	1824	Large, 18 lb/hr, 600 CFM		1.64	100	300	900	73.10	
	1830	Distributor, asphalt, trailer mtd, 2000 gal., 38 H.P. diesel		9.15	340	1,020	3,050	277.20	
	1840	3000 gal., 38 H.P. diesel		10.40	370	1,105	3,325	304.20	
	1850	Drill, rotary hammer, electric, 1-1/2" diameter		.70	22	66	198	18.80	
	1860	Carbide bit for above		.05	9	27	81	5.80	
	1865	Rotary, crawler, 250 H.P.		122.45	2,075	6,240	18,700	2,228	
	1870	Emulsion sprayer, 65 gal., 5 H.P. gas engine		3.20	99	297	890	85	
	1880	200 gal., 5 H.P. engine		6.85	165	495	1,475	153.80	
	1930	Floodlight, mercury vapor, or quartz, on tripod, 1000 watt		.34	13	39	117	10.50	
	1940	2000 watt		.59	24.50	73	219	19.30	
	1950	Floodlights, trailer mounted with generator, 1 - 300 watt light		3.00	70	210	630	66	
	1960	2 - 1000 watt lights		4.15	107	320	960	97.20	
	2000	4 - 300 watt lights		3.65	86.50	260	780	81.20	
	2005	Foam spray rig, incl. box trailer, compressor, generator, proportioner		24.90	480	1,445	4,325	488.20	
	2020	Forklift, straight mast, 12' lift, 5000 lb., 2 wheel drive, gas		22.00	198	595	1,775	295	
	2040	21' lift, 5000 lb., 4 wheel drive, diesel		16.65	247	740	2,225	281.20	
	2050	For rough terrain, 42' lift, 35' reach, 9000 lb., 110 HP		22.50	445	1,330	4,000	446	
	2060	For plant, 4 T. capacity, 80 H.P., 2 wheel drive, gas		13.55	105	315	945	171.40	
	2080	10 T. capacity, 120 H.P., 2 wheel drive, diesel		18.65	183	550	1,650	259.20	
	2100	Generator, electric, gas engine, 1.5 KW to 3 KW		3.10	10.35	31	93	31	
	2200	5 KW		3.95	14.35	43	129	40.20	
	2300	10 KW		7.45	33.50	100	300	79.60	
	2400	25 KW		9.10	65	195	585	111.80	
	2500	Diesel engine, 20 KW		9.25	73.50	220	660	118	
	2600	50 KW		16.60	98.50	295	885	191.80	
	2700	100 KW		31.90	120	360	1,075	327.20	
	2800	250 KW		63.05	232	695	2,075	643.40	

01 54 33 | Equipment Rental

			UNIT	HOURLY OPER. COST	RENT PER DAY	RENT PER WEEK	RENT PER MONTH	EQUIPMENT COST/DAY
40	2850	Hammer, hydraulic, for mounting on boom, to 500 ft.-lb.	Ea.	2.15	68.50	205	615	58.20
	2860	1000 ft.-lb.		3.65	112	335	1,000	96.20
	2900	Heaters, space, oil or electric, 50 MBH		2.50	6.65	20	60	24
	3000	100 MBH		4.52	9	27	81	41.55
	3100	300 MBH		14.52	36.50	110	330	138.15
	3150	500 MBH		29.33	45	135	405	261.65
	3200	Hose, water, suction with coupling, 20' long, 2" diameter		.02	3	9	27	1.95
	3210	3" diameter		.03	4.67	14	42	3.05
	3220	4" diameter		.03	5.35	16	48	3.45
	3230	6" diameter		.11	18.35	55	165	11.90
	3240	8" diameter		.20	33.50	100	300	21.60
	3250	Discharge hose with coupling, 50' long, 2" diameter		.01	1.67	5	15	1.10
	3260	3" diameter		.02	2.67	8	24	1.75
	3270	4" diameter		.02	3.67	11	33	2.35
	3280	6" diameter		.06	9.65	29	87	6.30
	3290	8" diameter		.20	33.50	100	300	21.60
	3295	Insulation blower		.22	6	18	54	5.35
	3300	Ladders, extension type, 16' to 36' long		.14	23	69	207	14.90
	3400	40' to 60' long		.20	33.50	100	300	21.60
	3405	Lance for cutting concrete		2.97	104	313	940	86.35
	3407	Lawn mower, rotary, 22", 5HP		2.48	65	195	585	58.85
	3408	48" self propelled		4.48	128	385	1,150	112.85
	3410	Level, laser type, for pipe and sewer leveling		1.46	97.50	292	875	70.10
	3430	Electronic		.75	50	150	450	36
	3440	Laser type, rotating beam for grade control		1.17	77.50	233	700	55.95
	3460	Builders level with tripod and rod		.08	14	42	126	9.05
	3500	Light towers, towable, with diesel generator, 2000 watt		3.65	86.50	260	780	81.20
	3600	4000 watt		4.15	107	320	960	97.20
	3700	Mixer, powered, plaster and mortar, 6 C.F., 7 H.P.		2.10	19.65	59	177	28.60
	3800	10 C.F., 9 H.P.		2.25	31.50	95	285	37
	3850	Nailer, pneumatic		.45	30	90	270	21.60
	3900	Paint sprayers complete, 8 CFM		.67	44.50	134	400	32.15
	4000	17 CFM		1.17	77.50	233	700	55.95
	4020	Pavers, bituminous, rubber tires, 8' wide, 50 H.P., diesel		35.70	1,050	3,155	9,475	916.60
	4030	10' wide, 150 H.P.		75.65	1,700	5,095	15,300	1,624
	4050	Crawler, 8' wide, 100 H.P., diesel		75.65	1,775	5,355	16,100	1,676
	4060	10' wide, 150 H.P.		84.80	2,175	6,505	19,500	1,979
	4070	Concrete paver, 12' to 24' wide, 250 H.P.		81.00	1,575	4,755	14,300	1,599
	4080	Placer-spreader-trimmer, 24' wide, 300 H.P.		116.90	2,600	7,825	23,500	2,500
	4100	Pump, centrifugal gas pump, 1-1/2" diam., 65 GPM		3.35	45	135	405	53.80
	4200	2" diameter, 130 GPM		4.55	50	150	450	66.40
	4300	3" diameter, 250 GPM		4.80	51.50	155	465	69.40
	4400	6" diameter, 1500 GPM		25.55	175	525	1,575	309.40
	4500	Submersible electric pump, 1-1/4" diameter, 55 GPM		.39	15.65	47	141	12.50
	4600	1-1/2" diameter, 83 GPM		.42	18	54	162	14.15
	4700	2" diameter, 120 GPM		1.20	22.50	67	201	23
	4800	3" diameter, 300 GPM		2.05	41.50	125	375	41.40
	4900	4" diameter, 560 GPM		8.90	163	490	1,475	169.20
	5000	6" diameter, 1590 GPM		13.05	222	665	2,000	237.40
	5100	Diaphragm pump, gas, single, 1-1/2" diameter		1.24	45.50	136	410	37.10
	5200	2" diameter		3.65	56.50	170	510	63.20
	5300	3" diameter		3.65	56.50	170	510	63.20
	5400	Double, 4" diameter		4.90	78.50	235	705	85.20
	5450	Pressure Washer 5 Ga.Pm,3000 PSI		4.00	35	105	315	53
	5500	Trash pump, self-priming, gas, 2" diameter		3.85	20	60	180	42.80
	5600	Diesel, 4" diameter		10.25	58.50	175	525	117
	5650	Diesel, 6" diameter		34.00	130	390	1,175	350
	5655	Grout Pump		13.80	93.50	280	840	166.40

01 54 33 | Equipment Rental

		01 54 33 Equipment Rental	UNIT	HOURLY OPER. COST	RENT PER DAY	RENT PER WEEK	RENT PER MONTH	EQUIPMENT COST/DAY	
40	5700	Salamanders, L.P. gas fired, 100,000 BTU	Ea.	4.56	15.35	46	138	45.70	40
	5705	50,000 BTU		3.40	8.65	26	78	32.40	
	5720	Sandblaster, portable, open top, 3 C.F. capacity		.55	27	81	243	20.60	
	5730	6 C.F. capacity		.85	40	120	360	30.80	
	5740	Accessories for above		.12	19.65	59	177	12.75	
	5750	Sander, floor		.81	18.65	56	168	17.70	
	5760	Edger		.75	25.50	77	231	21.40	
	5800	Saw, chain, gas engine, 18″ long		1.90	17.65	53	159	25.80	
	5900	Hydraulic powered, 36″ long		.60	55	165	495	37.80	
	5950	60″ long		.65	56.50	170	510	39.20	
	6000	Masonry, table mounted, 14″ diameter, 5 H.P.		1.25	53	159	475	41.80	
	6050	Portable cut-off, 8 H.P.		2.05	28	84	252	33.20	
	6100	Circular, hand held, electric, 7-1/4″ diameter		.19	4	12	36	3.90	
	6200	12″ diameter		.26	7.35	22	66	6.50	
	6250	Wall saw, w/hydraulic power, 10 H.P		7.50	60	180	540	96	
	6275	Shot blaster, walk behind, 20″ wide		4.60	300	905	2,725	217.80	
	6280	Sidewalk broom, walk-behind		2.31	57	171	515	52.70	
	6300	Steam cleaner, 100 gallons per hour		2.90	46.50	140	420	51.20	
	6310	200 gallons per hour		4.15	55	165	495	66.20	
	6340	Tar Kettle/Pot, 400 gallon		6.15	80	240	720	97.20	
	6350	Torch, cutting, acetylene-oxygen, 150′ hose		.50	21.50	65	195	17	
	6360	Hourly operating cost includes tips and gas		9.45				75.60	
	6410	Toilet, portable chemical		.11	19	57	171	12.30	
	6420	Recycle flush type		.14	23	69	207	14.90	
	6430	Toilet, fresh water flush, garden hose,		.16	26.50	79	237	17.10	
	6440	Hoisted, non-flush, for high rise		.14	22.50	68	204	14.70	
	6450	Toilet, trailers, minimum		.24	39.50	118	355	25.50	
	6460	Maximum		.72	119	358	1,075	77.35	
	6465	Tractor, farm with attachment		14.40	262	785	2,350	272.20	
	6500	Trailers, platform, flush deck, 2 axle, 25 ton capacity		4.75	107	320	960	102	
	6600	40 ton capacity		6.20	150	450	1,350	139.60	
	6700	3 axle, 50 ton capacity		6.70	165	495	1,475	152.60	
	6800	75 ton capacity		8.35	218	655	1,975	197.80	
	6810	Trailer mounted cable reel for H.V. line work		4.84	231	692	2,075	177.10	
	6820	Trailer mounted cable tensioning rig		9.59	455	1,370	4,100	350.70	
	6830	Cable pulling rig		69.62	2,575	7,710	23,100	2,099	
	6900	Water tank, engine driven discharge, 5000 gallons		6.25	143	430	1,300	136	
	6925	10,000 gallons		8.50	202	605	1,825	189	
	6950	Water truck, off highway, 6000 gallons		66.95	775	2,320	6,950	999.60	
	7010	Tram car for H.V. line work, powered, 2 conductor		6.48	125	375	1,125	126.85	
	7020	Transit (builder's level) with tripod		.08	14	42	126	9.05	
	7030	Trench box, 3000 lbs. 6′x8′		.56	93	279	835	60.30	
	7040	7200 lbs. 6′x20′		1.05	175	525	1,575	113.40	
	7050	8000 lbs., 8′ x 16′		.95	158	475	1,425	102.60	
	7060	9500 lbs., 8′x20′		1.16	194	581	1,750	125.50	
	7065	11,000 lbs., 8′x24′		1.27	212	637	1,900	137.55	
	7070	12,000 lbs., 10′ x 20′		1.71	285	855	2,575	184.70	
	7100	Truck, pickup, 3/4 ton, 2 wheel drive		10.35	56.50	170	510	116.80	
	7200	4 wheel drive		10.65	71.50	215	645	128.20	
	7250	Crew carrier, 9 passenger		14.70	86.50	260	780	169.60	
	7290	Flat bed truck, 20,000 G.V.W.		15.55	122	365	1,100	197.40	
	7300	Tractor, 4 x 2, 220 H.P.		21.80	190	570	1,700	288.40	
	7410	330 H.P.		32.15	262	785	2,350	414.20	
	7500	6 x 4, 380 H.P.		36.85	305	920	2,750	478.80	
	7600	450 H.P.		44.90	370	1,110	3,325	581.20	
	7620	Vacuum truck, hazardous material, 2500 gallon		10.80	310	925	2,775	271.40	
	7625	5,000 gallon		16.90	435	1,300	3,900	395.20	
40	7640	Tractor, with A frame, boom and winch, 225 H.P.	↓	24.35	267	800	2,400	354.80	

01 54 33 | Equipment Rental

			UNIT	HOURLY OPER. COST	RENT PER DAY	RENT PER WEEK	RENT PER MONTH	EQUIPMENT COST/DAY	
40	7650	Vacuum, H.E.P.A., 16 gal., wet/dry	Ea.	.82	18	54	162	17.35	40
	7655	55 gal, wet/dry		.83	27	81	243	22.85	
	7660	Water tank, portable		.17	28.50	85.50	257	18.45	
	7690	Sewer/catch basin vacuum, 14 CY, 1500 Gallon		21.45	650	1,950	5,850	561.60	
	7700	Welder, electric, 200 amp		3.63	17.65	53	159	39.65	
	7800	300 amp		5.36	21.50	64	192	55.70	
	7900	Gas engine, 200 amp		12.85	25.50	77	231	118.20	
	8000	300 amp		14.70	27.50	83	249	134.20	
	8100	Wheelbarrow, any size		.07	11.65	35	105	7.55	
	8200	Wrecking ball, 4000 lb.	↓	2.10	73.50	220	660	60.80	
50	0010	**HIGHWAY EQUIPMENT RENTAL** without operators R015433-10							50
	0050	Asphalt batch plant, portable drum mixer, 100 ton/hr.	Ea.	63.70	1,450	4,330	13,000	1,376	
	0060	200 ton/hr.		71.10	1,525	4,565	13,700	1,482	
	0070	300 ton/hr.		82.55	1,800	5,395	16,200	1,739	
	0100	Backhoe attachment, long stick, up to 185 HP, 10.5' long		.32	21.50	64	192	15.35	
	0140	Up to 250 HP, 12' long		.35	23	69	207	16.60	
	0180	Over 250 HP, 15' long		.47	31	93	279	22.35	
	0200	Special dipper arm, up to 100 HP, 32' long		.96	63.50	191	575	45.90	
	0240	Over 100 HP, 33' long		1.19	79.50	238	715	57.10	
	0280	Catch basin/sewer cleaning truck, 3 ton, 9 CY, 1000 Gal		34.00	405	1,210	3,625	514	
	0300	Concrete batch plant, portable, electric, 200 CY/Hr		17.20	510	1,530	4,600	443.60	
	0520	Grader/dozer attachment, ripper/scarifier, rear mounted, up to 135 HP		3.10	65	195	585	63.80	
	0540	Up to 180 HP		3.70	83.50	250	750	79.60	
	0580	Up to 250 HP		4.05	95	285	855	89.40	
	0700	Pvmt. removal bucket, for hyd. excavator, up to 90 HP		1.65	48.50	145	435	42.20	
	0740	Up to 200 HP		1.85	70	210	630	56.80	
	0780	Over 200 HP		1.95	83.50	250	750	65.60	
	0900	Aggregate spreader, self-propelled, 187 HP		54.90	695	2,080	6,250	855.20	
	1000	Chemical spreader, 3 C.Y.		2.75	43.50	130	390	48	
	1900	Hammermill, traveling, 250 HP		92.70	1,850	5,540	16,600	1,850	
	2000	Horizontal borer, 3" diam, 13 HP gas driven		5.45	56.50	170	510	77.60	
	2150	Horizontal directional drill, 20,000 lb. thrust, 78 H.P. diesel		26.15	700	2,095	6,275	628.20	
	2160	30,000 lb. thrust, 115 H.P. diesel		32.45	1,050	3,185	9,550	896.60	
	2170	50,000 lb. thrust, 170 H.P. diesel		45.85	1,350	4,080	12,200	1,183	
	2190	Mud trailer for HDD, 1500 gallon, 175 H.P., gas		24.40	147	440	1,325	283.20	
	2200	Hydromulchers, gas power, 3000 gal., for truck mounting		17.00	258	775	2,325	291	
	2400	Joint & crack cleaner, walk behind, 25 HP		3.10	50	150	450	54.80	
	2500	Filler, trailer mounted, 400 gal., 20 HP		7.90	218	655	1,975	194.20	
	3000	Paint striper, self propelled, double line, 30 HP		6.25	162	485	1,450	147	
	3200	Post drivers, 6" I-Beam frame, for truck mounting		13.40	435	1,305	3,925	368.20	
	3400	Road sweeper, self propelled, 8' wide, 90 HP		31.95	575	1,730	5,200	601.60	
	3450	Road sweeper, vacuum assisted, 4 CY, 220 Gal		56.60	630	1,895	5,675	831.80	
	4000	Road mixer, self-propelled, 130 HP		39.70	770	2,315	6,950	780.60	
	4100	310 HP		71.00	2,225	6,640	19,900	1,896	
	4220	Cold mix paver, incl pug mill and bitumen tank, 165 HP		86.60	2,175	6,515	19,500	1,996	
	4250	Paver, asphalt, wheel or crawler, 130 H.P., diesel		83.50	2,125	6,340	19,000	1,936	
	4300	Paver, road widener, gas 1' to 6', 67 HP		40.10	825	2,480	7,450	816.80	
	4400	Diesel, 2' to 14', 88 HP		52.30	1,050	3,175	9,525	1,053	
	4600	Slipform pavers, curb and gutter, 2 track, 75 HP		36.80	765	2,300	6,900	754.40	
	4700	4 track, 165 HP		45.85	815	2,445	7,325	855.80	
	4800	Median barrier, 215 HP		46.50	845	2,535	7,600	879	
	4901	Trailer, low bed, 75 ton capacity		9.00	218	655	1,975	203	
	5000	Road planer, walk behind, 10" cutting width, 10 HP		2.90	30.50	91	273	41.40	
	5100	Self propelled, 12" cutting width, 64 HP		8.25	127	380	1,150	142	
	5120	Traffic line remover, metal ball blaster, truck mounted, 115 HP		46.45	760	2,285	6,850	828.60	
	5140	Grinder, truck mounted, 115 HP		49.55	805	2,420	7,250	880.40	
	5160	Walk-behind, 11 HP		3.25	46.50	140	420	54	
	5200	Pavement profiler, 4' to 6' wide, 450 HP	↓	210.55	3,400	10,220	30,700	3,728	

01 54 33 | Equipment Rental

		Description	Unit	Hourly Oper. Cost	Rent Per Day	Rent Per Week	Rent Per Month	Equipment Cost/Day
50	5300	8' to 10' wide, 750 HP	Ea.	333.90	4,725	14,160	42,500	5,503
	5400	Roadway plate, steel, 1"x8'x20'		.07	11.35	34	102	7.35
	5600	Stabilizer,self-propelled, 150 HP		39.45	610	1,830	5,500	681.60
	5700	310 HP		66.45	1,350	4,060	12,200	1,344
	5800	Striper, thermal, truck mounted 120 gal. paint, 150 H.P.		49.90	505	1,520	4,550	703.20
	6000	Tar kettle, 330 gal., trailer mounted		5.77	61.50	185	555	83.15
	7000	Tunnel locomotive, diesel, 8 to 12 ton		26.70	585	1,750	5,250	563.60
	7005	Electric, 10 ton		23.25	665	2,000	6,000	586
	7010	Muck cars, 1/2 C.Y. capacity		1.75	23	69	207	27.80
	7020	1 C.Y. capacity		1.95	31.50	94	282	34.40
	7030	2 C.Y. capacity		2.10	36.50	110	330	38.80
	7040	Side dump, 2 C.Y. capacity		2.30	45	135	405	45.40
	7050	3 C.Y. capacity		3.10	51.50	155	465	55.80
	7060	5 C.Y. capacity		4.40	65	195	585	74.20
	7100	Ventilating blower for tunnel, 7-1/2 H.P.		1.35	51.50	155	465	41.80
	7110	10 H.P.		1.57	53.50	160	480	44.55
	7120	20 H.P.		2.58	69.50	208	625	62.25
	7140	40 H.P.		4.56	98.50	295	885	95.50
	7160	60 H.P.		6.90	152	455	1,375	146.20
	7175	75 H.P.		8.81	202	607	1,825	191.90
	7180	200 H.P.		19.95	305	910	2,725	341.60
	7800	Windrow loader, elevating		47.65	1,350	4,080	12,200	1,197
60	0010	**LIFTING AND HOISTING EQUIPMENT RENTAL** without operators (R015433-10)						
	0120	Aerial lift truck, 2 person, to 80'	Ea.	25.55	725	2,180	6,550	640.40
	0140	Boom work platform, 40' snorkel (R015433-15)		15.40	260	780	2,350	279.20
	0150	Crane, flatbed mntd, 3 ton cap.		13.50	195	585	1,750	225
	0200	Crane, climbing, 106' jib, 6000 lb. capacity, 410 FPM (R312316-45)		37.40	1,475	4,400	13,200	1,179
	0300	101' jib, 10,250 lb. capacity, 270 FPM		43.30	1,850	5,580	16,700	1,462
	0500	Tower, static, 130' high, 106' jib, 6200 lb. capacity at 400 FPM		40.85	1,700	5,090	15,300	1,345
	0600	Crawler mounted, lattice boom, 1/2 C.Y., 15 tons at 12' radius		30.66	615	1,850	5,550	615.30
	0700	3/4 C.Y., 20 tons at 12' radius		40.88	770	2,310	6,925	789.05
	0800	1 C.Y., 25 tons at 12' radius		54.50	1,025	3,080	9,250	1,052
	0900	1-1/2 C.Y., 40 tons at 12' radius		54.50	1,025	3,095	9,275	1,055
	1000	2 C.Y., 50 tons at 12' radius		54.20	1,125	3,380	10,100	1,110
	1100	3 C.Y., 75 tons at 12' radius		72.95	1,525	4,560	13,700	1,496
	1200	100 ton capacity, 60' boom		71.55	1,725	5,195	15,600	1,611
	1300	165 ton capacity, 60' boom		89.60	1,925	5,750	17,300	1,867
	1400	200 ton capacity, 70' boom		111.30	2,350	7,085	21,300	2,307
	1500	350 ton capacity, 80' boom		158.35	3,775	11,335	34,000	3,534
	1600	Truck mounted, lattice boom, 6 x 4, 20 tons at 10' radius		44.77	1,200	3,610	10,800	1,080
	1700	25 tons at 10' radius		48.31	1,300	3,930	11,800	1,172
	1800	30 tons at 10' radius		52.66	1,400	4,180	12,500	1,257
	1900	40 tons at 12' radius		56.59	1,450	4,370	13,100	1,327
	2000	60 tons at 15' radius		64.84	1,550	4,620	13,900	1,443
	2050	82 tons at 15' radius		73.58	1,650	4,940	14,800	1,577
	2100	90 tons at 15' radius		83.16	1,800	5,380	16,100	1,741
	2200	115 tons at 15' radius		94.14	2,000	6,020	18,100	1,957
	2300	150 tons at 18' radius		77.90	2,100	6,335	19,000	1,890
	2350	165 tons at 18' radius		112.04	2,250	6,720	20,200	2,240
	2400	Truck mounted, hydraulic, 12 ton capacity		49.10	625	1,880	5,650	768.80
	2500	25 ton capacity		49.30	660	1,975	5,925	789.40
	2550	33 ton capacity		50.35	695	2,080	6,250	818.80
	2560	40 ton capacity		48.50	670	2,015	6,050	791
	2600	55 ton capacity		72.00	915	2,745	8,225	1,125
	2700	80 ton capacity		85.50	1,025	3,060	9,175	1,296
	2720	100 ton capacity		104.90	1,550	4,685	14,100	1,776
	2740	120 ton capacity		83.95	2,000	5,975	17,900	1,867
	2760	150 ton capacity		89.85	2,175	6,490	19,500	2,017

01 54 33 | Equipment Rental

		01 54 33 Equipment Rental	UNIT	HOURLY OPER. COST	RENT PER DAY	RENT PER WEEK	RENT PER MONTH	EQUIPMENT COST/DAY	
60	2800	Self-propelled, 4 x 4, with telescoping boom, 5 ton	Ea.	14.65	235	705	2,125	258.20	60
	2900	12-1/2 ton capacity		34.60	540	1,625	4,875	601.80	
	3000	15 ton capacity		32.30	575	1,725	5,175	603.40	
	3050	20 ton capacity		37.95	660	1,985	5,950	700.60	
	3100	25 ton capacity		38.85	675	2,030	6,100	716.80	
	3150	40 ton capacity		50.15	895	2,680	8,050	937.20	
	3200	Derricks, guy, 20 ton capacity, 60' boom, 75' mast		33.55	360	1,078	3,225	484	
	3300	100' boom, 115' mast		51.95	615	1,850	5,550	785.60	
	3400	Stiffleg, 20 ton capacity, 70' boom, 37' mast		35.80	465	1,400	4,200	566.40	
	3500	100' boom, 47' mast		54.68	745	2,240	6,725	885.45	
	3550	Helicopter, small, lift to 1250 lbs. maximum, w/pilot		105.67	2,900	8,710	26,100	2,587	
	3600	Hoists, chain type, overhead, manual, 3/4 ton		.10	1	3	9	1.40	
	3900	10 ton		.70	9.65	29	87	11.40	
	4000	Hoist and tower, 5000 lb. cap., portable electric, 40' high		4.65	207	621	1,875	161.40	
	4100	For each added 10' section, add		.10	16.35	49	147	10.60	
	4200	Hoist and single tubular tower, 5000 lb. electric, 100' high		6.26	289	867	2,600	223.50	
	4300	For each added 6'-6" section, add		.17	27.50	83	249	17.95	
	4400	Hoist and double tubular tower, 5000 lb., 100' high		6.70	320	955	2,875	244.60	
	4500	For each added 6'-6" section, add		.19	31	93	279	20.10	
	4550	Hoist and tower, mast type, 6000 lb., 100' high		7.26	330	990	2,975	256.10	
	4570	For each added 10' section, add		.12	19.65	59	177	12.75	
	4600	Hoist and tower, personnel, electric, 2000 lb., 100' @ 125 FPM		14.74	880	2,640	7,925	645.90	
	4700	3000 lb., 100' @ 200 FPM		16.88	995	2,990	8,975	733.05	
	4800	3000 lb., 150' @ 300 FPM		18.63	1,125	3,340	10,000	817.05	
	4900	4000 lb., 100' @ 300 FPM		19.36	1,125	3,410	10,200	836.90	
	5000	6000 lb., 100' @ 275 FPM	▼	20.98	1,200	3,580	10,700	883.85	
	5100	For added heights up to 500', add	L.F.	.01	1.67	5	15	1.10	
	5200	Jacks, hydraulic, 20 ton	Ea.	.05	1.67	5	15	1.40	
	5500	100 ton		.35	10.65	32	96	9.20	
	6100	Jacks, hydraulic, climbing w/ 50' jackrods, control console, 30 ton cap		1.79	119	357	1,075	85.70	
	6150	For each added 10' jackrod section, add		.05	3.33	10	30	2.40	
	6300	50 ton capacity		2.87	191	574	1,725	137.75	
	6350	For each added 10' jackrod section, add		.06	4	12	36	2.90	
	6500	125 ton capacity		7.55	505	1,510	4,525	362.40	
	6550	For each added 10' jackrod section, add		.52	34.50	103	310	24.75	
	6600	Cable jack, 10 ton capacity with 200' cable		1.50	99.50	299	895	71.80	
	6650	For each added 50' of cable, add	▼	.17	11.35	34	102	8.15	
70	0010	**WELLPOINT EQUIPMENT RENTAL** without operators R015433-10							70
	0020	Based on 2 months rental							
	0100	Combination jetting & wellpoint pump, 60 H.P. diesel	Ea.	21.79	295	884	2,650	351.10	
	0200	High pressure gas jet pump, 200 H.P., 300 psi	"	57.29	252	756	2,275	609.50	
	0300	Discharge pipe, 8" diameter	L.F.	.01	.48	1.44	4.32	.35	
	0350	12" diameter		.01	.71	2.12	6.35	.50	
	0400	Header pipe, flows up to 150 G.P.M., 4" diameter		.01	.43	1.29	3.87	.35	
	0500	400 G.P.M., 6" diameter		.01	.51	1.54	4.62	.40	
	0600	800 G.P.M., 8" diameter		.01	.71	2.12	6.35	.50	
	0700	1500 G.P.M., 10" diameter		.01	.74	2.22	6.65	.50	
	0800	2500 G.P.M., 12" diameter		.02	1.40	4.21	12.65	1	
	0900	4500 G.P.M., 16" diameter		.03	1.80	5.39	16.15	1.30	
	0950	For quick coupling aluminum and plastic pipe, add	▼	.03	1.86	5.58	16.75	1.35	
	1100	Wellpoint, 25' long, with fittings & riser pipe, 1-1/2" or 2" diameter	Ea.	.06	3.71	11.13	33.50	2.70	
	1200	Wellpoint pump, diesel powered, 4" diameter, 20 H.P.		8.77	170	510	1,525	172.15	
	1300	6" diameter, 30 H.P.		12.22	211	632	1,900	224.15	
	1400	8" suction, 40 H.P.		16.47	289	867	2,600	305.15	
	1500	10" suction, 75 H.P.		26.59	340	1,013	3,050	415.30	
	1600	12" suction, 100 H.P.		37.34	540	1,620	4,850	622.70	
	1700	12" suction, 175 H.P.	▼	57.89	590	1,770	5,300	817.10	

01 54 33 | Equipment Rental

			UNIT	HOURLY OPER. COST	RENT PER DAY	RENT PER WEEK	RENT PER MONTH	EQUIPMENT COST/DAY	
80	0010	**MARINE EQUIPMENT RENTAL** without operators R015433-10							80
	0200	Barge, 400 Ton, 30' wide x 90' long	Ea.	14.80	1,050	3,140	9,425	746.40	
	0240	800 Ton, 45' wide x 90' long		17.95	1,275	3,800	11,400	903.60	
	2000	Tugboat, diesel, 100 HP		27.75	195	585	1,750	339	
	2040	250 HP		56.05	360	1,075	3,225	663.40	
	2080	380 HP	↓	115.60	1,075	3,195	9,575	1,564	

Crew A-1	Bare Costs Hr.	Bare Costs Daily	Incl. Subs O & P Hr.	Incl. Subs O & P Daily	Cost Per Labor-Hour Bare Costs	Cost Per Labor-Hour Incl. O&P
1 Building Laborer	$31.60	$252.80	$49.00	$392.00	$31.60	$49.00
1 Concrete saw, gas manual		63.20		69.52	7.90	8.69
8 L.H., Daily Totals		$316.00		$461.52	$39.50	$57.69

Crew A-1A	Hr.	Daily	Hr.	Daily	Bare Costs	Incl. O&P
1 Skilled Worker	$40.85	$326.80	$63.25	$506.00	$40.85	$63.25
1 Shot Blaster, 20"		217.80		239.58	27.23	29.95
8 L.H., Daily Totals		$544.60		$745.58	$68.08	$93.20

Crew A-1B	Hr.	Daily	Hr.	Daily	Bare Costs	Incl. O&P
1 Building Laborer	$31.60	$252.80	$49.00	$392.00	$31.60	$49.00
1 Concrete Saw		141.00		155.10	17.63	19.39
8 L.H., Daily Totals		$393.80		$547.10	$49.23	$68.39

Crew A-1C	Hr.	Daily	Hr.	Daily	Bare Costs	Incl. O&P
1 Building Laborer	$31.60	$252.80	$49.00	$392.00	$31.60	$49.00
1 Chain saw, gas, 18"		25.80		28.38	3.23	3.55
8 L.H., Daily Totals		$278.60		$420.38	$34.83	$52.55

Crew A-1D	Hr.	Daily	Hr.	Daily	Bare Costs	Incl. O&P
1 Building Laborer	$31.60	$252.80	$49.00	$392.00	$31.60	$49.00
1 Vibrating plate, gas, 18"		31.80		34.98	3.98	4.37
8 L.H., Daily Totals		$284.60		$426.98	$35.58	$53.37

Crew A-1E	Hr.	Daily	Hr.	Daily	Bare Costs	Incl. O&P
1 Building Laborer	$31.60	$252.80	$49.00	$392.00	$31.60	$49.00
1 Vibratory Plate, Gas, 21"		39.20		43.12	4.90	5.39
8 L.H., Daily Totals		$292.00		$435.12	$36.50	$54.39

Crew A-1F	Hr.	Daily	Hr.	Daily	Bare Costs	Incl. O&P
1 Building Laborer	$31.60	$252.80	$49.00	$392.00	$31.60	$49.00
1 Rammer/tamper, gas, 8"		43.40		47.74	5.42	5.97
8 L.H., Daily Totals		$296.20		$439.74	$37.02	$54.97

Crew A-1G	Hr.	Daily	Hr.	Daily	Bare Costs	Incl. O&P
1 Building Laborer	$31.60	$252.80	$49.00	$392.00	$31.60	$49.00
1 Rammer/tamper, gas, 15"		49.40		54.34	6.17	6.79
8 L.H., Daily Totals		$302.20		$446.34	$37.77	$55.79

Crew A-1H	Hr.	Daily	Hr.	Daily	Bare Costs	Incl. O&P
1 Building Laborer	$31.60	$252.80	$49.00	$392.00	$31.60	$49.00
1 Exterior Steam Cleaner		51.20		56.32	6.40	7.04
8 L.H., Daily Totals		$304.00		$448.32	$38.00	$56.04

Crew A-1J	Hr.	Daily	Hr.	Daily	Bare Costs	Incl. O&P
1 Building Laborer	$31.60	$252.80	$49.00	$392.00	$31.60	$49.00
1 Cultivator, Walk-Behind, 5 H.P.		57.05		62.76	7.13	7.84
8 L.H., Daily Totals		$309.85		$454.76	$38.73	$56.84

Crew A-1K	Hr.	Daily	Hr.	Daily	Bare Costs	Incl. O&P
1 Building Laborer	$31.60	$252.80	$49.00	$392.00	$31.60	$49.00
1 Cultivator, Walk-Behind, 8 H.P.		91.10		100.21	11.39	12.53
8 L.H., Daily Totals		$343.90		$492.21	$42.99	$61.53

Crew A-1M	Hr.	Daily	Hr.	Daily	Bare Costs	Incl. O&P
1 Building Laborer	$31.60	$252.80	$49.00	$392.00	$31.60	$49.00
1 Snow Blower, Walk-Behind		52.70		57.97	6.59	7.25
8 L.H., Daily Totals		$305.50		$449.97	$38.19	$56.25

Crew A-2	Bare Costs Hr.	Bare Costs Daily	Incl. Subs O & P Hr.	Incl. Subs O & P Daily	Cost Per Labor-Hour Bare Costs	Cost Per Labor-Hour Incl. O&P
2 Laborers	$31.60	$505.60	$49.00	$784.00	$31.38	$48.55
1 Truck Driver (light)	30.95	247.60	47.65	381.20		
1 Flatbed Truck, Gas, 1.5 Ton		194.40		213.84	8.10	8.91
24 L.H., Daily Totals		$947.60		$1379.04	$39.48	$57.46

Crew A-2A	Hr.	Daily	Hr.	Daily	Bare Costs	Incl. O&P
2 Laborers	$31.60	$505.60	$49.00	$784.00	$31.38	$48.55
1 Truck Driver (light)	30.95	247.60	47.65	381.20		
1 Flatbed Truck, Gas, 1.5 Ton		194.40		213.84		
1 Concrete Saw		141.00		155.10	13.98	15.37
24 L.H., Daily Totals		$1088.60		$1534.14	$45.36	$63.92

Crew A-2B	Hr.	Daily	Hr.	Daily	Bare Costs	Incl. O&P
1 Truck Driver (light)	$30.95	$247.60	$47.65	$381.20	$30.95	$47.65
1 Flatbed Truck, Gas, 1.5 Ton		194.40		213.84	24.30	26.73
8 L.H., Daily Totals		$442.00		$595.04	$55.25	$74.38

Crew A-3A	Hr.	Daily	Hr.	Daily	Bare Costs	Incl. O&P
1 Truck Driver (light)	$30.95	$247.60	$47.65	$381.20	$30.95	$47.65
1 Pickup truck, 4 x 4, 3/4 ton		128.20		141.02	16.02	17.63
8 L.H., Daily Totals		$375.80		$522.22	$46.98	$65.28

Crew A-3B	Hr.	Daily	Hr.	Daily	Bare Costs	Incl. O&P
1 Equip. Oper. (medium)	$41.35	$330.80	$62.15	$497.20	$36.65	$55.65
1 Truck Driver (heavy)	31.95	255.60	49.15	393.20		
1 Dump Truck, 12 C.Y., 400 H.P.		533.00		586.30		
1 F.E. Loader, W.M.,2.5 C.Y		395.80		435.38	58.05	63.85
16 L.H., Daily Totals		$1515.20		$1912.08	$94.70	$119.51

Crew A-3C	Hr.	Daily	Hr.	Daily	Bare Costs	Incl. O&P
1 Equip. Oper. (light)	$39.05	$312.40	$58.70	$469.60	$39.05	$58.70
1 Loader, Skid Steer, 78 H.P.		259.60		285.56	32.45	35.70
8 L.H., Daily Totals		$572.00		$755.16	$71.50	$94.39

Crew A-3D	Hr.	Daily	Hr.	Daily	Bare Costs	Incl. O&P
1 Truck Driver, Light	$30.95	$247.60	$47.65	$381.20	$30.95	$47.65
1 Pickup truck, 4 x 4, 3/4 ton		128.20		141.02		
1 Flatbed Trailer, 25 Ton		102.00		112.20	28.77	31.65
8 L.H., Daily Totals		$477.80		$634.42	$59.73	$79.30

Crew A-3E	Hr.	Daily	Hr.	Daily	Bare Costs	Incl. O&P
1 Equip. Oper. (crane)	$42.55	$340.40	$63.95	$511.60	$37.25	$56.55
1 Truck Driver (heavy)	31.95	255.60	49.15	393.20		
1 Pickup truck, 4 x 4, 3/4 ton		128.20		141.02	8.01	8.81
16 L.H., Daily Totals		$724.20		$1045.82	$45.26	$65.36

Crew A-3F	Hr.	Daily	Hr.	Daily	Bare Costs	Incl. O&P
1 Equip. Oper. (crane)	$42.55	$340.40	$63.95	$511.60	$37.25	$56.55
1 Truck Driver (heavy)	31.95	255.60	49.15	393.20		
1 Pickup truck, 4 x 4, 3/4 ton		128.20		141.02		
1 Truck Tractor, 6x4, 380 H.P.		478.80		526.68		
1 Lowbed Trailer, 75 Ton		203.00		223.30	50.63	55.69
16 L.H., Daily Totals		$1406.00		$1795.80	$87.88	$112.24

Crew No.	Bare Costs		Incl. Subs O & P		Cost Per Labor-Hour	
Crew A-3G	**Hr.**	**Daily**	**Hr.**	**Daily**	**Bare Costs**	**Incl. O&P**
1 Equip. Oper. (crane)	$42.55	$340.40	$63.95	$511.60	$37.25	$56.55
1 Truck Driver (heavy)	31.95	255.60	49.15	393.20		
1 Pickup truck, 4 x 4, 3/4 ton		128.20		141.02		
1 Truck Tractor, 6x4, 450 H.P.		581.20		639.32		
1 Lowbed Trailer, 75 Ton		203.00		223.30	57.02	62.73
16 L.H., Daily Totals		$1508.40		$1908.44	$94.28	$119.28

Crew A-3H	Hr.	Daily	Hr.	Daily	Bare Costs	Incl. O&P
1 Equip. Oper. (crane)	$42.55	$340.40	$63.95	$511.60	$42.55	$63.95
1 Hyd. crane, 12 Ton (daily)		1018.00		1119.80	127.25	139.97
8 L.H., Daily Totals		$1358.40		$1631.40	$169.80	$203.93

Crew A-3I	Hr.	Daily	Hr.	Daily	Bare Costs	Incl. O&P
1 Equip. Oper. (crane)	$42.55	$340.40	$63.95	$511.60	$42.55	$63.95
1 Hyd. crane, 25 Ton (daily)		1054.00		1159.40	131.75	144.93
8 L.H., Daily Totals		$1394.40		$1671.00	$174.30	$208.88

Crew A-3J	Hr.	Daily	Hr.	Daily	Bare Costs	Incl. O&P
1 Equip. Oper. (crane)	$42.55	$340.40	$63.95	$511.60	$42.55	$63.95
1 Hyd. crane, 40 Ton (daily)		1058.00		1163.80	132.25	145.47
8 L.H., Daily Totals		$1398.40		$1675.40	$174.80	$209.43

Crew A-3K	Hr.	Daily	Hr.	Daily	Bare Costs	Incl. O&P
1 Equip. Oper. (crane)	$42.55	$340.40	$63.95	$511.60	$39.67	$59.63
1 Equip. Oper. Oiler	36.80	294.40	55.30	442.40		
1 Hyd. crane, 55 Ton (daily)		1491.00		1640.10		
1 P/U Truck, 3/4 Ton (daily)		137.80		151.58	101.80	111.98
16 L.H., Daily Totals		$2263.60		$2745.68	$141.47	$171.60

Crew A-3L	Hr.	Daily	Hr.	Daily	Bare Costs	Incl. O&P
1 Equip. Oper. (crane)	$42.55	$340.40	$63.95	$511.60	$39.67	$59.63
1 Equip. Oper. Oiler	36.80	294.40	55.30	442.40		
1 Hyd. crane, 80 Ton (daily)		1704.00		1874.40		
1 P/U Truck, 3/4 Ton (daily)		137.80		151.58	115.11	126.62
16 L.H., Daily Totals		$2476.60		$2979.98	$154.79	$186.25

Crew A-3M	Hr.	Daily	Hr.	Daily	Bare Costs	Incl. O&P
1 Equip. Oper. (crane)	$42.55	$340.40	$63.95	$511.60	$39.67	$59.63
1 Equip. Oper. Oiler	36.80	294.40	55.30	442.40		
1 Hyd. crane, 100 Ton (daily)		2399.00		2638.90		
1 P/U Truck, 3/4 Ton (daily)		137.80		151.58	158.55	174.41
16 L.H., Daily Totals		$3171.60		$3744.48	$198.22	$234.03

Crew A-3N	Hr.	Daily	Hr.	Daily	Bare Costs	Incl. O&P
1 Equip. Oper. (crane)	$42.55	$340.40	$63.95	$511.60	$42.55	$63.95
1 Tower crane (monthly)		1022.00		1124.20	127.75	140.53
8 L.H., Daily Totals		$1362.40		$1635.80	$170.30	$204.47

Crew A-3P	Hr.	Daily	Hr.	Daily	Bare Costs	Incl. O&P
1 Equip. Oper., Light	$39.05	$312.40	$58.70	$469.60	$39.05	$58.70
1 A.T. Forklift, 42' lift		446.00		490.60	55.75	61.33
8 L.H., Daily Totals		$758.40		$960.20	$94.80	$120.03

Crew A-4	Hr.	Daily	Hr.	Daily	Bare Costs	Incl. O&P
2 Carpenters	$39.95	$639.20	$61.95	$991.20	$38.37	$58.88
1 Painter, Ordinary	35.20	281.60	52.75	422.00		
24 L.H., Daily Totals		$920.80		$1413.20	$38.37	$58.88

Crew No.	Bare Costs		Incl. Subs O & P		Cost Per Labor-Hour	
Crew A-5	**Hr.**	**Daily**	**Hr.**	**Daily**	**Bare Costs**	**Incl. O&P**
2 Laborers	$31.60	$505.60	$49.00	$784.00	$31.53	$48.85
.25 Truck Driver (light)	30.95	61.90	47.65	95.30		
.25 Flatbed Truck, Gas, 1.5 Ton		48.60		53.46	2.70	2.97
18 L.H., Daily Totals		$616.10		$932.76	$34.23	$51.82

Crew A-6	Hr.	Daily	Hr.	Daily	Bare Costs	Incl. O&P
1 Instrument Man	$40.85	$326.80	$63.25	$506.00	$39.73	$60.80
1 Rodman/Chainman	38.60	308.80	58.35	466.80		
1 Laser Transit/Level		70.10		77.11	4.38	4.82
16 L.H., Daily Totals		$705.70		$1049.91	$44.11	$65.62

Crew A-7	Hr.	Daily	Hr.	Daily	Bare Costs	Incl. O&P
1 Chief Of Party	$50.20	$401.60	$77.15	$617.20	$43.22	$66.25
1 Instrument Man	40.85	326.80	63.25	506.00		
1 Rodman/Chainman	38.60	308.80	58.35	466.80		
1 Laser Transit/Level		70.10		77.11	2.92	3.21
24 L.H., Daily Totals		$1107.30		$1667.11	$46.14	$69.46

Crew A-8	Hr.	Daily	Hr.	Daily	Bare Costs	Incl. O&P
1 Chief of Party	$50.20	$401.60	$77.15	$617.20	$42.06	$64.28
1 Instrument Man	40.85	326.80	63.25	506.00		
2 Rodmen/Chainmen	38.60	617.60	58.35	933.60		
1 Laser Transit/Level		70.10		77.11	2.19	2.41
32 L.H., Daily Totals		$1416.10		$2133.91	$44.25	$66.68

Crew A-9	Hr.	Daily	Hr.	Daily	Bare Costs	Incl. O&P
1 Asbestos Foreman	$44.60	$356.80	$69.90	$559.20	$44.16	$69.20
7 Asbestos Workers	44.10	2469.60	69.10	3869.60		
64 L.H., Daily Totals		$2826.40		$4428.80	$44.16	$69.20

Crew A-10A	Hr.	Daily	Hr.	Daily	Bare Costs	Incl. O&P
1 Asbestos Foreman	$44.60	$356.80	$69.90	$559.20	$44.27	$69.37
2 Asbestos Workers	44.10	705.60	69.10	1105.60		
24 L.H., Daily Totals		$1062.40		$1664.80	$44.27	$69.37

Crew A-10B	Hr.	Daily	Hr.	Daily	Bare Costs	Incl. O&P
1 Asbestos Foreman	$44.60	$356.80	$69.90	$559.20	$44.23	$69.30
3 Asbestos Workers	44.10	1058.40	69.10	1658.40		
32 L.H., Daily Totals		$1415.20		$2217.60	$44.23	$69.30

Crew A-10C	Hr.	Daily	Hr.	Daily	Bare Costs	Incl. O&P
3 Asbestos Workers	$44.10	$1058.40	$69.10	$1658.40	$44.10	$69.10
1 Flatbed Truck, Gas, 1.5 Ton		194.40		213.84	8.10	8.91
24 L.H., Daily Totals		$1252.80		$1872.24	$52.20	$78.01

Crew A-10D	Hr.	Daily	Hr.	Daily	Bare Costs	Incl. O&P
2 Asbestos Workers	$44.10	$705.60	$69.10	$1105.60	$41.89	$64.36
1 Equip. Oper. (crane)	42.55	340.40	63.95	511.60		
1 Equip. Oper. Oiler	36.80	294.40	55.30	442.40		
1 Hydraulic Crane, 33 Ton		818.80		900.68	25.59	28.15
32 L.H., Daily Totals		$2159.20		$2960.28	$67.47	$92.51

Crew A-11	Hr.	Daily	Hr.	Daily	Bare Costs	Incl. O&P
1 Asbestos Foreman	$44.60	$356.80	$69.90	$559.20	$44.16	$69.20
7 Asbestos Workers	44.10	2469.60	69.10	3869.60		
2 Chip. Hammers, 12 Lb., Elec.		34.00		37.40	.53	.58
64 L.H., Daily Totals		$2860.40		$4466.20	$44.69	$69.78

Crew No.	Bare Costs		Incl. Subs O & P		Cost Per Labor-Hour	
Crew A-12	**Hr.**	**Daily**	**Hr.**	**Daily**	**Bare Costs**	**Incl. O&P**
1 Asbestos Foreman	$44.60	$356.80	$69.90	$559.20	$44.16	$69.20
7 Asbestos Workers	44.10	2469.60	69.10	3869.60		
1 Trk-mtd vac, 14 CY, 1500 Gal.		561.60		617.76		
1 Flatbed Truck, 20,000 GVW		197.40		217.14	11.86	13.05
64 L.H., Daily Totals		$3585.40		$5263.70	$56.02	$82.25
Crew A-13	**Hr.**	**Daily**	**Hr.**	**Daily**	**Bare Costs**	**Incl. O&P**
1 Equip. Oper. (light)	$39.05	$312.40	$58.70	$469.60	$39.05	$58.70
1 Trk-mtd vac, 14 CY, 1500 Gal.		561.60		617.76		
1 Flatbed Truck, 20,000 GVW		197.40		217.14	94.88	104.36
8 L.H., Daily Totals		$1071.40		$1304.50	$133.93	$163.06
Crew B-1	**Hr.**	**Daily**	**Hr.**	**Daily**	**Bare Costs**	**Incl. O&P**
1 Labor Foreman (outside)	$33.60	$268.80	$52.10	$416.80	$32.27	$50.03
2 Laborers	31.60	505.60	49.00	784.00		
24 L.H., Daily Totals		$774.40		$1200.80	$32.27	$50.03
Crew B-1A	**Hr.**	**Daily**	**Hr.**	**Daily**	**Bare Costs**	**Incl. O&P**
1 Laborer Foreman	$33.60	$268.80	$52.10	$416.80	$32.27	$50.03
2 Laborers	31.60	505.60	49.00	784.00		
2 Cutting Torches		34.00		37.40		
2 Gases		151.20		166.32	7.72	8.49
24 L.H., Daily Totals		$959.60		$1404.52	$39.98	$58.52
Crew B-1B	**Hr.**	**Daily**	**Hr.**	**Daily**	**Bare Costs**	**Incl. O&P**
1 Laborer Foreman	$33.60	$268.80	$52.10	$416.80	$34.84	$53.51
2 Laborers	31.60	505.60	49.00	784.00		
1 Equip. Oper. (crane)	42.55	340.40	63.95	511.60		
2 Cutting Torches		34.00		37.40		
2 Gases		151.20		166.32		
1 Hyd. Crane, 12 Ton		768.80		845.68	29.81	32.79
32 L.H., Daily Totals		$2068.80		$2761.80	$64.65	$86.31
Crew B-2	**Hr.**	**Daily**	**Hr.**	**Daily**	**Bare Costs**	**Incl. O&P**
1 Labor Foreman (outside)	$33.60	$268.80	$52.10	$416.80	$32.00	$49.62
4 Laborers	31.60	1011.20	49.00	1568.00		
40 L.H., Daily Totals		$1280.00		$1984.80	$32.00	$49.62
Crew B-3	**Hr.**	**Daily**	**Hr.**	**Daily**	**Bare Costs**	**Incl. O&P**
1 Labor Foreman (outside)	$33.60	$268.80	$52.10	$416.80	$33.67	$51.76
2 Laborers	31.60	505.60	49.00	784.00		
1 Equip. Oper. (med.)	41.35	330.80	62.15	497.20		
2 Truck Drivers (heavy)	31.95	511.20	49.15	786.40		
1 Crawler Loader, 3 C.Y.		990.00		1089.00		
2 Dump Trucks 12 C.Y., 400 H.P.		1066.00		1172.60	42.83	47.12
48 L.H., Daily Totals		$3672.40		$4746.00	$76.51	$98.88
Crew B-3A	**Hr.**	**Daily**	**Hr.**	**Daily**	**Bare Costs**	**Incl. O&P**
4 Laborers	$31.60	$1011.20	$49.00	$1568.00	$33.55	$51.63
1 Equip. Oper. (med.)	41.35	330.80	62.15	497.20		
1 Hyd. Excavator, 1.5 C.Y.		865.80		952.38	21.65	23.81
40 L.H., Daily Totals		$2207.80		$3017.58	$55.20	$75.44
Crew B-3B	**Hr.**	**Daily**	**Hr.**	**Daily**	**Bare Costs**	**Incl. O&P**
2 Laborers	$31.60	$505.60	$49.00	$784.00	$34.13	$52.33
1 Equip. Oper. (med.)	41.35	330.80	62.15	497.20		
1 Truck Driver (heavy)	31.95	255.60	49.15	393.20		
1 Backhoe Loader, 80 H.P.		345.00		379.50		
1 Dump Truck, 12 C.Y., 400 H.P.		533.00		586.30	27.44	30.18
32 L.H., Daily Totals		$1970.00		$2640.20	$61.56	$82.51

Crew No.	Bare Costs		Incl. Subs O & P		Cost Per Labor-Hour	
Crew B-3C	**Hr.**	**Daily**	**Hr.**	**Daily**	**Bare Costs**	**Incl. O&P**
3 Laborers	$31.60	$758.40	$49.00	$1176.00	$34.04	$52.29
1 Equip. Oper. (med.)	41.35	330.80	62.15	497.20		
1 Crawler Loader, 4 C.Y.		1450.00		1595.00	45.31	49.84
32 L.H., Daily Totals		$2539.20		$3268.20	$79.35	$102.13
Crew B-4	**Hr.**	**Daily**	**Hr.**	**Daily**	**Bare Costs**	**Incl. O&P**
1 Labor Foreman (outside)	$33.60	$268.80	$52.10	$416.80	$31.99	$49.54
4 Laborers	31.60	1011.20	49.00	1568.00		
1 Truck Driver (heavy)	31.95	255.60	49.15	393.20		
1 Truck Tractor, 220 H.P.		288.40		317.24		
1 Flatbed Trailer, 40 Ton		139.60		153.56	8.92	9.81
48 L.H., Daily Totals		$1963.60		$2848.80	$40.91	$59.35
Crew B-5	**Hr.**	**Daily**	**Hr.**	**Daily**	**Bare Costs**	**Incl. O&P**
1 Labor Foreman (outside)	$33.60	$268.80	$52.10	$416.80	$34.67	$53.20
4 Laborers	31.60	1011.20	49.00	1568.00		
2 Equip. Oper. (med.)	41.35	661.60	62.15	994.40		
1 Air Compressor, 250 cfm		162.40		178.64		
2 Breakers, Pavement, 60 lb.		16.40		18.04		
2 -50' Air Hoses, 1.5"		12.60		13.86		
1 Crawler Loader, 3 C.Y.		990.00		1089.00	21.10	23.21
56 L.H., Daily Totals		$3123.00		$4278.74	$55.77	$76.41
Crew B-5A	**Hr.**	**Daily**	**Hr.**	**Daily**	**Bare Costs**	**Incl. O&P**
1 Foreman	$33.60	$268.80	$52.10	$416.80	$34.07	$52.28
6 Laborers	31.60	1516.80	49.00	2352.00		
2 Equip. Oper. (med.)	41.35	661.60	62.15	994.40		
1 Equip. Oper. (light)	39.05	312.40	58.70	469.60		
2 Truck Drivers (heavy)	31.95	511.20	49.15	786.40		
1 Air Compressor, 365 cfm		211.20		232.32		
2 Breakers, Pavement, 60 lb.		16.40		18.04		
8 -50' Air Hoses, 1"		32.80		36.08		
2 Dump Trucks, 8 C.Y., 220 H.P.		652.80		718.08	9.51	10.46
96 L.H., Daily Totals		$4184.00		$6023.72	$43.58	$62.75
Crew B-5B	**Hr.**	**Daily**	**Hr.**	**Daily**	**Bare Costs**	**Incl. O&P**
1 Powderman	$40.85	$326.80	$63.25	$506.00	$36.57	$55.83
2 Equip. Oper. (med.)	41.35	661.60	62.15	994.40		
3 Truck Drivers (heavy)	31.95	766.80	49.15	1179.60		
1 F.E. Loader, W.M.,2.5 C.Y.		395.80		435.38		
3 Dump Trucks, 12 C.Y., 400 H.P.		1599.00		1758.90		
1 Air Compressor, 365 cfm		211.20		232.32	45.96	50.55
48 L.H., Daily Totals		$3961.20		$5106.60	$82.53	$106.39
Crew B-5C	**Hr.**	**Daily**	**Hr.**	**Daily**	**Bare Costs**	**Incl. O&P**
3 Laborers	$31.60	$758.40	$49.00	$1176.00	$34.92	$53.34
1 Equip. Oper. (medium)	41.35	330.80	62.15	497.20		
2 Truck Drivers (heavy)	31.95	511.20	49.15	786.40		
1 Equip. Oper. (crane)	42.55	340.40	63.95	511.60		
1 Equip. Oper. Oiler	36.80	294.40	55.30	442.40		
2 Dump Trucks, 12 C.Y., 400 H.P.		1066.00		1172.60		
1 Crawler Loader, 4 C.Y.		1450.00		1595.00		
1 S.P. Crane, 4x4, 25 Ton		716.80		788.48	50.51	55.56
64 L.H., Daily Totals		$5468.00		$6969.68	$85.44	$108.90
Crew B-6	**Hr.**	**Daily**	**Hr.**	**Daily**	**Bare Costs**	**Incl. O&P**
2 Laborers	$31.60	$505.60	$49.00	$784.00	$34.08	$52.23
1 Equip. Oper. (light)	39.05	312.40	58.70	469.60		
1 Backhoe Loader, 48 H.P.		293.80		323.18	12.24	13.47
24 L.H., Daily Totals		$1111.80		$1576.78	$46.33	$65.70

Crew No.	Bare Costs		Incl. Subs O & P		Cost Per Labor-Hour	
Crew B-6A	**Hr.**	**Daily**	**Hr.**	**Daily**	**Bare Costs**	**Incl. O&P**
.5 Labor Foreman (outside)	$33.60	$134.40	$52.10	$208.40	$35.90	$54.88
1 Laborer	31.60	252.80	49.00	392.00		
1 Equip. Oper. (med.)	41.35	330.80	62.15	497.20		
1 Vacuum Trk.,5000 Gal.		395.20		434.72	19.76	21.74
20 L.H., Daily Totals		$1113.20		$1532.32	$55.66	$76.62

Crew B-6B	Hr.	Daily	Hr.	Daily	Bare Costs	Incl. O&P
2 Labor Foremen (out)	$33.60	$537.60	$52.10	$833.60	$32.27	$50.03
4 Laborers	31.60	1011.20	49.00	1568.00		
1 S.P. Crane, 4x4, 5 Ton		258.20		284.02		
1 Flatbed Truck, Gas, 1.5 Ton		194.40		213.84		
1 Butt Fusion Machine		199.80		219.78	13.59	14.95
48 L.H., Daily Totals		$2201.20		$3119.24	$45.86	$64.98

Crew B-7	Hr.	Daily	Hr.	Daily	Bare Costs	Incl. O&P
1 Labor Foreman (outside)	$33.60	$268.80	$52.10	$416.80	$33.56	$51.71
4 Laborers	31.60	1011.20	49.00	1568.00		
1 Equip. Oper. (med.)	41.35	330.80	62.15	497.20		
1 Brush Chipper, 12", 130 H.P.		225.60		248.16		
1 Crawler Loader, 3 C.Y.		990.00		1089.00		
2 Chain Saws, Gas, 36" Long		75.60		83.16	26.90	29.59
48 L.H., Daily Totals		$2902.00		$3902.32	$60.46	$81.30

Crew B-7A	Hr.	Daily	Hr.	Daily	Bare Costs	Incl. O&P
2 Laborers	$31.60	$505.60	$49.00	$784.00	$34.08	$52.23
1 Equip. Oper. (light)	39.05	312.40	58.70	469.60		
1 Rake w/Tractor		257.70		283.47		
2 Chain Saws, gas, 18"		51.60		56.76	12.89	14.18
24 L.H., Daily Totals		$1127.30		$1593.83	$46.97	$66.41

Crew B-8	Hr.	Daily	Hr.	Daily	Bare Costs	Incl. O&P
1 Labor Foreman (outside)	$33.60	$268.80	$52.10	$416.80	$35.02	$53.50
2 Laborers	31.60	505.60	49.00	784.00		
2 Equip. Oper. (med.)	41.35	661.60	62.15	994.40		
1 Equip. Oper. Oiler	36.80	294.40	55.30	442.40		
2 Truck Drivers (heavy)	31.95	511.20	49.15	786.40		
1 Hyd. Crane, 25 Ton		789.40		868.34		
1 Crawler Loader, 3 C.Y.		990.00		1089.00		
2 Dump Trucks, 12 C.Y., 400 H.P.		1066.00		1172.60	44.46	48.91
64 L.H., Daily Totals		$5087.00		$6553.94	$79.48	$102.41

Crew B-9	Hr.	Daily	Hr.	Daily	Bare Costs	Incl. O&P
1 Labor Foreman (outside)	$33.60	$268.80	$52.10	$416.80	$32.00	$49.62
4 Laborers	31.60	1011.20	49.00	1568.00		
1 Air Compressor, 250 cfm		162.40		178.64		
2 Breakers, Pavement, 60 lb.		16.40		18.04		
2 -50' Air Hoses, 1.5"		12.60		13.86	4.79	5.26
40 L.H., Daily Totals		$1471.40		$2195.34	$36.78	$54.88

Crew B-9A	Hr.	Daily	Hr.	Daily	Bare Costs	Incl. O&P
2 Laborers	$31.60	$505.60	$49.00	$784.00	$31.72	$49.05
1 Truck Driver (heavy)	31.95	255.60	49.15	393.20		
1 Water Tanker, 5000 Gal.		136.00		149.60		
1 Truck Tractor, 220 H.P.		288.40		317.24		
2 -50' Discharge Hoses, 3"		3.50		3.85	17.83	19.61
24 L.H., Daily Totals		$1189.10		$1647.89	$49.55	$68.66

Crew B-9B	Hr.	Daily	Hr.	Daily	Bare Costs	Incl. O&P
2 Laborers	$31.60	$505.60	$49.00	$784.00	$31.72	$49.05
1 Truck Driver (heavy)	31.95	255.60	49.15	393.20		
2 -50' Discharge Hoses, 3"		3.50		3.85		
1 Water Tanker, 5000 Gal.		136.00		149.60		
1 Truck Tractor, 220 H.P.		288.40		317.24		
1 Pressure Washer		53.00		58.30	20.04	22.04
24 L.H., Daily Totals		$1242.10		$1706.19	$51.75	$71.09

Crew B-9D	Hr.	Daily	Hr.	Daily	Bare Costs	Incl. O&P
1 Labor Foreman (Outside)	$33.60	$268.80	$52.10	$416.80	$32.00	$49.62
4 Common Laborers	31.60	1011.20	49.00	1568.00		
1 Air Compressor, 250 cfm		162.40		178.64		
2 -50' Air Hoses, 1.5"		12.60		13.86		
2 Air Powered Tampers		48.00		52.80	5.58	6.13
40 L.H., Daily Totals		$1503.00		$2230.10	$37.58	$55.75

Crew B-10	Hr.	Daily	Hr.	Daily	Bare Costs	Incl. O&P
1 Equip. Oper. (med.)	$41.35	$330.80	$62.15	$497.20	$38.10	$57.77
.5 Laborer	31.60	126.40	49.00	196.00		
12 L.H., Daily Totals		$457.20		$693.20	$38.10	$57.77

Crew B-10A	Hr.	Daily	Hr.	Daily	Bare Costs	Incl. O&P
1 Equip. Oper. (med.)	$41.35	$330.80	$62.15	$497.20	$38.10	$57.77
.5 Laborer	31.60	126.40	49.00	196.00		
1 Roller, 2-Drum, W.B., 7.5 H.P.		145.20		159.72	12.10	13.31
12 L.H., Daily Totals		$602.40		$852.92	$50.20	$71.08

Crew B-10B	Hr.	Daily	Hr.	Daily	Bare Costs	Incl. O&P
1 Equip. Oper. (med.)	$41.35	$330.80	$62.15	$497.20	$38.10	$57.77
.5 Laborer	31.60	126.40	49.00	196.00		
1 Dozer, 200 H.P.		1082.00		1190.20	90.17	99.18
12 L.H., Daily Totals		$1539.20		$1883.40	$128.27	$156.95

Crew B-10C	Hr.	Daily	Hr.	Daily	Bare Costs	Incl. O&P
1 Equip. Oper. (med.)	$41.35	$330.80	$62.15	$497.20	$38.10	$57.77
.5 Laborer	31.60	126.40	49.00	196.00		
1 Dozer, 200 H.P.		1082.00		1190.20		
1 Vibratory Roller, Towed, 23 Ton		411.00		452.10	124.42	136.86
12 L.H., Daily Totals		$1950.20		$2335.50	$162.52	$194.63

Crew B-10D	Hr.	Daily	Hr.	Daily	Bare Costs	Incl. O&P
1 Equip. Oper. (med.)	$41.35	$330.80	$62.15	$497.20	$38.10	$57.77
.5 Laborer	31.60	126.40	49.00	196.00		
1 Dozer, 200 H.P.		1082.00		1190.20		
1 Sheepsft. Roller, Towed		465.40		511.94	128.95	141.85
12 L.H., Daily Totals		$2004.60		$2395.34	$167.05	$199.61

Crew B-10E	Hr.	Daily	Hr.	Daily	Bare Costs	Incl. O&P
1 Equip. Oper. (med.)	$41.35	$330.80	$62.15	$497.20	$38.10	$57.77
.5 Laborer	31.60	126.40	49.00	196.00		
1 Tandem Roller, 5 Ton		137.80		151.58	11.48	12.63
12 L.H., Daily Totals		$595.00		$844.78	$49.58	$70.40

Crew B-10F	Hr.	Daily	Hr.	Daily	Bare Costs	Incl. O&P
1 Equip. Oper. (med.)	$41.35	$330.80	$62.15	$497.20	$38.10	$57.77
.5 Laborer	31.60	126.40	49.00	196.00		
1 Tandem Roller, 10 Ton		233.00		256.30	19.42	21.36
12 L.H., Daily Totals		$690.20		$949.50	$57.52	$79.13

Crew No.	Bare Costs		Incl. Subs O & P		Cost Per Labor-Hour	

Crew B-10G	Hr.	Daily	Hr.	Daily	Bare Costs	Incl. O&P
1 Equip. Oper. (med.)	$41.35	$330.80	$62.15	$497.20	$38.10	$57.77
.5 Laborer	31.60	126.40	49.00	196.00		
1 Sheepsft. Roll., 240 H.P.		1142.00		1256.20	95.17	104.68
12 L.H., Daily Totals		$1599.20		$1949.40	$133.27	$162.45

Crew B-10H	Hr.	Daily	Hr.	Daily	Bare Costs	Incl. O&P
1 Equip. Oper. (med.)	$41.35	$330.80	$62.15	$497.20	$38.10	$57.77
.5 Laborer	31.60	126.40	49.00	196.00		
1 Diaphragm Water Pump, 2"		63.20		69.52		
1 -20' Suction Hose, 2"		1.95		2.15		
2 -50' Discharge Hoses, 2"		2.20		2.42	5.61	6.17
12 L.H., Daily Totals		$524.55		$767.28	$43.71	$63.94

Crew B-10I	Hr.	Daily	Hr.	Daily	Bare Costs	Incl. O&P
1 Equip. Oper. (med.)	$41.35	$330.80	$62.15	$497.20	$38.10	$57.77
.5 Laborer	31.60	126.40	49.00	196.00		
1 Diaphragm Water Pump, 4"		86.20		94.82		
1 -20' Suction Hose, 4"		3.45		3.79		
2 -50' Discharge Hoses, 4"		4.70		5.17	7.86	8.65
12 L.H., Daily Totals		$551.55		$796.99	$45.96	$66.42

Crew B-10J	Hr.	Daily	Hr.	Daily	Bare Costs	Incl. O&P
1 Equip. Oper. (med.)	$41.35	$330.80	$62.15	$497.20	$38.10	$57.77
.5 Laborer	31.60	126.40	49.00	196.00		
1 Centrifugal Water Pump, 3"		69.40		76.34		
1 -20' Suction Hose, 3"		3.05		3.36		
2 -50' Discharge Hoses, 3"		3.50		3.85	6.33	6.96
12 L.H., Daily Totals		$533.15		$776.75	$44.43	$64.73

Crew B-10K	Hr.	Daily	Hr.	Daily	Bare Costs	Incl. O&P
1 Equip. Oper. (med.)	$41.35	$330.80	$62.15	$497.20	$38.10	$57.77
.5 Laborer	31.60	126.40	49.00	196.00		
1 Centr. Water Pump, 6"		309.40		340.34		
1 -20' Suction Hose, 6"		11.90		13.09		
2 -50' Discharge Hoses, 6"		12.60		13.86	27.82	30.61
12 L.H., Daily Totals		$791.10		$1060.49	$65.92	$88.37

Crew B-10L	Hr.	Daily	Hr.	Daily	Bare Costs	Incl. O&P
1 Equip. Oper. (med.)	$41.35	$330.80	$62.15	$497.20	$38.10	$57.77
.5 Laborer	31.60	126.40	49.00	196.00		
1 Dozer, 80 H.P.		399.20		439.12	33.27	36.59
12 L.H., Daily Totals		$856.40		$1132.32	$71.37	$94.36

Crew B-10M	Hr.	Daily	Hr.	Daily	Bare Costs	Incl. O&P
1 Equip. Oper. (med.)	$41.35	$330.80	$62.15	$497.20	$38.10	$57.77
.5 Laborer	31.60	126.40	49.00	196.00		
1 Dozer, 300 H.P.		1423.00		1565.30	118.58	130.44
12 L.H., Daily Totals		$1880.20		$2258.50	$156.68	$188.21

Crew B-10N	Hr.	Daily	Hr.	Daily	Bare Costs	Incl. O&P
1 Equip. Oper. (med.)	$41.35	$330.80	$62.15	$497.20	$38.10	$57.77
.5 Laborer	31.60	126.40	49.00	196.00		
1 F.E. Loader, T.M., 1.5 C.Y		367.00		403.70	30.58	33.64
12 L.H., Daily Totals		$824.20		$1096.90	$68.68	$91.41

Crew B-10O	Hr.	Daily	Hr.	Daily	Bare Costs	Incl. O&P
1 Equip. Oper. (med.)	$41.35	$330.80	$62.15	$497.20	$38.10	$57.77
.5 Laborer	31.60	126.40	49.00	196.00		
1 F.E. Loader, T.M., 2.25 C.Y.		767.40		844.14	63.95	70.34
12 L.H., Daily Totals		$1224.60		$1537.34	$102.05	$128.11

Crew B-10P	Hr.	Daily	Hr.	Daily	Bare Costs	Incl. O&P
1 Equip. Oper. (med.)	$41.35	$330.80	$62.15	$497.20	$38.10	$57.77
.5 Laborer	31.60	126.40	49.00	196.00		
1 Crawler Loader, 3 C.Y.		990.00		1089.00	82.50	90.75
12 L.H., Daily Totals		$1447.20		$1782.20	$120.60	$148.52

Crew B-10Q	Hr.	Daily	Hr.	Daily	Bare Costs	Incl. O&P
1 Equip. Oper. (med.)	$41.35	$330.80	$62.15	$497.20	$38.10	$57.77
.5 Laborer	31.60	126.40	49.00	196.00		
1 Crawler Loader, 4 C.Y.		1450.00		1595.00	120.83	132.92
12 L.H., Daily Totals		$1907.20		$2288.20	$158.93	$190.68

Crew B-10R	Hr.	Daily	Hr.	Daily	Bare Costs	Incl. O&P
1 Equip. Oper. (med.)	$41.35	$330.80	$62.15	$497.20	$38.10	$57.77
.5 Laborer	31.60	126.40	49.00	196.00		
1 F.E. Loader, W.M., 1 C.Y.		239.00		262.90	19.92	21.91
12 L.H., Daily Totals		$696.20		$956.10	$58.02	$79.67

Crew B-10S	Hr.	Daily	Hr.	Daily	Bare Costs	Incl. O&P
1 Equip. Oper. (med.)	$41.35	$330.80	$62.15	$497.20	$38.10	$57.77
.5 Laborer	31.60	126.40	49.00	196.00		
1 F.E. Loader, W.M., 1.5 C.Y.		328.20		361.02	27.35	30.09
12 L.H., Daily Totals		$785.40		$1054.22	$65.45	$87.85

Crew B-10T	Hr.	Daily	Hr.	Daily	Bare Costs	Incl. O&P
1 Equip. Oper. (med.)	$41.35	$330.80	$62.15	$497.20	$38.10	$57.77
.5 Laborer	31.60	126.40	49.00	196.00		
1 F.E. Loader, W.M.,2.5 C.Y.		395.80		435.38	32.98	36.28
12 L.H., Daily Totals		$853.00		$1128.58	$71.08	$94.05

Crew B-10U	Hr.	Daily	Hr.	Daily	Bare Costs	Incl. O&P
1 Equip. Oper. (med.)	$41.35	$330.80	$62.15	$497.20	$38.10	$57.77
.5 Laborer	31.60	126.40	49.00	196.00		
1 F.E. Loader, W.M., 5.5 C.Y.		854.40		939.84	71.20	78.32
12 L.H., Daily Totals		$1311.60		$1633.04	$109.30	$136.09

Crew B-10V	Hr.	Daily	Hr.	Daily	Bare Costs	Incl. O&P
1 Equip. Oper. (med.)	$41.35	$330.80	$62.15	$497.20	$38.10	$57.77
.5 Laborer	31.60	126.40	49.00	196.00		
1 Dozer, 700 H.P.		4232.00		4655.20	352.67	387.93
12 L.H., Daily Totals		$4689.20		$5348.40	$390.77	$445.70

Crew B-10W	Hr.	Daily	Hr.	Daily	Bare Costs	Incl. O&P
1 Equip. Oper. (med.)	$41.35	$330.80	$62.15	$497.20	$38.10	$57.77
.5 Laborer	31.60	126.40	49.00	196.00		
1 Dozer, 105 H.P.		600.60		660.66	50.05	55.06
12 L.H., Daily Totals		$1057.80		$1353.86	$88.15	$112.82

Crew B-10X	Hr.	Daily	Hr.	Daily	Bare Costs	Incl. O&P
1 Equip. Oper. (med.)	$41.35	$330.80	$62.15	$497.20	$38.10	$57.77
.5 Laborer	31.60	126.40	49.00	196.00		
1 Dozer, 410 H.P.		1849.00		2033.90	154.08	169.49
12 L.H., Daily Totals		$2306.20		$2727.10	$192.18	$227.26

Crew B-10Y	Hr.	Daily	Hr.	Daily	Bare Costs	Incl. O&P
1 Equip. Oper. (med.)	$41.35	$330.80	$62.15	$497.20	$38.10	$57.77
.5 Laborer	31.60	126.40	49.00	196.00		
1 Vibr. Roller, Towed, 12 Ton		469.60		516.56	39.13	43.05
12 L.H., Daily Totals		$926.80		$1209.76	$77.23	$100.81

Crew B-11A	Bare Costs Hr.	Bare Costs Daily	Incl. Subs O & P Hr.	Incl. Subs O & P Daily	Cost Per Labor-Hour Bare Costs	Cost Per Labor-Hour Incl. O&P
1 Equipment Oper. (med.)	$41.35	$330.80	$62.15	$497.20	$36.48	$55.58
1 Laborer	31.60	252.80	49.00	392.00		
1 Dozer, 200 H.P.		1082.00		1190.20	67.63	74.39
16 L.H., Daily Totals		$1665.60		$2079.40	$104.10	$129.96

Crew B-11B	Bare Costs Hr.	Bare Costs Daily	Incl. Subs O & P Hr.	Incl. Subs O & P Daily	Cost Per Labor-Hour Bare Costs	Cost Per Labor-Hour Incl. O&P
1 Equipment Oper. (light)	$39.05	$312.40	$58.70	$469.60	$35.33	$53.85
1 Laborer	31.60	252.80	49.00	392.00		
1 Air Powered Tamper		24.00		26.40		
1 Air Compressor, 365 cfm		211.20		232.32		
2 -50' Air Hoses, 1.5"		12.60		13.86	15.49	17.04
16 L.H., Daily Totals		$813.00		$1134.18	$50.81	$70.89

Crew B-11C	Bare Costs Hr.	Bare Costs Daily	Incl. Subs O & P Hr.	Incl. Subs O & P Daily	Cost Per Labor-Hour Bare Costs	Cost Per Labor-Hour Incl. O&P
1 Equipment Oper. (med.)	$41.35	$330.80	$62.15	$497.20	$36.48	$55.58
1 Laborer	31.60	252.80	49.00	392.00		
1 Backhoe Loader, 48 H.P.		293.80		323.18	18.36	20.20
16 L.H., Daily Totals		$877.40		$1212.38	$54.84	$75.77

Crew B-11J	Bare Costs Hr.	Bare Costs Daily	Incl. Subs O & P Hr.	Incl. Subs O & P Daily	Cost Per Labor-Hour Bare Costs	Cost Per Labor-Hour Incl. O&P
1 Equipment Oper. (med.)	$41.35	$330.80	$62.15	$497.20	$36.48	$55.58
1 Laborer	31.60	252.80	49.00	392.00		
1 Grader, 30,000 Lbs.		550.20		605.22		
1 Ripper, beam & 1 shank		79.60		87.56	39.36	43.30
16 L.H., Daily Totals		$1213.40		$1581.98	$75.84	$98.87

Crew B-11K	Bare Costs Hr.	Bare Costs Daily	Incl. Subs O & P Hr.	Incl. Subs O & P Daily	Cost Per Labor-Hour Bare Costs	Cost Per Labor-Hour Incl. O&P
1 Equipment Oper. (med.)	$41.35	$330.80	$62.15	$497.20	$36.48	$55.58
1 Laborer	31.60	252.80	49.00	392.00		
1 Trencher, Chain Type, 8' D		1777.00		1954.70	111.06	122.17
16 L.H., Daily Totals		$2360.60		$2843.90	$147.54	$177.74

Crew B-11L	Bare Costs Hr.	Bare Costs Daily	Incl. Subs O & P Hr.	Incl. Subs O & P Daily	Cost Per Labor-Hour Bare Costs	Cost Per Labor-Hour Incl. O&P
1 Equipment Oper. (med.)	$41.35	$330.80	$62.15	$497.20	$36.48	$55.58
1 Laborer	31.60	252.80	49.00	392.00		
1 Grader, 30,000 Lbs.		550.20		605.22	34.39	37.83
16 L.H., Daily Totals		$1133.80		$1494.42	$70.86	$93.40

Crew B-11M	Bare Costs Hr.	Bare Costs Daily	Incl. Subs O & P Hr.	Incl. Subs O & P Daily	Cost Per Labor-Hour Bare Costs	Cost Per Labor-Hour Incl. O&P
1 Equipment Oper. (med.)	$41.35	$330.80	$62.15	$497.20	$36.48	$55.58
1 Laborer	31.60	252.80	49.00	392.00		
1 Backhoe Loader, 80 H.P.		345.00		379.50	21.56	23.72
16 L.H., Daily Totals		$928.60		$1268.70	$58.04	$79.29

Crew B-11N	Bare Costs Hr.	Bare Costs Daily	Incl. Subs O & P Hr.	Incl. Subs O & P Daily	Cost Per Labor-Hour Bare Costs	Cost Per Labor-Hour Incl. O&P
1 Labor Foreman	$33.60	$268.80	$52.10	$416.80	$34.22	$52.37
2 Equipment Operators (med.)	41.35	661.60	62.15	994.40		
6 Truck Drivers (hvy.)	31.95	1533.60	49.15	2359.20		
1 F.E. Loader, W.M., 5.5 C.Y.		854.40		939.84		
1 Dozer, 410 H.P.		1849.00		2033.90		
6 Dump Trucks, Off Hwy., 50 Ton		9492.00		10441.20	169.38	186.32
72 L.H., Daily Totals		$14659.40		$17185.34	$203.60	$238.69

Crew B-11Q	Bare Costs Hr.	Bare Costs Daily	Incl. Subs O & P Hr.	Incl. Subs O & P Daily	Cost Per Labor-Hour Bare Costs	Cost Per Labor-Hour Incl. O&P
1 Equipment Operator (med.)	$41.35	$330.80	$62.15	$497.20	$38.10	$57.77
.5 Laborer	31.60	126.40	49.00	196.00		
1 Dozer, 140 H.P.		682.00		750.20	56.83	62.52
12 L.H., Daily Totals		$1139.20		$1443.40	$94.93	$120.28

Crew B-11R	Bare Costs Hr.	Bare Costs Daily	Incl. Subs O & P Hr.	Incl. Subs O & P Daily	Cost Per Labor-Hour Bare Costs	Cost Per Labor-Hour Incl. O&P
1 Equipment Operator (med.)	$41.35	$330.80	$62.15	$497.20	$38.10	$57.77
.5 Laborer	31.60	126.40	49.00	196.00		
1 Dozer, 200 H.P.		1082.00		1190.20	90.17	99.18
12 L.H., Daily Totals		$1539.20		$1883.40	$128.27	$156.95

Crew B-11S	Bare Costs Hr.	Bare Costs Daily	Incl. Subs O & P Hr.	Incl. Subs O & P Daily	Cost Per Labor-Hour Bare Costs	Cost Per Labor-Hour Incl. O&P
1 Equipment Operator (med.)	$41.35	$330.80	$62.15	$497.20	$38.10	$57.77
.5 Laborer	31.60	126.40	49.00	196.00		
1 Dozer, 300 H.P.		1423.00		1565.30		
1 Ripper, beam & 1 shank		79.60		87.56	125.22	137.74
12 L.H., Daily Totals		$1959.80		$2346.06	$163.32	$195.51

Crew B-11T	Bare Costs Hr.	Bare Costs Daily	Incl. Subs O & P Hr.	Incl. Subs O & P Daily	Cost Per Labor-Hour Bare Costs	Cost Per Labor-Hour Incl. O&P
1 Equipment Operator (med.)	$41.35	$330.80	$62.15	$497.20	$38.10	$57.77
.5 Laborer	31.60	126.40	49.00	196.00		
1 Dozer, 410 H.P.		1849.00		2033.90		
1 Ripper, beam & 2 shanks		89.40		98.34	161.53	177.69
12 L.H., Daily Totals		$2395.60		$2825.44	$199.63	$235.45

Crew B-11U	Bare Costs Hr.	Bare Costs Daily	Incl. Subs O & P Hr.	Incl. Subs O & P Daily	Cost Per Labor-Hour Bare Costs	Cost Per Labor-Hour Incl. O&P
1 Equipment Operator (med.)	$41.35	$330.80	$62.15	$497.20	$38.10	$57.77
.5 Laborer	31.60	126.40	49.00	196.00		
1 Dozer, 520 H.P.		2464.00		2710.40	205.33	225.87
12 L.H., Daily Totals		$2921.20		$3403.60	$243.43	$283.63

Crew B-11V	Bare Costs Hr.	Bare Costs Daily	Incl. Subs O & P Hr.	Incl. Subs O & P Daily	Cost Per Labor-Hour Bare Costs	Cost Per Labor-Hour Incl. O&P
3 Laborers	$31.60	$758.40	$49.00	$1176.00	$31.60	$49.00
1 Roller, 2-Drum, W.B., 7.5 H.P.		145.20		159.72	6.05	6.66
24 L.H., Daily Totals		$903.60		$1335.72	$37.65	$55.66

Crew B-11W	Bare Costs Hr.	Bare Costs Daily	Incl. Subs O & P Hr.	Incl. Subs O & P Daily	Cost Per Labor-Hour Bare Costs	Cost Per Labor-Hour Incl. O&P
1 Equipment Operator (med.)	$41.35	$330.80	$62.15	$497.20	$32.70	$50.22
1 Common Laborer	31.60	252.80	49.00	392.00		
10 Truck Drivers (hvy.)	31.95	2556.00	49.15	3932.00		
1 Dozer, 200 H.P.		1082.00		1190.20		
1 Vibratory Roller, Towed, 23 Ton		411.00		452.10		
10 Dump Trucks, 8 C.Y., 220 H.P.		3264.00		3590.40	49.55	54.51
96 L.H., Daily Totals		$7896.60		$10053.90	$82.26	$104.73

Crew B-11Y	Bare Costs Hr.	Bare Costs Daily	Incl. Subs O & P Hr.	Incl. Subs O & P Daily	Cost Per Labor-Hour Bare Costs	Cost Per Labor-Hour Incl. O&P
1 Labor Foreman (Outside)	$33.60	$268.80	$52.10	$416.80	$35.07	$53.73
5 Common Laborers	31.60	1264.00	49.00	1960.00		
3 Equipment Operators (med.)	41.35	992.40	62.15	1491.60		
1 Dozer, 80 H.P.		399.20		439.12		
2 Rollers, 2-Drum, W.B., 7.5 H.P.		290.40		319.44		
4 Vibratory Plates, Gas, 21"		156.80		172.48	11.76	12.93
72 L.H., Daily Totals		$3371.60		$4799.44	$46.83	$66.66

Crew B-12A	Bare Costs Hr.	Bare Costs Daily	Incl. Subs O & P Hr.	Incl. Subs O & P Daily	Cost Per Labor-Hour Bare Costs	Cost Per Labor-Hour Incl. O&P
1 Equip. Oper. (crane)	$42.55	$340.40	$63.95	$511.60	$37.08	$56.48
1 Laborer	31.60	252.80	49.00	392.00		
1 Hyd. Excavator, 1 C.Y.		669.20		736.12	41.83	46.01
16 L.H., Daily Totals		$1262.40		$1639.72	$78.90	$102.48

Crew B-12B	Bare Costs Hr.	Bare Costs Daily	Incl. Subs O & P Hr.	Incl. Subs O & P Daily	Cost Per Labor-Hour Bare Costs	Cost Per Labor-Hour Incl. O&P
1 Equip. Oper. (crane)	$42.55	$340.40	$63.95	$511.60	$37.08	$56.48
1 Laborer	31.60	252.80	49.00	392.00		
1 Hyd. Excavator, 1.5 C.Y.		865.80		952.38	54.11	59.52
16 L.H., Daily Totals		$1459.00		$1855.98	$91.19	$116.00

Crew No.	Bare Costs		Incl. Subs O & P		Cost Per Labor-Hour	
Crew B-12C	**Hr.**	**Daily**	**Hr.**	**Daily**	**Bare Costs**	**Incl. O&P**
1 Equip. Oper. (crane)	$42.55	$340.40	$63.95	$511.60	$37.08	$56.48
1 Laborer	31.60	252.80	49.00	392.00		
1 Hyd. Excavator, 2 C.Y.		1182.00		1300.20	73.88	81.26
16 L.H., Daily Totals		$1775.20		$2203.80	$110.95	$137.74
Crew B-12D	**Hr.**	**Daily**	**Hr.**	**Daily**	**Bare Costs**	**Incl. O&P**
1 Equip. Oper. (crane)	$42.55	$340.40	$63.95	$511.60	$37.08	$56.48
1 Laborer	31.60	252.80	49.00	392.00		
1 Hyd. Excavator, 3.5 C.Y.		2162.00		2378.20	135.13	148.64
16 L.H., Daily Totals		$2755.20		$3281.80	$172.20	$205.11
Crew B-12E	**Hr.**	**Daily**	**Hr.**	**Daily**	**Bare Costs**	**Incl. O&P**
1 Equip. Oper. (crane)	$42.55	$340.40	$63.95	$511.60	$37.08	$56.48
1 Laborer	31.60	252.80	49.00	392.00		
1 Hyd. Excavator, .5 C.Y.		394.60		434.06	24.66	27.13
16 L.H., Daily Totals		$987.80		$1337.66	$61.74	$83.60
Crew B-12F	**Hr.**	**Daily**	**Hr.**	**Daily**	**Bare Costs**	**Incl. O&P**
1 Equip. Oper. (crane)	$42.55	$340.40	$63.95	$511.60	$37.08	$56.48
1 Laborer	31.60	252.80	49.00	392.00		
1 Hyd. Excavator, .75 C.Y.		598.60		658.46	37.41	41.15
16 L.H., Daily Totals		$1191.80		$1562.06	$74.49	$97.63
Crew B-12G	**Hr.**	**Daily**	**Hr.**	**Daily**	**Bare Costs**	**Incl. O&P**
1 Equip. Oper. (crane)	$42.55	$340.40	$63.95	$511.60	$37.08	$56.48
1 Laborer	31.60	252.80	49.00	392.00		
1 Crawler Crane, 15 Ton		615.30		676.83		
1 Clamshell Bucket, .5 C.Y.		36.60		40.26	40.74	44.82
16 L.H., Daily Totals		$1245.10		$1620.69	$77.82	$101.29
Crew B-12H	**Hr.**	**Daily**	**Hr.**	**Daily**	**Bare Costs**	**Incl. O&P**
1 Equip. Oper. (crane)	$42.55	$340.40	$63.95	$511.60	$37.08	$56.48
1 Laborer	31.60	252.80	49.00	392.00		
1 Crawler Crane, 25 Ton		1052.00		1157.20		
1 Clamshell Bucket, 1 C.Y.		46.20		50.82	68.64	75.50
16 L.H., Daily Totals		$1691.40		$2111.62	$105.71	$131.98
Crew B-12I	**Hr.**	**Daily**	**Hr.**	**Daily**	**Bare Costs**	**Incl. O&P**
1 Equip. Oper. (crane)	$42.55	$340.40	$63.95	$511.60	$37.08	$56.48
1 Laborer	31.60	252.80	49.00	392.00		
1 Crawler Crane, 20 Ton		789.05		867.96		
1 Dragline Bucket, .75 C.Y.		20.00		22.00	50.57	55.62
16 L.H., Daily Totals		$1402.25		$1793.56	$87.64	$112.10
Crew B-12J	**Hr.**	**Daily**	**Hr.**	**Daily**	**Bare Costs**	**Incl. O&P**
1 Equip. Oper. (crane)	$42.55	$340.40	$63.95	$511.60	$37.08	$56.48
1 Laborer	31.60	252.80	49.00	392.00		
1 Gradall, 5/8 C.Y.		958.80		1054.68	59.92	65.92
16 L.H., Daily Totals		$1552.00		$1958.28	$97.00	$122.39
Crew B-12K	**Hr.**	**Daily**	**Hr.**	**Daily**	**Bare Costs**	**Incl. O&P**
1 Equip. Oper. (crane)	$42.55	$340.40	$63.95	$511.60	$37.08	$56.48
1 Laborer	31.60	252.80	49.00	392.00		
1 Gradall, 3 Ton, 1 C.Y.		1130.00		1243.00	70.63	77.69
16 L.H., Daily Totals		$1723.20		$2146.60	$107.70	$134.16

Crew No.	Bare Costs		Incl. Subs O & P		Cost Per Labor-Hour	
Crew B-12L	**Hr.**	**Daily**	**Hr.**	**Daily**	**Bare Costs**	**Incl. O&P**
1 Equip. Oper. (crane)	$42.55	$340.40	$63.95	$511.60	$37.08	$56.48
1 Laborer	31.60	252.80	49.00	392.00		
1 Crawler Crane, 15 Ton		615.30		676.83		
1 F.E. Attachment, .5 C.Y.		52.80		58.08	41.76	45.93
16 L.H., Daily Totals		$1261.30		$1638.51	$78.83	$102.41
Crew B-12M	**Hr.**	**Daily**	**Hr.**	**Daily**	**Bare Costs**	**Incl. O&P**
1 Equip. Oper. (crane)	$42.55	$340.40	$63.95	$511.60	$37.08	$56.48
1 Laborer	31.60	252.80	49.00	392.00		
1 Crawler Crane, 20 Ton		789.05		867.96		
1 F.E. Attachment, .75 C.Y.		58.20		64.02	52.95	58.25
16 L.H., Daily Totals		$1440.45		$1835.58	$90.03	$114.72
Crew B-12N	**Hr.**	**Daily**	**Hr.**	**Daily**	**Bare Costs**	**Incl. O&P**
1 Equip. Oper. (crane)	$42.55	$340.40	$63.95	$511.60	$37.08	$56.48
1 Laborer	31.60	252.80	49.00	392.00		
1 Crawler Crane, 25 Ton		1052.00		1157.20		
1 F.E. Attachment, 1 C.Y.		64.00		70.40	69.75	76.72
16 L.H., Daily Totals		$1709.20		$2131.20	$106.83	$133.20
Crew B-12O	**Hr.**	**Daily**	**Hr.**	**Daily**	**Bare Costs**	**Incl. O&P**
1 Equip. Oper. (crane)	$42.55	$340.40	$63.95	$511.60	$37.08	$56.48
1 Laborer	31.60	252.80	49.00	392.00		
1 Crawler Crane, 40 Ton		1055.00		1160.50		
1 F.E. Attachment, 1.5 C.Y.		74.20		81.62	70.58	77.63
16 L.H., Daily Totals		$1722.40		$2145.72	$107.65	$134.11
Crew B-12P	**Hr.**	**Daily**	**Hr.**	**Daily**	**Bare Costs**	**Incl. O&P**
1 Equip. Oper. (crane)	$42.55	$340.40	$63.95	$511.60	$37.08	$56.48
1 Laborer	31.60	252.80	49.00	392.00		
1 Crawler Crane, 40 Ton		1055.00		1160.50		
1 Dragline Bucket, 1.5 C.Y.		32.80		36.08	67.99	74.79
16 L.H., Daily Totals		$1681.00		$2100.18	$105.06	$131.26
Crew B-12Q	**Hr.**	**Daily**	**Hr.**	**Daily**	**Bare Costs**	**Incl. O&P**
1 Equip. Oper. (crane)	$42.55	$340.40	$63.95	$511.60	$37.08	$56.48
1 Laborer	31.60	252.80	49.00	392.00		
1 Hyd. Excavator, 5/8 C.Y.		517.20		568.92	32.33	35.56
16 L.H., Daily Totals		$1110.40		$1472.52	$69.40	$92.03
Crew B-12S	**Hr.**	**Daily**	**Hr.**	**Daily**	**Bare Costs**	**Incl. O&P**
1 Equip. Oper. (crane)	$42.55	$340.40	$63.95	$511.60	$37.08	$56.48
1 Laborer	31.60	252.80	49.00	392.00		
1 Hyd. Excavator, 2.5 C.Y.		1619.00		1780.90	101.19	111.31
16 L.H., Daily Totals		$2212.20		$2684.50	$138.26	$167.78
Crew B-12T	**Hr.**	**Daily**	**Hr.**	**Daily**	**Bare Costs**	**Incl. O&P**
1 Equip. Oper. (crane)	$42.55	$340.40	$63.95	$511.60	$37.08	$56.48
1 Laborer	31.60	252.80	49.00	392.00		
1 Crawler Crane, 75 Ton		1496.00		1645.60		
1 F.E. Attachment, 3 C.Y.		97.60		107.36	99.60	109.56
16 L.H., Daily Totals		$2186.80		$2656.56	$136.68	$166.04
Crew B-12V	**Hr.**	**Daily**	**Hr.**	**Daily**	**Bare Costs**	**Incl. O&P**
1 Equip. Oper. (crane)	$42.55	$340.40	$63.95	$511.60	$37.08	$56.48
1 Laborer	31.60	252.80	49.00	392.00		
1 Crawler Crane, 75 Ton		1496.00		1645.60		
1 Dragline Bucket, 3 C.Y.		46.20		50.82	96.39	106.03
16 L.H., Daily Totals		$2135.40		$2600.02	$133.46	$162.50

Crew No.	Bare Costs		Incl. Subs O & P		Cost Per Labor-Hour	
Crew B-13	**Hr.**	**Daily**	**Hr.**	**Daily**	**Bare Costs**	**Incl. O&P**
1 Labor Foreman (outside)	$33.60	$268.80	$52.10	$416.80	$34.19	$52.48
4 Laborers	31.60	1011.20	49.00	1568.00		
1 Equip. Oper. (crane)	42.55	340.40	63.95	511.60		
1 Equip. Oper. Oiler	36.80	294.40	55.30	442.40		
1 Hyd. Crane, 25 Ton		789.40		868.34	14.10	15.51
56 L.H., Daily Totals		$2704.20		$3807.14	$48.29	$67.98
Crew B-13A	**Hr.**	**Daily**	**Hr.**	**Daily**	**Bare Costs**	**Incl. O&P**
1 Foreman	$33.60	$268.80	$52.10	$416.80	$34.77	$53.24
2 Laborers	31.60	505.60	49.00	784.00		
2 Equipment Operators (med.)	41.35	661.60	62.15	994.40		
2 Truck Drivers (heavy)	31.95	511.20	49.15	786.40		
1 Crawler Crane, 75 Ton		1496.00		1645.60		
1 Crawler Loader, 4 C.Y.		1450.00		1595.00		
2 Dump Trucks, 8 C.Y., 220 H.P.		652.80		718.08	64.26	70.69
56 L.H., Daily Totals		$5546.00		$6940.28	$99.04	$123.93
Crew B-13B	**Hr.**	**Daily**	**Hr.**	**Daily**	**Bare Costs**	**Incl. O&P**
1 Labor Foreman (outside)	$33.60	$268.80	$52.10	$416.80	$34.19	$52.48
4 Laborers	31.60	1011.20	49.00	1568.00		
1 Equip. Oper. (crane)	42.55	340.40	63.95	511.60		
1 Equip. Oper. Oiler	36.80	294.40	55.30	442.40		
1 Hyd. Crane, 55 Ton		1125.00		1237.50	20.09	22.10
56 L.H., Daily Totals		$3039.80		$4176.30	$54.28	$74.58
Crew B-13C	**Hr.**	**Daily**	**Hr.**	**Daily**	**Bare Costs**	**Incl. O&P**
1 Labor Foreman (outside)	$33.60	$268.80	$52.10	$416.80	$34.19	$52.48
4 Laborers	31.60	1011.20	49.00	1568.00		
1 Equip. Oper. (crane)	42.55	340.40	63.95	511.60		
1 Equip. Oper. Oiler	36.80	294.40	55.30	442.40		
1 Crawler Crane, 100 Ton		1611.00		1772.10	28.77	31.64
56 L.H., Daily Totals		$3525.80		$4710.90	$62.96	$84.12
Crew B-13D	**Hr.**	**Daily**	**Hr.**	**Daily**	**Bare Costs**	**Incl. O&P**
1 Laborer	$31.60	$252.80	$49.00	$392.00	$37.08	$56.48
1 Equip. Oper. (crane)	42.55	340.40	63.95	511.60		
1 Hyd. Excavator, 1 C.Y.		669.20		736.12		
1 Trench Box		113.40		124.74	48.91	53.80
16 L.H., Daily Totals		$1375.80		$1764.46	$85.99	$110.28
Crew B-13E	**Hr.**	**Daily**	**Hr.**	**Daily**	**Bare Costs**	**Incl. O&P**
1 Laborer	$31.60	$252.80	$49.00	$392.00	$37.08	$56.48
1 Equip. Oper. (crane)	42.55	340.40	63.95	511.60		
1 Hyd. Excavator, 1.5 C.Y.		865.80		952.38		
1 Trench Box		113.40		124.74	61.20	67.32
16 L.H., Daily Totals		$1572.40		$1980.72	$98.28	$123.80
Crew B-13F	**Hr.**	**Daily**	**Hr.**	**Daily**	**Bare Costs**	**Incl. O&P**
1 Laborer	$31.60	$252.80	$49.00	$392.00	$37.08	$56.48
1 Equip. Oper. (crane)	42.55	340.40	63.95	511.60		
1 Hyd. Excavator, 3.5 C.Y.		2162.00		2378.20		
1 Trench Box		113.40		124.74	142.21	156.43
16 L.H., Daily Totals		$2868.60		$3406.54	$179.29	$212.91
Crew B-13G	**Hr.**	**Daily**	**Hr.**	**Daily**	**Bare Costs**	**Incl. O&P**
1 Laborer	$31.60	$252.80	$49.00	$392.00	$37.08	$56.48
1 Equip. Oper. (crane)	42.55	340.40	63.95	511.60		
1 Hyd. Excavator, .75 C.Y.		598.60		658.46		
1 Trench Box		113.40		124.74	44.50	48.95
16 L.H., Daily Totals		$1305.20		$1686.80	$81.58	$105.43

Crew No.	Bare Costs		Incl. Subs O & P		Cost Per Labor-Hour	
Crew B-13H	**Hr.**	**Daily**	**Hr.**	**Daily**	**Bare Costs**	**Incl. O&P**
1 Laborer	$31.60	$252.80	$49.00	$392.00	$37.08	$56.48
1 Equip. Oper. (crane)	42.55	340.40	63.95	511.60		
1 Gradall, 5/8 C.Y.		958.80		1054.68		
1 Trench Box		113.40		124.74	67.01	73.71
16 L.H., Daily Totals		$1665.40		$2083.02	$104.09	$130.19
Crew B-13I	**Hr.**	**Daily**	**Hr.**	**Daily**	**Bare Costs**	**Incl. O&P**
1 Laborer	$31.60	$252.80	$49.00	$392.00	$37.08	$56.48
1 Equip. Oper. (crane)	42.55	340.40	63.95	511.60		
1 Gradall, 3 Ton, 1 C.Y.		1130.00		1243.00		
1 Trench Box		113.40		124.74	77.71	85.48
16 L.H., Daily Totals		$1836.60		$2271.34	$114.79	$141.96
Crew B-13J	**Hr.**	**Daily**	**Hr.**	**Daily**	**Bare Costs**	**Incl. O&P**
1 Laborer	$31.60	$252.80	$49.00	$392.00	$37.08	$56.48
1 Equip. Oper. (crane)	42.55	340.40	63.95	511.60		
1 Hyd. Excavator, 2.5 C.Y.		1619.00		1780.90		
1 Trench Box		113.40		124.74	108.28	119.10
16 L.H., Daily Totals		$2325.60		$2809.24	$145.35	$175.58
Crew B-14	**Hr.**	**Daily**	**Hr.**	**Daily**	**Bare Costs**	**Incl. O&P**
1 Labor Foreman (outside)	$33.60	$268.80	$52.10	$416.80	$33.17	$51.13
4 Laborers	31.60	1011.20	49.00	1568.00		
1 Equip. Oper. (light)	39.05	312.40	58.70	469.60		
1 Backhoe Loader, 48 H.P.		293.80		323.18	6.12	6.73
48 L.H., Daily Totals		$1886.20		$2777.58	$39.30	$57.87
Crew B-14A	**Hr.**	**Daily**	**Hr.**	**Daily**	**Bare Costs**	**Incl. O&P**
1 Equip. Oper. (crane)	$42.55	$340.40	$63.95	$511.60	$38.90	$58.97
.5 Laborer	31.60	126.40	49.00	196.00		
1 Hyd. Excavator, 4.5 C.Y.		2635.00		2898.50	219.58	241.54
12 L.H., Daily Totals		$3101.80		$3606.10	$258.48	$300.51
Crew B-14B	**Hr.**	**Daily**	**Hr.**	**Daily**	**Bare Costs**	**Incl. O&P**
1 Equip. Oper. (crane)	$42.55	$340.40	$63.95	$511.60	$38.90	$58.97
.5 Laborer	31.60	126.40	49.00	196.00		
1 Hyd. Excavator, 6 C.Y.		3080.00		3388.00	256.67	282.33
12 L.H., Daily Totals		$3546.80		$4095.60	$295.57	$341.30
Crew B-14C	**Hr.**	**Daily**	**Hr.**	**Daily**	**Bare Costs**	**Incl. O&P**
1 Equip. Oper. (crane)	$42.55	$340.40	$63.95	$511.60	$38.90	$58.97
.5 Laborer	31.60	126.40	49.00	196.00		
1 Hyd. Excavator, 7 C.Y.		3256.00		3581.60	271.33	298.47
12 L.H., Daily Totals		$3722.80		$4289.20	$310.23	$357.43
Crew B-14F	**Hr.**	**Daily**	**Hr.**	**Daily**	**Bare Costs**	**Incl. O&P**
1 Equip. Oper. (crane)	$42.55	$340.40	$63.95	$511.60	$38.90	$58.97
.5 Laborer	31.60	126.40	49.00	196.00		
1 Hyd. Shovel, 7 C.Y.		3090.00		3399.00	257.50	283.25
12 L.H., Daily Totals		$3556.80		$4106.60	$296.40	$342.22
Crew B-14G	**Hr.**	**Daily**	**Hr.**	**Daily**	**Bare Costs**	**Incl. O&P**
1 Equip. Oper. (crane)	$42.55	$340.40	$63.95	$511.60	$38.90	$58.97
.5 Laborer	31.60	126.40	49.00	196.00		
1 Hyd. Shovel, 12 C.Y.		4148.00		4562.80	345.67	380.23
12 L.H., Daily Totals		$4614.80		$5270.40	$384.57	$439.20

Crew No.	Bare Costs		Incl. Subs O & P		Cost Per Labor-Hour	
Crew B-14J	**Hr.**	**Daily**	**Hr.**	**Daily**	**Bare Costs**	**Incl. O&P**
1 Equip. Oper. (med.)	$41.35	$330.80	$62.15	$497.20	$38.10	$57.77
.5 Laborer	31.60	126.40	49.00	196.00		
1 F.E. Loader, 8 C.Y.		1497.00		1646.70	124.75	137.22
12 L.H., Daily Totals		$1954.20		$2339.90	$162.85	$194.99

Crew B-14K	Hr.	Daily	Hr.	Daily	Bare Costs	Incl. O&P
1 Equip. Oper. (med.)	$41.35	$330.80	$62.15	$497.20	$38.10	$57.77
.5 Laborer	31.60	126.40	49.00	196.00		
1 F.E. Loader, 10 C.Y.		2225.00		2447.50	185.42	203.96
12 L.H., Daily Totals		$2682.20		$3140.70	$223.52	$261.73

Crew B-15	Hr.	Daily	Hr.	Daily	Bare Costs	Incl. O&P
1 Equipment Oper. (med)	$41.35	$330.80	$62.15	$497.20	$34.59	$52.84
.5 Laborer	31.60	126.40	49.00	196.00		
2 Truck Drivers (heavy)	31.95	511.20	49.15	786.40		
2 Dump Trucks, 12 C.Y., 400 H.P.		1066.00		1172.60		
1 Dozer, 200 H.P.		1082.00		1190.20	76.71	84.39
28 L.H., Daily Totals		$3116.40		$3842.40	$111.30	$137.23

Crew B-16	Hr.	Daily	Hr.	Daily	Bare Costs	Incl. O&P
1 Labor Foreman (outside)	$33.60	$268.80	$52.10	$416.80	$32.19	$49.81
2 Laborers	31.60	505.60	49.00	784.00		
1 Truck Driver (heavy)	31.95	255.60	49.15	393.20		
1 Dump Truck, 12 C.Y., 400 H.P.		533.00		586.30	16.66	18.32
32 L.H., Daily Totals		$1563.00		$2180.30	$48.84	$68.13

Crew B-17	Hr.	Daily	Hr.	Daily	Bare Costs	Incl. O&P
2 Laborers	$31.60	$505.60	$49.00	$784.00	$33.55	$51.46
1 Equip. Oper. (light)	39.05	312.40	58.70	469.60		
1 Truck Driver (heavy)	31.95	255.60	49.15	393.20		
1 Backhoe Loader, 48 H.P.		293.80		323.18		
1 Dump Truck, 8 C.Y., 220 H.P.		326.40		359.04	19.38	21.32
32 L.H., Daily Totals		$1693.80		$2329.02	$52.93	$72.78

Crew B-17A	Hr.	Daily	Hr.	Daily	Bare Costs	Incl. O&P
2 Laborer Foremen	$33.60	$537.60	$52.10	$833.60	$34.05	$52.78
6 Laborers	31.60	1516.80	49.00	2352.00		
1 Skilled Worker Foreman	42.85	342.80	66.35	530.80		
1 Skilled Worker	40.85	326.80	63.25	506.00		
80 L.H., Daily Totals		$2724.00		$4222.40	$34.05	$52.78

Crew B-18	Hr.	Daily	Hr.	Daily	Bare Costs	Incl. O&P
1 Labor Foreman (outside)	$33.60	$268.80	$52.10	$416.80	$32.27	$50.03
2 Laborers	31.60	505.60	49.00	784.00		
1 Vibratory Plate, Gas, 21"		39.20		43.12	1.63	1.80
24 L.H., Daily Totals		$813.60		$1243.92	$33.90	$51.83

Crew B-19	Hr.	Daily	Hr.	Daily	Bare Costs	Incl. O&P
1 Pile Driver Foreman	$40.50	$324.00	$65.75	$526.00	$39.55	$62.37
4 Pile Drivers	38.50	1232.00	62.50	2000.00		
2 Equip. Oper. (crane)	42.55	680.80	63.95	1023.20		
1 Equip. Oper. Oiler	36.80	294.40	55.30	442.40		
1 Crawler Crane, 40 Ton		1055.00		1160.50		
1 Lead, 90' high		118.80		130.68		
1 Hammer, Diesel, 22k Ft-Lb		590.00		649.00	27.56	30.32
64 L.H., Daily Totals		$4295.00		$5931.78	$67.11	$92.68

Crew B-19A	Hr.	Daily	Hr.	Daily	Bare Costs	Incl. O&P
1 Pile Driver Foreman	$40.50	$324.00	$65.75	$526.00	$39.55	$62.37
4 Pile Drivers	38.50	1232.00	62.50	2000.00		
2 Equip. Oper. (crane)	42.55	680.80	63.95	1023.20		
1 Equip. Oper. Oiler	36.80	294.40	55.30	442.40		
1 Crawler Crane, 75 Ton		1496.00		1645.60		
1 Lead, 90' high		118.80		130.68		
1 Hammer, Diesel, 41k Ft-Lb		677.60		745.36	35.82	39.40
64 L.H., Daily Totals		$4823.60		$6513.24	$75.37	$101.77

Crew B-20	Hr.	Daily	Hr.	Daily	Bare Costs	Incl. O&P
1 Labor Foreman (out)	$33.60	$268.80	$52.10	$416.80	$35.35	$54.78
1 Skilled Worker	40.85	326.80	63.25	506.00		
1 Laborer	31.60	252.80	49.00	392.00		
24 L.H., Daily Totals		$848.40		$1314.80	$35.35	$54.78

Crew B-20A	Hr.	Daily	Hr.	Daily	Bare Costs	Incl. O&P
1 Labor Foreman	$33.60	$268.80	$52.10	$416.80	$38.24	$58.20
1 Laborer	31.60	252.80	49.00	392.00		
1 Plumber	48.75	390.00	73.15	585.20		
1 Plumber Apprentice	39.00	312.00	58.55	468.40		
32 L.H., Daily Totals		$1223.60		$1862.40	$38.24	$58.20

Crew B-21	Hr.	Daily	Hr.	Daily	Bare Costs	Incl. O&P
1 Labor Foreman (out)	$33.60	$268.80	$52.10	$416.80	$36.38	$56.09
1 Skilled Worker	40.85	326.80	63.25	506.00		
1 Laborer	31.60	252.80	49.00	392.00		
.5 Equip. Oper. (crane)	42.55	170.20	63.95	255.80		
.5 S.P. Crane, 4x4, 5 Ton		129.10		142.01	4.61	5.07
28 L.H., Daily Totals		$1147.70		$1712.61	$40.99	$61.16

Crew B-21A	Hr.	Daily	Hr.	Daily	Bare Costs	Incl. O&P
1 Labor Foreman	$33.60	$268.80	$52.10	$416.80	$39.10	$59.35
1 Laborer	31.60	252.80	49.00	392.00		
1 Plumber	48.75	390.00	73.15	585.20		
1 Plumber Apprentice	39.00	312.00	58.55	468.40		
1 Equip. Oper. (crane)	42.55	340.40	63.95	511.60		
1 S.P. Crane, 4x4, 12 Ton		601.80		661.98	15.05	16.55
40 L.H., Daily Totals		$2165.80		$3035.98	$54.15	$75.90

Crew B-21B	Hr.	Daily	Hr.	Daily	Bare Costs	Incl. O&P
1 Laborer Foreman	$33.60	$268.80	$52.10	$416.80	$34.19	$52.61
3 Laborers	31.60	758.40	49.00	1176.00		
1 Equip. Oper. (crane)	42.55	340.40	63.95	511.60		
1 Hyd. Crane, 12 Ton		768.80		845.68	19.22	21.14
40 L.H., Daily Totals		$2136.40		$2950.08	$53.41	$73.75

Crew B-21C	Hr.	Daily	Hr.	Daily	Bare Costs	Incl. O&P
1 Laborer Foreman	$33.60	$268.80	$52.10	$416.80	$34.19	$52.48
4 Laborers	31.60	1011.20	49.00	1568.00		
1 Equip. Oper. (crane)	42.55	340.40	63.95	511.60		
1 Equip. Oper. Oiler	36.80	294.40	55.30	442.40		
2 Cutting Torches		34.00		37.40		
2 Gases		151.20		166.32		
1 Lattice Boom Crane, 90 Ton		1741.00		1915.10	34.40	37.84
56 L.H., Daily Totals		$3841.00		$5057.62	$68.59	$90.31

Crew No.	Bare Costs		Incl. Subs O & P		Cost Per Labor-Hour	
Crew B-22	**Hr.**	**Daily**	**Hr.**	**Daily**	**Bare Costs**	**Incl. O&P**
1 Labor Foreman (out)	$33.60	$268.80	$52.10	$416.80	$36.79	$56.62
1 Skilled Worker	40.85	326.80	63.25	506.00		
1 Laborer	31.60	252.80	49.00	392.00		
.75 Equip. Oper. (crane)	42.55	255.30	63.95	383.70		
.75 S.P. Crane, 4x4, 5 Ton		193.65		213.01	6.46	7.10
30 L.H., Daily Totals		$1297.35		$1911.52	$43.24	$63.72

Crew B-22A	Hr.	Daily	Hr.	Daily	Bare Costs	Incl. O&P
1 Labor Foreman (out)	$33.60	$268.80	$52.10	$416.80	$35.70	$55.01
1 Skilled Worker	40.85	326.80	63.25	506.00		
2 Laborers	31.60	505.60	49.00	784.00		
.75 Equipment Oper. (crane)	42.55	255.30	63.95	383.70		
.75 S.P. Crane, 4x4, 5 Ton		193.65		213.01		
1 Generator, 5 kW		40.20		44.22		
1 Butt Fusion Machine		199.80		219.78	11.41	12.55
38 L.H., Daily Totals		$1790.15		$2567.51	$47.11	$67.57

Crew B-22B	Hr.	Daily	Hr.	Daily	Bare Costs	Incl. O&P
1 Skilled Worker	$40.85	$326.80	$63.25	$506.00	$36.23	$56.13
1 Laborer	31.60	252.80	49.00	392.00		
1 Electro Fusion Machine		146.80		161.48	9.18	10.09
16 L.H., Daily Totals		$726.40		$1059.48	$45.40	$66.22

Crew B-23	Hr.	Daily	Hr.	Daily	Bare Costs	Incl. O&P
1 Labor Foreman (outside)	$33.60	$268.80	$52.10	$416.80	$32.00	$49.62
4 Laborers	31.60	1011.20	49.00	1568.00		
1 Drill Rig, Truck-Mounted		2533.00		2786.30		
1 Flatbed Truck, Gas, 3 Ton		242.60		266.86	69.39	76.33
40 L.H., Daily Totals		$4055.60		$5037.96	$101.39	$125.95

Crew B-23A	Hr.	Daily	Hr.	Daily	Bare Costs	Incl. O&P
1 Labor Foreman (outside)	$33.60	$268.80	$52.10	$416.80	$35.52	$54.42
1 Laborer	31.60	252.80	49.00	392.00		
1 Equip. Operator (medium)	41.35	330.80	62.15	497.20		
1 Drill Rig, Truck-Mounted		2533.00		2786.30		
1 Pickup Truck, 3/4 Ton		116.80		128.48	110.41	121.45
24 L.H., Daily Totals		$3502.20		$4220.78	$145.93	$175.87

Crew B-23B	Hr.	Daily	Hr.	Daily	Bare Costs	Incl. O&P
1 Labor Foreman (outside)	$33.60	$268.80	$52.10	$416.80	$35.52	$54.42
1 Laborer	31.60	252.80	49.00	392.00		
1 Equip. Operator (medium)	41.35	330.80	62.15	497.20		
1 Drill Rig, Truck-Mounted		2533.00		2786.30		
1 Pickup Truck, 3/4 Ton		116.80		128.48		
1 Centr. Water Pump, 6"		309.40		340.34	123.30	135.63
24 L.H., Daily Totals		$3811.60		$4561.12	$158.82	$190.05

Crew B-24	Hr.	Daily	Hr.	Daily	Bare Costs	Incl. O&P
1 Cement Finisher	$38.30	$306.40	$56.05	$448.40	$36.62	$55.67
1 Laborer	31.60	252.80	49.00	392.00		
1 Carpenter	39.95	319.60	61.95	495.60		
24 L.H., Daily Totals		$878.80		$1336.00	$36.62	$55.67

Crew B-25	Hr.	Daily	Hr.	Daily	Bare Costs	Incl. O&P
1 Labor Foreman	$33.60	$268.80	$52.10	$416.80	$34.44	$52.87
7 Laborers	31.60	1769.60	49.00	2744.00		
3 Equip. Oper. (med.)	41.35	992.40	62.15	1491.60		
1 Asphalt Paver, 130 H.P.		1936.00		2129.60		
1 Tandem Roller, 10 Ton		233.00		256.30		
1 Roller, Pneum. Whl, 12 Ton		307.80		338.58	28.15	30.96
88 L.H., Daily Totals		$5507.60		$7376.88	$62.59	$83.83

Crew No.	Bare Costs		Incl. Subs O & P		Cost Per Labor-Hour	
Crew B-25B	**Hr.**	**Daily**	**Hr.**	**Daily**	**Bare Costs**	**Incl. O&P**
1 Labor Foreman	$33.60	$268.80	$52.10	$416.80	$35.02	$53.64
7 Laborers	31.60	1769.60	49.00	2744.00		
4 Equip. Oper. (medium)	41.35	1323.20	62.15	1988.80		
1 Asphalt Paver, 130 H.P.		1936.00		2129.60		
2 Tandem Rollers, 10 Ton		466.00		512.60		
1 Roller, Pneum. Whl, 12 Ton		307.80		338.58	28.23	31.05
96 L.H., Daily Totals		$6071.40		$8130.38	$63.24	$84.69

Crew B-25C	Hr.	Daily	Hr.	Daily	Bare Costs	Incl. O&P
1 Labor Foreman	$33.60	$268.80	$52.10	$416.80	$35.18	$53.90
3 Laborers	31.60	758.40	49.00	1176.00		
2 Equip. Oper. (medium)	41.35	661.60	62.15	994.40		
1 Asphalt Paver, 130 H.P.		1936.00		2129.60		
1 Tandem Roller, 10 Ton		233.00		256.30	45.19	49.71
48 L.H., Daily Totals		$3857.80		$4973.10	$80.37	$103.61

Crew B-26	Hr.	Daily	Hr.	Daily	Bare Costs	Incl. O&P
1 Labor Foreman (outside)	$33.60	$268.80	$52.10	$416.80	$35.34	$54.48
6 Laborers	31.60	1516.80	49.00	2352.00		
2 Equip. Oper. (med.)	41.35	661.60	62.15	994.40		
1 Rodman (reinf.)	44.55	356.40	72.85	582.80		
1 Cement Finisher	38.30	306.40	56.05	448.40		
1 Grader, 30,000 Lbs.		550.20		605.22		
1 Paving Mach. & Equip.		2500.00		2750.00	34.66	38.13
88 L.H., Daily Totals		$6160.20		$8149.62	$70.00	$92.61

Crew B-26A	Hr.	Daily	Hr.	Daily	Bare Costs	Incl. O&P
1 Labor Foreman (outside)	$33.60	$268.80	$52.10	$416.80	$35.34	$54.48
6 Laborers	31.60	1516.80	49.00	2352.00		
2 Equip. Oper. (med.)	41.35	661.60	62.15	994.40		
1 Rodman (reinf.)	44.55	356.40	72.85	582.80		
1 Cement Finisher	38.30	306.40	56.05	448.40		
1 Grader, 30,000 Lbs.		550.20		605.22		
1 Paving Mach. & Equip.		2500.00		2750.00		
1 Concrete Saw		141.00		155.10	36.26	39.89
88 L.H., Daily Totals		$6301.20		$8304.72	$71.60	$94.37

Crew B-26B	Hr.	Daily	Hr.	Daily	Bare Costs	Incl. O&P
1 Labor Foreman (outside)	$33.60	$268.80	$52.10	$416.80	$35.84	$55.12
6 Laborers	31.60	1516.80	49.00	2352.00		
3 Equip. Oper. (med.)	41.35	992.40	62.15	1491.60		
1 Rodman (reinf.)	44.55	356.40	72.85	582.80		
1 Cement Finisher	38.30	306.40	56.05	448.40		
1 Grader, 30,000 Lbs.		550.20		605.22		
1 Paving Mach. & Equip.		2500.00		2750.00		
1 Concrete Pump, 110' Boom		977.00		1074.70	41.95	46.15
96 L.H., Daily Totals		$7468.00		$9721.52	$77.79	$101.27

Crew B-27	Hr.	Daily	Hr.	Daily	Bare Costs	Incl. O&P
1 Labor Foreman (outside)	$33.60	$268.80	$52.10	$416.80	$32.10	$49.77
3 Laborers	31.60	758.40	49.00	1176.00		
1 Berm Machine		271.40		298.54	8.48	9.33
32 L.H., Daily Totals		$1298.60		$1891.34	$40.58	$59.10

Crew B-28	Hr.	Daily	Hr.	Daily	Bare Costs	Incl. O&P
2 Carpenters	$39.95	$639.20	$61.95	$991.20	$37.17	$57.63
1 Laborer	31.60	252.80	49.00	392.00		
24 L.H., Daily Totals		$892.00		$1383.20	$37.17	$57.63

Crew No.	Bare Costs		Incl. Subs O & P		Cost Per Labor-Hour	
Crew B-29	**Hr.**	**Daily**	**Hr.**	**Daily**	**Bare Costs**	**Incl. O&P**
1 Labor Foreman (outside)	$33.60	$268.80	$52.10	$416.80	$34.19	$52.48
4 Laborers	31.60	1011.20	49.00	1568.00		
1 Equip. Oper. (crane)	42.55	340.40	63.95	511.60		
1 Equip. Oper. Oiler	36.80	294.40	55.30	442.40		
1 Gradall, 5/8 C.Y.		958.80		1054.68	17.12	18.83
56 L.H., Daily Totals		$2873.60		$3993.48	$51.31	$71.31
Crew B-30	**Hr.**	**Daily**	**Hr.**	**Daily**	**Bare Costs**	**Incl. O&P**
1 Equip. Oper. (med.)	$41.35	$330.80	$62.15	$497.20	$35.08	$53.48
2 Truck Drivers (heavy)	31.95	511.20	49.15	786.40		
1 Hyd. Excavator, 1.5 C.Y.		865.80		952.38		
2 Dump Trucks, 12 C.Y., 400 H.P.		1066.00		1172.60	80.49	88.54
24 L.H., Daily Totals		$2773.80		$3408.58	$115.58	$142.02
Crew B-31	**Hr.**	**Daily**	**Hr.**	**Daily**	**Bare Costs**	**Incl. O&P**
1 Labor Foreman (outside)	$33.60	$268.80	$52.10	$416.80	$33.67	$52.21
3 Laborers	31.60	758.40	49.00	1176.00		
1 Carpenter	39.95	319.60	61.95	495.60		
1 Air Compressor, 250 cfm		162.40		178.64		
1 Sheeting Driver		5.75		6.33		
2 -50' Air Hoses, 1.5"		12.60		13.86	4.52	4.97
40 L.H., Daily Totals		$1527.55		$2287.22	$38.19	$57.18
Crew B-32	**Hr.**	**Daily**	**Hr.**	**Daily**	**Bare Costs**	**Incl. O&P**
1 Laborer	$31.60	$252.80	$49.00	$392.00	$38.91	$58.86
3 Equip. Oper. (med.)	41.35	992.40	62.15	1491.60		
1 Grader, 30,000 Lbs.		550.20		605.22		
1 Tandem Roller, 10 Ton		233.00		256.30		
1 Dozer, 200 H.P.		1082.00		1190.20	58.29	64.12
32 L.H., Daily Totals		$3110.40		$3935.32	$97.20	$122.98
Crew B-32A	**Hr.**	**Daily**	**Hr.**	**Daily**	**Bare Costs**	**Incl. O&P**
1 Laborer	$31.60	$252.80	$49.00	$392.00	$38.10	$57.77
2 Equip. Oper. (medium)	41.35	661.60	62.15	994.40		
1 Grader, 30,000 Lbs.		550.20		605.22		
1 Roller, Vibratory, 25 Ton		594.60		654.06	47.70	52.47
24 L.H., Daily Totals		$2059.20		$2645.68	$85.80	$110.24
Crew B-32B	**Hr.**	**Daily**	**Hr.**	**Daily**	**Bare Costs**	**Incl. O&P**
1 Laborer	$31.60	$252.80	$49.00	$392.00	$38.10	$57.77
2 Equip. Oper. (medium)	41.35	661.60	62.15	994.40		
1 Dozer, 200 H.P.		1082.00		1190.20		
1 Roller, Vibratory, 25 Ton		594.60		654.06	69.86	76.84
24 L.H., Daily Totals		$2591.00		$3230.66	$107.96	$134.61
Crew B-32C	**Hr.**	**Daily**	**Hr.**	**Daily**	**Bare Costs**	**Incl. O&P**
1 Labor Foreman	$33.60	$268.80	$52.10	$416.80	$36.81	$56.09
2 Laborers	31.60	505.60	49.00	784.00		
3 Equip. Oper. (medium)	41.35	992.40	62.15	1491.60		
1 Grader, 30,000 Lbs.		550.20		605.22		
1 Tandem Roller, 10 Ton		233.00		256.30		
1 Dozer, 200 H.P.		1082.00		1190.20	38.86	42.74
48 L.H., Daily Totals		$3632.00		$4744.12	$75.67	$98.84
Crew B-33A	**Hr.**	**Daily**	**Hr.**	**Daily**	**Bare Costs**	**Incl. O&P**
1 Equip. Oper. (med.)	$41.35	$330.80	$62.15	$497.20	$38.56	$58.39
.5 Laborer	31.60	126.40	49.00	196.00		
.25 Equip. Oper. (med.)	41.35	82.70	62.15	124.30		
1 Scraper, Towed, 7 C.Y.		102.20		112.42		
1.25 Dozers, 300 H.P.		1778.75		1956.63	134.35	147.79
14 L.H., Daily Totals		$2420.85		$2886.55	$172.92	$206.18

Crew No.	Bare Costs		Incl. Subs O & P		Cost Per Labor-Hour	
Crew B-33B	**Hr.**	**Daily**	**Hr.**	**Daily**	**Bare Costs**	**Incl. O&P**
1 Equip. Oper. (med.)	$41.35	$330.80	$62.15	$497.20	$38.56	$58.39
.5 Laborer	31.60	126.40	49.00	196.00		
.25 Equip. Oper. (med.)	41.35	82.70	62.15	124.30		
1 Scraper, Towed, 10 C.Y.		153.60		168.96		
1.25 Dozers, 300 H.P.		1778.75		1956.63	138.03	151.83
14 L.H., Daily Totals		$2472.25		$2943.09	$176.59	$210.22
Crew B-33C	**Hr.**	**Daily**	**Hr.**	**Daily**	**Bare Costs**	**Incl. O&P**
1 Equip. Oper. (med.)	$41.35	$330.80	$62.15	$497.20	$38.56	$58.39
.5 Laborer	31.60	126.40	49.00	196.00		
.25 Equip. Oper. (med.)	41.35	82.70	62.15	124.30		
1 Scraper, Towed, 15 C.Y.		168.80		185.68		
1.25 Dozers, 300 H.P.		1778.75		1956.63	139.11	153.02
14 L.H., Daily Totals		$2487.45		$2959.80	$177.68	$211.41
Crew B-33D	**Hr.**	**Daily**	**Hr.**	**Daily**	**Bare Costs**	**Incl. O&P**
1 Equip. Oper. (med.)	$41.35	$330.80	$62.15	$497.20	$38.56	$58.39
.5 Laborer	31.60	126.40	49.00	196.00		
.25 Equip. Oper. (med.)	41.35	82.70	62.15	124.30		
1 S.P. Scraper, 14 C.Y.		1509.00		1659.90		
.25 Dozer, 300 H.P.		355.75		391.32	133.20	146.52
14 L.H., Daily Totals		$2404.65		$2868.72	$171.76	$204.91
Crew B-33E	**Hr.**	**Daily**	**Hr.**	**Daily**	**Bare Costs**	**Incl. O&P**
1 Equip. Oper. (med.)	$41.35	$330.80	$62.15	$497.20	$38.56	$58.39
.5 Laborer	31.60	126.40	49.00	196.00		
.25 Equip. Oper. (med.)	41.35	82.70	62.15	124.30		
1 S.P. Scraper, 21 C.Y.		1924.00		2116.40		
.25 Dozer, 300 H.P.		355.75		391.32	162.84	179.12
14 L.H., Daily Totals		$2819.65		$3325.22	$201.40	$237.52
Crew B-33F	**Hr.**	**Daily**	**Hr.**	**Daily**	**Bare Costs**	**Incl. O&P**
1 Equip. Oper. (med.)	$41.35	$330.80	$62.15	$497.20	$38.56	$58.39
.5 Laborer	31.60	126.40	49.00	196.00		
.25 Equip. Oper. (med.)	41.35	82.70	62.15	124.30		
1 Elev. Scraper, 11 C.Y.		1095.00		1204.50		
.25 Dozer, 300 H.P.		355.75		391.32	103.63	113.99
14 L.H., Daily Totals		$1990.65		$2413.32	$142.19	$172.38
Crew B-33G	**Hr.**	**Daily**	**Hr.**	**Daily**	**Bare Costs**	**Incl. O&P**
1 Equip. Oper. (med.)	$41.35	$330.80	$62.15	$497.20	$38.56	$58.39
.5 Laborer	31.60	126.40	49.00	196.00		
.25 Equip. Oper. (med.)	41.35	82.70	62.15	124.30		
1 Elev. Scraper, 22 C.Y.		2127.00		2339.70		
.25 Dozer, 300 H.P.		355.75		391.32	177.34	195.07
14 L.H., Daily Totals		$3022.65		$3548.53	$215.90	$253.47
Crew B-33H	**Hr.**	**Daily**	**Hr.**	**Daily**	**Bare Costs**	**Incl. O&P**
.5 Laborer	$31.60	$126.40	$49.00	$196.00	$38.56	$58.39
1 Equipment Operator (med.)	41.35	330.80	62.15	497.20		
.25 Equipment Operator (med.)	41.35	82.70	62.15	124.30		
1 S.P. Scraper, 44 C.Y.		3452.00		3797.20		
.25 Dozer, 410 H.P.		462.25		508.48	279.59	307.55
14 L.H., Daily Totals		$4454.15		$5123.18	$318.15	$365.94
Crew B-33J	**Hr.**	**Daily**	**Hr.**	**Daily**	**Bare Costs**	**Incl. O&P**
1 Equipment Operator (med.)	$41.35	$330.80	$62.15	$497.20	$41.35	$62.15
1 S.P. Scraper, 14 C.Y.		1509.00		1659.90	188.63	207.49
8 L.H., Daily Totals		$1839.80		$2157.10	$229.97	$269.64

Crew B-33K	Hr.	Daily	Hr.	Daily	Bare Costs	Incl. O&P
1 Equipment Operator (med.)	$41.35	$330.80	$62.15	$497.20	$38.56	$58.39
.25 Equipment Operator (med.)	41.35	82.70	62.15	124.30		
.5 Laborer	31.60	126.40	49.00	196.00		
1 S.P. Scraper, 31 C.Y.		2878.00		3165.80		
.25 Dozer, 410 H.P.		462.25		508.48	238.59	262.45
14 L.H., Daily Totals		$3880.15		$4491.77	$277.15	$320.84

Crew B-34A	Hr.	Daily	Hr.	Daily	Bare Costs	Incl. O&P
1 Truck Driver (heavy)	$31.95	$255.60	$49.15	$393.20	$31.95	$49.15
1 Dump Truck, 8 C.Y., 220 H.P.		326.40		359.04	40.80	44.88
8 L.H., Daily Totals		$582.00		$752.24	$72.75	$94.03

Crew B-34B	Hr.	Daily	Hr.	Daily	Bare Costs	Incl. O&P
1 Truck Driver (heavy)	$31.95	$255.60	$49.15	$393.20	$31.95	$49.15
1 Dump Truck, 12 C.Y., 400 H.P.		533.00		586.30	66.63	73.29
8 L.H., Daily Totals		$788.60		$979.50	$98.58	$122.44

Crew B-34C	Hr.	Daily	Hr.	Daily	Bare Costs	Incl. O&P
1 Truck Driver (heavy)	$31.95	$255.60	$49.15	$393.20	$31.95	$49.15
1 Truck Tractor, 6x4, 380 H.P.		478.80		526.68		
1 Dump Trailer, 16.5 C.Y.		115.60		127.16	74.30	81.73
8 L.H., Daily Totals		$850.00		$1047.04	$106.25	$130.88

Crew B-34D	Hr.	Daily	Hr.	Daily	Bare Costs	Incl. O&P
1 Truck Driver (heavy)	$31.95	$255.60	$49.15	$393.20	$31.95	$49.15
1 Truck Tractor, 6x4, 380 H.P.		478.80		526.68		
1 Dump Trailer, 20 C.Y.		130.20		143.22	76.13	83.74
8 L.H., Daily Totals		$864.60		$1063.10	$108.08	$132.89

Crew B-34E	Hr.	Daily	Hr.	Daily	Bare Costs	Incl. O&P
1 Truck Driver (heavy)	$31.95	$255.60	$49.15	$393.20	$31.95	$49.15
1 Dump Truck, Off Hwy., 25 Ton		1280.00		1408.00	160.00	176.00
8 L.H., Daily Totals		$1535.60		$1801.20	$191.95	$225.15

Crew B-34F	Hr.	Daily	Hr.	Daily	Bare Costs	Incl. O&P
1 Truck Driver (heavy)	$31.95	$255.60	$49.15	$393.20	$31.95	$49.15
1 Dump Truck, Off Hwy., 35 Ton		1144.00		1258.40	143.00	157.30
8 L.H., Daily Totals		$1399.60		$1651.60	$174.95	$206.45

Crew B-34G	Hr.	Daily	Hr.	Daily	Bare Costs	Incl. O&P
1 Truck Driver (heavy)	$31.95	$255.60	$49.15	$393.20	$31.95	$49.15
1 Dump Truck, Off Hwy., 50 Ton		1582.00		1740.20	197.75	217.53
8 L.H., Daily Totals		$1837.60		$2133.40	$229.70	$266.68

Crew B-34H	Hr.	Daily	Hr.	Daily	Bare Costs	Incl. O&P
1 Truck Driver (heavy)	$31.95	$255.60	$49.15	$393.20	$31.95	$49.15
1 Dump Truck, Off Hwy., 65 Ton		1594.00		1753.40	199.25	219.18
8 L.H., Daily Totals		$1849.60		$2146.60	$231.20	$268.32

Crew B-34J	Hr.	Daily	Hr.	Daily	Bare Costs	Incl. O&P
1 Truck Driver (heavy)	$31.95	$255.60	$49.15	$393.20	$31.95	$49.15
1 Dump Truck, Off Hwy., 100 Ton		2047.00		2251.70	255.88	281.46
8 L.H., Daily Totals		$2302.60		$2644.90	$287.82	$330.61

Crew B-34K	Hr.	Daily	Hr.	Daily	Bare Costs	Incl. O&P
1 Truck Driver (heavy)	$31.95	$255.60	$49.15	$393.20	$31.95	$49.15
1 Truck Tractor, 6x4, 450 H.P.		581.20		639.32		
1 Lowbed Trailer, 75 Ton		203.00		223.30	98.03	107.83
8 L.H., Daily Totals		$1039.80		$1255.82	$129.97	$156.98

Crew B-34L	Hr.	Daily	Hr.	Daily	Bare Costs	Incl. O&P
1 Equip. Oper. (light)	$39.05	$312.40	$58.70	$469.60	$39.05	$58.70
1 Flatbed Truck, Gas, 1.5 Ton		194.40		213.84	24.30	26.73
8 L.H., Daily Totals		$506.80		$683.44	$63.35	$85.43

Crew B-34N	Hr.	Daily	Hr.	Daily	Bare Costs	Incl. O&P
1 Truck Driver (heavy)	$31.95	$255.60	$49.15	$393.20	$31.95	$49.15
1 Dump Truck, 8 C.Y., 220 H.P.		326.40		359.04		
1 Flatbed Trailer, 40 Ton		139.60		153.56	58.25	64.08
8 L.H., Daily Totals		$721.60		$905.80	$90.20	$113.22

Crew B-34P	Hr.	Daily	Hr.	Daily	Bare Costs	Incl. O&P
1 Pipe Fitter	$49.35	$394.80	$74.05	$592.40	$40.55	$61.28
1 Truck Driver (light)	30.95	247.60	47.65	381.20		
1 Equip. Oper. (medium)	41.35	330.80	62.15	497.20		
1 Flatbed Truck, Gas, 3 Ton		242.60		266.86		
1 Backhoe Loader, 48 H.P.		293.80		323.18	22.35	24.59
24 L.H., Daily Totals		$1509.60		$2060.84	$62.90	$85.87

Crew B-34Q	Hr.	Daily	Hr.	Daily	Bare Costs	Incl. O&P
1 Pipe Fitter	$49.35	$394.80	$74.05	$592.40	$40.95	$61.88
1 Truck Driver (light)	30.95	247.60	47.65	381.20		
1 Eqip. Oper. (crane)	42.55	340.40	63.95	511.60		
1 Flatbed Trailer, 25 Ton		102.00		112.20		
1 Dump Truck, 8 C.Y., 220 H.P.		326.40		359.04		
1 Hyd. Crane, 25 Ton		789.40		868.34	50.74	55.82
24 L.H., Daily Totals		$2200.60		$2824.78	$91.69	$117.70

Crew B-34R	Hr.	Daily	Hr.	Daily	Bare Costs	Incl. O&P
1 Pipe Fitter	$49.35	$394.80	$74.05	$592.40	$40.95	$61.88
1 Truck Driver (light)	30.95	247.60	47.65	381.20		
1 Eqip. Oper. (crane)	42.55	340.40	63.95	511.60		
1 Flatbed Trailer, 25 Ton		102.00		112.20		
1 Dump Truck, 8 C.Y., 220 H.P.		326.40		359.04		
1 Hyd. Crane, 25 Ton		789.40		868.34		
1 Hyd. Excavator, 1 C.Y.		669.20		736.12	78.63	86.49
24 L.H., Daily Totals		$2869.80		$3560.90	$119.58	$148.37

Crew B-34S	Hr.	Daily	Hr.	Daily	Bare Costs	Incl. O&P
2 Pipe Fitters	$49.35	$789.60	$74.05	$1184.80	$43.30	$65.30
1 Truck Driver (heavy)	31.95	255.60	49.15	393.20		
1 Eqip. Oper. (crane)	42.55	340.40	63.95	511.60		
1 Flatbed Trailer, 40 Ton		139.60		153.56		
1 Truck Tractor, 6x4, 380 H.P.		478.80		526.68		
1 Hyd. Crane, 80 Ton		1296.00		1425.60		
1 Hyd. Excavator, 2 C.Y.		1182.00		1300.20	96.76	106.44
32 L.H., Daily Totals		$4482.00		$5495.64	$140.06	$171.74

Crew B-34T	Hr.	Daily	Hr.	Daily	Bare Costs	Incl. O&P
2 Pipe Fitters	$49.35	$789.60	$74.05	$1184.80	$43.30	$65.30
1 Truck Driver (heavy)	31.95	255.60	49.15	393.20		
1 Eqip. Oper. (crane)	42.55	340.40	63.95	511.60		
1 Flatbed Trailer, 40 Ton		139.60		153.56		
1 Truck Tractor, 6x4, 380 H.P.		478.80		526.68		
1 Hyd. Crane, 80 Ton		1296.00		1425.60	59.83	65.81
32 L.H., Daily Totals		$3300.00		$4195.44	$103.13	$131.11

Crew No.	Bare Costs		Incl. Subs O & P		Cost Per Labor-Hour	
Crew B-35	**Hr.**	**Daily**	**Hr.**	**Daily**	**Bare Costs**	**Incl. O&P**
1 Laborer Foreman (out)	$33.60	$268.80	$52.10	$416.80	$39.02	$59.46
1 Skilled Worker	40.85	326.80	63.25	506.00		
1 Welder (plumber)	48.75	390.00	73.15	585.20		
1 Laborer	31.60	252.80	49.00	392.00		
1 Equip. Oper. (crane)	42.55	340.40	63.95	511.60		
1 Equip. Oper. Oiler	36.80	294.40	55.30	442.40		
1 Welder, electric, 300 amp		55.70		61.27		
1 Hyd. Excavator, .75 C.Y.		598.60		658.46	13.63	14.99
48 L.H., Daily Totals		$2527.50		$3573.73	$52.66	$74.45

Crew B-35A	Hr.	Daily	Hr.	Daily	Bare Costs	Incl. O&P
1 Laborer Foreman (out)	$33.60	$268.80	$52.10	$416.80	$37.96	$57.96
2 Laborers	31.60	505.60	49.00	784.00		
1 Skilled Worker	40.85	326.80	63.25	506.00		
1 Welder (plumber)	48.75	390.00	73.15	585.20		
1 Equip. Oper. (crane)	42.55	340.40	63.95	511.60		
1 Equip. Oper. Oiler	36.80	294.40	55.30	442.40		
1 Welder, gas engine, 300 amp		134.20		147.62		
1 Crawler Crane, 75 Ton		1496.00		1645.60	29.11	32.02
56 L.H., Daily Totals		$3756.20		$5039.22	$67.08	$89.99

Crew B-36	Hr.	Daily	Hr.	Daily	Bare Costs	Incl. O&P
1 Labor Foreman (outside)	$33.60	$268.80	$52.10	$416.80	$35.90	$54.88
2 Laborers	31.60	505.60	49.00	784.00		
2 Equip. Oper. (med.)	41.35	661.60	62.15	994.40		
1 Dozer, 200 H.P.		1082.00		1190.20		
1 Aggregate Spreader		35.00		38.50		
1 Tandem Roller, 10 Ton		233.00		256.30	33.75	37.13
40 L.H., Daily Totals		$2786.00		$3680.20	$69.65	$92.00

Crew B-36A	Hr.	Daily	Hr.	Daily	Bare Costs	Incl. O&P
1 Labor Foreman (outside)	$33.60	$268.80	$52.10	$416.80	$37.46	$56.96
2 Laborers	31.60	505.60	49.00	784.00		
4 Equip. Oper. (med.)	41.35	1323.20	62.15	1988.80		
1 Dozer, 200 H.P.		1082.00		1190.20		
1 Aggregate Spreader		35.00		38.50		
1 Tandem Roller, 10 Ton		233.00		256.30		
1 Roller, Pneum. Whl, 12 Ton		307.80		338.58	29.60	32.56
56 L.H., Daily Totals		$3755.40		$5013.18	$67.06	$89.52

Crew B-36B	Hr.	Daily	Hr.	Daily	Bare Costs	Incl. O&P
1 Labor Foreman (outside)	$33.60	$268.80	$52.10	$416.80	$36.77	$55.98
2 Laborers	31.60	505.60	49.00	784.00		
4 Equip. Oper. (medium)	41.35	1323.20	62.15	1988.80		
1 Truck Driver, Heavy	31.95	255.60	49.15	393.20		
1 Grader, 30,000 Lbs.		550.20		605.22		
1 F.E. Loader, crl, 1.5 C.Y.		463.00		509.30		
1 Dozer, 300 H.P.		1423.00		1565.30		
1 Roller, Vibratory, 25 Ton		594.60		654.06		
1 Truck Tractor, 6x4, 450 H.P.		581.20		639.32		
1 Water Tanker, 5000 Gal.		136.00		149.60	58.56	64.42
64 L.H., Daily Totals		$6101.20		$7705.60	$95.33	$120.40

Crew B-36C	Hr.	Daily	Hr.	Daily	Bare Costs	Incl. O&P
1 Labor Foreman (outside)	$33.60	$268.80	$52.10	$416.80	$37.92	$57.54
3 Equip. Oper. (medium)	41.35	992.40	62.15	1491.60		
1 Truck Driver, Heavy	31.95	255.60	49.15	393.20		
1 Grader, 30,000 Lbs.		550.20		605.22		
1 Dozer, 300 H.P.		1423.00		1565.30		
1 Roller, Vibratory, 25 Ton		594.60		654.06		
1 Truck Tractor, 6x4, 450 H.P.		581.20		639.32		
1 Water Tanker, 5000 Gal.		136.00		149.60	82.13	90.34
40 L.H., Daily Totals		$4801.80		$5915.10	$120.05	$147.88

Crew B-37	Hr.	Daily	Hr.	Daily	Bare Costs	Incl. O&P
1 Labor Foreman (outside)	$33.60	$268.80	$52.10	$416.80	$33.17	$51.13
4 Laborers	31.60	1011.20	49.00	1568.00		
1 Equip. Oper. (light)	39.05	312.40	58.70	469.60		
1 Tandem Roller, 5 Ton		137.80		151.58	2.87	3.16
48 L.H., Daily Totals		$1730.20		$2605.98	$36.05	$54.29

Crew B-38	Hr.	Daily	Hr.	Daily	Bare Costs	Incl. O&P
1 Labor Foreman (outside)	$33.60	$268.80	$52.10	$416.80	$35.44	$54.19
2 Laborers	31.60	505.60	49.00	784.00		
1 Equip. Oper. (light)	39.05	312.40	58.70	469.60		
1 Equip. Oper. (medium)	41.35	330.80	62.15	497.20		
1 Backhoe Loader, 48 H.P.		293.80		323.18		
1 Hyd.Hammer, (1200 lb.)		153.00		168.30		
1 F.E. Loader, W.M., 4 C.Y.		541.00		595.10		
1 Pavt. Rem. Bucket		56.80		62.48	26.11	28.73
40 L.H., Daily Totals		$2462.20		$3316.66	$61.56	$82.92

Crew B-39	Hr.	Daily	Hr.	Daily	Bare Costs	Incl. O&P
1 Labor Foreman (outside)	$33.60	$268.80	$52.10	$416.80	$33.17	$51.13
4 Laborers	31.60	1011.20	49.00	1568.00		
1 Equip. Oper. (light)	39.05	312.40	58.70	469.60		
1 Air Compressor, 250 cfm		162.40		178.64		
2 Breakers, Pavement, 60 lb.		16.40		18.04		
2 -50' Air Hoses, 1.5"		12,60		13.86	3.99	4.39
48 L.H., Daily Totals		$1783.80		$2664.94	$37.16	$55.52

Crew B-40	Hr.	Daily	Hr.	Daily	Bare Costs	Incl. O&P
1 Pile Driver Foreman (out)	$40.50	$324.00	$65.75	$526.00	$39.55	$62.37
4 Pile Drivers	38.50	1232.00	62.50	2000.00		
2 Equip. Oper. (crane)	42.55	680.80	63.95	1023.20		
1 Equip. Oper. Oiler	36.80	294.40	55.30	442.40		
1 Crawler Crane, 40 Ton		1055.00		1160.50		
1 Vibratory Hammer & Gen.		2257.00		2482.70	51.75	56.92
64 L.H., Daily Totals		$5843.20		$7634.80	$91.30	$119.29

Crew B-40B	Hr.	Daily	Hr.	Daily	Bare Costs	Incl. O&P
1 Laborer Foreman	$33.60	$268.80	$52.10	$416.80	$34.63	$53.06
3 Laborers	31.60	758.40	49.00	1176.00		
1 Equip. Oper. (crane)	42.55	340.40	63.95	511.60		
1 Equip. Oper. Oiler	36.80	294.40	55.30	442.40		
1 Lattice Boom Crane, 40 Ton		1327.00		1459.70	27.65	30.41
48 L.H., Daily Totals		$2989.00		$4006.50	$62.27	$83.47

Crew B-41	Hr.	Daily	Hr.	Daily	Bare Costs	Incl. O&P
1 Labor Foreman (outside)	$33.60	$268.80	$52.10	$416.80	$32.70	$50.53
4 Laborers	31.60	1011.20	49.00	1568.00		
.25 Equip. Oper. (crane)	42.55	85.10	63.95	127.90		
.25 Equip. Oper. Oiler	36.80	73.60	55.30	110.60		
.25 Crawler Crane, 40 Ton		263.75		290.13	5.99	6.59
44 L.H., Daily Totals		$1702.45		$2513.43	$38.69	$57.12

Crew No.	Bare Costs		Incl. Subs O & P		Cost Per Labor-Hour	
Crew B-42	**Hr.**	**Daily**	**Hr.**	**Daily**	**Bare Costs**	**Incl. O&P**
1 Labor Foreman (outside)	$33.60	$268.80	$52.10	$416.80	$35.51	$55.88
4 Laborers	31.60	1011.20	49.00	1568.00		
1 Equip. Oper. (crane)	42.55	340.40	63.95	511.60		
1 Equip. Oper. Oiler	36.80	294.40	55.30	442.40		
1 Welder	44.70	357.60	79.65	637.20		
1 Hyd. Crane, 25 Ton		789.40		868.34		
1 Welder, gas engine, 300 amp		134.20		147.62		
1 Horz. Boring Csg. Mch.		444.20		488.62	21.37	23.51
64 L.H., Daily Totals		$3640.20		$5080.58	$56.88	$79.38

Crew B-43	Hr.	Daily	Hr.	Daily	Bare Costs	Incl. O&P
1 Labor Foreman (outside)	$33.60	$268.80	$52.10	$416.80	$34.63	$53.06
3 Laborers	31.60	758.40	49.00	1176.00		
1 Equip. Oper. (crane)	42.55	340.40	63.95	511.60		
1 Equip. Oper. Oiler	36.80	294.40	55.30	442.40		
1 Drill Rig, Truck-Mounted		2533.00		2786.30	52.77	58.05
48 L.H., Daily Totals		$4195.00		$5333.10	$87.40	$111.11

Crew B-44	Hr.	Daily	Hr.	Daily	Bare Costs	Incl. O&P
1 Pile Driver Foreman	$40.50	$324.00	$65.75	$526.00	$38.90	$61.58
4 Pile Drivers	38.50	1232.00	62.50	2000.00		
2 Equip. Oper. (crane)	42.55	680.80	63.95	1023.20		
1 Laborer	31.60	252.80	49.00	392.00		
1 Crawler Crane, 40 Ton		1055.00		1160.50		
1 Lead, 60' high		69.60		76.56		
1 Hammer, diesel, 15K Ft.-Lbs.		578.60		636.46	26.61	29.27
64 L.H., Daily Totals		$4192.80		$5814.72	$65.51	$90.86

Crew B-45	Hr.	Daily	Hr.	Daily	Bare Costs	Incl. O&P
1 Equip. Oper. (med.)	$41.35	$330.80	$62.15	$497.20	$36.65	$55.65
1 Truck Driver (heavy)	31.95	255.60	49.15	393.20		
1 Dist. Tanker, 3000 Gallon		304.20		334.62		
1 Truck Tractor, 6x4, 380 H.P.		478.80		526.68	48.94	53.83
16 L.H., Daily Totals		$1369.40		$1751.70	$85.59	$109.48

Crew B-46	Hr.	Daily	Hr.	Daily	Bare Costs	Incl. O&P
1 Pile Driver Foreman	$40.50	$324.00	$65.75	$526.00	$35.38	$56.29
2 Pile Drivers	38.50	616.00	62.50	1000.00		
3 Laborers	31.60	758.40	49.00	1176.00		
1 Chain Saw, Gas, 36" Long		37.80		41.58	.79	.87
48 L.H., Daily Totals		$1736.20		$2743.58	$36.17	$57.16

Crew B-47	Hr.	Daily	Hr.	Daily	Bare Costs	Incl. O&P
1 Blast Foreman	$33.60	$268.80	$52.10	$416.80	$34.75	$53.27
1 Driller	31.60	252.80	49.00	392.00		
1 Equip. Oper. (light)	39.05	312.40	58.70	469.60		
1 Air Track Drill, 4"		850.60		935.66		
1 Air Compressor, 600 cfm		434.40		477.84		
2 -50' Air Hoses, 3"		32.40		35.64	54.89	60.38
24 L.H., Daily Totals		$2151.40		$2727.54	$89.64	$113.65

Crew B-47A	Hr.	Daily	Hr.	Daily	Bare Costs	Incl. O&P
1 Drilling Foreman	$33.60	$268.80	$52.10	$416.80	$37.65	$57.12
1 Equip. Oper. (heavy)	42.55	340.40	63.95	511.60		
1 Oiler	36.80	294.40	55.30	442.40		
1 Air Track Drill, 5"		974.00		1071.40	40.58	44.64
24 L.H., Daily Totals		$1877.60		$2442.20	$78.23	$101.76

Crew B-47C	Hr.	Daily	Hr.	Daily	Bare Costs	Incl. O&P
1 Laborer	$31.60	$252.80	$49.00	$392.00	$35.33	$53.85
1 Equip. Oper. (light)	39.05	312.40	58.70	469.60		
1 Air Compressor, 750 cfm		439.60		483.56		
2 -50' Air Hoses, 3"		32.40		35.64		
1 Air Track Drill, 4"		850.60		935.66	82.66	90.93
16 L.H., Daily Totals		$1887.80		$2316.46	$117.99	$144.78

Crew B-47E	Hr.	Daily	Hr.	Daily	Bare Costs	Incl. O&P
1 Laborer Foreman	$33.60	$268.80	$52.10	$416.80	$32.10	$49.77
3 Laborers	31.60	758.40	49.00	1176.00		
1 Flatbed Truck, Gas, 3 Ton		242.60		266.86	7.58	8.34
32 L.H., Daily Totals		$1269.80		$1859.66	$39.68	$58.11

Crew B-47G	Hr.	Daily	Hr.	Daily	Bare Costs	Incl. O&P
1 Laborer Foreman	$33.60	$268.80	$52.10	$416.80	$33.96	$52.20
2 Laborers	31.60	505.60	49.00	784.00		
1 Equip. Oper. (light)	39.05	312.40	58.70	469.60		
1 Air Track Drill, 4"		850.60		935.66		
1 Air Compressor, 600 cfm		434.40		477.84		
2 -50' Air Hoses, 3"		32.40		35.64		
1 Grout Pump		166.40		183.04	46.37	51.01
32 L.H., Daily Totals		$2570.60		$3302.58	$80.33	$103.21

Crew B-48	Hr.	Daily	Hr.	Daily	Bare Costs	Incl. O&P
1 Labor Foreman (outside)	$33.60	$268.80	$52.10	$416.80	$35.26	$53.86
3 Laborers	31.60	758.40	49.00	1176.00		
1 Equip. Oper. (crane)	42.55	340.40	63.95	511.60		
1 Equip. Oper. Oiler	36.80	294.40	55.30	442.40		
1 Equip. Oper. (light)	39.05	312.40	58.70	469.60		
1 Centr. Water Pump, 6"		309.40		340.34		
1 -20' Suction Hose, 6"		11.90		13.09		
1 -50' Discharge Hose, 6"		6.30		6.93		
1 Drill Rig, Truck-Mounted		2533.00		2786.30	51.08	56.19
56 L.H., Daily Totals		$4835.00		$6163.06	$86.34	$110.05

Crew B-49	Hr.	Daily	Hr.	Daily	Bare Costs	Incl. O&P
1 Labor Foreman (outside)	$33.60	$268.80	$52.10	$416.80	$36.65	$56.48
3 Laborers	31.60	758.40	49.00	1176.00		
2 Equip. Oper. (crane)	42.55	680.80	63.95	1023.20		
2 Equip. Oper. Oilers	36.80	588.80	55.30	884.80		
1 Equip. Oper. (light)	39.05	312.40	58.70	469.60		
2 Pile Drivers	38.50	616.00	62.50	1000.00		
1 Hyd. Crane, 25 Ton		789.40		868.34		
1 Centr. Water Pump, 6"		309.40		340.34		
1 -20' Suction Hose, 6"		11.90		13.09		
1 -50' Discharge Hose, 6"		6.30		6.93		
1 Drill Rig, Truck-Mounted		2533.00		2786.30	41.48	45.63
88 L.H., Daily Totals		$6875.20		$8985.40	$78.13	$102.11

Crew B-50	Hr.	Daily	Hr.	Daily	Bare Costs	Incl. O&P
2 Pile Driver Foremen	$40.50	$648.00	$65.75	$1052.00	$37.76	$59.76
6 Pile Drivers	38.50	1848.00	62.50	3000.00		
2 Equip. Oper. (crane)	42.55	680.80	63.95	1023.20		
1 Equip. Oper. Oiler	36.80	294.40	55.30	442.40		
3 Laborers	31.60	758.40	49.00	1176.00		
1 Crawler Crane, 40 Ton		1055.00		1160.50		
1 Lead, 60' high		69.60		76.56		
1 Hammer, diesel, 15K Ft.-Lbs.		578.60		636.46		
1 Air Compressor, 600 cfm		434.40		477.84		
2 -50' Air Hoses, 3"		32.40		35.64		
1 Chain Saw, Gas, 36" Long		37.80		41.58	19.71	21.68
112 L.H., Daily Totals		$6437.40		$9122.18	$57.48	$81.45

Crew No.	Bare Costs		Incl. Subs O & P		Cost Per Labor-Hour	
Crew B-51	**Hr.**	**Daily**	**Hr.**	**Daily**	**Bare Costs**	**Incl. O&P**
1 Labor Foreman (outside)	$33.60	$268.80	$52.10	$416.80	$31.82	$49.29
4 Laborers	31.60	1011.20	49.00	1568.00		
1 Truck Driver (light)	30.95	247.60	47.65	381.20		
1 Flatbed Truck, Gas, 1.5 Ton		194.40		213.84	4.05	4.46
48 L.H., Daily Totals		$1722.00		$2579.84	$35.88	$53.75
Crew B-52	**Hr.**	**Daily**	**Hr.**	**Daily**	**Bare Costs**	**Incl. O&P**
1 Carpenter Foreman	$41.95	$335.60	$65.05	$520.40	$36.85	$56.79
1 Carpenter	39.95	319.60	61.95	495.60		
3 Laborers	31.60	758.40	49.00	1176.00		
1 Cement Finisher	38.30	306.40	56.05	448.40		
.5 Rodman (reinf.)	44.55	178.20	72.85	291.40		
.5 Equip. Oper. (med.)	41.35	165.40	62.15	248.60		
.5 Crawler Loader, 3 C.Y.		495.00		544.50	8.84	9.72
56 L.H., Daily Totals		$2558.60		$3724.90	$45.69	$66.52
Crew B-53	**Hr.**	**Daily**	**Hr.**	**Daily**	**Bare Costs**	**Incl. O&P**
1 Equip. Oper. (light)	$39.05	$312.40	$58.70	$469.60	$39.05	$58.70
1 Trencher, Chain, 12 H.P.		59.60		65.56	7.45	8.20
8 L.H., Daily Totals		$372.00		$535.16	$46.50	$66.89
Crew B-54	**Hr.**	**Daily**	**Hr.**	**Daily**	**Bare Costs**	**Incl. O&P**
1 Equip. Oper. (light)	$39.05	$312.40	$58.70	$469.60	$39.05	$58.70
1 Trencher, Chain, 40 H.P.		301.00		331.10	37.63	41.39
8 L.H., Daily Totals		$613.40		$800.70	$76.67	$100.09
Crew B-54A	**Hr.**	**Daily**	**Hr.**	**Daily**	**Bare Costs**	**Incl. O&P**
.17 Labor Foreman (outside)	$33.60	$45.70	$52.10	$70.86	$40.22	$60.69
1 Equipment Operator (med.)	41.35	330.80	62.15	497.20		
1 Wheel Trencher, 67 H.P.		1009.00		1109.90	107.80	118.58
9.36 L.H., Daily Totals		$1385.50		$1677.96	$148.02	$179.27
Crew B-54B	**Hr.**	**Daily**	**Hr.**	**Daily**	**Bare Costs**	**Incl. O&P**
.25 Labor Foreman (outside)	$33.60	$67.20	$52.10	$104.20	$39.80	$60.14
1 Equipment Operator (med.)	41.35	330.80	62.15	497.20		
1 Wheel Trencher, 150 H.P.		1753.00		1928.30	175.30	192.83
10 L.H., Daily Totals		$2151.00		$2529.70	$215.10	$252.97
Crew B-54C	**Hr.**	**Daily**	**Hr.**	**Daily**	**Bare Costs**	**Incl. O&P**
1 Laborer	$31.60	$252.80	$49.00	$392.00	$36.48	$55.58
1 Equipment Operator (med.)	41.35	330.80	62.15	497.20		
1 Wheel Trencher, 67 H.P.		1009.00		1109.90	63.06	69.37
16 L.H., Daily Totals		$1592.60		$1999.10	$99.54	$124.94
Crew B-55	**Hr.**	**Daily**	**Hr.**	**Daily**	**Bare Costs**	**Incl. O&P**
2 Laborers	$31.60	$505.60	$49.00	$784.00	$31.38	$48.55
1 Truck Driver (light)	30.95	247.60	47.65	381.20		
1 Truck-mounted earth auger		706.60		777.26		
1 Flatbed Truck, Gas, 3 Ton		242.60		266.86	39.55	43.51
24 L.H., Daily Totals		$1702.40		$2209.32	$70.93	$92.06
Crew B-56	**Hr.**	**Daily**	**Hr.**	**Daily**	**Bare Costs**	**Incl. O&P**
1 Laborer	$31.60	$252.80	$49.00	$392.00	$35.33	$53.85
1 Equip. Oper. (light)	39.05	312.40	58.70	469.60		
1 Air Track Drill, 4"		850.60		935.66		
1 Air Compressor, 600 cfm		434.40		477.84		
1 -50' Air Hose, 3"		16.20		17.82	81.33	89.46
16 L.H., Daily Totals		$1866.40		$2292.92	$116.65	$143.31

Crew No.	Bare Costs		Incl. Subs O & P		Cost Per Labor-Hour	
Crew B-57	**Hr.**	**Daily**	**Hr.**	**Daily**	**Bare Costs**	**Incl. O&P**
1 Labor Foreman (outside)	$33.60	$268.80	$52.10	$416.80	$35.87	$54.67
2 Laborers	31.60	505.60	49.00	784.00		
1 Equip. Oper. (crane)	42.55	340.40	63.95	511.60		
1 Equip. Oper. (light)	39.05	312.40	58.70	469.60		
1 Equip. Oper. Oiler	36.80	294.40	55.30	442.40		
1 Crawler Crane, 25 Ton		1052.00		1157.20		
1 Clamshell Bucket, 1 C.Y.		46.20		50.82		
1 Centr. Water Pump, 6"		309.40		340.34		
1 -20' Suction Hose, 6"		11.90		13.09		
20 -50' Discharge Hoses, 6"		126.00		138.60	32.20	35.42
48 L.H., Daily Totals		$3267.10		$4324.45	$68.06	$90.09
Crew B-58	**Hr.**	**Daily**	**Hr.**	**Daily**	**Bare Costs**	**Incl. O&P**
2 Laborers	$31.60	$505.60	$49.00	$784.00	$34.08	$52.23
1 Equip. Oper. (light)	39.05	312.40	58.70	469.60		
1 Backhoe Loader, 48 H.P.		293.80		323.18		
1 Small Helicopter, w/pilot		2587.00		2845.70	120.03	132.04
24 L.H., Daily Totals		$3698.80		$4422.48	$154.12	$184.27
Crew B-59	**Hr.**	**Daily**	**Hr.**	**Daily**	**Bare Costs**	**Incl. O&P**
1 Truck Driver (heavy)	$31.95	$255.60	$49.15	$393.20	$31.95	$49.15
1 Truck Tractor, 220 H.P.		288.40		317.24		
1 Water Tanker, 5000 Gal.		136.00		149.60	53.05	58.35
8 L.H., Daily Totals		$680.00		$860.04	$85.00	$107.51
Crew B-59A	**Hr.**	**Daily**	**Hr.**	**Daily**	**Bare Costs**	**Incl. O&P**
2 Laborers	$31.60	$505.60	$49.00	$784.00	$31.72	$49.05
1 Truck Driver (heavy)	31.95	255.60	49.15	393.20		
1 Water Tanker, 5000 Gal.		136.00		149.60		
1 Truck Tractor, 220 H.P.		288.40		317.24	17.68	19.45
24 L.H., Daily Totals		$1185.60		$1644.04	$49.40	$68.50
Crew B-60	**Hr.**	**Daily**	**Hr.**	**Daily**	**Bare Costs**	**Incl. O&P**
1 Labor Foreman (outside)	$33.60	$268.80	$52.10	$416.80	$36.32	$55.25
2 Laborers	31.60	505.60	49.00	784.00		
1 Equip. Oper. (crane)	42.55	340.40	63.95	511.60		
2 Equip. Oper. (light)	39.05	624.80	58.70	939.20		
1 Equip. Oper. Oiler	36.80	294.40	55.30	442.40		
1 Crawler Crane, 40 Ton		1055.00		1160.50		
1 Lead, 60' high		69.60		76.56		
1 Hammer, diesel, 15K Ft.-Lbs.		578.60		636.46		
1 Backhoe Loader, 48 H.P.		293.80		323.18	35.66	39.23
56 L.H., Daily Totals		$4031.00		$5290.70	$71.98	$94.48
Crew B-61	**Hr.**	**Daily**	**Hr.**	**Daily**	**Bare Costs**	**Incl. O&P**
1 Labor Foreman (outside)	$33.60	$268.80	$52.10	$416.80	$33.49	$51.56
3 Laborers	31.60	758.40	49.00	1176.00		
1 Equip. Oper. (light)	39.05	312.40	58.70	469.60		
1 Cement Mixer, 2 C.Y.		187.80		206.58		
1 Air Compressor, 160 cfm		138.80		152.68	8.16	8.98
40 L.H., Daily Totals		$1666.20		$2421.66	$41.66	$60.54
Crew B-62	**Hr.**	**Daily**	**Hr.**	**Daily**	**Bare Costs**	**Incl. O&P**
2 Laborers	$31.60	$505.60	$49.00	$784.00	$34.08	$52.23
1 Equip. Oper. (light)	39.05	312.40	58.70	469.60		
1 Loader, Skid Steer, 30 H.P., gas		149.20		164.12	6.22	6.84
24 L.H., Daily Totals		$967.20		$1417.72	$40.30	$59.07

Crew No.	Bare Costs		Incl. Subs O & P		Cost Per Labor-Hour	
Crew B-63	**Hr.**	**Daily**	**Hr.**	**Daily**	**Bare Costs**	**Incl. O&P**
4 Laborers	$31.60	$1011.20	$49.00	$1568.00	$33.09	$50.94
1 Equip. Oper. (light)	39.05	312.40	58.70	469.60		
1 Loader, Skid Steer, 30 H.P., gas		149.20		164.12	3.73	4.10
40 L.H., Daily Totals		$1472.80		$2201.72	$36.82	$55.04
Crew B-64	**Hr.**	**Daily**	**Hr.**	**Daily**	**Bare Costs**	**Incl. O&P**
1 Laborer	$31.60	$252.80	$49.00	$392.00	$31.27	$48.33
1 Truck Driver (light)	30.95	247.60	47.65	381.20		
1 Power Mulcher (small)		136.60		150.26		
1 Flatbed Truck, Gas, 1.5 Ton		194.40		213.84	20.69	22.76
16 L.H., Daily Totals		$831.40		$1137.30	$51.96	$71.08
Crew B-65	**Hr.**	**Daily**	**Hr.**	**Daily**	**Bare Costs**	**Incl. O&P**
1 Laborer	$31.60	$252.80	$49.00	$392.00	$31.27	$48.33
1 Truck Driver (light)	30.95	247.60	47.65	381.20		
1 Power Mulcher (large)		271.80		298.98		
1 Flatbed Truck, Gas, 1.5 Ton		194.40		213.84	29.14	32.05
16 L.H., Daily Totals		$966.60		$1286.02	$60.41	$80.38
Crew B-66	**Hr.**	**Daily**	**Hr.**	**Daily**	**Bare Costs**	**Incl. O&P**
1 Equip. Oper. (light)	$39.05	$312.40	$58.70	$469.60	$39.05	$58.70
1 Loader-Backhoe		231.20		254.32	28.90	31.79
8 L.H., Daily Totals		$543.60		$723.92	$67.95	$90.49
Crew B-67	**Hr.**	**Daily**	**Hr.**	**Daily**	**Bare Costs**	**Incl. O&P**
1 Millwright	$41.60	$332.80	$61.05	$488.40	$40.33	$59.88
1 Equip. Oper. (light)	39.05	312.40	58.70	469.60		
1 Forklift, R/T, 4,000 Lb.		281.20		309.32	17.57	19.33
16 L.H., Daily Totals		$926.40		$1267.32	$57.90	$79.21
Crew B-68	**Hr.**	**Daily**	**Hr.**	**Daily**	**Bare Costs**	**Incl. O&P**
2 Millwrights	$41.60	$665.60	$61.05	$976.80	$40.75	$60.27
1 Equip. Oper. (light)	39.05	312.40	58.70	469.60		
1 Forklift, R/T, 4,000 Lb.		281.20		309.32	11.72	12.89
24 L.H., Daily Totals		$1259.20		$1755.72	$52.47	$73.16
Crew B-69	**Hr.**	**Daily**	**Hr.**	**Daily**	**Bare Costs**	**Incl. O&P**
1 Labor Foreman (outside)	$33.60	$268.80	$52.10	$416.80	$34.63	$53.06
3 Laborers	31.60	758.40	49.00	1176.00		
1 Equip Oper. (crane)	42.55	340.40	63.95	511.60		
1 Equip Oper. Oiler	36.80	294.40	55.30	442.40		
1 Hyd. Crane, 80 Ton		1296.00		1425.60	27.00	29.70
48 L.H., Daily Totals		$2958.00		$3972.40	$61.63	$82.76
Crew B-69A	**Hr.**	**Daily**	**Hr.**	**Daily**	**Bare Costs**	**Incl. O&P**
1 Labor Foreman	$33.60	$268.80	$52.10	$416.80	$34.67	$52.88
3 Laborers	31.60	758.40	49.00	1176.00		
1 Equip. Oper. (medium)	41.35	330.80	62.15	497.20		
1 Concrete Finisher	38.30	306.40	56.05	448.40		
1 Curb/Gutter Paver, 2-Track		754.40		829.84	15.72	17.29
48 L.H., Daily Totals		$2418.80		$3368.24	$50.39	$70.17
Crew B-69B	**Hr.**	**Daily**	**Hr.**	**Daily**	**Bare Costs**	**Incl. O&P**
1 Labor Foreman	$33.60	$268.80	$52.10	$416.80	$34.67	$52.88
3 Laborers	31.60	758.40	49.00	1176.00		
1 Equip. Oper. (medium)	41.35	330.80	62.15	497.20		
1 Cement Finisher	38.30	306.40	56.05	448.40		
1 Curb/Gutter Paver, 4-Track		855.80		941.38	17.83	19.61
48 L.H., Daily Totals		$2520.20		$3479.78	$52.50	$72.50

Crew No.	Bare Costs		Incl. Subs O & P		Cost Per Labor-Hour	
Crew B-70	**Hr.**	**Daily**	**Hr.**	**Daily**	**Bare Costs**	**Incl. O&P**
1 Labor Foreman (outside)	$33.60	$268.80	$52.10	$416.80	$36.06	$55.08
3 Laborers	31.60	758.40	49.00	1176.00		
3 Equip. Oper. (med.)	41.35	992.40	62.15	1491.60		
1 Grader, 30,000 Lbs.		550.20		605.22		
1 Ripper, beam & 1 shank		79.60		87.56		
1 Road Sweeper, S.P., 8' wide		601.60		661.76		
1 F.E. Loader, W.M., 1.5 C.Y.		328.20		361.02	27.85	30.64
56 L.H., Daily Totals		$3579.20		$4799.96	$63.91	$85.71
Crew B-70A	**Hr.**	**Daily**	**Hr.**	**Daily**	**Bare Costs**	**Incl. O&P**
1 Laborer	$31.60	$252.80	$49.00	$392.00	$39.40	$59.52
4 Equip. Oper. (med.)	41.35	1323.20	62.15	1988.80		
1 Grader, 40,000 Lbs.		961.00		1057.10		
1 F.E. Loader, W.M.,2.5 C.Y.		395.80		435.38		
1 Dozer, 80 H.P.		399.20		439.12		
1 Roller, Pneum. Whl, 12 Ton		307.80		338.58	51.59	56.75
40 L.H., Daily Totals		$3639.80		$4650.98	$91.00	$116.27
Crew B-71	**Hr.**	**Daily**	**Hr.**	**Daily**	**Bare Costs**	**Incl. O&P**
1 Labor Foreman (outside)	$33.60	$268.80	$52.10	$416.80	$36.06	$55.08
3 Laborers	31.60	758.40	49.00	1176.00		
3 Equip. Oper. (med.)	41.35	992.40	62.15	1491.60		
1 Pvmt. Profiler, 750 H.P.		5503.00		6053.30		
1 Road Sweeper, S.P., 8' wide		601.60		661.76		
1 F.E. Loader, W.M., 1.5 C.Y.		328.20		361.02	114.87	126.36
56 L.H., Daily Totals		$8452.40		$10160.48	$150.94	$181.44
Crew B-72	**Hr.**	**Daily**	**Hr.**	**Daily**	**Bare Costs**	**Incl. O&P**
1 Labor Foreman (outside)	$33.60	$268.80	$52.10	$416.80	$36.73	$55.96
3 Laborers	31.60	758.40	49.00	1176.00		
4 Equip. Oper. (med.)	41.35	1323.20	62.15	1988.80		
1 Pvmt. Profiler, 750 H.P.		5503.00		6053.30		
1 Hammermill, 250 H.P.		1850.00		2035.00		
1 Windrow Loader		1197.00		1316.70		
1 Mix Paver 165 H.P.		1996.00		2195.60		
1 Roller, Pneum. Whl, 12 Ton		307.80		338.58	169.59	186.55
64 L.H., Daily Totals		$13204.20		$15520.78	$206.32	$242.51
Crew B-73	**Hr.**	**Daily**	**Hr.**	**Daily**	**Bare Costs**	**Incl. O&P**
1 Labor Foreman (outside)	$33.60	$268.80	$52.10	$416.80	$37.94	$57.61
2 Laborers	31.60	505.60	49.00	784.00		
5 Equip. Oper. (med.)	41.35	1654.00	62.15	2486.00		
1 Road Mixer, 310 H.P.		1896.00		2085.60		
1 Tandem Roller, 10 Ton		233.00		256.30		
1 Hammermill, 250 H.P.		1850.00		2035.00		
1 Grader, 30,000 Lbs.		550.20		605.22		
.5 F.E. Loader, W.M., 1.5 C.Y.		164.10		180.51		
.5 Truck Tractor, 220 H.P.		144.20		158.62		
.5 Water Tanker, 5000 Gal.		68.00		74.80	76.65	84.31
64 L.H., Daily Totals		$7333.90		$9082.85	$114.59	$141.92

Crews

Crew No.	Bare Costs		Incl. Subs O & P		Cost Per Labor-Hour	
Crew B-74	**Hr.**	**Daily**	**Hr.**	**Daily**	**Bare Costs**	**Incl. O&P**
1 Labor Foreman (outside)	$33.60	$268.80	$52.10	$416.80	$36.81	$56.00
1 Laborer	31.60	252.80	49.00	392.00		
4 Equip. Oper. (med.)	41.35	1323.20	62.15	1988.80		
2 Truck Drivers (heavy)	31.95	511.20	49.15	786.40		
1 Grader, 30,000 Lbs.		550.20		605.22		
1 Ripper, beam & 1 shank		79.60		87.56		
2 Stabilizers, 310 H.P.		2688.00		2956.80		
1 Flatbed Truck, Gas, 3 Ton		242.60		266.86		
1 Chem. Spreader, Towed		48.00		52.80		
1 Roller, Vibratory, 25 Ton		594.60		654.06		
1 Water Tanker, 5000 Gal.		136.00		149.60		
1 Truck Tractor, 220 H.P.		288.40		317.24	72.30	79.53
64 L.H., Daily Totals		$6983.40		$8674.14	$109.12	$135.53

Crew No.	Bare Costs		Incl. Subs O & P		Cost Per Labor-Hour	
Crew B-75	**Hr.**	**Daily**	**Hr.**	**Daily**	**Bare Costs**	**Incl. O&P**
1 Labor Foreman (outside)	$33.60	$268.80	$52.10	$416.80	$37.51	$56.98
1 Laborer	31.60	252.80	49.00	392.00		
4 Equip. Oper. (med.)	41.35	1323.20	62.15	1988.80		
1 Truck Driver (heavy)	31.95	255.60	49.15	393.20		
1 Grader, 30,000 Lbs.		550.20		605.22		
1 Ripper, beam & 1 shank		79.60		87.56		
2 Stabilizers, 310 H.P.		2688.00		2956.80		
1 Dist. Tanker, 3000 Gallon		304.20		334.62		
1 Truck Tractor, 6x4, 380 H.P.		478.80		526.68		
1 Roller, Vibratory, 25 Ton		594.60		654.06	83.85	92.23
56 L.H., Daily Totals		$6795.80		$8355.74	$121.35	$149.21

Crew No.	Bare Costs		Incl. Subs O & P		Cost Per Labor-Hour	
Crew B-76	**Hr.**	**Daily**	**Hr.**	**Daily**	**Bare Costs**	**Incl. O&P**
1 Dock Builder Foreman	$40.50	$324.00	$65.75	$526.00	$39.43	$62.38
5 Dock Builders	38.50	1540.00	62.50	2500.00		
2 Equip. Oper. (crane)	42.55	680.80	63.95	1023.20		
1 Equip. Oper. Oiler	36.80	294.40	55.30	442.40		
1 Crawler Crane, 50 Ton		1110.00		1221.00		
1 Barge, 400 Ton		746.40		821.04		
1 Hammer, diesel, 15K Ft.-Lbs.		578.60		636.46		
1 Lead, 60' high		69.60		76.56		
1 Air Compressor, 600 cfm		434.40		477.84		
2 -50' Air Hoses, 3"		32.40		35.64	41.27	45.40
72 L.H., Daily Totals		$5810.60		$7760.14	$80.70	$107.78

Crew No.	Bare Costs		Incl. Subs O & P		Cost Per Labor-Hour	
Crew B-76A	**Hr.**	**Daily**	**Hr.**	**Daily**	**Bare Costs**	**Incl. O&P**
1 Laborer Foreman	$33.60	$268.80	$52.10	$416.80	$33.87	$52.04
5 Laborers	31.60	1264.00	49.00	1960.00		
1 Equip. Oper. (crane)	42.55	340.40	63.95	511.60		
1 Equip. Oper. Oiler	36.80	294.40	55.30	442.40		
1 Crawler Crane, 50 Ton		1110.00		1221.00		
1 Barge, 400 Ton		746.40		821.04	29.01	31.91
64 L.H., Daily Totals		$4024.00		$5372.84	$62.88	$83.95

Crew No.	Bare Costs		Incl. Subs O & P		Cost Per Labor-Hour	
Crew B-77	**Hr.**	**Daily**	**Hr.**	**Daily**	**Bare Costs**	**Incl. O&P**
1 Labor Foreman	$33.60	$268.80	$52.10	$416.80	$31.87	$49.35
3 Laborers	31.60	758.40	49.00	1176.00		
1 Truck Driver (light)	30.95	247.60	47.65	381.20		
1 Crack Cleaner, 25 H.P.		54.80		60.28		
1 Crack Filler, Trailer Mtd.		194.20		213.62		
1 Flatbed Truck, Gas, 3 Ton		242.60		266.86	12.29	13.52
40 L.H., Daily Totals		$1766.40		$2514.76	$44.16	$62.87

Crew No.	Bare Costs		Incl. Subs O & P		Cost Per Labor-Hour	
Crew B-78	**Hr.**	**Daily**	**Hr.**	**Daily**	**Bare Costs**	**Incl. O&P**
1 Labor Foreman	$33.60	$268.80	$52.10	$416.80	$31.82	$49.29
4 Laborers	31.60	1011.20	49.00	1568.00		
1 Truck Driver (light)	30.95	247.60	47.65	381.20		
1 Paint Striper, S.P.		147.00		161.70		
1 Flatbed Truck, Gas, 3 Ton		242.60		266.86		
1 Pickup Truck, 3/4 Ton		116.80		128.48	10.55	11.61
48 L.H., Daily Totals		$2034.00		$2923.04	$42.38	$60.90

Crew No.	Bare Costs		Incl. Subs O & P		Cost Per Labor-Hour	
Crew B-78A	**Hr.**	**Daily**	**Hr.**	**Daily**	**Bare Costs**	**Incl. O&P**
1 Equip. Oper. (light)	$39.05	$312.40	$58.70	$469.60	$39.05	$58.70
1 Line Rem. (metal balls) 115 H.P.		828.60		911.46	103.58	113.93
8 L.H., Daily Totals		$1141.00		$1381.06	$142.63	$172.63

Crew No.	Bare Costs		Incl. Subs O & P		Cost Per Labor-Hour	
Crew B-78B	**Hr.**	**Daily**	**Hr.**	**Daily**	**Bare Costs**	**Incl. O&P**
2 Laborers	$31.60	$505.60	$49.00	$784.00	$32.43	$50.08
.25 Equip. Oper. (light)	39.05	78.10	58.70	117.40		
1 Pickup Truck, 3/4 Ton		116.80		128.48		
1 Line Rem., 11 H.P., walk behind		54.00		59.40		
.25 Road Sweeper, S.P., 8' wide		150.40		165.44	17.84	19.63
18 L.H., Daily Totals		$904.90		$1254.72	$50.27	$69.71

Crew No.	Bare Costs		Incl. Subs O & P		Cost Per Labor-Hour	
Crew B-79	**Hr.**	**Daily**	**Hr.**	**Daily**	**Bare Costs**	**Incl. O&P**
1 Labor Foreman	$33.60	$268.80	$52.10	$416.80	$31.87	$49.35
3 Laborers	31.60	758.40	49.00	1176.00		
1 Truck Driver (light)	30.95	247.60	47.65	381.20		
1 Thermo. Striper, T.M.		703.20		773.52		
1 Flatbed Truck, Gas, 3 Ton		242.60		266.86		
2 Pickup Trucks, 3/4 Ton		233.60		256.96	29.48	32.43
40 L.H., Daily Totals		$2454.20		$3271.34	$61.35	$81.78

Crew No.	Bare Costs		Incl. Subs O & P		Cost Per Labor-Hour	
Crew B-79A	**Hr.**	**Daily**	**Hr.**	**Daily**	**Bare Costs**	**Incl. O&P**
1.5 Equip. Oper. (light)	$39.05	$468.60	$58.70	$704.40	$39.05	$58.70
.5 Line Remov. (grinder) 115 H.P.		440.20		484.22		
1 Line Rem. (metal balls) 115 H.P.		828.60		911.46	105.73	116.31
12 L.H., Daily Totals		$1737.40		$2100.08	$144.78	$175.01

Crew No.	Bare Costs		Incl. Subs O & P		Cost Per Labor-Hour	
Crew B-80	**Hr.**	**Daily**	**Hr.**	**Daily**	**Bare Costs**	**Incl. O&P**
1 Labor Foreman	$33.60	$268.80	$52.10	$416.80	$33.80	$51.86
1 Laborer	31.60	252.80	49.00	392.00		
1 Truck Driver (light)	30.95	247.60	47.65	381.20		
1 Equip. Oper. (light)	39.05	312.40	58.70	469.60		
1 Flatbed Truck, Gas, 3 Ton		242.60		266.86		
1 Earth Auger, Truck-Mtd.		412.00		453.20	20.46	22.50
32 L.H., Daily Totals		$1736.20		$2379.66	$54.26	$74.36

Crew No.	Bare Costs		Incl. Subs O & P		Cost Per Labor-Hour	
Crew B-80A	**Hr.**	**Daily**	**Hr.**	**Daily**	**Bare Costs**	**Incl. O&P**
3 Laborers	$31.60	$758.40	$49.00	$1176.00	$31.60	$49.00
1 Flatbed Truck, Gas, 3 Ton		242.60		266.86	10.11	11.12
24 L.H., Daily Totals		$1001.00		$1442.86	$41.71	$60.12

Crew No.	Bare Costs		Incl. Subs O & P		Cost Per Labor-Hour	
Crew B-80B	**Hr.**	**Daily**	**Hr.**	**Daily**	**Bare Costs**	**Incl. O&P**
3 Laborers	$31.60	$758.40	$49.00	$1176.00	$33.46	$51.42
1 Equip. Oper. (light)	39.05	312.40	58.70	469.60		
1 Crane, Flatbed Mounted, 3 Ton		225.00		247.50	7.03	7.73
32 L.H., Daily Totals		$1295.80		$1893.10	$40.49	$59.16

Crews

Crew No.	Bare Costs		Incl. Subs O & P		Cost Per Labor-Hour	

Crew B-80C	Hr.	Daily	Hr.	Daily	Bare Costs	Incl. O&P
2 Laborers	$31.60	$505.60	$49.00	$784.00	$31.38	$48.55
1 Truck Driver (light)	30.95	247.60	47.65	381.20		
1 Flatbed Truck, Gas, 1.5 Ton		194.40		213.84		
1 Manual fence post auger, gas		7.40		8.14	8.41	9.25
24 L.H., Daily Totals		$955.00		$1387.18	$39.79	$57.80

Crew B-81	Hr.	Daily	Hr.	Daily	Bare Costs	Incl. O&P
1 Laborer	$31.60	$252.80	$49.00	$392.00	$34.97	$53.43
1 Equip. Oper. (med.)	41.35	330.80	62.15	497.20		
1 Truck Driver (heavy)	31.95	255.60	49.15	393.20		
1 Hydromulcher, T.M.		291.00		320.10		
1 Truck Tractor, 220 H.P.		288.40		317.24	24.14	26.56
24 L.H., Daily Totals		$1418.60		$1919.74	$59.11	$79.99

Crew B-82	Hr.	Daily	Hr.	Daily	Bare Costs	Incl. O&P
1 Laborer	$31.60	$252.80	$49.00	$392.00	$35.33	$53.85
1 Equip. Oper. (light)	39.05	312.40	58.70	469.60		
1 Horiz. Borer, 6 H.P.		77.60		85.36	4.85	5.34
16 L.H., Daily Totals		$642.80		$946.96	$40.17	$59.19

Crew B-82A	Hr.	Daily	Hr.	Daily	Bare Costs	Incl. O&P
1 Laborer	$31.60	$252.80	$49.00	$392.00	$35.33	$53.85
1 Equip. Oper. (light)	39.05	312.40	58.70	469.60		
1 Flatbed Truck, Gas, 3 Ton		242.60		266.86		
1 Flatbed Trailer, 25 Ton		102.00		112.20		
1 Horiz. Dir. Drill, 20k lb. thrust		628.20		691.02	60.80	66.88
16 L.H., Daily Totals		$1538.00		$1931.68	$96.13	$120.73

Crew B-82B	Hr.	Daily	Hr.	Daily	Bare Costs	Incl. O&P
2 Laborers	$31.60	$505.60	$49.00	$784.00	$34.08	$52.23
1 Equip. Oper. (light)	39.05	312.40	58.70	469.60		
1 Flatbed Truck, Gas, 3 Ton		242.60		266.86		
1 Flatbed Trailer, 25 Ton		102.00		112.20		
1 Horiz. Dir. Drill, 30k lb. thrust		896.60		986.26	51.72	56.89
24 L.H., Daily Totals		$2059.20		$2618.92	$85.80	$109.12

Crew B-82C	Hr.	Daily	Hr.	Daily	Bare Costs	Incl. O&P
2 Laborers	$31.60	$505.60	$49.00	$784.00	$34.08	$52.23
1 Equip. Oper. (light)	39.05	312.40	58.70	469.60		
1 Flatbed Truck, Gas, 3 Ton		242.60		266.86		
1 Flatbed Trailer, 25 Ton		102.00		112.20		
1 Horiz. Dir. Drill, 50k lb. thrust		1183.00		1301.30	63.65	70.02
24 L.H., Daily Totals		$2345.60		$2933.96	$97.73	$122.25

Crew B-82D	Hr.	Daily	Hr.	Daily	Bare Costs	Incl. O&P
1 Equip. Oper. (light)	$39.05	$312.40	$58.70	$469.60	$39.05	$58.70
1 Mud Trailer for HDD, 1500 gallon		283.20		311.52	35.40	38.94
8 L.H., Daily Totals		$595.60		$781.12	$74.45	$97.64

Crew B-83	Hr.	Daily	Hr.	Daily	Bare Costs	Incl. O&P
1 Tugboat Captain	$41.35	$330.80	$62.15	$497.20	$36.48	$55.58
1 Tugboat Hand	31.60	252.80	49.00	392.00		
1 Tugboat, 250 H.P.		663.40		729.74	41.46	45.61
16 L.H., Daily Totals		$1247.00		$1618.94	$77.94	$101.18

Crew B-84	Hr.	Daily	Hr.	Daily	Bare Costs	Incl. O&P
1 Equip. Oper. (med.)	$41.35	$330.80	$62.15	$497.20	$41.35	$62.15
1 Rotary Mower/Tractor		273.40		300.74	34.17	37.59
8 L.H., Daily Totals		$604.20		$797.94	$75.53	$99.74

Crew B-85	Hr.	Daily	Hr.	Daily	Bare Costs	Incl. O&P
3 Laborers	$31.60	$758.40	$49.00	$1176.00	$33.62	$51.66
1 Equip. Oper. (med.)	41.35	330.80	62.15	497.20		
1 Truck Driver (heavy)	31.95	255.60	49.15	393.20		
1 Aerial Lift Truck, 80'		640.40		704.44		
1 Brush Chipper, 12", 130 H.P.		225.60		248.16		
1 Pruning Saw, Rotary		6.50		7.15	21.81	23.99
40 L.H., Daily Totals		$2217.30		$3026.15	$55.43	$75.65

Crew B-86	Hr.	Daily	Hr.	Daily	Bare Costs	Incl. O&P
1 Equip. Oper. (med.)	$41.35	$330.80	$62.15	$497.20	$41.35	$62.15
1 Stump Chipper, S.P.		201.10		221.21	25.14	27.65
8 L.H., Daily Totals		$531.90		$718.41	$66.49	$89.80

Crew B-86A	Hr.	Daily	Hr.	Daily	Bare Costs	Incl. O&P
1 Equip. Oper. (medium)	$41.35	$330.80	$62.15	$497.20	$41.35	$62.15
1 Grader, 30,000 Lbs.		550.20		605.22	68.78	75.65
8 L.H., Daily Totals		$881.00		$1102.42	$110.13	$137.80

Crew B-86B	Hr.	Daily	Hr.	Daily	Bare Costs	Incl. O&P
1 Equip. Oper. (medium)	$41.35	$330.80	$62.15	$497.20	$41.35	$62.15
1 Dozer, 200 H.P.		1082.00		1190.20	135.25	148.78
8 L.H., Daily Totals		$1412.80		$1687.40	$176.60	$210.93

Crew B-87	Hr.	Daily	Hr.	Daily	Bare Costs	Incl. O&P
1 Laborer	$31.60	$252.80	$49.00	$392.00	$39.40	$59.52
4 Equip. Oper. (med.)	41.35	1323.20	62.15	1988.80		
2 Feller Bunchers, 100 H.P.		1122.80		1235.08		
1 Log Chipper, 22" Tree		588.40		647.24		
1 Dozer, 105 H.P.		600.60		660.66		
1 Chain Saw, Gas, 36" Long		37.80		41.58	58.74	64.61
40 L.H., Daily Totals		$3925.60		$4965.36	$98.14	$124.13

Crew B-88	Hr.	Daily	Hr.	Daily	Bare Costs	Incl. O&P
1 Laborer	$31.60	$252.80	$49.00	$392.00	$39.96	$60.27
6 Equip. Oper. (med.)	41.35	1984.80	62.15	2983.20		
2 Feller Bunchers, 100 H.P.		1122.80		1235.08		
1 Log Chipper, 22" Tree		588.40		647.24		
2 Log Skidders, 50 H.P.		1795.20		1974.72		
1 Dozer, 105 H.P.		600.60		660.66		
1 Chain Saw, Gas, 36" Long		37.80		41.58	74.01	81.42
56 L.H., Daily Totals		$6382.40		$7934.48	$113.97	$141.69

Crew B-89	Hr.	Daily	Hr.	Daily	Bare Costs	Incl. O&P
1 Equip. Oper. (light)	$39.05	$312.40	$58.70	$469.60	$35.00	$53.17
1 Truck Driver (light)	30.95	247.60	47.65	381.20		
1 Flatbed Truck, Gas, 3 Ton		242.60		266.86		
1 Concrete Saw		141.00		155.10		
1 Water Tank, 65 Gal.		18.45		20.30	25.13	27.64
16 L.H., Daily Totals		$962.05		$1293.06	$60.13	$80.82

Crew B-89A	Hr.	Daily	Hr.	Daily	Bare Costs	Incl. O&P
1 Skilled Worker	$40.85	$326.80	$63.25	$506.00	$36.23	$56.13
1 Laborer	31.60	252.80	49.00	392.00		
1 Core Drill (large)		110.20		121.22	6.89	7.58
16 L.H., Daily Totals		$689.80		$1019.22	$43.11	$63.70

Crew No.	Bare Costs		Incl. Subs O & P		Cost Per Labor-Hour	
Crew B-89B	**Hr.**	**Daily**	**Hr.**	**Daily**	**Bare Costs**	**Incl. O&P**
1 Equip. Oper. (light)	$39.05	$312.40	$58.70	$469.60	$35.00	$53.17
1 Truck Driver, Light	30.95	247.60	47.65	381.20		
1 Wall Saw, Hydraulic, 10 H.P.		96.00		105.60		
1 Generator, Diesel, 100 kW		327.20		359.92		
1 Water Tank, 65 Gal.		18.45		20.30		
1 Flatbed Truck, Gas, 3 Ton		242.60		266.86	42.77	47.04
16 L.H., Daily Totals		$1244.25		$1603.47	$77.77	$100.22

Crew B-90	**Hr.**	**Daily**	**Hr.**	**Daily**	**Bare Costs**	**Incl. O&P**
1 Labor Foreman (outside)	$33.60	$268.80	$52.10	$416.80	$33.80	$51.85
3 Laborers	31.60	758.40	49.00	1176.00		
2 Equip. Oper. (light)	39.05	624.80	58.70	939.20		
2 Truck Drivers (heavy)	31.95	511.20	49.15	786.40		
1 Road Mixer, 310 H.P.		1896.00		2085.60		
1 Dist. Truck, 2000 Gal.		277.20		304.92	33.96	37.35
64 L.H., Daily Totals		$4336.40		$5708.92	$67.76	$89.20

Crew B-90A	**Hr.**	**Daily**	**Hr.**	**Daily**	**Bare Costs**	**Incl. O&P**
1 Labor Foreman	$33.60	$268.80	$52.10	$416.80	$37.46	$56.96
2 Laborers	31.60	505.60	49.00	784.00		
4 Equip. Oper. (medium)	41.35	1323.20	62.15	1988.80		
2 Graders, 30,000 Lbs.		1100.40		1210.44		
1 Tandem Roller, 10 Ton		233.00		256.30		
1 Roller, Pneum. Whl, 12 Ton		307.80		338.58	29.31	32.24
56 L.H., Daily Totals		$3738.80		$4994.92	$66.76	$89.19

Crew B-90B	**Hr.**	**Daily**	**Hr.**	**Daily**	**Bare Costs**	**Incl. O&P**
1 Labor Foreman	$33.60	$268.80	$52.10	$416.80	$36.81	$56.09
2 Laborers	31.60	505.60	49.00	784.00		
3 Equip. Oper. (medium)	41.35	992.40	62.15	1491.60		
1 Tandem Roller, 10 Ton		233.00		256.30		
1 Roller, Pneum. Whl, 12 Ton		307.80		338.58		
1 Road Mixer, 310 H.P.		1896.00		2085.60	50.77	55.84
48 L.H., Daily Totals		$4203.60		$5372.88	$87.58	$111.94

Crew B-91	**Hr.**	**Daily**	**Hr.**	**Daily**	**Bare Costs**	**Incl. O&P**
1 Labor Foreman (outside)	$33.60	$268.80	$52.10	$416.80	$36.77	$55.98
2 Laborers	31.60	505.60	49.00	784.00		
4 Equip. Oper. (med.)	41.35	1323.20	62.15	1988.80		
1 Truck Driver (heavy)	31.95	255.60	49.15	393.20		
1 Dist. Tanker, 3000 Gallon		304.20		334.62		
1 Truck Tractor, 6x4, 380 H.P.		478.80		526.68		
1 Aggreg. Spreader, S.P.		855.20		940.72		
1 Roller, Pneum. Whl, 12 Ton		307.80		338.58		
1 Tandem Roller, 10 Ton		233.00		256.30	34.05	37.45
64 L.H., Daily Totals		$4532.20		$5979.70	$70.82	$93.43

Crew B-92	**Hr.**	**Daily**	**Hr.**	**Daily**	**Bare Costs**	**Incl. O&P**
1 Labor Foreman (outside)	$33.60	$268.80	$52.10	$416.80	$32.10	$49.77
3 Laborers	31.60	758.40	49.00	1176.00		
1 Crack Cleaner, 25 H.P.		54.80		60.28		
1 Air Compressor, 60 cfm		122.60		134.86		
1 Tar Kettle, T.M.		83.15		91.47		
1 Flatbed Truck, Gas, 3 Ton		242.60		266.86	15.72	17.30
32 L.H., Daily Totals		$1530.35		$2146.26	$47.82	$67.07

Crew B-93	**Hr.**	**Daily**	**Hr.**	**Daily**	**Bare Costs**	**Incl. O&P**
1 Equip. Oper. (med.)	$41.35	$330.80	$62.15	$497.20	$41.35	$62.15
1 Feller Buncher, 100 H.P.		561.40		617.54	70.17	77.19
8 L.H., Daily Totals		$892.20		$1114.74	$111.53	$139.34

Crew No.	Bare Costs		Incl. Subs O & P		Cost Per Labor-Hour	
Crew B-94A	**Hr.**	**Daily**	**Hr.**	**Daily**	**Bare Costs**	**Incl. O&P**
1 Laborer	$31.60	$252.80	$49.00	$392.00	$31.60	$49.00
1 Diaphragm Water Pump, 2″		63.20		69.52		
1 -20' Suction Hose, 2″		1.95		2.15		
2 -50' Discharge Hoses, 2″		2.20		2.42	8.42	9.26
8 L.H., Daily Totals		$320.15		$466.08	$40.02	$58.26

Crew B-94B	**Hr.**	**Daily**	**Hr.**	**Daily**	**Bare Costs**	**Incl. O&P**
1 Laborer	$31.60	$252.80	$49.00	$392.00	$31.60	$49.00
1 Diaphragm Water Pump, 4″		86.20		94.82		
1 -20' Suction Hose, 4″		3.45		3.79		
2 -50' Discharge Hoses, 4″		4.70		5.17	11.79	12.97
8 L.H., Daily Totals		$347.15		$495.79	$43.39	$61.97

Crew B-94C	**Hr.**	**Daily**	**Hr.**	**Daily**	**Bare Costs**	**Incl. O&P**
1 Laborer	$31.60	$252.80	$49.00	$392.00	$31.60	$49.00
1 Centrifugal Water Pump, 3″		69.40		76.34		
1 -20' Suction Hose, 3″		3.05		3.36		
2 -50' Discharge Hoses, 3″		3.50		3.85	9.49	10.44
8 L.H., Daily Totals		$328.75		$475.55	$41.09	$59.44

Crew B-94D	**Hr.**	**Daily**	**Hr.**	**Daily**	**Bare Costs**	**Incl. O&P**
1 Laborer	$31.60	$252.80	$49.00	$392.00	$31.60	$49.00
1 Centr. Water Pump, 6″		309.40		340.34		
1 -20' Suction Hose, 6″		11.90		13.09		
2 -50' Discharge Hoses, 6″		12.60		13.86	41.74	45.91
8 L.H., Daily Totals		$586.70		$759.29	$73.34	$94.91

Crew C-1	**Hr.**	**Daily**	**Hr.**	**Daily**	**Bare Costs**	**Incl. O&P**
3 Carpenters	$39.95	$958.80	$61.95	$1486.80	$37.86	$58.71
1 Laborer	31.60	252.80	49.00	392.00		
32 L.H., Daily Totals		$1211.60		$1878.80	$37.86	$58.71

Crew C-2	**Hr.**	**Daily**	**Hr.**	**Daily**	**Bare Costs**	**Incl. O&P**
1 Carpenter Foreman (out)	$41.95	$335.60	$65.05	$520.40	$38.89	$60.31
4 Carpenters	39.95	1278.40	61.95	1982.40		
1 Laborer	31.60	252.80	49.00	392.00		
48 L.H., Daily Totals		$1866.80		$2894.80	$38.89	$60.31

Crew C-2A	**Hr.**	**Daily**	**Hr.**	**Daily**	**Bare Costs**	**Incl. O&P**
1 Carpenter Foreman (out)	$41.95	$335.60	$65.05	$520.40	$38.62	$59.33
3 Carpenters	39.95	958.80	61.95	1486.80		
1 Cement Finisher	38.30	306.40	56.05	448.40		
1 Laborer	31.60	252.80	49.00	392.00		
48 L.H., Daily Totals		$1853.60		$2847.60	$38.62	$59.33

Crew C-3	**Hr.**	**Daily**	**Hr.**	**Daily**	**Bare Costs**	**Incl. O&P**
1 Rodman Foreman	$46.55	$372.40	$76.10	$608.80	$40.88	$65.53
4 Rodmen (reinf.)	44.55	1425.60	72.85	2331.20		
1 Equip. Oper. (light)	39.05	312.40	58.70	469.60		
2 Laborers	31.60	505.60	49.00	784.00		
3 Stressing Equipment		27.60		30.36		
.5 Grouting Equipment		75.80		83.38	1.62	1.78
64 L.H., Daily Totals		$2719.40		$4307.34	$42.49	$67.30

Crew C-4	**Hr.**	**Daily**	**Hr.**	**Daily**	**Bare Costs**	**Incl. O&P**
1 Rodman Foreman	$46.55	$372.40	$76.10	$608.80	$45.05	$73.66
3 Rodmen (reinf.)	44.55	1069.20	72.85	1748.40		
3 Stressing Equipment		27.60		30.36	.86	.95
32 L.H., Daily Totals		$1469.20		$2387.56	$45.91	$74.61

Crew No.	Bare Costs		Incl. Subs O & P		Cost Per Labor-Hour	
Crew C-5	**Hr.**	**Daily**	**Hr.**	**Daily**	**Bare Costs**	**Incl. O&P**
1 Rodman Foreman	$46.55	$372.40	$76.10	$608.80	$43.44	$69.54
4 Rodmen (reinf.)	44.55	1425.60	72.85	2331.20		
1 Equip. Oper. (crane)	42.55	340.40	63.95	511.60		
1 Equip. Oper. Oiler	36.80	294.40	55.30	442.40		
1 Hyd. Crane, 25 Ton		789.40		868.34	14.10	15.51
56 L.H., Daily Totals		$3222.20		$4762.34	$57.54	$85.04
Crew C-6	**Hr.**	**Daily**	**Hr.**	**Daily**	**Bare Costs**	**Incl. O&P**
1 Labor Foreman (outside)	$33.60	$268.80	$52.10	$416.80	$33.05	$50.69
4 Laborers	31.60	1011.20	49.00	1568.00		
1 Cement Finisher	38.30	306.40	56.05	448.40		
2 Gas Engine Vibrators		52.00		57.20	1.08	1.19
48 L.H., Daily Totals		$1638.40		$2490.40	$34.13	$51.88
Crew C-7	**Hr.**	**Daily**	**Hr.**	**Daily**	**Bare Costs**	**Incl. O&P**
1 Labor Foreman (outside)	$33.60	$268.80	$52.10	$416.80	$34.23	$52.29
5 Laborers	31.60	1264.00	49.00	1960.00		
1 Cement Finisher	38.30	306.40	56.05	448.40		
1 Equip. Oper. (med.)	41.35	330.80	62.15	497.20		
1 Equip. Oper. (oiler)	36.80	294.40	55.30	442.40		
2 Gas Engine Vibrators		52.00		57.20		
1 Concrete Bucket, 1 C.Y.		20.20		22.22		
1 Hyd. Crane, 55 Ton		1125.00		1237.50	16.63	18.29
72 L.H., Daily Totals		$3661.60		$5081.72	$50.86	$70.58
Crew C-7A	**Hr.**	**Daily**	**Hr.**	**Daily**	**Bare Costs**	**Incl. O&P**
1 Labor Foreman (outside)	$33.60	$268.80	$52.10	$416.80	$31.94	$49.42
5 Laborers	31.60	1264.00	49.00	1960.00		
2 Truck Drivers (Heavy)	31.95	511.20	49.15	786.40		
2 Conc. Transit Mixers		1772.80		1950.08	27.70	30.47
64 L.H., Daily Totals		$3816.80		$5113.28	$59.64	$79.89
Crew C-7B	**Hr.**	**Daily**	**Hr.**	**Daily**	**Bare Costs**	**Incl. O&P**
1 Labor Foreman (outside)	$33.60	$268.80	$52.10	$416.80	$33.87	$52.04
5 Laborers	31.60	1264.00	49.00	1960.00		
1 Equipment Oper. (crane)	42.55	340.40	63.95	511.60		
1 Equipment Oiler	36.80	294.40	55.30	442.40		
1 Conc. Bucket, 2 C.Y.		32.00		35.20		
1 Lattice Boom Crane, 165 Ton		2240.00		2464.00	35.50	39.05
64 L.H., Daily Totals		$4439.60		$5830.00	$69.37	$91.09
Crew C-7C	**Hr.**	**Daily**	**Hr.**	**Daily**	**Bare Costs**	**Incl. O&P**
1 Labor Foreman (outside)	$33.60	$268.80	$52.10	$416.80	$34.29	$52.67
5 Laborers	31.60	1264.00	49.00	1960.00		
2 Equipment Operators (medium)	41.35	661.60	62.15	994.40		
2 F.E. Loaders, W.M., 4 C.Y.		1082.00		1190.20	16.91	18.60
64 L.H., Daily Totals		$3276.40		$4561.40	$51.19	$71.27
Crew C-7D	**Hr.**	**Daily**	**Hr.**	**Daily**	**Bare Costs**	**Incl. O&P**
1 Labor Foreman (outside)	$33.60	$268.80	$52.10	$416.80	$33.28	$51.32
5 Laborers	31.60	1264.00	49.00	1960.00		
1 Equip. Oper. (med.)	41.35	330.80	62.15	497.20		
1 Concrete Conveyer		179.00		196.90	3.20	3.52
56 L.H., Daily Totals		$2042.60		$3070.90	$36.48	$54.84

Crew No.	Bare Costs		Incl. Subs O & P		Cost Per Labor-Hour	
Crew C-8	**Hr.**	**Daily**	**Hr.**	**Daily**	**Bare Costs**	**Incl. O&P**
1 Labor Foreman (outside)	$33.60	$268.80	$52.10	$416.80	$35.19	$53.34
3 Laborers	31.60	758.40	49.00	1176.00		
2 Cement Finishers	38.30	612.80	56.05	896.80		
1 Equip. Oper. (med.)	41.35	330.80	62.15	497.20		
1 Concrete Pump (small)		737.80		811.58	13.18	14.49
56 L.H., Daily Totals		$2708.60		$3798.38	$48.37	$67.83
Crew C-8A	**Hr.**	**Daily**	**Hr.**	**Daily**	**Bare Costs**	**Incl. O&P**
1 Labor Foreman (outside)	$33.60	$268.80	$52.10	$416.80	$34.17	$51.87
3 Laborers	31.60	758.40	49.00	1176.00		
2 Cement Finishers	38.30	612.80	56.05	896.80		
48 L.H., Daily Totals		$1640.00		$2489.60	$34.17	$51.87
Crew C-8B	**Hr.**	**Daily**	**Hr.**	**Daily**	**Bare Costs**	**Incl. O&P**
1 Labor Foreman (outside)	$33.60	$268.80	$52.10	$416.80	$33.95	$52.25
3 Laborers	31.60	758.40	49.00	1176.00		
1 Equip. Oper. (med.)	41.35	330.80	62.15	497.20		
1 Vibrating Power Screed		70.35		77.39		
1 Roller, Vibratory, 25 Ton		594.60		654.06		
1 Dozer, 200 H.P.		1082.00		1190.20	43.67	48.04
40 L.H., Daily Totals		$3104.95		$4011.65	$77.62	$100.29
Crew C-8C	**Hr.**	**Daily**	**Hr.**	**Daily**	**Bare Costs**	**Incl. O&P**
1 Labor Foreman (outside)	$33.60	$268.80	$52.10	$416.80	$34.67	$52.88
3 Laborers	31.60	758.40	49.00	1176.00		
1 Cement Finisher	38.30	306.40	56.05	448.40		
1 Equip. Oper. (med.)	41.35	330.80	62.15	497.20		
1 Shotcrete Rig, 12 C.Y./Hr		237.80		261.58	4.95	5.45
48 L.H., Daily Totals		$1902.20		$2799.98	$39.63	$58.33
Crew C-8D	**Hr.**	**Daily**	**Hr.**	**Daily**	**Bare Costs**	**Incl. O&P**
1 Labor Foreman (outside)	$33.60	$268.80	$52.10	$416.80	$35.64	$53.96
1 Laborer	31.60	252.80	49.00	392.00		
1 Cement Finisher	38.30	306.40	56.05	448.40		
1 Equipment Oper. (light)	39.05	312.40	58.70	469.60		
1 Air Compressor, 250 cfm		162.40		178.64		
2 -50' Air Hoses, 1"		8.20		9.02	5.33	5.86
32 L.H., Daily Totals		$1311.00		$1914.46	$40.97	$59.83
Crew C-8E	**Hr.**	**Daily**	**Hr.**	**Daily**	**Bare Costs**	**Incl. O&P**
1 Labor Foreman (outside)	$33.60	$268.80	$52.10	$416.80	$35.64	$53.96
1 Laborer	31.60	252.80	49.00	392.00		
1 Cement Finisher	38.30	306.40	56.05	448.40		
1 Equipment Oper. (light)	39.05	312.40	58.70	469.60		
1 Air Compressor, 250 cfm		162.40		178.64		
2 -50' Air Hoses, 1"		8.20		9.02		
1 Concrete Pump (small)		737.80		811.58	28.39	31.23
32 L.H., Daily Totals		$2048.80		$2726.04	$64.03	$85.19
Crew C-10	**Hr.**	**Daily**	**Hr.**	**Daily**	**Bare Costs**	**Incl. O&P**
1 Laborer	$31.60	$252.80	$49.00	$392.00	$36.07	$53.70
2 Cement Finishers	38.30	612.80	56.05	896.80		
24 L.H., Daily Totals		$865.60		$1288.80	$36.07	$53.70
Crew C-10B	**Hr.**	**Daily**	**Hr.**	**Daily**	**Bare Costs**	**Incl. O&P**
3 Laborers	$31.60	$758.40	$49.00	$1176.00	$34.28	$51.82
2 Cement Finishers	38.30	612.80	56.05	896.80		
1 Concrete Mixer, 10 C.F.		155.20		170.72		
2 Trowels, 48" Walk-Behind		84.80		93.28	6.00	6.60
40 L.H., Daily Totals		$1611.20		$2336.80	$40.28	$58.42

Crews

Crew C-10C	Bare Costs Hr.	Bare Costs Daily	Incl. Subs O & P Hr.	Incl. Subs O & P Daily	Cost Per Labor-Hour Bare Costs	Cost Per Labor-Hour Incl. O&P
1 Laborer	$31.60	$252.80	$49.00	$392.00	$36.07	$53.70
2 Cement Finishers	38.30	612.80	56.05	896.80		
1 Trowel, 48" Walk-Behind		42.40		46.64	1.77	1.94
24 L.H., Daily Totals		$908.00		$1335.44	$37.83	$55.64

Crew C-10D	Hr.	Daily	Hr.	Daily	Bare Costs	Incl. O&P
1 Laborer	$31.60	$252.80	$49.00	$392.00	$36.07	$53.70
2 Cement Finishers	38.30	612.80	56.05	896.80		
1 Vibrating Power Screed		70.35		77.39		
1 Trowel, 48" Walk-Behind		42.40		46.64	4.70	5.17
24 L.H., Daily Totals		$978.35		$1412.83	$40.76	$58.87

Crew C-10E	Hr.	Daily	Hr.	Daily	Bare Costs	Incl. O&P
1 Laborer	$31.60	$252.80	$49.00	$392.00	$36.07	$53.70
2 Cement Finishers	38.30	612.80	56.05	896.80		
1 Vibrating Power Screed		70.35		77.39		
1 Cement Trowel, 96" Ride-On		166.60		183.26	9.87	10.86
24 L.H., Daily Totals		$1102.55		$1549.44	$45.94	$64.56

Crew C-11	Hr.	Daily	Hr.	Daily	Bare Costs	Incl. O&P
1 Struc. Steel Foreman	$46.70	$373.60	$83.20	$665.60	$43.81	$75.59
6 Struc. Steel Workers	44.70	2145.60	79.65	3823.20		
1 Equip. Oper. (crane)	42.55	340.40	63.95	511.60		
1 Equip. Oper. Oiler	36.80	294.40	55.30	442.40		
1 Lattice Boom Crane, 150 Ton		1890.00		2079.00	26.25	28.88
72 L.H., Daily Totals		$5044.00		$7521.80	$70.06	$104.47

Crew C-12	Hr.	Daily	Hr.	Daily	Bare Costs	Incl. O&P
1 Carpenter Foreman (out)	$41.95	$335.60	$65.05	$520.40	$39.33	$60.64
3 Carpenters	39.95	958.80	61.95	1486.80		
1 Laborer	31.60	252.80	49.00	392.00		
1 Equip. Oper. (crane)	42.55	340.40	63.95	511.60		
1 Hyd. Crane, 12 Ton		768.80		845.68	16.02	17.62
48 L.H., Daily Totals		$2656.40		$3756.48	$55.34	$78.26

Crew C-13	Hr.	Daily	Hr.	Daily	Bare Costs	Incl. O&P
1 Struc. Steel Worker	$44.70	$357.60	$79.65	$637.20	$43.12	$73.75
1 Welder	44.70	357.60	79.65	637.20		
1 Carpenter	39.95	319.60	61.95	495.60		
1 Welder, gas engine, 300 amp		134.20		147.62	5.59	6.15
24 L.H., Daily Totals		$1169.00		$1917.62	$48.71	$79.90

Crew C-14	Hr.	Daily	Hr.	Daily	Bare Costs	Incl. O&P
1 Carpenter Foreman (out)	$41.95	$335.60	$65.05	$520.40	$39.01	$60.75
5 Carpenters	39.95	1598.00	61.95	2478.00		
4 Laborers	31.60	1011.20	49.00	1568.00		
4 Rodmen (reinf.)	44.55	1425.60	72.85	2331.20		
2 Cement Finishers	38.30	612.80	56.05	896.80		
1 Equip. Oper. (crane)	42.55	340.40	63.95	511.60		
1 Equip. Oper. Oiler	36.80	294.40	55.30	442.40		
1 Hyd. Crane, 80 Ton		1296.00		1425.60	9.00	9.90
144 L.H., Daily Totals		$6914.00		$10174.00	$48.01	$70.65

Crew C-14A	Hr.	Daily	Hr.	Daily	Bare Costs	Incl. O&P
1 Carpenter Foreman (out)	$41.95	$335.60	$65.05	$520.40	$40.09	$62.55
16 Carpenters	39.95	5113.60	61.95	7929.60		
4 Rodmen (reinf.)	44.55	1425.60	72.85	2331.20		
2 Laborers	31.60	505.60	49.00	784.00		
1 Cement Finisher	38.30	306.40	56.05	448.40		
1 Equip. Oper. (med.)	41.35	330.80	62.15	497.20		
1 Gas Engine Vibrator		26.00		28.60		
1 Concrete Pump (small)		737.80		811.58	3.82	4.20
200 L.H., Daily Totals		$8781.40		$13350.98	$43.91	$65.75

Crew C-14B	Hr.	Daily	Hr.	Daily	Bare Costs	Incl. O&P
1 Carpenter Foreman (out)	$41.95	$335.60	$65.05	$520.40	$40.02	$62.30
16 Carpenters	39.95	5113.60	61.95	7929.60		
4 Rodmen (reinf.)	44.55	1425.60	72.85	2331.20		
2 Laborers	31.60	505.60	49.00	784.00		
2 Cement Finishers	38.30	612.80	56.05	896.80		
1 Equip. Oper. (med.)	41.35	330.80	62.15	497.20		
1 Gas Engine Vibrator		26.00		28.60		
1 Concrete Pump (small)		737.80		811.58	3.67	4.04
208 L.H., Daily Totals		$9087.80		$13799.38	$43.69	$66.34

Crew C-14C	Hr.	Daily	Hr.	Daily	Bare Costs	Incl. O&P
1 Carpenter Foreman (out)	$41.95	$335.60	$65.05	$520.40	$38.25	$59.61
6 Carpenters	39.95	1917.60	61.95	2973.60		
2 Rodmen (reinf.)	44.55	712.80	72.85	1165.60		
4 Laborers	31.60	1011.20	49.00	1568.00		
1 Cement Finisher	38.30	306.40	56.05	448.40		
1 Gas Engine Vibrator		26.00		28.60	.23	.26
112 L.H., Daily Totals		$4309.60		$6704.60	$38.48	$59.36

Crew C-14D	Hr.	Daily	Hr.	Daily	Bare Costs	Incl. O&P
1 Carpenter Foreman (out)	$41.95	$335.60	$65.05	$520.40	$39.72	$61.68
18 Carpenters	39.95	5752.80	61.95	8920.80		
2 Rodmen (reinf.)	44.55	712.80	72.85	1165.60		
2 Laborers	31.60	505.60	49.00	784.00		
1 Cement Finisher	38.30	306.40	56.05	448.40		
1 Equip. Oper. (med.)	41.35	330.80	62.15	497.20		
1 Gas Engine Vibrator		26.00		28.60		
1 Concrete Pump (small)		737.80		811.58	3.82	4.20
200 L.H., Daily Totals		$8707.80		$13176.58	$43.54	$65.88

Crew C-14E	Hr.	Daily	Hr.	Daily	Bare Costs	Incl O&P
1 Carpenter Foreman (out)	$41.95	$335.60	$65.05	$520.40	$39.38	$62.13
2 Carpenters	39.95	639.20	61.95	991.20		
4 Rodmen (reinf.)	44.55	1425.60	72.85	2331.20		
3 Laborers	31.60	758.40	49.00	1176.00		
1 Cement Finisher	38.30	306.40	56.05	448.40		
1 Gas Engine Vibrator		26.00		28.60	.30	.33
88 L.H., Daily Totals		$3491.20		$5495.80	$39.67	$62.45

Crew C-14F	Hr.	Daily	Hr.	Daily	Bare Costs	Incl. O&P
1 Laborer Foreman (out)	$33.60	$268.80	$52.10	$416.80	$36.29	$54.04
2 Laborers	31.60	505.60	49.00	784.00		
6 Cement Finishers	38.30	1838.40	56.05	2690.40		
1 Gas Engine Vibrator		26.00		28.60	.36	.40
72 L.H., Daily Totals		$2638.80		$3919.80	$36.65	$54.44

Crew No.	Bare Costs		Incl. Subs O & P		Cost Per Labor-Hour	
Crew C-14G	**Hr.**	**Daily**	**Hr.**	**Daily**	**Bare Costs**	**Incl. O&P**
1 Laborer Foreman (out)	$33.60	$268.80	$52.10	$416.80	$35.71	$53.47
2 Laborers	31.60	505.60	49.00	784.00		
4 Cement Finishers	38.30	1225.60	56.05	1793.60		
1 Gas Engine Vibrator		26.00		28.60	.46	.51
56 L.H., Daily Totals		$2026.00		$3023.00	$36.18	$53.98
Crew C-14H	**Hr.**	**Daily**	**Hr.**	**Daily**	**Bare Costs**	**Incl. O&P**
1 Carpenter Foreman (out)	$41.95	$335.60	$65.05	$520.40	$39.38	$61.14
2 Carpenters	39.95	639.20	61.95	991.20		
1 Rodman (reinf.)	44.55	356.40	72.85	582.80		
1 Laborer	31.60	252.80	49.00	392.00		
1 Cement Finisher	38.30	306.40	56.05	448.40		
1 Gas Engine Vibrator		26.00		28.60	.54	.60
48 L.H., Daily Totals		$1916.40		$2963.40	$39.92	$61.74
Crew C-14L	**Hr.**	**Daily**	**Hr.**	**Daily**	**Bare Costs**	**Incl. O&P**
1 Carpenter Foreman (out)	$41.95	$335.60	$65.05	$520.40	$37.20	$57.40
6 Carpenters	39.95	1917.60	61.95	2973.60		
4 Laborers	31.60	1011.20	49.00	1568.00		
1 Cement Finisher	38.30	306.40	56.05	448.40		
1 Gas Engine Vibrator		26.00		28.60	.27	.30
96 L.H., Daily Totals		$3596.80		$5539.00	$37.47	$57.70
Crew C-15	**Hr.**	**Daily**	**Hr.**	**Daily**	**Bare Costs**	**Incl. O&P**
1 Carpenter Foreman (out)	$41.95	$335.60	$65.05	$520.40	$37.53	$57.88
2 Carpenters	39.95	639.20	61.95	991.20		
3 Laborers	31.60	758.40	49.00	1176.00		
2 Cement Finishers	38.30	612.80	56.05	896.80		
1 Rodman (reinf.)	44.55	356.40	72.85	582.80		
72 L.H., Daily Totals		$2702.40		$4167.20	$37.53	$57.88
Crew C-16	**Hr.**	**Daily**	**Hr.**	**Daily**	**Bare Costs**	**Incl. O&P**
1 Labor Foreman (outside)	$33.60	$268.80	$52.10	$416.80	$37.27	$57.67
3 Laborers	31.60	758.40	49.00	1176.00		
2 Cement Finishers	38.30	612.80	56.05	896.80		
1 Equip. Oper. (med.)	41.35	330.80	62.15	497.20		
2 Rodmen (reinf.)	44.55	712.80	72.85	1165.60		
1 Concrete Pump (small)		737.80		811.58	10.25	11.27
72 L.H., Daily Totals		$3421.40		$4963.98	$47.52	$68.94
Crew C-17	**Hr.**	**Daily**	**Hr.**	**Daily**	**Bare Costs**	**Incl. O&P**
2 Skilled Worker Foremen	$42.85	$685.60	$66.35	$1061.60	$41.25	$63.87
8 Skilled Workers	40.85	2614.40	63.25	4048.00		
80 L.H., Daily Totals		$3300.00		$5109.60	$41.25	$63.87
Crew C-17A	**Hr.**	**Daily**	**Hr.**	**Daily**	**Bare Costs**	**Incl. O&P**
2 Skilled Worker Foremen	$42.85	$685.60	$66.35	$1061.60	$41.27	$63.87
8 Skilled Workers	40.85	2614.40	63.25	4048.00		
.125 Equip. Oper. (crane)	42.55	42.55	63.95	63.95		
.125 Hyd. Crane, 80 Ton		162.00		178.20	2.00	2.20
81 L.H., Daily Totals		$3504.55		$5351.75	$43.27	$66.07
Crew C-17B	**Hr.**	**Daily**	**Hr.**	**Daily**	**Bare Costs**	**Incl. O&P**
2 Skilled Worker Foremen	$42.85	$685.60	$66.35	$1061.60	$41.28	$63.87
8 Skilled Workers	40.85	2614.40	63.25	4048.00		
.25 Equip. Oper. (crane)	42.55	85.10	63.95	127.90		
.25 Hyd. Crane, 80 Ton		324.00		356.40		
.25 Trowel, 48" Walk-Behind		10.60		11.66	4.08	4.49
82 L.H., Daily Totals		$3719.70		$5605.56	$45.36	$68.36

Crew No.	Bare Costs		Incl. Subs O & P		Cost Per Labor-Hour	
Crew C-17C	**Hr.**	**Daily**	**Hr.**	**Daily**	**Bare Costs**	**Incl. O&P**
2 Skilled Worker Foremen	$42.85	$685.60	$66.35	$1061.60	$41.30	$63.87
8 Skilled Workers	40.85	2614.40	63.25	4048.00		
.375 Equip. Oper. (crane)	42.55	127.65	63.95	191.85		
.375 Hyd. Crane, 80 Ton		486.00		534.60	5.86	6.44
83 L.H., Daily Totals		$3913.65		$5836.05	$47.15	$70.31
Crew C-17D	**Hr.**	**Daily**	**Hr.**	**Daily**	**Bare Costs**	**Incl. O&P**
2 Skilled Worker Foremen	$42.85	$685.60	$66.35	$1061.60	$41.31	$63.87
8 Skilled Workers	40.85	2614.40	63.25	4048.00		
.5 Equip. Oper. (crane)	42.55	170.20	63.95	255.80		
.5 Hyd. Crane, 80 Ton		648.00		712.80	7.71	8.49
84 L.H., Daily Totals		$4118.20		$6078.20	$49.03	$72.36
Crew C-17E	**Hr.**	**Daily**	**Hr.**	**Daily**	**Bare Costs**	**Incl. O&P**
2 Skilled Worker Foremen	$42.85	$685.60	$66.35	$1061.60	$41.25	$63.87
8 Skilled Workers	40.85	2614.40	63.25	4048.00		
1 Hyd. Jack with Rods		85.70		94.27	1.07	1.18
80 L.H., Daily Totals		$3385.70		$5203.87	$42.32	$65.05
Crew C-18	**Hr.**	**Daily**	**Hr.**	**Daily**	**Bare Costs**	**Incl. O&P**
.125 Labor Foreman (out)	$33.60	$33.60	$52.10	$52.10	$31.82	$49.34
1 Laborer	31.60	252.80	49.00	392.00		
1 Concrete Cart, 10 C.F.		55.20		60.72	6.13	6.75
9 L.H., Daily Totals		$341.60		$504.82	$37.96	$56.09
Crew C-19	**Hr.**	**Daily**	**Hr.**	**Daily**	**Bare Costs**	**Incl. O&P**
.125 Labor Foreman (out)	$33.60	$33.60	$52.10	$52.10	$31.82	$49.34
1 Laborer	31.60	252.80	49.00	392.00		
1 Concrete Cart, 18 C.F.		86.40		95.04	9.60	10.56
9 L.H., Daily Totals		$372.80		$539.14	$41.42	$59.90
Crew C-20	**Hr.**	**Daily**	**Hr.**	**Daily**	**Bare Costs**	**Incl. O&P**
1 Labor Foreman (outside)	$33.60	$268.80	$52.10	$416.80	$33.91	$51.91
5 Laborers	31.60	1264.00	49.00	1960.00		
1 Cement Finisher	38.30	306.40	56.05	448.40		
1 Equip. Oper. (med.)	41.35	330.80	62.15	497.20		
2 Gas Engine Vibrators		52.00		57.20		
1 Concrete Pump (small)		737.80		811.58	12.34	13.57
64 L.H., Daily Totals		$2959.80		$4191.18	$46.25	$65.49
Crew C-21	**Hr.**	**Daily**	**Hr.**	**Daily**	**Bare Costs**	**Incl. O&P**
1 Labor Foreman (outside)	$33.60	$268.80	$52.10	$416.80	$33.91	$51.91
5 Laborers	31.60	1264.00	49.00	1960.00		
1 Cement Finisher	38.30	306.40	56.05	448.40		
1 Equip. Oper. (med.)	41.35	330.80	62.15	497.20		
2 Gas Engine Vibrators		52.00		57.20		
1 Concrete Conveyer		179.00		196.90	3.61	3.97
64 L.H., Daily Totals		$2401.00		$3576.50	$37.52	$55.88
Crew C-22	**Hr.**	**Daily**	**Hr.**	**Daily**	**Bare Costs**	**Incl. O&P**
1 Rodman Foreman	$46.55	$372.40	$76.10	$608.80	$44.70	$72.84
4 Rodmen (reinf.)	44.55	1425.60	72.85	2331.20		
.125 Equip. Oper. (crane)	42.55	42.55	63.95	63.95		
.125 Equip. Oper. Oiler	36.80	36.80	55.30	55.30		
.125 Hyd. Crane, 25 Ton		98.67		108.54	2.35	2.58
42 L.H., Daily Totals		$1976.03		$3167.79	$47.05	$75.42

Crew No.	Bare Costs		Incl. Subs O & P		Cost Per Labor-Hour	

Crew C-23	Hr.	Daily	Hr.	Daily	Bare Costs	Incl. O&P
2 Skilled Worker Foremen	$42.85	$685.60	$66.35	$1061.60	$41.02	$63.15
6 Skilled Workers	40.85	1960.80	63.25	3036.00		
1 Equip. Oper. (crane)	42.55	340.40	63.95	511.60		
1 Equip. Oper. Oiler	36.80	294.40	55.30	442.40		
1 Lattice Boom Crane, 90 Ton		1741.00		1915.10	21.76	23.94
80 L.H., Daily Totals		$5022.20		$6966.70	$62.78	$87.08

Crew C-23A	Hr.	Daily	Hr.	Daily	Bare Costs	Incl. O&P
1 Labor Foreman (outside)	$33.60	$268.80	$52.10	$416.80	$35.23	$53.87
2 Laborers	31.60	505.60	49.00	784.00		
1 Equip. Oper. (crane)	42.55	340.40	63.95	511.60		
1 Equip. Oper. Oiler	36.80	294.40	55.30	442.40		
1 Crawler Crane, 100 Ton		1611.00		1772.10		
3 Conc. Buckets, 8 C.Y.		603.00		663.30	55.35	60.88
40 L.H., Daily Totals		$3623.20		$4590.20	$90.58	$114.76

Crew C-24	Hr.	Daily	Hr.	Daily	Bare Costs	Incl. O&P
2 Skilled Worker Foremen	$42.85	$685.60	$66.35	$1061.60	$41.02	$63.15
6 Skilled Workers	40.85	1960.80	63.25	3036.00		
1 Equip. Oper. (crane)	42.55	340.40	63.95	511.60		
1 Equip. Oper. Oiler	36.80	294.40	55.30	442.40		
1 Lattice Boom Crane, 150 Ton		1890.00		2079.00	23.63	25.99
80 L.H., Daily Totals		$5171.20		$7130.60	$64.64	$89.13

Crew C-25	Hr.	Daily	Hr.	Daily	Bare Costs	Incl. O&P
2 Rodmen (reinf.)	$44.55	$712.80	$72.85	$1165.60	$34.95	$57.77
2 Rodmen Helpers	25.35	405.60	42.70	683.20		
32 L.H., Daily Totals		$1118.40		$1848.80	$34.95	$57.77

Crew C-27	Hr.	Daily	Hr.	Daily	Bare Costs	Incl. O&P
2 Cement Finishers	$38.30	$612.80	$56.05	$896.80	$38.30	$56.05
1 Concrete Saw		141.00		155.10	8.81	9.69
16 L.H., Daily Totals		$753.80		$1051.90	$47.11	$65.74

Crew C-28	Hr.	Daily	Hr.	Daily	Bare Costs	Incl. O&P
1 Cement Finisher	$38.30	$306.40	$56.05	$448.40	$38.30	$56.05
1 Portable Air Compressor, Gas		19.10		21.01	2.39	2.63
8 L.H., Daily Totals		$325.50		$469.41	$40.69	$58.68

Crew D-1	Hr.	Daily	Hr.	Daily	Bare Costs	Incl. O&P
1 Bricklayer	$40.50	$324.00	$61.45	$491.60	$36.33	$55.10
1 Bricklayer Helper	32.15	257.20	48.75	390.00		
16 L.H., Daily Totals		$581.20		$881.60	$36.33	$55.10

Crew D-2	Hr.	Daily	Hr.	Daily	Bare Costs	Incl. O&P
3 Bricklayers	$40.50	$972.00	$61.45	$1474.80	$37.41	$56.88
2 Bricklayer Helpers	32.15	514.40	48.75	780.00		
.5 Carpenter	39.95	159.80	61.95	247.80		
44 L.H., Daily Totals		$1646.20		$2502.60	$37.41	$56.88

Crew D-3	Hr.	Daily	Hr.	Daily	Bare Costs	Incl. O&P
3 Bricklayers	$40.50	$972.00	$61.45	$1474.80	$37.29	$56.64
2 Bricklayer Helpers	32.15	514.40	48.75	780.00		
.25 Carpenter	39.95	79.90	61.95	123.90		
42 L.H., Daily Totals		$1566.30		$2378.70	$37.29	$56.64

Crew No.	Bare Costs		Incl. Subs O & P		Cost Per Labor-Hour	

Crew D-4	Hr.	Daily	Hr.	Daily	Bare Costs	Incl. O&P
1 Bricklayer	$40.50	$324.00	$61.45	$491.60	$35.96	$54.41
2 Bricklayer Helpers	32.15	514.40	48.75	780.00		
1 Equip. Oper. (light)	39.05	312.40	58.70	469.60		
1 Grout Pump, 50 C.F./Hr.		126.80		139.48	3.96	4.36
32 L.H., Daily Totals		$1277.60		$1880.68	$39.92	$58.77

Crew D-5	Hr.	Daily	Hr.	Daily	Bare Costs	Incl. O&P
1 Bricklayer	$40.50	$324.00	$61.45	$491.60	$40.50	$61.45
8 L.H., Daily Totals		$324.00		$491.60	$40.50	$61.45

Crew D-6	Hr.	Daily	Hr.	Daily	Bare Costs	Incl. O&P
3 Bricklayers	$40.50	$972.00	$61.45	$1474.80	$36.47	$55.37
3 Bricklayer Helpers	32.15	771.60	48.75	1170.00		
.25 Carpenter	39.95	79.90	61.95	123.90		
50 L.H., Daily Totals		$1823.50		$2768.70	$36.47	$55.37

Crew D-7	Hr.	Daily	Hr.	Daily	Bare Costs	Incl. O&P
1 Tile Layer	$38.10	$304.80	$55.80	$446.40	$34.08	$49.90
1 Tile Layer Helper	30.05	240.40	44.00	352.00		
16 L.H., Daily Totals		$545.20		$798.40	$34.08	$49.90

Crew D-8	Hr.	Daily	Hr.	Daily	Bare Costs	Incl. O&P
3 Bricklayers	$40.50	$972.00	$61.45	$1474.80	$37.16	$56.37
2 Bricklayer Helpers	32.15	514.40	48.75	780.00		
40 L.H., Daily Totals		$1486.40		$2254.80	$37.16	$56.37

Crew D-9	Hr.	Daily	Hr.	Daily	Bare Costs	Incl. O&P
3 Bricklayers	$40.50	$972.00	$61.45	$1474.80	$36.33	$55.10
3 Bricklayer Helpers	32.15	771.60	48.75	1170.00		
48 L.H., Daily Totals		$1743.60		$2644.80	$36.33	$55.10

Crew D-10	Hr.	Daily	Hr.	Daily	Bare Costs	Incl. O&P
1 Bricklayer Foreman	$42.50	$340.00	$64.45	$515.60	$39.42	$59.65
1 Bricklayer	40.50	324.00	61.45	491.60		
1 Bricklayer Helper	32.15	257.20	48.75	390.00		
1 Equip. Oper. (crane)	42.55	340.40	63.95	511.60		
1 S.P. Crane, 4x4, 12 Ton		601.80		661.98	18.81	20.69
32 L.H., Daily Totals		$1863.40		$2570.78	$58.23	$80.34

Crew D-11	Hr.	Daily	Hr.	Daily	Bare Costs	Incl. O&P
1 Bricklayer Foreman	$42.50	$340.00	$64.45	$515.60	$38.38	$58.22
1 Bricklayer	40.50	324.00	61.45	491.60		
1 Bricklayer Helper	32.15	257.20	48.75	390.00		
24 L.H., Daily Totals		$921.20		$1397.20	$38.38	$58.22

Crew D-12	Hr.	Daily	Hr.	Daily	Bare Costs	Incl. O&P
1 Bricklayer Foreman	$42.50	$340.00	$64.45	$515.60	$36.83	$55.85
1 Bricklayer	40.50	324.00	61.45	491.60		
2 Bricklayer Helpers	32.15	514.40	48.75	780.00		
32 L.H., Daily Totals		$1178.40		$1787.20	$36.83	$55.85

Crew D-13	Hr.	Daily	Hr.	Daily	Bare Costs	Incl. O&P
1 Bricklayer Foreman	$42.50	$340.00	$64.45	$515.60	$38.30	$58.22
1 Bricklayer	40.50	324.00	61.45	491.60		
2 Bricklayer Helpers	32.15	514.40	48.75	780.00		
1 Carpenter	39.95	319.60	61.95	495.60		
1 Equip. Oper. (crane)	42.55	340.40	63.95	511.60		
1 S.P. Crane, 4x4, 12 Ton		601.80		661.98	12.54	13.79
48 L.H., Daily Totals		$2440.20		$3456.38	$50.84	$72.01

Crew No.	Bare Costs		Incl. Subs O & P		Cost Per Labor-Hour	
Crew E-1	**Hr.**	**Daily**	**Hr.**	**Daily**	**Bare Costs**	**Incl. O&P**
1 Welder Foreman	$46.70	$373.60	$83.20	$665.60	$43.48	$73.85
1 Welder	44.70	357.60	79.65	637.20		
1 Equip. Oper. (light)	39.05	312.40	58.70	469.60		
1 Welder, gas engine, 300 amp		134.20		147.62	5.59	6.15
24 L.H., Daily Totals		$1177.80		$1920.02	$49.08	$80.00

Crew E-2	Hr.	Daily	Hr.	Daily	Bare Costs	Incl. O&P
1 Struc. Steel Foreman	$46.70	$373.60	$83.20	$665.60	$43.55	$74.44
4 Struc. Steel Workers	44.70	1430.40	79.65	2548.80		
1 Equip. Oper. (crane)	42.55	340.40	63.95	511.60		
1 Equip. Oper. Oiler	36.80	294.40	55.30	442.40		
1 Lattice Boom Crane, 90 Ton		1741.00		1915.10	31.09	34.20
56 L.H., Daily Totals		$4179.80		$6083.50	$74.64	$108.63

Crew E-3	Hr.	Daily	Hr.	Daily	Bare Costs	Incl. O&P
1 Struc. Steel Foreman	$46.70	$373.60	$83.20	$665.60	$45.37	$80.83
1 Struc. Steel Worker	44.70	357.60	79.65	637.20		
1 Welder	44.70	357.60	79.65	637.20		
1 Welder, gas engine, 300 amp		134.20		147.62	5.59	6.15
24 L.H., Daily Totals		$1223.00		$2087.62	$50.96	$86.98

Crew E-4	Hr.	Daily	Hr.	Daily	Bare Costs	Incl. O&P
1 Struc. Steel Foreman	$46.70	$373.60	$83.20	$665.60	$45.20	$80.54
3 Struc. Steel Workers	44.70	1072.80	79.65	1911.60		
1 Welder, gas engine, 300 amp		134.20		147.62	4.19	4.61
32 L.H., Daily Totals		$1580.60		$2724.82	$49.39	$85.15

Crew E-5	Hr.	Daily	Hr.	Daily	Bare Costs	Incl. O&P
2 Struc. Steel Foremen	$46.70	$747.20	$83.20	$1331.20	$44.09	$76.36
5 Struc. Steel Workers	44.70	1788.00	79.65	3186.00		
1 Equip. Oper. (crane)	42.55	340.40	63.95	511.60		
1 Welder	44.70	357.60	79.65	637.20		
1 Equip. Oper. Oiler	36.80	294.40	55.30	442.40		
1 Lattice Boom Crane, 90 Ton		1741.00		1915.10		
1 Welder, gas engine, 300 amp		134.20		147.62	23.44	25.78
80 L.H., Daily Totals		$5402.80		$8171.12	$67.53	$102.14

Crew E-6	Hr.	Daily	Hr.	Daily	Bare Costs	Incl. O&P
3 Struc. Steel Foremen	$46.70	$1120.80	$83.20	$1996.80	$44.09	$76.50
9 Struc. Steel Workers	44.70	3218.40	79.65	5734.80		
1 Equip. Oper. (crane)	42.55	340.40	63.95	511.60		
1 Welder	44.70	357.60	79.65	637.20		
1 Equip. Oper. Oiler	36.80	294.40	55.30	442.40		
1 Equip. Oper. (light)	39.05	312.40	58.70	469.60		
1 Lattice Boom Crane, 90 Ton		1741.00		1915.10		
1 Welder, gas engine, 300 amp		134.20		147.62		
1 Air Compressor, 160 cfm		138.80		152.68		
2 Impact Wrenches		32.40		35.64	15.99	17.59
128 L.H., Daily Totals		$7690.40		$12043.44	$60.08	$94.09

Crew E-7	Hr.	Daily	Hr.	Daily	Bare Costs	Incl. O&P
1 Struc. Steel Foreman	$46.70	$373.60	$83.20	$665.60	$44.09	$76.36
4 Struc. Steel Workers	44.70	1430.40	79.65	2548.80		
1 Equip. Oper. (crane)	42.55	340.40	63.95	511.60		
1 Equip. Oper. Oiler	36.80	294.40	55.30	442.40		
1 Welder Foreman	46.70	373.60	83.20	665.60		
2 Welders	44.70	715.20	79.65	1274.40		
1 Lattice Boom Crane, 90 Ton		1741.00		1915.10		
2 Welders, gas engine, 300 amp		268.40		295.24	25.12	27.63
80 L.H., Daily Totals		$5537.00		$8318.74	$69.21	$103.98

Crew E-8	Hr.	Daily	Hr.	Daily	Bare Costs	Incl. O&P
1 Struc. Steel Foreman	$46.70	$373.60	$83.20	$665.60	$43.80	$75.50
4 Struc. Steel Workers	44.70	1430.40	79.65	2548.80		
1 Welder Foreman	46.70	373.60	83.20	665.60		
4 Welders	44.70	1430.40	79.65	2548.80		
1 Equip. Oper. (crane)	42.55	340.40	63.95	511.60		
1 Equip. Oper. Oiler	36.80	294.40	55.30	442.40		
1 Equip. Oper. (light)	39.05	312.40	58.70	469.60		
1 Lattice Boom Crane, 90 Ton		1741.00		1915.10		
4 Welders, gas engine, 300 amp		536.80		590.48	21.90	24.09
104 L.H., Daily Totals		$6833.00		$10357.98	$65.70	$99.60

Crew E-9	Hr.	Daily	Hr.	Daily	Bare Costs	Incl. O&P
2 Struc. Steel Foremen	$46.70	$747.20	$83.20	$1331.20	$44.09	$76.50
5 Struc. Steel Workers	44.70	1788.00	79.65	3186.00		
1 Welder Foreman	46.70	373.60	83.20	665.60		
5 Welders	44.70	1788.00	79.65	3186.00		
1 Equip. Oper. (crane)	42.55	340.40	63.95	511.60		
1 Equip. Oper. Oiler	36.80	294.40	55.30	442.40		
1 Equip. Oper. (light)	39.05	312.40	58.70	469.60		
1 Lattice Boom Crane, 90 Ton		1741.00		1915.10		
5 Welders, gas engine, 300 amp		671.00		738.10	18.84	20.73
128 L.H., Daily Totals		$8056.00		$12445.60	$62.94	$97.23

Crew E-10	Hr.	Daily	Hr.	Daily	Bare Costs	Incl. O&P
1 Welder Foreman	$46.70	$373.60	$83.20	$665.60	$45.70	$81.42
1 Welder	44.70	357.60	79.65	637.20		
1 Welder, gas engine, 300 amp		134.20		147.62		
1 Flatbed Truck, Gas, 3 Ton		242.60		266.86	23.55	25.91
16 L.H., Daily Totals		$1108.00		$1717.28	$69.25	$107.33

Crew E-11	Hr.	Daily	Hr.	Daily	Bare Costs	Incl. O&P
2 Painters, Struc. Steel	$36.00	$576.00	$65.35	$1045.60	$35.66	$59.60
1 Building Laborer	31.60	252.80	49.00	392.00		
1 Equip. Oper. (light)	39.05	312.40	58.70	469.60		
1 Air Compressor, 250 cfm		162.40		178.64		
1 Sandblaster, portable, 3 C.F.		20.60		22.66		
1 Set Sand Blasting Accessories		12.75		14.03	6.12	6.73
32 L.H., Daily Totals		$1336.95		$2122.53	$41.78	$66.33

Crew E-12	Hr.	Daily	Hr.	Daily	Bare Costs	Incl. O&P
1 Welder Foreman	$46.70	$373.60	$83.20	$665.60	$42.88	$70.95
1 Equip. Oper. (light)	39.05	312.40	58.70	469.60		
1 Welder, gas engine, 300 amp		134.20		147.62	8.39	9.23
16 L.H., Daily Totals		$820.20		$1282.82	$51.26	$80.18

Crew E-13	Hr.	Daily	Hr.	Daily	Bare Costs	Incl. O&P
1 Welder Foreman	$46.70	$373.60	$83.20	$665.60	$44.15	$75.03
.5 Equip. Oper. (light)	39.05	156.20	58.70	234.80		
1 Welder, gas engine, 300 amp		134.20		147.62	11.18	12.30
12 L.H., Daily Totals		$664.00		$1048.02	$55.33	$87.33

Crew E-14	Hr.	Daily	Hr.	Daily	Bare Costs	Incl. O&P
1 Welder Foreman	$46.70	$373.60	$83.20	$665.60	$46.70	$83.20
1 Welder, gas engine, 300 amp		134.20		147.62	16.77	18.45
8 L.H., Daily Totals		$507.80		$813.22	$63.48	$101.65

Crew E-16	Hr.	Daily	Hr.	Daily	Bare Costs	Incl. O&P
1 Welder Foreman	$46.70	$373.60	$83.20	$665.60	$45.70	$81.42
1 Welder	44.70	357.60	79.65	637.20		
1 Welder, gas engine, 300 amp		134.20		147.62	8.39	9.23
16 L.H., Daily Totals		$865.40		$1450.42	$54.09	$90.65

Crew No.	Bare Costs		Incl. Subs O & P		Cost Per Labor-Hour	
Crew E-17	**Hr.**	**Daily**	**Hr.**	**Daily**	**Bare Costs**	**Incl. O&P**
1 Structural Steel Foreman	$46.70	$373.60	$83.20	$665.60	$45.70	$81.42
1 Structural Steel Worker	44.70	357.60	79.65	637.20		
16 L.H., Daily Totals		$731.20		$1302.80	$45.70	$81.42
Crew E-18	**Hr.**	**Daily**	**Hr.**	**Daily**	**Bare Costs**	**Incl. O&P**
1 Structural Steel Foreman	$46.70	$373.60	$83.20	$665.60	$44.43	$76.86
3 Structural Steel Workers	44.70	1072.80	79.65	1911.60		
1 Equipment Operator (med.)	41.35	330.80	62.15	497.20		
1 Lattice Boom Crane, 20 Ton		1080.00		1188.00	27.00	29.70
40 L.H., Daily Totals		$2857.20		$4262.40	$71.43	$106.56
Crew E-19	**Hr.**	**Daily**	**Hr.**	**Daily**	**Bare Costs**	**Incl. O&P**
1 Structural Steel Worker	$44.70	$357.60	$79.65	$637.20	$43.48	$73.85
1 Structural Steel Foreman	46.70	373.60	83.20	665.60		
1 Equip. Oper. (light)	39.05	312.40	58.70	469.60		
1 Lattice Boom Crane, 20 Ton		1080.00		1188.00	45.00	49.50
24 L.H., Daily Totals		$2123.60		$2960.40	$88.48	$123.35
Crew E-20	**Hr.**	**Daily**	**Hr.**	**Daily**	**Bare Costs**	**Incl. O&P**
1 Structural Steel Foreman	$46.70	$373.60	$83.20	$665.60	$43.69	$75.09
5 Structural Steel Workers	44.70	1788.00	79.65	3186.00		
1 Equip. Oper. (crane)	42.55	340.40	63.95	511.60		
1 Oiler	36.80	294.40	55.30	442.40		
1 Lattice Boom Crane, 40 Ton		1327.00		1459.70	20.73	22.81
64 L.H., Daily Totals		$4123.40		$6265.30	$64.43	$97.90
Crew E-22	**Hr.**	**Daily**	**Hr.**	**Daily**	**Bare Costs**	**Incl. O&P**
1 Skilled Worker Foreman	$42.85	$342.80	$66.35	$530.80	$41.52	$64.28
2 Skilled Workers	40.85	653.60	63.25	1012.00		
24 L.H., Daily Totals		$996.40		$1542.80	$41.52	$64.28
Crew E-24	**Hr.**	**Daily**	**Hr.**	**Daily**	**Bare Costs**	**Incl. O&P**
3 Structural Steel Workers	$44.70	$1072.80	$79.65	$1911.60	$43.86	$75.28
1 Equipment Operator (medium)	41.35	330.80	62.15	497.20		
1 Hyd. Crane, 25 Ton		789.40		868.34	24.67	27.14
32 L.H., Daily Totals		$2193.00		$3277.14	$68.53	$102.41
Crew E-25	**Hr.**	**Daily**	**Hr.**	**Daily**	**Bare Costs**	**Incl. O&P**
1 Welder Foreman	$46.70	$373.60	$83.20	$665.60	$46.70	$83.20
1 Cutting Torch		17.00		18.70		
1 Gases		75.60		83.16	11.57	12.73
8 L.H., Daily Totals		$466.20		$767.46	$58.27	$95.93
Crew F-3	**Hr.**	**Daily**	**Hr.**	**Daily**	**Bare Costs**	**Incl. O&P**
4 Carpenters	$39.95	$1278.40	$61.95	$1982.40	$40.47	$62.35
1 Equip. Oper. (crane)	42.55	340.40	63.95	511.60		
1 Hyd. Crane, 12 Ton		768.80		845.68	19.22	21.14
40 L.H., Daily Totals		$2387.60		$3339.68	$59.69	$83.49
Crew F-4	**Hr.**	**Daily**	**Hr.**	**Daily**	**Bare Costs**	**Incl. O&P**
4 Carpenters	$39.95	$1278.40	$61.95	$1982.40	$39.86	$61.17
1 Equip. Oper. (crane)	42.55	340.40	63.95	511.60		
1 Equip. Oper. Oiler	36.80	294.40	55.30	442.40		
1 Hyd. Crane, 55 Ton		1125.00		1237.50	23.44	25.78
48 L.H., Daily Totals		$3038.20		$4173.90	$63.30	$86.96
Crew F-5	**Hr.**	**Daily**	**Hr.**	**Daily**	**Bare Costs**	**Incl. O&P**
1 Carpenter Foreman	$41.95	$335.60	$65.05	$520.40	$40.45	$62.73
3 Carpenters	39.95	958.80	61.95	1486.80		
32 L.H., Daily Totals		$1294.40		$2007.20	$40.45	$62.73

Crew No.	Bare Costs		Incl. Subs O & P		Cost Per Labor-Hour	
Crew F-6	**Hr.**	**Daily**	**Hr.**	**Daily**	**Bare Costs**	**Incl. O&P**
2 Carpenters	$39.95	$639.20	$61.95	$991.20	$37.13	$57.17
2 Building Laborers	31.60	505.60	49.00	784.00		
1 Equip. Oper. (crane)	42.55	340.40	63.95	511.60		
1 Hyd. Crane, 12 Ton		768.80		845.68	19.22	21.14
40 L.H., Daily Totals		$2254.00		$3132.48	$56.35	$78.31
Crew F-7	**Hr.**	**Daily**	**Hr.**	**Daily**	**Bare Costs**	**Incl. O&P**
2 Carpenters	$39.95	$639.20	$61.95	$991.20	$35.77	$55.48
2 Building Laborers	31.60	505.60	49.00	784.00		
32 L.H., Daily Totals		$1144.80		$1775.20	$35.77	$55.48
Crew G-1	**Hr.**	**Daily**	**Hr.**	**Daily**	**Bare Costs**	**Incl. O&P**
1 Roofer Foreman	$36.25	$290.00	$61.10	$488.80	$31.99	$53.90
4 Roofers, Composition	34.25	1096.00	57.70	1846.40		
2 Roofer Helpers	25.35	405.60	42.70	683.20		
1 Application Equipment		173.20		190.52		
1 Tar Kettle/Pot		97.20		106.92		
1 Crew Truck		169.60		186.56	7.86	8.64
56 L.H., Daily Totals		$2231.60		$3502.40	$39.85	$62.54
Crew G-2	**Hr.**	**Daily**	**Hr.**	**Daily**	**Bare Costs**	**Incl. O&P**
1 Plasterer	$36.15	$289.20	$54.55	$436.40	$33.35	$50.77
1 Plasterer Helper	32.30	258.40	48.75	390.00		
1 Building Laborer	31.60	252.80	49.00	392.00		
1 Grout Pump, 50 C.F./Hr.		126.80		139.48	5.28	5.81
24 L.H., Daily Totals		$927.20		$1357.88	$38.63	$56.58
Crew G-2A	**Hr.**	**Daily**	**Hr.**	**Daily**	**Bare Costs**	**Incl. O&P**
1 Roofer, composition	$34.25	$274.00	$57.70	$461.60	$30.40	$49.80
1 Roofer Helper	25.35	202.80	42.70	341.60		
1 Building Laborer	31.60	252.80	49.00	392.00		
1 Foam spray rig, trailer-mtd.		488.20		537.02		
1 Pickup Truck, 3/4 Ton		116.80		128.48	25.21	27.73
24 L.H., Daily Totals		$1334.60		$1860.70	$55.61	$77.53
Crew G-3	**Hr.**	**Daily**	**Hr.**	**Daily**	**Bare Costs**	**Incl. O&P**
2 Sheet Metal Workers	$47.20	$755.20	$72.65	$1162.40	$39.40	$60.83
2 Building Laborers	31.60	505.60	49.00	784.00		
32 L.H., Daily Totals		$1260.80		$1946.40	$39.40	$60.83
Crew G-4	**Hr.**	**Daily**	**Hr.**	**Daily**	**Bare Costs**	**Incl. O&P**
1 Labor Foreman (outside)	$33.60	$268.80	$52.10	$416.80	$32.27	$50.03
2 Building Laborers	31.60	505.60	49.00	784.00		
1 Flatbed Truck, Gas, 1.5 Ton		194.40		213.84		
1 Air Compressor, 160 cfm		138.80		152.68	13.88	15.27
24 L.H., Daily Totals		$1107.60		$1567.32	$46.15	$65.31
Crew G-5	**Hr.**	**Daily**	**Hr.**	**Daily**	**Bare Costs**	**Incl. O&P**
1 Roofer Foreman	$36.25	$290.00	$61.10	$488.80	$31.09	$52.38
2 Roofers, Composition	34.25	548.00	57.70	923.20		
2 Roofer Helpers	25.35	405.60	42.70	683.20		
1 Application Equipment		173.20		190.52	4.33	4.76
40 L.H., Daily Totals		$1416.80		$2285.72	$35.42	$57.14
Crew G-6A	**Hr.**	**Daily**	**Hr.**	**Daily**	**Bare Costs**	**Incl. O&P**
2 Roofers Composition	$34.25	$548.00	$57.70	$923.20	$34.25	$57.70
1 Small Compressor, Electric		11.35		12.48		
2 Pneumatic Nailers		43.20		47.52	3.41	3.75
16 L.H., Daily Totals		$602.55		$983.21	$37.66	$61.45

Crew No.	Bare Costs Hr.	Bare Costs Daily	Incl. Subs O & P Hr.	Incl. Subs O & P Daily	Cost Per Labor-Hour Bare Costs	Cost Per Labor-Hour Incl. O&P
Crew G-7						
1 Carpenter	$39.95	$319.60	$61.95	$495.60	$39.95	$61.95
1 Small Compressor, Electric		11.35		12.48		
1 Pneumatic Nailer		21.60		23.76	4.12	4.53
8 L.H., Daily Totals		$352.55		$531.85	$44.07	$66.48

Crew H-1	Hr.	Daily	Hr.	Daily	Bare Costs	Incl. O&P
2 Glaziers	$38.60	$617.60	$58.35	$933.60	$41.65	$69.00
2 Struc. Steel Workers	44.70	715.20	79.65	1274.40		
32 L.H., Daily Totals		$1332.80		$2208.00	$41.65	$69.00

Crew H-2	Hr.	Daily	Hr.	Daily	Bare Costs	Incl. O&P
2 Glaziers	$38.60	$617.60	$58.35	$933.60	$36.27	$55.23
1 Building Laborer	31.60	252.80	49.00	392.00		
24 L.H., Daily Totals		$870.40		$1325.60	$36.27	$55.23

Crew H-3	Hr.	Daily	Hr.	Daily	Bare Costs	Incl. O&P
1 Glazier	$38.60	$308.80	$58.35	$466.80	$34.45	$52.58
1 Helper	30.30	242.40	46.80	374.40		
16 L.H., Daily Totals		$551.20		$841.20	$34.45	$52.58

Crew J-1	Hr.	Daily	Hr.	Daily	Bare Costs	Incl. O&P
3 Plasterers	$36.15	$867.60	$54.55	$1309.20	$34.61	$52.23
2 Plasterer Helpers	32.30	516.80	48.75	780.00		
1 Mixing Machine, 6 C.F.		126.60		139.26	3.17	3.48
40 L.H., Daily Totals		$1511.00		$2228.46	$37.77	$55.71

Crew J-2	Hr.	Daily	Hr.	Daily	Bare Costs	Incl. O&P
3 Plasterers	$36.15	$867.60	$54.55	$1309.20	$34.77	$52.29
2 Plasterer Helpers	32.30	516.80	48.75	780.00		
1 Lather	35.55	284.40	52.60	420.80		
1 Mixing Machine, 6 C.F.		126.60		139.26	2.64	2.90
48 L.H., Daily Totals		$1795.40		$2649.26	$37.40	$55.19

Crew J-3	Hr.	Daily	Hr.	Daily	Bare Costs	Incl. O&P
1 Terrazzo Worker	$37.70	$301.60	$55.20	$441.60	$34.27	$50.17
1 Terrazzo Helper	30.85	246.80	45.15	361.20		
1 Terrazzo Grinder, Electric		81.60		89.76		
1 Terrazzo Mixer		171.60		188.76	15.82	17.41
16 L.H., Daily Totals		$801.60		$1081.32	$50.10	$67.58

Crew K-1	Hr.	Daily	Hr.	Daily	Bare Costs	Incl. O&P
1 Carpenter	$39.95	$319.60	$61.95	$495.60	$35.45	$54.80
1 Truck Driver (light)	30.95	247.60	47.65	381.20		
1 Flatbed Truck, Gas, 3 Ton		242.60		266.86	15.16	16.68
16 L.H., Daily Totals		$809.80		$1143.66	$50.61	$71.48

Crew K-2	Hr.	Daily	Hr.	Daily	Bare Costs	Incl. O&P
1 Struc. Steel Foreman	$46.70	$373.60	$83.20	$665.60	$40.78	$70.17
1 Struc. Steel Worker	44.70	357.60	79.65	637.20		
1 Truck Driver (light)	30.95	247.60	47.65	381.20		
1 Flatbed Truck, Gas, 3 Ton		242.60		266.86	10.11	11.12
24 L.H., Daily Totals		$1221.40		$1950.86	$50.89	$81.29

Crew L-1	Hr.	Daily	Hr.	Daily	Bare Costs	Incl. O&P
1 Electrician	$47.00	$376.00	$69.95	$559.60	$47.88	$71.55
1 Plumber	48.75	390.00	73.15	585.20		
16 L.H., Daily Totals		$766.00		$1144.80	$47.88	$71.55

Crew L-2	Hr.	Daily	Hr.	Daily	Bare Costs	Incl. O&P
1 Carpenter	$39.95	$319.60	$61.95	$495.60	$35.13	$54.38
1 Carpenter Helper	30.30	242.40	46.80	374.40		
16 L.H., Daily Totals		$562.00		$870.00	$35.13	$54.38

Crew L-3	Hr.	Daily	Hr.	Daily	Bare Costs	Incl. O&P
1 Carpenter	$39.95	$319.60	$61.95	$495.60	$43.52	$66.63
.5 Electrician	47.00	188.00	69.95	279.80		
.5 Sheet Metal Worker	47.20	188.80	72.65	290.60		
16 L.H., Daily Totals		$696.40		$1066.00	$43.52	$66.63

Crew L-3A	Hr.	Daily	Hr.	Daily	Bare Costs	Incl. O&P
1 Carpenter Foreman (outside)	$41.95	$335.60	$65.05	$520.40	$43.70	$67.58
.5 Sheet Metal Worker	47.20	188.80	72.65	290.60		
12 L.H., Daily Totals		$524.40		$811.00	$43.70	$67.58

Crew L-4	Hr.	Daily	Hr.	Daily	Bare Costs	Incl. O&P
2 Skilled Workers	$40.85	$653.60	$63.25	$1012.00	$37.33	$57.77
1 Helper	30.30	242.40	46.80	374.40		
24 L.H., Daily Totals		$896.00		$1386.40	$37.33	$57.77

Crew L-5	Hr.	Daily	Hr.	Daily	Bare Costs	Incl. O&P
1 Struc. Steel Foreman	$46.70	$373.60	$83.20	$665.60	$44.68	$77.91
5 Struc. Steel Workers	44.70	1788.00	79.65	3186.00		
1 Equip. Oper. (crane)	42.55	340.40	63.95	511.60		
1 Hyd. Crane, 25 Ton		789.40		868.34	14.10	15.51
56 L.H., Daily Totals		$3291.40		$5231.54	$58.77	$93.42

Crew L-5A	Hr.	Daily	Hr.	Daily	Bare Costs	Incl. O&P
1 Structural Steel Foreman	$46.70	$373.60	$83.20	$665.60	$44.66	$76.61
2 Structural Steel Workers	44.70	715.20	79.65	1274.40		
1 Equip. Oper. (crane)	42.55	340.40	63.95	511.60		
1 S.P. Crane, 4x4, 25 Ton		716.80		788.48	22.40	24.64
32 L.H., Daily Totals		$2146.00		$3240.08	$67.06	$101.25

Crew L-5B	Hr.	Daily	Hr.	Daily	Bare Costs	Incl. O&P
1 Structural Steel Foreman	$46.70	$373.60	$83.20	$665.60	$45.35	$72.19
2 Structural Steel Workers	44.70	715.20	79.65	1274.40		
2 Electricians	47.00	752.00	69.95	1119.20		
2 Steamfitters/Pipefitters	49.35	789.60	74.05	1184.80		
1 Equip. Oper. (crane)	42.55	340.40	63.95	511.60		
1 Equip. Oper. Oiler	36.80	294.40	55.30	442.40		
1 Hyd. Crane, 80 Ton		1296.00		1425.60	18.00	19.80
72 L.H., Daily Totals		$4561.20		$6623.60	$63.35	$91.99

Crew L-6	Hr.	Daily	Hr.	Daily	Bare Costs	Incl. O&P
1 Plumber	$48.75	$390.00	$73.15	$585.20	$48.17	$72.08
.5 Electrician	47.00	188.00	69.95	279.80		
12 L.H., Daily Totals		$578.00		$865.00	$48.17	$72.08

Crew L-7	Hr.	Daily	Hr.	Daily	Bare Costs	Incl. O&P
2 Carpenters	$39.95	$639.20	$61.95	$991.20	$38.57	$59.39
1 Building Laborer	31.60	252.80	49.00	392.00		
.5 Electrician	47.00	188.00	69.95	279.80		
28 L.H., Daily Totals		$1080.00		$1663.00	$38.57	$59.39

Crew L-8	Hr.	Daily	Hr.	Daily	Bare Costs	Incl. O&P
2 Carpenters	$39.95	$639.20	$61.95	$991.20	$41.71	$64.19
.5 Plumber	48.75	195.00	73.15	292.60		
20 L.H., Daily Totals		$834.20		$1283.80	$41.71	$64.19

Crew L-9	Bare Costs Hr.	Bare Costs Daily	Incl. Subs O & P Hr.	Incl. Subs O & P Daily	Cost Per Labor-Hour Bare Costs	Cost Per Labor-Hour Incl. O&P
1 Labor Foreman (inside)	$32.10	$256.80	$49.80	$398.40	$36.33	$58.32
2 Building Laborers	31.60	505.60	49.00	784.00		
1 Struc. Steel Worker	44.70	357.60	79.65	637.20		
.5 Electrician	47.00	188.00	69.95	279.80		
36 L.H., Daily Totals		$1308.00		$2099.40	$36.33	$58.32

Crew L-10	Hr.	Daily	Hr.	Daily	Bare Costs	Incl. O&P
1 Structural Steel Foreman	$46.70	$373.60	$83.20	$665.60	$44.65	$75.60
1 Structural Steel Worker	44.70	357.60	79.65	637.20		
1 Equip. Oper. (crane)	42.55	340.40	63.95	511.60		
1 Hyd. Crane, 12 Ton		768.80		845.68	32.03	35.24
24 L.H., Daily Totals		$1840.40		$2660.08	$76.68	$110.84

Crew L-11	Hr.	Daily	Hr.	Daily	Bare Costs	Incl. O&P
2 Wreckers	$31.60	$505.60	$54.15	$866.40	$36.20	$57.74
1 Equip. Oper. (crane)	42.55	340.40	63.95	511.60		
1 Equip. Oper. (light)	39.05	312.40	58.70	469.60		
1 Hyd. Excavator, 2.5 C.Y.		1619.00		1780.90		
1 Loader, Skid Steer, 78 H.P.		259.60		285.56	58.71	64.58
32 L.H., Daily Totals		$3037.00		$3914.06	$94.91	$122.31

Crew M-1	Hr.	Daily	Hr.	Daily	Bare Costs	Incl. O&P
3 Elevator Constructors	$56.60	$1358.40	$84.30	$2023.20	$53.77	$80.09
1 Elevator Apprentice	45.30	362.40	67.45	539.60		
5 Hand Tools		57.00		62.70	1.78	1.96
32 L.H., Daily Totals		$1777.80		$2625.50	$55.56	$82.05

Crew M-3	Hr.	Daily	Hr.	Daily	Bare Costs	Incl. O&P
1 Electrician Foreman (out)	$49.00	$392.00	$72.90	$583.20	$45.37	$68.04
1 Common Laborer	31.60	252.80	49.00	392.00		
.25 Equipment Operator, Medium	41.35	82.70	62.15	124.30		
1 Elevator Constructor	56.60	452.80	84.30	674.40		
1 Elevator Apprentice	45.30	362.40	67.45	539.60		
.25 S.P. Crane, 4x4, 20 Ton		175.15		192.66	5.15	5.67
34 L.H., Daily Totals		$1717.85		$2506.17	$50.52	$73.71

Crew M-4	Hr.	Daily	Hr.	Daily	Bare Costs	Incl. O&P
1 Electrician Foreman (out)	$49.00	$392.00	$72.90	$583.20	$44.96	$67.44
1 Common Laborer	31.60	252.80	49.00	392.00		
.25 Equipment Operator, Crane	42.55	85.10	63.95	127.90		
.25 Equipment Operator, Oiler	36.80	73.60	55.30	110.60		
1 Elevator Constructor	56.60	452.80	84.30	674.40		
1 Elevator Apprentice	45.30	362.40	67.45	539.60		
.25 S.P. Crane, 4x4, 40 Ton		234.30		257.73	6.51	7.16
36 L.H., Daily Totals		$1853.00		$2685.43	$51.47	$74.60

Crew Q-1	Hr.	Daily	Hr.	Daily	Bare Costs	Incl. O&P
1 Plumber	$48.75	$390.00	$73.15	$585.20	$43.88	$65.85
1 Plumber Apprentice	39.00	312.00	58.55	468.40		
16 L.H., Daily Totals		$702.00		$1053.60	$43.88	$65.85

Crew Q-1A	Hr.	Daily	Hr.	Daily	Bare Costs	Incl. O&P
.25 Plumber Foreman (out)	$50.75	$101.50	$76.20	$152.40	$49.15	$73.76
1 Plumber	48.75	390.00	73.15	585.20		
10 L.H., Daily Totals		$491.50		$737.60	$49.15	$73.76

Crew Q-1C	Hr.	Daily	Hr.	Daily	Bare Costs	Incl. O&P
1 Plumber	$48.75	$390.00	$73.15	$585.20	$43.03	$64.62
1 Plumber Apprentice	39.00	312.00	58.55	468.40		
1 Equip. Oper. (medium)	41.35	330.80	62.15	497.20		
1 Trencher, Chain Type, 8' D		1777.00		1954.70	74.04	81.45
24 L.H., Daily Totals		$2809.80		$3505.50	$117.08	$146.06

Crew Q-2	Hr.	Daily	Hr.	Daily	Bare Costs	Incl. O&P
2 Plumbers	$48.75	$780.00	$73.15	$1170.40	$45.50	$68.28
1 Plumber Apprentice	39.00	312.00	58.55	468.40		
24 L.H., Daily Totals		$1092.00		$1638.80	$45.50	$68.28

Crew Q-3	Hr.	Daily	Hr.	Daily	Bare Costs	Incl. O&P
1 Plumber Foreman (inside)	$49.25	$394.00	$73.90	$591.20	$46.44	$69.69
2 Plumbers	48.75	780.00	73.15	1170.40		
1 Plumber Apprentice	39.00	312.00	58.55	468.40		
32 L.H., Daily Totals		$1486.00		$2230.00	$46.44	$69.69

Crew Q-4	Hr.	Daily	Hr.	Daily	Bare Costs	Incl. O&P
1 Plumber Foreman (inside)	$49.25	$394.00	$73.90	$591.20	$46.44	$69.69
1 Plumber	48.75	390.00	73.15	585.20		
1 Welder (plumber)	48.75	390.00	73.15	585.20		
1 Plumber Apprentice	39.00	312.00	58.55	468.40		
1 Welder, electric, 300 amp		55.70		61.27	1.74	1.91
32 L.H., Daily Totals		$1541.70		$2291.27	$48.18	$71.60

Crew Q-5	Hr.	Daily	Hr.	Daily	Bare Costs	Incl. O&P
1 Steamfitter	$49.35	$394.80	$74.05	$592.40	$44.42	$66.67
1 Steamfitter Apprentice	39.50	316.00	59.30	474.40		
16 L.H., Daily Totals		$710.80		$1066.80	$44.42	$66.67

Crew Q-6	Hr.	Daily	Hr.	Daily	Bare Costs	Incl. O&P
2 Steamfitters	$49.35	$789.60	$74.05	$1184.80	$46.07	$69.13
1 Steamfitter Apprentice	39.50	316.00	59.30	474.40		
24 L.H., Daily Totals		$1105.60		$1659.20	$46.07	$69.13

Crew Q-7	Hr.	Daily	Hr.	Daily	Bare Costs	Incl. O&P
1 Steamfitter Foreman (inside)	$49.85	$398.80	$74.80	$598.40	$47.01	$70.55
2 Steamfitters	49.35	789.60	74.05	1184.80		
1 Steamfitter Apprentice	39.50	316.00	59.30	474.40		
32 L.H., Daily Totals		$1504.40		$2257.60	$47.01	$70.55

Crew Q-8	Hr.	Daily	Hr.	Daily	Bare Costs	Incl. O&P
1 Steamfitter Foreman (inside)	$49.85	$398.80	$74.80	$598.40	$47.01	$70.55
1 Steamfitter	49.35	394.80	74.05	592.40		
1 Welder (steamfitter)	49.35	394.80	74.05	592.40		
1 Steamfitter Apprentice	39.50	316.00	59.30	474.40		
1 Welder, electric, 300 amp		55.70		61.27	1.74	1.91
32 L.H., Daily Totals		$1560.10		$2318.87	$48.75	$72.46

Crew Q-9	Hr.	Daily	Hr.	Daily	Bare Costs	Incl. O&P
1 Sheet Metal Worker	$47.20	$377.60	$72.65	$581.20	$42.48	$65.38
1 Sheet Metal Apprentice	37.75	302.00	58.10	464.80		
16 L.H., Daily Totals		$679.60		$1046.00	$42.48	$65.38

Crew Q-10	Hr.	Daily	Hr.	Daily	Bare Costs	Incl. O&P
2 Sheet Metal Workers	$47.20	$755.20	$72.65	$1162.40	$44.05	$67.80
1 Sheet Metal Apprentice	37.75	302.00	58.10	464.80		
24 L.H., Daily Totals		$1057.20		$1627.20	$44.05	$67.80

Crew No.	Bare Costs		Incl. Subs O & P		Cost Per Labor-Hour	
Crew Q-11	**Hr.**	**Daily**	**Hr.**	**Daily**	**Bare Costs**	**Incl. O&P**
1 Sheet Metal Foreman (inside)	$47.70	$381.60	$73.40	$587.20	$44.96	$69.20
2 Sheet Metal Workers	47.20	755.20	72.65	1162.40		
1 Sheet Metal Apprentice	37.75	302.00	58.10	464.80		
32 L.H., Daily Totals		$1438.80		$2214.40	$44.96	$69.20

Crew Q-12	Hr.	Daily	Hr.	Daily	Bare Costs	Incl. O&P
1 Sprinkler Installer	$48.15	$385.20	$72.25	$578.00	$43.33	$65.03
1 Sprinkler Apprentice	38.50	308.00	57.80	462.40		
16 L.H., Daily Totals		$693.20		$1040.40	$43.33	$65.03

Crew Q-13	Hr.	Daily	Hr.	Daily	Bare Costs	Incl. O&P
1 Sprinkler Foreman (inside)	$48.65	$389.20	$73.00	$584.00	$45.86	$68.83
2 Sprinkler Installers	48.15	770.40	72.25	1156.00		
1 Sprinkler Apprentice	38.50	308.00	57.80	462.40		
32 L.H., Daily Totals		$1467.60		$2202.40	$45.86	$68.83

Crew Q-14	Hr.	Daily	Hr.	Daily	Bare Costs	Incl. O&P
1 Asbestos Worker	$44.10	$352.80	$69.10	$552.80	$39.70	$62.20
1 Asbestos Apprentice	35.30	282.40	55.30	442.40		
16 L.H., Daily Totals		$635.20		$995.20	$39.70	$62.20

Crew Q-15	Hr.	Daily	Hr.	Daily	Bare Costs	Incl. O&P
1 Plumber	$48.75	$390.00	$73.15	$585.20	$43.88	$65.85
1 Plumber Apprentice	39.00	312.00	58.55	468.40		
1 Welder, electric, 300 amp		55.70		61.27	3.48	3.83
16 L.H., Daily Totals		$757.70		$1114.87	$47.36	$69.68

Crew Q-16	Hr.	Daily	Hr.	Daily	Bare Costs	Incl. O&P
2 Plumbers	$48.75	$780.00	$73.15	$1170.40	$45.50	$68.28
1 Plumber Apprentice	39.00	312.00	58.55	468.40		
1 Welder, electric, 300 amp		55.70		61.27	2.32	2.55
24 L.H., Daily Totals		$1147.70		$1700.07	$47.82	$70.84

Crew Q-17	Hr.	Daily	Hr.	Daily	Bare Costs	Incl. O&P
1 Steamfitter	$49.35	$394.80	$74.05	$592.40	$44.42	$66.67
1 Steamfitter Apprentice	39.50	316.00	59.30	474.40		
1 Welder, electric, 300 amp		55.70		61.27	3.48	3.83
16 L.H., Daily Totals		$766.50		$1128.07	$47.91	$70.50

Crew Q-17A	Hr.	Daily	Hr.	Daily	Bare Costs	Incl. O&P
1 Steamfitter	$49.35	$394.80	$74.05	$592.40	$43.80	$65.77
1 Steamfitter Apprentice	39.50	316.00	59.30	474.40		
1 Equip. Oper. (crane)	42.55	340.40	63.95	511.60		
1 Hyd. Crane, 12 Ton		768.80		845.68		
1 Welder, electric, 300 amp		55.70		61.27	34.35	37.79
24 L.H., Daily Totals		$1875.70		$2485.35	$78.15	$103.56

Crew Q-18	Hr.	Daily	Hr.	Daily	Bare Costs	Incl. O&P
2 Steamfitters	$49.35	$789.60	$74.05	$1184.80	$46.07	$69.13
1 Steamfitter Apprentice	39.50	316.00	59.30	474.40		
1 Welder, electric, 300 amp		55.70		61.27	2.32	2.55
24 L.H., Daily Totals		$1161.30		$1720.47	$48.39	$71.69

Crew Q-19	Hr.	Daily	Hr.	Daily	Bare Costs	Incl. O&P
1 Steamfitter	$49.35	$394.80	$74.05	$592.40	$45.28	$67.77
1 Steamfitter Apprentice	39.50	316.00	59.30	474.40		
1 Electrician	47.00	376.00	69.95	559.60		
24 L.H., Daily Totals		$1086.80		$1626.40	$45.28	$67.77

Crew No.	Bare Costs		Incl. Subs O & P		Cost Per Labor-Hour	
Crew Q-20	**Hr.**	**Daily**	**Hr.**	**Daily**	**Bare Costs**	**Incl. O&P**
1 Sheet Metal Worker	$47.20	$377.60	$72.65	$581.20	$43.38	$66.29
1 Sheet Metal Apprentice	37.75	302.00	58.10	464.80		
.5 Electrician	47.00	188.00	69.95	279.80		
20 L.H., Daily Totals		$867.60		$1325.80	$43.38	$66.29

Crew Q-21	Hr.	Daily	Hr.	Daily	Bare Costs	Incl. O&P
2 Steamfitters	$49.35	$789.60	$74.05	$1184.80	$46.30	$69.34
1 Steamfitter Apprentice	39.50	316.00	59.30	474.40		
1 Electrician	47.00	376.00	69.95	559.60		
32 L.H., Daily Totals		$1481.60		$2218.80	$46.30	$69.34

Crew Q-22	Hr.	Daily	Hr.	Daily	Bare Costs	Incl. O&P
1 Plumber	$48.75	$390.00	$73.15	$585.20	$43.88	$65.85
1 Plumber Apprentice	39.00	312.00	58.55	468.40		
1 Hyd. Crane, 12 Ton		768.80		845.68	48.05	52.85
16 L.H., Daily Totals		$1470.80		$1899.28	$91.92	$118.71

Crew Q-22A	Hr.	Daily	Hr.	Daily	Bare Costs	Incl. O&P
1 Plumber	$48.75	$390.00	$73.15	$585.20	$40.48	$61.16
1 Plumber Apprentice	39.00	312.00	58.55	468.40		
1 Laborer	31.60	252.80	49.00	392.00		
1 Equip. Oper. (crane)	42.55	340.40	63.95	511.60		
1 Hyd. Crane, 12 Ton		768.80		845.68	24.02	26.43
32 L.H., Daily Totals		$2064.00		$2802.88	$64.50	$87.59

Crew Q-23	Hr.	Daily	Hr.	Daily	Bare Costs	Incl. O&P
1 Plumber Foreman	$50.75	$406.00	$76.20	$609.60	$46.95	$70.50
1 Plumber	48.75	390.00	73.15	585.20		
1 Equip. Oper. (medium)	41.35	330.80	62.15	497.20		
1 Lattice Boom Crane, 20 Ton		1080.00		1188.00	45.00	49.50
24 L.H., Daily Totals		$2206.80		$2880.00	$91.95	$120.00

Crew R-1	Hr.	Daily	Hr.	Daily	Bare Costs	Incl. O&P
1 Electrician Foreman	$47.50	$380.00	$70.70	$565.60	$41.52	$62.36
3 Electricians	47.00	1128.00	69.95	1678.80		
2 Helpers	30.30	484.80	46.80	748.80		
48 L.H., Daily Totals		$1992.80		$2993.20	$41.52	$62.36

Crew R-1A	Hr.	Daily	Hr.	Daily	Bare Costs	Incl. O&P
1 Electrician	$47.00	$376.00	$69.95	$559.60	$38.65	$58.38
1 Helper	30.30	242.40	46.80	374.40		
16 L.H., Daily Totals		$618.40		$934.00	$38.65	$58.38

Crew R-2	Hr.	Daily	Hr.	Daily	Bare Costs	Incl. O&P
1 Electrician Foreman	$47.50	$380.00	$70.70	$565.60	$41.66	$62.59
3 Electricians	47.00	1128.00	69.95	1678.80		
2 Helpers	30.30	484.80	46.80	748.80		
1 Equip. Oper. (crane)	42.55	340.40	63.95	511.60		
1 S.P. Crane, 4x4, 5 Ton		258.20		284.02	4.61	5.07
56 L.H., Daily Totals		$2591.40		$3788.82	$46.27	$67.66

Crew R-3	Hr.	Daily	Hr.	Daily	Bare Costs	Incl. O&P
1 Electrician Foreman	$47.50	$380.00	$70.70	$565.60	$46.31	$69.05
1 Electrician	47.00	376.00	69.95	559.60		
.5 Equip. Oper. (crane)	42.55	170.20	63.95	255.80		
.5 S.P. Crane, 4x4, 5 Ton		129.10		142.01	6.46	7.10
20 L.H., Daily Totals		$1055.30		$1523.01	$52.77	$76.15

Crew No.	Bare Costs		Incl. Subs O & P		Cost Per Labor-Hour	
Crew R-4	**Hr.**	**Daily**	**Hr.**	**Daily**	**Bare Costs**	**Incl. O&P**
1 Struc. Steel Foreman	$46.70	$373.60	$83.20	$665.60	$45.56	$78.42
3 Struc. Steel Workers	44.70	1072.80	79.65	1911.60		
1 Electrician	47.00	376.00	69.95	559.60		
1 Welder, gas engine, 300 amp		134.20		147.62	3.36	3.69
40 L.H., Daily Totals		$1956.60		$3284.42	$48.91	$82.11

Crew R-5	Hr.	Daily	Hr.	Daily	Bare Costs	Incl. O&P
1 Electrician Foreman	$47.50	$380.00	$70.70	$565.60	$40.97	$61.60
4 Electrician Linemen	47.00	1504.00	69.95	2238.40		
2 Electrician Operators	47.00	752.00	69.95	1119.20		
4 Electrician Groundmen	30.30	969.60	46.80	1497.60		
1 Crew Truck		169.60		186.56		
1 Flatbed Truck, 20,000 GVW		197.40		217.14		
1 Pickup Truck, 3/4 Ton		116.80		128.48		
.2 Hyd. Crane, 55 Ton		225.00		247.50		
.2 Hyd. Crane, 12 Ton		153.76		169.14		
.2 Earth Auger, Truck-Mtd.		82.40		90.64		
1 Tractor w/Winch		354.80		390.28	14.77	16.25
88 L.H., Daily Totals		$4905.36		$6850.54	$55.74	$77.85

Crew R-6	Hr.	Daily	Hr.	Daily	Bare Costs	Incl. O&P
1 Electrician Foreman	$47.50	$380.00	$70.70	$565.60	$40.97	$61.60
4 Electrician Linemen	47.00	1504.00	69.95	2238.40		
2 Electrician Operators	47.00	752.00	69.95	1119.20		
4 Electrician Groundmen	30.30	969.60	46.80	1497.60		
1 Crew Truck		169.60		186.56		
1 Flatbed Truck, 20,000 GVW		197.40		217.14		
1 Pickup Truck, 3/4 Ton		116.80		128.48		
.2 Hyd. Crane, 55 Ton		225.00		247.50		
.2 Hyd. Crane, 12 Ton		153.76		169.14		
.2 Earth Auger, Truck-Mtd.		82.40		90.64		
1 Tractor w/Winch		354.80		390.28		
3 Cable Trailers		531.30		584.43		
.5 Tensioning Rig		175.35		192.88		
.5 Cable Pulling Rig		1049.50		1154.45	34.73	38.20
88 L.H., Daily Totals		$6661.51		$8782.30	$75.70	$99.80

Crew R-7	Hr.	Daily	Hr.	Daily	Bare Costs	Incl. O&P
1 Electrician Foreman	$47.50	$380.00	$70.70	$565.60	$33.17	$50.78
5 Electrician Groundmen	30.30	1212.00	46.80	1872.00		
1 Crew Truck		169.60		186.56	3.53	3.89
48 L.H., Daily Totals		$1761.60		$2624.16	$36.70	$54.67

Crew R-8	Hr.	Daily	Hr.	Daily	Bare Costs	Incl. O&P
1 Electrician Foreman	$47.50	$380.00	$70.70	$565.60	$41.52	$62.36
3 Electrician Linemen	47.00	1128.00	69.95	1678.80		
2 Electrician Groundmen	30.30	484.80	46.80	748.80		
1 Pickup Truck, 3/4 Ton		116.80		128.48		
1 Crew Truck		169.60		186.56	5.97	6.56
48 L.H., Daily Totals		$2279.20		$3308.24	$47.48	$68.92

Crew R-9	Hr.	Daily	Hr.	Daily	Bare Costs	Incl. O&P
1 Electrician Foreman	$47.50	$380.00	$70.70	$565.60	$38.71	$58.47
1 Electrician Lineman	47.00	376.00	69.95	559.60		
2 Electrician Operators	47.00	752.00	69.95	1119.20		
4 Electrician Groundmen	30.30	969.60	46.80	1497.60		
1 Pickup Truck, 3/4 Ton		116.80		128.48		
1 Crew Truck		169.60		186.56	4.47	4.92
64 L.H., Daily Totals		$2764.00		$4057.04	$43.19	$63.39

Crew No.	Bare Costs		Incl. Subs O & P		Cost Per Labor-Hour	
Crew R-10	**Hr.**	**Daily**	**Hr.**	**Daily**	**Bare Costs**	**Incl. O&P**
1 Electrician Foreman	$47.50	$380.00	$70.70	$565.60	$44.30	$66.22
4 Electrician Linemen	47.00	1504.00	69.95	2238.40		
1 Electrician Groundman	30.30	242.40	46.80	374.40		
1 Crew Truck		169.60		186.56		
3 Tram Cars		380.55		418.61	11.46	12.61
48 L.H., Daily Totals		$2676.55		$3783.57	$55.76	$78.82

Crew R-11	Hr.	Daily	Hr.	Daily	Bare Costs	Incl. O&P
1 Electrician Foreman	$47.50	$380.00	$70.70	$565.60	$44.24	$66.21
4 Electricians	47.00	1504.00	69.95	2238.40		
1 Equip. Oper. (crane)	42.55	340.40	63.95	511.60		
1 Common Laborer	31.60	252.80	49.00	392.00		
1 Crew Truck		169.60		186.56		
1 Hyd. Crane, 12 Ton		768.80		845.68	16.76	18.43
56 L.H., Daily Totals		$3415.60		$4739.84	$60.99	$84.64

Crew R-12	Hr.	Daily	Hr.	Daily	Bare Costs	Incl. O&P
1 Carpenter Foreman	$40.45	$323.60	$62.75	$502.00	$37.52	$58.94
4 Carpenters	39.95	1278.40	61.95	1982.40		
4 Common Laborers	31.60	1011.20	49.00	1568.00		
1 Equip. Oper. (med.)	41.35	330.80	62.15	497.20		
1 Steel Worker	44.70	357.60	79.65	637.20		
1 Dozer, 200 H.P.		1082.00		1190.20		
1 Pickup Truck, 3/4 Ton		116.80		128.48	13.62	14.98
88 L.H., Daily Totals		$4500.40		$6505.48	$51.14	$73.93

Crew R-13	Hr.	Daily	Hr.	Daily	Bare Costs	Incl. O&P
1 Electrician Foreman	$47.50	$380.00	$70.70	$565.60	$44.94	$67.02
3 Electricians	47.00	1128.00	69.95	1678.80		
.25 Equip. Oper. (crane)	42.55	85.10	63.95	127.90		
1 Equipment Oiler	36.80	294.40	55.30	442.40		
.25 Hydraulic Crane, 33 Ton		204.70		225.17	4.87	5.36
42 L.H., Daily Totals		$2092.20		$3039.87	$49.81	$72.38

Crew R-15	Hr.	Daily	Hr.	Daily	Bare Costs	Incl. O&P
1 Electrician Foreman	$47.50	$380.00	$70.70	$565.60	$45.76	$68.20
4 Electricians	47.00	1504.00	69.95	2238.40		
1 Equipment Oper. (light)	39.05	312.40	58.70	469.60		
1 Aerial Lift Truck		328.80		361.68	6.85	7.54
48 L.H., Daily Totals		$2525.20		$3635.28	$52.61	$75.73

Crew R-18	Hr.	Daily	Hr.	Daily	Bare Costs	Incl. O&P
.25 Electrician Foreman	$47.50	$95.00	$70.70	$141.40	$36.76	$55.76
1 Electrician	47.00	376.00	69.95	559.60		
2 Helpers	30.30	484.80	46.80	748.80		
26 L.H., Daily Totals		$955.80		$1449.80	$36.76	$55.76

Crew R-19	Hr.	Daily	Hr.	Daily	Bare Costs	Incl. O&P
.5 Electrician Foreman	$47.50	$190.00	$70.70	$282.80	$47.10	$70.10
2 Electricians	47.00	752.00	69.95	1119.20		
20 L.H., Daily Totals		$942.00		$1402.00	$47.10	$70.10

Crew R-21	Hr.	Daily	Hr.	Daily	Bare Costs	Incl. O&P
1 Electrician Foreman	$47.50	$380.00	$70.70	$565.60	$46.98	$69.94
3 Electricians	47.00	1128.00	69.95	1678.80		
.1 Equip. Oper. (med.)	41.35	33.08	62.15	49.72		
.1 S.P. Crane, 4x4, 25 Ton		71.68		78.85	2.19	2.40
32.8 L.H., Daily Totals		$1612.76		$2372.97	$49.17	$72.35

Crew No.	Bare Costs		Incl. Subs O & P		Cost Per Labor-Hour	
Crew R-22	**Hr.**	**Daily**	**Hr.**	**Daily**	**Bare Costs**	**Incl. O&P**
.66 Electrician Foreman	$47.50	$250.80	$70.70	$373.30	$39.90	$60.12
2 Helpers	30.30	484.80	46.80	748.80		
2 Electricians	47.00	752.00	69.95	1119.20		
37.28 L.H., Daily Totals		$1487.60		$2241.30	$39.90	$60.12
Crew R-30	**Hr.**	**Daily**	**Hr.**	**Daily**	**Bare Costs**	**Incl. O&P**
.25 Electrician Foreman (out)	$49.00	$98.00	$72.90	$145.80	$37.68	$57.28
1 Electrician	47.00	376.00	69.95	559.60		
2 Laborers, (Semi-Skilled)	31.60	505.60	49.00	784.00		
26 L.H., Daily Totals		$979.60		$1489.40	$37.68	$57.28
Crew R-31	**Hr.**	**Daily**	**Hr.**	**Daily**	**Bare Costs**	**Incl. O&P**
1 Electrician	$47.00	$376.00	$69.95	$559.60	$47.00	$69.95
1 Core Drill, Electric, 2.5 H.P.		52.95		58.24	6.62	7.28
8 L.H., Daily Totals		$428.95		$617.85	$53.62	$77.23
Crew W-41E	**Hr.**	**Daily**	**Hr.**	**Daily**	**Bare Costs**	**Incl. O&P**
.5 Plumber Foreman (out)	$50.75	$203.00	$76.20	$304.80	$42.29	$64.10
1 Plumber	48.75	390.00	73.15	585.20		
1 Laborer	31.60	252.80	49.00	392.00		
20 L.H., Daily Totals		$845.80		$1282.00	$42.29	$64.10

Historical Cost Indexes

The table below lists both the RSMeans Historical Cost Index based on Jan. 1, 1993 = 100 as well as the computed value of an index based on Jan. 1, 2009 costs. Since the Jan. 1, 2009 figure is estimated, space is left to write in the actual index figures as they become available through either the quarterly "RSMeans Construction Cost Indexes" or as printed in the "Engineering News-Record." To compute the actual index based on Jan. 1, 2009 = 100, divide the Historical Cost Index for a particular year by the actual Jan. 1, 2009 Construction Cost Index. Space has been left to advance the index figures as the year progresses.

Year	Historical Cost Index Jan. 1, 1993 = 100		Current Index Based on Jan. 1, 2009 = 100		Year	Historical Cost Index Jan. 1, 1993 = 100	Current Index Based on Jan. 1, 2009 = 100		Year	Historical Cost Index Jan. 1, 1993 = 100	Current Index Based on Jan. 1, 2009 = 100	
	Est.	Actual	Est.	Actual		Actual	Est.	Actual		Actual	Est.	Actual
Oct 2009					July 1994	104.4	60.3		July 1976	46.9	27.1	
July 2009					1993	101.7	54.7		1975	44.8	24.1	
April 2009					1992	99.4	53.5		1974	41.4	22.3	
Jan 2009	185.9		100.0	100.0	1991	96.8	52.1		1973	37.7	20.3	
July 2008		180.4	97.0		1990	94.3	50.7		1972	34.8	18.7	
2007		169.4	91.1		1989	92.1	49.6		1971	32.1	17.3	
2006		162.0	87.1		1988	89.9	48.3		1970	28.7	15.4	
2005		151.6	81.5		1987	87.7	47.2		1969	26.9	14.5	
2004		143.7	77.3		1986	84.2	45.3		1968	24.9	13.4	
2003		132.0	71.0		1985	82.6	44.4		1967	23.5	12.6	
2002		128.7	69.2		1984	82.0	44.1		1966	22.7	12.2	
2001		125.1	67.3		1983	80.2	43.1		1965	21.7	11.7	
2000		120.9	65.0		1982	76.1	41.0		1964	21.2	11.4	
1999		117.6	63.3		1981	70.0	37.6		1963	20.7	11.1	
1998		115.1	61.9		1980	62.9	33.8		1962	20.2	10.9	
1997		112.8	60.7		1979	57.8	31.1		1961	19.8	10.7	
1996		110.2	59.3		1978	53.5	28.8		1960	19.7	10.6	
1995		107.6	57.9		1977	49.5	26.6		1959	19.3	10.4	

Adjustments to Costs

The Historical Cost Index can be used to convert National Average building costs at a particular time to the approximate building costs for some other time.

Example:

Estimate and compare construction costs for different years in the same city.

To estimate the National Average construction cost of a building in 1970, knowing that it cost $900,000 in 2009:

INDEX in 1970 = 28.7

INDEX in 2009 = 185.9

Time Adjustment using the Historical Cost Indexes:

$$\frac{\text{Index for Year A}}{\text{Index for Year B}} \text{ x Cost in Year B = Cost in Year A}$$

$$\frac{\text{INDEX 1970}}{\text{INDEX 2009}} \text{ x Cost 2009 = Cost 1970}$$

$$\frac{28.7}{185.9} \text{ x } \$900{,}000 = .154 \text{ x } \$900{,}000 = \$138{,}600$$

The construction cost of the building in 1970 is $138,600.

Note: The City Cost Indexes for Canada can be used to convert U.S. National averages to local costs in Canadian dollars.

How to Use the City Cost Indexes

What you should know before you begin

RSMeans City Cost Indexes (CCI) are an extremely useful tool to use when you want to compare costs from city to city and region to region.

This publication contains average construction cost indexes for 316 U.S. and Canadian cities covering over 930 three-digit zip code locations.

Keep in mind that a City Cost Index number is a percentage ratio of a specific city's cost to the national average cost of the same item at a stated time period.

In other words, these index figures represent relative construction factors (or, if you prefer, multipliers) for Material and Installation costs, as well as the weighted average for Total In Place costs for each CSI MasterFormat division. Installation costs include both labor and equipment rental costs. When estimating equipment rental rates only, for a specific location, use 01543 CONTRACTOR EQUIPMENT index.

The 30 City Average Index is the average of 30 major U.S. cities and serves as a National Average.

Index figures for both material and installation are based on the 30 major city average of 100 and represent the cost relationship as of July 1, 2008. The index for each division is computed from representative material and labor quantities for that division. The weighted average for each city is a weighted total of the components listed above it, but does not include relative productivity between trades or cities.

As changes occur in local material prices, labor rates, and equipment rental rates, (including fuel costs) the impact of these changes should be accurately measured by the change in the City Cost Index for each particular city (as compared to the 30 City Average).

Therefore, if you know (or have estimated) building costs in one city today, you can easily convert those costs to expected building costs in another city.

In addition, by using the Historical Cost Index, you can easily convert National Average building costs at a particular time to the approximate building costs for some other time. The City Cost Indexes can then be applied to calculate the costs for a particular city.

Quick Calculations

Location Adjustment Using the City Cost Indexes:

$$\frac{\text{Index for City A}}{\text{Index for City B}} \text{ x Cost in City B = Cost in City A}$$

Time Adjustment for the National Average Using the Historical Cost Index:

$$\frac{\text{Index for Year A}}{\text{Index for Year B}} \text{ x Cost in Year B = Cost in Year A}$$

Adjustment from the National Average:

$$\frac{\text{Index for City A}}{100} \text{ x National Average Cost = Cost in City A}$$

Since each of the other RSMeans publications contains many different items, any *one* item multiplied by the particular city index may give incorrect results. However, the larger the number of items compiled, the closer the results should be to actual costs for that particular city.

The City Cost Indexes for Canadian cities are calculated using Canadian material and equipment prices and labor rates, in Canadian dollars. Therefore, indexes for Canadian cities can be used to convert U.S. National Average prices to local costs in Canadian dollars.

How to use this section

1. Compare costs from city to city.

In using the RSMeans Indexes, remember that an index number is not a fixed number but a ratio: It's a percentage ratio of a building component's cost at any stated time to the National Average cost of that same component at the same time period. Put in the form of an equation:

$$\frac{\text{Specific City Cost}}{\text{National Average Cost}} \text{ x 100 = City Index Number}$$

Therefore, when making cost comparisons between cities, do not subtract one city's index number from the index number of another city and read the result as a percentage difference. Instead, divide one city's index number by that of the other city. The resulting number may then be used as a multiplier to calculate cost differences from city to city.

The formula used to find cost differences between cities for the purpose of comparison is as follows:

$$\frac{\text{City A Index}}{\text{City B Index}} \text{ x City B Cost (Known) = City A Cost (Unknown)}$$

In addition, you can use RSMeans CCI to calculate and compare costs division by division between cities using the same basic formula. (Just be sure that you're comparing similar divisions.)

2. Compare a specific city's construction costs with the National Average.

When you're studying construction location feasibility, it's advisable to compare a prospective project's cost index with an index of the National Average cost.

For example, divide the weighted average index of construction costs of a specific city by that of the 30 City Average, which = 100.

$$\frac{\text{City Index}}{100} = \text{\% of National Average}$$

As a result, you get a ratio that indicates the relative cost of construction in that city in comparison with the National Average.

3. Convert U.S. National Average to actual costs in Canadian City.

$$\frac{\text{Index for Canadian City}}{100} \times \text{National Average Cost = Cost in Canadian City in \$ CAN}$$

4. **Adjust construction cost data based on a National Average.**

When you use a source of construction cost data which is based on a National Average (such as RSMeans cost data publications), it is necessary to adjust those costs to a specific location.

$$\frac{\text{City Index}}{100} \times \begin{array}{c}\text{"Book" Cost Based on}\\ \text{National Average Costs}\end{array} = \begin{array}{c}\text{City Cost}\\ \text{(Unknown)}\end{array}$$

5. **When applying the City Cost Indexes to demolition projects, use the appropriate division installation index.** For example, for removal of existing doors and windows, use Division 8 (Openings) index.

What you might like to know about how we developed the Indexes

The information presented in the CCI is organized according to the Construction Specifications Institute (CSI) MasterFormat 2004.

To create a reliable index, RSMeans researched the building type most often constructed in the United States and Canada. Because it was concluded that no one type of building completely represented the building construction industry, nine different types of buildings were combined to create a composite model.

The exact material, labor and equipment quantities are based on detailed analysis of these nine building types, then each quantity is weighted in proportion to expected usage. These various material items, labor hours, and equipment rental rates are thus combined to form a composite building representing as closely as possible the actual usage of materials, labor and equipment used in the North American Building Construction Industry.

The following structures were chosen to make up that composite model:

1. Factory, 1 story
2. Office, 2–4 story
3. Store, Retail
4. Town Hall, 2–3 story
5. High School, 2–3 story
6. Hospital, 4–8 story
7. Garage, Parking
8. Apartment, 1–3 story
9. Hotel/Motel, 2–3 story

For the purposes of ensuring the timeliness of the data, the components of the index for the composite model have been streamlined. They currently consist of:

- specific quantities of 66 commonly used construction materials;
- specific labor-hours for 21 building construction trades; and
- specific days of equipment rental for 6 types of construction equipment (normally used to install the 66 material items by the 21 trades.) Fuel costs and routine maintenance costs are included in the equipment costs.

A sophisticated computer program handles the updating of all costs for each city on a quarterly basis. Material and equipment price quotations are gathered quarterly from 316 cities in the United States and Canada. These prices and the latest negotiated labor wage rates for 21 different building trades are used to compile the quarterly update of the City Cost Index.

The 30 major U.S. cities used to calculate the National Average are:

Atlanta, GA	Memphis, TN
Baltimore, MD	Milwaukee, WI
Boston, MA	Minneapolis, MN
Buffalo, NY	Nashville, TN
Chicago, IL	New Orleans, LA
Cincinnati, OH	New York, NY
Cleveland, OH	Philadelphia, PA
Columbus, OH	Phoenix, AZ
Dallas, TX	Pittsburgh, PA
Denver, CO	St. Louis, MO
Detroit, MI	San Antonio, TX
Houston, TX	San Diego, CA
Indianapolis, IN	San Francisco, CA
Kansas City, MO	Seattle, WA
Los Angeles, CA	Washington, DC

What the CCI does not indicate

The weighted average for each city is a total of the divisional components weighted to reflect typical usage, but it does not include the productivity variations between trades or cities.

In addition, the CCI does not take into consideration factors such as the following:

- managerial efficiency
- competitive conditions
- automation
- restrictive union practices
- unique local requirements
- regional variations due to specific building codes

	DIVISION	UNITED STATES			ALABAMA															
		ANYTOWN			BIRMINGHAM			HUNTSVILLE			MOBILE			MONTGOMERY			TUSCALOOSA			
		MAT.	INST.	TOTAL	MAT.	INST.	TOTAL	MAT.	INST.	TOTAL	MAT.	INST.	TOTAL	MAT.	INST.	TOTAL	MAT.	INST.	TOTAL	
015433	CONTRACTOR EQUIPMENT		100.0	100.0		100.5	100.5		100.4	100.4		98.9	98.9		98.9	98.9		100.4	100.4	
0241, 31 - 34	SITE & INFRASTRUCTURE, DEMOLITION	100.0	100.0	100.0	86.5	91.9	90.2	82.2	91.4	88.5	95.7	87.7	90.2	93.7	87.8	89.6	82.7	90.8	88.2	
0310	Concrete Forming & Accessories	100.0	100.0	100.0	91.0	72.9	75.3	94.9	67.1	70.9	95.0	59.2	64.1	92.7	49.9	55.7	94.8	53.8	59.4	
0320	Concrete Reinforcing	100.0	100.0	100.0	89.7	80.3	85.8	89.7	76.3	84.2	92.5	52.8	76.3	92.5	78.9	87.0	89.7	79.3	85.4	
0330	Cast-in-Place Concrete	100.0	100.0	100.0	99.2	71.4	88.6	89.7	68.5	81.6	94.7	64.3	83.1	97.0	52.0	79.8	93.4	54.5	78.6	
03	CONCRETE	100.0	100.0	100.0	92.8	74.8	84.6	88.6	70.6	80.4	91.5	61.3	77.8	92.5	57.8	76.7	90.4	60.4	76.8	
04	MASONRY	100.0	100.0	100.0	87.3	81.1	83.6	86.8	66.0	74.3	87.5	60.7	71.4	86.0	41.3	59.2	85.9	52.9	66.1	
05	METALS	100.0	100.0	100.0	99.9	91.5	97.4	101.3	90.4	98.1	100.1	80.0	94.2	99.9	88.7	96.6	100.5	89.2	97.2	
06	WOOD, PLASTICS & COMPOSITES	100.0	100.0	100.0	91.7	70.1	79.9	94.6	67.4	79.7	94.8	59.1	75.2	91.3	49.6	68.4	94.6	52.0	71.2	
07	THERMAL & MOISTURE PROTECTION	100.0	100.0	100.0	94.8	85.8	91.3	92.4	78.2	86.9	92.4	77.4	86.6	91.8	68.2	82.6	92.7	72.1	84.7	
08	OPENINGS	100.0	100.0	100.0	96.8	73.7	91.1	97.1	64.9	89.1	97.1	56.5	87.0	97.1	57.2	87.2	97.1	63.8	88.8	
0920	Plaster & Gypsum Board	100.0	100.0	100.0	103.6	69.6	81.5	98.3	66.7	77.8	100.7	58.1	73.0	103.6	48.3	67.7	100.7	50.8	68.3	
0950, 0980	Ceilings & Acoustic Treatment	100.0	100.0	100.0	97.1	69.6	80.0	100.4	66.7	79.6	100.4	58.1	74.2	97.1	48.3	66.9	100.4	50.8	69.7	
0960	Flooring	100.0	100.0	100.0	103.5	67.6	93.8	103.5	60.6	91.9	113.4	62.8	99.7	111.6	33.2	90.4	103.5	42.1	86.9	
0970, 0990	Wall Finishes & Painting/Coating	100.0	100.0	100.0	100.1	63.9	78.4	100.1	60.5	76.4	104.2	62.4	79.1	100.1	61.3	76.8	100.1	52.8	71.7	
09	FINISHES	100.0	100.0	100.0	97.7	70.6	83.4	97.4	64.8	80.2	102.5	59.9	79.9	101.0	46.9	72.4	97.7	50.8	72.9	
COVERS	DIVS. 10 - 14, 25, 28, 41, 43, 44	100.0	100.0	100.0	100.0	87.4	97.5	100.0	83.7	96.8	100.0	81.4	96.3	100.0	78.7	95.8	100.0	80.9	96.2	
21, 22, 23	FIRE SUPPRESSION, PLUMBING & HVAC	100.0	100.0	100.0	99.9	67.3	86.9	99.8	62.1	84.8	99.8	63.7	85.4	99.9	40.6	76.3	99.9	40.4	76.2	
26, 27, 3370	ELECTRICAL, COMMUNICATIONS & UTIL.	100.0	100.0	100.0	108.6	60.1	85.0	100.5	67.3	84.3	100.5	58.4	80.0	103.4	71.0	87.6	100.0	60.1	80.5	
MF2004	WEIGHTED AVERAGE	100.0	100.0	100.0	98.3	74.9	88.5	97.0	71.1	86.1	97.9	65.9	84.5	98.0	59.2	81.7	97.0	60.5	81.7	

	DIVISION	ALASKA									ARIZONA								
		ANCHORAGE			FAIRBANKS			JUNEAU			FLAGSTAFF			MESA/TEMPE			PHOENIX		
		MAT.	INST.	TOTAL	MAT.	INST.	TOTAL	MAT.	INST.	TOTAL	MAT.	INST.	TOTAL	MAT.	INST.	TOTAL	MAT.	INST.	TOTAL
015433	CONTRACTOR EQUIPMENT		113.0	113.0		113.0	113.0		113.0	113.0		96.2	96.2		99.2	99.2		99.8	99.8
0241, 31 - 34	SITE & INFRASTRUCTURE, DEMOLITION	140.1	126.1	130.5	126.3	126.2	126.3	139.2	126.1	130.2	81.1	102.1	95.4	86.2	104.8	98.9	86.7	106.0	99.9
0310	Concrete Forming & Accessories	131.5	114.1	116.5	137.2	116.5	119.3	133.2	114.0	116.6	103.2	65.4	70.6	103.8	60.0	66.0	104.8	69.2	74.0
0320	Concrete Reinforcing	136.8	103.8	123.3	139.4	103.8	124.9	104.9	103.7	104.4	99.9	79.4	91.5	98.8	79.4	90.9	97.4	79.7	90.2
0330	Cast-in-Place Concrete	168.8	111.3	146.8	148.0	112.6	134.5	169.7	111.2	147.3	96.0	77.4	88.9	104.8	65.6	89.8	104.9	77.9	94.6
03	CONCRETE	138.1	110.7	125.6	125.3	112.1	119.3	132.2	110.6	122.4	121.0	72.3	98.9	108.5	65.9	89.1	108.1	74.3	92.7
04	MASONRY	207.3	116.1	152.7	216.8	117.7	157.5	198.8	116.1	149.3	97.8	61.8	76.3	110.7	47.6	72.9	97.8	64.9	78.1
05	METALS	124.0	100.9	117.3	124.1	101.1	117.4	124.1	100.8	117.3	95.3	74.9	89.3	93.5	75.4	88.2	95.1	76.6	89.7
06	WOOD, PLASTICS & COMPOSITES	114.3	113.6	113.9	118.5	115.7	117.0	114.3	113.6	113.9	105.5	65.2	83.4	96.2	65.3	79.3	97.1	70.0	82.2
07	THERMAL & MOISTURE PROTECTION	181.2	111.1	153.9	178.9	113.9	153.6	179.5	111.1	152.9	98.9	67.0	86.5	107.4	58.4	88.3	107.3	68.1	92.1
08	OPENINGS	125.0	109.8	121.3	122.2	111.0	119.4	122.2	109.8	119.1	100.8	68.9	92.9	96.3	66.7	89.0	97.3	71.5	90.9
0920	Plaster & Gypsum Board	139.3	113.8	122.7	156.6	116.0	130.2	139.3	113.8	122.7	85.7	64.0	71.6	84.9	64.0	71.3	87.3	68.8	75.2
0950, 0980	Ceilings & Acoustic Treatment	125.5	113.8	118.3	131.4	116.0	121.8	125.5	113.8	118.3	112.0	64.0	82.3	101.2	64.0	78.2	108.7	68.8	84.0
0960	Flooring	158.9	120.7	148.6	158.9	120.7	148.6	158.9	120.7	148.6	98.3	44.5	83.7	102.9	55.5	90.1	103.2	58.0	91.0
0970, 0990	Wall Finishes & Painting/Coating	155.7	112.9	130.0	155.7	121.0	134.8	155.7	112.9	130.0	90.4	55.1	69.2	94.0	55.1	70.7	94.0	57.2	71.9
09	FINISHES	148.5	115.6	131.1	150.3	118.2	133.3	147.3	115.6	130.5	101.0	59.8	79.2	100.2	58.1	78.0	102.4	65.3	82.8
COVERS	DIVS. 10 - 14, 25, 28, 41, 43, 44	100.0	110.2	102.0	100.0	111.0	102.2	100.0	110.2	102.0	100.0	81.5	96.3	100.0	76.7	95.4	100.0	82.4	96.5
21, 22, 23	FIRE SUPPRESSION, PLUMBING & HVAC	100.5	103.0	101.5	100.4	109.8	104.2	100.5	94.4	98.1	100.2	76.8	90.9	100.1	66.6	86.8	100.1	76.9	90.9
26, 27, 3370	ELECTRICAL, COMMUNICATIONS & UTIL.	140.0	109.5	125.1	148.2	109.5	129.4	141.5	109.5	125.9	101.6	61.3	81.9	92.5	61.3	77.3	100.3	64.5	82.9
MF2004	WEIGHTED AVERAGE	128.3	110.3	120.8	127.6	112.6	121.3	127.0	108.5	119.2	101.4	71.6	88.9	99.2	66.7	85.6	99.9	74.0	89.0

	DIVISION	ARIZONA						ARKANSAS											
		PRESCOTT			TUCSON			FORT SMITH			JONESBORO			LITTLE ROCK			PINE BLUFF		
		MAT.	INST.	TOTAL	MAT.	INST.	TOTAL	MAT.	INST.	TOTAL	MAT.	INST.	TOTAL	MAT.	INST.	TOTAL	MAT.	INST.	TOTAL
015433	CONTRACTOR EQUIPMENT		96.2	96.2		99.2	99.2		92.0	92.0		105.6	105.6		92.0	92.0		92.0	92.0
0241, 31 - 34	SITE & INFRASTRUCTURE, DEMOLITION	69.8	100.8	91.0	82.0	105.6	98.1	78.5	90.7	86.9	101.1	96.1	97.7	87.0	90.8	89.6	80.6	90.7	87.5
0310	Concrete Forming & Accessories	99.4	49.6	56.4	104.1	68.8	73.6	99.9	40.9	48.9	88.5	55.8	60.3	92.2	68.5	71.7	81.9	68.2	70.1
0320	Concrete Reinforcing	99.9	74.3	89.4	81.4	79.4	80.6	99.6	69.8	87.4	95.3	71.7	85.7	99.7	66.5	86.2	99.7	66.4	86.1
0330	Cast-in-Place Concrete	95.9	59.1	81.8	107.9	77.8	96.4	89.9	68.1	81.5	85.6	62.4	76.7	90.3	68.3	81.9	82.4	68.2	77.0
03	CONCRETE	106.7	57.9	84.5	106.3	74.0	91.6	89.0	56.8	74.4	85.8	62.5	75.2	88.8	68.4	79.5	86.9	68.2	78.4
04	MASONRY	98.1	50.9	69.8	95.8	61.8	75.4	97.4	57.0	73.2	93.6	48.6	66.6	94.5	57.0	72.0	118.5	57.0	81.6
05	METALS	95.3	69.2	87.7	94.3	75.5	88.8	102.8	72.3	93.9	96.8	83.8	93.0	98.6	71.7	90.7	101.3	71.4	92.6
06	WOOD, PLASTICS & COMPOSITES	101.1	47.9	71.9	96.4	70.0	81.9	103.9	36.4	66.9	90.3	56.5	71.8	98.0	73.2	84.4	83.1	73.2	77.6
07	THERMAL & MOISTURE PROTECTION	97.6	55.0	81.0	108.7	64.4	91.5	99.2	52.7	81.1	102.5	56.3	84.5	98.5	56.6	82.2	98.0	56.6	81.9
08	OPENINGS	100.8	53.1	89.0	92.6	71.5	87.4	97.2	44.6	84.2	99.1	60.4	89.5	97.2	65.1	89.3	92.5	65.1	85.7
0920	Plaster & Gypsum Board	83.1	46.0	59.0	90.4	68.8	76.3	86.9	34.7	53.0	98.2	55.1	70.2	83.1	72.7	76.4	80.6	72.7	75.5
0950, 0980	Ceilings & Acoustic Treatment	110.3	46.0	70.6	102.1	68.8	81.5	92.8	34.7	56.9	91.6	55.1	69.1	90.3	72.7	79.4	88.6	72.7	78.8
0960	Flooring	96.7	44.3	82.5	93.7	44.5	80.4	112.0	68.2	100.1	74.9	59.9	70.8	113.1	68.2	101.0	101.8	68.2	92.7
0970, 0990	Wall Finishes & Painting/Coating	90.4	42.4	61.6	94.0	55.1	70.7	98.5	61.9	76.5	87.4	55.6	68.3	98.5	63.5	77.5	98.5	63.5	77.5
09	FINISHES	98.6	47.0	71.3	98.3	62.6	79.4	95.6	47.1	70.0	88.2	56.3	71.4	95.0	68.9	81.2	90.9	68.9	79.3
COVERS	DIVS. 10 - 14, 25, 28, 41, 43, 44	100.0	77.8	95.6	100.0	82.4	96.5	100.0	69.7	94.0	100.0	56.6	91.4	100.0	74.0	94.9	100.0	74.0	94.9
21, 22, 23	FIRE SUPPRESSION, PLUMBING & HVAC	100.2	71.2	88.7	100.1	68.8	87.7	100.1	48.1	79.5	100.3	48.1	79.6	100.1	66.4	86.7	100.1	50.8	80.5
26, 27, 3370	ELECTRICAL, COMMUNICATIONS & UTIL.	101.3	61.2	81.8	94.8	58.8	77.3	95.8	75.3	85.8	103.8	52.2	78.7	100.5	76.6	88.9	95.1	76.6	86.1
MF2004	WEIGHTED AVERAGE	99.2	63.7	84.3	98.0	70.6	86.5	97.4	60.1	81.8	96.9	60.2	81.5	97.1	69.9	85.7	96.9	66.6	84.2

DIVISION		ARKANSAS			CALIFORNIA															
		TEXARKANA			ANAHEIM			BAKERSFIELD			FRESNO			LOS ANGELES			OAKLAND			
		MAT.	INST.	TOTAL	MAT.	INST.	TOTAL	MAT.	INST.	TOTAL	MAT.	INST.	TOTAL	MAT.	INST.	TOTAL	MAT.	INST.	TOTAL	
015433	CONTRACTOR EQUIPMENT		90.7	90.7		102.2	102.2		100.7	100.7		100.7	100.7		100.3	100.3		104.6	104.6	
0241, 31 - 34	SITE & INFRASTRUCTURE, DEMOLITION	95.9	87.8	90.4	101.8	109.4	107.0	108.0	107.8	107.9	107.6	107.2	107.4	98.5	110.1	106.4	141.8	105.6	117.1	
0310	Concrete Forming & Accessories	88.6	46.1	51.9	106.2	126.7	123.9	102.1	126.1	122.8	102.5	126.4	123.2	104.2	126.7	123.6	108.4	142.5	137.8	
0320	Concrete Reinforcing	99.2	69.7	87.2	93.2	118.2	103.4	105.0	118.3	110.4	88.9	118.5	101.0	104.4	118.6	110.2	98.2	119.2	106.8	
0330	Cast-in-Place Concrete	89.9	49.0	74.2	101.2	123.5	109.7	108.3	122.4	113.7	106.7	117.0	110.6	108.8	122.0	113.9	132.9	120.7	128.2	
03	CONCRETE	86.1	52.5	70.8	104.4	122.9	112.8	107.6	122.3	114.3	105.9	120.5	112.5	108.7	122.4	115.0	122.1	129.1	125.3	
04	MASONRY	97.4	33.2	59.0	88.8	115.3	104.7	109.8	117.4	114.3	112.8	118.2	116.1	95.5	121.8	111.2	156.6	127.6	139.2	
05	METALS	93.7	69.9	86.8	106.4	103.8	105.6	101.3	103.1	101.8	106.5	102.7	105.4	104.0	103.2	103.8	98.8	106.9	101.2	
06	WOOD, PLASTICS & COMPOSITES	91.9	49.2	68.4	94.5	126.5	112.1	87.6	126.6	109.0	99.6	127.0	114.6	86.8	126.3	108.5	106.4	146.2	128.2	
07	THERMAL & MOISTURE PROTECTION	98.9	45.9	78.3	102.5	116.7	108.1	100.5	109.8	104.1	96.4	112.5	102.7	104.2	118.8	109.9	107.3	125.6	114.4	
08	OPENINGS	97.7	52.0	86.4	104.5	121.5	108.7	102.0	119.4	106.3	103.8	118.1	107.3	98.3	121.4	104.0	104.2	133.2	111.4	
0920	Plaster & Gypsum Board	84.8	47.9	60.8	105.6	127.3	119.7	100.8	127.3	118.0	101.3	127.8	118.5	104.4	127.3	119.3	108.3	147.3	133.6	
0950, 0980	Ceilings & Acoustic Treatment	94.5	47.9	65.7	114.8	127.3	122.6	116.4	127.3	123.2	119.5	127.8	124.6	126.1	127.3	126.9	120.5	147.3	137.0	
0960	Flooring	104.4	57.2	91.6	114.2	113.6	114.1	114.2	87.9	107.1	121.1	133.7	124.5	102.3	113.6	105.3	117.6	116.2	117.2	
0970, 0990	Wall Finishes & Painting/Coating	98.5	32.6	59.0	104.7	105.8	105.3	106.2	101.6	103.5	124.4	100.4	110.0	97.5	111.7	106.0	108.1	144.5	129.9	
09	FINISHES	94.2	46.6	69.0	109.0	122.4	116.1	110.5	117.3	114.1	114.9	126.0	120.7	109.6	122.9	116.6	115.6	139.4	128.2	
COVERS	DIVS. 10 - 14, 25, 28, 41, 43, 44	100.0	41.1	88.3	100.0	114.8	102.9	100.0	115.2	103.0	100.0	126.3	105.2	100.0	114.3	102.8	100.0	130.1	106.0	
21, 22, 23	FIRE SUPPRESSION, PLUMBING & HVAC	100.1	35.6	74.5	100.1	115.5	106.3	100.1	99.9	100.0	100.3	111.6	104.8	100.3	115.5	106.3	100.3	143.0	117.3	
26, 27, 3370	ELECTRICAL, COMMUNICATIONS & UTIL.	97.0	42.0	70.2	91.5	105.2	98.2	104.2	97.6	101.0	90.7	94.5	92.5	100.4	114.5	107.3	104.9	130.0	117.1	
MF2004	WEIGHTED AVERAGE	95.9	48.8	76.1	101.5	114.7	107.0	103.3	109.4	105.8	103.2	112.6	107.2	102.2	116.7	108.3	108.8	129.5	117.5	

DIVISION		CALIFORNIA																	
		OXNARD			REDDING			RIVERSIDE			SACRAMENTO			SAN DIEGO			SAN FRANCISCO		
		MAT.	INST.	TOTAL	MAT.	INST.	TOTAL	MAT.	INST.	TOTAL	MAT.	INST.	TOTAL	MAT.	INST.	TOTAL	MAT.	INST.	TOTAL
015433	CONTRACTOR EQUIPMENT		100.0	100.0		100.6	100.6		101.4	101.4		102.9	102.9		99.6	99.6		108.1	108.1
0241, 31 - 34	SITE & INFRASTRUCTURE, DEMOLITION	108.0	106.3	106.9	112.2	106.2	108.1	100.1	107.8	105.4	111.5	110.7	110.9	105.6	104.1	104.6	144.3	111.2	121.7
0310	Concrete Forming & Accessories	105.7	126.6	123.8	105.4	126.0	123.2	106.4	126.5	123.7	106.9	127.0	124.2	108.1	111.0	110.6	108.2	144.2	139.3
0320	Concrete Reinforcing	105.0	117.9	110.3	101.9	118.0	108.5	104.1	118.0	109.8	90.8	118.6	102.1	99.3	118.2	107.0	111.8	119.9	115.1
0330	Cast-in-Place Concrete	107.5	122.7	113.3	121.8	116.7	119.8	105.0	123.4	112.0	112.3	118.0	114.5	107.0	105.8	106.6	136.3	122.7	131.1
03	CONCRETE	107.5	122.6	114.3	116.4	120.1	118.1	106.2	122.8	113.7	110.9	121.1	115.6	108.5	109.9	109.1	126.4	130.8	128.4
04	MASONRY	114.7	109.9	111.8	117.9	110.1	113.3	87.8	110.8	101.6	125.0	118.3	121.0	98.7	111.5	106.4	157.2	136.0	144.5
05	METALS	101.1	103.1	101.7	105.9	101.6	104.7	106.6	103.3	105.6	94.5	103.4	97.1	103.5	103.1	103.4	104.9	110.5	106.5
06	WOOD, PLASTICS & COMPOSITES	94.3	126.6	112.0	94.5	127.0	112.3	94.5	126.5	112.1	100.3	127.2	115.1	95.3	107.6	102.0	106.4	146.4	128.3
07	THERMAL & MOISTURE PROTECTION	105.8	112.5	108.4	106.2	108.9	107.3	102.8	111.8	106.3	112.8	114.5	113.4	108.4	105.6	107.3	109.9	138.3	120.9
08	OPENINGS	100.9	121.6	106.0	103.5	118.9	107.3	103.0	121.5	107.6	117.4	120.2	118.1	99.9	109.8	102.3	108.6	135.8	115.3
0920	Plaster & Gypsum Board	105.6	127.3	119.7	104.0	127.8	119.4	105.0	127.3	119.5	103.4	127.8	119.2	117.3	107.6	111.0	110.7	147.3	134.5
0950, 0980	Ceilings & Acoustic Treatment	121.4	127.3	125.1	126.4	127.8	127.2	118.9	127.3	124.1	118.6	127.8	124.3	107.2	107.6	107.4	127.8	147.3	139.8
0960	Flooring	112.7	113.6	113.0	113.5	109.6	112.4	118.4	113.6	117.1	117.0	114.3	116.3	105.4	107.6	106.0	117.6	126.2	119.9
0970, 0990	Wall Finishes & Painting/Coating	105.5	96.5	100.1	105.5	104.7	105.0	102.7	105.8	104.6	105.2	114.3	110.7	103.8	111.7	108.5	108.1	156.9	137.4
09	FINISHES	111.8	121.4	116.9	113.4	122.1	118.0	110.7	122.4	116.9	112.2	124.3	118.6	107.6	110.1	108.9	117.8	142.9	131.1
COVERS	DIVS. 10 - 14, 25, 28, 41, 43, 44	100.0	115.0	103.0	100.0	126.3	105.2	100.0	114.7	102.9	100.0	126.9	105.3	100.0	111.7	102.3	100.0	130.7	106.1
21, 22, 23	FIRE SUPPRESSION, PLUMBING & HVAC	100.2	115.6	106.3	100.2	103.5	101.5	100.1	115.5	106.2	100.1	112.9	105.2	100.3	113.8	105.6	100.3	164.9	126.0
26, 27, 3370	ELECTRICAL, COMMUNICATIONS & UTIL.	96.4	99.5	97.9	99.4	98.1	98.8	91.6	96.3	93.9	99.6	103.7	101.6	97.7	104.1	100.8	104.9	162.2	132.8
MF2004	WEIGHTED AVERAGE	102.9	112.8	107.0	105.8	109.8	107.5	101.6	112.7	106.3	104.9	114.6	108.9	102.4	108.8	105.1	111.2	141.3	123.8

DIVISION		CALIFORNIA												COLORADO					
		SAN JOSE			SANTA BARBARA			STOCKTON			VALLEJO			COLORADO SPRINGS			DENVER		
		MAT.	INST.	TOTAL	MAT.	INST.	TOTAL	MAT.	INST.	TOTAL	MAT.	INST.	TOTAL	MAT.	INST.	TOTAL	MAT.	INST.	TOTAL
015433	CONTRACTOR EQUIPMENT		102.4	102.4		100.7	100.7		100.6	100.6		103.0	103.0		94.5	94.5		98.3	98.3
0241, 31 - 34	SITE & INFRASTRUCTURE, DEMOLITION	144.1	103.4	116.3	107.9	107.8	107.9	105.7	107.1	106.7	110.6	110.1	110.2	89.9	94.0	92.7	89.0	102.2	98.1
0310	Concrete Forming & Accessories	107.9	143.1	138.3	106.5	126.4	123.7	106.0	126.3	123.6	107.3	141.4	136.8	93.6	80.7	82.5	99.9	82.1	84.5
0320	Concrete Reinforcing	92.9	119.3	103.7	105.0	118.2	110.4	105.6	118.1	110.7	97.8	118.8	106.4	106.9	79.6	95.8	106.9	79.7	95.8
0330	Cast-in-Place Concrete	127.8	120.7	125.1	107.1	122.6	113.0	107.0	117.2	110.9	122.3	119.1	121.1	101.8	82.9	94.5	96.3	84.2	91.7
03	CONCRETE	114.6	129.5	121.4	107.3	122.4	114.2	107.4	120.5	113.4	117.0	127.9	122.0	108.0	81.5	96.0	102.5	82.6	93.5
04	MASONRY	148.2	128.9	136.6	111.1	114.7	113.2	114.3	110.2	111.9	88.9	121.4	108.3	101.4	75.9	86.1	103.3	79.0	88.7
05	METALS	100.5	109.6	103.2	101.5	103.2	102.0	102.9	102.9	102.9	97.7	104.4	99.6	101.7	86.0	97.1	104.2	86.1	98.9
06	WOOD, PLASTICS & COMPOSITES	101.9	145.9	126.1	94.3	126.6	112.0	96.1	127.0	113.1	98.0	145.9	124.3	94.4	84.4	88.9	102.1	84.2	92.3
07	THERMAL & MOISTURE PROTECTION	102.1	135.8	115.2	101.9	109.6	104.9	105.8	110.1	107.4	110.0	122.1	114.7	100.0	80.2	92.3	98.9	76.6	90.2
08	OPENINGS	93.9	135.5	104.2	102.0	121.6	106.8	101.2	120.1	105.9	117.0	135.5	121.6	99.5	85.2	96.0	99.4	85.1	95.9
0920	Plaster & Gypsum Board	100.9	147.3	131.0	105.6	127.3	119.7	106.0	127.8	120.1	108.2	147.3	133.6	82.4	83.9	83.4	95.8	83.9	88.1
0950, 0980	Ceilings & Acoustic Treatment	110.5	147.3	133.3	121.4	127.3	125.1	121.4	127.8	125.3	128.0	147.3	139.9	107.2	83.9	92.8	106.3	83.9	92.4
0960	Flooring	106.2	126.2	111.6	114.2	87.9	107.1	114.2	105.4	111.8	125.9	126.2	126.0	104.0	76.5	96.6	109.2	88.8	103.7
0970, 0990	Wall Finishes & Painting/Coating	106.2	146.2	130.2	105.5	96.5	100.1	105.5	105.1	105.3	105.1	128.2	119.0	105.7	46.6	70.2	106.0	76.5	88.3
09	FINISHES	108.6	141.4	126.0	112.5	117.0	114.8	112.3	121.4	117.2	115.6	139.1	128.0	100.0	76.7	87.7	103.1	83.0	92.4
COVERS	DIVS. 10 - 14, 25, 28, 41, 43, 44	100.0	129.5	105.8	100.0	115.0	103.0	100.0	126.3	105.2	100.0	128.9	105.7	100.0	87.8	97.6	100.0	88.6	97.7
21, 22, 23	FIRE SUPPRESSION, PLUMBING & HVAC	100.2	145.6	118.3	100.2	115.5	106.3	100.2	103.7	101.6	100.2	117.5	107.1	100.1	75.1	90.2	100.0	86.7	94.7
26, 27, 3370	ELECTRICAL, COMMUNICATIONS & UTIL.	103.1	141.0	121.6	88.4	100.3	94.2	99.1	111.7	105.2	95.6	116.5	105.8	104.9	83.3	94.3	106.7	85.4	96.3
MF2004	WEIGHTED AVERAGE	106.0	132.5	117.1	101.9	112.9	106.5	103.5	112.0	107.0	104.1	121.6	111.4	101.5	81.0	92.9	101.8	85.5	95.0

		COLORADO												CONNECTICUT					
	DIVISION	FORT COLLINS			GRAND JUNCTION			GREELEY			PUEBLO			BRIDGEPORT			BRISTOL		
		MAT.	INST.	TOTAL	MAT.	INST.	TOTAL	MAT.	INST.	TOTAL	MAT.	INST.	TOTAL	MAT.	INST.	TOTAL	MAT.	INST.	TOTAL
015433	CONTRACTOR EQUIPMENT		96.0	96.0		98.9	98.9		96.0	96.0		95.8	95.8		100.4	100.4		100.4	100.4
0241, 31 - 34	SITE & INFRASTRUCTURE, DEMOLITION	99.8	96.8	97.8	119.3	99.5	105.7	87.3	95.5	92.9	111.1	93.0	98.7	98.6	104.2	102.5	97.8	104.2	102.2
0310	Concrete Forming & Accessories	100.5	74.6	78.2	110.1	75.0	79.8	98.5	47.2	54.2	107.5	81.0	84.6	96.4	121.9	118.4	96.4	121.6	118.2
0320	Concrete Reinforcing	107.9	77.7	95.6	107.5	77.6	95.3	107.8	77.2	95.3	103.6	79.6	93.8	100.4	127.1	111.3	100.4	127.1	111.3
0330	Cast-in-Place Concrete	111.9	78.1	98.9	118.5	79.0	103.4	93.5	56.1	79.2	106.2	84.0	97.7	100.0	120.2	107.7	93.7	120.1	103.8
03	CONCRETE	114.9	76.8	97.6	119.9	77.2	100.5	100.2	57.0	80.6	109.6	82.1	97.1	107.4	122.0	114.0	104.5	121.8	112.4
04	MASONRY	112.1	52.7	76.5	139.9	54.8	88.9	106.6	35.0	63.7	103.5	74.2	86.0	106.6	132.0	121.8	99.2	132.0	118.8
05	METALS	99.8	81.5	94.5	98.1	80.9	93.1	99.8	80.1	94.0	99.7	86.8	96.0	94.9	123.3	103.2	94.9	123.1	103.1
06	WOOD, PLASTICS & COMPOSITES	101.9	78.6	89.1	103.9	78.7	90.1	99.4	46.9	70.6	101.2	84.7	92.2	101.4	119.6	111.4	101.4	119.6	111.4
07	THERMAL & MOISTURE PROTECTION	99.7	64.4	86.0	103.6	61.1	87.1	99.1	54.1	81.6	103.2	79.6	94.0	94.5	127.7	107.4	94.6	124.1	106.1
08	OPENINGS	95.3	81.8	92.0	99.3	81.9	95.0	95.3	64.6	87.7	93.8	85.4	91.7	104.9	129.4	110.9	104.9	127.4	110.4
0920	Plaster & Gypsum Board	94.3	78.0	83.7	105.1	78.0	87.5	93.1	45.2	62.0	80.6	83.9	82.7	104.8	119.5	114.4	104.8	119.5	114.4
0950, 0980	Ceilings & Acoustic Treatment	100.5	78.0	86.6	106.2	78.0	88.8	100.5	45.2	66.3	117.8	83.9	96.8	99.1	119.5	111.8	99.1	119.5	111.8
0960	Flooring	109.0	61.4	96.2	119.4	61.4	103.7	107.8	61.4	95.2	115.8	88.8	108.5	96.8	125.6	104.6	96.8	111.9	100.9
0970, 0990	Wall Finishes & Painting/Coating	106.0	46.2	70.1	108.9	76.5	89.5	106.0	28.8	59.7	108.9	43.2	69.5	91.7	116.7	106.7	91.7	116.7	106.7
09	FINISHES	102.0	69.4	84.8	110.7	73.4	91.0	100.8	46.5	72.1	108.6	78.8	92.9	100.5	121.9	111.8	100.5	119.5	110.5
COVERS	DIVS. 10 - 14, 25, 28, 41, 43, 44	100.0	85.9	97.2	100.0	87.0	97.4	100.0	78.8	95.8	100.0	88.6	97.8	100.0	111.7	102.3	100.0	111.7	102.3
21, 22, 23	FIRE SUPPRESSION, PLUMBING & HVAC	100.0	79.1	91.7	99.9	65.2	86.1	100.0	74.1	89.7	99.9	68.0	87.2	100.2	114.4	105.9	100.2	114.4	105.8
26, 27, 3370	ELECTRICAL, COMMUNICATIONS & UTIL.	101.2	85.3	93.5	94.7	61.7	78.7	101.2	85.3	93.5	95.7	77.6	86.9	97.0	109.9	103.3	97.0	109.6	103.2
MF2004	WEIGHTED AVERAGE	102.2	77.3	91.7	104.9	72.0	91.1	99.7	67.0	86.0	101.3	79.0	91.9	100.3	118.6	108.0	99.6	118.0	107.3

		CONNECTICUT																	
	DIVISION	HARTFORD			NEW BRITAIN			NEW HAVEN			NORWALK			STAMFORD			WATERBURY		
		MAT.	INST.	TOTAL	MAT.	INST.	TOTAL	MAT.	INST.	TOTAL	MAT.	INST.	TOTAL	MAT.	INST.	TOTAL	MAT.	INST.	TOTAL
015433	CONTRACTOR EQUIPMENT		100.4	100.4		100.4	100.4		100.6	100.6		100.4	100.4		100.4	100.4		100.4	100.4
0241, 31 - 34	SITE & INFRASTRUCTURE, DEMOLITION	95.1	104.2	101.3	98.0	104.2	102.2	97.9	104.6	102.5	98.4	104.2	102.4	99.0	104.3	102.6	98.2	104.2	102.3
0310	Concrete Forming & Accessories	94.5	121.6	117.9	96.6	121.6	118.2	96.1	121.6	118.1	96.4	122.1	118.6	96.4	122.4	118.8	96.4	121.8	118.3
0320	Concrete Reinforcing	100.4	127.1	111.3	100.4	127.1	111.3	100.4	127.1	111.2	100.4	127.2	111.3	100.4	127.3	111.3	100.4	127.1	111.3
0330	Cast-in-Place Concrete	96.1	120.1	105.3	95.2	120.1	104.7	96.9	111.0	102.3	98.4	128.8	110.0	100.0	128.9	111.1	100.0	120.1	107.7
03	CONCRETE	105.5	121.8	112.9	105.2	121.8	112.7	120.8	118.7	119.8	106.7	125.1	115.0	107.4	125.3	115.5	107.4	121.9	114.0
04	MASONRY	99.4	132.0	118.9	101.2	132.0	119.6	99.4	132.0	118.9	98.8	133.7	119.7	99.7	133.7	120.0	99.7	132.0	119.0
05	METALS	99.5	123.1	106.4	91.6	123.1	100.8	91.8	123.0	100.9	94.9	123.6	103.3	94.9	123.9	103.4	94.9	123.2	103.2
06	WOOD, PLASTICS & COMPOSITES	99.9	119.6	110.7	101.4	119.6	111.4	101.4	119.6	111.4	101.4	119.6	111.4	101.4	119.6	111.4	101.4	119.6	111.4
07	THERMAL & MOISTURE PROTECTION	95.1	124.1	106.4	94.6	124.5	106.2	94.7	123.0	105.7	94.7	129.5	108.2	94.6	129.5	108.2	94.6	124.5	106.2
08	OPENINGS	105.6	127.4	111.0	104.9	127.4	110.4	104.9	129.4	110.9	104.9	129.4	110.9	104.9	129.4	110.9	104.9	129.4	110.9
0920	Plaster & Gypsum Board	106.1	119.5	114.8	104.8	119.5	114.4	104.8	119.5	114.4	104.8	119.5	114.4	104.8	119.5	114.4	104.8	119.5	114.4
0950, 0980	Ceilings & Acoustic Treatment	99.1	119.5	111.8	99.1	119.5	111.8	99.1	119.5	111.8	99.1	119.5	111.8	99.1	119.5	111.8	99.1	119.5	111.8
0960	Flooring	96.8	111.9	100.9	96.8	111.9	100.9	96.8	125.6	104.6	96.8	120.6	103.2	96.8	120.6	103.2	96.8	125.6	104.6
0970, 0990	Wall Finishes & Painting/Coating	91.7	116.7	106.7	91.7	116.7	106.7	91.7	116.7	106.7	91.7	116.7	106.7	91.7	116.7	106.7	91.7	116.7	106.7
09	FINISHES	100.2	119.5	110.4	100.5	119.5	110.5	100.5	121.9	111.8	100.5	121.0	111.3	100.6	121.0	111.4	100.4	121.9	111.7
COVERS	DIVS. 10 - 14, 25, 28, 41, 43, 44	100.0	111.7	102.3	100.0	111.7	102.3	100.0	111.7	102.3	100.0	111.7	102.3	100.0	111.9	102.3	100.0	111.7	102.3
21, 22, 23	FIRE SUPPRESSION, PLUMBING & HVAC	100.1	114.4	105.8	100.2	114.4	105.8	100.2	114.4	105.8	100.2	114.5	105.9	100.2	114.5	105.9	100.2	114.4	105.9
26, 27, 3370	ELECTRICAL, COMMUNICATIONS & UTIL.	96.9	110.5	103.5	97.1	109.6	103.2	97.0	109.6	103.1	97.0	108.3	102.5	97.0	154.5	125.0	96.6	109.9	103.1
MF2004	WEIGHTED AVERAGE	100.5	118.1	107.9	99.2	118.0	107.1	101.0	118.0	108.2	99.9	119.0	107.9	100.0	125.4	110.7	99.9	118.5	107.7

		D.C.			DELAWARE			FLORIDA											
	DIVISION	WASHINGTON			WILMINGTON			DAYTONA BEACH			FORT LAUDERDALE			JACKSONVILLE			MELBOURNE		
		MAT.	INST.	TOTAL	MAT.	INST.	TOTAL	MAT.	INST.	TOTAL	MAT.	INST.	TOTAL	MAT.	INST.	TOTAL	MAT.	INST.	TOTAL
015433	CONTRACTOR EQUIPMENT		103.6	103.6		114.7	114.7		98.9	98.9		93.2	93.2		98.9	98.9		98.9	98.9
0241, 31 - 34	SITE & INFRASTRUCTURE, DEMOLITION	110.3	91.8	97.7	91.3	109.0	103.4	116.1	89.5	97.9	101.4	78.7	85.9	116.2	88.7	97.4	124.6	88.7	100.0
0310	Concrete Forming & Accessories	101.3	81.9	84.5	99.5	102.8	102.4	95.8	77.8	80.2	94.0	68.1	71.6	95.5	59.1	64.1	92.0	79.7	81.4
0320	Concrete Reinforcing	113.5	91.1	104.3	99.5	100.1	99.7	92.5	82.2	88.3	92.5	80.4	87.6	92.5	51.9	75.9	93.5	82.2	88.9
0330	Cast-in-Place Concrete	127.6	88.5	112.7	98.3	89.1	94.8	91.9	77.2	86.3	96.4	74.4	88.0	92.8	66.0	82.5	110.8	79.5	98.8
03	CONCRETE	121.1	87.3	105.8	102.8	98.5	100.8	90.4	79.4	85.4	92.4	74.0	84.0	90.8	61.8	77.6	101.5	81.1	92.2
04	MASONRY	101.7	81.9	89.8	107.6	93.8	99.3	86.3	72.3	77.9	86.6	68.2	75.6	86.2	58.6	69.7	83.7	76.3	79.2
05	METALS	98.0	109.3	101.3	99.6	114.7	104.0	101.7	94.5	99.6	101.6	94.4	99.5	100.2	81.2	94.7	110.7	94.9	106.1
06	WOOD, PLASTICS & COMPOSITES	100.9	81.2	90.1	99.8	104.2	102.2	95.7	81.0	87.6	91.6	65.6	77.4	95.7	57.9	74.9	91.4	81.0	85.7
07	THERMAL & MOISTURE PROTECTION	105.3	84.7	97.3	99.8	102.9	101.0	95.1	79.8	89.1	95.1	82.3	90.1	95.3	64.7	83.4	95.5	85.9	91.7
08	OPENINGS	107.0	92.7	103.4	95.2	108.6	98.5	99.7	74.9	93.6	97.1	66.1	89.4	99.7	52.1	88.0	98.9	80.9	94.5
0920	Plaster & Gypsum Board	111.0	80.6	91.2	111.0	104.3	106.6	100.7	80.8	87.8	100.3	64.9	77.3	100.7	56.9	72.2	97.0	80.8	86.5
0950, 0980	Ceilings & Acoustic Treatment	107.4	80.6	90.8	94.7	104.3	100.6	100.4	80.8	88.3	100.4	64.9	78.4	100.4	56.9	73.5	96.2	80.8	86.7
0960	Flooring	127.7	100.2	120.3	82.1	98.2	86.5	117.2	79.9	107.1	117.2	65.4	103.2	117.2	55.9	100.6	114.0	79.9	104.7
0970, 0990	Wall Finishes & Painting/Coating	125.2	87.4	102.5	95.8	98.0	97.1	115.3	76.7	92.1	110.4	54.2	76.7	115.3	49.6	75.9	115.3	105.7	109.5
09	FINISHES	111.0	85.3	97.5	100.5	101.4	101.0	106.3	78.4	91.6	103.9	65.1	83.4	106.3	57.2	80.4	104.5	82.7	93.0
COVERS	DIVS. 10 - 14, 25, 28, 41, 43, 44	100.0	98.9	99.8	100.0	105.5	101.1	100.0	83.5	96.7	100.0	86.3	97.3	100.0	79.0	95.8	100.0	85.0	97.0
21, 22, 23	FIRE SUPPRESSION, PLUMBING & HVAC	100.1	93.8	97.6	100.2	112.4	105.1	99.9	69.7	87.8	99.9	67.2	86.9	99.9	48.1	79.3	99.9	71.9	88.7
26, 27, 3370	ELECTRICAL, COMMUNICATIONS & UTIL.	103.4	101.5	102.5	105.3	108.1	106.6	99.2	62.9	81.6	99.2	75.5	87.7	98.9	68.2	84.0	100.5	72.9	87.1
MF2004	WEIGHTED AVERAGE	104.6	92.8	99.7	100.6	105.9	102.8	99.1	76.5	89.6	98.5	73.6	88.0	98.8	63.1	83.8	102.0	80.0	92.7

DIVISION		FLORIDA																	
		MIAMI			ORLANDO			PANAMA CITY			PENSACOLA			ST. PETERSBURG			TALLAHASSEE		
		MAT.	INST.	TOTAL	MAT.	INST.	TOTAL	MAT.	INST.	TOTAL	MAT.	INST.	TOTAL	MAT.	INST.	TOTAL	MAT.	INST.	TOTAL
015433	CONTRACTOR EQUIPMENT		93.2	93.2		98.9	98.9		98.9	98.9		98.9	98.9		98.9	98.9		98.9	98.9
0241, 31 - 34	SITE & INFRASTRUCTURE, DEMOLITION	104.8	79.0	87.2	118.7	88.3	98.0	131.2	86.4	100.6	130.8	87.9	101.5	117.0	87.8	97.0	114.3	87.6	96.0
0310	Concrete Forming & Accessories	98.8	68.8	72.9	95.5	79.4	81.6	94.8	43.6	50.6	92.8	55.9	60.9	93.6	49.9	55.9	97.3	45.9	52.9
0320	Concrete Reinforcing	92.5	80.6	87.6	92.5	79.0	87.0	96.6	48.8	77.1	99.0	48.7	78.5	95.8	55.4	79.3	92.5	51.6	75.8
0330	Cast-in-Place Concrete	104.6	75.8	93.6	121.1	76.9	104.2	97.5	46.3	77.9	120.5	62.6	98.4	104.0	61.8	87.8	102.3	54.3	83.9
03	CONCRETE	96.5	74.8	86.6	105.3	79.5	93.5	98.7	47.1	75.2	109.1	58.6	86.1	96.6	56.9	78.5	95.3	51.9	75.6
04	MASONRY	85.9	72.0	77.6	90.3	72.3	79.5	90.5	44.5	62.9	109.4	53.0	75.7	130.7	48.6	81.5	89.0	47.1	63.9
05	METALS	107.0	94.5	103.4	107.8	92.8	103.4	101.5	69.5	92.1	102.6	80.0	96.0	105.0	81.7	98.2	94.7	80.7	90.6
06	WOOD, PLASTICS & COMPOSITES	97.6	65.6	80.1	95.4	83.9	89.1	94.6	43.7	66.7	93.0	57.6	73.5	93.1	49.4	69.1	93.5	44.4	66.6
07	THERMAL & MOISTURE PROTECTION	103.2	79.5	94.0	96.2	80.1	89.9	95.5	47.4	76.8	95.5	59.2	81.3	95.0	55.6	79.7	100.5	74.5	90.4
08	OPENINGS	97.1	66.4	89.5	99.7	73.0	93.1	97.4	36.6	82.3	97.3	53.8	86.6	98.3	50.1	86.4	100.2	45.6	86.7
0920	Plaster & Gypsum Board	96.3	64.9	75.9	109.7	83.8	92.9	98.8	42.2	62.0	100.0	56.5	71.8	98.0	48.0	65.6	114.3	42.9	67.9
0950, 0980	Ceilings & Acoustic Treatment	97.1	64.9	77.2	100.4	83.8	90.2	94.6	42.2	62.2	96.2	56.5	71.7	94.6	48.0	65.8	100.4	42.9	64.8
0960	Flooring	124.4	68.9	109.4	117.4	79.9	107.2	116.6	46.9	97.8	110.9	59.4	97.0	115.6	60.3	100.7	117.2	50.3	99.1
0970, 0990	Wall Finishes & Painting/Coating	110.4	54.2	76.7	115.3	58.7	81.4	115.3	41.0	70.7	115.3	49.0	75.5	115.3	43.3	72.1	115.3	43.8	72.4
09	FINISHES	105.0	66.2	84.5	107.6	78.1	92.0	105.9	44.0	73.2	104.4	55.7	78.7	104.1	50.7	75.9	107.8	45.5	74.9
COVERS	DIVS. 10 - 14, 25, 28, 41, 43, 44	100.0	86.9	97.4	100.0	83.7	96.8	100.0	46.9	89.5	100.0	48.0	89.7	100.0	56.8	91.4	100.0	71.7	94.4
21, 22, 23	FIRE SUPPRESSION, PLUMBING & HVAC	99.9	72.1	88.8	99.8	64.3	85.7	99.9	37.8	75.2	99.9	57.1	82.8	99.9	55.6	82.3	99.8	42.2	76.9
26, 27, 3370	ELECTRICAL, COMMUNICATIONS & UTIL.	107.1	76.3	92.1	107.0	40.4	74.6	97.9	41.8	70.6	101.9	58.2	80.6	99.5	47.0	73.9	106.8	36.7	72.6
MF2004	WEIGHTED AVERAGE	101.2	75.4	90.3	103.1	72.0	90.0	100.3	48.6	78.6	102.9	61.3	85.4	102.3	58.1	83.7	99.6	53.0	80.1

DIVISION		FLORIDA			GEORGIA														
		TAMPA			ALBANY			ATLANTA			AUGUSTA			COLUMBUS			MACON		
		MAT.	INST.	TOTAL	MAT.	INST.	TOTAL	MAT.	INST.	TOTAL	MAT.	INST.	TOTAL	MAT.	INST.	TOTAL	MAT.	INST.	TOTAL
015433	CONTRACTOR EQUIPMENT		98.9	98.9		94.3	94.3		93.4	93.4		92.7	92.7		94.3	94.3		102.3	102.3
0241, 31 - 34	SITE & INFRASTRUCTURE, DEMOLITION	117.3	88.5	97.6	101.7	81.2	87.6	104.1	94.1	97.3	100.4	90.9	93.9	101.6	81.9	88.1	102.1	94.0	96.6
0310	Concrete Forming & Accessories	96.7	82.8	84.7	94.9	49.5	55.7	97.6	76.1	79.0	95.1	55.1	60.6	94.8	62.4	66.8	94.2	59.4	64.1
0320	Concrete Reinforcing	92.5	86.2	89.9	92.1	89.8	91.2	98.4	91.7	95.6	99.3	69.3	87.0	92.5	90.9	91.9	93.7	90.2	92.2
0330	Cast-in-Place Concrete	101.6	67.4	88.5	98.1	51.4	80.3	112.0	74.5	97.7	105.7	53.6	85.8	97.8	61.1	83.7	96.4	53.5	80.0
03	CONCRETE	94.9	79.0	87.7	93.1	59.2	77.7	101.6	78.6	91.1	97.3	57.9	79.4	93.0	68.6	81.9	92.6	64.5	79.8
04	MASONRY	87.5	73.0	78.8	89.4	45.4	63.1	87.7	73.7	79.3	87.9	45.1	62.3	88.4	64.0	73.8	101.5	43.2	66.6
05	METALS	103.9	95.5	101.5	101.0	89.0	97.5	93.6	83.5	90.7	92.4	71.5	86.3	100.5	92.8	98.2	96.1	92.0	94.9
06	WOOD, PLASTICS & COMPOSITES	97.0	86.4	91.2	94.6	44.5	67.1	97.3	77.4	86.4	94.5	55.6	73.1	94.6	61.0	76.2	103.4	59.9	79.5
07	THERMAL & MOISTURE PROTECTION	95.3	90.3	93.3	92.3	60.5	80.0	94.3	78.2	88.0	93.9	56.1	79.2	92.5	66.4	82.3	91.1	66.2	81.4
08	OPENINGS	99.7	77.8	94.3	97.1	51.0	85.7	99.1	75.4	93.2	93.3	54.8	83.8	97.1	62.1	88.4	95.2	60.5	86.6
0920	Plaster & Gypsum Board	100.7	86.4	91.4	98.0	43.0	62.3	118.8	77.0	91.7	117.7	54.3	76.6	100.5	60.1	74.3	108.5	58.9	76.3
0950, 0980	Ceilings & Acoustic Treatment	100.4	86.4	91.8	99.6	43.0	64.6	109.8	77.0	89.5	110.7	54.3	75.8	99.6	60.1	75.2	95.7	58.9	73.0
0960	Flooring	117.2	60.3	101.8	117.2	34.6	94.9	89.4	71.8	84.6	88.3	44.1	76.3	117.2	56.5	100.8	91.2	40.7	77.5
0970, 0990	Wall Finishes & Painting/Coating	115.3	61.6	83.1	110.4	46.5	72.1	102.8	86.2	92.9	102.8	42.0	66.3	110.4	54.9	77.1	112.7	51.8	76.1
09	FINISHES	106.4	76.7	90.7	103.3	44.8	72.4	100.7	76.2	87.7	100.2	51.2	74.3	103.6	59.7	80.4	92.4	54.1	72.2
COVERS	DIVS. 10 - 14, 25, 28, 41, 43, 44	100.0	84.7	97.0	100.0	81.4	96.3	100.0	85.7	97.2	100.0	59.5	92.0	100.0	83.3	96.7	100.0	81.1	96.3
21, 22, 23	FIRE SUPPRESSION, PLUMBING & HVAC	99.9	89.3	95.7	99.8	65.6	86.2	100.0	78.0	91.2	100.0	73.6	89.5	99.9	55.9	82.4	99.9	65.7	86.3
26, 27, 3370	ELECTRICAL, COMMUNICATIONS & UTIL.	99.1	47.0	73.8	94.5	58.3	76.9	98.7	76.7	88.0	99.6	68.7	84.5	94.6	61.4	78.4	93.1	63.8	78.8
MF2004	WEIGHTED AVERAGE	100.1	78.7	91.1	97.9	62.0	82.9	98.2	79.2	90.2	96.9	64.3	83.2	97.8	66.8	84.8	96.5	66.6	83.9

DIVISION		GEORGIA						HAWAII						IDAHO					
		SAVANNAH			VALDOSTA			HONOLULU			STATES & POSS., GUAM			BOISE			LEWISTON		
		MAT.	INST.	TOTAL	MAT.	INST.	TOTAL	MAT.	INST.	TOTAL	MAT.	INST.	TOTAL	MAT.	INST.	TOTAL	MAT.	INST.	TOTAL
015433	CONTRACTOR EQUIPMENT		95.0	95.0		94.3	94.3		101.7	101.7		166.8	166.8		101.1	101.1		97.4	97.4
0241, 31 - 34	SITE & INFRASTRUCTURE, DEMOLITION	104.1	81.4	88.6	112.2	81.1	90.9	137.0	107.0	116.5	175.5	104.4	126.9	77.2	101.1	93.6	84.6	100.5	95.4
0310	Concrete Forming & Accessories	95.0	54.7	60.2	84.2	47.4	52.4	116.5	134.3	131.9	110.8	65.8	71.9	101.1	75.5	79.0	119.6	63.6	71.2
0320	Concrete Reinforcing	93.4	69.5	83.6	94.4	48.1	75.5	105.0	112.4	108.1	185.6	35.3	124.3	102.2	72.7	90.2	109.4	95.6	103.8
0330	Cast-in-Place Concrete	114.0	52.5	90.5	96.0	54.8	80.3	174.6	120.7	154.0	177.7	117.2	154.6	96.8	81.3	90.9	108.8	90.1	101.7
03	CONCRETE	100.7	58.2	81.4	96.9	51.9	76.4	139.4	124.4	132.6	156.7	78.9	121.3	105.8	77.2	92.8	116.7	79.3	99.7
04	MASONRY	88.4	55.8	68.9	94.2	50.6	68.1	126.5	125.4	125.9	193.1	50.3	107.6	129.1	64.4	90.4	131.3	85.4	103.8
05	METALS	96.5	82.9	92.5	100.1	75.6	92.9	137.8	105.6	128.4	147.9	85.3	129.6	100.9	76.3	93.7	94.3	87.3	92.2
06	WOOD, PLASTICS & COMPOSITES	109.0	52.4	77.9	81.9	43.4	60.8	109.8	139.0	125.8	109.7	67.2	86.4	93.5	76.2	84.0	103.4	55.7	77.2
07	THERMAL & MOISTURE PROTECTION	95.1	57.8	80.6	92.7	63.3	81.2	115.3	122.7	118.2	123.5	73.1	103.9	95.4	71.0	85.9	146.2	78.1	119.7
08	OPENINGS	98.2	50.6	86.4	92.6	40.8	79.8	106.1	129.4	111.9	105.6	56.9	93.6	92.6	71.5	87.4	113.1	62.9	100.7
0920	Plaster & Gypsum Board	97.1	51.2	67.3	92.8	41.9	59.7	136.6	140.2	138.9	209.0	53.9	108.2	81.2	75.2	77.3	136.2	54.1	82.8
0950, 0980	Ceilings & Acoustic Treatment	100.4	51.2	70.0	94.6	41.9	62.0	119.7	140.2	132.3	244.1	53.9	126.5	112.2	75.2	89.3	143.8	54.1	88.3
0960	Flooring	118.0	52.0	100.1	109.3	43.4	91.5	164.8	125.5	154.2	164.5	52.0	134.1	100.3	71.3	92.5	142.5	89.7	128.2
0970, 0990	Wall Finishes & Painting/Coating	112.3	52.6	76.5	110.4	37.0	66.4	104.0	139.2	125.1	105.5	41.3	67.0	98.1	45.4	66.5	119.6	67.5	88.3
09	FINISHES	103.9	53.6	77.3	99.9	44.9	70.8	133.9	134.7	134.3	200.4	61.4	126.9	100.8	72.2	85.7	168.1	67.5	114.9
COVERS	DIVS. 10 - 14, 25, 28, 41, 43, 44	100.0	61.0	92.3	100.0	59.8	92.0	100.0	116.4	103.2	100.0	91.3	98.3	100.0	71.5	94.4	100.0	74.7	95.0
21, 22, 23	FIRE SUPPRESSION, PLUMBING & HVAC	99.9	59.6	83.9	99.9	40.8	76.4	100.2	107.3	103.0	102.8	44.9	79.8	99.9	69.2	87.7	101.0	83.4	94.0
26, 27, 3370	ELECTRICAL, COMMUNICATIONS & UTIL.	97.4	60.5	79.4	92.7	36.0	65.1	111.7	121.3	116.4	159.7	48.1	105.4	97.4	77.1	87.5	84.8	78.0	81.5
MF2004	WEIGHTED AVERAGE	98.8	62.0	83.3	97.8	51.0	78.2	118.7	118.7	118.7	138.1	64.3	107.1	100.6	74.8	89.8	109.0	80.4	97.0

DIVISION		IDAHO						ILLINOIS											
		POCATELLO			TWIN FALLS			CHICAGO			DECATUR			EAST ST. LOUIS			JOLIET		
		MAT.	INST.	TOTAL	MAT.	INST.	TOTAL	MAT.	INST.	TOTAL	MAT.	INST.	TOTAL	MAT.	INST.	TOTAL	MAT.	INST.	TOTAL
015433	CONTRACTOR EQUIPMENT		101.1	101.1		101.1	101.1		94.1	94.1		101.3	101.3		108.1	108.1		92.0	92.0
0241, 31 - 34	SITE & INFRASTRUCTURE, DEMOLITION	78.5	101.1	94.0	84.9	99.1	94.6	98.1	95.3	96.2	88.3	95.1	93.0	104.6	98.0	100.1	97.9	93.6	94.9
0310	Concrete Forming & Accessories	101.2	75.2	78.8	102.0	32.3	41.8	100.5	153.2	146.1	98.1	108.8	107.4	93.2	110.3	107.9	102.5	134.2	129.9
0320	Concrete Reinforcing	102.6	73.0	90.5	104.4	71.2	90.8	106.3	157.6	127.2	99.0	96.9	98.1	101.0	105.9	103.0	106.3	146.9	122.9
0330	Cast-in-Place Concrete	99.6	81.1	92.5	102.1	43.4	79.7	105.9	145.2	120.9	99.4	108.5	102.9	93.0	107.7	98.6	105.8	138.0	118.1
03	CONCRETE	105.2	77.0	92.4	112.9	44.8	81.9	104.5	149.8	125.1	97.3	106.6	101.5	89.1	109.2	98.2	104.6	136.9	119.3
04	MASONRY	126.4	70.6	93.0	129.2	33.8	72.1	94.4	152.2	129.0	72.1	109.4	94.5	75.6	113.5	98.3	97.2	139.5	122.5
05	METALS	108.9	76.4	99.4	108.9	73.0	98.4	90.1	130.7	102.0	95.0	104.6	97.8	92.7	117.4	99.9	88.2	124.1	98.7
06	WOOD, PLASTICS & COMPOSITES	93.5	76.2	84.0	94.1	31.3	59.6	105.2	152.3	131.1	102.9	108.2	105.8	100.2	109.1	105.1	107.0	131.5	120.5
07	THERMAL & MOISTURE PROTECTION	95.5	66.3	84.1	96.2	41.7	75.0	100.7	142.7	117.1	99.2	102.3	100.4	93.6	108.4	99.3	100.6	134.6	113.8
08	OPENINGS	93.3	65.7	86.5	96.6	38.8	82.3	101.6	155.3	114.8	96.9	105.3	99.0	86.6	113.9	93.3	99.4	141.2	109.8
0920	Plaster & Gypsum Board	78.7	75.2	76.4	78.4	28.6	46.1	92.0	154.1	132.3	98.5	108.4	104.9	95.1	109.4	104.4	89.4	132.6	117.5
0950, 0980	Ceilings & Acoustic Treatment	117.8	75.2	91.5	111.2	28.6	60.1	101.5	154.1	134.0	93.4	108.4	102.7	87.6	109.4	101.1	101.5	132.6	120.8
0960	Flooring	103.6	71.3	94.8	104.9	48.3	89.6	92.0	149.5	107.5	100.2	111.2	103.2	109.6	118.3	111.9	91.6	138.4	104.2
0970, 0990	Wall Finishes & Painting/Coating	98.0	47.2	67.5	98.1	27.8	55.9	86.5	148.0	123.4	93.0	106.6	101.2	100.6	103.3	102.2	84.6	135.3	115.0
09	FINISHES	102.8	72.4	86.7	102.2	33.8	66.0	93.8	152.7	124.9	97.1	109.9	103.8	96.2	110.9	104.0	93.2	135.2	115.4
COVERS	DIVS. 10 - 14, 25, 28, 41, 43, 44	100.0	71.5	94.4	100.0	41.3	88.4	100.0	126.8	105.3	100.0	95.4	99.1	100.0	96.9	99.4	100.0	121.3	104.2
21, 22, 23	FIRE SUPPRESSION, PLUMBING & HVAC	99.9	69.2	87.7	99.9	57.6	83.1	99.9	133.8	113.4	99.9	101.2	100.4	99.9	96.5	98.5	100.0	125.8	110.3
26, 27, 3370	ELECTRICAL, COMMUNICATIONS & UTIL.	93.8	72.8	83.6	87.9	63.6	76.1	97.5	135.8	116.1	96.5	91.2	93.9	93.5	100.0	96.7	96.3	118.0	106.9
MF2004	WEIGHTED AVERAGE	101.6	74.5	90.2	102.5	54.0	82.1	97.9	138.3	114.9	96.2	102.5	98.8	93.9	105.7	98.8	97.4	126.9	109.8

DIVISION		ILLINOIS									INDIANA								
		PEORIA			ROCKFORD			SPRINGFIELD			ANDERSON			BLOOMINGTON			EVANSVILLE		
		MAT.	INST.	TOTAL	MAT.	INST.	TOTAL	MAT.	INST.	TOTAL	MAT.	INST.	TOTAL	MAT.	INST.	TOTAL	MAT.	INST.	TOTAL
015433	CONTRACTOR EQUIPMENT		100.5	100.5		100.5	100.5		101.3	101.3		96.9	96.9		87.5	87.5		113.5	113.5
0241, 31 - 34	SITE & INFRASTRUCTURE, DEMOLITION	97.2	94.2	95.2	96.3	95.6	95.8	95.5	95.2	95.3	83.4	94.7	91.1	74.8	94.8	88.5	79.4	121.3	108.0
0310	Concrete Forming & Accessories	96.5	109.8	108.0	100.6	119.9	117.3	98.7	108.7	107.4	94.9	78.7	80.9	99.9	76.1	79.4	93.9	81.8	83.4
0320	Concrete Reinforcing	99.0	107.5	102.4	93.5	127.0	107.2	99.0	100.1	99.4	95.1	76.5	87.5	86.1	76.5	82.2	94.4	78.7	88.0
0330	Cast-in-Place Concrete	97.1	103.1	99.4	99.4	116.6	106.0	88.4	103.8	94.3	95.0	79.6	89.1	95.9	76.5	88.5	91.7	89.6	90.9
03	CONCRETE	96.1	107.2	101.1	96.3	120.0	107.1	92.1	105.6	98.2	90.2	79.3	85.2	97.9	76.3	88.1	98.6	84.2	92.1
04	MASONRY	112.5	109.2	110.5	87.6	110.3	101.2	76.8	110.3	96.8	85.8	80.4	82.6	86.5	74.2	79.1	82.8	84.6	83.9
05	METALS	95.0	109.6	99.2	95.0	119.6	102.2	97.4	106.0	99.9	100.2	87.8	96.6	94.3	76.1	89.0	87.0	86.1	86.8
06	WOOD, PLASTICS & COMPOSITES	102.9	106.1	104.7	102.9	117.7	111.1	104.0	107.4	105.8	106.6	78.4	91.1	115.7	75.4	93.6	95.2	80.5	87.1
07	THERMAL & MOISTURE PROTECTION	99.3	107.5	102.5	102.0	116.7	107.7	99.0	105.8	101.6	96.2	78.8	89.4	92.8	79.0	87.4	96.8	85.3	92.3
08	OPENINGS	97.1	107.5	99.6	97.1	119.8	102.7	98.1	105.7	100.0	95.6	78.1	91.3	103.1	76.5	96.5	95.5	79.1	91.5
0920	Plaster & Gypsum Board	98.5	106.2	103.5	98.5	118.3	111.4	94.9	107.6	103.1	95.7	78.1	84.3	88.8	75.2	80.0	84.7	79.1	81.1
0950, 0980	Ceilings & Acoustic Treatment	93.4	106.2	101.4	93.4	118.3	108.8	91.8	107.6	101.6	91.1	78.1	83.1	79.9	75.2	77.0	85.3	79.1	81.5
0960	Flooring	100.2	110.2	102.9	100.2	113.6	103.8	100.2	102.7	100.9	92.9	87.0	91.3	104.8	84.1	99.2	98.3	82.6	94.0
0970, 0990	Wall Finishes & Painting/Coating	93.0	112.8	104.9	93.0	122.6	110.8	93.0	106.2	100.9	95.1	69.4	79.7	91.6	85.5	87.9	97.7	82.0	88.3
09	FINISHES	97.2	110.3	104.1	97.2	119.2	108.8	96.0	107.8	102.3	91.4	79.3	85.0	93.7	78.6	85.7	92.4	81.9	86.9
COVERS	DIVS. 10 - 14, 25, 28, 41, 43, 44	100.0	102.0	100.4	100.0	102.4	100.5	100.0	95.6	99.1	100.0	89.9	98.0	100.0	80.3	96.1	100.0	85.6	97.1
21, 22, 23	FIRE SUPPRESSION, PLUMBING & HVAC	99.9	105.9	102.3	100.0	111.9	104.7	99.8	102.1	100.7	99.9	74.0	89.6	99.7	78.3	91.2	99.9	80.6	92.2
26, 27, 3370	ELECTRICAL, COMMUNICATIONS & UTIL.	95.6	95.1	95.4	96.0	122.8	109.0	101.8	96.9	99.4	83.2	93.0	88.0	98.5	78.8	89.0	93.8	87.3	90.6
MF2004	WEIGHTED AVERAGE	98.2	104.9	101.0	97.1	115.0	104.7	97.0	103.4	99.7	94.8	82.5	89.6	96.3	78.8	89.3	94.4	86.5	91.1

DIVISION		INDIANA																	
		FORT WAYNE			GARY			INDIANAPOLIS			MUNCIE			SOUTH BEND			TERRE HAUTE		
		MAT.	INST.	TOTAL	MAT.	INST.	TOTAL	MAT.	INST.	TOTAL	MAT.	INST.	TOTAL	MAT.	INST.	TOTAL	MAT.	INST.	TOTAL
015433	CONTRACTOR EQUIPMENT		96.9	96.9		96.9	96.9		92.0	92.0		96.5	96.5		105.8	105.8		113.5	113.5
0241, 31 - 34	SITE & INFRASTRUCTURE, DEMOLITION	84.5	94.5	91.3	83.9	97.6	93.2	83.0	98.0	93.2	74.9	95.1	88.7	86.0	95.0	92.1	81.1	121.4	108.7
0310	Concrete Forming & Accessories	93.6	74.1	76.7	95.0	106.7	105.1	95.5	83.7	85.3	92.2	78.4	80.3	96.5	77.7	80.3	95.0	80.8	82.7
0320	Concrete Reinforcing	95.1	75.8	87.2	95.1	91.8	93.8	94.5	80.9	88.9	95.3	76.5	87.6	95.1	72.9	86.1	94.4	83.8	90.0
0330	Cast-in-Place Concrete	100.9	78.3	92.3	99.2	107.2	102.3	91.4	85.9	89.3	100.8	80.8	93.2	96.2	82.3	90.9	88.8	87.6	88.3
03	CONCRETE	92.9	76.6	85.5	92.2	104.0	97.6	91.7	83.8	88.1	97.5	79.5	89.3	87.4	79.9	84.0	101.0	84.0	93.3
04	MASONRY	89.4	77.8	82.5	87.0	104.4	97.4	93.5	83.6	87.6	88.5	80.4	83.6	86.0	79.0	81.8	89.4	81.2	84.5
05	METALS	100.2	86.9	96.3	100.2	98.8	99.8	102.2	80.7	95.9	95.9	87.5	93.5	100.2	98.0	99.5	87.7	88.6	88.0
06	WOOD, PLASTICS & COMPOSITES	106.3	73.1	88.1	104.5	107.0	105.9	103.5	83.3	92.4	109.8	78.2	92.5	103.7	77.1	89.1	97.3	80.2	87.9
07	THERMAL & MOISTURE PROTECTION	96.0	81.7	90.4	95.2	104.4	98.8	94.4	83.9	90.3	95.2	78.9	88.9	91.2	83.8	88.3	96.8	80.9	90.7
08	OPENINGS	95.6	72.7	90.0	95.6	104.7	97.9	103.6	83.4	98.6	97.9	78.0	93.0	88.9	76.2	85.8	96.1	80.9	92.3
0920	Plaster & Gypsum Board	94.9	72.6	80.4	90.5	107.8	101.7	94.0	83.0	86.8	84.5	78.1	80.4	88.6	76.7	80.9	84.7	78.8	80.9
0950, 0980	Ceilings & Acoustic Treatment	91.1	72.6	79.7	91.1	107.8	101.4	95.3	83.0	87.7	80.8	78.1	79.1	87.0	76.7	80.6	85.3	78.8	81.3
0960	Flooring	92.9	76.2	88.4	92.9	112.5	98.2	92.7	89.6	91.9	97.7	87.0	94.8	91.3	83.4	89.2	98.3	87.3	95.3
0970, 0990	Wall Finishes & Painting/Coating	95.1	76.1	83.7	95.1	104.9	100.9	95.1	90.8	92.5	91.6	69.4	78.3	94.6	81.4	86.7	97.7	84.1	89.6
09	FINISHES	91.2	74.6	82.5	90.7	108.0	99.8	92.5	85.6	88.9	90.9	79.1	84.7	89.0	79.1	83.8	92.4	82.3	87.1
COVERS	DIVS. 10 - 14, 25, 28, 41, 43, 44	100.0	79.4	95.9	100.0	86.2	97.3	100.0	91.7	98.4	100.0	89.3	97.9	100.0	80.2	96.1	100.0	92.8	98.6
21, 22, 23	FIRE SUPPRESSION, PLUMBING & HVAC	99.9	73.7	89.5	99.9	100.5	100.2	99.9	85.7	94.2	99.7	74.0	89.4	99.8	80.4	92.1	99.9	82.7	93.1
26, 27, 3370	ELECTRICAL, COMMUNICATIONS & UTIL.	84.0	77.4	80.8	96.0	97.4	96.7	98.3	93.0	95.7	89.1	79.7	84.5	103.7	82.7	93.5	92.1	83.7	88.0
MF2004	WEIGHTED AVERAGE	95.3	78.4	88.2	96.3	101.5	98.5	98.0	86.7	93.2	95.5	80.6	89.3	95.6	83.1	90.4	95.0	86.6	91.5

	DIVISION	IOWA																	
		CEDAR RAPIDS			COUNCIL BLUFFS			DAVENPORT			DES MOINES			DUBUQUE			SIOUX CITY		
		MAT.	INST.	TOTAL	MAT.	INST.	TOTAL	MAT.	INST.	TOTAL	MAT.	INST.	TOTAL	MAT.	INST.	TOTAL	MAT.	INST.	TOTAL
015433	CONTRACTOR EQUIPMENT		96.8	96.8		96.1	96.1		99.9	99.9		101.2	101.2		95.9	95.9		99.9	99.9
0241, 31 - 34	SITE & INFRASTRUCTURE, DEMOLITION	94.5	96.5	95.9	101.6	92.4	95.3	92.6	99.8	97.5	85.1	99.7	95.1	92.5	93.6	93.3	103.8	95.9	98.4
0310	Concrete Forming & Accessories	100.0	77.3	80.4	82.1	54.4	58.2	99.7	87.9	89.5	100.7	72.6	76.4	83.5	68.8	70.8	100.0	62.1	67.3
0320	Concrete Reinforcing	102.3	80.6	93.4	104.3	75.4	92.5	102.3	92.5	98.3	102.3	77.1	92.0	100.9	80.4	92.6	102.3	60.8	85.4
0330	Cast-in-Place Concrete	105.3	79.0	95.3	109.5	77.0	97.1	101.4	91.7	97.7	97.2	92.5	95.4	103.1	89.1	97.8	104.4	54.0	85.2
03	CONCRETE	100.6	79.3	90.9	103.0	67.6	86.9	98.8	90.7	95.1	95.8	81.3	89.2	97.6	79.0	89.1	100.0	60.3	81.9
04	MASONRY	102.5	75.8	86.5	102.9	76.2	86.9	98.6	84.9	90.4	94.1	69.4	79.3	103.4	67.3	81.8	96.2	53.0	70.3
05	METALS	87.7	91.0	88.7	93.1	88.5	91.8	87.7	99.9	91.3	87.9	90.6	88.7	86.5	90.6	87.7	87.7	81.0	85.8
06	WOOD, PLASTICS & COMPOSITES	108.4	76.4	90.8	87.1	47.8	65.5	108.4	86.6	96.4	108.2	72.1	88.4	88.9	68.4	77.7	108.4	62.4	83.1
07	THERMAL & MOISTURE PROTECTION	101.7	77.6	92.3	101.1	64.4	86.9	101.2	86.7	95.6	102.4	74.4	91.6	101.4	68.4	88.6	101.2	52.2	82.2
08	OPENINGS	99.0	79.8	94.2	98.0	63.1	89.4	99.0	89.5	96.6	99.0	77.0	93.6	98.0	77.4	92.9	99.0	59.9	89.3
0920	Plaster & Gypsum Board	98.5	75.8	83.8	89.9	46.2	61.5	98.5	86.2	90.5	88.9	71.2	77.4	90.3	67.6	75.6	98.5	61.1	74.2
0950, 0980	Ceilings & Acoustic Treatment	113.2	75.8	90.1	109.1	46.2	70.2	113.2	86.2	96.5	109.1	71.2	85.6	109.1	67.6	83.4	113.2	61.1	81.0
0960	Flooring	132.2	50.0	110.0	105.8	40.8	88.2	116.8	93.1	110.4	116.5	37.3	95.1	121.3	34.8	97.9	117.3	52.7	99.9
0970, 0990	Wall Finishes & Painting/Coating	105.6	73.3	86.2	98.4	70.2	81.5	102.0	88.5	93.9	102.0	76.4	86.6	104.6	73.3	85.8	103.0	61.0	77.8
09	FINISHES	114.3	71.0	91.4	103.7	51.5	76.1	109.5	88.8	98.5	106.0	65.6	84.6	108.7	62.0	84.0	111.2	60.2	84.2
COVERS	DIVS. 10 - 14, 25, 28, 41, 43, 44	100.0	84.7	97.0	100.0	83.5	96.7	100.0	88.0	97.6	100.0	83.3	96.7	100.0	81.5	96.3	100.0	80.5	96.1
21, 22, 23	FIRE SUPPRESSION, PLUMBING & HVAC	100.2	81.3	92.7	100.2	77.8	91.3	100.2	88.0	95.4	100.1	70.5	88.4	100.2	73.0	89.4	100.2	72.4	89.1
26, 27, 3370	ELECTRICAL, COMMUNICATIONS & UTIL.	96.5	78.6	87.7	101.6	82.4	92.3	93.9	87.3	90.7	109.1	72.8	91.4	100.2	75.0	87.9	96.5	71.5	84.3
MF2004	WEIGHTED AVERAGE	98.8	80.8	91.2	99.6	74.6	89.1	97.7	90.2	94.5	98.2	76.7	89.2	97.9	75.8	88.6	98.4	68.8	86.0

	DIVISION	IOWA			KANSAS														
		WATERLOO			DODGE CITY			KANSAS CITY			SALINA			TOPEKA			WICHITA		
		MAT.	INST.	TOTAL	MAT.	INST.	TOTAL	MAT.	INST.	TOTAL	MAT.	INST.	TOTAL	MAT.	INST.	TOTAL	MAT.	INST.	TOTAL
015433	CONTRACTOR EQUIPMENT		99.9	99.9		101.9	101.9		96.4	96.4		101.9	101.9		100.5	100.5		101.9	101.9
0241, 31 - 34	SITE & INFRASTRUCTURE, DEMOLITION	93.1	95.0	94.4	111.7	90.9	97.5	91.9	85.4	87.5	101.4	91.3	94.5	95.7	89.1	91.2	94.2	91.1	92.0
0310	Concrete Forming & Accessories	100.4	43.7	51.5	95.3	61.1	65.8	100.0	99.0	99.1	91.8	49.4	55.2	99.0	46.2	53.4	98.7	51.0	57.4
0320	Concrete Reinforcing	102.3	80.2	93.3	107.2	54.8	85.8	101.8	88.5	96.4	106.5	70.6	91.9	99.0	95.2	97.4	99.0	78.0	90.4
0330	Cast-in-Place Concrete	105.3	45.9	82.6	115.4	53.9	91.9	90.5	96.2	92.7	100.1	46.8	79.7	91.5	51.2	76.1	87.3	55.0	75.0
03	CONCRETE	100.6	53.0	79.0	113.8	58.8	88.8	95.7	96.3	96.0	101.7	54.2	80.1	93.6	58.8	77.8	91.6	58.9	76.8
04	MASONRY	97.6	55.6	72.5	104.7	51.3	72.8	105.5	94.8	99.1	119.4	45.0	74.9	99.6	58.9	75.2	93.5	58.3	72.4
05	METALS	87.7	89.0	88.1	92.4	79.3	88.6	97.2	96.9	97.1	92.3	85.6	90.3	97.4	96.0	97.0	97.4	87.2	94.4
06	WOOD, PLASTICS & COMPOSITES	109.0	43.4	73.0	96.3	67.8	80.6	102.4	100.4	101.3	92.8	52.0	70.4	98.2	43.7	68.3	99.8	49.8	72.4
07	THERMAL & MOISTURE PROTECTION	101.0	48.4	80.5	99.9	57.5	83.4	98.5	97.3	98.0	99.5	53.1	81.4	100.4	64.6	86.5	99.0	59.4	83.6
08	OPENINGS	94.7	56.2	85.2	96.8	57.9	87.2	95.8	92.5	95.0	96.7	51.9	85.6	97.1	59.6	87.8	97.1	57.1	87.2
0920	Plaster & Gypsum Board	98.5	41.4	61.4	95.6	66.5	76.7	92.1	100.3	97.4	94.9	50.2	65.8	93.8	41.5	59.8	90.9	47.9	62.9
0950, 0980	Ceilings & Acoustic Treatment	113.2	41.4	68.8	86.8	66.5	74.2	87.6	100.3	95.5	86.8	50.2	64.1	87.6	41.5	59.1	85.9	47.9	62.4
0960	Flooring	118.7	52.9	100.9	99.5	43.9	84.5	89.2	97.8	91.5	97.6	32.1	79.9	101.1	40.8	84.8	100.0	70.6	92.0
0970, 0990	Wall Finishes & Painting/Coating	103.0	32.0	60.4	93.0	44.7	64.0	100.6	77.8	86.9	93.0	44.7	64.0	93.0	60.4	73.4	93.0	55.1	70.3
09	FINISHES	110.1	43.3	74.8	96.1	57.0	75.4	94.0	95.9	95.0	94.7	45.1	68.5	95.3	44.8	68.6	94.2	54.6	73.3
COVERS	DIVS. 10 - 14, 25, 28, 41, 43, 44	100.0	74.6	95.0	100.0	46.0	89.3	100.0	74.5	94.9	100.0	56.1	91.3	100.0	59.3	91.9	100.0	58.5	91.8
21, 22, 23	FIRE SUPPRESSION, PLUMBING & HVAC	100.2	39.9	76.2	99.9	59.4	83.8	99.9	90.2	96.0	99.9	59.3	83.8	99.9	63.9	85.6	99.8	57.5	83.0
26, 27, 3370	ELECTRICAL, COMMUNICATIONS & UTIL.	96.5	64.9	81.1	96.7	71.8	84.5	101.8	98.7	100.3	96.4	71.8	84.4	108.4	72.2	90.7	103.2	71.8	87.9
MF2004	WEIGHTED AVERAGE	97.8	58.4	81.2	99.9	64.0	84.8	98.3	93.7	96.4	98.7	61.6	83.1	98.9	66.1	85.1	97.6	64.9	83.9

	DIVISION	KENTUCKY												LOUISIANA					
		BOWLING GREEN			LEXINGTON			LOUISVILLE			OWENSBORO			ALEXANDRIA			BATON ROUGE		
		MAT.	INST.	TOTAL	MAT.	INST.	TOTAL	MAT.	INST.	TOTAL	MAT.	INST.	TOTAL	MAT.	INST.	TOTAL	MAT.	INST.	TOTAL
015433	CONTRACTOR EQUIPMENT		94.2	94.2		101.0	101.0		94.2	94.2		113.5	113.5		90.7	90.7		89.6	89.6
0241, 31 - 34	SITE & INFRASTRUCTURE, DEMOLITION	67.9	96.2	87.2	75.3	99.3	91.7	65.8	95.9	86.4	79.4	121.8	108.4	103.3	88.4	93.1	117.8	86.5	96.4
0310	Concrete Forming & Accessories	85.2	83.5	83.7	97.8	75.8	78.8	93.5	82.0	83.6	90.6	75.0	77.1	84.2	39.7	45.8	100.0	63.7	68.7
0320	Concrete Reinforcing	85.5	88.2	86.6	94.4	93.4	94.0	94.4	93.8	94.1	85.6	91.5	88.0	101.1	63.1	85.6	100.9	63.3	85.6
0330	Cast-in-Place Concrete	83.7	106.4	92.4	90.4	84.0	87.9	94.3	78.8	88.4	86.3	103.0	92.7	93.9	42.6	74.3	91.5	62.6	80.4
03	CONCRETE	90.5	92.4	91.4	93.1	82.3	88.2	94.7	83.3	89.5	98.8	88.2	93.9	90.9	46.6	70.7	93.7	63.9	80.1
04	MASONRY	89.1	76.0	81.3	86.6	68.3	75.6	87.8	80.4	83.4	86.1	74.3	79.0	115.2	49.6	75.9	106.9	54.3	75.4
05	METALS	92.0	87.2	90.6	93.4	90.4	92.5	100.8	89.6	97.5	83.9	90.0	85.7	92.4	75.4	87.4	105.1	73.7	95.9
06	WOOD, PLASTICS & COMPOSITES	91.0	82.5	86.4	101.2	75.1	86.9	98.6	82.5	89.8	91.4	71.3	80.4	86.6	38.8	60.3	104.1	69.2	84.9
07	THERMAL & MOISTURE PROTECTION	85.3	86.6	85.8	97.1	89.0	94.0	91.1	80.8	87.1	96.8	78.5	89.7	99.3	49.9	80.1	101.3	59.7	85.1
08	OPENINGS	95.1	80.4	91.5	96.1	80.0	92.1	96.1	85.4	93.4	93.3	78.1	89.6	99.3	45.9	86.1	102.6	62.3	92.6
0920	Plaster & Gypsum Board	80.3	82.2	81.5	87.3	73.5	78.4	83.8	82.2	82.8	80.3	69.6	73.4	81.9	37.1	52.8	99.4	68.4	79.3
0950, 0980	Ceilings & Acoustic Treatment	85.3	82.2	83.4	89.5	73.5	79.6	86.1	82.2	83.7	77.0	69.6	72.4	92.0	37.1	58.0	98.6	68.4	80.0
0960	Flooring	94.1	85.0	91.7	99.0	62.2	89.1	97.5	71.4	90.5	96.6	85.0	93.5	102.0	64.0	91.7	107.2	68.7	96.8
0970, 0990	Wall Finishes & Painting/Coating	97.7	72.4	82.5	97.7	72.2	82.4	97.7	79.8	87.0	97.7	92.7	94.7	98.5	36.6	61.4	106.6	42.0	67.8
09	FINISHES	90.7	82.5	86.3	94.1	72.1	82.5	91.9	79.6	85.4	89.4	78.0	83.4	93.0	43.1	66.6	102.0	63.1	81.4
COVERS	DIVS. 10 - 14, 25, 28, 41, 43, 44	100.0	69.7	94.0	100.0	93.4	98.7	100.0	92.9	98.6	100.0	89.3	97.9	100.0	59.3	91.9	100.0	77.9	95.6
21, 22, 23	FIRE SUPPRESSION, PLUMBING & HVAC	99.9	85.3	94.1	99.9	68.9	87.6	99.9	83.4	93.4	99.9	73.0	89.2	100.1	55.4	82.3	99.9	54.4	81.8
26, 27, 3370	ELECTRICAL, COMMUNICATIONS & UTIL.	92.8	79.4	86.2	93.2	76.6	85.1	104.3	79.4	92.2	92.2	80.4	86.5	93.4	55.6	75.0	107.6	62.9	85.8
MF2004	WEIGHTED AVERAGE	93.6	84.6	89.8	95.1	78.5	88.1	97.1	83.9	91.6	93.4	83.2	89.1	97.0	56.0	79.8	102.2	63.8	86.1

DIVISION		LOUISIANA												MAINE					
		LAKE CHARLES			MONROE			NEW ORLEANS			SHREVEPORT			AUGUSTA			BANGOR		
		MAT.	INST.	TOTAL	MAT.	INST.	TOTAL	MAT.	INST.	TOTAL	MAT.	INST.	TOTAL	MAT.	INST.	TOTAL	MAT.	INST.	TOTAL
015433	**CONTRACTOR EQUIPMENT**		89.6	89.6		90.7	90.7		91.0	91.0		90.7	90.7		100.4	100.4		100.4	100.4
0241, 31 - 34	**SITE & INFRASTRUCTURE, DEMOLITION**	119.6	86.2	96.8	103.3	87.9	92.8	122.9	89.8	100.2	105.0	87.8	93.2	81.6	100.5	94.5	80.8	98.1	92.6
0310	Concrete Forming & Accessories	103.4	54.1	60.8	83.6	40.4	46.3	104.8	66.4	71.7	99.2	42.6	50.3	95.0	88.6	89.5	90.4	83.8	84.7
0320	Concrete Reinforcing	100.9	60.4	84.4	100.0	63.1	84.9	100.9	64.9	86.2	99.6	63.2	84.8	82.0	99.0	88.9	81.7	99.2	88.8
0330	Cast-in-Place Concrete	93.4	62.6	81.6	93.9	60.7	81.2	92.5	71.9	84.6	93.3	48.9	76.3	83.5	58.8	74.0	77.5	59.5	70.6
03	**CONCRETE**	94.8	59.1	78.6	90.6	53.0	73.5	94.4	68.7	82.7	90.6	49.9	72.1	100.6	79.8	91.1	97.0	77.8	88.3
04	**MASONRY**	105.0	51.8	73.1	109.9	50.2	74.2	106.3	59.7	78.4	101.0	51.9	71.6	97.6	45.5	66.4	111.9	52.0	76.0
05	**METALS**	99.3	72.2	91.4	92.4	74.3	87.1	113.3	76.7	102.6	88.6	74.2	84.4	82.7	83.3	82.9	82.4	79.5	81.6
06	**WOOD, PLASTICS & COMPOSITES**	108.8	56.3	80.0	85.9	39.1	60.2	105.0	69.2	85.3	103.1	42.5	69.8	99.9	97.7	98.7	94.8	90.4	92.4
07	**THERMAL & MOISTURE PROTECTION**	102.9	57.1	85.1	99.3	53.5	81.5	103.7	62.8	87.8	98.6	52.4	80.7	94.4	54.1	78.7	94.8	54.4	79.1
08	**OPENINGS**	102.6	55.8	91.0	99.3	50.2	87.2	103.4	68.1	94.7	97.2	48.0	85.0	101.7	77.4	95.7	101.6	74.9	95.0
0920	Plaster & Gypsum Board	105.2	55.1	72.6	81.5	37.5	52.9	104.0	68.4	80.9	86.3	40.9	56.8	106.9	96.8	100.4	100.4	89.3	93.2
0950, 0980	Ceilings & Acoustic Treatment	97.8	55.1	71.4	92.0	37.5	58.3	96.9	68.4	79.3	90.3	40.9	59.8	90.0	96.8	94.2	90.8	89.3	89.9
0960	Flooring	107.4	64.0	95.6	101.5	64.0	91.4	107.7	68.7	97.2	111.4	68.7	99.9	96.8	37.7	80.8	94.6	47.0	81.7
0970, 0990	Wall Finishes & Painting/Coating	106.6	38.7	65.8	98.5	43.7	65.6	108.1	63.4	81.3	98.5	36.6	61.4	91.7	31.7	55.7	91.7	29.2	54.2
09	**FINISHES**	102.5	54.2	77.0	92.8	44.2	67.1	102.9	66.8	83.8	96.3	46.2	69.8	97.3	74.7	85.3	96.0	72.5	83.6
COVERS	**DIVS. 10 - 14, 25, 28, 41, 43, 44**	100.0	76.4	95.3	100.0	54.5	91.0	100.0	79.8	96.0	100.0	69.0	93.9	100.0	58.2	91.7	100.0	70.4	94.1
21, 22, 23	**FIRE SUPPRESSION, PLUMBING & HVAC**	100.0	55.9	82.5	100.1	49.1	79.8	100.0	66.1	86.5	100.0	55.3	82.3	100.0	69.1	87.7	100.2	67.9	87.3
26, 27, 3370	**ELECTRICAL, COMMUNICATIONS & UTIL.**	95.3	61.1	78.7	95.2	56.7	76.5	96.3	68.6	82.8	102.6	68.7	86.1	98.3	80.1	89.5	95.3	80.0	87.9
MF2004	**WEIGHTED AVERAGE**	100.1	61.1	83.7	96.9	56.0	79.7	102.9	69.6	88.9	96.8	59.2	81.0	96.1	74.3	86.9	95.8	73.7	86.5

DIVISION		MAINE						MARYLAND						MASSACHUSETTS					
		LEWISTON			PORTLAND			BALTIMORE			HAGERSTOWN			BOSTON			BROCKTON		
		MAT.	INST.	TOTAL	MAT.	INST.	TOTAL	MAT.	INST.	TOTAL	MAT.	INST.	TOTAL	MAT.	INST.	TOTAL	MAT.	INST.	TOTAL
015433	**CONTRACTOR EQUIPMENT**		100.4	100.4		100.4	100.4		104.8	104.8		101.9	101.9		105.7	105.7		102.3	102.3
0241, 31 - 34	**SITE & INFRASTRUCTURE, DEMOLITION**	78.2	98.1	91.8	79.6	98.1	92.2	96.0	97.2	96.8	85.7	94.6	91.8	85.3	106.8	100.0	82.9	103.8	97.2
0310	Concrete Forming & Accessories	95.3	83.8	85.3	94.4	83.8	85.2	101.7	77.9	81.1	91.6	79.4	81.1	99.3	137.7	132.5	99.5	122.8	119.6
0320	Concrete Reinforcing	100.4	99.2	99.9	100.4	99.2	99.9	100.9	86.7	95.1	89.2	73.9	83.0	99.3	146.1	118.4	100.4	145.7	118.9
0330	Cast-in-Place Concrete	79.1	59.5	71.6	100.6	59.5	84.9	95.2	79.8	89.3	81.0	60.1	73.0	100.0	144.7	117.1	95.2	142.8	113.4
03	**CONCRETE**	97.6	77.8	88.6	107.6	77.8	94.0	101.3	81.3	92.2	87.0	72.8	80.5	110.3	140.6	124.1	107.9	133.2	119.4
04	**MASONRY**	96.5	52.0	69.8	96.8	52.0	70.0	96.9	71.5	81.7	100.6	73.3	84.3	114.9	155.6	139.3	111.4	146.4	132.4
05	**METALS**	85.3	79.5	83.6	86.5	79.5	84.5	97.3	97.9	97.5	93.8	91.7	93.2	95.8	126.2	105.4	93.8	123.2	102.4
06	**WOOD, PLASTICS & COMPOSITES**	100.6	90.4	95.0	99.8	90.4	94.6	97.9	79.9	88.0	86.7	79.8	82.9	104.1	136.6	121.9	102.9	120.1	112.3
07	**THERMAL & MOISTURE PROTECTION**	94.6	54.4	79.0	97.2	54.4	80.5	97.6	80.2	90.8	96.9	68.5	85.8	94.8	148.7	115.8	94.9	142.3	113.3
08	**OPENINGS**	104.9	74.9	97.5	104.9	74.9	97.5	93.7	85.8	91.8	90.9	78.9	88.0	100.9	141.0	110.8	98.9	129.0	106.3
0920	Plaster & Gypsum Board	105.5	89.3	95.0	111.1	89.3	96.9	104.6	79.4	88.2	99.8	79.4	86.5	103.0	137.2	127.0	103.1	120.0	114.1
0950, 0980	Ceilings & Acoustic Treatment	99.1	89.3	93.0	96.6	89.3	92.1	95.4	79.4	85.5	96.3	79.4	85.9	98.2	137.2	122.3	100.1	120.0	112.4
0960	Flooring	97.5	47.0	83.9	96.8	47.0	83.3	89.9	78.2	86.7	85.4	72.9	82.0	97.3	167.4	116.2	97.8	167.4	116.6
0970, 0990	Wall Finishes & Painting/Coating	91.7	29.2	54.2	91.7	29.2	54.2	94.6	81.0	86.4	94.6	35.7	59.3	94.4	158.2	132.7	93.9	130.3	115.7
09	**FINISHES**	99.1	72.5	85.1	99.0	72.5	85.0	96.0	77.8	86.4	93.4	73.7	83.0	99.1	145.7	123.7	99.0	131.6	116.3
COVERS	**DIVS. 10 - 14, 25, 28, 41, 43, 44**	100.0	70.4	94.1	100.0	70.4	94.1	100.0	86.0	97.2	100.0	88.7	97.8	100.0	121.9	104.3	100.0	118.1	103.6
21, 22, 23	**FIRE SUPPRESSION, PLUMBING & HVAC**	100.2	67.9	87.3	100.0	67.9	87.2	99.9	88.5	95.4	99.9	87.9	95.1	100.2	131.5	112.6	100.2	108.5	103.5
26, 27, 3370	**ELECTRICAL, COMMUNICATIONS & UTIL.**	97.2	80.0	88.8	103.7	80.0	92.2	98.2	91.7	95.0	95.4	81.4	88.6	96.3	137.1	116.2	95.4	99.5	97.4
MF2004	**WEIGHTED AVERAGE**	96.4	73.7	86.9	98.5	73.7	88.1	98.2	86.0	93.1	94.9	81.6	89.3	100.5	135.9	115.4	99.2	120.7	108.2

DIVISION		MASSACHUSETTS																	
		FALL RIVER			HYANNIS			LAWRENCE			LOWELL			NEW BEDFORD			PITTSFIELD		
		MAT.	INST.	TOTAL	MAT.	INST.	TOTAL	MAT.	INST.	TOTAL	MAT.	INST.	TOTAL	MAT.	INST.	TOTAL	MAT.	INST.	TOTAL
015433	**CONTRACTOR EQUIPMENT**		103.2	103.2		102.3	102.3		102.3	102.3		100.4	100.4		103.2	103.2		100.4	100.4
0241, 31 - 34	**SITE & INFRASTRUCTURE, DEMOLITION**	82.0	103.7	96.8	79.8	103.8	96.2	83.6	103.8	97.4	82.5	103.8	97.0	80.5	103.7	96.4	83.6	101.8	96.0
0310	Concrete Forming & Accessories	99.5	122.5	119.4	91.6	122.2	118.0	99.3	122.8	119.6	96.5	122.9	119.3	99.5	122.5	119.4	96.5	103.9	102.9
0320	Concrete Reinforcing	100.4	122.5	109.4	80.5	122.4	97.6	99.5	131.4	112.5	100.4	131.6	113.1	100.4	122.5	109.4	83.3	111.1	94.7
0330	Cast-in-Place Concrete	92.1	144.9	112.3	86.9	144.3	108.8	96.0	140.7	113.1	87.3	140.8	107.7	81.1	144.9	105.5	95.2	117.3	103.7
03	**CONCRETE**	106.4	129.5	116.9	97.5	129.2	111.9	108.0	129.9	118.0	99.1	129.8	113.1	101.3	129.5	114.1	99.4	109.4	103.9
04	**MASONRY**	111.2	148.3	133.4	110.2	148.4	133.0	110.8	143.8	130.6	97.1	146.5	126.7	110.3	148.3	133.0	97.7	114.9	108.0
05	**METALS**	93.8	114.5	99.9	90.5	114.1	97.4	91.4	117.7	99.1	91.4	115.6	98.5	93.8	114.5	99.9	91.2	102.0	94.4
06	**WOOD, PLASTICS & COMPOSITES**	102.9	120.3	112.5	93.7	120.0	108.2	102.9	120.1	112.3	102.3	120.1	112.1	102.9	120.3	112.5	102.3	104.7	103.6
07	**THERMAL & MOISTURE PROTECTION**	94.9	137.4	111.4	94.5	138.2	111.5	94.9	141.3	112.9	94.7	141.9	113.0	94.9	137.4	111.4	94.7	111.7	101.3
08	**OPENINGS**	98.9	117.9	103.6	95.0	117.7	100.6	98.9	122.4	104.7	104.9	122.4	109.2	98.9	117.9	103.6	104.9	106.7	105.3
0920	Plaster & Gypsum Board	103.1	120.0	114.1	96.1	120.0	111.6	105.5	120.0	115.0	105.5	120.0	115.0	103.1	120.0	114.1	105.5	104.1	104.6
0950, 0980	Ceilings & Acoustic Treatment	100.1	120.0	112.4	91.7	120.0	109.2	99.1	120.0	112.1	99.1	120.0	112.1	100.1	120.0	112.4	99.1	104.1	102.2
0960	Flooring	97.6	167.4	116.5	94.1	167.4	113.9	96.8	167.4	115.9	96.8	167.4	115.9	97.6	167.4	116.5	97.0	131.4	106.3
0970, 0990	Wall Finishes & Painting/Coating	93.9	130.3	115.7	93.9	127.1	113.8	91.8	130.3	114.9	91.7	127.1	113.0	93.9	130.3	115.7	91.7	104.5	99.4
09	**FINISHES**	99.0	131.8	116.4	94.8	131.3	114.1	98.7	131.6	116.1	98.7	131.3	115.9	98.9	131.8	116.3	98.7	109.4	104.4
COVERS	**DIVS. 10 - 14, 25, 28, 41, 43, 44**	100.0	118.8	103.7	100.0	118.1	103.6	100.0	118.2	103.6	100.0	118.2	103.6	100.0	118.8	103.7	100.0	94.7	98.9
21, 22, 23	**FIRE SUPPRESSION, PLUMBING & HVAC**	100.2	108.3	103.4	100.2	108.2	103.4	100.2	112.9	105.3	100.2	122.9	109.2	100.2	108.3	103.4	100.2	87.5	95.1
26, 27, 3370	**ELECTRICAL, COMMUNICATIONS & UTIL.**	95.2	99.4	97.3	93.5	99.4	96.4	96.7	125.4	110.7	97.1	125.4	110.9	96.3	99.4	97.8	97.1	90.8	94.1
MF2004	**WEIGHTED AVERAGE**	99.0	118.9	107.3	96.2	118.7	105.7	98.9	123.6	109.3	97.7	125.7	109.5	98.4	118.9	107.0	97.8	101.0	99.1

DIVISION		MASSACHUSETTS						MICHIGAN											
		SPRINGFIELD			WORCESTER			ANN ARBOR			DEARBORN			DETROIT			FLINT		
		MAT.	INST.	TOTAL	MAT.	INST.	TOTAL	MAT.	INST.	TOTAL	MAT.	INST.	TOTAL	MAT.	INST.	TOTAL	MAT.	INST.	TOTAL
015433	CONTRACTOR EQUIPMENT		100.4	100.4		100.4	100.4		111.3	111.3		111.3	111.3		98.5	98.5		111.3	111.3
0241, 31 - 34	SITE & INFRASTRUCTURE, DEMOLITION	83.0	102.1	96.0	82.9	103.7	97.1	77.1	96.5	90.4	76.9	96.7	90.4	90.9	98.3	95.9	67.3	95.7	86.7
0310	Concrete Forming & Accessories	96.7	105.4	104.2	97.0	126.0	122.0	98.2	110.1	108.5	98.1	119.5	116.6	99.1	119.6	116.8	101.2	95.0	95.8
0320	Concrete Reinforcing	100.4	109.0	103.9	100.4	141.6	117.2	96.1	128.4	109.3	96.1	128.7	109.4	95.4	128.7	109.0	96.1	127.9	109.1
0330	Cast-in-Place Concrete	90.8	119.2	101.7	90.3	142.6	110.3	85.0	114.5	96.3	83.1	118.0	96.4	91.3	118.0	101.5	85.6	100.4	91.3
03	CONCRETE	100.8	110.3	105.1	100.5	133.6	115.6	89.6	115.3	101.3	88.7	120.7	103.3	92.5	119.5	104.7	90.1	103.6	96.2
04	MASONRY	97.4	118.2	109.9	96.9	146.5	126.6	92.7	113.2	105.0	92.5	119.5	108.7	91.8	119.5	108.4	92.7	101.3	97.8
05	METALS	93.8	101.3	96.0	93.8	118.6	101.1	86.1	121.4	96.4	86.1	122.4	96.7	89.9	104.6	94.2	86.1	119.0	95.7
06	WOOD, PLASTICS & COMPOSITES	102.3	104.7	103.6	102.8	125.4	115.2	96.7	109.5	103.7	96.7	119.9	109.4	97.2	119.9	109.7	99.9	93.6	96.4
07	THERMAL & MOISTURE PROTECTION	94.7	113.1	101.8	94.7	134.8	110.3	102.5	111.9	106.2	101.1	121.3	109.0	99.3	121.3	107.8	100.2	96.9	99.0
08	OPENINGS	104.9	106.1	105.2	104.9	130.8	111.3	94.5	109.6	98.2	94.5	115.3	99.6	96.2	117.4	101.4	94.5	97.0	95.1
0920	Plaster & Gypsum Board	105.5	104.1	104.6	105.5	125.5	118.5	85.4	108.7	100.5	85.4	119.6	107.6	85.4	119.6	107.6	86.9	92.2	90.3
0950, 0980	Ceilings & Acoustic Treatment	99.1	104.1	102.2	99.1	125.5	115.4	88.7	108.7	101.1	88.7	119.6	107.8	89.7	119.6	108.2	88.7	92.2	90.9
0960	Flooring	96.7	131.4	106.1	96.8	163.3	114.8	92.7	120.3	100.2	92.4	126.2	101.5	92.4	126.2	101.6	92.5	82.5	89.8
0970, 0990	Wall Finishes & Painting/Coating	93.3	104.5	100.0	91.7	130.3	114.9	92.9	102.4	98.6	92.9	108.3	102.1	94.4	108.3	102.7	92.9	88.8	90.4
09	FINISHES	98.8	110.3	104.8	98.7	133.9	117.3	90.8	111.2	101.6	90.7	120.1	106.2	92.0	120.1	106.8	90.2	91.4	90.8
COVERS	DIVS. 10 - 14, 25, 28, 41, 43, 44	100.0	95.9	99.2	100.0	101.5	100.3	100.0	106.7	101.3	100.0	109.0	101.8	100.0	109.0	101.8	100.0	94.3	98.9
21, 22, 23	FIRE SUPPRESSION, PLUMBING & HVAC	100.2	101.9	100.9	100.2	105.4	102.3	99.9	99.5	99.7	99.9	110.7	104.2	99.9	112.8	105.0	99.9	101.4	100.5
26, 27, 3370	ELECTRICAL, COMMUNICATIONS & UTIL.	97.2	90.8	94.1	97.2	95.7	96.5	95.2	76.9	86.3	95.2	116.0	105.3	96.5	115.9	106.0	95.2	99.3	97.2
MF2004	WEIGHTED AVERAGE	98.4	104.5	101.0	98.3	118.9	107.0	93.7	104.5	98.2	93.5	115.5	102.7	95.3	114.2	103.2	93.4	100.8	96.5

DIVISION		MICHIGAN															MINNESOTA		
		GRAND RAPIDS			KALAMAZOO			LANSING			MUSKEGON			SAGINAW			DULUTH		
		MAT.	INST.	TOTAL	MAT.	INST.	TOTAL	MAT.	INST.	TOTAL	MAT.	INST.	TOTAL	MAT.	INST.	TOTAL	MAT.	INST.	TOTAL
015433	CONTRACTOR EQUIPMENT		101.4	101.4		101.4	101.4		111.3	111.3		101.4	101.4		111.3	111.3		101.1	101.1
0241, 31 - 34	SITE & INFRASTRUCTURE, DEMOLITION	81.3	81.7	81.6	80.0	82.4	81.7	85.4	95.2	92.1	77.9	82.3	80.9	70.4	94.8	87.1	93.3	102.7	99.7
0310	Concrete Forming & Accessories	98.1	72.0	75.5	97.7	89.5	90.7	101.8	95.1	96.0	98.3	81.1	83.4	98.3	86.8	88.4	97.1	114.3	112.0
0320	Concrete Reinforcing	94.4	83.7	90.0	94.4	85.7	90.8	96.1	127.6	108.9	95.0	85.0	90.9	96.1	127.4	108.9	102.0	101.7	101.9
0330	Cast-in-Place Concrete	91.4	90.4	91.0	92.1	104.6	96.9	87.4	99.4	92.0	90.0	99.0	93.4	84.0	93.6	87.7	107.7	104.9	106.6
03	CONCRETE	93.6	80.4	87.6	96.0	93.5	94.9	90.9	103.3	96.5	91.6	87.7	89.8	89.2	97.6	93.0	100.8	109.3	104.7
04	MASONRY	88.8	47.0	63.8	91.9	86.7	88.8	86.5	96.5	92.5	90.5	73.9	80.6	94.0	89.5	91.3	102.0	116.9	110.9
05	METALS	93.8	80.7	90.0	92.3	85.1	90.2	84.7	118.2	94.5	90.1	83.7	88.2	86.2	117.5	95.3	89.3	119.0	98.0
06	WOOD, PLASTICS & COMPOSITES	94.9	73.7	83.2	97.3	89.9	93.2	99.3	94.1	96.4	93.7	79.6	86.0	92.6	86.2	89.1	111.8	114.3	113.2
07	THERMAL & MOISTURE PROTECTION	94.4	56.5	79.7	92.7	87.5	90.7	101.4	93.8	98.4	91.9	76.2	85.8	101.0	88.2	96.0	102.3	115.3	107.3
08	OPENINGS	94.1	66.1	87.2	90.7	83.4	88.9	94.5	97.3	95.2	89.9	76.2	86.5	92.4	93.0	92.6	94.7	116.1	100.0
0920	Plaster & Gypsum Board	84.3	68.4	74.0	85.8	85.2	85.4	83.8	92.8	89.6	72.6	74.6	73.9	85.4	84.6	84.9	94.5	115.3	108.0
0950, 0980	Ceilings & Acoustic Treatment	87.8	68.4	75.8	89.5	85.2	86.8	87.1	92.8	90.6	91.1	74.6	80.9	88.7	84.6	86.2	96.3	115.3	108.1
0960	Flooring	98.1	38.0	81.8	97.9	74.2	91.5	102.6	87.9	98.6	97.1	70.0	89.8	92.7	79.9	89.3	107.5	123.6	111.9
0970, 0990	Wall Finishes & Painting/Coating	97.7	37.1	61.4	97.7	75.3	84.3	105.2	91.7	97.1	97.0	53.8	71.1	92.9	83.8	87.4	91.3	100.9	97.1
09	FINISHES	93.1	61.8	76.5	93.4	85.2	89.1	94.8	93.3	94.0	91.2	75.6	83.0	90.5	84.8	87.5	101.7	115.2	108.9
COVERS	DIVS. 10 - 14, 25, 28, 41, 43, 44	100.0	93.8	98.8	100.0	98.9	99.8	100.0	94.5	98.9	100.0	97.5	99.5	100.0	91.6	98.3	100.0	95.1	99.0
21, 22, 23	FIRE SUPPRESSION, PLUMBING & HVAC	99.9	56.3	82.6	99.9	82.3	92.9	99.8	99.8	99.8	99.8	84.2	93.6	99.9	84.2	93.6	99.8	102.4	100.9
26, 27, 3370	ELECTRICAL, COMMUNICATIONS & UTIL.	103.2	53.6	79.1	92.6	81.8	87.3	101.2	77.2	89.5	93.2	79.1	86.3	92.9	72.8	83.1	102.5	96.9	99.7
MF2004	WEIGHTED AVERAGE	96.1	65.2	83.1	94.8	85.6	91.0	94.4	96.9	95.5	93.5	81.4	88.4	93.0	89.9	91.7	98.2	108.2	102.4

DIVISION		MINNESOTA												MISSISSIPPI					
		MINNEAPOLIS			ROCHESTER			SAINT PAUL			ST. CLOUD			BILOXI			GREENVILLE		
		MAT.	INST.	TOTAL	MAT.	INST.	TOTAL	MAT.	INST.	TOTAL	MAT.	INST.	TOTAL	MAT.	INST.	TOTAL	MAT.	INST.	TOTAL
015433	CONTRACTOR EQUIPMENT		104.1	104.1		101.1	101.1		101.1	101.1		100.8	100.8		99.5	99.5		99.5	99.5
0241, 31 - 34	SITE & INFRASTRUCTURE, DEMOLITION	91.9	107.7	102.7	90.9	101.7	98.3	95.7	103.5	101.0	83.7	105.3	98.5	103.7	89.6	94.1	107.6	89.7	95.4
0310	Concrete Forming & Accessories	99.4	135.9	131.0	99.8	104.9	104.2	93.5	125.6	121.2	85.1	123.3	118.1	93.6	61.8	66.1	81.0	74.8	75.6
0320	Concrete Reinforcing	102.2	122.7	110.6	102.0	122.2	110.3	98.4	122.6	108.3	103.3	122.2	111.0	92.5	63.2	80.6	100.3	70.6	88.1
0330	Cast-in-Place Concrete	111.9	121.2	115.4	109.7	101.3	106.5	113.9	120.4	116.4	95.6	118.8	104.5	106.8	60.3	89.0	107.2	63.7	90.6
03	CONCRETE	104.4	128.4	115.3	101.9	107.8	104.6	106.4	123.6	114.2	91.9	121.9	105.5	97.1	63.2	81.7	100.6	71.4	87.3
04	MASONRY	100.9	131.8	119.4	100.1	115.7	109.5	109.8	131.8	123.0	105.0	124.3	116.5	88.7	50.1	65.6	129.1	56.0	85.3
05	METALS	91.8	131.8	103.5	89.2	129.7	101.0	88.8	131.3	101.2	90.7	129.1	101.9	97.6	87.9	94.7	95.7	91.2	94.4
06	WOOD, PLASTICS & COMPOSITES	115.2	135.9	126.6	115.2	102.8	108.4	107.8	122.3	115.7	94.4	120.8	108.9	94.9	65.7	78.8	78.9	79.7	79.4
07	THERMAL & MOISTURE PROTECTION	102.5	133.8	114.7	102.6	102.3	102.5	103.6	131.8	114.6	101.1	113.7	106.0	92.6	62.2	80.7	92.7	68.8	83.4
08	OPENINGS	98.0	142.0	108.9	94.7	124.0	102.0	92.3	134.5	102.8	91.2	133.8	101.8	97.1	66.2	89.4	96.4	71.8	90.3
0920	Plaster & Gypsum Board	98.9	137.5	124.0	98.5	103.4	101.7	91.8	123.6	112.4	83.9	122.2	108.8	104.0	64.9	78.6	92.8	79.5	84.2
0950, 0980	Ceilings & Acoustic Treatment	99.7	137.5	123.1	98.0	103.4	101.4	93.0	123.6	111.9	69.9	122.2	102.2	100.4	64.9	78.5	94.6	79.5	85.2
0960	Flooring	105.3	126.6	111.0	107.7	92.5	103.6	101.2	126.6	108.0	105.0	126.6	110.9	117.2	59.2	101.5	108.0	59.2	94.8
0970, 0990	Wall Finishes & Painting/Coating	98.9	128.0	116.4	94.2	101.5	98.6	98.9	119.8	111.4	103.2	128.0	118.1	110.4	49.9	74.1	110.4	75.7	89.6
09	FINISHES	102.7	134.3	119.4	102.6	102.8	102.7	99.4	125.3	113.1	93.2	124.8	109.9	104.3	60.3	81.1	99.1	72.9	85.3
COVERS	DIVS. 10 - 14, 25, 28, 41, 43, 44	100.0	106.0	101.2	100.0	96.9	99.4	100.0	104.0	100.8	100.0	98.8	99.8	100.0	69.6	94.0	100.0	73.5	94.8
21, 22, 23	FIRE SUPPRESSION, PLUMBING & HVAC	99.8	117.1	106.7	99.8	98.0	99.1	99.8	112.5	104.8	99.5	115.8	106.0	99.9	50.0	80.0	99.9	57.4	83.0
26, 27, 3370	ELECTRICAL, COMMUNICATIONS & UTIL.	101.9	115.3	108.5	100.6	85.7	93.4	100.1	102.4	101.2	100.8	102.4	101.6	99.9	55.5	78.3	99.2	73.2	86.5
MF2004	WEIGHTED AVERAGE	99.3	124.2	109.8	98.1	104.7	100.9	98.6	118.7	107.0	95.8	117.5	104.9	98.5	62.6	83.4	100.0	70.9	87.8

DIVISION		MISSISSIPPI						MISSOURI											
		JACKSON			MERIDIAN			CAPE GIRARDEAU			COLUMBIA			JOPLIN			KANSAS CITY		
		MAT.	INST.	TOTAL	MAT.	INST.	TOTAL	MAT.	INST.	TOTAL	MAT.	INST.	TOTAL	MAT.	INST.	TOTAL	MAT.	INST.	TOTAL
015433	CONTRACTOR EQUIPMENT		99.5	99.5		99.5	99.5		107.1	107.1		108.1	108.1		105.9	105.9		100.9	100.9
0241, 31 - 34	SITE & INFRASTRUCTURE, DEMOLITION	98.9	89.7	92.6	98.7	90.0	92.7	93.4	93.9	93.7	102.8	95.7	97.9	104.2	98.8	100.5	96.7	93.5	94.5
0310	Concrete Forming & Accessories	93.0	67.4	70.9	79.8	74.1	74.9	86.6	74.2	75.9	88.2	66.7	69.6	104.3	69.2	74.0	103.3	107.7	107.1
0320	Concrete Reinforcing	92.5	61.7	79.9	99.0	61.7	83.8	105.1	83.1	96.1	102.9	113.6	107.3	100.8	70.9	88.6	95.8	114.7	103.5
0330	Cast-in-Place Concrete	99.2	63.7	85.6	101.1	66.3	87.8	94.2	83.8	90.2	89.0	85.3	87.6	100.3	69.9	88.7	93.2	107.9	98.8
03	CONCRETE	93.5	66.5	81.3	93.6	70.4	83.1	93.6	80.9	87.8	86.6	83.8	85.3	97.9	71.0	85.6	93.9	109.3	100.9
04	MASONRY	89.8	56.0	69.6	88.4	60.6	71.8	119.3	77.3	94.1	131.0	84.4	103.1	93.1	58.8	72.6	98.8	107.6	104.1
05	METALS	97.5	87.6	94.6	95.6	87.8	93.3	94.2	104.9	97.4	90.0	118.4	98.3	94.9	87.6	92.8	102.4	113.4	105.6
06	WOOD, PLASTICS & COMPOSITES	93.7	69.8	80.6	77.8	76.3	77.0	88.5	72.1	79.5	94.8	59.6	75.5	104.2	68.2	84.4	103.5	107.6	105.8
07	THERMAL & MOISTURE PROTECTION	94.4	67.7	84.0	92.3	70.4	83.8	102.0	78.1	92.7	93.7	81.5	88.9	95.6	64.8	84.2	96.3	106.2	100.1
08	OPENINGS	97.4	63.9	89.2	96.4	73.9	90.8	98.8	72.5	92.3	91.4	86.9	90.3	91.5	70.4	86.3	97.5	110.3	100.7
0920	Plaster & Gypsum Board	96.8	69.2	78.9	92.8	75.9	81.8	95.2	71.0	79.5	93.6	58.1	70.5	97.0	66.8	77.4	93.7	107.7	102.8
0950, 0980	Ceilings & Acoustic Treatment	97.1	69.2	79.8	94.6	75.9	83.0	86.1	71.0	76.8	87.6	58.1	69.3	85.0	66.8	73.8	90.8	107.7	101.3
0960	Flooring	117.6	59.2	101.8	106.9	59.2	94.0	86.9	67.2	81.5	106.5	88.5	101.7	122.7	45.2	101.7	98.9	99.9	99.2
0970, 0990	Wall Finishes & Painting/Coating	110.4	75.7	89.6	110.4	87.5	96.7	98.3	77.8	86.0	100.6	75.5	85.5	94.6	40.3	62.0	99.5	120.1	111.9
09	FINISHES	102.7	67.1	83.9	98.1	73.4	85.0	90.8	72.3	81.1	94.9	68.1	80.7	103.1	61.8	81.3	98.3	107.5	103.2
COVERS	DIVS. 10 - 14, 25, 28, 41, 43, 44	100.0	72.3	94.5	100.0	74.7	95.0	100.0	60.5	92.2	100.0	94.1	98.8	100.0	69.5	94.0	100.0	99.4	99.9
21, 22, 23	FIRE SUPPRESSION, PLUMBING & HVAC	99.9	63.4	85.4	99.9	65.9	86.4	99.9	89.9	95.9	99.9	91.3	96.5	100.1	52.6	81.2	100.0	104.1	101.6
26, 27, 3370	ELECTRICAL, COMMUNICATIONS & UTIL.	104.6	73.2	89.3	98.9	58.9	79.4	100.9	108.3	104.5	96.0	85.3	90.8	90.9	68.8	80.1	101.2	105.2	103.1
MF2004	WEIGHTED AVERAGE	98.5	69.9	86.5	96.8	70.9	85.9	98.2	87.4	93.7	96.3	88.1	92.9	97.1	68.0	84.9	99.2	106.1	102.1

DIVISION		MISSOURI									MONTANA								
		SPRINGFIELD			ST. JOSEPH			ST. LOUIS			BILLINGS			BUTTE			GREAT FALLS		
		MAT.	INST.	TOTAL	MAT.	INST.	TOTAL	MAT.	INST.	TOTAL	MAT.	INST.	TOTAL	MAT.	INST.	TOTAL	MAT.	INST.	TOTAL
015433	CONTRACTOR EQUIPMENT		101.2	101.2		99.5	99.5		108.0	108.0		99.1	99.1		98.7	98.7		98.7	98.7
0241, 31 - 34	SITE & INFRASTRUCTURE, DEMOLITION	95.9	91.6	92.9	98.2	88.6	91.7	93.2	97.5	96.1	91.0	97.7	95.6	99.9	96.7	97.7	103.6	97.5	99.4
0310	Concrete Forming & Accessories	101.2	71.7	75.7	103.1	79.4	82.6	98.8	105.3	104.4	96.9	65.3	69.6	85.4	69.7	71.8	96.8	66.7	70.8
0320	Concrete Reinforcing	99.0	113.3	104.8	94.6	104.7	98.7	96.1	109.4	101.5	102.3	70.6	89.4	111.0	70.7	94.6	102.3	70.5	89.3
0330	Cast-in-Place Concrete	96.6	62.9	83.8	93.2	103.3	97.0	94.2	112.3	101.1	108.7	69.2	93.6	119.7	70.9	101.0	126.5	68.4	104.3
03	CONCRETE	96.1	77.4	87.6	93.6	93.4	93.5	92.5	109.3	100.2	102.3	68.8	87.1	106.2	71.3	90.4	110.3	69.1	91.6
04	MASONRY	88.7	78.1	82.4	98.4	90.3	93.6	99.5	110.3	106.0	120.5	71.1	90.9	116.3	74.4	91.2	120.3	73.2	92.1
05	METALS	93.8	105.3	97.2	98.6	107.2	101.1	98.9	119.5	104.9	106.1	86.5	100.3	98.9	86.8	95.4	102.8	86.5	98.1
06	WOOD, PLASTICS & COMPOSITES	103.2	72.6	86.4	104.2	74.6	88.0	102.4	102.5	102.5	102.6	65.0	82.0	90.4	70.0	79.2	103.8	65.0	82.5
07	THERMAL & MOISTURE PROTECTION	98.2	73.3	88.5	96.6	92.3	94.9	101.9	107.8	104.2	101.3	69.3	88.9	101.2	73.8	90.5	101.7	68.2	88.7
08	OPENINGS	97.1	81.0	93.1	96.2	89.9	94.7	97.7	109.7	100.7	97.7	62.6	89.0	95.6	65.8	88.2	99.0	63.1	90.1
0920	Plaster & Gypsum Board	96.9	71.5	80.4	98.4	73.5	82.2	103.2	102.6	102.8	93.6	64.1	74.4	91.0	69.3	76.9	98.5	64.1	76.1
0950, 0980	Ceilings & Acoustic Treatment	87.6	71.5	77.7	90.0	73.5	79.8	91.1	102.6	98.2	106.7	64.1	80.3	111.6	69.3	85.4	113.2	64.1	82.8
0960	Flooring	109.2	45.2	91.9	102.0	97.0	100.7	94.0	102.4	96.3	110.1	61.4	96.9	106.9	83.2	100.5	115.5	83.2	106.7
0970, 0990	Wall Finishes & Painting/Coating	95.0	57.7	72.6	95.1	85.4	89.3	98.3	109.2	104.8	100.5	53.8	72.5	98.4	58.9	74.7	98.4	55.9	72.9
09	FINISHES	98.4	64.9	80.7	99.4	81.8	90.1	95.3	104.3	100.1	105.0	62.9	82.8	104.5	71.1	86.9	108.8	68.2	87.3
COVERS	DIVS. 10 - 14, 25, 28, 41, 43, 44	100.0	89.1	97.8	100.0	92.8	98.6	100.0	103.9	100.8	100.0	73.6	94.8	100.0	74.7	95.0	100.0	74.9	95.0
21, 22, 23	FIRE SUPPRESSION, PLUMBING & HVAC	99.9	68.3	87.4	100.1	90.1	96.1	99.9	108.7	103.4	100.1	76.1	90.6	100.2	69.9	88.2	100.2	74.3	89.9
26, 27, 3370	ELECTRICAL, COMMUNICATIONS & UTIL.	99.7	73.1	86.7	99.0	84.0	91.7	103.6	111.7	107.6	95.1	75.5	85.6	102.4	72.6	87.9	95.2	72.5	84.1
MF2004	WEIGHTED AVERAGE	97.3	77.6	89.1	98.3	90.2	94.9	98.6	108.8	102.9	101.9	74.6	90.4	101.5	75.0	90.4	103.1	74.8	91.2

DIVISION		MONTANA						NEBRASKA											
		HELENA			MISSOULA			GRAND ISLAND			LINCOLN			NORTH PLATTE			OMAHA		
		MAT.	INST.	TOTAL	MAT.	INST.	TOTAL	MAT.	INST.	TOTAL	MAT.	INST.	TOTAL	MAT.	INST.	TOTAL	MAT.	INST.	TOTAL
015433	CONTRACTOR EQUIPMENT		98.7	98.7		98.7	98.7		100.5	100.5		100.5	100.5		100.5	100.5		91.9	91.9
0241, 31 - 34	SITE & INFRASTRUCTURE, DEMOLITION	100.7	96.6	97.9	81.3	96.6	91.8	99.7	90.7	93.6	89.3	90.7	90.3	100.6	90.2	93.5	80.1	90.7	87.3
0310	Concrete Forming & Accessories	97.1	66.1	70.3	88.2	61.0	64.7	97.9	82.8	84.8	102.5	65.5	70.5	97.8	81.6	83.8	94.4	72.4	75.4
0320	Concrete Reinforcing	105.8	70.5	91.4	112.9	80.2	99.6	107.8	73.6	93.8	99.0	73.0	88.4	109.2	74.3	95.0	103.6	73.2	91.2
0330	Cast-in-Place Concrete	106.7	67.0	91.5	88.1	66.0	79.7	116.1	74.6	100.2	92.6	77.3	86.7	116.1	71.2	98.9	99.3	79.2	91.6
03	CONCRETE	101.8	68.4	86.6	86.0	67.5	77.6	108.2	78.8	94.8	94.3	72.1	84.2	108.5	77.2	94.3	97.1	75.3	87.2
04	MASONRY	113.0	71.6	88.2	139.5	69.1	97.3	107.5	84.8	93.9	96.6	74.6	83.4	92.2	97.1	95.2	101.8	80.2	88.9
05	METALS	102.0	86.4	97.5	95.9	90.8	94.4	90.1	87.7	89.4	95.0	86.8	92.6	90.3	88.4	89.8	94.3	78.7	89.8
06	WOOD, PLASTICS & COMPOSITES	103.9	65.0	82.6	94.0	60.6	75.6	99.8	84.6	91.4	102.7	61.5	80.1	99.6	84.6	91.4	93.7	71.3	81.4
07	THERMAL & MOISTURE PROTECTION	101.4	73.7	90.6	100.5	70.8	89.0	99.0	80.5	91.8	98.5	75.3	89.5	99.0	87.9	94.6	94.4	82.9	89.9
08	OPENINGS	98.5	61.8	89.4	95.6	63.1	87.6	90.7	77.9	87.5	96.2	61.0	87.5	90.0	77.2	86.8	99.3	71.2	92.4
0920	Plaster & Gypsum Board	91.2	64.1	73.6	91.7	59.4	70.8	94.9	83.9	87.8	91.4	60.0	71.0	95.1	83.9	87.8	93.7	70.7	78.8
0950, 0980	Ceilings & Acoustic Treatment	108.2	64.1	80.9	111.6	59.4	79.3	86.8	83.9	85.0	89.3	60.0	71.2	87.6	83.9	85.3	102.1	70.7	82.7
0960	Flooring	115.5	66.3	102.2	109.0	83.2	102.0	98.5	120.2	104.3	100.2	93.7	98.4	98.4	112.5	102.2	128.1	54.5	108.2
0970, 0990	Wall Finishes & Painting/Coating	98.4	49.7	69.2	98.4	54.7	72.2	93.0	77.7	83.8	93.0	89.5	90.9	93.0	73.3	81.2	158.4	71.7	106.4
09	FINISHES	106.4	64.2	84.1	104.2	63.8	82.8	95.2	88.6	91.7	95.5	72.6	83.4	95.4	86.1	90.5	116.0	68.3	90.8
COVERS	DIVS. 10 - 14, 25, 28, 41, 43, 44	100.0	86.3	97.3	100.0	83.7	96.8	100.0	88.2	97.7	100.0	85.5	97.1	100.0	82.2	96.5	100.0	85.3	97.1
21, 22, 23	FIRE SUPPRESSION, PLUMBING & HVAC	100.1	70.3	88.3	100.2	66.6	86.8	99.9	80.1	92.1	99.8	80.1	92.0	99.9	78.2	91.3	99.7	80.0	91.8
26, 27, 3370	ELECTRICAL, COMMUNICATIONS & UTIL.	101.0	72.5	87.1	100.3	76.7	88.8	89.9	74.0	82.1	103.8	74.0	89.3	91.8	89.9	90.9	103.6	83.2	93.7
MF2004	WEIGHTED AVERAGE	101.8	73.5	89.9	99.0	73.2	88.2	97.3	82.4	91.0	97.6	77.1	89.0	96.8	85.0	91.8	99.5	78.8	90.8

DIVISION		NEVADA									NEW HAMPSHIRE								
		CARSON CITY			LAS VEGAS			RENO			MANCHESTER			NASHUA			PORTSMOUTH		
		MAT.	INST.	TOTAL	MAT.	INST.	TOTAL	MAT.	INST.	TOTAL	MAT.	INST.	TOTAL	MAT.	INST.	TOTAL	MAT.	INST.	TOTAL
015433	CONTRACTOR EQUIPMENT		101.1	101.1		101.1	101.1		101.1	101.1		100.4	100.4		100.4	100.4		100.4	100.4
0241, 31 - 34	SITE & INFRASTRUCTURE, DEMOLITION	68.7	103.0	92.2	63.5	106.4	92.9	61.8	103.0	90.0	82.1	98.4	93.2	84.7	98.4	94.1	79.1	99.7	93.2
0310	Concrete Forming & Accessories	101.7	92.8	94.0	103.7	110.5	109.6	98.7	92.9	93.7	94.5	81.3	83.1	96.7	81.3	83.4	86.0	83.7	84.0
0320	Concrete Reinforcing	107.5	113.2	109.8	99.6	125.4	110.2	102.1	124.4	111.2	100.4	93.2	97.4	100.4	93.2	97.4	80.3	93.3	85.6
0330	Cast-in-Place Concrete	109.5	91.2	102.5	106.6	115.0	109.8	116.4	91.2	106.8	101.0	117.1	107.2	88.2	117.1	99.3	83.6	120.9	97.9
03	CONCRETE	110.8	96.1	104.1	108.0	114.4	110.9	112.8	98.2	106.1	107.8	96.0	102.4	101.9	96.0	99.2	92.7	98.5	95.4
04	MASONRY	123.0	78.0	96.0	117.2	103.7	109.1	123.8	78.0	96.4	98.0	87.9	91.9	97.8	87.9	91.8	93.3	93.1	93.2
05	METALS	94.9	101.3	96.7	101.1	110.0	103.7	95.3	105.6	98.3	93.8	90.6	92.9	93.8	90.6	92.8	90.4	93.6	91.4
06	WOOD, PLASTICS & COMPOSITES	93.3	93.6	93.5	93.3	108.8	101.8	89.4	93.6	91.7	98.9	80.4	88.7	102.3	80.4	90.3	89.6	80.4	84.6
07	THERMAL & MOISTURE PROTECTION	99.4	87.1	94.6	115.4	102.4	110.4	100.8	87.1	95.5	94.2	92.3	93.4	94.8	92.3	93.8	94.5	113.5	101.9
08	OPENINGS	93.3	106.4	96.5	94.8	117.7	100.5	93.6	106.6	96.8	104.9	77.3	98.1	104.9	77.3	98.1	105.5	72.2	97.3
0920	Plaster & Gypsum Board	78.5	93.2	88.1	84.5	109.0	100.4	78.7	93.2	88.2	108.7	78.9	89.4	105.5	78.9	88.3	97.0	78.9	85.3
0950, 0980	Ceilings & Acoustic Treatment	107.8	93.2	98.8	121.2	109.0	113.6	117.8	93.2	102.6	96.6	78.9	85.7	99.1	78.9	86.6	90.8	78.9	83.5
0960	Flooring	104.9	57.3	92.0	98.8	98.1	98.6	104.5	57.3	91.8	97.9	100.4	98.5	96.8	100.4	97.8	91.8	100.4	94.1
0970, 0990	Wall Finishes & Painting/Coating	99.2	87.7	92.3	100.4	118.4	111.2	98.1	87.7	91.9	91.7	96.7	94.7	91.7	96.7	94.7	91.7	96.7	94.7
09	FINISHES	101.0	85.1	92.6	102.4	108.7	105.7	102.4	85.1	93.2	98.7	86.0	92.0	99.1	86.0	92.2	94.1	87.2	90.5
COVERS	DIVS. 10 - 14, 25, 28, 41, 43, 44	100.0	109.7	101.9	100.0	104.5	100.9	100.0	109.7	101.9	100.0	69.9	94.0	100.0	69.9	94.0	100.0	71.7	94.4
21, 22, 23	FIRE SUPPRESSION, PLUMBING & HVAC	99.9	87.3	94.9	100.0	110.2	104.1	99.9	87.5	95.0	100.0	85.0	94.1	100.2	85.0	94.2	100.2	87.7	95.2
26, 27, 3370	ELECTRICAL, COMMUNICATIONS & UTIL.	98.8	99.4	99.1	99.0	128.0	113.1	94.5	99.4	96.9	96.7	83.4	90.2	97.0	83.4	90.4	95.2	83.4	89.5
MF2004	WEIGHTED AVERAGE	99.9	93.1	97.1	101.1	112.0	105.7	99.8	93.9	97.3	99.1	87.7	94.3	98.6	87.7	94.0	95.9	90.1	93.5

DIVISION		NEW JERSEY																	
		CAMDEN			ELIZABETH			JERSEY CITY			NEWARK			PATERSON			TRENTON		
		MAT.	INST.	TOTAL	MAT.	INST.	TOTAL	MAT.	INST.	TOTAL	MAT.	INST.	TOTAL	MAT.	INST.	TOTAL	MAT.	INST.	TOTAL
015433	CONTRACTOR EQUIPMENT		98.5	98.5		100.4	100.4		98.5	98.5		100.4	100.4		100.4	100.4		98.1	98.1
0241, 31 - 34	SITE & INFRASTRUCTURE, DEMOLITION	87.9	103.7	98.7	100.8	104.6	103.4	87.9	104.4	99.2	103.4	104.6	104.2	99.1	104.5	102.8	87.6	104.3	99.0
0310	Concrete Forming & Accessories	96.7	120.5	117.2	104.7	121.1	118.9	96.7	121.2	117.9	93.3	121.2	117.4	95.5	121.1	117.6	93.9	120.8	117.1
0320	Concrete Reinforcing	100.4	114.2	106.0	77.3	119.5	94.5	100.4	119.5	108.2	100.4	119.5	108.2	100.4	119.5	108.2	100.4	113.2	105.6
0330	Cast-in-Place Concrete	76.7	123.8	94.7	84.5	125.1	100.0	76.7	125.2	95.2	98.1	125.1	108.4	97.8	125.1	108.2	87.3	125.0	101.7
03	CONCRETE	96.5	119.4	106.9	97.8	121.4	108.5	96.5	121.2	107.8	106.3	121.4	113.2	106.3	121.3	113.2	101.4	119.8	109.7
04	MASONRY	91.2	120.3	108.6	110.7	120.3	116.4	88.6	120.3	107.6	99.9	120.3	112.1	95.4	120.3	110.3	87.2	120.3	107.0
05	METALS	93.6	102.4	96.2	90.2	108.8	95.6	93.7	106.5	97.4	93.7	108.9	98.1	88.9	108.8	94.7	88.9	102.3	92.8
06	WOOD, PLASTICS & COMPOSITES	102.3	120.8	112.5	116.2	120.8	118.8	102.3	120.8	112.5	100.9	120.8	111.8	104.1	120.8	113.3	99.3	120.8	111.1
07	THERMAL & MOISTURE PROTECTION	94.4	120.6	104.6	95.1	125.8	107.0	94.4	125.8	106.6	94.6	125.8	106.7	95.0	120.7	105.0	90.6	120.7	102.3
08	OPENINGS	104.9	117.4	108.0	105.9	119.9	109.4	104.9	119.9	108.6	110.9	119.9	113.1	110.9	119.9	113.1	105.4	117.1	108.3
0920	Plaster & Gypsum Board	105.5	120.8	115.5	109.3	120.8	116.8	105.5	120.8	115.5	108.7	120.8	116.6	105.5	120.8	115.5	108.7	120.8	116.6
0950, 0980	Ceilings & Acoustic Treatment	99.1	120.8	112.5	90.8	120.8	109.4	99.1	120.8	112.5	96.6	120.8	111.6	99.1	120.8	112.5	96.6	120.8	111.6
0960	Flooring	96.8	136.7	107.6	101.1	151.8	114.8	96.8	151.8	111.7	97.0	151.8	111.8	96.8	151.8	111.7	97.2	151.8	111.9
0970, 0990	Wall Finishes & Painting/Coating	91.7	124.2	111.2	91.6	126.9	112.8	91.7	126.9	112.8	91.6	126.9	112.8	91.6	126.9	112.8	91.7	124.2	111.2
09	FINISHES	99.5	124.6	112.8	100.1	127.5	114.6	99.5	127.5	114.3	99.6	127.5	114.4	99.6	127.5	114.4	99.3	127.2	114.0
COVERS	DIVS. 10 - 14, 25, 28, 41, 43, 44	100.0	110.2	102.0	100.0	118.0	103.6	100.0	118.0	103.6	100.0	118.0	103.6	100.0	118.0	103.6	100.0	110.1	102.0
21, 22, 23	FIRE SUPPRESSION, PLUMBING & HVAC	100.2	113.9	105.6	100.2	120.6	108.3	100.2	122.8	109.2	100.2	120.6	108.3	100.2	122.8	109.2	100.3	121.8	108.8
26, 27, 3370	ELECTRICAL, COMMUNICATIONS & UTIL.	97.5	134.0	115.3	95.0	137.5	115.6	98.8	141.1	119.4	99.0	141.1	119.5	98.8	137.5	117.6	101.4	139.2	119.8
MF2004	WEIGHTED AVERAGE	97.8	117.8	106.2	98.6	121.5	108.2	97.8	122.2	108.0	100.5	122.0	109.5	99.3	121.8	108.8	97.6	120.6	107.3

DIVISION		NEW MEXICO															NEW YORK		
		ALBUQUERQUE			FARMINGTON			LAS CRUCES			ROSWELL			SANTA FE			ALBANY		
		MAT.	INST.	TOTAL	MAT.	INST.	TOTAL	MAT.	INST.	TOTAL	MAT.	INST.	TOTAL	MAT.	INST.	TOTAL	MAT.	INST.	TOTAL
015433	CONTRACTOR EQUIPMENT		112.1	112.1		112.1	112.1		90.9	90.9		112.1	112.1		112.1	112.1		112.9	112.9
0241, 31 - 34	SITE & INFRASTRUCTURE, DEMOLITION	78.2	108.0	98.6	84.5	108.0	100.6	89.2	91.1	90.5	89.0	108.0	102.0	84.0	108.0	100.4	73.4	105.3	95.2
0310	Concrete Forming & Accessories	99.8	69.1	73.3	99.8	69.1	73.3	97.0	67.4	71.4	99.8	68.8	73.1	98.9	69.1	73.2	96.1	87.4	88.6
0320	Concrete Reinforcing	102.8	70.9	89.8	112.3	70.9	95.4	106.4	50.6	83.7	111.4	50.9	86.7	110.2	70.9	94.2	99.1	94.3	97.1
0330	Cast-in-Place Concrete	108.9	77.6	96.9	109.9	77.6	97.6	95.6	69.2	85.5	101.3	77.5	92.2	125.5	77.6	107.2	83.4	101.9	90.5
03	CONCRETE	109.5	73.5	93.1	113.8	73.5	95.5	91.2	65.7	79.6	114.0	69.6	93.8	118.7	73.5	98.1	99.0	94.8	97.1
04	MASONRY	113.9	65.3	84.8	120.3	65.3	87.3	107.2	62.7	80.6	121.7	65.3	87.9	116.6	65.3	85.9	91.8	99.1	96.2
05	METALS	101.4	89.8	98.0	99.1	89.8	96.4	99.0	73.7	91.6	99.1	80.8	93.7	99.1	89.8	96.4	94.7	105.7	97.9
06	WOOD, PLASTICS & COMPOSITES	93.4	69.5	80.3	93.5	69.5	80.3	84.5	68.3	75.6	93.5	69.5	80.3	92.8	69.5	80.0	95.1	84.5	89.3
07	THERMAL & MOISTURE PROTECTION	100.3	74.8	90.4	100.6	74.8	90.6	86.5	69.1	79.7	100.9	74.8	90.7	100.2	74.8	90.3	89.5	91.8	90.4
08	OPENINGS	93.4	71.8	88.1	96.2	71.8	90.1	86.0	64.9	80.8	92.2	65.6	85.6	92.4	71.8	87.3	96.0	82.3	92.7
0920	Plaster & Gypsum Board	83.2	68.0	73.3	76.1	68.0	70.8	78.3	68.0	71.6	76.1	68.0	70.8	84.5	68.0	73.8	107.9	83.7	92.2
0950, 0980	Ceilings & Acoustic Treatment	112.4	68.0	84.9	107.8	68.0	83.2	101.3	68.0	80.7	107.8	68.0	83.2	107.8	68.0	83.2	92.4	83.7	87.0
0960	Flooring	102.1	69.0	93.2	104.5	69.0	94.9	138.2	65.2	118.4	104.5	69.0	94.9	104.5	69.0	94.9	83.8	86.7	84.6
0970, 0990	Wall Finishes & Painting/Coating	102.6	54.9	74.0	98.1	54.9	72.2	90.8	54.9	69.3	98.1	54.9	72.2	98.1	54.9	72.2	82.4	79.3	80.5
09	FINISHES	101.7	67.5	83.6	100.4	67.5	83.0	115.2	65.8	89.0	100.8	67.5	83.2	101.6	67.5	83.6	94.3	85.8	89.8
COVERS	DIVS. 10 - 14, 25, 28, 41, 43, 44	100.0	75.1	95.1	100.0	75.1	95.1	100.0	71.8	94.4	100.0	75.1	95.1	100.0	75.1	95.1	100.0	93.2	98.7
21, 22, 23	FIRE SUPPRESSION, PLUMBING & HVAC	100.0	72.3	89.0	99.9	72.3	88.9	100.2	71.7	88.9	99.9	72.0	88.8	99.9	72.3	88.9	100.0	93.0	97.3
26, 27, 3370	ELECTRICAL, COMMUNICATIONS & UTIL.	88.6	73.8	81.4	86.8	73.8	80.5	88.0	57.1	72.9	87.5	73.8	80.9	101.6	73.8	88.1	106.9	93.9	100.6
MF2004	WEIGHTED AVERAGE	99.8	76.0	89.8	100.3	76.0	90.1	97.0	68.5	85.0	100.3	74.3	89.3	102.0	76.0	91.1	97.4	94.7	96.3

	DIVISION	NEW YORK																	
		BINGHAMTON			BUFFALO			HICKSVILLE			NEW YORK			RIVERHEAD			ROCHESTER		
		MAT.	INST.	TOTAL	MAT.	INST.	TOTAL	MAT.	INST.	TOTAL	MAT.	INST.	TOTAL	MAT.	INST.	TOTAL	MAT.	INST.	TOTAL
015433	CONTRACTOR EQUIPMENT		117.7	117.7		98.1	98.1		109.6	109.6		110.2	110.2		109.6	109.6		114.6	114.6
0241, 31 - 34	SITE & INFRASTRUCTURE, DEMOLITION	95.6	96.0	95.9	95.5	99.3	98.1	114.6	121.4	119.3	130.7	121.2	124.2	115.6	121.4	119.6	72.0	105.3	94.8
0310	Concrete Forming & Accessories	101.0	79.0	82.0	97.1	113.7	111.4	90.2	144.0	136.7	107.0	184.8	174.2	94.9	144.0	137.3	98.0	92.4	93.2
0320	Concrete Reinforcing	98.1	94.4	96.6	98.4	102.3	100.0	99.4	184.3	134.0	106.7	190.7	140.9	101.2	184.3	135.2	99.2	89.2	95.1
0330	Cast-in-Place Concrete	102.9	90.2	98.1	110.7	116.6	112.9	96.3	156.1	119.1	108.8	168.8	131.7	97.9	156.1	120.1	99.5	99.1	99.4
03	CONCRETE	99.7	87.7	94.2	107.4	111.8	109.4	103.5	154.3	126.6	112.0	177.7	141.9	104.3	154.3	127.1	111.1	95.2	103.9
04	MASONRY	107.0	81.0	91.5	109.3	118.4	114.8	110.9	159.6	140.0	107.8	171.5	145.9	116.4	159.6	142.3	105.3	95.8	99.6
05	METALS	91.3	114.7	98.1	94.1	95.5	94.5	101.1	138.4	112.0	108.4	140.9	117.9	101.5	138.4	112.3	90.8	105.9	95.2
06	WOOD, PLASTICS & COMPOSITES	105.3	79.7	91.3	98.4	113.4	106.6	87.5	140.7	116.7	110.5	188.2	153.2	92.9	140.7	119.2	95.4	92.8	94.0
07	THERMAL & MOISTURE PROTECTION	103.1	80.5	94.3	99.1	109.0	102.9	106.4	150.4	123.5	109.3	167.8	132.1	107.0	150.4	123.9	94.9	95.6	95.2
08	OPENINGS	91.0	78.8	88.0	95.2	102.2	96.9	88.8	150.6	104.1	98.0	177.2	117.5	88.8	150.6	104.1	98.8	87.9	96.1
0920	Plaster & Gypsum Board	119.9	78.5	93.0	99.0	113.6	108.5	107.3	142.2	129.9	128.3	191.1	169.1	109.4	142.2	130.7	98.2	92.5	94.5
0950, 0980	Ceilings & Acoustic Treatment	96.6	78.5	85.4	98.5	113.6	107.8	79.1	142.2	118.1	109.1	191.1	159.8	81.6	142.2	119.1	97.8	92.5	94.5
0960	Flooring	94.4	81.9	91.0	89.1	117.2	96.7	94.2	160.0	112.0	93.3	169.6	113.9	95.5	160.0	112.9	79.3	101.1	85.2
0970, 0990	Wall Finishes & Painting/Coating	86.7	85.2	85.8	86.1	112.6	102.0	111.0	154.5	137.1	90.3	156.7	130.2	111.0	154.5	137.1	84.7	96.0	91.5
09	FINISHES	97.3	79.8	88.1	92.5	115.0	104.4	104.7	147.1	127.1	109.2	180.1	146.7	105.9	147.1	127.7	92.1	94.6	93.5
COVERS	DIVS. 10 - 14, 25, 28, 41, 43, 44	100.0	93.3	98.7	100.0	107.5	101.5	100.0	128.9	105.7	100.0	139.6	107.8	100.0	128.9	105.7	100.0	88.4	97.7
21, 22, 23	FIRE SUPPRESSION, PLUMBING & HVAC	100.4	79.0	91.9	99.9	97.1	98.8	99.8	146.9	118.5	100.2	164.6	125.8	99.9	146.9	118.6	99.9	89.6	95.8
26, 27, 3370	ELECTRICAL, COMMUNICATIONS & UTIL.	102.2	86.8	94.7	99.8	98.5	99.1	102.2	149.3	125.1	109.4	173.1	140.4	103.8	149.3	126.0	101.0	86.2	93.8
MF2004	WEIGHTED AVERAGE	98.1	86.9	93.4	99.0	104.8	101.5	101.2	146.4	120.1	106.1	164.6	130.7	102.0	146.4	120.6	98.4	94.1	96.6

	DIVISION	NEW YORK																	
		SCHENECTADY			SYRACUSE			UTICA			WATERTOWN			WHITE PLAINS			YONKERS		
		MAT.	INST.	TOTAL	MAT.	INST.	TOTAL	MAT.	INST.	TOTAL	MAT.	INST.	TOTAL	MAT.	INST.	TOTAL	MAT.	INST.	TOTAL
015433	CONTRACTOR EQUIPMENT		112.9	112.9		112.9	112.9		112.9	112.9		112.9	112.9		109.7	109.7		109.7	109.7
0241, 31 - 34	SITE & INFRASTRUCTURE, DEMOLITION	73.4	105.4	95.3	94.4	104.9	101.6	71.6	102.8	92.9	80.1	105.4	97.4	120.5	114.7	116.5	130.0	114.5	119.4
0310	Concrete Forming & Accessories	100.7	88.1	89.8	99.9	88.9	90.4	101.2	85.6	87.8	86.2	92.4	91.6	106.4	128.8	125.7	106.7	128.8	125.8
0320	Concrete Reinforcing	97.7	94.3	96.3	99.1	94.7	97.3	99.1	91.4	96.0	99.7	94.6	97.6	99.8	183.6	134.0	103.8	183.6	136.3
0330	Cast-in-Place Concrete	94.5	102.9	97.7	95.5	96.9	96.0	87.4	90.1	88.4	101.8	101.0	101.5	95.9	133.8	110.4	107.3	133.8	117.4
03	CONCRETE	104.1	95.3	100.1	103.0	93.7	98.8	101.1	89.3	95.7	113.8	96.6	106.0	100.8	139.6	118.4	110.8	139.6	123.9
04	MASONRY	91.2	100.7	96.9	98.0	96.0	96.8	89.8	85.7	87.4	91.0	98.5	95.5	101.8	133.1	120.6	107.3	133.1	122.7
05	METALS	94.8	105.7	98.0	94.6	104.8	97.6	92.9	103.5	96.0	92.9	104.8	96.4	95.9	132.6	106.6	104.7	132.6	112.8
06	WOOD, PLASTICS & COMPOSITES	101.7	84.5	92.3	101.7	87.2	93.7	101.7	87.7	94.0	84.0	91.2	87.9	111.6	129.2	121.3	111.4	129.2	121.2
07	THERMAL & MOISTURE PROTECTION	90.1	92.6	91.1	100.2	94.7	98.0	89.9	90.2	90.0	90.2	96.4	92.6	109.7	141.0	121.9	110.0	141.0	122.1
08	OPENINGS	96.0	82.8	92.7	94.0	82.1	91.0	96.0	83.9	93.0	96.0	88.3	94.1	92.5	142.9	105.0	96.0	143.5	107.7
0920	Plaster & Gypsum Board	115.0	83.7	94.7	115.0	86.4	96.4	115.0	87.0	96.8	107.9	90.6	96.7	121.9	129.9	127.1	127.9	129.9	129.2
0950, 0980	Ceilings & Acoustic Treatment	96.6	83.7	88.6	96.6	86.4	90.3	96.6	87.0	90.7	96.6	90.6	92.9	83.3	129.9	112.1	107.5	129.9	121.3
0960	Flooring	83.4	86.7	84.3	85.1	88.4	86.0	83.4	93.3	86.1	76.1	93.3	80.8	89.7	164.6	110.0	89.3	164.6	109.6
0970, 0990	Wall Finishes & Painting/Coating	82.4	79.3	80.5	88.3	91.0	89.9	82.4	80.1	81.0	82.4	89.0	86.4	88.2	154.5	128.0	88.2	154.5	128.0
09	FINISHES	95.9	86.2	90.8	97.5	88.7	92.9	96.0	86.6	91.0	93.5	92.3	92.8	100.3	137.7	120.1	107.4	137.7	123.4
COVERS	DIVS. 10 - 14, 25, 28, 41, 43, 44	100.0	93.9	98.8	100.0	97.5	99.5	100.0	90.0	98.0	100.0	94.9	99.0	100.0	122.9	104.5	100.0	122.9	104.5
21, 22, 23	FIRE SUPPRESSION, PLUMBING & HVAC	100.2	93.9	97.7	100.2	86.9	94.9	100.2	81.7	92.9	100.2	76.0	90.6	100.4	127.3	111.1	100.4	127.3	111.1
26, 27, 3370	ELECTRICAL, COMMUNICATIONS & UTIL.	102.2	93.9	98.2	102.2	91.0	96.8	99.7	90.4	95.2	102.2	90.9	96.7	99.0	137.5	117.8	106.8	137.5	121.7
MF2004	WEIGHTED AVERAGE	97.7	95.3	96.7	98.7	93.1	96.4	96.7	89.5	93.6	98.4	92.2	95.8	99.3	132.8	113.6	104.7	132.8	116.5

	DIVISION	NORTH CAROLINA																	
		ASHEVILLE			CHARLOTTE			DURHAM			FAYETTEVILLE			GREENSBORO			RALEIGH		
		MAT.	INST.	TOTAL	MAT.	INST.	TOTAL	MAT.	INST.	TOTAL	MAT.	INST.	TOTAL	MAT.	INST.	TOTAL	MAT.	INST.	TOTAL
015433	CONTRACTOR EQUIPMENT		98.0	98.0		98.0	98.0		102.1	102.1		102.1	102.1		102.1	102.1		102.1	102.1
0241, 31 - 34	SITE & INFRASTRUCTURE, DEMOLITION	106.5	78.0	87.0	110.4	78.0	88.3	106.6	85.3	92.0	105.4	85.5	91.8	106.5	85.5	92.1	110.6	85.5	93.5
0310	Concrete Forming & Accessories	96.6	41.4	48.9	101.2	43.4	51.3	99.0	41.9	49.7	96.1	42.7	49.9	98.8	42.6	50.2	99.3	43.5	51.1
0320	Concrete Reinforcing	101.1	45.2	78.3	101.6	42.2	77.4	102.7	52.8	82.3	104.9	53.0	83.7	101.6	52.7	81.6	101.6	53.0	81.7
0330	Cast-in-Place Concrete	103.2	49.5	82.7	116.0	51.6	91.4	104.9	49.7	83.8	108.2	52.9	87.1	104.1	51.1	83.8	118.9	56.5	95.1
03	CONCRETE	106.2	47.1	79.3	111.9	48.1	82.9	106.8	48.8	80.4	108.3	50.2	81.9	106.2	49.5	80.4	113.1	51.8	85.2
04	MASONRY	85.3	41.9	59.3	92.1	45.9	64.4	90.9	37.7	59.0	89.0	42.9	61.4	89.0	38.4	58.7	91.3	43.9	62.9
05	METALS	88.3	77.6	85.2	92.5	76.7	87.9	102.1	80.7	95.8	106.4	80.7	98.9	95.7	80.3	91.2	93.3	80.7	89.6
06	WOOD, PLASTICS & COMPOSITES	96.4	40.8	65.9	101.7	43.0	69.5	98.9	42.3	67.8	95.2	42.3	66.1	98.6	42.2	67.7	98.6	43.0	68.1
07	THERMAL & MOISTURE PROTECTION	102.3	45.2	80.1	99.8	46.8	79.2	103.1	47.3	81.4	102.2	45.8	80.2	102.8	44.8	80.3	102.0	46.1	80.2
08	OPENINGS	94.2	40.5	80.9	98.3	42.1	84.4	98.3	44.8	85.1	94.3	44.8	82.0	98.3	44.7	85.0	95.1	44.9	82.6
0920	Plaster & Gypsum Board	106.6	38.4	62.3	108.1	40.7	64.4	110.6	40.0	64.8	110.0	40.0	64.5	112.3	39.9	65.3	108.1	40.7	64.3
0950, 0980	Ceilings & Acoustic Treatment	92.6	38.4	59.1	95.1	40.7	61.5	97.6	40.0	62.0	93.4	40.0	60.4	97.6	39.9	61.9	95.1	40.7	61.5
0960	Flooring	107.3	46.7	90.9	110.9	46.9	93.6	111.5	46.7	94.0	107.4	46.7	91.0	111.5	43.2	93.0	111.5	46.7	94.0
0970, 0990	Wall Finishes & Painting/Coating	114.4	36.8	67.8	114.4	42.0	71.0	114.4	38.6	68.9	114.4	36.8	67.8	114.4	35.8	67.2	114.4	40.6	70.1
09	FINISHES	101.8	41.6	70.0	103.5	43.7	71.9	104.8	42.1	71.6	102.5	42.5	70.7	105.1	41.7	71.6	104.0	43.5	72.0
COVERS	DIVS. 10 - 14, 25, 28, 41, 43, 44	100.0	75.0	95.1	100.0	75.6	95.2	100.0	70.2	94.1	100.0	71.1	94.3	100.0	75.3	95.1	100.0	71.5	94.4
21, 22, 23	FIRE SUPPRESSION, PLUMBING & HVAC	100.2	41.5	76.9	99.9	41.7	76.8	100.3	40.2	76.4	100.1	41.4	76.8	100.2	41.5	76.8	100.0	39.6	76.0
26, 27, 3370	ELECTRICAL, COMMUNICATIONS & UTIL.	97.7	40.2	69.7	100.3	52.3	76.9	98.6	50.9	75.4	97.2	46.2	72.4	97.5	40.4	69.7	104.9	40.1	73.3
MF2004	WEIGHTED AVERAGE	97.6	49.6	77.5	100.2	52.2	80.0	101.1	51.7	80.4	101.1	52.1	80.5	99.8	50.7	79.2	100.7	51.3	80.0

	DIVISION	NORTH CAROLINA						NORTH DAKOTA											
		WILMINGTON			WINSTON-SALEM			BISMARCK			FARGO			GRAND FORKS			MINOT		
		MAT.	INST.	TOTAL	MAT.	INST.	TOTAL	MAT.	INST.	TOTAL	MAT.	INST.	TOTAL	MAT.	INST.	TOTAL	MAT.	INST.	TOTAL
015433	CONTRACTOR EQUIPMENT		98.0	98.0		102.1	102.1		98.7	98.7		98.7	98.7		98.7	98.7		98.7	98.7
0241, 31 - 34	SITE & INFRASTRUCTURE, DEMOLITION	107.6	78.0	87.3	106.8	85.4	92.2	96.1	97.3	96.9	95.6	97.3	96.8	104.5	94.4	97.6	101.8	97.3	98.7
0310	Concrete Forming & Accessories	98.0	41.9	49.5	100.0	41.7	49.6	94.7	45.5	52.2	95.6	45.8	52.6	91.7	39.5	46.6	87.6	60.6	64.3
0320	Concrete Reinforcing	101.9	51.1	81.2	101.6	41.9	77.2	111.0	69.7	94.1	102.3	68.9	88.7	109.3	70.1	93.3	113.0	70.1	95.5
0330	Cast-in-Place Concrete	102.8	50.8	82.9	106.7	48.5	84.4	101.0	53.0	82.7	97.2	54.8	81.0	104.9	50.4	84.1	104.9	51.9	84.6
03	CONCRETE	106.2	48.8	80.1	107.5	46.2	79.6	100.1	54.1	79.2	105.6	54.8	82.5	106.5	50.4	81.0	105.5	60.4	85.0
04	MASONRY	75.1	39.7	53.9	89.2	39.5	59.5	100.2	58.0	74.9	101.6	43.4	66.8	102.7	65.2	80.2	103.0	65.2	80.4
05	METALS	88.0	79.8	85.6	93.3	76.2	88.3	92.7	80.8	89.2	94.9	79.7	90.4	92.7	77.0	88.1	92.9	81.6	89.6
06	WOOD, PLASTICS & COMPOSITES	98.1	41.6	67.1	98.6	41.5	67.3	87.0	40.7	61.6	86.9	41.3	61.8	83.7	37.5	58.3	79.2	60.8	69.1
07	THERMAL & MOISTURE PROTECTION	102.3	45.2	80.1	102.8	44.4	80.1	101.6	50.6	81.8	100.2	48.3	80.0	102.5	52.8	83.2	102.3	54.4	83.7
08	OPENINGS	94.3	43.3	81.7	98.3	41.4	84.2	99.1	46.4	86.1	99.1	46.8	86.1	99.1	40.9	84.7	99.3	57.4	88.9
0920	Plaster & Gypsum Board	107.5	39.3	63.2	112.3	39.2	64.8	99.2	38.8	60.0	101.6	39.4	61.2	106.4	35.5	60.4	104.9	59.7	75.5
0950, 0980	Ceilings & Acoustic Treatment	93.4	39.3	59.9	97.6	39.2	61.5	139.6	38.8	77.3	139.6	39.4	77.7	142.1	35.5	76.2	142.1	59.7	91.1
0960	Flooring	108.2	46.7	91.6	111.5	46.7	94.0	115.9	67.4	102.8	115.6	40.2	95.2	114.2	40.2	94.2	111.5	80.8	103.2
0970, 0990	Wall Finishes & Painting/Coating	114.4	35.7	67.2	114.4	37.5	68.3	98.4	33.8	59.7	98.4	69.7	81.2	98.4	25.8	54.9	98.4	28.3	56.4
09	FINISHES	102.4	41.8	70.4	105.1	42.0	71.7	115.4	46.9	79.2	115.7	45.8	78.7	117.1	37.4	75.0	115.9	60.8	86.8
COVERS	DIVS. 10 - 14, 25, 28, 41, 43, 44	100.0	70.8	94.2	100.0	74.9	95.0	100.0	80.0	96.0	100.0	80.1	96.1	100.0	40.9	88.3	100.0	82.4	96.5
21, 22, 23	FIRE SUPPRESSION, PLUMBING & HVAC	100.2	41.2	76.7	100.2	40.1	76.3	100.3	60.9	84.6	100.3	64.6	86.1	100.4	39.8	76.3	100.4	60.4	84.5
26, 27, 3370	ELECTRICAL, COMMUNICATIONS & UTIL.	98.0	38.8	69.2	97.5	44.0	71.5	103.0	66.2	85.1	104.9	66.3	86.1	97.4	62.5	80.4	100.6	67.4	84.4
MF2004	WEIGHTED AVERAGE	97.2	49.6	77.2	99.5	50.0	78.7	100.1	62.9	84.5	101.4	62.0	84.9	100.8	55.1	81.6	100.9	67.3	86.8

	DIVISION	OHIO																	
		AKRON			CANTON			CINCINNATI			CLEVELAND			COLUMBUS			DAYTON		
		MAT.	INST.	TOTAL	MAT.	INST.	TOTAL	MAT.	INST.	TOTAL	MAT.	INST.	TOTAL	MAT.	INST.	TOTAL	MAT.	INST.	TOTAL
015433	CONTRACTOR EQUIPMENT		95.4	95.4		95.4	95.4		100.2	100.2		96.1	96.1		95.3	95.3		95.1	95.1
0241, 31 - 34	SITE & INFRASTRUCTURE, DEMOLITION	89.0	102.8	98.4	89.1	102.8	98.5	73.9	105.2	95.3	89.0	104.5	99.6	88.5	100.1	96.4	73.0	104.9	94.8
0310	Concrete Forming & Accessories	101.0	96.3	96.9	101.0	85.4	87.5	96.2	85.4	86.8	101.1	103.3	103.0	98.3	83.6	85.6	96.1	80.6	82.7
0320	Concrete Reinforcing	98.0	93.3	96.1	98.0	77.1	89.5	89.7	84.8	87.7	98.6	93.8	96.6	92.8	86.2	90.1	89.7	80.5	85.9
0330	Cast-in-Place Concrete	93.2	100.7	96.1	94.1	98.3	95.7	81.6	87.3	83.8	91.4	109.9	98.5	89.5	93.0	90.8	75.9	90.4	81.5
03	CONCRETE	95.0	96.4	95.6	95.4	87.8	92.0	87.8	86.1	87.0	94.3	102.9	98.2	92.6	87.1	90.1	85.2	83.8	84.6
04	MASONRY	90.0	94.7	92.8	90.7	87.0	88.4	75.7	92.4	85.7	94.4	107.2	102.1	93.1	93.2	93.1	75.2	88.1	82.9
05	METALS	92.1	82.3	89.2	92.1	75.5	87.2	97.5	88.1	94.8	93.5	85.7	91.3	91.9	83.4	89.4	96.8	79.1	91.6
06	WOOD, PLASTICS & COMPOSITES	96.9	96.8	96.9	97.2	85.5	90.8	93.8	82.8	87.8	96.0	100.8	98.6	98.1	81.2	88.8	95.0	78.1	85.7
07	THERMAL & MOISTURE PROTECTION	104.4	97.5	101.7	105.3	93.7	100.8	98.8	93.5	96.8	103.2	110.2	106.0	100.2	93.8	97.7	104.3	89.3	98.5
08	OPENINGS	107.6	96.0	104.7	101.7	75.2	95.1	97.8	82.9	94.1	97.8	98.2	97.9	100.2	81.4	95.6	98.0	78.3	93.1
0920	Plaster & Gypsum Board	90.7	96.4	94.4	91.5	84.6	87.0	92.6	82.5	86.0	89.9	100.5	96.8	88.0	80.5	83.1	92.6	77.6	82.9
0950, 0980	Ceilings & Acoustic Treatment	92.3	96.4	94.8	92.3	84.6	87.5	99.2	82.5	88.9	90.6	100.5	96.8	91.1	80.5	84.5	100.2	77.6	86.2
0960	Flooring	102.6	88.2	98.7	102.8	80.9	96.9	106.6	93.6	103.1	102.4	103.3	102.6	93.6	91.5	93.0	109.3	86.2	103.0
0970, 0990	Wall Finishes & Painting/Coating	105.7	108.0	107.1	105.7	81.4	91.1	106.9	88.9	96.1	105.7	112.3	109.7	97.3	94.2	95.4	106.9	85.5	94.1
09	FINISHES	99.9	96.3	98.0	100.1	84.2	91.7	99.4	86.8	92.7	99.5	104.2	102.0	93.6	85.6	89.4	100.4	81.6	90.5
COVERS	DIVS. 10 - 14, 25, 28, 41, 43, 44	100.0	88.8	97.8	100.0	68.1	93.7	100.0	91.9	98.4	100.0	103.5	100.7	100.0	93.2	98.6	100.0	90.5	98.1
21, 22, 23	FIRE SUPPRESSION, PLUMBING & HVAC	99.9	95.5	98.2	99.9	81.4	92.6	100.0	87.0	94.9	99.9	105.5	102.1	99.7	89.8	95.8	101.0	85.2	94.7
26, 27, 3370	ELECTRICAL, COMMUNICATIONS & UTIL.	97.5	87.7	92.7	96.8	88.2	92.6	94.7	80.0	87.5	97.3	104.2	100.7	98.1	85.1	91.8	93.1	83.3	88.3
MF2004	WEIGHTED AVERAGE	97.8	93.7	96.1	97.3	85.0	92.1	95.3	88.1	92.3	97.2	102.6	99.5	96.3	88.6	93.0	95.2	85.5	91.1

	DIVISION	OHIO												OKLAHOMA					
		LORAIN			SPRINGFIELD			TOLEDO			YOUNGSTOWN			ENID			LAWTON		
		MAT.	INST.	TOTAL	MAT.	INST.	TOTAL	MAT.	INST.	TOTAL	MAT.	INST.	TOTAL	MAT.	INST.	TOTAL	MAT.	INST.	TOTAL
015433	CONTRACTOR EQUIPMENT		95.4	95.4		95.1	95.1		97.4	97.4		95.4	95.4		81.5	81.5		82.1	82.1
0241, 31 - 34	SITE & INFRASTRUCTURE, DEMOLITION	88.4	102.3	97.9	73.3	103.2	93.8	87.7	99.5	95.8	88.9	103.4	98.8	109.5	91.1	96.9	104.6	92.2	96.1
0310	Concrete Forming & Accessories	101.1	95.1	95.9	96.1	82.2	84.1	98.3	102.8	102.2	101.1	89.9	91.4	96.3	36.5	44.6	99.8	49.3	56.1
0320	Concrete Reinforcing	98.0	93.7	96.2	89.7	85.8	88.1	92.8	93.1	93.0	98.0	88.2	94.0	99.3	77.2	90.3	99.6	77.2	90.4
0330	Cast-in-Place Concrete	88.7	105.6	95.2	78.0	89.9	82.5	89.5	106.1	95.8	92.3	102.3	96.1	92.6	49.2	76.0	89.6	49.2	74.2
03	CONCRETE	92.9	97.7	95.1	86.1	85.3	85.8	92.6	101.5	96.7	94.6	93.2	94.0	92.1	49.2	72.6	88.9	54.8	73.4
04	MASONRY	86.7	100.9	95.2	75.4	87.3	82.5	102.1	100.3	101.0	90.2	93.6	92.3	103.5	55.6	74.8	97.4	55.6	72.3
05	METALS	92.6	83.9	90.1	96.8	80.8	92.1	91.7	89.1	91.0	92.1	80.6	88.8	95.9	70.0	88.3	99.5	70.1	90.9
06	WOOD, PLASTICS & COMPOSITES	96.9	93.9	95.2	96.3	80.9	87.9	98.1	103.6	101.1	96.9	88.6	92.4	100.9	33.3	63.8	103.9	50.5	74.6
07	THERMAL & MOISTURE PROTECTION	105.2	104.2	104.8	104.2	89.3	98.4	101.8	107.2	103.9	105.5	94.8	101.3	99.5	57.9	83.3	99.3	59.7	83.9
08	OPENINGS	101.7	94.4	99.9	95.9	81.1	92.2	98.1	98.5	98.2	101.7	89.7	98.7	95.5	46.5	83.4	97.2	55.8	87.0
0920	Plaster & Gypsum Board	90.7	93.3	92.4	92.6	80.5	84.7	88.0	103.7	98.2	90.7	87.9	88.9	84.5	31.6	50.2	86.9	49.4	62.6
0950, 0980	Ceilings & Acoustic Treatment	92.3	93.3	92.9	100.2	80.5	88.0	91.1	103.7	98.9	92.3	87.9	89.5	83.6	31.6	51.5	92.8	49.4	66.0
0960	Flooring	102.8	108.5	104.3	109.3	86.2	103.0	92.8	98.8	94.4	102.8	93.1	100.2	109.3	46.1	92.2	112.0	46.1	94.2
0970, 0990	Wall Finishes & Painting/Coating	105.7	112.3	109.7	106.9	85.5	94.1	97.3	104.5	101.6	105.7	93.8	98.5	98.5	58.9	74.7	98.5	58.3	74.7
09	FINISHES	100.0	99.4	99.7	100.4	83.0	91.2	93.4	102.6	98.2	100.1	90.7	95.1	94.4	38.6	64.9	97.2	48.7	71.5
COVERS	DIVS. 10 - 14, 25, 28, 41, 43, 44	100.0	100.2	100.0	100.0	90.5	98.1	100.0	91.3	98.3	100.0	87.7	97.6	100.0	65.0	93.1	100.0	67.0	93.5
21, 22, 23	FIRE SUPPRESSION, PLUMBING & HVAC	99.9	84.1	93.6	101.0	84.8	94.6	99.7	98.4	99.2	99.9	86.1	94.5	100.1	63.6	85.6	100.1	63.6	85.6
26, 27, 3370	ELECTRICAL, COMMUNICATIONS & UTIL.	97.0	91.6	94.4	93.1	82.2	87.8	98.0	105.0	101.4	97.0	86.0	91.6	95.5	66.6	81.4	97.1	66.6	82.3
MF2004	WEIGHTED AVERAGE	96.9	93.6	95.5	95.1	85.7	91.2	96.5	99.7	97.9	97.2	89.7	94.1	97.4	59.7	81.6	97.8	62.5	83.0

DIVISION		OKLAHOMA									OREGON								
		MUSKOGEE			OKLAHOMA CITY			TULSA			EUGENE			MEDFORD			PORTLAND		
		MAT.	INST.	TOTAL	MAT.	INST.	TOTAL	MAT.	INST.	TOTAL	MAT.	INST.	TOTAL	MAT.	INST.	TOTAL	MAT.	INST.	TOTAL
015433	CONTRACTOR EQUIPMENT		90.7	90.7		82.3	82.3		90.7	90.7		100.7	100.7		100.7	100.7		100.7	100.7
0241, 31 - 34	SITE & INFRASTRUCTURE, DEMOLITION	95.0	87.8	90.1	104.0	92.7	96.3	101.6	89.3	93.2	106.9	104.9	105.5	116.1	104.9	108.4	109.7	104.9	106.5
0310	Concrete Forming & Accessories	101.0	32.8	42.0	98.9	42.7	50.4	100.8	42.8	50.7	105.7	100.0	100.8	101.4	100.0	100.2	107.1	100.4	101.3
0320	Concrete Reinforcing	99.3	34.3	72.8	99.6	77.2	90.4	99.6	77.1	90.4	102.3	95.5	99.5	99.8	95.5	98.0	103.1	95.8	100.1
0330	Cast-in-Place Concrete	83.3	38.8	66.3	92.3	52.0	76.9	90.8	48.1	74.5	104.9	102.8	104.1	108.3	102.8	106.2	107.8	103.0	106.0
03	CONCRETE	84.8	36.7	62.9	90.1	52.9	73.2	89.5	52.4	72.6	105.7	100.0	103.1	111.8	100.0	106.4	107.3	100.2	104.1
04	MASONRY	113.5	48.9	74.8	97.1	57.5	73.4	96.8	57.5	73.3	114.5	98.1	104.7	111.6	98.1	103.5	115.5	103.2	108.1
05	METALS	95.8	60.3	85.4	101.5	70.0	92.3	98.9	81.5	93.8	91.3	95.9	92.6	90.9	95.9	92.3	92.3	96.6	93.6
06	WOOD, PLASTICS & COMPOSITES	106.0	33.0	65.9	103.1	40.7	68.8	104.5	41.9	70.2	94.2	100.0	97.3	88.4	100.0	94.7	95.5	100.0	97.9
07	THERMAL & MOISTURE PROTECTION	99.1	43.7	77.6	96.8	59.5	82.3	99.1	56.8	82.7	105.4	91.5	100.0	106.0	93.5	101.1	105.2	99.9	103.1
08	OPENINGS	95.5	32.0	79.8	97.2	50.5	85.7	97.2	50.8	85.7	99.2	101.3	99.7	101.8	101.3	101.7	96.8	101.3	97.9
0920	Plaster & Gypsum Board	88.4	31.1	51.2	83.1	39.3	54.6	86.9	40.4	56.7	102.3	99.7	100.7	100.9	99.7	100.1	107.8	99.7	102.6
0950, 0980	Ceilings & Acoustic Treatment	92.8	31.1	54.7	90.3	39.3	58.7	92.8	40.4	60.4	103.3	99.7	101.1	118.9	99.7	107.0	107.0	99.7	102.5
0960	Flooring	113.2	37.2	92.6	112.0	46.1	94.2	111.8	48.0	94.6	112.4	99.0	108.8	110.0	99.0	107.0	109.9	99.0	106.9
0970, 0990	Wall Finishes & Painting/Coating	98.5	31.6	58.3	98.5	58.9	74.7	98.5	46.6	67.3	110.1	67.7	84.7	110.1	62.1	81.3	109.6	72.7	87.5
09	FINISHES	97.0	33.7	63.5	95.9	43.4	68.2	96.6	43.1	68.3	107.4	96.5	101.6	110.8	95.9	102.9	108.3	97.1	102.3
COVERS	DIVS. 10 - 14, 25, 28, 41, 43, 44	100.0	64.2	92.9	100.0	66.6	93.4	100.0	66.9	93.5	100.0	101.3	100.3	100.0	101.3	100.3	100.0	101.4	100.3
21, 22, 23	FIRE SUPPRESSION, PLUMBING & HVAC	100.1	25.8	70.6	100.0	64.6	86.0	100.1	53.1	81.4	100.2	99.5	99.9	100.2	99.5	99.9	100.1	104.3	101.8
26, 27, 3370	ELECTRICAL, COMMUNICATIONS & UTIL.	95.2	30.7	63.8	103.0	66.6	85.3	97.1	40.1	69.4	99.3	95.1	97.2	102.8	85.0	94.1	99.6	99.8	99.7
MF2004	WEIGHTED AVERAGE	96.9	41.6	73.7	98.7	61.7	83.1	97.6	56.3	80.3	100.7	98.4	99.7	102.3	97.0	100.1	101.1	101.0	101.0

DIVISION		OREGON			PENNSYLVANIA														
		SALEM			ALLENTOWN			ALTOONA			ERIE			HARRISBURG			PHILADELPHIA		
		MAT.	INST.	TOTAL	MAT.	INST.	TOTAL	MAT.	INST.	TOTAL	MAT.	INST.	TOTAL	MAT.	INST.	TOTAL	MAT.	INST.	TOTAL
015433	CONTRACTOR EQUIPMENT		100.7	100.7		112.9	112.9		112.9	112.9		112.9	112.9		112.1	112.1		96.8	96.8
0241, 31 - 34	SITE & INFRASTRUCTURE, DEMOLITION	103.5	104.9	104.5	93.2	104.4	100.9	98.1	104.1	102.2	94.4	104.8	101.5	83.6	102.8	96.7	103.0	97.5	99.3
0310	Concrete Forming & Accessories	104.5	100.2	100.7	99.3	112.3	110.5	84.5	80.5	81.0	98.8	88.7	90.0	92.5	84.3	85.4	100.8	135.8	131.0
0320	Concrete Reinforcing	103.2	95.8	100.2	99.1	103.5	100.9	96.1	92.7	94.7	98.1	93.8	96.4	99.1	100.9	99.8	101.3	140.6	117.3
0330	Cast-in-Place Concrete	94.7	102.9	97.9	86.5	100.6	91.9	96.4	84.4	91.8	94.8	86.3	91.6	96.1	93.8	95.2	109.1	127.5	116.1
03	CONCRETE	101.1	100.1	100.7	97.8	107.3	102.1	93.1	85.7	89.7	92.1	90.2	91.2	100.1	92.2	96.5	108.2	133.1	119.6
04	MASONRY	119.7	103.2	109.8	94.9	99.5	97.7	97.0	74.6	83.6	87.2	90.9	89.4	95.0	87.8	90.7	96.6	134.8	119.5
05	METALS	91.7	96.4	93.1	95.2	119.3	102.2	89.1	112.0	95.8	89.3	112.6	96.1	97.0	117.6	103.0	102.3	127.7	109.7
06	WOOD, PLASTICS & COMPOSITES	91.6	100.0	96.2	101.1	116.0	109.3	80.0	82.7	81.5	97.3	87.8	92.1	94.9	83.0	88.4	100.0	137.2	120.5
07	THERMAL & MOISTURE PROTECTION	102.8	95.6	100.0	100.2	114.4	105.7	99.4	88.0	95.0	99.6	94.5	97.6	105.4	105.4	105.4	99.5	133.9	112.9
08	OPENINGS	98.7	101.3	99.3	94.0	110.9	98.2	88.4	88.3	88.3	88.5	89.5	88.7	94.0	90.8	93.2	97.7	141.5	108.5
0920	Plaster & Gypsum Board	95.4	99.7	98.2	112.6	116.3	115.0	106.3	81.8	90.4	112.6	87.1	96.0	109.7	82.1	91.8	103.8	138.5	126.3
0950, 0980	Ceilings & Acoustic Treatment	104.7	99.7	101.6	87.4	116.3	105.3	91.6	81.8	85.5	87.4	87.1	87.2	89.9	82.1	85.1	96.4	138.5	122.4
0960	Flooring	112.6	99.0	108.9	85.1	94.7	87.7	79.2	51.1	71.6	86.5	76.1	83.7	85.7	90.0	86.8	84.3	136.6	98.5
0970, 0990	Wall Finishes & Painting/Coating	110.1	67.7	84.7	88.3	81.1	84.0	84.2	105.7	97.1	94.5	89.4	91.4	88.3	87.7	88.0	97.3	144.5	125.6
09	FINISHES	106.5	96.5	101.2	95.0	105.5	100.6	93.7	76.8	84.8	96.4	86.0	90.9	94.1	85.2	89.4	100.4	137.3	119.9
COVERS	DIVS. 10 - 14, 25, 28, 41, 43, 44	100.0	101.3	100.3	100.0	103.4	100.7	100.0	94.0	98.8	100.0	98.7	99.8	100.0	88.9	97.8	100.0	120.3	104.0
21, 22, 23	FIRE SUPPRESSION, PLUMBING & HVAC	100.0	99.6	99.9	100.2	109.1	103.8	99.8	84.8	93.8	99.8	86.4	94.5	100.1	91.6	96.7	100.0	127.6	111.0
26, 27, 3370	ELECTRICAL, COMMUNICATIONS & UTIL.	107.7	95.1	101.6	101.3	96.3	98.9	90.6	98.0	94.2	92.4	85.6	89.1	100.0	84.4	92.4	98.7	137.5	117.6
MF2004	WEIGHTED AVERAGE	101.0	99.1	100.2	97.7	106.5	101.4	94.3	89.4	92.2	94.2	92.0	93.3	97.9	93.1	95.9	101.0	130.0	113.2

DIVISION		PENNSYLVANIA												PUERTO RICO			RHODE ISLAND		
		PITTSBURGH			READING			SCRANTON			YORK			SAN JUAN			PROVIDENCE		
		MAT.	INST.	TOTAL	MAT.	INST.	TOTAL	MAT.	INST.	TOTAL	MAT.	INST.	TOTAL	MAT.	INST.	TOTAL	MAT.	INST.	TOTAL
015433	CONTRACTOR EQUIPMENT		111.7	111.7		114.5	114.5		112.9	112.9		112.1	112.1		85.9	85.9		102.0	102.0
0241, 31 - 34	SITE & INFRASTRUCTURE, DEMOLITION	108.6	105.1	106.2	100.1	107.9	105.4	93.7	104.3	100.9	82.4	102.8	96.4	118.7	82.9	94.2	81.6	103.3	96.4
0310	Concrete Forming & Accessories	99.7	93.0	93.9	100.2	84.7	86.8	99.4	82.7	85.0	82.3	84.5	84.2	90.3	18.4	28.2	98.4	112.2	110.3
0320	Concrete Reinforcing	96.1	104.8	99.7	99.5	97.5	98.7	99.1	104.7	101.4	98.1	100.9	99.3	180.4	10.8	111.2	100.4	113.9	105.9
0330	Cast-in-Place Concrete	96.2	91.5	94.4	79.7	96.6	86.1	90.4	88.9	89.8	85.2	94.1	88.6	100.0	31.3	73.7	94.2	117.2	103.0
03	CONCRETE	97.7	95.9	96.9	94.1	92.8	93.5	99.6	90.5	95.5	98.1	92.5	95.5	114.5	22.7	72.8	107.3	113.8	110.3
04	MASONRY	86.6	93.2	90.5	97.8	91.1	93.8	95.2	94.8	95.0	95.1	88.4	91.1	88.1	15.1	44.4	107.4	123.6	117.1
05	METALS	93.9	118.6	101.1	98.8	116.2	103.9	97.1	119.1	103.5	94.7	117.6	101.4	107.3	37.5	86.9	93.8	108.6	98.1
06	WOOD, PLASTICS & COMPOSITES	96.0	93.2	94.5	100.2	82.0	90.2	101.1	78.9	88.9	86.7	83.0	84.7	95.1	18.6	53.1	102.0	110.1	106.4
07	THERMAL & MOISTURE PROTECTION	102.6	96.6	100.3	99.1	107.7	102.5	100.1	98.2	99.4	99.3	105.7	101.8	131.3	21.7	88.6	93.0	114.7	101.4
08	OPENINGS	91.8	102.6	94.5	95.2	90.6	94.1	94.0	89.9	93.0	90.8	90.8	90.8	152.2	15.3	118.4	99.5	112.2	102.6
0920	Plaster & Gypsum Board	99.1	92.8	95.0	107.4	81.2	90.4	115.0	77.8	90.9	108.9	82.1	91.5	119.4	15.5	51.9	103.3	109.7	107.5
0950, 0980	Ceilings & Acoustic Treatment	85.5	92.8	90.0	86.3	81.2	83.2	96.6	77.8	85.0	89.1	82.1	84.8	219.3	15.5	93.2	95.9	109.7	104.4
0960	Flooring	97.3	99.5	97.9	82.5	91.3	84.9	85.1	107.4	91.1	81.1	90.0	83.5	199.5	14.7	149.6	97.2	126.8	105.2
0970, 0990	Wall Finishes & Painting/Coating	97.0	105.7	102.2	95.8	99.0	97.7	88.3	98.0	94.1	88.3	87.7	88.0	198.5	13.7	87.6	93.9	120.7	110.0
09	FINISHES	96.3	94.8	95.5	98.0	86.1	91.7	97.5	88.0	92.5	92.5	85.3	88.7	198.7	17.4	102.9	98.0	115.9	107.5
COVERS	DIVS. 10 - 14, 25, 28, 41, 43, 44	100.0	98.5	99.7	100.0	96.7	99.4	100.0	97.5	99.5	100.0	89.1	97.8	100.0	19.5	84.1	100.0	98.2	99.6
21, 22, 23	FIRE SUPPRESSION, PLUMBING & HVAC	99.9	93.8	97.5	100.2	105.6	102.3	100.2	88.8	95.7	100.2	91.9	96.9	104.1	13.7	68.1	100.0	107.2	102.9
26, 27, 3370	ELECTRICAL, COMMUNICATIONS & UTIL.	96.7	98.0	97.3	100.9	88.0	94.6	101.3	95.8	98.7	94.9	84.4	89.8	122.6	13.8	69.6	95.9	94.5	95.2
MF2004	WEIGHTED AVERAGE	96.9	98.7	97.6	98.5	97.6	98.1	98.5	95.2	97.1	96.1	93.3	94.9	120.1	23.9	79.7	98.8	109.1	103.2

	DIVISION	SOUTH CAROLINA															SOUTH DAKOTA		
		CHARLESTON			COLUMBIA			FLORENCE			GREENVILLE			SPARTANBURG			ABERDEEN		
		MAT.	INST.	TOTAL	MAT.	INST.	TOTAL	MAT.	INST.	TOTAL	MAT.	INST.	TOTAL	MAT.	INST.	TOTAL	MAT.	INST.	TOTAL
015433	CONTRACTOR EQUIPMENT		101.7	101.7		101.7	101.7		101.7	101.7		101.7	101.7		101.7	101.7		98.7	98.7
0241, 31 - 34	SITE & INFRASTRUCTURE, DEMOLITION	101.6	85.5	90.5	102.8	85.4	90.9	113.3	85.4	94.2	107.6	85.0	92.2	107.4	85.0	92.1	90.7	95.3	93.8
0310	Concrete Forming & Accessories	98.5	45.4	52.6	100.5	44.3	52.0	86.2	44.5	50.2	98.2	44.3	51.7	101.5	44.3	52.1	94.9	43.0	50.0
0320	Concrete Reinforcing	101.6	64.9	86.6	101.6	64.7	86.5	101.2	64.8	86.4	101.1	51.7	81.0	101.1	51.7	81.0	109.5	44.4	82.9
0330	Cast-in-Place Concrete	95.5	54.1	79.7	102.3	54.7	84.1	81.4	53.9	70.9	81.4	53.8	70.8	81.4	53.8	70.8	100.0	49.9	80.8
03	CONCRETE	102.2	53.8	80.2	105.4	53.6	81.8	102.5	53.4	80.1	100.8	50.9	78.1	101.0	50.9	78.2	99.9	47.2	75.9
04	MASONRY	96.8	42.6	64.4	91.5	40.3	60.9	79.6	42.6	57.4	77.6	42.6	56.6	79.6	42.6	57.4	104.3	60.4	78.0
05	METALS	88.9	82.3	87.0	88.9	81.5	86.7	87.9	81.8	86.1	87.9	77.3	84.8	87.9	77.3	84.8	99.1	70.6	90.8
06	WOOD, PLASTICS & COMPOSITES	98.6	44.3	68.8	100.9	43.8	69.5	83.8	43.8	61.8	98.4	43.8	68.4	102.5	43.8	70.2	101.5	42.2	68.9
07	THERMAL & MOISTURE PROTECTION	102.5	46.4	80.6	101.6	45.0	79.6	102.7	46.3	80.8	102.7	46.3	80.8	102.7	46.3	80.8	101.1	53.0	82.4
08	OPENINGS	98.3	48.3	85.9	98.3	48.1	85.9	94.3	48.1	82.8	94.2	45.0	82.0	94.2	45.0	82.0	94.8	43.0	82.0
0920	Plaster & Gypsum Board	114.8	42.1	67.5	108.1	41.5	64.9	102.7	41.5	63.0	108.1	41.5	64.8	109.9	41.5	65.5	100.3	40.4	61.4
0950, 0980	Ceilings & Acoustic Treatment	97.6	42.1	63.2	95.1	41.5	62.0	93.4	41.5	61.3	92.6	41.5	61.0	92.6	41.5	61.0	112.0	40.4	67.7
0960	Flooring	111.5	65.3	99.0	107.9	48.9	92.0	101.4	48.9	87.2	108.7	64.1	96.6	110.4	64.1	97.9	116.2	56.4	100.1
0970, 0990	Wall Finishes & Painting/Coating	114.4	45.0	72.8	111.9	45.0	71.7	114.4	45.0	72.8	114.4	45.0	72.8	114.4	45.0	72.8	98.4	42.7	65.0
09	FINISHES	105.8	48.6	75.6	102.8	45.0	72.3	101.3	45.5	71.8	103.6	48.1	74.3	104.4	48.1	74.6	108.6	45.0	75.0
COVERS	DIVS. 10 - 14, 25, 28, 41, 43, 44	100.0	68.9	93.8	100.0	68.6	93.8	100.0	68.6	93.8	100.0	68.6	93.8	100.0	68.6	93.8	100.0	51.0	90.3
21, 22, 23	FIRE SUPPRESSION, PLUMBING & HVAC	100.2	44.8	78.1	99.9	39.6	76.0	100.2	39.7	76.1	100.2	38.0	75.5	100.2	38.0	75.5	100.1	47.9	79.4
26, 27, 3370	ELECTRICAL, COMMUNICATIONS & UTIL.	97.2	103.6	100.3	101.3	53.0	77.8	95.2	50.4	73.4	97.3	54.5	76.4	97.3	54.5	76.4	97.9	59.7	79.3
MF2004	WEIGHTED AVERAGE	98.4	62.2	83.2	98.6	53.3	79.6	96.6	53.3	78.4	96.7	52.9	78.3	96.9	52.9	78.4	99.9	56.3	81.6

	DIVISION	SOUTH DAKOTA									TENNESSEE								
		PIERRE			RAPID CITY			SIOUX FALLS			CHATTANOOGA			JACKSON			JOHNSON CITY		
		MAT.	INST.	TOTAL	MAT.	INST.	TOTAL	MAT.	INST.	TOTAL	MAT.	INST.	TOTAL	MAT.	INST.	TOTAL	MAT.	INST.	TOTAL
015433	CONTRACTOR EQUIPMENT		98.7	98.7		98.7	98.7		99.4	99.4		103.7	103.7		103.8	103.8		98.8	98.8
0241, 31 - 34	SITE & INFRASTRUCTURE, DEMOLITION	89.8	95.3	93.6	88.9	95.4	93.3	90.5	96.6	94.7	102.0	95.5	97.6	100.7	93.9	96.1	111.5	86.6	94.5
0310	Concrete Forming & Accessories	93.5	43.9	50.7	101.2	41.7	49.8	94.3	46.7	53.2	94.6	42.0	49.2	88.1	34.0	41.4	82.1	42.7	48.1
0320	Concrete Reinforcing	109.0	55.9	87.3	102.3	55.7	83.3	102.3	55.9	83.4	84.7	59.6	74.5	87.3	37.9	67.2	85.3	58.6	74.4
0330	Cast-in-Place Concrete	92.0	48.0	75.2	96.4	48.6	78.1	91.9	49.0	75.5	97.3	46.6	78.0	102.7	40.9	79.1	78.3	50.5	67.7
03	CONCRETE	95.4	49.1	74.3	96.5	48.3	74.6	93.0	50.7	73.8	90.6	49.1	71.7	93.9	39.5	69.2	95.8	50.5	75.2
04	MASONRY	99.8	60.0	76.0	101.1	55.9	74.0	97.2	58.1	73.8	97.7	38.4	62.2	109.4	29.9	61.8	108.7	38.3	66.5
05	METALS	99.1	75.7	92.3	100.8	75.7	93.4	100.9	76.2	93.7	103.4	83.8	97.7	98.1	72.8	90.7	100.4	83.4	95.4
06	WOOD, PLASTICS & COMPOSITES	99.3	43.0	68.4	105.2	39.1	68.9	100.2	46.0	70.5	99.3	42.3	68.0	86.0	33.9	57.4	74.0	43.3	57.1
07	THERMAL & MOISTURE PROTECTION	100.9	51.3	81.6	101.5	52.9	82.6	102.6	54.7	84.0	95.4	54.6	79.6	92.0	36.3	70.3	90.2	54.7	76.4
08	OPENINGS	98.0	47.0	85.4	99.0	44.9	85.6	99.2	48.6	86.7	99.4	48.4	86.8	100.5	35.4	84.4	95.3	49.3	83.9
0920	Plaster & Gypsum Board	94.3	41.3	59.9	100.2	37.2	59.3	99.8	44.4	63.8	82.6	40.8	55.4	90.6	32.0	52.5	96.2	41.8	60.8
0950, 0980	Ceilings & Acoustic Treatment	107.0	41.3	66.4	116.2	37.2	67.3	109.5	44.4	69.2	97.1	40.8	62.3	95.1	32.0	56.1	93.5	41.8	61.5
0960	Flooring	115.5	40.2	95.1	115.5	71.0	103.5	115.5	76.7	105.0	99.5	44.3	84.6	91.7	21.5	72.7	93.8	45.6	80.7
0970, 0990	Wall Finishes & Painting/Coating	98.4	51.2	70.1	98.4	51.2	70.1	98.4	51.2	70.1	108.8	41.3	68.3	100.4	27.9	56.9	106.1	48.4	71.5
09	FINISHES	106.2	43.3	73.0	109.2	47.4	76.5	107.7	52.4	78.5	93.8	41.9	66.4	94.0	31.1	60.7	95.6	43.3	68.0
COVERS	DIVS. 10 - 14, 25, 28, 41, 43, 44	100.0	75.8	95.2	100.0	75.7	95.2	100.0	76.2	95.3	100.0	46.6	89.4	100.0	50.4	90.2	100.0	53.0	90.7
21, 22, 23	FIRE SUPPRESSION, PLUMBING & HVAC	100.0	47.5	79.2	100.1	48.0	79.4	100.0	46.2	78.6	100.0	39.7	76.0	99.9	48.8	79.6	99.7	52.2	80.8
26, 27, 3370	ELECTRICAL, COMMUNICATIONS & UTIL.	100.9	51.8	77.0	94.6	51.8	73.8	102.9	69.1	86.5	106.5	69.3	88.4	104.1	45.8	75.7	95.0	53.9	75.0
MF2004	WEIGHTED AVERAGE	99.5	56.5	81.4	99.7	56.5	81.5	99.9	60.1	83.2	99.4	55.0	80.8	99.0	47.9	77.5	98.4	55.4	80.3

	DIVISION	TENNESSEE									TEXAS								
		KNOXVILLE			MEMPHIS			NASHVILLE			ABILENE			AMARILLO			AUSTIN		
		MAT.	INST.	TOTAL	MAT.	INST.	TOTAL	MAT.	INST.	TOTAL	MAT.	INST.	TOTAL	MAT.	INST.	TOTAL	MAT.	INST.	TOTAL
015433	CONTRACTOR EQUIPMENT		98.8	98.8		101.9	101.9		104.9	104.9		90.7	90.7		90.7	90.7		90.0	90.0
0241, 31 - 34	SITE & INFRASTRUCTURE, DEMOLITION	88.0	86.6	87.1	90.4	91.5	91.1	96.9	98.6	98.1	102.0	88.4	92.7	103.2	89.3	93.7	91.2	88.8	89.5
0310	Concrete Forming & Accessories	93.7	42.7	49.6	94.8	60.5	65.2	94.0	64.0	68.1	97.9	39.2	47.2	98.2	49.2	55.9	91.9	54.1	59.3
0320	Concrete Reinforcing	84.7	58.6	74.1	91.0	65.6	80.6	91.0	64.5	80.2	96.6	53.3	78.9	96.6	48.0	76.8	97.4	46.2	76.5
0330	Cast-in-Place Concrete	91.4	46.8	74.4	86.6	66.5	78.9	90.5	67.4	81.7	94.2	42.3	74.4	93.3	47.5	75.8	91.5	48.0	74.9
03	CONCRETE	88.1	49.2	70.4	87.1	65.3	77.2	89.3	66.9	79.1	90.3	44.4	69.4	90.0	49.6	71.6	84.5	51.4	69.5
04	MASONRY	75.1	38.3	53.0	87.8	68.0	76.0	83.4	62.6	71.0	100.4	50.3	70.4	100.1	45.8	67.6	94.0	48.8	66.9
05	METALS	104.0	83.6	98.1	105.1	91.0	101.0	105.1	89.4	100.5	102.8	70.5	93.4	102.8	68.4	92.8	98.1	65.3	88.5
06	WOOD, PLASTICS & COMPOSITES	87.6	43.3	63.3	91.7	61.1	74.9	97.1	64.9	79.4	100.6	38.3	66.4	99.0	51.7	73.0	96.5	53.1	75.4
07	THERMAL & MOISTURE PROTECTION	88.4	54.2	75.1	90.1	67.5	81.3	93.4	64.6	82.2	99.2	46.2	78.6	99.4	45.1	78.3	87.0	51.3	73.1
08	OPENINGS	92.2	49.3	81.6	98.4	63.3	89.7	97.6	65.6	89.7	93.2	42.3	80.6	93.2	46.2	81.6	101.2	54.7	89.7
0920	Plaster & Gypsum Board	102.7	41.8	63.1	91.4	60.1	71.1	98.7	64.0	76.2	86.9	36.6	54.3	82.7	50.5	61.8	89.9	57.1	68.6
0950, 0980	Ceilings & Acoustic Treatment	94.3	41.8	61.8	98.7	60.1	74.8	96.1	64.0	76.3	92.8	36.6	58.1	90.3	50.5	65.7	86.3	57.1	68.3
0960	Flooring	99.2	45.6	84.7	94.2	40.5	79.6	101.6	71.6	93.5	112.0	64.7	99.2	111.8	57.3	97.1	94.6	43.9	80.9
0970, 0990	Wall Finishes & Painting/Coating	106.1	48.4	71.5	101.3	57.0	74.7	108.4	66.4	83.2	97.1	49.2	68.4	97.1	34.1	59.3	92.7	38.3	60.0
09	FINISHES	90.7	43.3	65.6	90.8	55.8	72.3	99.3	65.7	81.5	96.6	44.5	69.0	95.7	49.1	71.1	89.5	50.6	68.9
COVERS	DIVS. 10 - 14, 25, 28, 41, 43, 44	100.0	53.0	90.7	100.0	79.0	95.9	100.0	79.9	96.0	100.0	73.3	94.7	100.0	66.5	93.4	100.0	67.0	93.5
21, 22, 23	FIRE SUPPRESSION, PLUMBING & HVAC	99.7	52.2	80.8	99.9	66.6	86.6	99.9	78.9	91.5	100.1	39.9	76.2	100.0	48.3	79.5	100.0	55.5	82.3
26, 27, 3370	ELECTRICAL, COMMUNICATIONS & UTIL.	100.8	58.3	80.1	103.2	68.7	86.4	103.0	65.4	84.7	97.3	43.8	71.2	101.9	68.6	85.7	104.2	57.6	86.4
MF2004	WEIGHTED AVERAGE	95.8	55.8	79.0	97.5	70.1	86.0	98.5	73.6	88.0	98.2	50.8	78.3	98.6	56.8	81.0	96.5	59.1	80.8

	DIVISION	TEXAS																	
		BEAUMONT			CORPUS CHRISTI			DALLAS			EL PASO			FORT WORTH			HOUSTON		
		MAT.	INST.	TOTAL	MAT.	INST.	TOTAL	MAT.	INST.	TOTAL	MAT.	INST.	TOTAL	MAT.	INST.	TOTAL	MAT.	INST.	TOTAL
015433	**CONTRACTOR EQUIPMENT**		92.0	92.0		96.8	96.8		99.4	99.4		90.7	90.7		90.7	90.7		100.1	100.1
0241, 31 - 34	**SITE & INFRASTRUCTURE, DEMOLITION**	98.8	89.0	92.1	122.7	82.5	95.2	119.7	88.1	98.1	100.2	87.7	91.7	104.7	89.0	93.9	123.0	85.7	97.5
0310	Concrete Forming & Accessories	104.7	48.2	55.9	98.0	35.7	44.1	94.8	59.3	64.2	97.3	43.2	50.6	93.7	58.5	63.3	94.1	65.9	69.8
0320	Concrete Reinforcing	96.4	39.4	73.2	96.7	44.8	75.5	96.7	52.9	78.8	96.6	43.5	75.0	96.6	52.7	78.7	98.0	58.2	81.8
0330	Cast-in-Place Concrete	94.3	52.5	78.3	98.9	45.0	78.3	92.3	55.4	78.2	86.4	41.0	69.1	98.2	51.5	80.3	97.2	68.9	86.4
03	**CONCRETE**	92.7	49.3	72.9	90.0	42.8	68.5	88.7	58.4	75.0	86.7	43.7	67.1	92.0	55.9	75.6	92.2	67.2	80.8
04	**MASONRY**	101.3	56.1	74.2	83.8	48.5	62.7	102.2	58.6	76.1	98.0	46.2	67.0	94.7	58.5	73.0	96.1	60.8	75.0
05	**METALS**	102.3	65.8	91.6	97.6	76.1	91.3	102.5	82.2	96.6	102.6	63.5	91.1	99.5	70.4	91.0	106.9	87.5	101.2
06	**WOOD, PLASTICS & COMPOSITES**	111.4	48.3	76.8	115.1	34.9	71.0	98.7	60.2	77.6	100.2	45.8	70.3	101.1	60.1	78.6	96.8	66.7	80.3
07	**THERMAL & MOISTURE PROTECTION**	104.5	55.7	85.5	95.3	44.3	75.5	93.2	63.4	81.6	95.7	52.4	78.8	100.4	52.1	81.6	96.8	67.4	85.4
08	**OPENINGS**	94.9	44.3	82.4	108.6	36.1	90.7	103.4	54.0	91.2	93.2	41.3	80.3	87.1	53.9	78.9	104.5	63.6	94.4
0920	Plaster & Gypsum Board	96.3	47.1	64.3	93.9	32.9	54.3	90.5	59.2	70.1	85.2	44.4	58.7	86.2	59.2	68.6	94.8	65.9	76.0
0950, 0980	Ceilings & Acoustic Treatment	104.4	47.1	69.0	87.9	32.9	53.9	94.2	59.2	72.5	90.3	44.4	61.9	92.8	59.2	72.0	105.8	65.9	81.1
0960	Flooring	111.1	70.9	100.2	107.1	43.4	89.9	100.6	53.2	87.8	112.0	62.4	98.6	145.6	42.6	117.7	98.5	58.7	87.7
0970, 0990	Wall Finishes & Painting/Coating	93.9	46.0	65.1	105.5	53.5	74.3	103.5	53.1	73.2	97.1	32.7	58.5	98.5	48.6	68.6	101.4	61.2	77.3
09	**FINISHES**	95.0	51.8	72.2	96.3	38.2	65.6	98.4	57.5	76.7	95.7	46.0	69.4	107.1	54.7	79.4	99.9	64.0	80.9
COVERS	**DIVS. 10 - 14, 25, 28, 41, 43, 44**	100.0	80.4	96.1	100.0	77.8	95.6	100.0	79.2	95.9	100.0	64.9	93.0	100.0	78.9	95.8	100.0	86.0	97.2
21, 22, 23	**FIRE SUPPRESSION, PLUMBING & HVAC**	100.0	59.1	83.7	99.9	40.6	76.3	99.9	63.8	85.5	100.0	34.9	74.1	100.0	53.6	81.6	99.9	71.9	88.8
26, 27, 3370	**ELECTRICAL, COMMUNICATIONS & UTIL.**	95.0	65.0	80.4	97.6	50.1	74.5	97.3	69.2	83.6	98.6	51.9	75.9	98.1	60.3	79.7	97.2	68.5	83.2
MF2004	**WEIGHTED AVERAGE**	98.4	60.1	82.3	98.4	50.4	78.3	99.4	66.2	85.4	97.5	49.8	77.5	98.0	60.7	82.4	100.7	71.2	88.3

	DIVISION	TEXAS																	
		LAREDO			LUBBOCK			ODESSA			SAN ANTONIO			WACO			WICHITA FALLS		
		MAT.	INST.	TOTAL	MAT.	INST.	TOTAL	MAT.	INST.	TOTAL	MAT.	INST.	TOTAL	MAT.	INST.	TOTAL	MAT.	INST.	TOTAL
015433	**CONTRACTOR EQUIPMENT**		90.0	90.0		98.4	98.4		90.7	90.7		91.8	91.8		90.7	90.7		90.7	90.7
0241, 31 - 34	**SITE & INFRASTRUCTURE, DEMOLITION**	90.5	88.6	89.2	130.8	84.3	99.0	102.2	88.8	93.0	90.2	92.3	91.6	100.8	88.9	92.7	101.6	88.5	92.6
0310	Concrete Forming & Accessories	92.6	35.9	43.6	96.9	38.6	46.5	97.9	36.7	45.0	92.7	55.6	60.6	98.5	38.8	46.9	98.5	39.6	47.6
0320	Concrete Reinforcing	97.4	44.9	75.9	97.7	53.1	79.5	96.6	53.0	78.8	104.4	49.3	81.9	96.6	46.1	76.0	96.6	47.3	76.5
0330	Cast-in-Place Concrete	76.2	59.8	69.9	94.3	47.7	76.5	94.2	44.1	75.0	74.7	67.5	71.9	85.0	52.9	72.7	90.8	47.3	74.2
03	**CONCRETE**	81.2	47.3	65.8	89.5	46.7	70.0	90.3	43.8	69.2	82.0	59.5	71.7	86.1	46.5	68.1	88.8	45.3	69.0
04	**MASONRY**	91.7	49.6	66.5	99.7	44.4	66.6	100.4	44.7	67.0	91.5	60.6	73.0	97.0	55.3	72.0	97.5	56.3	72.8
05	**METALS**	99.1	64.5	89.0	106.5	82.3	99.4	102.1	69.5	92.6	99.7	69.4	90.9	102.7	67.4	92.4	102.6	71.3	93.5
06	**WOOD, PLASTICS & COMPOSITES**	95.8	34.7	62.3	100.4	38.4	66.4	100.6	36.5	65.4	95.8	54.0	72.9	106.9	34.0	66.9	106.9	38.3	69.2
07	**THERMAL & MOISTURE PROTECTION**	91.4	49.4	75.1	89.9	46.8	73.1	99.2	42.5	77.2	91.4	64.9	81.1	99.7	47.2	79.3	99.7	51.0	80.8
08	**OPENINGS**	101.9	36.6	85.7	104.2	40.7	88.5	93.2	39.3	79.9	103.9	53.6	91.4	87.1	35.0	74.2	87.1	42.2	76.0
0920	Plaster & Gypsum Board	90.9	32.9	53.3	87.4	36.6	54.4	86.9	34.7	53.0	90.9	52.9	66.2	86.9	32.1	51.3	86.9	36.6	54.2
0950, 0980	Ceilings & Acoustic Treatment	87.9	32.9	53.9	94.5	36.6	58.7	92.8	34.7	56.9	87.9	52.9	66.3	92.8	32.1	55.3	92.8	36.6	58.0
0960	Flooring	91.6	43.4	78.6	104.0	36.1	85.6	112.0	35.7	91.4	91.6	66.5	84.8	145.6	33.1	115.2	146.7	75.0	127.3
0970, 0990	Wall Finishes & Painting/Coating	92.7	50.0	67.1	108.4	30.6	61.7	97.1	30.6	57.2	92.7	50.0	67.1	98.5	31.6	58.3	102.1	50.0	70.8
09	**FINISHES**	89.2	37.7	62.0	98.8	36.3	65.7	96.6	35.1	64.1	89.2	55.9	71.6	106.9	35.5	69.1	107.5	46.6	75.3
COVERS	**DIVS. 10 - 14, 25, 28, 41, 43, 44**	100.0	71.7	94.4	100.0	73.1	94.7	100.0	64.1	92.9	100.0	78.0	95.7	100.0	75.8	95.2	100.0	64.9	93.0
21, 22, 23	**FIRE SUPPRESSION, PLUMBING & HVAC**	99.9	36.6	74.7	99.6	45.9	78.2	100.1	33.4	73.6	99.9	67.0	86.8	100.1	54.1	81.9	100.1	45.6	78.5
26, 27, 3370	**ELECTRICAL, COMMUNICATIONS & UTIL.**	99.3	64.9	82.6	95.8	41.2	69.2	97.4	38.3	68.7	99.8	64.9	82.8	97.4	71.9	85.0	99.7	60.7	80.7
MF2004	**WEIGHTED AVERAGE**	95.8	51.7	77.3	100.1	51.1	79.5	98.1	46.3	76.3	96.2	65.4	83.3	97.8	56.8	80.6	98.5	55.3	80.3

	DIVISION	UTAH												VERMONT					
		LOGAN			OGDEN			PROVO			SALT LAKE CITY			BURLINGTON			RUTLAND		
		MAT.	INST.	TOTAL	MAT.	INST.	TOTAL	MAT.	INST.	TOTAL	MAT.	INST.	TOTAL	MAT.	INST.	TOTAL	MAT.	INST.	TOTAL
015433	**CONTRACTOR EQUIPMENT**		101.4	101.4		101.4	101.4		100.7	100.7		101.4	101.4		100.4	100.4		100.4	100.4
0241, 31 - 34	**SITE & INFRASTRUCTURE, DEMOLITION**	88.6	101.7	97.5	77.5	101.7	94.0	85.2	100.4	95.6	77.3	101.7	93.9	76.5	96.8	90.3	76.6	96.8	90.4
0310	Concrete Forming & Accessories	105.2	55.4	62.2	105.3	55.4	62.2	106.5	55.5	62.5	104.7	55.5	62.2	93.5	55.8	60.9	97.0	56.3	61.8
0320	Concrete Reinforcing	102.6	74.0	90.9	102.2	74.0	90.7	110.9	74.0	95.9	104.6	74.0	92.1	100.4	81.1	92.5	100.4	81.2	92.5
0330	Cast-in-Place Concrete	92.3	71.4	84.3	93.7	71.4	85.2	92.4	71.4	84.4	102.4	71.4	90.6	94.2	107.0	99.1	88.1	107.2	95.4
03	**CONCRETE**	112.9	65.1	91.2	102.6	65.1	85.6	113.1	65.2	91.3	122.1	65.2	96.2	104.5	78.9	92.9	101.9	79.2	91.6
04	**MASONRY**	113.2	57.9	80.1	106.9	57.9	77.6	118.6	57.9	82.3	120.8	57.9	83.2	107.1	53.8	75.2	85.9	53.8	66.7
05	**METALS**	100.8	75.6	93.5	101.3	75.6	93.8	99.0	75.7	92.2	106.0	75.7	97.1	95.4	79.7	90.8	93.8	80.3	89.8
06	**WOOD, PLASTICS & COMPOSITES**	85.2	52.3	67.1	85.2	52.3	67.1	86.6	52.3	67.7	86.9	52.3	67.9	97.3	55.9	74.6	102.8	55.9	77.1
07	**THERMAL & MOISTURE PROTECTION**	99.5	63.9	85.7	98.3	63.9	84.9	101.3	63.9	86.8	101.8	63.9	87.1	94.1	54.6	78.7	94.1	63.0	82.0
08	**OPENINGS**	88.2	54.8	79.9	88.2	54.8	79.9	92.5	54.8	83.1	90.1	54.8	81.3	104.9	55.5	92.6	104.9	55.5	92.6
0920	Plaster & Gypsum Board	76.3	50.4	59.5	76.3	50.4	59.5	76.7	50.4	59.6	78.7	50.4	60.3	105.1	53.6	71.6	104.0	53.6	71.2
0950, 0980	Ceilings & Acoustic Treatment	108.7	50.4	72.6	108.7	50.4	72.6	108.7	50.4	72.6	103.1	50.4	70.5	92.5	53.6	68.4	93.3	53.6	68.7
0960	Flooring	104.5	53.5	90.7	102.2	53.5	89.0	105.3	53.5	91.3	103.3	53.5	89.9	97.0	61.8	87.5	96.8	61.8	87.4
0970, 0990	Wall Finishes & Painting/Coating	98.1	43.1	65.1	98.1	43.1	65.1	98.1	60.7	75.7	100.7	60.7	76.7	91.7	35.7	58.1	91.7	35.7	58.1
09	**FINISHES**	101.5	52.9	75.8	99.7	52.9	75.0	102.2	54.9	77.2	99.7	54.9	76.0	96.4	54.0	74.0	96.3	54.0	74.0
COVERS	**DIVS. 10 - 14, 25, 28, 41, 43, 44**	100.0	64.0	92.9	100.0	64.0	92.9	100.0	64.0	92.9	100.0	64.0	92.9	100.0	81.9	96.4	100.0	81.9	96.4
21, 22, 23	**FIRE SUPPRESSION, PLUMBING & HVAC**	99.9	69.5	87.8	99.9	69.5	87.8	99.9	69.5	87.8	100.0	69.5	87.9	100.0	64.9	86.1	100.2	65.0	86.2
26, 27, 3370	**ELECTRICAL, COMMUNICATIONS & UTIL.**	94.5	66.8	81.0	94.9	66.8	81.2	95.1	69.7	82.8	98.2	69.7	84.3	99.2	62.8	81.5	96.9	62.8	80.3
MF2004	**WEIGHTED AVERAGE**	100.3	67.3	86.4	98.4	67.3	85.3	100.8	67.8	87.0	102.9	67.9	88.2	99.3	67.7	86.1	97.5	68.1	85.2

DIVISION		VIRGINIA																	
		ALEXANDRIA			ARLINGTON			NEWPORT NEWS			NORFOLK			PORTSMOUTH			RICHMOND		
		MAT.	INST.	TOTAL	MAT.	INST.	TOTAL	MAT.	INST.	TOTAL	MAT.	INST.	TOTAL	MAT.	INST.	TOTAL	MAT.	INST.	TOTAL
015433	CONTRACTOR EQUIPMENT		102.9	102.9		102.1	102.1		105.8	105.8		106.2	106.2		105.8	105.8		105.8	105.8
0241, 31 - 34	SITE & INFRASTRUCTURE, DEMOLITION	116.5	89.1	97.8	128.2	86.7	99.8	110.7	87.7	95.0	111.7	88.4	95.8	109.2	87.0	94.0	111.5	88.1	95.5
0310	Concrete Forming & Accessories	94.1	76.1	78.6	93.4	74.1	76.7	98.7	72.5	76.1	100.4	72.6	76.4	88.3	59.5	63.4	98.2	65.5	70.0
0320	Concrete Reinforcing	90.2	82.2	86.9	101.8	78.7	92.4	101.6	71.3	89.2	101.6	71.4	89.3	101.2	71.3	89.0	101.6	71.8	89.4
0330	Cast-in-Place Concrete	106.8	82.9	97.7	103.9	80.3	94.9	104.6	65.6	89.7	115.0	65.6	96.1	103.6	65.0	88.8	106.2	60.2	88.6
03	CONCRETE	108.2	80.8	95.7	113.7	77.9	97.4	106.4	71.2	90.4	111.4	71.2	93.1	105.3	65.1	87.0	107.1	66.3	88.5
04	MASONRY	91.6	72.3	80.1	104.3	69.2	83.3	97.7	58.4	74.2	103.1	58.4	76.3	102.5	58.0	75.9	96.0	60.9	75.0
05	METALS	97.2	96.7	97.0	95.9	86.6	93.2	97.2	91.7	95.6	96.2	91.7	94.9	96.2	89.8	94.3	99.3	92.4	97.3
06	WOOD, PLASTICS & COMPOSITES	97.7	75.7	85.6	94.4	75.7	84.2	98.6	77.4	87.0	100.5	77.4	87.8	86.4	60.1	72.0	97.4	68.4	81.5
07	THERMAL & MOISTURE PROTECTION	101.5	80.6	93.4	103.2	72.5	91.3	102.5	61.3	86.4	99.9	61.3	84.8	102.5	58.7	85.5	101.5	61.8	86.0
08	OPENINGS	98.3	78.3	93.3	96.2	72.3	90.3	98.3	69.5	91.2	98.3	69.5	91.2	98.4	60.1	88.9	98.3	65.7	90.2
0920	Plaster & Gypsum Board	112.3	74.7	87.9	109.0	74.7	86.7	112.3	75.6	88.5	108.8	75.6	87.2	105.3	57.7	74.4	117.0	66.4	84.1
0950, 0980	Ceilings & Acoustic Treatment	97.6	74.7	83.4	93.4	74.7	81.8	97.6	75.6	84.0	97.6	75.6	84.0	97.6	57.7	72.9	100.9	66.4	79.5
0960	Flooring	111.5	88.9	105.4	109.5	61.7	96.6	111.5	56.7	96.7	110.9	56.7	96.3	102.5	71.7	94.2	110.9	79.8	102.5
0970, 0990	Wall Finishes & Painting/Coating	127.6	83.3	101.0	127.6	83.3	101.0	114.4	48.3	74.8	114.4	64.6	84.5	114.4	64.6	84.5	114.4	67.6	86.3
09	FINISHES	105.9	78.9	91.6	105.1	73.4	88.4	105.1	67.6	85.3	104.4	69.4	85.9	101.5	61.9	80.6	106.2	68.7	86.4
COVERS	DIVS. 10 - 14, 25, 28, 41, 43, 44	100.0	89.2	97.9	100.0	80.8	96.2	100.0	80.8	96.2	100.0	80.7	96.2	100.0	71.9	94.4	100.0	79.0	95.8
21, 22, 23	FIRE SUPPRESSION, PLUMBING & HVAC	100.2	88.5	95.5	100.2	82.4	93.1	100.2	66.9	86.9	100.0	65.9	86.4	100.2	65.9	86.5	100.0	61.7	84.8
26, 27, 3370	ELECTRICAL, COMMUNICATIONS & UTIL.	97.6	98.2	97.9	95.1	93.4	94.3	97.2	63.8	80.9	105.5	60.4	83.5	95.4	60.4	78.4	104.2	69.8	87.5
MF2004	WEIGHTED AVERAGE	100.7	86.0	94.5	101.5	80.8	92.8	100.5	70.9	88.1	101.9	70.5	88.7	99.8	67.4	86.2	101.7	70.1	88.4

DIVISION		VIRGINIA			WASHINGTON														
		ROANOKE			EVERETT			RICHLAND			SEATTLE			SPOKANE			TACOMA		
		MAT.	INST.	TOTAL	MAT.	INST.	TOTAL	MAT.	INST.	TOTAL	MAT.	INST.	TOTAL	MAT.	INST.	TOTAL	MAT.	INST.	TOTAL
015433	CONTRACTOR EQUIPMENT		102.1	102.1		101.3	101.3		95.9	95.9		101.8	101.8		95.9	95.9		101.3	101.3
0241, 31 - 34	SITE & INFRASTRUCTURE, DEMOLITION	108.2	86.2	93.2	98.0	110.8	106.7	104.1	96.0	98.5	102.3	110.1	107.6	103.4	96.0	98.3	101.4	111.1	108.0
0310	Concrete Forming & Accessories	98.5	69.0	73.0	111.8	100.6	102.1	117.8	80.6	85.7	103.4	102.7	102.8	123.3	80.6	86.4	103.4	102.5	102.6
0320	Concrete Reinforcing	101.6	65.7	86.9	113.0	93.3	105.0	99.5	86.1	94.0	111.7	93.3	104.2	100.2	86.2	94.5	111.7	93.3	104.2
0330	Cast-in-Place Concrete	117.6	65.2	97.6	101.0	100.9	101.0	111.2	86.8	101.9	106.1	108.7	107.1	115.6	86.8	104.6	103.9	108.6	105.7
03	CONCRETE	112.5	68.5	92.4	98.0	98.8	98.3	103.1	83.8	94.3	100.7	102.5	101.5	105.6	83.8	95.7	99.7	102.3	100.9
04	MASONRY	98.2	61.2	76.1	138.6	96.3	113.3	115.6	81.3	95.1	133.2	100.8	113.8	116.3	83.7	96.8	133.0	100.8	113.7
05	METALS	97.0	88.0	94.3	109.5	88.6	103.4	91.2	83.8	89.0	111.2	90.3	105.1	93.7	83.9	90.8	111.2	88.8	104.7
06	WOOD, PLASTICS & COMPOSITES	98.6	73.3	84.7	103.9	102.1	102.9	96.8	79.9	87.5	94.8	102.1	98.8	105.7	79.9	91.5	93.7	102.1	98.3
07	THERMAL & MOISTURE PROTECTION	102.4	60.0	85.9	105.1	93.7	100.6	151.3	80.0	123.6	105.0	97.3	102.0	148.0	80.9	121.9	104.7	96.4	101.5
08	OPENINGS	98.3	65.2	90.1	103.7	97.9	102.3	111.6	74.8	102.5	106.0	97.9	104.0	112.2	74.8	103.0	104.4	97.9	102.8
0920	Plaster & Gypsum Board	112.3	72.1	86.2	112.7	102.1	105.8	140.1	78.9	100.4	107.3	102.1	103.9	137.2	78.9	99.4	109.7	102.1	104.8
0950, 0980	Ceilings & Acoustic Treatment	97.6	72.1	81.8	105.1	102.1	103.3	111.4	78.9	91.3	108.2	102.1	104.4	106.7	78.9	89.5	108.4	102.1	104.5
0960	Flooring	111.5	43.9	93.2	122.4	102.7	117.1	107.6	38.8	89.0	114.7	102.9	111.5	108.0	70.2	97.8	115.3	102.9	111.9
0970, 0990	Wall Finishes & Painting/Coating	114.4	47.8	74.4	109.3	81.6	92.7	109.0	67.3	84.0	109.3	89.6	97.5	108.9	67.3	83.9	109.3	89.6	97.5
09	FINISHES	105.0	62.9	82.7	110.2	99.4	104.5	122.6	70.8	95.2	108.1	101.4	104.5	121.3	77.2	98.0	108.5	101.4	104.8
COVERS	DIVS. 10 - 14, 25, 28, 41, 43, 44	100.0	71.9	94.4	100.0	98.2	99.6	100.0	76.1	95.3	100.0	101.4	100.3	100.0	76.0	95.3	100.0	101.4	100.3
21, 22, 23	FIRE SUPPRESSION, PLUMBING & HVAC	100.2	53.2	81.5	100.2	96.4	98.7	100.6	96.4	98.9	100.2	107.0	102.9	100.5	82.1	93.2	100.2	98.8	99.7
26, 27, 3370	ELECTRICAL, COMMUNICATIONS & UTIL.	97.2	44.4	71.5	103.3	90.3	97.0	97.0	90.8	93.9	103.1	99.4	101.3	95.3	76.3	86.1	103.1	97.1	100.2
MF2004	WEIGHTED AVERAGE	101.2	63.7	85.4	105.0	96.7	101.5	104.1	85.7	96.4	105.4	101.7	103.9	104.5	81.8	95.0	105.1	99.6	102.8

DIVISION		WASHINGTON						WEST VIRGINIA											
		VANCOUVER			YAKIMA			CHARLESTON			HUNTINGTON			PARKERSBURG			WHEELING		
		MAT.	INST.	TOTAL	MAT.	INST.	TOTAL	MAT.	INST.	TOTAL	MAT.	INST.	TOTAL	MAT.	INST.	TOTAL	MAT.	INST.	TOTAL
015433	CONTRACTOR EQUIPMENT		98.2	98.2		101.3	101.3		102.1	102.1		102.1	102.1		102.1	102.1		102.1	102.1
0241, 31 - 34	SITE & INFRASTRUCTURE, DEMOLITION	113.5	100.9	104.9	104.4	109.4	107.8	103.9	89.4	94.0	107.3	90.3	95.7	112.4	89.4	96.7	113.1	88.9	96.6
0310	Concrete Forming & Accessories	104.4	91.6	93.4	103.7	94.9	96.1	104.3	84.5	87.2	99.7	89.4	90.8	90.5	83.3	84.3	92.2	83.6	84.8
0320	Concrete Reinforcing	112.6	92.9	104.6	112.2	85.7	101.4	101.6	81.1	93.2	101.6	86.3	95.4	100.2	88.1	95.2	99.5	84.0	93.1
0330	Cast-in-Place Concrete	116.1	98.0	109.2	111.1	82.4	100.1	99.4	99.8	99.5	110.4	104.1	108.0	103.1	95.7	100.3	103.1	96.4	100.6
03	CONCRETE	109.1	94.0	102.2	104.2	88.5	97.1	104.3	90.0	97.8	109.2	94.6	102.6	109.2	89.4	100.2	109.2	89.0	100.0
04	MASONRY	133.4	93.0	109.2	125.2	63.4	88.2	94.7	89.4	91.5	97.6	86.6	91.0	81.7	83.9	83.1	105.3	81.5	91.0
05	METALS	108.4	89.5	102.9	109.4	84.2	102.0	97.2	96.7	97.1	97.3	98.2	97.5	95.8	98.7	96.6	95.9	97.4	96.4
06	WOOD, PLASTICS & COMPOSITES	86.4	91.6	89.3	94.0	102.1	98.4	104.8	82.8	92.7	98.6	87.8	92.7	88.3	81.2	84.4	90.1	82.8	86.1
07	THERMAL & MOISTURE PROTECTION	105.3	86.7	98.1	104.8	76.6	93.9	99.7	87.7	95.0	102.7	88.3	97.1	102.5	84.7	95.6	102.9	83.8	95.5
08	OPENINGS	102.2	91.4	99.6	103.8	84.5	99.1	99.5	77.5	94.1	98.3	81.5	94.1	98.9	78.3	93.8	99.8	82.1	95.4
0920	Plaster & Gypsum Board	107.8	91.5	97.2	109.1	102.1	104.6	107.9	82.0	91.1	111.0	87.1	95.5	105.2	80.4	89.1	105.6	82.0	90.3
0950, 0980	Ceilings & Acoustic Treatment	103.5	91.5	96.1	103.4	102.1	102.6	94.3	82.0	86.7	92.6	87.1	89.2	91.8	80.4	84.7	91.8	82.0	85.7
0960	Flooring	121.2	71.4	107.7	116.5	58.3	100.8	111.5	105.0	109.7	111.3	93.6	106.5	105.7	95.6	103.0	106.8	94.5	103.5
0970, 0990	Wall Finishes & Painting/Coating	115.7	64.1	84.7	109.3	67.3	84.1	114.4	87.1	98.0	114.4	87.1	98.0	114.4	87.2	98.1	114.4	91.3	100.5
09	FINISHES	107.0	84.7	95.2	107.8	86.0	96.3	103.5	88.5	95.6	103.5	89.9	96.3	101.5	85.7	93.1	101.9	86.2	93.6
COVERS	DIVS. 10 - 14, 25, 28, 41, 43, 44	100.0	72.3	94.5	100.0	95.7	99.2	100.0	83.9	96.8	100.0	85.7	97.2	100.0	83.8	96.8	100.0	93.5	98.7
21, 22, 23	FIRE SUPPRESSION, PLUMBING & HVAC	100.3	100.9	100.6	100.2	93.0	97.4	100.0	83.0	93.3	100.2	84.9	94.1	100.1	87.3	95.0	100.2	90.6	96.4
26, 27, 3370	ELECTRICAL, COMMUNICATIONS & UTIL.	109.7	97.6	103.8	106.1	90.8	98.6	102.5	88.0	95.4	97.2	95.8	96.5	97.7	94.3	96.0	94.8	92.8	93.8
MF2004	WEIGHTED AVERAGE	106.5	93.8	101.1	105.3	88.0	98.0	100.4	87.8	95.1	100.6	90.3	96.3	99.6	88.7	95.0	100.6	89.2	95.8

	DIVISION	WISCONSIN																	
		EAU CLAIRE			GREEN BAY			KENOSHA			LA CROSSE			MADISON			MILWAUKEE		
		MAT.	INST.	TOTAL	MAT.	INST.	TOTAL	MAT.	INST.	TOTAL	MAT.	INST.	TOTAL	MAT.	INST.	TOTAL	MAT.	INST.	TOTAL
015433	**CONTRACTOR EQUIPMENT**		101.0	101.0		98.7	98.7		98.3	98.3		101.0	101.0		100.4	100.4		89.4	89.4
0241, 31 - 34	**SITE & INFRASTRUCTURE, DEMOLITION**	90.8	104.0	99.8	94.1	99.5	97.8	97.0	103.1	101.1	84.4	104.1	97.9	91.0	106.5	101.6	91.9	97.4	95.7
0310	Concrete Forming & Accessories	97.0	94.4	94.7	105.0	93.6	95.2	104.1	104.4	104.4	84.4	94.9	93.5	95.9	95.7	95.7	99.6	113.8	111.8
0320	Concrete Reinforcing	100.8	100.0	100.5	98.9	85.6	93.5	100.4	102.3	101.1	100.5	88.2	95.5	100.6	88.6	95.7	100.6	102.7	101.5
0330	Cast-in-Place Concrete	97.8	94.4	96.5	101.2	96.2	99.3	108.2	100.4	105.2	87.9	94.5	90.4	97.3	98.2	97.6	97.1	108.5	101.5
03	**CONCRETE**	97.7	95.8	96.8	98.3	93.4	96.1	102.5	102.6	102.5	89.7	93.8	91.6	96.9	95.4	96.2	97.2	109.0	102.6
04	**MASONRY**	89.5	95.1	92.8	119.3	94.2	104.3	100.4	107.7	104.8	88.7	98.0	94.2	101.2	104.0	102.9	103.2	117.5	111.8
05	**METALS**	91.9	101.6	94.8	94.1	95.7	94.5	99.3	102.6	100.3	91.8	96.5	93.2	101.1	95.7	99.5	101.8	96.2	100.2
06	**WOOD, PLASTICS & COMPOSITES**	113.5	94.8	103.3	115.8	94.8	104.3	112.2	103.9	107.6	98.3	94.8	96.4	107.0	94.7	100.3	112.3	113.6	113.0
07	**THERMAL & MOISTURE PROTECTION**	100.1	85.2	94.3	102.0	83.5	94.8	99.8	97.3	98.8	99.6	85.2	94.0	95.5	92.9	94.5	97.8	112.1	103.4
08	**OPENINGS**	100.7	93.5	98.9	98.8	91.8	97.1	98.4	105.9	100.3	100.7	82.9	96.3	103.3	95.2	101.3	105.6	111.3	107.0
0920	Plaster & Gypsum Board	100.4	95.0	96.9	92.7	95.0	94.2	85.0	104.4	97.6	94.8	95.0	94.9	92.1	95.0	94.0	96.5	114.3	108.1
0950, 0980	Ceilings & Acoustic Treatment	99.9	95.0	96.9	92.5	95.0	94.0	82.6	104.4	96.1	98.3	95.0	96.2	85.3	95.0	91.3	86.8	114.3	103.8
0960	Flooring	97.0	106.7	99.6	116.7	106.7	114.0	122.8	112.2	119.9	89.5	108.7	94.7	98.8	105.0	100.5	107.4	120.5	110.9
0970, 0990	Wall Finishes & Painting/Coating	89.2	91.8	90.8	101.0	74.5	85.1	102.8	101.8	102.2	89.2	71.5	78.6	93.6	93.5	93.5	95.6	113.1	106.1
09	**FINISHES**	99.4	96.8	98.0	103.8	94.6	99.0	101.3	106.3	103.9	95.6	95.6	95.6	94.8	97.6	96.3	98.6	115.5	107.5
COVERS	**DIVS. 10 - 14, 25, 28, 41, 43, 44**	100.0	87.3	97.5	100.0	86.6	97.3	100.0	102.3	100.5	100.0	88.3	97.7	100.0	87.9	97.6	100.0	105.1	101.0
21, 22, 23	**FIRE SUPPRESSION, PLUMBING & HVAC**	100.1	80.1	92.2	100.4	83.8	93.8	100.1	88.5	95.5	100.1	81.3	92.7	99.7	89.3	95.6	99.9	101.4	100.5
26, 27, 3370	**ELECTRICAL, COMMUNICATIONS & UTIL.**	101.9	87.7	94.9	96.6	83.4	90.2	98.6	92.9	95.8	102.1	87.7	95.1	101.3	88.9	95.3	99.6	104.3	101.9
MF2004	**WEIGHTED AVERAGE**	97.9	92.0	95.4	99.7	90.4	95.8	100.0	99.3	99.7	96.3	91.3	94.2	99.5	95.0	97.6	100.3	106.4	102.9

	DIVISION	WISCONSIN			WYOMING									CANADA					
		RACINE			CASPER			CHEYENNE			ROCK SPRINGS			CALGARY, ALBERTA			EDMONTON, ALBERTA		
		MAT.	INST.	TOTAL	MAT.	INST.	TOTAL	MAT.	INST.	TOTAL	MAT.	INST.	TOTAL	MAT.	INST.	TOTAL	MAT.	INST.	TOTAL
015433	**CONTRACTOR EQUIPMENT**		100.4	100.4		101.1	101.1		101.1	101.1		101.1	101.1		106.4	106.4		106.4	106.4
0241, 31 - 34	**SITE & INFRASTRUCTURE, DEMOLITION**	91.0	107.3	102.1	92.6	100.1	97.7	85.8	100.2	95.6	82.2	98.9	93.6	124.3	104.8	111.0	154.6	104.8	120.6
0310	Concrete Forming & Accessories	97.4	104.6	103.6	103.9	45.0	53.0	105.5	61.3	67.3	101.8	40.4	48.7	127.0	90.2	95.2	128.4	90.2	95.4
0320	Concrete Reinforcing	100.6	102.3	101.3	109.3	46.0	83.5	103.0	46.9	80.1	111.4	46.7	85.0	147.9	63.2	113.4	147.9	63.2	113.4
0330	Cast-in-Place Concrete	97.5	100.2	98.5	102.7	76.1	92.5	98.0	76.3	89.7	99.0	56.8	82.9	186.7	101.0	153.9	208.9	101.0	167.6
03	**CONCRETE**	97.1	102.6	99.6	108.1	57.1	84.9	105.3	64.5	86.7	106.7	48.4	80.2	157.0	89.4	126.3	167.5	89.4	131.9
04	**MASONRY**	103.3	107.7	105.9	107.3	39.3	66.6	104.0	54.2	74.2	162.7	48.6	94.4	213.5	91.4	140.4	202.0	91.4	135.8
05	**METALS**	100.2	102.8	100.9	98.3	64.7	88.5	99.5	66.3	89.8	95.5	64.1	86.3	151.1	88.1	132.7	151.6	88.0	133.0
06	**WOOD, PLASTICS & COMPOSITES**	110.3	103.9	106.8	97.7	42.4	67.3	99.3	63.7	79.8	96.1	39.1	64.8	114.3	89.5	100.7	110.9	89.5	99.2
07	**THERMAL & MOISTURE PROTECTION**	99.5	98.6	99.1	97.6	51.9	79.8	99.4	58.1	83.3	100.7	49.9	80.9	123.0	87.9	109.3	123.7	87.9	109.8
08	**OPENINGS**	103.3	105.9	103.9	92.7	41.9	80.2	94.8	53.5	84.6	99.1	40.1	84.5	92.0	79.9	89.0	92.0	79.9	89.0
0920	Plaster & Gypsum Board	95.6	104.4	101.3	86.7	40.2	56.5	82.1	62.2	69.2	84.1	36.7	53.3	162.0	88.6	114.3	158.2	88.6	113.0
0950, 0980	Ceilings & Acoustic Treatment	83.4	104.4	96.4	112.1	40.2	67.6	106.0	62.2	78.9	108.8	36.7	64.2	144.8	88.6	110.0	148.1	88.6	111.3
0960	Flooring	104.5	112.2	106.6	103.5	37.0	85.5	108.0	59.4	94.9	107.0	52.3	92.2	134.4	88.8	122.1	135.6	88.8	122.9
0970, 0990	Wall Finishes & Painting/Coating	93.5	101.8	98.5	98.0	49.5	68.9	101.2	49.5	70.2	97.3	31.7	57.9	109.6	100.7	104.3	109.7	92.2	99.2
09	**FINISHES**	96.7	106.3	101.8	104.7	42.9	72.1	104.8	60.0	81.1	102.3	40.4	69.6	132.7	91.6	111.0	135.2	90.7	111.6
COVERS	**DIVS. 10 - 14, 25, 28, 41, 43, 44**	100.0	102.3	100.5	100.0	79.4	95.9	100.0	82.0	96.4	100.0	65.3	93.1	140.0	102.3	132.5	140.0	102.3	132.5
21, 22, 23	**FIRE SUPPRESSION, PLUMBING & HVAC**	99.9	93.7	97.5	100.0	60.7	84.4	99.9	59.9	84.0	99.9	57.3	83.0	100.1	87.6	95.1	100.1	87.6	95.2
26, 27, 3370	**ELECTRICAL, COMMUNICATIONS & UTIL.**	98.0	96.2	97.1	101.2	59.3	80.8	97.6	72.7	85.5	93.3	60.4	77.3	125.2	84.7	105.5	118.5	84.7	102.1
MF2004	**WEIGHTED AVERAGE**	99.5	101.3	100.2	100.6	58.4	82.9	100.0	65.9	85.7	102.0	56.5	82.9	130.7	89.9	113.6	131.8	89.8	114.1

	DIVISION	CANADA																	
		HALIFAX, NOVA SCOTIA			HAMILTON, ONTARIO			KITCHENER, ONTARIO			LAVAL, QUEBEC			LONDON, ONTARIO			MONTREAL, QUEBEC		
		MAT.	INST.	TOTAL	MAT.	INST.	TOTAL	MAT.	INST.	TOTAL	MAT.	INST.	TOTAL	MAT.	INST.	TOTAL	MAT.	INST.	TOTAL
015433	**CONTRACTOR EQUIPMENT**		102.9	102.9		107.3	107.3		103.8	103.8		104.0	104.0		104.1	104.1		105.6	105.6
0241, 31 - 34	**SITE & INFRASTRUCTURE, DEMOLITION**	99.3	99.2	99.2	114.9	111.1	112.3	101.3	104.7	103.6	94.4	101.1	99.0	114.5	104.9	108.0	101.1	101.0	101.0
0310	Concrete Forming & Accessories	96.4	76.4	79.1	129.6	95.4	100.1	117.4	87.8	91.8	129.2	86.1	91.9	129.7	89.0	94.5	135.9	93.7	99.5
0320	Concrete Reinforcing	154.5	61.2	116.4	161.2	90.3	132.3	110.2	90.2	102.0	156.6	79.7	125.2	130.4	88.9	113.5	146.2	93.9	124.9
0330	Cast-in-Place Concrete	171.3	75.7	134.7	151.9	101.4	132.6	140.8	82.6	118.6	134.5	95.5	119.6	149.3	99.4	130.2	174.7	101.9	146.9
03	**CONCRETE**	150.7	74.0	115.8	143.6	96.6	122.2	121.6	86.7	105.7	134.5	88.5	113.6	136.2	92.8	116.5	151.6	96.7	126.6
04	**MASONRY**	173.5	82.9	119.2	181.0	101.5	133.4	161.9	97.8	123.5	160.9	85.4	115.7	182.4	99.1	132.5	167.3	93.7	123.2
05	**METALS**	120.4	81.4	109.0	127.9	94.2	118.1	116.1	93.8	109.6	105.5	89.3	100.8	125.3	92.1	115.6	123.5	94.8	115.1
06	**WOOD, PLASTICS & COMPOSITES**	86.1	75.5	80.3	115.5	94.6	104.0	109.3	86.1	96.6	127.9	85.7	104.8	115.5	87.0	99.9	127.9	93.9	109.3
07	**THERMAL & MOISTURE PROTECTION**	106.0	76.2	94.4	120.1	96.5	110.9	110.0	93.2	103.5	105.1	88.3	98.5	120.9	93.2	110.1	117.1	96.9	109.3
08	**OPENINGS**	82.8	68.9	79.4	92.0	91.4	91.9	83.3	85.2	83.7	92.0	74.7	87.7	93.2	85.4	91.2	92.0	82.1	89.6
0920	Plaster & Gypsum Board	163.1	74.4	105.5	195.8	94.3	129.9	152.4	85.5	109.0	145.7	84.9	106.2	199.5	86.4	126.1	152.5	93.2	114.0
0950, 0980	Ceilings & Acoustic Treatment	105.4	74.4	86.2	120.6	94.3	104.3	105.4	85.5	93.1	95.4	84.9	88.9	138.5	86.4	106.3	102.0	93.2	96.5
0960	Flooring	110.1	64.5	97.8	134.4	94.8	123.7	127.8	94.8	118.9	132.0	95.4	122.1	134.6	94.8	123.9	133.9	95.4	123.5
0970, 0990	Wall Finishes & Painting/Coating	109.6	74.4	88.5	109.6	100.5	104.2	109.6	89.7	97.7	109.6	88.6	97.0	109.6	97.0	102.0	109.6	104.2	106.4
09	**FINISHES**	114.2	74.2	93.1	130.8	95.8	112.3	118.0	89.0	102.6	116.0	88.2	101.3	135.6	90.6	111.8	119.2	95.8	106.9
COVERS	**DIVS. 10 - 14, 25, 28, 41, 43, 44**	140.0	73.8	126.9	140.0	106.6	133.4	140.0	104.6	133.0	140.0	91.6	130.4	140.0	105.2	133.1	140.0	94.1	130.9
21, 22, 23	**FIRE SUPPRESSION, PLUMBING & HVAC**	99.2	76.1	90.0	100.0	91.9	96.8	99.2	88.9	95.1	99.2	92.7	96.6	100.0	89.0	95.6	100.1	94.4	97.8
26, 27, 3370	**ELECTRICAL, COMMUNICATIONS & UTIL.**	128.4	74.6	102.2	128.8	96.1	112.9	124.1	93.4	109.1	123.5	74.4	99.6	123.3	93.4	108.8	129.8	86.1	108.5
MF2004	**WEIGHTED AVERAGE**	119.0	78.0	101.8	123.4	96.9	112.3	114.5	92.2	105.1	114.5	87.6	103.2	122.0	93.3	110.0	121.7	93.8	110.0

DIVISION		CANADA																	
		OSHAWA, ONTARIO			OTTAWA, ONTARIO			QUEBEC, QUEBEC			REGINA, SASKATCHEWAN			SASKATOON, SASKATCHEWAN			ST CATHARINES, ONTARIO		
		MAT.	INST.	TOTAL	MAT.	INST.	TOTAL	MAT.	INST.	TOTAL	MAT.	INST.	TOTAL	MAT.	INST.	TOTAL	MAT.	INST.	TOTAL
015433	**CONTRACTOR EQUIPMENT**		103.8	103.8		103.8	103.8		106.0	106.0		101.2	101.2		101.2	101.2		102.2	102.2
0241, 31 - 34	**SITE & INFRASTRUCTURE, DEMOLITION**	114.1	104.0	107.2	111.7	104.6	106.9	99.7	101.1	100.7	113.0	96.8	101.9	107.4	97.0	100.3	102.1	101.9	102.0
0310	Concrete Forming & Accessories	123.3	87.7	92.6	130.2	91.7	97.0	137.1	94.0	99.8	106.0	59.5	65.8	106.0	59.4	65.7	115.3	92.6	95.7
0320	Concrete Reinforcing	174.5	85.3	138.1	160.1	88.8	131.0	140.1	94.0	121.3	125.7	62.6	100.0	118.8	62.6	95.9	111.1	90.2	102.6
0330	Cast-in-Place Concrete	162.9	87.0	133.9	153.0	98.5	132.2	148.3	102.3	130.7	156.3	70.3	123.4	142.0	70.3	114.6	134.4	101.2	121.7
03	**CONCRETE**	151.0	87.4	122.1	143.9	93.7	121.1	138.1	96.9	119.4	131.1	64.8	100.9	123.1	64.7	96.5	118.6	95.3	108.0
04	**MASONRY**	165.4	93.8	122.5	179.7	97.2	130.3	174.3	93.7	126.0	166.4	63.6	104.8	165.9	63.6	104.6	161.4	101.5	125.5
05	**METALS**	107.1	92.7	102.9	126.2	93.7	116.7	125.2	95.1	116.4	105.5	76.4	97.0	105.5	76.3	97.0	106.4	93.7	102.7
06	**WOOD, PLASTICS & COMPOSITES**	116.4	85.9	99.6	115.2	90.9	101.9	128.6	94.0	109.6	95.8	57.8	74.9	94.2	57.8	74.2	106.9	91.5	98.4
07	**THERMAL & MOISTURE PROTECTION**	110.9	86.4	101.4	123.4	92.7	111.5	118.6	97.1	110.2	106.0	63.5	89.5	104.9	62.5	88.4	110.0	93.4	103.5
08	**OPENINGS**	90.8	85.8	89.6	92.0	88.4	91.1	92.0	89.6	91.4	87.0	55.6	79.2	86.0	55.6	78.5	82.7	88.5	84.2
0920	Plaster & Gypsum Board	155.3	85.3	109.8	234.7	90.5	141.1	191.0	93.2	127.4	165.1	56.2	94.3	145.5	56.2	87.5	137.2	91.1	107.2
0950, 0980	Ceilings & Acoustic Treatment	99.6	85.3	90.7	131.4	90.5	106.1	93.6	93.2	93.3	119.5	56.2	80.3	119.5	56.2	80.3	99.6	91.1	94.3
0960	Flooring	132.0	97.2	122.6	134.4	93.4	123.4	134.4	95.4	123.9	119.8	60.7	103.8	119.8	60.7	103.8	126.0	94.8	117.6
0970, 0990	Wall Finishes & Painting/Coating	109.6	104.9	106.8	109.6	91.4	98.7	110.2	104.2	106.6	109.6	64.5	82.6	109.6	55.0	76.9	109.6	100.5	104.2
09	**FINISHES**	119.0	91.1	104.2	138.2	92.0	113.8	123.1	95.9	108.7	122.0	59.8	89.1	119.1	58.7	87.2	114.0	94.0	103.4
COVERS	**DIVS. 10 - 14, 25, 28, 41, 43, 44**	140.0	104.7	133.0	140.0	103.1	132.7	140.0	94.3	130.9	140.0	69.7	126.1	140.0	69.7	126.1	140.0	81.2	128.4
21, 22, 23	**FIRE SUPPRESSION, PLUMBING & HVAC**	99.2	103.2	100.8	100.2	89.4	95.9	99.9	94.4	97.7	99.4	77.6	90.8	99.3	77.6	90.7	99.2	90.6	95.8
26, 27, 3370	**ELECTRICAL, COMMUNICATIONS & UTIL.**	125.3	92.7	109.4	119.7	94.4	107.4	121.7	86.1	104.4	126.7	64.4	96.4	127.0	64.4	96.5	126.3	94.5	110.8
MF2004	**WEIGHTED AVERAGE**	117.9	94.7	108.2	122.7	93.9	110.6	120.1	94.2	109.2	115.1	69.9	96.1	113.6	69.8	95.2	112.3	94.2	104.7

DIVISION		CANADA																	
		ST JOHNS, NEWFOUNDLAND			THUNDER BAY, ONTARIO			TORONTO, ONTARIO			VANCOUVER, BRITISH COLUMBIA			WINDSOR, ONTARIO			WINNIPEG, MANITOBA		
		MAT.	INST.	TOTAL	MAT.	INST.	TOTAL	MAT.	INST.	TOTAL	MAT.	INST.	TOTAL	MAT.	INST.	TOTAL	MAT.	INST.	TOTAL
015433	**CONTRACTOR EQUIPMENT**		104.8	104.8		102.2	102.2		104.1	104.1		112.7	112.7		102.2	102.2		106.4	106.4
0241, 31 - 34	**SITE & INFRASTRUCTURE, DEMOLITION**	115.7	99.5	104.7	107.6	101.8	103.7	139.0	105.5	116.1	121.8	107.4	111.9	97.2	101.5	100.1	119.3	100.9	106.7
0310	Concrete Forming & Accessories	103.8	64.3	69.7	123.3	93.0	97.1	131.1	100.7	104.8	124.8	78.0	84.4	123.3	90.0	94.5	130.3	65.5	74.3
0320	Concrete Reinforcing	163.8	57.7	120.5	99.3	89.6	95.4	157.0	91.0	130.1	156.7	72.3	122.3	108.9	88.9	100.7	147.9	56.0	110.4
0330	Cast-in-Place Concrete	172.4	77.6	136.2	148.0	99.8	129.6	147.4	109.5	132.9	163.0	92.2	135.9	137.7	100.8	123.6	183.5	72.3	141.0
03	**CONCRETE**	164.6	68.7	121.0	126.3	94.8	112.0	140.8	101.8	123.1	153.2	82.7	121.1	120.2	93.7	108.2	155.5	67.1	115.3
04	**MASONRY**	164.6	65.1	105.1	162.1	101.3	125.7	193.0	106.7	141.4	177.4	81.1	119.8	161.5	99.9	124.6	180.0	62.1	109.4
05	**METALS**	108.2	79.1	99.7	106.2	92.7	102.3	127.7	94.9	118.2	165.9	89.2	143.5	106.3	93.4	102.5	159.9	78.6	136.1
06	**WOOD, PLASTICS & COMPOSITES**	95.2	64.3	78.3	116.4	91.6	102.8	116.4	99.4	107.1	114.3	75.1	92.8	116.4	88.3	101.0	114.2	66.0	87.7
07	**THERMAL & MOISTURE PROTECTION**	109.3	63.4	91.5	110.2	94.8	104.2	121.0	101.2	113.3	132.4	82.1	112.8	110.0	93.6	103.6	114.1	67.3	95.9
08	**OPENINGS**	98.2	60.2	88.8	81.7	87.9	83.3	90.8	96.0	92.1	93.8	75.8	89.3	81.5	86.6	82.7	92.0	60.7	84.3
0920	Plaster & Gypsum Board	168.5	62.7	99.8	166.5	91.2	117.6	177.3	99.3	126.6	154.9	73.5	102.0	156.7	87.8	111.9	157.0	64.1	96.7
0950, 0980	Ceilings & Acoustic Treatment	104.6	62.7	78.7	95.4	91.2	92.8	132.6	99.3	112.0	141.0	73.5	99.3	95.4	87.8	90.7	121.9	64.1	86.2
0960	Flooring	114.8	54.9	98.6	132.0	54.6	111.1	135.6	100.5	126.1	134.8	92.0	123.2	132.0	95.5	122.1	133.3	67.3	115.5
0970, 0990	Wall Finishes & Painting/Coating	109.6	64.5	82.6	109.6	93.2	99.8	111.5	104.9	107.5	109.6	88.7	97.1	109.6	92.8	99.5	109.7	48.1	72.7
09	**FINISHES**	117.4	62.5	88.3	119.2	86.7	102.0	133.2	101.2	116.3	130.7	81.2	104.5	117.5	91.3	103.6	125.4	64.2	93.1
COVERS	**DIVS. 10 - 14, 25, 28, 41, 43, 44**	140.0	71.2	126.4	140.0	81.6	128.4	140.0	108.5	133.8	140.0	99.9	132.1	140.0	80.6	128.2	140.0	71.7	126.5
21, 22, 23	**FIRE SUPPRESSION, PLUMBING & HVAC**	99.6	64.2	85.5	99.2	90.7	95.8	99.9	95.9	98.3	99.9	73.3	89.3	99.2	90.2	95.6	99.9	65.6	86.3
26, 27, 3370	**ELECTRICAL, COMMUNICATIONS & UTIL.**	122.9	65.8	95.1	124.1	92.7	108.8	122.0	96.8	109.8	124.7	75.0	100.5	130.2	94.4	112.8	127.9	67.0	98.3
MF2004	**WEIGHTED AVERAGE**	119.9	69.2	98.6	113.5	92.9	104.9	123.6	99.8	113.6	131.2	82.0	110.5	113.0	93.2	104.7	129.6	69.6	104.4

Location Factors

Costs shown in RSMeans cost data publications are based on national averages for materials and installation. To adjust these costs to a specific location, simply multiply the base cost by the factor and divide by 100 for that city. The data is arranged alphabetically by state and postal zip code numbers. For a city not listed, use the factor for a nearby city with similar economic characteristics.

STATE/ZIP	CITY	MAT.	INST.	TOTAL
ALABAMA				
350-352	Birmingham	98.3	74.9	88.5
354	Tuscaloosa	97.0	60.5	81.7
355	Jasper	97.2	54.0	79.0
356	Decatur	97.0	59.9	81.4
357-358	Huntsville	97.0	71.1	86.1
359	Gadsden	96.9	59.4	81.2
360-361	Montgomery	98.0	59.2	81.7
362	Anniston	96.3	52.9	78.1
363	Dothan	96.9	52.5	78.3
364	Evergreen	96.4	57.3	80.0
365-366	Mobile	97.9	65.9	84.5
367	Selma	96.6	54.4	78.9
368	Phenix City	97.3	59.4	81.4
369	Butler	96.8	55.7	79.5
ALASKA				
995-996	Anchorage	128.3	110.3	120.8
997	Fairbanks	127.6	112.6	121.3
998	Juneau	127.0	108.5	119.2
999	Ketchikan	139.6	108.1	126.4
ARIZONA				
850,853	Phoenix	99.9	74.0	89.0
852	Mesa/Tempe	99.2	66.7	85.6
855	Globe	99.5	63.1	84.2
856-857	Tucson	98.0	70.6	86.5
859	Show Low	99.5	64.2	84.7
860	Flagstaff	101.4	71.6	88.9
863	Prescott	99.2	63.7	84.3
864	Kingman	97.5	69.6	85.8
865	Chambers	97.5	64.4	83.6
ARKANSAS				
716	Pine Bluff	96.9	66.6	84.2
717	Camden	94.6	44.6	73.6
718	Texarkana	95.9	48.8	76.1
719	Hot Springs	93.8	47.9	74.5
720-722	Little Rock	97.1	69.9	85.7
723	West Memphis	96.0	59.6	80.8
724	Jonesboro	96.9	60.2	81.5
725	Batesville	94.7	54.3	77.7
726	Harrison	95.9	55.0	78.8
727	Fayetteville	93.4	54.7	77.1
728	Russellville	94.5	56.7	78.6
729	Fort Smith	97.4	60.1	81.8
CALIFORNIA				
900-902	Los Angeles	102.2	116.7	108.3
903-905	Inglewood	97.6	107.9	101.9
906-908	Long Beach	99.3	108.0	102.9
910-912	Pasadena	98.0	107.9	102.2
913-916	Van Nuys	101.2	107.9	104.0
917-918	Alhambra	100.2	107.9	103.4
919-921	San Diego	102.4	108.8	105.1
922	Palm Springs	99.2	106.8	102.4
923-924	San Bernardino	96.9	107.2	101.2
925	Riverside	101.6	112.7	106.3
926-927	Santa Ana	98.9	107.5	102.5
928	Anaheim	101.5	114.7	107.0
930	Oxnard	102.9	112.8	107.0
931	Santa Barbara	101.9	112.9	106.5
932-933	Bakersfield	103.3	109.4	105.8
934	San Luis Obispo	102.5	105.3	103.7
935	Mojave	99.6	104.3	101.6
936-938	Fresno	103.2	112.6	107.2
939	Salinas	103.3	116.4	108.8
940-941	San Francisco	111.2	141.3	123.8
942,956-958	Sacramento	104.9	114.6	108.9
943	Palo Alto	103.5	123.7	112.0
944	San Mateo	106.4	129.3	116.0
945	Vallejo	104.1	121.6	111.4
946	Oakland	108.8	129.5	117.5
947	Berkeley	108.3	123.8	114.8
948	Richmond	107.4	123.9	114.3
949	San Rafael	108.7	124.5	115.3
950	Santa Cruz	108.4	116.6	111.8
CALIFORNIA (CONT'D)				
951	San Jose	106.0	132.5	117.1
952	Stockton	103.5	112.0	107.0
953	Modesto	103.4	111.8	106.9
954	Santa Rosa	103.3	126.1	112.9
955	Eureka	104.7	109.6	106.7
959	Marysville	104.0	111.2	107.0
960	Redding	105.8	109.8	107.5
961	Susanville	104.6	110.0	106.8
COLORADO				
800-802	Denver	101.8	85.5	95.0
803	Boulder	98.6	82.0	91.6
804	Golden	100.7	80.8	92.4
805	Fort Collins	102.2	77.3	91.7
806	Greeley	99.7	67.0	86.0
807	Fort Morgan	99.2	80.6	91.4
808-809	Colorado Springs	101.5	81.0	92.9
810	Pueblo	101.3	79.0	91.9
811	Alamosa	102.4	75.3	91.0
812	Salida	102.2	76.2	91.3
813	Durango	102.8	76.0	91.6
814	Montrose	101.4	74.4	90.1
815	Grand Junction	104.9	72.0	91.1
816	Glenwood Springs	102.4	77.7	92.1
CONNECTICUT				
060	New Britain	99.2	118.0	107.1
061	Hartford	100.5	118.1	107.9
062	Willimantic	99.8	117.5	107.2
063	New London	95.8	117.1	104.8
064	Meriden	97.9	118.5	106.6
065	New Haven	101.0	118.0	108.2
066	Bridgeport	100.3	118.6	108.0
067	Waterbury	99.9	118.5	107.7
068	Norwalk	99.9	119.0	107.9
069	Stamford	100.0	125.4	110.7
D.C.				
200-205	Washington	104.6	92.8	99.7
DELAWARE				
197	Newark	100.5	105.8	102.7
198	Wilmington	100.6	105.9	102.8
199	Dover	101.2	105.8	103.1
FLORIDA				
320,322	Jacksonville	98.8	63.1	83.8
321	Daytona Beach	99.1	76.5	89.6
323	Tallahassee	99.6	53.0	80.1
324	Panama City	100.3	48.6	78.6
325	Pensacola	102.9	61.3	85.4
326,344	Gainesville	100.5	66.0	86.0
327-328,347	Orlando	103.1	72.0	90.0
329	Melbourne	102.0	80.0	92.7
330-332,340	Miami	101.2	75.4	90.3
333	Fort Lauderdale	98.5	73.6	88.0
334,349	West Palm Beach	97.2	69.8	85.7
335-336,346	Tampa	100.1	78.7	91.1
337	St. Petersburg	102.3	58.1	83.7
338	Lakeland	99.2	77.8	90.2
339,341	Fort Myers	98.6	70.1	86.6
342	Sarasota	100.4	71.1	88.1
GEORGIA				
300-303,399	Atlanta	98.2	79.2	90.2
304	Statesboro	97.9	48.1	77.0
305	Gainesville	96.5	63.2	82.5
306	Athens	95.9	66.9	83.7
307	Dalton	98.0	52.3	78.8
308-309	Augusta	96.9	64.3	83.2
310-312	Macon	96.5	66.6	83.9
313-314	Savannah	98.8	62.0	83.3
315	Waycross	97.7	57.2	80.7
316	Valdosta	97.8	51.0	78.2
317,398	Albany	97.9	62.0	82.9
318-319	Columbus	97.8	66.8	84.8

Location Factors

STATE/ZIP	CITY	MAT.	INST.	TOTAL
HAWAII				
967	Hilo	114.2	118.7	116.1
968	Honolulu	118.7	118.7	118.7
STATES & POSS.				
969	Guam	138.1	64.3	107.1
IDAHO				
832	Pocatello	101.6	74.5	90.2
833	Twin Falls	102.5	54.0	82.1
834	Idaho Falls	99.9	59.8	83.0
835	Lewiston	109.0	80.4	97.0
836-837	Boise	100.6	74.8	89.8
838	Coeur d'Alene	108.3	77.4	95.3
ILLINOIS				
600-603	North Suburban	97.4	123.7	108.5
604	Joliet	97.4	126.9	109.8
605	South Suburban	97.4	124.5	108.8
606-608	Chicago	97.9	138.3	114.9
609	Kankakee	93.6	107.0	99.2
610-611	Rockford	97.1	115.0	104.7
612	Rock Island	94.7	99.5	96.7
613	La Salle	95.9	106.3	100.3
614	Galesburg	95.7	99.0	97.1
615-616	Peoria	98.2	104.9	101.0
617	Bloomington	95.1	106.4	99.8
618-619	Champaign	98.4	105.0	101.2
620-622	East St. Louis	93.9	105.7	98.8
623	Quincy	94.9	94.8	94.9
624	Effingham	94.3	95.9	95.0
625	Decatur	96.2	102.5	98.8
626-627	Springfield	97.0	103.4	99.7
628	Centralia	92.4	100.5	95.8
629	Carbondale	92.1	94.1	92.9
INDIANA				
460	Anderson	94.8	82.5	89.6
461-462	Indianapolis	98.0	86.7	93.2
463-464	Gary	96.3	101.5	98.5
465-466	South Bend	95.6	83.1	90.4
467-468	Fort Wayne	95.3	78.4	88.2
469	Kokomo	92.5	80.8	87.6
470	Lawrenceburg	91.1	76.9	85.1
471	New Albany	92.4	74.4	84.9
472	Columbus	94.7	79.5	88.3
473	Muncie	95.5	80.6	89.3
474	Bloomington	96.9	78.8	89.3
475	Washington	92.7	82.4	88.4
476-477	Evansville	94.4	86.5	91.1
478	Terre Haute	95.0	86.6	91.5
479	Lafayette	94.3	82.1	89.2
IOWA				
500-503,509	Des Moines	98.2	76.7	89.2
504	Mason City	95.7	61.4	81.3
505	Fort Dodge	96.0	57.1	79.7
506-507	Waterloo	97.8	58.4	81.2
508	Creston	96.2	61.2	81.5
510-511	Sioux City	98.4	68.8	86.0
512	Sibley	96.9	47.6	76.2
513	Spencer	98.5	46.3	76.6
514	Carroll	95.6	51.4	77.0
515	Council Bluffs	99.6	74.6	89.1
516	Shenandoah	96.3	49.3	76.6
520	Dubuque	97.9	75.8	88.6
521	Decorah	96.5	48.8	76.5
522-524	Cedar Rapids	98.8	80.8	91.2
525	Ottumwa	96.6	69.3	85.1
526	Burlington	95.8	70.7	85.2
527-528	Davenport	97.7	90.2	94.5
KANSAS				
660-662	Kansas City	98.3	93.7	96.4
664-666	Topeka	98.9	66.1	85.1
667	Fort Scott	96.6	71.4	86.0
668	Emporia	96.7	59.7	81.2
669	Belleville	98.4	61.0	82.7
670-672	Wichita	97.6	64.9	83.9
673	Independence	98.1	63.3	83.5
674	Salina	98.7	61.6	83.1
675	Hutchinson	93.8	61.2	80.1
676	Hays	97.6	62.5	82.9
677	Colby	98.4	62.3	83.3
KANSAS (CONT'D)				
678	Dodge City	99.9	64.0	84.8
679	Liberal	97.5	62.2	82.7
KENTUCKY				
400-402	Louisville	97.1	83.9	91.6
403-405	Lexington	95.1	78.5	88.1
406	Frankfort	96.3	78.9	89.0
407-409	Corbin	92.3	65.8	81.2
410	Covington	92.8	99.3	95.5
411-412	Ashland	91.7	96.7	93.8
413-414	Campton	93.1	66.2	81.8
415-416	Pikeville	94.0	81.7	88.8
417-418	Hazard	92.4	58.7	78.3
420	Paducah	90.9	87.2	89.4
421-422	Bowling Green	93.6	84.6	89.8
423	Owensboro	93.4	83.2	89.1
424	Henderson	90.7	86.8	89.0
425-426	Somerset	90.6	70.3	82.1
427	Elizabethtown	90.1	83.6	87.3
LOUISIANA				
700-701	New Orleans	102.9	69.6	88.9
703	Thibodaux	100.4	63.7	85.0
704	Hammond	97.6	58.4	81.2
705	Lafayette	100.0	58.5	82.6
706	Lake Charles	100.1	61.1	83.7
707-708	Baton Rouge	102.2	63.8	86.1
710-711	Shreveport	96.8	59.2	81.0
712	Monroe	96.9	56.0	79.7
713-714	Alexandria	97.0	56.0	79.8
MAINE				
039	Kittery	93.0	74.4	85.2
040-041	Portland	98.5	73.7	88.1
042	Lewiston	96.4	73.7	86.9
043	Augusta	96.1	74.3	86.9
044	Bangor	95.8	73.7	86.5
045	Bath	94.2	74.4	85.9
046	Machias	93.7	73.9	85.4
047	Houlton	93.9	74.4	85.7
048	Rockland	92.9	74.5	85.2
049	Waterville	94.2	74.4	85.9
MARYLAND				
206	Waldorf	100.5	69.5	87.5
207-208	College Park	100.4	79.4	91.6
209	Silver Spring	99.7	75.5	89.5
210-212	Baltimore	98.2	86.0	93.1
214	Annapolis	98.7	80.2	90.9
215	Cumberland	94.0	80.4	88.3
216	Easton	95.5	42.8	73.4
217	Hagerstown	94.9	81.6	89.3
218	Salisbury	95.8	50.3	76.7
219	Elkton	93.1	62.9	80.4
MASSACHUSETTS				
010-011	Springfield	98.4	104.5	101.0
012	Pittsfield	97.8	101.0	99.1
013	Greenfield	95.7	102.3	98.4
014	Fitchburg	94.4	117.6	104.1
015-016	Worcester	98.3	118.9	107.0
017	Framingham	93.9	126.1	107.4
018	Lowell	97.7	125.7	109.5
019	Lawrence	98.9	123.6	109.3
020-022, 024	Boston	100.5	135.9	115.4
023	Brockton	99.2	120.7	108.2
025	Buzzards Bay	93.3	118.2	103.8
026	Hyannis	96.2	118.7	105.7
027	New Bedford	98.4	118.9	107.0
MICHIGAN				
480,483	Royal Oak	91.4	103.9	96.6
481	Ann Arbor	93.7	104.5	98.2
482	Detroit	95.3	114.2	103.2
484-485	Flint	93.4	100.8	96.5
486	Saginaw	93.0	89.9	91.7
487	Bay City	93.1	90.0	91.8
488-489	Lansing	94.4	96.9	95.5
490	Battle Creek	94.5	88.2	91.9
491	Kalamazoo	94.8	85.6	91.0
492	Jackson	92.7	91.6	92.3
493,495	Grand Rapids	96.1	65.2	83.1
494	Muskegon	93.5	81.4	88.4

STATE/ZIP	CITY	MAT.	INST.	TOTAL
MICHIGAN (CONT'D)				
496	Traverse City	92.3	69.1	82.5
497	Gaylord	93.3	72.1	84.4
498-499	Iron Mountain	95.3	83.2	90.2
MINNESOTA				
550-551	Saint Paul	98.6	118.7	107.0
553-555	Minneapolis	99.3	124.2	109.8
556-558	Duluth	98.2	108.2	102.4
MINNESOTA (CONT'd)				
559	Rochester	98.1	104.7	100.9
560	Mankato	96.0	103.2	99.0
561	Windom	94.8	77.6	87.6
562	Willmar	94.3	82.8	89.5
563	St. Cloud	95.8	117.5	104.9
564	Brainerd	95.9	99.5	97.4
565	Detroit Lakes	97.7	94.6	96.4
566	Bemidji	97.1	95.9	96.6
567	Thief River Falls	96.7	91.7	94.6
MISSISSIPPI				
386	Clarksdale	96.6	60.4	81.4
387	Greenville	100.0	70.9	87.8
388	Tupelo	97.9	63.1	83.2
389	Greenwood	97.8	59.9	81.9
390-392	Jackson	98.5	69.9	86.5
393	Meridian	96.8	70.9	85.9
394	Laurel	97.8	63.7	83.5
395	Biloxi	98.5	62.6	83.4
396	Mccomb	96.3	59.3	80.8
397	Columbus	97.7	61.0	82.3
MISSOURI				
630-631	St. Louis	98.6	108.8	102.9
633	Bowling Green	97.1	90.3	94.2
634	Hannibal	95.9	80.1	89.3
635	Kirksville	98.7	73.3	88.0
636	Flat River	98.1	91.1	95.1
637	Cape Girardeau	98.2	87.4	93.7
638	Sikeston	96.1	77.2	88.2
639	Poplar Bluff	95.6	77.7	88.1
640-641	Kansas City	99.2	106.1	102.1
644-645	St. Joseph	98.3	90.2	94.9
646	Chillicothe	95.1	68.5	83.9
647	Harrisonville	94.7	98.4	96.2
648	Joplin	97.1	68.0	84.9
650-651	Jefferson City	95.4	86.0	91.5
652	Columbia	96.3	88.1	92.9
653	Sedalia	95.4	81.7	89.6
654-655	Rolla	94.1	73.0	85.2
656-658	Springfield	97.3	77.6	89.1
MONTANA				
590-591	Billings	101.9	74.6	90.4
592	Wolf Point	101.1	71.5	88.7
593	Miles City	98.9	72.8	88.0
594	Great Falls	103.1	74.8	91.2
595	Havre	100.0	72.8	88.6
596	Helena	101.8	73.5	89.9
597	Butte	101.6	75.0	90.4
598	Missoula	99.0	73.2	88.2
599	Kalispell	97.9	72.0	87.0
NEBRASKA				
680-681	Omaha	99.5	78.8	90.8
683-685	Lincoln	97.6	77.1	89.0
686	Columbus	95.5	76.4	87.5
687	Norfolk	97.2	80.4	90.1
688	Grand Island	97.3	82.4	91.0
689	Hastings	96.5	84.8	91.6
690	Mccook	96.4	75.1	87.5
691	North Platte	96.8	85.0	91.8
692	Valentine	98.7	73.6	88.2
693	Alliance	98.6	71.6	87.3
NEVADA				
889-891	Las Vegas	101.1	112.0	105.7
893	Ely	99.5	72.0	88.0
894-895	Reno	99.8	93.9	97.3
897	Carson City	99.9	93.1	97.1
898	Elko	98.2	77.6	89.6
NEW HAMPSHIRE				
030	Nashua	98.6	87.7	94.0
031	Manchester	99.1	87.7	94.3

STATE/ZIP	CITY	MAT.	INST.	TOTAL
NEW HAMPSHIRE (CONT'D)				
032-033	Concord	97.0	83.8	91.5
034	Keene	95.1	53.4	77.6
035	Littleton	95.0	60.6	80.5
036	Charleston	94.6	50.8	76.2
037	Claremont	93.7	50.8	75.7
038	Portsmouth	95.9	90.1	93.5
NEW JERSEY				
070-071	Newark	100.5	122.0	109.5
072	Elizabeth	98.6	121.5	108.2
073	Jersey City	97.8	122.2	108.0
074-075	Paterson	99.3	121.8	108.8
076	Hackensack	97.3	122.3	107.8
077	Long Branch	97.1	121.2	107.2
078	Dover	97.6	122.0	107.8
079	Summit	97.6	121.5	107.6
080,083	Vineland	95.7	117.6	104.9
081	Camden	97.8	117.8	106.2
082,084	Atlantic City	96.3	117.6	105.2
085-086	Trenton	97.6	120.6	107.3
087	Point Pleasant	97.7	120.8	107.4
088-089	New Brunswick	98.1	121.4	107.9
NEW MEXICO				
870-872	Albuquerque	99.8	76.0	89.8
873	Gallup	99.9	76.0	89.9
874	Farmington	100.3	76.0	90.1
875	Santa Fe	102.0	76.0	91.1
877	Las Vegas	98.4	75.8	88.9
878	Socorro	97.9	76.0	88.7
879	Truth/Consequences	98.1	70.9	86.7
880	Las Cruces	97.0	68.5	85.0
881	Clovis	98.5	74.2	88.3
882	Roswell	100.3	74.3	89.3
883	Carrizozo	100.7	76.0	90.3
884	Tucumcari	99.3	74.2	88.8
NEW YORK				
100-102	New York	106.1	164.6	130.7
103	Staten Island	101.9	162.0	127.1
104	Bronx	100.0	162.0	126.0
105	Mount Vernon	99.8	132.8	113.6
106	White Plains	99.8	132.8	113.6
107	Yonkers	104.7	132.8	116.5
108	New Rochelle	100.1	132.8	113.8
109	Suffern	99.9	121.8	109.1
110	Queens	101.3	161.9	126.7
111	Long Island City	102.9	161.9	127.6
112	Brooklyn	103.1	161.9	127.8
113	Flushing	103.0	161.9	127.7
114	Jamaica	101.2	161.9	126.7
115,117,118	Hicksville	101.2	146.4	120.1
116	Far Rockaway	103.1	161.9	127.8
119	Riverhead	102.0	146.4	120.6
120-122	Albany	97.4	94.7	96.3
123	Schenectady	97.7	95.3	96.7
124	Kingston	100.8	112.5	105.7
125-126	Poughkeepsie	100.0	127.3	111.5
127	Monticello	99.3	115.7	106.2
128	Glens Falls	92.2	91.0	91.7
129	Plattsburgh	96.7	85.9	92.2
130-132	Syracuse	98.7	93.1	96.4
133-135	Utica	96.7	89.5	93.6
136	Watertown	98.4	92.2	95.8
137-139	Binghamton	98.1	86.9	93.4
140-142	Buffalo	99.0	104.8	101.5
143	Niagara Falls	96.7	101.5	98.7
144-146	Rochester	98.4	94.1	96.6
147	Jamestown	95.6	82.5	90.1
148-149	Elmira	95.5	84.4	90.8
NORTH CAROLINA				
270,272-274	Greensboro	99.8	50.7	79.2
271	Winston-Salem	99.5	50.0	78.7
275-276	Raleigh	100.7	51.3	80.0
277	Durham	101.1	51.7	80.4
278	Rocky Mount	96.9	41.6	73.7
279	Elizabeth City	97.7	43.0	74.7
280	Gastonia	98.5	49.6	78.0
281-282	Charlotte	100.2	52.2	80.0
283	Fayetteville	101.1	52.1	80.5
284	Wilmington	97.2	49.6	77.2
285	Kinston	95.2	43.3	73.4

Location Factors

STATE/ZIP	CITY	MAT.	INST.	TOTAL
NORTH CAROLINA (CONT'D)				
286	Hickory	95.5	45.9	74.7
287-288	Asheville	97.6	49.6	77.5
289	Murphy	96.3	35.7	70.9
NORTH DAKOTA				
580-581	Fargo	101.4	62.0	84.9
582	Grand Forks	100.8	55.1	81.6
583	Devils Lake	100.0	57.3	82.0
584	Jamestown	100.0	49.1	78.7
585	Bismarck	100.1	62.9	84.5
586	Dickinson	100.9	60.1	83.7
587	Minot	100.9	67.3	86.8
588	Williston	99.4	60.1	82.9
OHIO				
430-432	Columbus	96.3	88.6	93.0
433	Marion	92.6	83.5	88.8
434-436	Toledo	96.5	99.7	97.9
437-438	Zanesville	93.1	82.5	88.7
439	Steubenville	94.1	91.6	93.0
440	Lorain	96.9	93.6	95.5
441	Cleveland	97.2	102.6	99.5
442-443	Akron	97.8	93.7	96.1
444-445	Youngstown	97.2	89.7	94.1
446-447	Canton	97.3	85.0	92.1
448-449	Mansfield	94.4	88.3	91.9
450	Hamilton	95.0	85.8	91.1
451-452	Cincinnati	95.3	88.1	92.3
453-454	Dayton	95.2	85.5	91.1
455	Springfield	95.1	85.7	91.2
456	Chillicothe	93.8	91.4	92.8
457	Athens	96.7	75.9	88.0
458	Lima	97.0	85.3	92.1
OKLAHOMA				
730-731	Oklahoma City	98.7	61.7	83.1
734	Ardmore	95.2	61.1	80.9
735	Lawton	97.8	62.5	83.0
736	Clinton	96.6	59.7	81.1
737	Enid	97.4	59.7	81.6
738	Woodward	95.4	59.8	80.4
739	Guymon	96.4	32.0	69.4
740-741	Tulsa	97.6	56.3	80.3
743	Miami	94.1	64.5	81.7
744	Muskogee	96.9	41.6	73.7
745	Mcalester	93.8	52.4	76.5
746	Ponca City	94.4	60.4	80.1
747	Durant	94.4	60.4	80.2
748	Shawnee	95.9	57.8	79.9
749	Poteau	93.5	62.8	80.6
OREGON				
970-972	Portland	101.1	101.0	101.0
973	Salem	101.0	99.1	100.2
974	Eugene	100.7	98.4	99.7
975	Medford	102.3	97.0	100.1
976	Klamath Falls	102.2	96.7	99.9
977	Bend	101.0	98.5	100.0
978	Pendleton	96.1	99.0	97.3
979	Vale	93.8	90.3	92.3
PENNSYLVANIA				
150-152	Pittsburgh	96.9	98.7	97.6
153	Washington	93.8	98.4	95.7
154	Uniontown	94.1	95.5	94.7
155	Bedford	95.1	89.3	92.6
156	Greensburg	95.0	96.8	95.8
157	Indiana	93.9	95.6	94.6
158	Dubois	95.5	93.2	94.5
159	Johnstown	95.1	93.2	94.3
160	Butler	91.9	96.3	93.8
161	New Castle	91.9	94.7	93.1
162	Kittanning	92.4	98.6	95.0
163	Oil City	91.9	92.9	92.3
164-165	Erie	94.2	92.0	93.3
166	Altoona	94.3	89.4	92.2
167	Bradford	95.4	90.7	93.4
168	State College	94.9	90.1	92.9
169	Wellsboro	96.0	90.7	93.8
170-171	Harrisburg	97.9	93.1	95.9
172	Chambersburg	95.5	88.9	92.7
173-174	York	96.1	93.3	94.9
175-176	Lancaster	94.2	87.9	91.6

STATE/ZIP	CITY	MAT.	INST.	TOTAL
PENNSYLVANIA (CONT'D)				
177	Williamsport	92.9	81.1	87.9
178	Sunbury	95.0	92.2	93.8
179	Pottsville	94.1	92.3	93.4
180	Lehigh Valley	95.5	111.4	102.2
181	Allentown	97.7	106.5	101.4
182	Hazleton	94.9	92.4	93.9
183	Stroudsburg	94.8	98.8	96.5
184-185	Scranton	98.5	95.2	97.1
186-187	Wilkes-Barre	94.7	92.7	93.8
188	Montrose	94.4	93.7	94.1
189	Doylestown	94.4	118.6	104.5
190-191	Philadelphia	101.0	130.0	113.2
193	Westchester	97.2	119.9	106.8
194	Norristown	96.2	127.4	109.3
195-196	Reading	98.5	97.6	98.1
PUERTO RICO				
009	San Juan	120.1	23.9	79.7
RHODE ISLAND				
028	Newport	97.7	109.1	102.5
029	Providence	98.8	109.1	103.2
SOUTH CAROLINA				
290-292	Columbia	98.6	53.3	79.6
293	Spartanburg	96.9	52.9	78.4
294	Charleston	98.4	62.2	83.2
295	Florence	96.6	53.3	78.4
296	Greenville	96.7	52.9	78.3
297	Rock Hill	96.1	50.9	77.1
298	Aiken	97.1	70.0	85.7
299	Beaufort	97.9	46.2	76.2
SOUTH DAKOTA				
570-571	Sioux Falls	99.9	60.1	83.2
572	Watertown	98.1	55.3	80.1
573	Mitchell	97.1	55.2	79.5
574	Aberdeen	99.9	56.3	81.6
575	Pierre	99.5	56.5	81.4
576	Mobridge	97.7	55.0	79.8
577	Rapid City	99.7	56.5	81.5
TENNESSEE				
370-372	Nashville	98.5	73.6	88.0
373-374	Chattanooga	99.4	55.0	80.8
375,380-381	Memphis	97.5	70.1	86.0
376	Johnson City	98.4	55.4	80.3
377-379	Knoxville	95.8	55.8	79.0
382	McKenzie	97.1	55.2	79.5
383	Jackson	99.0	47.9	77.5
384	Columbia	95.6	57.0	79.4
385	Cookeville	97.0	58.1	80.6
TEXAS				
750	McKinney	98.7	51.5	78.8
751	Waxahackie	98.6	54.2	80.0
752-753	Dallas	99.4	66.2	85.4
754	Greenville	98.7	38.4	73.4
755	Texarkana	97.6	50.8	77.9
756	Longview	98.0	39.9	73.6
757	Tyler	98.8	54.2	80.1
758	Palestine	95.0	40.3	72.0
759	Lufkin	95.7	43.9	74.0
760-761	Fort Worth	98.0	60.7	82.4
762	Denton	97.6	48.1	76.8
763	Wichita Falls	98.5	55.3	80.3
764	Eastland	96.9	39.4	72.8
765	Temple	95.6	48.7	75.9
766-767	Waco	97.8	56.8	80.6
768	Brownwood	98.0	37.2	72.5
769	San Angelo	97.6	45.3	75.7
770-772	Houston	100.7	71.2	88.3
773	Huntsville	98.8	37.9	73.2
774	Wharton	100.2	42.2	75.8
775	Galveston	98.2	69.1	86.0
776-777	Beaumont	98.4	60.1	82.3
778	Bryan	95.4	62.6	81.6
779	Victoria	100.1	44.3	76.7
780	Laredo	95.8	51.7	77.3
781-782	San Antonio	96.2	65.4	83.3
783-784	Corpus Christi	98.4	50.4	78.3
785	McAllen	98.0	45.1	75.3
786-787	Austin	96.5	59.1	80.3

Location Factors

STATE/ZIP	CITY	MAT.	INST.	TOTAL
TEXAS (CONT'D)				
788	Del Rio	97.6	31.9	70.1
789	Giddings	94.9	39.5	71.7
790-791	Amarillo	98.6	56.8	81.0
792	Childress	97.6	49.6	77.4
793-794	Lubbock	100.1	51.1	79.5
795-796	Abilene	98.2	50.8	78.3
797	Midland	100.0	47.4	77.9
798-799,885	El Paso	97.5	49.8	77.5
UTAH				
840-841	Salt Lake City	102.9	67.9	88.2
842,844	Ogden	98.4	67.3	85.3
843	Logan	100.3	67.3	86.4
845	Price	100.7	46.5	78.0
846-847	Provo	100.8	67.8	87.0
VERMONT				
050	White River Jct.	96.3	56.6	79.7
051	Bellows Falls	94.8	64.0	81.9
052	Bennington	95.2	66.8	83.2
053	Brattleboro	95.6	69.0	84.4
054	Burlington	99.3	67.7	86.1
056	Montpelier	95.9	68.1	84.2
057	Rutland	97.5	68.1	85.2
058	St. Johnsbury	96.3	57.1	79.9
059	Guildhall	95.1	56.9	79.0
VIRGINIA				
220-221	Fairfax	100.2	82.2	92.6
222	Arlington	101.5	80.8	92.8
223	Alexandria	100.7	86.0	94.5
224-225	Fredericksburg	98.8	73.8	88.3
226	Winchester	99.4	66.5	85.6
227	Culpeper	99.3	72.0	87.9
228	Harrisonburg	99.6	68.2	86.4
229	Charlottesville	100.0	66.6	85.9
230-232	Richmond	101.7	70.1	88.4
233-235	Norfolk	101.9	70.5	88.7
236	Newport News	100.5	70.9	88.1
237	Portsmouth	99.8	67.4	86.2
238	Petersburg	99.9	70.1	87.4
239	Farmville	99.0	56.9	81.3
240-241	Roanoke	101.2	63.7	85.4
242	Bristol	98.7	57.4	81.4
243	Pulaski	98.4	54.7	80.1
244	Staunton	99.3	62.8	84.0
245	Lynchburg	99.4	68.1	86.2
246	Grundy	98.8	54.1	80.0
WASHINGTON				
980-981,987	Seattle	105.4	101.7	103.9
982	Everett	105.0	96.7	101.5
983-984	Tacoma	105.1	99.6	102.8
985	Olympia	103.1	99.5	101.6
986	Vancouver	106.5	93.8	101.1
988	Wenatchee	104.8	81.4	95.0
989	Yakima	105.3	88.0	98.0
990-992	Spokane	104.5	81.8	95.0
993	Richland	104.1	85.7	96.4
994	Clarkston	102.9	82.2	94.2
WEST VIRGINIA				
247-248	Bluefield	97.5	77.3	89.0
249	Lewisburg	99.3	82.0	92.0
250-253	Charleston	100.4	87.8	95.1
254	Martinsburg	98.9	76.5	89.5
255-257	Huntington	100.6	90.3	96.3
258-259	Beckley	97.3	86.0	92.6
260	Wheeling	100.6	89.2	95.8
261	Parkersburg	99.6	88.7	95.0
262	Buckhannon	98.9	88.9	94.7
263-264	Clarksburg	99.3	87.8	94.5
265	Morgantown	99.4	88.5	94.8
266	Gassaway	98.7	88.8	94.5
267	Romney	98.8	82.3	91.9
268	Petersburg	98.6	84.6	92.7
WISCONSIN				
530,532	Milwaukee	100.3	106.4	102.9
531	Kenosha	100.0	99.3	99.7
534	Racine	99.5	101.3	100.2
535	Beloit	99.3	94.6	97.3
537	Madison	99.5	95.0	97.6

STATE/ZIP	CITY	MAT.	INST.	TOTAL
WISCONSIN (CONT'D)				
538	Lancaster	96.9	90.3	94.1
539	Portage	95.6	94.3	95.0
540	New Richmond	95.5	93.6	94.7
541-543	Green Bay	99.7	90.4	95.8
544	Wausau	94.9	88.5	92.2
545	Rhinelander	98.0	88.7	94.1
546	La Crosse	96.3	91.3	94.2
547	Eau Claire	97.9	92.0	95.4
548	Superior	95.4	95.7	95.5
549	Oshkosh	95.5	89.9	93.2
WYOMING				
820	Cheyenne	100.0	65.9	85.7
821	Yellowstone Nat'l Park	97.5	58.6	81.2
822	Wheatland	98.6	58.6	81.8
823	Rawlins	100.2	58.4	82.6
824	Worland	98.1	56.5	80.6
825	Riverton	99.2	56.5	81.3
826	Casper	100.6	58.4	82.9
827	Newcastle	97.9	58.4	81.3
828	Sheridan	100.7	60.5	83.8
829-831	Rock Springs	102.0	56.5	82.9
CANADIAN FACTORS (reflect Canadian currency)				
ALBERTA				
	Calgary	130.7	89.9	113.6
	Edmonton	131.8	89.8	114.1
	Fort McMurray	127.7	92.6	112.9
	Lethbridge	121.9	92.0	109.3
	Lloydminster	116.4	88.8	104.8
	Medicine Hat	116.4	88.1	104.5
	Red Deer	116.9	88.1	104.8
BRITISH COLUMBIA				
	Kamloops	117.2	90.8	106.1
	Prince George	118.3	90.8	106.7
	Vancouver	131.2	82.0	110.5
	Victoria	118.3	78.8	101.7
MANITOBA				
	Brandon	116.4	76.1	99.5
	Portage la Prairie	116.5	74.7	98.9
	Winnipeg	129.6	69.6	104.4
NEW BRUNSWICK				
	Bathurst	114.7	67.8	95.0
	Dalhousie	114.6	67.8	95.0
	Fredericton	117.1	72.4	98.3
	Moncton	115.0	68.7	95.5
	Newcastle	114.7	67.8	95.0
	St. John	117.3	72.4	98.4
NEWFOUNDLAND				
	Corner Brook	119.7	68.2	98.0
	St. Johns	119.9	69.2	98.6
NORTHWEST TERRITORIES				
	Yellowknife	121.4	85.7	106.4
NOVA SCOTIA				
	Bridgewater	116.1	75.0	98.9
	Dartmouth	117.5	75.0	99.7
	Halifax	119.0	78.0	101.8
	New Glasgow	115.5	75.0	98.5
	Sydney	113.1	75.0	97.1
	Truro	115.5	75.0	98.5
	Yarmouth	115.4	75.0	98.5
ONTARIO				
	Barrie	119.2	93.0	108.2
	Brantford	118.4	96.9	109.4
	Cornwall	118.5	93.3	107.9
	Hamilton	123.4	96.9	112.3
	Kingston	119.4	93.4	108.5
	Kitchener	114.5	92.2	105.1
	London	122.0	93.3	110.0
	North Bay	118.4	91.2	107.0
	Oshawa	117.9	94.7	108.2
	Ottawa	122.7	93.9	110.6
	Owen Sound	119.4	91.2	107.6
	Peterborough	118.4	93.2	107.8
	Sarnia	117.9	97.5	109.3
ONTARIO (CONT'D)				

STATE/ZIP	CITY	MAT.	INST.	TOTAL
	Sault Ste Marie	112.9	92.6	104.4
	St. Catharines	112.3	94.2	104.7
	Sudbury	112.7	92.9	104.4
	Thunder Bay	113.5	92.9	104.9
	Timmins	118.6	91.2	107.1
	Toronto	123.6	99.8	113.6
	Windsor	113.0	93.2	104.7
PRINCE EDWARD ISLAND				
	Charlottetown	117.8	63.3	94.9
	Summerside	117.2	63.3	94.6
QUEBEC				
	Cap-de-la-Madeleine	115.3	87.8	103.8
	Charlesbourg	115.3	87.8	103.8
	Chicoutimi	114.0	93.5	105.4
	Gatineau	114.9	87.6	103.4
	Granby	115.1	87.6	103.5
	Hull	115.0	87.6	103.5
	Joliette	115.5	87.8	103.9
	Laval	114.5	87.6	103.2
	Montreal	121.7	93.8	110.0
	Quebec	120.1	94.2	109.2
	Rimouski	114.4	93.5	105.6
	Rouyn-Noranda	114.8	87.6	103.4
	Saint Hyacinthe	114.2	87.6	103.0
	Sherbrooke	115.2	87.6	103.6
	Sorel	115.4	87.8	103.8
	St. Jerome	114.8	87.6	103.4
	Trois Rivieres	115.4	87.8	103.8
SASKATCHEWAN				
	Moose Jaw	113.5	69.9	95.2
	Prince Albert	112.5	68.1	93.9
	Regina	115.1	69.9	96.1
	Saskatoon	113.6	69.8	95.2
YUKON				
	Whitehorse	113.2	68.9	94.6

R011105-05 Tips for Accurate Estimating

1. Use pre-printed or columnar forms for orderly sequence of dimensions and locations and for recording telephone quotations.
2. Use only the front side of each paper or form except for certain pre-printed summary forms.
3. Be consistent in listing dimensions: For example, length x width x height. This helps in rechecking to ensure that, the total length of partitions is appropriate for the building area.
4. Use printed (rather than measured) dimensions where given.
5. Add up multiple printed dimensions for a single entry where possible.
6. Measure all other dimensions carefully.
7. Use each set of dimensions to calculate multiple related quantities.
8. Convert foot and inch measurements to decimal feet when listing. Memorize decimal equivalents to .01 parts of a foot (1/8″ equals approximately .01′).
9. Do not "round off" quantities until the final summary.
10. Mark drawings with different colors as items are taken off.
11. Keep similar items together, different items separate.
12. Identify location and drawing numbers to aid in future checking for completeness.
13. Measure or list everything on the drawings or mentioned in the specifications.
14. It may be necessary to list items not called for to make the job complete.
15. Be alert for: Notes on plans such as N.T.S. (not to scale); changes in scale throughout the drawings; reduced size drawings; discrepancies between the specifications and the drawings.
16. Develop a consistent pattern of performing an estimate. For example:
    a. Start the quantity takeoff at the lower floor and move to the next higher floor.
    b. Proceed from the main section of the building to the wings.
    c. Proceed from south to north or vice versa, clockwise or counterclockwise.
    d. Take off floor plan quantities first, elevations next, then detail drawings.
17. List all gross dimensions that can be either used again for different quantities, or used as a rough check of other quantities for verification (exterior perimeter, gross floor area, individual floor areas, etc.).
18. Utilize design symmetry or repetition (repetitive floors, repetitive wings, symmetrical design around a center line, similar room layouts, etc.). Note: Extreme caution is needed here so as not to omit or duplicate an area.
19. Do not convert units until the final total is obtained. For instance, when estimating concrete work, keep all units to the nearest cubic foot, then summarize and convert to cubic yards.
20. When figuring alternatives, it is best to total all items involved in the basic system, then total all items involved in the alternates. Therefore you work with positive numbers in all cases. When adds and deducts are used, it is often confusing whether to add or subtract a portion of an item; especially on a complicated or involved alternate.

R011105-10 Unit Gross Area Requirements

The figures in the table below indicate typical ranges in square feet as a function of the "occupant" unit. This table is best used in the preliminary design stages to help determine the probable size requirement for the total project.

Building Type	Unit	Gross Area in S.F.		
		1/4	Median	3/4
Apartments	Unit	660	860	1,100
Auditorium & Play Theaters	Seat	18	25	38
Bowling Alleys	Lane		940	
Churches & Synagogues	Seat	20	28	39
Dormitories	Bed	200	230	275
Fraternity & Sorority Houses	Bed	220	315	370
Garages, Parking	Car	325	355	385
Hospitals	Bed	685	850	1,075
Hotels	Rental Unit	475	600	710
Housing for the elderly	Unit	515	635	755
Housing, Public	Unit	700	875	1,030
Ice Skating Rinks	Total	27,000	30,000	36,000
Motels	Rental Unit	360	465	620
Nursing Homes	Bed	290	350	450
Restaurants	Seat	23	29	39
Schools, Elementary	Pupil	65	77	90
Junior High & Middle	↓	85	110	129
Senior High	↓	102	130	145
Vocational	↓	110	135	195
Shooting Ranges	Point		450	
Theaters & Movies	Seat		15	

R011105-20 Floor Area Ratios

Table below lists commonly used gross to net area and net to gross area ratios expressed in % for various building types.

Building Type	Gross to Net Ratio	Net to Gross Ratio	Building Type	Gross to Net Ratio	Net to Gross Ratio
Apartment	156	64	School Buildings (campus type)		
Bank	140	72	Administrative	150	67
Church	142	70	Auditorium	142	70
Courthouse	162	61	Biology	161	62
Department Store	123	81	Chemistry	170	59
Garage	118	85	Classroom	152	66
Hospital	183	55	Dining Hall	138	72
Hotel	158	63	Dormitory	154	65
Laboratory	171	58	Engineering	164	61
Library	132	76	Fraternity	160	63
Office	135	75	Gymnasium	142	70
Restaurant	141	70	Science	167	60
Warehouse	108	93	Service	120	83
			Student Union	172	59

The gross area of a building is the total floor area based on outside dimensions.

The net area of a building is the usable floor area for the function intended and excludes such items as stairways, corridors and mechanical rooms. In the case of a commercial building, it might be considered as the "leasable area."

R011105-30 Occupancy Determinations

Description		S.F. Required per Person		
		BOCA	SBC	UBC
Assembly Areas	Fixed Seats	**	6	7
	Movable Seats		15	15
	Concentrated	7		
	Unconcentrated	15		
	Standing Space	3		
Educational	Unclassified			
	Classrooms	20	40	20
	Shop Areas	50	100	50
Institutional	Unclassified		125	
	In-Patient Areas	240		
	Sleeping Areas	120		
Mercantile	Basement	30	30	20
	Ground Floor	30	30	30
	Upper Floors	60	60	50
Office		100	100	100

BOCA=Building Officials & Code Administrators
SBC=Southern Building Code
UBC=Uniform Building Code

** The occupancy load for assembly area with fixed seats shall be determined by the number of fixed seats installed.

R011110-10 Architectural Fees

Tabulated below are typical percentage fees by project size, for good professional architectural service. Fees may vary from those listed depending upon degree of design difficulty and economic conditions in any particular area.

Rates can be interpolated horizontally and vertically. Various portions of the same project requiring different rates should be adjusted proportionately. For alterations, add 50% to the fee for the first $500,000 of project cost and add 25% to the fee for project cost over $500,000.

Architectural fees tabulated below include Structural, Mechanical and Electrical Engineering Fees. They do not include the fees for special consultants such as kitchen planning, security, acoustical, interior design, etc.

Civil Engineering fees are included in the Architectural fee for project sites requiring minimal design such as city sites. However, separate Civil Engineering fees must be added when utility connections require design, drainage calculations are needed, stepped foundations are required, or provisions are required to protect adjacent wetlands.

Building Types	Total Project Size in Thousands of Dollars						
	100	250	500	1,000	5,000	10,000	50,000
Factories, garages, warehouses, repetitive housing	9.0%	8.0%	7.0%	6.2%	5.3%	4.9%	4.5%
Apartments, banks, schools, libraries, offices, municipal buildings	12.2	12.3	9.2	8.0	7.0	6.6	6.2
Churches, hospitals, homes, laboratories, museums, research	15.0	13.6	12.7	11.9	9.5	8.8	8.0
Memorials, monumental work, decorative furnishings	—	16.0	14.5	13.1	10.0	9.0	8.3

R011110-30 Engineering Fees

Typical **Structural Engineering Fees** based on type of construction and total project size. These fees are included in Architectural Fees.

Type of Construction	Total Project Size (in thousands of dollars)			
	$500	$500-$1,000	$1,000-$5,000	Over $5000
Industrial buildings, factories & warehouses	Technical payroll times 2.0 to 2.5	1.60%	1.25%	1.00%
Hotels, apartments, offices, dormitories, hospitals, public buildings, food stores	↓	2.00%	1.70%	1.20%
Museums, banks, churches and cathedrals	↓	2.00%	1.75%	1.25%
Thin shells, prestressed concrete, earthquake resistive	↓	2.00%	1.75%	1.50%
Parking ramps, auditoriums, stadiums, convention halls, hangars & boiler houses	↓	2.50%	2.00%	1.75%
Special buildings, major alterations, underpinning & future expansion	↓	Add to above 0.5%	Add to above 0.5%	Add to above 0.5%

For complex reinforced concrete or unusually complicated structures, add 20% to 50%.

Typical **Mechanical and Electrical Engineering Fees** are based on the size of the subcontract. The fee structure for both are shown below. These fees are included in Architectural Fees.

Type of Construction	Subcontract Size							
	$25,000	$50,000	$100,000	$225,000	$350,000	$500,000	$750,000	$1,000,000
Simple structures	6.4%	5.7%	4.8%	4.5%	4.4%	4.3%	4.2%	4.1%
Intermediate structures	8.0	7.3	6.5	5.6	5.1	5.0	4.9	4.8
Complex structures	10.1	9.0	9.0	8.0	7.5	7.5	7.0	7.0

For renovations, add 15% to 25% to applicable fee.

R012153-10 Repair and Remodeling

Cost figures are based on new construction utilizing the most cost-effective combination of labor, equipment and material with the work scheduled in proper sequence to allow the various trades to accomplish their work in an efficient manner.

The costs for repair and remodeling work must be modified due to the following factors that may be present in any given repair and remodeling project.

1. Equipment usage curtailment due to the physical limitations of the project, with only hand-operated equipment being used.
2. Increased requirement for shoring and bracing to hold up the building while structural changes are being made and to allow for temporary storage of construction materials on above-grade floors.
3. Material handling becomes more costly due to having to move within the confines of an enclosed building. For multi-story construction, low capacity elevators and stairwells may be the only access to the upper floors.
4. Large amount of cutting and patching and attempting to match the existing construction is required. It is often more economical to remove entire walls rather than create many new door and window openings. This sort of trade-off has to be carefully analyzed.
5. Cost of protection of completed work is increased since the usual sequence of construction usually cannot be accomplished.
6. Economies of scale usually associated with new construction may not be present. If small quantities of components must be custom fabricated due to job requirements, unit costs will naturally increase. Also, if only small work areas are available at a given time, job scheduling between trades becomes difficult and subcontractor quotations may reflect the excessive start-up and shut-down phases of the job.
7. Work may have to be done on other than normal shifts and may have to be done around an existing production facility which has to stay in production during the course of the repair and remodeling.
8. Dust and noise protection of adjoining non-construction areas can involve substantial special protection and alter usual construction methods.
9. Job may be delayed due to unexpected conditions discovered during demolition or removal. These delays ultimately increase construction costs.
10. Piping and ductwork runs may not be as simple as for new construction. Wiring may have to be snaked through walls and floors.
11. Matching "existing construction" may be impossible because materials may no longer be manufactured. Substitutions may be expensive.
12. Weather protection of existing structure requires additional temporary structures to protect building at openings.
13. On small projects, because of local conditions, it may be necessary to pay a tradesman for a minimum of four hours for a task that is completed in one hour.

All of the above areas can contribute to increased costs for a repair and remodeling project. Each of the above factors should be considered in the planning, bidding and construction stage in order to minimize the increased costs associated with repair and remodeling jobs.

R012909-80 Sales Tax by State

State sales tax on materials is tabulated below (5 states have no sales tax). Many states allow local jurisdictions, such as a county or city, to levy additional sales tax.

Some projects may be sales tax exempt, particularly those constructed with public funds.

State	Tax (%)	State	Tax (%)	State	Tax (%)	State	Tax (%)
Alabama	4	Illinois	6.25	Montana	0	Rhode Island	7
Alaska	0	Indiana	6	Nebraska	5.5	South Carolina	6
Arizona	5.6	Iowa	5	Nevada	6.5	South Dakota	4
Arkansas	6	Kansas	5.3	New Hampshire	0	Tennessee	7
California	7.25	Kentucky	6	New Jersey	7	Texas	6.25
Colorado	2.9	Louisiana	4	New Mexico	5	Utah	4.65
Connecticut	6	Maine	5	New York	4	Vermont	6
Delaware	0	Maryland	6	North Carolina	4.25	Virginia	5
District of Columbia	5.75	Massachusetts	5	North Dakota	5	Washington	6.5
Florida	6	Michigan	6	Ohio	5.5	West Virginia	6
Georgia	4	Minnesota	6.5	Oklahoma	4.5	Wisconsin	5
Hawaii	4	Mississippi	7	Oregon	0	Wyoming	4
Idaho	6	Missouri	4.225	Pennsylvania	6	Average	4.91 %

Sales Tax by Province (Canada)

GST - a value-added tax, which the government imposes on most goods and services provided in or imported into Canada. PST - a retail sales tax, which five of the provinces impose on the price of most goods and some services. QST - a value-added tax, similar to the federal GST, which Quebec imposes. HST - Three provinces have combined their retail sales tax with the federal GST into one harmonized tax.

Province	PST (%)	QST (%)	GST(%)	HST(%)
Alberta	0	0	6	0
British Columbia	7	0	6	0
Manitoba	7	0	6	0
New Brunswick	0	0	0	13
Newfoundland	0	0	0	13
Northwest Territories	0	0	6	0
Nova Scotia	0	0	0	13
Ontario	8	0	6	0
Prince Edward Island	10	0	6	0
Quebec	0	7.5	6	0
Saskatchewan	5	0	6	0
Yukon	0	0	6	0

R012909-85 Unemployment Taxes and Social Security Taxes

State Unemployment Tax rates vary not only from state to state, but also with the experience rating of the contractor. The Federal Unemployment Tax rate is 6.2% of the first $7,000 of wages. This is reduced by a credit of up to 5.4% for timely payment to the state. The minimum Federal Unemployment Tax is 0.8% after all credits.

Social Security (FICA) for 2009 is estimated at time of publication to be 7.65% of wages up to $102,000.

R012909-90 Overtime

One way to improve the completion date of a project or eliminate negative float from a schedule is to compress activity duration times. This can be achieved by increasing the crew size or working overtime with the proposed crew.

To determine the costs of working overtime to compress activity duration times, consider the following examples. Below is an overtime efficiency and cost chart based on a five, six, or seven day week with an eight through twelve hour day. Payroll percentage increases for time and one half and double time are shown for the various working days.

Days per Week	Hours per Day	Production Efficiency					Payroll Cost Factors	
		1st Week	2nd Week	3rd Week	4th Week	Average 4 Weeks	@ 1-1/2 Times	@ 2 Times
5	8	100%	100%	100%	100%	100 %	100 %	100 %
	9	100	100	95	90	96.25	105.6	111.1
	10	100	95	90	85	91.25	110.0	120.0
	11	95	90	75	65	81.25	113.6	127.3
	12	90	85	70	60	76.25	116.7	133.3
6	8	100	100	95	90	96.25	108.3	116.7
	9	100	95	90	85	92.50	113.0	125.9
	10	95	90	85	80	87.50	116.7	133.3
	11	95	85	70	65	78.75	119.7	139.4
	12	90	80	65	60	73.75	122.2	144.4
7	8	100	95	85	75	88.75	114.3	128.6
	9	95	90	80	70	83.75	118.3	136.5
	10	90	85	75	65	78.75	121.4	142.9
	11	85	80	65	60	72.50	124.0	148.1
	12	85	75	60	55	68.75	126.2	152.4

R013113-40 Builder's Risk Insurance

Builder's Risk Insurance is insurance on a building during construction. Premiums are paid by the owner or the contractor. Blasting, collapse and underground insurance would raise total insurance costs above those listed. Floater policy for materials delivered to the job runs $.75 to $1.25 per $100 value. Contractor equipment insurance runs $.50 to $1.50 per $100 value. Insurance for miscellaneous tools to $1,500 value runs from $3.00 to $7.50 per $100 value.

Tabulated below are New England Builder's Risk insurance rates in dollars per $100 value for $1,000 deductible. For $25,000 deductible, rates can be reduced 13% to 34%. On contracts over $1,000,000, rates may be lower than those tabulated. Policies are written annually for the total completed value in place. For "all risk" insurance (excluding flood, earthquake and certain other perils) add $.025 to total rates below.

Coverage	Frame Construction (Class 1)		Brick Construction (Class 4)		Fire Resistive (Class 6)	
	Range	Average	Range	Average	Range	Average
Fire Insurance	$.350 to $ .850	$.600	$.158 to $.189	$.174	$.052 to $.080	$.070
Extended Coverage	.115 to .200	.158	.080 to .105	.101	.081 to .105	.100
Vandalism	.012 to .016	.014	.008 to .011	.011	.008 to .011	.010
Total Annual Rate	$.477 to $1.066	$.772	$.246 to $.305	$.286	$.141 to $.196	$.180

R013113-50 General Contractor's Overhead

There are two distinct types of overhead on a construction project: Project Overhead and Main Office Overhead. Project Overhead includes those costs at a construction site not directly associated with the installation of construction materials. Examples of Project Overhead costs include the following:

1. Superintendent
2. Construction office and storage trailers
3. Temporary sanitary facilities
4. Temporary utilities
5. Security fencing
6. Photographs
7. Clean up
8. Performance and payment bonds

The above Project Overhead items are also referred to as General Requirements and therefore are estimated in Division 1. Division 1 is the first division listed in the CSI MasterFormat but it is usually the last division estimated. The sum of the costs in Divisions 1 through 49 is referred to as the sum of the direct costs.

All construction projects also include indirect costs. The primary components of indirect costs are the contractor's Main Office Overhead and profit. The amount of the Main Office Overhead expense varies depending on the following:

1. Owner's compensation
2. Project managers and estimator's wages
3. Clerical support wages
4. Office rent and utilities
5. Corporate legal and accounting costs
6. Advertising
7. Automobile expenses
8. Association dues
9. Travel and entertainment expenses

These costs are usually calculated as a percentage of annual sales volume. This percentage can range from 35% for a small contractor doing less than $500,000 to 5% for a large contractor with sales in excess of $100 million.

R013113-60 Workers' Compensation Insurance Rates by Trade

The table below tabulates the national averages for Workers' Compensation insurance rates by trade and type of building. The average "Insurance Rate" is multiplied by the "% of Building Cost" for each trade. This produces the "Workers' Compensation Cost" by % of total labor cost, to be added for each trade by building type to determine the weighted average Workers' Compensation rate for the building types analyzed.

Trade	Insurance Rate (% Labor Cost)		% of Building Cost			Workers' Compensation		
	Range	Average	Office Bldgs.	Schools & Apts.	Mfg.	Office Bldgs.	Schools & Apts.	Mfg.
Excavation, Grading, etc.	4.2 % to 17.5%	10.0%	4.8%	4.9%	4.5%	0.48%	0.49%	0.45%
Piles & Foundations	5.9 to 40.3	20.1	7.1	5.2	8.7	1.43	1.05	1.75
Concrete	5.1 to 26.8	14.6	5.0	14.8	3.7	0.73	2.16	0.54
Masonry	4.8 to 43.3	14.4	6.9	7.5	1.9	0.99	1.08	0.27
Structural Steel	5.9 to 104.1	37.9	10.7	3.9	17.6	4.06	1.48	6.67
Miscellaneous & Ornamental Metals	3.4 to 22.8	10.7	2.8	4.0	3.6	0.30	0.43	0.39
Carpentry & Millwork	5.9 to 53.2	17.8	3.7	4.0	0.5	0.66	0.71	0.09
Metal or Composition Siding	5.9 to 38	16.6	2.3	0.3	4.3	0.38	0.05	0.71
Roofing	5.9 to 77.1	31.2	2.3	2.6	3.1	0.72	0.81	0.97
Doors & Hardware	4.9 to 32	11.6	0.9	1.4	0.4	0.10	0.16	0.05
Sash & Glazing	5.9 to 32.4	13.9	3.5	4.0	1.0	0.49	0.56	0.14
Lath & Plaster	3.3 to 43.9	13.6	3.3	6.9	0.8	0.45	0.94	0.11
Tile, Marble & Floors	3.1 to 17.9	9.1	2.6	3.0	0.5	0.24	0.27	0.05
Acoustical Ceilings	2.6 to 45.6	10.6	2.4	0.2	0.3	0.25	0.02	0.03
Painting	4.7 to 29.6	12.5	1.5	1.6	1.6	0.19	0.20	0.20
Interior Partitions	5.9 to 53.2	17.8	3.9	4.3	4.4	0.69	0.77	0.78
Miscellaneous Items	2.1 to 168.2	16.0	5.2	3.7	9.7	0.83	0.59	1.55
Elevators	2.8 to 11.8	6.6	2.1	1.1	2.2	0.14	0.07	0.15
Sprinklers	2.5 to 14.3	7.8	0.5	—	2.0	0.04	—	0.16
Plumbing	2.9 to 12.4	7.8	4.9	7.2	5.2	0.38	0.56	0.41
Heat., Vent., Air Conditioning	4.3 to 23	11.6	13.5	11.0	12.9	1.57	1.28	1.50
Electrical	2.8 to 11.5	6.5	10.1	8.4	11.1	0.66	0.55	0.72
Total	2.1 % to 168.2%	—	100.0%	100.0%	100.0%	15.78%	14.23%	17.69%
		Overall Weighted Average	15.90%					

Workers' Compensation Insurance Rates by States

The table below lists the weighted average Workers' Compensation base rate for each state with a factor comparing this with the national average of 15.5%.

State	Weighted Average	Factor	State	Weighted Average	Factor	State	Weighted Average	Factor
Alabama	24.2%	156	Kentucky	18.4%	119	North Dakota	13.7%	88
Alaska	21.7	140	Louisiana	28.3	183	Ohio	14.8	95
Arizona	9.5	61	Maine	15.4	99	Oklahoma	14.6	94
Arkansas	13.2	85	Maryland	16.7	108	Oregon	13.5	87
California	19.6	126	Massachusetts	12.5	81	Pennsylvania	13.8	89
Colorado	13.2	85	Michigan	17.3	112	Rhode Island	21.2	137
Connecticut	21.0	135	Minnesota	25.7	166	South Carolina	18.3	118
Delaware	15.1	97	Mississippi	19.1	123	South Dakota	19.1	123
District of Columbia	13.7	88	Missouri	17.2	111	Tennessee	15.1	97
Florida	13.8	89	Montana	16.6	107	Texas	12.2	79
Georgia	26.2	169	Nebraska	23.0	148	Utah	12.3	79
Hawaii	17.0	110	Nevada	11.6	75	Vermont	18.0	116
Idaho	10.8	70	New Hampshire	20.0	129	Virginia	11.7	75
Illinois	21.6	139	New Jersey	13.4	86	Washington	9.8	63
Indiana	5.9	38	New Mexico	18.5	119	West Virginia	9.2	59
Iowa	11.5	74	New York	12.6	81	Wisconsin	15.0	97
Kansas	8.9	57	North Carolina	18.9	122	Wyoming	6.5	42
				Weighted Average for U.S. is	15.9% of payroll = 100%			

Rates in the following table are the base or manual costs per $100 of payroll for Workers' Compensation in each state. Rates are usually applied to straight time wages only and not to premium time wages and bonuses.

The weighted average skilled worker rate for 35 trades is 15.5%. For bidding purposes, apply the full value of Workers' Compensation directly to total labor costs, or if labor is 38%, materials 42% and overhead and profit 20% of total cost, carry 38/80 x 15.5% =7.4% of cost (before overhead and profit) into overhead. Rates vary not only from state to state but also with the experience rating of the contractor.

Rates are the most current available at the time of publication.

R013113-060 Workers' Compensation Insurance Rates by Trade and State (cont.)

State	Carpentry — 3 stories or less 5651	Carpentry — interior cab. work 5437	Carpentry — general 5403	Concrete Work — NOC 5213	Concrete Work — flat (flr., sdwk.) 5221	Electrical Wiring — inside 5190	Excavation — earth NOC 6217	Excavation — rock 6217	Glaziers 5462	Insulation Work 5479	Lathing 5443	Masonry 5022	Painting & Decorating 5474	Pile Driving 6003	Plastering 5480	Plumbing 5183	Roofing 5551	Sheet Metal Work (HVAC) 5538	Steel Erection — door & sash 5102	Steel Erection — inter., ornam. 5102	Steel Erection — structure 5040	Steel Erection — NOC 5057	Tile Work — (interior ceramic) 5348	Waterproofing 9014	Wrecking 5701
AL	37.98	22.82	32.46	16.20	11.27	10.68	11.16	11.16	21.26	16.28	11.93	27.00	25.81	32.65	18.85	12.27	58.14	22.96	11.01	11.01	55.68	25.37	13.31	6.56	55.68
AK	14.73	12.27	17.44	11.07	9.71	8.72	12.14	12.14	32.36	27.78	9.74	43.27	21.95	32.40	43.85	9.97	35.12	8.61	9.39	9.39	43.01	33.00	6.76	6.61	43.01
AZ	11.60	6.45	15.40	8.56	4.45	4.59	4.90	4.90	6.93	12.48	6.61	6.51	6.33	11.45	5.78	5.24	13.38	7.38	10.66	10.66	21.10	15.34	3.17	2.59	21.10
AR	14.82	7.45	16.26	12.17	6.48	5.02	7.76	7.76	9.82	16.49	5.93	9.93	11.50	16.48	16.10	5.15	23.09	13.09	6.77	6.77	32.38	25.50	6.08	3.74	32.38
CA	31.95	31.95	31.95	11.89	11.89	7.98	13.79	13.79	15.74	12.50	11.24	16.21	17.08	15.27	20.53	12.03	44.43	15.60	12.46	12.46	17.17	20.88	8.54	17.08	20.88
CO	13.57	8.92	12.33	11.57	7.87	5.28	10.34	10.34	9.75	16.84	6.02	14.01	9.02	14.31	9.58	7.43	26.09	11.61	9.31	9.31	32.62	17.58	7.06	5.03	17.58
CT	15.67	14.66	28.76	24.79	10.09	8.01	12.91	12.91	22.98	15.94	21.94	22.24	17.61	22.88	14.48	11.16	41.95	15.61	18.57	18.57	42.72	23.45	11.58	5.41	42.72
DE	15.17	15.17	11.46	11.51	9.49	5.89	9.32	9.32	13.11	11.46	13.11	12.77	26.77	18.75	13.11	7.87	26.71	9.78	12.28	12.28	26.77	12.28	9.29	12.77	26.77
DC	11.01	12.78	10.19	9.66	9.14	6.01	12.54	12.54	17.02	8.18	9.40	11.97	7.51	12.11	12.48	10.38	19.21	8.93	10.61	10.61	37.21	16.49	17.86	4.00	37.21
FL	13.05	10.47	14.39	15.31	6.97	6.64	8.30	8.30	9.91	9.24	5.98	11.32	9.73	38.24	18.70	6.75	22.19	12.68	8.85	8.85	27.89	14.56	6.55	5.17	27.89
GA	35.68	20.54	25.67	16.85	13.26	10.82	17.38	17.38	18.77	23.64	15.09	21.06	27.77	27.36	22.83	11.4	54.71	20.76	19.93	19.93	64.78	42.21	12.42	8.50	64.78
HI	17.38	11.62	28.20	13.74	12.27	7.10	7.73	7.73	20.55	21.19	10.94	17.87	11.33	20.15	15.61	6.06	33.24	8.02	11.11	11.11	31.71	21.70	9.78	11.54	31.71
ID	9.16	6.04	10.35	11.25	6.13	3.70	5.72	5.72	8.70	7.67	8.89	7.59	8.45	12.13	13.37	5.36	30.27	8.89	5.72	5.72	29.16	10.28	9.79	4.75	29.16
IL	24.93	14.46	20.78	26.59	12.10	9.62	10.06	10.06	19.18	17.48	15.59	19.27	11.62	34.85	15.88	11.07	31.06	14.96	16.05	16.05	72.39	25.69	14.02	4.84	72.39
IN	9.65	4.92	7.29	5.10	3.22	2.84	4.87	4.87	6.28	7.44	2.62	4.84	4.72	6.44	3.72	2.91	10.88	4.45	3.43	3.43	12.12	7.04	3.09	2.47	12.12
IA	10.43	10.02	10.34	12.12	7.00	4.96	6.27	6.27	9.42	7.06	5.58	9.61	6.65	9.78	7.33	6.46	22.58	7.52	6.15	6.15	30.21	35.21	7.42	3.87	35.21
KS	10.05	7.85	9.89	7.19	6.42	3.37	4.41	4.41	9.22	7.45	4.68	6.91	7.57	11.31	6.25	4.85	15.06	7.14	4.82	4.82	26.19	13.21	6.35	3.46	13.21
KY	18.72	14.07	25.30	13.39	8.00	5.43	9.15	9.15	17.37	17.59	11.24	9.12	13.18	27.72	14.41	7.06	38.25	19.58	12.32	12.32	54.03	22.56	13.67	5.74	56.16
LA	22.02	24.93	53.17	26.43	15.61	9.88	17.49	17.49	20.14	21.43	24.51	28.33	29.58	31.25	22.37	8.64	77.12	21.20	18.98	18.98	51.76	24.21	13.77	13.78	66.41
ME	13.79	10.40	28.55	20.94	8.74	6.31	9.05	9.05	18.14	11.82	7.90	14.76	12.73	16.70	11.79	10.56	22.87	11.61	9.73	9.73	31.67	23.11	7.00	7.12	31.67
MD	15.05	11.06	14.11	15.67	6.67	9.24	9.20	9.20	20.03	15.02	8.83	11.43	7.48	20.09	15.02	6.11	34.74	13.54	9.51	9.51	59.03	28.60	7.60	5.36	28.60
MA	6.80	5.60	11.46	19.51	6.57	3.20	4.19	4.19	8.96	10.06	6.35	10.81	4.79	14.68	5.08	3.98	32.80	5.15	7.61	7.61	45.71	36.69	6.21	2.09	25.88
MI	18.40	11.39	20.64	19.45	8.95	4.72	10.20	10.20	12.41	14.75	13.59	15.54	13.64	40.27	15.18	6.84	30.67	10.00	10.16	10.16	40.27	22.21	11.11	4.87	40.27
MN	19.53	19.80	41.48	14.65	15.73	7.53	14.95	14.95	15.68	11.90	22.05	19.18	16.95	24.63	22.05	10.88	70.18	14.23	10.58	10.58	104.13	33.30	14.55	6.68	36.03
MS	21.07	11.11	22.29	15.49	9.26	6.28	11.91	11.91	13.07	14.85	7.91	14.09	13.14	35.34	31.72	9.80	44.11	16.91	14.23	14.23	36.11	25.03	9.84	4.18	36.11
MO	29.99	11.16	13.98	18.91	10.39	7.64	9.17	9.17	11.13	15.17	7.93	15.01	12.53	19.12	17.02	10.05	29.98	13.19	11.31	11.31	34.56	35.84	11.18	6.51	34.56
MT	17.12	11.15	22.57	12.37	12.79	6.39	15.54	15.54	10.67	32.03	11.06	13.75	9.61	25.18	10.02	8.75	37.61	11.53	8.70	8.70	27.45	15.73	7.53	7.46	15.73
NE	24.65	17.47	21.80	25.07	15.13	10.38	16.90	16.90	23.70	31.50	12.20	22.22	20.07	22.50	21.67	12.35	35.22	19.42	16.05	16.05	45.40	35.27	11.10	6.38	47.05
NV	15.25	7.14	11.58	10.76	8.23	7.17	9.77	9.77	10.19	8.12	5.27	7.50	7.97	13.33	7.30	5.98	14.08	18.10	8.67	8.67	25.10	21.54	5.39	5.55	21.54
NH	25.92	13.36	18.90	26.79	14.17	6.94	14.52	14.52	13.07	21.36	9.63	24.34	7.77	18.99	11.71	10.13	49.23	13.59	15.27	15.27	52.24	18.83	11.37	6.09	52.24
NJ	14.34	8.94	14.34	16.01	9.99	4.76	8.35	8.35	8.55	12.65	11.64	13.09	11.92	16.03	11.64	6.17	37.77	6.49	9.46	9.46	23.72	13.59	6.99	5.72	22.31
NM	18.91	7.15	20.11	17.36	10.05	7.51	11.20	11.20	21.26	13.72	8.39	15.34	11.64	21.86	13.44	8.13	36.37	13.47	13.69	13.69	55.78	38.81	6.59	6.29	55.78
NY	12.72	6.45	12.67	16.02	11.64	6.03	8.29	8.29	10.97	7.98	12.10	16.02	9.89	12.77	9.03	6.88	27.98	10.46	8.84	8.84	23.45	12.72	6.66	6.04	8.73
NC	18.76	14.56	15.36	18.15	6.88	11.50	11.70	11.70	15.21	16.00	15.28	12.51	12.90	17.50	15.68	9.16	29.70	15.90	10.44	10.44	79.66	26.21	9.28	5.55	79.66
ND	11.17	11.17	11.17	6.42	6.42	3.63	6.64	6.64	11.17	11.17	7.51	7.71	6.61	22.78	7.51	5.22	23.53	5.22	22.78	22.78	22.78	22.78	11.17	23.53	10.98
OH	11.04	8.52	11.68	12.33	9.99	6.14	8.70	8.70	7.86	16.94	45.64	12.71	12.37	19.06	3.28	6.99	31.63	9.71	7.59	7.59	29.23	16.59	9.60	6.42	16.59
OK	13.42	9.17	11.76	11.76	6.34	5.73	10.83	10.83	16.07	19.47	8.45	10.02	8.47	20.25	12.19	7.09	21.13	9.05	13.76	13.76	39.72	23.97	6.50	5.71	39.72
OR	15.43	8.10	15.69	14.22	8.97	5.14	9.33	9.33	15.22	11.80	7.60	14.07	10.63	16.17	12.73	5.50	26.30	9.41	7.93	7.93	26.94	18.78	11.04	4.54	26.94
PA	13.18	13.18	11.38	14.21	10.46	5.95	8.27	8.27	11.47	11.38	11.47	12.22	13.31	15.78	11.47	7.33	27.42	7.84	14.30	14.30	21.31	14.30	8.02	12.22	21.37
RI	19.53	11.65	18.07	18.23	16.24	4.43	10.38	10.38	12.85	22.78	11.97	25.11	24.13	37.66	17.25	8.52	33.92	10.29	14.07	14.07	59.49	37.50	14.36	7.70	59.49
SC	22.54	15.41	20.91	14.80	7.92	10.92	13.16	13.16	16.12	14.69	11.32	12.88	16.52	19.48	16.02	11.47	45.24	12.95	12.99	12.99	32.57	26.39	9.62	6.68	47.05
SD	27.77	8.13	29.58	25.56	9.55	5.52	11.01	11.01	12.16	12.05	7.70	11.31	9.53	21.32	12.44	11.75	20.73	12.48	8.98	8.98	80.86	40.98	8.74	4.98	80.86
TN	15.68	13.11	16.06	13.89	7.97	7.35	13.68	13.68	13.09	11.80	7.30	15.60	12.63	24.89	16.02	7.66	22.25	12.77	7.17	7.17	29.18	23.47	10.15	4.65	29.18
TX	11.25	8.89	11.25	9.00	7.26	6.15	8.54	8.54	9.65	12.87	6.36	10.91	8.98	22.43	8.98	6.75	19.59	12.84	9.09	9.09	29.68	13.37	7.07	6.69	9.96
UT	15.74	8.88	12.97	9.81	8.47	4.57	11.59	11.59	14.70	12.49	6.35	11.25	8.37	12.47	7.26	6.18	24.99	12.64	7.16	7.16	25.67	14.67	6.80	4.27	18.37
VT	15.70	10.75	17.37	17.27	7.21	6.98	12.59	12.59	19.75	17.78	8.33	15.82	11.62	16.74	10.78	10.44	36.43	11.59	12.51	12.51	44.95	42.80	11.24	7.28	44.95
VA	11.05	8.02	9.16	11.05	5.31	5.52	7.57	7.57	10.25	9.90	14.34	9.62	8.99	10.75	10.22	5.83	22.52	7.65	8.37	8.37	33.52	19.66	4.74	3.04	33.52
WA	8.48	8.48	8.48	7.32	7.32	2.81	6.69	6.69	12.27	9.77	8.43	9.09	10.93	16.12	9.95	4.17	16.36	4.27	7.60	7.60	7.60	7.60	8.45	16.36	7.60
WV	10.65	7.30	10.23	8.24	4.41	4.70	6.22	6.22	7.89	7.25	6.11	7.98	7.90	11.90	7.93	4.90	18.05	5.74	5.58	5.58	24.17	13.20	5.65	2.66	17.34
WI	9.99	11.82	14.45	11.15	9.69	6.02	8.19	8.19	10.44	13.58	6.25	15.31	13.03	16.47	9.96	6.24	37.36	7.21	9.36	9.36	28.23	44.93	13.80	4.86	28.23
WY	5.87	5.87	5.87	5.87	5.87	5.87	5.87	5.87	5.87	5.87	5.87	5.87	5.87	5.87	5.87	5.87	5.87	5.87	5.87	5.87	5.87	5.87	5.87	5.87	5.87
AVG.	16.63	11.62	17.80	14.58	9.14	6.46	10.01	10.01	13.89	14.44	10.63	14.37	12.49	20.09	13.60	7.84	31.18	11.57	10.74	10.74	37.94	23.15	9.13	6.69	34.13

R013113-060 Workers' Compensation (cont.) (Canada in Canadian dollars)

Province		Alberta	British Columbia	Manitoba	Ontario	New Brunswick	Newfndld. & Labrador	Northwest Territories	Nova Scotia	Prince Edward Island	Quebec	Saskat-chewan	Yukon
Carpentry—3 stories or less	Rate	7.32	4.74	4.36	4.35	3.79	10.08	3.84	7.82	5.96	15.05	6.23	8.89
	Code	42143	721028	40102	723	4226	4226	4-41	4226	401	80110	B1317	202
Carpentry—interior cab. work	Rate	2.18	4.90	4.36	4.35	4.40	4.69	3.84	4.52	3.71	15.05	3.51	8.89
	Code	42133	721021	40102	723	4279	4270	4-41	4274	402	80110	B11-27	202
CARPENTRY—general	Rate	7.32	4.74	4.36	4.35	3.79	4.69	3.84	7.82	5.96	15.05	6.23	8.89
	Code	42143	721028	40102	723	4226	4299	4-41	4226	401	80110	B1317	202
CONCRETE WORK—NOC	Rate	4.41	4.81	7.46	16.02	3.79	10.08	3.84	4.52	5.96	17.99	6.24	4.67
	Code	42104	721010	40110	748	4224	4224	4-41	4224	401	80100	B13-14	203
CONCRETE WORK—flat (flr. sidewalk)	Rate	4.41	4.81	7.46	16.02	3.79	10.08	3.84	4.52	5.96	17.99	6.24	4.67
	Code	42104	721010	40110	748	4224	4224	4-41	4224	401	80100	B13-14	203
ELECTRICAL Wiring—inside	Rate	2.13	1.77	2.51	3.79	2.17	3.01	3.52	2.31	3.71	6.12	3.51	4.67
	Code	42124	721019	40203	704	4261	4261	4-46	4261	402	80170	B11-05	206
EXCAVATION—earth NOC	Rate	2.51	3.56	3.76	4.55	2.62	4.37	3.64	3.34	3.82	7.13	3.93	4.67
	Code	40604	721031	40706	711	4214	4214	4-43	4214	404	80030	R11-06	207
EXCAVATION—rock	Rate	2.51	3.56	3.76	4.55	2.62	4.37	3.64	3.34	3.82	7.13	3.93	4.67
	Code	40604	721031	40706	711	4214	4214	4-43	4214	404	80030	R11-06	207
GLAZIERS	Rate	3.16	3.62	4.36	8.90	4.55	6.81	3.84	5.21	3.71	17.08	6.23	4.67
	Code	42121	715020	40109	751	4233	4233	4-41	4233	402	80150	B13-04	212
INSULATION WORK	Rate	2.57	8.42	4.36	8.90	4.55	6.81	3.84	5.21	5.96	15.05	5.98	8.89
	Code	42184	721029	40102	751	4234	4234	4-41	4234	401	80110	B12-07	202
LATHING	Rate	5.59	6.78	4.36	4.35	4.40	4.69	3.84	4.52	3.71	15.05	6.23	8.89
	Code	42135	721033	40102	723	4273	4279	4-41	4271	402	80110	B13-16	202
MASONRY	Rate	4.41	6.78	4.36	11.15	4.55	6.81	3.84	5.21	5.96	17.99	6.23	8.89
	Code	42102	721037	40102	741	4231	4231	4-41	4231	401	80100	B13-18	202
PAINTING & DECORATING	Rate	4.35	4.94	3.18	6.75	4.40	4.69	3.84	4.52	3.71	15.05	5.98	8.89
	Code	42111	721041	40105	719	4275	4275	4-41	4275	402	80110	B12-01	202
PILE DRIVING	Rate	4.41	5.18	3.76	6.34	3.79	9.78	3.64	4.64	5.96	7.13	6.23	8.89
	Code	42159	722004	40706	732	4221	4221	4-43	4221	401	80030	B13-10	202
PLASTERING	Rate	5.59	6.78	5.08	6.75	4.40	4.69	3.84	4.52	3.71	15.05	5.98	8.89
	Code	42135	721042	40108	719	4271	4271	4-41	4271	402	80110	B12-21	202
PLUMBING	Rate	2.13	3.11	3.05	4.02	2.32	3.37	3.52	2.43	3.71	6.93	3.51	4.67
	Code	42122	721043	40204	707	4241	4241	4-46	4241	402	80160	B11-01	214
ROOFING	Rate	8.12	8.68	7.02	12.98	4.55	10.08	3.84	10.57	5.96	23.01	6.23	8.89
	Code	42118	721036	40403	728	4236	4236	4-41	4236	401	80130	B13-20	202
SHEET METAL WORK (HVAC)	Rate	2.13	3.11	7.02	4.02	2.32	3.37	3.52	2.43	3.71	6.93	3.51	4.67
	Code	42117	721043	40402	707	4244	4244	4-46	4244	402	80160	B11-07	208
STEEL ERECTION—door & sash	Rate	2.57	14.07	11.32	16.02	3.79	10.08	3.84	5.21	5.96	25.32	6.23	8.89
	Code	42106	722005	40502	748	4227	4227	4-41	4227	401	80080	B13-22	202
STEEL ERECTION—inter., ornam.	Rate	2.57	14.07	11.32	16.02	3.79	10.08	3.84	5.21	5.96	25.32	6.23	8.89
	Code	42106	722005	40502	748	4227	4227	4-41	4227	401	80080	B13-22	202
STEEL ERECTION—structure	Rate	2.57	14.07	11.32	16.02	3.79	10.08	3.84	5.21	5.96	25.32	6.23	8.89
	Code	42106	722005	40502	748	4227	4227	4-41	4227	401	80080	B13-22	202
STEEL ERECTION—NOC	Rate	2.57	14.07	11.32	16.02	3.79	10.08	3.84	5.21	5.96	25.32	6.23	8.89
	Code	42106	722005	40502	748	4227	4227	4-41	4227	401	80080	B13-22	202
TILE WORK—inter. (ceramic)	Rate	3.63	5.62	1.97	6.75	4.40	10.08	3.84	4.52	3.71	15.05	6.23	8.89
	Code	42113	721054	40103	719	4276	4276	4-41	4276	402	80110	B13-01	202
WATERPROOFING	Rate	4.35	4.93	4.36	4.35	4.55	4.69	3.84	5.21	3.71	23.01	5.98	8.89
	Code	42139	721016	40102	723	4239	4299	4-41	4239	402	80130	B12-17	202
WRECKING	Rate	2.51	4.59	6.59	16.02	2.62	4.39	3.64	3.34	5.96	15.05	6.23	8.89
	Code	40604	721005	40106	748	4211	4211	4-43	4211	401	80110	B13-09	202

General Requirements — R0131 Project Management & Coordination

R013113-80 Performance Bond

This table shows the cost of a Performance Bond for a construction job scheduled to be completed in 12 months. Add 1% of the premium cost per month for jobs requiring more than 12 months to complete. The rates are "standard" rates offered to contractors that the bonding company considers financially sound and capable of doing the work. Preferred rates are offered by some bonding companies based upon financial strength of the contractor. Actual rates vary from contractor to contractor and from bonding company to bonding company. Contractors should prequalify through a bonding agency before submitting a bid on a contract that requires a bond.

Contract Amount	Building Construction Class B Projects	Highways & Bridges: Class A New Construction	Highways & Bridges: Class A-1 Highway Resurfacing
First $ 100,000 bid	$25.00 per M	$15.00 per M	$9.40 per M
Next 400,000 bid	$ 2,500 plus $15.00 per M	$ 1,500 plus $10.00 per M	$ 940 plus $7.20 per M
Next 2,000,000 bid	8,500 plus 10.00 per M	5,500 plus 7.00 per M	3,820 plus 5.00 per M
Next 2,500,000 bid	28,500 plus 7.50 per M	19,500 plus 5.50 per M	15,820 plus 4.50 per M
Next 2,500,000 bid	47,250 plus 7.00 per M	33,250 plus 5.00 per M	28,320 plus 4.50 per M
Over 7,500,000 bid	64,750 plus 6.00 per M	45,750 plus 4.50 per M	39,570 plus 4.00 per M

General Requirements — R0154 Construction Aids

R015423-10 Steel Tubular Scaffolding

On new construction, tubular scaffolding is efficient up to 60′ high or five stories. Above this it is usually better to use a hung scaffolding if construction permits. Swing scaffolding operations may interfere with tenants. In this case, the tubular is more practical at all heights.

In repairing or cleaning the front of an existing building the cost of tubular scaffolding per S.F. of building front increases as the height increases above the first tier. The first tier cost is relatively high due to leveling and alignment.

The minimum efficient crew for erecting and dismantling is three workers. They can set up and remove 18 frame sections per day up to 5 stories high. For 6 to 12 stories high, a crew of four is most efficient. Use two or more on top and two on the bottom for handing up or hoisting. They can also set up and remove 18 frame sections per day. At 7' horizontal spacing, this will run about 800 S.F. per day of erecting and dismantling. Time for placing and removing planks must be added to the above. A crew of three can place and remove 72 planks per day up to 5 stories. For over 5 stories, a crew of four can place and remove 80 planks per day.

The table below shows the number of pieces required to erect tubular steel scaffolding for 1000 S.F. of building frontage. This area is made up of a scaffolding system that is 12 frames (11 bays) long by 2 frames high.

For jobs under twenty-five frames, add 50% to rental cost. Rental rates will be lower for jobs over three months duration. Large quantities for long periods can reduce rental rates by 20%.

Description of Component	Number of Pieces for 1000 S.F. of Building Front	Unit
5′ Wide Standard Frame, 6′-4″ High	24	Ea.
Leveling Jack & Plate	24	↓
Cross Brace	44	↓
Side Arm Bracket, 21″	12	↓
Guardrail Post	12	↓
Guardrail, 7′ section	22	↓
Stairway Section	2	↓
Stairway Starter Bar	1	↓
Stairway Inside Handrail	2	↓
Stairway Outside Handrail	2	↓
Walk-Thru Frame Guardrail	2	↓

Scaffolding is often used as falsework over 15′ high during construction of cast-in-place concrete beams and slabs. Two foot wide scaffolding is generally used for heavy beam construction. The span between frames depends upon the load to be carried with a maximum span of 5′.

Heavy duty shoring frames with a capacity of 10,000#/leg can be spaced up to 10′ O.C. depending upon form support design and loading.

Scaffolding used as horizontal shoring requires less than half the material required with conventional shoring.

On new construction, erection is done by carpenters.

Rolling towers supporting horizontal shores can reduce labor and speed the job. For maintenance work, catwalks with spans up to 70′ can be supported by the rolling towers.

R015433-10 Contractor Equipment

Rental Rates shown elsewhere in the book pertain to late model high quality machines in excellent working condition, rented from equipment dealers. Rental rates from contractors may be substantially lower than the rental rates from equipment dealers depending upon economic conditions; for older, less productive machines, reduce rates by a maximum of 15%. Any overtime must be added to the base rates. For shift work, rates are lower. Usual rule of thumb is 150% of one shift rate for two shifts; 200% for three shifts.

For periods of less than one week, operated equipment is usually more economical to rent than renting bare equipment and hiring an operator.

Costs to move equipment to a job site (mobilization) or from a job site (demobilization) are not included in rental rates, nor in any Equipment costs on any Unit Price line items or crew listings. These costs can be found elsewhere. If a piece of equipment is already at a job site, it is not appropriate to utilize mob/demob costs in an estimate again.

Rental rates vary throughout the country with larger cities generally having lower rates. Lease plans for new equipment are available for periods in excess of six months with a percentage of payments applying toward purchase.

Monthly rental rates vary from 2% to 5% of the cost of the equipment depending on the anticipated life of the equipment and its wearing parts. Weekly rates are about 1/3 the monthly rates and daily rental rates about 1/3 the weekly rate.

The hourly operating costs for each piece of equipment include costs to the user such as fuel, oil, lubrication, normal expendables for the equipment, and a percentage of mechanic's wages chargeable to maintenance. The hourly operating costs listed do not include the operator's wages.

The daily cost for equipment used in the standard crews is figured by dividing the weekly rate by five, then adding eight times the hourly operating cost to give the total daily equipment cost, not including the operator. This figure is in the right hand column of the Equipment listings under Equipment Cost/Day.

Pile Driving rates shown for pile hammer and extractor do not include leads, crane, boiler or compressor. Vibratory pile driving requires an added field specialist during set-up and pile driving operation for the electric model. The hydraulic model requires a field specialist for set-up only. Up to 125 reuses of sheet piling are possible using vibratory drivers. For normal conditions, crane capacity for hammer type and size are as follows.

Crane Capacity	Hammer Type and Size		
	Air or Steam	Diesel	Vibratory
25 ton	to 8,750 ft.-lb.		70 H.P.
40 ton	15,000 ft.-lb.	to 32,000 ft.-lb.	170 H.P.
60 ton	25,000 ft.-lb.		300 H.P.
100 ton		112,000 ft.-lb.	

Cranes should be specified for the job by size, building and site characteristics, availability, performance characteristics, and duration of time required.

Backhoes & Shovels rent for about the same as equivalent size cranes but maintenance and operating expense is higher. Crane operators rate must be adjusted for high boom heights. Average adjustments: for 150′ boom add 2% per hour; over 185′, add 4% per hour; over 210′, add 6% per hour; over 250′, add 8% per hour and over 295′, add 12% per hour.

Tower Cranes of the climbing or static type have jibs from 50′ to 200′ and capacities at maximum reach range from 4,000 to 14,000 pounds. Lifting capacities increase up to maximum load as the hook radius decreases.

Typical rental rates, based on purchase price are about 2% to 3% per month.

Erection and dismantling runs between 500 and 2000 labor hours. Climbing operation takes 10 labor hours per 20′ climb. Crane dead time is about 5 hours per 40′ climb. If crane is bolted to side of the building add cost of ties and extra mast sections. Climbing cranes have from 80′ to 180′ of mast while static cranes have 80′ to 800′ of mast.

Truck Cranes can be converted to tower cranes by using tower attachments. Mast heights over 400′ have been used.

A single 100′ high material **Hoist and Tower** can be erected and dismantled in about 400 labor hours; a double 100′ high hoist and tower in about 600 labor hours. Erection times for additional heights are 3 and 4 labor hours per vertical foot respectively up to 150′, and 4 to 5 labor hours per vertical foot over 150′ high. A 40′ high portable Buck hoist takes about 160 labor hours to erect and dismantle. Additional heights take 2 labor hours per vertical foot to 80′ and 3 labor hours per vertical foot for the next 100′. Most material hoists do not meet local code requirements for carrying personnel.

A 150′ high **Personnel Hoist** requires about 500 to 800 labor hours to erect and dismantle. Budget erection time at 5 labor hours per vertical foot for all trades. Local code requirements or labor scarcity requiring overtime can add up to 50% to any of the above erection costs.

Earthmoving Equipment: The selection of earthmoving equipment depends upon the type and quantity of material, moisture content, haul distance, haul road, time available, and equipment available. Short haul cut and fill operations may require dozers only, while another operation may require excavators, a fleet of trucks, and spreading and compaction equipment. Stockpiled material and granular material are easily excavated with front end loaders. Scrapers are most economically used with hauls between 300′ and 1-1/2 miles if adequate haul roads can be maintained. Shovels are often used for blasted rock and any material where a vertical face of 8′ or more can be excavated. Special conditions may dictate the use of draglines, clamshells, or backhoes. Spreading and compaction equipment must be matched to the soil characteristics, the compaction required and the rate the fill is being supplied.

Existing Conditions — R0241 Demolition

R024119-10 Demolition Defined

Whole Building Demolition - Demolition of the whole building with no concern for any particular building element, component, or material type being demolished. This type of demolition is accomplished with large pieces of construction equipment that break up the structure, load it into trucks and haul it to a disposal site, but disposal or dump fees are not included. Demolition of below-grade foundation elements, such as footings, foundation walls, grade beams, slabs on grade, etc., is not included. Certain mechanical equipment containing flammable liquids or ozone-depleting refrigerants, electric lighting elements, communication equipment components, and other building elements may contain hazardous waste, and must be removed, either selectively or carefully, as hazardous waste before the building can be demolished.

Foundation Demolition - Demolition of below-grade foundation footings, foundation walls, grade beams, and slabs on grade. This type of demolition is accomplished by hand or pneumatic hand tools, and does not include saw cutting, or handling, loading, hauling, or disposal of the debris.

Gutting - Removal of building interior finishes and electrical/mechanical systems down to the load-bearing and sub-floor elements of the rough building frame, with no concern for any particular building element, component, or material type being demolished. This type of demolition is accomplished by hand or pneumatic hand tools, and includes loading into trucks, but not hauling, disposal or dump fees, scaffolding, or shoring. Certain mechanical equipment containing flammable liquids or ozone-depleting refrigerants, electric lighting elements, communication equipment components, and other building elements may contain hazardous waste, and must be removed, either selectively or carefully, as hazardous waste, before the building is gutted.

Selective Demolition - Demolition of a selected building element, component, or finish, with some concern for surrounding or adjacent elements, components, or finishes (see the first Subdivision (s) at the beginning of appropriate Divisions). This type of demolition is accomplished by hand or pneumatic hand tools, and does not include handling, loading, storing, hauling, or disposal of the debris, scaffolding, or shoring. "Gutting" methods may be used in order to save time, but damage that is caused to surrounding or adjacent elements, components, or finishes may have to be repaired at a later time.

Careful Removal - Removal of a piece of service equipment, building element or component, or material type, with great concern for both the removed item and surrounding or adjacent elements, components or finishes. The purpose of careful removal may be to protect the removed item for later re-use, preserve a higher salvage value of the removed item, or replace an item while taking care to protect surrounding or adjacent elements, components, connections, or finishes from cosmetic and/or structural damage. An approximation of the time required to perform this type of removal is 1/3 to 1/2 the time it would take to install a new item of like kind (see Reference Number R220105-10). This type of removal is accomplished by hand or pneumatic hand tools, and does not include loading, hauling, or storing the removed item, scaffolding, shoring, or lifting equipment.

Cutout Demolition - Demolition of a small quantity of floor, wall, roof, or other assembly, with concern for the appearance and structural integrity of the surrounding materials. This type of demolition is accomplished by hand or pneumatic hand tools, and does not include saw cutting, handling, loading, hauling, or disposal of debris, scaffolding, or shoring.

Rubbish Handling - Work activities that involve handling, loading or hauling of debris. Generally, the cost of rubbish handling must be added to the cost of all types of demolition, with the exception of whole building demolition.

Minor Site Demolition - Demolition of site elements outside the footprint of a building. This type of demolition is accomplished by hand or pneumatic hand tools, or with larger pieces of construction equipment, and may include loading a removed item onto a truck (check the Crew for equipment used). It does not include saw cutting, hauling or disposal of debris, and, sometimes, handling or loading.

R024119-20 Dumpsters

Dumpster rental costs on construction sites are presented in two ways.

The cost per week rental includes the delivery of the dumpster; its pulling or emptying once per week, and its final removal. The assumption is made that the dumpster contractor could choose to empty a dumpster by simply bringing in an empty unit and removing the full one. These costs also include the disposal of the materials in the dumpster.

The Alternate Pricing can be used when actual planned conditions are not approximated by the weekly numbers. For example, these lines can be used when a dumpster is needed for 4 weeks and will need to be emptied 2 or 3 times per week. Conversely the Alternate Pricing lines can be used when a dumpster will be rented for several weeks or months but needs to be emptied only a few times over this period.

Existing Conditions — R0265 Underground Storage Tank Removal

R026510-20 Underground Storage Tank Removal

Underground Storage Tank Removal can be divided into two categories: Non-Leaking and Leaking. Prior to removing an underground storage tank, tests should be made, with the proper authorities present, to determine whether a tank has been leaking or the surrounding soil has been contaminated.

To safely remove Liquid Underground Storage Tanks:

1. Excavate to the top of the tank.
2. Disconnect all piping.
3. Open all tank vents and access ports.
4. Remove all liquids and/or sludge.
5. Purge the tank with an inert gas.
6. Provide access to the inside of the tank and clean out the interior using proper personal protective equipment (PPE).
7. Excavate soil surrounding the tank using proper PPE for on-site personnel.
8. Pull and properly dispose of the tank.
9. Clean up the site of all contaminated material.
10. Install new tanks or close the excavation.

Existing Conditions — R0282 Asbestos Remediation

R028213-20 Asbestos Removal Process

Asbestos removal is accomplished by a specialty contractor who understands the federal and state regulations regarding the handling and disposal of the material. The process of asbestos removal is divided into many individual steps. An accurate estimate can be calculated only after all the steps have been priced.

The steps are generally as follows:

1. Obtain an asbestos abatement plan from an industrial hygienist.
2. Monitor the air quality in and around the removal area and along the path of travel between the removal area and transport area. This establishes the background contamination.
3. Construct a two part decontamination chamber at entrance to removal area.
4. Install a HEPA filter to create a negative pressure in the removal area.
5. Install wall, floor and ceiling protection as required by the plan, usually 2 layers of fireproof 6 mil polyethylene.
6. Industrial hygienist visually inspects work area to verify compliance with plan.
7. Provide temporary supports for conduit and piping affected by the removal process.
8. Proceed with asbestos removal and bagging process. Monitor air quality as described in Step #2. Discontinue operations when contaminate levels exceed applicable standards.
9. Document the legal disposal of materials in accordance with EPA standards.
10. Thoroughly clean removal area including all ledges, crevices and surfaces.
11. Post abatement inspection by industrial hygienist to verify plan compliance.
12. Provide a certificate from a licensed industrial hygienist attesting that contaminate levels are within acceptable standards before returning area to regular use.

Thermal & Moist. Protec. — R0784 Firestopping

R078413-30 Firestopping

Firestopping is the sealing of structural, mechanical, electrical and other penetrations through fire-rated assemblies. The basic components of firestop systems are safing insulation and firestop sealant on both sides of wall penetrations and the top side of floor penetrations.

Pipe penetrations are assumed to be through concrete, grout, or joint compound and can be sleeved or unsleeved. Costs for the penetrations and sleeves are not included. An annular space of 1″ is assumed. Escutcheons are not included.

Metallic pipe is assumed to be copper, aluminum, cast iron or similar metallic material. Insulated metallic pipe is assumed to be covered with a thermal insulating jacket of varying thickness and materials.

Non-metallic pipe is assumed to be PVC, CPVC, FR Polypropylene or similar plastic piping material. Intumescent firestop sealant or wrap strips are included. Collars on both sides of wall penetrations and a sheet metal plate on the underside of floor penetrations are included.

Ductwork is assumed to be sheet metal, stainless steel or similar metallic material. Duct penetrations are assumed to be through concrete, grout or joint compound. Costs for penetrations and sleeves are not included. An annular space of 1/2″ is assumed.

Multi-trade openings include costs for sheet metal forms, firestop mortar, wrap strips, collars and sealants as necessary.

Structural penetrations joints are assumed to be 1/2″ or less. CMU walls are assumed to be within 1-1/2″ of metal deck. Drywall walls are assumed to be tight to the underside of metal decking.

Metal panel, glass or curtain wall systems include a spandrel area of 5' filled with mineral wool foil-faced insulation. Fasteners and stiffeners are included.

Finishes — R0991 Painting

R099100-10 Painting Estimating Techniques

Proper estimating methodology is needed to obtain an accurate painting estimate. There is no known reliable shortcut or square foot method. The following steps should be followed:

- List all surfaces to be painted, with an accurate quantity (area) of each. Items having similar surface condition, finish, application method and accessibility may be grouped together.
- List all the tasks required for each surface to be painted, including surface preparation, masking, and protection of adjacent surfaces. Surface preparation may include minor repairs, washing, sanding and puttying.
- Select the proper Means line for each task. Review and consider all adjustments to labor and materials for type of paint and location of work. Apply the height adjustment carefully. For instance, when applying the adjustment for work over 8' high to a wall that is 12' high, apply the adjustment only to the area between 8' and 12' high, and not to the entire wall.

When applying more than one percent (%) adjustment, apply each to the base cost of the data, rather than applying one percentage adjustment on top of the other.

When estimating the cost of painting walls and ceilings remember to add the brushwork for all cut-ins at inside corners and around windows and doors as a LF measure. One linear foot of cut-in with brush equals one square foot of painting.

All items for spray painting include the labor for roll-back.

Deduct for openings greater than 100 SF, or openings that extend from floor to ceiling and are greater than 5' wide. Do not deduct small openings.

The cost of brushes, rollers, ladders and spray equipment are considered to be part of a painting contractor's overhead, and should not be added to the estimate. The cost of rented equipment such as scaffolding and swing staging should be added to the estimate.

Plumbing — R2201 Operation & Maintenance of Plumbing

R220105-10 Demolition (Selective vs. Removal for Replacement)

Demolition can be divided into two basic categories.

One type of demolition involves the removal of material with no concern for its replacement. The labor-hours to estimate this work are found under "Selective Demolition" in the Fire Protection, Plumbing and HVAC Divisions. It is selective in that individual items or all the material installed as a system or trade grouping such as plumbing or heating systems are removed. This may be accomplished by the easiest way possible, such as sawing, torch cutting, or sledge hammer as well as simple unbolting.

The second type of demolition is the removal of some item for repair or replacement. This removal may involve careful draining, opening of unions, disconnecting and tagging of electrical connections, capping of pipes/ducts to prevent entry of debris or leakage of the material contained as well as transport of the item away from its in-place location to a truck/dumpster. An approximation of the time required to accomplish this type of demolition is to use half of the time indicated as necessary to install a new unit. For example; installation of a new pump might be listed as requiring 6 labor-hours so if we had to estimate the removal of the old pump we would allow an additional 3 hours for a total of 9 hours. That is, the complete replacement of a defective pump with a new pump would be estimated to take 9 labor-hours.

Plumbing — R2205 Common Work Results for Plumbing

R220523-80 Valve Materials

VALVE MATERIALS

Bronze:
Bronze is one of the oldest materials used to make valves. It is most commonly used in hot and cold water systems and other non-corrosive services. It is often used as a seating surface in larger iron body valves to ensure tight closure.

Carbon Steel:
Carbon steel is a high strength material. Therefore, valves made from this metal are used in higher pressure services, such as steam lines up to 600 psi at 850° F. Many steel valves are available with butt-weld ends for economy and are generally used in high pressure steam service as well as other higher pressure non-corrosive services.

Forged Steel:
Valves from tough carbon steel are used in service up to 2000 psi and temperatures up to 1000° F in Gate, Globe and Check valves.

Iron:
Valves are normally used in medium to large pipe lines to control non-corrosive fluid and gases, where pressures do not exceed 250 psi at 450° F or 500 psi cold water, oil or gas.

Stainless Steel:
Developed steel alloys can be used in over 90% corrosive services.

Plastic PVC:
This is used in a great variety of valves generally in high corrosive service with lower temperatures and pressures.

VALVE SERVICE PRESSURES

Pressure ratings on valves provide an indication of the safe operating pressure for a valve at some elevated temperature. This temperature is dependent upon the materials used and the fabrication of the valve. When specific data is not available, a good "rule-of-thumb" to follow is the temperature of saturated steam on the primary rating indicated on the valve body. Example: The valve has the number 150S printed on the side indicating 150 psi and hence, a maximum operating temperature of 367° F (temperature of saturated steam and 150 psi).

DEFINITIONS

1. "WOG" – Water, oil, gas (cold working pressures).
2. "SWP" – Steam working pressure.
3. 100% area (full port) – means the area through the valve is equal to or greater than the area of standard pipe.
4. "Standard Opening" – means that the area through the valve is less than the area of standard pipe and therefore these valves should be used only where restriction of flow is unimportant.
5. "Round Port" – means the valve has a full round opening through the plug and body, of the same size and area as standard pipe.
6. "Rectangular Port" – valves have rectangular shaped ports through the plug body. The area of the port is either equal to 100% of the area of standard pipe, or restricted (standard opening). In either case it is clearly marked.
7. "ANSI" – American National Standards Institute.

R220523-90 Valve Selection Considerations

INTRODUCTION: In any piping application, valve performance is critical. Valves should be selected to give the best performance at the lowest cost.

The following is a list of performance characteristics generally expected of valves.

1. Stopping flow or starting it.
2. Throttling flow (Modulation).
3. Flow direction changing.
4. Checking backflow (Permitting flow in only one direction).
5. Relieving or regulating pressure.

In order to properly select the right valve, some facts must be determined.

A. What liquid or gas will flow through the valve?

B. Does the fluid contain suspended particles?

C. Does the fluid remain in liquid form at all times?

D. Which metals does fluid corrode?

E. What are the pressure and temperature limits? (As temperature and pressure rise, so will the price of the valve.)

F. Is there constant line pressure?

G. Is the valve merely an on-off valve?

H. Will checking of backflow be required?

I. Will the valve operate frequently or infrequently?

Valves are classified by design type into such classifications as Gate, Globe, Angle, Check, Ball, Butterfly and Plug. They are also classified by end connection, stem, pressure restrictions and material such as bronze, cast iron, etc. Each valve has a specific use. A quality valve used correctly will provide a lifetime of trouble- free service, but a high quality valve installed in the wrong service may require frequent attention.

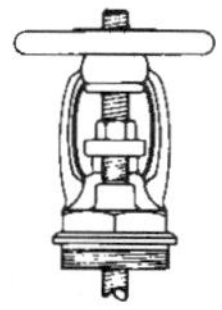

STEM TYPES
(OS & Y)—Rising Stem-Outside Screw and Yoke

Offers a visual indication of whether the valve is open or closed. Recommended where high temperatures, corrosives, and solids in the line might cause damage to inside-valve stem threads. The stem threads are engaged by the yoke bushing so the stem rises through the hand wheel as it is turned.

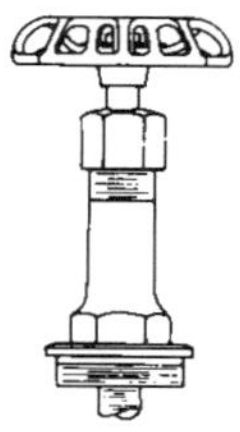

(R.S.)—Rising Stem-Inside Screw

Adequate clearance for operation must be provided because both the hand wheel and the stem rise.
The valve wedge position is indicated by the position of the stem and hand wheel.

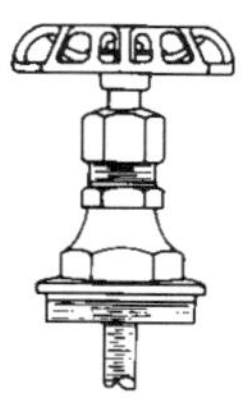

(N.R.S.)—Non-Rising Stem-Inside Screw

A minimum clearance is required for operating this type of valve. Excessive wear or damage to stem threads inside the valve may be caused by heat, corrosion, and solids. Because the hand wheel and stem do not rise, wedge position cannot be visually determined.

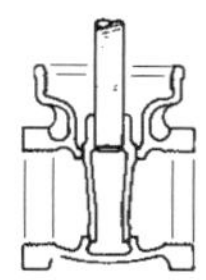

VALVE TYPES
Gate Valves

Provide full flow, minute pressure drop, minimum turbulence and minimum fluid trapped in the line.
They are normally used where operation is infrequent.

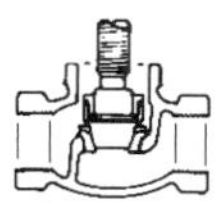

Globe Valves

Globe valves are designed for throttling and/or frequent operation with positive shut-off. Particular attention must be paid to the several types of seating materials available to avoid unnecessary wear. The seats must be compatible with the fluid in service and may be composition or metal. The configuration of the Globe valve opening causes turbulence which results in increased resistance. Most bronze Globe valves are rising stem-inside screw, but they are also available on O.S. & Y.

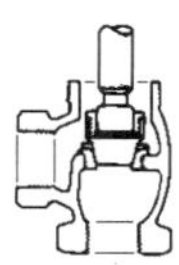

Angle Valves

The fundamental difference between the Angle valve and the Globe valve is the fluid flow through the Angle valve. It makes a 90° turn and offers less resistance to flow than the Globe valve while replacing an elbow. An Angle valve thus reduces the number of joints and installation time.

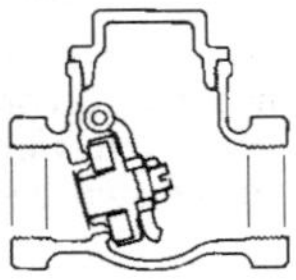

Check Valves

Check valves are designed to prevent backflow by automatically seating when the direction of fluid is reversed.
Swing Check valves are generally installed with Gate-valves, as they provide comparable full flow. Usually recommended for lines where flow velocities are low and should not be used on lines with pulsating flow. Recommended for horizontal installation, or in vertical lines only where flow is upward.

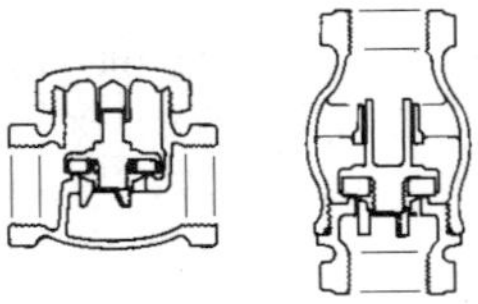

Lift Check Valves

These are commonly used with Globe and Angle valves since they have similar diaphragm seating arrangements and are recommended for preventing backflow of steam, air, gas and water, and on vapor lines with high flow velocities. For horizontal lines, horizontal lift checks should be used and vertical lift checks for vertical lines.

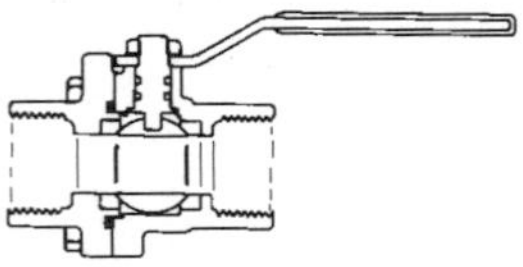

Ball Valves

Ball valves are light and easily installed, yet because of modern elastomeric seats, provide tight closure. Flow is controlled by rotating up to 90° a drilled ball which fits tightly against resilient seals. This ball seats with flow in either direction, and valve handle indicates the degree of opening. Recommended for frequent operation readily adaptable to automation, ideal for installation where space is limited.

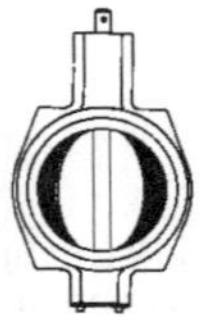

Butterfly Valves

Butterfly valves provide bubble-tight closure with excellent throttling characteristics. They can be used for full-open, closed and for throttling applications.

The Butterfly valve consists of a disc within the valve body which is controlled by a shaft. In its closed position, the valve disc seals against a resilient seat. The disc position throughout the full 90° rotation is visually indicated by the position of the operator.

A Butterfly valve is only a fraction of the weight of a Gate valve and requires no gaskets between flanges in most cases. Recommended for frequent operation and adaptable to automation where space is limited.

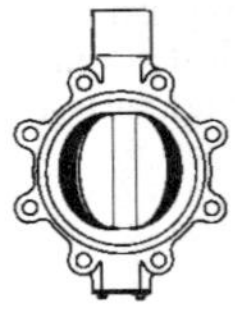

Wafer and Lug type bodies when installed between two pipe flanges, can be easily removed from the line. The pressure of the bolted flanges holds the valve in place.
Locating lugs makes installation easier.

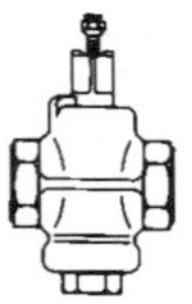

Plug Valves

Lubricated plug valves, because of the wide range of service to which they are adapted, may be classified as all purpose valves. They can be safely used at all pressure and vacuums, and at all temperatures up to the limits of available lubricants. They are the most satisfactory valves for the handling of gritty suspensions and many other destructive, erosive, corrosive and chemical solutions.

R221113-50 Pipe Material Considerations

1. Malleable fittings should be used for gas service.
2. Malleable fittings are used where there are stresses/strains due to expansion and vibration.
3. Cast fittings may be broken as an aid to disassembling of heating lines frozen by long use, temperature and minerals.
4. Cast iron pipe is extensively used for underground and submerged service.
5. Type M (light wall) copper tubing is available in hard temper only and is used for nonpressure and less severe applications than K and L.
6. Type L (medium wall) copper tubing, available hard or soft for interior service.
7. Type K (heavy wall) copper tubing, available in hard or soft temper for use where conditions are severe. For underground and interior service.
8. Hard drawn tubing requires fewer hangers or supports but should not be bent. Silver brazed fittings are recommended, however soft solder is normally used.
9. Type DMV (very light wall) copper tubing designed for drainage, waste and vent plus other non-critical pressure services.

Domestic/Imported Pipe and Fittings Cost

The prices shown in this publication for steel/cast iron pipe and steel, cast iron, malleable iron fittings are based on domestic production sold at the normal trade discounts. The above listed items of foreign manufacture may be available at prices of 1/3 to 1/2 those shown. Some imported items after minor machining or finishing operations are being sold as domestic to further complicate the system.

Caution: Most pipe prices in this book also include a coupling and pipe hangers which for the larger sizes can add significantly to the per foot cost and should be taken into account when comparing "book cost" with quoted supplier's cost.

R221113-70 Piping to 10′ High

When taking off pipe, it is important to identify the different material types and joining procedures, as well as distances between supports and components required for proper support.

During the takeoff, measure through all fittings. Do not subtract the lengths of the fittings, valves, or strainers, etc. This added length plus the final rounding of the totals will compensate for nipples and waste.

When rounding off totals always increase the actual amount to correspond with manufacturer's shipping lengths.

A. Both red brass and yellow brass pipe are normally furnished in 12′ lengths, plain end. The Unit Price section includes in the linear foot costs two field threads and one coupling per 10′ length. A carbon steel clevis type hanger assembly every 10′ is also prorated into the linear foot costs, including both material and labor.

B. Cast iron soil pipe is furnished in either 5′ or 10′ lengths. For pricing purposes, the Unit Price section features 10′ lengths with a joint and a carbon steel clevis hanger assembly every 5′ prorated into the per foot costs of both material and labor.

Three methods of joining are considered: lead and oakum poured joints or push-on gasket type joints for the bell and spigot pipe and a joint clamp for the no-hub soil pipe. The labor and material costs for each of these individual joining procedures are also prorated into the linear costs per foot.

C. Copper tubing covers types K, L, M, and DWV which are furnished in 20′ lengths. Means pricing data is based on a tubing cut each length and a coupling and two soft soldered joints every 10′. A carbon steel, clevis type hanger assembly every 10′ is also prorated into the per foot costs. The prices for refrigeration tubing are for materials only. Labor for full lengths may be based on the type L labor but short cut measures in tight areas can increase the installation labor-hours from 20 to 40%.

D. Corrosion-resistant piping does not lend itself to one particular standard of hanging or support assembly due to its diversity of application and placement. The several varieties of corrosion-resistant piping do not include any material or labor costs for hanger assemblies (See the Unit Price section for appropriate selection).

E. Glass pipe is furnished in standard lengths either 5′ or 10′ long, beaded on one end. Special orders for diverse lengths beaded on both ends are also available. For pricing purposes, R.S. Means features 10′ lengths with a coupling and a carbon steel band hanger assembly every 10′ prorated into the per foot linear costs.

Glass pipe is also available with conical ends and standard lengths ranging from 6″ through 3′ in 6″ increments, then up to 10′ in 12″ increments. Special lengths can be customized for particular installation requirements.

For pricing purposes, Means has based the labor and material pricing on 10′ lengths. Included in these costs per linear foot are the prorated costs for a flanged assembly every 10′ consisting of two flanges, a gasket, two insertable seals, and the required number of bolts and nuts. A carbon steel band hanger assembly based on 10′ center lines has also been prorated into the costs per foot for labor and materials.

F. Plastic pipe of several compositions and joining methods are considered. Fiberglass reinforced pipe (FRP) is priced, based on 10′ lengths (20′ lengths are also available), with coupling and epoxy joints every 10′. FRP is furnished in both "General Service" and "High Strength." A carbon steel clevis hanger assembly, 3 for every 10′, is built into the prorated labor and material costs on a per foot basis.

The PVC and CPVC pipe schedules 40, 80 and 120, plus SDR ratings are all based on 20′ lengths with a coupling installed every 10′, as well as a carbon steel clevis hanger assembly every 3′. The PVC and ABS type DWV piping is based on 10′ lengths with solvent weld couplings every 10′, and with carbon steel clevis hanger assemblies, 3 for every 10′. The rest of the plastic piping in this section is based on flexible 100′ coils and does not include any coupling or supports.

This section ends with PVC drain and sewer piping based on 10′ lengths with bell and spigot ends and 0-ring type, push-on joints.

G. Stainless steel piping includes both weld end and threaded piping, both in the type 304 and 316 specification and in the following schedules, 5, 10, 40, 80, and 160. Although this piping is usually furnished in 20′ lengths, this cost grouping has a joint (either heli-arc butt-welded or threads and coupling) every 10′. A carbon steel clevis type hanger assembly is also included at 10′ intervals and prorated into the linear foot costs.

H. Carbon steel pipe includes both black and galvanized. This section encompasses schedules 40 (standard) and 80 (extra heavy).

Several common methods of joining steel pipe — such as thread and coupled, butt welded, and flanged (150 lb. weld neck flanges) are also included.

For estimating purposes, it is assumed that the piping is purchased in 20′ lengths and that a compatible joint is made up every 10′. These joints are prorated into the labor and material costs per linear foot. The following hanger and support assemblies every 10′ are also included: carbon steel clevis for the T & C pipe, and single rod roll type for both the welded and flanged piping. All of these hangers are oversized to accommodate pipe insulation 3/4″ thick through 5″ pipe size and 1-1/2″ thick from 6″ through 12″ pipe size.

I. Grooved joint steel pipe is priced both black and galvanized, in schedules 10, 40, and 80, furnished in 20′ lengths. This section describes two joining methods: cut groove and roll groove. The schedule 10 piping is roll-grooved, while the heavier schedules are cut-grooved. The labor and material costs are prorated into per linear foot prices, including a coupled joint every 10′, as well as a carbon steel clevis hanger assembly.

Notes:

The pipe hanger assemblies mentioned in the preceding paragraphs include the described hanger; appropriately sized steel, box-type insert and nut; plus 18″ of threaded hanger rod.

C clamps are used when the pipe is to be supported from steel shapes rather than anchored in the slab. C clamps are slightly less costly than inserts. However, to save time in estimating, it is advisable to use the given line number cost, rather than substituting a C clamp for the insert.

Add to piping labor for elevated installation:

10′ to 14.5′ high	10%	30′ to 34.5′ high	40%
15′ to 19.5′ high	20%	35′ to 39.5′ high	50%
20′ to 24.5′ high	25%	Over 40′ and higher	55%
25′ to 29.5′ high	35%		

When using the percentage adds for elevated piping installations as shown above, bear in mind that the given heights are for the pipe supports, even though the insert, anchor, or clamp may be several feet higher than the pipe itself.

An allowance has been included in the piping installation time for testing and minor tightening of leaking joints, fittings, stuffing boxes, packing glands, etc. For extraordinary test requirements such as x-rays, prolonged pressure or demonstration tests, a percentage of the piping labor, based on the estimator's experience, must be added to the labor total. A testing service specializing in weld x-rays should be consulted for pricing if it is an estimate requirement. Equipment installation time includes start-up with associated adjustments.

R230500-10 Subcontractors

On the unit cost pages of the R.S. Means Cost Data books, the last column is entitled "Total Incl. O&P". This is normally the cost of the installing contractor. In the HVAC Division, this is the cost of the mechanical contractor. If the particular work being estimated is to be performed by a sub to the mechanical contractor, the mechanical's profit and handling charge (usually 10%) is added to the total of the last column.

R233100-10 Loudness Levels for Moving Air thru Fans, Diffusers, Register, Etc. (Measured in Sones)

Recommended Loudness Levels (Sones)											
Area	Very Quiet	Quiet	Noisy	Area	Very Quiet	Quiet	Noisy	Area	Very Quiet	Quiet	Noisy
Auditoriums				**Hotels**				**Offices**			
Auditorium lobbies	3	4	6	Banquet Rooms	1.5	3	6	Conference rooms	1	1.7	3
Concert and opera halls	0.8	1	1.5	Individual rooms, suites	1	2	4	Drafting	2	4	8
Courtrooms	2	3	4	Kitchens and laundries	4	7	10	General open offices	2	4	8
Lecture halls	1.5	2	3	Lobbies	2	4	8	Halls and corridors	2.5	5	10
Movie theaters	1.5	2	3	**Indoor Sports**				Professional offices	1.5	3	6
Churches and Schools				Bowling alleys	3	4	6	Reception room	1.5	3	6
Kitchens	4	6	8	Gymnasiums	2	4	6	Tabulation & computation	3	6	12
Laboratories	2	4	6	Swimming pools	4	7	10	**Public Buildings**			
Libraries	1.5	2	3	**Manufacturing Areas**				Banks	2	4	6
Recreation halls	2	4	8	Assembly lines	5	12	30	Court houses	2.5	4	6
Sanctuaries	1	1.7	3	Foreman's office	3	5	8	Museums	2	3	4
Schools and classrooms	1.5	2.5	4	Foundries	10	20	40	Planetariums	1.5	2	3
Hospital and Clinics				General storage	5	10	20	Post offices	2.5	4	6
Laboratories	2	4	6	Heavy machinery	10	25	60	Public libraries	1.5	2	4
Lobbies, waiting rooms	2	4	6	Light machinery	5	12	30	Waiting rooms	3	5	8
Halls and corridors	2	4	6	Tool maintenance	4	7	10	**Restaurants**			
Operating rooms	1.5	2.5	4	**Stores**				Cafeterias	3	6	10
Private rooms	1	1.7	3	Department stores	3	6	8	Night clubs	2.5	4	6
Wards	1.5	2.5	4	Supermarkets	4	7	10	Restaurants	2	4	8

R233100-20 Ductwork

Duct Weight in Pounds per L.F., Straight Runs

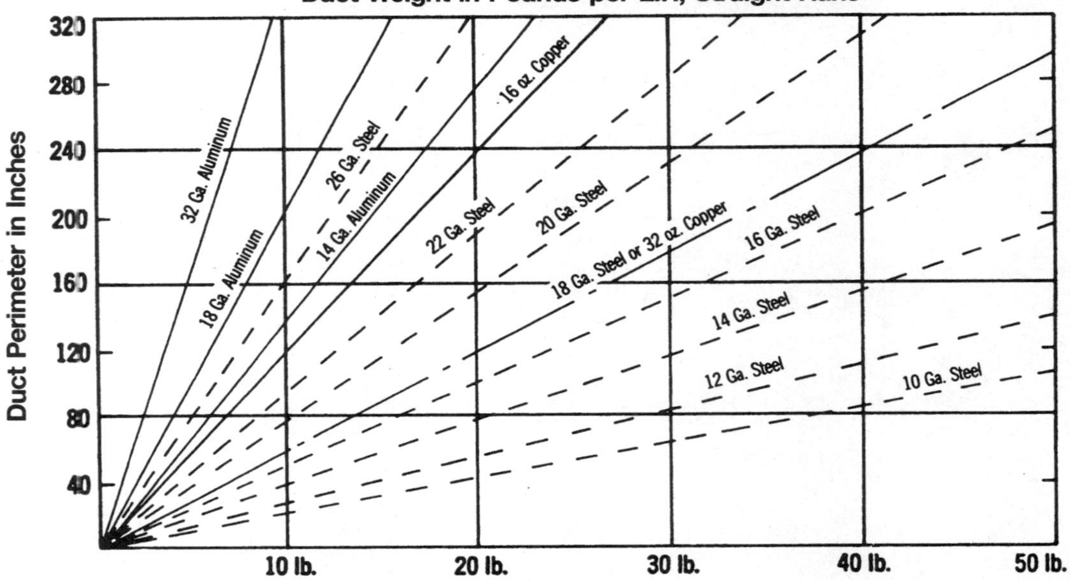

Add to the above for fittings; 90° elbow is 3 L.F.; 45° elbow is 2.5 L.F.; offset is 4 L.F.; transition offset is 6 L.F.; square-to-round transition is 4 L.F.; 90° reducing elbow is 5 L.F. For bracing and waste, add 20% to aluminum and copper, 15% to steel.

R233100-30 Duct Fabrication/Installation

The labor cost for fabricated sheet metal duct includes both the cost of fabrication and installation of the duct. The split is approximately 40% for fabrication, 60% for installation. It is for this reason that the percentage add for elevated installation of fabricated duct is less than the percentage add for prefabricated/preformed duct.

Example: assume a piece of duct cost $100 installed (labor only)

Sheet Metal Fabrication = 40% = $40
Installation = 60% = $60

The add for elevated installation is:

Based on total labor (fabrication & installation) = $100 x 6% = $6.00

Based on installation cost only (Material purchased prefabricated) = $60 x 10% = $6.00

The $6.00 markup (10′ to 15′ high) is the same.

R233100-40 Sheet Metal Calculator (Weight in Lb./Ft. of Length)

Gauge	26	24	22	20	18	16	Gauge	26	24	22	20	18	16
Wt.-Lb./S.F.	.906	1.156	1.406	1.656	2.156	2.656	Wt.-Lb./S.F.	.906	1.156	1.406	1.656	2.156	2.656
SMACNA Max. Dimension – Long Side		30″	54″	84″	85″ Up		SMACNA Max. Dimension – Long Side		30″	54″	84″	85″ Up	
Sum-2 sides							Sum-2 Sides						
2	.3	.40	.50	.60	.80	.90	56	9.3	12.0	14.0	16.2	21.3	25.2
3	.5	.65	.80	.90	1.1	1.4	57	9.5	12.3	14.3	16.5	21.7	25.7
4	.7	.85	1.0	1.2	1.5	1.8	58	9.7	12.5	14.5	16.8	22.0	26.1
5	.8	1.1	1.3	1.5	1.9	2.3	59	9.8	12.7	14.8	17.1	22.4	26.6
6	1.0	1.3	1.5	1.7	2.3	2.7	60	10.0	12.9	15.0	17.4	22.8	27.0
7	1.2	1.5	1.8	2.0	2.7	3.2	61	10.2	13.1	15.3	17.7	23.2	27.5
8	1.3	1.7	2.0	2.3	3.0	3.6	62	10.3	13.3	15.5	18.0	23.6	27.9
9	1.5	1.9	2.3	2.6	3.4	4.1	63	10.5	13.5	15.8	18.3	24.0	28.4
10	1.7	2.2	2.5	2.9	3.8	4.5	64	10.7	13.7	16.0	18.6	24.3	28.8
11	1.8	2.4	2.8	3.2	4.2	5.0	65	10.8	13.9	16.3	18.9	24.7	29.3
12	2.0	2.6	3.0	3.5	4.6	5.4	66	11.0	14.1	16.5	19.1	25.1	29.7
13	2.2	2.8	3.3	3.8	4.9	5.9	67	11.2	14.3	16.8	19.4	25.5	30.2
14	2.3	3.0	3.5	4.1	5.3	6.3	68	11.3	14.6	17.0	19.7	25.8	30.6
15	2.5	3.2	3.8	4.4	5.7	6.8	69	11.5	14.8	17.3	20.0	26.2	31.1
16	2.7	3.4	4.0	4.6	6.1	7.2	70	11.7	15.0	17.5	20.3	26.6	31.5
17	2.8	3.7	4.3	4.9	6.5	7.7	71	11.8	15.2	17.8	20.6	27.0	32.0
18	3.0	3.9	4.5	5.2	6.8	8.1	72	12.0	15.4	18.0	20.9	27.4	32.4
19	3.2	4.1	4.8	5.5	7.2	8.6	73	12.2	15.6	18.3	21.2	27.7	32.9
20	3.3	4.3	5.0	5.8	7.6	9.0	74	12.3	15.8	18.5	21.5	28.1	33.3
21	3.5	4.5	5.3	6.1	8.0	9.5	75	12.5	16.1	18.8	21.8	28.5	33.8
22	3.7	4.7	5.5	6.4	8.4	9.9	76	12.7	16.3	19.0	22.0	28.9	34.2
23	3.8	5.0	5.8	6.7	8.7	10.4	77	12.8	16.5	19.3	22.3	29.3	34.7
24	4.0	5.2	6.0	7.0	9.1	10.8	78	13.0	16.7	19.5	22.6	29.6	35.1
25	4.2	5.4	6.3	7.3	9.5	11.3	79	13.2	16.9	19.8	22.9	30.0	35.6
26	4.3	5.6	6.5	7.5	9.9	11.7	80	13.3	17.1	20.0	23.2	30.4	36.0
27	4.5	5.8	6.8	7.8	10.3	12.2	81	13.5	17.3	20.3	23.5	30.8	36.5
28	4.7	6.0	7.0	8.1	10.6	12.6	82	13.7	17.5	20.5	23.8	31.2	36.9
29	4.8	6.2	7.3	8.4	11.0	13.1	83	13.8	17.8	20.8	24.1	31.5	37.4
30	5.0	6.5	7.5	8.7	11.4	13.5	84	14.0	18.0	21.0	24.4	31.9	37.8
31	5.2	6.7	7.8	9.0	11.8	14.0	85	14.2	18.2	21.3	24.7	32.3	38.3
32	5.3	6.9	8.0	9.3	12.2	14.4	86	14.3	18.4	21.5	24.9	32.7	38.7
33	5.5	7.1	8.3	9.6	12.5	14.9	87	14.5	18.6	21.8	25.2	33.1	39.2
34	5.7	7.3	8.5	9.9	12.9	15.3	88	14.7	18.8	22.0	25.5	33.4	39.6
35	5.8	7.5	8.8	10.2	13.3	15.8	89	14.8	19.0	22.3	25.8	33.8	40.1
36	6.0	7.8	9.0	10.4	13.7	16.2	90	15.0	19.3	22.5	26.1	34.2	40.5
37	6.2	8.0	9.3	10.7	14.1	16.7	91	15.2	19.5	22.8	26.4	34.6	41.0
38	6.3	8.2	9.5	11.0	14.4	17.1	92	15.3	19.7	23.0	26.7	35.0	41.4
39	6.5	8.4	9.8	11.3	14.8	17.6	93	15.5	19.9	23.3	27.0	35.3	41.9
40	6.7	8.6	10.0	11.6	15.2	18.0	94	15.7	20.1	23.5	27.3	35.7	42.3
41	6.8	8.8	10.3	11.9	15.6	18.5	95	15.8	20.3	23.8	27.6	36.1	42.8
42	7.0	9.0	10.5	12.2	16.0	18.9	96	16.0	20.5	24.0	27.8	36.5	43.2
43	7.2	9.2	10.8	12.5	16.3	19.4	97	16.2	20.8	24.3	28.1	36.9	43.7
44	7.3	9.5	11.0	12.8	16.7	19.8	98	16.3	21.0	24.5	28.4	37.2	44.1
45	7.5	9.7	11.3	13.1	17.1	20.3	99	16.5	21.2	24.8	28.7	37.6	44.6
46	7.7	9.9	11.5	13.3	17.5	20.7	100	16.7	21.4	25.0	29.0	38.0	45.0
47	7.8	10.1	11.8	13.6	17.9	21.2	101	16.8	21.6	25.3	29.3	38.4	45.5
48	8.0	10.3	12.0	13.9	18.2	21.6	102	17.0	21.8	25.5	29.6	38.8	45.9
49	8.2	10.5	12.3	14.2	18.6	22.1	103	17.2	22.0	25.8	29.9	39.1	46.4
50	8.3	10.7	12.5	14.5	19.0	22.5	104	17.3	22.3	26.0	30.2	39.5	46.8
51	8.5	11.0	12.8	14.8	19.4	23.0	105	17.5	22.5	26.3	30.5	39.9	47.3
52	8.7	11.2	13.0	15.1	19.8	23.4	106	17.7	22.7	26.5	30.7	40.3	47.7
53	8.8	11.4	13.3	15.4	20.1	23.9	107	17.8	22.9	26.8	31.0	40.7	48.2
54	9.0	11.6	13.5	15.7	20.5	24.3	108	18.0	23.1	27.0	31.3	41.0	48.6
55	9.2	11.8	13.8	16.0	20.9	24.8	109	18.2	23.3	27.3	31.6	41.4	49.1
							110	18.3	23.5	27.5	31.9	41.8	49.5

Example: If duct is 34″ x 20″ x 15′ long, 34″ is greater than 30″ maximum, for 24 ga. so must be 22 ga. 34″ + 20″ = 54″ going across from 54″ find 13.5 lb. per foot. 13.5 x 15′ = 202.5 lbs. For S.F. of surface area 202.5 ÷ 1.406 = 144 S.F.

Note: Figures include an allowance for scrap.

Heating, Ventilating & A.C. R2331 HVAC Ducts & Casings

R233100-50 Ductwork Packages (per Ton of Cooling)

System	Sheet Metal	Insulation	Diffusers	Return Register
Roof Top Unit Single Zone	120 Lbs.	52 S.F.	1	1
Roof Top Unit Multizone	240 Lbs.	104 S.F.	2	1
Self-contained Air or Water Cooled	108 Lbs.	—	2	—
Split System Air Cooled	102 Lbs.	—	2	—

Systems reflect most common usage.
Refer to system graphics for duct layout.

Heating, Ventilating & A.C. R2334 HVAC Fans

R233400-10 Recommended Ventilation Air Changes

Table below lists range of time in minutes per change for various types of facilities.

Facility	Minutes	Facility	Minutes	Facility	Minutes
Assembly Halls	2-10	Dance Halls	2-10	Laundries	1-3
Auditoriums	2-10	Dining Rooms	3-10	Markets	2-10
Bakeries	2-3	Dry Cleaners	1-5	Offices	2-10
Banks	3-10	Factories	2-5	Pool Rooms	2-5
Bars	2-5	Garages	2-10	Recreation Rooms	2-10
Beauty Parlors	2-5	Generator Rooms	2-5	Sales Rooms	2-10
Boiler Rooms	1-5	Gymnasiums	2-10	Theaters	2-8
Bowling Alleys	2-10	Kitchens-Hospitals	2-5	Toilets	2-5
Churches	5-10	Kitchens-Restaurant	1-3	Transformer Rooms	1-5

CFM air required for changes = Volume of room in cubic feet ÷ Minutes per change.

Heating, Ventilating & A.C. R2337 Air Outlets & Inlets

R233700-60 Diffuser Evaluation

CFM = V × An × K where V = Outlet velocity in feet per minute. An = Neck area in square feet and K = Diffuser delivery factor. An undersized diffuser for a desired CFM will produce a high velocity and noise level. When air moves past people at a velocity in excess of 25 FPM, an annoying draft is felt. An oversized diffuser will result in low velocity with poor mixing. Consideration must be given to avoid vertical stratification or horizontal areas of stagnation.

R235000-10 Heating Systems

Heating Systems

The basic function of a heating system is to bring an enclosed volume up to a desired temperature and then maintain that temperature within a reasonable range. To accomplish this, the selected system must have sufficient capacity to offset transmission losses resulting from the temperature difference on the interior and exterior of the enclosing walls in addition to losses due to cold air infiltration through cracks, crevices and around doors and windows. The amount of heat to be furnished is dependent upon the building size, construction, temperature difference, air leakage, use, shape, orientation and exposure. Air circulation is also an important consideration. Circulation will prevent stratification which could result in heat losses through uneven temperatures at various levels. For example, the most efficient use of unit heaters can usually be achieved by circulating the space volume through the total number of units once every 20 minutes or 3 times an hour. This general rule must, of course, be adapted for special cases such as large buildings with low ratios of heat transmitting surface to cubical volume. The type of occupancy of a building will have considerable bearing on the number of heat transmitting units and the location selected. It is axiomatic, however, that the basis of any successful heating system is to provide the maximum amount of heat at the points of maximum heat loss such as exposed walls, windows, and doors. Large roof areas, wind direction, and wide doorways create problems of excessive heat loss and require special consideration and treatment.

Heat Transmission

Heat transfer is an important parameter to be considered during selection of the exterior wall style, material and window area. A high rate of transfer will permit greater heat loss during the wintertime with the resultant increase in heating energy costs and a greater rate of heat gain in the summer with proportionally greater cooling cost. Several terms are used to describe various aspects of heat transfer. However, for general estimating purposes this book lists U values for systems of construction materials. U is the "overall heat transfer coefficient." It is defined as the heat flow per hour through one square foot when the temperature difference in the air on either side of the structure wall, roof, ceiling or floor is one degree Fahrenheit. The structural segment may be a single homogeneous material or a composite.

Total heat transfer is found using the following equation:

$\mathbf{Q = AU(T_2 - T_1)}$ where

Q = Heat flow, BTU per hour
A = Area, square feet
U = Overall heat transfer coefficient
$\mathbf{(T_2 - T_1)}$ = Difference in temperature of air on each side of the construction component. (Also abbreviated TD)

Note that heat can flow through all surfaces of any building and this flow is in addition to heat gain or loss due to ventilation, infiltration and generation (appliances, machinery, people).

R235000-20 Heating Approximations for Quick Estimating

Oil Piping & Boiler Room Piping

Small System	20 to 30% of Boiler
Complex System with Pumps, Headers, Etc.	80 to 110% of Boiler
Breeching With Insulation:	
Small	10 to 15% of Boiler
Large	15 to 25% of Boiler
Coils:	15 to 30% of Containing Unit
Balancing (Independent)	1/2% of H.V.A.C. Estimate

Quality/Complexity Adjustment: For all heating installations add these adjustments to the estimate to more closely allow for the equipment and conditions of the particular job under consideration.

Economy installation, add	0 to 5% of System
Good quality, medium complexity, add	5 to 15% of System
Above average quality and complexity, add	15 to 25% of System

R235000-30 The Basics of a Heating System

The function of a heating system is to achieve and maintain a desired temperature in a room or building by replacing the amount of heat being dissipated. There are four kinds of heating systems: hot-water, steam, warm-air and electric resistance. Each has certain essential and similar elements with the exception of electric resistance heating.

The basic elements of a heating system are:

A. A **combustion chamber** in which fuel is burned and heat transferred to a conveying medium.

B. The **"fluid"** used for conveying the heat (water, steam or air).

C. **Conductors** or pipes for transporting the fluid to specific desired locations.

D. A means of disseminating the heat, sometimes called **terminal units.**

A. The **combustion chamber** in a furnace heats air which is then distributed. This is called a warm-air system.

The combustion chamber in a boiler heats water which is either distributed as hot water or steam and this is termed a hydronic system.

The maximum allowable working pressures are limited by ASME "Code for Heating Boilers" to 15 PSI for steam and 160 PSI for hot water heating boilers, with a maximum temperature limitation of 250° F. Hot water boilers are generally rated for a working pressure of 30 PSI. High pressure boilers are governed by the ASME "Code for Power Boilers" which is used almost universally for boilers operating over 15 PSIG. High pressure boilers used for a combination of heating/process loads are usually designed for 150 PSIG.

Boiler ratings are usually indicated as either Gross or Net Output. The Gross Load is equal to the Net Load plus a piping and pickup allowance. When this allowance cannot be determined, divide the gross output rating by 1.25 for a value equal to or greater than the net heat loss requirement of the building.

B. Of the three **fluids** used, steam carries the greatest amount of heat per unit volume. This is due to the fact that it gives up its latent heat of vaporization at a temperature considerably above room temperature. Another advantage is that the pressure to produce a positive circulation is readily available. Piping conducts the steam to terminal units and returns condensate to the boiler.

The **steam system** is well adapted to large buildings because of its positive circulation, its comparatively economical installation and its ability to deliver large quantities of heat. Nearly all large office buildings, stores, hotels, and industrial buildings are so heated, in addition to many residences.

Hot water, when used as the heat carrying fluid, gives up a portion of its sensible heat and then returns to the boiler or heating apparatus for reheating. As the heat conveyed by each pound of water is about one-fiftieth of the heat conveyed by a pound of steam, it is necessary to circulate about fifty times as much water as steam by weight (although only one-thirtieth as much by volume). The hot water system is usually, although not necessarily, designed to operate at temperatures below that of the ordinary steam system and so the amount of heat transfer surface must be correspondingly greater. A temperature of 190° F to 200° F is normally the maximum. Circulation in small buildings may depend on the difference in density between hot water and the cool water returning to the boiler; circulating pumps are normally used to maintain a desired rate of flow. Pumps permit a greater degree of flexibility and better control.

In **warm-air** furnace systems, cool air is taken from one or more points in the building, passed over the combustion chamber and flue gas passages and then distributed through a duct system. A disadvantage of this system is that the ducts take up much more building volume than steam or hot water pipes. Advantages of this system are the relative ease with which humidification can be accomplished by the evaporation of water as the air circulates through the heater, and the lack of need for expensive disseminating units as the warm air simply becomes part of the interior atmosphere of the building.

C. Conductors (pipes and ducts) have been lightly treated in the discussion of conveying fluids. For more detailed information such as sizing and distribution methods, the reader is referred to technical publications such as the American Society of Heating, Refrigerating and Air-Conditioning Engineers "Handbook of Fundamentals."

D. Terminal units come in an almost infinite variety of sizes and styles, but the basic principles of operation are very limited. As previously mentioned, warm-air systems require only a simple register or diffuser to mix heated air with that present in the room. Special application items such as radiant coils and infrared heaters are available to meet particular conditions but are not usually considered for general heating needs. Most heating is accomplished by having air flow over coils or pipes containing the heat transporting medium (steam, hot-water, electricity). These units, while varied, may be separated into two general types, (1) radiator/convectors and (2) unit heaters.

Radiator/convectors may be cast, fin-tube or pipe assemblies. They may be direct, indirect, exposed, concealed or mounted within a cabinet enclosure, upright or baseboard style. These units are often collectively referred to as "radiators" or "radiation" although none gives off heat either entirely by radiation or by convection but rather a combination of both. The air flows over the units as a gravity "current." It is necessary to have one or more heat-emitting units in each room. The most efficient placement is low along an outside wall or under a window to counteract the cold coming into the room and achieve an even distribution.

In contrast to radiator/convectors which operate most effectively against the walls of smaller rooms, **unit heaters** utilize a fan to move air over heating coils and are very effective in locations of relatively large volume. Unit heaters, while usually suspended overhead, may be floor mounted. They also may take in fresh outside air for ventilation. The heat distributed by unit heaters may be from a remote source and conveyed by a fluid or it may be from the combustion of fuel in each individual heater. In the latter case the only piping required would be for fuel, however, a vent for the products of combustion would be necessary.

The following list gives may of the advantages of unit heaters for applications other than office or residential:

a. Large capacity so smaller number of units are required,
b. Piping system simplified, **c.** Space saved where they are located overhead out of the way, **d.** Rapid heating directed where needed with effective wide distribution, **e.** Difference between floor and ceiling temperature reduced, **f.** Circulation of air obtained, and ventilation with introduction of fresh air possible, **g.** Heat output flexible and easily controlled.

R235000-50 Factor for Determining Heat Loss for Various Types of Buildings

General: While the most accurate estimates of heating requirements would naturally be based on detailed information about the building being considered, it is possible to arrive at a reasonable approximation using the following procedure:

1. Calculate the cubic volume of the room or building.
2. Select the appropriate factor from Table 1 below. Note that the factors apply only to inside temperatures listed in the first column and to 0° F outside temperature.
3. If the building has bad north and west exposures, multiply the heat loss factor by 1.1.
4. If the outside design temperature is other than 0° F, multiply the factor from Table 1 by the factor from Table 2.
5. Multiply the cubic volume by the factor selected from Table 1. This will give the estimated BTUH heat loss which must be made up to maintain inside temperature.

Table 1 — Building Type	Conditions	Qualifications	Loss Factor*
Factories & Industrial Plants General Office Areas at 70° F	One Story	Skylight in Roof	6.2
		No Skylight in Roof	5.7
	Multiple Story	Two Story	4.6
		Three Story	4.3
		Four Story	4.1
		Five Story	3.9
		Six Story	3.6
	All Walls Exposed	Flat Roof	6.9
		Heated Space Above	5.2
	One Long Warm Common Wall	Flat Roof	6.3
		Heated Space Above	4.7
	Warm Common Walls on Both Long Sides	Flat Roof	5.8
		Heated Space Above	4.1
Warehouses at 60° F	All Walls Exposed	Skylights in Roof	5.5
		No Skylight in Roof	5.1
		Heated Space Above	4.0
	One Long Warm Common Wall	Skylight in Roof	5.0
		No Skylight in Roof	4.9
		Heated Space Above	3.4
	Warm Common Walls on Both Long Sides	Skylight in Roof	4.7
		No Skylight in Roof	4.4
		Heated Space Above	3.0

*Note: This table tends to be conservative particularly for new buildings designed for minimum energy consumption.

Table 2 — Outside Design Temperature Correction Factor (for Degrees Fahrenheit)									
Outside Design Temperature	50	40	30	20	10	0	-10	-20	-30
Correction Factor	.29	.43	.57	.72	.86	1.00	1.14	1.28	1.43

R235000-70 Transmission of Heat

R235000-70 and R235000-80 provide a way to calculate heat transmission of various construction materials from their U values and the TD (Temperature Difference).

1. From the Exterior Enclosure Division or elsewhere, determine U values for the construction desired.
2. Determine the coldest design temperature. The difference between this temperature and the desired interior temperature is the TD (temperature difference).
3. Enter R235000-70 or R235000-80 at correct U Value. Cross horizontally to the intersection with appropriate TD. Read transmission per square foot from bottom of figure.
4. Multiply this value of BTU per hour transmission per square foot of area by the total surface area of that type of construction.

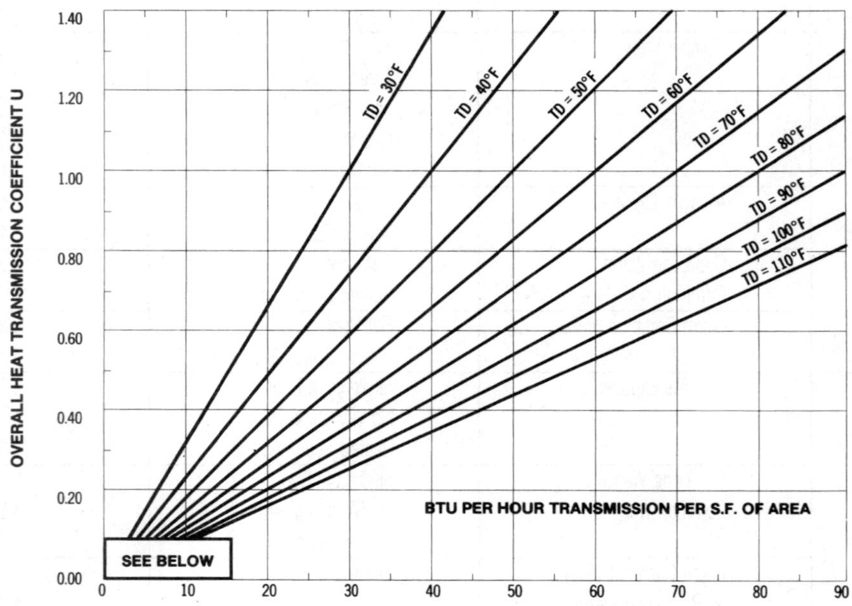

R235000-80 Transmission of Heat (Low Rate)

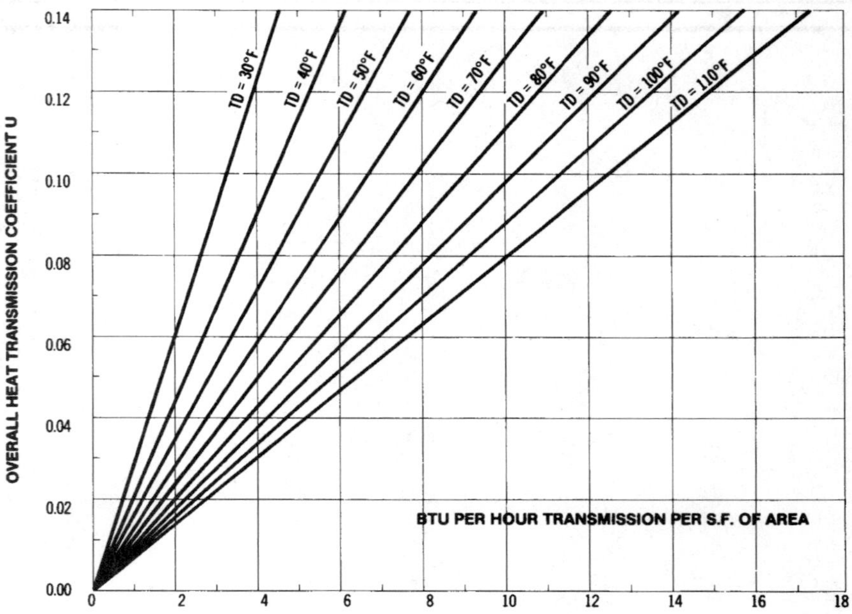

Heating, Ventilating & A.C. R2356 Solar Energy Heating Equipment

R235616-60 Solar Heating (Space and Hot Water)

Collectors should face as close to due South as possible, however, variations of up to 20 degrees on either side of true South are acceptable. Local climate and collector type may influence the choice between east or west deviations. Obviously they should be located so they are not shaded from the sun's rays. Incline collectors at a slope of latitude minus 5 degrees for domestic hot water and latitude plus 15 degrees for space heating.

Flat plate collectors consist of a number of components as follows: Insulation to reduce heat loss through the bottom and sides of the collector. The enclosure which contains all the components in this assembly is usually weatherproof and prevents dust, wind and water from coming in contact with the absorber plate. The cover plate usually consists of one or more layers of a variety of glass or plastic and reduces the reradiation by creating an air space which traps the heat between the cover and the absorber plates.

The absorber plate must have a good thermal bond with the fluid passages. The absorber plate is usually metallic and treated with a surface coating which improves absorptivity. Black or dark paints or selective coatings are used for this purpose, and the design of this passage and plate combination helps determine a solar system's effectiveness.

Heat transfer fluid passage tubes are attached above and below or integral with an absorber plate for the purpose of transferring thermal energy from the absorber plate to a heat transfer medium. The heat exchanger is a device for transferring thermal energy from one fluid to another.

Piping and storage tanks should be well insulated to minimize heat losses.

Size domestic water heating storage tanks to hold 20 gallons of water per user, minimum, plus 10 gallons per dishwasher or washing machine. For domestic water heating an optimum collector size is approximately 3/4 square foot of area per gallon of water storage. For space heating of residences and small commercial applications the collector is commonly sized between 30% and 50% of the internal floor area. For space heating of large commercial applications, collector areas less than 30% of the internal floor area can still provide significant heat reductions.

A supplementary heat source is recommended for Northern states for December through February.

The solar energy transmission per square foot of collector surface varies greatly with the material used. Initial cost, heat transmittance and useful life are obviously interrelated.

Heating, Ventilating & A.C. R2360 Central Cooling Equipment

R236000-10 Air Conditioning

General: The purpose of air conditioning is to control the environment of a space so that comfort is provided for the occupants and/or conditions are suitable for the processes or equipment contained therein. The several items which should be evaluated to define system objectives are:

- Temperature Control
- Humidity Control
- Cleanliness
- Odor, smoke and fumes
- Ventilation

Efforts to control the above parameters must also include consideration of the degree or tolerance of variation, the noise level introduced, the velocity of air motion and the energy requirements to accomplish the desired results.

The variation in **temperature** and **humidity** is a function of the sensor and the controller. The controller reacts to a signal from the sensor and produces the appropriate suitable response in either the terminal unit, the conductor of the transporting medium (air, steam, chilled water, etc.), or the source (boiler, evaporating coils, etc.).

The **noise level** is a by-product of the energy supplied to moving components of the system. Those items which usually contribute the most noise are pumps, blowers, fans, compressors and diffusers. The level of noise can be partially controlled through use of vibration pads, isolators, proper sizing, shields, baffles and sound absorbing liners.

Some **air motion** is necessary to prevent stagnation and stratification. The maximum acceptable velocity varies with the degree of heating or cooling which is taking place. Most people feel air moving past them at velocities in excess of 25 FPM as an annoying draft. However, velocities up to 45 FPM may be acceptable in certain cases. Ventilation, expressed as air changes per hour and percentage of fresh air, is usually an item regulated by local codes.

Selection of the system to be used for a particular application is usually a trade-off. In some cases the building size, style, or room available for mechanical use limits the range of possibilities. Prime factors influencing the decision are first cost and total life (operating, maintenance and replacement costs). The accuracy with which each parameter is determined will be an important measure of the reliability of the decision and subsequent satisfactory operation of the installed system.

Heat delivery may be desired from an air conditioning system. Heating capability usually is added as follows: A gas fired burner or hot water/steam/electric coils may be added to the air handling unit directly and heat all air equally. For limited or localized heat requirements the water/steam/electric coils may be inserted into the duct branch supplying the cold areas. Gas fired duct furnaces are also available.

Note: When water or steam coils are used the cost of the piping and boiler must also be added. For a rough estimate use the cost per square foot of the appropriate sized hydronic system with unit heaters. This will provide a cost for the boiler and piping, and the unit heaters of the system would equate to the approximate cost of the heating coils.

R236000-20 Air Conditioning Requirements

BTU's per hour per S.F. of floor area and S.F. per ton of air conditioning.

Type of Building	BTU/Hr per S.F.	S.F. per Ton	Type of Building	BTU/Hr per S.F.	S.F. per Ton	Type of Building	BTU/Hr per S.F.	S.F. per Ton
Apartments, Individual	26	450	Dormitory, Rooms	40	300	Libraries	50	240
Corridors	22	550	Corridors	30	400	Low Rise Office, Exterior	38	320
Auditoriums & Theaters	40	300/18*	Dress Shops	43	280	Interior	33	360
Banks	50	240	Drug Stores	80	150	Medical Centers	28	425
Barber Shops	48	250	Factories	40	300	Motels	28	425
Bars & Taverns	133	90	High Rise Office—Ext. Rms.	46	263	Office (small suite)	43	280
Beauty Parlors	66	180	Interior Rooms	37	325	Post Office, Individual Office	42	285
Bowling Alleys	68	175	Hospitals, Core	43	280	Central Area	46	260
Churches	36	330/20*	Perimeter	46	260	Residences	20	600
Cocktail Lounges	68	175	Hotel, Guest Rooms	44	275	Restaurants	60	200
Computer Rooms	141	85	Corridors	30	400	Schools & Colleges	46	260
Dental Offices	52	230	Public Spaces	55	220	Shoe Stores	55	220
Dept. Stores, Basement	34	350	Industrial Plants, Offices	38	320	Shop'g. Ctrs., Supermarkets	34	350
Main Floor	40	300	General Offices	34	350	Retail Stores	48	250
Upper Floor	30	400	Plant Areas	40	300	Specialty	60	200

*Persons per ton

12,000 BTU = 1 ton of air conditioning

R236000-30 Psychrometric Table

Dewpoint or Saturation Temperature (F)

Relative humidity (%)															
100	32	35	40	45	50	55	60	65	70	75	80	85	90	95	100
90	30	33	37	42	47	52	57	62	67	72	77	82	87	92	97
80	27	30	34	39	44	49	54	58	64	68	73	78	83	88	93
70	24	27	31	36	40	45	50	55	60	64	69	74	79	84	88
60	20	24	28	32	36	41	46	51	55	60	65	69	74	79	83
50	16	20	24	28	33	36	41	46	50	55	60	64	69	73	78
40	12	15	18	23	27	31	35	40	45	49	53	58	62	67	71
30	8	10	14	18	21	25	29	33	37	42	46	50	54	59	62
20	6	7	8	9	13	16	20	24	28	31	35	40	43	48	52
10	4	4	5	5	6	8	9	10	13	17	20	24	27	30	34
	32	**35**	**40**	**45**	**50**	**55**	**60**	**65**	**70**	**75**	**80**	**85**	**90**	**95**	**100**

Dry bulb temperature (F)

This table shows the relationship between RELATIVE HUMIDITY, DRY BULB TEMPERATURE AND DEWPOINT.

As an example, assume that the thermometer in a room reads 75° F, and we know that the relative humidity is 50%. The chart shows the dewpoint temperature to be 55° F. That is, any surface colder than 55° F will "sweat" or collect condensing moisture. This surface could be the outside of an uninsulated chilled water pipe in the summertime, or the inside surface of a wall or deck in the wintertime. After determining the extreme ambient parameters, the table at the left is useful in determining which surfaces need insulation or vapor barrier protection.

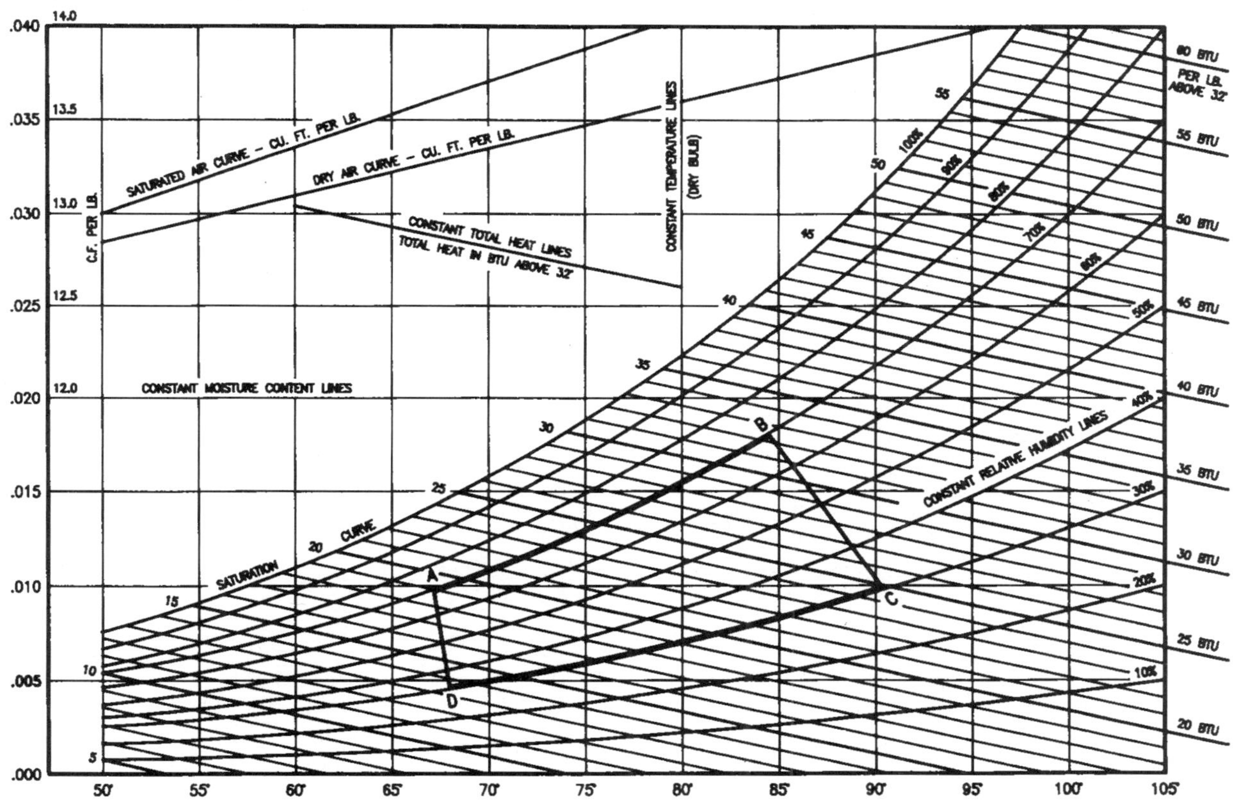

TEMPERATURE DEGREES FAHRENHEIT
TOTAL PRESSURE = 14.696 LB. PER SQ. IN. ABS.

Psychrometric chart showing different variables based on one pound of dry air. Space marked A B C D is temperature-humidity range which is most comfortable for majority of people.

R236000-90 Quality/Complexity Adjustment for Air Conditioning Systems

Economy installation, add	0 to 5%
Good quality, medium complexity, add	5 to 15%
Above average quality and complexity, add	15 to 25%

Add the above adjustments to the estimate to more closely allow for the equipment and conditions of the particular job under consideration.

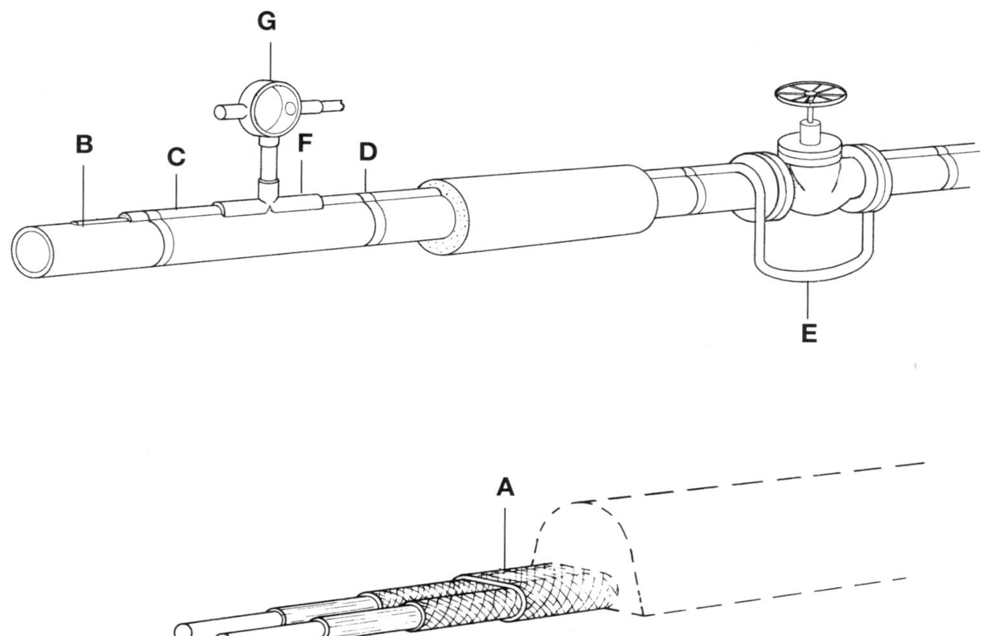

Fig. R238313-11

R238313-10 Heat Trace Systems

Before you can determine the cost of a HEAT TRACE installation, the method of attachment must be established. There are (4) common methods:

1. Cable is simply attached to the pipe with polyester tape every 12′.
2. Cable is attached with a continuous cover of 2″ wide aluminum tape.
3. Cable is attached with factory extruded heat transfer cement and covered with metallic raceway with clips every 10′.
4. Cable is attached between layers of pipe insulation using either clips or polyester tape.

Example: Components for method 3 must include:

A. Heat trace cable by voltage and watts per linear foot.
B. Heat transfer cement, 1 gallon per 60 linear feet of cover.
C. Metallic raceway by size and type.
D. Raceway clips by size of pipe.

When taking off linear foot lengths of cable add the following for each valve in the system. (E)

In all of the above methods each component of the system must be priced individually.

SCREWED OR WELDED VALVE:

1/2″	=	6″
3/4″	=	9″
1″	=	1′ -0″
1-1/2″	=	1′ -6″
2″	=	2′
2-1/2″	=	2′ -6″
3″	=	2′ -6″
4″	=	4′ -0″
6″	=	7′ -0″
8″	=	9′ -6″
10″	=	12′ -6″
12″	=	15′ -0″
14″	=	18′ -0″
16″	=	21′ -6″
18″	=	25′ -6″
20″	=	28′ -6″
24″	=	34′ -0″
30″	=	40′ -0″

FLANGED VALVE:

1/2″	=	1′ -0″
3/4″	=	1′ -6″
1″	=	2′ -0″
1-1/2″	=	2′ -6″
2″	=	2′ -6″
2-1/2″	=	3′ -0″
3″	=	3′ -6″
4″	=	4′ -0″
6″	=	8′ -0″
8″	=	11′ -0″
10″	=	14′ -0″
12″	=	16′ -6″
14″	=	19′ -6″
16″	=	23′ -0″
18″	=	27′ -0″
20″	=	30′ -0″
24″	=	36′ -0″
30″	=	42′ -0″

BUTTERFLY VALVES:

1/2″	=	0′
3/4″	=	0′
1″	=	1′ -0″
1-1/2″	=	1′ -6″
2″	=	2′ -0″
2-1/2″	=	2′ -6″
3″	=	2′ -6″
4″	=	3′ -0″
6″	=	3′ -6″
8″	=	4′ -0″
10″	=	4′ -0″
12″	=	5′ -0″
14″	=	5′ -6″
16″	=	6′ -0″
18″	=	6′ -6″
20″	=	7′ -0″
24″	=	8′ -0″
30″	=	10′ -0″

R238313-10 Heat Trace Systems (cont.)

Add the following quantities of heat transfer cement to linear foot totals for each valve:

Nominal Valve Size	Gallons of Cement per Valve
1/2″	0.14
3/4″	0.21
1″	0.29
1-1/2″	0.36
2″	0.43
2-1/2″	0.70
3″	0.71
4″	1.00
6″	1.43
8″	1.48
10″	1.50
12″	1.60
14″	1.75
16″	2.00
18″	2.25
20″	2.50
24″	3.00
30″	3.75

The following must be added to the list of components to accurately price HEAT TRACE systems:

1. Expediter fitting and clamp fasteners (F)
2. Junction box and nipple connected to expediter fitting (G)
3. Field installed terminal blocks within junction box
4. Ground lugs
5. Piping from power source to expediter fitting
6. Controls
7. Thermostats
8. Branch wiring
9. Cable splices
10. End of cable terminations
11. Branch piping fittings and boxes

Deduct the following percentages from labor if cable lengths in the same area exceed:

150′ to 250′	10%	351′ to 500′	20%
251′ to 350′	15%	Over 500′	25%

Add the following percentages to labor for elevated installations:

15′ to 20′ high	10%	31′ to 35′ high	40%
21′ to 25′ high	20%	36′ to 40′ high	50%
26′ to 30′ high	30%	Over 40′ high	60%

R238313-20 Spiral-Wrapped Heat Trace Cable (Pitch Table)

In order to increase the amount of heat, occasionally heat trace cable is wrapped in a spiral fashion around a pipe; increasing the number of feet of heater cable per linear foot of pipe.

Engineers first determine the heat loss per foot of pipe (based on the insulating material, its thickness, and the temperature differential across it). A ratio is then calculated by the formula:

$$\text{Feet of Heat Trace per Foot of Pipe} = \frac{\text{Watts/Foot of Heat Loss}}{\text{Watts/Foot of the Cable}}$$

The linear distance between wraps (pitch) is then taken from a chart or table. Generally, the pitch is listed on a drawing leaving the estimator to calculate the total length of heat tape required. An approximation may be taken from this table.

Feet of Heat Trace Per Foot of Pipe																
	Nominal Pipe Size in Inches															
Pitch In Inches	1	1¼	1½	2	2½	3	4	6	8	10	12	14	16	18	20	24
3.5	1.80															
4	1.65															
5	1.46	1.60	1.80													
6	1.34	1.45	1.55	1.75												
7	1.25	1.35	1.43	1.57	1.75											
8	1.20	1.28	1.34	1.45	1.60	1.80										
9	1.16	1.23	1.28	1.37	1.51	1.68										
10	1.13	1.19	1.24	1.32	1.44	1.57	1.82									
15	1.06	1.08	1.10	1.15	1.21	1.29	1.42	1.78								
20	1.04	1.05	1.06	1.08	1.13	1.17	1.25	1.49	1.73							
25		1.04	1.04	1.06	1.08	1.11	1.17	1.33	1.51	1.72						
30				1.04	1.05	1.07	1.12	1.24	1.37	1.54	1.70	1.80				
35					1.06	1.06	1.09	1.17	1.28	1.42	1.54	1.64	1.78			
40						1.05	1.07	1.14	1.22	1.33	1.44	1.52	1.64	1.75		
50							1.05	1.09	1.15	1.22	1.29	1.35	1.44	1.53	1.64	1.83
60								1.06	1.11	1.16	1.21	1.25	1.31	1.39	1.46	1.62
70								1.05	1.08	1.12	1.17	1.19	1.24	1.30	1.35	1.47
80									1.06	1.09	1.13	1.15	1.19	1.24	1.28	1.38
90									1.04	1.06	1.10	1.13	1.16	1.19	1.23	1.32
100										1.05	1.08	1.10	1.13	1.15	1.19	1.23

Note: Common practice would normally limit the lower end of the table to 5% of additional heat and above 80% an engineer would likely opt for two (2) parallel cables.

R312319-90 Wellpoints

A single stage wellpoint system is usually limited to dewatering an average 15′ depth below normal ground water level. Multi-stage systems are employed for greater depth with the pumping equipment installed only at the lowest header level. Ejectors with unlimited lift capacity can be economical when two or more stages of wellpoints can be replaced or when horizontal clearance is restricted, such as in deep trenches or tunneling projects, and where low water flows are expected. Wellpoints are usually spaced on 2-1/2′ to 10′ centers along a header pipe. Wellpoint spacing, header size, and pump size are all determined by the expected flow as dictated by soil conditions.

In almost all soils encountered in wellpoint dewatering, the wellpoints may be jetted into place. Cemented soils and stiff clays may require sand wicks about 12″ in diameter around each wellpoint to increase efficiency and eliminate weeping into the excavation. These sand wicks require 1/2 to 3 C.Y. of washed filter sand and are installed by using a 12″ diameter steel casing and hole puncher jetted into the ground 2′ deeper than the wellpoint. Rock may require predrilled holes.

Labor required for the complete installation and removal of a single stage wellpoint system is in the range of 3/4 to 2 labor-hours per linear foot of header, depending upon jetting conditions, wellpoint spacing, etc.

Continuous pumping is necessary except in some free draining soil where temporary flooding is permissible (as in trenches which are backfilled after each day's work). Good practice requires provision of a stand-by pump during the continuous pumping operation.

Systems for continuous trenching below the water table should be installed three to four times the length of expected daily progress to ensure uninterrupted digging, and header pipe size should not be changed during the job.

For pervious free draining soils, deep wells in place of wellpoints may be economical because of lower installation and maintenance costs. Daily production ranges between two to three wells per day, for 25′ to 40′ depths, to one well per day for depths over 50′.

Detailed analysis and estimating for any dewatering problem is available at no cost from wellpoint manufacturers. Major firms will quote "sufficient equipment" quotes or their affiliates offer lump sum proposals to cover complete dewatering responsibility.

Description for 200′ System with 8″ Header		Quantities
Equipment & Material	Wellpoints 25′ long, 2″ diameter @ 5′ O.C.	40 Each
	Header pipe, 8″ diameter	200 L.F.
	Discharge pipe, 8″ diameter	100 L.F.
	8″ valves	3 Each
	Combination jetting & wellpoint pump (standby)	1 Each
	Wellpoint pump, 8″ diameter	1 Each
	Transportation to and from site	1 Day
	Fuel for 30 days x 60 gal./day	1800 Gallons
	Lubricants for 30 days x 16 lbs./day	480 Lbs.
	Sand for points	40 C.Y.
Labor	Technician to supervise installation	1 Week
	Labor for installation and removal of system	300 Labor-hours
	4 Operators straight time 40 hrs./wk. for 4.33 wks.	693 Hrs.
	4 Operators overtime 2 hrs./wk. for 4.33 wks.	35 Hrs.

Earthwork — R3141 Shoring

R314116-40 Wood Sheet Piling

Wood sheet piling may be used for depths to 20′ where there is no ground water. If moderate ground water is encountered Tongue & Groove sheeting will help to keep it out. When considerable ground water is present, steel sheeting must be used.

For estimating purposes on trench excavation, sizes are as follows:

Depth	Sheeting	Wales	Braces	B.F. per S.F.
To 8′	3 x 12′s	6 x 8′s, 2 line	6 x 8′s, @ 10′	4.0 @ 8′
8′ x 12′	3 x 12′s	10 x 10′s, 2 line	10 x 10′s, @ 9′	5.0 average
12′ to 20′	3 x 12′s	12 x 12′s, 3 line	12 x 12′s, @ 8′	7.0 average

Sheeting to be toed in at least 2′ depending upon soil conditions. A five person crew with an air compressor and sheeting driver can drive and brace 440 SF/day at 8′ deep, 360 SF/day at 12′ deep, and 320 SF/day at 16′ deep. For normal soils, piling can be pulled in 1/3 the time to install. Pulling difficulty increases with the time in the ground. Production can be increased by high pressure jetting.

R314116-45 Steel Sheet Piling

Limiting weights are 22 to 38#/S.F. of wall surface with 27#/S.F. average for usual types and sizes. (Weights of piles themselves are from 30.7#/L.F. to 57#/L.F. but they are 15″ to 21″ wide.) Lightweight sections 12″ to 28″ wide from 3 ga. to 12 ga. thick are also available for shallow excavations. Piles may be driven two at a time with an impact or vibratory hammer (use vibratory to pull) hung from a crane without leads. A reasonable estimate of the life of steel sheet piling is 10 uses with up to 125 uses possible if a vibratory hammer is used. Used piling costs from 50% to 80% of new piling depending on location and market conditions. Sheet piling and H piles can be rented for about 30% of the delivered mill price for the first month and 5% per month thereafter. Allow 1 labor-hour per pile for cleaning and trimming after driving. These costs increase with depth and hydrostatic head. Vibratory drivers are faster in wet granular soils and are excellent for pile extraction. Pulling difficulty increases with the time in the ground and may cost more than driving. It is often economical to abandon the sheet piling, especially if it can be used as the outer wall form. Allow about 1/3 additional length or more for toeing into ground. Add bracing, waler and strut costs. Waler costs can equal the cost per ton of sheeting.

Utilities — R3311 Water Utility Distribution Piping

R331113-80 Piping Designations

There are several systems currently in use to describe pipe and fittings. The following paragraphs will help to identify and clarify classifications of piping systems used for water distribution.

Piping may be classified by schedule. Piping schedules include 5S, 10S, 10, 20, 30, Standard, 40, 60, Extra Strong, 80, 100, 120, 140, 160 and Double Extra Strong. These schedules are dependent upon the pipe wall thickness. The wall thickness of a particular schedule may vary with pipe size.

Ductile iron pipe for water distribution is classified by Pressure Classes such as Class 150, 200, 250, 300 and 350. These classes are actually the rated water working pressure of the pipe in pounds per square inch (psi). The pipe in these pressure classes is designed to withstand the rated water working pressure plus a surge allowance of 100 psi.

The American Water Works Association (AWWA) provides standards for various types of **plastic pipe.** C-900 is the specification for polyvinyl chloride (PVC) piping used for water distribution in sizes ranging from 4″ through 12″. C-901 is the specification for polyethylene (PE) pressure pipe, tubing and fittings used for water distribution in sizes ranging from 1/2″ through 3″. C-905 is the specification for PVC piping sizes 14″ and greater.

PVC pressure-rated pipe is identified using the standard dimensional ratio (SDR) method. This method is defined by the American Society for Testing and Materials (ASTM) Standard D 2241. This pipe is available in SDR numbers 64, 41, 32.5, 26, 21, 17, and 13.5. Pipe with an SDR of 64 will have the thinnest wall while pipe with an SDR of 13.5 will have the thickest wall. When the pressure rating (PR) of a pipe is given in psi, it is based on a line supplying water at 73 degrees F.

The National Sanitation Foundation (NSF) seal of approval is applied to products that can be used with potable water. These products have been tested to ANSI/NSF Standard 14.

Valves and strainers are classified by American National Standards Institute (ANSI) Classes. These Classes are 125, 150, 200, 250, 300, 400, 600, 900, 1500 and 2500. Within each class there is an operating pressure range dependent upon temperature. Design parameters should be compared to the appropriate material dependent, pressure-temperature rating chart for accurate valve selection.

Change Orders

Change Order Considerations

A Change Order is a written document, usually prepared by the design professional, and signed by the owner, the architect/ engineer and the contractor. A change order states the agreement of the parties to: an addition, deletion, or revision in the work; an adjustment in the contract sum, if any; or an adjustment in the contract time, if any. Change orders, or "extras" in the construction process occur after execution of the construction contract and impact architects/ engineers, contractors and owners.

Change orders that are properly recognized and managed can ensure orderly, professional and profitable progress for all who are involved in the project. There are many causes for change orders and change order requests. In all cases, change orders or change order requests should be addressed promptly and in a precise and prescribed manner. The following paragraphs include information regarding change order pricing and procedures.

The Causes of Change Orders

Reasons for issuing change orders include:

- Unforeseen field conditions that require a change in the work
- Correction of design discrepancies, errors or omissions in the contract documents
- Owner-requested changes, either by design criteria, scope of work, or project objectives
- Completion date changes for reasons unrelated to the construction process
- Changes in building code interpretations, or other public authority requirements that require a change in the work
- Changes in availability of existing or new materials and products

Procedures

Properly written contract documents must include the correct change order procedures for all parties—owners, design professionals and contractors—to follow in order to avoid costly delays and litigation.

Being "in the right" is not always a sufficient or acceptable defense. The contract provisions requiring notification and documentation must be adhered to within a defined or reasonable time frame.

The appropriate method of handling change orders is by a written proposal and acceptance by all parties involved. Prior to starting work on a project, all parties should identify their authorized agents who may sign and accept change orders, as well as any limits placed on their authority.

Time may be a critical factor when the need for a change arises. For such cases, the contractor might be directed to proceed on a "time and materials" basis, rather than wait for all paperwork to be processed—a delay that could impede progress. In this situation, the contractor must still follow the prescribed change order procedures, including but not limited to, notification and documentation.

All forms used for change orders should be dated and signed by the proper authority. Lack of documentation can be very costly, especially if legal judgments are to be made and if certain field personnel are no longer available. For time and material change orders, the contractor should keep accurate daily records of all labor and material allocated to the change. Forms that can be used to document change order work are available in *Means Forms for Building Construction Professionals.*

Owners or awarding authorities who do considerable and continual building construction (such as the federal government) realize the inevitability of change orders for numerous reasons, both predictable and unpredictable. As a result, the federal government, the American Institute of Architects (AIA), the Engineers Joint Contract Documents Committee (EJCDC) and other contractor, legal and technical organizations have developed standards and procedures to be followed by all parties to achieve contract continuance and timely completion, while being financially fair to all concerned.

In addition to the change order standards put forth by industry associations, there are also many books available on the subject.

Pricing Change Orders

When pricing change orders, regardless of their cause, the most significant factor is when the change occurs. The need for a change may be perceived in the field or requested by the architect/engineer *before* any of the actual installation has begun, or may evolve or appear *during* construction when the item of work in question is partially installed. In the latter cases, the original sequence of construction is disrupted, along with all contiguous and supporting systems. Change orders cause the greatest impact when they occur *after* the installation has been completed and must be uncovered, or even replaced. Post-completion changes may be caused by necessary design changes, product failure, or changes in the owner's requirements that are not discovered until the building or the systems begin to function.

Specified procedures of notification and record keeping must be adhered to and enforced regardless of the stage of construction: *before, during,* or *after* installation. Some bidding documents anticipate change orders by requiring that unit prices including overhead and profit percentages—for additional as well as deductible changes—be listed. Generally these unit prices do not fully take into account the ripple effect, or impact on other trades, and should be used for general guidance only.

When pricing change orders, it is important to classify the time frame in which the change occurs. There are two basic time frames for change orders: *pre-installation change orders,* which occur before the start of construction, and *post-installation change orders,* which involve reworking after the original installation. Change orders that occur between these stages may be priced according to the extent of work completed using a combination of techniques developed for pricing *pre-* and *post-installation* changes.

The following factors are the basis for a check list to use when preparing a change order estimate.

Factors To Consider When Pricing Change Orders

As an estimator begins to prepare a change order, the following questions should be reviewed to determine their impact on the final price.

General

- Is the change order work *pre-installation* or *post-installation?*

  Change order work costs vary according to how much of the installation has been completed. Once workers have the project scoped in their mind, even though they have not started, it can be difficult to refocus. Consequently they may spend more than the normal amount of time understanding the change. Also, modifications to work in place such as trimming or refitting usually take more time than was initially estimated. The greater the amount of work in place, the more reluctant workers are to change it. Psychologically they may resent the change and as a result the rework takes longer than normal. Post-installation change order estimates must include demolition of existing work as required to accomplish the change. If the work is performed at a later time, additional obstacles such as building finishes may be present which must be protected. Regardless of whether the change occurs pre-installation or post-installation, attempt to isolate the identifiable factors and price them separately. For example, add shipping costs that may be required pre-installation or any demolition required post-installation. Then analyze the potential impact on productivity of psychological and/or learning curve factors and adjust the output rates accordingly. One approach is to break down the typical workday into segments and quantify the impact on each segment. The following chart may be useful as a guide:

Activities (Productivity) Expressed as Percentages of a Workday			
Task	Means Mechanical Cost Data (for New Construction)	Pre-Installation Change Orders	Post-Installation Change Orders
1. Study plans	3%	6%	6%
2. Material procurement	3%	3%	3%
3. Receiving and storing	3%	3%	3%
4. Mobilization	5%	5%	5%
5. Site movement	5%	5%	8%
6. Layout and marking	8%	10%	12%
7. Actual installation	64%	59%	54%
8. Clean-up	3%	3%	3%
9. Breaks—non-productive	6%	6%	6%
Total	100%	100%	100%

Change Order Installation Efficiency

The labor-hours expressed (for new construction) are based on average installation time, using an efficiency level of approximately 60-65%. For change order situations, adjustments to this efficiency level should reflect the daily labor-hour allocation for that particular occurrence.

If any of the specific percentages expressed in the above chart do not apply to a particular project situation, then those percentage points should be reallocated to the appropriate task(s). Example: Using data for new construction, assume there is no new material being utilized. The percentages for Tasks 2 and 3 would therefore be reallocated to other tasks. If the time required for Tasks 2 and 3 can now be applied to installation, we can add the time allocated for *Material Procurement* and *Receiving and Storing* to the *Actual Installation* time for new construction, thereby increasing the Actual Installation percentage.

This chart shows that, due to reduced productivity, labor costs will be higher than those for new construction by 5% to 15% for pre-installation change orders and by 15% to 25% for post-installation change orders. Each job and change order is unique and must be examined individually. Many factors, covered elsewhere in this section, can each have a significant impact on productivity and change order costs. All such factors should be considered in every case.

- Will the change substantially delay the original completion date?

  A significant change in the project may cause the original completion date to be extended. The extended schedule may subject the contractor to new wage rates dictated by relevant labor contracts. Project supervision and other project overhead must also be extended beyond the original completion date. The schedule extension may also put installation into a new weather season. For example, underground piping scheduled for October installation was delayed until January. As a result, frost penetrated the trench area, thereby changing the degree of difficulty of the task. Changes and delays may have a ripple effect throughout the project. This effect must be analyzed and negotiated with the owner.

- What is the net effect of a deduct change order?

  In most cases, change orders resulting in a deduction or credit reflect only bare costs. The contractor may retain the overhead and profit based on the original bid.

Materials

- Will you have to pay more or less for the new material, required by the change order, than you paid for the original purchase?

  The same material prices or discounts will usually apply to materials purchased for change orders as new construction. In some instances, however, the contractor may forfeit the advantages of competitive pricing for change orders. Consider the following example:

  A contractor purchased over $20,000 worth of fan coil units for an installation, and obtained the maximum discount. Some time later it was determined the project required an additional matching unit. The contractor has to purchase this unit from the original supplier to ensure a match. The supplier at this time may not discount the unit because of the small quantity, and the fact that he is no longer in a competitive situation. The impact of quantity on purchase can add between 0% and 25% to material prices and/or subcontractor quotes.

- If materials have been ordered or delivered to the job site, will they be subject to a cancellation charge or restocking fee?

  Check with the supplier to determine if ordered materials are subject to a cancellation charge. Delivered materials not used as result of a change order may be subject to a restocking fee if returned to the supplier. Common restocking charges run between 20% and 40%. Also, delivery charges to return the goods to the supplier must be added.

Labor

- How efficient is the existing crew at the actual installation?

  Is the same crew that performed the initial work going to do the change order? Possibly the change consists of the installation of a unit identical to one already installed; therefore the change should take less time. Be sure to consider this potential productivity increase and modify the productivity rates accordingly.

- If the crew size is increased, what impact will that have on supervision requirements?

  Under most bargaining agreements or management practices, there is a point at which a working foreman is replaced by a nonworking foreman. This replacement increases project overhead by adding a nonproductive worker. If additional workers are added to accelerate the project or to perform changes while maintaining the schedule, be sure to add additional supervision time if warranted. Calculate the hours involved and the additional cost directly if possible.

- What are the other impacts of increased crew size?

  The larger the crew, the greater the potential for productivity to decrease. Some of the factors that cause this productivity loss are: overcrowding (producing restrictive conditions in the working space), and possibly a shortage of any special tools and equipment required. Such factors affect not only the crew working on the elements directly involved in the change order, but other crews whose movement may also be hampered.

  As the crew increases, check its basic composition for changes by the addition or deletion of apprentices or nonworking foreman and quantify the potential effects of equipment shortages or other logistical factors.

- As new crews, unfamiliar with the project, are brought onto the site, how long will it take them to become oriented to the project requirements?

  The orientation time for a new crew to become 100% effective varies with the site and type of project. Orientation is easiest at a new construction site, and most difficult at existing, very restrictive renovation sites. The type of work also affects orientation time. When all elements of the work are exposed, such as concrete or masonry work, orientation is decreased. When the work is concealed or less visible, such as existing electrical systems, orientation takes longer. Usually orientation can be accomplished in one day or less. Costs for added orientation should be itemized and added to the total estimated cost.

- How much actual production can be gained by working overtime?

  Short term overtime can be used effectively to accomplish more work in a day. However, as overtime is scheduled to run beyond several weeks, studies have shown marked decreases in output. The following chart shows the effect of long term overtime on worker efficiency. If the anticipated change requires extended overtime to keep the job on schedule, these factors can be used as a guide to predict the impact on time and cost. Add project overhead, particularly supervision, that may also be incurred.

		Production Efficiency					Payroll Cost Factors	
Days per Week	Hours per Day	1 Week	2 Weeks	3 Weeks	4 Weeks	Average 4 Weeks	@ 1-1/2 Times	@ 2 Times
	8	100%	100%	100%	100%	100%	100%	100%
	9	100	100	95	90	96.25	105.6	111.1
5	10	100	95	90	85	91.25	110.0	120.0
	11	95	90	75	65	81.25	113.6	127.3
	12	90	85	70	60	76.25	116.7	133.3
	8	100	100	95	90	96.25	108.3	116.7
	9	100	95	90	85	92.50	113.0	125.9
6	10	95	90	85	80	87.50	116.7	133.3
	11	95	85	70	65	78.75	119.7	139.4
	12	90	80	65	60	73.75	122.2	144.4
	8	100	95	85	75	88.75	114.3	128.6
	9	95	90	80	70	83.75	118.3	136.5
7	10	90	85	75	65	78.75	121.4	142.9
	11	85	80	65	60	72.50	124.0	148.1
	12	85	75	60	55	68.75	126.2	152.4

Effects of Overtime

Caution: Under many labor agreements, Sundays and holidays are paid at a higher premium than the normal overtime rate.

The use of long-term overtime is counterproductive on almost any construction job; that is, the longer the period of overtime, the lower the actual production rate. Numerous studies have been conducted, and while they have resulted in slightly different numbers, all reach the same conclusion. The figure above tabulates the effects of overtime work on efficiency.

As illustrated, there can be a difference between the *actual* payroll cost per hour and the *effective* cost per hour for overtime work. This is due to the reduced production efficiency with the increase in weekly hours beyond 40. This difference between actual and effective cost results from overtime work over a prolonged period. Short-term overtime work does not result in as great a reduction in efficiency, and in such cases, effective cost may not vary significantly from the actual payroll cost. As the total hours per week are increased on a regular basis, more time is lost because of fatigue, lowered morale, and an increased accident rate.

As an example, assume a project where workers are working 6 days a week, 10 hours per day. From the figure above (based on productivity studies), the average effective productive hours over a four-week period are:

$$0.875 \times 60 = 52.5$$

Depending upon the locale and day of week, overtime hours may be paid at time and a half or double time. For time and a half, the overall (average) *actual* payroll cost (including regular and overtime hours) is determined as follows:

$$\frac{40 \text{ reg. hrs.} + (20 \text{ overtime hrs.} \times 1.5)}{60 \text{ hrs.}} = 1.167$$

Based on 60 hours, the payroll cost per hour will be 116.7% of the normal rate at 40 hours per week. However, because the effective production (efficiency) for 60 hours is reduced to the equivalent of 52.5 hours, the effective cost of overtime is calculated as follows:

For time and a half:

$$\frac{40 \text{ reg. hrs.} + (20 \text{ overtime hrs.} \times 1.5)}{52.5 \text{ hrs.}} = 1.33$$

Installed cost will be 133% of the normal rate (for labor).

Thus, when figuring overtime, the actual cost per unit of work will be higher than the apparent overtime payroll dollar increase, due to the reduced productivity of the longer workweek. These efficiency calculations are true only for those cost factors determined by hours worked. Costs that are applied weekly or monthly, such as equipment rentals, will not be similarly affected.

Equipment

- What equipment is required to complete the change order?

  Change orders may require extending the rental period of equipment already on the job site, or the addition of special equipment brought in to accomplish the change work. In either case, the additional rental charges and operator labor charges must be added.

Summary

The preceding considerations and others you deem appropriate should be analyzed and applied to a change order estimate. The impact of each should be quantified and listed on the estimate to form an audit trail.

Change orders that are properly identified, documented, and managed help to ensure the orderly, professional and profitable progress of the work. They also minimize potential claims or disputes at the end of the project.

Estimating Tips

- The cost figures in this Square Foot Cost section were derived from approximately 11,200 projects contained in the RSMeans database of completed construction projects. They include the contractor's overhead and profit, but do not generally include architectural fees or land costs. The figures have been adjusted to January of the current year. New projects are added to our files each year, and outdated projects are discarded. For this reason, certain costs may not show a uniform annual progression. In no case are all subdivisions of a project listed.
- These projects were located throughout the U.S. and reflect a tremendous variation in square foot (S.F.) and cubic foot (C.F.) costs. This is due to differences, not only in labor and material costs, but also in individual owners' requirements. For instance, a bank in a large city would have different features than one in a rural area. This is true of all the different types of buildings analyzed. Therefore, caution should be exercised when using these Square Foot costs. For example, for court houses, costs in the database are local court house costs and will not apply to the larger, more elaborate federal court houses. As a general rule, the projects in the 1/4 column do not include any site work or equipment, while the projects in the 3/4 column may include both equipment and site work. The median figures do not generally include site work.
- None of the figures "go with" any others. All individual cost items were computed and tabulated separately. Thus, the sum of the median figures for Plumbing, HVAC and Electrical will not normally total up to the total Mechanical and Electrical costs arrived at by separate analysis and tabulation of the projects.
- Each building was analyzed as to total and component costs and percentages. The figures were arranged in ascending order with the results tabulated as shown. The 1/4 column shows that 25% of the projects had lower costs and 75% had higher. The 3/4 column shows that 75% of the projects had lower costs and 25% had higher. The median column shows that 50% of the projects had lower costs and 50% had higher.
- There are two times when square foot costs are useful. The first is in the conceptual stage when no details are available. Then square foot costs make a useful starting point. The second is after the bids are in and the costs can be worked back into their appropriate units for information purposes. As soon as details become available in the project design, the square foot approach should be discontinued and the project priced as to its particular components. When more precision is required, or for estimating the replacement cost of specific buildings, the current edition of *RSMeans Square Foot Costs* should be used.
- In using the figures in this section, it is recommended that the median column be used for preliminary figures if no additional information is available. The median figures, when multiplied by the total city construction cost index figures (see City Cost Indexes) and then multiplied by the project size modifier at the end of this section, should present a fairly accurate base figure, which would then have to be adjusted in view of the estimator's experience, local economic conditions, code requirements, and the owner's particular requirements. There is no need to factor the percentage figures, as these should remain constant from city to city. All tabulations mentioning air conditioning had at least partial air conditioning.
- The editors of this book would greatly appreciate receiving cost figures on one or more of your recent projects, which would then be included in the averages for next year. All cost figures received will be kept confidential, except that they will be averaged with other similar projects to arrive at Square Foot cost figures for next year's book. See the last page of the book for details and the discount available for submitting one or more of your projects.

50 17 00 | S.F. Costs

		50 17 00 | S.F. Costs	UNIT	UNIT COSTS 1/4	UNIT COSTS MEDIAN	UNIT COSTS 3/4	% OF TOTAL 1/4	% OF TOTAL MEDIAN	% OF TOTAL 3/4	
01	0010	**APARTMENTS Low Rise (1 to 3 story)**	S.F.	67	84.50	112				01
	0020	Total project cost	C.F.	6	8	9.85				
	0100	Site work	S.F.	5.75	7.85	13.75	6.05%	10.55%	14.05%	
	0500	Masonry		1.32	3.07	5.30	1.54%	3.67%	6.35%	
	1500	Finishes		7.10	9.75	12.10	9.05%	10.75%	12.85%	
	1800	Equipment		2.19	3.32	4.94	2.73%	4.03%	5.95%	
	2720	Plumbing		5.20	6.70	8.55	6.65%	8.95%	10.05%	
	2770	Heating, ventilating, air conditioning		3.33	4.10	6.05	4.20%	5.60%	7.60%	
	2900	Electrical		3.88	5.15	6.90	5.20%	6.65%	8.40%	
	3100	Total: Mechanical & Electrical		13.45	17.10	21.50	15.90%	18.05%	23%	
	9000	Per apartment unit, total cost	Apt.	62,500	95,500	140,500				
	9500	Total: Mechanical & Electrical	"	11,800	18,600	24,300				
02	0010	**APARTMENTS Mid Rise (4 to 7 story)**	S.F.	89	107	133				02
	0020	Total project costs	C.F.	6.95	9.60	13.10				
	0100	Site work	S.F.	3.56	7.05	12.65	5.25%	6.70%	9.15%	
	0500	Masonry		5.90	8.15	11.15	5.10%	7.25%	10.50%	
	1500	Finishes		11.20	14.60	18.40	10.55%	13.45%	17.70%	
	1800	Equipment		2.58	3.90	5.05	2.54%	3.48%	4.31%	
	2500	Conveying equipment		1.91	2.44	2.96	1.94%	2.27%	2.69%	
	2720	Plumbing		5.20	8.35	8.85	5.70%	7.20%	8.95%	
	2900	Electrical		5.85	7.95	9.65	6.65%	7.20%	8.95%	
	3100	Total: Mechanical & Electrical		18.75	23.50	28.50	18.50%	21%	23%	
	9000	Per apartment unit, total cost	Apt.	100,500	118,500	196,500				
	9500	Total: Mechanical & Electrical	"	19,000	22,000	27,800				
03	0010	**APARTMENTS High Rise (8 to 24 story)**	S.F.	101	116	139				03
	0020	Total project costs	C.F.	9.80	11.40	14.55				
	0100	Site work	S.F.	3.66	5.90	8.30	2.58%	4.84%	6.15%	
	0500	Masonry		5.85	10.60	13.20	4.74%	9.65%	11.05%	
	1500	Finishes		11.20	14	16.50	9.75%	11.80%	13.70%	
	1800	Equipment		3.24	3.99	5.30	2.78%	3.49%	4.35%	
	2500	Conveying equipment		2.29	3.48	4.73	2.23%	2.78%	3.37%	
	2720	Plumbing		7.40	8.75	12.25	6.80%	7.20%	10.45%	
	2900	Electrical		6.95	8.75	11.85	6.45%	7.65%	8.80%	
	3100	Total: Mechanical & Electrical		21	26.50	32	17.95%	22.50%	24.50%	
	9000	Per apartment unit, total cost	Apt.	105,000	115,500	160,000				
	9500	Total: Mechanical & Electrical	"	22,700	25,900	27,400				
04	0010	**AUDITORIUMS**	S.F.	104	141	203				04
	0020	Total project costs	C.F.	6.55	9.15	13.10				
	2720	Plumbing	S.F.	6.35	9.20	11.15	5.85%	7.20%	8.70%	
	2900	Electrical		8.10	11.70	15.75	6.80%	8.95%	11.30%	
	3100	Total: Mechanical & Electrical		23	46	56	24.50%	30.50%	31.50%	
05	0010	**AUTOMOTIVE SALES**	S.F.	77.50	105	130				05
	0020	Total project costs	C.F.	5.10	6.15	7.95				
	2720	Plumbing	S.F.	3.54	6.15	6.70	2.89%	6.05%	6.50%	
	2770	Heating, ventilating, air conditioning		5.45	8.35	9	4.61%	10%	10.35%	
	2900	Electrical		6.25	9.80	13.30	7.25%	9.80%	12.15%	
	3100	Total: Mechanical & Electrical		19.60	28	33.50	19.15%	20.50%	22%	
06	0010	**BANKS**	S.F.	151	189	240				06
	0020	Total project costs	C.F.	10.85	14.75	19.50				
	0100	Site work	S.F.	17.35	26.50	38	7.75%	12.45%	17%	
	0500	Masonry		7.90	15.15	26.50	3.36%	6.95%	10.35%	
	1500	Finishes		13.45	19.50	24	5.80%	8.45%	11.25%	
	1800	Equipment		5.70	12.60	26	1.34%	5.95%	10.65%	
	2720	Plumbing		4.76	6.80	9.95	2.82%	3.90%	4.93%	
	2770	Heating, ventilating, air conditioning		9.05	12.10	16.10	4.86%	7.15%	8.50%	
	2900	Electrical		14.35	19.20	25	8.20%	10.20%	12.15%	
	3100	Total: Mechanical & Electrical		33.50	45.50	53	16.25%	19.40%	23%	
	3500	See also Divisions 11020 & 11030 (MF2004 11 16 00 & 11 17 00)								

	50 17 00 | S.F. Costs		UNIT	UNIT COSTS 1/4	UNIT COSTS MEDIAN	UNIT COSTS 3/4	% OF TOTAL 1/4	% OF TOTAL MEDIAN	% OF TOTAL 3/4	
13	0010	**CHURCHES**	S.F.	102	130	168				13
	0020	Total project costs	C.F.	6.35	8	10.55				
	1800	Equipment	S.F.	1.22	2.93	6.25	.95%	2.11%	4.50%	
	2720	Plumbing		3.99	5.55	8.20	3.51%	4.96%	6.25%	
	2770	Heating, ventilating, air conditioning		9.30	12.15	17.20	7.50%	10%	12%	
	2900	Electrical		8.60	11.80	15.90	7.35%	8.80%	11%	
	3100	Total: Mechanical & Electrical	↓	26.50	34.50	46	18.30%	22%	24.50%	
	3500	See also Division 11040 (MF2004 11 91 00)								
15	0010	**CLUBS, COUNTRY**	S.F.	110	132	167				15
	0020	Total project costs	C.F.	8.85	10.80	14.90				
	2720	Plumbing	S.F.	6.65	9.85	22.50	5.60%	7.90%	10%	
	2900	Electrical		8.65	11.85	15.45	7%	8.95%	11%	
	3100	Total: Mechanical & Electrical	↓	46	57.50	60.50	19%	26.50%	29.50%	
17	0010	**CLUBS, SOCIAL Fraternal**	S.F.	87.50	126	169				17
	0020	Total project costs	C.F.	5.50	8.30	9.90				
	2720	Plumbing	S.F.	5.50	6.85	10.40	5.60%	6.90%	8.55%	
	2770	Heating, ventilating, air conditioning		7.95	9.60	12.35	8.20%	9.25%	14.40%	
	2900	Electrical		6.60	10.85	12.40	6.50%	9.50%	10.55%	
	3100	Total: Mechanical & Electrical	↓	19.45	37	47	21%	23%	23.50%	
18	0010	**CLUBS, Y.M.C.A.**	S.F.	110	143	179				18
	0020	Total project costs	C.F.	5.05	8.50	12.65				
	2720	Plumbing	S.F.	6.95	13.85	15.55	5.65%	7.60%	10.85%	
	2900	Electrical		8.80	11	15.15	6.05%	7.60%	9.25%	
	3100	Total: Mechanical & Electrical	↓	33.50	37.50	51.50	18.40%	21.50%	28.50%	
19	0010	**COLLEGES Classrooms & Administration**	S.F.	111	152	201				19
	0020	Total project costs	C.F.	8.10	11.80	18.20				
	0500	Masonry	S.F.	8.15	15.35	18.85	5.65%	8.25%	10.50%	
	2720	Plumbing		5.65	11.60	20.50	5.10%	6.60%	8.95%	
	2900	Electrical		9.25	14.05	19.20	7.70%	9.85%	12%	
	3100	Total: Mechanical & Electrical	↓	37	51.50	61.50	24%	28%	31.50%	
21	0010	**COLLEGES Science, Engineering, Laboratories**	S.F.	207	242	280				21
	0020	Total project costs	C.F.	11.85	17.30	19.65				
	1800	Equipment	S.F.	11.50	26	28.50	2%	6.45%	12.65%	
	2900	Electrical		17.05	23.50	37	7.10%	9.40%	12.10%	
	3100	Total: Mechanical & Electrical	↓	63.50	75	116	28.50%	31.50%	41%	
	3500	See also Division 11600 (MF2004 11 53 00)								
23	0010	**COLLEGES Student Unions**	S.F.	132	179	216				23
	0020	Total project costs	C.F.	7.35	9.65	11.90				
	3100	Total: Mechanical & Electrical	S.F.	49.50	53.50	63.50	23.50%	26%	29%	
25	0010	**COMMUNITY CENTERS**	S.F.	109	134	181				25
	0020	Total project costs	C.F.	7.10	10.15	13.15				
	1800	Equipment	S.F.	2.64	4.46	7.10	1.48%	3.01%	5.45%	
	2720	Plumbing		5.15	9	12.30	4.85%	7%	8.95%	
	2770	Heating, ventilating, air conditioning		8.25	12.10	17.20	6.80%	10.35%	12.90%	
	2900	Electrical		8.90	11.90	17.05	7.30%	9%	10.45%	
	3100	Total: Mechanical & Electrical	↓	31	38	54	20%	25%	31%	
28	0010	**COURT HOUSES**	S.F.	157	180	233				28
	0020	Total project costs	C.F.	12.05	14.40	18.15				
	2720	Plumbing	S.F.	7.45	10.45	15	5.95%	7.45%	8.20%	
	2900	Electrical		16.70	18.50	27	8.90%	10.45%	11.55%	
	3100	Total: Mechanical & Electrical	↓	42.50	58	64	22.50%	27.50%	30.50%	
30	0010	**DEPARTMENT STORES**	S.F.	58	78.50	99				30
	0020	Total project costs	C.F.	3.11	4.03	5.50				
	2720	Plumbing	S.F.	1.81	2.28	3.47	1.82%	4.21%	5.90%	
	2770	Heating, ventilating, air conditioning	↓	5.30	8.15	12.25	8.20%	9.10%	14.80%	

50 17 00 | S.F. Costs

			UNIT	UNIT COSTS 1/4	UNIT COSTS MEDIAN	UNIT COSTS 3/4	% OF TOTAL 1/4	% OF TOTAL MEDIAN	% OF TOTAL 3/4	
30	2900	Electrical	S.F.	6.65	9.15	10.80	9.05%	12.15%	14.95%	30
	3100	Total: Mechanical & Electrical		11.75	15	26.50	13.20%	21.50%	50%	
31	0010	**DORMITORIES Low Rise (1 to 3 story)**	S.F.	109	143	185				31
	0020	Total project costs	C.F.	6.20	10.05	15.05				
	2720	Plumbing	S.F.	6.60	8.85	11.20	8.05%	9%	9.65%	
	2770	Heating, ventilating, air conditioning		7	8.40	11.15	4.61%	8.05%	10%	
	2900	Electrical		7.15	10.95	14.85	6.25%	8.65%	9.50%	
	3100	Total: Mechanical & Electrical		36.50	40.50	63.50	21.50%	24%	27%	
	9000	Per bed, total cost	Bed	46,700	52,000	111,000				
32	0010	**DORMITORIES Mid Rise (4 to 8 story)**	S.F.	135	176	218				32
	0020	Total project costs	C.F.	14.90	16.35	19.60				
	2900	Electrical	S.F.	14.35	16.30	22	8.20%	10.20%	11.95%	
	3100	Total: Mechanical & Electrical	″	40	41.50	81.50	25.50%	34.50%	37.50%	
	9000	Per bed, total cost	Bed	19,200	43,900	111,000				
34	0010	**FACTORIES**	S.F.	51	76.50	117				34
	0020	Total project costs	C.F.	3.28	4.89	8.10				
	0100	Site work	S.F.	5.85	10.65	16.85	6.95%	11.45%	17.95%	
	2720	Plumbing		2.76	5.15	8.50	3.73%	6.05%	8.10%	
	2770	Heating, ventilating, air conditioning		5.35	7.70	10.40	5.25%	8.45%	11.35%	
	2900	Electrical		6.35	10.05	15.35	8.10%	10.50%	14.20%	
	3100	Total: Mechanical & Electrical		18.15	24.50	37	21%	28.50%	35.50%	
36	0010	**FIRE STATIONS**	S.F.	102	139	182				36
	0020	Total project costs	C.F.	5.95	8.15	10.85				
	0500	Masonry	S.F.	15	27	34.50	8.60%	11.65%	16.45%	
	1140	Roofing		3.31	8.95	10.20	1.90%	4.94%	5.05%	
	1580	Painting		2.57	3.84	3.93	1.37%	1.57%	2.07%	
	1800	Equipment		1.31	2.57	4.47	.74%	1.98%	3.54%	
	2720	Plumbing		5.70	9.15	13.25	5.85%	7.35%	9.45%	
	2770	Heating, ventilating, air conditioning		5.60	9.10	14.10	5.15%	7.40%	9.40%	
	2900	Electrical		7.20	12.55	16.95	6.85%	8.65%	10.60%	
	3100	Total: Mechanical & Electrical		35.50	45.50	52.50	18.75%	23%	27%	
37	0010	**FRATERNITY HOUSES & Sorority Houses**	S.F.	102	131	179				37
	0020	Total project costs	C.F.	10.10	10.55	12.70				
	2720	Plumbing	S.F.	7.65	8.80	16.10	6.80%	8%	10.85%	
	2900	Electrical		6.70	14.45	17.70	6.60%	9.90%	10.65%	
	3100	Total: Mechanical & Electrical		17.90	26	31		15.10%	15.90%	
38	0010	**FUNERAL HOMES**	S.F.	107	146	265				38
	0020	Total project costs	C.F.	10.95	12.15	23.50				
	2900	Electrical	S.F.	4.73	8.65	9.50	3.58%	4.44%	5.95%	
	3100	Total: Mechanical & Electrical		16.75	24.50	34	12.90%	12.90%	12.90%	
39	0010	**GARAGES, COMMERCIAL (Service)**	S.F.	60.50	93.50	129				39
	0020	Total project costs	C.F.	3.99	5.90	8.55				
	1800	Equipment	S.F.	3.42	7.70	12	2.21%	4.62%	6.80%	
	2720	Plumbing		4.20	6.45	11.80	5.45%	7.85%	10.65%	
	2730	Heating & ventilating		5.50	7.30	9.90	5.25%	6.85%	8.20%	
	2900	Electrical		5.75	8.80	12.65	7.15%	9.25%	10.85%	
	3100	Total: Mechanical & Electrical		12.70	24.50	36	12.35%	17.40%	26%	
40	0010	**GARAGES, MUNICIPAL (Repair)**	S.F.	89	119	167				40
	0020	Total project costs	C.F.	5.55	7.05	12.15				
	0500	Masonry	S.F.	8.35	16.35	25.50	5.60%	9.15%	12.50%	
	2720	Plumbing		4	7.65	14.45	3.59%	6.70%	7.95%	
	2730	Heating & ventilating		6.85	9.90	19.10	6.15%	7.45%	13.50%	
	2900	Electrical		6.60	10.35	14.90	6.65%	8.15%	11.15%	
	3100	Total: Mechanical & Electrical		30.50	42	61	21.50%	25.50%	28.50%	

	50 17 00 | S.F. Costs		UNIT	UNIT COSTS 1/4	UNIT COSTS MEDIAN	UNIT COSTS 3/4	% OF TOTAL 1/4	% OF TOTAL MEDIAN	% OF TOTAL 3/4	
41	0010	**GARAGES, PARKING**	S.F.	34.50	50.50	87				41
	0020	Total project costs	C.F.	3.26	4.42	6.45				
	2720	Plumbing	S.F.	.98	1.52	2.35	1.72%	2.70%	3.85%	
	2900	Electrical		1.90	2.33	3.67	4.33%	5.20%	6.30%	
	3100	Total: Mechanical & Electrical	↓	3.89	5.40	6.75	7%	8.90%	11.05%	
	3200									
	9000	Per car, total cost	Car	14,600	18,400	23,400				
43	0010	**GYMNASIUMS**	S.F.	96.50	129	174				43
	0020	Total project costs	C.F.	4.81	6.50	8				
	1800	Equipment	S.F.	2.29	4.30	8.25	1.81%	3.30%	6.70%	
	2720	Plumbing		6.10	7.25	9.35	4.65%	6.40%	7.75%	
	2770	Heating, ventilating, air conditioning		6.55	10	20	5.15%	9.05%	11.10%	
	2900	Electrical		7.40	10.10	13.30	6.60%	8.30%	10.30%	
	3100	Total: Mechanical & Electrical	↓	27	36.50	43.50	19.75%	23.50%	29%	
	3500	See also Division 11480 (MF2004 11 67 00)								
46	0010	**HOSPITALS**	S.F.	185	228	315				46
	0020	Total project costs	C.F.	14.05	17.45	25				
	1800	Equipment	S.F.	4.70	9.05	15.60	.96%	2.63%	5%	
	2720	Plumbing		15.95	22.50	28.50	7.60%	9.10%	10.85%	
	2770	Heating, ventilating, air conditioning		23.50	30	40.50	7.80%	12.95%	16.65%	
	2900	Electrical		20	26.50	39	9.85%	11.55%	13.90%	
	3100	Total: Mechanical & Electrical	↓	57	81.50	123	27%	33.50%	36.50%	
	9000	Per bed or person, total cost	Bed	214,500	295,500	340,500				
	9900	See also Division 11700 (MF2004 11 71 00)								
48	0010	**HOUSING For the Elderly**	S.F.	91	115	142				48
	0020	Total project costs	C.F.	6.50	9	11.50				
	0100	Site work	S.F.	6.35	9.85	14.45	5.05%	7.90%	12.10%	
	0500	Masonry		2.77	10.35	15.15	1.30%	6.05%	11%	
	1800	Equipment		2.20	3.03	4.82	1.88%	3.23%	4.43%	
	2510	Conveying systems		2.21	2.98	4.03	1.78%	2.20%	2.81%	
	2720	Plumbing		6.75	8.60	10.85	8.15%	9.55%	10.50%	
	2730	Heating, ventilating, air conditioning		3.47	4.91	7.35	3.30%	5.60%	7.25%	
	2900	Electrical		6.80	9.20	11.80	7.30%	8.50%	10.25%	
	3100	Total: Mechanical & Electrical	↓	23.50	28	37	18.10%	22.50%	29%	
	9000	Per rental unit, total cost	Unit	84,500	99,000	110,500				
	9500	Total: Mechanical & Electrical	"	18,800	21,700	25,300				
50	0010	**HOUSING Public (Low Rise)**	S.F.	76.50	106	138				50
	0020	Total project costs	C.F.	6.80	8.50	10.55				
	0100	Site work	S.F.	9.75	14.05	22.50	8.35%	11.75%	16.50%	
	1800	Equipment		2.08	3.39	5.15	2.26%	3.03%	4.24%	
	2720	Plumbing		5.50	7.30	9.25	7.15%	9.05%	11.60%	
	2730	Heating, ventilating, air conditioning		2.77	5.40	5.90	4.26%	6.05%	6.45%	
	2900	Electrical		4.63	6.90	9.55	5.10%	6.55%	8.25%	
	3100	Total: Mechanical & Electrical	↓	22	28.50	31.50	14.50%	17.55%	26.50%	
	9000	Per apartment, total cost	Apt.	84,000	95,500	120,000				
	9500	Total: Mechanical & Electrical	"	17,900	22,100	24,500				
51	0010	**ICE SKATING RINKS**	S.F.	65.50	153	168				51
	0020	Total project costs	C.F.	4.81	4.92	5.65				
	2720	Plumbing	S.F.	2.44	4.58	4.69	3.12%	3.23%	5.65%	
	2900	Electrical		7	10.75	11.35	6.30%	10.15%	15.05%	
	3100	Total: Mechanical & Electrical	↓	11.65	16.50	20.50	18.95%	18.95%	18.95%	
52	0010	**JAILS**	S.F.	199	257	330				52
	0020	Total project costs	C.F.	17.95	25	29.50				
	1800	Equipment	S.F.	7.75	23	39	2.80%	5.55%	10.35%	
	2720	Plumbing		19.15	25.50	34	7%	8.90%	13.35%	
	2770	Heating, ventilating, air conditioning		17.95	24	46.50	7.50%	9.45%	17.75%	
	2900	Electrical	↓	20.50	27.50	35	8.20%	11.55%	14.95%	

		50 17 00 | S.F. Costs	UNIT	UNIT COSTS			% OF TOTAL			
				1/4	MEDIAN	3/4	1/4	MEDIAN	3/4	
52	3100	Total: Mechanical & Electrical	S.F.	54	99	117	28%	30%	34%	52
53	0010	**LIBRARIES**	S.F.	126	161	207				53
	0020	Total project costs	C.F.	8.60	10.80	13.75				
	0500	Masonry	S.F.	9.85	17.40	29	5.80%	7.60%	11.80%	
	1800	Equipment		1.72	4.63	7	.37%	1.50%	4.16%	
	2720	Plumbing		4.65	6.75	9.15	3.38%	4.60%	5.70%	
	2770	Heating, ventilating, air conditioning		10.30	17.40	22.50	7.80%	10.95%	12.80%	
	2900	Electrical		12.85	16.65	21	8.30%	10.30%	11.95%	
	3100	Total: Mechanical & Electrical		38	48	58.50	19.65%	23%	26.50%	
54	0010	**LIVING, ASSISTED**	S.F.	116	137	162				54
	0020	Total project costs	C.F.	9.80	11.45	13				
	0500	Masonry	S.F.	3.42	4.08	4.79	2.37%	3.16%	3.86%	
	1800	Equipment		2.65	3.08	3.94	2.12%	2.45%	2.66%	
	2720	Plumbing		9.75	13.05	13.55	6.05%	8.15%	10.60%	
	2770	Heating, ventilating, air conditioning		11.60	12.10	13.25	7.95%	9.35%	9.70%	
	2900	Electrical		11.40	12.60	14.55	9%	10%	10.70%	
	3100	Total: Mechanical & Electrical		32	37.50	43	26%	29%	31.50%	
55	0010	**MEDICAL CLINICS**	S.F.	118	146	185				55
	0020	Total project costs	C.F.	8.70	11.25	14.95				
	1800	Equipment	S.F.	3.20	6.70	10.45	1.05%	2.94%	6.35%	
	2720	Plumbing		7.85	11.10	14.80	6.15%	8.40%	10.10%	
	2770	Heating, ventilating, air conditioning		9.40	12.30	18.10	6.65%	8.85%	11.35%	
	2900	Electrical		10.20	14.45	18.90	8.10%	10%	12.25%	
	3100	Total: Mechanical & Electrical		32.50	44	60.50	22.50%	27%	33.50%	
	3500	See also Division 11700 (MF2004 11 71 00)								
57	0010	**MEDICAL OFFICES**	S.F.	112	138	169				57
	0020	Total project costs	C.F.	8.30	11.25	15.25				
	1800	Equipment	S.F.	3.68	7.25	10.35	.70%	5.10%	7.05%	
	2720	Plumbing		6.15	9.50	12.80	5.60%	6.80%	8.50%	
	2770	Heating, ventilating, air conditioning		7.45	10.75	14.20	6.10%	8%	9.70%	
	2900	Electrical		9.10	13.05	18.10	7.65%	9.80%	11.70%	
	3100	Total: Mechanical & Electrical		26	35.50	51.50	19.35%	23%	29%	
59	0010	**MOTELS**	S.F.	70.50	102	133				59
	0020	Total project costs	C.F.	6.25	8.40	13.70				
	2720	Plumbing	S.F.	7.15	9.05	10.80	9.45%	10.60%	12.55%	
	2770	Heating, ventilating, air conditioning		4.34	6.50	11.60	5.60%	5.60%	10%	
	2900	Electrical		6.65	8.40	10.45	7.45%	9.05%	10.45%	
	3100	Total: Mechanical & Electrical		22.50	28.50	48.50	18.50%	24%	25.50%	
	5000									
	9000	Per rental unit, total cost	Unit	35,800	68,000	73,500				
	9500	Total: Mechanical & Electrical	″	6,975	10,600	12,300				
60	0010	**NURSING HOMES**	S.F.	110	142	174				60
	0020	Total project costs	C.F.	8.65	10.80	14.80				
	1800	Equipment	S.F.	3.47	4.60	7.70	2.02%	3.62%	4.99%	
	2720	Plumbing		9.45	14.30	17.25	8.75%	10.10%	12.70%	
	2770	Heating, ventilating, air conditioning		9.95	15.10	20	9.70%	11.45%	11.80%	
	2900	Electrical		10.90	13.60	18.55	9.40%	10.55%	12.45%	
	3100	Total: Mechanical & Electrical		26	36.50	61	26%	29.50%	30.50%	
	9000	Per bed or person, total cost	Bed	48,900	61,000	79,000				
61	0010	**OFFICES Low Rise (1 to 4 story)**	S.F.	93	120	155				61
	0020	Total project costs	C.F.	6.65	9.15	12.05				
	0100	Site work	S.F.	7.20	12.60	18.85	5.90%	9.70%	13.60%	
	0500	Masonry		3.46	7.25	13.15	2.62%	5.50%	8.45%	
	1800	Equipment		.99	1.93	5.25	.69%	1.50%	3.63%	
	2720	Plumbing		3.32	5.15	7.50	3.66%	4.50%	6.10%	
	2770	Heating, ventilating, air conditioning		7.35	10.25	15	7.20%	10.30%	11.70%	
	2900	Electrical		7.55	10.75	15.20	7.45%	9.60%	11.35%	

50 17 00 | S.F. Costs

		50 17 00 | S.F. Costs	UNIT	UNIT COSTS 1/4	UNIT COSTS MEDIAN	UNIT COSTS 3/4	% OF TOTAL 1/4	% OF TOTAL MEDIAN	% OF TOTAL 3/4	
61	3100	Total: Mechanical & Electrical	S.F.	20.50	27.50	41	18%	22%	27%	61
62	0010	**OFFICES Mid Rise (5 to 10 story)**	S.F.	98.50	119	157				62
	0020	Total project costs	C.F.	6.95	8.90	12.60				
	2720	Plumbing	S.F.	2.97	4.61	6.60	2.83%	3.74%	4.50%	
	2770	Heating, ventilating, air conditioning	↓	7.45	10.65	17.05	7.65%	9.40%	11%	
	2900	Electrical	↓	7.30	9.35	12.95	6.35%	7.80%	10%	
	3100	Total: Mechanical & Electrical	↓	18.90	24.50	46.50	19.15%	21.50%	27.50%	
63	0010	**OFFICES High Rise (11 to 20 story)**	S.F.	121	152	187				63
	0020	Total project costs	C.F.	8.45	10.55	15.15				
	2900	Electrical	S.F.	7.35	8.95	13.30	5.80%	7.85%	10.50%	
	3100	Total: Mechanical & Electrical	↓	23.50	31.50	53.50	16.90%	23.50%	34%	
64	0010	**POLICE STATIONS**	S.F.	145	190	240				64
	0020	Total project costs	C.F.	11.55	14.15	19.35				
	0500	Masonry	S.F.	13.60	24	30	6.70%	9.10%	11.35%	
	1800	Equipment	↓	2.30	10.10	16.05	.98%	3.35%	6.70%	
	2720	Plumbing	↓	8.10	16.15	20	5.65%	6.90%	10.75%	
	2770	Heating, ventilating, air conditioning	↓	12.65	16.80	23	5.85%	10.55%	11.70%	
	2900	Electrical	↓	15.85	22.50	30	9.80%	11.85%	14.80%	
	3100	Total: Mechanical & Electrical	↓	52	62.50	84.50	25%	31.50%	32.50%	
65	0010	**POST OFFICES**	S.F.	114	141	180				65
	0020	Total project costs	C.F.	6.90	8.70	9.90				
	2720	Plumbing	S.F.	5.15	6.40	8.05	4.24%	5.30%	5.60%	
	2770	Heating, ventilating, air conditioning	↓	8.05	9.95	11.05	6.65%	7.15%	9.35%	
	2900	Electrical	↓	9.45	13.30	15.75	7.25%	9%	11%	
	3100	Total: Mechanical & Electrical	↓	27.50	35.50	40.50	16.25%	18.80%	22%	
66	0010	**POWER PLANTS**	S.F.	795	1,000	1,925				66
	0020	Total project costs	C.F.	22	47.50	102				
	2900	Electrical	S.F.	56	119	177	9.30%	12.75%	21.50%	
	8100	Total: Mechanical & Electrical	↓	140	455	1,025	32.50%	32.50%	52.50%	
67	0010	**RELIGIOUS EDUCATION**	S.F.	92	120	148				67
	0020	Total project costs	C.F.	5.10	7.35	9.15				
	2720	Plumbing	S.F.	3.85	5.45	7.70	4.40%	5.30%	7.10%	
	2770	Heating, ventilating, air conditioning	↓	9.75	11	15.55	10.05%	11.45%	12.35%	
	2900	Electrical	↓	7.30	10.30	13.65	7.60%	9.05%	10.35%	
	3100	Total: Mechanical & Electrical	↓	29.50	39	47.50	22%	23%	27%	
69	0010	**RESEARCH Laboratories & Facilities**	S.F.	138	198	288				69
	0020	Total project costs	C.F.	10.40	20	24				
	1800	Equipment	S.F.	6.10	12.05	29.50	.94%	4.58%	8.80%	
	2720	Plumbing	↓	13.85	17.70	28.50	6.15%	8.30%	10.80%	
	2770	Heating, ventilating, air conditioning	↓	12.40	41.50	49.50	7.25%	16.50%	17.50%	
	2900	Electrical	↓	16.15	26.50	43.50	9.55%	11.50%	15.40%	
	3100	Total: Mechanical & Electrical	↓	49	92	132	29.50%	37%	42%	
70	0010	**RESTAURANTS**	S.F.	134	172	224				70
	0020	Total project costs	C.F.	11.25	14.75	19.35				
	1800	Equipment	S.F.	8.65	21	32	6.10%	13%	15.65%	
	2720	Plumbing	↓	10.60	12.85	16.85	6.10%	8.15%	9%	
	2770	Heating, ventilating, air conditioning	↓	13.45	18.60	24	9.20%	12%	12.40%	
	2900	Electrical	↓	14.15	17.45	22.50	8.35%	10.55%	11.55%	
	3100	Total: Mechanical & Electrical	↓	44	47	60.50	21%	24.50%	29.50%	
	9000	Per seat unit, total cost	Seat	4,900	6,550	7,750				
	9500	Total: Mechanical & Electrical	"	1,225	1,625	1,925				
72	0010	**RETAIL STORES**	S.F.	62.50	84	111				72
	0020	Total project costs	C.F.	4.23	6.05	8.40				
	2720	Plumbing	S.F.	2.26	3.78	6.45	3.26%	4.60%	6.80%	
	2770	Heating, ventilating, air conditioning	↓	4.89	6.70	10.05	6.75%	8.75%	10.15%	
	2900	Electrical	↓	5.65	7.70	11.10	7.25%	9.90%	11.65%	
	3100	Total: Mechanical & Electrical	↓	14.95	19.20	26.50	17.05%	21%	23.50%	

		50 17 00 | S.F. Costs	UNIT	UNIT COSTS 1/4	UNIT COSTS MEDIAN	UNIT COSTS 3/4	% OF TOTAL 1/4	% OF TOTAL MEDIAN	% OF TOTAL 3/4	
74	0010	**SCHOOLS Elementary**	S.F.	101	125	155				74
	0020	Total project costs	C.F.	6.65	8.55	11.05				
	0500	Masonry	S.F.	9.10	15.70	23.50	5.80%	11%	14.95%	
	1800	Equipment		2.78	4.70	8.75	1.89%	3.32%	4.71%	
	2720	Plumbing		5.85	8.30	11.05	5.70%	7.15%	9.35%	
	2730	Heating, ventilating, air conditioning		8.80	14	19.55	8.15%	10.80%	14.90%	
	2900	Electrical		9.60	12.70	15.95	8.40%	10.05%	11.70%	
	3100	Total: Mechanical & Electrical	↓	34	43	52.50	25%	27.50%	30%	
	9000	Per pupil, total cost	Ea.	11,700	17,400	50,000				
	9500	Total: Mechanical & Electrical	″	3,300	4,175	14,500				
76	0010	**SCHOOLS Junior High & Middle**	S.F.	103	129	157				76
	0020	Total project costs	C.F.	6.65	8.65	9.70				
	0500	Masonry	S.F.	13.10	16.90	19.75	8%	11.40%	14.30%	
	1800	Equipment		3.36	5.40	8.15	1.81%	3.26%	4.86%	
	2720	Plumbing		6.10	7.55	9.35	5.30%	6.80%	7.25%	
	2770	Heating, ventilating, air conditioning		12.20	14.85	26	8.90%	11.55%	14.20%	
	2900	Electrical		10.30	12.40	15.95	7.90%	9.35%	10.60%	
	3100	Total: Mechanical & Electrical	↓	33.50	43	53	23.50%	27%	29.50%	
	9000	Per pupil, total cost	Ea.	13,300	17,500	23,500				
78	0010	**SCHOOLS Senior High**	S.F.	108	133	167				78
	0020	Total project costs	C.F.	6.60	9.70	15.65				
	1800	Equipment	S.F.	2.87	6.75	9.50	1.88%	2.98%	4.80%	
	2720	Plumbing		6.15	9.20	16.80	5.60%	6.90%	8.30%	
	2770	Heating, ventilating, air conditioning		12.50	14.35	27.50	8.95%	11.60%	15%	
	2900	Electrical		10.95	14.25	21	8.65%	10.15%	11.95%	
	3100	Total: Mechanical & Electrical	↓	36.50	42.50	71.50	23.50%	26.50%	28.50%	
	9000	Per pupil, total cost	Ea.	10,300	21,000	26,200				
80	0010	**SCHOOLS Vocational**	S.F.	88.50	128	159				80
	0020	Total project costs	C.F.	5.50	7.90	10.90				
	0500	Masonry	S.F.	5.20	12.80	19.60	3.53%	6.70%	10.95%	
	1800	Equipment		2.77	6.90	9.60	1.24%	3.10%	4.26%	
	2720	Plumbing		5.65	8.45	12.40	5.40%	6.90%	8.55%	
	2770	Heating, ventilating, air conditioning		7.90	14.75	24.50	8.60%	11.90%	14.65%	
	2900	Electrical		9.20	12.05	16.60	8.45%	10.95%	13.20%	
	3100	Total: Mechanical & Electrical	↓	32	35.50	61	23.50%	27.50%	31%	
	9000	Per pupil, total cost	Ea.	12,300	33,000	49,200				
83	0010	**SPORTS ARENAS**	S.F.	77.50	103	159				83
	0020	Total project costs	C.F.	4.20	7.50	9.70				
	2720	Plumbing	S.F.	4.49	6.80	14.35	4.35%	6.35%	9.40%	
	2770	Heating, ventilating, air conditioning		9.65	11.40	15.85	8.80%	10.20%	13.55%	
	2900	Electrical		8.05	10.95	14.15	8.60%	9.90%	12.25%	
	3100	Total: Mechanical & Electrical	↓	20	35	47	21.50%	25%	27.50%	
85	0010	**SUPERMARKETS**	S.F.	71.50	83	97				85
	0020	Total project costs	C.F.	3.98	4.81	7.30				
	2720	Plumbing	S.F.	3.99	5.05	5.85	5.40%	6%	7.45%	
	2770	Heating, ventilating, air conditioning		5.85	7.80	9.50	8.60%	8.65%	9.60%	
	2900	Electrical		8.95	10.30	12.15	10.40%	12.45%	13.60%	
	3100	Total: Mechanical & Electrical	↓	23	25	32.50	20.50%	26.50%	31%	
86	0010	**SWIMMING POOLS**	S.F.	116	194	415				86
	0020	Total project costs	C.F.	9.25	11.55	12.60				
	2720	Plumbing	S.F.	10.70	12.20	16.60	4.80%	9.70%	20.50%	
	2900	Electrical		8.70	14.10	20.50	5.75%	6.95%	7.60%	
	3100	Total: Mechanical & Electrical	↓	21	55	73.50	11.15%	14.10%	23.50%	
87	0010	**TELEPHONE EXCHANGES**	S.F.	154	225	285				87
	0020	Total project costs	C.F.	9.55	15.35	21				
	2720	Plumbing	S.F.	6.50	10	14.65	4.52%	5.80%	6.90%	
	2770	Heating, ventilating, air conditioning	↓	15.05	30.50	37.50	11.80%	16.05%	18.40%	

	50 17 00 | S.F. Costs		UNIT	UNIT COSTS 1/4	UNIT COSTS MEDIAN	UNIT COSTS 3/4	% OF TOTAL 1/4	% OF TOTAL MEDIAN	% OF TOTAL 3/4
87	2900	Electrical	S.F.	15.65	25	44	10.90%	14%	17.85%
	3100	Total: Mechanical & Electrical	↓	46	87.50	124	29.50%	33.50%	44.50%
91	0010	**THEATERS**	S.F.	96.50	120	183			
	0020	Total project costs	C.F.	4.46	6.60	9.70			
	2720	Plumbing	S.F.	3.22	3.49	14.25	2.92%	4.70%	6.80%
	2770	Heating, ventilating, air conditioning		9.40	11.35	14.05	8%	12.25%	13.40%
	2900	Electrical		8.45	11.40	23	8.05%	9.95%	12.25%
	3100	Total: Mechanical & Electrical	↓	21.50	32.50	66.50	23%	26.50%	27.50%
94	0010	**TOWN HALLS City Halls & Municipal Buildings**	S.F.	108	137	179			
	0020	Total project costs	C.F.	9.85	11.80	16.60			
	2720	Plumbing	S.F.	4.50	8.40	15.50	4.31%	5.95%	7.95%
	2770	Heating, ventilating, air conditioning		8.15	16.15	23.50	7.05%	9.05%	13.45%
	2900	Electrical		10.25	14.60	19.95	8.05%	9.45%	11.65%
	3100	Total: Mechanical & Electrical	↓	35.50	45	69	22%	26.50%	31%
97	0010	**WAREHOUSES & Storage Buildings**	S.F.	40.50	60	86			
	0020	Total project costs	C.F.	2.11	3.30	5.45			
	0100	Site work	S.F.	4.15	8.25	12.45	6.05%	12.95%	19.85%
	0500	Masonry		2.42	5.70	12.35	3.73%	7.40%	12.30%
	1800	Equipment		.65	1.39	7.80	.91%	1.82%	5.55%
	2720	Plumbing		1.34	2.41	4.50	2.90%	4.80%	6.55%
	2730	Heating, ventilating, air conditioning		1.53	4.32	5.80	2.41%	5%	8.90%
	2900	Electrical		2.38	4.48	7.40	5.15%	7.20%	10.10%
	3100	Total: Mechanical & Electrical	↓	6.65	10.20	20	12.75%	18.90%	26%
99	0010	**WAREHOUSE & OFFICES Combination**	S.F.	49.50	66	90.50			
	0020	Total project costs	C.F.	2.53	3.67	5.45			
	1800	Equipment	S.F.	.86	1.66	2.47	.52%	1.20%	2.40%
	2720	Plumbing		1.91	3.39	4.96	3.74%	4.76%	6.30%
	2770	Heating, ventilating, air conditioning		3.02	4.72	6.60	5%	5.65%	10.05%
	2900	Electrical		3.31	4.93	7.75	5.75%	8%	10%
	3100	Total: Mechanical & Electrical	↓	9.25	14.25	22.50	14.40%	19.95%	24.50%

Square Foot Project Size Modifier

One factor that affects the S.F. cost of a particular building is the size. In general, for buildings built to the same specifications in the same locality, the larger building will have the lower S.F. cost. This is due mainly to the decreasing contribution of the exterior walls plus the economy of scale usually achievable in larger buildings. The Area Conversion Scale shown below will give a factor to convert costs for the typical size building to an adjusted cost for the particular project.

The Square Foot Base Size lists the median costs, most typical project size in our accumulated data, and the range in size of the projects.

The Size Factor for your project is determined by dividing your project area in S.F. by the typical project size for the particular Building Type. With this factor, enter the Area Conversion Scale at the appropriate Size Factor and determine the appropriate cost multiplier for your building size.

Example: Determine the cost per S.F. for a 100,000 S.F. Mid-rise apartment building.

$$\frac{\text{Proposed building area} = 100{,}000 \text{ S.F.}}{\text{Typical size from below} = 50{,}000 \text{ S.F.}} = 2.00$$

Enter Area Conversion scale at 2.0, intersect curve, read horizontally the appropriate cost multiplier of .94. Size adjusted cost becomes .94 x $107.00 = $101.00 based on national average costs.

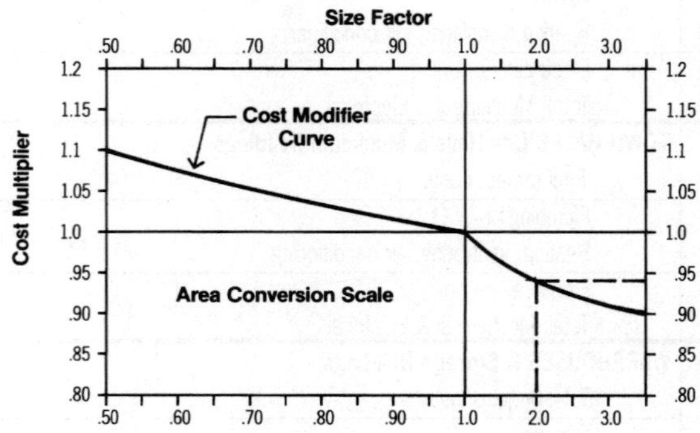

Note: For Size Factors less than .50, the Cost Multiplier is 1.1
For Size Factors greater than 3.5, the Cost Multiplier is .90

Square Foot Base Size							
Building Type	Median Cost per S.F.	Typical Size Gross S.F.	Typical Range Gross S.F.	Building Type	Median Cost per S.F.	Typical Size Gross S.F.	Typical Range Gross S.F.
Apartments, Low Rise	$ 84.50	21,000	9,700 - 37,200	Jails	$ 257.00	40,000	5,500 - 145,000
Apartments, Mid Rise	107.00	50,000	32,000 - 100,000	Libraries	161.00	12,000	7,000 - 31,000
Apartments, High Rise	116.00	145,000	95,000 - 600,000	Living, Assisted	137.00	32,300	23,500 - 50,300
Auditoriums	141.00	25,000	7,600 - 39,000	Medical Clinics	146.00	7,200	4,200 - 15,700
Auto Sales	105.00	20,000	10,800 - 28,600	Medical Offices	138.00	6,000	4,000 - 15,000
Banks	189.00	4,200	2,500 - 7,500	Motels	102.00	40,000	15,800 - 120,000
Churches	130.00	17,000	2,000 - 42,000	Nursing Homes	142.00	23,000	15,000 - 37,000
Clubs, Country	132.00	6,500	4,500 - 15,000	Offices, Low Rise	120.00	20,000	5,000 - 80,000
Clubs, Social	126.00	10,000	6,000 - 13,500	Offices, Mid Rise	119.00	120,000	20,000 - 300,000
Clubs, YMCA	143.00	28,300	12,800 - 39,400	Offices, High Rise	152.00	260,000	120,000 - 800,000
Colleges (Class)	152.00	50,000	15,000 - 150,000	Police Stations	190.00	10,500	4,000 - 19,000
Colleges (Science Lab)	242.00	45,600	16,600 - 80,000	Post Offices	141.00	12,400	6,800 - 30,000
College (Student Union)	179.00	33,400	16,000 - 85,000	Power Plants	1000.00	7,500	1,000 - 20,000
Community Center	134.00	9,400	5,300 - 16,700	Religious Education	120.00	9,000	6,000 - 12,000
Court Houses	180.00	32,400	17,800 - 106,000	Research	198.00	19,000	6,300 - 45,000
Dept. Stores	78.50	90,000	44,000 - 122,000	Restaurants	172.00	4,400	2,800 - 6,000
Dormitories, Low Rise	143.00	25,000	10,000 - 95,000	Retail Stores	84.00	7,200	4,000 - 17,600
Dormitories, Mid Rise	176.00	85,000	20,000 - 200,000	Schools, Elementary	125.00	41,000	24,500 - 55,000
Factories	76.50	26,400	12,900 - 50,000	Schools, Jr. High	129.00	92,000	52,000 - 119,000
Fire Stations	139.00	5,800	4,000 - 8,700	Schools, Sr. High	133.00	101,000	50,500 - 175,000
Fraternity Houses	131.00	12,500	8,200 - 14,800	Schools, Vocational	128.00	37,000	20,500 - 82,000
Funeral Homes	146.00	10,000	4,000 - 20,000	Sports Arenas	103.00	15,000	5,000 - 40,000
Garages, Commercial	93.50	9,300	5,000 - 13,600	Supermarkets	83.00	44,000	12,000 - 60,000
Garages, Municipal	119.00	8,300	4,500 - 12,600	Swimming Pools	194.00	20,000	10,000 - 32,000
Garages, Parking	50.50	163,000	76,400 - 225,300	Telephone Exchange	225.00	4,500	1,200 - 10,600
Gymnasiums	129.00	19,200	11,600 - 41,000	Theaters	120.00	10,500	8,800 - 17,500
Hospitals	228.00	55,000	27,200 - 125,000	Town Halls	137.00	10,800	4,800 - 23,400
House (Elderly)	115.00	37,000	21,000 - 66,000	Warehouses	60.00	25,000	8,000 - 72,000
Housing (Public)	106.00	36,000	14,400 - 74,400	Warehouse & Office	66.00	25,000	8,000 - 72,000
Ice Rinks	153.00	29,000	27,200 - 33,600				

Abbreviations

A	Area Square Feet; Ampere
ABS	Acrylonitrile Butadiene Stryrene; Asbestos Bonded Steel
A.C.	Alternating Current; Air-Conditioning; Asbestos Cement; Plywood Grade A & C
A.C.I.	American Concrete Institute
AD	Plywood, Grade A & D
Addit.	Additional
Adj.	Adjustable
af	Audio-frequency
A.G.A.	American Gas Association
Agg.	Aggregate
A.H.	Ampere Hours
A hr.	Ampere-hour
A.H.U.	Air Handling Unit
A.I.A.	American Institute of Architects
AIC	Ampere Interrupting Capacity
Allow.	Allowance
alt.	Altitude
Alum.	Aluminum
a.m.	Ante Meridiem
Amp.	Ampere
Anod.	Anodized
Approx.	Approximate
Apt.	Apartment
Asb.	Asbestos
A.S.B.C.	American Standard Building Code
Asbe.	Asbestos Worker
ASCE.	American Society of Civil Engineers
A.S.H.R.A.E.	American Society of Heating, Refrig. & AC Engineers
A.S.M.E.	American Society of Mechanical Engineers
A.S.T.M.	American Society for Testing and Materials
Attchmt.	Attachment
Avg., Ave.	Average
A.W.G.	American Wire Gauge
AWWA	American Water Works Assoc.
Bbl.	Barrel
B&B	Grade B and Better; Balled & Burlapped
B.&S.	Bell and Spigot
B.&W.	Black and White
b.c.c.	Body-centered Cubic
B.C.Y.	Bank Cubic Yards
BE	Bevel End
B.F.	Board Feet
Bg. cem.	Bag of Cement
BHP	Boiler Horsepower; Brake Horsepower
B.I.	Black Iron
Bit., Bitum.	Bituminous
Bit., Conc.	Bituminous Concrete
Bk.	Backed
Bkrs.	Breakers
Bldg.	Building
Blk.	Block
Bm.	Beam
Boil.	Boilermaker
B.P.M.	Blows per Minute
BR	Bedroom
Brg.	Bearing
Brhe.	Bricklayer Helper
Bric.	Bricklayer
Brk.	Brick
Brng.	Bearing
Brs.	Brass
Brz.	Bronze
Bsn.	Basin
Btr.	Better
Btu	British Thermal Unit
BTUH	BTU per Hour
B.U.R.	Built-up Roofing
BX	Interlocked Armored Cable
°C	degree centegrade
c	Conductivity, Copper Sweat
C	Hundred; Centigrade
C/C	Center to Center, Cedar on Cedar
C-C	Center to Center
Cab.	Cabinet
Cair.	Air Tool Laborer
Calc	Calculated
Cap.	Capacity
Carp.	Carpenter
C.B.	Circuit Breaker
C.C.A.	Chromate Copper Arsenate
C.C.F.	Hundred Cubic Feet
cd	Candela
cd/sf	Candela per Square Foot
CD	Grade of Plywood Face & Back
CDX	Plywood, Grade C & D, exterior glue
Cefi.	Cement Finisher
Cem.	Cement
CF	Hundred Feet
C.F.	Cubic Feet
CFM	Cubic Feet per Minute
c.g.	Center of Gravity
CHW	Chilled Water; Commercial Hot Water
C.I.	Cast Iron
C.I.P.	Cast in Place
Circ.	Circuit
C.L.	Carload Lot
Clab.	Common Laborer
Clam	Common maintenance laborer
C.L.F.	Hundred Linear Feet
CLF	Current Limiting Fuse
CLP	Cross Linked Polyethylene
cm	Centimeter
CMP	Corr. Metal Pipe
C.M.U.	Concrete Masonry Unit
CN	Change Notice
Col.	Column
CO_2	Carbon Dioxide
Comb.	Combination
Compr.	Compressor
Conc.	Concrete
Cont.	Continuous; Continued, Container
Corr.	Corrugated
Cos	Cosine
Cot	Cotangent
Cov.	Cover
C/P	Cedar on Paneling
CPA	Control Point Adjustment
Cplg.	Coupling
C.P.M.	Critical Path Method
CPVC	Chlorinated Polyvinyl Chloride
C.Pr.	Hundred Pair
CRC	Cold Rolled Channel
Creos.	Creosote
Crpt.	Carpet & Linoleum Layer
CRT	Cathode-ray Tube
CS	Carbon Steel, Constant Shear Bar Joist
Csc	Cosecant
C.S.F.	Hundred Square Feet
CSI	Construction Specifications Institute
C.T.	Current Transformer
CTS	Copper Tube Size
Cu	Copper, Cubic
Cu. Ft.	Cubic Foot
cw	Continuous Wave
C.W.	Cool White; Cold Water
Cwt.	100 Pounds
C.W.X.	Cool White Deluxe
C.Y.	Cubic Yard (27 cubic feet)
C.Y./Hr.	Cubic Yard per Hour
Cyl.	Cylinder
d	Penny (nail size)
D	Deep; Depth; Discharge
Dis., Disch.	Discharge
Db.	Decibel
Dbl.	Double
DC	Direct Current
DDC	Direct Digital Control
Demob.	Demobilization
d.f.u.	Drainage Fixture Units
D.H.	Double Hung
DHW	Domestic Hot Water
DI	Ductile Iron
Diag.	Diagonal
Diam., Dia	Diameter
Distrib.	Distribution
Div.	Division
Dk.	Deck
D.L.	Dead Load; Diesel
DLH	Deep Long Span Bar Joist
Do.	Ditto
Dp.	Depth
D.P.S.T.	Double Pole, Single Throw
Dr.	Drive
Drink.	Drinking
D.S.	Double Strength
D.S.A.	Double Strength A Grade
D.S.B.	Double Strength B Grade
Dty.	Duty
DWV	Drain Waste Vent
DX	Deluxe White, Direct Expansion
dyn	Dyne
e	Eccentricity
E	Equipment Only; East
Ea.	Each
E.B.	Encased Burial
Econ.	Economy
E.C.Y	Embankment Cubic Yards
EDP	Electronic Data Processing
EIFS	Exterior Insulation Finish System
E.D.R.	Equiv. Direct Radiation
Eq.	Equation
EL	elevation
Elec.	Electrician; Electrical
Elev.	Elevator; Elevating
EMT	Electrical Metallic Conduit; Thin Wall Conduit
Eng.	Engine, Engineered
EPDM	Ethylene Propylene Diene Monomer
EPS	Expanded Polystyrene
Eqhv.	Equip. Oper., Heavy
Eqlt.	Equip. Oper., Light
Eqmd.	Equip. Oper., Medium
Eqmm.	Equip. Oper., Master Mechanic
Eqol.	Equip. Oper., Oilers
Equip.	Equipment
ERW	Electric Resistance Welded
E.S.	Energy Saver
Est.	Estimated
esu	Electrostatic Units
E.W.	Each Way
EWT	Entering Water Temperature
Excav.	Excavation
Exp.	Expansion, Exposure
Ext.	Exterior
Extru.	Extrusion
f.	Fiber stress
F	Fahrenheit; Female; Fill
Fab.	Fabricated

FBGS	Fiberglass
F.C.	Footcandles
f.c.c.	Face-centered Cubic
f'c.	Compressive Stress in Concrete; Extreme Compressive Stress
F.E.	Front End
FEP	Fluorinated Ethylene Propylene (Teflon)
F.G.	Flat Grain
F.H.A.	Federal Housing Administration
Fig.	Figure
Fin.	Finished
Fixt.	Fixture
Fl. Oz.	Fluid Ounces
Flr.	Floor
F.M.	Frequency Modulation; Factory Mutual
Fmg.	Framing
Fdn.	Foundation
Fori.	Foreman, Inside
Foro.	Foreman, Outside
Fount.	Fountain
fpm	Feet per Minute
FPT	Female Pipe Thread
Fr.	Frame
F.R.	Fire Rating
FRK	Foil Reinforced Kraft
FRP	Fiberglass Reinforced Plastic
FS	Forged Steel
FSC	Cast Body; Cast Switch Box
Ft.	Foot; Feet
Ftng.	Fitting
Ftg.	Footing
Ft lb.	Foot Pound
Furn.	Furniture
FVNR	Full Voltage Non-Reversing
FXM	Female by Male
Fy.	Minimum Yield Stress of Steel
g	Gram
G	Gauss
Ga.	Gauge
Gal., gal.	Gallon
gpm, GPM	Gallon per Minute
Galv.	Galvanized
Gen.	General
G.F.I.	Ground Fault Interrupter
Glaz.	Glazier
GPD	Gallons per Day
GPH	Gallons per Hour
GPM	Gallons per Minute
GR	Grade
Gran.	Granular
Grnd.	Ground
H	High Henry
H.C.	High Capacity
H.D.	Heavy Duty; High Density
H.D.O.	High Density Overlaid
H.D.P.E.	high density polyethelene
Hdr.	Header
Hdwe.	Hardware
Help.	Helper Average
HEPA	High Efficiency Particulate Air Filter
Hg	Mercury
HIC	High Interrupting Capacity
HM	Hollow Metal
HMWPE	high molecular weight polyethylene
H.O.	High Output
Horiz.	Horizontal

H.P.	Horsepower; High Pressure
H.P.F.	High Power Factor
Hr.	Hour
Hrs./Day	Hours per Day
HSC	High Short Circuit
Ht.	Height
Htg.	Heating
Htrs.	Heaters
HVAC	Heating, Ventilation & Air-Conditioning
Hvy.	Heavy
HW	Hot Water
Hyd.;Hydr.	Hydraulic
Hz.	Hertz (cycles)
I.	Moment of Inertia
IBC	International Building Code
I.C.	Interrupting Capacity
ID	Inside Diameter
I.D.	Inside Dimension; Identification
I.F.	Inside Frosted
I.M.C.	Intermediate Metal Conduit
In.	Inch
Incan.	Incandescent
Incl.	Included; Including
Int.	Interior
Inst.	Installation
Insul.	Insulation/Insulated
I.P.	Iron Pipe
I.P.S.	Iron Pipe Size
I.P.T.	Iron Pipe Threaded
I.W.	Indirect Waste
J	Joule
J.I.C.	Joint Industrial Council
K	Thousand; Thousand Pounds; Heavy Wall Copper Tubing, Kelvin
K.A.H.	Thousand Amp. Hours
KCMIL	Thousand Circular Mils
KD	Knock Down
K.D.A.T.	Kiln Dried After Treatment
kg	Kilogram
kG	Kilogauss
kgf	Kilogram Force
kHz	Kilohertz
Kip.	1000 Pounds
KJ	Kiljoule
K.L.	Effective Length Factor
K.L.F.	Kips per Linear Foot
Km	Kilometer
K.S.F.	Kips per Square Foot
K.S.I.	Kips per Square Inch
kV	Kilovolt
kVA	Kilovolt Ampere
K.V.A.R.	Kilovar (Reactance)
KW	Kilowatt
KWh	Kilowatt-hour
L	Labor Only; Length; Long; Medium Wall Copper Tubing
Lab.	Labor
lat	Latitude
Lath.	Lather
Lav.	Lavatory
lb.; #	Pound
L.B.	Load Bearing; L Conduit Body
L. & E.	Labor & Equipment
lb./hr.	Pounds per Hour
lb./L.F.	Pounds per Linear Foot
lbf/sq.in.	Pound-force per Square Inch
L.C.L.	Less than Carload Lot
L.C.Y.	Loose Cubic Yard
Ld.	Load

LE	Lead Equivalent
LED	Light Emitting Diode
L.F.	Linear Foot
L.F. Nose	Linear Foot of Stair Nosing
L.F. Rsr	Linear Foot of Stair Riser
Lg.	Long; Length; Large
L & H	Light and Heat
LH	Long Span Bar Joist
L.H.	Labor Hours
L.L.	Live Load
L.L.D.	Lamp Lumen Depreciation
lm	Lumen
lm/sf	Lumen per Square Foot
lm/W	Lumen per Watt
L.O.A.	Length Over All
log	Logarithm
L-O-L	Lateralolet
long.	longitude
L.P.	Liquefied Petroleum; Low Pressure
L.P.F.	Low Power Factor
LR	Long Radius
L.S.	Lump Sum
Lt.	Light
Lt. Ga.	Light Gauge
L.T.L.	Less than Truckload Lot
Lt. Wt.	Lightweight
L.V.	Low Voltage
M	Thousand; Material; Male; Light Wall Copper Tubing
M^2CA	Meters Squared Contact Area
m/hr.; M.H.	Man-hour
mA	Milliampere
Mach.	Machine
Mag. Str.	Magnetic Starter
Maint.	Maintenance
Marb.	Marble Setter
Mat; Mat'l.	Material
Max.	Maximum
MBF	Thousand Board Feet
MBH	Thousand BTU's per hr.
MC	Metal Clad Cable
M.C.F.	Thousand Cubic Feet
M.C.F.M.	Thousand Cubic Feet per Minute
M.C.M.	Thousand Circular Mils
M.C.P.	Motor Circuit Protector
MD	Medium Duty
M.D.O.	Medium Density Overlaid
Med.	Medium
MF	Thousand Feet
M.F.B.M.	Thousand Feet Board Measure
Mfg.	Manufacturing
Mfrs.	Manufacturers
mg	Milligram
MGD	Million Gallons per Day
MGPH	Thousand Gallons per Hour
MH, M.H.	Manhole; Metal Halide; Man-Hour
MHz	Megahertz
Mi.	Mile
MI	Malleable Iron; Mineral Insulated
mm	Millimeter
Mill.	Millwright
Min., min.	Minimum, minute
Misc.	Miscellaneous
ml	Milliliter, Mainline
M.L.F.	Thousand Linear Feet
Mo.	Month
Mobil.	Mobilization
Mog.	Mogul Base
MPH	Miles per Hour
MPT	Male Pipe Thread

Abbreviations

MRT	Mile Round Trip
ms	Millisecond
M.S.F.	Thousand Square Feet
Mstz.	Mosaic & Terrazzo Worker
M.S.Y.	Thousand Square Yards
Mtd., mtd.	Mounted
Mthe.	Mosaic & Terrazzo Helper
Mtng.	Mounting
Mult.	Multi; Multiply
M.V.A.	Million Volt Amperes
M.V.A.R.	Million Volt Amperes Reactance
MV	Megavolt
MW	Megawatt
MXM	Male by Male
MYD	Thousand Yards
N	Natural; North
nA	Nanoampere
NA	Not Available; Not Applicable
N.B.C.	National Building Code
NC	Normally Closed
N.E.M.A.	National Electrical Manufacturers Assoc.
NEHB	Bolted Circuit Breaker to 600V.
N.L.B.	Non-Load-Bearing
NM	Non-Metallic Cable
nm	Nanometer
No.	Number
NO	Normally Open
N.O.C.	Not Otherwise Classified
Nose.	Nosing
N.P.T.	National Pipe Thread
NQOD	Combination Plug-on/Bolt on Circuit Breaker to 240V.
N.R.C.	Noise Reduction Coefficient/ Nuclear Regulator Commission
N.R.S.	Non Rising Stem
ns	Nanosecond
nW	Nanowatt
OB	Opposing Blade
OC	On Center
OD	Outside Diameter
O.D.	Outside Dimension
ODS	Overhead Distribution System
O.G.	Ogee
O.H.	Overhead
O&P	Overhead and Profit
Oper.	Operator
Opng.	Opening
Orna.	Ornamental
OSB	Oriented Strand Board
O.S.&Y.	Outside Screw and Yoke
Ovhd.	Overhead
OWG	Oil, Water or Gas
Oz.	Ounce
P.	Pole; Applied Load; Projection
p.	Page
Pape.	Paperhanger
P.A.P.R.	Powered Air Purifying Respirator
PAR	Parabolic Reflector
Pc., Pcs.	Piece, Pieces
P.C.	Portland Cement; Power Connector
P.C.F.	Pounds per Cubic Foot
P.C.M.	Phase Contrast Microscopy
P.E.	Professional Engineer; Porcelain Enamel; Polyethylene; Plain End
Perf.	Perforated
PEX	Cross linked polyethylene
Ph.	Phase
P.I.	Pressure Injected
Pile.	Pile Driver
Pkg.	Package
Pl.	Plate
Plah.	Plasterer Helper
Plas.	Plasterer
Pluh.	Plumbers Helper
Plum.	Plumber
Ply.	Plywood
p.m.	Post Meridiem
Pntd.	Painted
Pord.	Painter, Ordinary
pp	Pages
PP, PPL	Polypropylene
P.P.M.	Parts per Million
Pr.	Pair
P.E.S.B.	Pre-engineered Steel Building
Prefab.	Prefabricated
Prefin.	Prefinished
Prop.	Propelled
PSF, psf	Pounds per Square Foot
PSI, psi	Pounds per Square Inch
PSIG	Pounds per Square Inch Gauge
PSP	Plastic Sewer Pipe
Pspr.	Painter, Spray
Psst.	Painter, Structural Steel
P.T.	Potential Transformer
P. & T.	Pressure & Temperature
Ptd.	Painted
Ptns.	Partitions
Pu	Ultimate Load
PVC	Polyvinyl Chloride
Pvmt.	Pavement
Pwr.	Power
Q	Quantity Heat Flow
Qt.	Quart
Quan., Qty.	Quantity
Q.C.	Quick Coupling
r	Radius of Gyration
R	Resistance
R.C.P.	Reinforced Concrete Pipe
Rect.	Rectangle
Reg.	Regular
Reinf.	Reinforced
Req'd.	Required
Res.	Resistant
Resi.	Residential
Rgh.	Rough
RGS	Rigid Galvanized Steel
R.H.W.	Rubber, Heat & Water Resistant; Residential Hot Water
rms	Root Mean Square
Rnd.	Round
Rodm.	Rodman
Rofc.	Roofer, Composition
Rofp.	Roofer, Precast
Rohe.	Roofer Helpers (Composition)
Rots.	Roofer, Tile & Slate
R.O.W.	Right of Way
RPM	Revolutions per Minute
R.S.	Rapid Start
Rsr	Riser
RT	Round Trip
S.	Suction; Single Entrance; South
SC	Screw Cover
SCFM	Standard Cubic Feet per Minute
Scaf.	Scaffold
Sch., Sched.	Schedule
S.C.R.	Modular Brick
S.D.	Sound Deadening
S.D.R.	Standard Dimension Ratio
S.E.	Surfaced Edge
Sel.	Select
S.E.R., S.E.U.	Service Entrance Cable
S.F.	Square Foot
S.F.C.A.	Square Foot Contact Area
S.F. Flr.	Square Foot of Floor
S.F.G.	Square Foot of Ground
S.F. Hor.	Square Foot Horizontal
S.F.R.	Square Feet of Radiation
S.F. Shlf.	Square Foot of Shelf
S4S	Surface 4 Sides
Shee.	Sheet Metal Worker
Sin.	Sine
Skwk.	Skilled Worker
SL	Saran Lined
S.L.	Slimline
Sldr.	Solder
SLH	Super Long Span Bar Joist
S.N.	Solid Neutral
S-O-L	Socketolet
sp	Standpipe
S.P.	Static Pressure; Single Pole; Self-Propelled
Spri.	Sprinkler Installer
spwg	Static Pressure Water Gauge
S.P.D.T.	Single Pole, Double Throw
SPF	Spruce Pine Fir
S.P.S.T.	Single Pole, Single Throw
SPT	Standard Pipe Thread
Sq.	Square; 100 Square Feet
Sq. Hd.	Square Head
Sq. In.	Square Inch
S.S.	Single Strength; Stainless Steel
S.S.B.	Single Strength B Grade
sst, ss	Stainless Steel
Sswk.	Structural Steel Worker
Sswl.	Structural Steel Welder
St.;Stl.	Steel
S.T.C.	Sound Transmission Coefficient
Std.	Standard
Stg.	Staging
STK	Select Tight Knot
STP	Standard Temperature & Pressure
Stpi.	Steamfitter, Pipefitter
Str.	Strength; Starter; Straight
Strd.	Stranded
Struct.	Structural
Sty.	Story
Subj.	Subject
Subs.	Subcontractors
Surf.	Surface
Sw.	Switch
Swbd.	Switchboard
S.Y.	Square Yard
Syn.	Synthetic
S.Y.P.	Southern Yellow Pine
Sys.	System
t.	Thickness
T	Temperature; Ton
Tan	Tangent
T.C.	Terra Cotta
T & C	Threaded and Coupled
T.D.	Temperature Difference
Tdd	Telecommunications Device for the Deaf
T.E.M.	Transmission Electron Microscopy
TFE	Tetrafluoroethylene (Teflon)
T. & G.	Tongue & Groove; Tar & Gravel
Th., Thk.	Thick
Thn.	Thin
Thrded	Threaded
Tilf.	Tile Layer, Floor
Tilh.	Tile Layer, Helper
THHN	Nylon Jacketed Wire
THW.	Insulated Strand Wire

Abbreviations

THWN	Nylon Jacketed Wire
T.L.	Truckload
T.M.	Track Mounted
Tot.	Total
T-O-L	Threadolet
T.S.	Trigger Start
Tr.	Trade
Transf.	Transformer
Trhv.	Truck Driver, Heavy
Trlr	Trailer
Trlt.	Truck Driver, Light
TTY	Teletypewriter
TV	Television
T.W.	Thermoplastic Water Resistant Wire
UCI	Uniform Construction Index
UF	Underground Feeder
UGND	Underground Feeder
U.H.F.	Ultra High Frequency
U.I.	United Inch
U.L.	Underwriters Laboratory
Uld.	unloading
Unfin.	Unfinished
URD	Underground Residential Distribution
US	United States
USP	United States Primed
UTP	Unshielded Twisted Pair
V	Volt
V.A.	Volt Amperes
V.C.T.	Vinyl Composition Tile
VAV	Variable Air Volume
VC	Veneer Core
Vent.	Ventilation
Vert.	Vertical
V.F.	Vinyl Faced
V.G.	Vertical Grain
V.H.F.	Very High Frequency
VHO	Very High Output
Vib.	Vibrating
V.L.F.	Vertical Linear Foot
Vol.	Volume
VRP	Vinyl Reinforced Polyester
W	Wire; Watt; Wide; West
w/	With
W.C.	Water Column; Water Closet
W.F.	Wide Flange
W.G.	Water Gauge
Wldg.	Welding
W. Mile	Wire Mile
W-O-L	Weldolet
W.R.	Water Resistant
Wrck.	Wrecker
W.S.P.	Water, Steam, Petroleum
WT., Wt.	Weight
WWF	Welded Wire Fabric
XFER	Transfer
XFMR	Transformer
XHD	Extra Heavy Duty
XHHW, XLPE	Cross-Linked Polyethylene Wire Insulation
XLP	Cross-linked Polyethylene
Y	Wye
yd	Yard
yr	Year
Δ	Delta
%	Percent
~	Approximately
Ø	Phase; diameter
@	At
#	Pound; Number
<	Less Than
>	Greater Than
Z	zone

A

B

C

D

G

H

I

J

K

L

T

U

V

W

X

Y

Z

Notes

Notes

Notes

Notes

Notes

Notes

Notes

Division Notes

I	CREW	DAILY OUTPUT	LABOR-HOURS	UNIT	2000 BARE COSTS MAT.	LABOR	EQUIP.	TOTAL	TOTAL INCL O&P

Division Notes

		CREW	DAILY OUTPUT	LABOR-HOURS	UNIT	2000 BARE COSTS MAT.	LABOR	EQUIP.	TOTAL	TOTAL INCL O&P

Division Notes

		CREW	DAILY OUTPUT	LABOR-HOURS	UNIT	2000 BARE COSTS MAT.	LABOR	EQUIP.	TOTAL	TOTAL INCL O&P

Division Notes

	CREW	DAILY OUTPUT	LABOR-HOURS	UNIT	2000 BARE COSTS MAT.	LABOR	EQUIP.	TOTAL	TOTAL INCL O&P

Division Notes

		CREW	DAILY OUTPUT	LABOR-HOURS	UNIT	2000 BARE COSTS MAT.	LABOR	EQUIP.	TOTAL	TOTAL INCL O&P

Division Notes

	CREW	DAILY OUTPUT	LABOR-HOURS	UNIT	2000 BARE COSTS MAT.	LABOR	EQUIP.	TOTAL	TOTAL INCL O&P

Division Notes

		CREW	DAILY OUTPUT	LABOR-HOURS	UNIT	2000 BARE COSTS MAT.	LABOR	EQUIP.	TOTAL	TOTAL INCL O&P

Division Notes

	CREW	DAILY OUTPUT	LABOR-HOURS	UNIT	2000 BARE COSTS MAT.	LABOR	EQUIP.	TOTAL	TOTAL INCL O&P

Reed Construction Data/RSMeans—a tradition of excellence in construction cost information and services since 1942

Table of Contents

For more information visit the RSMeans website at www.rsmeans.com

Book Selection Guide

The following table provides definitive information on the content of each cost data publication. The number of lines of data provided in each unit price or assemblies division, as well as the number of reference tables and crews, is listed for each book. The presence of other elements such as equipment rental costs, historical cost indexes, city cost indexes, square foot models, or cross-referenced indexes is also indicated. You can use the table to help select the RSMeans book that has the quantity and type of information you most need in your work.

Unit Cost Divisions	Building Construction Costs	Mechanical	Electrical	Repair & Remodel	Square Foot	Site Work Landsc.	Assemblies	Interior	Concrete Masonry	Open Shop	Heavy Construction	Light Commercial	Facilities Construction	Plumbing	Residential
1	553	372	385	474		489		301	437	552	483	220	967	386	157
2	555	239	60	544		782		333	189	553	680	409	1022	262	217
3	1471	193	179	830		1321		266	1803	1467	1459	304	1386	162	309
4	878	24	0	677		683		557	1082	854	576	469	1125	0	385
5	1847	197	152	951		796		1023	751	1816	1045	809	1837	275	724
6	1870	78	80	1458		461		1191	310	1849	395	1562	1484	22	2076
7	1395	164	71	1418		506		513	439	1393	0	1091	1451	175	852
8	1941	61	10	1969		298		1677	673	1906	0	1389	2134	0	1319
9	1800	72	26	1606		292		1899	392	1761	0	1513	2035	54	1357
10	978	17	10	582		226		800	156	978	29	465	1072	233	212
11	998	209	160	476		126		848	28	983	0	223	1012	170	110
12	547	0	2	303		248		1691	12	536	19	349	1736	23	317
13	730	119	111	247		341		255	69	702	253	80	740	74	79
14	281	36	0	230		28		260	0	280	0	12	300	21	6
21	78	0	16	31		0		252	0	78	0	60	370	375	0
22	1090	6955	154	1049		1409		596	20	1069	1634	751	6824	8656	572
23	1189	7103	609	776		159		682	38	1103	110	649	5143	1860	425
26	1246	493	9825	991		743		1053	55	1234	527	1060	9761	438	557
27	75	0	200	34		9		67	0	75	35	48	198	0	4
28	87	59	94	64		0		69	0	72	0	42	109	45	27
31	1904	1162	1031	1225		2731		7	1635	1848	2804	1027	1964	1087	1033
32	737	46	0	603		3849		360	271	710	1366	353	1370	158	412
33	514	1020	429	199		1957		0	225	512	1883	114	1547	1189	112
34	99	0	20	4		134		0	31	55	113	0	119	0	0
35	18	0	0	0		166		0	0	18	281	0	83	0	0
41	52	0	0	24		7		22	0	52	30	0	59	14	0
44	98	90	0	0		32		0	0	23	32	0	98	105	0
Totals	23031	18709	13264	16765		17793		14722	8616	22479	13754	12999	45946	15784	11262

Assembly Divisions	Building Construction Costs	Mechanical	Elec-trical	Repair & Remodel	Square Foot	Site Work Landscape	Assemblies	Interior	Concrete Masonry	Open Shop	Heavy Construction	Light Commercial	Facilities Construction	Plumbing	Asm Div	Residential
A		15	0	182	150	530	598	0	536		570	154	24	0	1	374
B		0	0	808	2471	0	5577	329	1914		368	2014	143	0	2	217
C		0	0	630	864	0	1229	1567	146		0	765	242	0	3	588
D		1066	810	694	1779	0	2408	765	0		0	1270	1011	1068	4	861
E		0	0	85	260	0	297	5	0		0	257	5	0	5	392
F		0	0	0	123	0	124	0	0		0	123	3	0	6	358
G		528	294	311	202	2206	668	0	535		1194	200	293	685	7	301
															8	760
															9	80
															10	0
															11	0
															12	0
Totals		1609	1104	2710	5849	2736	10901	2666	3131		2132	4783	1721	1753		3931

Reference Section	Building Construction Costs	Mechanical	Electrical	Repair & Remodel	Square Foot	Site Work Landscape	Assemblies	Interior	Concrete Masonry	Open Shop	Heavy Construction	Light Commercial	Facilities Construction	Plumbing	Residential
Reference Tables	yes	yes	yes	yes	no	yes	yes	yes	yes	yes	yes	yes	yes	yes	yes
Models					105							46			32
Crews	482	482	482	462		482		482	482	460	482	460	462	482	460
Equipment Rental Costs	yes	yes	yes	yes		yes		yes	yes	yes	yes	yes	yes	yes	yes
Historical Cost Indexes	yes	yes	yes	yes	yes	yes	yes	yes	yes	yes	yes	yes	yes	yes	no
City Cost Indexes	yes	yes	yes	yes	yes	yes	yes	yes	yes	yes	yes	yes	yes	yes	yes

Annual Cost Guides

For more information visit the RSMeans website at www.rsmeans.com

RSMeans Building Construction Cost Data 2009

Offers you unchallenged unit price reliability in an easy-to-use arrangement. Whether used for complete, finished estimates or for periodic checks, it supplies more cost facts better and faster than any comparable source. Over 20,000 unit prices for 2009. The City Cost Indexes and Location Factors cover over 930 areas, for indexing to any project location in North America. Order and get *RSMeans Quarterly Update Service* FREE. You'll have year-long access to the RSMeans Estimating **HOTLINE** FREE with your subscription. Expert assistance when using RSMeans data is just a phone call away.

$154.95 per copy
Available Oct. 2008
Catalog no. 60019

Unit prices organized according to MasterFormat 2004!

RSMeans Mechanical Cost Data 2009

- **HVAC**
- **Controls**

Total unit and systems price guidance for mechanical construction. . . materials, parts, fittings, and complete labor cost information. Includes prices for piping, heating, air conditioning, ventilation, and all related construction.

Plus new 2009 unit costs for:

- Over 2,500 installed HVAC/controls assemblies components
- "On-site" Location Factors for over 930 cities and towns in the U.S. and Canada
- Crews, labor, and equipment

$154.95 per copy
Available Oct. 2008
Catalog no. 60029

RSMeans Plumbing Cost Data 2009

Comprehensive unit prices and assemblies for plumbing, irrigation systems, commercial and residential fire protection, point-of-use water heaters, and the latest approved materials. This publication and its companion, *RSMeans Mechanical Cost Data*, provide full-range cost estimating coverage for all the mechanical trades.

Now contains more lines of no-hub CI soil pipe fittings, more flange-type escutcheons, fiberglass pipe insulation in a full range of sizes for 2-1/2" and 3" wall thicknesses, 220 lines of grease duct, and much more.

$154.95 per copy
Available Oct. 2008
Catalog no. 60219

Unit prices organized according to MasterFormat 2004!

RSMeans Facilities Construction Cost Data 2009

For the maintenance and construction of commercial, industrial, municipal, and institutional properties. Costs are shown for new and remodeling construction and are broken down into materials, labor, equipment, and overhead and profit. Special emphasis is given to sections on mechanical, electrical, furnishings, site work, building maintenance, finish work, and demolition.

More than 43,000 unit costs, plus assemblies costs and a comprehensive Reference Section are included.

$368.95 per copy
Available Nov. 2008
Catalog no. 60209

For more information
visit the RSMeans website
at www.rsmeans.com

Annual Cost Guides

RSMeans Square Foot Costs 2009

It's Accurate and Easy To Use!

- **Updated price information** based on nationwide figures from suppliers, estimators, labor experts, and contractors
- "How-to-Use" sections, with **clear examples** of commercial, residential, industrial, and institutional structures
- Realistic graphics, offering true-to-life illustrations of building projects
- Extensive information on using square foot cost data, including sample estimates and alternate pricing methods

$168.95 per copy
Available Oct. 2008
Catalog no. 60059

RSMeans Repair & Remodeling Cost Data 2009

Commercial/Residential

Use this valuable tool to estimate commercial and residential renovation and remodeling.

Includes: New costs for hundreds of unique methods, materials, and conditions that only come up in repair and remodeling, PLUS:

- Unit costs for over 15,000 construction components
- Installed costs for over 90 assemblies
- Over 930 "on-site" localization factors for the U.S. and Canada.

Unit prices organized according to MasterFormat 2004!

$132.95 per copy
Available Nov. 2008
Catalog no. 60049

RSMeans Electrical Cost Data 2009

Pricing information for every part of electrical cost planning. More than 13,000 unit and systems costs with design tables; clear specifications and drawings; engineering guides; illustrated estimating procedures; complete labor-hour and materials costs for better scheduling and procurement; and the latest electrical products and construction methods.

- A variety of special electrical systems including cathodic protection
- Costs for maintenance, demolition, HVAC/mechanical, specialties, equipment, and more

$154.95 per copy
Available Oct. 2008
Catalog no. 60039

Unit prices organized according to MasterFormat 2004!

RSMeans Electrical Change Order Cost Data 2009

RSMeans Electrical Change Order Cost Data provides you with electrical unit prices exclusively for pricing change orders—based on the recent, direct experience of contractors and suppliers. Analyze and check your own change order estimates against the experience others have had doing the same work. It also covers productivity analysis and change order cost justifications. With useful information for calculating the effects of change orders and dealing with their administration.

$154.95 per copy
Available Dec. 2008
Catalog no. 60239

Annual Cost Guides

For more information visit the RSMeans website at www.rsmeans.com

RSMeans Assemblies Cost Data 2009

RSMeans Assemblies Cost Data takes the guesswork out of preliminary or conceptual estimates. Now you don't have to try to calculate the assembled cost by working up individual component costs. We've done all the work for you.

Presents detailed illustrations, descriptions, specifications, and costs for every conceivable building assembly—240 types in all—arranged in the easy-to-use UNIFORMAT II system. Each illustrated "assembled" cost includes a complete grouping of materials and associated installation costs, including the installing contractor's overhead and profit.

$245.95 per copy
Available Oct. 2008
Catalog no. 60069

RSMeans Open Shop Building Construction Cost Data 2009

The latest costs for accurate budgeting and estimating of new commercial and residential construction. . . renovation work. . . change orders. . . cost engineering.

RSMeans Open Shop "BCCD" will assist you to:

- Develop benchmark prices for change orders
- Plug gaps in preliminary estimates and budgets
- Estimate complex projects
- Substantiate invoices on contracts
- Price ADA-related renovations

Unit prices organized according to MasterFormat 2004!

$154.95 per copy
Available Dec. 2008
Catalog no. 60159

RSMeans Residential Cost Data 2009

Contains square foot costs for 30 basic home models with the look of today, plus hundreds of custom additions and modifications you can quote right off the page. With costs for the 100 residential systems you're most likely to use in the year ahead. Complete with blank estimating forms, sample estimates, and step-by-step instructions.

Now contains line items for cultured stone and brick, PVC trim lumber, and TPO roofing.

$132.95 per copy
Available Oct. 2008
Catalog no. 60179

Unit prices organized according to MasterFormat 2004!

RSMeans Site Work & Landscape Cost Data 2009

Includes unit and assemblies costs for earthwork, sewerage, piped utilities, site improvements, drainage, paving, trees & shrubs, street openings/repairs, underground tanks, and more. Contains 57 tables of assemblies costs for accurate conceptual estimates.

Includes:

- Estimating for infrastructure improvements
- Environmentally-oriented construction
- ADA-mandated handicapped access
- Hazardous waste line items

$154.95 per copy
Available Nov. 2008
Catalog no. 60289

RSMeans Facilities Maintenance & Repair Cost Data 2009

RSMeans Facilities Maintenance & Repair Cost Data gives you a complete system to manage and plan your facility repair and maintenance costs and budget efficiently. Guidelines for auditing a facility and developing an annual maintenance plan. Budgeting is included, along with reference tables on cost and management, and information on frequency and productivity of maintenance operations.

The only nationally recognized source of maintenance and repair costs. Developed in cooperation with the Civil Engineering Research Laboratory (CERL) of the Army Corps of Engineers.

$336.95 per copy
Available Dec. 2008
Catalog no. 60309

RSMeans Light Commercial Cost Data 2009

Specifically addresses the light commercial market, which is a specialized niche in the construction industry. Aids you, the owner/designer/contractor, in preparing all types of estimates—from budgets to detailed bids. Includes new advances in methods and materials.

Assemblies Section allows you to evaluate alternatives in the early stages of design/planning.

Over 11,000 unit costs ensure that you have the prices you need. . . when you need them.

Unit prices organized according to MasterFormat 2004!

$132.95 per copy
Available Nov. 2008
Catalog no. 60189

For more information
visit the RSMeans website
at www.rsmeans.com

Annual Cost Guides

RSMeans Concrete & Masonry Cost Data 2009

Provides you with cost facts for virtually all concrete/masonry estimating needs, from complicated formwork to various sizes and face finishes of brick and block—all in great detail. The comprehensive Unit Price Section contains more than 7,500 selected entries. Also contains an Assemblies [Cost] Section, and a detailed Reference Section that supplements the cost data.

Unit prices organized according to MasterFormat 2004!

$139.95 per copy
Available Dec. 2008
Catalog no. 60119

RSMeans Labor Rates for the Construction Industry 2009

Complete information for estimating labor costs, making comparisons, and negotiating wage rates by trade for over 300 U.S. and Canadian cities. With 46 construction trades listed by local union number in each city, and historical wage rates included for comparison. Each city chart lists the county and is alphabetically arranged with handy visual flip tabs for quick reference.

$336.95 per copy
Available Dec. 2008
Catalog no. 60129

RSMeans Construction Cost Indexes 2009

What materials and labor costs will change unexpectedly this year? By how much?

- Breakdowns for 316 major cities
- National averages for 30 key cities
- Expanded five major city indexes
- Historical construction cost indexes

$335.00 per year (subscription)
Catalog no. 50149

$83.25 individual quarters
Catalog no. 60149 A,B,C,D

RSMeans Interior Cost Data 2009

Provides you with prices and guidance needed to make accurate interior work estimates. Contains costs on materials, equipment, hardware, custom installations, furnishings, and labor costs. . . for new and remodel commercial and industrial interior construction, including updated information on office furnishings, and reference information.

Unit prices organized according to MasterFormat 2004!

$154.95 per copy
Available Nov. 2008
Catalog no. 60099

RSMeans Heavy Construction Cost Data 2009

A comprehensive guide to heavy construction costs. Includes costs for highly specialized projects such as tunnels, dams, highways, airports, and waterways. Information on labor rates, equipment, and materials costs is included. Features unit price costs, systems costs, and numerous reference tables for costs and design.

$154.95 per copy
Available Dec. 2008
Catalog no. 60169

Unit prices organized according to MasterFormat 2004!

Reference Books

For more information visit the RSMeans website at www.rsmeans.com

Value Engineering: Practical Applications

For Design, Construction, Maintenance & Operations

by Alphonse Dell'Isola, PE

A tool for immediate application—for engineers, architects, facility managers, owners, and contractors. Includes making the case for VE—the management briefing; integrating VE into planning, budgeting, and design; conducting life cycle costing; using VE methodology in design review and consultant selection; case studies; VE workbook; and a life cycle costing program on disk.

$79.95 per copy
Over 450 pages, illustrated, softcover
Catalog no. 67319A

Facilities Operations & Engineering Reference

by the Association for Facilities Engineering and RSMeans

An all-in-one technical reference for planning and managing facility projects and solving day-to-day operations problems. Selected as the official certified plant engineer reference, this handbook covers financial analysis, maintenance, HVAC and energy efficiency, and more.

$54.98 per copy
Over 700 pages, illustrated, hardcover
Catalog no. 67318

The Building Professional's Guide to Contract Documents

3rd Edition

by Waller S. Poage, AIA, CSI, CVS

A comprehensive reference for owners, design professionals, contractors, and students

- Structure your documents for maximum efficiency.
- Effectively communicate construction requirements.
- Understand the roles and responsibilities of construction professionals.
- Improve methods of project delivery.

$64.95 per copy, 400 pages
Diagrams and construction forms, hardcover
Catalog no. 67261A

Complete Book of Framing

by Scot Simpson

This straightforward, easy-to-learn method will help framers, carpenters, and handy homeowners build their skills in rough carpentry and framing. Shows how to frame all the parts of a house: floors, walls, roofs, door & window openings, and stairs—with hundreds of color photographs and drawings that show every detail.

The book gives beginners all the basics they need to go from zero framing knowledge to a journeyman level.

$29.95 per copy
352 pages, softcover
Catalog no. 67353

Cost Planning & Estimating for Facilities Maintenance

In this unique book, a team of facilities management authorities shares their expertise on:

- Evaluating and budgeting maintenance operations
- Maintaining and repairing key building components
- Applying *RSMeans Facilities Maintenance & Repair Cost Data* to your estimating

Covers special maintenance requirements of the ten major building types.

$89.95 per copy
Over 475 pages, hardcover
Catalog no. 67314

Life Cycle Costing for Facilities

by Alphonse Dell'Isola and Dr. Steven Kirk

Guidance for achieving higher quality design and construction projects at lower costs! Cost-cutting efforts often sacrifice quality to yield the cheapest product. Life cycle costing enables building designers and owners to achieve both. The authors of this book show how LCC can work for a variety of projects — from roads to HVAC upgrades to different types of buildings.

$99.95 per copy
450 pages, hardcover
Catalog no. 67341

Planning & Managing Interior Projects 2nd Edition

by Carol E. Farren, CFM

Expert guidance on managing renovation & relocation projects

This book guides you through every step in relocating to a new space or renovating an old one. From initial meeting through design and construction, to post-project administration, it helps you get the most for your company or client. Includes sample forms, spec lists, agreements, drawings, and much more!

$34.98 per copy
200 pages, softcover
Catalog no. 67245A

Construction Business Management

by Nick Ganaway

Only 43% of construction firms stay in business after four years. Make sure your company thrives with valuable guidance from a pro with 25 years of success as a commercial contractor. Find out what it takes to build all aspects of a business that is profitable, enjoyable, and enduring. With a bonus chapter on retail construction.

$49.95 per copy
200 pages, softcover
Catalog no. 67352

Interior Home Improvement Costs 9th Edition

Updated estimates for the most popular remodeling and repair projects—from small, do-it-yourself jobs to major renovations and new construction. Includes: kitchens & baths; new living space from your attic, basement, or garage; new floors, paint, and wallpaper; tearing out or building new walls; closets, stairs, and fireplaces; new energy-saving improvements, home theaters, and more!

$24.95 per copy
250 pages, illustrated, softcover
Catalog no. 67308E

Exterior Home Improvement Costs 9th Edition

Updated estimates for the most popular remodeling and repair projects—from small, do-it-yourself jobs, to major renovations and new construction. Includes: curb appeal projects—landscaping, patios, porches, driveways, and walkways; new windows and doors; decks, greenhouses, and sunrooms; room additions and garages; roofing, siding, and painting; "green" improvements to save energy & water.

$24.95 per copy
Over 275 pages, illustrated, softcover
Catalog no. 67309E

Builder's Essentials: Plan Reading & Material Takeoff

For Residential and Light Commercial Construction

by Wayne J. DelPico

A valuable tool for understanding plans and specs, and accurately calculating material quantities. Step-by-step instructions and takeoff procedures based on a full set of working drawings.

$35.95 per copy
Over 420 pages, softcover
Catalog no. 67307

Means Unit Price Estimating Methods

New 4th Edition

This new edition includes up-to-date cost data and estimating examples, updated to reflect changes to the CSI numbering system and new features of RSMeans cost data. It describes the most productive, universally accepted ways to estimate, and uses checklists and forms to illustrate shortcuts and timesavers. A model estimate demonstrates procedures. A new chapter explores computer estimating alternatives.

$59.95 per copy
Over 350 pages, illustrated, softcover
Catalog no. 67303B

Total Productive Facilities Management

by Richard W. Sievert, Jr.

Today, facilities are viewed as strategic resources. . . elevating the facility manager to the role of asset manager supporting the organization's overall business goals. Now, Richard Sievert Jr., in this well-articulated guidebook, sets forth a new operational standard for the facility manager's emerging role. . . a comprehensive program for managing facilities as a true profit center.

$29.98 per copy
275 pages, softcover
Catalog no. 67321

Green Building: Project Planning & Cost Estimating 2nd Edition

This new edition has been completely updated with the latest in green building technologies, design concepts, standards, and costs. Now includes a 2009 Green Building *CostWorks* CD with more than 300 green building assemblies and over 5,000 unit price line items for sustainable building. The new edition is also full-color with all new case studies—plus a new chapter on deconstruction, a key aspect of green building.

$129.95 per copy
350 pages, softcover
Catalog no. 67338A

Concrete Repair and Maintenance Illustrated

by Peter Emmons

Hundreds of illustrations show users how to analyze, repair, clean, and maintain concrete structures for optimal performance and cost effectiveness. From parking garages to roads and bridges to structural concrete, this comprehensive book describes the causes, effects, and remedies for concrete wear and failure. Invaluable for planning jobs, selecting materials, and training employees, this book is a must-have for concrete specialists, general contractors, facility managers, civil and structural engineers, and architects.

$69.95 per copy
300 pages, illustrated, softcover
Catalog no. 67146

Reference Books

For more information visit the RSMeans website at www.rsmeans.com

Means Illustrated Construction Dictionary Condensed, 2nd Edition

The best portable dictionary for office or field use—an essential tool for contractors, architects, insurance and real estate personnel, facility managers, homeowners, and anyone who needs quick, clear definitions for construction terms. The second edition has been further enhanced with updates and hundreds of new terms and illustrations . . . in keeping with the most recent developments in the construction industry.

Now with a quick-reference Spanish section. Includes tools and equipment, materials, tasks, and more!

$59.95 per copy
Over 500 pages, softcover
Catalog no. 67282A

Means Repair & Remodeling Estimating 4th Edition

By Edward B. Wetherill and RSMeans

This important reference focuses on the unique problems of estimating renovations of existing structures, and helps you determine the true costs of remodeling through careful evaluation of architectural details and a site visit.

New section on disaster restoration costs.

$69.95 per copy
Over 450 pages, illustrated, hardcover
Catalog no. 67265B

Facilities Planning & Relocation

by David D. Owen

A complete system for planning space needs and managing relocations. Includes step-by-step manual, over 50 forms, and extensive reference section on materials and furnishings.

New lower price and user-friendly format.

$49.98 per copy
Over 450 pages, softcover
Catalog no. 67301

Means Square Foot & UNIFORMAT Assemblies Estimating Methods 3rd Edition

Develop realistic square foot and assemblies costs for budgeting and construction funding. The new edition features updated guidance on square foot and assemblies estimating using UNIFORMAT II. An essential reference for anyone who performs conceptual estimates.

$69.95 per copy
Over 300 pages, illustrated, softcover
Catalog no. 67145B

Means Electrical Estimating Methods 3rd Edition

Expanded edition includes sample estimates and cost information in keeping with the latest version of the CSI MasterFormat and UNIFORMAT II. Complete coverage of fiber optic and uninterruptible power supply electrical systems, broken down by components, and explained in detail. Includes a new chapter on computerized estimating methods. A practical companion to *RSMeans Electrical Cost Data.*

$64.95 per copy
Over 325 pages, hardcover
Catalog no. 67230B

Means Mechanical Estimating Methods 4th Edition

Completely updated, this guide assists you in making a review of plans, specs, and bid packages, with suggestions for takeoff procedures, listings, substitutions, and pre-bid scheduling for all components of HVAC. Includes suggestions for budgeting labor and equipment usage. Compares materials and construction methods to allow you to select the best option.

$64.95 per copy
Over 350 pages, illustrated, softcover
Catalog no. 67294B

Means ADA Compliance Pricing Guide 2nd Edition

by Adaptive Environments and RSMeans

Completely updated and revised to the new 2004 *Americans with Disabilities Act Accessibility Guidelines,* this book features more than 70 of the most commonly needed modifications for ADA compliance. Projects range from installing ramps and walkways, widening doorways and entryways, and installing and refitting elevators, to relocating light switches and signage.

$79.95 per copy
Over 350 pages, illustrated, softcover
Catalog no. 67310A

Project Scheduling & Management for Construction 3rd Edition

by David R. Pierce, Jr.

A comprehensive yet easy-to-follow guide to construction project scheduling and control—from vital project management principles through the latest scheduling, tracking, and controlling techniques. The author is a leading authority on scheduling, with years of field and teaching experience at leading academic institutions. Spend a few hours with this book and come away with a solid understanding of this essential management topic.

$64.95 per copy
Over 300 pages, illustrated, hardcover
Catalog no. 67247B

The Practice of Cost Segregation Analysis

by Bruce A. Desrosiers and Wayne J. DelPico

This expert guide walks you through the practice of cost segregation analysis, which enables property owners to defer taxes and benefit from "accelerated cost recovery" through depreciation deductions on assets that are properly identified and classified.

With a glossary of terms, sample cost segregation estimates for various building types, key information resources, and updates via a dedicated website, this book is a critical resource for anyone involved in cost segregation analysis.

$99.95 per copy
Over 225 pages
Catalog no. 67345

Preventive Maintenance for Multi-Family Housing

by John C. Maciha

Prepared by one of the nation's leading experts on multi-family housing.

This complete PM system for apartment and condominium communities features expert guidance, checklists for buildings and grounds maintenance tasks and their frequencies, a reusable wall chart to track maintenance, and a dedicated website featuring customizable electronic forms. A must-have for anyone involved with multi-family housing maintenance and upkeep.

$89.95 per copy
225 pages
Catalog no. 67346

How to Estimate with Means Data & CostWorks

New 3rd Edition

by RSMeans and Saleh A. Mubarak, Ph.D.

New 3rd Edition—fully updated with new chapters, plus new CD with updated *CostWorks* cost data and MasterFormat organization. Includes all major construction items—with more than 300 exercises and two sets of plans that show how to estimate for a broad range of construction items and systems—including general conditions and equipment costs.

$59.95 per copy
272 pages, softcover
Includes CostWorks CD
Catalog no. 67324B

Job Order Contracting

Expediting Construction Project Delivery

by Allen Henderson

Expert guidance to help you implement JOC—fast becoming the preferred project delivery method for repair and renovation, minor new construction, and maintenance projects in the public sector and in many states and municipalities. The author, a leading JOC expert and practitioner, shows how to:

- Establish a JOC program
- Evaluate proposals and award contracts
- Handle general requirements and estimating
- Partner for maximum benefits

$89.95 per copy
192 pages, illustrated, hardcover
Catalog no. 67348

Builder's Essentials: Estimating Building Costs

For the Residential & Light Commercial Contractor

by Wayne J. DelPico

Step-by-step estimating methods for residential and light commercial contractors. Includes a detailed look at every construction specialty—explaining all the components, takeoff units, and labor needed for well-organized, complete estimates. Covers correctly interpreting plans and specifications, and developing accurate and complete labor and material costs.

$29.95 per copy
Over 400 pages, illustrated, softcover
Catalog no. 67343

Building & Renovating Schools

This all-inclusive guide covers every step of the school construction process—from initial planning, needs assessment, and design, right through moving into the new facility. A must-have resource for anyone concerned with new school construction or renovation. With square foot cost models for elementary, middle, and high school facilities, and real-life case studies of recently completed school projects.

The contributors to this book—architects, construction project managers, contractors, and estimators who specialize in school construction—provide start-to-finish, expert guidance on the process.

$69.98 per copy
Over 425 pages, hardcover
Catalog no. 67342

Reference Books

The Homeowner's Guide to Mold By Michael Pugliese

Expert guidance to protect your health and your home.

Mold, whether caused by leaks, humidity or flooding, is a real health and financial issue—for homeowners and contractors. This full-color book explains:
- Construction and maintenance practices to prevent mold
- How to inspect for and remove mold
- Mold remediation procedures and costs
- What to do after a flood
- How to deal with insurance companies

$21.95 per copy
144 pages, softcover
Catalog no. 67344

Means Illustrated Construction Dictionary

Unabridged 3rd Edition, with CD-ROM

Long regarded as the industry's finest, *Means Illustrated Construction Dictionary* is now even better. With the addition of over 1,000 new terms and hundreds of new illustrations, it is the clear choice for the most comprehensive and current information. The companion CD-ROM that comes with this new edition adds many extra features: larger graphics, expanded definitions, and links to both CSI MasterFormat numbers and product information.

$99.95 per copy
Over 790 pages, illustrated, hardcover
Catalog no. 67292A

Means Landscape Estimating Methods

New 5th Edition

Answers questions about preparing competitive landscape construction estimates, with up-to-date cost estimates and the new MasterFormat classification system. Expanded and revised to address the latest materials and methods, including new coverage on approaches to green building. Includes:
- Step-by-step explanation of the estimating process
- Sample forms and worksheets that save time and prevent errors

$64.95 per copy
Over 350 pages, softcover
Catalog no. 67295C

Means Plumbing Estimating Methods 3rd Edition

by Joseph Galeno and Sheldon Greene

Updated and revised! This practical guide walks you through a plumbing estimate, from basic materials and installation methods through change order analysis. *Plumbing Estimating Methods* covers residential, commercial, industrial, and medical systems, and features sample takeoff and estimate forms and detailed illustrations of systems and components.

$29.98 per copy
330+ pages, softcover
Catalog no. 67283B

Understanding & Negotiating Construction Contracts

by Kit Werremeyer

Take advantage of the author's 30 years' experience in small-to-large (including international) construction projects. Learn how to identify, understand, and evaluate high risk terms and conditions typically found in all construction contracts—then negotiate to lower or eliminate the risk, improve terms of payment, and reduce exposure to claims and disputes. The author avoids "legalese" and gives real-life examples from actual projects.

$69.95 per copy
300 pages, softcover
Catalog no. 67350

Means Estimating Handbook

2nd Edition

Updated Second Edition answers virtually any estimating technical question—all organized by CSI MasterFormat. This comprehensive reference covers the full spectrum of technical data required to estimate construction costs. The book includes information on sizing, productivity, equipment requirements, code-mandated specifications, design standards, and engineering factors.

$99.95 per copy
Over 900 pages, hardcover
Catalog no. 67276A

Means Spanish/English Construction Dictionary 2nd Edition

by RSMeans and the International Code Council

This expanded edition features thousands of the most common words and useful phrases in the construction industry with easy-to-follow pronunciations in both Spanish and English. Over 800 new terms, phrases, and illustrations have been added. It also features a new stand-alone "Safety & Emergencies" section, with colored pages for quick access.

Unique to this dictionary are the systems illustrations showing the relationship of components in the most common building systems for all major trades.

$23.95 per copy
Over 400 pages
Catalog no. 67327A

How Your House Works

by Charlie Wing

A must-have reference for every homeowner, handyman, and contractor—for repair, remodeling, and new construction. This book uncovers the mysteries behind just about every major appliance and building element in your house—from electrical, heating and AC, to plumbing, framing, foundations and appliances. Clear, "exploded" drawings show exactly how things should be put together and how they function—what to check if they don't work, and what you can do that might save you having to call in a professional.

$21.95 per copy
160 pages, softcover
Catalog no. 67351

For more information visit the RSMeans website at www.rsmeans.com

Professional Development

Means CostWorks® Training

This one-day course helps users become more familiar with the functionality of *Means CostWorks* program. Each menu, icon, screen, and function found in the program is explained in depth. Time is devoted to hands-on estimating exercises.

Some of what you'll learn:

- Search the database utilizing all navigation methods
- Export RSMeans Data to your preferred spreadsheet format
- View crews, assembly components, and much more!
- Automatically regionalize the database

You are required to bring your own laptop computer for this course.

When you register for this course you will receive an outline for your laptop requirements.

Unit Price Estimating

This interactive *two-day* seminar teaches attendees how to interpret project information and process it into final, detailed estimates with the greatest accuracy level.

The most important credential an estimator can take to the job is the ability to visualize construction, and estimate accurately.

Some of what you'll learn:

- Interpreting the design in terms of cost
- The most detailed, time-tested methodology for accurate pricing
- Key cost drivers—material, labor, equipment, staging, and subcontracts
- Understanding direct and indirect costs for accurate job cost accounting and change order management

Who should attend: Corporate and government estimators and purchasers, architects, engineers… and others needing to produce accurate project estimates.

Square Foot & Assemblies Estimating

This *two-day* course teaches attendees how to quickly deliver accurate square foot estimates using limited budget and design information.

Some of what you'll learn:

- How square foot costing gets the estimate done faster
- Taking advantage of a "systems" or "assemblies" format
- The RSMeans "building assemblies/square foot cost approach"
- How to create a reliable preliminary systems estimate using bare-bones design information

Who should attend: Facilities managers, facilities engineers, estimators, planners, developers, construction finance professionals… and others needing to make quick, accurate construction cost estimates at commercial, government, educational, and medical facilities.

Repair & Remodeling Estimating

This *two-day* seminar emphasizes all the underlying considerations unique to repair/remodeling estimating and presents the correct methods for generating accurate, reliable R&R project costs using the unit price and assemblies methods.

Some of what you'll learn:

- Estimating considerations—like labor-hours, building code compliance, working within existing structures, purchasing materials in smaller quantities, unforeseen deficiencies
- Identifying problems and providing solutions to estimating building alterations
- Rules for factoring in minimum labor costs, accurate productivity estimates, and allowances for project contingencies
- R&R estimating examples calculated using unit price and assemblies data

Who should attend: Facilities managers, plant engineers, architects, contractors, estimators, builders… and others who are concerned with the proper preparation and/or evaluation of repair and remodeling estimates.

Mechanical & Electrical Estimating

This *two-day* course teaches attendees how to prepare more accurate and complete mechanical/electrical estimates, avoiding the pitfalls of omission and double-counting, while understanding the composition and rationale within the RSMeans Mechanical/Electrical database.

Some of what you'll learn:

- The unique way mechanical and electrical systems are interrelated
- M&E estimates–conceptual, planning, budgeting, and bidding stages
- Order of magnitude, square foot, assemblies, and unit price estimating
- Comparative cost analysis of equipment and design alternatives

Who should attend: Architects, engineers, facilities managers, mechanical and electrical contractors… and others needing a highly reliable method for developing, understanding, and evaluating mechanical and electrical contracts.

Green Building Planning & Construction

In this *two-day* course, learn about tailoring a building and its placement on the site to the local climate, site conditions, culture, and community. Includes information to reduce resource consumption and augment resource supply.

Some of what you'll learn:

- Green technologies, materials, systems, and standards
- Energy efficiencies with energy modeling tools
- Cost vs. value of green products over their life cycle
- Low-cost strategies and economic incentives and funding
- Health, comfort, and productivity goals and techniques

Who should attend: Contractors, project managers, building owners, building officials, healthcare and insurance professionals.

Facilities Maintenance & Repair Estimating

This *two-day* course teaches attendees how to plan, budget, and estimate the cost of ongoing and preventive maintenance and repair for existing buildings and grounds.

Some of what you'll learn:

- The most financially favorable maintenance, repair, and replacement scheduling and estimating
- Auditing and value engineering facilities
- Preventive planning and facilities upgrading
- Determining both in-house and contract-out service costs
- Annual, asset-protecting M&R plan

Who should attend: Facility managers, maintenance supervisors, buildings and grounds superintendents, plant managers, planners, estimators… and others involved in facilities planning and budgeting.

Practical Project Management for Construction Professionals

In this *two-day* course, acquire the essential knowledge and develop the skills to effectively and efficiently execute the day-to-day responsibilities of the construction project manager.

Covers:

- General conditions of the construction contract
- Contract modifications: change orders and construction change directives
- Negotiations with subcontractors and vendors
- Effective writing: notification and communications
- Dispute resolution: claims and liens

Who should attend: Architects, engineers, owner's representatives, project managers.

Facilities Repair & Remodeling Estimating

In this *two-day* course, professionals working in facilities management can get help with their daily challenges to establish budgets for all phase of a project

Some of what you'll learn:

- Determine the full scope of a project
- Identify the scope of risks & opportunities
- Creative solutions to estimating issues
- Organizing estimates for presentation & discussion
- Special techniques for repair/remodel and maintenance projects
- Negotiating project change orders

Who should attend: Facility managers, engineers, contractors, facility trades-people, planners & project managers

Professional Development

For more information visit the RSMeans website at www.rsmeans.com

Scheduling and Project Management

This *two-day* course teaches attendees the most current and proven scheduling and management techniques needed to bring projects in on time and on budget.

Some of what you'll learn:

- Crucial phases of planning and scheduling
- How to establish project priorities and develop realistic schedules and management techniques
- Critical Path and Precedence Methods
- Special emphasis on cost control

Who should attend: Construction project managers, supervisors, engineers, estimators, contractors… and others who want to improve their project planning, scheduling, and management skills.

Understanding & Negotiating Construction Contracts

In this *two-day* course, learn how to protect the assets of your company by justifying or eliminating commercial risk through negotiation of a contract's terms and conditions.

Some of what you'll learn:

- Myths and paradigms of contracts
- Scope of work, terms of payment, and scheduling
- Dispute resolution
- Why clients love CLAIMS!
- Negotiating issues and much more

Who should attend: Contractors & subcontractors, material suppliers, project managers, risk & insurance managers, procurement managers, owners and facility managers, corporate executives.

Site Work & Heavy Construction Estimating

This *two-day* course teaches attendees how to estimate earthwork, site utilities, foundations, and site improvements, using the assemblies and the unit price methods.

Some of what you'll learn:

- Basic site work and heavy construction estimating skills
- Estimating foundations, utilities, earthwork, and site improvements
- Correct equipment usage, quality control, and site investigation for estimating purposes

Who should attend: Project managers, design engineers, estimators, and contractors doing site work and heavy construction.

Assessing Scope of Work for Facility Construction Estimating

This *two-day* course is a practical training program that addresses the vital importance of understanding the SCOPE of projects in order to produce accurate cost estimates in a facility repair and remodeling environment.

Some of what you'll learn:

- Discussions of site visits, plans/specs, record drawings of facilities, and site-specific lists
- Review of CSI divisions, including means, methods, materials, and the challenges of scoping each topic
- Exercises in SCOPE identification and SCOPE writing for accurate estimating of projects
- Hands-on exercises that require SCOPE, take-off, and pricing

Who should attend: Corporate and government estimators, planners, facility managers, and others who need to produce accurate project estimates.

2009 RSMeans Seminar Schedule

Note: call for exact dates and details.

Location	Dates
Las Vegas, NV	March
Washington, DC	April
Denver, CO	May
San Francisco, CA	June
Washington, DC	September
Dallas, TX	September
Las Vegas, NV	October
Atlantic City, NJ	October
Orlando, FL	November
San Diego, CA	December

1-800-334-3509, ext. 5115

Registration Information

Register early... Save up to $100! Register 30 days before the start date of a seminar and save $100 off your total fee. *Note: This discount can be applied only once per order. It cannot be applied to team discount registrations or any other special offer.*

How to register Register by phone today! The RSMeans toll-free number for making reservations is **1-800-334-3509, ext. 5115.**

Individual seminar registration fee - $935. *Means CostWorks®* training registration fee - $375. To register by mail, complete the registration form and return, with your full fee, to: RSMeans Seminars, 63 Smiths Lane, Kingston, MA 02364.

Government pricing All federal government employees save off the regular seminar price. Other promotional discounts cannot be combined with the government discount.

Team discount program for two to four seminar registrations. Call for pricing: 1-800-334-3509, ext. 5115

Multiple course discounts When signing up for two or more courses, call for pricing.

Refund policy Cancellations will be accepted up to ten business days prior to the seminar start. There are no refunds for cancellations received later than ten working days prior to the first day of the seminar. A $150 processing fee will be applied for all cancellations. Written notice of cancellation is required. Substitutions can be made at any time before the session starts. **No-shows are subject to the full seminar fee.**

AACE approved courses Many seminars described and offered here have been approved for 14 hours (1.4 recertification credits) of credit by the AACE International Certification Board toward meeting the continuing education requirements for recertification as a Certified Cost Engineer/Certified Cost Consultant.

AIA Continuing Education We are registered with the AIA Continuing Education System (AIA/CES) and are committed to developing quality learning activities in accordance with the CES criteria. Many seminars meet the AIA/CES criteria for Quality Level 2. AIA members may receive (14) learning units (LUs) for each two-day RSMeans course.

NASBA CPE sponsor credits We are part of the National Registry of CPE sponsors. Attendees may be eligible for (16) CPE credits.

Daily course schedule The first day of each seminar session begins at 8:30 a.m. and ends at 4:30 p.m. The second day begins at 8:00 a.m. and ends at 4:00 p.m. Participants are urged to bring a hand-held calculator, since many actual problems will be worked out in each session.

Continental breakfast Your registration includes the cost of a continental breakfast, and a morning and afternoon refreshment break. These informal segments allow you to discuss topics of mutual interest with other seminar attendees. (You are free to make your own lunch and dinner arrangements.)

Hotel/transportation arrangements RSMeans arranges to hold a block of rooms at most host hotels. To take advantage of special group rates when making your reservation, be sure to mention that you are attending the RSMeans seminar. You are, of course, free to stay at the lodging place of your choice. **(Hotel reservations and transportation arrangements should be made directly by seminar attendees.)**

Important Class sizes are limited, so please register as soon as possible.

Note: *Pricing subject to change.*

Registration Form

ADDS-1000

Call 1-800-334-3509, ext. 5115 to register or FAX this form 1-800-632-6732. Visit our Web site: www.rsmeans.com

Please register the following people for the RSMeans construction seminars as shown here. We understand that we must make our own hotel reservations if overnight stays are necessary.

☐ Full payment of $____________________ enclosed.

☐ Bill me

Please print name of registrant(s)

(To appear on certificate of completion)

____

____

____

____

P.O. #: ______________________________

GOVERNMENT AGENCIES MUST SUPPLY PURCHASE ORDER NUMBER OR TRAINING FORM.

Firm name______________________________

Address______________________________

City/State/Zip______________________________

Telephone no.____________________ Fax no.____________________

E-mail address______________________________

Charge registration(s) to: ☐ MasterCard ☐ VISA ☐ American Express

Account no.____________________ Exp. date____________________

Cardholder's signature______________________________

Seminar name______________________________

Seminar City ______________________________

Please mail check to: RSMeans Seminars, 63 Smiths Lane, P.O. Box 800, Kingston, MA 02364 USA

Harness the full power of RSMeans online with MeansCostWorks.com

MeansCostWorks.com

Manage your estimates in a secure environment... available 24/7 from a web-enabled computer.

To Learn MORE call 1.800.334.3509 or go to: www.MeansCostWorks.com
View the online demo...Get a FREE TRIAL Subscription!

MeansCostWorks.com from RSMeans is web-based software for estimating construction costs on commercial and residential construction projects. It provides fast and accurate construction estimating using the latest construction cost data available from RSMeans. *It's an invaluable tool for construction, engineering and design professionals.*

Introducing *MeansCostsWorks.com*

Now you can access and use the world's largest construction cost database for estimating interior projects... verifying bids, managing change orders, budgeting, designing, and more.

MeansCostWorks.com is current, comprehensive, and very easy to use. It combines exhaustive cost data for virtually every construction component and task with state-of-the-art estimating and robust customization.

- Estimate faster and more accurately than you ever thought possible.
- Collaborate seamlessly with authorized team members.
- Easily print, e-mail, or export your work product to your Microsoft EXCEL program.

MeansData™

Software Integration

CONSTRUCTION COSTS FOR SOFTWARE APPLICATIONS

Your construction estimating software is only as good as your cost data.

BECK TECHNOLOGY

D Profiler

Have you ever wondered why so many projects are over budget, over scope, and take longer to design and build than expected?

DProfiler with RSMeans is the solution that reduces your costs, the risks to the owner, and gives you a competitive advantage.

Use DProfiler with RSMeans cost data at the start of a project to establish the project scope and budget.

www.beck-technology.com

SECURE INTERNET ORDER SITE **www.rsmeans.com** RSMeans

The Gypsum Construction Handbook by USG

An invaluable reference of construction procedures for gypsum drywall, cement board, veneer plaster, and conventional plaster. This book includes the newest product developments, installation methods, fire- and sound-rated construction information, illustrated framing-to-finish application instructions, estimating and planning information, and more. Great for architects and engineers; contractors, builders, and dealers; apprentices and training programs; building inspectors and code officials; and anyone interested in gypsum construction. Features information on tools and safety practices, a glossary of construction terms, and a list of important agencies and associations. Also available in Spanish.

$29.95
Over 500 pages, illustrated, softcover
Catalog no. 67357

BIM Handbook

A Guide to Building Information Modeling for Owners, Managers, Designers, Engineers and Contractors

by Chuck Eastman, Paul Teicholz, Rafael Sacks, Kathleen Liston

Building Information Modeling (BIM) is a new approach to design, construction, and facility management where a digital depiction of the building process is used to facilitate the exchange of information in digital format.

The BIM Handbook provides an in-depth understanding of BIM technologies, the business and organizational issues associated with its implementation, and the profound advantages that effective use of BIM can provide to all members of a project team. Also includes a rich set of informative BIM case studies.

$85.00
504 pages
Catalog no. 67356

Universal Design Ideas for Style, Comfort & Safety

by RSMeans and Lexicon Consulting, Inc.

Incorporating universal design when building or remodeling helps people of any age and physical ability more fully and safely enjoy their living spaces. This book shows how universal design can be artfully blended into the most attractive homes. It discusses specialized products like adjustable countertops and chair lifts, as well as simple ways to enhance a home's safety and comfort. With color photos and expert guidance, every area of the home is covered. Includes budget estimates that give an idea how much projects will cost.

$21.95
160 pages, illustrated, softcover
Catalog no. 67354

Green Home Improvement

by Daniel D. Chiras, PhD

With energy costs rising and environmental awareness increasing, many people are looking to make their homes greener. This book makes it easy, with 65 projects and actual costs and projected savings that will help homeowners prioritize their green improvements.

Projects range from simple water savers that cost only a few dollars, to bigger-ticket items such as efficient appliances and HVAC systems, green roofing, flooring, landscaping, and much more. With color photos and cost estimates, each project compares options and describes the work involved, the benefits, and the savings.

$24.95
320 pages, illustrated, softcover
Catalog no. 67355

Residential & Light Commercial Construction Standards, 3rd Edition, Updated

by RSMeans and contributing authors

This book provides authoritative requirements and recommendations compiled from leading professional associations, industry publications, and building code organizations. This all-in-one reference helps establish a standard for workmanship, quickly resolve disputes, and avoid defect claims. Updated third edition includes new coverage of green building, seismic, hurricane, and mold-resistant construction, plus updated national building code requirements.

Covers all areas of building construction, including paving; concrete; framing; finish carpentry; insulation; roofing and siding; doors and windows; drywall; ceilings and floors; wall coverings; plumbing; HVAC; electrical; and more!

$59.95
Over 550 pages, illustrated, softcover
Catalog no. 67322B

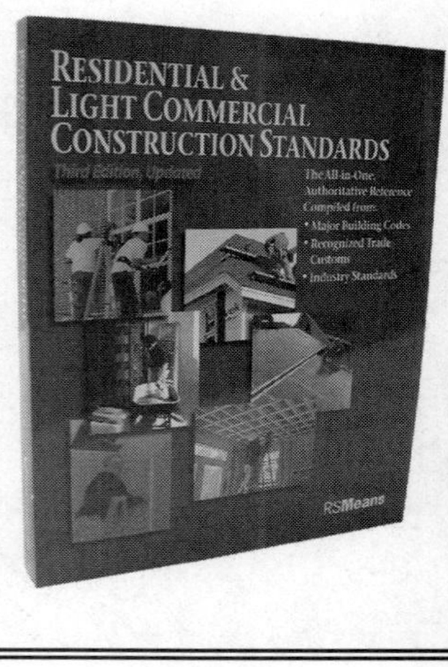

ORDER TOLL FREE 1-800-334-3509
OR FAX 1-800-632-6732

2009 Order Form

Qty.	Book no.	COST ESTIMATING BOOKS	Unit Price	Total
	60069	Assemblies Cost Data 2009	$245.95	
	60019	Building Construction Cost Data 2009	154.95	
	60119	Concrete & Masonry Cost Data 2009	136.95	
	50149	Construction Cost Indexes 2009 (subscription)	335.00	
	60149A	Construction Cost Index–January 2009	83.75	
	60149B	Construction Cost Index–April 2009	83.75	
	60149C	Construction Cost Index–July 2009	83.75	
	60149D	Construction Cost Index–October 2009	83.75	
	60349	Contr. Pricing Guide: Resid. R & R Costs 2009	39.95	
	60339	Contr. Pricing Guide: Resid. Detailed 2009	39.95	
	60329	Contr. Pricing Guide: Resid. Sq. Ft. 2009	39.95	
	60239	Electrical Change Order Cost Data 2009	154.95	
	60039	Electrical Cost Data 2009	154.95	
	60209	Facilities Construction Cost Data 2009	368.95	
	60309	Facilities Maintenance & Repair Cost Data 2009	336.95	
	60169	Heavy Construction Cost Data 2009	154.95	
	60099	Interior Cost Data 2009	154.95	
	60129	Labor Rates for the Const. Industry 2009	336.95	
	60189	Light Commercial Cost Data 2009	132.95	
	60029	Mechanical Cost Data 2009	154.95	
	60159	Open Shop Building Const. Cost Data 2009	154.95	
	60219	Plumbing Cost Data 2009	154.95	
	60049	Repair & Remodeling Cost Data 2009	132.95	
	60179	Residential Cost Data 2009	132.95	
	60289	Site Work & Landscape Cost Data 2009	154.95	
	60059	Square Foot Costs 2009	168.95	
	62018	Yardsticks for Costing (2008)	149.95	
	62019	Yardsticks for Costing (2009)	154.95	
		REFERENCE BOOKS		
	67310A	ADA Compliance Pricing Guide, 2nd Ed.	79.95	
	67330	Bldrs Essentials: Adv. Framing Methods	12.98	
	67329	Bldrs Essentials: Best Bus. Practices for Bldrs	29.95	
	67298A	Bldrs Essentials: Framing/Carpentry 2nd Ed.	15.98	
	67298AS	Bldrs Essentials: Framing/Carpentry Spanish	15.98	
	67307	Bldrs Essentials: Plan Reading & Takeoff	35.95	
	67342	Building & Renovating Schools	69.98	
	67339	Building Security: Strategies & Costs	44.98	
	67353	Complete Book of Framing	29.95	
	67146	Concrete Repair & Maintenance Illustrated	69.95	
	67352	Construction Business Management	49.95	
	67314	Cost Planning & Est. for Facil. Maint.	89.95	
	67328A	Designing & Building with the IBC, 2nd Ed.	99.95	
	67230B	Electrical Estimating Methods, 3rd Ed.	64.95	
	64777A	Environmental Remediation Est. Methods, 2nd Ed.	49.98	
	67343	Estimating Bldg. Costs for Resi. & Lt. Comm.	29.95	
	67276A	Estimating Handbook, 2nd Ed.	$99.95	
	67318	Facilities Operations & Engineering Reference	54.98	
	67301	Facilities Planning & Relocation	49.98	
	67338A	Green Building: Proj. Planning & Cost Est., 2nd Ed.	129.95	
	67355	Green Home Improvement	24.95	

Qty.	Book no.	REFERENCE BOOKS (Cont.)	Unit Price	Total
	67357	The Gypsum Construction Handbook	29.95	
	67357S	The Gypsum Construction Handbook (Spanish)	29.95	
	67323	Historic Preservation: Proj. Planning & Est.	49.98	
	67349	Home Addition & Renovation Project Costs	19.98	
	67308E	Home Improvement Costs–Int. Projects, 9th Ed.	24.95	
	67309E	Home Improvement Costs–Ext. Projects, 9th Ed.	24.95	
	67344	Homeowner's Guide to Mold	21.95	
	67324B	How to Est.w/Means Data & CostWorks, 3rd Ed.	59.95	
	67351	How Your House Works	21.95	
	67282A	Illustrated Const. Dictionary, Condensed, 2nd Ed.	59.95	
	67292A	Illustrated Const. Dictionary, w/CD-ROM, 3rd Ed.	99.95	
	67348	Job Order Contracting	89.95	
	67347	Kitchen & Bath Project Costs	19.98	
	67295C	Landscape Estimating Methods, 5th Ed.	64.95	
	67341	Life Cycle Costing for Facilities	99.95	
	67294B	Mechanical Estimating Methods, 4th Ed.	64.95	
	67245A	Planning & Managing Interior Projects, 2nd Ed.	34.98	
	67283B	Plumbing Estimating Methods, 3rd Ed.	29.98	
	67345	Practice of Cost Segregation Analysis	99.95	
	67337	Preventive Maint. for Higher Education Facilities	149.95	
	67346	Preventive Maint. for Multi-Family Housing	89.95	
	67326	Preventive Maint. Guidelines for School Facil.	149.95	
	67247B	Project Scheduling & Management for Constr. 3rd Ed.	64.95	
	67265B	Repair & Remodeling Estimating Methods, 4th Ed.	69.95	
	67322B	Resi. & Light Commercial Const. Stds., 3rd Ed.	59.95	
	67327A	Spanish/English Construction Dictionary, 2nd Ed.	23.95	
	67145B	Sq. Ft. & Assem. Estimating Methods, 3rd Ed.	69.95	
	67321	Total Productive Facilities Management	29.98	
	67350	Understanding and Negotiating Const. Contracts	69.95	
	67303B	Unit Price Estimating Methods, 4th Ed.	59.95	
	67354	Universal Design	21.95	
	67319A	Value Engineering: Practical Applications	79.95	
		MA residents add 5% state sales tax		
		Shipping & Handling**		
		Total (U.S. Funds)*		

Prices are subject to change and are for U.S. delivery only. *Canadian customers may call for current prices. **Shipping & handling charges: Add 7% of total order for check and credit card payments. Add 9% of total order for invoiced orders.

Send order to: **ADDV-1000**

Name (please print) ____________________

Company ____________________

☐ **Company**

☐ **Home** **Address** ____________________

____

City/State/Zip ____________________

Phone # ____________________ P.O. # ____________________

(Must accompany all orders being billed)

Mail to: RSMeans, P.O. Box 800, Kingston, MA 02364

RSMeans Project Cost Report

By filling out this report, your project data will contribute to the database that supports the RSMeans Project Cost Square Foot Data. When you fill out this form, RSMeans will provide a $30 discount off one of the RSMeans products advertised in the preceding pages. Please complete the form including all items where you have cost data, and all the items marked (✔).

$30.00 Discount per product for each report you submit.

Project Description (No remodeling projects, please.)

✔ Building Use (Office, School...) ____________________

✔ Address (City, State) ____________________

____

✔ Building Area (SF) ____________________

✔ Frame (Wood, Steel...) ____________________

✔ Exterior Wall (Brick, Tilt-up...) ____________________

✔ Basement: (check one) ☐ Full ☐ Partial ☐ None

✔ Number Stories ____________________

✔ Floor-to-Floor Height ____________________

✔ Volume (C.F.) ____________________

% Air Conditioned ____________ Tons ________

Total Project Cost $ ____________

Owner ____________________

Architect ____________________

General Contractor ____________________

✔ Bid Date ____________________

✔ Typical Bay Size ____________________

✔ Capacity ____________________

✔ Labor Force: _____ % Union _____ % Non-Union

✔ Project Description (Circle one number in each line)

1. Economy 2. Average 3. Custom 4. Luxury
1. Square 2. Rectangular 3. Irregular 4. Very Irregular

Comments ____________________

____

A	✔	General Conditions	$
B	✔	Site Work	$
C	✔	Concrete	$
D	✔	Masonry	$
E	✔	Metals	$
F	✔	Wood & Plastics	$
G	✔	Thermal & Moisture Protection	$
GR		Roofing & Flashing	$
H	✔	Doors and Windows	$
J	✔	Finishes	$
JP		Painting & Wall Covering	$

K	✔	Specialties	$
L	✔	Equipment	$
M	✔	Furnishings	$
N	✔	Special Construction	$
P	✔	Conveying Systems	$
Q	✔	Mechanical	$
QP		Plumbing	$
QB		HVAC	$
R	✔	Electrical	$
S	✔	Mech./Elec. Combined	$

Please specify the RSMeans product you wish to receive. Complete the address information.

Product Name ____________________

Product Number ____________________

Your Name ____________________

Title ____________________

Company ____________________

☐ Company
☐ Home Street Address ____________________

City, State, Zip ____________________

Return by mail or Fax 888-492-6770.

Method of Payment:

Credit Card # ____________________

Expiration Date ____________________

Check ____________________

Purchase Order ____________________

R.S. Means Company, Inc.
Square Foot Costs Department
P.O. Box 800
Kingston, MA 02364-9988